The System Engineer's Handbook

A guide to building VMEbus and VXIbus systems

Edited by

John Black

602-966-5936

ACADEMIC PRESS, INC.
Harcourt Brace Jovanovich, Publishers
San Diego New York Boston London Sydney Tokyo Toronto

To the man from Galilee who,
when I was 19 years old,
changed the course of my life forever.

Design and typesetting by Four Peaks Publishing, Inc.
Fountain Hills, AZ 85268 602-837-3756

This book is printed on acid-free paper. ∞

Academic Press, Inc.
1250 Sixth Avenue, San Diego, California 92101-4311

United Kingdom Edition published by
Academic Press Limited
24–28 Oval Road, London NW1 7DX

Library of Congress Cataloging-in-Publication Data

The system engineer's handbook : a guide to building VMEbus and VXIbus systems / John Black, editor.
p. cm.
ISBN 0-12-102820-8
1. VME (Computer bus) 2. VXI (Computer bus) 3. Multiprocessors-Design and construction. I. Black, John (John Arthur)
TK7895.B87S97 1992
621.39'16–dc20 92-7422
CIP

PRINTED IN THE UNITED STATES OF AMERICA

92 93 94 95 96 97 QW 9 8 7 6 5 4 3 2 1

List of Contributors

Tom Barrett .... Chapter 18
Robert I. Birenbaum .... Chapters 35 & 37
Ken Boyette .... Chapter 8
Bruce Brunette .... Chapter 25
Laurie Burger .... Chapter 4
Alfred Chao .... Chapter 13
Tim Coombs .... Chapter 1
Dick Eue .... Chapter 34
Jim Fainer .... Chapter 2
Mark Fedde .... Chapter 34
Jerry Fiddler .... Chapter 33
Marco Flagg .... Chapter 14
Tom Foley .... Chapter 19
David Forbes .... Chapter 20
Steve Gibson .... Chapter 31
Ernest E. Godsey .... Chapters 27, 30, & 32
Frank G. Goss .... Chapter 24
Willliam A. Gowan .... Chapter 38
Robert Greenfield, Jr. .. Chapter 30
David A. Haworth .... Chapter 21
Walter S. Heath .... Chapter 7
Jim Jonas .... Chapter 20
David R. King .... Chapter 35
Leslie Kirby .... Chapter 33
Larry Larsen .... Chapter 16
Roger Larson .... Chapter 34
Joel Libove .... Chapter 3
Kevin J. Lynch .... Chapter 17
Bruce C. MacDougall .. Chapter 36
Joe Mannas .... Chapter 31
James M. Mooney ...... Chapter 29
Bill Moren .... Chapter 28
Glenford J. Myers .... Chapters 11 & 23
Gregory M.Nielson ...... Chapter 19
Donald Peterson .... Chapter 26
Manos Pinacoulakis .... Chapter 9
Paul Schur .... Chapter 2
Terry Shankland .... Chapter 6
John Simpson .... Chapter 5
Mark Stewart .... Chapter 25
Jacob Tal .... Chapter 39
Tom Thawley .... Chapter 27
Ed Theobald .... Chapter 34
Mike Tobias .... Chapter 10
Don Turrell .... Chapter 15
Dennis Waldenmayer .. Chapter 10
Ron Wolfe .... Chapter 22

Preface

Today's computer systems incorporate a wide variety of specialized technologies, such as VLSI chips, multilayer printed circuit boards, connectors, serial communication protocols, Winchester and floppy disk technologies, keyboard technology, CRT technology, local area network protocols, operating systems, and high-level languages.

Each of these technologies is constantly advancing. To keep yourself up-to-date, you must commit yourself to a life-long process of re-education. This is no simple task. Even engineering schools are having difficulty keeping up with these rapidly evolving technologies. Even if you're fortunate enough to live within commuting distance of a university that offers up-to-date courses, you may still find it difficult to find the time for school. If you're like most engineers, you'll struggle with this problem as long as you remain in the profession.

This book is my attempt to help you update yourself on the key technologies and standards that you'll need to build state-of-the-art multiprocessor systems. In putting this book together I've invited some of the most knowledgeable experts in the computer industry to be my "guest lecturers". You're invited to attend these "lectures" at your leisure, and on your own schedule. You may attend them in any order. You can skip those that are of less interest, while attending those you may need for your current design project.

As the editor of this book I have attended these "lectures" previously, asking questions of the speakers, and suggesting changes that I think will clarify some difficult topics. In that sense, my job is largely done. While I will be giving some of the lectures myself, my primary role in this book will be to introduce the speakers, and to give you some idea what they will be discussing. I hope you'll enjoy (and learn as much from) these speakers as I have.

John Black

Acknowledgements

I would like to thank all of the contributing authors, not only for the time and effort they expended in writing the chapters in this book, but also for the patience they showed in answering my seemingly endless questions. As the editor, I made a commitment to myself that I would include nothing that I could not understand myself, and the contributing authors were the unfortunate ones who often got stuck with the job of educating me.

I would also like to thank Harry Helms. Had it not been for his encouragement and enthusiasm, I would probably never have undertaken the job of putting this book together.

Special thanks to my beloved wife Beverly, who proofread this book (picking the lint from the garment when the tailoring was done) and to my parents, John and Polly, who instilled in me a love for learning that continues to this day.

A few ~~verbalizations~~ words from the ~~redactor~~ editor

This book could never have been written by one person. In that respect it reflects the proliferation and specialization of computer technology over the last 2 decades.

The microprocessor chip has produced far more than just lower cost computers. Prior to its invention, a "computer system" consisted of a single Central Processing Unit (CPU) surrounded by memory and peripheral controllers. With the appearance of the microprocessor, the meaning of this term has been transformed. Today the term "computer system" often refers to a collection of several "intelligent" plug-in boards, each containing its own onboard microprocessor, memory, and peripheral interfaces.

The term "computer architecture" has also taken on a new meaning. At one time the term was used to describe the register set and the instruction set of the CPU. However, that same term is now being used to describe the data paths between the various computing elements in the system. These data paths often include backplane buses, FIFO buffers, shared memory, DMA controllers, and pipelined buses.

It takes teamwork

In fact, the complexity of some new computer systems has reached the point where no single person can possibly understand all of the software and hardware technologies that they incorporate. In short, computer technology can no longer be mastered by a mortal. As a result, these complex systems must now be designed by *teams* of engineers, each of whom is a specialist in one (or a few) particular hardware or software technologies.

Whenever a team of people work together on a design project, there will always be the human factor. To ensure that the various subsystems they design will function together properly, the designers of these various subsystems must *communicate* with each other.

For example, if I'm designing an MPU board and you're designing a disk controller board, we must agree upon an *interface*, through which they will communicate. This interface might involve both hardware and software. The hardware interface might be defined by a list of signal lines and a set of timing diagrams, which define the operation of a backplane. The software interface might be defined as a parameter block, which holds numerical values (or possibly ASCII strings) that gets passed between the MPU board and the disk controller board through shared memory.

If you and I work in the same design group, we might define this interface in a rather informal manner. For example, we might exchange copies of our schematics and code listings. Any remaining questions could be resolved verbally. However, if we work at separate locations we might need to *formalize* the interface definition by creating a document. By describing the interface in writing we reduce our dependence on verbal communication to ensure compatibility.

This formalized interface definition (like all other design documents) must be debugged during the design process. Each incompatibility found between the designs must be traced back to an inadequacy in the formal interface definition. That document must then be modified to ensure that the problem does not recur.

Trust is important

A carefully defined interface, such as a backplane standard, permits many engineers to contribute their specialized expertise to a complex system, even when they can't communicate because of the distance or time between them. However, there still remains a human factor. Designers who communicate with each other can develop a level of *trust* in each other. As this trust develops, each designer becomes more confident that the other designers understand the interface specification, and will comply with it in all respects.

This level of confidence is hard to develop when the various designers can't communicate. The resulting system typically reflects that lack of trust. For example, hardware engineers may leave extra timing margins in their interface designs, "just in case the *other* engineer doesn't fully comply with the interface timing." Software engineers may add extra code "just to verify that all the parameters passed are within the allowed limits." The resulting inefficiencies can cripple a system's performance.

These "just in case" inefficiencies are less common when the designers can communicate readily and have confidence in each other's abilities. As a result, the components designed within a tightly knit group of engineers typically function very efficiently.

These observations have led me to formulate the first axiom of system design:

The first axiom

Axiom 1:

> The structure and the function of a system reflects the structure and function of the organization that designed it.

This leads naturally to the following corollary:

Corollary 1.1:

> When organizing a group of engineers to design a complex system, you should structure the design groups to match the system architecture that you envision. Thus, each subsystem should be designed by a tightly knit group of engineers. The organizational boundaries between design groups should correspond to the software interfaces and the hardware interfaces between the subsystems.

In structuring the organization you should also take into account how that organization will *function*. To do this you must consider the individual personalities of each engineer. Engineers who are better communicators should be assigned the task of defining the hardware and software interfaces between the groups.

This project management technique works well where all of the engineers work for a single large organization. However, as computer technologies have become more specialized, it has become impossible for any one company to do everything. For example, very few companies have the technical know-how to design their own VLSI chips, multilayer printed circuit boards, disk drives, keyboards, CRT monitors, operating systems and high-level language compilers. As a result, today's computer industry consists of thousands of specialized manufacturers, each selling components that incorporate their own technological expertise.

When taken together, these thousands of companies can be viewed as an enormous *distributed design* organization, consisting of engineers located thousands of miles apart. It's the loosely coupled nature of this design organization that has spurred the development of the open interfacing standards we see today, such as VMEbus, SCSI bus, and Ethernet. These open standards allow the construction of larger and more complex *distributed computer systems*.

In general, this interconnection is progressing from the bottom upward. For example, "islands of automation" are typically seen in a manufacturing facility as it moves toward full automation. In the early stages, individual machines are automated. Later, these machines are interconnected, to allow for coordination and data gathering. In a similar fashion, an office may be automated by first purchasing several desktop systems, and later interconnecting them with a local area network. *Connectivity* is becoming a major factor in the purchase of new computer systems.

The second axiom

This observation has led me to formulate the 2nd axiom of system design:

Axiom 2:

> Today's systems are tomorrow's subsystems.

Corollary 2:

> When designing any system, you should always include some interface for connecting it into a distributed system.

While open interface standards can ensure *compatibility* between the interconnected systems, the *performance* of a distributed system (as well as the individual systems within it) depends upon how well the individual system interfaces are designed, and upon how they're interconnected. Data buffering and the prioritization of requests for system resources can have a major impact on system performance. Sometimes a few minor adjustments to a distributed system can improve its performance dramatically.

Compatibility and performance is what systems engineering is all about. And system engineering is what this book is all about. In it you'll learn about the most popular open interface standards, and about the characteristics of the various hardware and software technologies you can use to construct multiprocessor systems. You'll also learn how to choose your system components wisely, how to assemble and debug your system hardware and software, and how to evaluate and optimize the resulting system performance.

TABLE OF CONTENTS

CHAPTER 1

How To Use Special-Purpose Tools To Debug Your VMEbus Systems

By Tim Coombs

VMEbus users are fortunate to have available a wide range of special-purpose debugging tools, in addition to the general-purpose tools which have been available since the mid-1970s. In fact, there are so many specialized tools for VMEbus that you may have some difficulty figuring out which tool is the most appropriate one for the job at hand. In this chapter Tim describes the various type of VMEbus-specific debugging aids, and explains when each is appropriate.

In the beginning there was the multimeter and the oscilloscope. While it was possible to debug complex boards and systems with these tools alone, you needed lots of time and patience.

Then came general-purpose logic analyzers and emulators. These instruments were very helpful in tracking down subtle problems in complex systems. However, they took a long time to set up, and the information they provided still required considerable study and interpretation.

Today, with the widespread use of standardized backplane buses like VMEbus, it has become practical to design more specialized diagnostic tools. These tools greatly reduce debugging time, and improve user confidence in system operation. While some of these tools are fairly simple, others are complex instruments that even include their own user interface.

Debugging a backplane-based system
Because of the modular nature of VMEbus systems, debugging can be split into 3 stages:

- board-level debugging
- system integration
- incorporation of the application software

Each of these 3 stages requires different debugging techniques and, consequently, different tools. In this chapter, we'll survey the various instruments available, and discuss when each is useful during the 3 debugging stages shown above.

Board-level debugging
Most engineers working with VMEbus products should not have to worry about debugging at the board level. After all, one of the advantages to using VMEbus is the wide range of off-the-shelf boards that have already been designed and (hopefully) debugged!

Nevertheless, there may be instances where you need to design a special-purpose board yourself. Most of the debugging at the board level can be carried out using traditional tools. However, you will likely find VMEbus-specific tools helpful for confirming the correct operation of the board's VMEbus interface logic.

System integration
If you can buy all the boards you need off the shelf, you'll be able to start with the system integration phase. During this phase you will likely spend a lot of time tracking down subtle bugs. It's during this phase that VMEbus-specific tools are most helpful.

The process of system integration involves the assembly of the various VMEbus boards, a power supply, a backplane, and a chassis, and the verification that all the boards are interacting properly. In many cases you will also need to install an operating system and I/O driver software, since many VMEbus boards don't function normally until the operating system is in place. For example, intelligent peripheral boards often have the bulk of their software loaded from disk, during system initialization.

Problems are usually caused by simple things, such as:

- a missing DC power level
- a missing bus terminator
- incorrectly configured boards

Another common problem is a mismatch between the I/O driver and the I/O port's hardware configuration, such as control line usage on a serial port.

When the assembled system fails to work, there is often very little information to indicate what is happening (or NOT happening). Faced with this situation, many engineers have a tendency to assume that the cause of the problem is far more serious than it actually is. However, it's fairly unusual for the cause to be faulty products, or nonconformance to specifications: 2 culprits that often (incorrectly) get the blame. More often than not, the problem is just due to incorrect hardware configuration, and can be quickly remedied, once it has been identified.

Incorporation of application software
Once the operating system is running properly, application software can be developed in a fashion similar to that used on any other computer. On the whole, traditional editors, compilers and debuggers can be used. However, there are some situations in which specialized VMEbus diagnostic and debugging tools might be useful.

When a system contains several MPU boards (or several intelligent peripheral controllers) you may find that a few of the masters use far more than their share of bus bandwidth. To solve this problem may need special tools to help you monitor and reconfigure the system hardware and software, for balanced bus usage.

Another situation is when you need to be able to verify that exceptional events are being handled properly. For example, systems used for industrial control (or process control) invariably have software and/or hardware features to protect them against unusual events, such as spurious interrupts. Without special debugging tools, you may have a very difficult time testing these features in a reliable or repeatable way.

Using conventional instruments
An obvious questions arises. How are the available tools suited to solving these problems?

Multimeters
A multimeter is an absolutely essential piece of equipment for system integration. Once the power supply has been connected to the backplane (but *prior* to the insertion of any conventional VMEbus boards) you *must* check the voltages at the backplane for correct DC levels. Failure to do this can be *very* expensive. It's easy to assume that someone else has checked the system (perhaps the technician who assembled the chassis?) but there is simply no substitute for checking the voltages yourself.

Oscilloscopes
Oscilloscopes are extremely versatile general-purpose tools. They're most useful for:

- observing repetitive waveforms.
- analyzing the precise timing relationship between a small number of signals.

Most of the VMEbus signal lines (address, data, and address modifier lines) don't carry repetitive waveforms during normal operation. To capture the non-repetitive levels of these lines during a specific data transfer cycle, a storage scope may be helpful. However, this still leaves the problem of triggering the storage scope, during that particular bus cycle. To do that you may need to trigger the scope, based on the address of the location being accessed. The simple edge triggering provided by the oscilloscope does not allow this.

However, an oscilloscope can still be useful for verifying set-up times and hold times between the strobe lines (AS*, DS0*, DS1*, DTACK* and BERR*) and the corresponding address and data bus lines. This is particularly helpful when you're debugging a new board design.

Logic analyzers
General-purpose logic analyzers are well suited to the simultaneous display of large numbers of waveforms: just the thing you need to analyze address and data bus transfers. Their probes can be connected to simultaneously display onboard signals and VMEbus signals. This is very helpful when you want to determine whether your board's VMEbus interface is working properly.

However, you may find yourself spending considerable time making all the necessary probe connections. After you connect all the probes, you must also configure the analyzer to display the signal name of each VMEbus waveform. That allows you to more easily interpret the display visually.

Even after the analyzer has been connected and configured, it may not provide a very efficient way to debug a VMEbus system. That's because the VMEbus consists of 4 loosely-coupled buses:

- The Data Transfer Bus
- The Arbitration bus
- The Interrupt bus
- The Utility bus

The semi-independent protocols on these 4 buses complicate the selection of trigger conditions. Unless you choose the right trigger conditions, you may get swamped with too much useless information. For example, it would be pointless to capture the bus request (BR0*-BR3*) signals on the falling edge of the data strobes, and then try to use that information to determine whether an arbiter is operating properly. Since the levels of the bus request lines don't change in synchronism with the data transfers, you can't properly evaluate the operation of the arbiter with this fragmentary information.

Even within the data transfer bus, you would want to sample the data bus at a different time during different cycle types. For example, during a *write* cycle you would sample the data bus at the falling edge of the data strobes, and during a *read* cycle, at the falling edge of DTACK*.

Finally, the captured information must be *interpreted*. Not all VMEbus signal levels are significant at all times. For example, the 2 most significant bytes of the address bus carry no information during A16 address broadcasts. A general-purpose logic analyzer has no provision for recognizing the address width (based on the address modifier code) and modifying the display (on a cycle-by-cycle basis) to show only the valid address levels.

The value of logic analyzers for software debugging is also debatable, especially in a VMEbus system with multiple masters. Arbitration takes place on a different set of lines from the data transfers, and the levels on the arbitration lines don't necessarily remain static during each master's use of the backplane. Thus, the analyzer can't distinguish between a transaction initiated by one master and a transaction initiated by another.

Emulators

Microprocessor emulators can be very useful when debugging new board designs. Once the MPU board's internal logic has been verified, the debugging software within the emulator can then be used to exercise the VMEbus interface logic on the board. If the emulator includes hardware and software for timing analysis, it can be used to measure delays over the VMEbus. This may suggest ways to improve the board design.

The timing analysis provided by the emulator might also allow you to measure the performance of different system configurations, allowing you to *tune* the system for maximum performance.

Using VMEbus-specific tools

What is a VMEbus-specific tool? As the name suggests, it's a tool or instrument that's designed specifically to simplify the debugging of VMEbus boards and systems. Because it is so specialized, it is typically of little (or no) value for other purposes, although similar tools may exist for other backplane standards. VMEbus-specific tools are characterized by:

- ease of use, since they connect directly to the bus signal lines by plugging into the backplane, like any other VMEbus board.
- better timing characteristics, since long conductors aren't needed between the bus lines and the analyzer logic.
- more effective capture of the bus state, since their design is based specifically on the VMEbus protocols.
- more effective display of capture state information, since they are designed specifically to display the results in an easily understood way.

A summary of VMEbus-specific tools

The following is a list of the types of VMEbus-specific tools available, and their main characteristics:

- VMEbus systems
- Power supervisors
- Extender boards
- Firmware debug monitors
- Signal breakout boards
- VMEbus monitors
- Anomaly triggers
- VMEbus analyzers
- VMEbus stimulators
- VMEbus ownership monitors

VMEbus systems

When debugging new board designs, one of the most valuable bus-specific tools is a working VMEbus system. This system should include at least one of each type of VMEbus functional module (master, slave, requester, arbiter, etc). Recently, a number of VMEbus card cage manufacturers have announced special open-frame card cages, which make debugging easier.

Power supervisors

A *power supervisor* should be the first board you plug into any newly-wired chassis. It is designed to withstand power supply output levels that are out of adjustment. Indicators on its front panel show, at a glance, if DC power is within specifications. It also applies dummy loads to the power supply outputs, allowing you to adjust the supplies to provide the correct levels.

Using a power supervisor is much safer (and cheaper) than using thousands of dollars worth of VMEbus boards for the same purpose! The dummy loads can be designed to protect themselves (and the backplane connector pins) if the DC levels are too high. In addition, they may also be used as a heat source for testing the system cooling. A version without the dummy loads can be used where permanent display of power supply status is desired. A VMEbus power supervisor is available from Concise Technology.

Extender boards

Extender boards are essential for debugging new board designs, or for finding failed components on faulty boards. In fact, extender boards have proven so useful that they're available for virtually every backplane bus standard. A well-designed extender board provides:

- access to the components on a board, even when the board is connected to the VMEbus.
- tie points for connecting probes to individual VMEbus signals.
- jumpers or switches to disconnect individual onboard signal lines from the backplane.
- jumpers or switches to disconnect DC power from the board. (This permits insertion of an ammeter, to measure the board's DC power consumption.)

VMEbus extender boards are available from many different sources.

Firmware debug monitors
A *firmware debug monitor* is essential during system integration, since it allows you to generate read and write cycles on the VMEbus. Without this ability you can't verify operation of the backplane or the slave boards in the system.

Over the years firmware debug monitors have become more and more sophisticated, and hence more useful. Many of the current generation of monitors are derived from the original Motorola "Tutor", and include single-line assembly/disassembly. This feature is very useful for setting up loops to test new hardware. Each firmware monitor is customized to the I/O ports on its own particular MPU board, and many also have special commands that allow you to interact with the VMEbus interface logic on the board.

Signal breakout boards
Breakout boards bring the individual VMEbus signals out to eyelets, making it easier to connect logic analyzer probes. Buffering is typically provided to prevent the analyzer from overloading the VMEbus signal lines. However, because of the propagation delays of these buffers, the timing displayed by the logic analyzer will inevitably be a little different from the timing on the actual backplane signals.

VMEbus monitors
Special-purpose VMEbus analyzers solve many of the problems associated with more general-purpose analyzers. However, they don't necessarily capture the ancillary signals at the correct time. This limits their usefulness when investigating arbitration problems.

A bus *monitor* plugs directly into the backplane, and has an LED for each of the VMEbus signals, as shown in Figure 1. Thus, no external CRT terminal is required.

The various signals are grouped logically on the display. In the relevant capture mode, each signal is captured at the most auspicious moment. Special capture modes are available, such as pulse stretching (for displaying signals with low duty cycles) or latching (for permanently capturing spurious signals).

Capture of the address and data (which are displayed according to the width of the transaction) may be qualified by cycle type. For example, the capture of a data transfer or an interrupt acknowledge cycle can be qualified by either a DTACK* or a BERR* and/or by an external qualifier.

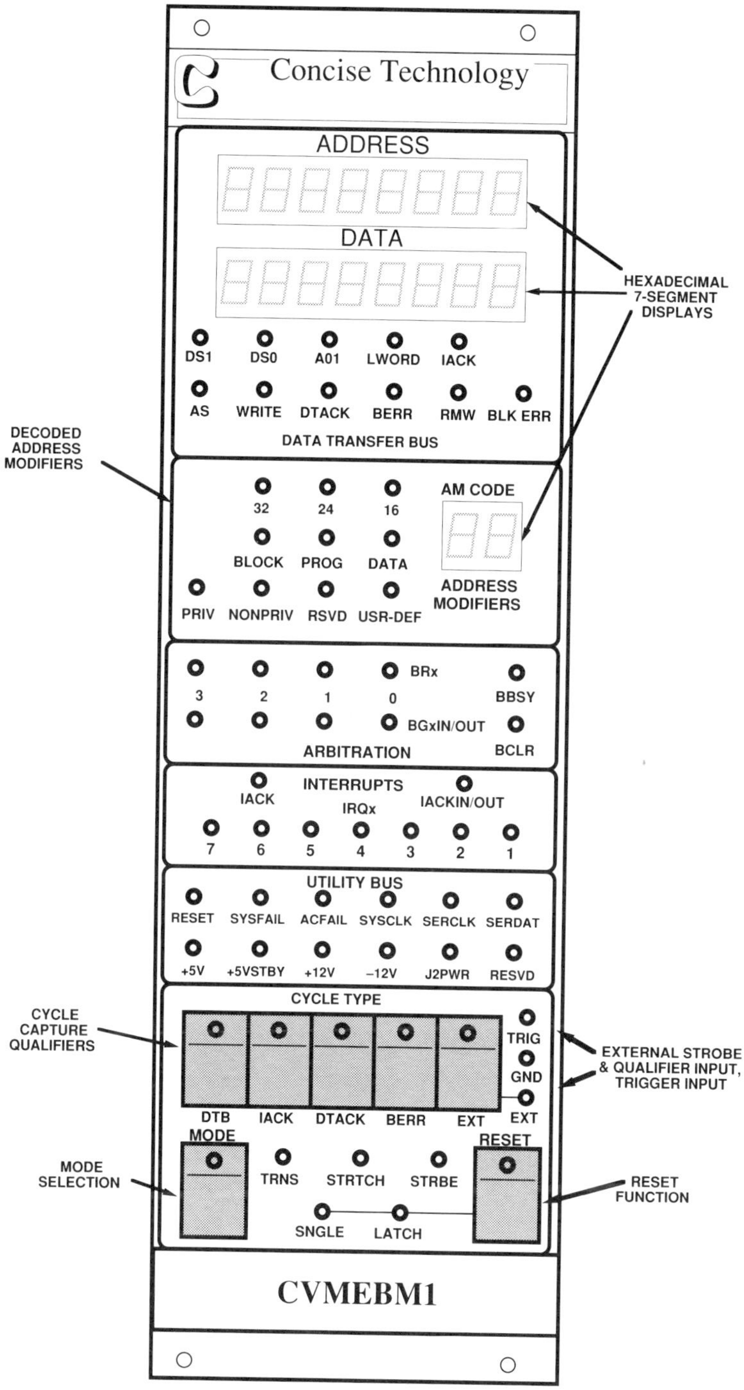
Concise Technology
ADDRESS
DATA
HEXADECIMAL 7-SEGMENT DISPLAYS
DS1 DS0 A01 LWORD IACK
AS WRITE DTACK BERR RMW BLK ERR
DATA TRANSFER BUS
DECODED ADDRESS MODIFIERS
32 24 16
AM CODE
BLOCK PROG DATA
PRIV NONPRIV RSVD USR-DEF
ADDRESS MODIFIERS
3 2 1 0 BRx
BBSY
BGxIN/OUT
ARBITRATION
BCLR
IACK
INTERRUPTS
IRQx
IACKIN/OUT
7 6 5 4 3 2 1
UTILITY BUS
RESET SYSFAIL ACFAIL SYSCLK SERCLK SERDAT
+5V +5VSTBY +12V –12V J2PWR RESVD
CYCLE TYPE
CYCLE CAPTURE QUALIFIERS
TRIG
GND
EXT
EXTERNAL STROBE & QUALIFIER INPUT, TRIGGER INPUT
DTB IACK DTACK BERR EXT
MODE
RESET
MODE SELECTION
TRNS STRTCH STRBE
SNGLE LATCH
RESET FUNCTION
CVMEBM1

Figure 1

In addition, a bus monitor can be designed to provide a *trigger* output. It can also be designed to detect certain specific problems, such as the absence of a bus terminator to pull up a VMEbus signal line.

A VMEbus monitor is most useful as a system integration tool. As previously discussed, most of the problems that develop during system integration have simple causes, but may take a long time to locate. A VMEbus monitor provides a simple and fast way to locate these problems.

VMEbus monitors also provide a simple way to verify that the correct byte lanes are being used to write data to (and read data from) memory. This is particularly important when designing CPU boards with processors other than the 68000 family. Since the processor on a master board writes data to (and reads data from) RAM in the same byte-lane order, a byte-lane problem might not otherwise be discovered until a processor with a different byte ordering is also plugged into the backplane.

Anomaly triggers

An *anomaly trigger* plugs directly into the VMEbus backplane, and continuously monitors the VMEbus traffic to detect certain specified VMEbus timing violations. When one of these violations occurs, the board gives a visual indication, and generates a trigger signal. This trigger signal can be used to trigger a logic analyzer, which could be configured to retain a detailed history of the bus timing leading up to the anomaly.

Since it is hard-wired to detect a number of important VMEbus trigger conditions, an anomaly trigger greatly enhances the usefulness of a logic analyzer. However, it still has its limitations. Because of the timing tolerances of the components used to build an anomaly trigger, it's not possible to guarantee that every bus timing violation will be detected without also generating "false alarms". For example, the documentation for the Ultraview/VMETRO anomaly trigger board (called the VBAT) says that the board detects "significant violations".

VMEbus analyzers

A *VMEbus analyzer* is essentially a special-purpose logic analyzer. Since it is integrated onto a VMEbus board, it avoids many of the problems associated with general-purpose analyzers. Connection to the backplane is no longer a problem (providing there's a spare slot available) and the state capturing and presentation are specifically geared to VMEbus.

These features save considerable time. Because of the compact nature of the products, an external ASCII terminal is typically used. However, the character-oriented nature of such a terminal limits the display of histograms and waveform displays to some degree.

VMEbus stimulators

Some features in a VMEbus system are designed to handle exceptional events. For example, the interrupt handler might be used to respond to events occurring external to the system. Since these external events may not be totally under your control (and may not even be repeatable) debugging the interrupt handling hardware and software can be a chore.

Other exceptional events might include glitches on the arbitration bus, or spurious interrupts, due to high levels of noise in the operating environment. Simulation of spurious signals is particularly important for testing hardware and software designed to protect process control systems from these real-world events.

A bus stimulator can be used to exercise the Arbitration bus, Interrupt bus, and Utility bus, either within the constraints of the VMEbus specification, or to generate spurious events. It's useful for testing existing (as well as new) boards, and for ensuring that interrupt handlers operate properly.

VMEbus ownership monitors
An ownership monitor shows which master has control of the VMEbus at any given time. It does this by monitoring (and remembering) the extent to which bus grant signals have propagated along the daisy chain. In a similar way, it monitors the IACK daisy-chain, to determine which interrupt handler initiates each interrupt acknowledge cycle, and which interrupter responds to each interrupt acknowledge cycle.

In addition to an LED display (See Figure 2 on the next page), the ownership monitor can generate real-time output signals to indicate which master is using the backplane at any given time. The LEDs provide a *qualitative* indication of the degree of bus usage, and the real-time output signals provide a *quantitative* one. By making incremental changes to the system configuration, while observing the effects in real time, you can improve overall system performance and/or solve bus utilization problems.

The real-time output signals from the ownership monitor can also be used to qualify sampling by some other instrument, such as a bus analyzer. This permits capture and display of only those data transfers initiated by a particular master, solving one major problem with using analyzers in multimaster systems.

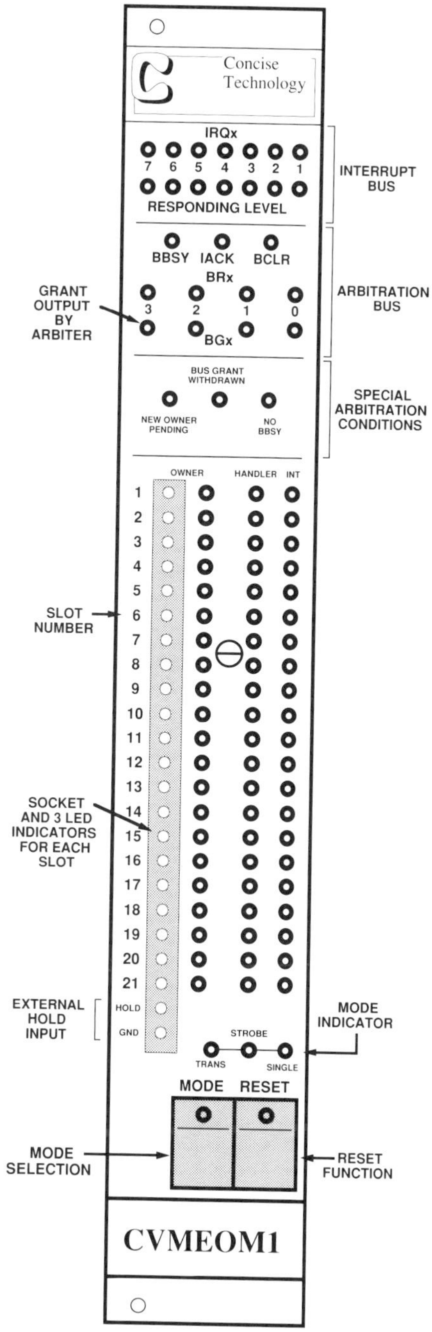

Figure 2

The ownership monitor may also be used to check new arbiter, requester, and interrupter designs for glitches.

Summary
Traditional general-purpose instruments are not always the best tool to use when debugging VMEbus systems. Since they often take a lot of time to connect and use, they don't provide a very efficient way to locate simple configuration problems. However, VMEbus users are fortunate to have available a wide range of VMEbus-specific tools, which provide a more effective and efficient solution in these problems.

In fact, it's perhaps a reflection of the popularity of VMEbus that so many different bus-specific tools exist for it: probably more than for any other backplane bus. A cynic might say this is because VMEbus is difficult to use. A more reasonable view is that the large VMEbus user base has made it possible for many companies to successfully develop and market these specialized tools.

Tim Coombs formed Concise Technology in 1986 to develop a range of VMEbus-specific tools and instruments. Previously, he was the U.K. manager of MPU field application engineering for Motorola, and was responsible for a team supporting Motorola's processor chips, as well as its VMEbus hardware and software products. More recently, he has been involved with interfacing transputers to VMEbus, and with developing a range of fiber optic interfaces for VMEbus. Tim has a B.Sc. in electronics and an M.Sc. in digital electronics from the University of Manchester Institute of Science and Technology.

CHAPTER 2

How To Use A VMEbus Analyzer To Integrate Your System

By Jim Fainer and Paul Schur

The integration of a multiprocessing VMEbus system can sometimes be long and arduous. You can spend a lot of your time tracking down problems that turn out to be as simple as incorrect board jumpering, or as complex as a time-dependent bug in the real-time application code. While traditional logic analyzers can be used with some success, they have important limitations. You may also sometimes spend more time setting them up than you spend finding the problem. In this chapter Jim and Paul describe a faster way to zero in on problems during the system integration process, through the use of VMEbus analyzers.

Suppose you're integrating a disk-based system comprised of 5 VMEbus boards. Three of these boards, a processor, memory, and a disk controller have been purchased from 3 different companies. The other 2 boards, a robotic arm controller and a motion sensor, have been designed by your own company. Your company has also purchased a real-time operating system, and written a library of software routines to control movement of the robotic arm, based on inputs from the motion sensor. It is your job to integrate all of these pieces into a working system.

In theory, by using a backplane standard like VMEbus, you have simplified the process of system integration. You just insert the boards into a card rack and verify that each board works correctly. If the system doesn't work, you determine which board is at fault, and then either contact the engineer in your company that designed the board, or contact the company from which you bought the board.

That all sounds very simple, but it's not always so easy to determine which board is at fault. In some cases, the fault may even be in the software, instead of in the hardware. As a result, you may be asked to provide some hard evidence that you have localized the problem before others will commit their time to helping you.

Using a bus analyzer

A bus analyzer is a specialized tool designed to quickly pinpoint board-level and system-level problems. VMEbus analyzers are typically built onto a single-slot Eurocard, which can be plugged directly into a VMEbus system slot. They monitor (and in some cases they even stimulate) activity on the VMEbus. Let's take the system described above as an example, to show how you might use an analyzer to find problems in your system.

You would probably begin by looking at the most basic elements of the system: the processor board and the memory board. We will assume that the processor has a firmware-resident debug monitor, which allows you to perform block transfers over the VMEbus to (and from) the memory board. You first enter a monitor command that performs a block transfer write operation to the memory board. Then you issue another command to read the block of data back, for verification. If that verification fails, there may be a number of things causing the problem. For example:

- The processor board may be writing to an incorrect address.
- The processor board may be writing incorrect data.
- The memory board may be storing the data at an incorrect location.
- The data may not be valid on the bus when the memory board captures it.
- The backplane may have shorts between its signal lines.

A bus analyzer can be used to quickly determine if any of these is causing the problem. The analyzer samples the VMEbus signal lines, and stores the samples in a *trace buffer*. During *synchronous sampling*, bus signals are clocked into the trace buffer on the falling edge of DTACK* or BERR*.

After the analyzer has finished capturing this information, you can examine the contents of the trace buffer on a CRT screen. In this example, the trace display would reveal if the processor board's address, data, or any other signals involved in the transfer, are carrying incorrect levels. You are easily able to detect stuck address lines or data lines.

If the analyzer detected no backplane activity, the problem is likely to be incorrect jumpering of the local bus addressing logic, which has caused the processor to access onboard resources, instead of going offboard to the VMEbus system memory.

Adding the disk drive

Suppose you find that the processor board and the memory board are performing data transfers correctly. You can now add the disk controller to the system. We will assume that this particular controller can do DMA transfers to (and from) system memory, over the VMEbus. We will also assume that the MPU board's monitor firmware has routines that allow you to pass *read sector* commands and *write sector* commands to the disk controller. The disk controller board then acknowledges completion of sector read or sector write by interrupting the processor over the VMEbus.

To write a sector to the disk drive, the MPU first writes the sector data into the memory board over the VMEbus. Then it commands the controller to read that sector data from memory, and write it to the disk, at a specified track and sector location.

Both the processor board and the disk controller board access the system memory during a sector read (or a sector write) operation. In fact, bus cycles initiated by the 2 boards may get interleaved in such a way as to make it difficult to distinguish one from the other. To verify that the disk controller board is working correctly, it's helpful to isolate the cycles initiated by the controller board from other cycles occurring on the bus. Bus analyzers allow you to do just that, by providing trigger and trace qualifiers that control when (and how much) information is captured in the trace buffer.

Looking in the right place
A *trigger* is a bus event around which you want to study bus activity. Ideally, you would like to have an analyzer that allows you to store this trigger event anywhere in the trace buffer. For example, if the analyzer lets you store the trigger event in the middle of the trace buffer, you will be able to see bus activity both *before* and *after* that event. When the trace buffer contents are displayed, the trigger event can be highlighted, to provide a reference for all the other trace data in the buffer.

In this example, a trigger condition could be set to look for the MPU operation that initiates the disk controller's sector write. (For example, this might be a VMEbus write cycle to the disk controller's onboard interrupt register.) By using this trigger event, you can start capturing bus activity at the beginning of the sector write operation. Activity immediately after the trigger should show the controller board doing a DMA transfer of the sector data from the VMEbus memory board into the controller board.

Since the MPU in this system can be transferring data at the same time as the disk controller, more qualification is needed if you want to avoid capturing the MPU activity that's interleaved with the disk controller activity. *Trace qualifiers* provide you with this capability. A trace qualifier specifies what *type* of information to store in the trace buffer. In this case, you can define a trace qualifier that allows the analyzer to capture data only when the accesses are within the address range of the sector buffer on the VMEbus memory board.

Using this set-up, you can verify that the disk controller is reading the correct data from VMEbus memory. When the disk controller is commanded to read a sector from the disk, and write it into memory, you can also verify that operation in a similar fashion.

Analyzing a sequence of events
Once you have verified proper operation of the MPU board, the memory board, and the disk controller board, you can install the operating system. The first step in bringing up the operating system is the bootstrap procedure. In some cases a simple firmware-resident bootstrap loader routine reads a more sophisticated bootstrap loader routine from the first several tracks of the disk drive. Once the second bootstrap loader has been read into memory and started, it finds the disk file containing the operating system code, loads that code into memory, and then executes it.

Obviously, if this 3-step process fails to complete, it may not be easy to determine where the failure occurred. However, if your bus analyzer allows you to specify more than one trigger or trace qualifier in sequence (or in logical combination) you'll be able to verify each step in the process. For example, to determine whether the secondary bootstrap loader was able to find the operating system file in the disk directory, you might set up a sequential trigger like the one shown in Figure 1.

Another feature of a bus analyzer, *occurrence counting*, lets you count how many times a given event has occurred, and then trigger on a specified count. For example, if you want to look at the data from the 10th disk sector read, you could generate a trigger when the MPU writes the 10th *sector read* command into the disk controller's interrupt register.

Putting it all together
You now have 3 of the 5 VMEbus boards verified, and you have your operating system running. Your last task is to integrate your application code with your company's robotic

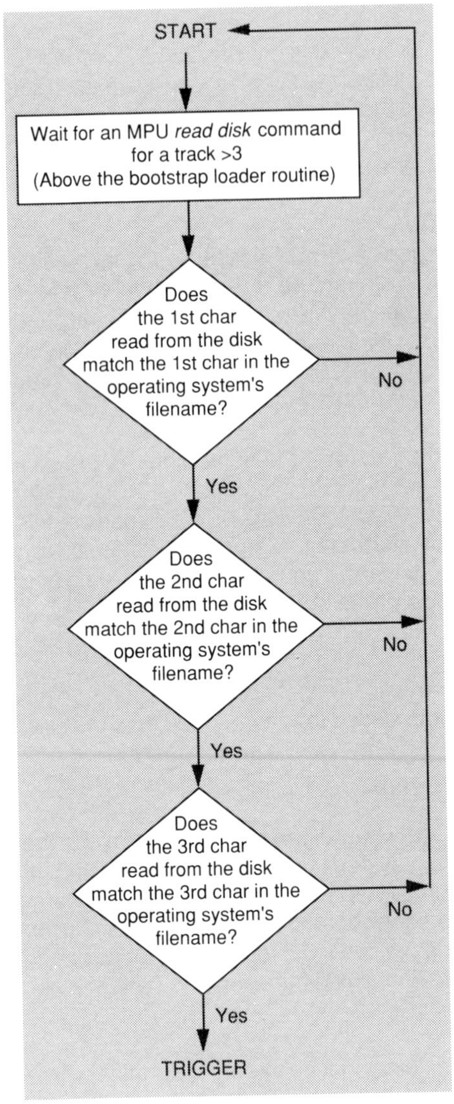

Figure 1

arm controller and motion sensor boards, and with the rest of the software in the system. Since these boards are a new design, you may want to verify their bus timing more carefully than the others. You can get a detailed picture of their VMEbus timing by configuring the analyzer to sample the bus *asynchronously*. In this mode you set a free-running onboard clock to the sampling rate that you need, and then allow the analyzer to sample the bus signal levels at fixed intervals. The relationships between bus signals can then be displayed with waveform representations on a CRT display.

Tuning your system

Suppose you have verified the timing of the new boards to be within the VMEbus specifications. However, you find that the system response is slower than you expected. A bus analyzer can then be used to determine where, in the application code, the processor is spending its time. By displaying histograms, the analyzer can help identify hot spots, allowing you to modify the application code, to speed things up.

Typically you would start by histogramming a wide address range, and then dividing up that range into smaller ranges. Once you identify the hot spots, you can use the normal trace and trigger features to scrutinize the activity occurring within that address range. For example, suppose you progressively narrow your search for a hot spot until you find that most of the processor's time is spent assessing locations between addresses 1000_{16} and 2000_{16}. Knowing this, you set a trace qualifier for this address range, and start capturing bus transactions. The trace buffer collects all bus transactions in that address range. The trace display would then reveal what type of transfers are occurring.

What to look for in a VMEbus analyzer

There are several features to look for when you're choosing a VMEbus analyzer:

- Trace buffer size
- Trace buffer expandability
- Maximum sampling speed

- Trigger/trace qualifiers
- VMEbus stimulation and mastership
- Temperature and voltage monitoring

Trace buffer size
VMEbus analyzers are available with trace buffer capacities as small as 2 Ksamples, and as large as 128 Ksamples. The amount of trace buffer that you will need obviously depends upon your application. However, as a rule of thumb, if you always use trigger and trace qualifiers to restrict what gets captured, you probably won't need more than 16 Ksamples of storage.

However, there may come a day when you're faced with an intermittent problem that hangs your system up, and you may have no clue as to where to begin looking. To find that kind of a problem, you may need to do unqualified traces. That requires more memory.

It's also important to ensure that the trace buffer can capture all of the bus signals that you're interested in. For example, if you're working with 32-bit addresses and 32-bit data, your analyzer's trace buffer should be wide enough to capture all these bits.

Trace buffer expandability
If you buy an analyzer with less than 16 Ksamples of trace buffer, you should consider the possibility that you may want to increase its buffer size at a later date. If the analyzer is designed properly this can be done through a simple memory chip upgrade.

Maximum sampling speed
Theoretically, the VMEbus can transfer data at rates in excess of 10 MHz. Current VMEbus boards don't transfer data that fast, but your analyzer should have a maximum synchronous sampling frequency at least that high. For asynchronous sampling, you should keep in mind that your ability to measure the timing difference between signal transitions on the bus is limited by the sampling rate. A 20 MHz sampling rate provides only 50 nsec resolution. A 100 MHz sampling rate provides a 10 nsec resolution. In some cases, even a 10 nsec resolution may not be adequate to measure skews between signals, or narrow glitches on signal lines. That's when it's important to have an *external trigger output* on the analyzer. This trigger output can be used to trigger an oscilloscope, so you can get a better look at the signal waveforms.

Trigger/trace qualifiers
Trigger and trace qualifiers allow you to zero in on a problem, but only if they provide you with enough flexibility. As a minimum, you should be able to:

- *specify* address and data ranges as a condition.
- *specify* more than one bus condition simultaneously.
- *combine* conditions and occurrence counts logically.
- *sequence* conditions and occurrence counts.

By logically combining several conditions to generate a trigger, or by requiring a specified sequence of conditions before a trigger occurs, you can zero in on a problem.

VMEbus stimulation and mastership
In addition to sampling the bus, some bus analyzers can also drive the bus signal lines either high or low. The analyzer can activate individual bus signals to:

- generate an interrupt
- reset the system
- perform a backplane shorts test

By driving the signal lines high and low in the appropriate sequence, these line drivers can also be used to generate data transfer cycles. This can prove very useful when you find that just monitoring the bus doesn't provide you with enough information. For example, you might configure the analyzer to generate VMEbus write cycles and read cycles, in order to verify a memory board's operation.

Temperature and voltage monitoring
Temperature and voltage monitoring can be used to continuously monitor system operation. The purpose behind this monitoring is to *prevent* failures, rather than repair them. A rising temperature might indicate that a fan filter is dirty, or that some board in the system is overheating. Voltage monitoring also indicates how stable the power supply outputs are as the system is warmed up, and as the loads change.

The N9210 analyzer
The Nissho N9210 bus analyzer is an example of a dedicated VMEbus analyzer. (See Figure 2) It is built on a single Eurocard, which plugs directly into a VMEbus system. Any ASCII terminal can be connected through a front panel connector, to provide the user interface. No other connections are required.

Figure 2

The N9210 powers up with all qualifiers *don't cared*, so if you just want to capture all activity, no set-up is required. You just type GO, and the analyzer captures bus activity (by sampling the backplane signals on each bus cycle at the falling edge of DTACK* or BERR*) and stores this captured information in its 64 Ksample trace buffer.

Each sample is stored in the trace buffer with a time tag, which permits interval computation between samples at a later time. The board can synchronously sample VMEbus cycles at rates up to 25 MHz, which is much higher than the maximum speed of VMEbus transfers. It can also sample asynchronously at speeds up to 50 MHz, giving a 20 nsec sampling interval.

The N9210 can store up to 4 trigger/trace qualifiers simultaneously in its nonvolatile memory, and can combine those qualifiers either logically or sequentially, to control when tracing is started and stopped. Each of these conditions defines a bus event, comprised of up to 95 bus signal levels, plus one external trigger signal.

In addition to the analyzer features listed above, the N9210 also includes some useful extras. For example, it can act as a bus master, and it allows direct read-out of the DC power supply voltages, as well as the temperature. It also has a 2nd serial port, which can be used to provide hardcopy output of the screen displays.

After data is captured, the contents of the trace buffer can be presented on the screen as waveforms or in hexadecimal form. Hexadecimal display permits quick checking of addresses and data values.

Histograms are provided to show address and signal line usage. The board also accepts commands for generating signal transitions on the VMEbus. The 9210's trace buffer capacity can also be expanded from 4 Kbytes to 64 Kbytes with the simple replacement of socketed parts.

A case history
VMEbus systems are increasingly being used to provide high-performance peripheral support for larger systems. One example is a multi-spindle large-capacity disk server, for a VAX minicomputer. Such a disk server can provide functions that would otherwise burden the VAX, or that may not even be available on the VAX. By reducing the overhead imposed on the host, and by increasing the number of I/O operations, it allows more system throughput.

Such a system was recently developed. (See Figure 3) The server consisted of a special interface controller board that managed the physical and logical link to the VAX host. This management was done with a 16-bit microprocessor, and a bit-slice engine. This interface controller board received command packets and status packets over the VMEbus from a 68020-based Single-Board Computer (SBC). This SBC had:

- a block of dual-ported memory
- a block of local memory
- hardware to handle VMEbus interrupts
- 2 serial interfaces
- a control program for the system

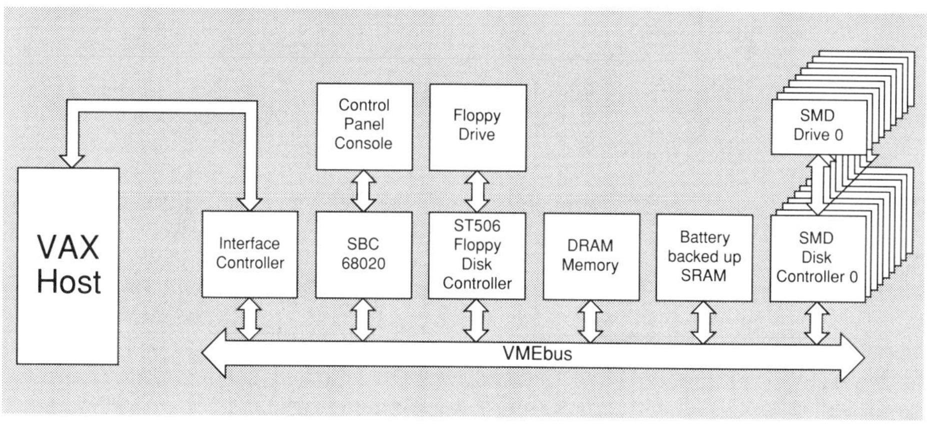

Figure 3

The 68020-based SBC managed the system resources, which resided on other VMEbus boards. These boards included:

- 9 intelligent disk controllers (packet driven)
- 1 dynamic memory board
- 1 nonvolatile memory board
- 1 controller for a small winchester disk, and floppy drive
- A front panel

Each SMD disk controller board was equipped with a full track buffer, and could queue up to 10 command packets. Each could also perform scatter/gather operations. The memory boards were used for caching data. The dynamic memory was used to provide a read cache, and the non-volatile memory was used for the write buffer. The disk controller for the small winchester and the floppy provided a way to download the control program to the 68020-based SBC.

However, during the development process, the engineers encountered a perplexing problem. The VAX operating system included disk diagnostics that, over time, sent out every command to the disk server that the system would ever use to read and write to the disk. When these diagnostics were enabled, the disk server would consistently hang up after 18 hours of operation. The only way to clear this hung-up condition was to perform a cold boot operation. The problem was not an unpredictable one. It could be repeated, but it took 18 hours of operation to get the system to hang up again.

Even though the error was repeatable, the fact that it occurred only once ever 18 hours made is practically impossible to isolate, with any conventional logic analyzers. Faced with this, the engineers configured the N9210 analyzer to continuously capture all VMEbus activity in the disk server, and to store the captured information in the trace buffer. Hopefully, when the system hung up, examination of the trace buffer would reveal the command packet of the most recently issued command from the VAX to the disk server. These command packets contained:

- the command
- the number of sectors to be read from (or written to) the disk
- the VMEbus memory address that the sectors would be written to (or read from)
- the VMEbus memory address of the next command packet in the queue

By studying the information captured in the trace buffer of the N9210, the system designers were able to determine that the command being processed when the system hung up was a *read-data* command.

The VAX had issued a read data command to the disk server. (There was certainly nothing unusual about this, since the VAX diagnostic routines had issued this command many millions of times before.) The interface controller board in the VMEbus chassis had received this command, and then sent a command packet over the backplane to the SBC. The last bus transaction was when the SBC wrote a status word into the interface controller's mailbox. That word (which was captured in the trace buffer) indicated that the SBC had received the read data command from the interface controller. However, the SBC never returned from processing that read data command.

At this point the question was "What is the 68020 doing?" A 68020 in-circuit emulator was hooked up to the SBC, and the problem was found to be in the cache routine. The 68020 (in processing the read data command) had called the *read cache routine* (to determine if the requested data was already in the cache) and had never returned.

Summary

When a complex VMEbus system fails, you may need specialized test equipment to identify the problem. This type of equipment can provide valuable aid in troubleshooting a complex system. Because dedicated logic analyzers require a minimal amount of hook-up time, and a small amount of set-up time, they are the most efficient way to identify where the problem is.

Jim Fainer is a senior technical support engineer for Nissho Electronics. Jim has been involved in the design of high performance disk and tape drive controllers for the past 10 years. He is also been involved in the development for high performance multi-spindle disk systems. His hobbies include camping, hiking, fishing, and target shooting.

Paul Schur is president and founder of Silicon Control, which specializes in the design and manufacture of bus analyzers. He holds a BSEE degree from the University of Illinois at Urbana. Paul has been involved in the design of digital systems for the past 15 years. Prior to starting Silicon Control, Paul worked at Northrop Defense Systems, designing airborne equipment for various government contracts.

CHAPTER 3

How To Use Anomaly Triggers To Detect Subtle Bus Timing Violations

By Joel Libove

High speed backplane buses like VMEbus provide a great deal of performance and flexibility. However, along with these benefits come increased design complexity. The stringent timing requirements, and the large number of possible bus cycle types, provide a challenge to even the best hardware designers. As a result, some VMEbus boards contain subtle design errors which may eventually cause an unexpected system failure. In this chapter Joel describes how bus anomaly triggers can be used to detect these subtle design errors ***before*** *they cause a system failure.*

Some design errors are quickly found during early testing of a new VMEbus board. However, others may remain hidden, and may not manifest themselves until months (or even years) later when, perhaps, a new VMEbus board is integrated into the system.

When this happens, the result may be intermittent problems that are difficult to localize. Many different bugs can produce the same gross symptom: a system which "hangs up" occasionally. It's difficult to know where to begin debugging.

Even when a logic analyzer is available, you often have no clue as to how you should set up the triggering on the logic analyzer to capture the offending bus cycles. In some cases, the gross symptom doesn't even show itself for some time after the bad cycle, that precipitated the problem. For example, if a master fails to meet the data set-up time during a write to memory, the contents of that memory location may be bad, but no symptom is produced until that bad location is read.

These bus timing violations may be difficult (if not impossible) to catch with a traditional logic analyzer. However, a device with built-in knowledge of the VMEbus timing specs can spot a violation immediately, by systematically checking the timing of each signal transition, as it occurs on the backplane. Such a device is called a *Bus Anomaly Detector.*

This chapter discusses how a VMEbus anomaly detector is designed, and how it can be used to detect a variety of common bus timing errors.

Common VMEbus timing violations

There are an infinite number of ways to violate the timing constraints given in the VMEbus specification. However, some violations are much more common than others. These common violations can be grouped into 4 classes:

- Control signal interlock violations.
- Control signal duration errors.
- Glitches and bus signal degradation.
- Violations of stability requirements for address and data signals.

Detecting control signal interlock violations

The VMEbus spec defines the protocols that VMEbus boards must abide by when communicating with each other. Each data transfer cycle consists of an interlocked sequence of transitions on strobe lines. For example, during a read cycle the data strobe line(s) are asserted by the master. The selected slave responds by putting data on the data bus and then asserting the DTACK* line. The master then rescinds the data strobe(s), after which the slave releases DTACK*. (See Figure 1)

Detection of a Rule 2.35 violation

This interlock is ensured by several rules in the VMEbus specification. One of these rules is Rule 2.35, which states that a master must not drive the first data strobe low until the slave that was addressed in the previous cycle releases both DTACK* and BERR*.

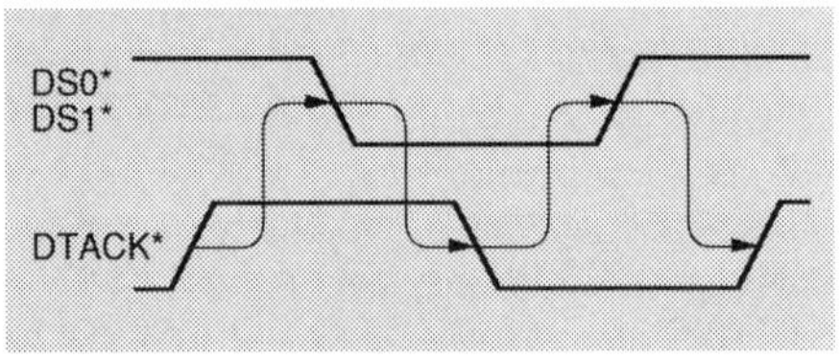

Figure 1

To detect a violation of Rule 2.35 you can use a fast JK flip-flop and some gates, as shown in Figure 2. (This circuit is implemented using a PLD.) The positive-edge-triggered clock input of this flip-flop (U1A) is sensitive to the rising (leading) edge of signal

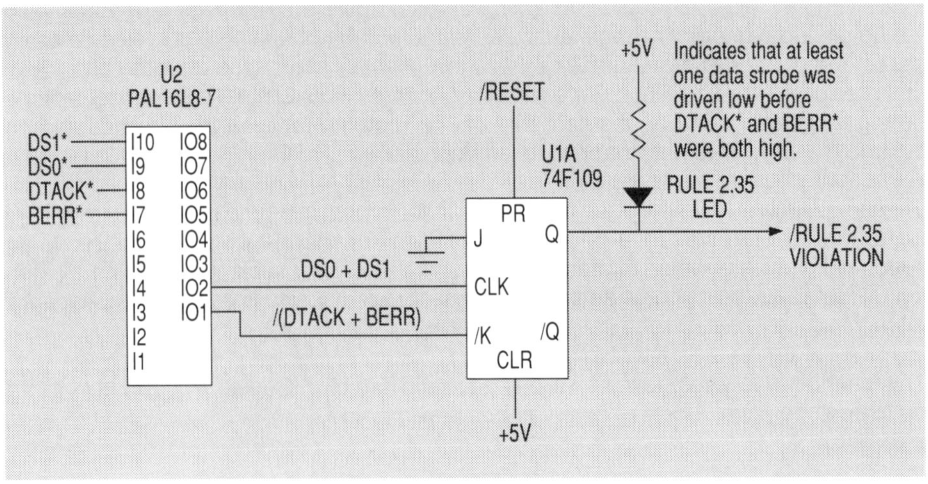

Figure 2

DS0+DS1, which goes high when either DS0* or DS1* goes low (active) on the VMEbus backplane. The flip-flop's /K input is connected to the signal /(DTACK + BERR), which goes low whenever DTACK* or BERR* is low (active) on the VMEbus backplane.

As Figure 2 shows, /(DTACK+BERR) is sampled by flip-flop U1A on the leading edge of the first data strobe. If either DTACK* or BERR* is active on the VMEbus, /(DTACK+BERR) will be low. This low level indicates that the first data strobe has gone active illegally, before DTACK* (or BERR*) has been rescinded from the previous cycle. When this happens, the Q output of U1A will go low, signalling an error condition, which is labelled */RULE 2.35 VIOLATION*. The timing diagram in Figure 3 illustrates a normal cycle, followed by a 2nd cycle, in which this violation occurs.

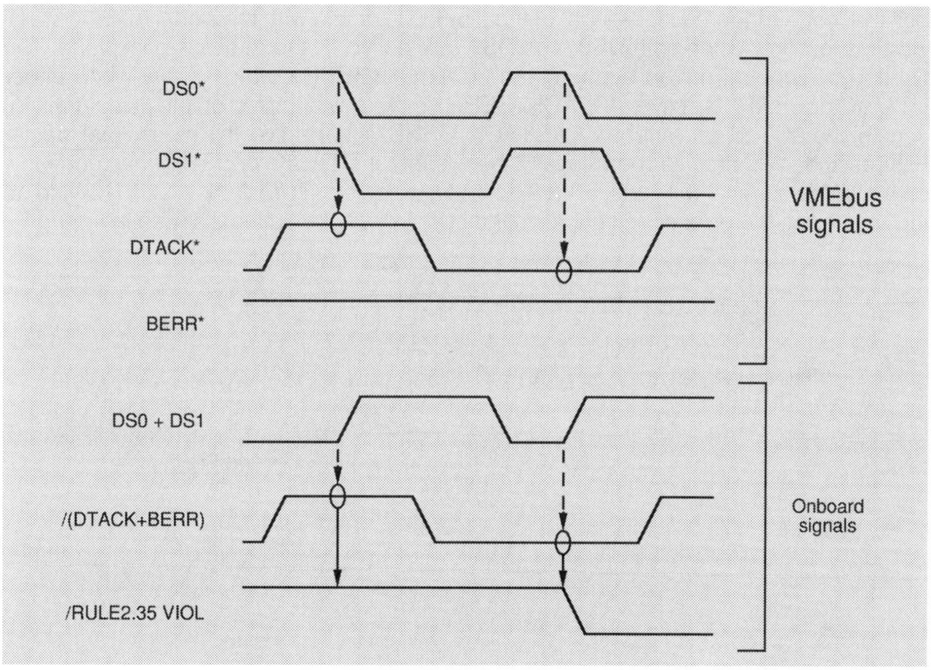

Figure 3

The error signal */RULE 2.35 VIOLATION* can be sent to an LED indicator. More important, this fast trigger signal (which is generated less than 20 nsecs after the first sign of the violation) can also be used to trigger a logic analyzer or bus timing analyzer, to stop it from continuously sampling the bus, and preserve the illegal bus cycle for later study.

Even a logic analyzer with relatively simple triggering capabilities can easily capture the cycle, since it only needs to monitor the level of the */RULE2.35 VIOLATION* signal from the anomaly detector. Its other probes can be connected to VMEbus lines DS0*, DS1*, DTACK*, and BERR* to capture the violation and display it.

In this way, the anomaly detector provides you with a direct lead to this violation, which you might not have suspected previously. Once you have observed the violation, you can attach additional logic analyzer probes to appropriate points on the board that's generating the problem, and then track the problem down to the gate level.

Detection of Rule 2.17 violation

Another possible violation of the interlocked timing relationship (previously shown in Figure 1) would be a violation of Rule 2.17. It states that, once a master drives its data strobe(s) low, it must not drive them high again without first seeing the addressed slave drive either DTACK* or BERR* low. An anomaly detector should therefore illuminate its *RULE 2.17 VIOLATION* LED (and generate a fast trigger) whenever it detects that either data strobe is rescinded before DTACK* or BERR* goes low.

This violation may be detected with a fast JK flip-flop, shown in Figure 4 as U1B. The positive-edge-triggered clock input of the flip-flop is driven with the signal /(DS0 * DS1), which is low only when both DS0* and DS1* are active on the VMEbus backplane. The flip-flop's J input is driven by the signal /(DTACK+BERR), which is low when either DTACK* or BERR* is active. /(DTACK+BERR) is thereby sampled on the rising (trailing) edge of the ANDed data strobes. If /(DTACK+BERR) is not active, this means that master has rescinded its data strobes illegally, before either DTACK* or BERR* has gone low on the VMEbus. This will cause U1B's /Q output (labeled */RULE2.17 VIOLATION*) to go low, signalling the error condition. (See the bus cycle in the timing diagram in Figure 5.)

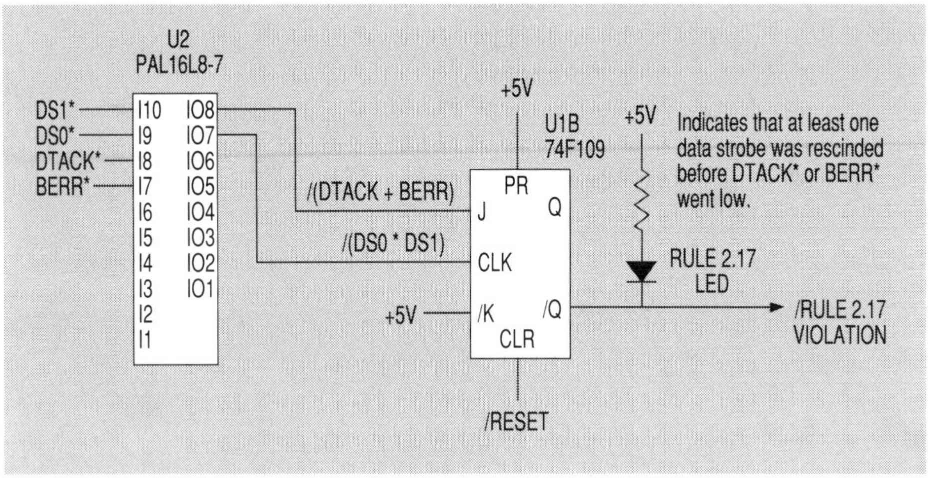

Figure 4

Many other interlocked timing violations can also be detected with circuitry similar to that described above. These include violations of Rules 2.20, 2.28, 2.44, 3.6, 3.10, 4.5, 4.6, and 4.45.

Detection of strobe-duration violations

The VMEbus spec prescribes minimum durations for assertion and de-assertion of various strobe lines, as well as minimum times between certain interlocked strobe edges. Violations of these rules can easily be detected, using fast JK flip-flops, precision delay lines, and fast PLDs, as shown below.

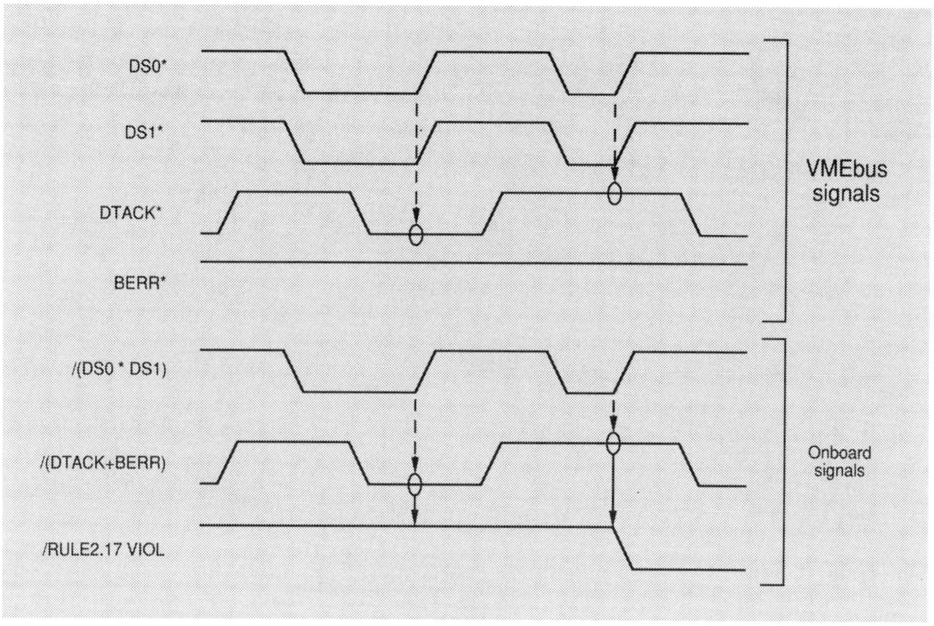

Figure 5

Detection of a Rule 3.7 violation
Rule 3.7 requires each requester to hold BBSY* active for at least 90 nsecs. Failure to follow this rule can cause the arbitration process to fail. This might result in a hung up system, or a cycle in which 2 masters erroneously drive the bus at the same time.

Even if the requester on each master board drives BBSY* correctly, the BBSY* signal on the backplane may still momentarily spike high if the onboard master simultaneously drives all of its address lines from high to low with high-current drivers. (This problem is sometimes called the *BBSY* glitch*.)

The severity of the BBSY* glitch depends upon the board design. In extreme cases it can be severe enough to cause the arbiter to think that BBSY* has gone high, indicating that the current master is relinquishing the VMEbus. The arbiter then issues a new bus grant to another master. This can be fatal, since the new master will then turn on its bus drivers and superimpose its own cycle on top of the current cycle.

A Rule 3.7 violation can be detected by a fast JK flip-flop (U3A in Figure 6). The signal /BSY (which is merely a buffered version of BBSY*) is fed into the positive-edge-triggered clock input of U3A, and is also fed into an 85-nsec delay line. The delayed signal (called /BSYDLY) is then routed to the J input of U3A.

When BBSY* goes low on the VMEbus, the clock input to U3A goes low. However, since U3A is clocked by the rising edge, it doesn't yet latch the level on its J input. Meanwhile, the same falling edge propagates through the delay line, and emerges after 85 nsec as /BSYDLY, and is then routed to the J input of the flip-flop.

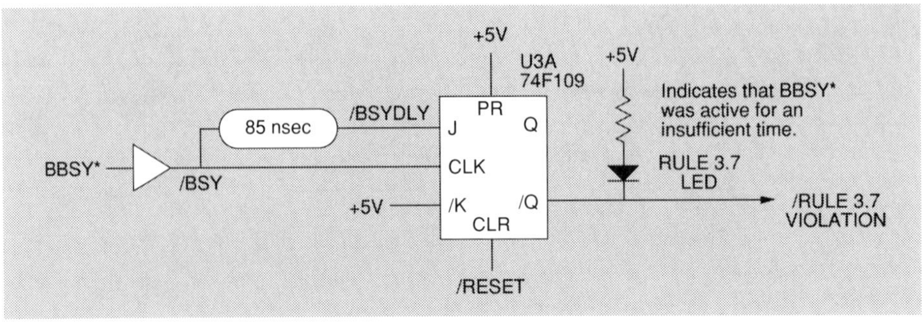

Figure 6

When BBSY* eventually goes high on the VMEbus, U3A latches the /BSYDLY signal on its J input. If this happens *before* the BBSY* falling edge has propagated entirely through the delay line, the flip-flop will latch a high level from its J input, and the /Q output of U3A (*/RULE3.7 VIOLATION*) will go low, signalling the violation.

The timing diagram in Figure 7 illustrates first a normal cycle, and then a cycle in which an error occurs due to improper design of the master's BBSY* generating circuitry. Figure 8 shows another illegal cycle, in which BBSY* is correctly generated, but suffers a *BBSY* glitch*, as described earlier. If this glitch occurs less than 85 nsec after the start of BBSY*, the anomaly trigger will recognize this glitch as a Rule 3.7 violation.

Note: Although Rule 3.7 specifies a minimum BBSY* duration at 90 nsec, the above-described anomaly detection circuitry (which is TTL-based) doesn't have enough precision to detect a duration of *exactly* 90 nsec. Several factors, (including delay line tolerance) introduce aggregate errors of approximately 5 nsec. Therefore, an 85 nsec delay line is used to make the anomaly detector slightly more lenient. It is guaranteed

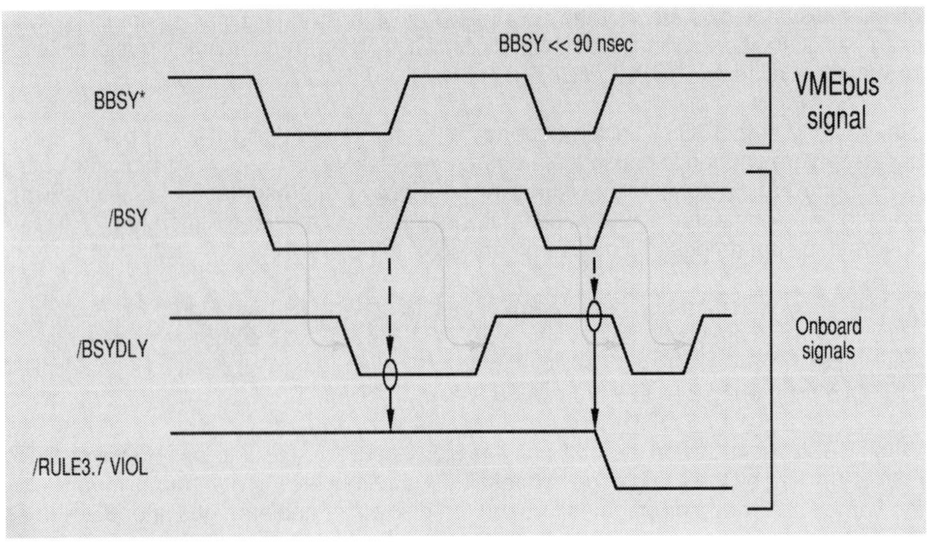

Figure 7

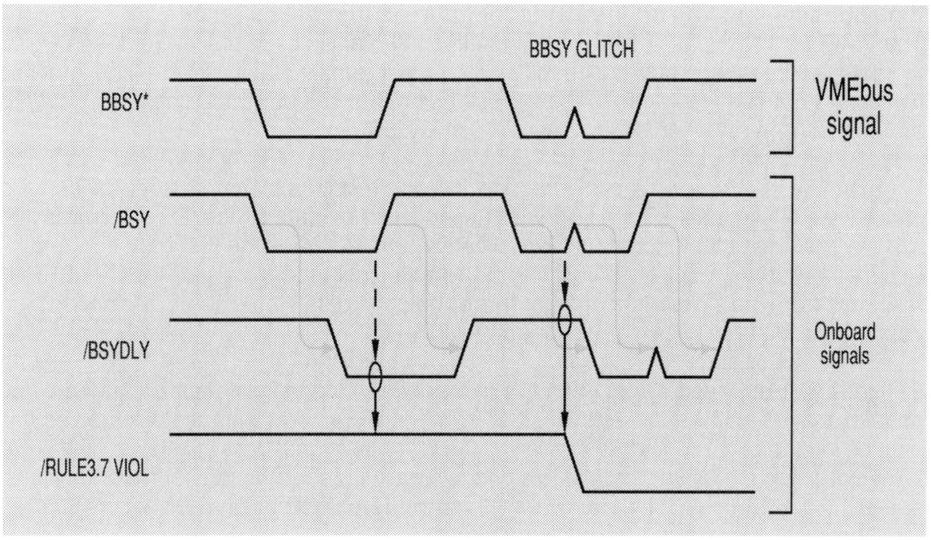

Figure 8

to detect an error only if BBSY* is low for less than 80 nsecs. Because of this design approach, no false error indications should ever occur. Admittedly, a VMEbus signal that is out of spec by a few nsecs might be missed by this anomaly trigger, but fortunately most VMEbus violations that are severe enough to cause system failures are more than just a few nsecs out of spec. Therefore anomaly detectors are still highly effective in detecting such errors.

Many other violations of specified strobe durations can be detected with circuitry similar to that shown above, including violations of Rules 2.31, 2.37, and 2.45.

In addition, violations of minimum delays between various strobe edges can be detected with similar circuitry, including those specified by Rule 2.39, 4.41 and 4.46.

Detecting glitches and bus signal degradation
The type of violation-detection circuitry described above will not only trigger on violations of the VMEbus specification rules. It might also trigger on certain noise conditions, such as severe ringing on strobe edges. This is because the bus anomaly circuitry looks for incorrect sequences of edges, and if ringing is severe enough, it will create extra transitions. For example, extreme ringing at the leading edge of the data strobes might generate a Rule 2.17 violation, since the extra rising edge in the initial ring looks much like a premature rescission of the data strobes, before the slave has driven DTACK* or BERR* low.

Other noise-induced problems (similiar to the BBSY* glitch described earlier) can also masquerade as timing violations, appearing as premature signal transitions. Glitches like this should **not** be ignored. If they can trigger the anomaly detector, they might also be seen by other VMEbus boards in the system. Extra transitions (due to ringing) on either the leading or trailing edges of strobes can cause malfunctions in boards that respond to these strobes.

Violations of stability requirements for address and data signals
The VMEbus specification clearly defines time intervals in each bus cycle during which the data, address, and address modifier lines must remain stable. These time intervals (which are defined in relation to the transitions on the various strobe lines) are covered by Rules 2.30, 2.34, 2.40, 2.41, 2.42, 2.48 and 2.56.

For example, during a write cycle the specification requires that a master provide valid data on the bus at least 35 nsec *before* it asserts the first data strobe. (This ensures that a slave sees this stable data at least 10 nsecs before it sees the first data strobe fall.) The master is then required to *hold* this data stable until it receives a DTACK* or BERR* response from the slave. If the master fails to hold the data stable during the above time interval, the slave may capture erroneous data.

During a VMEbus read cycle, the slave is required to place valid data on the data bus no later than the time it asserts DTACK*. (This ensures that the master receives this data no later than 25 nsec *after* it receives the DTACK* signal from the slave.) The slave is then required to *hold* this data stable on the bus until it sees the master rescind at least one of the data strobes. If the slave fails to hold data stable during the above time interval, the master might capture erroneous data.

Note: One of the most common errors in early VMEbus slave board designs was the failure to hold data stable until the master's data strobe disappeared. Instead, the slave would remove its data shortly after the address (supplied by the master) changed.

When you're using a traditional logic analyzer, its difficult to trigger on data instability errors. That's because the analyzer must be configured to trigger on either a rising edge, or on a falling edge, and you don't know which edge will be generated by the unstable data. To detect instability, you need to be able to detect any *changes* in data bus levels, without regard to the level of these lines. The same applies to detection of instabilities on the address bus and the address modifier bus.

Detection of instability on data Lines D00-D07
Figure 9 is a simplified diagram of the circuitry used to check for data line instability in a commercially available bus anomaly trigger (Ultraview's VME Bus Anomaly Trigger).

The error signal /LDER goes low (and also lights an LED) if any of the data lines (D00-D07) change during an interval when they should be stable. The portion of the VMEbus cycle during which the data lines are checked for stability depends upon whether the cycle is a read cycle or a write cycle.

During a write cycle, the VMEbus specification ensures that an addressed slave will receive stable data from 10 nsecs before the first data strobe goes low until it drives either DTACK* or BERR* low. The bus anomaly detector circuit shown in Figure 9 is slightly more lenient, due to the accuracy limitations of its own components. It merely requires that the data be stable from 0 nsec before the time DS0* goes low until the time DTACK* goes low.

PLD U34 generates a low level on signal /DST8, which indicates the period during which stability is to be checked. The timing diagram in Figure 10 shows that the signal /DST8 is low from the time DS0* goes low until DTACK* goes low.

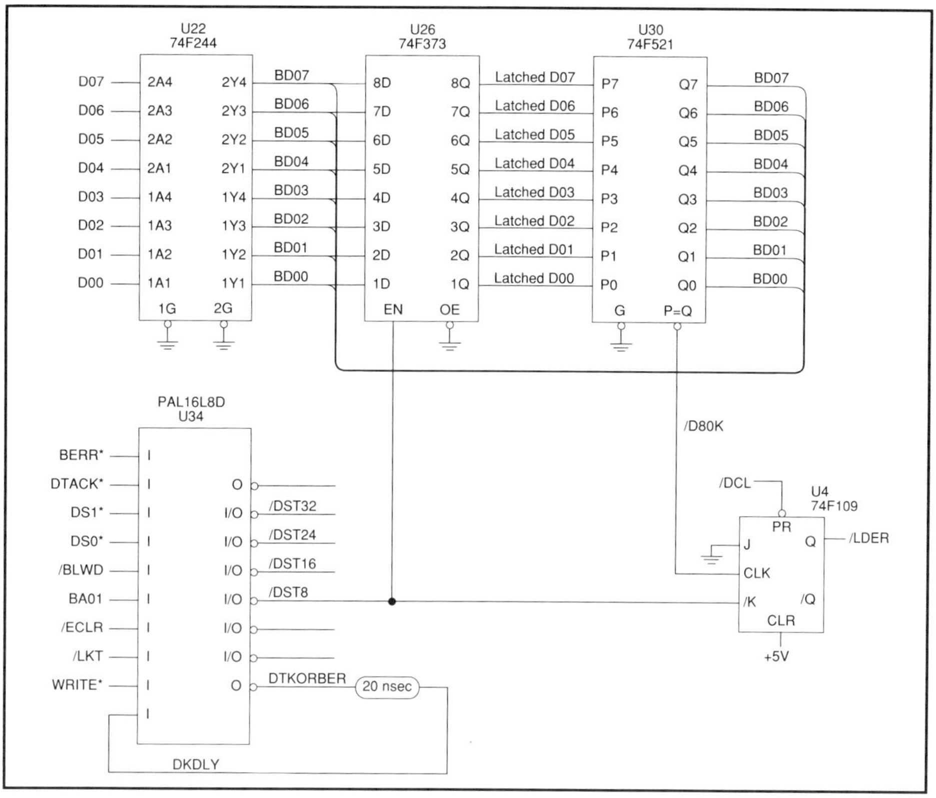

Figure 9

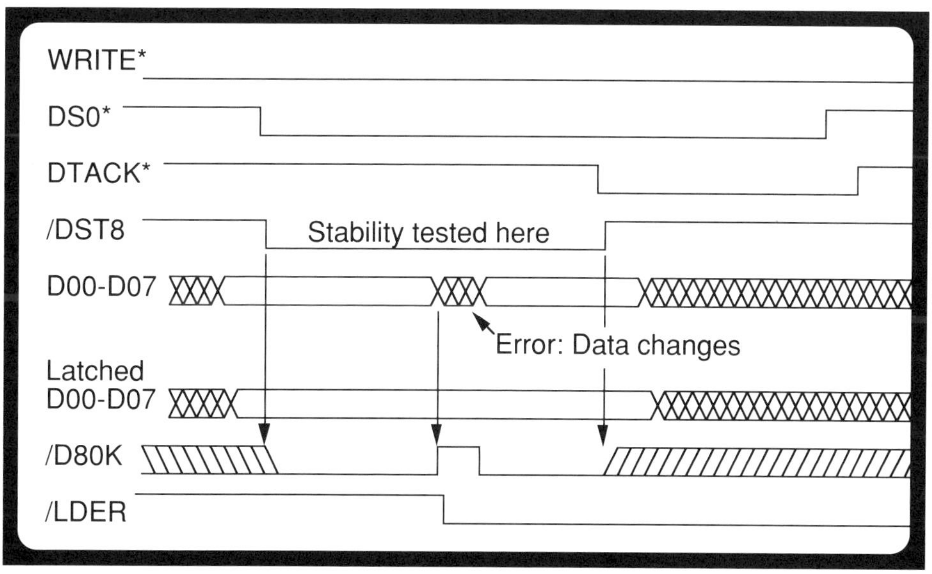

Figure 10

During a read cycle, the VMEbus specification ensures that a master will receive stable data no later than 25 nsecs after DTACK* goes low, and should see this data remain stable through the time that it drives its data strobe high again. The timing diagram in Figure 11 shows that the signal /DST8 is low from approximately 25 nsecs after DTACK* goes low until the time when DS0* goes high. The PLD logic equations (in U34) for generating /DST8 are shown in Figure 12.

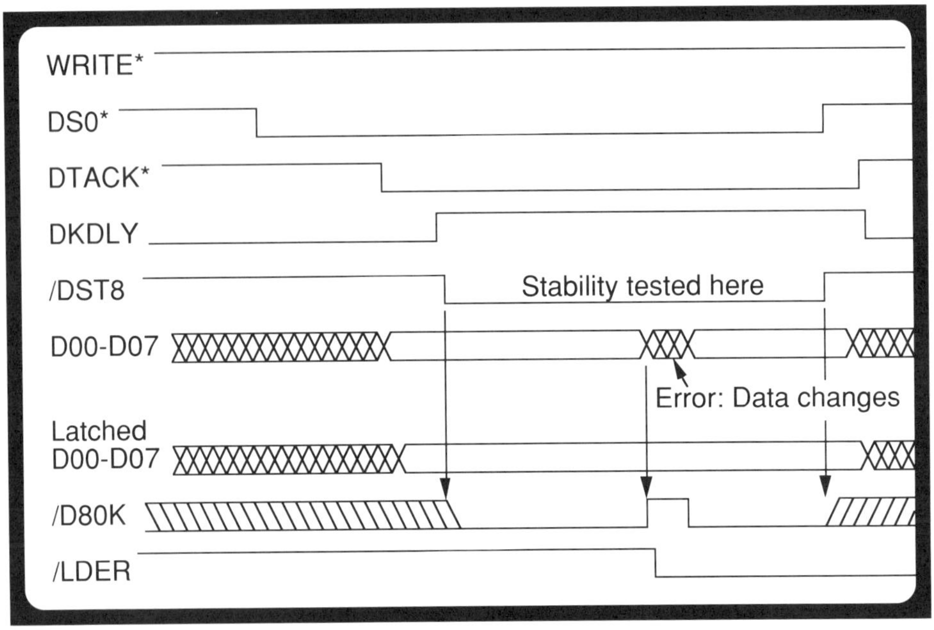

Figure 11

```
DST8 =      ;LOW WHEN DATA BITS D0-7 SHOULD BE STABLE
  ;WRITE CYCLE:
  WRITE *         DS0 * /A01 * LWORD * /DTACK * /BERR ;QUAD BYTE WRITE
+ WRITE * /DS1 * DS0 * /A01 * LWORD * /DTACK * /BERR ;UAT1-3 WRITE
+ WRITE *         DS0       * /LWORD * /DTACK * /BERR ;WORD, ODD BYTE WRITE
  ;READ CYCLE:
+ /WRITE *         DS0 * /A01 * LWORD * DTACK * DKDLY ;QUAD BYTE READ
+ /WRITE * /DS1 * DS0 * /A01 * LWORD * DTACK * DKDLY ;UAT1-3 READ
+ /WRITE * DS0                * /LWORD * DTACK * DKDLY ;WORD, ODD BYTE READ
```

Figure 12

You can see, from these PLD equations, that stability is required on D00-D07 for:

- quad byte (32-bit) data transfers
- word (16-bit) data transfers
- odd byte data transfers
- unaligned byte 1-3 data transfers

These cycle types are described in detail in the VMEbus specification.

Referring to Figure 9, U22 merely buffers VMEbus data lines D00-D07. The buffered data is fed into the data inputs of latch U26 (74F373). Comparator U30 (74F521) compares the buffered data inputs with the outputs of latch U26. During the time when /DST8 is high, the buffered data flows through latch U26, which is then in transparent mode. However, as soon as /DST8 goes low (meaning that stability is to be tested), the outputs of latch U26 are frozen.

As long as the incoming data does not change from the values latched in U26, the comparator output /D8OK will remain low, indicating equality between the incoming data and the latched data. However, should any of the 8 VMEbus data lines change (even for only a few nsecs) the buffered incoming data will momentarily differ from the latched data, and /D8OK will momentarily go high. This rising edge will clock flip-flop U4. If this high-going transition occurs while /DST8 is low (that is, while the data lines are supposed to remain stable) U4's Q output (/LDER) will go low, indicating a VMEbus timing violation.

Circuitry similar to the above is used to detect instability in lines D08-D15, D16-D23, D24-D31. Using somewhat different PLD equations, similar circuitry can also be used to detect instabilities in lines LWORD*, A01-A07, A08-A15, A16-A23, A24-A31 and AM0-AM5.

The above circuit descriptions were necessarily brief. There are some other factors that you need to take into account when designing a bus anomaly trigger. For example, when measuring the minimum pulse width of VMEbus strobe line signals, you need to take into account the fact that rising and falling edges may have different propagation delays through the onboard logic. In the case of VMEbus, the minimum pulse widths are long enough that this differential delay is not a problem. However, to use circuits like those shown above for a faster bus (such as Futurebus+) the differential delays will have to be reduced to a maximum of 1 to 2 nsecs, making design of the anomaly detector more challenging.

The Ultraview VBAT

The Ultraview VBAT, (shown in Figure 13) is a double-height VMEbus-compatible Bus Anomaly Trigger, which can be installed in a VMEbus system by simply plugging it into a spare VMEbus slot. It then monitors bus operation to detect VMEbus timing violations as they happen during actual system operation.

Figure 13

In addition to the circuits described earlier in this chapter, the VBAT has over 100 individual violation detector circuits. Since all of these circuits operate in parallel, they can provide a fast, effective screen for catching most VMEbus timing violations.

Figure 14 shows a trace from a storage scope, which was triggered with the Ultraview VME Bus Anomaly Trigger. Using the VBAT, the customer was able to detect (and subsequently correct) an instability in the D01 data bus line, which occurred during some VMEbus write cycles.

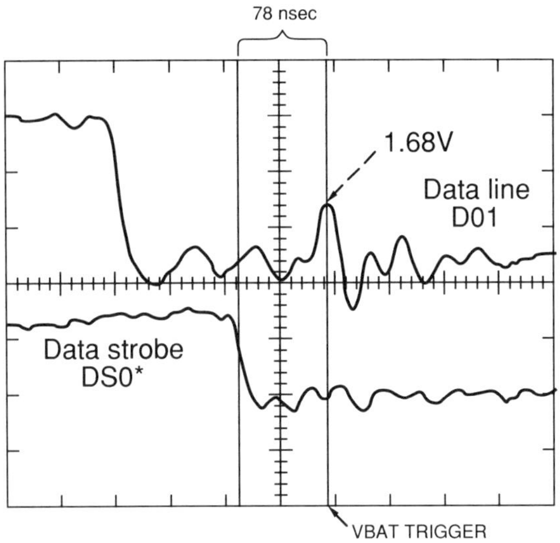

Figure 14

Conclusion

Bus Anomaly Detectors are useful for screening new designs for serious violations. They also speed up system debugging by automatically recognizing cycles in which timing violations occur. Of course, anomaly detectors cannot check for every type of violation, so they're not a substitute for the careful design of new boards. Nevertheless, they are remarkably proficient at ferreting out timing violations that the designer might other-wise overlook.

The anomaly detection methods described here may be extended to newer buses, and should be especially valuable for designers tackling the new, powerful, and more complex Futurebus+.

Joel Libove is president of Ultraview. He holds a BSEE from Cornell University, and an MSEE and Ph.D. in Electrical Engineering from the University of California at Berkeley. Prior to founding Ultraview Corporation, Joel was a consultant to several companies, where he specialized primarily in biomedical electronics. He has several patents, and has won 3 product design awards.

CHAPTER 4

How To Use Soft Architectures To Build Versatile Diagnostic Tools

By Laurie Burger

Systems based on VMEbus are getting more complex, as engineers take full advantage of its multiprocessing capabilities. These systems generally make better use of the VMEbus bandwidth than earlier systems. However, in many cases the traffic is too great for a single bus to handle, and secondary buses (such as VSB) are being used. With the recent introduction of VMEbus "extensions" (such as VXIbus and VME64) bus architectures are getting more elaborate every day. In addition, with several processors in the system, the protocols needed to carry messages between them add another layer of software complexity. To deal with all this complexity, system engineers need more versatile diagnostic tools. In this chapter Laurie describes a versatile bus analyzer with a "soft architecture", which allows it to be restructured to suit the job at hand.

Board-level VMEbus analyzers have been available for several years. They allow you to see the backplane signal levels in an easy-to-read format. These analyzers include 4 basic functions:

- a *bus interface*, to sample signal levels from the backplane
- a *trigger/trace qualifier*, to select when and what samples to store
- a *trace memory*, to store the information
- a *user interface*, typically provided through a serial port

Early board-level analyzers were designed mainly to let you see what information was being passed over the backplane. To accomplish this, the address lines and data lines were sampled once each bus cycle, at a time where they were guaranteed to be valid. By knowing what information was being passed, you could determine what was going on in the higher layers of system protocol, which were usually implemented in software.

These early analyzers were designed to capture a "stream of consciousness" when a simple trigger condition was met. However, the recent trend is toward much more elaborate features such as:

- high-speed sampling of waveforms.
- increased trace memory size.
- sophisticated triggering functions, to allow more selective capture of information.
- waveform displays.
- backplane short circuit testing.
- a state machine, for generating signal sequences on the backplane.

Early VMEbus analyzers sampled the VMEbus signals at 16 to 20 Msamples/sec, and could retain about 2 Ksamples. The new analyzers can sample at rates of 50 Msamples/sec: almost 3 times as fast. This higher sampling rate obviously demands more trace memory. It would seem that a memory 4 times as large would be adequate. However, if the trace memory is made even larger, you can use more sophisticated triggering capabilities to answer more difficult questions.

For example, you may want to examine bus activity between 2 widely separated events. (One event starts the trace, and the other event stops it.) Another case when you need plenty of trace memory is when the system is behaving erratically. You may want to trace several occurrences of a particular event, and then look for any differences. Yet another case is when some external event is occurring at unpredictable intervals. You might want to capture several events, and then compute the time between these events to look for a pattern.

To provide these capabilities, recent analyzers have been designed to retain up to 256 Ksamples. This increased trace memory size (along with the higher sampling rates) also allows you to analyze the timing of VMEbus signals much more precisely. You need this kind of precision to determine signal duration and signal stability: both of which determine whether the VMEbus will operate properly. This relatively high resolution can also help you analyze event duration, interval timing, and the timing of interrupt routines.

Driving VMEbus signal lines

An analyzer can be equipped with VMEbus line drivers and with a state machine to control them. The state machine can then be loaded with microcode that specifies:

- which backplane lines are to be asserted.
- the duration of each assertion.
- how long until their next assertion.

The resulting sequence of signal transitions on the bus can be used to test other VMEbus boards. For example, the state machine might be coded to generate write cycles, with a particular set of timing parameters. By varying the timing parameters within the limits prescribed by the VMEbus spec, the analyzer can determine whether or not a slave board has been designed properly.

All of these features have pushed the capabilities of board-level analyzers far beyond those of traditional general-purpose logic analyzers. In fact, they have become versatile interactive testers. These testers can be used during system debug, or they can be permanently installed in a VMEbus system, where they permit periodic automatic testing, under computer control.

Using a VMEbus analyzer

Plug-in VMEbus analyzers are easy to install, and easy to use. You no longer need to know which backplane pin carries the read/write signal, or the various address bits, since the board is already hardwired to connect to each of these signals. The levels and transitions of each bus signal line are displayed on a CRT screen, alongside its signal name.

VMEbus analyzers are now being used not just by hardware engineers, but also by software and system engineers, who would normally cringe at the thought of connecting and configuring a traditional general-purpose logic analyzer.

Figure 1, shows how you might configure a VMEbus analyzer to trigger whenever location 80000 is read. You simply type the address into the appropriate field on the screen. When some master on the VMEbus initiates a read cycle to that location, the analyzer will:

- *detect* the high level on the WRITE* line
- *decode* the address
- *generate* the trigger
- *begin sampling* the backplane signal levels, and storing them in its onboard trace buffer.

```
*A
MATCH SPECIFICATION A
                                   TRANSFER          INTERRUPT                 BUS
      ADDRESS   DATA      AM AS LW DS ER AK      REQ    ACK ACK    GRANT  REQ  BSY
                                   10            7654321            3210  3210
----------------------------------------------------------------------------------
R  >xx080000 <xxxxxxxx XX  X  X XX  X  X     XXXXXXX X    X        XXXX  XXXX  X
```

Figure 1

Because this customized user interface makes it so easy to select trace and trigger qualifiers, you can quickly narrow down your trace to include only the information you need.

Data Collection

VMEbus board-level analyzers typically collect data in either of 2 modes:

- synchronous
- asynchronous

In the *synchronous* mode the analyzer samples the address bus and the data bus each time DTACK* or BERR* goes true at the end of the cycle. This mode is useful when you only need one sample per data transfer cycle.

In the *asynchronous* mode the analyzer samples the bus repetitively, at a fixed time interval. For example, it might sample the bus at a 50 MHz rate, with one sample taken every 20 nsec. The synchronous mode is useful when you need to know the *precise signal timing*. For example, you might use it to determine:

- whether the address strobe is staying low for the required minimum time.
- whether the data bus becomes stable before or after the slave drives DTACK* low.
- whether the new master is waiting for BBSY* to go high before taking control of the bus.

There are 2 popular formats for displaying the timing information captured by the analyzer:

- The *state display* format
- The *waveform display* format

Figure 2 shows a state display. It presents the address and data bus levels in hexadecimal form, and is most useful for studying the address and data values being passed over the backplane.

```
**********************TRACE STARTED AT 10:23:02 ON 11/10/89**************
*    CYCLES BEFORE TRIGGER = 2048          CYCLES AFTER TRIGGER = 0000
**************************************************************************
                              TRANSFER             INTERRUPT             BU
FRAME R  ADDRESS    DATA     AM AS LW DS ER AK   REQ  ACK ACK  GRANT REQ BS
      W (INTERVAL) (SEC.MS.US.NS)               7654321    I/O  3210  3210
---------------------------------------------------------------------------
-0007 R 024E5782  E5E5E5E5  3F  0  1  0  1  0 ------- 1   1    1011  1011 1
        (35.2US)  (000.023.000.000)
-0006 W 00001000  0000567A  3E  0  0  0  1  0 ------- 1   1    0111  0111 1
        (35.9US)  (000.023.035.200)
-0005 R 01000010  44444444  2E  0  1  0  1  1 -----2- 0   1    1101  1101 1
        (40.8US)  (000.023.071.100)
-0004 R 01000014  5FEA3400  3F  0  1  0  1  1 ------- 1   1    1101  0101 1
        (41.1US)  (000.023.111.900)
-0003 W 182000FE  FE021109  1F  0  0  0  1  0 ------- 1   1    0111  0111 1
        (43.2US)  (000.023.153.000)
-0002 R 004E2222  D4E66F53  2F  0  0  0  1  0 -6---2- 1   0    1110  1110 1
        (45.0US)  (000.023.196.200)
-0001 R 0043222A  5E632F76  2F  0  0  0  1  0 -6---2- 0   0    1110  1110 1
        (46.1US)  (000.023.241.200)
 TRIG R 00432232  C4A55267  2F  0  0  0  1  0 -----2- 0   0    1110  1110 1
        (46.7US)  (000.023.287.300)
 **********************TRACE STOPPED AT 10:24:05 ON 11/10/89*************
```

Figure 2

Figure 3 shows a *waveform display*. It's most useful for determining the relative timing between signal transitions on the backplane.

Active versus passive bus analyzers

VMEbus analyzers have one huge advantage over general-purpose analyzers. Since they are *plugged into* the VMEbus backplane, they can *assert* signals on the bus. By making the analyzer board an *active* element in the VMEbus system, a whole new range of possibilities opens up. For example, in addition to monitoring bus traffic, the analyzer can generate a system reset, and can then load a boot routine.

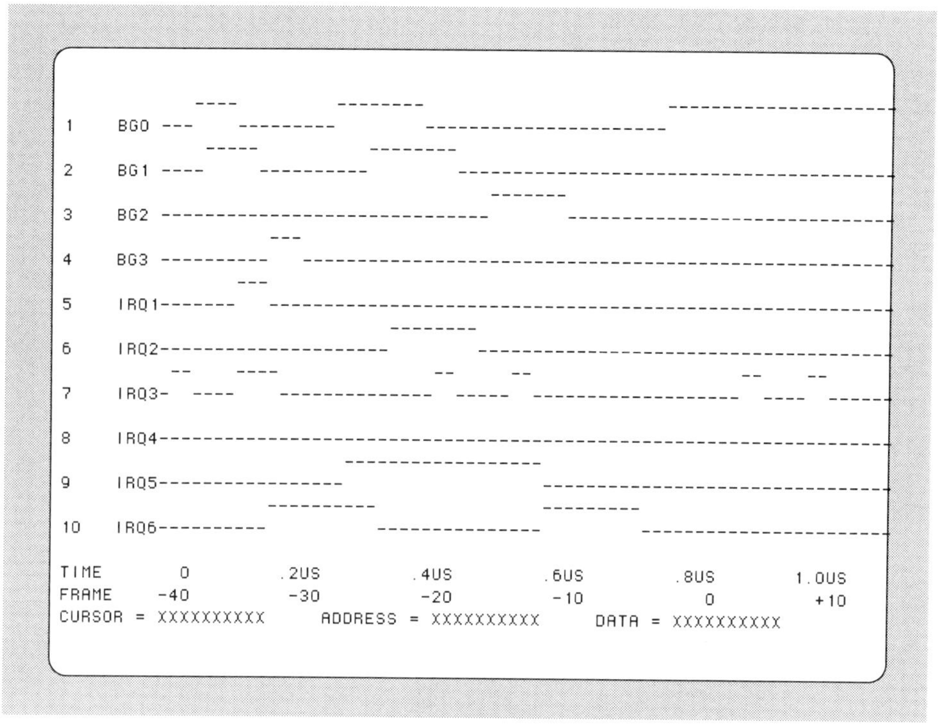

Figure 3

This ability to reinitialize a system is invaluable for remote field service support. A field engineer can remotely access a system, for diagnostic purposes, and not only determine what's wrong, but possibly even remedy the problem. Consider the configuration shown in Figure 4. Here, a VMEbus analyzer with 2 serial ports is installed in a VMEbus chassis. One of its serial ports is connected to a modem (which can be called by a remote system) and its other port is connected to the serial port on the system's MPU board.

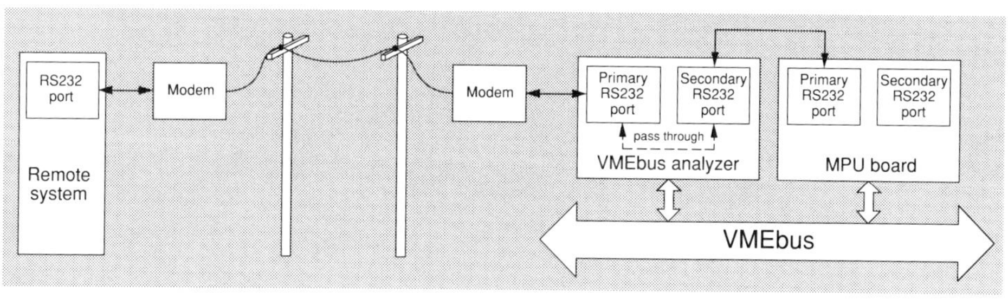

Figure 4

In a typical scenario, the remote system:

- *calls* the analyzer, via the modem.
- *resets* the local system through the analyzer, if necessary.
- *configures* the analyzer to collect data.
- *switches* the analyzer to *transparent mode*. (In the transparent mode the analyzer passes through the serial character stream from the modem to the serial port on the MPU board.)
- *commands* the firmware monitor on the MPU board, to run a diagnostic routine. (This routine initiates VMEbus cycles that trigger the analyzer.)
- *switches* the analyzer out of the transparent mode. (This allows the remote system to examine the contents of the trace memory on the analyzer board.)
- *performs* any other necessary follow-up diagnostic procedures, such as using the analyzer to generate an interrupt, and monitoring the system's response.

Clearly, this technique provides a very powerful tool for remote field support.

Computer versus human interface

One of the obstacles to using most analyzers for remote field support is that the user interface was designed primarily for *human* operation. This requires that all communication between the analyzer and the operator be encoded in ASCII.

When an analyzer is being accessed by a remote *computer*, it's possible to communicate in a more compact fashion. For example, the remote computer could download a parameter block, instead of setting up the analyzer with a series of ASCII commands. In addition, samples taken by the analyzer (and stored in the trace memory) could be uploaded (over the serial link) to the remote system as binary data.

Signature analysis

Using these features, a remote system could do signature analysis on a failed system. The remote system:

- configures the analyzer in the ailing system to collect patterns during several different diagnostic routines.
- initiates execution of these diagnostic routines.
- uploads the signature data from the trace memory.
- compares the uploaded signature data to signature data previously captured from a properly functioning system.

Hard versus soft hardware architecture

In early VMEbus analyzers the triggering and data capture logic was implemented with TTL comparators or PALs. However, with the advent of RAM-programmable gate arrays, analyzer boards can now be designed to have a *soft architecture*. This allows you to dynamically restructure the *hardware architecture* of the analyzer by simply changing the pattern stored in the RAM of the gate array. If the board is designed to copy the pattern into the gate array's RAM from an onboard ROM, the analyzer's architecture can be restructured by simply replacing the ROM.

The 5010 analyzer

Mupac's 5010 system analyzer uses this soft architecture approach. Figure 5 is a block diagram of the board.

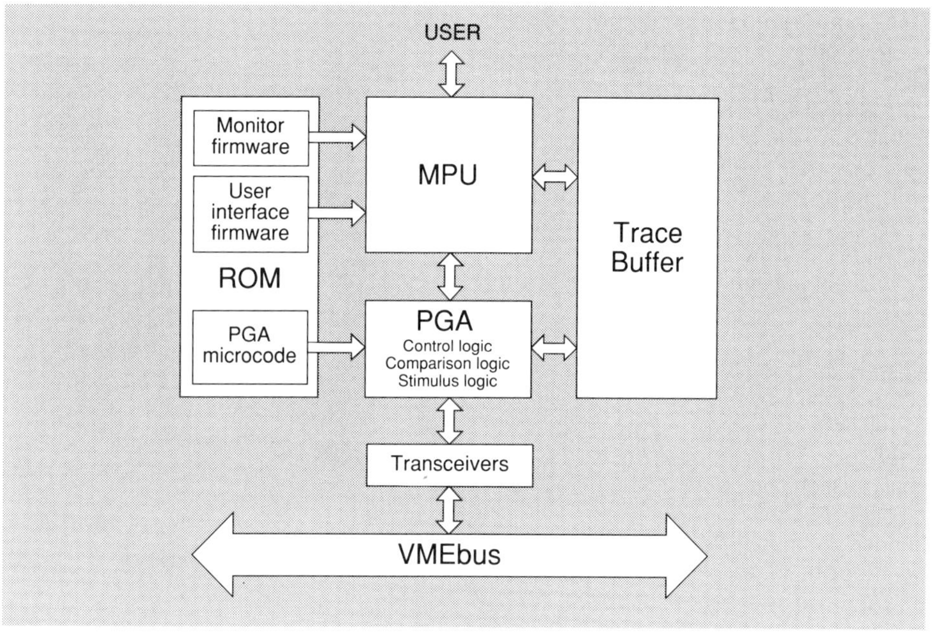

Figure 5

Upon power-up the onboard processor executes *monitor firmware* in the ROM, which initializes the serial ports and copies the programmable gate array's pattern (microcode) from the ROM into the gate array's RAM.

The *user interface firmware* also resides in the ROM. It provides both a human interface and an interface for a remote host computer. The remote host computer can set up the analyzer by downloading parameter blocks, and can upload the trace memory contents in binary form. Additional space in the ROM permits the inclusion of customized code. This code might be used to provide special routines, designed specifically to diagnose your system from a remote location.

Using a soft architecture to analyze 2 buses

A full-featured analyzer for a bus like VMEbus can cost as much as $7,000. Many VMEbus systems use a dual bus VME/VSB architecture. While VSB is not as elaborate as VMEbus, a VSB analyzer could cost in excess of $3,000. That adds up to over $10,000: not a very cost effective way to provide bus analysis for your system. You could save some money by using less capable bus analyzers. However, an analyzer with a soft architecture can be used to support both buses. All you need is a *bus adapter board*, and new firmware.

For example, Figure 6 shows a VSB adapter board. The VMEbus analyzer board is plugged into the adapter board, and then the entire assembly is plugged into a slot in the VMEbus chassis. The adapter board includes hardware that can be configured to route either the VMEbus or the VSB signals to the analyzer board. Because the 5010 analyzer board has a soft architecture, it can be configured to analyze either bus, without any hardware changes.

This same approach could be used to support other buses, such as VXIbus, Multibus II, Futurebus+, and SUNbus. In the case of VXIbus, Multibus II and Futurebus+ the adapter board serves a dual purpose: it routes the backplane signals to the analyzer board, and it provides some additional depth to position the front panel of the analyzer board flush with the front panel of the card rack.

Figure 6

Note: In the case of VXIbus, the adapter board would also need to have additional register hardware, if you want the analyzer to participate in the auto-configuration procedure during power-up.

In each case, the VMEbus analyzer can be thought of as the basic analyzer board, and the adapter board as a customized backplane interface. Firmware defines the analyzer board's soft architecture, and provides a customized user interface for each bus. Once you have a VMEbus analyzer, you can customize it for other backplanes for under $1,000.

Summary

Today's VMEbus systems are much more complex than those just a few years ago. The VMEbus backplane is being used to move ever-increasing amounts of data at faster and faster rates. In many cases secondary buses are needed to carry excess data traffic. When such a system fails, you need specialized tools to find and correct the problem efficiently. This is true not only during the development phase, but also after the system has been installed in the field.

In our increasingly computer-dependent society, the ability to promptly restore a failed system to normal operation is often vital. Many users are quite willing to pay the additional cost of an embedded analyzer, if it allows prompt diagnosis. An analyzer with a soft architecture is ideally suited for this application.

This accounts for the fact that VMEbus analyzer sales are growing rapidly. This trend toward the use of embedded analyzers will continue, and will accelerate as even more complex systems (based on Futurebus+) are developed.

Laurie Burger is the product marketing manager for Mupac. Prior to joining Mupac she had 10 years experience as a systems design engineer at Texas Instruments and Cheseborough Ponds. She has a BSEE and an MSEE from University of Rhode Island and is a member of IEEE and SME.

CHAPTER 5

How To Use The Right Tools To Integrate Your VMEbus System

By John Simpson

Integration of a VMEbus system composed of boards from many different vendors can be a tedious job without the right tools. In the early stages, you may spend a lot of time trying to track down board configuration errors, such as a misplaced jumper. Later, you may need to be able to see what addresses are being broadcast over the bus. Even after the system seems to be running properly, you may encounter transient or intermittent failures. In this chapter, John describes the various types of bus analysis tools, and shows where each fits into the system integration process.

Hardware and software engineers developing and troubleshooting VMEbus systems frequently need to review the address, data, and control information passed between boards over the backplane. General-purpose logic analyzers are, to a limited degree, able to capture this information. However, to integrate and debug your system efficiently, you need dedicated tools, designed specifically for VMEbus.

Dedicated VMEbus analyzers are typically designed as double-height boards, which can be inserted directly into the VMEbus backplane. This approach eliminates the time-consuming task of physically connecting dozens of probes, as you would have to do with a general-purpose logic analyzer.

These dedicated analyzers are microprocessor-controlled instruments, and include onboard firmware that translates the captured address, data, and control signal levels into formatted summaries of bus cycles, which are then displayed on an ASCII terminal.

Unlike a general-purpose logic analyzer, a dedicated VMEbus analyzer can also be designed to understand the VMEbus protocols, and can automatically detect violations of the VMEbus timing specifications.

The 3 phases of debugging

When you design a VMEbus board, you will typically go through 3 debugging phases, each requiring a different kind of tool:

Phase 1: In the 1st phase, you must verify that the board is able to drive and respond to the handshake lines properly. To do this, you need a high-speed *VMEbus timing analyzer.*

Phase 2: In the 2nd phase, you need to get the low-level software working on the board. To do this, you need to be able to view the address and data transfers over the VMEbus. For this purpose, you need a *VMEbus state analyzer.*

Phase 3: In the 3rd phase, you take your board through a "shake-down cruise" to verify that it works with a variety of other VMEbus boards. For this purpose you need a *VMEbus anomaly trigger*, which acts as a violation detector.

VMEbus timing analyzers

A timing analyzer samples the VMEbus signals asynchronously, based on an internal high-speed free-running clock. These samples are then presented on a CRT screen as digital waveforms, allowing you to determine the timing relationship between different signals. (Figure 1 shows how these waveforms might be displayed on an enhanced ASCII terminal.)

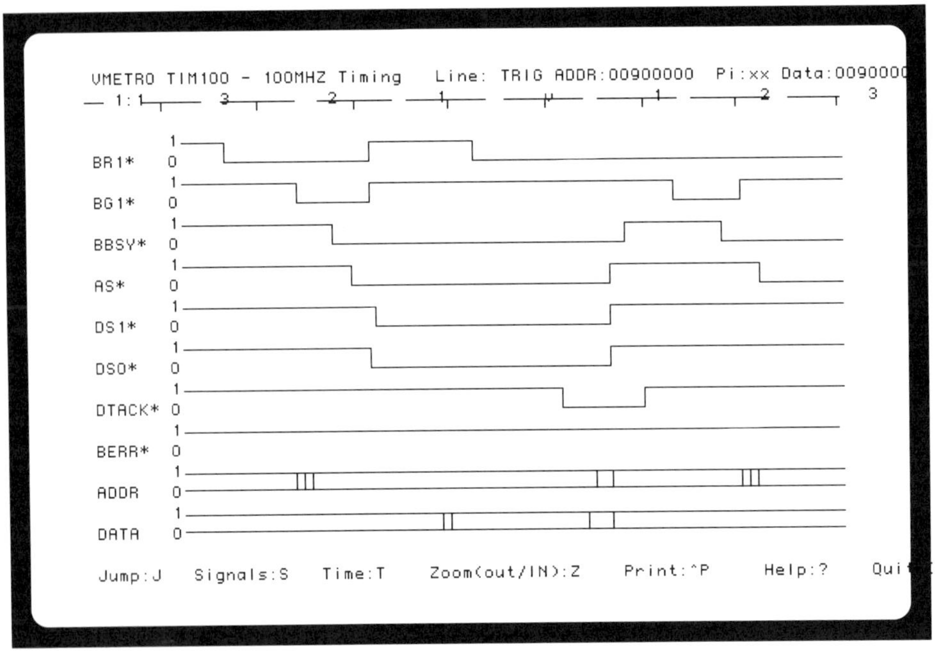

Figure 1

To guarantee that every transition of every sampled signal is captured, the analyzer must sample at least twice as fast as the fastest switching signals in the system.

The fastest switching signals on the VMEbus are usually the bus grant lines (BGxIN*/ BGxOUT*) and the cycle termination signals (DTACK* and BERR*). Between 2 neighboring boards in a short, lightly-loaded backplane, these handshake signals might have a duration as short as 20 nsec. This demands a 10 nsec (100 MHz) sampling rate for proper detection.

Of course, a sampling rate of 100 MHz on a 50 MHz signal doesn't provide a very accurate representation of the original waveform. (For example, the high time and low time on the displayed waveform are always multiples of 10 nsec.) However, the signals in actual VMEbus systems normally switch at much slower rates. For example, the BGxIN*/BGxOUT* and the DTACK* and BERR* signals are likely to have high times and low times that range from 150 to 1000 nsecs. For this reason, a 100 MHz timing analyzer is usually adequate for debugging the bus interface logic on VMEbus boards.

Metastability effects
Because the timing analyzer samples the backplane signals asynchronously, it may occasionally sample a signal just as that signal makes a transition. This might lead to metastability of an input latch on the analyzer board. When this happens, the output of the latch hangs midway between a high and a low level, or may even *oscillate* back and forth between a high and a low level, until it finally settles.

Since the latch output might still be unstable when it gets routed into the trace memory, it might be stored as either the level prior to the transition, or as the level subsequent to the transition. Whenever the *previous* level is stored, the analyzer will display the signal transition *one clock interval later* than the time at which it actually took place.

VMEbus state analyzers
A *VMEbus state analyzer* differs from a *VMEbus timing analyzer* in the way that it derives its sampling clock. Instead of using an internal free-running clock, a bus state analyzer uses the transitions of the VMEbus handshake lines. In doing this, the analyzer synchronizes itself to VMEbus backplane activity, as opposed to the asynchronous operation of a VMEbus timing analyzer.

A VMEbus state analyzer samples all of the address lines, the data lines, and the control signals once during each data transfer cycle on the VMEbus. It then uses these samples to produce an alphanumeric display of the bus cycles on an ASCII CRT terminal. (See Figure 2) Because this display presents the address and data, without cluttering the display with protocol details, it's ideal for software debugging, and for hardware/software integration.

In fact, state analyzers are indispensable for getting VMEbus hardware and software working together. In addition to giving you insight into backplane activity, a state analyzer permits you to easily verify that your jumper configurations provide the desired base address, interrupt levels, and arbitration levels.

The signal transitions used to sample data from the VMEbus must be carefully chosen. Not all of the VMEbus signals can be sampled at the same time. For example, the levels of the bus grant lines change, after the new master drives BBSY* low, to take control of the bus. Therefore, the bus grant levels must be latched on the falling edge of BBSY*. Later when the address bus, the data bus, and the rest of the control signals on the backplane are sampled (on the falling edge of DTACK* or BERR*) the contents of these bus grant latches are stored with them in the trace buffer.

Filtering
To preserve space in the trace memory, VMEbus state analyzers support highly selective cycle storage. This is done with a *filter* (sometimes called a *qualifier*). A filter allows the

```
VMETRO VMEbus Trace Group 1 of 3  Synchronous sampling  All lines displaye

       ! TIME      BUS  ADDRESS     DATA      R/W  SIZE  STAT   IRQ*    IACK*
       ! rel.    LEVEL                                       7654321 OC I/O
  -----!-------------------------------------------------------------------
==>TRIG ! 00.00      3   12060000  00004E75   R   LWORD  OK    1111111  1  1
   0001 ! 2.046 us 3   12060004  0000xxxx   R   WORD   OK    1111111  1  1
   0002 ! 1.798 us 3   12050078  00654326   R   LWORD  OK    1111111  1  1
   0003 ! 0.558 us 3   1205007C  xxxx43xx   R   UBYTE  OK    1111111  1  1
   0004 ! 0.558 us 3   1205007D  xxxxxx26   R   LBYTE  OK    1111111  1  1
   0005 ! 0.558 us 3   1205007E  xxxx0328   R   WORD   OK    1111111  1  1
   0006 ! 0.558 us 3   12050088  8765432A   R   LWORD  OK    1111111  1  1
   0007 ! 0.558 us 3   1205008C  8700002C   R   LWORD  OK    1111111  1  1
   0008 ! 0.558 us 3   12050002  xx6542xx   R   UNAL2  OK    1111111  1  1
   0009 ! 50.58 ms 2   40070000  FD654326   W   LWORD  BERR  1111111  1  1
   0010 ! 0.558 us 1   32050001  xxxxxx42   R   LBYTE  OK    1111111  1  1
   0011 ! 0.558 us 1   32050004  576545xx   R   UNAL3  OK    1111111  1  1
   0012 ! 1.054 us 1   32050008  3765422A   R   LWORD  OK    1111111  1  1
   0013 ! 0.558 us 1   3205000B  xx6E442A   R   UNAL3  OK    1111111  1  1
   0014 ! 0.558 us 1   32050010  8765462B   R   LWORD  OK    1111111  1  1
   0015 ! 1.054 us 1   32050014  77654328   R   LWORD  OK    1111111  1  1

 Time:T  Group:1,2,3  Jump:X  Search:S  Extract:E  Help:?  Print:^P  Quit:(
```

Figure 2

analyzer to store some cycles, while discarding others. For example, a filter might only allow storage of write cycles at a given bus grant level, and to a given address.

Filtering reduces the number of cycles you need to inspect in order to follow the execution of a program. The filters are also protocol sensitive, by virtue of their ability to distinguish between cycle types, such as read-modify-write or block transfer cycles.

Filtering is accomplished by comparing each sample with a user-defined pattern (called an *event*) to decide whether a sample is to be stored or discarded.

Software debugging
While *VMEbus timing analyzers* are designed mainly for *hardware* debugging, *VMEbus state analyzers* are designed mainly for *software* debugging. Hardware events are defined in terms of *signal transitions*, while software events are typically defined in terms of *bus cycles*. As a result, the triggering schemes of VMEbus state analyzers are much more complex than for VMEbus timing analyzers.

A VMEbus state analyzer should have several *word recognizers*, that can be configured to generate a trigger on a certain *sequence* of events. This feature is very helpful when you want to trigger on a call to a particular subroutine, but *only* when it's called from a particular address, instead of on every call.

How fast should a VMEbus analyzer be?
Since the sampling intervals of a VMEbus state analyzer are not based on a periodic clock, you can't determine time intervals directly, from a display of the cycle-by-cycle

samples. However, there are many situations when you want to know how much time elapsed between 2 events when they were displayed. To allow this, VMEbus state analyzers often store a *time tag* with each sample. An onboard timer provides a time value (for storage with each sample) that indicates the elapsed time since a reference event. These time tags can later be used to compute time intervals.

To capture a sample of each bus cycle, a bus state analyzer must be able to sample at least as fast as the cycle rate of the backplane. For the VMEbus, the propagation delays in the handshake sequence give a minimum cycle time of at least 80 nsec, for a cycle rate of 12.5 MHz. A VMEbus state analyzer must be able to sample the bus at speeds up to 12.5 MHz.

Providing bus analysis for VSB

Many of today's VMEbus systems employ the user-defined pins on the **a** and **c** rows of the P2 connector to provide a VSB (VME Subsystem Bus). To really see what's going on in the system, you will need to have another state analyzer for these VSB signals.

Obviously, this VSB state analyzer should derive its sampling clock from the VSB. However, since VSB activity is often closely tied to VMEbus activity, the VMEbus state analyzer should be designed to trigger the VSB state analyzer, and vice versa.

Since the VSB multiplexes address and data on the same set of signal lines, a VSB state analyzer should have 2 sets of input latches for each of the multiplexed VSB signal lines. By clocking one latch during the address phase, and the other latch during the data phase, a demultiplexing is achieved. The resulting captured samples can then be displayed, as shown in Figure 3.

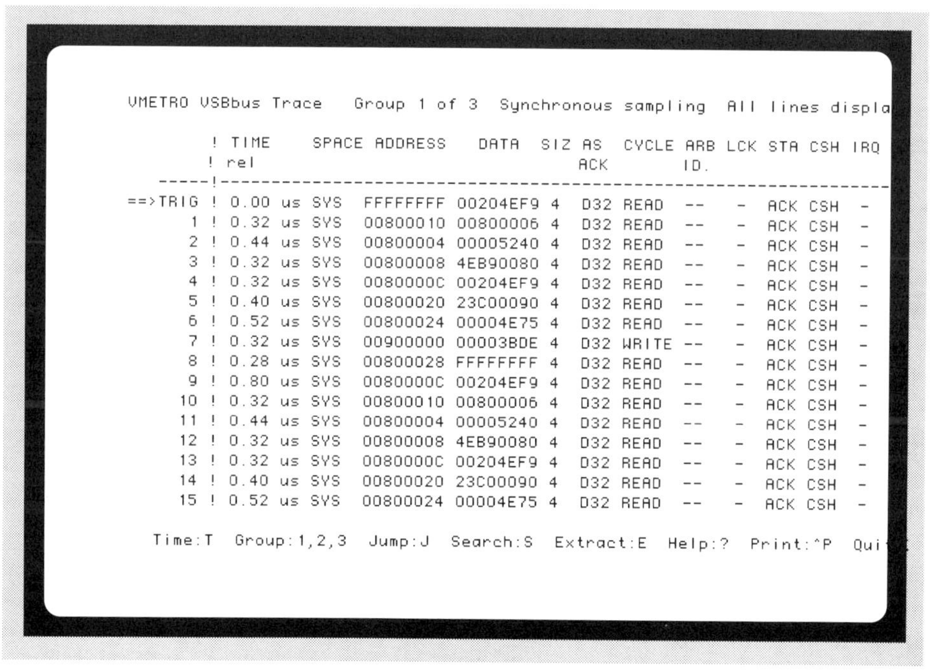

VMETRO VSBbus Trace   Group 1 of 3   Synchronous sampling   All lines displa

	! TIME rel	SPACE	ADDRESS	DATA	SIZ	AS ACK	CYCLE	ARB ID.	LCK	STA	CSH	IRQ
==>TRIG	! 0.00 us	SYS	FFFFFFFF	00204EF9	4	D32	READ	--	-	ACK	CSH	-
1	! 0.32 us	SYS	00800010	00800006	4	D32	READ	--	-	ACK	CSH	-
2	! 0.44 us	SYS	00800004	00005240	4	D32	READ	--	-	ACK	CSH	-
3	! 0.32 us	SYS	00800008	4EB90080	4	D32	READ	--	-	ACK	CSH	-
4	! 0.32 us	SYS	0080000C	00204EF9	4	D32	READ	--	-	ACK	CSH	-
5	! 0.40 us	SYS	00800020	23C00090	4	D32	READ	--	-	ACK	CSH	-
6	! 0.52 us	SYS	00800024	00004E75	4	D32	READ	--	-	ACK	CSH	-
7	! 0.32 us	SYS	00900000	00003BDE	4	D32	WRITE	--	-	ACK	CSH	-
8	! 0.28 us	SYS	00800028	FFFFFFFF	4	D32	READ	--	-	ACK	CSH	-
9	! 0.80 us	SYS	0080000C	00204EF9	4	D32	READ	--	-	ACK	CSH	-
10	! 0.32 us	SYS	00800010	00800006	4	D32	READ	--	-	ACK	CSH	-
11	! 0.44 us	SYS	00800004	00005240	4	D32	READ	--	-	ACK	CSH	-
12	! 0.32 us	SYS	00800008	4EB90080	4	D32	READ	--	-	ACK	CSH	-
13	! 0.32 us	SYS	0080000C	00204EF9	4	D32	READ	--	-	ACK	CSH	-
14	! 0.40 us	SYS	00800020	23C00090	4	D32	READ	--	-	ACK	CSH	-
15	! 0.52 us	SYS	00800024	00004E75	4	D32	READ	--	-	ACK	CSH	-

Time:T   Group:1,2,3   Jump:J   Search:S   Extract:E   Help:?   Print:^P   Qui

Figure 3

The bus anomaly trigger

If you are adding a new VMEbus board to your system, you should verify that it is performing within the VMEbus timing specifications. However, because of the asynchronous handshaking protocols used on the VMEbus, each board's signal timing is dependent (to some degree) upon the timing of others in the system. Even if the board works correctly when interacting with some boards, that's no guarantee that it will not violate the VMEbus timing specifications when interacting with others.

To locate this kind of problem, you need a board called a *bus anomaly trigger*. This board checks the timing of the backplane signal transitions against the VMEbus timing specifications. The anomaly trigger board actually consists of a collection of *violation detectors*. Each detector is specially designed to detect violations of a particular rule in the VMEbus specification.

Anomaly triggers are useful for detecting problems such as bus ringing, glitches, illegal sequences of control signals, address or data instabilities, and multiple simultaneous bus grants.

When using an anomaly trigger you don't need to know the cause of a system failure. The anomaly trigger detects any bus timing violation, and generates a trigger signal. This trigger signal can be fed into a VMEbus timing analyzer (or into an oscilloscope) to reveal the actual cause of the problem. Many violations detected by an anomaly trigger would be impossible to localize with a conventional analyzer, since you would have no idea what to trigger on.

Performance and statistical analysis

When the hardware debugging has been completed, you will often want to measure and "tune" the system for optimal performance. This may involve measurement of:

- interrupt response times
- program execution times
- memory access times

These times are easily measured by means of the time tag discussed above. Software utilities can read these time tags from the trace memory, and use them to do performance computations. For example, a *transfer rate utility* could calculate a data transfer rate (either in Mtransfers/sec or in Mbytes/sec) for each bus arbitration level.

You may also want to do statistical analysis on the sampled data. This is best done with histograms. For example, you might want to display histograms that show:

- the statistical distribution of accesses, within a specified address range.
- the statistical distribution of bus traffic among the 4 bus grant levels.
- the relative frequency of certain user-defined events.

To be useful, these statistics should be based on samples taken over an extended period of time. There are 2 different methods for collecting and analyzing the data:

- the software method
- the hardware method

The software method
The software method uses a 2-step, repetitive process. In step 1 the analyzer continuously stores samples, until the trace memory is full. Then, in step 2, the onboard processor reduces the captured samples to statistical values. This 2-step process is then repeated again and again, with the trace memory getting overwritten each time.

Since all the analysis is done in software, this technique offers a great deal of flexibility. However, the sampling process must be periodically suspended, to allow the processor to analyze the data in the full trace memory.

This means that the statistics reflect the bus activity *only* during those times that the trace memory is being filled. If bus activity tends to be transient (as it often is in real-time systems) the resulting statistics might not be representative of overall bus activity.

The hardware method
The hardware method uses *word recognizers*. A word recognizer is a circuit that compares each bus sample to a user-specified word. If a match occurs, it increments a hardware counter. Since the sampling and comparison can be done continuously, all bus activity is reflected in the statistics. The counters accurately reflect bus activity, even when that activity is very transient.

The VBT-321
Figure 4 shows VMETRO's VMEbus Tracer (VBT-321), which is a double-height VMEbus state analyzer that can be customized by the installation of any 1 of 4 piggyback modules:

- A VMEbus anomaly trigger, called the VBAT-PB
- A 100 MHz VMEbus timing analyzer, called the TIM100-PB
- A VSB state analyzer, called the VSB-PB
- A 256 Kbyte trace memory expansion, called the XMEM-PB

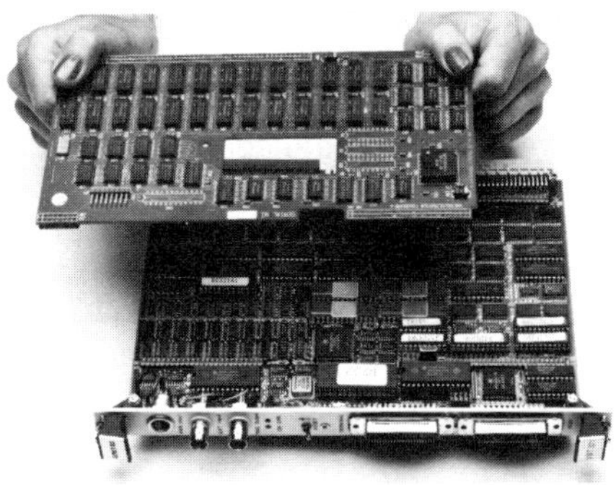

Figure 4

A typical scenario
Suppose your lab is going to integrate a newly designed I/O board of their own into an already-functioning VMEbus system, consisting of a number of off-the-shelf MPU boards, memory boards, and I/O boards.

Suppose you install the board in the system, bring the system up, and then attempt to initialize the board from one of the MPUs in the system. You get no response when accessing the registers on the new board. Where do you go from here?

To find the problem, you plug in a VBT-321 VMEbus state analyzer with a TIM100-PB piggyback VMEbus timing analyzer.

You want to capture a detailed view of the write cycle to the new board's control registers. To do that, you set up the VBT-321 to trigger whenever you address a control register on the new I/O board. Since you aren't sure whether your new board is responding with a DTACK*, you configure the VBT-321 to sample asynchronously.

You then write to the I/O board again, triggering the VBT-321. The VBT-321 compares and displays a timing waveform that indicates that the new board is not responding with DTACK*. An inspection of the board reveals that the DTACK* jumper (which is configured for the RAM access time) has not been installed. You then install the jumper, and the I/O board responds to VMEbus cycles in a normal manner.

You now write to the board's control registers, setting up the board to do a block transfer to VMEbus memory. However, when you then read that memory, you find that the block transfer did not write data into the specified locations. You suspect that the data was written to the wrong address. To investigate that problem, you configure the VBT-321 to take a bus trace in its *synchronous* sampling mode.

Since you have no idea what addresses the board is actually generating, you set the VBT-321 to trigger on any longword block transfers on bus grant level 2. The VBT-321 then captures information which shows that the address the I/O board is placing on the VMEbus is the *complement* of the desired address! You solve this problem by modifying the software driver to complement the address before writing it into the I/O board's control register.

Having eliminated all obvious bugs, you take the new board on a "shakedown cruise". Under heavy load, and with several active bus masters, you notice that the I/O board occasionally fails to do a block transfer to system memory when commanded to do so. Since you have no idea what is going on, you install the VMEbus anomaly trigger (VBAT) on the VBT-321. If the abnormal operation of the system is caused by violations of the VMEbus timing specifications, the VBAT-PB will display a message about this on the screen, as well as illuminating an LED on its front panel.

To capture any cycles with timing violations, you configure the VBT-321 state analyzer to trigger whenever the piggyback VBAT-PB detects a violation. The resulting trace indicates that the failure always occurs on write cycles to the I/O board's control register, when the bus grant 2 level is active. To further narrow down the problem, you replace the VBAT piggyback board with the TIM100 piggyback board. By selectively sampling the write cycles to the I/O board's control registers on bus grant 2, you determine that the master is failing to provide adequate data set up time.

This example shows how the different analyzers described in this chapter can be used to find a solution to all the problems that arise during board debugging, and during integration of your VMEbus system.

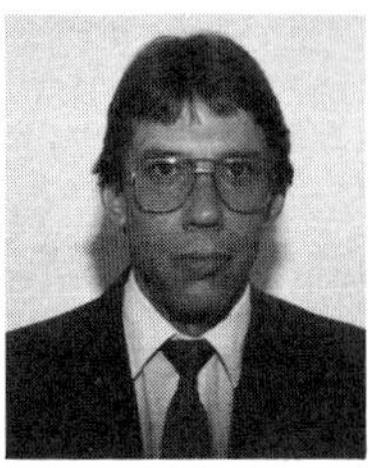

John Simpson is vice president of VMETRO. He came to VMETRO from Planning Research in February of 1990. His experience includes system architecture design, distributed operating system design, and development of several large-scale multiple-processor systems for the Department of Defense. John got his first bus analyzer experience in 1982, while working on PRC's proprietary backplane system. He also managed the design and development of the 1st Multibus II analyzer, which was introduced by PRC in 1989.

CHAPTER 6

How To Use C To Debug Your VMEbus Hardware

By Terry Shankland and John Black

Hardware standardization has now progressed to the point that you can purchase plug-in printed circuit boards (and assemble a very powerful multiprocessing system) without having to design any hardware. Software standardization is also progressing toward a similar level. For example, if you purchase one of the widely-used real-time kernels or operating systems, you can typically get the software drivers you need to incorporate an I/O board into your system. If all goes well, you can bring the system up quickly, and begin developing your application code. However, if the system fails to work, you may be forced to write assembly code, in order to fiddle with the status and control register bits on the I/O boards. In this chapter Terry and John suggest another alternative: the use of a high-level language to debug your hardware.

Interface standards are getting a lot of attention these days. Two of the earliest interface standards in our industry (both of which we take for granted) were the DIP packaging concept and the TTL signal levels. These hardware standards allowed components manufactured by different companies to be easily interfaced to PC boards, and to each other. Other hardware interface standards that have also facilitated the growth in computer technology are the serial RS-232 standard, the various backplane standards (such as STDbus, Multibus, and VMEbus), and local area network standards, such as Ethernet.

With all of this hardware standardization, you can now do your hardware designs at a much higher level. For example, you can assemble very complex microcomputer systems (or even networks of microcomputer systems) from off-the-shelf hardware components. However, if you assemble a complex system and it fails to work (which it often does), you will need some powerful debugging tools to debug that system.

The tools for debugging software have improved a lot in recent years. However tools for debugging hardware have lagged far behind. Logic analyzers were once quite useful, but, because of the prefetching and caching of the new high-performance 32-bit microprocessors, you may have a difficult time figuring out what a processor is doing, just by watching its bus interface. In addition, the behavior of complex I/O controller chips (such as Ethernet controller chips and SCSI controller chips) are largely dependent upon their stored internal states, and you can't always deduce those states from recent bus activity.

Using assembly language to create diagnostic programs
Because of the lack of good tools, you may be forced to debug complex systems with only a ROM-based debugger, and a simple assembler/disassembler. If you're under schedule pressure to "just get it working", the assembly language diagnostic programs that you develop will likely be poorly documented. This (coupled with the fact that it takes a lot of code to track down bugs in complex systems) results in the creation of throw-away code, which can only be understood by its creator, and gets discarded as soon as the source of the problem is found.

There's another problem with writing your diagnostic code in assembly language: it can only be run on the type of microprocessor for which it was intended. For example, if I write an assembly language diagnostic routine to test an intelligent VMEbus hard disk controller board in my 68020 system, you can't use that routine to test the same hard disk controller board in your 80386-based system. That discourages VMEbus companies from providing comprehensive diagnostic software with their VMEbus boards.

Another problem with assembly language is that it forces you to work at a low level. For example, if you want an intelligent disk controller to read sector 4 from the disk, you may have to build a command block specifying all sorts of parameters. Failure to initialize some critical parameter in the command block may cause a crash, diverting your attention away from the real bug that you're trying to find.

The logical way to hide all of this detail is to put it into a subroutine. However, passing lots of parameters to a subroutine in assembly language can be tedious and error prone. As a result, you may be tempted to use jumps instead of subroutine calls, and just to make all variables global. This leads to the creation of spaghetti code. As the debug process continues (and as your diagnostic program grows) you find the code increasingly difficult to modify. When you make a change to one portion, another portion stops working. As the code grows larger, these interactions may become so severe that you may have to scrap the whole program, and start over with a more structured approach.

Features of an idealized debugging language
If assembly language isn't the answer, then what is? Let's start by listing the features that an ideal debugging language would provide. This ideal language would...

- allow you to express your diagnostic algorithm in a processor-independent fashion. This feature would encourage VMEbus board manufacturers to provide processor-independent diagnostic software for their boards, which you could run in your VMEbus system, regardless of what type of microprocessor you have.
- allow you to write programs which are self-documenting. For example, you would like to be able to use meaningful mnemonics for variables, instead of register numbers. You would also like to be able to see what parameters get passed to subroutines. These features would allow you to easily understand (and add to) your debugging software, as you pursue a particular bug.
- allow you to make subroutine calls (and pass parameters to those subroutines) easily, so you can hide the details of low-level code, and see the high-level structure of your program.
- allow you to handle strings efficiently, instead of writing loops of assembly code to handle them character-by-character. (This is useful for displaying messages on the screen, and accepting commands from the keyboard.)

- allow you to do decimal and hexadecimal arithmetic operations efficiently, so that you don't have to add, subtract, multiply, or divide parameter values, byte-by-byte.
- allow you to use powerful control structures (such as loops) so you can expend your effort on exercising the system hardware, instead of debugging loop counters (Oh... I see what the problem is... `loop_count` should have been initialized to 1 instead of 0!) or exit conditions (I guess it just never occurred to me that `buffer_count` might go negative!).
- allow you to assign names to the bits and the fields of control and status registers, so you can access them with easily understood variable names.
- allow you to define, access, and pass custom data structures efficiently. For example, if you need to pass a command block to an I/O controller board, or if you need to interpret (and possibly display) a response block from the controller board, you would like to access each field in the command block, using meaningful mnemonics.

High Level Languages
High Level Languages (HLLs) like Fortran, Basic, Pascal, and Cobol have been designed for programmers who want to write processor-independent programs. In general, these languages assume that the programmer doesn't know very much about the internal resources (registers, memory map, I/O controller logic, etc) of the computer on which the program will run, and (to a large extent) doesn't even want to know. The processor-independent source code file of a HLL program is translated into the native machine language of the processor upon which it runs by either a compiler or an interpreter.

HLL compilers
A compiler is a *language translator* that turns your HLL program into machine language before the program is to be executed. It does this by substituting predefined blocks of machine code instructions for each HLL statement. These blocks of machine code instructions perform the operations specified by the HLL statements. For example, the statement...

```
string_a = string_b
```

...might cause the compiler to insert a block of machine language instructions that copy an ASCII string from an area of memory (designated as string_b) into another area of memory (designated string_a).

In some cases, the predefined block of machine code may be large. One good example is the multiplication of 2 floating point numbers. In that case, the compiler simply inserts a subroutine call to a floating-point-multiply routine. At the same time, it records the fact that the floating point multiply routine is needed, so the routine will later get copied from a library of routines on disk into the final machine code program file. There are 2 reasons for doing it this way:

- Floating point multiply routines take up too much memory to be inserted everywhere a multiply operation is done.
- If a given HLL program doesn't ever multiply any floating point numbers, the floating point multiply subroutine will not be included. This results in a much smaller machine language program file.

HLL interpreters
An interpreter, on the other hand, keeps a copy of the HLL program (or a compacted form of it) in memory while executing it. Instead of inserting blocks of machine code into a file for later execution, an interpreter calls a subroutine for each HLL operation that it encounters, as it steps through the program. Because the program is executed as it is being translated (instead of first being translated by a compiler, saved in a file, and later executed) we call this *immediate execution*.

Using High Level Languages to create diagnostic programs
Looking back at the list of features we would like to have in an idealized debugging language, we find that many of them are provided by high level languages. Obviously, HLLs allow you to represent your program's algorithm in a processor-independent fashion: that's what they are designed to do. Most HLLs also allow you to write programs which are largely self-documenting, by using meaningful mnemonics for variables and for subroutines. (One exception to this is Basic, which allows only single character variable names, and uses a cryptic GOSUB XXX statement to call a subroutine. However, newer versions of Basic have improved on this.)

All high-level languages allow you to call subroutines, and the more recently developed HLLs also allow you to formally declare what variables should be passed to a subroutine. This is very helpful when you're trying to understand what a subroutine does. For example, if you're familiar with Pascal, you can probably figure out that a Pascal function which starts with...

```
FUNCTION average(X, Y: real): real;
```

...should be passed 2 real number variables, and will return a real number result, which is the average of the 2 numbers.

By using subroutines properly, you can hide the low-level details of your program code. That allows you to see the high-level structure of your program much more quickly. For example, Listing 1 shows a subroutine that performs a complicated activity, but still allows you to see how its algorithm is structured at a high level. Each line of the program calls a subroutine. Because of the names given to each of these subroutines, you can readily figure out what they do.

All HLLs allow you to input and output character strings, and do arithmetic operations in some fashion. All of them allow looping, based upon a loop count, or based upon some exit condition.

Given all of this, you might think that HLLs are ideally suited to writing diagnostic software. However, they aren't. As we said earlier, HLLs are designed to allow the programmer to write processor-independent (and hardware-independent) programs. Since the address maps of different systems look quite different, HLL compilers and interpreters go to great lengths to ensure that the programmer doesn't access memory locations, status registers, or control registers based upon their address. For example, if you define a variable `number_of_turnips`, the compiler doesn't allow you to specify where the value of that variable will be stored in the memory map.

There are valid reasons for doing this. In large multiuser systems it would be suicidal to leave the assignment of the memory map to the individual programmers.

Listing 1

```
PROCEDURE buy_milk(amount: gallons)
BEGIN
  get_car_key(key);
  start_car(key);
  drive_to(store);
  enter_store;
  acquire_milk(amount, price);
  pay_for_milk(price);
  exit_store;
  start_car(key);
  drive_to(home);
END;
```

If 2 programmers happened to choose the same location for storage of some variable, their programs would interact in an unpredictable way.

Since HLLs insulate programmers from the address map, they make it impossible to access status and control registers. For example, if you need to build and pass a command block to an intelligent peripheral controller, you will find that you can build the block in memory, but you can't write the address of that block into the appropriate register on the controller board.

We are led to the conclusion that what we really need is a quasi-HLL, which provides the conveniences of a HLL, but allows us to read from (and write to) any location that we choose within the address map. It turns out that there is such a language: it's called C.

The C programming language

C is a unique programming language. You are probably aware that the UNIX operating system is written in C, but you may not know that the original version of UNIX was actually written in assembly language! In the early 1970s, Dennis Ritchie designed C to allow the people at Bell Laboratories to rewrite the UNIX operating system in a form that could be more easily understood.

The power of the C language allowed the UNIX systems programmers to think at a much higher level. This, in turn permitted them to add major new features to UNIX, the most important being the inclusion of multiprogramming. (UNIX, up to that point, could support only one user program in core memory at a time!) But the most important result was that the structure of the UNIX software became much easier to understand.

Expressing your algorithm in a processor-independent fashion

When C was designed, portability was not the prime concern: both the assembly language version of UNIX, and the transliterated C version of UNIX ran on a PDP-11. However, portability has turned out to be one of C's most important features, and UNIX's success has largely been due to the fact that its C source code can be easily recompiled to run on machines other than the original PDP-11.

The fact that...

- C was developed primarily to make it easier for systems programmers to express and understand their algorithms, and...
- C turned out to provide processor independence...

...tells us that using assembly language to write large programs is a flawed strategy. It intermixes the programmer's attempt to express an algorithm with the process of adapting it to the machine on which it will run. While the use of assembly language may be the optimal strategy for small throw-away programs, a better long-term strategy is to provide programmers with a means to express their algorithms in a more natural way. They will then produce more easily understood (and more machine-independent) programs. The ability to separate the expression of the algorithm from the architecture of the machine is becoming increasingly important, as emphasis shifts from the optimal use of machine resources to optimal use of human resources.

Writing programs which are self-documenting.
Assembly language allows the incorporation of comments into the program to aid in its documentation. However, this is only partly successful because:

- The format, content, and quantity of comments in a program are strictly left to the programmer. Some programmers are meticulous about including helpful comments, while others regard comments as largely unnecessary.

- When the program code is changed during debugging, the comments are often not updated immediately. If the debug session lasts a long time, some of the early changes may be forgotten.

Because C allows you to express algorithms in a more understandable way, it is (at least to a greater degree than assembly language) self-documenting. It allows you to use meaningful mnemonics for variables and for function calls. (Listing 1 demonstrated that you can write easily understood programs in HLLs with no comments at all.)

Making subroutine calls, and passing parameters to those subroutines
Subroutines were originally invented to allow the generation of more compact programs. (This was a major concern when machines had less than 64 Kbytes of memory.) Subroutines allow you to take a code section that appears in many different places in a program, express it as a single subroutine, and then call it wherever it's needed. However, when you call such a subroutine, you often need to pass parameters to it. If your programming language provides no formal way to pass parameters, you may forget what variables it changes. Then you may find yourself *debugging* your diagnostic program, instead of using it.

C provides a formal way to pass parameters to a called function. For example, suppose the C function `mid_elem()` (See Listing 2) is called to get the middle element in an array, based upon the first and last index of the array. When `mid_elem()` is called, it gets passed 2 parameters, as follows...

```
mid_elem(lo_index, hi_index)
```

Listing 2

```
int midpoint(low, high)
int low, high;
{
   int middle;

   middle = (low + high + 1) / 2;    /* round to nearest int */
   return(middle);
}
```

The function then returns the array index number of the middle element (rounded upward). We could use this index to access the array and store the contents of the middle element as follows...

```
middle_element = array[mid_elem(lo_index, hi_index)];
```

Handling strings efficiently

Diagnostic programs generally do a lot of string input and string output operations. Input operations are needed when accepting commands from the user, and output operations are needed to indicate the status of the test, as it proceeds. C compilers provide standard `gets()` and `puts()` functions for inputting and outputting strings, respectively.

Listing 3 shows how the `gets()` function is used. The code calls `gets()` to get a character string from the console. Then it checks to see if the first character of that string is an `end_of_string` (NULL) character. If it is not, then the code calls the string-comparison

Listing 3

```
{
   char *gets(),    /* declares the type of value that     */
                    /*  gets() returns                     */
        s[100],     /* a buffer into which to place string */
        *standard;  /* a pointer to a character string     */
   ...
   standard = "Standard by which others are judged";
   gets(s);
   if (*s)
      {
      if (strcmp(s, standard) == 0)
         printf("Input string matches the standard string\n");
      else
         printf("Input doesn't match the standard string\n");
      }
   else
      printf("Operator entered only Return\n");
   ...
}
```

function `strcmp()` to determine whether or not the input string matches the standard string.

Doing decimal and hexadecimal arithmetic operations

When you are writing diagnostic programs, you often want to be able to express an address in hexadecimal form. You also want to be able to add and subtract offsets in hexadecimal. If you're forced to do all your arithmetic in decimal, some simple operations can be difficult to follow. Unlike some HLLs, C allows you to express numbers in hexadecimal, binary, octal, or decimal form.

You may also need to do add, subtract, multiply, and divide operations on 8-bit, 16-bit, or even 32-bit signed and unsigned integers. C provides you with a way to express what type of integer a variable represents, and takes this into account automatically when doing arithmetic operations. Listing 4 shows how you declare these various types of variables.

Listing 4

```
char              8_bit_signed_integer;
unsigned char     8_bit_unsigned_integer;
int               16_bit_signed_integer;
unsigned          16_bit_unsigned_integer;
long              32_bit_signed_integer;
unsigned long     32_bit_unsigned_integer;
float             32_bit_real;
double            64_bit_real;

Note: The definition of the C language does not specify
the length or precision of any of these types.  These
are the most commonly found values in the minicomputer
and microcomputer industries.
```

Using powerful control structures

C provides some very powerful and compact control structures for looping. Listing 5 shows a loop that sets the index `ii` to zero, and then repeats the print statement over and over, incrementing `ii` each time, until `ii` is equal to the number of elements in the array. The `printf()` function generates formatted output, where each line contains the decimal representation of one array element (`%d`) followed by a single character (`%c`) which is either a space or a newline. The decision as to whether to print the space or the newline is specified by the...

```
(ii%5 == 4) ? '\n' : ' '
```

...at the end of the print statement. It's a conditional statement that chooses between one of 2 options, based upon whether the expression in parenthesis is true or false. The end result is that the print statement prints 5 decimal numbers per line, each separated by a space.

Listing 5

```
#define NUM_ELEMENTS 10

{
    int ii,
        array[NUM_ELEMENTS];

    ...
    for (ii = 0; ii < NUM_ELEMENTS; ii++)
        printf("%d%c", array[ii], (ii%5 == 4) ? '\n' : '
');
    ...
}
```

Accessing status and control registers with easily understood variable names. C lets you interact with hardware status and control registers on the bit and byte level. It was designed that way because the UNIX programmers had to be able to access the status and control registers of the PDP-11, on which UNIX ran.

Figure 1 shows the registers of a hypothetical RS-232 interface chip. In order to verify that the RS-232 port works, you need to:

- write to the control register
- write to the output data register
- read the status register
- read the data input register

STATUS	DSR	BRK	CTS	OVR	PER	TRDY	RRDY	
CONTROL	DTR	SBR	RTS	BAUD		PARITY		RST
DATA IN								
DATA OUT								

Figure 1

You can do that in C by first defining a *structure*, which serves as a kind of template that can be overlayed upon a designated portion of the memory map. Listing 6 defines a template designed for use with our hypothetical serial RS-232 interface chip.

Once you have defined the template, you can overlay it onto the registers of the serial chip (at location $012345 of the address map) by a 2-step process, as shown in Listing 7.

Listing 6

```
/***********************************************************/
/* Define the status register for the serial port.  No storage */
/* is allocated, this is only a definition                  */
/***********************************************************/

typedef
struct {
   unsigned        : 1;  /* bit 0   - UNUSED          */
   unsigned rrdy   : 1;  /*     1   - rx ready        */
   unsigned trdy   : 1;  /*     2   - tx ready        */
   unsigned per    : 1;  /*     3   - parity error    */
   unsigned ovr    : 1;  /*     4   - overrun error   */
   unsigned cts    : 1;  /*     5   - clear to send   */
   unsigned brk    : 1;  /*     6   - break received  */
   unsigned dsr    : 1;  /*     7   - data set ready  */
} STATUSREG;

/****************************************************/
/* Define the control register for the serial port.  No */
/* storage is allocated, this is only a definition    */
/****************************************************/

typedef
struct {
   unsigned rst    : 1;  /* bit 0   - reset           */
   unsigned parity : 2;  /*     1-2 - parity          */
   unsigned baud   : 2;  /*     3-4 - baud rate       */
   unsigned rts    : 1;  /*     5   - request to send */
   unsigned sbr    : 1;  /*     6   - send break      */
   unsigned dtr    : 1;  /*     7   - data term. ready */
} CTRLREG;

/***************************************************/
/* Define the format of a serial port.  No storage is */
/* allocated, this is only a definition              */
/***************************************************/

typedef
struct {
   STATUSREG       sreg;
   CTRLREG         creg;
   unsigned char   data_in;
   unsigned char   data_out;
} SERIAL_PORT;

/***********************************************************/
/* Declare a serial port.  Storage is allocated according to */
/* the definition of SERIAL_PORT                             */
/***********************************************************/

SERIAL_PORT  serial_port_1;
```

Listing 7

```
/****************************/
/* Define a pointer variable */
/****************************/

SERIAL_PORT *spptr1;

/*************************************************/
/* Set the pointer to point to the base address of   */
/* the serial port's registers at hex address 012345 */
/*************************************************/

spptr1 = (SERIAL_PORT *)0x12345;
```

Step 1:
Define a pointer variable (called `spptr1`) to point to serial port number 1. (A pointer variable holds the address of some location in the address map.)

Step 2:
Set the pointer variable to point to the base address of the serial port's registers. (The assignment statement in Listing 7 is a bit cryptic, but it does the job.)

Once you have declared the pointer, and assigned an address to it, you use it to write to the bits and fields of the serial port's registers, just by using the names declared in the template as variable names. For example, if you want to initialize the serial port to transmit (and receive) at 1200 baud, with even parity, you could execute the code shown in Listing 8.

Listing 8

```
/*****************************************************************/
/* Symbolic names for the values used to configure the serial port */
/*****************************************************************/

#define RUN  0
#define LOAD 1

/*******************************************************/
/* Symbolic names for the possible parity condition to be */
/* passed to the setup function                          */
/*******************************************************/

#define EVEN 0
#define ODD  1
#define NONE 2
#define MARK 3

/*******************************************/
/* Permissable baud rates for a serial port */
/*******************************************/

unsigned baud_rate_table[] = {110, 300, 1200, 9600};
```

(continued)

Listing 8 *(continued)*

```
*****************************************************************/
/* A portion of a function that calls the port-setup function */
/****************************************************************/

{
   SERIAL_PORT *spptr1;
   void        serial_setup();
   unsigned    baud_rate, parity;

   ...
   spptr1 = (SERIAL_PORT *)0x12345;
   baud_rate = 1200;
   parity = EVEN;
   serial_setup(spptr1, baud_rate, parity);
   ...
}

/***********************************/
/* Function to set up a serial port */
/***********************************/

void
serial_setup(spptr, baud_rate, parity)
SERIAL_PORT *spptr;

unsigned    baud_rate, parity;
{
   void     wait();
   unsigned b;

   /* Test the requested baud rate against the permissable values
      found in the table. The index into the table for the matching
      rate is the value to pass to initialize the port */

   for (b = 0; baud_rate_table[b] != baud_rate; b++)
      ;
   spptr->creg.baud = b;

   /* Set the parity condition.  The calling function will have
      used one of the symbolic values defined above, so we can
      put it directly into the port */

   spptr->creg.parity = parity;

   /* Initialize the port */

   spptr->creg.rst = LOAD;
   wait(10);
   spptr->creg.rst = RUN;
}
```

Listing 9 shows the Intel 8086 assembly code that might be generated when the port setup function in Listing 8 is compiled. (Note: The lines in Listing 9 that begin with semicolons are the C source code, retained as comments in the assembly source for clarity.)

Defining, accessing, and passing custom data structures efficiently

While simple I/O chips are generally accessed through status and control registers, the trend in systems today is to equip the I/O controllers with DMA hardware, and at least some rudimentary intelligence. That allows the I/O controller to copy strings of bytes into (or out of) memory, without intervention by the system's CPU. For example, many VMEbus disk controller boards, local area network controller boards, and graphics controller boards include a dedicated onboard processor, as well as firmware that allows them to accept command blocks.

Listing 9

```
;void
;serial_setup(spptr, baud_rate, parity)
;SERIAL_PORT *spptr;
;unsigned    baud_rate, parity;
;{
;   void      wait();
;   unsigned b;
;
        push    bp
        mov     bp,sp
;
;   for (b = 0; baud_rate_table[b] != baud_rate; b++)
;      ;
;
        mov     word ptr -2[bp],00H
        jmp     $2
$1:     inc     word ptr -2[bp]
$2:     mov     bx,word ptr -2[bp]
        shl     bx,1
        mov     ax,word ptr baud_rate_table_[bx]
        cmp     ax,word ptr 6[bp]
        bne     $1
;
;   spptr->creg.baud = b;
;
        mov     ax,word ptr -2[bp]
        mov     bx,word ptr 4[bp]
        and     ax,03H
        shl     ax,1
        shl     ax,1
        shl     ax,1
        and     word ptr 2[bx],0ffe7H
        or      word ptr 2[bx],ax
```

(continued)

Listing 9 *(continued)*

```
;
;    spptr->creg.parity = parity;
;
        mov     ax,word ptr 8[bp]
        mov     bx,word ptr 4[bp]
        and     ax,03H
        shl     ax,1
        and     word ptr 2[bx],0fff9H
        or      word ptr 2[bx],ax
;
;    spptr->creg.rst = LOAD;
;
        mov     bx,word ptr 4[bp]
        or      word ptr 2[bx],01H
;


;    wait(10);
;
        mov     ax,0aH
        push    ax
        call    near ptr wait_
        pop     cx
;
;    spptr->creg.rst = RUN;
;
        mov     bx,word ptr 4[bp]
        and     word ptr 2[bx],0fffeH
;}
        mov     sp,bp
        pop     bp
        ret
```

Suppose we have an 8-port serial I/O board that accepts a command block, like the one shown in Figure 2. The most efficient way to pass a command block to this board is to build the block in global VMEbus memory, and then write the base address of the command block into a local interrupt register on the I/O controller board. Creating a command block in C is a 2-step process, as shown in Listing 10:

- Define a structure, in a fashion similar to the serial port example.
- Declare a command block variable, called `cmd_block`.

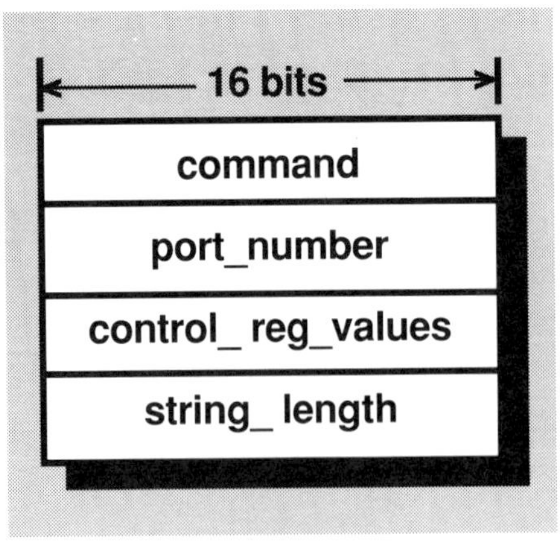

Figure 2

Listing 10

```
/********************************************/
/* Define the control register.  No storage is */
/* allocated, this is only a definition       */
/********************************************/

typedef
struct {
   unsigned rst    : 1;  /* bit 0   - reset            */
   unsigned parity : 2;  /*     1-2 - parity           */
   unsigned baud   : 2;  /*     3-4 - baud rate        */
   unsigned rts    : 1;  /*     5   - request to send  */
   unsigned sbr    : 1;  /*     6   - send break       */
   unsigned dtr    : 1;  /*     7   - data term. ready */
} CTRLREG;

/****************************************************/
/* Define the format of a command block.  No storage is */
/* allocated, this is only a definition                */
/****************************************************/

typedef
struct {

   unsigned command;
   unsigned port_number;
   CTRLREG  control_reg_values;
   char     *string_begin;
   unsigned string_length;

} CMD_BLOCK;

/***********************************************************/
/* Declare a command block.  Storage is allocated according to */
/* the definition of CMD_BLOCK                                 */
/***********************************************************/

CMD_BLOCK cmd_block;

/********************************************************/
/* A portion of a function that fills the command block and */
/* sends its address to the serial I/O board                */
/********************************************************/

{
   unsigned command, port_number, string_length;
   char     *string_begin;
   CTRLREG  control_reg_values;
   void     interrupt_serial_board();
```

(continued)

Listing 10 *(continued)*

```
    ...
    port_number = 2;
    /* Fill other command-block locations here */
    ...
    interrupt_serial_board(&cmd_block);
    ...
}

void
interrupt_serial_board(addr)
CMD_BLOCK *addr;
{
    /* Code implementing the interrupt goes here */
}
```

Once you have declared the command block variable, you can write to (or read from) the variables within the block, just as you did with the serial port registers. However, when you finish entering all of the parameters, you must pass the base address of the command block to the serial I/O board. The board can then read the command block, and execute the command that it specifies.

Suppose you pass the base address to the controller board by simply writing it into the controller board's interrupt register, which appears to the VMEbus as a memory location. When you write to this location, the controller board's onboard processor gets interrupted. It then gets the command block's address out of the interrupt register, and uses it to access the command block in global memory.

Not all controller boards work the same way. For example, some boards might require you to write the address of the command block into one register, and then access another register to interrupt the controller board's processor. The code in Listing 10 hides the implementation of the actual interrupt mechanism in a function called `interrupt_serial_board()`, which takes as an argument the address of the command block. The function's descriptive name makes it easy to see what it does, but the process by which it interrupts the controller board is hidden.

Interpreter or compiler?
Assuming that you want to use C to write your diagnostic software, you will need to choose a *translator* to convert your C source code to machine language for the particular processor type on which it will run. But that raises a question: should you choose a compiler or an interpreter?

Compilers have one main advantage over interpreters: once they finish compiling a program, the resulting machine language file can be executed very quickly. In fact, some compilers include optimizers, which scan the finished machine language program looking for inefficiencies, and correcting them.

Inefficiencies occur because a block of code derived from a template doesn't know the state of the processor just prior to its execution. For example, a block of code might load the value of a variable into one of the CPU's registers, even though the preceding block has already loaded that same value into that same register. A good optimizer would recognize this and delete the extra load instruction.

Some compilers carry optimization to an extreme. They spend a considerable amount of time getting rid of any excess machine code. The result is a machine language program that runs very fast: sometimes as much as 10 times faster than an interpreted program. However, when you're debugging hardware, you're typically more concerned about how quickly you can make changes in your program and then rerun it. A compiler that takes a lot of time optimizing keeps you waiting, and since your program's execution speed is usually not of great concern (at least at this stage of the development) you may not get much benefit from the optimization.

Another problem with using a compiler is that you aren't quite sure what portion of the program code is executing when your diagnostic program hangs up in a loop, or crashes. The only way to find out may be to leave a trail of breadcrumbs as your program executes. For example, you might print on the terminal screen the value of a significant variable each time through a loop, or you might just print "I successfully reached point A in the program". That may not tell you exactly where it crashed, but you will be able to determine that you got to point A, but never reached point B. Then you progressively reduce the execution distance between the print statements until you find the source code line with the error in it. Of course, every time you insert or move a print statement, you have to recompile the source code, before you can run the program again.

Diagnostic software should be written to accept changes readily. You want to be able to:

- find the bug in your program quickly.
- change the program quickly.
- run the new version of the program immediately.

That's where an interpreter really shines. Since an interpreter executes your program as it translates it, you can have it indicate where it is in the program as it executes. Then, if the program crashes, you can figure out what line of your diagnostic program was being executed when it crashed. You could also have the interpreter single-step through the program, so you can check the current value of any variable. That helps you figure out why your program is hung up in an infinite loop.

Using canned diagnostic software

Hopefully, if C becomes popular as a language for writing VMEbus diagnostic software, manufacturers of VMEbus products will be encouraged to provide canned diagnostic routines. For example, the company selling a VMEbus disk controller board might provide useful diagnostic routines (read a sector, write a sector, etc) in C. Your diagnostic programs would then consist of these canned diagnostic routines interspersed with some code of your own.

Terry Shankland is a consulting engineer. Since earning his BSEE in 1971 at Brown University, Terry has spent most of his time doing real-time programming. He has designed real-time graphics demos and benchmarks for Megatek graphics systems and a graphics subsystem for an Air Force fighter weapons training system. More recently, he has designed user interfaces for ultrasound instrumentation and an image processing system. His hobbies include listening to classical music, hiking, and the creation of freeware.

John Black is the president of Micrology pbt. He previously worked for Motorola, where he was responsible for the development of the VMEbus standard. He has authored numerous articles on advanced microcomputer architectures. He has been the editor of VMEbus System Magazine since 1986, and is also the editor of this book. John earned his BSEE and MSE degrees at Arizona State University, and is currently working toward a PhD in computer science.

CHAPTER 7

Downloading To VMEbus: Some Software Utilities

By Walter S. Heath

UNIX systems have now become so widely available that they are available to virtually every VMEbus engineer. This fact (combined with the fact that there are lots of very good UNIX-compatible software development tools, such as editors, compilers, and debugging utilities) makes UNIX systems very attractive for developing your application code. Typically, you can enter, compile, run, debug and edit your basic algorithms, all on the UNIX system. However, the final stages of the debugging (which often require I/O interaction) will still need to be done in your VMEbus system. That raises a question. How do you download the compiled object code from the UNIX system into your VMEbus system? In this chapter, Walter describes the utility software that he uses to download object files from his UNIX system. He uses Motorola's S-Record format, which represents binary files as ASCII character strings. Since his utilities are written in C, you can easily modify them for use with other operating systems.

Most VMEbus single board computers are equipped with ROM-resident monitors that accept downloaded object code files through an RS-232 port. Since the most significant bit of each byte sent over RS-232 links has traditionally been used to provide a parity bit, the monitor's download algorithm typically assumes that the sender has converted each 8-bit byte of the object code file into a pair of 7-bit ASCII characters. As the monitor receives each ASCII character pair, it converts the pair back into the original 8-bit byte value, and then writes that byte value into memory.

The two most popular ASCII formats have been popularized by Motorola and Intel. In both formats each 8-bit byte of the downloaded object code is represented by a pair of ASCII characters, chosen from the following set:

0, 1, 2, 3, 4, 5, 6, 7, 8, 9, A, B, C, D, E, F

Each pair of ASCII characters represents the 2-digit hexadecimal value of the 8-bit byte from which it was derived.

The reason for converting the file to ASCII, is that it can be sent through any serial port, whether or not that port uses the 8th bit for parity. The disadvantage is that, since each byte value of the memory image is represented by a pair of ASCII characters, the ASCII file must be at least twice as large as the original binary file.

Record formats

The ASCII character pairs are typically grouped into strings, which are each terminated with a carriage return and a line feed to form *records*. Each record typically also includes a destination memory address, a byte count, and a check sum, in addition to the hexadecimal ASCII representation of the object code.

Motorola S-Records

The records used by Motorola are called S-Records. When an object code file is to be sent over a serial link, the sender can convert the file to S-Records, as it sends the file. However, the 2 utility programs in this chapter accomplish the file transfer in two steps, as shown in Figure 1.

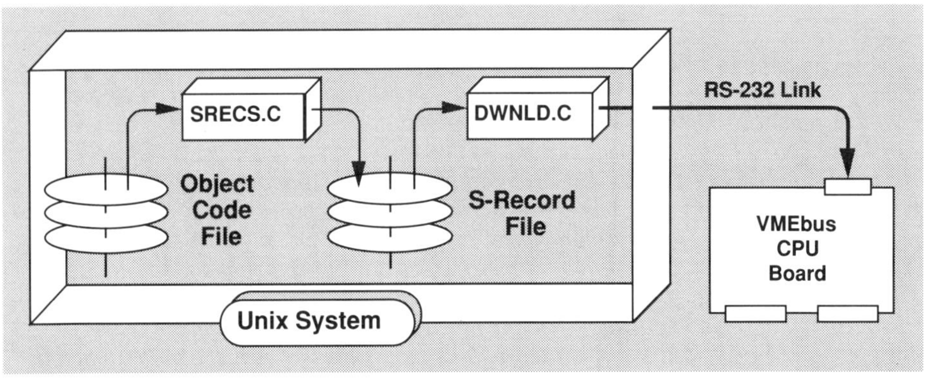

Figure 1

In the first step, a utility called SRECS.C reads the entire UNIX binary file, and generates an S-Record file. Then a utility called DWNLD.C sends the S-Record file over an RS-232 serial link to a VMEbus system.

An example of an S-Record file

Figure 2 shows a typical S-Record file, created by SRECS.C. S-Record files are composed entirely of printable characters, except for the CR/LF at the end of each record. That makes it easy to visually monitor. the file transmission with an RS-232 terminal.

```
L01
S00600004844521B
S2141000004E56000033FC0001001040F242790010FA
S21410001040EA4279001040EC33FC000100103C48E6
•
•
•
S214103C2000000000000000000000000000000000007F
S214103C3000000000000000000000000000000000006F
S208103C4000103C40DF
S9030000FC
```

Figure 2

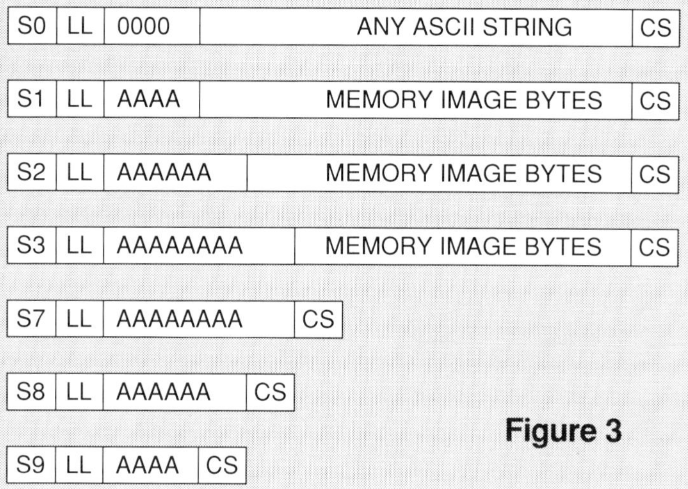

Figure 3

Figure 3 shows 7 types of S-Records, as defined by Motorola. The first record in an S-Record file is typically a header record. It begins with the 2 ASCII characters S0. The 2 hexadecimal ASCII characters following the S0 (represented in Figure 3 by LL) tell how many byte values there are in the remainder of the S-Record. The next 4 ASCII characters after the LL field typically represent the destination address for the data being transferred, but the data transferred by an S0 record is not meant to be stored in the destination system, so its field is typically 0000. (S0 records contain any ASCII character string that the sender wants to pass along with the file: a program name, version number, etc. This character string is typically displayed on a terminal at the destination system, for viewing by the user.) At the end of the S0 record (as in all S-Records) there is a checksum byte, represented by 2 hexadecimal ASCII characters.

Depending upon the type of processor receiving the records, the main body of the S-Record file consists of either S1, S2, or S3 records. S1 records are used to pass bytes to processors that have a 16-bit addressing space. S2 records are used to pass bytes to processors with a 24-bit addressing space, and S3 records pass bytes to processors with a 32-bit addressing space. The string of hexadecimal ASCII pairs in S1, S2, and S3 records represent a block of 8-bit bytes that will be stored in the receiving processor's memory map. The receiving processor converts each ASCII pair into an 8-bit byte value, and writes that byte into its memory, starting at the address indicated in the address field (AAA...) of the S-Record.

The end of the file is indicated with a termination record, which begins with S7, S8, or S9. The optional address field (AAA...) of the termination record tells the receiving monitor the starting address of the downloaded object code program. If the receiving monitor retains this starting address, it can either start executing the program immediately (upon completion of the download) or it can allow the user to start execution with a single key stroke. (Note: Not all ROM-resident monitors retain this starting address.) An S9 record is used to terminate a block of S1 records. S8 is used to terminate a block of S2 records, and S7 is used for S3 records. (Some ROM-resident monitors are very tolerant about this. For example, a monitor might accept an S9 record as a terminator for a block of S2 records.)

Looking back at Figure 2, if we disregard the first line of the file for the moment, we see a header record (S0) with the ASCII string "HDR (484452)", and a checksum byte (1B). Since the file shown in Figure 2 is being sent to a 68000 (which has a 24-bit address map) S2 records are used to send the object code bytes. The first S2 record has the hexadecimal base address of 100000, so its 16 memory image bytes will be deposited in the receiving processor's memory map, starting at that address. The last S2 record deposits its memory image from 103C40 to 103C47. The S9 record indicates the end of the block of records, and normally provides a starting address. However, in this case the receiver does not retain the starting address, so a 0000 is used as a dummy value.

The SRECS utility

Listing 1 (at the end of this chapter) shows the C language utility program SRECS. This program reads a binary file generated by the UNIX linker, and writes to an S-Record file. This particular version of SRECS is designed to produce S-Record files for transmission to a single board computer whose monitor accepts S-Records through the same serial port that it accepts monitor commands. SRECS inserts a load command (LO1) at the beginning of the S-Record file. When this load command is sent to the single board computer over the serial link, it commands the board's ROM-resident monitor to load the subsequent S-Record file. (This explains the L01 on the first line of Figure 2.)

Some monitors input the S-Record file through a different serial port from the one used to enter the load command. When using SRECS with such a monitor, you will want to remove the code that inserts the load command.

The SRECS program listing is thoroughly commented. One interesting feature is the use of a union construct...

```
union {
    struct{
    .
    .
    .
    }head
    unsigned char inbuf[32]
}u
```

This overlays the structure `head` over the 32 byte array `inbuf[32]`. Defining the array `inbuf[32]` makes it easy to read in the file header...

```
      read(fl,u.inbuf,32)
```

Once the file header has been read from the disk into the `inbuf[32]` array, the overlaid structure `head` serves as a template for accessing the various fields of the file header. For example, a test of the "magic number" in the file header is done as follows...

```
     if(u.head.fmagic != 0407) {...}
```

Once the file header has been read (and its values extracted) the `inbuf[32]` array is then used to read the memory image bytes from the disk file, to allow construction of the S2 records.

Each S2 record ends with a checksum. The checksum is computed by summing all byte values (after the S2), taking the ones complement of the sum, and then logically ANDing off all but the lower 8 bits.

The function `intasc()` accepts a binary integer `(intgr)`, and a number of characters `(nchrs)`, and then generates a hexadecimal ASCII representation of the integer. Then it concatenates that ASCII string onto the end of the current contents of the output buffer, `outbuf[]`.

The DWNLD utility

Once an S-Record file has been *produced* by SRECS, the DWNLD utility can be used to *send* the S-Records to the VMEbus processor board. The C language source code for DWNLD is shown in Listing 2. DWNLD opens the S-Record file (`userfile.sr`) and the output device port (`/dev/tty1`), reads blocks of S-Records from the file, and writes them to the port.

Note: It's important that the serial port used by DWNLD be "configured out" of the UNIX system, so UNIX doesn't set it up for use as a user port when the system is powered up.

Flow control protocols

While Motorola defined the S-Record *format*, they didn't specify any particular *flow control*, for use during the download of the S-Record files.

Some senders use an XON/XOFF protocol. That allows the receiver to temporarily stop the sender, by transmitting an XOFF character (13 hex). Then (when it has digested the previously sent ASCII characters) it sends an XON character (11 hex) to allow the transmission to resume.

Other senders use an ACK/NAK protocol. The sender sends one record. Then it waits for the receiver to verify the checksum, and send an ACK character (06 Hex). When it receives the ACK character, it sends the next record. If, at any point, the receiver gets an S-Record with a bad checksum, it sends a NAK character (15 hex) to the sender, and the sender retransmits the record.

The ACK/NAK protocol allows for retransmission when an error occurs, but there is a price to pay. Waiting for the ACK slows things down a lot. The DWNLD program presented here just sends the entire S-Record file, without checking for any response from the receiver. That speeds up the transmission considerably. Experience has shown this approach to be satisfactory.

Using DWNLD on non-UNIX systems

DWNLD was specifically created to download S-Record files from a UNIX system. For operating systems other than UNIX (such as MS-DOS) you may be able to use a standard copy utility to copy the S-Record file from the disk to an output serial port. Obviously you need to make sure that the serial port is configured for the proper baud rate, parity, number of stop bits, etc.

Downloading via diskette

S-Record downloading is a reliable and practical method for downloading programs of modest size. However, when programs get large, the time required to download them can be annoying. For example, at 9600 baud, it takes about a minute to download a 16 Kbyte program. Since downloading is part of the debug cycle, this slows development.

A faster way to transfer large programs is to have the UNIX operating system write the object file onto a floppy disk. Then you insert the diskette into the disk drive of your VMEbus system. Some single board computers include floppy disk controllers that can interface directly to a disk drive through the board's P2 connector.

To make this work, the VMEbus processor board's monitor firmware must contain routines for accessing the files placed on the disk by the UNIX operating system. These utilities must be able to do 2 things:

- They must be able to configure the board's floppy disk controller chip to read the format used by the UNIX system.
- They must be able to interpret the UNIX file structure on the disk, which involves reading the directories and the file headers.

Loading object files into a multiprocessor system

We could even carry this one step farther. When you're working with a multiprocessor system you may need to load object code files into the onboard memories of several processors within your VMEbus system, before you can begin debugging. To load these files from a floppy disk, you could equip the ROM-resident monitor on one of the processor boards to read a particular script file from the disk, which contains the filenames of all the object code files that must be loaded from the disk. The monitor could then automatically read each of these files from the diskette, and write them into the dual-ported memories of the other processor boards in the system.

Once the monitor is equipped with subroutines to access disk files, it would be fairly easy to provide monitor system calls that allow your VMEbus application software to also access these files.

A call for action

Routines to do all of these things can be (and have been) implemented by patching existing ROM-resident monitors. But why can't these capabilities be standardized and included in the monitors delivered with the processor board? If the various VMEbus manufacturers got together and agreed on a standard disk format and file structure, programs could be developed on any operating system, written to diskette, and then loaded into any VMEbus system. It would then be possible to sell software on diskette, for use on a wide variety of VMEbus systems.

Walter Heath is a computer systems designer with 23 years experience in designing and building real-time embedded computer systems. He is currently working at Lincoln Laboratory, Lexington, MA, on a contract basis, building research prototype systems. He holds an MSEE from the University of Michigan.

Some comments from the editor

Walter's system consists of a 68000 running UNIX System III. The files that he downloads are generated with the resident 68000 assembler. Once linked, the resulting executable files can be run on the UNIX system, or downloaded to a 68000-based VMEbus system.

SRECS assumes that the file it's converting to S-Records is an executable file: one that has already been linked by the UNIX linker to form an `a.out` binary memory image file. Executable files in UNIX System III contain a file header (consisting of eight 32-bit words) followed by memory image bytes, which consist of 3 parts:

- text (instruction code)
- initialized data
- zeroed data.

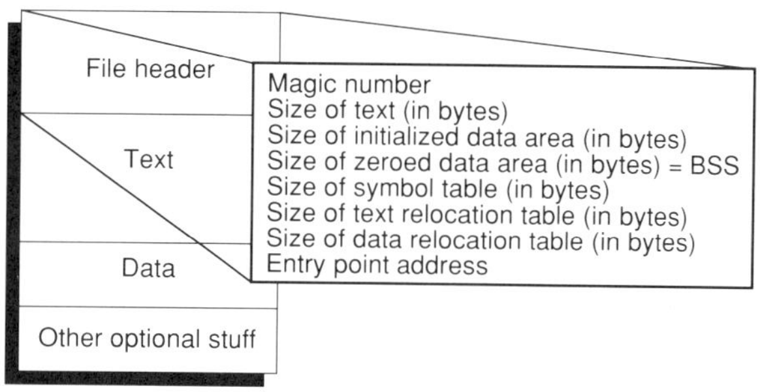

The first 32-bit entry in the file header is called the "magic number". (The origin of the magic number is best left a mystery.) For executable files, the magic number is either 407 (octal) or 410 (octal). The magic number indicates whether an executable file can or can't be shared by 2 or more UNIX users. For example, a file containing self-modifying (non-shareable) code would have a 407.

The memory image files generated by Walter's 68000 assembler typically have a magic number of 407, so SRECS checks to make sure that the first entry in the file is a 407. (SRECS warns the caller if it has been asked to convert a non-executable file.)

The file header also contains other information. SRECS reads:

- the size of the text
- the size of the initialized data area
- the base address where the text segment is to be loaded into memory.

Unix System III assumes that the entry point for the code is the same as its base address.

If you want to use this SRECS utility to download binary files from other versions of UNIX you may need to change it, since the file headers on other versions are different. For example, UNIX System V file headers contain both a base address for the text, and an entry point address. (The base address is the sixth 32-bit word, and the entry point is the fifth 32-bit word.)

Intel's ASCII format

The ASCII format used by Intel to download binary files consists of 4 basic record types:

00 is a *data* record.
01 is an *end* record.
02 is an *extended address* record.
03 is a *start* record.

A file transmission is begun with the *start* record. Then an *extended address* record is typically sent. It provides a 16-bit offset value (represented by 4 ASCII characters) which must be added to the 16-bit address (AAAA) of subsequent data records. The sum provides the actual 16-bit memory address at which the memory image bytes are to be written into the receiving processor's memory map. The end record indicates that the transmission has been completed.

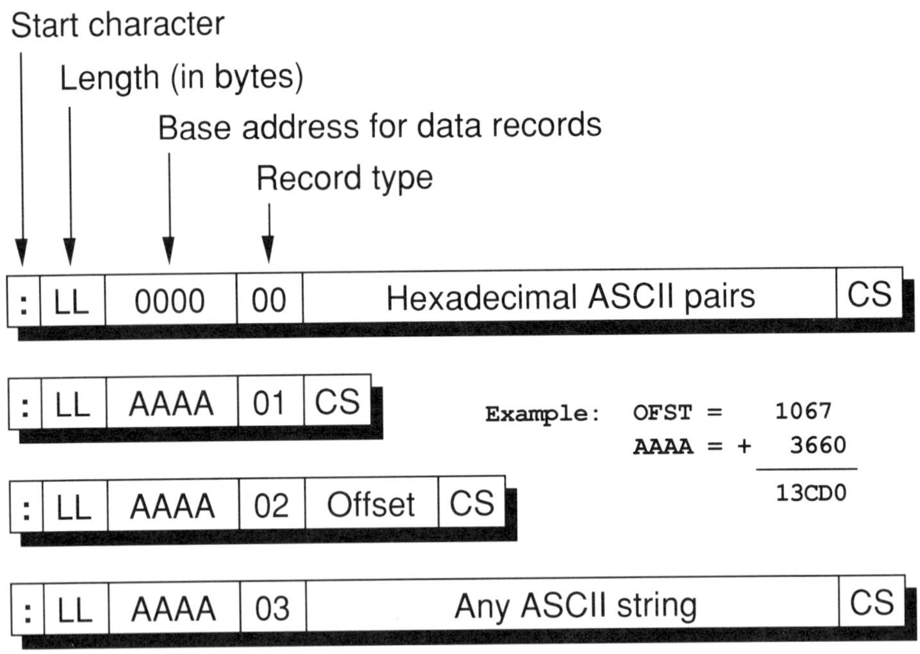

Listing 1

```
/***********************************************************************/
/* FILE SRECS.C                                                        */
/*                                                                     */
/* This program reads the load module file named in the first argument */
/* of its command line and generates the "S-record" output file named  */
/* in the second argument of its command line.  The output file is     */
/* suitable for downloading to a Force Co. SYS68K/CPU-2 computer.  The */
/* input file load module MUST have the format of a UNIX "a.out" file, */
/* as produced by the UNIX "ld" linker.  An example linker command is: */
/*                                                                     */
/*      ld -X -LT 100000 -o userfile.a userfile.o -lc -lm              */
/*                                                                     */
/* An example srecs calling syntax is:                                 */
/*                                                                     */
/*      srecs userfile.a userfile.sr                                   */
/*                                                                     */
/* Note that the SYS68K/CPU-2 computer accepts S-records over the same */
/* serial port as it inputs monitor commands.  This program first puts */
/* an 'LO1' command string in the output file, to get the monitor ready */
/* to accept S-records.  For SBCs that accept S-records over a separate */
/* serial port, this code should be removed.                           */
/***********************************************************************/
#define PMODE    0644                  /* RW for owner, R for others   */
#define BUFSIZ    16                   /* input buffer size (bytes)    */

int f1,f2                             ;/* in & out file descriptors    */
union{
    struct{                            /* Unix a.out file header struct */
        long fmagic                   ;/* Unix "magic number"          */
        long tsize                    ;/* text size (bytes, even)      */
        long dsize                    ;/* data size (  "      "  )     */
        long bsize                    ;/* bss size                     */
        long ssize                    ;
        long rtsize                   ;
        long rdsize                   ;
        long entry                    ;/* program-start address        */
    }head                             ;
    unsigned char inbuf[32]           ;/* struct overlay input buffer  */
}u                                    ;
int nr                                ;/* read byte count              */
long filsiz                           ;/* object file size (bytes)     */
char load1[] = {'L','O','1','\r','\n',0} ;/* the LO1 monitor command   */
char strec[] = "S00600004844521B"     ;/* start record (S0)            */
char serec[] = "S9030000FC"           ;/* end record (S9)              */
char recend[] = {'\r','\n',0}         ;/* end of a record (CR,LF,NUL)  */
long ldaddr                           ;/* load address                 */
char outbuf[100]                      ;/* S-record output buffer       */
long outidx                           ;/* outbuf[] index               */
long count                            ;/*  # of bytes read from input  */
long i                                ;/* working index                */
int chksum                            ;/* checksum calculation variable */
unsigned int inchr                    ;/* input character              */
unsigned int digit                    ;/* output digit                 */
```

(continued)

Listing 1 *(continued)*

```
/****************/
/* main program */
/****************/
main(argc,argv)
int argc           ;
char *argv[]       ;
{
/**********************************/
/*Check number of input arguments. */
/**********************************/
if(argc != 3){
    printf("One input file and one output file needed\n") ;
    return                                               ;
}
/******************************************/
/* Open the input file; create output file. */
/******************************************/
if((f1 = open(argv[1],0)) == -1){
    printf("Can't open %s\n",argv[1])     ;
    return                                ;
}
if((f2 = creat(argv[2],PMODE)) == -1){
    printf("Can't create %s\n",argv[2])   ;
    return                                ;
}
/*********************************************************************/
/* Output LO1 command (directs Force PROM monitor to accept S-records).*/
/*********************************************************************/
if(write(f2,load1,sizeof(load1)) != sizeof(load1)){
    printf("Can't output LO1 command\n")              ;
    return                                            ;
}
/********************************/
/* Output the S0 header record. */
/********************************/
if(write(f2,strec,16) != 16){
    printf("Can't write S0 header record to %s\n",argv[2]) ;
    return                                                 ;
}
/****************************/
/* Output record-end string. */
/****************************/
if(write(f2,recend,3) != 3){
    printf("Can't output record-end string(1)\n") ;
    return                                        ;
}
/*****************************************************************/
/* Read file header; check it and get file size (text & data only) */
/* and program entry point.                                      */
/*****************************************************************/
if((nr = read(f1,u.inbuf,32)) <= 0){
    printf("Can't read %s header\n",argv[1]) ;
    return                                   ;
}
else{
    if(u.head.fmagic != 0407){
        printf("Illegal input file header: fmagic = %o\n",u.head.fmagic) ;
        return                                       ;
    }
    filsiz = u.head.tsize + u.head.dsize     ;/* get file size          */
    ldaddr = u.head.entry                    ;/* get program entry point */
}
```

(continued)

Listing 1 *(continued)*

```
/*****************************************************/
/* Read bytes from input; form S2 records and output them.*/
/*****************************************************/
do{
    if(filsiz > BUFSIZ){                         /* choose record size    */
        count = BUFSIZ                          ;
        filsiz -= BUFSIZ                        ;
    }
    else{
        count = filsiz                          ;
        filsiz = 0                              ;
    }
    if((nr = read(f1,u.inbuf,count)) <= 0){
        printf("Can't read %s\n",argv[1])    ;
        return                                  ;
    }
/*****************************************************/
/* Build the S2 header, the count and the load address; */
/* start checksum calc.                                 */
/*****************************************************/
    outidx = 0                                  ;/* start the index           */
    outbuf[outidx++] = 'S'                      ;/* store 'S2'                */
    outbuf[outidx++] = '2'                      ;
    chksum = count + 4                          ;/* start checksum calculation */
                                                 /*  +4 is for length & addr.  */
    intasc(count + 4,2)                         ;/* convert length; store it   */
    chksum += ((ldaddr >> 16) & 0xFF)           ;/* add 3-byte load address    */
    chksum += ((ldaddr >> 8) & 0xFF)            ;
    chksum += (ldaddr & 0xFF)                   ;
    intasc(ldaddr,6)                            ;/* convert & store load addr. */
    ldaddr += count                             ;/* advance load address       */
/*****************************************************/
/* Now read each byte from inbuf[]; add it to checksum; */
/* convert & store it.                                  */
/*****************************************************/
    for(i = 0; i < count; i++){
        inchr = u.inbuf[i]                      ;
        chksum += inchr                         ;
        intasc(inchr,2)                         ;
    }
/*******************************************************************/
/* 1s complement the checksum; AND off all but lower byte; store it. */
/*******************************************************************/
    chksum = ~chksum & 0xFF                     ;
    intasc(chksum,2)                            ;
/*****************************************/
/* Now output the contents of outbuf[]. */
/*****************************************/
    if(write(f2,outbuf,outidx) != outidx){
        printf("Can't write an S2 record\n") ;
        return                                  ;
    }
/*****************************/
/* Output record-end string. */
/*****************************/
    if(write(f2,recend,3) != 3){
        printf("Can't output record-end string(2)\n") ;
        return                                         ;
    }
}while(filsiz > 0)                              ;/* repeat while bytes remain */
```

(continued)

Listing 1 *(continued)*

```
/***************************************/
/* Output file-termination S9 record. */
/***************************************/
if(write(f2,serec,10) != 10){
    printf("Can't write final S9 record\n") ;
    return                                  ;
}
/****************************************************/
/* Output record-end string to terminate S9 record. */
/****************************************************/
if(write(f2,recend,3) != 3){
    printf("Can't write record-end string(3)\n") ;
    return                                       ;
}
}                                      /* end of main() function  */

/*********************************************************************/
/* intasc(intgr,nchrs)                                               */
/* This function converts the binary number in 'intgr' to 'nchrs'    */
/* of ASCII hex digits and places them in array outbuf[].  If more than */
/* 'nchrs' hex digits exist in 'intgr' or if 'nchrs' is out of range */
/* (1 to 6), it informs the operator and exits the program.          */
/*********************************************************************/
intasc(intgr,nchrs)
unsigned int intgr                ;
int nchrs                         ;
{

switch(nchrs){
    case 6:
        digit = intgr/0x100000           ;/* get sixth digit       */
        intgr %= 0x100000                ;/* save remainder        */
        if(digit > 0xF){
            printf("intasc() input integer too large\n") ;
            exit(0)             ;
        }
        if(digit > 9)                     /* convert to ASCII      */
            digit += ('A' - 10)     ;
        else
            digit += '0'          ;
        outbuf[outidx++] = digit         ;/* store it              */
                                        /* Note: Falls thru!!      */
    case 5:
        digit = intgr/0x10000            ;/* get fifth digit       */
        intgr %= 0x10000                 ;
        if(digit > 9)                     /* convert it to ASCII   */
            digit += ('A' - 10)          ;
        else
            digit += '0'                 ;
        outbuf[outidx++] = digit         ;

    case 4:
        digit = intgr/0x1000             ;/* get fourth digit      */
        intgr %= 0x1000                  ;
        if(digit > 9)                     /* convert it to ASCII   */
            digit += ('A' - 10)          ;
        else
            digit += '0'                 ;
        outbuf[outidx++] = digit         ;
```

(continued)

Listing 1 *(continued)*

```
    case 3:
        digit = intgr/0x100                        ;/* get third digit       */
        intgr %= 0x100                             ;
        if(digit > 9)                               /* convert it to ASCII   */
            digit += ('A' - 10)                    ;
        else
            digit += '0'                           ;
        outbuf[outidx++] = digit                   ;

    case 2:                                     /* etc.                      */
        digit = intgr/0x10                         ;
        intgr %= 0x10                              ;
        if(digit > 9)
            digit += ('A' - 10)                    ;
        else
            digit += '0'                           ;
        outbuf[outidx++] = digit                   ;

    case 1:
        if(intgr > 9)
            digit = intgr + ('A' - 10)             ;
        else
            digit = intgr + '0'                    ;
        outbuf[outidx++] = digit                   ;
        break                                      ;

    default:
        printf("intasc() illegal nchrs: nchrs = %s\n",nchrs) ;
        exit(0)                                               ;
}                                               /* end of switch           */
}                                               /* end of intasc() function */
```

Listing 2

```
/**********************************************************************/
/* FILE DWNLD.C                                                       */
/*                                                                    */
/* This program downloads the S-records file specified by arg1 to a   */
/* single board computer which is connected to the serial port        */
/* specified by arg2.                                                 */
/*                                                                    */
/*     An example calling syntax is:                                  */
/*                                                                    */
/*     dwnld userfile.sr /dev/tty1                                    */
/*                                                                    */
/* Note that the tty port MUST NOT be configured into the Unix system */
/* (i. e., the /etc/inittab file entry for tty1 must contain :o:, then */
/* re-boot Unix).                                                     */
/**********************************************************************/

#define BUFSIZ    2048                   /*transfer buffer size       */

int f1,f2                               ;/*I/O file descriptors       */
char buf[BUFSIZ]                        ;/*transfer buffer            */
char inchr                              ;/*received char              */
int n                                   ;/* # of bytes received       */

/****************/
/* main program */
/****************/
main(argc,argv)
int argc                                ;
char *argv[]                            ;
{
/************************************/
/* Check number of input arguments. */
/************************************/
if(argc != 3){
    printf("Two arguments required\n")  ;
    return                              ;
}
/***********************/
/* Open the input file. */
/***********************/
if((f1 = open(argv[1],0)) == -1){
    printf("Can't open %s\n",argv[1])   ;
    return                              ;
}
/*******************************/
/*Open the output file (port). */
/*******************************/
if((f2 = open(argv[2],1)) == -1){
    printf("Can't open %s\n",argv[2])   ;
    return                              ;
}
/***************************************************/
/* Send contents of input file to output file (port). */
/***************************************************/
while((n = read(f1,buf,BUFSIZ)) > 0){
    write(f2,buf,n)                     ;
}
printf("Download completed\n")          ;

}                                         /*end of main() function    */
```

CHAPTER 8

Adding A PC Slave To VMEbus

By Ken Boyette

When IBM PCs and VMEbus systems were introduced in the early 1980s, it was assumed that the PC-based systems would be most successful in office applications, while VMEbus systems would be used for high-performance real-time applications, such as industrial automation. Clearly VMEbus has followed those expectations, but the use of IBM PCs and PC-compatibles has expanded far beyond the confines of the office environment. Today PCs are often seen in laboratory systems, data acquisition systems, instrumentation systems, and even in embedded control systems. But using PCs and VMEbus-based systems need not be an either/or proposition. In this chapter Ken describes how you can merge the PC with VMEbus, to get the benefits of both.

The PC was designed to fit the needs of the individual or the small business. Since it was intended for single users, its MS-DOS operating system was not designed to run multiple programs concurrently. The peripheral I/O needs were easy to anticipate (1 or 2 serial ports, a parallel printer port, 1 or 2 floppy disks, and a hard disk) so the hardware and the operating system were designed specifically to support these.

Low cost was important. To keep manufacturing and test costs down, the PC was designed around a motherboard concept, which allowed all of the standard hardware components to be mounted on a single large printed circuit board. Since the processing power needed by a single user is fairly modest, no hardware provisions were made to support multiple microprocessors.

Contrast this with the VMEbus. VMEbus was designed for use in multiuser (or in real-time control) applications. Because of this, data transfer rates were important, so provisions were made for fast 32-bit data transfers. Another major strength of the VMEbus is its ability to accommodate multiple processors. As a general purpose bus, VMEbus provides the versatility needed to build a wide variety of system architectures, with varying degrees of complexity and cost.

Using a PC for industrial control

IBM did include a feature in the IBM PC to facilitate expansion; a feature they called the *I/O channel*. Their intention was apparently to allow the installation of additional memory and communication boards. However, the I/O channel (now widely known as the PC bus) has been used for much more. In fact the PC bus has accounted for much of the PC's ongoing success. Third party vendors have designed a wide range of add-in I/O boards, for interfacing the PC to many things never anticipated by the original PC designers.

With all of these specialized interface boards available, engineers have attempted to use PCs in all sorts of *real world* applications. However, having the right interface boards is not enough. You must also consider how *reliable* the PC will be when it's used in harsh industrial environments. The standard desk-top PC was simply not designed for hostile factory conditions.

Of course, you could purchase a PC motherboard, and package it in an environmentally secure enclosure, such as a NEMA-compatible box. (This is typically done by mounting the motherboard against the back wall of the enclosure, with its slots facing the front door, as shown in Figure 1.) NEMA enclosures provide protection against mechanical damage, while providing adequate ventilation. However, that still leaves another reliability problem: the PC bus edge connectors. Because market pressures have forced makers of PC-clone motherboards to use very inexpensive connectors, the connector reliability may be poor.

Servicing a NEMA-enclosed system can also be a problem. Once the motherboard slots have been filled with expansion cards, you may find it impossible to access the motherboard for diagnostic purposes. If you determine that there is a problem in the motherboard, replacing it can be both difficult and time consuming.

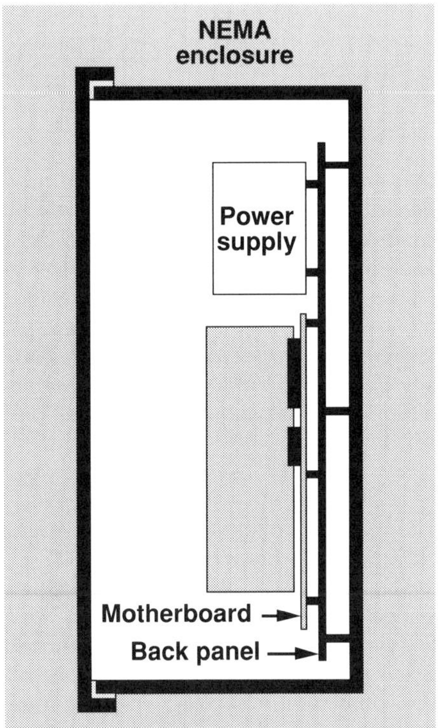

Figure 1

You can simplify motherboard replacement by designing a PC-bus-based system with a *passive backplane*, which contains no active components. You can then provide the active motherboard functions on a *plug-in board*. However, this still leaves the problem of supporting the boards in a high vibration environment.

PC bus expansion boards have been designed in a wide variety of physical sizes. That wasn't really a problem in office environments, where the PC wasn't subject to high levels of shock and vibration. But, where vibration is a consideration, this variable board size can make it difficult to clamp these boards into their edge connectors.

Using MS-DOS for real-time applications

All of these mechanical problems are important but, by far, the biggest problem in applying the PC to real-world applications is in the area of real-time response. How can MS-DOS (which is not a real-time operating system) be used?

If the real-time requirements aren't too demanding, you could develop a real-time operating system, and run it as an application program under MS-DOS. In that case,

MS-DOS system calls could be used to provide disk file access, and to manage the operator interface (the keyboard and CRT). However, if the response time provided by this approach is inadequate, you'll be forced to find some other solution.

While the PC was not designed to support multiple processors, it does support DMA masters. Some board manufacturers offer intelligent add-in boards for data acquisition or image processing. These boards can be plugged into the PC bus, where they do DMA transfers under the direction of the PC's processor. While these DMA boards can respond to external events promptly (without burdening the main processor) their data transfers are ultimately limited by the response time of the main processor's MS-DOS operating system.

Merging the PC and the VMEbus

Without a doubt, the driving force behind the acceptance of the PC has been the growing availability of low-cost ready-to-run software packages. Many of these packages could be very useful to VMEbus users, if there is a convenient way of using them.

Evidence of this can be seen by the fact that many VMEbus-based data acquisition systems today are connected (via RS232 links) to IBM PCs. The PC provides the operator interface, and the disk storage needed to store the data that's collected by the VMEbus system. In time-critical applications (where the data must be acquired quickly) the real-time data acquisition is handled by the VMEbus processor(s). Then (after the data has been collected and stored on the disk) the PC is used to post-process the data, using standard PC-compatible application programs, such as DBASE-III, Lotus 123, or laboratory plotting and graphics packages.

This configuration offers the advantages of both VMEbus and the PC. However, the obvious disadvantage is the slow serial data transfer rate between the systems.

Using shared memory for data transfers

How can we speed up the data transfer? Why not design a PC-compatible VMEbus board that shares memory with the rest of the VMEbus system? This board could be designed to plug directly into the VMEbus card cage, right alongside the VMEbus data acquisition boards. This VMEbus-compatible PC could be designed to withstand relatively high vibration and shock, just like other VMEbus boards. Since it gets its power directly from the VMEbus, it needs no separate power supply, and it can be removed and serviced just like any other card in the system.

To ensure that any standard PC software package will run on it, you must also provide all of the standard I/O devices: serial and parallel ports, a standard keyboard port, a calendar clock, a video controller, and a floppy/hard disk controller.

This PC-compatible VMEbus board could *send* data to other VMEbus boards by writing the data into global VMEbus memory. Likewise, it can *receive* data by reading from global memory the data written there by other boards. But how should this writing and reading be done? There are 2 possible methods:

- The PC can be a VMEbus master
- The PC can be a dual-ported-memory, intelligent VMEbus slave

Making the PC a VMEbus master
The tightest coupling between the PC and the VMEbus can be accomplished if the PC is a VMEbus master. One way to do this is to provide all of the PC's I/O ports onboard, and have it request the VMEbus whenever it needs to read or write to memory. Several manufacturers offer products that do just that.

However, that approach can add a lot of expense, which may be hard to justify in some applications. For example, the PC's processor chip runs from its own clock, and the chip expects incoming signals to be stable when it samples them on its clock edge. However, the VMEbus is asynchronous, and knows nothing about this clock, so its transitions can occur at any time. To avoid metastability problems inside the chip, you need to latch the incoming signals. However, that introduces delays that may lead to wait states for the processor.

Making the PC a dual-ported-memory, intelligent VMEbus slave
Fortunately, there is a simpler alternative: couple the PC to the VMEbus as an intelligent, dual-ported slave. The use of intelligent dual-ported-memory slaves is quite common in VMEbus systems. These intelligent slave boards are capable of independent operation, and are used for such demanding duties as data acquisition, network communication, and disk storage control. Once the slave's operating parameters have been set up by the main processor, the slave can operate in parallel with VMEbus activity.

This independent operation simplifies operating system software, and reduces the amount of competing traffic on the VMEbus. Since the intelligent slave can focus all of its onboard resources on its I/O operations, it can perform those operations more rapidly (and more thoroughly) than the main system processor, which shares its resources among several tasks.

A dual-ported-memory intelligent slave PC can be thought of as an extremely flexible intelligent slave board. Its internal programming can either be fixed in firmware, or can be dynamically loaded with any one of hundreds of off-the-shelf programs. A VMEbus system could even be equipped with several of these PC-compatible slave boards, each running a dedicated program.

Interrupts provide synchronization
A VMEbus-based PC can pass blocks of data to another VMEbus processor through its own dual-ported RAM. However, when the PC writes into its RAM, it must notify the other VMEbus processor that the data is available. To allow for this, you need to provide some kind of *synchronization mechanism*.

One way you can provide for synchronization is to equip the PC with a VMEbus *interrupter* and an *interrupt handler*. The PC's interrupter allows it to *send* interrupts to other VMEbus processor boards, and its interrupt handler allows it to *receive* interrupts from other boards.

But, how do you interface the interrupter hardware and the interrupt handler hardware to the MS-DOS software, and to the real-time operating system on the VMEbus processor board? One simple method is to treat the interrupter and the interrupt handler as I/O ports. You do this by providing a software driver that writes to the control registers (and reads from the status registers) of the interrupter and interrupt handler. With these drivers in place, application programs that run on the 2 operating systems can communicate with each other, just as if they were interconnected with an I/O link.

Using off-the-shelf PC software
By interfacing the VMEbus processor board to the VMEbus-compatible PC through this *simulated I/O port*, you permit the use of a number of very sophisticated industrial monitoring and control programs, that have been written to run on the PC. These programs treat the entire VMEbus system as an intelligent subsystem. The VMEbus system does all the real-time data acquisition and control, under the supervision of the PC's program.

For example, you might use Labtech's *LT/Control*. It allows you to define an interactive graphics interface, which permits an operator to adjust parameters dynamically. It also provides algorithms for the control of PID loops, provides mathematical routines, and allows the generation of reports, all without the need for programming. By using these off-the-shelf PC-compatible applications programs, you can greatly reduce your system development time.

Using the PC as an embedded development system
The versatile PC has largely replaced the relatively expensive dedicated development systems of the past. Hardware engineers use the PC to do schematic capture, circuit simulation, device programming, and documentation. Software engineers do their editing, cross-compiling, linking, simulation, and debugging, on PCs. With the addition of add-in emulators and logic analyzers, the need for breadboarding has also been largely eliminated.

Even after a VMEbus system has been completely integrated, and the operating system is running, the PC continues to play an important role in the development of application software. Many real-time VMEbus operating systems (such as Microware's OS-9) are sold with PC tool kits that provide specialized editors, compilers, device libraries, file transfer utilities, and debuggers, which greatly simplify the development of the target application software.

Even after the development process is complete, an embedded PC in the VMEbus system provides added convenience. If the PC is left in the system it provides an operator interface, and a standardized floppy disk format. When one of the serial ports is interfaced to a modem, the PC can also provide a way to remotely configure, control, and even diagnose a VMEbus system, after the system has been installed in the field.

The VME-1030/SXT
Logical Design Group has developed an IBM PC/XT compatible VMEbus board, based on the concepts presented in this chapter. (See Figure 2) The VME-1030/SXT provides all of the standard motherboard XT functions on a double-height, single-slot VMEbus board:

- 1 Mbyte of 16-bit DRAM
- 2 serial ports
- 1 parallel port
- an EGA-compatible video controller
- a floppy disk controller

With its NEC V30 processor chip (running at 10 MHz) and its 16-bit memory data path, the board has a performance level approaching that of an 8 MHz IBM PC/AT.

The board uses the A and C rows of the VMEbus P2 connector to provide PC bus signals for expansion. From the VMEbus point of view, the board is an A32/D16, 512-Kbyte slave. The board also provides 16 mailbox registers, which can be used (in conjunction with its 7-level interrupter) to facilitate communication and task synchronization between it and other VMEbus boards. Companion products include a dual floppy disk controller board, a hard disk controller board, and RAM-disk drive.

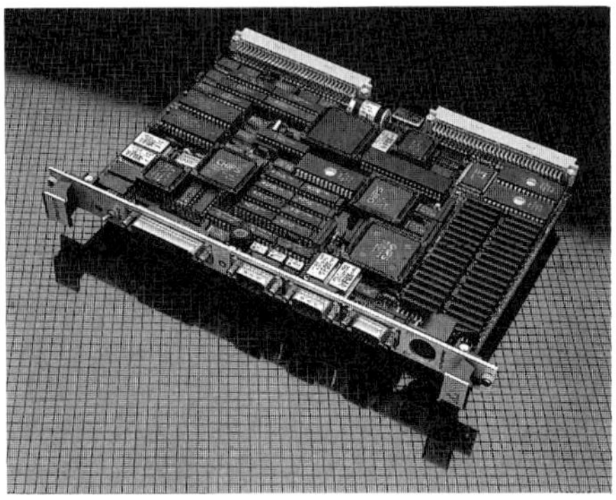

Figure 2

Summary

By embedding a VMEbus-compatible PC in your VMEbus system, you can solve many of the packaging, service, and reliability problems that accompany the use of an off-the-shelf PC clone. This PC can be either a VMEbus master or slave. While a master interface allows the PC's processor to access the global VMEbus memory directly, it can be expensive. In addition, data acquisition software for PCs typically assumes that the PC is connected to the data acquisition system through an I/O port. By designing the PC as a dual-ported VMEbus slave, you can use this off-the-shelf software while keeping costs low, and still benefitting from the close coupling between the PC and the VMEbus system.

Ken Boyette is the president of Logical Design Group in Raleigh, North Carolina. He has been active in board-level product design since 1978. Ken founded Logical Design in 1984 to design and manufacture VMEbus boards and custom microprocessor products.

CHAPTER 9

Running MS-DOS Without A Disk

By Manos Pinacoulakis

By using the large base of development software available under MS-DOS, you can write and debug your application programs in high level languages quickly and cheaply. At the same time, the wide selection of VMEbus I/O boards and the rugged nature of the VMEbus backplane standard makes it very attractive for use in industrial automation. To take full advantage of both of these standards you need to have a rugged PC-compatible VMEbus processor board. In this article Manos describes the problems you're likely to face in designing and using a PC-compatible processor board. He also shows how you can handle the most difficult problem of all: running MS-DOS without a fragile, failure-prone disk drive.

It's easy to see the many advantages of running MS-DOS in your VMEbus system. However, there are also some pitfalls. In implementing a PC/AT architecture on the VMEbus you should be aware of problems stemming from:

- Byte ordering
- 24-bit VMEbus addressing
- Expanded memory
- Short (I/O) address space
- 80286 vs. 80386

Byte ordering
Assembly language programmers who have spent their entire careers programming for the Motorola 68000 family architecture are surprised when they see a memory dump from an Intel 80X86-based system. Motorola processors store the *most significant* byte of each 16-bit or 32-bit word at the *lowest* byte address, while Intel processors store it at the *highest* byte address.

Since each processor reads bytes from memory in the same order that it stores them, that's usually not a problem. However, when you build a closely-coupled multiprocessing system that includes both a Motorola processor and an Intel processor, things get complicated. The problem can't be solved by simply reversing the order of the bytes whenever they're passed from one processor to another, because byte strings and multiple-byte words are stored differently.

24-bit VMEbus addressing
The first generation IBM-PCs were based upon the Intel 8088 processor chip, which is essentially an 8086 with an 8-bit data bus. Its internal address registers are 16-bits wide, which would normally provide access to 64 Kbytes of memory. To extend the addressing

range to 1 Mbyte, the 8088 was also equipped with 16-bit *segment registers*. The 16-bit values in these segment registers are shifted left 4 bits, and then added to the address registers, to produce a 20-bit address.

The 8088 and 8086 were not designed to support multiprocessing. While some attempts have been made to do multiprocessing on these processors, there is no way to keep one task from accidentally writing into the address space of some other task.

More recent Intel processors (the 80286 and the 80386) include memory management units (MMUs) which prevent each task from accessing any portion of the address space except its own. This MMU also allows each task running on the 286/386 processors to address up to 4 Gbytes of *virtual address space*. Unfortunately, IBM-PC users have already invested hundreds of millions of dollars in software that knows nothing about the 286/386 memory management hardware. To permit MS-DOS (and its application programs) to run on these processors, Intel also incorporated a *real mode* into these new processors, which essentially disables the memory management unit, and provides an 8086/8088 architecture. When running in real mode, the 80286 and 80386 processors run faster than the 8086, but they can still only generate 20-bit addresses.

The real mode does a good job of emulating the 8086/8088. However, it can create problems for the unwary. The danger lies in the fact that the 80286 processor has 24 address lines: A00-A23. Since the VMEbus has 23 address lines (A01-A23) plus 2 data strobes, you might assume that you can simply route the 80286's A01 to the VMEbus A01, the 80286's A02 to the VMEbus A02, etc.

That approach will work as long as the 80286 is running in its *virtual mode*, and is generating 24-bit addresses. However, to run MS-DOS, the processor must be in the *real mode*, and that means that only the lower 20 address lines are valid. The Intel 80286 specifications warn that address bits A20-A23 may not always be zero in real mode. To prevent problems, you must provide hardware to *force* the VMEbus address lines to the desired level, whenever the 80286 is in the real mode.

Expanded memory

The original IBM-PC design reserved 640 Kbytes of the 1-Mbyte address space for dynamic RAM. However, as new application programs were introduced, this was found to be inadequate. In order to allow the 8088/8086 to access more memory with its 20-bit addresses, Lotus, Intel, and Microsoft defined an expanded memory standard, now called LIM-EMS. This expanded memory standard uses *bank switching*. Each bank of memory is equipped with *bank switching hardware*, which allows that bank to be *selected* or *deselected*.

Before the processor accesses memory it must select one of the memory banks. This bank then responds every time the processor accesses memory. To access additional memory, the processor must first select another bank. (This automatically deselects the current bank.)

However, since the VMEbus standard provides 32 address lines, it does not include any provision for bank switching. VMEbus memory boards don't have the necessary bank switching hardware to allow selection and deselection of memory banks. Thus, a standard VMEbus memory board can't be used to provide LIM-EMS-compatible expanded memory.

Short (I/O) address space
VMEbus I/O boards with 8-bit peripheral chips appear as D08(O) VMEbus Slaves. This means that each 8-bit I/O register appears at an odd address, with vacant locations at the interleaved even addresses. However, MS-DOS I/O drivers (which were originally written for the 8-bit 8088) assume that the I/O registers occupy *contiguous* byte locations.

This problem is not unique to VMEbus. It also arose when the 8-bit PC bus was expanded to the 16-bit AT bus. Fortunately Intel processors indicate when they're doing I/O accesses. This allows you to include hardware between the processor chip and the bus that reroutes all 8-bit I/O transfers from the upper byte lane of the processor chip (D08-D15) to the lower byte lane (D00-D07) of the bus. This makes the 8-bit registers of the peripheral chips appear (to the processor) as contiguous byte locations. If done properly, this remapping is automatic, and is transparent to the software.

You *must* provide this onboard rerouting hardware, so that the 8-bit registers for serial ports and parallel ports will be contiguous, as they are in a PC-compatible computer. That allows you to use standard MS-DOS I/O drivers, and ensures compatibility even when application programs access the I/O registers directly.

80286 vs. 80386
Although the 80386 has generally been considered to be the processor of choice for high performance in AT designs, recently released benchmarks are telling a different story. 80286-based PCs execute slightly faster than 80386 machines running at the same clock speed, with the exception of integer and floating point math. Unless an 80386 board includes cache memory (or has a higher clock speed) you might be better off using a fast 80286 design.

Running MS-DOS without a disk
The IBM-PC is a disk-based machine. In fact its MS-DOS operating system can be considered simply a collection of disk services. Disks are very convenient, and they work well in stationary installations. However, they're also electromechanical devices, and are susceptible to wear and tear. When subjected to high temperatures, high vibration, contamination, and even physical abuse (typical of a factory floor) disk failure rates are unacceptable to industrial users.

However, it's hard to pass up the conveniences offered by MS-DOS. As a result, engineers have devised several methods for using MS-DOS software without a disk.

Creating an MS-DOS application program
To understand how you can build a diskless PC, you first need to understand how application programs are built and how they are run on a PC.

When you compile the source code of your application program on a PC, the result is a *relocatable object file*, with an .OBJ suffix. However, this is NOT an executable file. In order to generate an *executable file*, you use the MS-DOS LINK command. The LINK command converts the .OBJ file into an executable file, with an .EXE suffix.

You might think that MS-DOS could then simply load the .EXE file into memory and execute it, but that's not the case. The .EXE file actually contains a *program section* and a *data section*. MS-DOS loads these 2 sections into memory independently. Since some portion of the system's memory will be occupied by the MS-DOS operating system (and possibly by some additional *resident* programs) MS-DOS must first determine *where* the system has available memory, and *how much* memory is available.

If MS-DOS finds enough memory, then it:

- *reserves* a program area in memory
- *loads* the program segment from the disk into memory
- *reserves* a data area in memory
- *loads* the data segment from the disk into memory
- *reserves* a stack area in memory
- *tells* the program where its stack area is
- *tells* the program where to begin execution
- *initiates* execution of the program

Since MS-DOS is used to load and execute your .EXE file, the operating system must be resident in memory before the .EXE file can be run.

Running your .EXE file without a disk
But suppose you don't want to use a disk drive. How could you load and run your application program without a disk?

One method is to burn MS-DOS into EPROM, and then figure out a way to start it up when the system is turned on. That allows you to boot the system without a disk. However, you still need some way to load your .EXE application program into the system. You could put your application program in EPROMs, but that makes the process of updating your program a tedious one. A more flexible approach is to add some additional firmware to MS-DOS that allows you to *download* your .EXE file from a remote disk-based system, over a serial link.

Maybe your application program can do without MS-DOS
If you're willing to write your application program in such a way that it doesn't call upon any of the MS-DOS services, you can use a special kind of MS-DOS utility program (often called LOCATE) to convert your .EXE file to a *memory image file*.

When you invoke LOCATE (and give it the name of an .EXE file) it creates a memory image file. Since this memory image file is completely self contained, it can be run even when MS-DOS is NOT resident in memory. It is simply loaded into memory from disk, and then executed. Once you have the memory image file, you can burn it into a set of EPROMs, and then plug those EPROMs into your system.

However, one of the reasons for using MS-DOS in the first place is to take advantage of the services that it offers. Unless you are willing to provide all of these services yourself, you'll probably want to take a 3rd approach: use a solid state disk.

Solid state disks
Solid state disks are the most popular way to run MS-DOS in a harsh environment. The solid state disk might be anything from bubble memories to EEPROMs with battery-backed-up static RAMs. Since the hardware interface for these solid state disks will not look exactly like a standard PC disk controller board, you'll also need to have a software driver to interface the solid state disk to MS-DOS.

Assuming that you have a solid state disk in your VMEbus system, how would you get your application program developed and running on the system? The first step is to use a disk-based PC to develop and debug your application code. If your application code depends heavily on VMEbus I/O hardware, you'll probably want to do your development on a PC-compatible VMEbus processor board that resides in the VMEbus card rack.

After your program is thoroughly debugged and tested, you need to create a diskette with the files that you want to put onto the solid state disk. To do that, you format the diskette, using the standard MS-DOS FORMAT utility. (You invoke the /S option, so the MS-DOS files will be copied onto the diskette.) Then you copy your application program to the diskette. You'll probably also want to copy an AUTOEXEC.BAT file to the diskette, to automatically start your application program when the system is powered up.

Once you've created the diskette, all you need to do is to verify that it works properly, and then use a utility program to copy the diskette file to your solid state disk.

The VRG-286 processor board

The Intergon VRG-286 is a single-slot 20 MHz, 286-based processor board, designed around the NEAT CHIPset from Chips & Technologies. (See Figure 1) The board offers:

Figure 1

- a low-power 80C286 CMOS processor from Harris Semiconductor
- a real-time clock
- 2 serial ports
- a parallel port
- a coprocessor socket
- a keyboard controller
- 1 Mbyte of DRAM
- memory interleaving, for fast RAM access
- fast execution of the EPROM BIOS from DRAM
- an LIM-EMS memory controller
- VMEbus system controller functions

The VRG-286 occupies a single slot, and has all of its connectors accessible from the front panel. A graphics controller board (EGA, VGA or Monochrome) can be piggybacked, without violating the VMEbus single-slot dimensional constraints. Because its system controller functions can be enabled or disabled, it can be used either as the sole processor (in slot 1) or as part of a multiple processor system.

Although 80X86 processors generate 16-bit I/O addresses (allowing access to 64 Kbytes of I/O address space) the I/O boards designed for use in the IBM-PC only decode the lower 10 address lines (A0-A9). Thus, while those I/O boards can all be accessed within a range of 1,024 locations, they also appear redundantly at the higher addresses. Since the standard set of IBM-PC peripherals occupy most of this 1 Kbyte I/O space, there are only a handful of locations left for additional peripheral boards. To provide for more expansion options, the VRG-286 decodes all 16 of the onboard address bits, and accesses the 64 Kbyte VMEbus short I/O map whenever the onboard address is above the bottom 1024 locations. Whenever the processor accesses memory, the VMEbus standard (24-bit) addressing is used.

There are 15 interrupts in the PC/AT. All but 3 are assigned to standard peripherals such as video, keyboard, serial ports, etc. The VRG-286 assigns the 3 remaining interrupts to its VMEbus interrupt handler. This allows the VMEbus interrupt lines IRQ5*-IRQ7* to interrupt the onboard 80C286 processor. The processor then uses the status/ID byte (received from the VMEbus in the ensuing interrupt acknowledge cycle) to vector to any of 256 possible interrupt service routines.

The VMEbus IRQ1* and IRQ2* lines also interrupt the processor, which responds to them as floppy disk controller interrupts. However, the VRG-286 board doesn't generate a VMEbus interrupt acknowledge cycle in response to interrupts on these 2 VMEbus lines.

The VRG-202 solid state disk
The VRG-202 is a single slot VMEbus board with 12 JEDEC 28-pin sockets. By installing 27256-type or 27512-type EPROMs, or by installing 43256-type static RAMs (or any combination) the board can be configured to serve as either a read-only or a read/write solid state disk. The board provides a battery back-up for the RAMs. Its maximum storage capacity is 720 Kbytes. Additional VRG-202 boards can be used to provide increased capacity.

The VRG-202 is supplied with a ROMDISK utility that converts executable object files into binary files, based on the type of EPROMs used (27256 or 27512). This utility converts these binary files into the various hexadecimal ASCII formats used by different PROM programmers. Once the PROMs have been programmed, they can be inserted into the appropriate sockets on the VRG-202 card, allowing the system to run diskless. The firmware on the VRG-286 processor board contains the drivers needed to use the VRG-202 as drive A.

Interrupting VMEbus processors
The VRG-202 also includes a *general purpose interrupter*, for use by processor boards that would not otherwise be able to generate VMEbus interrupts. When any VMEbus master writes to a specific location on the VRG-202, the VRG-202's interrupter generates an interrupt on a jumper-selected IRQX* line.

For example, if the VRG-286 processor board wants to pass data to some other processor on the VMEbus it can do so by first writing the data into VMEbus memory, and then generating an interrupt to the destination processor by writing to the VRG-202.

The VRG-303
The VRG-303 *Floppy Card* provides the disk storage needed to develop application software in a VMEbus chassis. The VRG-303 is a double-slot VMEbus board, containing a floppy disk controller and a 3.5 inch, 720 Kbyte floppy drive. The board includes a ribbon connector on its front panel that allows connection of a 2nd floppy drive: preferably a 5.25 inch drive, to allow loading of software from 5.25 inch floppies.

The VRG-302
The VRG-302 *Hard Card* provides hard disk storage. This double-slot board contains a hard disk controller and a 3.5 inch hard disk, with a capacity of either 20 Mbytes or

40 Mbytes. Both the Floppy Card and the Hard Card are VMEbus slaves, and both communicate with the VRG-286 processor board over the VMEbus, via the P1 connector.

$50Network

Intergon's $50Network can be used in systems that don't have a slot for a Floppy Card or for a Hard Card. This software connects the VRG-286 board to a desktop PC over an RS-232 link, and allows the VRG-286 to access drives A, B, and C of your desktop machine, just as if they were its own disks. This allows editing, compiling, and downloading of any file for execution in the VMEbus system. In fact, either computer can gain access to the other's resources at any time.

Manos Pinacoulakis is a founder and president of marketing at Intergon Inc. He received his BSEE from the University of South Florida, his MSEE from Georgia Tech and his MBA degree from the University of Houston. Prior to founding Intergon in 1983, Manos held various design, management, and strategic marketing positions within the Texas Instruments OEM board products group in Houston.

CHAPTER 10

High Performance Hardware For Running MS-DOS On VMEbus

By Mike Tobias and Dennis Waldenmayer

The main reason for wanting to run MS-DOS in a VMEbus system is that it allows you to use a wide range of inexpensive, and useful software. However, you should consider more than just software compatibility. In this article, Mike and Dennis discuss some of the other features you need to have on your processor board.

IBM PC/AT-compatible processors are becoming very popular on several different backplane bus architectures. The driving force behind this is the wide range of application software available for PC compatible systems. As a VMEbus system user, you can take advantage of this software to greatly reduce the time it takes you to integrate and develop a VMEbus system.

However, you need to concern yourself with more than just PC/AT compatibility. For example, you should consider:

- how easy it is to access VMEbus memory and peripherals from your PC/AT board, and from MS-DOS application programs.
- how you should handle the byte ordering differences between Motorola and Intel processors.
- whether or not you need an onboard cache to deliver the performance you need.
- how you can coordinate message passing between the PC/AT and other VMEbus boards.

Access to VMEbus memory and peripherals
The 80386 microprocessor operates in 1 of 2 modes:

- 8086 real mode
- Protected virtual address mode

Since MS-DOS was originally written to run on the 8086/8088, it (and most MS-DOS compatible application software) runs on the 80386 in real mode. However, real mode provides only 1 Mbyte of addressing space, so your board must provide the processor with some way to access the much larger 16-Mbyte and 4-Gbyte VMEbus address spaces.

Using a real-mode window
To allow the processor to access more than 1 Mbyte in real mode, you can equip the board with a *real-mode window*. This window can be positioned over any portion of the

VMEbus address space, allowing direct reads and writes of VMEbus memory or peripheral registers within the window. One convenient area of the PC/AT memory map is the 64-Kbyte section from E0000 to EFFFF (hex). This is the area that IBM reserved for ROM BIOS expansion.

Through this window you can dynamically access:

- the 64 Kbyte VMEbus *Short I/O* address space
- any 64 Kbyte section of the 16-Mbyte VMEbus standard map
- any 64 Kbyte section of the 4 Gbyte VMEbus extended map.

To allow the 80386 to position the window over the desired area of VMEbus memory, you'll need to have control registers that can be accessed by the 80386 as I/O ports. These registers would allow the 80386 to access any portion of the VMEbus address map (including both memory and peripherals) without switching the processor into protected mode.

Solving the byte ordering problem

Most VMEbus systems have been designed around a Motorola 680X0 processor. When you introduce a PC/AT processor board into a 680X0 VMEbus environment, you must provide some way to handle the difference in byte ordering between Motorola and Intel processors.

Motorola processors store the Most Significant (MS) byte of a word or longword in the lowest memory address, and the Least Significant (LS) byte at the highest memory address. The Intel family of processors store their data in the opposite fashion: the MS byte is at the high memory address, and the LS byte is at the low memory address.

When you pass data between a Motorola processor and Intel processor, the bytes must be rearranged. If this is done in software it will slow the system down considerably. To avoid this slowdown, you can provide byte-swapping hardware. You do that by providing additional buffers to automatically reroute the data byte lanes between the 80386 processor and the VMEbus.

Byte-swapping hardware clearly outperforms software. But, there are some situations where you don't want to do any byte swapping. For example, if an Intel processor communicates with another Intel processor, there's no need to swap the bytes. To allow for this, you need to design the byte swapping hardware so that an application program can dynamically enable or disable the byte swapping as needed.

Using a cache

To run an 80386 processor without wait states at 25 MHz, you MUST have cache memory. The best way to construct this cache is with an Intel 82385 cache controller, which supports an 8K X 32-bit cache. As long as the processor executes code (or fetches data) that has been previously stored in the cache memory (a cache hit), it runs without wait states.

Avoiding stale data problems

If you make the local memory on your 80386 processor board dual ported, you permit other VMEbus masters to write directly into its memory. That can often speed things up.

But it also demands that you provide some mechanism to ensure that the local processor doesn't continue using the stale data from its cache after the dual-ported memory contents have been overwritten. There are 2 possible ways to do this:

- Flush the entire cache memory every time a VMEbus master accesses the onboard dual-ported RAM.
- Use a *bus snoop*.

It's easy to design hardware that flushes the entire cache whenever local memory is accessed. However, the results are not very satisfactory. Even an occasional access to the dual-ported memory drastically degrades performance.

However, you can design a bus snoop that watches for accesses to the dual-ported memory. When the snoop detects a write to a cached memory location, it only invalidates the small portion of the cache that contains that entry. If the write is to a non-cached location, the snoop takes no action, and the content of the cache is preserved.

Message passing
It's quite likely that you'll want the PC/AT processor to be able to pass information to some other VMEbus board. There are 2 good methods for passing data:

- Using dual-ported memory
- Using a virtual MS-DOS RAM disk

Using dual-ported memory
You can pass information from an MS-DOS application to a VMEbus processor by simply having the 80386 write the data to the onboard dual-ported memory. A VMEbus master could then read that data. The same method could be used to pass information from the VMEbus master to the 80386.

Virtual MS-DOS RAM disk
Reading and writing directly to dual-ported memory is the fastest way to pass information. However, if you write to the wrong area of memory, you might crash the operating system. For that reason, you may want to use a slower, but safer method: pass information through the MS-DOS operating system. You can do that by creating an MS-DOS *RAM disk I/O driver*, which looks to MS-DOS just like a standard disk drive, but actually stores its information in dual-ported memory.

Besides being safer, this RAM disk method has another big advantage: it allows any off-the-shelf MS-DOS application program to easily send information to (or receive information from) any VMEbus master. For example, an intelligent VMEbus data acquisition board could accumulate data, and then write that data to the MS-DOS RAM disk. Then an MS-DOS-compatible application program could read this data from the RAM disk, and generate a pen plot.

If your VMEbus system includes a real-time operating system, there's another good reason for using the RAM disk method. Several of the popular real-time operating systems are packaged with drivers that allow them to read from (and write to) MS-DOS-formatted disks. If you're using one of these operating systems, you can easily use this driver to interface it to an MS-DOS RAM disk.

Synchronization by interrupts

When passing data between a VMEbus master and the 80386 board, you need to provide some way for the 2 processors to interrupt each other. There are 2 useful mechanisms for doing that:

- Dual-ported memory interrupts
- VMEbus interrupts

Dual-ported memory interrupts

To provide for dual-ported memory interrupts, you need to equip your 80386 board with hardware that decodes the address of all VMEbus accesses to its onboard dual-ported memory. This decoding hardware generates an 80386 interrupt whenever a VMEbus master accesses a designated portion of the 80386 board's memory.

VMEbus interrupts

Dual-ported memory interrupts allow a VMEbus master to interrupt the 80386 processor, but they don't allow the 80386 to interrupt VMEbus processors. To provide for that, you need a VMEbus interrupter. It does 3 things:

- It generates an interrupt on a VMEbus IRQx line.
- It provides a vector when the interrupted VMEbus processor board responds with an interrupt acknowledge cycle.
- It informs its onboard 80836 processor when the VMEbus processor board has acknowledged the interrupt, and the vector has been passed to that board.

To allow the 80386 to interact with the interrupter, you need to provide 3 registers:

- a control register
- a vector register
- a status register

The 80386 first writes a status/ID vector into the vector register. (This status/ID vector will be passed to the interrupted VMEbus board, when it acknowledges the interrupt.) Note: This vector register must be loaded BEFORE the 80386 board generates any VMEbus interrupts.

AFTER the 80386 writes to the vector register, it generates the VMEbus interrupt, by writing to the control register.

The interrupted VMEbus board may take some time to acknowledge the interrupt, and to read the status/ID vector. The status register should have a bit to indicate the board's interrupt status, so the 80386 can determine whether or not its pending interrupt has been acknowledged.

If you have already provided hardware that allows the 80386 to access the VMEbus through a real-mode window, it takes very little additional hardware to provide a VMEbus interrupt handler.

When the 80386 processor detects a VMEbus interrupt, its interrupt service routine simply enables the real-mode window for the *IACK space*, instead of the usual memory space. Then (by reading a location within the real-mode window) the 80386 generates a VMEbus interrupt acknowledge cycle (which looks very much like a VMEbus memory read cycle). The 8-bit value that it reads from the interrupting VMEbus board will be the status/ID vector.

By using dual-ported memory and these interrupt techniques, your 80386 processor board can provide very efficient message passing.

The XVME-683

The Xycom XVME-683 is a double-height, 2-board, 80386-based processor module, which runs at a clock speed of 25 MHz. Its block diagram is shown in Figure 1.

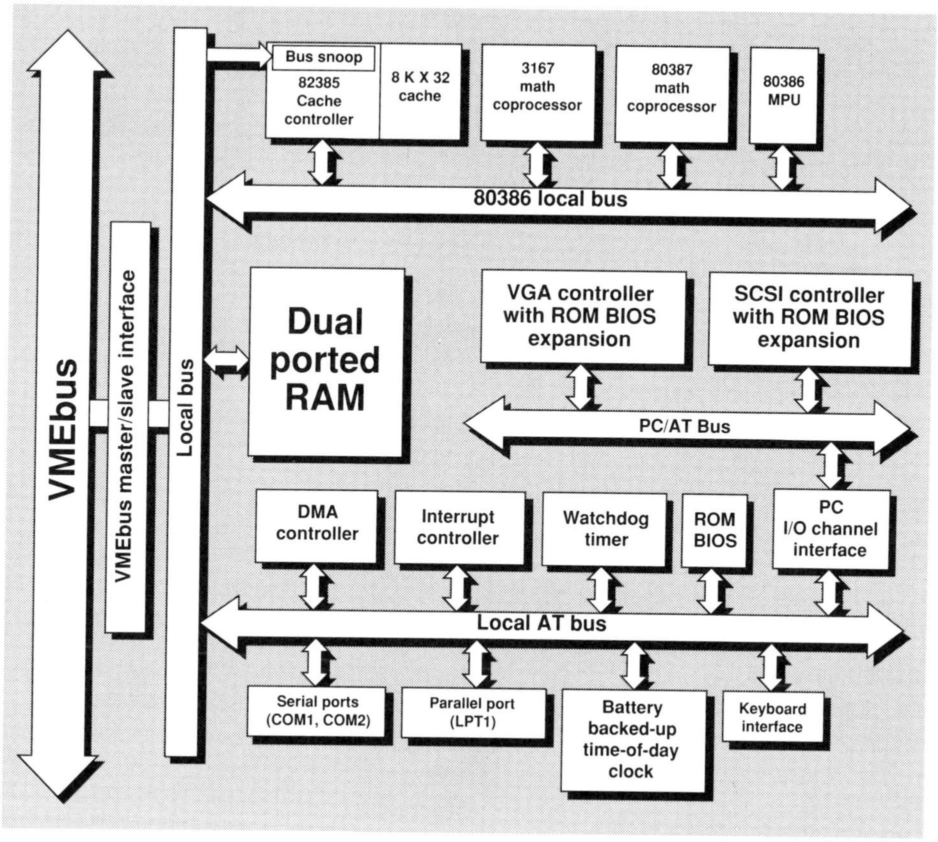

Figure 1

Figure 2 shows the local bus memory map. Except for the area from D0000 to EFFFF, the lowest 1 Mbyte of the XVME-683's address space looks like a standard MS-DOS computer. The area from E0000 to EFFFF provides a window that permits the 80386 to access the VMEbus address spaces in real mode.

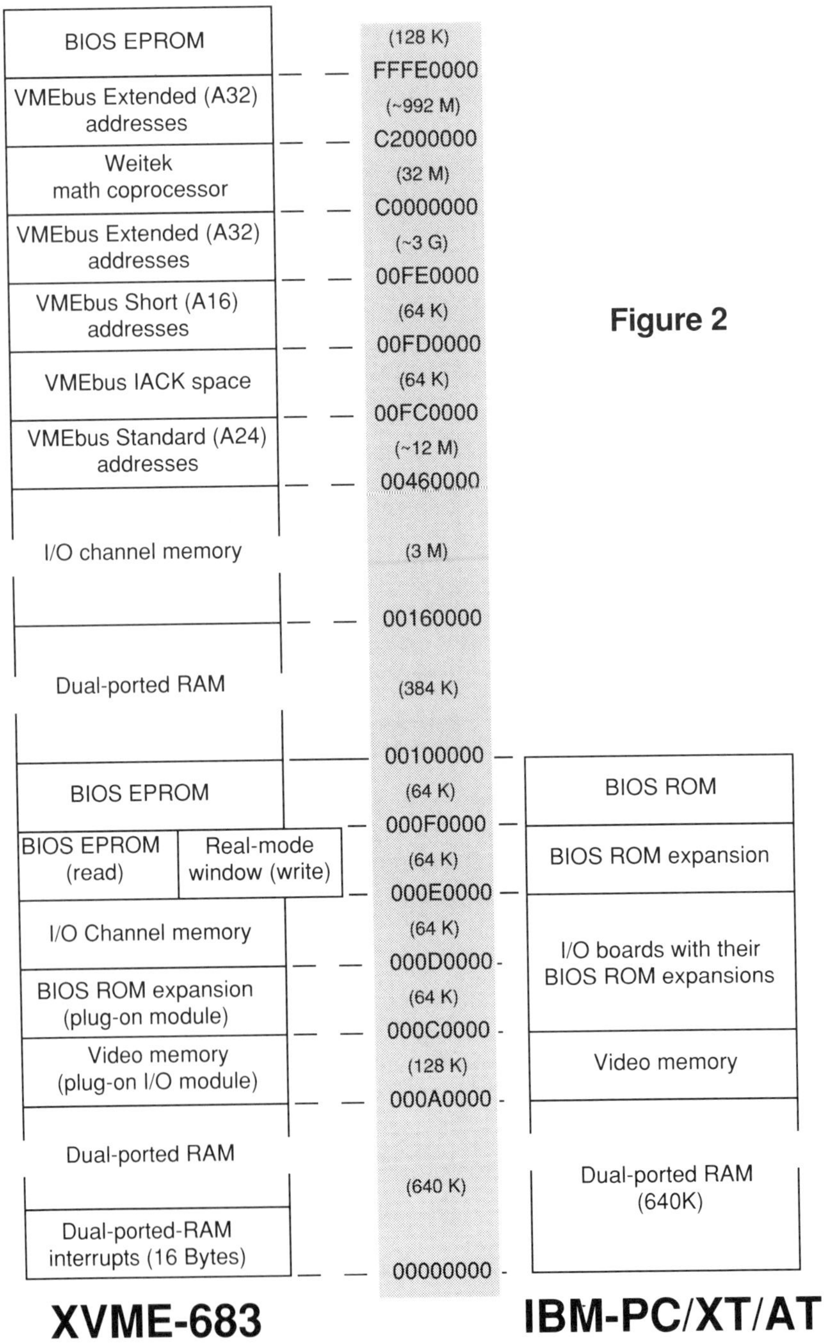

Figure 2

To allow byte swapping, the XVME-683 is equipped with special hardware, as shown in Figure 3. Before the 80386 writes a 32-bit longword to memory (for consumption by a 680X0 processor) or reads a 32-bit longword (provided by a 680X0 processor) it first enables the "A" buffers. This causes the 4 bytes of each 32-bit data transfer to be swapped.

When the 80386 reads or writes 16-bit data, only the bytes on the lower 2 byte lanes need to be swapped. The 80386 accomplishes this by enabling the "C" buffers.

When no byte swapping is needed, the software enables the "B" buffers, causing the data to be passed directly to the VMEbus.

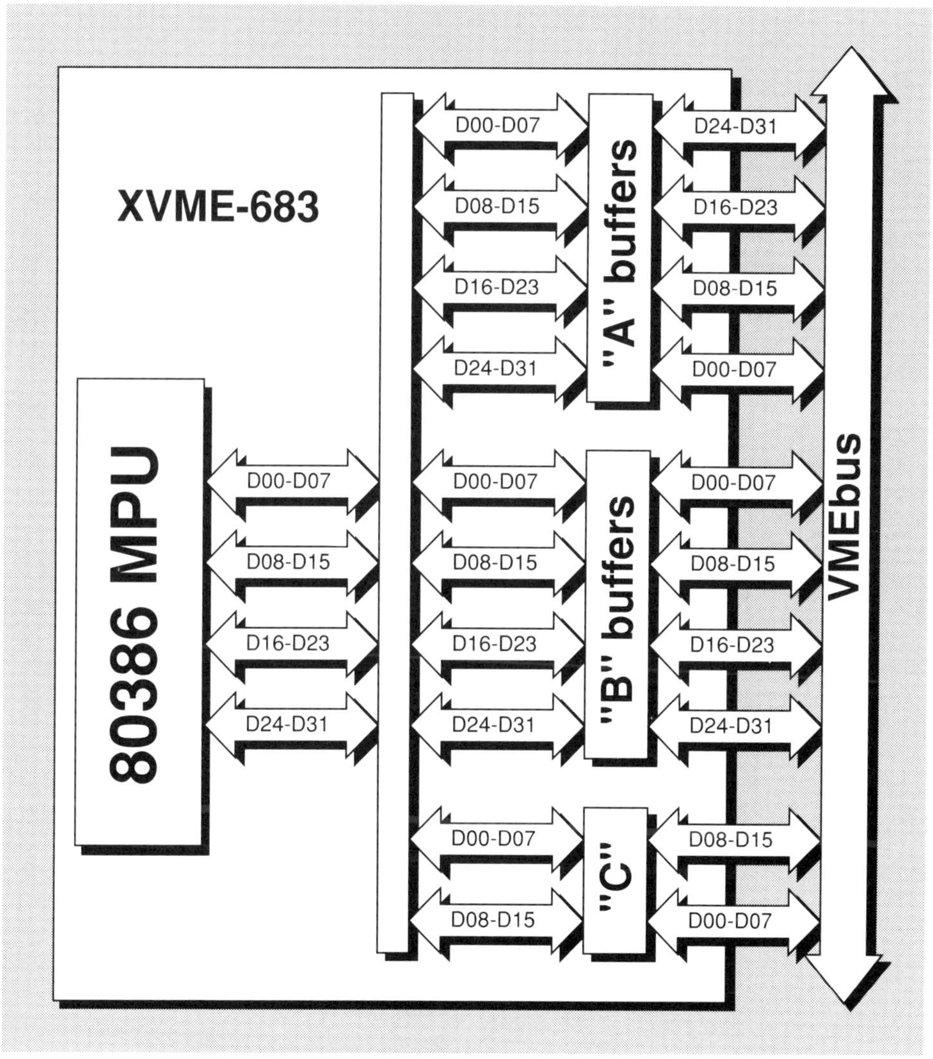

Figure 3

To allow the board to run the 80386 processor at 25 MHz, the XVME-683 is equipped with an Intel 82385 cache controller and an 8K by 32-bit cache memory. To guard against the use of stale data from its dual-ported RAM, the board is equipped with a bus snoop, which detects writes to the dual-ported memory, and invalidates any stale entries.

To support efficient message passing, the XVME-683 provides dual-ported memory interrupts, as well as a VMEbus interrupter, and a VMEbus interrupt handler. Any VMEbus master can generate a dual-ported memory interrupt to the onboard 80386 by simply reading 1 of the lowest 16 bytes of the dual-ported memory. The interrupter can be configured to generate a VMEbus interrupt on interrupt levels 1-7. The status/ID vector that it provides to the VMEbus is from the onboard vector register, which gets initialized by the 80386. An onboard status register allows the 80386 to detect when its pending interrupt has been acknowledged. The 80386 acknowledges VMEbus interrupts by mapping the real-mode window to the IACK space, and then reading the status/ID vector from the VMEbus interrupter through that window.

Mike Tobias is a senior design engineer with Xycom, where he has worked since 1979. His primary responsibilities include the design and development of VMEbus boards. In addition to these design responsibilities, Mike organizes and leads project teams within Xycom. Mike holds a BSEET from the University of Lake Superior. In his spare time he enjoys woodworking, and radio-control planes, cars, and boats.

Dennis Waldenmayer is the VMEbus product manager for Xycom, where he has worked since 1981. Prior to his current position, Dennis was Xycom's application engineering manager. Dennis has a BSEE in computer engineering from the University of Michigan. His hobbies include baseball and bowling.

CHAPTER 11

Coupling MS-DOS To VMEbus: It Takes More Than Just Hardware

By Glen Myers

Many VMEbus users have begun to use IBM-PC-compatible VMEbus modules in their systems. In this article Glen talks about why this is happening, and explains how to design a PC-compatible module so that it takes fullest possible advantage of the enormous quantity of off-the-shelf PC-compatible software.

Software written to run on IBM-PC compatibles includes far more than just MS-DOS and its compatible application programs. In addition to MS-DOS, you can purchase many other off-the-shelf operating systems that run on the PC family, such as:

- OS/2
- UNIX
- Xenix
- VRTX
- QNX

Along with these operating systems, you can also purchase:

- compilers
- expert systems
- graphics packages
- tool kits
- database systems
- networks
- utilities

There's also a large number of application-specific programs for such things as:

- data acquisition
- process control
- ATE program generation
- chromatographic data analysis
- signal processing

Because of the enormous size of the PC industry, the PC-compatible software business is intensely competitive. That results in a wide variety of choices, low prices, and high quality.

For instance, if your interest is in developing your own C application programs, there's a huge number of PC-based products, including:

- compilers (some with incremental compilation, some with excellent code optimization)
- source-language debuggers (such as Microsoft's impressive CodeView)
- C++ compilers
- Screen managers
- Subroutine libraries that go far beyond those provided with the compilers
- Source-code version control systems
- Execution profilers
- Translators (Fortran to C, etc)
- Cross-reference tools

Because of all of this software, PC-compatible VMEbus modules are replacing traditional single-board computers in an increasing number of VMEbus systems. However, interfacing the PC hardware architecture to the VMEbus is not completely straightforward. To make things work, you must:

- interface the PC hardware architecture to the VMEbus without compromising any of the rigid (but largely undocumented) PC compatibility constraints.
- include in the PC-compatible VMEbus board the features normally included on intelligent VMEbus boards: master and slave interfaces, an interrupter, an interrupt handler, and system controller functions.
- provide some way for MS-DOS application programs to access the VMEbus address spaces.
- provide useful software utilities, to help users take full advantage of VMEbus facilities.

Mating the 386 with VMEbus

The main problem with using an 80386 processor on the VMEbus is the difference in byte ordering between Intel's 80X86-family processors, and Motorola's 680X0-family processors. For example, suppose you store the hexadecimal byte values 01, 02, 03, and 04 in memory locations A000-A004 respectively. If an 80386 reads a 4-byte value from location A000, it will load the hexadecimal value 04030201 into the 386's register. A 68020 processor (reading from the same address) will load the value 01020304.

Byte ordering problems with I/O devices

Of course each of these processors writes data to memory, and then reads from memory, using the same byte ordering. So, as long as each processor only reads data that it has previously written to memory, you won't see this byte ordering difference.

However, the problem will surface when a 386-based processor board accesses a VMEbus peripheral board. These peripheral boards are designed with the 680X0's byte ordering in mind.

For example, suppose you just ignore the byte ordering problem, and simply design your 386 processor board such that its 32 data bus lines are connected directly to the corresponding 32 VMEbus data lines. Then, you write and compile the following C code, to read a 16-bit value from a VMEbus A/D converter board, using your 386 processor board.

```
/************************************************/
/* p points to the 16-bit channel 0 data register  */
/* which is located in A16 space                   */
/************************************************/

sample = *p;

if (sample >= previous) ...
```

Because the assumed byte ordering built into the 386 is the reverse of the byte ordering built into the VMEbus A/D converter board, the upper and lower bytes will be reversed. To use this value, the 386 must first call a subroutine that swaps the 2 bytes.

Byte dependent code causes problems
If you were only concerned about accessing I/O devices, you could include this byte-swapping routine in the I/O driver. That would hide the problem in the system code, where it could be forgotten.

However things aren't that simple. Many C-language programs contain code that assumes a particular byte ordering. For example, suppose you want to write a routine that tests an integer, to determine whether or not it's an exact multiple of 256. You can do such a test by reading the least significant byte of the 16-bit value, and then testing it for zero, as shown below...

```
union {integer whole; char part [2];} myvalue;
myvalue.whole = i;
if (myvalue.part[0] == 0) /* i is multiple of 256 */
```

Note: The `union` defines the integer storage location in such a way that it can be accessed as either a single 16-bit value, or as two 8-bit values.

But notice that the 3rd line in the C code shown above *assumes* that the *lowest* byte address (myvalue.part[0]) contains the *least significant* byte of the integer. That would be the correct assumption for most 80X86-based systems, but the VMEbus standard demands that 16-bit data be stored in VMEbus memory with the least significant byte in the *highest* address. If an 80X86 processor board reads and tests the lowest byte of a 16-bit value over the VMEbus, this algorithm will fail.

Solving the byte ordering problem with hardware
It would seem that you could solve this problem by connecting the 386's byte lanes to those of VMEbus, in a criss-cross manner. For example, you would connect the 386's D24-D31 lines to the VMEbus D00-D07 lines. Unfortunately, this approach leads to another problem.

While the 80X86 and the 680X0 processors *disagree* about byte ordering within integer values, programs (and library functions) written for the 80X86 and 680X0 processors *agree* on string ordering. Many of these programs, and library functions, access strings 2 or 4 bytes at a time, for higher performance.

Suppose we call a 386 library function to copy a string from VMEbus memory (where the byte lanes are swapped) to local memory (where the byte lanes are not swapped). If the function reads and writes the characters 4 bytes at a time, this high-speed copy operation would read the following string...

```
My dog has fleas
```

...from VMEbus memory, and write the following string...

```
d yMh gof sasael
```

...into local memory.

So what's the answer? You need to design the data path between the 386 and the VMEbus to either swap or not swap the byte lanes, depending upon what *type of access* is being done (integer, byte string, etc). Since the type of access is only visible in the application program, you must allow the application program to enable or disable the byte lane swapping dynamically, each time it accesses the VMEbus address space.

Mating the PC Hardware Architecture with VMEbus
In order to properly interface a PC-compatible processor board to the VMEbus, you need to allow the PC to:

- access each of the VMEbus address spaces
- monitor and respond to VMEbus interrupts
- monitor and respond to SYSFAIL
- respond to BERR in an appropriate fashion

All of this must be done without violating the rigid (and partially undocumented) rules concerning strict PC compatibility. If you fail to accomplish the latter, you will produce a PC-compatible board that runs some percentage of PC software, but will fail when running many other programs (such as Windows/386 and some networking software) that demand absolute compatibility.

The original IBM-PC was based on the Intel 8088, with its 1 Mbyte address space limitation. MS-DOS, and its application programs were designed with this in mind. While the newer 80X86 processors allow you to place memory (and memory-mapped registers) above the 1-Mbyte upper limit, you must provide additional software support to access those resources, and that overhead extracts a performance penalty. A better approach is to provide a movable *mapping window* somewhere within the PC's 1 Mbyte address space.

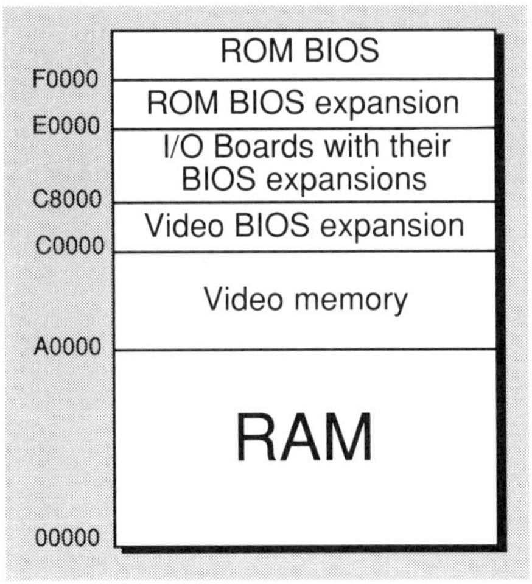

Figure 1

IBM defined the 1-Mbyte address map to look like Figure 1. Based on this map, it seems logical to use the area from D0000 to DFFFF to provide a 64-Kbyte VMEbus window. However, using that area would be a mistake, because many Ethernet, Arcnet, and token-ring network cards map their dual-ported RAM into this area.

That would seem to leave us with no 64-Kbyte window. But fortunately, there's one other area that you can use. IBM originally reserved the EXXXX area for extensions to the system BIOS, but that area has never been used.

Supporting VMEbus interrupts

The PC/AT architecture supports 15 hardware interrupts. Some of these are dedicated for specific purposes. Some additional ones are reserved for the normal complement of peripherals:

- serial ports
- parallel ports
- floppy disks
- hard disk

That leaves only 4 unassigned interrupt lines: fewer than the 7 VMEbus interrupt lines. To provide support for all 7 of the VMEbus interrupts, you need an *interrupt controller* that accepts interrupts from all 7 VMEbus interrupt lines, and routes them to one of these 4 available PC interrupt inputs. If you design this interrupt controller properly, it will also allow you to treat other time-critical VMEbus signals (such as SYSFAIL, and ACFAIL) as interrupts.

Miscellaneous status and control registers

To allow a programmer to position the VMEbus window over the desired VMEbus address range, you'll need to provide onboard *status and control registers* somewhere in the PC's 1-Mbyte map. You'll also need to provide onboard registers to support the VMEbus interrupt lines, and register bits to indicate the level of the VMEbus SYSFAIL* and ACFAIL* signals. It's best to address all of these registers in the PC's I/O space. However, you need to be careful where you put them, since some specialized PC application programs test certain I/O locations to detect the presence of I/O boards.

Mating MS-DOS with VMEbus

If you take into account all of the concerns listed above, you should be able to run off-the-shelf PC software without modifications. However, you may want to provide extensions to support VMEbus functions.

For example, you might want to expand the PC's ROM BIOS functions. Most BIOS's perform a power-on self test on the hardware. By expanding this self-test routine, you can also ensure that the VMEbus interface is working correctly.

Using an MS-DOS device driver

You'll also find it helpful to have special software to initialize and control:

- the window mapping
- the interrupt handling
- the VMEbus control registers.

Without this software, your application program development will be tedious, and error-prone.

MS-DOS includes a mechanism for automatically loading device drivers into system memory during power-up. These device drivers then become memory-resident extensions of the MS-DOS services. They are typically called by MS-DOS application programs to do I/O operations. By calling the device drivers the applications can accomplish their I/O without needing to know the details of the I/O hardware registers.

You'll probably want to create an MS-DOS device driver to:

- position the address window over the desired area of the VMEbus address space.
- allow application software to handle VMEbus interrupts.
- allow application software to send VMEbus interrupts.

With this driver, you can dramatically reduce the amount of time it takes to develop application programs.

The EPC-1
The Radisys EPC-1 is an example of a PC-compatible module for the VMEbus. It consists of a 2-slot-wide module which provides:

- an 80386-based PC-compatible computer
- a floppy disk
- a SCSI interface
- a graphics controller

Another 2-slot companion module contains:

- a 1.44 Mbyte floppy disk drive
- a 40-Mbyte SCSI hard-disk drive
- an adapter, which allows you to use a *short* PC/AT bus add-on card

The EPC-1 allows its 80386 processor to access the VMEbus through a 64-Kbyte window, beginning at address E0000. This window allows access to any location, in any of the VMEbus address spaces (A32, A24, A16), while using any address modifier, and any data width. The registers used to control this window are located in the PC's local I/O space, and are shown in Figure 2.

The first two registers in Figure 2 control the mapping. The address bits written into them are used to drive the upper VMEbus address lines. Additional bits in the 2nd register are used to drive the address modifier lines during the VMEbus access, and to specify whether or not the byte lane swapping hardware is enabled.

Since the 64-Kbyte window allows access to the entire VMEbus A16 short I/O space, the application program could simply initialize these registers to do VMEbus I/O accesses, and then just leave them alone. However, if the program also needs to access memory (in the VMEbus A24 and A32 spaces) it must include software to move the window around as needed, by writing into onboard control registers.

I/O address									
8150	A31	A30	A29	A28	A27	A26	A25	A24	Window address register
8151	A23	A22	Byte order	IACK	AM5	AM4	AM2	AM1	Window modifer register
8152	IRQ7	IRQ6	IRQ5	IRQ4	IRQ3	IRQ2	IRQ1	MSG	Interrupt state register
8153	IRQ7	IRQ6	IRQ5	IRQ4	IRQ3	IRQ2	IRQ1	MSG	Interrupt enable register
8154					Watch dog	AC-FAIL	BERR	SYS-FAIL	Event state register
8155					Watch dog	AC-FAIL	BERR	SYS-FAIL	Event enable register

Figure 2

When OS/2 or UNIX are run on a 386 the memory map is not limited to 1 Mbyte. For this reason, the EPC-1 automatically maps all local 386 addresses above the 16-Mbyte boundary directly into the VMEbus A32 space.

Software for the EPC-1
The EPC-1 is provided with a substantial amount of MS-DOS-based software, including:

- a bus manager
- a configurator
- a bus monitor
- a bus probe

The *Bus Manager* is an MS-DOS device driver that provides a hardware-independent interface to the VMEbus. Its allows an application program to position the window at any address in any specified VMEbus address space. It also allows the program to attach onboard interrupt handlers to VMEbus interrupt lines, and to perform high-speed copying of memory blocks between PC memory and VMEbus memory. To support application programs written in C or in Basic, interface libraries to this device driver are also provided.

To access particular VMEbus boards the user would write *interface functions* that call these VMEbus driver functions. In fact, a library of interface functions has already been created for a series of VMEbus products from Burr-Brown and Allen-Bradley.

The Configurator
The *Configurator* is an MS-DOS application program that maintains a database of descriptive information about the VMEbus boards in the system, including such things as where they're mapped in VMEbus address spaces, and what interrupts they generate.

This information allows it to display maps of each of the VMEbus address spaces, showing where VMEbus devices are located.

The Configurator can audit this database to find problems, such as overlapping address ranges. It can also access the VMEbus boards, to determine whether the database description of the system matches the actual physical makeup of the system.

In the MS-DOS world, Microsoft Windows is the premier software environment, and user interface; particularly the multitasking Windows/386. But Windows/386 is important for 2 other reasons:

- It allows multiple MS-DOS application programs to be run concurrently, using the 386's virtual 8086 mode.
- It provides an easy upgrade path to OS/2, both in terms of the user interface and the programming model.

For these reasons, several software packages provided with the EPC-1 are Windows/386 application programs.

The Bus Monitor
The *Bus Monitor* runs as a background task, sampling the VMEbus at a specified rate. It displays a simulated panel of LEDs on the PC screen, indicating the current state of the VMEbus, and the state of the PC's onboard hardware interface to the VMEbus.

The Bus Probe
The *Bus Probe* is the active counterpart of the Bus Monitor. It allows you to:

- perform VMEbus accesses (of any size, in any address space, using any address modifier, in either byte order)
- search VMEbus memory for matching patterns
- generate VMEbus interrupts
- reset the system

Another Windows application called the Auxiliary Processor Interface (AXPI) allows VMEbus processors to use the resources of the PC. For instance, a 680X0-family processor board on the VMEbus can:

- access the PC's disk directly
- read and write DOS files
- exchange messages with Windows application programs
- use the PC's keyboard as a virtual terminal
- display information in Windows on the PC's monitor

The AXPI package also includes C source code that allows a VMEbus processor to remotely call a procedure that runs on the EPC-1. For example, if your application program runs under pSOS, on a 68020 single board computer in the VMEbus chassis you can compile the AXPI source code, link the resulting object code into your 68020 application software, which could then call procedures that execute on the EPC-1.

Summary

VMEbus has provided an environment rich in hardware features, but poor in end-user and development software. The availability of PC-compatible plug-in modules for the VMEbus is a major factor in correcting this. However, these modules must be carefully designed to embrace VMEbus without compromising PC compatibility. Once this is accomplished (and a basic set of VMEbus-oriented software tools are available) PC-compatible modules will play a prominent role in new VMEbus-based designs.

Glenford J. Myers is president of Radisys in Beaverton, Oregon. Prior to founding Radisys in 1987, Glen spent 13 years at IBM, and 7 years at Intel, where he worked on the 80386 and 80960 products. He holds BSEE, MS and PhD degrees from Clarkson, Syracuse, and Polytechnic University.

CHAPTER 12

An Introduction To Multitasking

By John Black

A real-time system is a system that responds to at least one external event within a certain prescribed maximum time limit. Such a system can be as simple as a dedicated processor that controls a kitchen appliance, or as complex as a massive system that controls the continuous flow through a glass manufacturing plant. However, the key to designing any responsive real-time system is the proper design of the software. In this chapter, John describes several methods for writing this real-time software, and shows why real-time executives can be very useful.

A simple real-time system might consist of a microprocessor executing a program loop, and repeatedly reading a status register bit that indicates the occurrence of an external event. When the processor finally detects the event, it responds by executing a routine. The flow chart of the program for such a system would look like Figure 1.

A simple system like this is easy to design, since the processor is totally dedicated to servicing a single event. You can easily predict how fast the system will respond by simply counting the instructions in the program loop.

However, things are rarely this simple. Usually there will be several different types of events occurring, and you won't be able to justify dedicating a processor to service each event. However, if these events always occur in a cyclical pattern, the solution is still simple. For example, suppose events A, B, and C indicate 3 different angular positions through which a shaft rotates, as it makes a complete revolution. Figure 2 shows how you might program a single-processor real-time system to respond to these 3 events.

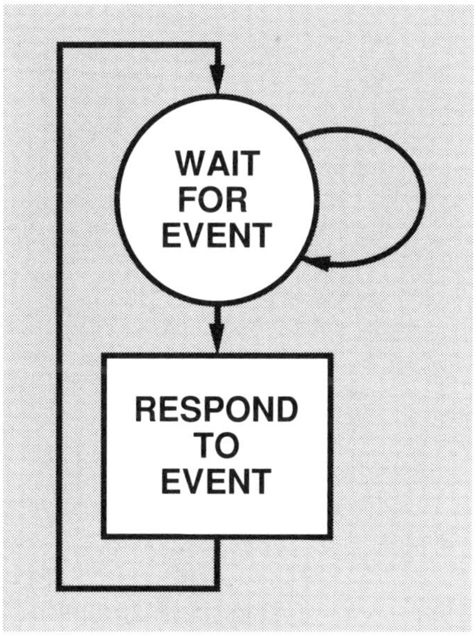

Figure 1

Unfortunately, events don't always occur in such a well behaved fashion. Suppose you have 3 photosensors positioned at 3 different places along a conveyer belt. Each sensor is used to detect the arrival of an object. Since there may be many objects moving along the conveyer belt at any given time, you can't be sure that photosensors A, B, and C will generate their respective events in a well-behaved cyclical fashion.

If you ignore this fact, and try to use a program like the one shown in Figure 2, you'll have serious problems. You might be waiting for an object to arrive at photosensor A and miss the fact that some other object has already arrived at (and possibly even passed) photosensor B.

Polling allows one processor to respond to many unrelated events
One solution is to modify the program slightly, as shown in Figure 3. Instead of waiting for event A, the processor just checks to see if event A has occurred. If it has, then the processor executes a routine to respond in the appropriate fashion. Then it checks to see whether event B has occurred. If event B has not occurred, then it checks to see whether event C has occurred, etc.

When you loop like this, checking to see whether any event has occurred, it's called *polling*. With a polled system, its still easy to calculate the worst-case time that the system might take to respond to an event. For example, suppose it takes 1 msec to check for each event, it takes 40 milliseconds to respond to event A, 35 milliseconds to respond to event B, and 30 milliseconds to respond to event C, as shown in Figure 3.

If you assume that any event can occur at any time, then the worst case response time would be...

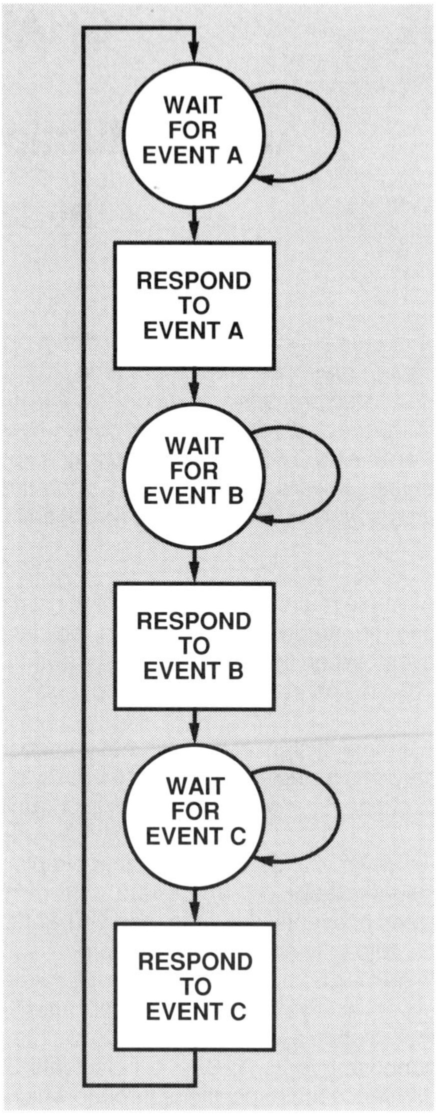

Figure 2

1 + 40 + 1 + 35 + 1 + 30 = 108 msec

In other words, the worst-case response time is the time it takes to go once around the polling loop, executing code for every event.

A modified version of polling can be used to handle urgent events
We defined a real time system as a system that responds to events within prescribed time limits. But different events may have different time limits. We call an event with

a very short time limit a very *urgent* event. An event with a longer time limit is less urgent.

A polled system works well when you need to respond to several equally urgent events. But suppose this is not the case. Suppose event A is more urgent than event B or C. For example, suppose event A has a time limit of 100 milliseconds, while events B and C can each wait as long as 150 msec. Since our loop time (as shown in Figure 3) is 108 msec, it appears that polling just isn't fast enough. However, you can still use polling if you *restructure* the program in Figure 3, to look like Figure 4.

In Figure 4, a test for event A has been inserted into the loop in 2 places. Because event A is checked twice each time around the loop, it will never have to wait the full loop delay. It will have to wait either...

40 + 1 + 35 + 1 = 77 msec
OR
40 + 1 + 30 + 1 = 72 msec

That satisfies event A's 100 msec time limit.

However, by inserting a test for A twice in the loop, you have increased the total loop time to 149 msec, so events B and C might have to wait up to 149 msec for a response from the system. Fortunately, this is adequate to meet their 150 msec time limit.

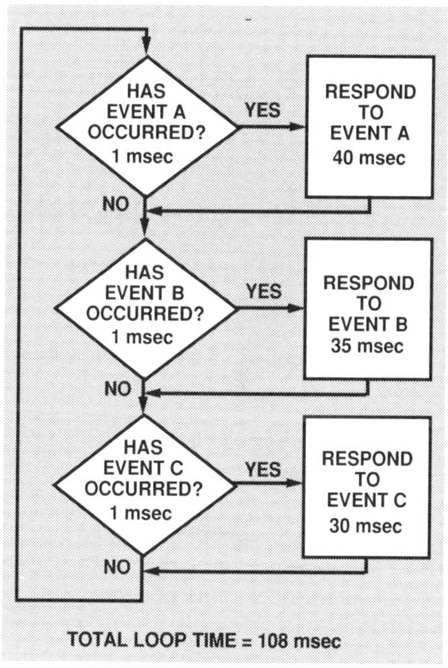

Figure 3

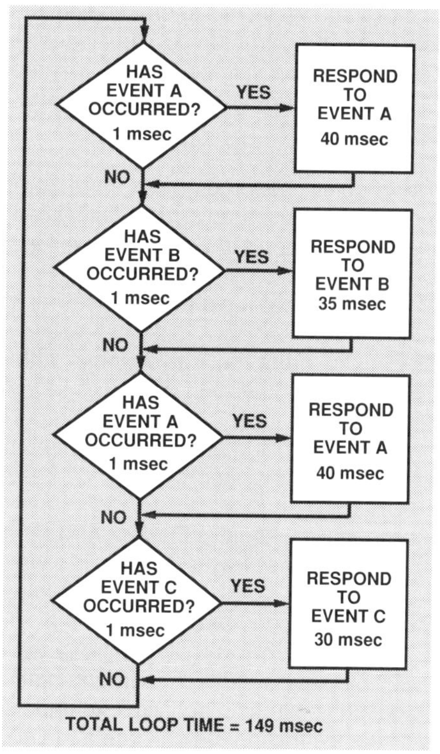

Figure 4

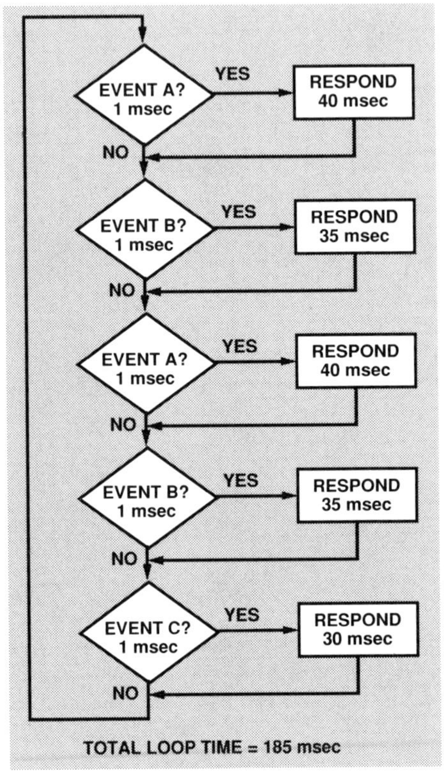

Figure 5

In some situations a polled approach doesn't get the job done

But now suppose we complicate the situation further. Suppose event A can wait 100 msec (as before) and C can wait 150 msec (as before) but event B can only wait 130 msec. If you try to restructure the loop as shown in Figure 5 (to satisfy B's response time) you find that the total loop time is 185 msec, and event C does not get serviced quickly enough. (If you then insert the test for C into the loop twice, you end up with timing identical to Figure 3, which means that A will not get serviced quickly enough.)

Suppose you get a little more creative, and restructure your software into 2 loops, as shown in Figure 6. Since the maximum loop time is 149 msec, event C now will get serviced fast enough. If you work out the maximum times for events A and B, you will see that they are also satisfactory. While this approach gets the job done, the structure of your code is getting pretty convoluted, and it's not hard to imagine where this would take us with a dozen events.

The problem becomes even more complicated when the routine for one event takes longer to execute than another event can wait. For example, suppose you have a system that must respond to 2 events as follows:

- Event A must have a response within 700 msec, and its routine takes 100 msec to execute.
- Event B must have a response within 100 msec, and its routine takes 20 msec to execute.

If you use a polling approach, you get the loop shown in Figure 7. The maximum loop time is 122 msec: clearly too long to satisfy event B.

Of course, you could split event A's responding routine into two 50 msec parts, as shown in Figure 8. That way, event B will always see a response within 72 msec.

But suppose you have a system that must respond to many events, instead of just 2. You might have to break up the larger response routines into many fragments. The resulting code, with all of its fragmentation (and possible branching, as shown in Figure 6) becomes almost impossible to understand and to debug. In addition, the insertion of a single instruction may dictate a major restructuring of the entire program.

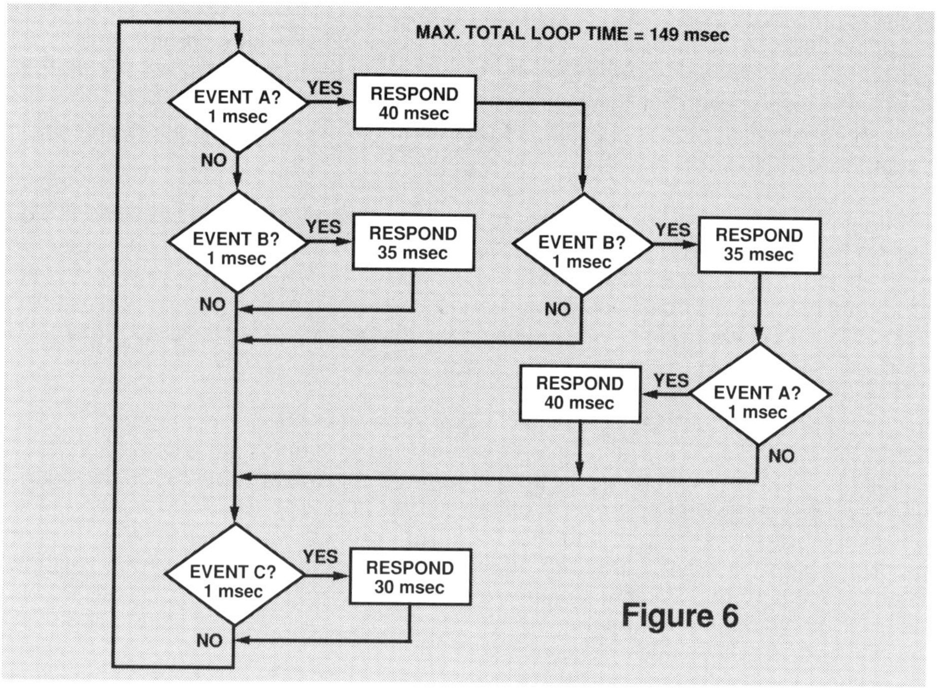

Figure 6

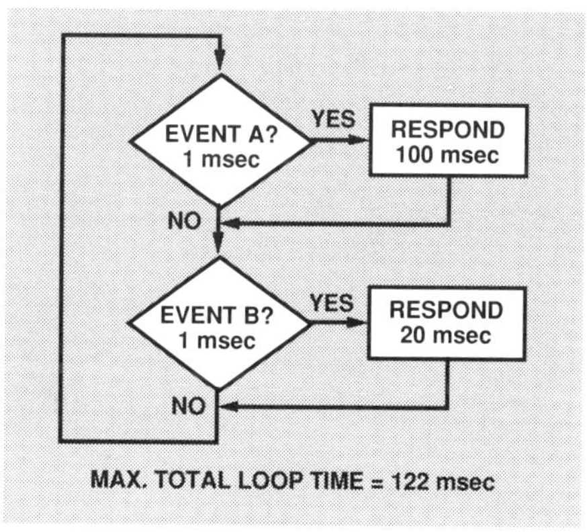

Figure 7

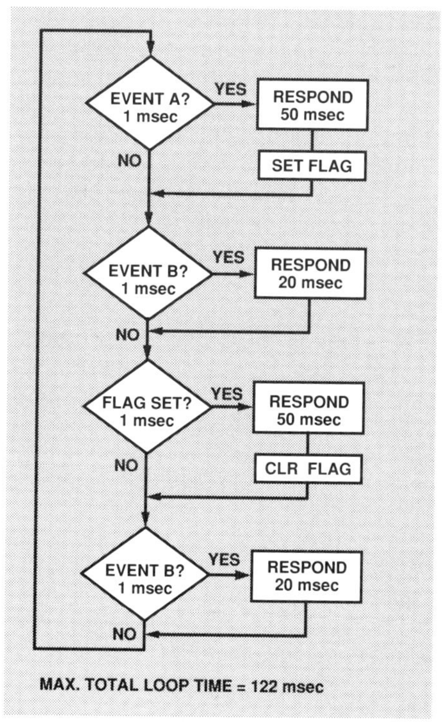

Figure 8

Multitasking provides an alternative to polling

Faced with all this complication, you might ask yourself, "Isn't there a better way?" The answer is "yes". Its called *multitasking* or *multiprogramming*. When we started this discussion we observed that things are very simple if you can dedicate a microprocessor to each event: each processor's job is to watch for its own event, and respond within the prescribed time.

However, dedicating a microprocessor to each possible event can be *expensive*. You end up paying for a lot of processing power that's wasted, executing non-productive program loops. A better solution is to design a kind of *multiplexer*, which allows several programs to share a single microprocessor. Such a multiplexer is called a *multitasking executive*, and each program which runs on the shared processor is called a *task*, or a *process*.

In order to create a multitasking executive, your processor must have interrupt handling hardware, which virtually every microprocessor has. This hardware allows an external signal (an interrupt) to temporarily take the processor away from a running program and give it to a special subroutine called an Interrupt Service Routine (ISR). The hardware does this by pushing the program counter contents (for the running program) onto the stack, and then loading the entry point address of the ISR into the program counter (This pushing and loading of the program counter is essentially a hardware-directed subroutine jump.)

Once this jump has taken place, the processor fetches and executes the ISR instructions until it fetches a *return from interrupt* instruction. At that point, the interrupt handling hardware returns control of the processor to the previously interrupted program, by *popping* the program counter value off the top of the stack and into the program counter. (This is essentially a return from subroutine.)

A simple multitasking executive

You can create a simple multitasking executive by writing a special ISR. This simple executive (like all multitasking executives) has a special data storage area in system memory, composed of task control blocks: 1 block for each task that's sharing the processor. Figure 9 shows a collection of 50 task control blocks, as they might appear in memory. A multitasking executive with 50 task control blocks would be able to run anywhere from 1 to 50 tasks, one at a time.

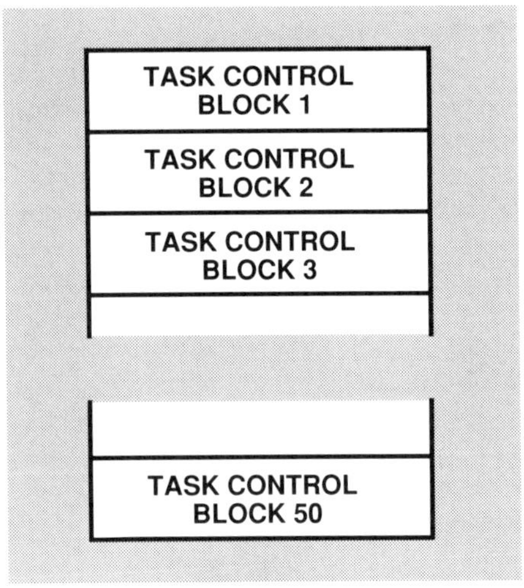

Figure 9

Each time the real-time clock interrupts the processor, your simple executive is entered. It does 5 things:

- It gets the interrupted task's *program counter value* (which was just pushed onto the stack) and stores it in the *task control block* of the interrupted task.

- It writes the contents of the processor's other registers (including the stack pointer register) into that same task control block. (Once all of these registers have been saved, the interrupted task is said to be *asleep*.)
- It chooses the task control block of some sleeping task, and copies the register values stored in that task control block into the processor's registers.
- It gets the *program counter value* from the same task control block, and *writes* (not pushes) it onto the top of the stack. Note: This overwrites the program counter value that was pushed onto the stack by the task that just went to sleep.
- It executes a *return from interrupt* instruction.

Upon execution of the return from interrupt instruction, the interrupt handling hardware *pops* the new program counter value from the stack, into the program counter. Since the executive has overwritten the stack's previous program counter value, control of the processor does not return to the previously interrupted task (which is now asleep) but returns to the task that the executive decided to awaken.

Notice that when a task is put to sleep, and later awakened, it is "unaware" that anything unusual has happened. As far as it can tell, it's the only program using the processor. That allows you to write each task like the one in Figure 1, just as if it were going to be run on a dedicated processor. The processor's executive then runs each of these tasks, one at a time.

A multitasking executive creates many *virtual processors*

By multiplexing the processor between all of the tasks, the executive is essentially creating a whole bunch of *virtual processors*. Since each of the programs running on these virtual processors is written to run independently, you can easily create and debug software that responds to dozens of different events.

Let's take a simple example. Suppose you need to design a system to handle 3 equally urgent events: A, B, and C. You write 3 tasks, like the one in Figure 1. Each task consists of an infinite loop that reads a status register bit over and over, until it detects an event, and then executes a response routine.

In order to make the system service all of the events, you want the executive to run task A for a short period, then put task A to sleep and run task B, then put task B to sleep and run task C, etc.

Time slicing with round-robin scheduling

You can accomplish this by driving the processor's interrupt input with a periodic signal from a clock circuit. The executive can then switch tasks when each clock interrupt occurs. When a processor is switched from task to task by a periodic clock signal, we call the procedure *time slicing*. Each periodic interrupt indicates the end of another *time slice*.

The executive can run each task for 1 time slice, before putting it back to sleep. Once a task has been awakened, run, and put back to sleep, it must then wait for all of the other tasks to run before the executive will awaken it again. Thus, the tasks would be run in a rotational fashion. This is called *round-robin* scheduling.

The executive keeps track of which task to wake up next by keeping a linked list of all the active task control blocks. For example, suppose there are 6 tasks

(A-F) being run by the executive. Figure 10 shows how the 6 task control blocks would be linked to a special *header block*.

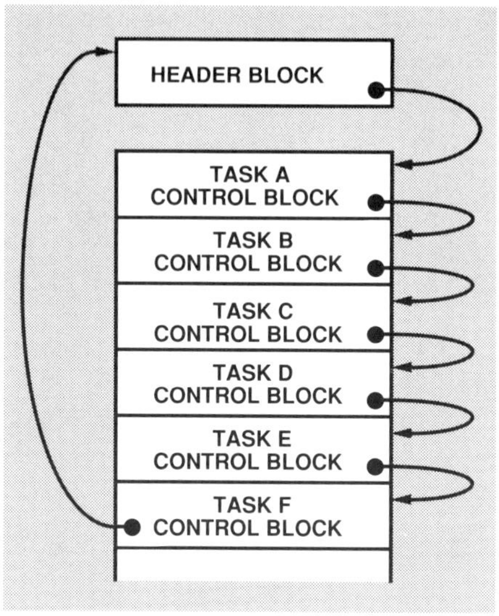

Figure 10

The executive first removes A's task control block from the list and awakens task A. After A has run for one time slice, the executive puts it back to sleep, and inserts A's task control block at the end of the linked list. (When the executive has run tasks A, B, and C, the linked list would look like Figure 11.) The executive continues stepping through the linked list, awakening tasks in the order in which they are linked.

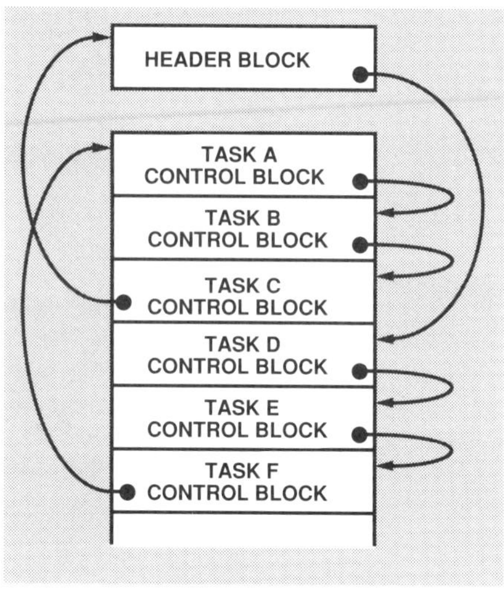

Figure 11

Since the processor is assigned in a round-robin fashion, each task is guaranteed the use of the processor within a certain number of time slices. (The number of active tasks.) However, the actual *time* each task waits will depend on the length of the time slice. How long should the time slice be? Let's take an example.

Suppose it takes the executive 1 msec to put one task to sleep and awaken another. (Most executives can accomplish this in 100 microseconds, but we'll assume 1 msec here, just to keep the arithmetic simple.) We will take the case described earlier, in Figure 3:

- Task A must finish execution within 100 msec after the event, and requires 40 msec to execute.
- Task B must finish execution within 150 msec after the event, and requires 35 msec to execute.
- Task C must finish execution within 150 msec after the event, and requires 30 msec to execute.

When a task is awakened, you want to guarantee that it has enough time to finish executing its response routine, before it's put back to sleep. Since task A takes the longest to execute, you need to choose a time slice of at least 41 msec. But with a 41 msec time slice, each task will be awakened only once every 123 msec, and that's not often enough to satisfy A's 100 msec response time!

In some cases, simple round-robin time slicing isn't fast enough

This situation appears to be analogous to Figure 3, and you might be tempted to apply an approach similar to Figure 4, by putting 2 copies of A's task control block in the linked list. (See Figure 12) Unfortunately, this doesn't work. With 4 task control blocks, B and C only get awakened once ever 164 msec, which doesn't permit them to respond quickly enough.

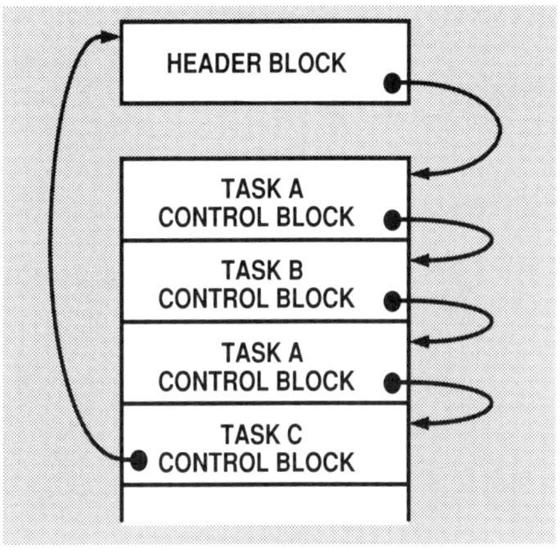

Figure 12

Figure 13 shows the difference between this situation and the one shown in Figure 4. The 41 msec time slice is more than tasks B and C need to execute but, since the executive allows them to continue executing until the end of the 41 msec time slice, they just execute a tight loop, testing their status register bit until the end of the time slice.

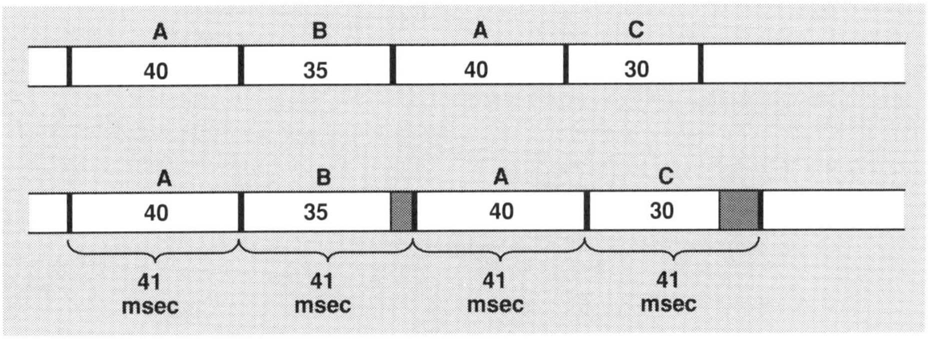

Figure 13

We chose the 41 msec time slice to guarantee that all of the tasks would finish execution of their respective response routines within one time slice. That works well when all of the tasks take about the same time to execute. But it doesn't work well when the various tasks take very different times. To take an extreme case, suppose you have 2 tasks:

1. Task X takes 100 msec to execute its response routine.
2. Task Y takes 20 msec to execute its response routine.

In this case, you might want to choose a 50 msec time slice, so task Y can respond quicker. (This is analogous to what we did with our polling software in Figure 8.)

But, suppose you use the task control block list shown in Figure 12 and vary the time slice length. Could you improve the worst-case response times by choosing a shorter time slice? If so, what's the optimum length? Since the response time depends upon the execution time of the specific tasks you're running, you must calculate the worst-case execution time of the tasks for each time slice length, and then choose the best length.

Figure 14 shows the total worst-case execution times for tasks B and C, as the time slice length is varied. There are clearly tradeoffs involved in choosing the time slice length. With very short time slices, the *overhead* goes up drastically. In fact, with a time slice of 2 msec (the shortest possible) the processor spends 50% of its time switching tasks! On the other hand, as the time slice gets longer, the idle time (See Figure 13) at the end of each time slice gets longer. (We are assuming here that the task doesn't detect another event near the end of its time slice: an assumption that might not be valid with very long time slices.)

According to Figure 14, the optimum length slice is 41 msec. (That's because the execution time of the 3 tasks (30, 35, and 40 msec) are very similar. Unfortunately, as we said before, even this optimum size is not good enough to provide adequate response times.

Putting each task to sleep as soon as it finishes executing
Suppose we try to solve the problem by finding unproductive uses of the processor's time, and eliminating them.

One place where our round-robin system is wasting processor time is at the end of each time slice. For example, with a 41 msec time slice, task C finishes execution of its response routine after only 30 msec, and then spends the last 10 msec of its time slice in a tight loop, checking its status register bit for another event.

You can largely eliminate this waste of time if you allow each task to return control of the processor to the executive when it finishes executing its response routine. Then, the executive can awaken another task, without waiting for the end of the time slice.

The task can give the processor back to the executive by executing a special *trap* instruction, which has the same effect as a time-slice interrupt. The executive then acts as if a time slice has just expired: it puts the current task to sleep and wakes up the next task.

To make use of this approach, you need to rewrite the code of tasks A, B and C to look like Figure 15. This approach eliminates wasted processor time in 2 ways:

- If a task determines that its event has not occurred since it was last put to sleep, it *immediately* executes a trap instruction, and is put to sleep again, releasing the processor.
- If a task determines that its event has occurred, it executes a trap instruction upon completion of its response routine. This allows the executive to awaken the next task before the end of the time slice.

Awakening tasks only when they have pending events
There's another place where a lot of processor time is wasted. The scheduler must still awaken each task so the task can check its status register for a pending event. As shown in Figure 15, if the executive awakens a task which has no event pending, that task simply executes its trap instruction, returning control of the processor to the executive. That

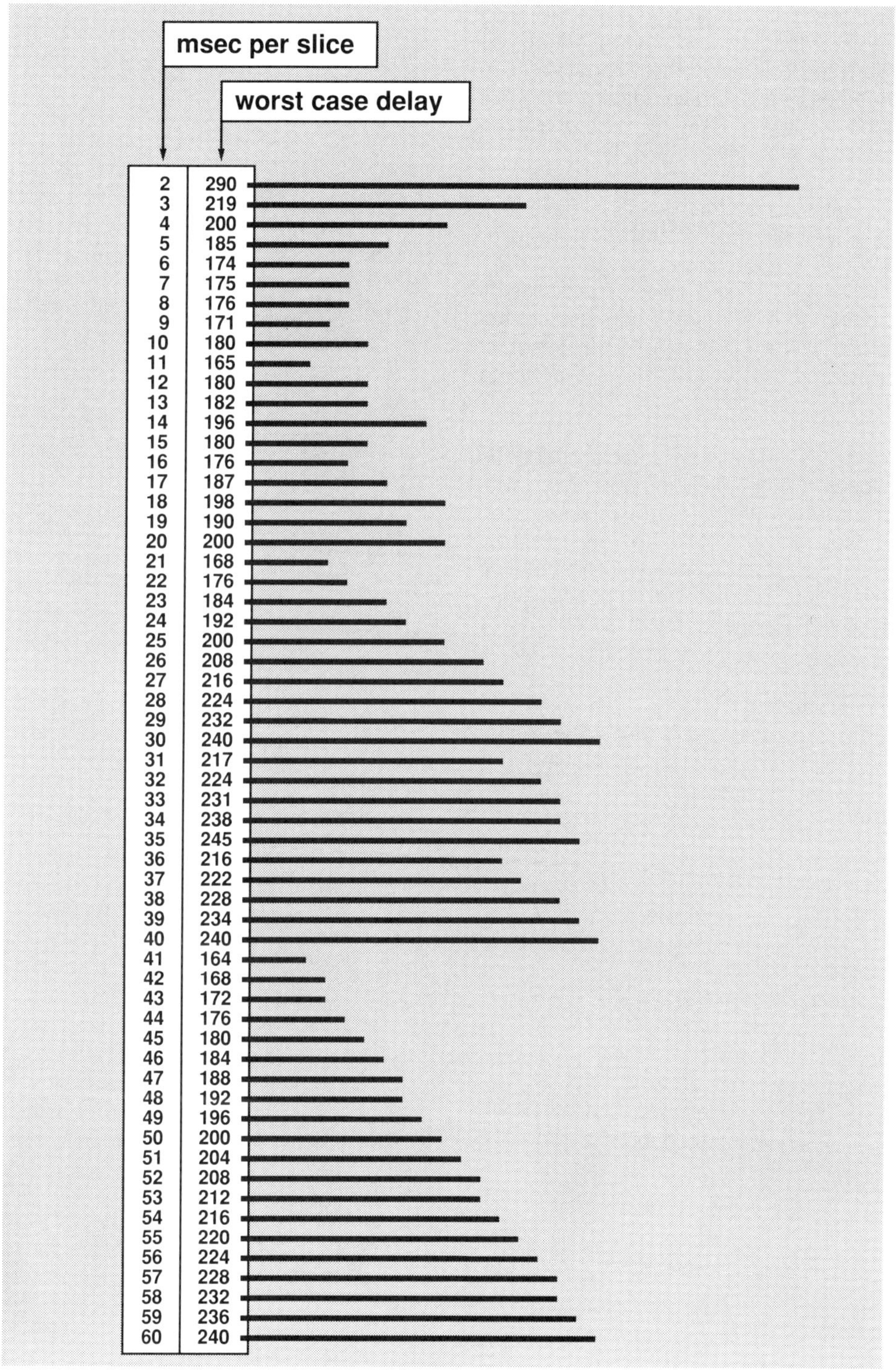
msec per slice
worst case delay
2 290
3 219
4 200
5 185
6 174
7 175
8 176
9 171
10 180
11 165
12 180
13 182
14 196
15 180
16 176
17 187
18 198
19 190
20 200
21 168
22 176
23 184
24 192
25 200
26 208
27 216
28 224
29 232
30 240
31 217
32 224
33 231
34 238
35 245
36 216
37 222
38 228
39 234
40 240
41 164
42 168
43 172
44 176
45 180
46 184
47 188
48 192
49 196
50 200
51 204
52 208
53 212
54 216
55 220
56 224
57 228
58 232
59 236
60 240

Figure 14

wastes a lot of processor time with register loading and saving. What you want to do is keep the executive from needlessly awakening a task. To do this the executive must know whether a given task has a pending event.

Up until now we have said that each task is always in 1 of 2 states: asleep or awake. But suppose you go one step farther and define 3 states. You do this by adding an *event bit* to each task control block. When this event bit is set, it indicates that the task has a pending event.

Once you add this bit, you can divide the sleeping tasks into 2 groups: those with a pending event (event bits set) and those with no pending event (event bit reset). A sleeping task with no pending event is in the *blocked* state. A sleeping task with a pending event is in the *ready to run* or *ready* state. (See Figure 16.)

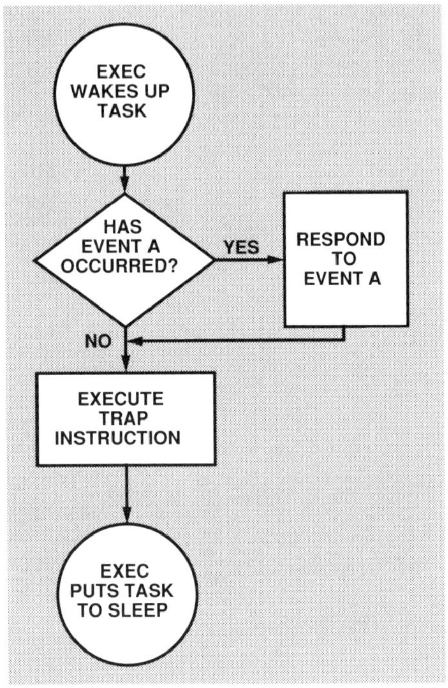

Figure 15

Once you add this bit to each of the task control blocks, you can rewrite the executive's round-robin scheduling routine to ignore any blocked task (any task with its event bit cleared). You must also rewrite each of the tasks, as shown in Figure 17. Since each task is now awakened only when it has a pending event, it simply responds

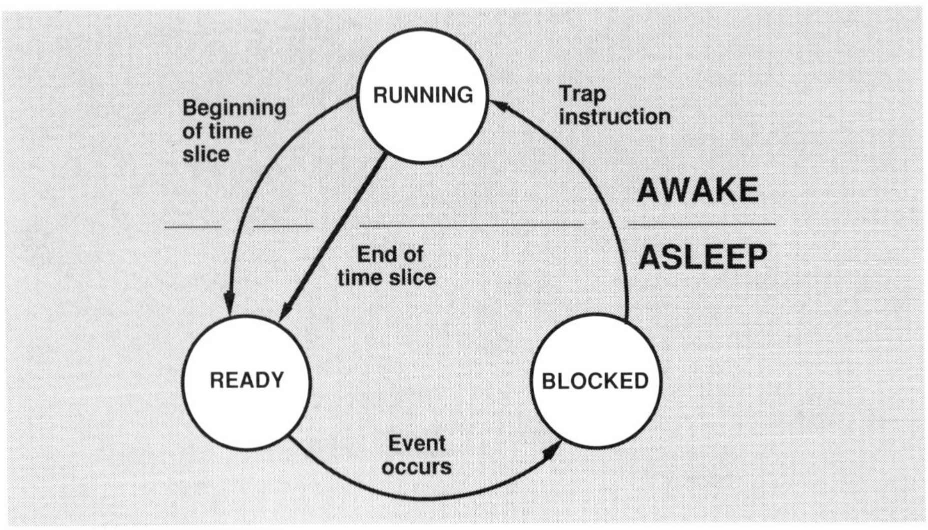

Figure 16

to the event, and then executes a *system call* instruction to transfer control of the processor back to the executive. The executive then puts the task back to sleep in the *blocked* state. (That is, it clears the event bit in the task's control block.) Once the executive clears the event bit, the task will not be reawakened until the executive's round-robin scheduling routine finds the event bit set again, indicating that a new event is pending for that task.

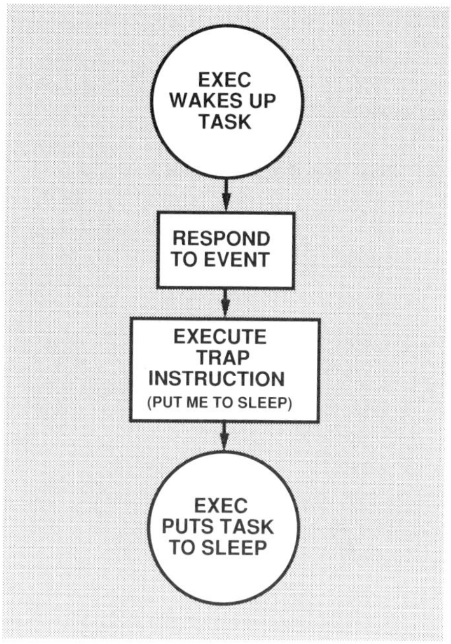

Figure 17

Now the question arises "Who sets this event bit?". To set the bit, you take advantage of the *vectored interrupt hardware* included on most microprocessors. This vectored interrupt hardware allows each different type of event to temporarily transfer control of the processor to its own unique ISR. The ISR executes a *system call* instruction that tells the executive to *set* the event bit in the task control block of the task that's supposed to respond to the event. To avoid problems, the time slice interrupt should be temporarily disabled, while the ISR executes. Otherwise the executive might put the ISR to sleep in the task control block of the task that the ISR interrupted, with unpredictable results.

Note: It would be simpler (and quicker) to just have the ISR set the event bit directly, instead of calling the executive, and asking it to set the bit. However, you need to keep the *application* code separate from the *executive* code. That way, once the executive is completely debugged, you never have to worry about its code again. If you let the *user* create an ISR that writes directly into the task control blocks, the user's ISR might overwrite the wrong thing. Then, when the executive reawakens the task, the results would be unpredictable.

In some cases the methods we have just discussed will improve the system response time enough to satisfy the needs of the more urgent events. (You might want to take the time at this point to figure out whether they would improve system response time enough to satisfy events A, B and C above.) However, if urgencies of the events differ enough, you *still* might not be able to do the job with round-robin, time sliced scheduling.

Prioritized scheduling algorithms provide improved response time for urgent events

Suppose you're designing a system to respond to 3 events, as follows:

- Event A must have a response within 100 msec, and its responding routine takes 40 msec to execute.

- Event B must have a response within 130 msec, and its responding routine takes 35 msec to execute.
- Event C must have a response within 150 msec, and its responding routine takes 30 msec to execute.

We have already seen that a round-robin time-sliced scheme cannot respond fast enough to satisfy these events. But suppose you assign a *priority level* to each of these 3 events. You then store the priority level of each event in its respective task control block. Then, at the end of each time slice, the executive awakens the ready task with the highest priority.

Figure 18 shows how prioritized, time-sliced, scheduling might work with these 3 tasks. (It assumes a 26 msec time slice.) Since task A is the most urgent (100 msec) you assign it the highest priority. You then assign event B (130 msec) a middle priority and event C (150 msec) the lowest priority. Now, suppose task C is executing when both event A and event B occur. As a result, ISR A and ISR B are run, setting the event bits in their respective task control blocks. At the end of the time slice, the executive regains control of the processor and runs task A. (Task B is forced to wait.) Task A continues to run for as many time slices as it needs to finish execution, since it has the highest priority.

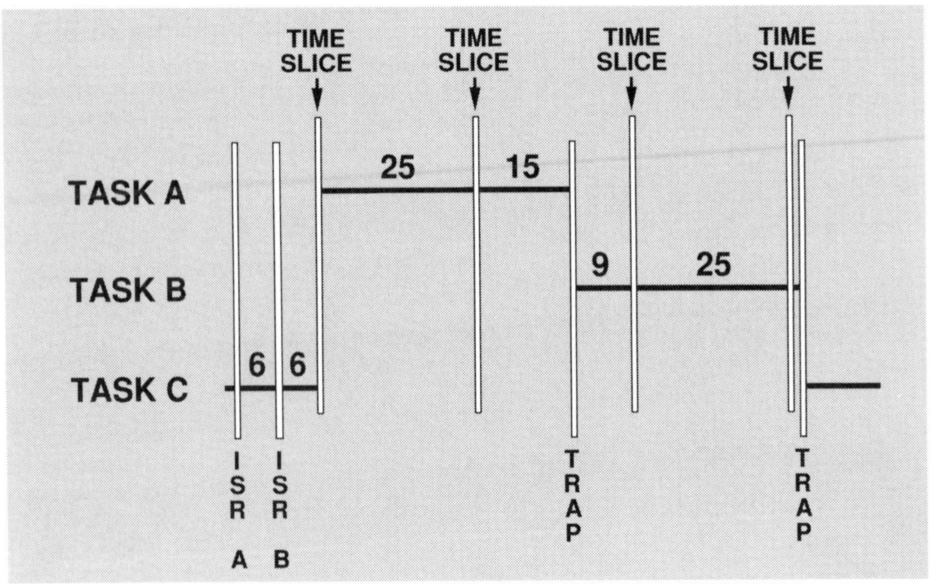

Figure 18

When task A finishes execution, it executes a trap instruction. This gives the processor back to the executive, and tells the executive to put Task A in the *blocked* state (event bit reset). The executive then looks for the highest priority ready task, (event bit set) and runs task B. When task B completes (and executes a trap instruction to block itself) the executive awakens task C to finish execution.

When a low priority task's execution is temporarily suspended (involuntarily) in order to let a higher priority task run, we say that the lower priority task has been *preempted*. By allowing tasks to be preempted, you can provide a better response time to urgent events. In this case, the worst case response time for events A and B will depend upon the length of the time slice. In general, the shorter the time slice, the quicker the response time. If you reduce the time slice to 3 msec, task C will be put to sleep (and task A awakened) within 3 msec after event A occurs. However, as we discussed earlier, shortening the time slice increases the executive overhead, so care must be taken not to shorten it too much.

But there is another way to improve the worst case response time for A and B. Suppose you define a special trap instruction that says, "Look for the highest priority task, and run it". When this trap instruction is executed, the executive acts just like it would at the end of a time slice: it puts the current task to sleep and then awakens the highest priority task. You could then put one of the trap instructions at the end of ISR A, ISR B and ISR C. Then, whenever an event occurs (and one of these ISR routines gets run) the ISR first executes a system call to set the appropriate event bit, and then executes another system call to tell the executive to run the highest priority task.

Figure 19 shows how this new scheme would work. As before, you assume that task C is executing when event A occurs. When ISR A has changed task A from the *blocked* state to the *ready* state (event bit set) it finishes by executing a system call that causes the executive to put task C to sleep and wake up task A. Then, when event B occurs, ISR B runs, changing task B's state from *blocked* to *ready*. Like ISR A, it then executes a trap instruction which causes the executive to put task A to sleep. But when the executive then searches for the highest priority task to run, it finds task A, and reawakens it.

Notice that, by using this scheme, task A completes execution 10 msec earlier, and task B completes execution 9 msec earlier.

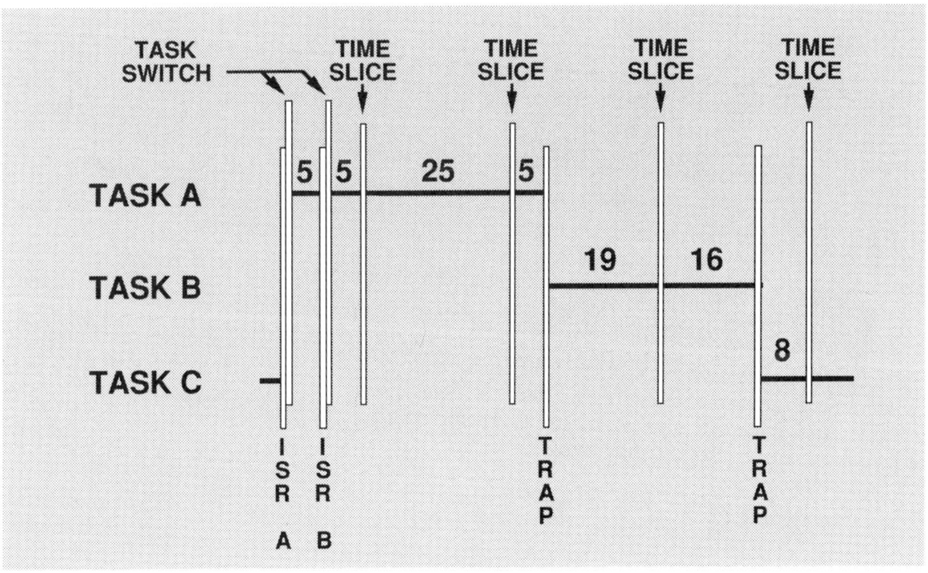

Figure 19

If you study Figure 19, you notice that the time slices serve no purpose! Each time a task is put to sleep by a time slice interrupt, it's immediately reawakened. In fact, you can disable the time slice interrupt and the executive still schedules the tasks in exactly the same order. The only difference is that you save 4 time slice delays: task A finishes 2 msec earlier, task B finishes 3 msec earlier, and task C finishes 4 msec earlier.

From this example, you see that it's possible to design a real-time multitasking system which is entirely event driven (i.e., with no time slicing). However, there are 2 reasons to use time slicing anyway.

The first reason to use time slicing is that the event driven system will only operate properly if each task ends with a trap instruction, which tells the executive to put it to sleep. If any programmer ever writes even a single task consisting of a loop (as shown in Figure 1) the system will *hang up* the first time that task is awakened. That means that the addition of a single new task to a fully debugged real-time system can bring the whole system to its knees.

The second reason to use time slicing is that you sometimes want to execute *background tasks.*

Background processing

Up to this point we have been talking about how a multitasking executive schedules real-time tasks. However, it's quite common to also have some other non-real-time tasks also being run by the executive. For example, the system may update a CRT terminal with current status, whenever real time events aren't demanding full use of the available processor time. These tasks are called *background tasks.*

Background tasks are typically infinite loops that use up as much processor time as they're allowed, so they're assigned the lowest possible priority level. If there are several background tasks, they're typically all given the same priority level. The idea is to share the available processor time among them, so a round-robin scheduler is used. It's also quite common to include a special entry in each task control block to indicate how many time slices a task is permitted to run before it's put to sleep. By putting a different number of time slices in the various background task control blocks, you can specify what portion of the available processor time will be made available to each background task.

John Black is the president of Micrology pbt. He previously worked for Motorola, where he was responsible for the development of the VMEbus standard. He has authored numerous articles on advanced microcomputer architectures. He has been the editor of VMEbus System Magazine since 1986, and is also the editor of this book. John earned his BSEE and MSE degrees at Arizona State University, and is currently working toward a PhD in computer science.

CHAPTER 13

Silicon Software Components: A Better Way To Build Real-Time Systems

By Alfred Chao

Engineers no longer build computers from discrete transistors: they use integrated circuits. Why should programmers continue to create software one instruction at a time? In this chapter, Alfred describes a method for building software from modular building blocks.

Over the years, as technology has advanced, hardware engineers have been able to work with higher levels of modular circuitry. During the 60's they had to build gates, counters and flip-flops from discrete transistors. During the 70's they were able to purchase these modules as inexpensive TTL integrated circuits. Today, the job of designing microcomputers has become a matter of selecting the right combination of VLSI chips, and interconnecting them to meet the needs of the application. This *building block* approach has allowed hardware engineers to work at much higher conceptual levels, allowing them to be more productive. At the same time, it has made systems much more reliable, since these VLSI chips have been thoroughly debugged through years of use.

Why not apply the same approach to the development of software? Why not identify commonly used software modules, package them in some convenient form, and make them available to programmers? For example, why not offer standardized software in ROMs? Since they could be sold in volume, the vendor could afford to expend considerable time and money developing and optimizing these software modules. Programmers who use these modules could spend their time working at a higher conceptual level, instead of reinventing wheels. We could call these software modules *silicon software components*.

If the components are offered in ROM, the programmer doesn't even need to have special equipment for compiling, linking or downloading the code into the target system. Since programmers aren't making "slight modifications" to the code stored in the ROMs (for their own application) each system works exactly like the next. That promotes compatibility, and minimizes the likelihood of subtle bugs, which can't be duplicated on similar (but not identical) systems.

How do you develop standardized silicon software components?
The first step in developing standardized components is to take the software used in typical systems and subdivide it into a hierarchy of structured software modules. Then you need to sort these modules into 2 groups:

- Modules that are independent of all system hardware (except the CPU)
- Modules that are not independent of all system hardware

The idea behind this is to figure out which modules can be mass produced, and which will have to be created by the engineer customizing the system for a specific application. The modules that you mass produce must not rely upon the unique hardware of any particular system. They provide hardware-independent services, which can be called upon by user-generated, hardware-dependent software.

Standardized silicon software components should contain only position-independent code. That allows the user to simply plug them into any convenient ROM socket. This is important for 2 reasons:

- When creating a software component you have no way of knowing where the user's ROM sockets will appear in the memory map. If the code in your ROM is position dependent, the user may not have a ROM socket at the required address.
- If the user purchases silicon software components from several sources, overlapping position-dependent code might make the ROMs incompatible.

Note: In many cases, the user might want to load the software component's code into RAM, instead of plugging in ROMs. This allows quick reconfiguration of the system memory map, as the system is debugged. For example, during the debugging process a user's position-dependent application code might grow until it eventually overlaps the area occupied by the software component. If the software component's code is stored in RAM, it can be moved out of the way, into another area of RAM. However, in the following discussion, we'll assume that the software component resides in ROM.

An example of a silicon software component: a multitasking executive
One complex type of software that lends itself very well to a silicon software component approach is a *multitasking executive.* By its very nature, executive code can be difficult to understand and debug. By encapsulating all of the executive's task switching functions inside a silicon software component, you can greatly simplify the application programmer's job. The programmer (when using such an executive) can write the application code as several independent *tasks*. Each task calls upon the executive (through system calls) to obtain the services it needs.

Silicon software components are, in essence, black boxes. In this case, the idea behind the black box approach is to *hide* the inner workings of the executive, while offering the user a simple, yet adequate, set of services.

What's the optimal number of services for a multitasking executive?
How many services should an executive provide? The answer to that question is debatable. There are 2 extremes:

- Have the executive provide lots of very simple services that have no options, and require minimal parameter passing.
- Have the executive provide just a few services that each have lots of options, and require considerable parameter passing.

The best approach is somewhere between these 2 extremes. Reducing the number of services makes it easier for the user to remember what's available. However, if you carry this too far, it requires too many calls to get the job done. For a multitasking executive, 30 to 40 services satisfy experienced users, while first time users can still learn the services quickly.

To make silicon software components useful in a broad range of applications, they should be designed to run as *fast* as possible. Again, this is best achieved by keeping things simple. A well designed executive should acknowledge interrupts in less than 30 microseconds, and should switch tasks within 80 microseconds.

In some cases, the user might want to adjust a certain critical parameter to "tune" the system's performance. For example, the user might want to specify a *time slice length* to optimize the execution speed of a certain set of tasks. To allow for this, the executive ROM should have a *set-up routine*. This routine reads certain key parameters from a *configuration table* in RAM.

How does the executive ROM find this configuration table? To provide a way, you must have an *anchor location:* a memory location that holds the base address of the configuration table.

The anchor location can be anywhere in memory, but its address is "wired into" the code in the executive ROM. The executive ROM's set-up routine looks at the address stored in the anchor location to find out where its own configuration table is.

This set-up routine also performs another important function. It writes the entry point addresses of its various *exception processing routines* into a *vector table* in RAM. Then, when an interrupt occurs (or a trap instruction gets executed) the processor will load the program counter from this vector table.

As mentioned earlier, the executive should be written to be entirely independent of the system hardware, except for the processor type. (The executive is, of necessity, dependent on the processor type, since its code consists of machine language.) The idea is to have all of the other hardware resources accessed through customized *drivers*, which are provided by the user.

Ideally, the executive should be fully functional, regardless of what hardware resources are available in the system. However, some standard executive services can only be provided if certain critical resources are available. One such critical resource is a periodic clock, which generates interrupts for time slicing. While an executive can be written to run in an entirely event driven fashion (with no periodic clock) there are some very useful system services that can only be provided if a clock is available. For example:

- Put me to sleep, and wake me after X microseconds.
- Put me to sleep until event N happens, OR until X microseconds have passed.
- What is the time of day?

In addition, *round-robin scheduling* (which is useful in balancing processor usage between several tasks) can only be implemented if there's a periodic clock signal.

A well-designed executive should provide these services when a clock is available, but should still operate (minus these services) when it is not.

What if the services provided by an executive aren't adequate?
It's important to provide the user with a way to expand the number of system services, as the need arises. For example, the user might want to add *file handling* or *debugging* services, by plugging in *auxiliary ROMs*. These auxiliary ROMs would then provide additional system services, which the user could access via additional system calls.

Each of these auxiliary ROMs would also need to have its own set-up routine, and its own configuration table in RAM. As in the case of the executive ROM, the auxiliary set-up routine sets things up according to the parameters in its own auxiliary configuration table. It also writes the entry point addresses of its own trap servicing routines into the interrupt vector table in RAM. This allows the user's software to call upon the new system services provided by the auxiliary ROM.

If the system includes several ROMs, the set-up subroutine in each ROM should be executed before any system calls are made. You accomplish this by first executing the executive ROM set-up routine (to initialize the system, based on its configuration table) and then by having the executive set-up routine call the set-up routines in each of the auxiliary ROMs.

Figure 1 shows what the executive configuration table might look like. In addition to a list of parameters for initializing the executive, the table also includes pointers to the base addresses of each of the auxiliary configuration tables. After initializing itself, the executive ROM set-up routine checks these pointers, to see whether they contain non-null values. If so, the executive's set-up routine looks in each auxiliary configuration table to find the address of the auxiliary set-up routine. Only after the executive has called the set-up routine in each auxiliary ROM does it commence normal operation.

When each auxiliary set-up routine is called, it must have the base address of its own configuration table. It finds this out by first reading the *anchor location*, to get the base address of the executive configuration table. Then it looks in the executive configuration table to find the base address of its own configuration table.

How does application software see a multitasking executive?
Up to this point we have discussed the design of the real-time executive. Let's now shift our focus to how you, as a user, would use a real-time executive.

As we mentioned earlier, the executive code knows nothing about the hardware configuration of your system. For example, it doesn't know what kinds of I/O devices your system has, or what interrupt vectors each I/O device generates. As a result, you must provide 2 kinds of software, in addition to the *tasks* that you will run under the executive:

- *I/O drivers*, which are collections of subroutines. The executive calls these subroutines to do operations on I/O devices.
- *Interrupt service routines*, which are routines automatically called by the interrupt hardware when interrupts occur. Since most of your I/O devices probably generate interrupts, you will typically write at least 1 interrupt service routine for each I/O device.

If an interrupt occurs while the executive is running, the interrupt handling hardware in the processor chip temporarily suspends execution of the executive code, until the appropriate interrupt service routine finishes executing.

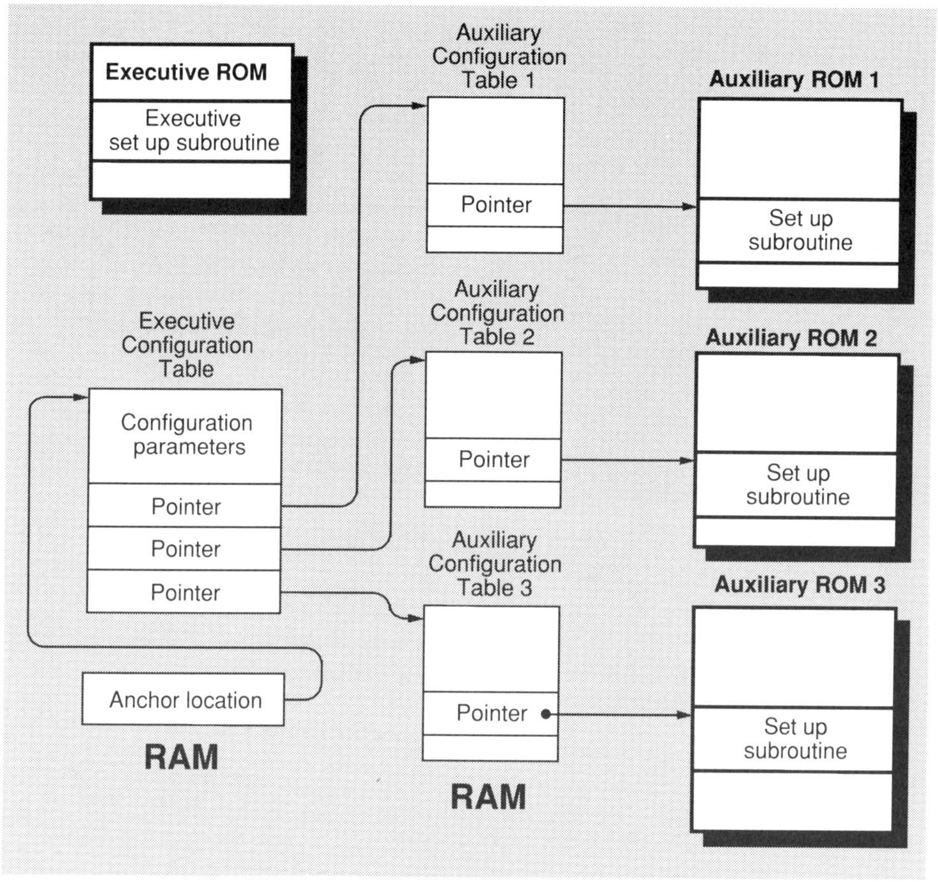

Figure 1

Since interrupt service routines are called by the hardware, the executive code doesn't "know" they exist. However, since the hardware determines the entry point to the interrupt service routine by *reading* it from the interrupt vector table in memory, you must include code in the executive's start-up routine that initializes the interrupt vector table, before any interrupts can occur.

When a task requests the executive to do an I/O operation (via a system call) the executive calls the appropriate *I/O driver subroutines*, as shown in Figure 2. Tasks must request the *executive* to do I/O operations, instead of accessing the I/O device registers directly. That allows the executive to schedule the use of *shared* I/O devices, and prevents 2 tasks from using the same device simultaneously. In effect, the executive serves as a *multiplexer* for the I/O devices.

When a task needs to do some I/O, it first requests the executive to *open* the I/O device. Depending upon the type of device, different things might happen, but the end result is that the executive grants the task exclusive use of the device, until the task tells the executive to *close* the device.

If a 2nd task then tries to open the I/O device while the 1st task is still using it, the executive puts the 2nd task to sleep. When the 1st task finishes using the device (and the executive reawakens the 2nd task) the 2nd task is unaware that anything unusual has happened. As far as it's concerned, it has simply been granted exclusive use of the I/O device.

Because of the way the executive handles conflicting I/O requests (without tasks being aware of it) you could say that the executive creates a set of *virtual I/O devices* for each task. Each task is written as if it has a set of dedicated I/O devices.

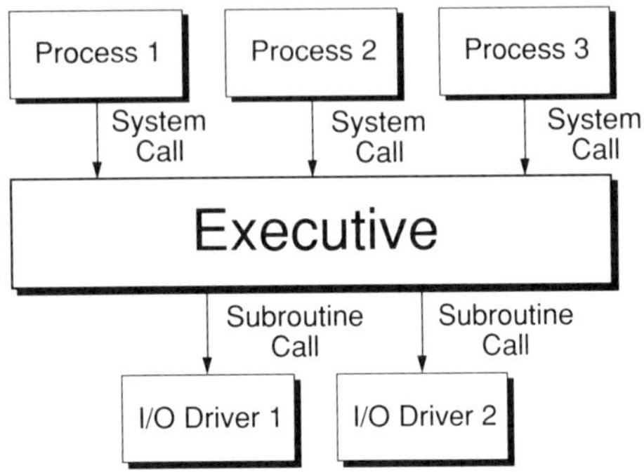

Figure 2

How do you partition application code in a multitasking system?
When some external event generates an interrupt, the interrupt handling hardware calls the appropriate interrupt service routine. If the executive (or one of the tasks) is executing when the interrupt occurs, its execution is temporarily suspended, until the interrupt service routine finishes execution.

What should your interrupt service routine do? Whatever you want it to do! There are no set rules. It might just generate a system call that asks the executive to wake up a certain task, or it might do something more elaborate, like reading information from an I/O device. However, you typically want to keep an interrupt service routine as short as possible, since normal task scheduling is deactivated as long as the interrupt service routine executes.

It all boils down to this: as a user you must make decisions about how to *partition* your application code. You must decide how much should be done in the interrupt service routines, and how much should be left for the awakening task to do.

You must also decide *how many tasks* to divide your application code into. Your real-time system will likely need to respond to several different kinds of external events. If the required response is fairly simple, you may be able to provide that response with a single task. However, if a complex response is needed, you may need to create several cooperating tasks.

Where several cooperating tasks are needed, you may find it difficult to decide how many tasks to create. If you create too many tasks, you'll use up too much processor time switching between tasks. On the other hand, if you create too few tasks, you may be holding up one action needlessly, while you wait for the right condition to initiate some other action.

There are no fixed rules for partitioning the application code into tasks. It depends upon the nature of your application. In general, if the system is supporting several completely independent activities (such as controlling several robots) then there should be at least one task dedicated to each activity. However, the decision as to whether each activity needs more than one task is not so simple. For example, should there be a single task for positioning the robot's arm, or should there be 2 separate tasks: one for controlling the elbow, and another for controlling the wrist?

Subdividing tasks
When should 1 task be divided up into 2 or more tasks? Here are some guidelines:

Divide up a task if it responds to 2 or more different events, and if...

- both of those events occur cyclically, but with different frequency.
- the occurrence of one event doesn't help predict the occurrence of the other event.
- the urgency of one event is different from the urgency of the other event.
- the events have no logical connection to each other in your mind.

In some cases, you will need a very large number of tasks to comply with these guidelines. However, a well-written multitasking executive doesn't set an upper limit on the number of tasks that it can run. It simply requires that you, as a user, specify the number of task you will be running.

When do you specify the number of tasks? If you're using a simple real-time multitasking executive, you may be required to specify the number of tasks when the executive is started, and then you can't change that number until it's shut down. This simple approach is adequate for many real-time applications, such as machine control, where the *structure* of the machine doesn't change during normal operation.

However, in some cases, you may want to allow a task to *spawn* another task. The result is typically 2 identical tasks that execute the same code, but have separate data areas. To allow this, the executive must provide special system calls, which a task can call to spawn, activate, or delete another task.

pSOS-68K: An example of a silicon software component
pSOS is a silicon software component which embodies a real-time multitasking executive like the one described above. It makes no hardware assumptions, other than the fact that it will be run on a 68000 microprocessor. All pSOS code is *position independent*, so it can be placed anywhere in the address map to suit the needs of the system it runs on. It can reside in RAM (which is especially convenient for debugging) or it can be programmed into a ROM/PROM/EPROM device.

In line with the concept of silicon software components, pSOS is used as a *black box*. A user simply provides an I/O driver (and possibly an interrupt service routine) for each of the system I/O devices, and then writes the application code in the form of tasks, which call upon pSOS services.

pSOS provides all of the basic executive functions, such as task switching, task scheduling, and I/O handling, as well as more elaborate ones, like task spawning. Figure 3 shows the basic services that it provides, via system calls.

Because it's simple, and because it's written in 68000 assembly language, pSOS is fast. On a 10 MHz 68000, it switches tasks in 80 microseconds, and allows interrupt responses within 28 microseconds.

Return to me the version no. of pSOS being used

Spawn a new process
Activate (allow execution of) spawned process X

Give me the process no. for the process with this name
Delete process no. X and reclaim its resources
Suspend execution of process no. X until a Resume call
Resume execution of process no. X
Change the priority of process no. X to Y
Change process no. X's mode bits (preemptable, round-robin)
Put process X into the supervisory state

Give me a memory segment
I'm through using memory segment no. X
Give memory segment X to process no. Y
Steal memory segment no. X from process Y and give it to me

Create a set of X buffers with a size of Y, and call it Z
Give me a buffer from set Z
I'm through using buffer set Z
I'm through using buffer X of set Z

Create a msg. queue, and call it X
Attach me to msg. queue X
Give me the msg. at the head of msg. queue X
Add this msg. to the end of msg. queue X
Put this msg. at the head of msg. queue X

Figure 3

Each silicon software component has a *jump table* at its base address. There's an address in the jump table for each system call provided by that component. Figure 4 shows the jump table for pSOS. Note that all of the basic executive services are obtained through one system call, and all I/O services through another system call.

The 3rd jump table entry has the address of a routine that's used to initialize pSOS. Like the system services routine and the I/O services routine, the set up routine can be called with a system call. However, this set up routine is usually called only once, when system operation begins.

The 4th jump table entry is used to provide a special system call. It is typically called by interrupt service routines at the end of their execution. It tells the executive to put the previously interrupted task to sleep, and then awaken the highest priority task. In many cases, the executive will awaken the task that it just put to sleep but, in some cases, the interrupt service routine changes the state of some high priority task from blocked to ready, and the executive will then awaken that task instead.

If the system includes a periodic clock (which can be used to generate a time slice interrupt) pSOS provides some additional services:

- An *alarm clock* to wake-up sleeping tasks after a certain delay
- A time-of-day service.

To keep the interface to the I/O simple, pSOS uses an *I/O device jump table*, much like Unix. (See Figure 5) Each I/O driver is a collection of 6 subroutines, as follows:

- *Initialize* the I/O device.
- *Open* the I/O device.
- *Close* the I/O device.
- *Read* from the I/O device.
- *Write* to the I/O device.
- *Perform a control function* on the I/O device.

The executive determines the entry points to these subroutines via the *I/O device jump table*, which is at the base address indicated in the executive's *configuration table*.

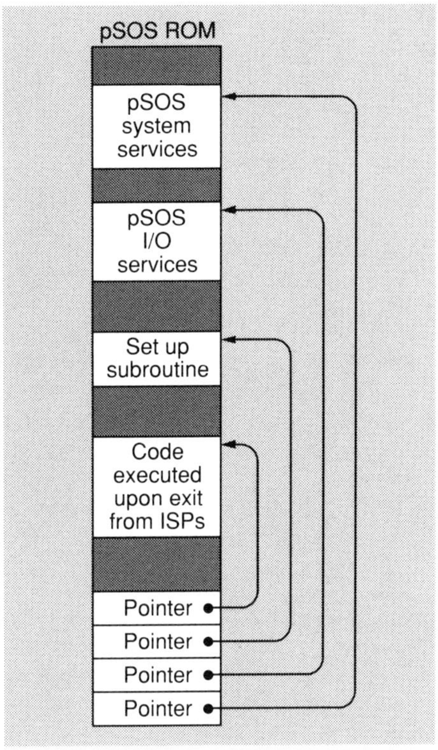

Figure 4

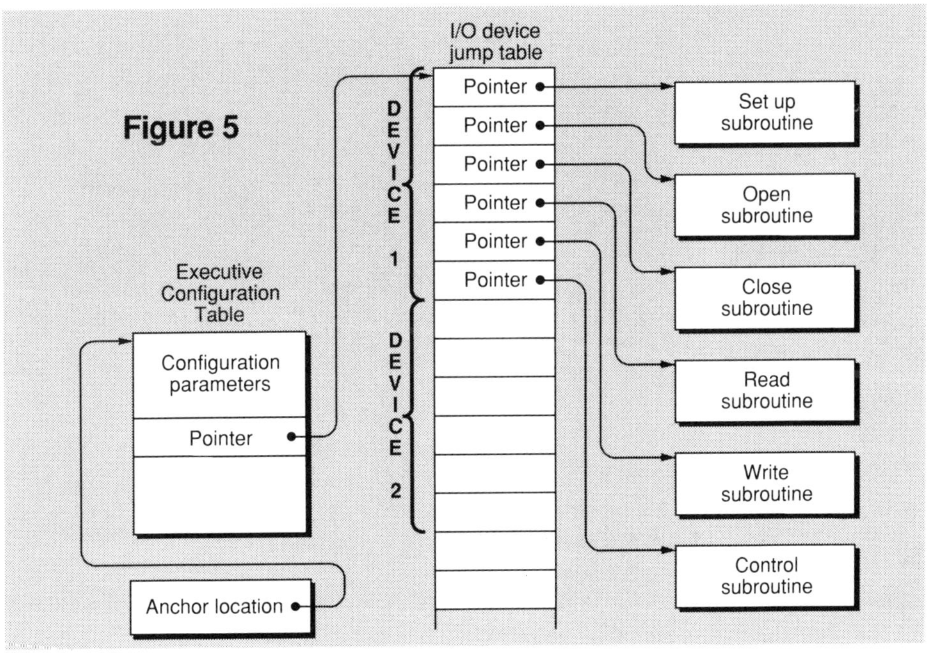

Figure 5

When a task needs to do some I/O, it generates a *system call*, and passes a *logical device number*. The executive then uses the logical device number as an *offset* into the I/O device jump table, to get the entry point address of the appropriate driver subroutine.

Expanding the services offered by pSOS
There are 3 other silicon software components available for use with pSOS:

- pHILE is a file system manager. It provides the services needed to build a Unix-style hierarchical directory tree, and allows additional file storage devices to be attached (as leaves) to that directory tree. It also allows any portion of a file to be *locked*, to prevent 2 tasks from simultaneously accessing the same data. It supports the IBM/PC-compatible floppy disk format, allowing a user to move files between pSOS-based systems and IBM-PC compatibles.
- pROBE is a system debugger and performance monitor. It provides services that let a user upload, download, assemble, and disassemble files. It also provides services that allow the user to see into (and control) the operation of a pSOS-based system. Since it takes full advantage of its knowledge of how pSOS works, its services allow the user to create *break points*, which are triggered by significant system events. pROBE's services also allow the user to monitor system activity, and identify performance bottlenecks.
- pRISM provides a set of services which allow users to build and debug multiple processor, pSOS-based systems. Each processor can have its own copy of pSOS, or all processors can share a single copy in global memory. pRISM allows any task in the system to call upon the services of its own processor's pSOS executive, or upon those of some other processor in the system. It also allows locally created message queues to be declared global, so tasks running on one processor can pass messages to tasks running on another. When used in conjunction with pROBE, pRISM also provides services that allow a user to do system-wide debugging from a single terminal.

Each of these components (like pSOS) is position independent, so it can be executed at any location in the address map. Each also includes its own set-up routine, which gets called by the pSOS set-up routine.

Alfred Chao is the founder and president of Software Components Group, a supplier of real-time system software. He received both his BS and MS degrees from MIT in 1975, and has held system design and management positions with Gould, Honeywell, and Intermetrics.

CHAPTER 14

Building Data Flow Machines On VMEbus

By Marco Flagg

Backplane buses like VMEbus have certainly changed the way systems are being designed. Only a few years ago engineers were designing with chip components. However, with the appearance of backplane buses, they began designing with components such as CPU boards, memory boards, and I/O boards. Then came single board computers. Recently, some companies have even introduced boards with several microprocessors. With these new boards you can now build VMEbus systems with dozens of processors in a single chassis. This raises a new question: What's the best way to interconnect and use these processors? In this chapter Marco explains the concepts of data flow architectures, and describes a VMEbus-based multiple processor data flow system that has been used to do real-time image processing.

Because VMEbus allows close coupling of many microprocessors, it has been widely used to build multiple processor systems. The availability of real-time kernels and operating systems with multiprocessing services has made it even easier to build multiprocessor systems.

Many of these multiprocessing systems are designed to take advantage of parallelism in the application software that they execute. For example, if all of the computations inside a loop rely solely on the loop count, each iteration of the loop can be executed in parallel by a separate processor.

Unfortunately, not every program is well suited to this approach. Even when a program can be subdivided into several programs that run in parallel, there may be one that takes much longer to execute than the others, creating a *critical path* that limits the overall execution speed.

Backplane-based multiprocessing systems also face another limitation, when it comes to parallel processing. As you plug more and more processor boards into the backplane, the bus traffic increases. If the bus bandwidth saturates, the backplane becomes a bottleneck. Faced with this limitation, you may find yourself looking for an alternative architecture. One good alternative is a *data flow architecture*.

Data flow architectures

Many applications require continuous processing of a stream of input data, to provide real-time output. These applications are well served by data flow architectures. A data flow machine consists of a number of *nodes,* interconnected by communication *channels*, as shown in Figure 1.

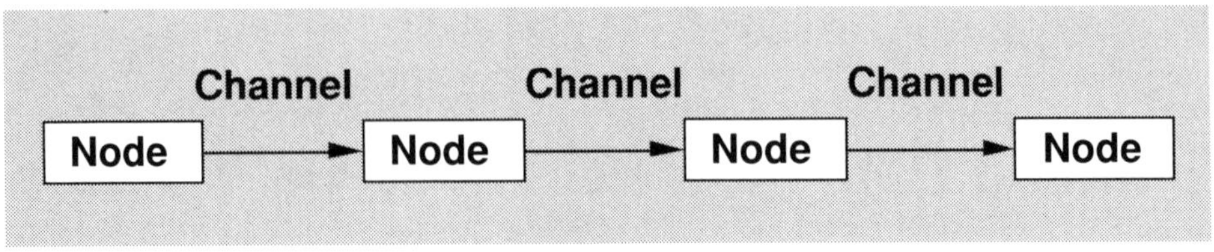

Figure 1

Data flows through each channel to a node. It is then processed by that node and passed through another channel to the next node. This concept is similar to that of a high volume manufacturing line, where a continuous stream of components (data) are moved by conveyer belts (channels) between assembly stations (nodes) as they are assembled into the final product.

Designing a data flow machine

Like a manufacturing line, each data flow architecture is tailored to do a particular kind of processing. To design an efficient data flow architecture, you need to first study your application to determine:

- where data enters the system.
- where data exits the system.
- what operations need to be performed on the data, as it flows through the system.

You then summarize the results in a block diagram, like the one shown in Figure 2. Each rectangular block represents an *operation* performed on the data, and each arrow indicates the *flow of data* between operations.

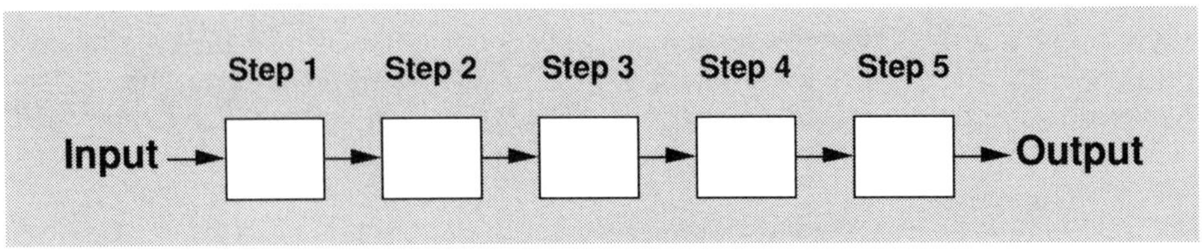

Figure 2

Since data flows continuously through the system, each node should take about the same amount of time to perform its operation as each of the other nodes. If some operation takes less time, its node will be idle part of the time, wasting a valuable processing resource. On the other hand, if an operation takes longer than the operations preceding it, a bottleneck will develop, and data will back up in the pipeline.

Proper partitioning is not always easy to do. You may not know (until you implement the various operations in hardware or software) how long each operation will take. If some operation takes much longer than you anticipated, it may prove to be a bottleneck.

Fortunately, you can eliminate bottlenecks through the use of *multiplexers* and *demultiplexors*. For example, suppose you have a 5-step data flow machine, like the one shown in Figure 2. After implementing this architecture, you find that step 3 takes twice as long as any of the other steps. By inserting a demultiplexor ahead of step 3, and a multiplexor after step 3, you can use 2 nodes in an alternating fashion to double the effective throughput of step 3. (See Figure 3)

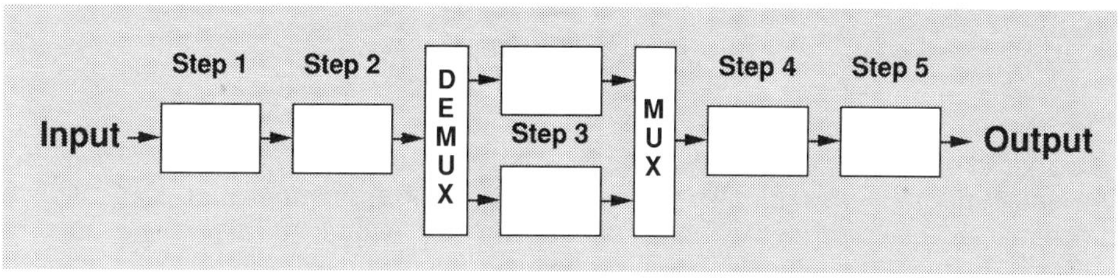

Figure 3

Of course, once you resolve one bottleneck, you may find another. By identifying and resolving the bottlenecks one by one, you can raise the performance of your system to any desired level. (If there are no bottlenecks in the system, but performance is still inadequate, you could use a demultiplexor at the front of the system, and a multiplexor at the end of the system to double the overall throughput, as shown in Figure 4.) This is the real advantage of data flow machines: you can multiplex the processing operations to greater and greater levels, to attain any required performance level.

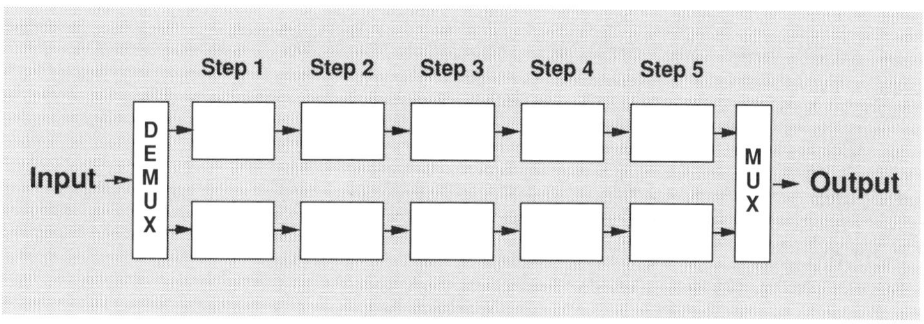

Figure 4

You need versatility
It's clear from the previous example that, to design data flow systems efficiently, you need to have the flexibility to *restructure* the system. In order to do that, you need to have a *flexible* channel implementation.

Suppose you implement each operation as a software module, running on its own dedicated microprocessor. You could then use dual-ported memory to provide a channel interconnection between the processors, as shown in Figure 5. Since you don't have to actually move the data, to pass it from one processor to another, this shared memory channel provides high throughput. One processor simply writes data into the shared memory, and the next processor *reads* the data from that same memory.

Figure 5

The shared memory approach also simplifies the job of demultiplexing and multiplexing the data stream. For example, demultiplexing can be accomplished by having a processor write consecutive elements of its output data stream into *alternating* memory buffers. These 2 buffers could then be read by 2 processors. However, to accomplish that, you need to have a triple-ported memory that's accessible to 3 processors, as shown in Figure 6.

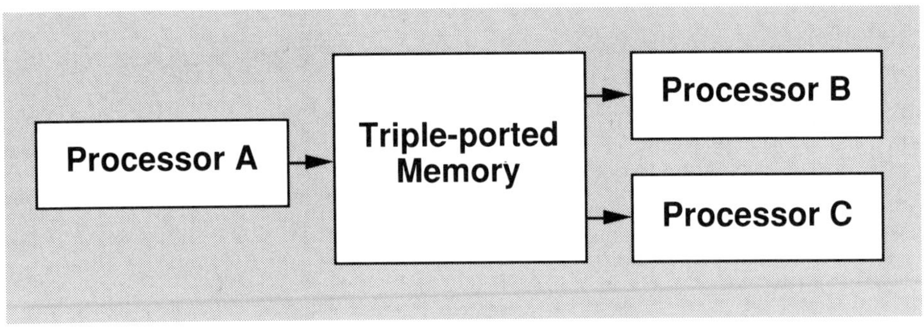

Figure 6

Obviously, shared memory communication channels provide a lot of flexibility. For example, if *all* the processors in the system share a common global memory, you can interconnect any 2 processors with a channel. However, you might pay a heavy performance price for that level of flexibility. For example, if you have a dozen processors (each communicating with its neighbors through VMEbus global memory) competition for access to the shared memory produces a bottleneck, even if the individual processors have adequate computing power.

A better solution is to have a *hierarchy* of shared memories, as shown in Figure 7. Each processor has a *private bus*, where it can fetch its instructions and perform the data operations assigned to it.

Several of these processors are then clustered around a *local bus*, which allows them to pass data efficiently from processor to processor through local memory, as well as to demultiplex and multiplex the data stream.

The last processor to process the data stream in each cluster writes its results (over the global bus) into *global* memory. That data is then accessible to the next cluster in the data flow.

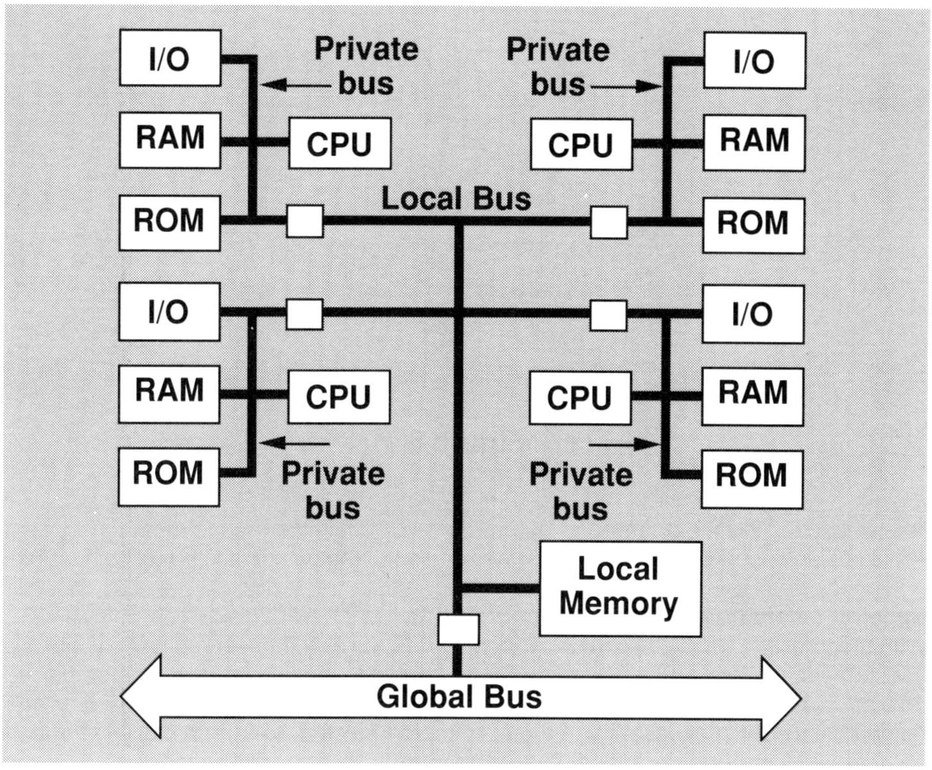

Figure 7

Response time
The *response time* of a data flow system is defined as the time from receipt of its input data to the time at which the needed response appears at its output. Response time is the sum of 2 times:

- Data processing time
- Data transfer time

Data processing time
The *data processing time* is the sum of all processing time expended on an input stream of data, from the point where it enters the system to the point where it exits the system. If the input data is routed in parallel through several chains of modules (as shown in Figure 8) then the system *response time* is determined by the slowest chain, which is called the *critical path*. In a data flow system, the critical path's processing time can be shortened arbitrarily, by demultiplexing and multiplexing the data stream, to allow several processors to do computations concurrently.

Data transfer time
The *data transfer time* is the total time needed to move the data into the system, from module to module, and back out of the system. The most significant portion of this

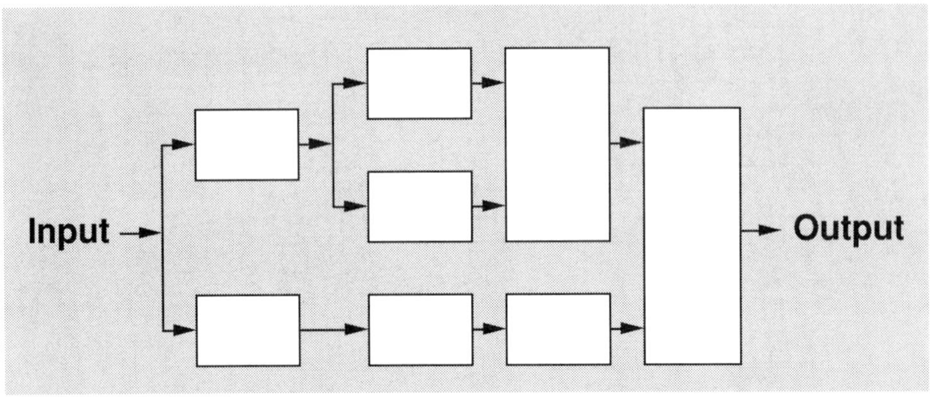

Figure 8

time is usually the input time and output time. To minimize these times, the system should be equipped with I/O interrupts, to make input and output as efficient as possible.

Building a data flow architecture on VMEbus

With high density VLSI chips you can design data flow systems that fit nicely into the VMEbus architecture. (See Figure 9) For example, several processors can be placed on a double-height VMEbus board; each equipped with its own private bus and private resources. These processors can then be interconnected (to form a cluster) through an onboard local bus, and local shared memory. This shared memory permits the passing of data between processors without use of the VMEbus, and also facilitates demultiplexing and multiplexing of data.

Any of the processors on the VMEbus board can access global VMEbus memory. This allows a processor on one board to pass data to a processor on another board. By plugging several of these boards into a VMEbus backplane, you can build a data flow architecture of arbitrary length.

In order to minimize data transfer times, between processors (and reduce the response time of the system) each processor should be equipped with one or more *location monitors*. Then, whenever some VMEbus master writes to the onboard memory, the location monitor assigned to watch that area interrupts it's onboard processor.

System input and output can be accomplished by either of 2 methods:

- VMEbus I/O
- VSB I/O

VMEbus I/O

Standard VMEbus I/O controller boards can be used to do DMA transfers between system I/O ports and global VMEbus memory. The advantage of this approach is that it allows any processor in the data flow system to initiate I/O. The disadvantage is that it uses some of the available VMEbus bandwidth, and competes with processors that are attempting to communicate with each other over the VMEbus.

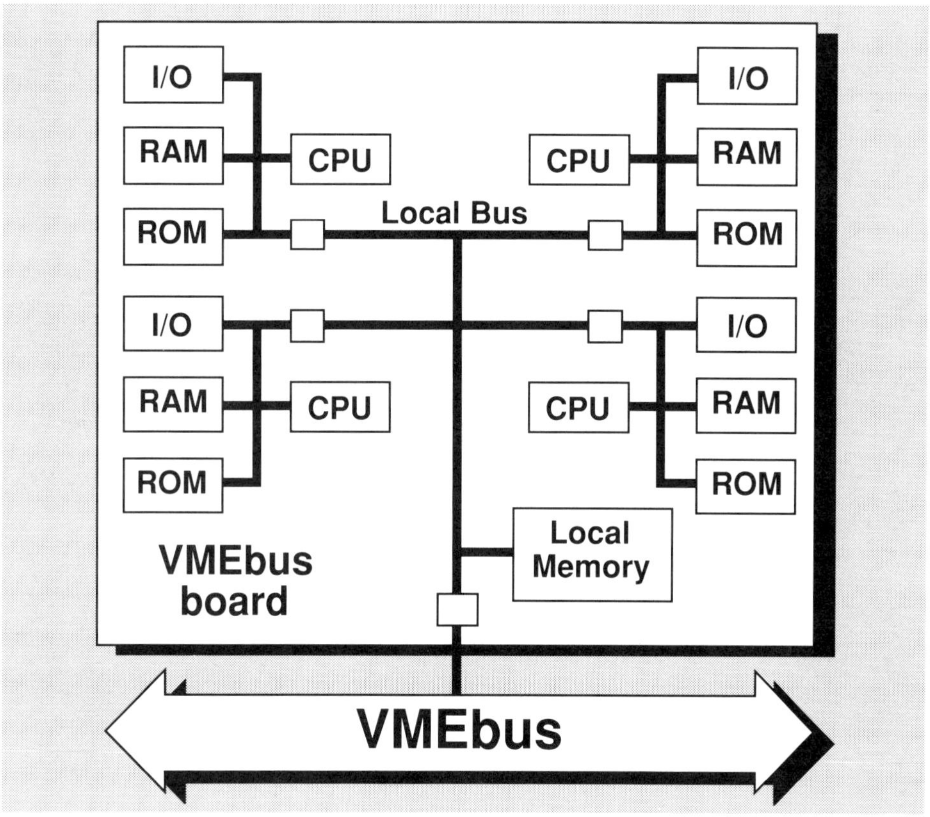

Figure 9

VSB I/O

If each processor board is equipped with a VSB port, you can use VSB I/O boards to provide I/O transfers between each board's local bus and the outside world. The advantage to this approach is that it doesn't use any of the VMEbus bandwidth. However, the VSB I/O board can only provide I/O to the local processors on the board.

Minimizing channel transit times

To minimize channel transit times, each processor should be equipped with interrupt routing logic that allows it to be configured to detect any VMEbus or VSB interrupts.

Building the data flow software

Suppose you have a block diagram with 8 nodes, like the one in Figure 10, and suppose you implement this architecture with a VMEbus system consisting of a VSB input board, 2 processor boards (4 processors on each) and a VSB output board, as shown in Figure 11. To complete this data flow architecture, you must create a software module to run on each of the 8 processors that implements the operation assigned to that node.

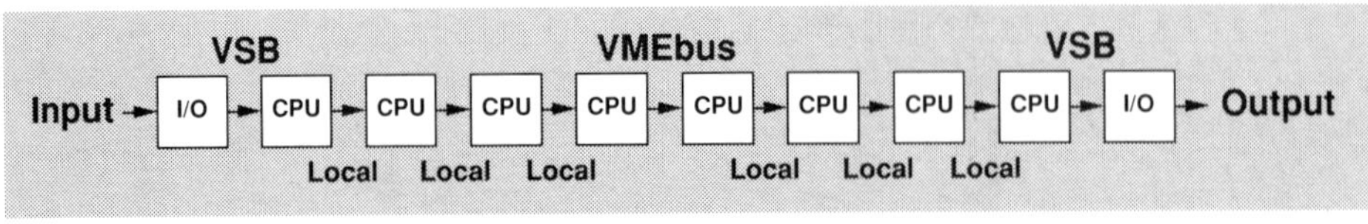

Figure 10

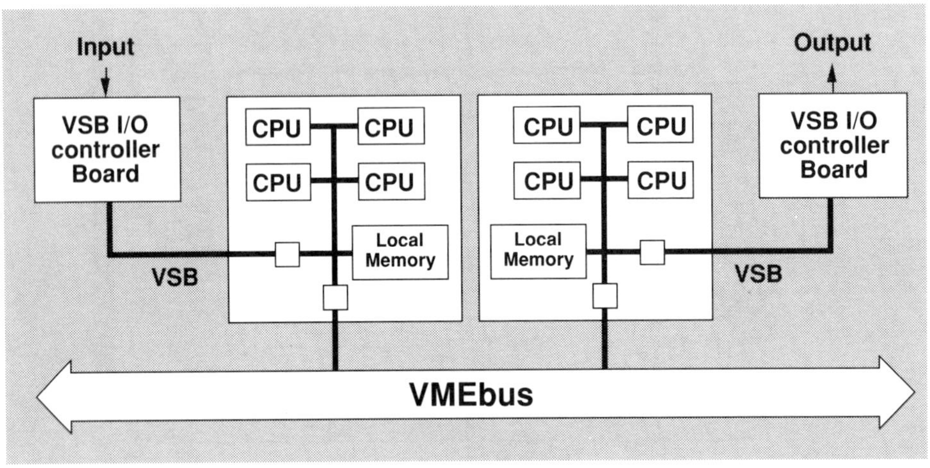

Figure 11

These software modules are typically small programs, written in C, or Fortran, or ADA, or assembly language, or whatever... To facilitate use of the communication channels (and any other system resources) you'll probably want to have a *function library* for the language in which you're writing.

Once you create the software modules, you'll need to assign a module to each processor. (To keep things simple, we'll assume that each processor executes only one software module.) Then you need to establish communication channels (in shared memory) between the processors, and configure the hardware to route the appropriate VMEbus interrupts to each processor.

It's not too difficult to do all of this manually, especially if you only have to do it once. However, you may need to make *changes* during the development phase. For example, you may need to add processors, and then redistribute the software among the processors. It would be helpful to have at least a rudimentary operating system to take care of the details. When the VMEbus system powers up, this operating system could:

- inventory the available processors
- determine which processors are clustered around a common local memory
- assign software modules to the various processors
- set up communication channels in the appropriate shared memory
- configure the interrupts

However, in order to have the operating system do that, the file containing each software module must include the following descriptive information:

- Where are the module's entry points?
- Which (if any) VMEbus interrupts does it service?
- How many input channel ports does it have?
- How many output channel ports does it have?

This descriptive information can be thought of as a *frame*, in which the *core* program is installed. To create the frame (in a standard format) and to install it over the core program, you'll need to have special utility programs.

Interconnection of processors
In addition to providing the software modules, you must also specify how these modules are to be *interconnected* with communication channels. You do that by creating a *network map*. The network map is a *file* which the operating system reads to determine how to set up the communication channels in shared memory.

For the same reasons mentioned earlier, you'll want to have an interactive utility program to help you build (or rebuild) this file. This utility allows you to make changes to the system structure, without having to rebuild the file manually. If the software modules are on the system disk while you're using this utility to build the file, the utility can also perform *consistency checks*. This ensures that you don't create channel connections between incompatible modules, and that you don't leave any channel ports unconnected in the final network.

The network map specifies where channels are needed between modules. However, it doesn't specify how those channels will be implemented. When the system is powered up, the operating system:

- examines the network map
- examines the module frames listed in the network map
- assigns the modules to the available processors
- implements the required data channels in shared memory
- configures the interrupt routing hardware.

By using an operating system that can determine the system resources, and assign the software modules to the processors in an optimal fashion, you can reconfigure the system without having to modify the software.

Multitasking can improve efficiency
In some cases, you may find that dedicating a processor to a single software module leaves the processor idle part of the time. To make full use of the processor's computing power, you may want to assign more than one software module to it. To do that, you'll need to provide *multitasking* capability.

The HyperFlo system
The HyperFlo system, manufactured by Pacific Cyber/Metrix, is a multiprocessor system, based on the data-flow principles described above. The system consists of a set of double-height VMEbus boards and supporting software, which includes the FLOS

operating system, as well as support utilities that run on UNIX or OS-9 development stations.

HyperFlo processor boards
There are 3 types of HyperFlo processor boards:

- the MPU-2 provides four 68020 processors
- the MPU-3 provides five 68030 processors
- the DSP-2 provides three TMS320C30 digital signal processors, plus one 68030

Each of these double-height VMEbus HyperFlo processor boards can be populated with several processors. Because of the design of the operating system, you can mix different processor boards: a system might include several of each processor board type.

The MPU-2 processor board
Figure 12 shows an MPU-2 processor board. Figure 13 shows the general architecture of HyperFlo processor boards, as exemplified by the MPU-2. In addition to the processors, this board contains floating-point coprocessors, and 1.5 Mbytes of static memory. Each processor has its own 68882 coprocessor, and 256 Kbytes of RAM for its exclusive use. An additional 5-ported block of local memory is also available for interprocessor communication.

Figure 12

Each processor has a private bus, which allows it exclusive access to its private resources. The 4 onboard processors are linked through a series of driver chips to a common local bus, which is regulated by a round-robin arbiter. The local bus permits each processor to access the 512 Kbytes of local static RAM, which can be used for communication between the processors on the board.

Any of the onboard processors can access the VMEbus through the local bus. The local RAM can also be accessed by other masters on the VMEbus. This allows it to be used for board-to-board communication channels.

The GMS-1 memory board
The companion GMS-1 memory board provides 6 Mbytes of static RAM, and supports data transfers to (or from) HyperFlo processor boards at up to 80 Mbytes/sec.

Note: Standard VMEbus protocols only provide speeds up to about 40 Mbytes/sec. To achieve this very high data transfer rate, a *synchronous* protocol is employed, using a user-defined address modifier code.

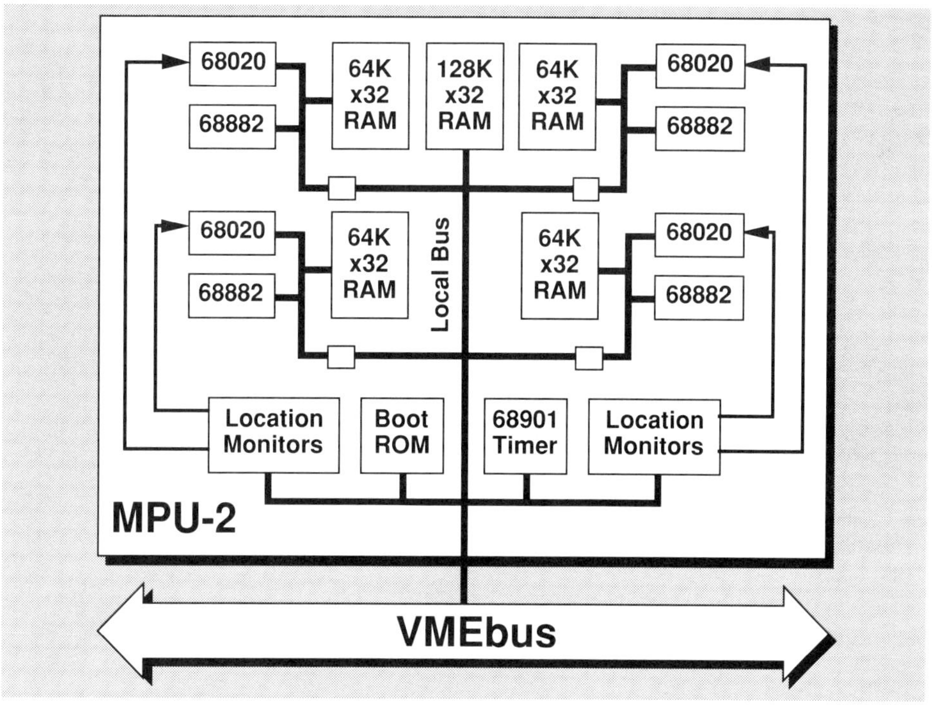

Figure 13

The SRB-2 system resource board
The 3rd board required for a minimal HyperFlo system is the SRB-2 *system resource board*. This board provides:

- a VMEbus arbiter
- a SCSI interface
- a time-of-day clock
- a real-time clock
- a power supply
- a temperature monitor
- ports for a system console
- several hardware mechanisms that are required in every HyperFlo system
- exception logging

System I/O
System I/O for a HyperFlo system can be provided with standard VMEbus and VSB I/O boards.

HyperFlo software
HyperFlo software support consists of:

- the FLOS run-time operating system
- a software-development toolkit

FLOS is a distributed data flow operating system, a portion of which runs on every HyperFlo processor. It takes care of all the administrative tasks in a HyperFlo system, including:

- task scheduling
- task distribution
- hardware performance checks
- resource allocation
- message passing
- system failure detection and reconfiguration

FLOS is a run-time-only operating system: it's used only to *execute* the application software.

All software *development* is done on another system (under UNIX or OS-9) using the software toolkit. The software toolkit includes software for development, debugging and operation of the application code.

Creation of HyperFlo software is a 2-step process:

- Creation of the software modules
- Creation of the application network map

Creating the module core

To facilitate use of system-specific resources and mechanisms, the HyperFlo toolkit includes a function library for C, FORTRAN, and assembly language. This library contains functions which allow a task to:

- use input and output data channels
- acquire the system time
- put itself to sleep
- communicate with the user, at the system console
- access the resources of a UNIX or OS-9 system

The application program calls UNIX-like functions in this library to communicate with a user terminal, or to access the resources of a UNIX or OS-9 computer. These function calls are accomplished using HyperFlo's built-in message passing capabilities.

Creating the module frame

The *module frame* provides information about the *module core* to the FLOS operating system. The structure of all module frames are the same, regardless of the type processor the module was written for, or the language it was written in.

The module frame can be generated with an interactive development tool, called the *Module Frame Editor*, which runs under UNIX or OS-9 on the front-end development system. Figure 14 shows the main screen of the Module Frame Editor for the module *volume*. This module has one read port (r0) and one write port (w0). The processor type has been specified as a 68020. Other information includes the desired stack size, and the interrupt levels used by the module.

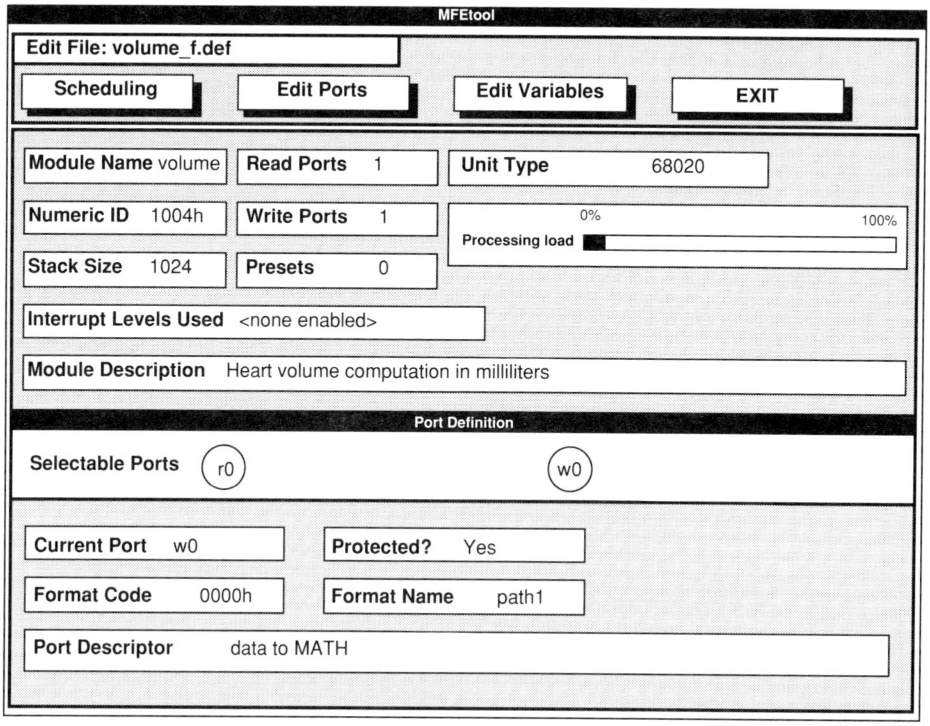

Figure 14

Once all the necessary information has been filled in on the screen, the editor builds the module frame automatically. An *executable module* can then be created by compiling the core, and linking it to the frame.

Creation of the application network map

The application network map is a disk file that specifies how the modules are to be linked together. It's created using HyperFlo's *application network configurator* utility.

Figure 15 shows how the configurator displays an application network map. The configurator looks at the modules in the library, and then displays a circle on the screen for each module. You then specify the interconnections by drawing lines on the screen, from circle to circle. The configurator then uses these lines to build the network map.

Once you have specified connections for each of the module inputs and outputs, you click the *Build* command on the right side of the screen. The configurator then constructs the *executable application software*, from the executable modules in the library. This software consists of 2 files:

- The module block
- The application network map

FLOS start-up

When you *execute* the application software, the FLOS operating system automatically takes an inventory of the available processors, and determines which ones reside on

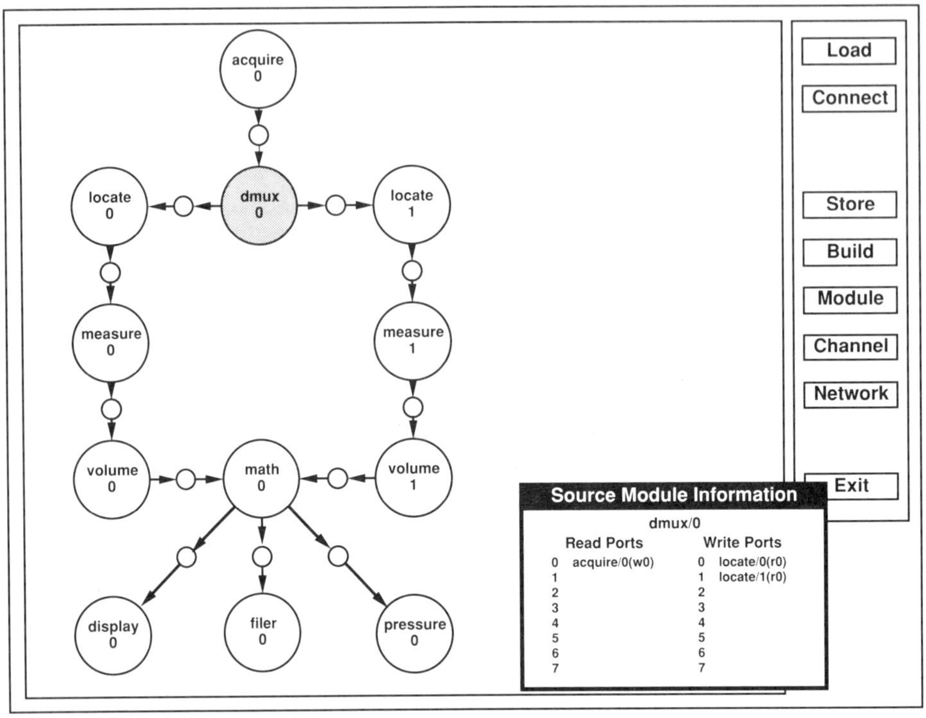

Figure 15

a common HyperFlo processor board, sharing local memory. Then FLOS assigns the software modules to the various processors in the system, and implements the channels between them in the appropriate shared memory. Each channel is accessible to both the module *sending* data to it and the module *receiving* data from it. To both modules, the channel appears to be a large FIFO.

If there are more modules than processors, FLOS will assign 2 or more modules to some processors, as necessary. In doing so, FLOS tries to minimize the use of shared buses, to keep them from becoming bottlenecks.

If you know that there will be a lot of traffic between a certain pair of modules, you can guide FLOS in its selections by setting distribution *restriction* parameters in the application network map. These parameters force FLOS to put certain software modules on a single processor board. That keeps their communication on that board's local bus, instead of on the VMEbus.

HyperFlo operation modes

A HyperFlo system is *not* designed to be a software development station. It assumes that you have already developed your application software on a separate platform. However, software is provided with HyperFlo to allow the use of a SUN-3 workstation, or an Eltec E-5 VMEbus processor board for software development.

When using the Eltec E-5 processor board (and the OS-9 operating system) the Hyperflo system communicates with the processor board through global VMEbus memory.

When using a SUN-3 workstation (and the UNIX operating system) the HyperFlo system communicates with the workstation through an Ethernet link.

HyperFlo stand-alone mode
When you're using the HyperFlo system in the stand-alone mode, you can retrieve application code from:

- the 2-Mbit mass-storage EPROM, located on the SRB-2 board.
- a floppy disk or a hard disk, connected to the SCSI port on the SRB-2.
- some other user-defined I/O device.

You communicate with FLOS in the stand-alone mode through the system console; a terminal connected to the RS-232C port on the SRB-2. You can start an application program and interact with it by entering commands on this console. You can also use commands to find out how the software modules have been assigned to the various processors in the system, and to measure system performance.

HyperFlo UNIX or OS-9 front end mode
When you operate the HyperFlo system through a front-end computer, you can retrieve an application program from a mass storage device on the front-end computer (hard disk, floppy disk, cartridge drive, etc.) and automatically load it into the HyperFlo system. The same commands are available as those on HyperFlo's system console, but in a more convenient menu/window format. Through this front-end system, several users can interact with HyperFlo at the same time.

You can also create a program that runs on the front-end computer, and coordinates its activities with a HyperFlo application program. For example, you might create a high-speed data acquisition and preprocessing program to run in the HyperFlo system, and use a UNIX-based off-the-shelf data base program to analyze and store the acquired data. HyperFlo would support communication and interaction between these two independent programs.

HyperFlo debugging tools
The HyperFlo package includes several useful debugging tools.

The *input data generator* is a module that can be incorporated into your application program to provide simulated data streams to other modules. It allows you to test your modules, as well as the application network that interconnects them.

The *data channel monitor* allows you to detect processing bottlenecks. It monitors the flow of data through the data channel network.

The *multiprocessor debugger* allows you to trace execution on any of the individual processors, examine their memory contents, and monitor any messages passed to them, or by them.

HyperFlo systems provide fault tolerance
One bonus provided by data flow systems is a degree of *fault tolerance*. Every processor in a HyperFlo system monitors the operation of every other processor. If any processor (or its associated hardware) fails, its demise is almost immediately detected by the other processors. FLOS then reassigns the failed unit's software module to

another processor, and the application continues to run. In many cases, a slight degradation in overall system performance is the only discernible consequence.

Calculating Hyperflo performance

Each HyperFlo processor devotes somewhere between 80% and 95% of its time to module processing. The remaining 5% to 20% of its time is used for the administrative processing of the FLOS operating system (module scheduling, message passing, command interpretation, etc.) .

If, for example a software module is written to run on an MPU-3 board, that module must not require performance of more than about 4.8 MIPS (80% of the 6 MIPS performance of each 68030 processor on that board).

An example of a VMEbus-based data flow system

DLR, the West German Space Agency has, for some time, been using a HyperFlo-based system (named SPACEMED) to analyze (at faster than real-time rates) physiologic functions (ECG, ICG, blood pressure, etc.) of astronauts in flight. The system is used to study the effects of a microgravity environment on human physiology.

This original SPACEMED system is now being expanded to include real-time analysis of ultrasonic heart-muscle images. A special ultrasound imaging machine acquires black-and-white heart-muscle images from the astronaut. (The heart appears as a dark area in the image.) These images (whose resolutions are comparable to that of a standard video camera) are then fed into the SPACEMED system. After some initial image filtering and enhancement, the system must determine the outlines of the heart chambers in each frame. Then it must use this data to compute the current *volume* of the heart, as well as other parameters of interest.

The image acquisition hardware must provide some initial *filtering* and *enhancement* of the raw images. For these functions, DLR selected a VMEbus board set from Imaging Technology. This board set performs the functions indicated by the emphasized boxes in Figure 16. The final processing step performed by the board set is a *feature extraction* function, which determines areas in the frame which fall within a user-defined brightness range. The data within these areas (some of which represent the heart muscle) are routed to the HyperFlo system for further processing.

On the output side, there is:

- a graphics controller board, for displaying the results
- a mass storage controller board, for data archiving
- an interface to some control hardware, by which real-time stimuli are provided to the astronaut

It's important that the computer perform all of its calculations in real time, since this allows the experimenters to provide various stimuli to the astronaut immediately.

In order to deliver the needed performance, the architecture shown in Figure 16 was envisioned. However, as development proceeded, it became clear that there would be a bottleneck at the *Heart chamber location and measurement* step. 10 MIPS was too much to demand from a single microprocessor. To resolve this bottleneck, 2 techniques

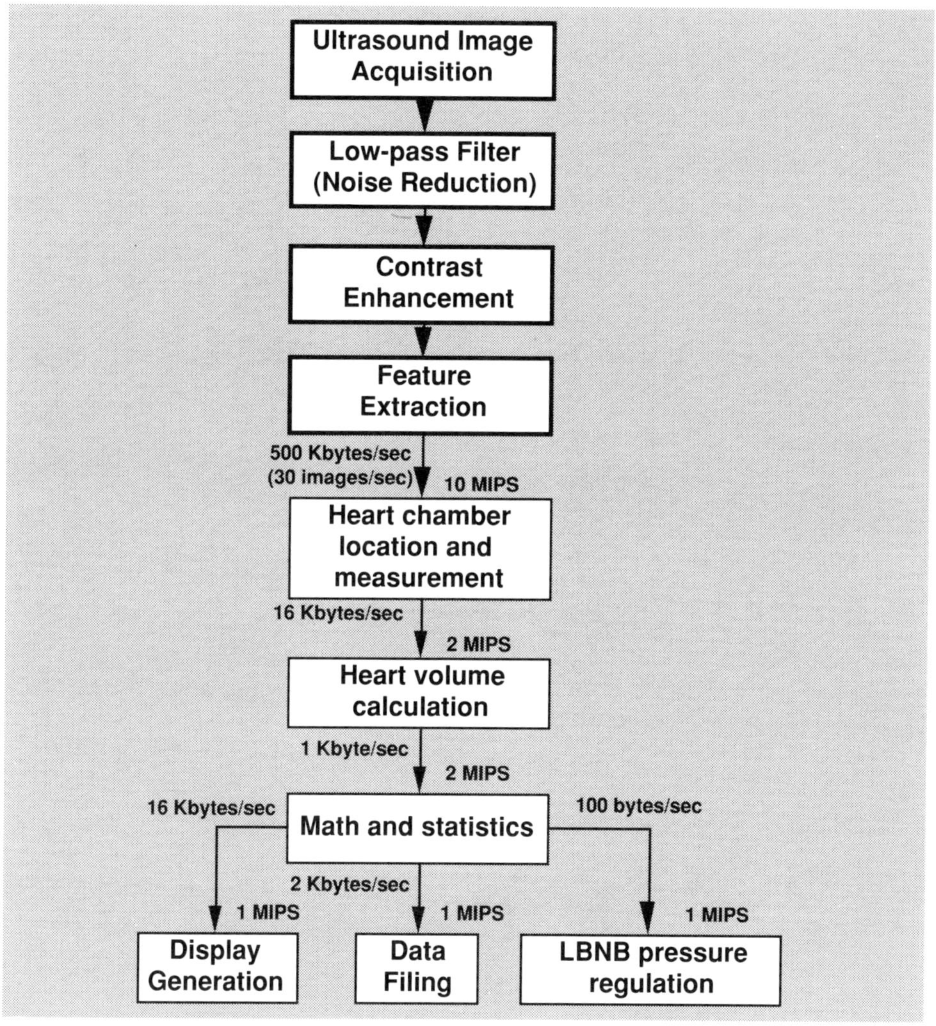

Figure 16

were then applied to split that step into multiple modules, allowing the use of more processors:

- Task splitting
- Data stream splitting

Task splitting
First, the heart chamber location and measurement operation was split into 2 serial steps, as shown in Figure 17:

- Heart chamber location
- Heart chamber measurement

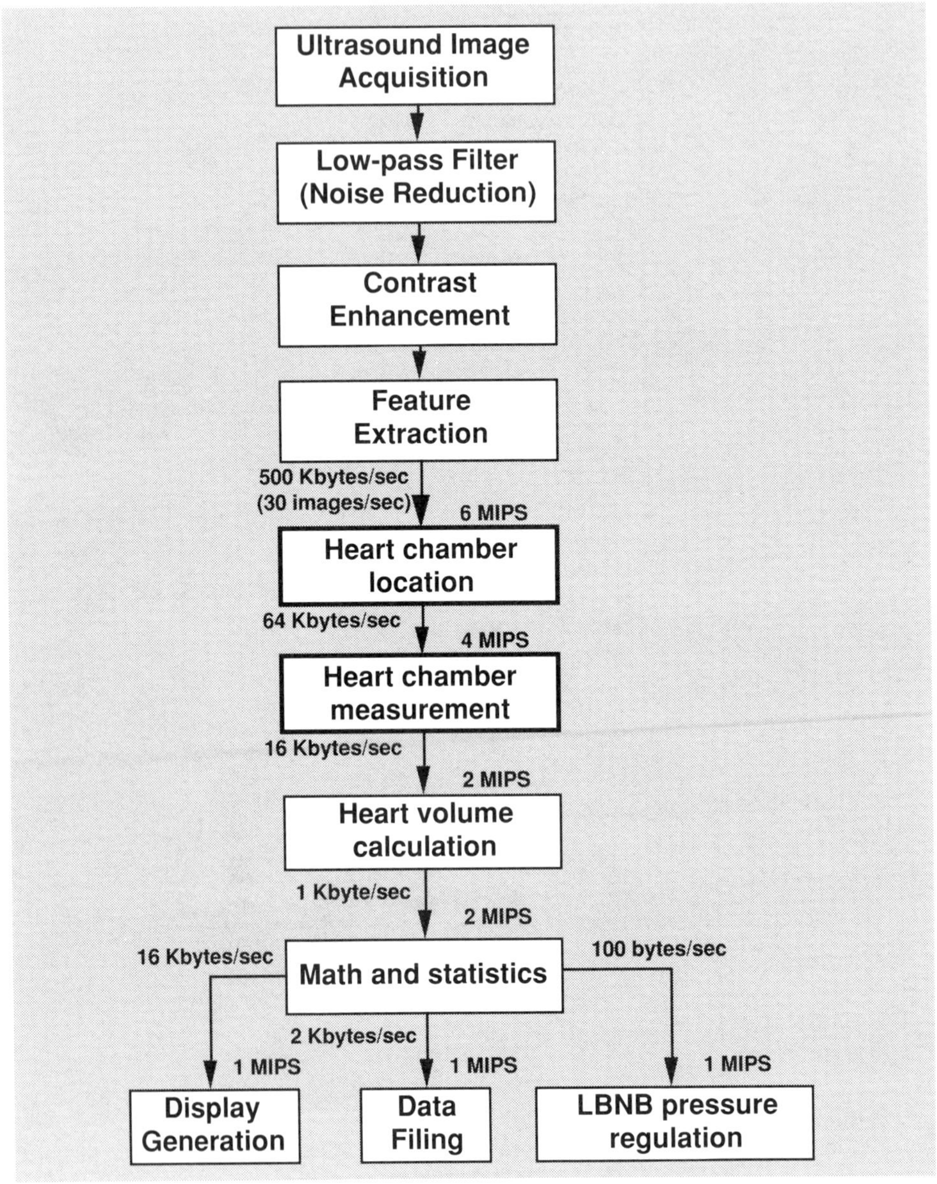

Figure 17

Splitting a task in this fashion allowed the use of 2 dedicated processors: 1 for each of the 2 steps. With each processor expending its full effort on half the job, the throughput increased.

This shows one of the unique advantages to data flow architectures: the total processing job can be broken down into smaller and smaller steps. With each processor in the chain doing a simpler job, its throughput goes up.

While this improved the situation, the *Heart chamber location* step still required 6 MIPS, and the *Heart chamber measuring* step still required 4 MIPS. This was still too much processing for a single processor, so *data stream splitting* was also required.

Data stream splitting

To provide data stream splitting, *demultiplexing* software was added. This software alternately provides image-frames to 2 identical module chains, as shown in Figure 18. Obviously, the number of images processed by each chain is half the previous number, and the throughput demanded of each chain is only half of the total. Combined with

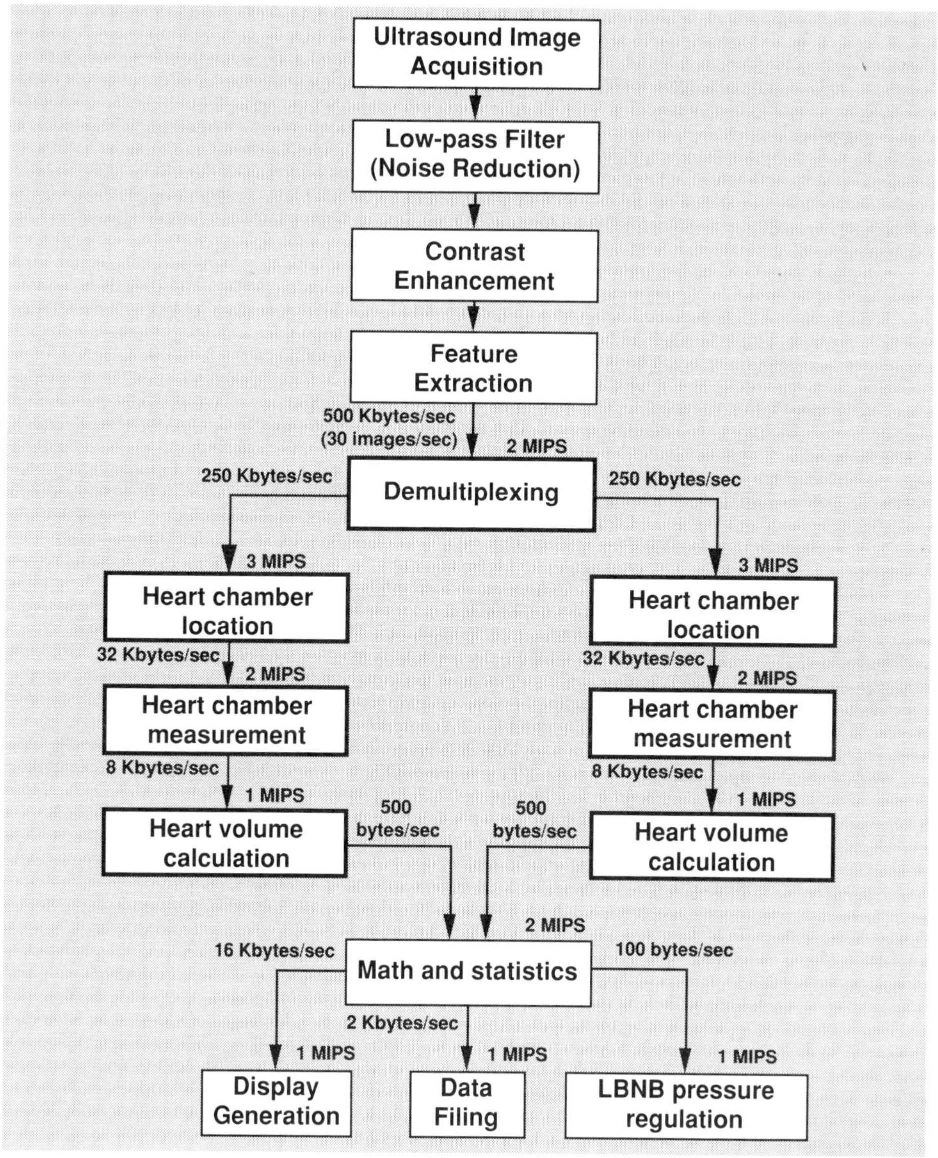

Figure 18

the task splitting above, this proved to be sufficient. However, if it had not been sufficient, additional module chains could have been provided, by splitting the data stream into 3 or more streams.

Using these data flow techniques, we reduced the demand for a 10 MIPS processor down to a demand for 4 processors, whose individual capabilities need not exceed 3 MIPS: well within the capability of a 68020.

Listing 1 (on the following page) shows the core code for the module that computes the heart volume in the SPACEMED system. This module communicates through the 2 data channels: R0 is an input port, and W0 is an output port. The function `c_init()` is called to initialize the data channels and the function `c_wacs()` is used to receive and transmit data through the channels.

Conclusion

Data flow architectures lend themselves well to applications where you need to process very high volumes of data quickly and continuously. Unlike many other multiprocessor architectures, the structure of a data flow architecture can be tailored to the topology of the application. Additional processing power can be added to the system, either by splitting the data stream and running it through separate chains of modules, or by splitting the processing steps into smaller increments. The number of processors that can be used in a VMEbus-based data flow system is limited only by the number of slots in the card rack, and even that number can be extended by the use of VMEbus rack-to-rack repeaters.

Marco Flagg has been project leader for the HyperFlo project at Pacific Cyber/Metrix since its inception in 1985. Marco was educated in West Germany, and emigrated to the US in 1984. He has 8 years experience in the design of real-time and multiprocessor machines for government and military agencies.

Listing 1

```
/*
 *  heart volume computation module
 */

#define REPEAT  while (1)
#define R0         0     /* first read port */
#define W0         8     /* first write port */

extern unsigned EstimateCrossSection();

typedef struct
  { unsigned pid;              /* picture id code */
     unsigned short count_y,  /* count of lines in image */
                    first_y;  /* first y position (not used) */
  } HEADER ;

typedef struct
  {  unsigned short first_x,  /* first and last points on a scan */
                    last_x;   /* line that are in the image     */
  } LINEDATA ;

typedef struct
  {  unsigned pid;            /* the copied picture id code */
     unsigned volume;         /* the computed volume */
  } OUTPUT;

static HEADER   hdr;          /* storage for structures */
static LINEDATA ldata;
static OUTPUT   odata;

module()
  {  register unsigned short i,j; register unsigned volume;

     c_init(R0);  c_init(W0);             /* initialize ports */
     REPEAT
     {   c_wacs(R0,sizeof(HEADER),&hdr);  /* get line count info */
         odata.pid = hdr.pid;             /* copy 'picture id'   */
         volume = 0;
         for (i=hdr.count_y; —i;          /* loop for each scan line */
         {   c_wacs(R0,sizeof(LINEDATA),&ldata);
             volume += EstimateCrossSection(ldata.first_x,ldata.last_x);
         }
         odata.volume = volume;           /* copy to output structure */
         c_wacs(W0,sizeof(OUTPUT),&odata);/* output to MATH module */
     }
  }
```

CHAPTER 15

Using VMEbus Repeaters To Build High Performance Distributed Processing Systems

By Don Turrell

As system designers tackle more challenging problems, they sometimes find that the 21-slot physical limitation of the VMEbus card rack prevents them from building the systems they need. In this chapter Don describes how engineers have attempted unsuccessfully to overcome this problem. Then he shows how a carefully designed bus repeater permits you to construct VMEbus systems of virtually unlimited size and processing power.

When the Revision A VMEbus spec was first released in 1981, most engineers thought the 21 board slots it provided were more than adequate. However, when more powerful processors, (like the 68020 and the 80386) were introduced, engineers found that these MPUs could service many more I/O devices than earlier 8-bit MPUs. In I/O intensive applications (such as industrial control or data acquisition) the number of I/O boards sometimes exceeded the 21 slots available in the VMEbus chassis.

Some users (in an attempt to provide more slots) have built bus extenders consisting of a cable (or cables) with a VMEbus-compatible connector at each end. The idea is to extend the signal conductors from a host VMEbus chassis into a 2nd VMEbus chassis. The 2nd chassis could then be used to house additional VMEbus boards.

However, the thought of using a crude *bus extender* of this type sends chills up and down the spine of any engineer who has experience designing high performance backplanes like VMEbus. Once this bus extender is plugged into the backplanes, all of the assumptions about the transmission line characteristics of the backplane signals become invalid. The resulting slow (and irregular-looking) signal transitions on both backplanes can be very ugly. In fact, ringing on the signal lines may even cause multiple transitions, where there should only be one.

This signal degradation would be bad enough by itself, but bus extenders are usually used in systems that have all 21 slots filled with boards. These systems already have a heavily loaded backplane, with the slowest permissible signal transitions. Thus, by the time a VMEbus system grows large enough to need a bus extender, its signal waveforms are already at their worst.

Bus expanders
Engineers, who better understand the transmission line theory behind backplane design, have devised a more sophisticated interconnect approach called a *bus*

expander. A bus expander consists of a pair of printed circuit boards, with driver chips and receiver chips that provide buffering between the cable conductors and the VMEbus backplane conductors, as shown in Figure 1.

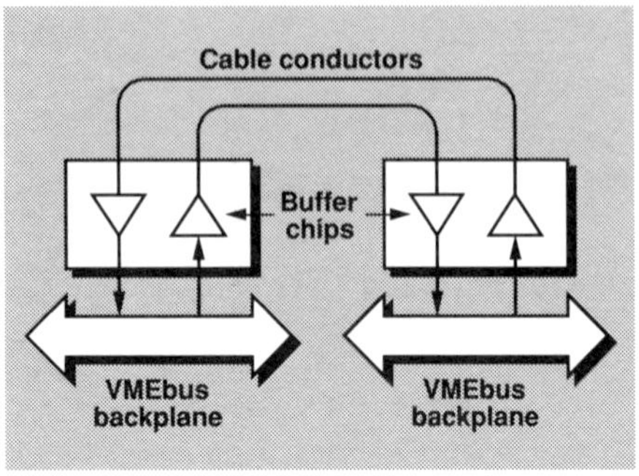

Figure 1

A bus expander is simple, easy to design and build, fairly inexpensive, and (because it takes into account the transmission line characteristics of the VMEbus backplane signal lines) it provides clean signal transitions. However, it may not preserve the relative timing between signal transitions from one backplane to the other. For this reason, bus expanders can produce systems with unreliable operation.

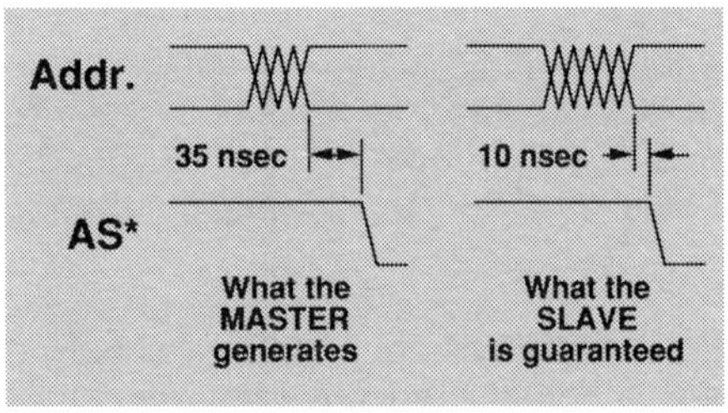

Figure 2

For example, the Rev C.1 VMEbus spec requires a master to drive the VMEbus address lines to a valid level 35 nsec prior to driving the address strobe low. (See Figure 2) Taking into account the worst case transmission line characteristics of the backplane, the spec then guarantees to the slave board that the address will be valid for at least 10 nsec before the slave sees the address strobe go low. The difference (25 nsec) provides time for the signal levels on the address bus to stabilize before the slave samples them.

When using expanders, the chief timing problem is that the address and data setup times in the destination chassis are essentially a degraded version of the setup times in the source chassis. These setup times may either be increased or decreased by the propagation and loading characteristics of the cable between the 2 systems. For example, if the address strobe from the source chassis is delayed more than the address bits, in its propagation to the destination chassis, the setup time in the destination chassis will be less than in the source chassis. As a result, the VMEbus slave board in the destination chassis might latch the address bits before those bits are valid on its backplane.

This can result in intermittent failures or (even worse) a failure that can be "cured" by unplugging some board that loads the address bus heavily. The latter situation, may start you on a futile search for a malfunction on the unplugged board.

Bus repeaters

However, you can design an interconnection device that takes into account both the transmission line characteristics and the setup times of the VMEbus by including:

- signal line drivers that regenerate the edges of the VMEbus signal transitions
- timing logic that regenerates the 35 nsec setup times required by the VMEbus spec

By including this timing logic you can avoid the situation that we saw earlier, where marginal operation leads to elusive bugs that seem to migrate throughout the system as you try to track them down.

Unfortunately, this advantage has its cost: the additional overhead incurred in regenerating the address and data setup times can add 50 to 100 nsec of delay to each data transfer from chassis to chassis. This overhead does not occur when using bus expanders. This is a good example of the maxim that there's no such thing as a *free lunch*.

Multiprocessing with bus repeaters
While the basic function of a repeater is to allow more than 21 boards in a VMEbus system, you can make a few simple additions that allow some spectacular performance increases.

For example, suppose you want to design a VMEbus system that captures and processes video images in real time. This image processing system consists of a video digitizer board and an image processing board. The video digitizer board digitizes an incoming video signal and then stores the resulting data in global VMEbus memory. The image processing board then accesses this digitized image.

Suppose the image processing board is not able to keep up with the incoming frame rate, partly because it must compete with the digitizer board for the VMEbus, and partly because it processes the images at a rate slower than the incoming frame rate.

By using a repeater, you can remove the image processing board (and a VMEbus memory board) from the digitizer's chassis, and insert it into a 2nd VMEbus chassis. If you then provide this 2nd chassis with basic system controller functions (like a system clock, and bus arbiter) it can then operate independently of the host chassis. Since the image processor board no longer has to compete for access to its VMEbus backplane, it can process the images faster.

However, even with this faster processing, the image processing board still might not be able to process the images as fast as they're digitized. To solve that problem, you could build several additional image processor chassis, and connect them to the host chassis through repeaters in a star configuration, as shown in Figure 3. With this configuration, the digitizer board could write the first digitized image into image processor chassis number 1, through one of the bus repeater links. Then, it could disconnect from that chassis, making available the full bandwidth of backplane number 1 to its image processing board. Then the digitizer board could write the 2nd image into chassis number 2, etc.

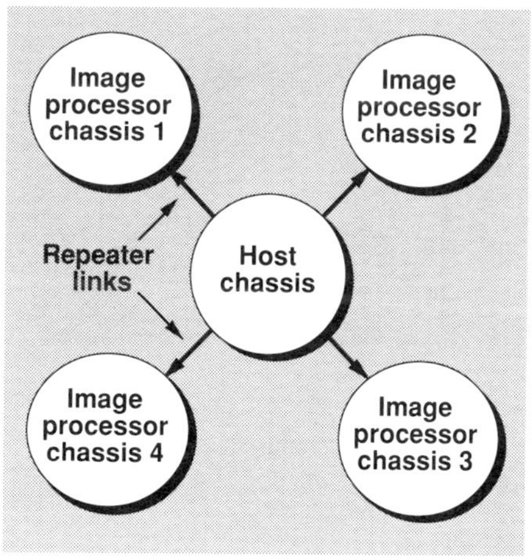

Figure 3

In using this approach, you solve 2 problems:

- You eliminate competition between the image processor board and the digitizer board, since the digitizer writes subsequent images into other chassis.
- You eliminate competition between image processor boards, since each resides in an independent chassis.

This is an example of an actual VMEbus system that has been built. It illustrates the essentially unlimited processing power that can be assembled using VME-to-VME repeaters. The key to this approach is that it allows connection or disconnection of the various chassis in the system through the repeater link. This permits any 2 chassis to work as a single tightly-coupled system (while transferring data) or as 2 independent systems (when processing in parallel).

Note: You may want to allow the auxiliary chassis to initiate host interrupts, even when the 2 systems are disconnected. For example, in the image processing system that we discussed above, you may want to have the image processor chassis interrupt the host chassis each time it finishes processing an image.

How a repeater-based system works

Now that we have seen how a connect/disconnect protocol can be used to build high-performance, loosely-coupled systems, let's take a closer look at how the connect/disconnect process would work.

A repeater that's designed to provide connection and disconnection must include hardware that provides all of the system controller functions the auxiliary chassis will need to operate independently of the host chassis. (See Figure 4) This includes a

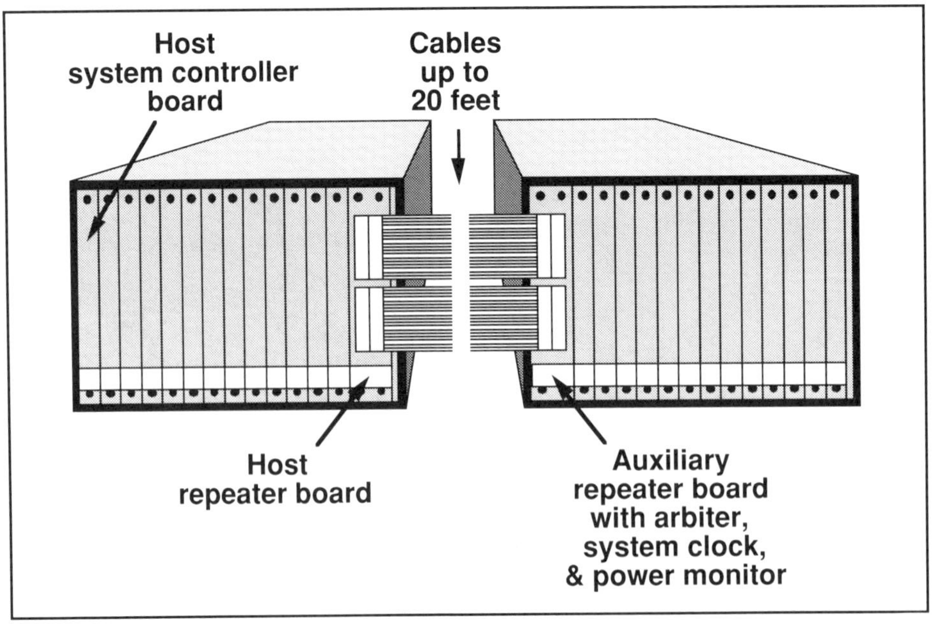

Figure 4

VMEbus arbiter, a system clock, and possibly a power monitor to detect power failures and initiate a shutdown sequence.

As long as the repeater provides a connection between 2 systems they function as one system, allowing any master in the auxiliary chassis to access slaves in the host chassis and any master in the host chassis to access any slave in the auxiliary chassis. While connected to the host chassis through the repeater, the VMEbus in the auxiliary chassis relies on the host's bus arbiter and system clock. In other words, the repeater effectively wires the auxiliary chassis backplane to the backplane in the host chassis.

The reason for using the host arbiter is to prevent deadlocks. For example, suppose a master in the auxiliary chassis tries to access the host memory at the same time a master in the host chassis tries to access memory in the auxiliary chassis. If the 2 VMEbus backplanes each had their own arbiter, each of these masters might gain control of its respective VMEbus. Then both masters would wait (possibly forever) for access to the other VMEbus repeater board.

However, by using a single arbiter for both the host chassis and the auxiliary chassis, you can ensure that this will never happen. (This one arbiter will never issue bus grants to 2 masters at the same time.)

Independent power down
A properly designed repeater link also provides you with another very useful feature: independent power down of the auxiliary chassis. When the repeater disconnects the 2 systems, the auxiliary chassis can be powered down without having to also power down the host chassis. This feature is especially useful when you're debugging the system, and you need to remove a board from the auxiliary chassis. Without this feature, you would have to power down the host chassis, which may involve the shutdown of a hard disk drive, or possibly even the shutdown of a multiuser operating system.

Shielding and RFI concerns
If the repeater's cabling will exit the system enclosure, you need to be concerned about EMI/RFI problems. It's much easier to design a repeater to conform to EMI/RFI radiation standards than it is to add them on later. To keep the radiation low, you should use a fully shielded cable (one with a foil shield and a drain wire).

The greatest source of EMI/RFI leakage (and the greatest reason for non-compliance) is not the cable: it's in the connectors. You should use shielded connectors (with conductive shells) and conductive mating connectors, and you should fully ground those shields to the front panel of the VMEbus repeater board.

Shielding works both ways. In addition to reducing radiation from the repeater's cables, it also reduces noise pickup on those same cables. Externally generated noise can come from other cables in the same cable tray, or even from the repeater's other cables.

Designing a bus repeater
The VMEbus spec provides rules that govern the physical and electrical design of VMEbus equipment. These rules ensure compatibility, between VMEbus boards purchased from different manufacturers. While some of these rules (like the maximum number of 21 board slots in a backplane) may seem arbitrary, they are based on worst

case mechanical and electrical computations. Failure to comply with them may result in mechanical incompatibilities or unreliable system operation.

Before designing a repeater, you must fully understand the reasons for the following 3 VMEbus requirements:

- the limits on physical length of the backplane
- the source current and sink current specifications for bus driver and receiver logic on VMEbus boards
- the VMEbus system timing requirements

It takes a finite time for the VMEbus signal lines to stabilize, after they have been driven high or low. That finite time depends upon the propagation time from one end of the backplane to the other. Since the propagation time is directly proportional to the physical length of the backplane, the physical length of the backplane is important.

The electrical loading on the VMEbus depends upon how many VMEbus boards are plugged into the backplane. As you insert additional boards into the backplane, the electrical loading on each signal line increases, changing the characteristics of that line. To ensure proper timing (even with a worst-case configuration of 21 boards) the VMEbus spec dictates the minimum drive capability, and the maximum loading, per board.

To ensure that the signal timing in both the host chassis and the auxiliary chassis will meet the VMEbus spec, the repeater board's drivers and receivers must also meet the drive and loading specifications, just like every other VMEbus board.

Meeting the address setup and data setup times

The VMEbus address and data setup timing is based upon the physical length of the VMEbus backplane, and upon the driver and receiver characteristics of the VMEbus boards. These setup times have been carefully calculated to take into account the effects of worst case bus loading. However there is no extra margin to compensate for the timing degradation that occurs over the cable of a repeater link.

While the use of high quality cabling can reduce the problem, impedance discontinuities and imperfect terminations will still introduce some signal waveform distortion. When these waveform distortions alter the timing of threshold crossings, they produce different propagation delays on the various signals passing through the cable. Since a certain amount of distortion is unavoidable, the repeater in the destination chassis should include circuitry that regenerates the proper setup times before using the signals to drive the backplane.

The PT-VME902A

In 1983 Performance Technologies introduced its first VME-to-VME link. Since that time MPU clock speeds have increased from 8 MHz to many times that speed. In addition, the early A24/D16 board products have evolved into much higher performance A32/D32 products.

As a result, we designed and introduced a 2nd generation VMEbus repeater called the PT-VME902A. (See Figure 5) It consists of a 2-board set, with cables that can be anywhere from 6 inches to 20 feet long, depending upon the needs of the application.

The loss of only a few nsec can mean additional wait states on today's very fast MPUs, so all delays through the PT-VME902A have been reduced to an absolute minimum, consistent with good engineering practice.

The PT-VME902A allows any master in the host chassis to access any slave in the auxiliary chassis, and any master in the auxiliary chassis to access any slave in the host chassis. All accesses by host masters over the link are completely transparent: there is no software involved.

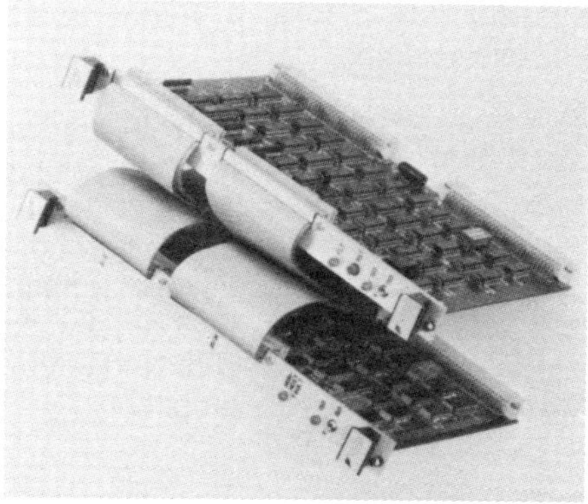

Figure 5

Through the use of gate arrays and custom chips, board designers have pushed bus timing much closer to the limits of the VMEbus spec. The tighter timing tolerances provided by these chips have reduced the VMEbus setup times and hold times that these newer boards generate. To ensure that the setup times of the signals in the destination chassis are adequate, but not excessive, the PT-VME902A has also been designed with tighter timing tolerances.

Using a VME-to-VME link during software development

A VME-to-VME link can be very useful when you're developing your application code on a UNIX system, and then running it in a VMEbus system. For example, most of the VMEbus-based real-time executives are sold with cross compilers and cross assemblers that run on the UNIX operating system. UNIX is great for software development, but once you have compiled, assembled, and linked your application program, you need some way to download it from the UNIX system into your target VMEbus chassis.

These downloads could be accomplished in the traditional manner, through an RS-232 link. However, the debugging process typically involves repeated compilations and downloads. If the downloaded files are large, you may spend a large percentage of your time waiting for the downloads to complete. By using a VME-to-VME link, instead of an RS-232 link, you can greatly speed up the downloads.

A good example of this situation can be seen at Los Alamos Laboratories, where engineers use the PT-VME902A to download their object files from a UNIX-based SUN workstation. Once the download has been accomplished, they run and debug the code from the SUN workstation, by communicating with the VMEbus chassis through an RS-232 link.

Summary

Expanding your VMEbus system with a VMEbus repeater allows you to stretch the 21-slot limit of the VMEbus. If you give careful attention to the VMEbus mechanical and electrical specifications, you can provide an unlimited number of slots, with direct addressing of all map locations, without the overhead of complex software drivers or networking software. If you go one step further, and provide some means to dynamically isolate and reconnect the auxiliary VMEbus chassis, you can build a loosely coupled system with a total processing power that far surpasses that of a single chassis system.

Don Turrell is the vice president of sales and marketing at Performance Technologies. He has done application engineering in real time computerized process control both in the US and abroad. He earned his BSEE at Pennsylvania State University, and his MBA from the University of Wisconsin at Madison. Don has authored and coauthored numerous articles on the subject of bus-to-bus adapters, ASIC interfacing to VMEbus, and the use of cache memory techniques in VMEbus processors.

CHAPTER 16

Building Large-scale Tightly-coupled Systems

By Larry Larsen

Process control systems typically control equipment that's widely dispersed throughout a manufacturing plant. The typical solution is to have several subsystems, which communicate with each other to keep things synchronized. However, if you use serial communication techniques to interconnect these subsystems, you may find it difficult to keep things coordinated, especially when events are occurring at a rapid pace. In this chapter Larry describes how you can interconnect widely dispersed VMEbus systems in such a way that all of the VMEbus resources appear (and can be accessed) as if they reside in a single VMEbus chassis.

Suppose you are asked to design a real-time process control system which coordinates the actions of equipment located hundreds of feet apart. Because you need fast response times, you may find a local area network (with its elaborate software overhead) is impractical. What you really need is a way to tightly couple VMEbus systems over long distances: a configuration that permits an MPU in one chassis to access the memory in another chassis, just as if that remote memory was local. By simply writing to the remote memory, an MPU could send information to a remote MPU quickly, without the software overhead of a local area network, or even a serial port driver.

The obvious method: extending the backplane signals

The most obvious way to provide such a link is to design a pair of VMEbus repeater boards. You could then install one of these boards in each of two chassis. You would then interconnect these boards with ribbon cables, which would carry the VMEbus backplane signals from chassis to chassis. (See Figure 1)

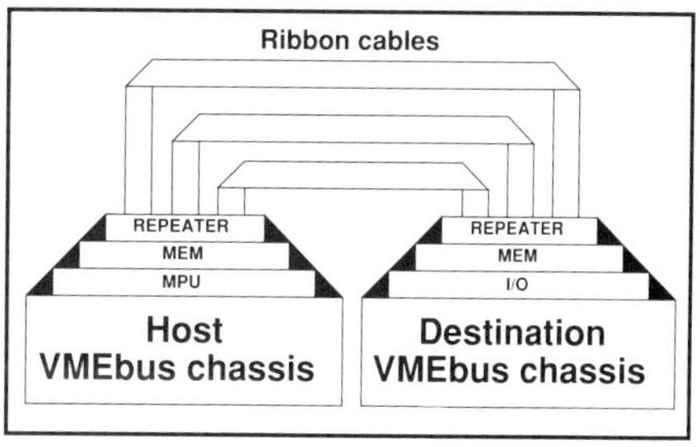

Figure 1

Of course, the repeater boards would have to include signal line buffers to keep the VMEbus backplane loading within the specifications. They would also need to have timing regeneration circuitry, to ensure that the address strobe setup time and the data strobe setup times are still legal in the destination chassis. If you design these boards properly, you end up with what amounts to a very long VMEbus backplane.

However, there are several problems with this approach:

- VMEbus has a very large number of signals, so you need several ribbon cables to route all of these signals from chassis to chassis. The large number of cables is especially awkward when the two chassis aren't mounted within the same enclosure. If the chassis-to-chassis distance is hundreds of feet, you may find that the cabling costs are exorbitant.
- You need to use high current (64 mA) drivers, in order to ensure that the signal transitions take place cleanly.
- You have to interleave the signal and ground lines on the ribbon cable, in order to keep crosstalk within acceptable levels. (This doubles the already-large number of conductors needed to carry the VMEbus signals.)
- Even with the interleaved ground and signal conductors, crosstalk can be a problem. For example, if all of the address lines change state simultaneously from FFFFFFFF to 00000000, there will probably be an unacceptable level of crosstalk in the cable. To keep this from becoming a problem, you have to limit the cable length to a few feet.

A better method: a multiplexed cable bus

These problems provide you with some compelling reasons to search for a better interconnect technique. In doing so, you want to reduce the number of signal conductors between the two chassis, without a corresponding reduction in throughput.

You could multiplex the VMEbus signals over the cable. This greatly reduces the total number of conductors you need. In addition, multiplexing permits higher reliability operation, since the smaller number of connector contact points reduces the likelihood of contact failures. It also allows you to include a parity line, for verification of the data as it is being sent.

However, if you multiplex the signal lines, you need to maintain a high data transfer rate. You can do that by:

- making optimum use of the signal line bandwidth, by taking into account transmission line effects.
- using a synchronous protocol to allow pipelined data transfers.

Taking into account transmission line effects

In order to get the maximum data throughput per wire, you need to use a controlled impedance conductor. If you use ribbon cable or twisted-pair cable, this will typically be about 100 ohms. To keep the signal transitions clean, you need to avoid discontinuities in the impedance. This rules out any stubs along the cable, and permits only one driver on each line. This, in turn, dictates that the line be used unidirectionally.

The chassis-to-chassis cable distance dictates the type of signal line drivers, and the type of cable, you should use. For distances up to 25 feet, you can use high current (64 mA) TTL drivers and twisted pair cable. Each TTL signal is carried by a twisted pair of conductors, with one conductor carrying the signal and the other carrying the return. To ensure reliable signal transmissions, you need to use high-quality controlled-impedance twisted-pair cable.

For distances between 25 and 50 feet, you need to use micro-coax cables, driven by high-current TTL drivers. Micro-coax cables provide excellent signal characteristics, but they're considerably more expensive than twisted pair cables.

To connect two VMEbus chassis that are more than 50 feet apart, you'll need to use differential signal line drivers and receivers. Differential drivers can drive much longer cable lengths than TTL drivers.

Using a synchronous protocol
In order to send data through the cable at a very high rate, you want to send it synchronously. (With reasonable care, you can send synchronous data at a rate of about 32 Mbits per second per wire). The high speed logic necessary to do this is more expensive (and takes more board real estate) per signal line. However, that additional expense is offset by the smaller number of wires, and the cheaper cabling.

To send data synchronously, you send the data down the data bus, as you generate edges on a clock signal. The clock signal edges indicate when data is changing, and when it's valid. For example, Figure 2 shows how the sender might send bytes over a cable with an 8-bit data bus. Even with high quality cables, there will be some signal waveform degradation, and some skewing of the clock signal with respect to the data. The worst case degradation and skew increases with cable length, so you will need to use lower clock frequencies for longer cable lengths.

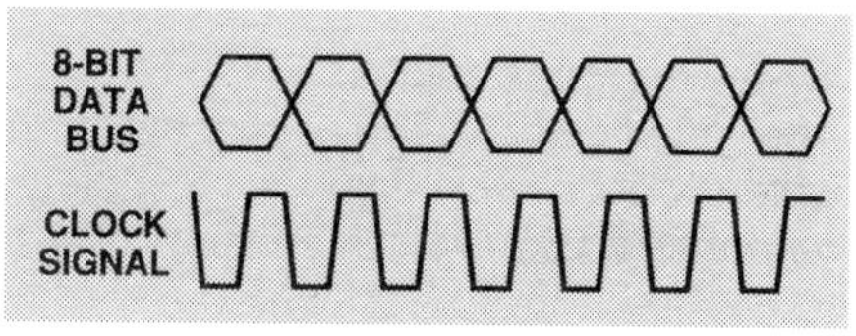

Figure 2

Minimizing cable crosstalk
It's impossible to completely eliminate cable crosstalk. However, synchronous data transmission provides a large degree of tolerance for crosstalk, by ensuring that the crosstalk occurs only when the receiver is not sampling the signal levels. When the receiving system detects the clock edge that indicates valid data, it waits for a prescribed delay, which is long enough to ensure that all of the switching transients in the cable have settled out. Since crosstalk occurs only during the signal transitions, crosstalk problems are avoided. When the cable is shorter, this settling process takes less time, so the clock frequency can be increased.

Handling externally generated noise
When cable lengths are very long (or when the cables are routed through noisy environments) you may have problems with externally generated noise. However, by adding a parity line to the cable bus, you permit the receiving adapter board to check the parity of each incoming byte parity against the level of the line.

Sending information over the cable bus
You can accomplish data transfers between the two adapter boards by sending *packets*. Each packet consists of a string of bytes. To indicate the beginning of a packet, you can use a packet synchronizing line. Since this line indicates the beginning and the end of the packet, we'll call it the *message envelope* signal. The meaning of each byte within the packet depends upon its position in the byte string.

In order to read and write to a destination chassis, we'll define 4 types of message packets:

- Write command packets
- Write response packets
- Read command packets
- Read response packets

Write command packets
The write command packet consists of 3 fields:

- A command field
- An address field
- A data field

Bits in the command field identify the type of operation, such as a *byte write*, a *word write*, a *longword write*, etc. The command field also specifies whether the data is being written to the destination's VMEbus global memory, or to the local dual-ported RAM on the destination's adapter board. (More about this dual-ported RAM later.)

Write response packet
After a *write command packet* has been sent to the destination chassis, and after the write VMEbus cycle has been completed, the destination chassis responds with a *write response packet*, which consists of a single field:

- A response field

The bits in this response field identify the message as a *write response message*, and indicate whether the write was successful (DTACK*) or unsuccessful (BERR*).

Read command packets
The *read command packet* consists of 2 fields:

- A command field
- An address field

As in the case of the write command packet, the bits in the command field identify the type of operation (such as byte read, word read, longword read, etc) and whether the data is being read from the destination VMEbus global memory, or from the local dual-ported RAM on the destination's adapter board.

The read command packet contains enough information to permit the adapter in the destination chassis to initiate a VMEbus read cycle. The data that's read from the memory must then be returned over the cable bus. This is accomplished with a re*ad response packet*.

Read response packet
The *read response packet* consists of 2 fields:

- A response field
- An optional data field

The bits in the response field identify the message as a *read response message*, and indicate whether the read operation was successful (DTACK*) or unsuccessful (BERR*). If the read was successful, the data field returns the data that was read from memory.

Address windowing
In order to avoid sending a packet over the cable every time a bus cycle occurs in the host chassis, you need to designate some address subrange that will cause an access to the destination chassis. You can do this by designating an *address window*, within which the host's adapter board will respond. The adapter board will ignore all data transfers to memory locations outside this window. Unless a reference is made to some location within the window, the adapter card is just a passive element in the system, and there is no data traffic between the two VMEbus chassis.

When a VMEbus master reads or writes to a location within this window, the adapter card latches the levels of the address bus, the data bus, and the control signals. Using these latched signal levels, the adapter board:

- constructs a command message packet, consisting of a string of bytes.
- sends this command packet over the cable (one byte at a time) to the adapter board in the destination chassis.

Since the construction and the transmission of this command packet is done entirely by hardware, the whole operation is transparent to the master in the host chassis that initiates the data transfer.

The adapter board in the destination chassis:

- requests the VMEbus.
- drives the address bus and data bus signal levels, based upon the bits in the command packet.
- initiates the data transfer cycle in the destination chassis.

When it completes the data transfer cycle, the destination adapter constructs a response packet, and then sends it back to the adapter board in the host chassis. The end result is that a block of memory in the destination chassis gets mapped to the address window in the host chassis, as shown in Figure 3.

Minimizing the size of command packets
Since the data transfer rate over the cable bus depends upon how many bytes there are in each command packet, you want to reduce the byte count wherever possible. One way to do that is to equip the destination adapter

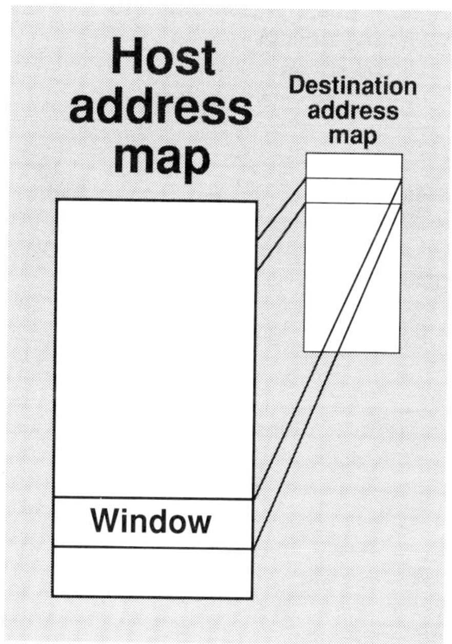

Figure 3

board with a register that holds the upper address bits. For example, if the adapter board has an 8-bit register to hold A16-A23, then you only need to send address bits A01-A15 over the cable bus to access an A24 slave location.

A similar technique can be used to handle address modifiers: The 6-bit address modifier code is first written into a register on the destination adapter board. Then, the subsequent accesses can be done without sending address modifier codes over the cable.

Speeding up data transfers with dual-ported RAM

Whenever a master in the host chassis accesses a memory location in the destination chassis, all other bus activity in both chassis must wait until the data transfer is complete. Since the data transfer rate of each of the two VMEbus backplanes is 40 Mbytes/sec, while the data transfer rate of the cable bus is only about 6 Mbytes/sec for writes, and 2.5 Mbytes/sec for reads, this is not a very efficient use of the backplanes.

However, you can greatly improve the backplane utilization by equipping the destination adapter board with dual-ported memory, as shown in Figure 4. This dual-ported memory appears as additional global memory to the masters in both chassis. A master in either chassis can write data to (or read data from) the dual-ported memory without using any of the bandwidth of the other VMEbus.

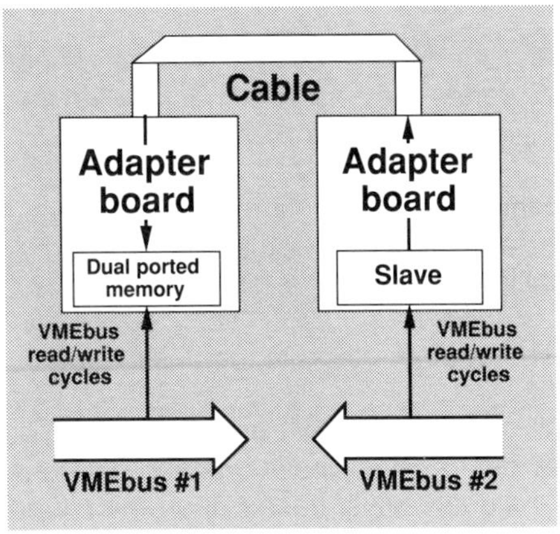

Figure 4

Providing a chassis-to-chassis interrupt mechanism

If you're going to pass data through a dual-ported memory, you'll need to provide some kind of chassis-to-chassis interrupt mechanism. This allows a master in one chassis to notify a master in the other chassis that updated information has been placed in the dual-ported RAM.

One way to provide chassis-to-chassis interrupts is to equip each adapter board with a VMEbus interrupter and interrupt handler, as shown in Figure 5. When adapter board A detects an interrupt request in its chassis, it responds to that interrupt, and captures the status/ID byte. Then it sends an *interrupt packet* over the cable bus to adapter board B. B then uses its onboard interrupter to generate an interrupt in its own chassis, and provides the status/ID byte when that interrupt is acknowledged.

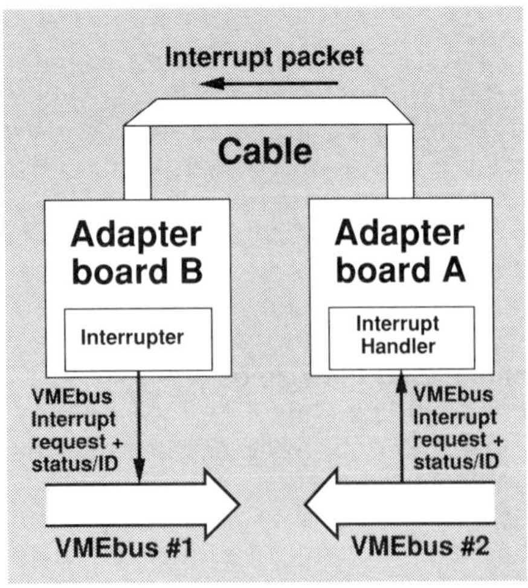

Figure 5

A description of the Bit 3 VME-to-VME link

The Bit 3 model 411 VME-to-VME link has been developed using the concepts described above. It consists of 2 VMEbus adapter boards: one for each of the two card racks that you need to interconnect (See Figures 6 and 7). These 2 boards are interconnected through two shielded cables, which provide two unidirectional bus ports to each adapter: a *data out* bus, and a *data in* bus. Each of these buses consist of 12 signals:

- 8 *data* lines
- 1 *parity* line
- 1 *message envelope* line
- 1 *message coming* line
- 1 *clock* line

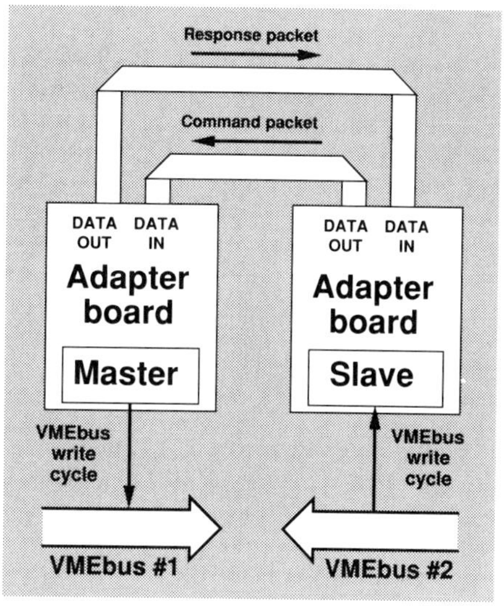

Figure 6

The transitions on all the other bus signals are synchronized to the edges on the clock line. The receiving adapter board samples the signal levels after a time delay from the clock edge, which ensures that the switching transients have settled out.

Note: Each of the Bit 3 adapter boards is equipped with both master and slave interfaces. This allows masters in either chassis to access VMEbus memory in the other chassis.

Figure 7

Cable bus bandwidth

The maximum bandwidth for the cable bus is 32 Mbytes/sec in each direction, for a total bandwidth of 64 Mbytes/sec. A complete command packet for a longword write to a 24-bit-address location can be sent over the cable in 288 nsec. The response message packet can be sent in 64 nsec. When all of the timing delays are taken into account, the entire process can be accomplished within 500 nsec.

In the 24-bit-address mode, all 24 of the address bits from the host chassis are sent over the cable bus to the destination chassis. That allows a VMEbus master in the host chassis to directly access 16 Mbytes of memory in the destination chassis, via read cycles and write cycles. (A VMEbus MPU in one chassis can even execute code stored in the memory of the other chassis.) A master in the host chassis can read 32-bit values from the memory in the destination chassis at a rate of 2.5 Mbytes/sec, and it can write 32-bit values at a rate of 6.0 Mbytes/sec.

These boards also provide a faster access mode that allows a master in the host chassis to access the memory in the destination chassis through a 64 Kbyte window. The master in the host chassis first writes an 8-bit value (over the cable) into a local register on the destination adapter. Then, when the master in the host chassis accesses a location within the 64 Kbyte window, only the lower 16 bits of the address are sent. The register on the destination adapter provides the upper 8 address bits. (The MPU in the host chassis can access any location in the destination chassis by writing the appropriate 8-bit value into the register.)

The parity line allows the adapter boards to check each byte of each incoming command packet against the parity line level. If it detects a parity error anywhere in the incoming packet, the adapter board aborts the command.

To allow transfers via dual-ported memory, the Bit-3 Model 411 adapter can be equipped with a plug-on 128-Kbyte piggyback RAM card. This dual-ported memory has two independent address decoders, allowing it to be jumper configured, to place the memory in a vacant address space in both VMEbus systems.

When connecting 2 chassis that are more than 50 feet apart, a companion VMEbus board can be used to convert the single ended TTL adapter board outputs to differential levels.

Summary

The techniques discussed in this chapter allow you to build tightly-coupled multiple-chassis VMEbus systems. By using these techniques, you can build very responsive distributed real-time systems, without the software overhead typical of local area networks.

Larry Larsen is the engineering manager at Bit 3 Computer Corp in Minneapolis, MN. He received his BSEE from the University of North Dakota in 1968. His design experience includes work on midrange computers and supercomputers.

CHAPTER 17

Using Data Transporters To Build Multiprocessing Systems

By Kevin J. Lynch

As engineers build multiprocessing VMEbus systems, they're finding that bus bandwidth limitations sometimes prevent all of the processors from working simultaneously. In this chapter Kevin describes how you can use data transporters to build pipelined VMEbus architectures that permit any degree of parallel processing.

Microprocessors are much faster today than they were only a few years ago. Data bus widths have gone from 16 bits to 32 bits. Meanwhile, clock speeds have been constantly pushed higher. But, despite all of this, engineers often need more performance than a single microprocessor can deliver. One solution is to build multiple processor systems. VMEbus provides an excellent platform for constructing multiprocessing systems.

In theory, the addition of each new processor adds a fixed amount of processing power to the total system. In order to provide the required processing power, you would simply keep adding processors until the performance is adequate. Of course, this assumes that you can coordinate the work done by each processor with the work done by the others.

The VMEbus is designed to allow you to build systems in this way. However, if you attempt to do so you'll probably find that it's not so simple. After a certain point, adding another processor may not result in a significant performance increase.

The problem is that VMEbus (like all buses) has a finite bandwidth. Each additional processor that you add to the bus, increases the usage of the bus bandwidth. Eventually, all of that bandwidth is used up. If you add additional processors beyond that point, some of them will be waiting to use the bus at any given time.

Using the VMEbus bandwidth more efficiently
Much of this bandwidth is used up by interprocessor message traffic, which involves a lot of data movement from place to place. The VMEbus specification includes several features that allow you to move data from place to place very efficiently:

- 32-bit data transfers
- block transfers
- address pipelining

32-bit data transfers
If you transfer data only 16 bits at a time, you're only using half of the available VMEbus bandwidth. Obviously, moving data 32 bits at a time is much more efficient.

Block transfers
Block transfer cycles allow a master to broadcast a single address over the VMEbus, and then write (or read) that location, and a string of adjacent locations, without having to broadcast another address. However, current microprocessors are not designed to write or read data in this manner.

Address pipelining
The VMEbus address pipelining feature is often used instead of block transfers. It offers about the same level of performance, with much less hardware on the slave boards. The slaves are required to latch the address from the bus immediately after the AS* signal goes low on each cycle. This allows the master to remove the address while the data transfer is still taking place and then broadcast the address for the next data transfer, while the current data transfer cycle finishes.

By using these VMEbus features you can improve the efficiency of data transfers. However, moving data from one VMEbus location to another with a general-purpose microprocessor is very inefficient.

The solution: a data transporter
A *data transporter* is a VMEbus board that is used to move data from one VMEbus location to another. It does so by reading data from one location, and then writing it to another location. An ideal data transporter should be able to do 32-bit reads and 32-bit writes, in order to move data in the most efficient manner. However, it should also be able to do 16-bit reads and writes, so it can also be used to transfer data to and from D16 boards.

Since you might want to execute some kind of operations on the data stream while you're moving it, you would want the data transporter to have some built-in intelligence. Ideally, these operations could be done without slowing down the data movement.

The data transporter should be designed so that it's not the limiting factor in the speed of data movement. The speed should depend solely upon how fast data can be read from the source board, and how fast data can be written to the destination board.

A data transporter should include a high speed FIFO
In order to move data efficiently, the data transporter should include a high-speed FIFO for intermediate data storage. The transporter first reads data from source VMEbus locations into the FIFO until it's full. Then it writes the FIFO contents to the destination VMEbus locations.

Depending upon the size of the data block being transported, moving data within a chassis may require one or more cycles of reading data from the source location, and then writing that data to the destination location. For example, to move 10 longwords from location A to location B, the transporter would have to initiate 20 longword data transfers over the VMEbus: 10 to read from location A into the FIFO, and 10 more to write from the FIFO to location B.

Increasing the VMEbus bandwidth

By using a data transporter to move data within a VMEbus system, you can greatly improve the efficiency of bus usage. But even with this higher efficiency, if you continue to add processor boards to the VMEbus, you will eventually run out of bandwidth.

An obvious way to add VMEbus bandwidth to a system is to add a 2nd VMEbus backplane. In theory, this doubles the available bandwidth. Of course, you need to do this in such a way that the 2 backplanes can operate independently. At the same time you want to have some way to move data efficiently from one VMEbus system to the other.

A data transporter should provide a cable interface

To transport data from one VMEbus chassis to another, the data transporter should provide a hardware interface to a cable. Ideally, this cable interface should be done through a plug-on board, so you can select the type of interface that's appropriate for the link you want to use. For example, you might use an RS-232 link, an RS-422 link, a parallel link, or even a fiber optic link.

A plug-on interface board also allows you to design a customized interface for connecting your VMEbus system to a VAX, a SUN workstation, an IBM PC, or some other non-VMEbus equipment. As long as you design the data transporter's software to hide the hardware details of the cable interface, you can use the same system software to move data to (or from) any of these remote computers.

Chassis-to-chassis transfers: a simple example

Assume you have installed a data transporter board in each of 2 VMEbus chassis, and you have connected them with a cable link, as shown in Figure 1. Further, assume that each chassis has a single processor board, which we will call the *controlling master.*

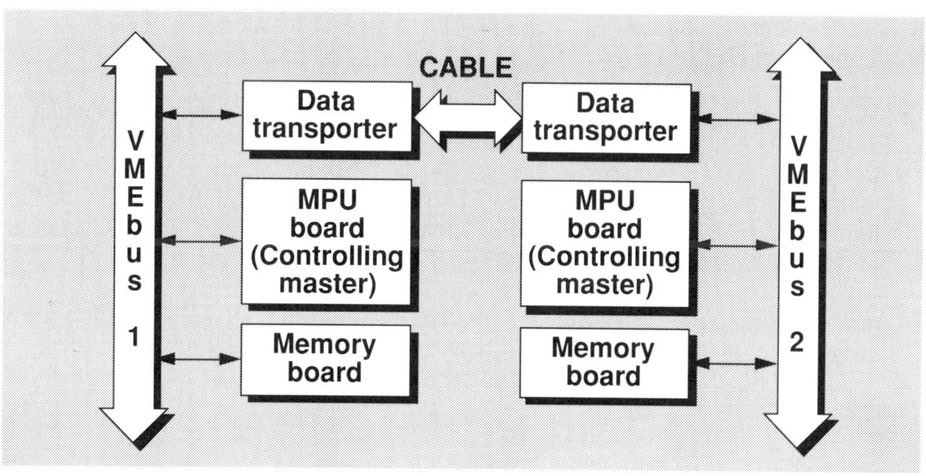

Figure 1

When the 2 chassis are powered up, the 2 controlling masters each initialize their respective transporter boards. Once the transporters have been initialized, one controlling master commands its transporter to read a block of data from a local VMEbus address, and send that data out over the cable link. The transporter executes this command by:

- reading from the local VMEbus memory, until its FIFO is full
- asserting a request signal that propagates over the cable to the remote transporter board in the other VMEbus system

If the remote transporter board is ready to receive, it asserts an acknowledge signal, and the transfer commences.

Because the data is being read into (and written out from) high-speed FIFOs in both transporter boards, the VMEbus block transfers proceed at the maximum speed permitted by each of the 2 VMEbus memory boards. For example, if the source memory location is in a bank of high-speed static RAM, and the destination memory location is in a bank of lower-speed dynamic RAM, the source read can take place in shorter bursts of bus traffic than the destination write: the reads from the high-speed static RAM need not be slowed down just to synchronize with the lower-speed writes to the dynamic RAM.

Transfer parameter blocks

Ideally, you want the command protocol to be independent of the type of hardware interface used to send data over the cable. This allows you to use the same protocols to initiate chassis-to-chassis data transfers, whether the transfer is done over an RS-485 differential driver link, a fiber optic link, a parallel link, or any other method of physical connection.

You can accomplish that by defining a Transfer Parameter Block (TPB) format. This TPB would include:

- a source location
- a destination location
- a transfer length

To pass a command to the data transporter, a controlling master builds a TPB in global VMEbus memory, and then writes the address of the TPB into one of the transporter's onboard registers. The transporter then reads the TBP from global VMEbus memory, and executes the command. The controlling master could then determine the transporter board's status (as well as any error conditions) by reading status registers on the transporter board.

To allow the queuing of commands for the data transporter, each TPB should have a field for a pointer to another TPB. This allows the controlling master to build a *linked list* of commands in global VMEbus memory. (The pointer of the last TPB in the list could be set to zero.)

The transporter should also have hardware that permits it to interrupt the controlling master when it completes the data transfer operation.

Chassis-to-chassis transfers in multiprocessing systems
The simple example cited above (for moving data from chassis to chassis) assumes that each chassis has a single central processor. However, VMEbus systems often have several processors in each chassis. To allow each of these processors to make use of the data transporter, you need to use a *client-server* approach.

The hardware configuration would still look the same as above. However, the program responsible for issuing commands to the data transporter would now be a *server task*, running under a multitasking kernel.

A program running on any VMEbus MPU board within the chassis could initiate data transfers to (or from) the other chassis by calling a *client subroutine*, and passing it a set of parameters that describe the kind of transfer needed. This client subroutine would typically be provided as part of a *software library*. Whenever you compile an application program that includes a call to the client subroutine, that subroutine would be copied from the library, into the executable code of your application program.

When the client subroutine is called, it notifies the server task that it has received a data transport request. The server task then builds an appropriate TPB in global memory and links this TPB into the linked list, for later execution. When the data transfer has completed, the server task returns status to the client subroutine, indicating whether or not the transfer was successful.

A similar procedure can be used to command the data transporter to move data from location to location within its own chassis. If the system includes a special communications protocol (for use between the 2 transporter boards) your application program could even command the movement of data between 2 VMEbus locations in a remote chassis.

Whether data is being transported from location to location within a chassis, or from chassis to chassis, the data being transported would still be buffered by FIFOs. The major difference is that, during chassis-to-chassis transfers, the FIFO can be filled with data from the source at the same time that it's being emptied to the destination.

The IV-3272
The *IV-3272 Full Speed Data Transporter* is an intelligent data transporter, with an onboard TMS-320 processor. (See Figure 2) Its design is based upon the principles discussed in this chapter. It transfers data over the bus at 10 million transfers/sec: near the maximum theoretical rate for a VMEbus system.

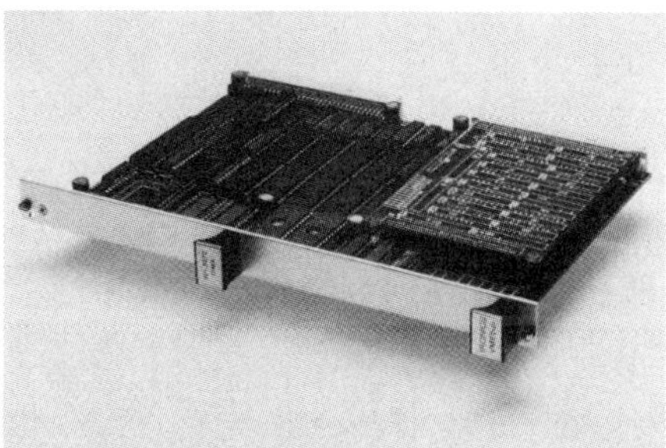

Figure 2

In a single-chassis system, the IV-3272 can be used to transfer data between any 2 VMEbus locations. The firmware supplied with the board has been designed to move data as fast as possible. In doing so, 2 restrictions have been placed on VME-to-VME transfers:

- You can't transfer data 8 bits at a time.
- The data width of the source must match the data width of the destination.

These are firmware restrictions only. The firmware can be modified to be more flexible, but performance will be lower.

The IV-3272 includes a general purpose hardware interface, which allows you to plug on a customized 4 X 6 inch interface board. This hardware interface consists of two 100-pin headers. The 200 signals provided on these headers include:

- all 64 user-defined pins of the VMEbus P2 connector.
- all signal lines from the onboard TMS-320 digital signal processor.
- 8 of the 32 VMEbus data lines.
- 32 I/O data lines.
- various status and control lines.
- power and ground.
- clocks.

The headers are mounted on the IV-3272 in a lightly populated section of the board, allowing a plug-on board to mount close enough to the IV-3272 that both will fit into single VMEbus slot. Alternatively, these header signals can be brought out (via ribbon cables) to a full-size circuit board.

The Ironics IV-3272-PIO is a general-purpose parallel interface board which drives 32 data lines and several handshake lines. It can be plugged onto the IV-3272, and used to drive cables for chassis-to-chassis data transfers. The cables are run between the P2 VMEbus connectors of the slots occupied by the two IV-3272s. With this configuration, the two IV-3272s can transfer data between their respective FIFOs with a cycle time of 150 nsec. If the data is moved over the cable 32 bits at a time, that's equivalent to about a 30 Mbyte/sec transfer rate.

However, data will typically be transferred over the cable at a slower rate, since the overall data transfer rate (from chassis to chassis) is limited by the memory access times.

Since the data that's read from the source memory goes directly into a very fast FIFO, the VMEbus data transfer rate from the source memory to the FIFO is determined by the source's memory access time. This transfer rate will be sustained as long as the FIFO doesn't fill up. However, if the VMEbus transfer rate exceeds the cable transfer rate (30 Mbytes/sec) the FIFO will eventually fill up, and further transfers have to wait until the FIFO empties.

As the source FIFO is filled with data from the source memory, it sends the data out over the cable. As that data arrives at the other end of the cable, the destination FIFO writes it into the destination VMEbus memory. However, if the destination memory is slower than the 30 Mbyte/sec cable transfer rate, the destination FIFO fills up, and further data transfers over the cable must be temporarily suspended.

Since the chassis-to-chassis data transfer rate is limited only by the memory access times, the total throughput for chassis-to-chassis transfers is nearly equivalent to the same transfers between a DMA device and memory within the same chassis.

The IV-3272 firmware

The IV-3272 is equipped with firmware that allows its onboard TMS-320 processor to interpret TPBs. You can use this firmware in 2 different modes:

- The first (and simplest) mode interprets the predefined fields in the TPB to initiate data transfers within the chassis, or from chassis to chassis.
- The second mode is similar to the first mode, except that it also provides a 16-bit status and control register, which can be accessed by VMEbus masters as 2 adjacent byte locations.

Customizing the firmware

The IV-3272 firmware assumes that, when a chassis-to-chassis transfer takes place, the destination chassis is expecting the transfer. (In other words, it assumes that some master in the destination chassis has passed a TPB to the destination IV-3272 board, commanding it to accept the transfer.) An alternative approach is to provide firmware that configures each IV-3272 board to accept a transfer over the cable whenever its linked list of commands is empty. This is possible because the TMS-320 processor chip directly controls the data line and control line levels of the parallel cable interface.

A typical application for a data transporter

Data transporters are ideally suited to systems which are partitioned by function, and which require a one-way, high-speed flow of data between chassis. For example, a radar system might consist of one VMEbus chassis which does the signal acquisition and peak detection. The resulting data is then passed (via the IV-3272 hardware interconnect) to a 2nd VMEbus chassis, which transforms and displays the data. A 3rd chassis could be connected to the 2nd chassis, and dedicated to data storage. Thus, the 2-chassis limitation is overcome by pipelining the data from one chassis to the next.

In the above radar example, you could also take advantage of the IV-3272's onboard TMS-320 digital signal processor by executing a Fourier transform on the data stream, as it passes between systems. (However, this would reduce the data transfer rate.)

Summary

Data transporters are useful in systems where you need to move a lot of data from chassis to chassis, or even from one memory area to another inside a single VMEbus chassis. Since the data transporter is optimized for that job, it can move the data much faster than a general purpose processor. Data transporters are especially useful in pipelined systems, where several VMEbus chassis are arranged in a linear fashion, with the data flowing from each system to the next system in the pipeline. This type of architecture permits any degree of parallel processing, limited only by the number of chassis in the pipeline.

Kevin J. Lynch is director of product engineering at Ironics in Ithaca, New York. Prior to joining Ironics, he wrote real time software for Universal Instruments in Binghamton, New York. He has a BSCS from SUNY, Binghamton.

CHAPTER 18

The Basic Principles Of 2D Graphics

By Tom Barrett

Graphics hardware has been evolving at an ever-accelerating pace in recent years. The falling price of RAM, and the higher clock speeds of VLSI chips, have made possible graphics systems with resolutions impossible only a few years ago. The benefits of these new systems are obvious, but it can be a painful process to adapt old graphics software to run on a new system. The solution to this problem is to structure your graphics software for portability. This can be done by putting all of the device-specific code in a standard set of library functions, which can be tailored to the hardware characteristics of each particular system. In this chapter Tom describes how this can be done, and uses the GKS standard as an illustration.

Computer graphics is the process of generating and presenting projections of objects on a 2-dimensional (2D) viewing surface. The objects themselves might be 2D (as in the case of an X-Y plot) or they might be 3D (such as a Space Shuttle). But, no matter how many dimensions the original object has, you need to reduce it to a 2D rendering, so it can be displayed on a 2D display surface.

To render a 3D object in 2 dimensions, you have to take into account such things as:

- the (x,y,z) coordinates of the *light source* (or sources)
- the (x,y,z) coordinates of the the *object*
- the (x,y,z) coordinates of the *viewer*
- the *orientation* of the object in (x,y,z) space

Using these numbers (as well as a mathematical description of the 3D object) you can compute the brightness and the angle of the light striking each surface of the object. This, in turn, allows you to render individual *pixels* on the screen.

Obviously, all of this computation can get quite involved, and it becomes difficult to see the forest for the trees. The best approach is to first understand the principles involved in rendering 2D objects. After you have mastered these principles, you will be better prepared to take on the additional challenge of rendering 3D objects.

CRT graphics
The most common graphics output device is the *Cathode Ray Tube* (CRT). The electron beam of the CRT scans the screen from left to right, then retraces to the left edge and repeats the scan just a little lower on the screen. This process repeats until the beam scans the entire screen surface, from top to bottom, and arrives at the lower-right corner

of the screen. The beam is then repositioned to the upper-left corner of the screen, and the entire scanning process repeats.

Note: In a color monitor, there are 3 electron beams. These 3 beams scan the screen simultaneously. However, just to keep things simple, let's assume we're talking about a monochrome monitor, which has only 1 beam.

As the beam scans the screen, the *control grids* in the neck of the CRT tube gate the beam on and off. When the beam is turned on, it illuminates a spot (pixel) on the face of the tube. The physics of the CRT tube, and the electronic circuitry driving the grids set an upper limit on how fast the beam can be turned on and off, as it scans across the screen. That sets an upper limit on how many pixels you can display on each scan line. The number of pixels per scan line is the *horizontal resolution* of the display. The number of scan lines determines the *vertical resolution* of the display.

Suppose you have a display with a vertical resolution of M, and a horizontal resolution of N. The ratio of the horizontal resolution to the vertical resolution (N/M) is called the *aspect ratio* of the CRT. Since most CRT tubes are about 33% wider than they are high, an aspect ratio of about 4/3 is considered ideal, since it produces a *square pixel.* Displays with non-square pixels are harder to work with. For example, if you try to draw a square with 100 pixels per side on a display with non-square pixels, you get a rectangle. This problem is not insurmountable. However, your software must compensate, in order to make the square look square.

The frame buffer

The control grid of the CRT is driven by a *CRT controller.* Inside the CRT controller is some RAM, which remembers which pixels are to be illuminated. This RAM is called the *frame buffer.* This frame buffer stores the 2D electronic image that will appear on the screen.

Each storage location in the frame buffer corresponds to a pixel on the display. In fact the term pixel is often used interchangeably to mean an illuminated point on the CRT screen, or the contents of the frame buffer location associated with that point.

There are 2 popular ways to organize the locations within a frame buffer:

- Multiplane
- Packed pixel

Multiplane frame buffers

Figure 1 shows how a single pixel is stored in a *multiplane frame buffer*, consisting of 4 planes. Each pixel is represented by a *stack* of 4 bits, where each bit in the stack lies above or below the others, in the stack of planes.

In the case of a 4-plane buffer (as described above) you would usually be able to access the 4 bits of each pixel simultaneously. In fact, since most microprocessors access 8, 16, or 32 bits at a time, you would probably be able to access several pixels simultaneously. This would permit you to rapidly fill the screen with a background color. In addition to the *pixel access mode*, some multiplane buffers also allow you to access bits *within a single plane*, without accessing any bits in the other planes. For example, you might be able to write simultaneously to 8 adjacent bits in one plane by doing a single byte write.

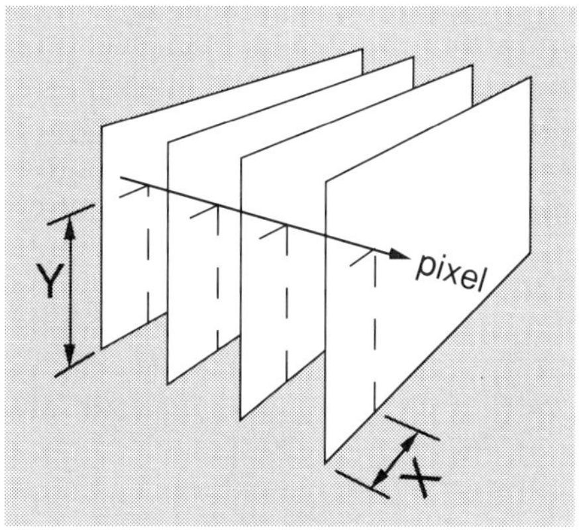

Figure 1

Packed pixel frame buffers

In a *packed pixel* frame buffer, each pixel is represented by a single, data storage location; usually a byte or a word. Each of these locations is individually addressable. This organization is depicted in Figure 2. For example, if each pixel storage location holds 8-bits, each pixel can display 256 colors.

A little about color

CRT monitors can be designed to display monochrome images, or to display color images. Since graphics are a lot more interesting when they're in color, let's talk about how that's done.

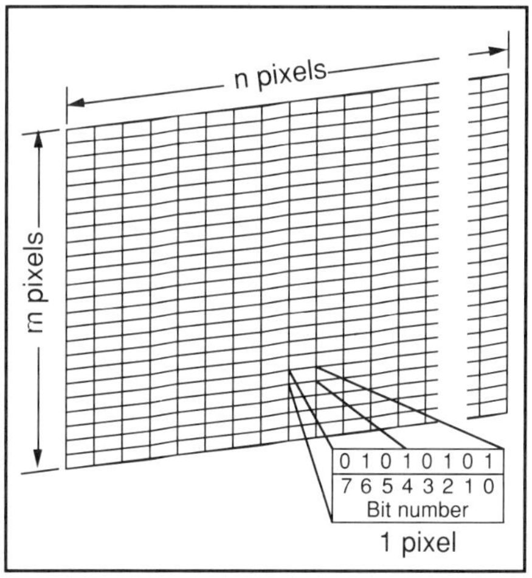

Figure 2

Color monitors generally have 3 input signals:

- Red
- Green
- Blue

This is known as the RGB color system. (There's another color system based on *intensity* and *hue* but, again, I'll be merciful, and spare you those details.)

The RGB input signals to the monitor are usually analog voltages, generated by 3 Digital-to-Analog converters (DACs) in the CRT controller: one per color. The resolution of the DAC determines how many shades of each primary color can be displayed. For example, if each DAC has 8 bits of resolution, 256 shades of red, 256 shades of blue, and 256 shades of green can be displayed. A 6-bit DAC will yield 64 shades of each primary color.

The palette

The total number of possible colors available in the *palette* depends upon the resolution of the DACs. Given three 8-bit DACs, you can generate a total of 16,777,216 color combinations (2^{24}). However, if the DACs have only 6-bit resolution, your palette is 18 bits wide, and you can generate only 262,144 colors (2^{18}).

Obviously, you would like to have the greater color resolution provided by storing 24 bits for each pixel. However, that takes a lot of frame buffer memory. (To display a

1024 X 1024 screen image, you would need to have 3 Mbytes of frame buffer memory.) While 18-bit color takes less memory, the frame buffer is still quite large.

Fortunately, if you don't need to use every one of the 16 million colors in your rendering, you can *substantially* reduce the amount of frame buffer memory you need. You do this by using a *Color Look-up Table* to define a *subset* of the total 24-bit color set.

Each entry in the table consists of 3 binary numbers. Each binary number specifies the brightness of red, green, and blue necessary to produce a particular composite color. The number of entries in the table determines the total number of unique colors that the table can display. For example, a table with 256 entries will be able to display 256 unique RGB colors.

How many entries should your table have? That depends upon the number of bits per pixel in your frame buffer. If your frame buffer has 8 bits per pixel, then the table must have 256 entries (2^8), as shown in Figure 3. If the frame buffer has 6 bits per pixel, then the table must have 64 entries (2^6).

The number of bits stored in the table for each primary RGB color depends on the resolution of the DACs. The binary value represented by these bits is the percentage of the maximum output voltage that the DAC can produce.

For example, if you use an 8-bit DAC, a bit pattern of 01010101 represents a 33% output on the DAC. See Figure 4 for additional examples.

Figure 5 shows how the RGB analog levels are generated for each color pixel on the screen. If you want to produce a brown pixel you need to have an entry for brown in the table. In this case, the entry for brown is found in entry number 6. Brown is formed by a combination of 66% red, and 33% green. (There is no blue component in brown.) To select this table entry, you need to store the number 6 in the frame buffer location associated with the screen pixel that you want to be brown.

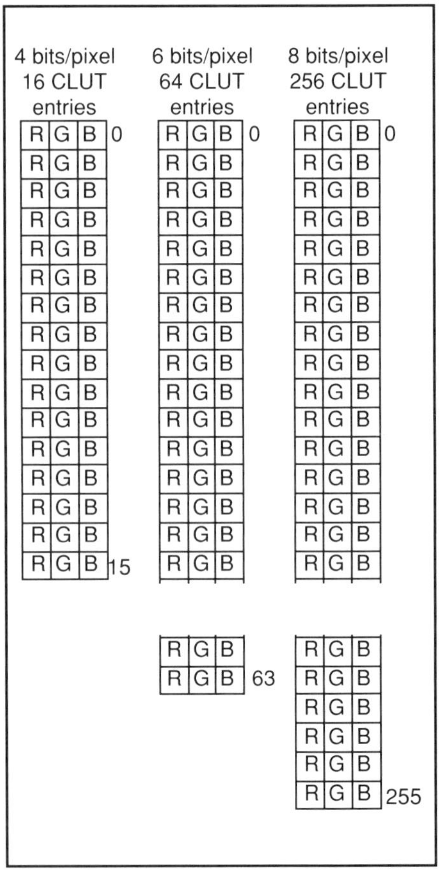

Figure 3

Now that you know how to illuminate each screen pixel with the color you want, you need to figure out *which* pixels to illuminate, to get the graphic that you want. To figure that out, you need to understand coordinate systems.

RGB value	Percentage
0000 0000	0
0011 1111	25
0101 0101	33
0111 1111	50
1011 1111	75
1111 1111	100

Figure 4

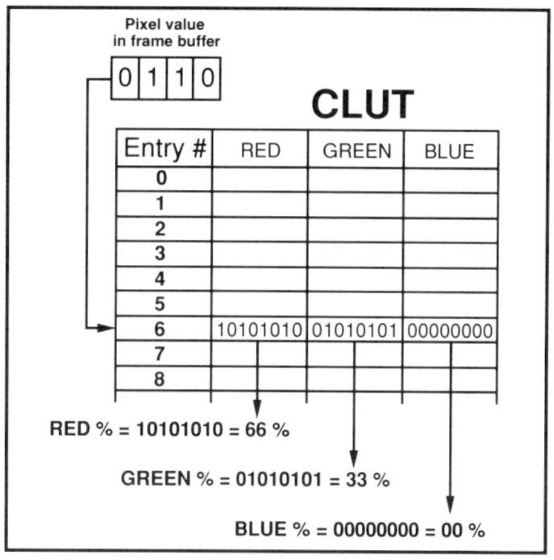

Figure 5

Coordinate systems
Remember that computer graphics involves the *rendering* of a real object as a 2D image. Real objects have:

- a size
- a location
- an orientation

To specify these characteristics in a way that a computer can handle them, you need to define a *coordinate system*. The essence of graphics is to take *points* (which might describe the vertices of an object) in a real world coordinate system, and *map* them into the coordinate space of a 2D viewing surface. That mapping is done through *transformations*, which describe the relationships between the coordinate values in one space and coordinate values in the other.

Most graphics packages use the following 3 coordinate systems:

- Device Coordinates
- World Coordinates
- Normalized Device Coordinates

Note: There are several different kinds of coordinate systems. (Cartesian, polar, spherical, etc) However, since a Cartesian coordinate system maps nicely into a rectangular frame buffer, it's the type most commonly used. In some cases you might want to use something else. For example a football is best described in spherical coordinates. However, if you use non-Cartesian coordinates to describe the objects you are rendering, your software will still have to convert these non-Cartesian coordinates to Cartesian somewhere in the process of generating the screen image.

Device coordinates
Device coordinates represent points on the 2D physical viewing surface. While a viewing surface (like the face of a CRT display) may appear to be continuous, it actually has a finite horizontal and vertical resolution. That means the number of (X,Y) coordinates

where you can actually illuminate a pixel is finite. This allows you to represent each pixel on the display surface as a pair of *integer* coordinates.

World coordinates

Most users want to be able to describe the object they're rendering in terms of familiar *units of measure*. For example, a cup might be 6 centimeters high, and 7 centimeters in diameter. To enter these values into the system, you need to define a *world coordinate* (WC) system.

A world coordinate system generally represents a physical system in the real world. Its units of measure might be seconds, meters, or degrees Celsius. Since world coordinate values are typically real numbers, standard graphics functions are written to accept *floating point* values from the application program that calls them.

Mapping world coordinates to device coordinates

Real number world coordinate values can be infinitely large. However a graphics display surface is finite, so you need to tell the graphics system what subset of the real world you want it to map onto the display surface.

Since the display surface is usually rectangular, you specify the subset of the real world (within which the object of interest lies) as a rectangular *window* in WC space. It's called a window because the viewer (who sits in front of the display surface) seems to be looking through a window at the objects in WC space. The viewer can't see any object which lies outside the window frame, since that portion of the view is masked by the "wall" around the window.

You tell the graphics package what portion of the WC space you want the viewer to see through the window by specifying the world coordinates of the 4 corners of the window. The simplest way to do that is to require that the sides and top of the window be parallel to the world coordinate axes. Then you only have to specify the world coordinates of 2 opposite corners of the window.

That is, in fact, how a *World Coordinate Window* is typically defined. Two coordinate pairs define the upper-left-hand corner (ULHC) and the lower-right-hand corner (LRHC) of the window. For example, if you're using a graphics library, you might define a window by executing the following function call:

```
Window(ULHCx,ULHCy,LRHCx,LRHCy)
```

The corner coordinates might be explicitly defined (with real numbers), as shown above, or you might define them indirectly by passing a pointer to the Window function...

```
Window(*coordinates),
```

...where *coordinates* is a data structure containing the (X,Y) world coordinates of the two corners.

To illustrate this, let's take an example. Suppose you have a CRT/frame buffer with a horizontal resolution of 640 pixels and a vertical resolution of 480 pixels. (This gives you the "ideal" aspect ratio of 4/3 for a CRT screen, producing a "square" pixel.)

Suppose you also have a description of an object, in world coordinates. To produce a graphic of the object you have to transform the *floating point* world coordinate values into *integer* device (screen) coordinate values.

But how do you do this transformation? You might first observe that the world coordinates of the upper-left-hand corner of the WC Window will be transformed into the pixel at row 0, column 0 of the frame buffer. Correspondingly, the world coordinates of the lower-right-hand corner of the WC Window will be transformed into the pixel at row 479, column 639. So, if you find a *linear transformation* that transforms these 2 points correctly, you can use it to transform all the other world coordinate values, as shown in Figure 6.

This certainly will produce the desired graphic on the 640x480 CRT. However, it's not very portable. It depends upon the particular horizontal and vertical resolution of the CRT you're using. If you decide to run your application program on another system (with a different horizontal or vertical resolution) you'll have to change the software.

So what's the answer? To keep things portable, you need to transform the real world coordinates into some kind of 2D device coordinates that don't depend upon the resolution of any particular display device. The answer is to define a 3rd type of coordinate system, called a *Normalized Device Coordinate (NDC)* system.

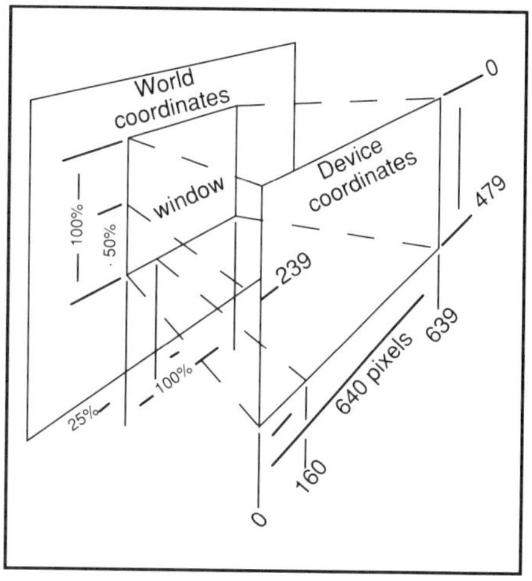

Figure 6

Normalized device coordinates

Normalized device coordinates are *real number* 2D (X,Y) coordinates, which can be infinitely large, and can be either positive or negative. Within this infinitely large 2D space, you define an imaginary finite viewing surface, with...

- its left edge at X = 0
- its right edge at X = 1
- its bottom edge at Y = 0
- its top edge at Y = 1

This square area can be thought of as a *virtual viewing surface*. In order to display a graphic on this viewing surface, you must map the world coordinates of your object into this viewing surface. Any portion of the object that gets mapped outside this square area will not be visible on the viewing surface.

When you want to display an object on the viewing surface, you need to specify *where*, on the surface, the graphic should appear. For example, you might want it to appear in the upper-left quadrant of the viewing surface.

You can specify where (on the viewing surface) you want to display a graphic by defining a *viewport*. When using a graphics library, you might define a viewport by executing the following function call:

Viewport(ULHCx,ULHCy,LRHCx,LRHCy)

The corner coordinates might be explicitly defined with real-number values, as shown above, or you might define them indirectly by passing a pointer to the Viewport function...

Viewport(*coordinates),

...where *coordinates* is a data structure containing the (X,Y) NDC coordinates of the 2 corners.

Note: The viewport need not be square. It might be rectangular. However, if you map a *square* window into a *rectangular* viewport, the graphic will be *stretched* along one axis.

Given the world coordinates of a window, and the normalized device coordinates of a viewport, you can derive a *normalization transform*, which maps each world coordinate to a normalized device coordinate. The mapping of points in WC space to points in NDC space is called *normalizing*.

If you define the viewport to be the upper-left quadrant of the viewing surface, then any object visible to an observer looking through the world coordinate window will appear in the NDC viewport in the upper-left quadrant of the viewing surface.

The user can reposition the graphic anywhere on the viewing surface by choosing a new set of (X,Y) coordinates for the 2 corners of the viewport rectangle. In Figure 7 the NDC coordinates of the viewport corners are defined to be...

- (0, 1) = upper-left corner
- (1, 1) = upper-right corner
- (1, 0) = lower-right corner
- (0, 0) = lower-left corner

Since this encompasses the entire virtual viewing surface, the world coordinate window view will occupy the entire viewing surface.

Contrast this with Figure 8, where the NDC coordinates of the viewport corners are...

- (.2, .8) = upper-left corner
- (.8, .8) = upper-right corner
- (.8, .5) = lower-right corner
- (.2, .5) = lower-left corner

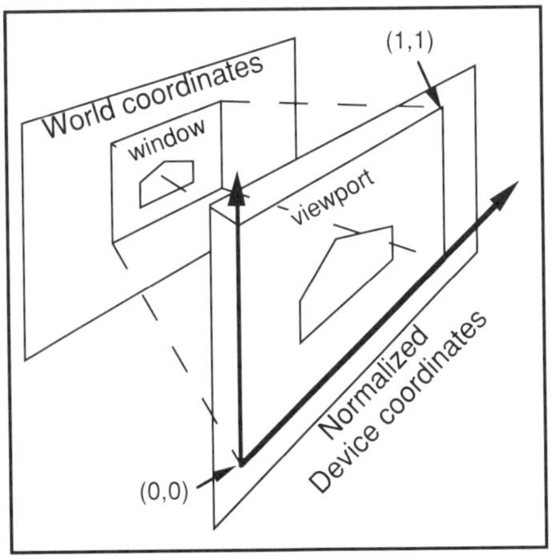

Figure 7

As you can see, the Window is now mapped to a much smaller area of the viewing surface than in Figure 7. By changing the viewport, it's easy to make a graphic smaller or larger, as well as to change its location on the viewing surface.

Putting it all together
Using the above technique, you can now transform points from the WC Window into NDC space. However, to display a graphic on an actual CRT display, you need to convert the *real number* NDC coordinates to *integer* device coordinates, based upon the horizontal and vertical resolution of the display. You do that by using another transform from NDC space to the device coordinate space. This transform is called the *device transform.*

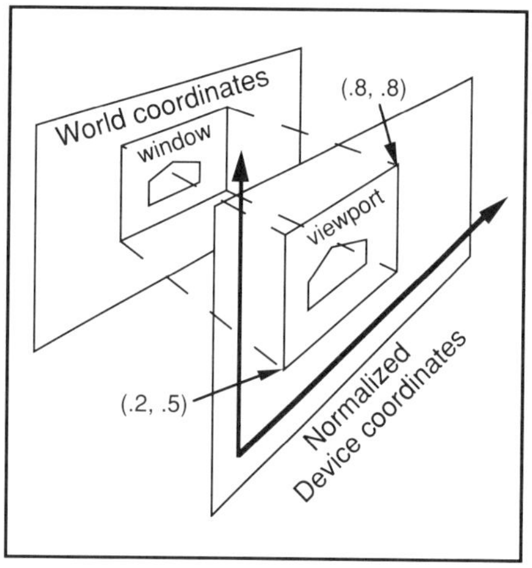

Figure 8

The device transform is fairly straightforward. You just map the virtual display surface to the physical display surface. For example, the upper-left corner of the virtual display surface is mapped to the upper-left corner of the physical display, the upper-right to the upper-right, etc.

Using the 640x480 CRT from the previous example, NDC (0,1) maps to the pixel (0,0) of the CRT. (See Figure 9) In clockwise order, the other 3 corners of the virtual display surface map to the CRT coordinates (639,0), (639,479), and (0,479) respectively. Since the viewport fills the entire virtual display surface, it will also fill the entire CRT screen.

But, suppose we go back to the example in Figure 8. Since the viewport occupies only a portion of the virtual display surface, it will also occupy only a portion of the physical display. The corners of the viewport now map to the CRT device coordinates (128, 96), (512, 96), (512, 240), and (128, 240), as shown in Figure 10.

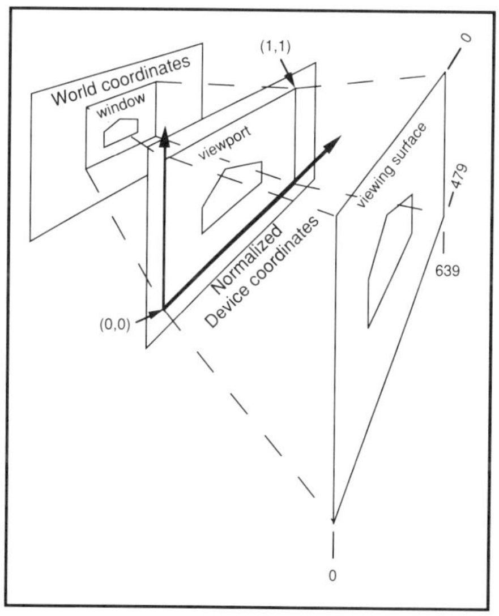

Figure 9

Why go to all this trouble?
It may seem like a lot of computational overhead, to go through 2 transformations. However, this technique allows you to either write or purchase device-independent graphics software. In fact this approach has made

graphics standards like GKS possible. By purchasing software written to the GKS standard you can save yourself a lot of time and effort. All you need to do is create a driver for your particular display device. This driver provides the needed device transform, while the GKS software permits you to enter data in world coordinates, which keeps your application code portable.

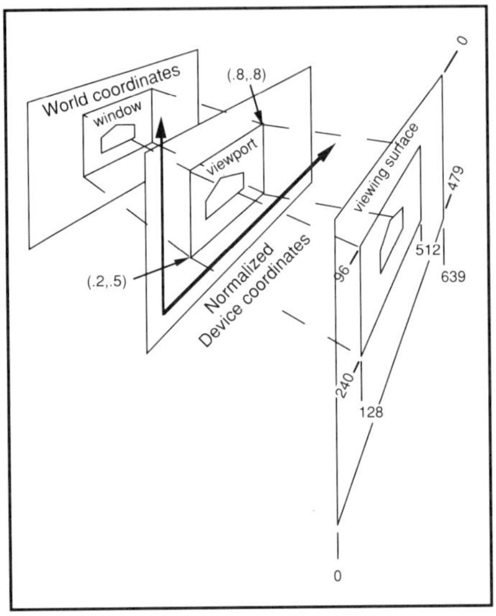

Figure 10

Drawing lines - How do you get from A to B?

There's probably no routine in a graphics library that gets more use than the line drawing routine. In view of this, it's important that you understand how lines are drawn on graphical display surfaces.

Note: In your analytical geometry course, you were taught that a *line* is infinitely long, while a *line segment* is bounded. I'm going to be less formal. Since we're working in bounded spaces, I'll use the word line interchangeably with line segment.

In a Cartesian system, a straight line between two points (x1, y1) and (x2, y2) (See Figure 11) can be expressed by the function...

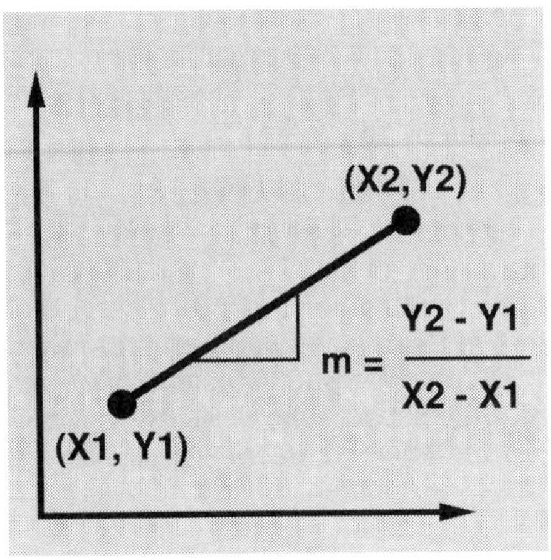

Figure 11

Equ (1) $y = mx + b$

...where m is the *slope* of the line, and b is the y-axis *intercept*. The slope of the line can be determined by the two-point slope formula...

Equ (2) $m = (y2 - y1) / (x2 - x1)$

...where (x1, y1) and (x2, y2) are the 2 endpoints of the line.

Note: In Equ (1) and Equ (2), all values are real numbers.

Since World Coordinates (WC) and Normalized Device Coordinates (NDC) are real numbers, there are an infinite number of points on the line between its endpoints. However, in Device Coordinate (DC) space there is a finite number of points, the number being determined by the horizontal and vertical resolution of the display device.

Whether a space has real number coordinates or integer coordinates, its Cartesian system is typically represented with a grid. Figure 12 depicts the case when the slope of the line is unity, and the points of the line lie on grid intersections. However, as shown in Figure 13, things aren't so simple when the line has a slope other than unity. The line doesn't necessarily pass through any grid intersection. That's when drawing a line on a graphics output device gets interesting.

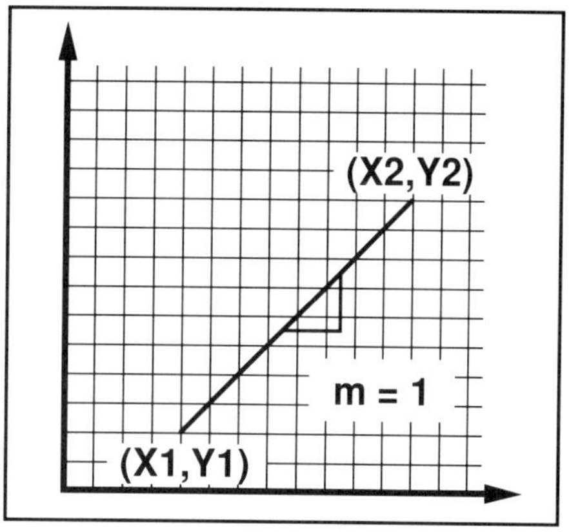

Figure 12

In DC space, the grid intersections may correspond to pixels on a CRT, dots on a matrix printer, or pen increments on a plotter. To keep things simple, let's assume we're going to display the line on a CRT screen, with a horizontal resolution and a vertical resolution that provide square pixels.

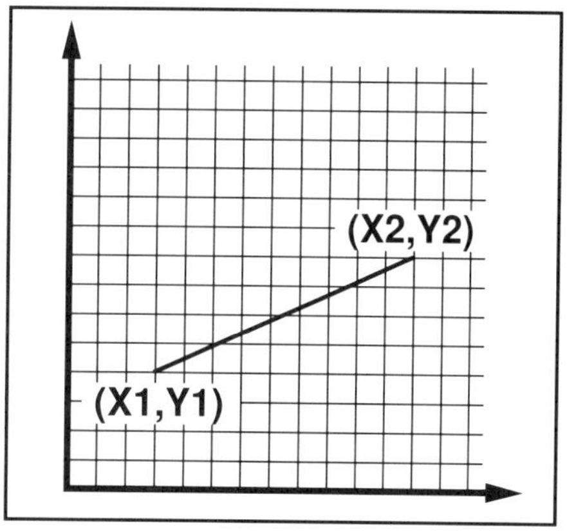

Figure 13

To draw a line on a CRT you simply illuminate the pixels that best represent the line. That sounds easy enough. However, before we proceed, let's consider things that will affect how the line is drawn.

Straightness

We certainly want the line to be straight. Since a line on a CRT is represented by a set of discrete pixels, you can only approximate the ideal line in WC space (except when the line is horizontal, or vertical, or when its slope is +1 or -1). In general, you want to illuminate the pixels along the path that makes the line appear as straight as possible.

Line thickness

You obviously want your line to have a consistent thickness along its entire length. To ensure that, you need to consider *pixel density*. When one interval of the line has more illuminated pixels than the next, that interval is perceived to be *brighter* than the rest of the line. (Note: When a line is displayed as dark pixels viewed against a bright background, an interval of greater pixel density is perceived as an *increased line width*.)

To avoid this undesirable effect, you need to illuminate the pixels in a way that gives a uniform appearance to the viewer.

Closure

The pentagram and the square in Figure 14 show another problem you want to avoid. Each is composed of a set of several lines, where each line shares its 2 endpoints with 2 adjacent lines. If the graphics software is not properly designed, lines which approach an endpoint from different directions may not end at exactly the same pixel. That produces a gap, which can be very irritating to the viewer. To ensure closure, you want to make sure that your line drawing routine terminates at the same pixel, no matter which direction the line is pointing.

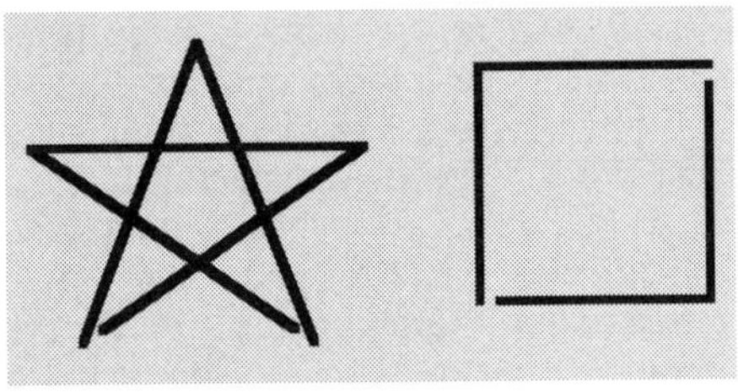

Figure 14

Speed

Lastly, your line-drawing routine should be as *fast* as possible. Since the line drawing routine is called by many other routines in the graphics library, it's often in the innermost of several nested loops. Any extra computations (or extra decision steps) in the line drawing routine greatly slow the entire graphics package.

Taking all of these things into account, let's get on with drawing a line.

A simple method for drawing a line

Since DC space is discrete (rather than continuous) line drawing is usually done in an incremental fashion. Let's take a look at how an incremental method works.

We start by choosing the axis whose pixel numbering will be used as the *independent* variable in our algorithm. In doing this, it's best to choose the axis with the *larger* component of the line's length. (This axis is sometimes called the *axis of greatest motion.*) For example, if |(x2 - x1)| is greater than |(y2 - y1)| you choose the x-coordinate as the independent variable. Otherwise, you choose the y-coordinate. From that point on, the line-drawing procedure is very straightforward. (Note: If the two are equal, it doesn't matter which axis you choose.)

Assume that you have chosen the x axis, as the axis of greatest motion. You could then draw the line as follows...

1. Initialize the *current coordinates* (x, y) to the real number *starting coordinates* of the line: (x1, y1).
2. Using the *integer portions* of the coordinates (x, y), illuminate the pixel at the starting coordinate.
3. Increment the independent variable (x) by 1. (That gives you the x coordinate of the next point on the line.)
4. Using Equ (1), calculate the corresponding value of the dependent variable (y). (This completes the (x, y) current coordinate pair.)
5. Using the *integer portions* of the coordinates (x, y), illuminate the pixel at the current coordinate.
6. Repeat steps 3, 4, and 5, until the independent variable (x) reaches its limit, as specified by the endpoint of the line, (x2, y2).

The resulting line will appear on your CRT as shown in Figure 15. However, notice that the center of the line appears to be lower than the ideal line, and the pixel density is not uniform (3,3,2,3,2,1). (The line is much brighter at its left end.) We can do better.

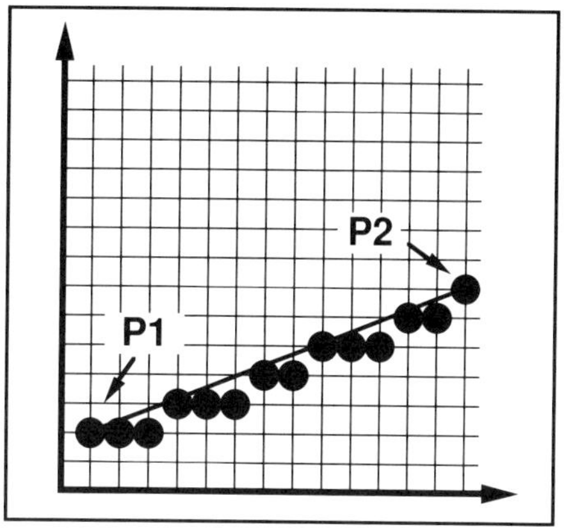

Figure 15

We can improve the positioning of the line by *half adjusting* the result of Equ (1), before using it to illuminate each pixel. To do that, you compute a new value (c) such that...

Equ (3) $c = b + 0.5$

You can then use the value c (in place of b) each time you use Equ (1) to calculate a new y value...

Equ (4) $y = mx + c$

Now, when you perform the line drawing algorithm again, using Equ (4), you get the line shown in Figure 16...not a bad line. The line is straight, the pixel density is more consistent (2,2,3,2,3,2), and it terminates properly.

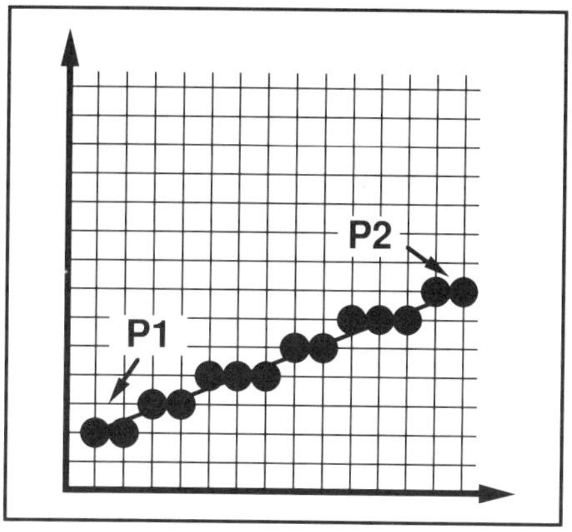

Figure 16

That would seem to settle the question, "How do you draw a line?" However there's still a problem. You have to do a real-number multiplication and a real-number addition to plot each pixel. Floating point operations are slow, so it takes a lot of time to draw the line. In fact, if you're using a software floating point arithmetic package, this method can be REAL slow. The situation is better if your computer has floating point hardware, but clearly the fewer floating point operations the better.

The DDA

There is a better method, which was used a lot in early computer graphics software. It's called the Digital Differential Analyzer (DDA). Our example above is similar to a simple DDA, but there are some differences.

As in the above example, the simple DDA generates the line along the axis of greatest motion. However, unlike the example, the DDA performs a one-time calculation of an increment for each axis...

Equ (5a) $xinc = (x2 - x1) / maxlength$
Equ (5b) $yinc = (y2 - y1) / maxlength$

...where maxlength is set equal to the larger of |(x2 - x1)| and |(y2 - y1)|. Note: Using this method, the increment corresponding to the axis of greatest motion is always 1.

The DDA initializes the current coordinate (x, y) to the values of (x1 + 0.5, y1 + 0.5). It then generates the line by the following procedure:

For zero to maxlength, repeat steps 1 through 3 below.

1. Using the *integer portions* of the coordinates (x, y), illuminate the pixel at the current coordinate.
2. Add xinc to the value of x.
3. Add yinc to the value of y.

The resulting line again looks like Figure 16, so the line quality hasn't suffered. But, by doing 2 real number divisions in the beginning, you have eliminated all of the real number multiplications in Equ (4).

However there are still 2 operations that take a lot of time:

- The floating-point-to-fixed-point operation in step 1, which is used to get the integer parts of x and y for pixel coordinates.
- The real number additions in steps 2 and 3. (Both xinc and yinc are reals.)

Given the fact that you still have to do these floating point operations, this algorithm may still be a little slow. But take heart! The crew at Big Blue came up with an algorithm that requires only integer operations!

Integer-based line drawing

In 1965, J.E. Bresenham (of IBM) presented a paper in which he described a new method for drawing a line on a digital plotter. In many respects it's like the DDA described above. However, it has one very unique feature. It produces an *error term*, which indicates how far a pixel is from the actual line.

Suppose that, instead of using Equ (1) to define your line, you use the following function...

Equ (6) $e(x, y) = (x)(y2) - (y)(x2)$

If you call e(x, y) the *error value*, you can define a line to be all of the points (x, y) which produce an error value of zero. (This actually defines a line of infinite length, but that doesn't turn out to be a problem.) Notice that Equ (6) includes the (x2, y2) coordinates of the *end point*, but not the *starting point* (x1, y1). That's because this simple equation can only define lines that pass through the *origin*. A more general equation would be needed for lines which don't pass through the origin, but the principles are the same. (We'll keep things simple by using this equation for the time being.)

If a point (x, y) is *exactly* on the line, the error value will be zero. However, if the point does not lie on the line, the error value will be non-zero. If the error value is greater than zero, then the point is *below* the actual line. If the error value is less than zero, the point is *above* the line.

As in earlier examples, you first select the axis of greatest motion. Let's assume that the line we want to plot has a slope between 0 and 1. That means that x is the axis of greatest motion. For any integer pixel coordinate (x, y), Bresenham's error term indicates the difference between the y coordinate of the pixel, and the y coordinate of the actual line.

Let's assume that you're going to plot the line starting at the origin. Obviously, you must first illuminate the pixel at the origin. Since the slope of the line is between 0 and 1, you know that next pixel you'll need to illuminate is either the pixel on its right, or the pixel on its upper right. In fact, each time you illuminate another pixel, there will be exactly 2 candidates for the next illumination: the pixel to the right, and the pixel to the upper right. Those two pixels are at (x+1, y) and (x+1, y+1), as depicted in Figure 17.

So how do you decide, which pixel to illuminate? Obviously, you illuminate the one closest to the actual line. How do you know which pixel is closest to the line? By using Bresenham's algorithm to determine which has the smallest error term.

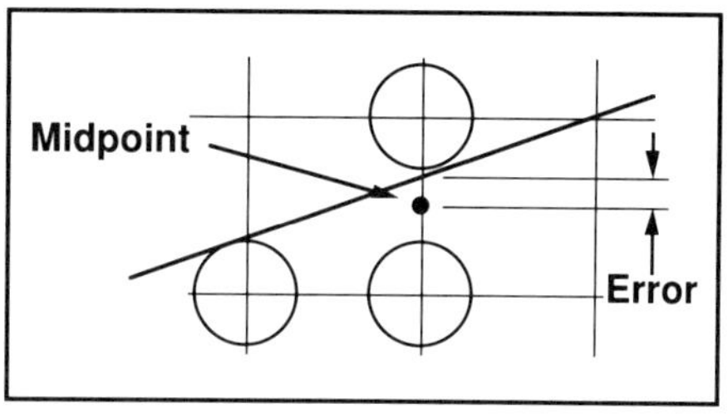

Figure 17

Bresenham's algorithm
Bresenham's algorithm evaluates Equ (6) at a y-axis point midway between the 2 candidate pixels, (x+1, y+0.5). If that midpoint is below the line (as shown in Figure 17) the error value is greater than zero, indicating that the pixel at (x+1, y+1) is closer to the line. Conversely, if the error value is less than zero, the midpoint lies above the line, as shown in Figure 18, and the pixel at (x+1, y) is closer.

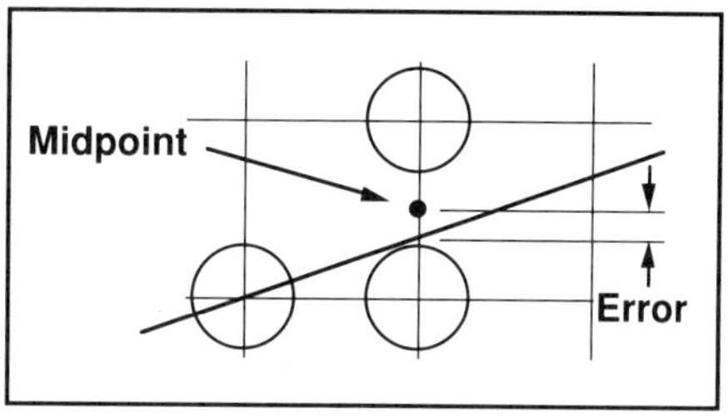

Figure 18

With Bresenham's algorithm you can avoid the use of real numbers completely. In fact, you don't even need to do any multiplies or divides. Only nice, fast, integer adds and subtracts are needed. Given the length of motion along the x and y axes (as dx and dy respectively) Bresenham's algorithm can be written in pseudo code, as shown in Listing 1.

Since the terms (`dy + dy`) and (`dy + dy - dx - dx`) are constant for any given line, you only need to compute them once. By doing that (and by placing the resulting values into 2 integer variables called `nondiag_inc` and `diag_inc` respectively) you can restate the algorithm as shown in Listing 2.

It can't get much simpler or faster than that. But remember, this was done for a line with a slope between 0 and 1, where both x2 and y2 were positive. There are seven other possibilities to deal with, as shown in Figure 19. In addition, not all lines are kind enough to start at the origin. The general algorithm essentially involves handling the initial conditions, the direction of line generation, and the diagonal and non-diagonal

Listing 1

```
begin
{
    e = dy + dy - dx
    x = dx
    y = dy
    for i from 0 to dx
    {
        SetPixel at (x,y)           /* turn on pixel */
        if e <= 0 then
            e = e + dy + dy         /* non-diagonal  */
        else
        {
            e = e + dy + dy - dx - dx /* diagonal */
            y = y + 1
        }
        x = x + 1
    }
}
end
```

Listing 2

```
begin
{
    nondiag_inc = dy + dy
    e = nondiag_inc - dx
    diag_inc = e - dx
    x = dx
    y = dy
    for i from 0 to dx
    {
        SetPixel at (x,y)               /* turn on pixel */
        if e <= 0 then
            e = e + nondiag_inc         /* non-diagonal  */
        else
        {
            e = e + diag_inc            /* diagonal      */
            y = y + 1
        }
        x = x + 1
    }
}
end
```

increment calculation. I'll leave its development as an exercise for the reader. (I've always wanted to say that!)

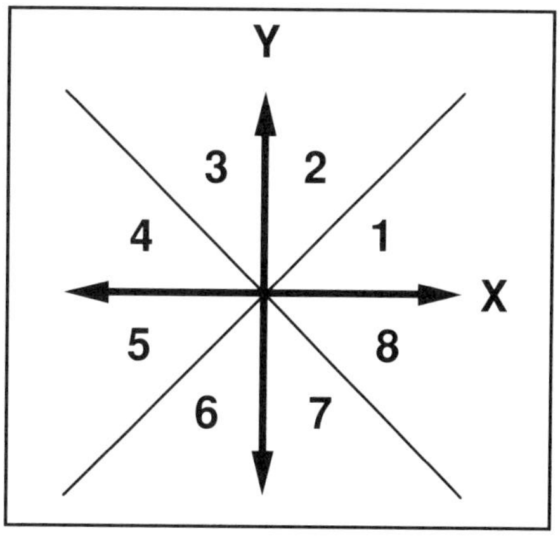

Figure 19

Now you know how to draw a line. However, most graphics packages (be they standards or not) provide the line drawing routine, in a more general form, called a *polyline* routine. For example, when you call the polyline routine in GKS, you pass the routine a pointer to an array of real-number (x, y) coordinate pairs in WC space. Each (x, y) coordinate pair in the array represents a point. The array of points defines a string of line segments, as shown in Figure 20. Note that each point (with the exception of the first and last point) is used to draw 2 line segments. The ending point of one line segment is the starting point of the next line segment.

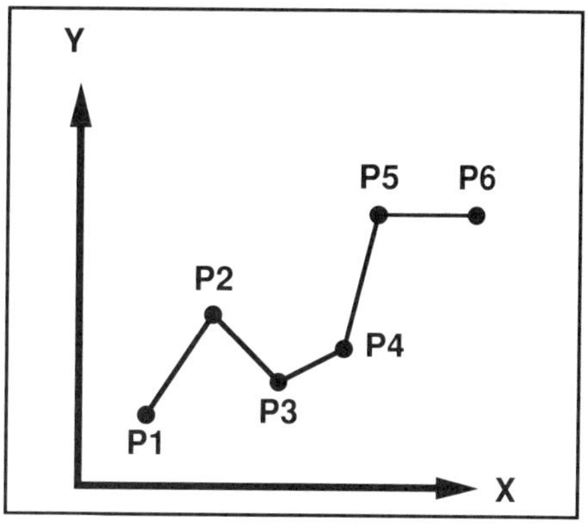

Figure 20

Figure 21 shows the processing stages that a typical polyline routine goes through, to generate the polyline. When you call polyline, you provide several attributes, which describe how you want the lines to look on the screen: the line color, the line width, and the line style. (I'll have more to say about these later.)

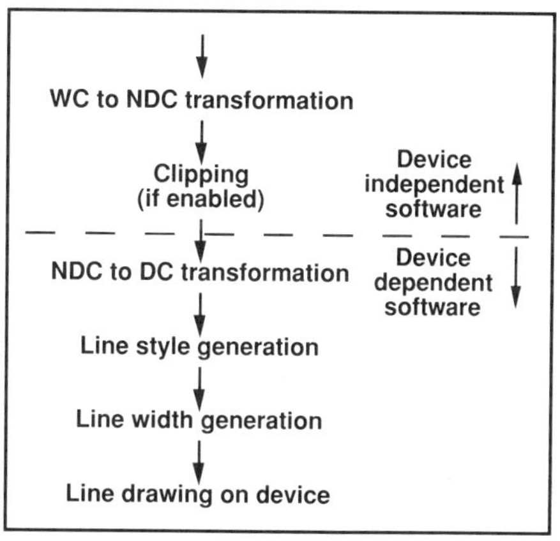

Figure 21

First, the (x,y) coordinates of the points are transformed from WC space to NDC. Then comes clipping...

Clipping

In some cases a line may actually lie outside the field of view. (It might be above the top of the screen, or below the bottom of the screen, etc.) If your graphic consists of a lot of lines, you don't want to waste time calculating the coordinates of each pixel in a line segment, just to determine whether it's within the screen boundaries or not.

Obviously, if both endpoints are above the top of the screen, or if both are below the bottom of the screen, the line won't be visible. But what if one endpoint is above the top, and the other is to the left of the screen? The line might be partially visible. If the line is only partially visible, you don't want to have to calculate the pixel coordinates of each point on the line, just to see whether or not it's on the screen.

To avoid all of those wasteful calculations, a technique called *clipping* is used. (I'll talk more about clipping later, but I mention it here just so you know where it fits into the scheme of things.) A clipping routine determines which (if any) portion of the line is visible. If only a portion of the line is visible, the clipping routine calculates new endpoints at the boundaries of the screen.

After clipping, the clipped (x, y) NDC coordinates of the visible portion of each line are transformed into the DC space of the display device. Next, the line styles are generated.

Line styles
Line styles describe whether the line is dotted, dashed, solid, etc. Software routines usually subdivide the line into visible and invisible sections, using a table of *run lengths* dictated by the line style.

Figure 22 shows 4 lines styles. When you want to draw a polyline, you could specify one of these 4 styles by passing a *style attribute*. The polyline routine could then use that style attribute to index into a table containing the necessary information about run lengths. The run lengths define how long each visible and invisible section is to be drawn, to produce the particular line style.

Line style index no.	Line style
1	————————
2	— — — — —
3	....
4	—·—·—·—·—

Figure 22

When drawing a dashed line, the invisible sections are skipped, to save time. Each visible section of the line is represented by a pair of endpoint coordinates. Those coordinates are then passed to the line width generation routine...

Line width
For a given visible section, the line width routine makes one or more calls to a line drawing routine, like the one we developed above. The number of calls depends on the *line width* attribute. A factor of 2 would double the line width, while a factor of 3 would triple it.

When a wide line is specified, the line width routine adjusts the starting and ending coordinates of the line (by a simple translation to an adjacent pixel) each time it calls the line drawing routine. Each additional call to the line drawing routine draws another line, adjacent and parallel to the previous line. Figure 23 shows a line having a width attribute of 3.0, and a type attribute of 4.

There are other techniques for handling line width and line style. In fact, it's probably fair to say that there are as many techniques as there are graphics software packages. There's no one "right" way to do it.

Line clipping: Now you see it...now you don't

Earlier, I briefly described a technique called line clipping, which is used to avoid the drawing of unnecessary lines. For example, Figure 24 shows a 2D object called BANISTER. This graphical object is represented by a set of line segments, which lie in an infinitely large 2D *world coordinate system*. The endpoints of those line segments are each specified by a set of (x,y) coordinates.

To display this graphical object, you must first specify what portion of this infinitely large 2D world coordinate space you want to map to your display screen. You do that by specifying the (x,y) coordinate values of the lower-left and upper-right corners of a *window* in world coordinate space. These corners define a rectangular area in world coordinate space that gets mapped to a rectangular area on your display surface. (See Figure 25)

The rectangular area on the display surface need not be the entire screen. A *viewport* is defined (using 2 pairs of (x,y) coordinates) to indicate where (on the screen) the lower-left and upper-right corners of the rectangular area are located. (See Figure 26)

Once you define the corners of the window and the corners of the viewport, you can figure out a *linear transformation* for mapping the world coordinate endpoints into device coordinates. Once you have the endpoints in device coordinates, all you need to do is draw lines connecting the endpoints, using a line drawing algorithm.

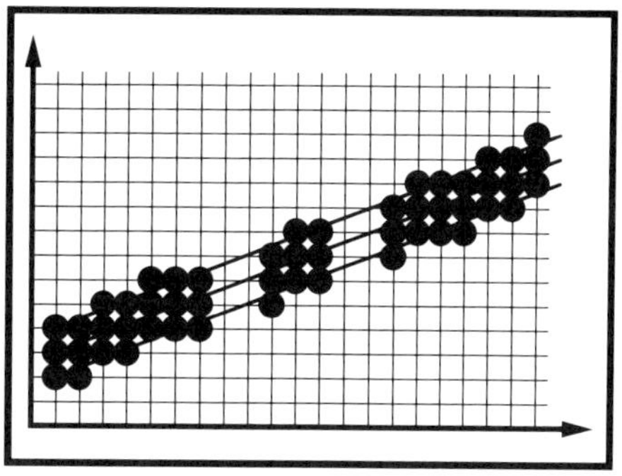

Figure 23

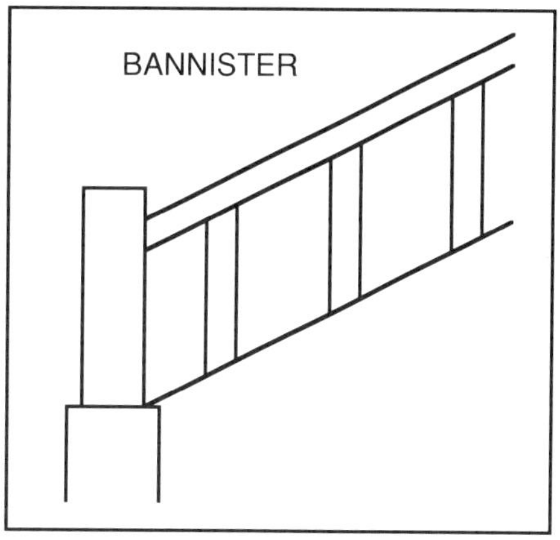

Figure 24

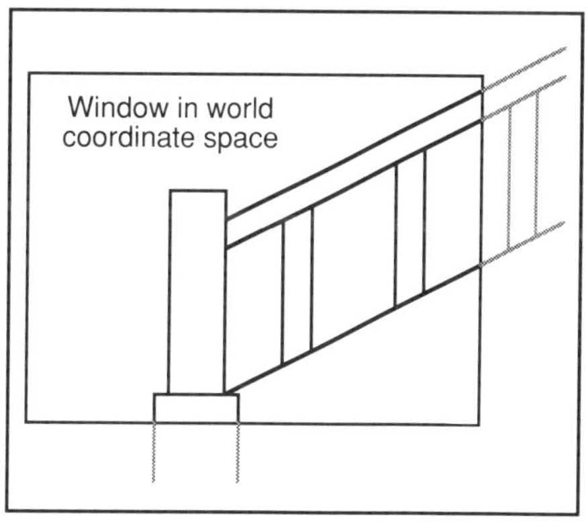

Figure 25

But now suppose you keep the linear transformation the same, and reduce the size of the world coordinate window. This might display an image like the one shown in Figure 27. Notice that some of the lines are no longer visible. If you reduce the size of the window further, additional lines disappear, as shown in Figure 28.

These lines still exist in the "real world". They just aren't visible inside the viewport of the device coordinate space.

A simple-minded algorithm would:

- transform all of the endpoint coordinates from world coordinate space to device coordinate space.
- call the line drawing algorithm, to draw all the lines connecting those endpoints.

But (as shown in Figure 27 and Figure 28) some of the lines may not be visible inside the viewport. When that happens, the line drawing algorithm will proceed to calculate the (x,y) pixel coordinates of each point along the line, in preparation for illuminating each pixel. However, If the pixel coordinates are beyond the limits of the screen, no pixel will be illuminated.

This is a waste of time. Drawing lines takes long enough as it is. Why waste time calculating the pixel coordinates of pixels that won't be displayed? In order to avoid these wasteful calculations, you need some way to determine whether or not a line is visible.

Obviously, if *both* endpoints are *inside* the viewing area, the line between them is visible. It's also easy to see that, if both endpoints lie to the *right* of the viewport, no point on the line is visible. By simply checking the (x,y) coordinates of the endpoints against the left, right, top and bottom limits of the viewport, you can avoid a lot of unnecessary calculations.

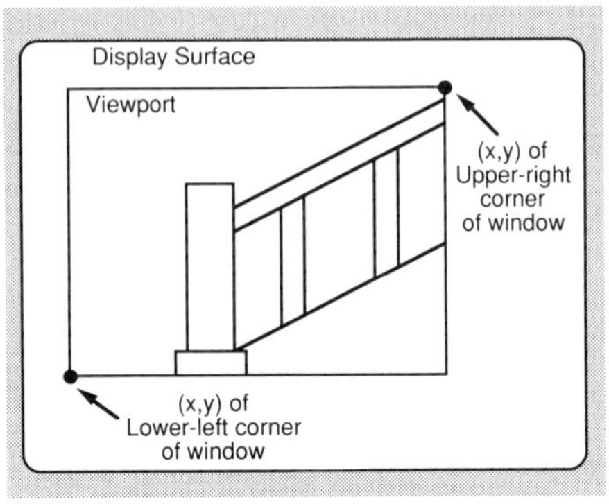

Figure 26

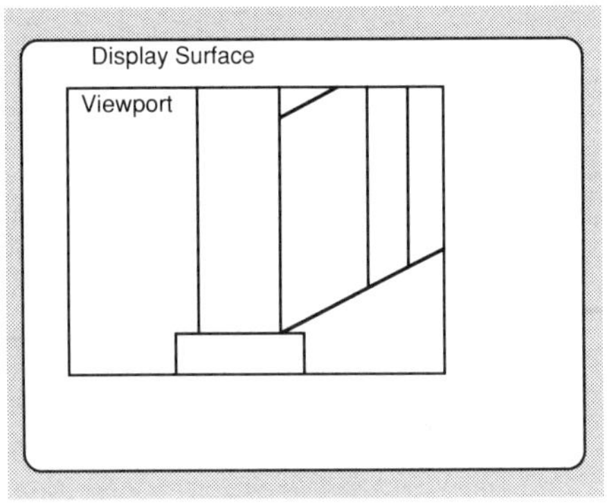

Figure 27

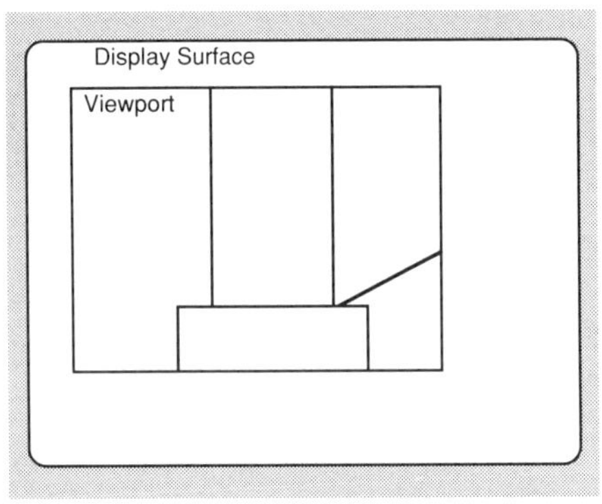

Figure 28

But how do you handle *partially* visible lines? Figure 29 shows 2 possibilities:

Line A: One of the endpoints lies inside the window.
Line B: Neither of the endpoints lies inside the window.

In both cases, you could just call the line drawing algorithm, and let it check each pixel coordinate, as it prepares to illuminate the pixel. However, this would add an additional step to each pixel plotting operation: a step that could slow the line drawing algorithm down considerably. A better approach is to use a technique called *clipping*.

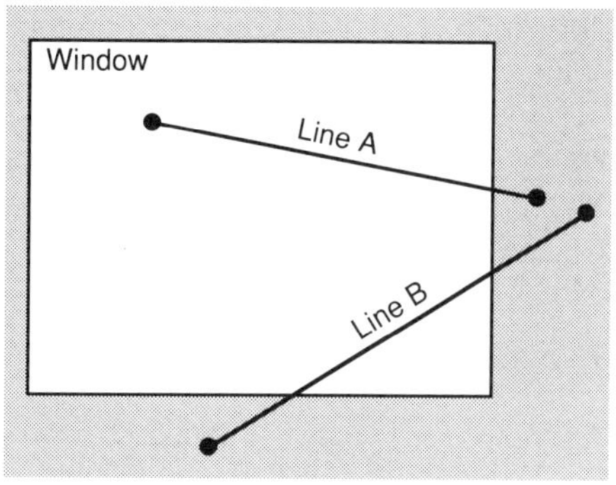

Figure 29

In order to execute a clipping algorithm, you must first answer 2 questions:

- Which viewport boundaries does the line intersect?
- What are the (x,y) coordinates of each intersection?

Using the answers to these 2 questions, we can *clip off* the invisible portion(s) of the line, and calculate new endpoints at each viewport boundary intersection.

In order to answer these 2 questions, we use the viewport boundaries to define a *clipping region*. This region is defined (as shown in Figure 30) by 4 values:

- $X_l = X_{left}$
- $Y_t = Y_{top}$
- $X_r = X_{right}$
- $Y_b = Y_{bottom}$

In order to determine the visibility of each line segment, you compare the x coordinates of its endpoints to X_l and X_r...

Equ (7) $X_l \leq x \leq X_r$

...and then compare the y coordinates of its endpoints to Y_t and Y_b...

$Y_b \leq y \leq Y_t$

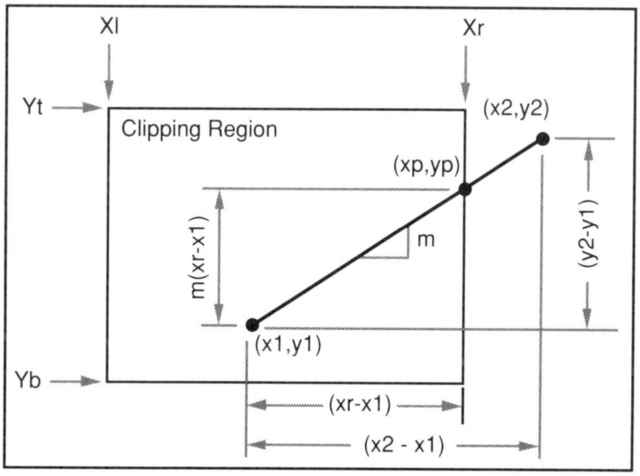

Figure 30

These are simple inequalities, and can be easily tested. In fact, the line drawing algorithm could test each pair of pixel coordinates *before* plotting the pixel. However, this would still take a lot of processing time.

But suppose we calculate the intersection of the line segment with the 4 boundaries of the viewport...

$x = X_r$
$x = X_l$
$y = Y_t$
$y = Y_b$

Those intersections could then be used as *new endpoints* for the line.

As shown in Figure 30, the point of intersection with the right edge of the clipping region (X_p, Y_p), can be computed as follows...

EQU (8) $Y_p = \{(Y2 - Y1)/(X2 - X1)\} \times \{(X_r - X1) + Y1\}$
$X_p = X_r$

...and the point of intersection within the left edge can be computed by a similar equation...

Equ (9) $Y_p = \{(Y2 - Y1)/(X2 - X1)\} \times \{(X_l - X1) + Y1\}$
$X_p = X_l$

The intersection with the top is similarly found by...

Equ (10) $X_p = \{(X2 - X1)/(Y2 - Y1)\} \times \{(Y_t - Y1) + X1\}$
$Y_p = Y_t$

For the bottom, the equation is...

Equ (11) $X_p = \{(X2 - X1)/(Y2 - Y1)\} \times \{(Y_b - Y1) + X1\}$
$Y_p = Y_b$

Note: The slope term is inverted in Equ (10) and Equ (11) because the intersecting edge is *horizontal*, instead of vertical.

Depending upon which edge of the clipping region the line intersects, one of these 4 equations is used to clip the line, and to calculate a new endpoint. If the line intersects 2 edges (instead of just 1) you need to apply 2 of the 4 equations, to get 2 new endpoints.

These equations are straightforward and are a great improvement over the testing of (x,y) coordinates of every pixel, prior to plotting it. However, you must first know which edge (or edges) the line intersects, in order to choose which of the 4 equations to apply.

It seems that you could just apply this clipping algorithm to every line in your graphic. However, that's not a good idea. First of all, if a lot of lines lie outside the boundaries of the clipping region, you're going to go through a lot of needless number crunching. But, even worse, when the clipping equations are applied to *invisible* lines, they produce new endpoints that lie *outside* the clipping region, as shown in Figure 31.

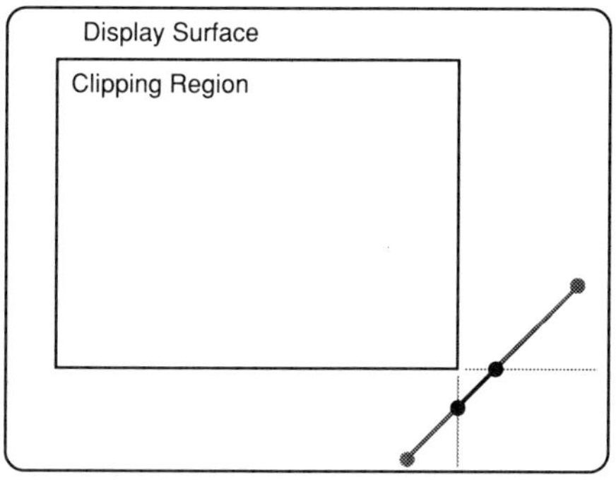

Figure 31

You need a quick way to determine:

- which lines are totally visible.
- which lines are totally invisible.
- which lines are partially visible, and which edges of the clipping region each partially visible line intersects.

Dan Cohen and Ivan Sutherland published a nifty solution to both of these problems. In this technique, the edges of the clipping region are used to define 9 regions, as depicted in Figure 32.

Each of the 9 regions is assigned a unique 4-bit code. Each bit of this code represents the answers to 1 of 4 questions about the point's position, with respect to the clipping region:

- Bit 0: Is the point to the left of X_l?
- Bit 1: Is the point to the right of X_r?
- Bit 2: Is the point below Y_b?
- Bit 3: Is the point above Y_t?

Note: Bit 0 is the most significant bit of the 4-bit code, and an answer of "yes" is represented by a 1 in each bit position.

Figure 33 shows the 9 regions, with the corresponding code for each. Since the answer to all 4 questions is "no" for any point within the clipping region, its 4-bit code is 0000.

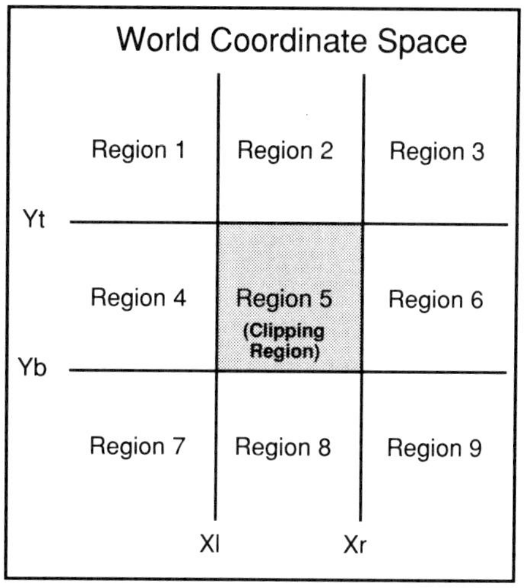

Figure 32

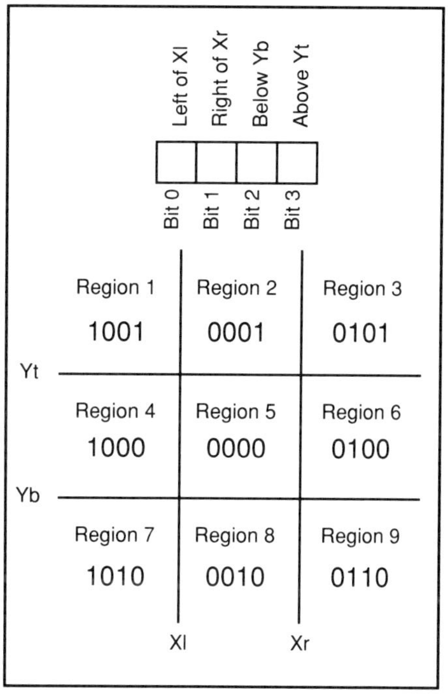

Figure 33

The 1st step in the Cohen-Sutherland algorithm is to determine the 4-bit codes for each of the endpoints. Once that's done, some interesting relationships appear:

- If both endpoint codes are zero, the line lies totally *inside* the clipping region, and may be drawn without further ado.
- If the bit-wise logical AND of the 2 endpoint codes is non-zero, both endpoints are on the same side of the clipping region. Such a line is *outside* the clipping region, and is discarded by the algorithm.

If the bit-wise logical AND of the 2 endpoint codes is zero, the line *might* pass through the clipping region and might have a visible segment. However, the algorithm doesn't *guarantee* this. For example, a line passing from Region 6 to Region 8 may (or may not) pass through the clipping region.

To summarize, when the 2 endpoint codes for each line are logically ANDed together, the result will be one of 3 possible conclusions:

- completely visible
- completely invisible
- perhaps partially visible

Since no clipping is needed in the 1st and 2nd case, the algorithm's job is done. In the 3rd case, the algorithm...

- chooses one of the endpoint's codes (non-zero)
- uses the 1s in that endpoint code to choose an appropriate intersection equation.
- calculates a new endpoint, at the intersection of the line and the edge of the clipping region.
- generates a 4-bit endpoint code for this new endpoint.
- logically ANDs the new endpoint code with the endpoint code for the other endpoint.

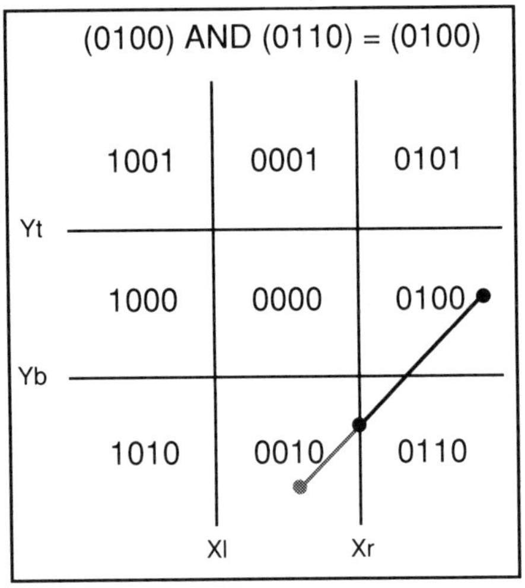

Figure 34

At this point, there are 3 possibilities:

- Both endpoint descriptors are zero, indicating that the line lies totally *inside* the clipping region. (This can only happen if one of the original endpoints was inside the clipping region.)
- The logical AND of the 2 codes is non-zero, indicating that both endpoints are on the same side of the clipping region. (See Figure 34) In this case, the line is completely *outside* the clipping region, and need not be drawn.
- The logical AND of the 2 codes is zero, indicating that an additional clipping operation is needed. (See Figure 35) This clipping process is repeated for each of the 1s in each of the 2 endpoint codes, until the clipped line is found to be either completely visible, or completely invisible.

The Cohen-Sutherland clipping algorithm is shown in pseudo code form in Listing 3. It provides a reasonable routine for the clipping operation, but it relies on a lot of real-number arithmetic, when done in the device-independent layer of a line drawing routine. If you have a math coprocessor in your system, this isn't much of a concern. However, if you're one of those poor souls who doesn't have a coprocessor, take heart. Ivan Sutherland has something for you too. He and Robert Sproull presented a paper on a *clipping divider* at the 1968 Fall Joint Computer Conference.

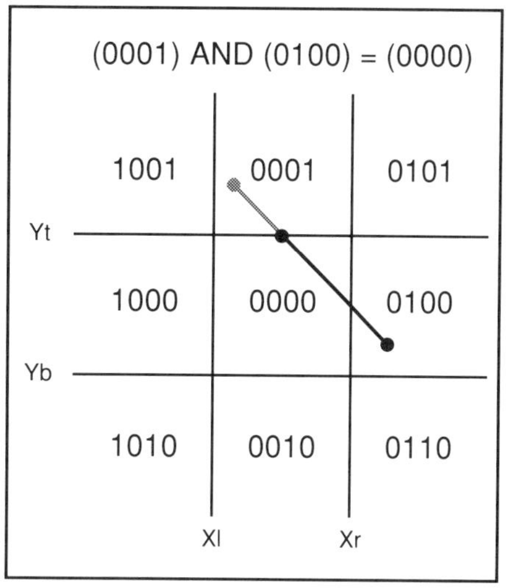

Figure 35

The clipping divider algorithm (also known as the *midpoint division algorithm*) determines the (x,y) coordinates of the visible point on a line that's farthest away from each endpoint. The visible portion of the line thus lies between these 2 points.

Like the Cohen-Sutherland algorithm, the midpoint division algorithm uses 4-bit endpoint codes to determine if the line is totally visible or totally invisible. If a line might be partially visible, the algorithm:

- calculates the (x,y) coordinates of its midpoint, M. (See Figure 36)
- generates a 4-bit code for M
- tests segment MB for visibility

If MB is found to be outside the clipping region, the coordinates of endpoint B are replaced with those of M. (Otherwise, the coordinates of A are replaced with M.) The remaining segment (either AM or MB) is again subdivided at its midpoint, and the process repeats until the visible point farthest from A is determined.

Once the algorithm finds the farthest visible point from A, it is repeated to find the farthest visible point from endpoint B. Once it finds this 2nd endpoint, the visible segment is defined as the segment connecting these 2 endpoints.

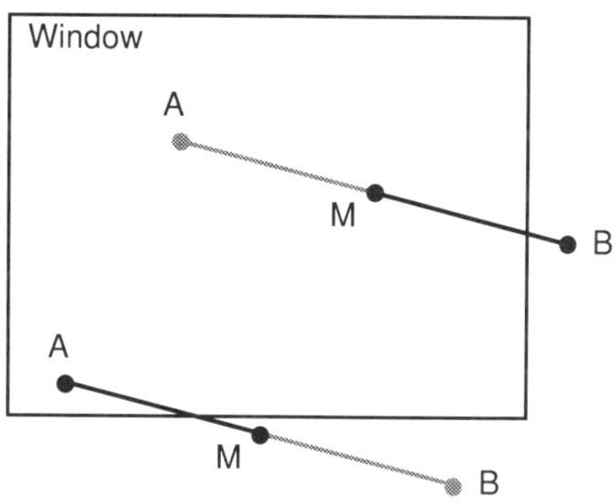

Figure 36

If your (x,y) coordinate values are represented by floating numbers, you might go on subdividing a line for a very long time before finding the farthest point to the precision allowed by the floating point representation. Fortunately, you

Listing 3

```
/* Cohen-Sutherland Clipping */

/*******************************************************
 * Given a line with endpoints at (X1,Y1) and (X2,Y2),  *
 * accept the line without clipping if it lies within   *
 * the Clipping Region.  Reject it if it is totally     *
 * outside the Clipping Region.  Otherwise, clip the    *
 * line to new visible endpoints.  In all cases,        *
 * return a code to indicate if the line is rejected or *
 * is visible.                                          *
 *******************************************************/

Clip (X1,Y1,X2,Y2)

variables: code1, code2, code, x, y, lim
externals: Xl, Xr, Yb, Yt

begin
{
        code1 = GetCode(X1,Y1)    /* get endpoint codes */
        code2 = GetCode(X2,Y2)

        while (code1 != 0 .OR. code2 != 0)
        {
                if ((code1 .AND. code2) != 0
                        return(reject)  /* line is rejected */

                else   /* line may have visible portion */
                {
                        code = code1
                        if (code1 == 0)
                                code = code2 /* (X1,Y1) visible */

                        if ((code .AND. 0xA) != 0)
                        {
                            /*********************************/
                            /* Line lies either to the left  */
                            /* of Xl or to the right of Xr   */
                            /*********************************/

                                lim = Xl
                                if (code .AND. 0x2)
                                        lim = Xr
                                y = (Y2-Y1)*(lim-X1)/(X2-X1) + Y1
                                x = lim
                        }
```

(continued)

Listing 3 *(continued)*

```
                else
                {
                        /********************************/
                        /* Line lies either below Yb or */
                        /* above Yt                     */
                        /********************************/

                        lim = Yb
                        if (code .AND. 0x4)
                                lim = Yt
                        x = (X2-X1)*(lim-Y1)/(Y2-Y1) + X1
                        y = lim
                }

        if ( code == code1)
        {
                /*************************************/
                /* Update endpoint (X1,Y1) and get   */
                /* new descriptor code               */
                /*************************************/

                code1 = GetCode(x,y)
                X1 = x
                Y1 = y
        }
        else
        {
                /*************************************/
                /* Update endpoint (X2,Y2) and get   */
                /* new descriptor code               */
                /*************************************/

                code2 = GetCode(x,y)
                X2 = x
                Y2 = y
        }
    }

return(visible)     /* (X1,Y1) to (X2,Y2) is visible */

}
end
```

(continued)

Listing 3 *(continued)*

```
/* Function to set up descriptor codes for a point */

GetCode(X,Y)

variables: c
externals: Xl, Xr, Yb, Yt

begin
{
        c = 0
        if (X < Xl)
                c = 8           /* Left of left edge   */
        if (X > Xr)
                c = 2           /* Right of right edge */
        if (Y < Yb)
                c = c + 1       /* Below bottom edge   */
        if (Y > Yt)
                c = c + 4       /* Above top edge      */
        return (c)
}
```

don't need a very high precision to draw the line on a typical display. This allows you to choose a lower limit (based on the resolution of your display device) and terminate the algorithm when the length of the subdivided segment becomes shorter than this limit.

Listing 4 shows the clipping divider algorithm in pseudo code.

Even though the clipping divider algorithm is iterative, it converges to the needed accuracy as a function of $\log_2(N)$, where N is the resolution of display device.

Since the divisions are always by 2, the divider algorithm can be implemented *without* a math coprocessor. (To divide by 2, you simply do a right shift.) Even better, the algorithm can be implemented with *integer* division, where you throw away the remainder each time. That allows you to incorporate it into your *device-dependent* line drawing algorithm. You do that by first defining the clipping region (and the line's endpoints) as *integer* coordinates in device coordinate space.

Graphics standards

In order to create portable graphics software, you need to write it in such a way that it calls upon a standard set of library graphics functions. There are several standards in use today: GKS, PHIGS and NAPLPS are the most widely used. In the following discussion I will focus on the GKS (Graphical Kernel System). GKS supports 2-dimensional (x,y) graphics, which is adequate for over 90% of today's graphics applications.

Listing 4

```
/* Clipping by Midpoint Division */

/********************************************************
 * This is the Sutherland-Sproull algorithm for the      *
 * clipping divider.  The function makes two calls to a  *
 * common function, FarPoint, which returns the          *
 * coordinates of the visible point farthest from the    *
 * first coordinate pair.  Return a code to indicate if  *
 * the line is rejected or is visible.                   *
 ********************************************************/

Clip (Ax,Ay,Bx,By)

variables: bx, by, code1, code2
externals: Xl, Xr, Yb, Yt

begin
{
      code1 = GetCode(Ax,Ay)
      code2 = GetCode(Bx,By)

      if (code1 = 0 .AND. code2 = 0)
            return(visible)    /* line is all visible */

      if ((code1 .AND. code2) != 0
            return(reject)     /* line is rejected */

      else          /* line may have visible portion */
      {
            /*******************************************/
            /* Find visible point farthest from A and  */
            /* return its coordinates in bx and by.    */
            /*******************************************/

            bx = Bx
            by = By
            FarPoint(Ax,Ay,bx,by,code2)
```

(continued)

Listing 4 *(continued)*

```
            /*******************************************/
            /* Find visible point farthest from B and  */
            /* return its coordinates in Ax and Ay.    */
            /*******************************************/

            FarPoint(Bx,By,Ax,Ay,code1)

            /*******************************************/
            /* (Ax,Ay) and (bx,by) define the ends of  */
            /* the visible part of original line AB.   */
            /* Return the two new endpoints as (Ax,Ay) */
            /* and (Bx,By).                            */
            /*******************************************/

            Bx = by
            By = by

            return(visible) /* (X1,Y1) to (X2,Y2) is visible */
      }
}
end

/*******************************************/
/* Find farthest visible point from (Px,Py) */
/*******************************************/

FarPoint(Px,Py,Qx,Qy,c)

variables: Mx, My, cm
external: cutoff   /* cutoff is tolerance for length */

begin
{
      while (abs(Qx-Px) > cutoff)
      {
            Mx = (Qx-Px)/2 + Px  /* calculate midpoint */
            My = (Qy-Py)/2 + Py
            cm = GetCode(Mx,My)  /* get descriptor code */

            if ((cm .AND. c) != 0)
            {
                  Qx = Mx    /* segment MQ is invisible */
                  Qy = My
                  c = cm
            }
```

(continued)

Listing 4 *(continued)*

```
                else
                {
                        Px = Mx /* segment MQ has visible part */
                        Py = My
                }
                }
}
end
```

An introduction to GKS

GKS is typically implemented as a library of functions, which range from very simple graphical functions that draw lines, all the way up to high level animation support. Ideally, you would like to be able to write all the GKS library functions to be independent of the system hardware. (If written in a high-level language, you could then simply recompile the library for each new system.) However, since some GKS functions interact directly with input and output hardware, they can't be hardware independent.

The GKS standard is currently designed to produce 2-dimensional pictures. In producing these pictures, it uses three 2-dimensional coordinate systems:

- *World coordinates* allow you to describe an object in familiar units like feet, amperes, seconds, or kilograms. The X and Y axis of the world coordinate system can extend to positive or negative infinity.
- *Normalized device coordinates* are finite: they range between 0 and 1. NDC space is dimensionless.
- *Device coordinates* are based upon the specific video RAM configuration of the system on which GKS is running.

GKS allows your application program to define a *window* in world coordinates, and a *viewport* in normalized device coordinates. It then uses these definitions to compute a *normalization transformation* that maps the window into the viewport. Some GKS implementations allow an application program to define several window/viewport pairs, for use in building a single drawing. Each window/viewport pair has its own transformation, which your application program then selects by number. Once your program selects a transformation, GKS uses that transformation until your program selects another transformation.

GKS also includes a special parameter, called the *clipping indicator*. Your application program can set this parameter to CLIP or to NOCLIP. If the clipping indicator is set to CLIP, then any (X,Y) world coordinate that gets transformed to an NDC point outside of the bounds of the viewport is not displayed. When the clipping indicator is set to NOCLIP, all (X,Y) coordinates are transformed to NDC space, and eventually passed along to the output device. Once your program sets the clipping indicator, the indicator retains the same value until you change it.

GKS output functions

Your application software draws pictures on a CRT by calling a set of basic GKS output functions. The 6 most important output functions are:

- polyline - accepts an array of real number NDC (X,Y) coordinate pairs, and writes into the video RAM, to draw a series of connected *line segments* on the CRT screen, in a connect-the-dots fashion.
- polymarker - accepts an array of real number NDC (X,Y) coordinate pairs, and writes into the video RAM to draw *symbols* on the CRT screen, positioning each at the coordinate indicated.
- text - accepts a real number NDC (X,Y) coordinate pair, and writes into the video RAM to display a *character string* on the CRT screen, starting at the coordinate indicated.
- fill area - accepts an array of real number NDC (X,Y) coordinate pairs, and writes into the video RAM to draw a *polygon*. Then it fills the polygon with a pattern, a hatch style, a solid color, or possibly nothing at all.
- cell array - copies an array of pixel values into the video RAM, to display a *cell* on the CRT screen.
- generalized drawing primitive (GDP) - writes pixel values to the video RAM to draw arcs, circles, ellipses, curves, etc.

Figure 37 shows the output generated by GKS when each of these graphics output functions is called.

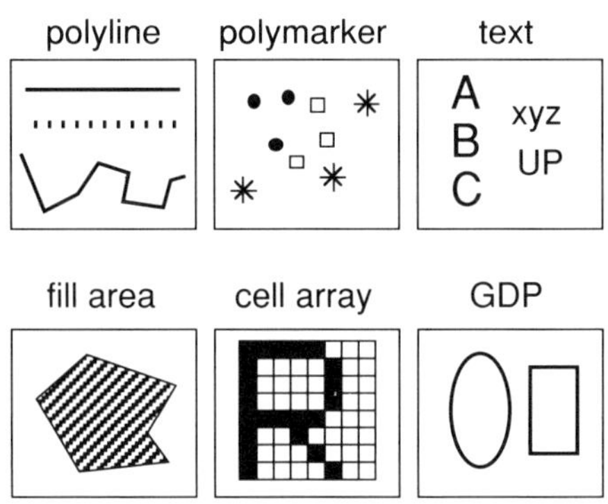

Figure 37

Some of the output functions are passed special parameters (called attributes) that further define what their output should look like. For example, the *polyline* function needs parameters to indicate:

- what *type* of line to draw (solid, dashed, etc)
- what *color* to make the line
- how *wide* to make the line

The *text* function is an even better example. It gets passed 8 attributes:

- *Character height*, which controls the size of text output
- *Text path*, which defines a line along which the text is to be written
- *Character up vector*, which is used to specify the orientation of the characters
- *Text alignment*, which allows the characters to be written side-by-side, or stacked vertically
- *Character expansion*, which governs the aspect ratio (height vs. width) of the text
- *Color*
- *Character spacing*
- *Font*
- *Text precision*

Text precision is a GKS concept which categorizes text into 3 classes or *precisions*:

- STROKE produces the highest quality text output. The characters are drawn onto the screen as an outline, and then filled. As a result they can be drawn at any angle, and can be any color, in order to satisfy all of the text attributes *precisely*.
- CHAR copies a *pixel pattern* of the character onto the screen. Depending upon what patterns are available, the characters may or may not satisfy all of the text attributes.
- STRING uses only the character height attribute (and only if the physical output device can produce variable height characters). It outputs the characters horizontally from left to right, beginning at the NDC (X,Y) coordinate provided by the application software.

With all of these attributes, it can be rather cumbersome to call GKS functions. To make things easier, some GKS implementations allow your application software to declare *bundles* of attributes. GKS assigns a *bundle ID number* to each declared bundle. Once your application software has declared several bundles, it can select any one of them for use with subsequent GKS output function calls.

GKS input functions

Up to this point we've been talking about the application software as if it runs independent of any outside influence. But, in many applications, the software must be able to *interact* with a user through an input device such as a mouse, a digitizer tablet, or a keyboard. Obviously, different input devices have fundamentally different characteristics, and must be handled differently.

In order to provide the application software with hardware-independent input functions, GKS classifies each type of input device into 1 of 6 basic categories, called *logical input device types*. Then it provides an *input function* for each, to allow the application software to interface with one of each type.

The 6 logical input device types are:

- LOCATOR inputs an (X,Y) world coordinate.
- PICK selects some displayed object.
- CHOICE inputs an integer, indicating a menu choice.
- VALUATOR inputs a real number from a keyboard or a potentiometer.
- STRING inputs a text string from a keyboard.
- STROKE inputs a sequence of (X,Y) world coordinate positions, typically from a digitizer tablet.

Each of these logical devices may operate in any of 3 modes:

- REQUEST MODE: Whenever the application program requests an input, it must then wait for the logical input device to respond. When the logical input device captures an input value, it returns that value to the program, and then waits for another request from the program before capturing any additional input values.

- SAMPLE MODE: The logical input device *periodically* samples (and holds) new input data, just in case the application program requests that data. Thus, when the application program requests input in this mode, it gets the most recently sampled data from the logical input device.
- EVENT MODE: The logical input device waits until input data is *presented* to it. When data is presented, the device sends the data to the system (unsolicited) and then *waits* for the application program to accept the data. When the application program accepts the data, the input device resumes its wait for more input data.

What is a segment?

A segment is a *sequence* of output function calls (along with their attributes) that describe a single graphical object. Each segment has a unique *identifier*. Creation of a segment is a 4-step process. The application program:

- *sets* the viewport and clipping indicator
- *opens* a new segment under a given identifier (much like it would open a new file)
- *calls* any number of output functions (polyline, polymarker, etc)
- *closes* the segment

Once the application program closes the segment, it can no longer *modify* it. However, it can reopen the segment and *delete* it.

Note: The application program is permitted to change the viewport anytime during the *creation* of the segment. Each time an output function call is added to the segment, the active viewport is also added to the segment. (This active viewport is later used, when the segment is displayed.)

Segments have 2 attributes that you may find useful:

- visibility
- priority

The *visibility* attribute allows the application software to make major changes in the image on the CRT screen by simply toggling the visibility attributes of the segments displayed. For example, you might want to have the application software display one of 2 menus, depending upon the situation. You can do that by including 2 segments in the screen painting function, and then toggling their visibility attributes. One segment can be made *invisible*, while the other is made *visible*.

The segment *priority* attribute allows the application program to specify the *order* in which overlapping segments will be drawn on the screen. The last segment drawn will be the one on top, allowing the application program to specify which segment will be seen, and which will be hidden.

When GKS is called upon to display a segment, it can be told to *scale*, *rotate*, and *shift* the segment. This is called *segment transformation*. Segment transformation is done on NDC coordinate values, instead of on world coordinates. As each output function listed in the segment is executed (to output the scaled/rotated/shifted result) it uses the *NDC viewport* and the *clipping indicator* that were stored in the segment at the time the function call was added to the segment. This allows several outputs to be independently positioned, clipped and drawn on the screen.

GKS workstations

A GKS system might have anywhere from 1 to 6 input devices, so GKS must be told which input devices are actually available. The user does that by creating a GKS *workstation description table.*

Many GKS implementations allow the system to have several different GKS workstation tables: each describing a different set of I/O devices. The application software can *open, activate, deactivate, clear,* or *close* any workstation table. Once the application software opens and activates a workstation, it may call output functions to draw on that workstation's screen.

While a workstation is open, GKS can be used to create segments for display on that workstation. However, once that workstation is closed, and another opened, those segments cannot be displayed on the new workstation. That can be an inconvenience, since you may want to create segments for display on several different workstations. GKS provides a special feature for that purpose, called Workstation Independent Segment Storage (WISS). WISS acts like a workstation. The application software opens it just like it would any workstation, and then creates segments. Once these segments have been closed (and the WISS has been closed) the application software can open some other physical workstation, and output the WISS segments to that new workstation's CRT screen.

Metafiles allow you to store segments

When a segment is created, it resides in RAM. If the system is powered down, the segment is lost. However, GKS also provides a facility called *metafiles*, which allow an application program to write segments to disk. Metafiles (like WISS) are opened and closed in the same way as a physical workstation.

Metafiles are especially useful when you want to develop a graphical image on one system, and then display it on another. However, while the GKS standard specifies the *content* of metafiles, it doesn't prescribe a *format.* Because of this, you will probably not be able to write a metafile with one company's implementation of GKS, and then read it with another company's implementation.

GKS Implementation Levels

Not all systems implement the full capabilities of the GKS standard. In order to provide a standard way to describe the implementation level of a given system, the GKS standard defines 12 levels of implementation. However, 3 of the 12 are very primitive, and are rarely implemented. The 9 levels you're likely to see implemented are shown in Figure 38.

	a0	a1	a2	b0	b1	b2	c0	c1	c2
LOCATOR modes									
request				•	•	•	•	•	•
sample							•	•	•
event							•	•	•
PICK modes									
request					•	•		•	•
sample								•	•
event								•	•
CHOICE modes									
request				•	•	•	•	•	•
sample							•	•	•
event							•	•	•
VALUATOR modes									
request				•	•	•	•	•	•
sample							•	•	•
event							•	•	•
STRING modes									
request				•	•	•	•	•	•
sample							•	•	•
event							•	•	•
STROKE modes									
request				•	•	•	•	•	•
sample							•	•	•
event							•	•	•
Predefined bundling	•			•			•		
Full bundling		•	•		•	•		•	•
Normalization transforms	M	M	M	M	M	M	M	M	M
Workstations	1	M	M	1	M	M	1	M	
Set viewport input priority				•	•	•	•	•	
Output functions									
polyline	•	•	•	•	•	•	•	•	
polymarkers	•	•	•	•	•	•	•	•	
text	•	•	•	•	•	•	•	•	
fill area	•	•	•	•	•	•	•	•	
cell array	•	•	•	•	•	•	•	•	
GDP	•	•	•	•	•	•	•	•	
Metafiles	?	•	•	?	•	•	?	•	
Segments		•	•		•	•		•	
WISS			•			•			

1 = Single M = Multiple ? = Optional

Figure 38

Formal standardization of GKS

The International Standards Organization (ISO) published the first formal standard for GKS in 1985. The American National Standards Institute (ANSI) put its stamp of approval on the standard soon after. These two groups have been the leaders in the evolution of the GKS standard.

Language bindings for GKS

The original ISO/ANSI GKS standard was developed as a language independent standard. However, in order to write portable application software, users needed some standard method for calling GKS functions from the programming language they were using. To use the popular terminology, GKS needed to have a *binding* to a programming language, before it could be used.

In the United States, the National Bureau of Standards (NBS) has supported ANSI and ISO in their efforts to standardize GKS. NBS has also developed validation programs for GKS software packages. By running these validation programs, suppliers and users can determine whether or not a particular GKS implementation (including its language binding) conforms with the ANSI/ISO standard.

Tom Barrett is the owner of A.T. Barrett and Associates, a software consulting firm specializing in real-time data acquisition, control systems, and graphics applications. He holds a BA in Math from the University of Texas, and has 31 years of software development experience.

CHAPTER 19

Practical Techniques For Producing 3D Graphical Images

By Thomas A. Foley and Gregory M. Nielson

In this chapter Tom and Greg show how you can use a pinhole camera model to produce 3D wire-frame images on a CRT screen. Then they discuss a special technique, called the Z-buffer algorithm, which is built into the hardware of many graphics display terminals, and is used to remove the hidden surfaces from parallel projections of 3D shaded objects. They also introduce the magic M transformation, which allows you to use Z-buffer hardware to remove hidden surfaces from a perspective projection. Since an understanding of the behavior of light is very important in rendering realistic images, Tom and Greg describe a popular lighting model, and show how to do lighting model calculations efficiently.

Although there are several widely read books describing techniques for the graphical display of 3D objects, (3 are listed in the references at the end of this chapter) you are typically forced to spend a considerable amount of time trying to piece together what you need to implement a program. Some of the computational details are left to the reader, some are in error, and some are done in a cumbersome manner. For example, the general approach taken by most authors to produce a perspective view of an object involves multiplication by several 4 x 4 matrices. However, there is a much simpler way. In fact, using the approach described in this article, you can achieve the same result using a subtraction, and a multiplication by a single 3 x 3 matrix.

A model for perspective projection

To develop a mathematical procedure for producing perspective projections, we must first invent a *model*. Our model will be that of a pinhole camera with zoom capability, which takes a square photograph. The CRT screen image produced by this model is the photograph taken by the pinhole camera.

The parameters that define the position and attitude of the camera are given below, and illustrated in Figure 1.

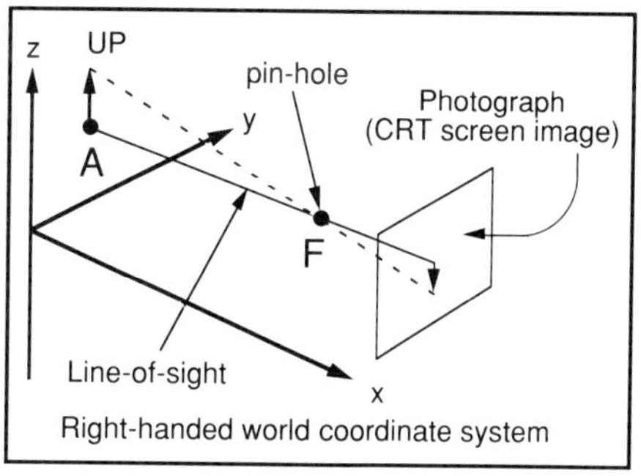

Figure 1

F is called the *FROM point*. It is a 3D world coordinate point (representing the location of the camera pinhole) which is used when doing *perspective* projections.

A is called the *AT point*. It is a 3D world coordinate point which lies at the center of the camera's field of view. Thus, the vector A-F is the *line-of-sight*. The *film* (or plane of projection) is *perpendicular* to this line-of-sight vector.

The FROM point fixes the location of the camera pinhole, and the line-of-sight specifies an axis through that pinhole, perpendicular to the film. However, the camera's *rotational* position about the line-of-sight axis has not yet been fixed. To specify the camera's rotational position, we need another vector...

U is called the *UP vector*. It is a vector in 3D world coordinates that defines which direction is *up* in world coordinate space. The camera is positioned so that the image of the UP vector on the film will be pointing up on the photograph.

v is called the *view angle*. It determines how wide an angle of view will be mapped to the square film area. In effect, the view angle specifies how close the square film will be placed to the pinhole of the camera.

In addition to the above values (which serve to position and orient the camera) we need to specify a rectangular region of the CRT screen, within which the image will be displayed. This region is called the *viewport*, and is specified by two (X,Y) coordinates: its lower-left corner (V_{left}, V_{bottom}) and its upper right-hand corner (V_{right}, V_{top}).

The coordinates for specifying the viewport location are called *normalized device coordinates*. An example of a viewport is shown in Figure 2. The viewport determines where the pinhole camera's "photograph" is to be displayed on the CRT screen. Even though the camera's photograph is square, the viewport need not be. If it isn't square, then the image will be *stretched* or *compressed* to fit the dimensions of the viewport rectangle. Note: This will result in some distortion of the image.

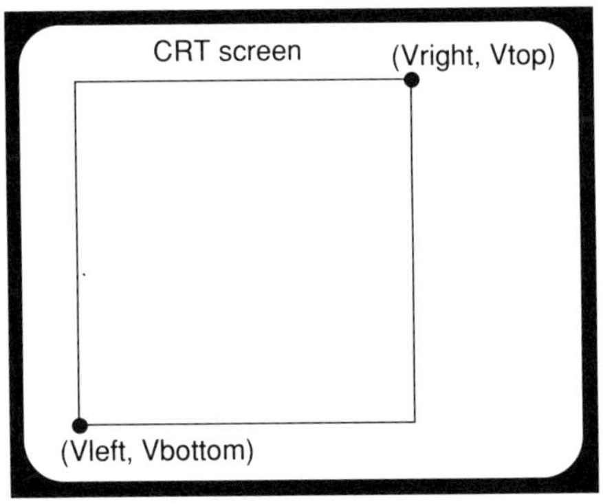

Figure 2

In specifying F, A and U, there are some restrictions:

- F and A may not be the *same* point. This would obviously cause a problem since 2 identical points would not define a unique line-of-sight.
- U cannot be a *null* vector, since a null vector would not specify a unique up-direction.
- U cannot be parallel to the line-of-sight (A-F), since this would not specify a unique rotational position for the camera, with respect to the line-of-sight.
- v must be greater than zero degrees, and less than 180 degrees.

Objects viewed on the CRT screen can be enlarged in 2 different ways:

- You can move F (the camera pinhole position) closer to the object, while maintaining the same distance between the pinhole and the film.
- You can move the film farther from the pinhole, while maintaining the position of the pinhole. (By doing this, you are reducing the view angle.)

The latter approach is called *zooming*. Many people assume that the effect on the resulting image will be the same; that *zooming in* on an object produces the same effect as *moving the camera* closer. However this is not true. For example, suppose we start with a wire-frame image of a cube. When we enlarge the image of the cube by moving the camera closer, the front face of the cube is much larger than the back face. As shown in Figure 3, this results in an exaggerated depth perception.

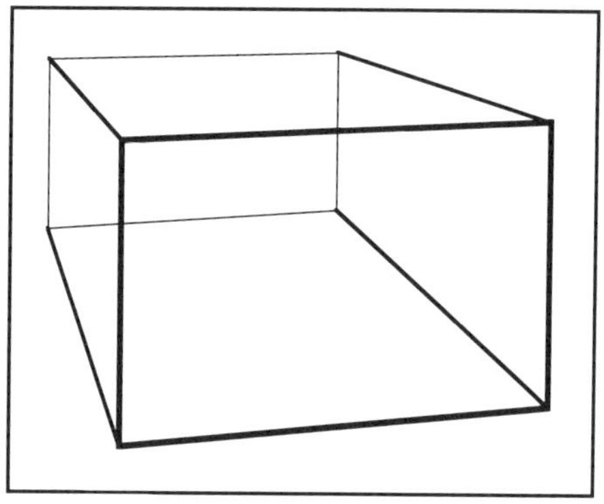

Figure 3

In contrast, if we enlarge the image by narrowing the angle of view (moving the film *farther* from the pinhole) the front and back faces of the cube are much nearer the same size, as shown in Figure 4.

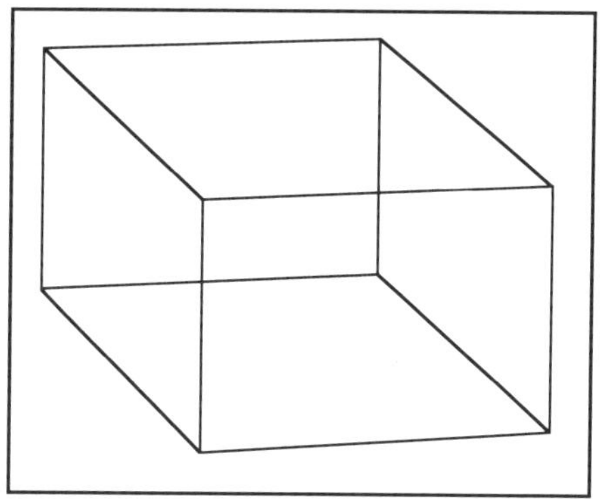

Figure 4

If the film is moved *very close* to the pinhole, you get a very wide angle view, much like the view through a *fish-eye* camera lens. This view can be quite distorted. Viewing angles between 40 degrees and 60 degrees give a much more realistic view.

An object in world coordinate space is typically specified as a set of 3D (X, Y, Z) *vertices*, and a set of *edges* connecting those vertices. In order to produce a 2D screen image of the object, the 3D coordinates of each vertex must first be *transformed* to 2D integer screen coordinates. This is called the *viewing transformation*.

The viewing transformation is usually done as a 2-step process:

- The 3D world coordinates of each vertex (X_W, Y_W, Z_W) are transformed into another set of 3D coordinates (X_e, Y_e, Z_e) which is based upon an *eye coordinate system*.
- The 3D eye coordinates of each vertex (X_e, Y_e, Z_e) are transformed to a 2D space called the *normalized device coordinate system* (X_{NDC}, Y_{NDC})

When drawing the eye coordinate system (on a 2D piece of paper) we usually show the positive X axis pointing to the right, the positive Y axis pointing upward, and the positive Z axis pointing toward the horizon. Note (for later discussion) that this is called a *left-handed* coordinate system.

We said earlier that F and A can be anywhere in world coordinate space, as long as they are not the same point. In many cases, placing A at the origin makes it easier to generate the vertex coordinates for the object you want to define. For example, if you want the object to be a torus, you might use the 8 (X, Y) point coordinates shown in Figure 5, and then rotate them about the Z axis to generate a torus.

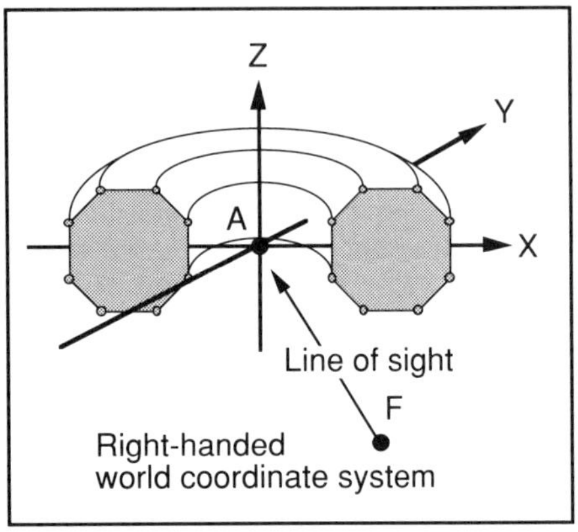

Figure 5

However, if you want to do a *perspective* projection of the object, it's more convenient to have the F point at the origin. That's why we transform the *world* vertex coordinates to eye vertex coordinates. After each of the vertex coordinates has been transformed to eye coordinates, the camera pinhole (F) is at the origin, A lies on the positive Z axis, and the UP vector has a positive Y component, and is parallel to the Y-Z plane. (See Figure 6)

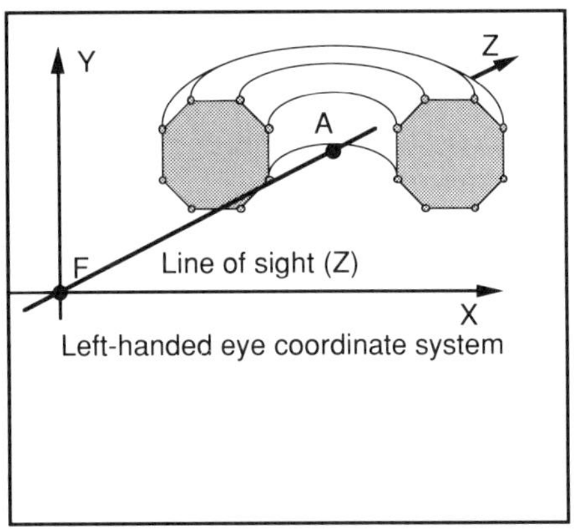

Figure 6

Once you have transformed all of the vertices to eye coordinates, it's easy to do a perspective projection. Let's assume that the plane of projection (the camera's film) lies a distance D from the origin (the camera's pin hole). Let's further assume that the film is square, with a height and width of $F_{height} = F_{width}$. (See Figure 7) The actual value chosen for D and for F_{height} isn't really important here. What we want is a ratio. For example, you would get the same affect by choosing D to be twice as large and, at the same time choosing F_{height} to be twice as large. For convenience we will choose D = 1.

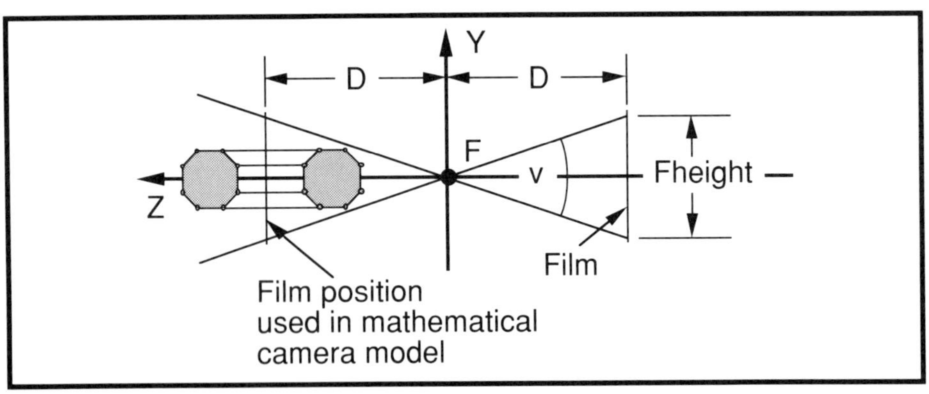

Figure 7

We can express F_{height} in terms of the viewing angle (v) as follows...

$(0.5 * F_{height}) / D = \tan(v / 2)$

Substituting 1 for D, and solving for F_{height}, we get equation **(1)...**

(1) $F_{height} = 2 * \tan(v / 2)$

Since $F_{width} = F_{height}$ we also get equation **(2)**...

(2) $F_{width} = 2 * \tan(v / 2)$

Note: You can see from Figure 7 that the X and Y vertex coordinates of the image on the pinhole camera's film will be negated from the actual vertex coordinates. To simplify the mathematics (eliminating this negation) we will assume that the film is on the same side of the pinhole as the object begin viewed.

Now suppose we have a vertex in eye coordinates at (X_e, Y_e, Z_e). We need to compute the coordinates of the point on the film where this vertex will be projected through the pinhole (X_s, Y_s, Z_s). Using similar triangles, it's easy to show that...

$X_s / D = X_e / Z_e$

Substituting 1 for D, we get equation **(3)...**

(3) $X_s = X_e / Z_e$

Similarly...

(4) $Y_s = Y_e / Z_e$

In order to plot this vertex on the CRT screen, we need to convert its 3D eye coordinates to 2D screen coordinates. Looking back at Figure 2, we see that we can calculate the viewport's height and width, as follows...

$V_{height} = V_{top} - V_{bottom}$
$V_{width} = V_{right} - V_{left}$

Then we can calculate the viewport's center-point coordinates, as follows...

$X_{vcenter} = (V_{right} + V_{left}) / 2$
$Y_{vcenter} = (V_{top} + V_{bottom}) / 2$

We can use these values to map the eye coordinates of the vertex's projection on the film to the viewport as follows...

$X_n = X_{vcenter} + (X_s * V_{width}) / F_{width}$
$Y_n = Y_{vcenter} + (Y_s * V_{height}) / F_{height}$

Substituting per equations **(1)**, **(2)**, **(3)**, and **(4)** for F_{width}, F_{height}, X_s, and Y_s we get...

(5) $X_n = X_{vcenter} + (X_e / Z_e) \{V_{width} / (2 * \tan (v / 2))\}$

(6) $Y_n = Y_{vcenter} + (Y_e / Z_e) \{V_{height} / (2 * \tan (v / 2))\}$

Since $X_{vcenter}$, $Y_{vcenter}$, V_{width}, V_{height}, and tan (v / 2) are all constants, we can calculate 4 constants, and then use these same 4 constants to do each of the eye-coordinate-to-screen-coordinate transformations, as follows...

$X_n = A + (X_e / Z_e) \{B\}$

$Y_n = C + (Y_e / Z_e) \{D\}$

These 2 equations provide us with the 2nd transformation we needed: from eye coordinates to screen coordinates. However, we still need to know how to do the first transformation: from world coordinates to eye coordinates.

Remember that F in world coordinates might be anywhere in world coordinate space, and that the object might also be anywhere. We want to transform the world coordinates to eye coordinates in such a way that F ends up at the origin of the eye coordinate system, A ends up on the positive Z axis, and the UP vector ends up in the positive Y-Z plane.

We could accomplish this transformation in 5 steps:

- *Translate* all vertex world coordinates by an amount that moves F to the origin. For example, if F is at (3, 5, 7) we would subtract 3 from the X component of each vertex, 5 from the Y component, and 7 from the Z component.
- *Rotate* all of the vertices around the world coordinate X axis until A lies in the X-Z plane.
- *Rotate* all of the vertices around the world coordinate Y axis, until A lies on the Z axis.
- *Rotate* all of the vertices around the world coordinate Z axis, until the UP vector is parallel to the Y-Z plane, and has a positive Y component.
- Invert the sign of the Z coordinate of each vertex.

The last step is needed because it's *traditional* to use a *right-handed* coordinate system for world coordinates, and a *left-handed* coordinate system for eye coordinates.

Each of the 3 rotations can be accomplished by multiplying the (X, Y, Z) coordinates of each vertex by a 3 x 3 matrix, followed by the final negation, to convert right-handed to left-handed coordinates. However, we can *drastically* reduce the amount of arithmetic by first *combining* these 3 matrices (and the negation) into a single 3 x 3 matrix, which we will call V. Once we have matrix V, we can transform the world coordinates (X_w, Y_w, Z_w) of each vertex to eye coordinates (X_e, Y_e, Z_e) by the following operations...

(7) $[\, X_e \; Y_e \; Z_e \,] = [\, (X_w - F_x)(Y_w - F_y)(Z_w - F_z) \,] \begin{bmatrix} V_{11} & V_{12} & V_{13} \\ V_{21} & V_{22} & V_{23} \\ V_{31} & V_{32} & V_{33} \end{bmatrix}$

As shown in the above equation, the first step is to subtract F's coordinate values from those of each vertex. Then we simply multiply each of the vertex coordinates by matrix V, to accomplish the 3 rotations and the transformation from right-handed to left-handed coordinates (the negation of the z coordinate).

Since matrix V transforms a world coordinate system with mutually perpendicular X, Y and Z axes to an eye coordinate system with mutually perpendicular X, Y and Z axes, we call it an *orthogonal matrix*. Because V is orthogonal, it has a very nice property that will help us derive it: its inverse (V^{-1}) is equal to its transpose (V^T). In other words...

$$V^{-1} = \begin{bmatrix} V_{11} & V_{21} & V_{31} \\ V_{12} & V_{22} & V_{32} \\ V_{13} & V_{23} & V_{33} \end{bmatrix}$$

Since this matrix is the inverse of matrix V, it transforms point coordinates from the eye coordinate system back to the world coordinate system. So, if we take unit vectors along the X, Y, and Z axes of the eye coordinate system, and apply matrix V^{-1} to them, it transforms them to unit vectors in world coordinates.

For example, when we multiply a unit vector along the X axis of the eye coordinate system by V^{-1}, we get...

$$[1\ 0\ 0] \begin{bmatrix} V_{11} & V_{21} & V_{31} \\ V_{12} & V_{22} & V_{32} \\ V_{13} & V_{23} & V_{33} \end{bmatrix} = [V_{11}\ V_{21}\ V_{31}]$$

When we multiply a unit vector along the Y axis of the eye coordinate system by V^{-1}, we get...

$$[0\ 1\ 0] \begin{bmatrix} V_{11} & V_{21} & V_{31} \\ V_{12} & V_{22} & V_{32} \\ V_{13} & V_{23} & V_{33} \end{bmatrix} = [V_{12}\ V_{22}\ V_{32}]$$

And, when we multiply a unit vector along the Z axis of the eye coordinate system by V^{-1}, we get...

$$[0\ 0\ 1] \begin{bmatrix} V_{11} & V_{21} & V_{31} \\ V_{12} & V_{22} & V_{32} \\ V_{13} & V_{23} & V_{33} \end{bmatrix} = [V_{13}\ V_{23}\ V_{33}]$$

These 3 equations give us a way to calculate the 9 terms of the matrix V. For example, if we can find numeric values for the X, Y, and Z components of the world coordinate line-of-sight vector, we will have values for V_{13}, V_{23}, and V_{33}.

We already said that the Z axis in the eye coordinate system is the line-of-sight. So what is the line-of-sight vector in the world coordinate system? It's just the vector from F to A. Let's call this world coordinate line-of-sight vector c...

$$c_X = A_X - F_X$$
$$c_Y = A_Y - F_Y$$
$$c_Z = A_Z - F_Z$$

We can *normalize* this vector (scale its X, Y, and Z components, so that its vector length is one) by dividing each component of c by the *length of c* (which we write as || c ||). We'll call the normalized line-of-sight vector c'.

$|| c || = SQRT [c_x^2 + c_y^2 + c_z^2]$

$c'_x = c_x / || c ||$
$c'_y = c_y / || c ||$
$c'_z = c_z / || c ||$

Since we previously showed that matrix V^{-1} transforms the eye coordinate Z axis unit vector to this unit line-of-sight vector, we now know how to calculate 3 terms of V...

$V_{13} = c'_x$
$V_{23} = c'_y$
$V_{33} = c'_z$

To get values for V_{11}, V_{21}, and V_{31}, we need to calculate the world coordinate transformation of the eye coordinate X axis unit vector. If you place your eye at the origin of the eye coordinate system and look along that positive Z axis, you want:

- the positive X axis to be perpendicular to it, and pointing to the right.
- the eye coordinate system's positive X axis to be perpendicular to the transformed UP vector.

From this, we can say that the world coordinate transformation of the eye coordinate X axis must be:

- perpendicular to the world line-of-sight vector.
- perpendicular to the world coordinate UP vector.

We can generate a vector that is perpendicular to 2 other vectors by calculating the cross-product of those 2 vectors. In this case, we calculate the cross product of the line-of-sight vector and the UP vector.

$a' = (c' X u) / || c' X u ||$

We can now calculate 3 more terms of V...

$V_{11} = a'_x$
$V_{21} = a'_y$
$V_{31} = a'_z$

Since we have calculated a normalized vector (c') that transforms to the eye coordinate Z axis, and another normalized vector (a') that transforms to the eye coordinate X axis, the calculation of the vector which transforms to the Y axis is simple: it's just the cross-product of a' and c'. Since a' and c' are perpendicular, and since the length of both a' and c' are 1, the cross product vector (b') will already be normalized to a length of 1.

$b' = a' X c'$

This gives us the final 3 terms of V...

$V_{12} = b'_x$
$V_{22} = b'_y$
$V_{32} = b'_z$

Once we have computed matrix V, we can use it to convert the world coordinates of each vertex to eye coordinates, using equation (7), and then use equations (5) and (6) to convert the eye coordinates to screen coordinates.

The magic M transformation

Using the previous formulation, you can transform vertex coordinates and then plot a *wire-frame* image by simply drawing 2D line segments connecting the projected points. However, you might want to render a more realistic image by shading the surfaces of the object. In order to do that, you must know which surfaces are hidden and which are not. Many raster scan graphics display terminals and workstations provide special hardware which implements a *Z-buffer algorithm.* The Z-buffer algorithm is discussed in detail in the 3 references listed at the end of this article. Z-buffer hardware associates an additional memory location with each pixel on the screen. When you pass a point's (X,Y) screen address to the graphics display terminal, you also specify a Z coordinate, which corresponds to the *depth* of that point. The hardware saves the depth value, along with the pixel value. Then, if another value is passed to the terminal for the same pixel, the Z-buffer hardware compares its depth to the one already stored at that pixel location. If the new pixel value is deeper, the hardware discards it. If it's not as deep, the hardware replaces the previous pixel value with the new one. The end result is that only the closest surfaces of the object appear on the screen.

Since the Z-buffer algorithm implements a *parallel* projection (perpendicular to the screen face) it doesn't do the job if you want to get a *perspective* view. However, you can still use Z-buffer hardware to get a perspective view if you do one additional transformation of the point coordinates, before converting them to integer pixel coordinates for the graphics display terminal.

This transformation is done by first defining the *view volume* E, as shown in Figure 8. It's a truncated pyramid, consisting of all of the eye coordinates (X_e, Y_e, Z_e) within the volume...

$-Z * \tan(v/2) \leq Xe \leq Z * \tan(v/2)$
$-Z * \tan(v/2) \leq Y_e \leq Z * \tan(v/2)$
$zmin \leq Z_e \leq zmax$

....where v is the view angle.

We want to transform the coordinates within this truncated pyramid to a rectangular box B, also shown in Figure 9...

$-1 \leq X_b \leq 1$
$-1 \leq Y_b \leq 1$
$0 \leq Z_b \leq 1$

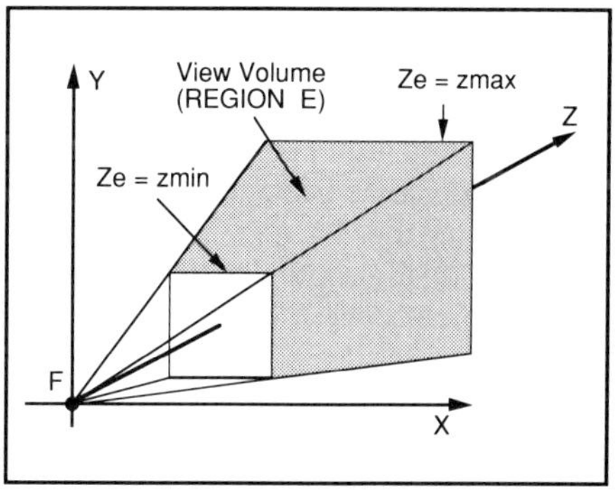

Figure 8

For a point $P_e = (X_e, Y_e, Z_e)$ in the eye coordinate pyramid E, define the transformation M to be...

$X_b = X_e / [Z_e * \tan (v / 2)]$
$Y_b = Y_e / [Ze * \tan (v / 2)]$
$Z_b = C * [l - zmin / Z_e])$

....where C = zmax/(zmax – zmin).

This transformation has 3 desirable properties:

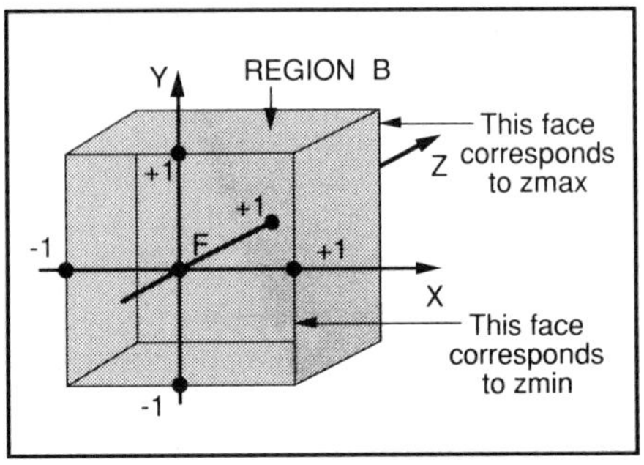

Figure 9

- It maps all of region E onto region B, in a one-to-one manner.
- It maps lines (and planes) in E to lines (and planes) in B.
- It preserves the relative depth in the Z-coordinate direction.

It's somewhat surprising that there exists a transformation with all of these desirable properties. For this reason, this transformation is sometimes called the "magic" M transformation.

If we apply this magic M transformation to each of the vertices of our object, prior to pixelizing them, we will get a perspective view of the object when the Z-buffer algorithm does its orthographic parallel projection of the (X_d, Y_d, Z_d) pixel coordinates.

The preservation of depth is important to ensure that the Z-buffer hardware does its job properly. The fact that the magic M transformation maps line segments to line segments allows us to simply pass the endpoints to the graphics display terminal. The terminal can then interpolate between those point coordinates to plot the line segment between them. Likewise, because the transformation maps planes to planes, when we want the terminal to plot a planar polygonal face of the object, we can simply pass the 3D vertices of that polygon, and the terminal can interpolate to get the internal points.

Keep in mind that the magic M transform maps real number coordinates (X_e, Y_e, Z_e) to real number coordinates (X_b, Y_b, Z_b). The resulting real number coordinates must then be mapped to the integer (pixel) coordinates (X_d, Y_d, Z_d) needed by the graphics display terminal. The pixel coordinates lie within a rectangular box, similar to the one shown in Figure 9, except that the box dimensions are based upon the number of vertical and horizontal pixels, and upon the depth of the Z-buffer.

Typical device coordinates for a high resolution display are 1024 by 1024 with 16 bits per pixel for the Z-buffer. If the entire view volume is to be displayed using the entire screen, then the width and height of the box would be 1024 and the depth would be 64K. We could calculate the pixel coordinates as follows...

$X_d = (512)(X_b) + 512$
$Y_d = (512)(X_b) + 512$
$Z_d = (64K)(Z_b)$

Lighting Model Computation

In addition to Z-buffer hardware, many raster scan graphics display terminals provide a large number of colors in RGB mode. Typically, the terminal is passed 3 integer (0 to 255) values, representing the red, green, and blue intensities at each pixel. This permits the display of 16.7 million different colors, with a storage requirement of 24 bits for each pixel. However, not all graphics display terminals provide 8 bits for red, green, and blue. To keep your software portable, it's a good idea to use *real number* RGB intensity values between 0 and 1, and then scale them to the range needed for each particular device. (A value of 1 is typically chosen to mean full intensity.)

In order to render a realistic image of an object, you need to calculate the light intensity that it emits at the various points across its surface. These intensity values are calculated using a mathematical *lighting model*. There are several well known lighting models. We will discuss one of the most popular (and simplest) which is known as the *Phong* lighting model.

Consider Figure 10, where P is a point on a surface, N is the outward normal vector of unit length, L is a unit vector directed towards the light source, R is the unit reflectance vector, which indicates the direction of the light reflecting off the surface, S is a unit vector pointing toward the viewer, and alpha and theta are the angles indicated. The Phong lighting model computes the light intensity (I_p) that the observer sees at the point P as follows...

$$
\begin{aligned}
I_P = \; & (K_a)\,(I_a) + \\
& (K_d)\,(\cos(\text{theta}))\,(I_L) + \\
& (K_s)\,(\cos^n(\text{alpha}))\,(I_L)
\end{aligned}
$$

...where I_L is the incident intensity of the *light source* at point P, I_a is the intensity of the *ambient* light, K_a, K_d, and K_s are real constants (between 0 and 1) and n is a non-negative number; usually an integer.

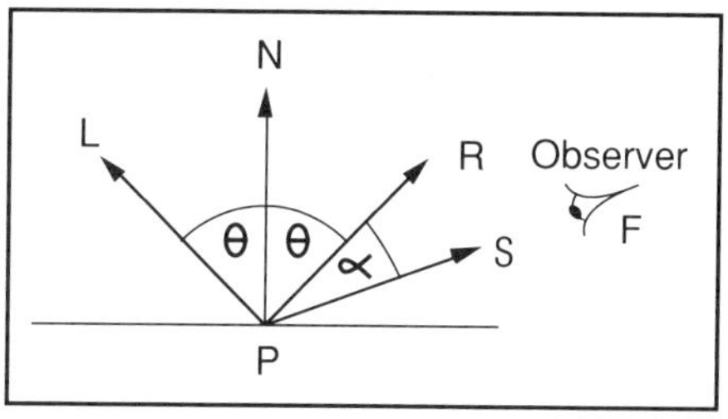

Figure 10

The 1st term calculates how much *ambient* light is emitted toward the observer. (The Phong model treats ambient light as a kind of uniform *glow* that's emitted by every point on the object's surface.)

The 2nd term calculates how much *diffused* light is reflected toward the observer. (Diffused light is the light that gets reflected from the object's surface in all directions, when a light source illuminates its surface). This diffuse term is based on Lambert's law for reflected light off a matte surface.

The 3rd term calculates the *specular* light that gets reflected from the object. (Specular light is the *glare* seen when a light source is mirrored in the surface of a shiny object.)

If we want to render just a black and white image, we only need to do this calculation once. However, if we want a color rendering, we must do it 3 times: once for red, once for green, and once for blue.

By varying the lighting parameters (K_a, K_b, K_s and n) we can change the appearance of an object dramatically. For example, by lowering the ambient level to zero, we can make the side of the object opposite the light source invisible. By increasing the value of n, we can sharpen the specular reflection, to give the appearance of a shiny object. For a more complete discussion of the Phong lighting model, see the references listed at the end of this article.

3D objects are typically specified to a graphics system as a collection of adjacent planar polygons. For example, a cube could be represented as 6 square polygons. In order to render the cube on a CRT screen, we would first calculate the *centroid* of each of its square faces (by averaging the vertex coordinates for that face). Then we would apply the lighting model to each of the 6 centroids (to get a light intensity for each) and pass these intensities to the graphics display terminal, along with the four 2D pixel coordinates of the vertices of each square face. The display terminal fills each 4-sided polygon with pixels of the specified intensity. The result is a *faceted* rendering, with sharp intensity changes at the boundaries between the faces of the cube.

In some cases, you may want to render a smoothly rounded object, such as a cylinder. Suppose we define the surface of the cylinder with a set of square faces, and then compute an intensity for each face, based on the location of the face's centroid. When we use these intensity values to display each face we will see what appears to be a faceted surface, instead of a smooth-looking cylinder.

To avoid the facets, you can apply the lighting model to each of the 4 *vertices* of each face, instead of just to the *centroid* of each face. (The surface normal vector used for each vertex is the *average* of the surface normals of the square faces around it.) This gives a different light intensity value for each vertex. Then, as you pass each pair of 2D vertex coordinates to the graphics display terminal, you also pass the corresponding pixel intensity for that vertex. The terminal then *interpolates* the intensities across each polygon, from vertex to vertex.

Since adjacent polygons share vertices, their shared vertex intensities will be identical, and the interpolated shading of each pixel along their common edge is the same. As a result, the common edge will not be visible, giving a smooth appearance. This technique is called *Gouraud shading.*

Notice that to use the lighting model described above, we must know the cosines of the angles theta and alpha. These angles depend upon the position of the light source, the position of the observer, and the orientation of the planar polygonal face, from which the light is being reflected.

To calculate the cosine of theta, we start by calculating the vector L, previously shown in Figure 10. It is simply the normalized *difference* between the light source coordinates (L_f) and the coordinates of point P, on the surface of the object...

$$L = (L_f - P) / || L_f - P ||$$

We then calculate the surface normal, at the point P, which might be the centroid of one of the planar polygonal faces. Since the surface normal at any point on the planar polygon is the same as the normal at any other point on the polygon, we can just calculate the normal at one of its vertices, and use that. We do that by taking the *cross product* of 2 edges of the polygon, and then normalizing the resulting vector to a length of 1. (The resulting vector is perpendicular to both edges, and is therefore perpendicular to the polygon.) Note: The order of the cross product is important, since you want to calculate the *outward-pointing* normal vector, not the inward-pointing one.

Once we have this surface normal, we can calculate the cosine of theta, as the dot product of N and L...

cos(theta) = N • L

However, there is a special case that we must take care of. If the surface normal points to the *opposite* side of the planar polygon from the light source, we *don't* want the surface to be illuminated. Thus...

IF (N • L ≤ 0) THEN cos(theta) = 0

To calculate cosine of alpha, we first calculate the vector S, which points to the observer. It is the normalized *difference* between the observer's coordinates and the coordinates of point P, on the surface of the object...

S = (F – P) / || F – P ||

We also need to calculate the direction vector (R) for the specular reflection from the surface of the object. Since the specular light will be reflected off the surface of the object at the same angle that it arrived, we can take the vector from the surface point to the light source (L), negate it (so it's pointing *toward* the surface, instead of away from it) and then add 2 times its normal component [(N • L) * N] to itself. (The first time cancels the inward component, and the second creates a outward component of equal magnitude, as shown in Figure 11.)

R = – L + 2(N • L) * N

Once we have S and R, we can calculate the cosine of alpha, as the dot product of S and R...

cos(alpha) = S • R

However, there are 2 special cases that we must take care of. If the surface normal points to the *opposite* side of the polygon from the light source, (N • L ≤ 0) then we don't want any reflectance. Also, if the observer and the reflection are on opposite sides of the normal vector, or on opposite sides of the polygon (S • R ≤ 0) we don't want any reflectance. Thus...

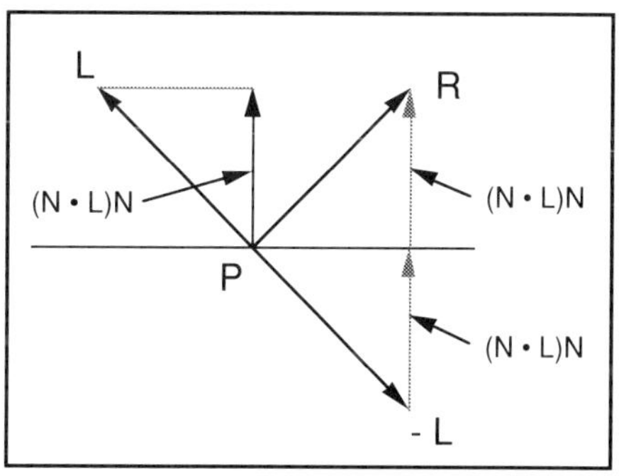

Figure 11

IF (N • L ≤ 0) OR (S • R ≤ 0) THEN cos(alpha) = 0

Courtesy of Heinrich Müller, Universität Karlsruhe, Germany

Figure 12

Courtesy of Heinrich Müller, Universität Karlsruhe, Germany

Figure 13

Note: There is one other special case. When you use this lighting model to calculate light intensities, you will sometimes find that the computed light intensity is greater than one. When that happens, just set the intensity to one.

While the computation of these vectors and angles is typically done with world coordinates, they could be calculated using eye coordinates. However, you should NEVER attempt to do the calculations *after* the magic M transformation has been applied.

There are many extensions to this simple lighting model which attempt to produce more realistic renderings, or to introduce more sophisticated light sources. For example, you can create light sources that give a spotlight effect. However, if you want to generate very realistic renderings of a scene (with shadows, reflections, and refractions) you would use a computationally expensive *ray tracing* algorithm. Figures 12 and 13 (on the following page) show the quality of the 3D realism obtained using ray tracing. Lighting model techniques similar to those discussed in this chapter are typically employed when using ray tracing algorithms.

References

1. J.D. Foley and A. Van Dam, Fundamentals of Interactive Computer Graphics, Addison Wesley, Reading, Massachusetts, 1982.

2. W.M. Newman and R.F. Sproull, Principles of Interactive Computer Graphics, McGraw-Hill, New York, 1979.

3. D.F. Rogers, Procedural Elements for Computer Graphics, McGraw-Hill, New York, 1985.

Tom Foley *is an associate professor in the Computer Science department at Arizona State University. He is a consultant for Lawrence Livermore National Laboratory, and has taught at California Polytechnical Institute, San Luis Obispo and at the University of Nevada at Las Vegas. He has a Ph.D. in Mathematics from ASU and has published numerous research articles in the area of computer graphics and computer-aided geometric design. He is currently a member of the executive board of the IEEE technical committee on computer graphics.*

Gregory M. Nielson *is a professor of computer science at Arizona State University, where he teaches courses in computer graphics and computer-aided geometric design. He has lectured in Italy, France, England, Chile, and Germany, and has published widely in the areas of computer graphics and geometric modeling. Greg has a Ph.D.from the University of Utah, is an associate editor of ACM Transactions on Graphics, Computer Aided Geometric Design, and Rocky Mountain Journal of Mathematics. He currently chairs the IEEE technical committee on computer graphics.*

CHAPTER 20

Pattern Transformers: A New Class Of Graphics Engine

By Jim Jonas and David Forbes

The cost of high resolution graphics hardware has dropped greatly in the last few years. With the falling price of memory, and the rising speed of VLSI graphics chips, you can now build a high resolution (1024 x 1024) graphics systems that cost an order of magnitude less than those only a few years ago. These graphics systems produce very high quality stationary images. However, with over a million pixels on the screen, it's hard to update the screen fast enough for real time animation applications. Dave and Jim feel that just pushing silicon speeds yet a little higher won't get the job done. You need a new type of processor for these high resolution graphics systems. They've developed a new processor that they call a pattern transformer.

While general purpose microprocessors have been around since the early 1970s, there is now a trend toward special purpose processors, such as floating point processors, digital signal processors, and graphics processors. Because each of these special purpose processors is designed to do just one job very efficiently, they can usually outperform a general purpose processor.

One type of special purpose processor that's getting a lot of attention these days is the graphics processor, often called a *graphics engine.* Graphics engines are optimized for drawing images onto video screens. To understand the various architectures used in graphics engines, you first need to understand how graphics systems display images.

Vector graphics displays
Vector graphics displays are most useful for drawing wire-frame objects. They have analog deflection circuitry that steers the electron beam across the face of the CRT, while the beam is turned on and off. Since the beam can be steered in any direction, sloped lines are just as easy to draw as horizontal or vertical lines. Since the beam need not traverse the entire face of the CRT, wire-frame objects can be drawn very quickly. That makes vector graphic displays very useful for real-time animation applications, especially where the depiction of motion is more important than realism.

Raster graphics displays
Raster graphics displays provide a much more realistic image, but the 2-dimensional nature of their pixel arrays makes it awkward to display rotating objects. As a line is rotated away from a horizontal or vertical position, its endpoints may no longer lie at the center of a pixel. The calculations needed to choose the appropriate pixels for the endpoints (as well as for all the intermediate pixels on the line between them) take

time. If the system is drawing a complex object, these calculations limit the speed at which rotations can be done without causing annoying jitter.

An introduction to graphics engines

In recent years raster graphics systems have become the dominant form of computer graphics, mainly for 2 reasons:

- Memory prices have been plunging.
- The *digital* technology used in raster graphics has advanced more than the *analog* technology used in vector graphics.

Raster graphics systems typically fall into 1 of 3 general categories:

- Tiling engines
- Geometric engines
- BITBLT engines

Tiling engines are very specialized and expensive machines that draw objects as thousands of triangular or rectangular patches, called *tiles*. Tiling engines don't keep the pixel values in video RAM. Instead, they calculate the pixel intensities from a *predigested* draw list, as the pixels are being scanned onto the CRT screen. Tiling engines are used for applications that require real time animation, such as out-the-window displays for flight simulators. Since they typically cost millions of dollars, they are in a completely different class than the other 2 types.

Geometric engines draw objects at the *pixel* level. They calculate and draw (in the video RAM) the boundaries of simple geometric figures (such as triangles, squares, and circles) one pixel at a time. Then they fill the area inside these boundaries, pixel by pixel, with small (16 x 16 or smaller) stipple patterns.

Since geometric engines write pixels one at a time, their speed is limited by 2 things:

- the rate at which the pixel coordinates of successive pixels can be computed.
- the speed at which the video RAM can be accessed.

Calculation of the pixel coordinates may involve more than just incrementing an address. For example, if each pixel corresponds to a single bit of video RAM (and if the video RAM is accessed 8 bits at a time) the graphics processor may need to calculate an 8-bit *mask*, so that adjacent pixels aren't accessed simultaneously.

Note: The uPD7270 graphics chip from NEC and the HD63484 graphics chip from Hitachi are geometric engines.

BIT boundary, block transfer engines (often called *bit blitters or BITBLT engines*) are able to fill video RAM faster than geometric engines because they write to several pixels simultaneously. For example, if 16 consecutive pixels on the screen are all in one 16-bit-wide video RAM location, a bit blitter could write to all 16 pixels with a single memory access.

Bit blitters can fill a screen fast. They're most useful for copying solid polygons or text from pattern storage to the video RAM. However, if you need to scale or rotate the pattern as you copy it to the screen, a bit blitter isn't very useful. In fact, when it comes to scaling or rotation, bit blitters revert to drawing pixel-by-pixel, and they often do so more slowly than geometric engines.

So, geometric engines and bit blitter engines each have their own unique advantages and disadvantages. Geometric engines draw lines and arcs very well, but they're much slower at drawing characters. On the other hand, bit blitter engines copy characters very well (because they can write to several pixels at a time) but only if both source and destination bit maps are aligned, and have the same scale factor.

Some applications demand a new kind of graphics engine
Graphics engines are being very widely used today to display symbolic images. For example, a large portion of the graphics job in CAD, CAE, and drafting applications involves the repeated drawing of complex symbols, like those shown in Figure 1.

These symbols are typically small, and are drawn as a collection of simple geometric objects, such as lines, arcs, and ellipses. Graphics systems in wide use today typically build these symbols on the screen by drawing several primitive geometric objects. (This building of the symbol is repeated over and over, at each location where the symbol is needed on the screen.) For example, the IC pads of a printed circuit board (like the one shown in Figure 1) may be built from a rectangle, with a half-round at each end. To give another example, an architectural drawing might differentiate wood from steel beams by their fill patterns. (The wood might have a sinuous grain pattern, while the steel might be filled with a stipple pattern.)

A bit blitter engine can copy patterns like half-rounds and stipple patterns to the screen quickly. However, they don't work so fast when none of the available half-round patterns

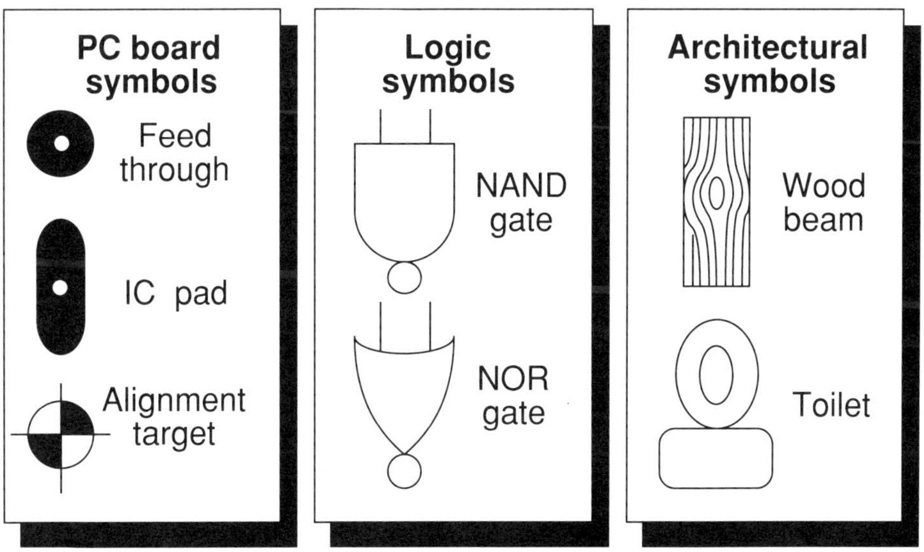

Figure 1

are the right diameter. They also have a problem with drawing a wood grain pattern on a sloping beam.

A geometric engine can draw half-rounds to any size you need, and can draw a wood grain pattern on a sloping beam, but it requires a lot of time to fill the screen.

What you need is a graphics engine that can not only copy pixel patterns onto the screen, several pixels at a time, but can also *rotate* that pixel pattern to any angle, and can *scale* the pattern in the process. We call this *pattern transformation*.

The pattern transformation algorithm

In order to understand how a pattern transformation engine works, you first need to understand what we mean by a *pattern*. For our purposes here, a pattern is a 2-dimensional array of integer pixel intensities. It doesn't matter how these intensity values are stored, but you need some way to access them, so we'll define a procedure:

```
read(x_coord, y_coord, pixel_value)
```

In order to acquire the pixel intensity at a particular point on the pattern, you call the procedure read, passing it the integer (X,Y) coordinates of the pixel.

Once you acquire a pixel intensity from the pattern, you want to write that intensity to a particular pixel location on the CRT screen. You do that by calling a 2nd procedure:

```
plot(x_coord, y_coord, pixel_value)
```

Procedures `read` and `plot` accept only *integer* (X,Y) coordinates. If you're using real number (X,Y) coordinates (as is often the case) you must *truncate* the real numbers before passing them to these procedures. (As shown in Figure 2, both `read` and `plot` access the pixel whose lower-left corner lies on the integer (X,Y) coordinates passed to it.) For example, if you want to illuminate the point (2.2, 1.7) with an intensity of 7, you first truncate the X and Y values to get (2,1), and then pass these integer values to `plot`, along with the intensity:

```
plot(2, 1, 7);
```

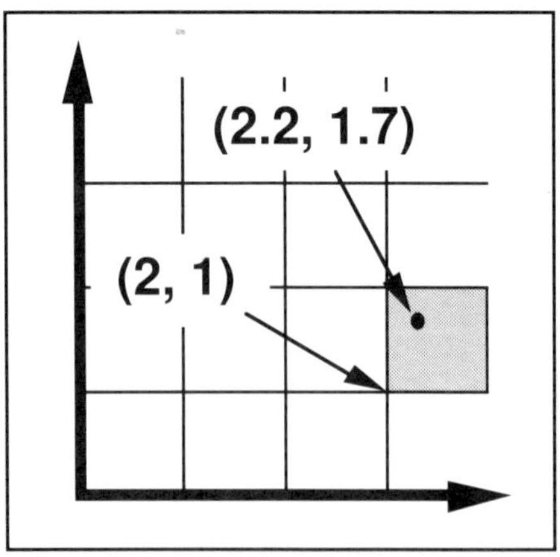

Figure 2

To perform a pattern transformation, you need to have some function that *maps* the (X,Y) coordinates of the pattern to the (X,Y) coordinates on the screen. While it's possible to map these coordinates in many different ways, the most useful mappings combine *translation, rotation*, and *scaling*, all of which maintain the *shape* of the pattern, while transforming it. For example, a rectangular set of pattern pixels will always map to a rectangular area on the screen.

Let's take a simple case in which you want to copy a pattern onto the CRT screen with a translation (sx, sy) but no rotation, and no scaling. If our pattern is a 7 x 14 array of pixel intensities, you would get a result like the one shown in Figure 3.

Since the boundaries of the rectangle are vertical and horizontal, it's easy to write a procedure that does this simple transformation. The procedure `copy_pattern` (shown in Listing 1) copies the pixel intensities from a rectangular pattern onto a rectangular area of the screen. You call it as follows:

```
copy_pattern(sx, sy);
```

When called, it first copies the left-most column of pixels, from the pattern to the screen, starting at the bottom and moving upward. Then it copies the next column of pixels, from bottom to top. It repeats this process (stepping one column to the right each time) until it reaches coordinates (sx2, sy), just to the right of the lower-right corner, as shown in Figure 4. At that point it stops, without writing to (sx2, sy).

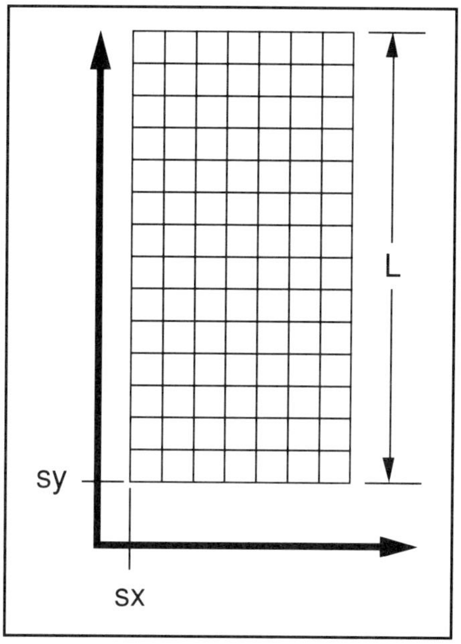

Figure 3

This simple procedure can only *translate* the pattern: it can't *rotate* or *scale* the pattern. But, suppose you also want to scale the pattern. For example, suppose you want to display a pattern at 3 times its unscaled size. You now face a problem. The pattern might be only 7 x 14 pixels, but now you need to copy 21 x 42 pixel intensities to the screen.

The obvious solution is to simply plot each pixel intensity 3 times. But suppose you want fractional scaling factors. For example, suppose you want to scale the pattern by 1.333333. The procedure `copy_and_scale_pattern` (shown in Listing 2) allows this. It's called as follows:

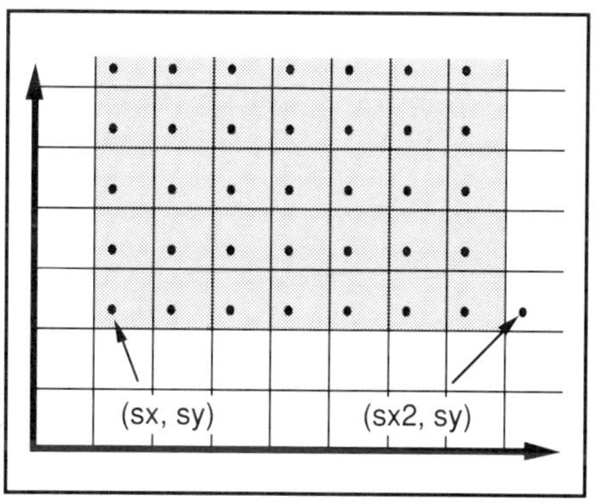

Figure 4

```
copy_and_scale_pattern(sx1, sy, sx2);
```

This procedure is similar to `copy_pattern`. It copies the first column of pixels from the pattern, starting at screen coordinates (sx1, sy). Then it copies each adjacent column of pixels, until it reaches coordinates (sx2, sy).

However, before the procedure starts copying pixels, it calculates a `scale_factor`, based upon the distance between sx1 and sx2. This scaling factor tells it how far apart to read the pattern pixels.

To read the pixel values from the pattern properly, `copy_and_scale_pattern` calculates the pattern's (X,Y) coordinates using real numbers, and then truncates those real number coordinates to get integer coordinates. (That's why `copy_and_scale_pattern` defines the variables pdx and pdy, and then uses them to step the pattern (X,Y) coordinates, instead of just using a step of 1.) In the example above (where we had a scaling factor of 1.333333) *`copy_and_scale_pattern`* reads the pixel intensities from the pattern as shown in Figure 5.

Listing 1

```
PROCEDURE copy_pattern(sx, sy: real);

VAR
sx2: real           (* Screen coordinate of lower-right corner *)
sy_temp: real;
spx, spy: integer;  (* Screen pixel coordinates *)
spy_temp: integer;

px, py: real;       (* Pattern coordinates *)
py_temp: real;

ppx, ppy: integer;  (* Pattern pixel coordinates *)
ppy_temp: integer;

pixel_no: integer;  (* Pixel number *)
srect_width: real;  (* Width of screen rectangle in pixels *)

BEGIN
(******************************************)
(* Set sx2 to the screen coordinate of the *)
(* lower-right corner of the rectangle.    *)
(******************************************)
sx2 := sx + PAT_WIDTH;

(**********************************************)
(* Set px and py to the pattern pixel coordinates *)
(* of the lower-left corner of the pattern.       *)
(**********************************************)
px := 0;
py := 0;
```

(continued)

Listing 1 *(continued)*

```
(***********************************************)
(* Starting with the left-most column, copy each  *)
(* column of pixel values from the pattern into   *)
(* the screen rectangle.                           *)
(***********************************************)
WHILE sx <> sx2
      BEGIN
      (***********************************************)
      (* Starting at the bottom pixel of the column,  *)
      (* copy the pixel values from the pattern into  *)
      (* the screen rectangle.                          *)
      (***********************************************)
      sy_temp := sy;
      py_temp := py;
      FOR pixel_no := 1 to PAT_LENGTH DO
            BEGIN
            (**************************************)
            (* Copy a pixel value onto the screen *)
            (**************************************)
            ppx := TRUNC(px);
            spx := TRUNC(sx);
            ppy_temp := TRUNC(py_temp);
            spy_temp := TRUNC(sy_temp);

            read(ppx, ppy_temp, pixel_value);
            plot(spx, spy_temp, pixel_value);

            (*******************************)
            (* Increment the y coordinates *)
            (*******************************)
            sy_temp := sy_temp + 1;
            py_temp := py_temp + 1;

            END; (* FOR pixel_no := ... *)

      (***************************)
      (* Step to the next column *)
      (***************************)
      sx := sx + 1;
      px := px + 1;

      END (* WHILE sx <> sx2 *)
END; (* PROCEDURE copy_pattern *)
```

Listing 2

```
PROCEDURE copy_and_scale_pattern(sx1, sy, sx2: real);

VAR
sx: real             (* Screen coordinates (sy declared above) *)
sy_temp: real;
spx, spy: integer;   (* Screen pixel coordinates *)
spy_temp: integer;

px, py: real;        (* Pattern coordinates *)
py_temp: real;
pdx, pdy: real;

ppx, ppy: integer;   (* Pattern pixel coordinates *)
ppy_temp: integer;

pixel_no: integer;   (* Pixel number *)
srect_width: real;   (* Width of screen rectangle in pixels  *)
srect_length: real;  (* Length of screen rectangle in pixels *)
scale_factor: real;

BEGIN
(*****************************************************)
(* Set sx to the screen coordinate of the lower-left  *)
(* corner of the rectangle.  Note: sy is already set. *)
(*****************************************************)
sx := sx1;

(***********************************************)
(* Set px and py to the pattern pixel coordinates *)
(* of the lower-left corner of the pattern.       *)
(***********************************************)
px := 0;
py := 0;

(*****************************************************)
(* Calculate the scale factor, and then calculate the   *)
(* horizontal step size and the vertical step size that *)
(* will be used for reading the pattern.  Note: This    *)
(* step size may be a fractional number of pixels.      *)
(*****************************************************)
srect_width := sx2 - sx1;
scale_factor := srect_width/PAT_WIDTH
pdx := 1/scale_factor;
pdy := 1/scale_factor;
srect_length := scale_factor * PAT_LENGTH;
```

(continued)

Listing 2 *(continued)*

```
(**********************************************)
(* Starting with the left-most column, copy each  *)
(* column of pixel values from the pattern into   *)
(* the screen rectangle.                          *)
(**********************************************)
WHILE TRUNC(sx) <> TRUNC(sx2) DO
       BEGIN
       (**********************************************)
       (* Starting at the bottom pixel of the column,  *)
       (* copy the pixel values from the pattern into  *)
       (* the screen rectangle.                        *)
       (* NOTE: Since srect_length may be a fractional *)
       (* number of pixels, we must truncate it.       *)
       (**********************************************)
       sy_temp := sy;
       py_temp := py;
       FOR pixel_no := 1 to TRUNC(srect_length) DO
              BEGIN
              (*************************************)
              (* Copy a pixel value onto the screen *)
              (*************************************)
              ppx := TRUNC(px);
              spx := TRUNC(sx);
              ppy_temp := TRUNC(py_temp);
              spy_temp := TRUNC(sy_temp);

              read(ppx, ppy_temp, pixel_value);
              plot(spx, spy_temp, pixel_value);

              (******************************)
              (* Increment the y coordinates *)
              (******************************)
              py_temp := py_temp + pdy;
              sy_temp := sy_temp + 1;

              END; (* FOR pixel_no := ... *)

       (*************************)
       (* Step to the next column *)
       (*************************)
       px := px + pdx;
       sx := sx + 1;

       END (* WHILE TRUNC(sx) <> ... *)
END; (* PROCEDURE copy_and_scale_pattern *)
```

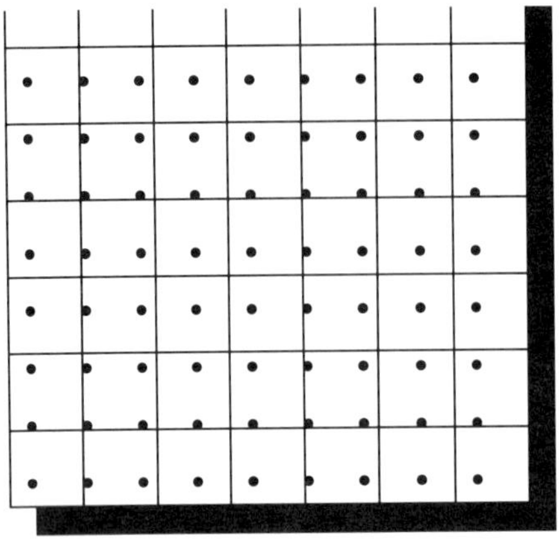

Figure 5

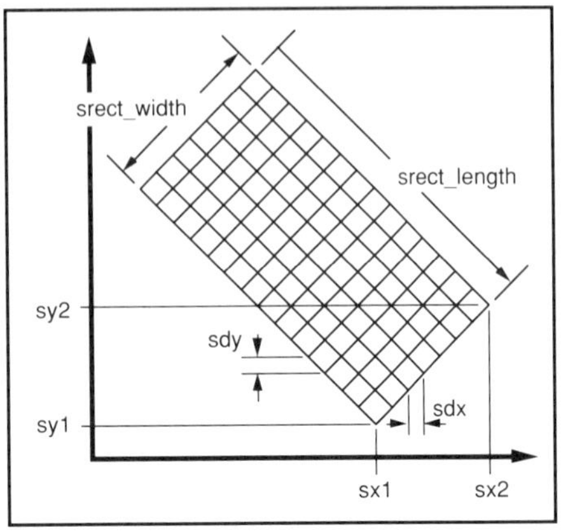

Figure 6

By copying the pattern to the screen over and over at different scales, you can display motion toward or away from the observer. This (in conjunction with the (X, Y) translation) permits you to display 3D motion. But suppose you also wanted to show the pattern *spinning* as it moves?

The procedure `copy_scale_and_rotate_pattern` (shown in Listing 3) does this. It's called as follows:

```
copy_scale_and_rotate_pattern(x1, y1, x2, y2);
```

Figure 6 shows how the 7 x 14 pattern might be plotted on the screen. Since you pass two Y-coordinate values, the procedure can determine the *length* and the *slope* of the *baseline*, and then copy the pattern at an angle. You can display rotational motion by copying the pattern to the screen, over and over, at different angles.

If you want the pattern to spin around a stationary point in the (X,Y) plane of the screen, you choose a length for the baseline, and then pass pairs of (X,Y) coordinates that always define a segment of that length. The procedure calculates the scaling factor from the length of this baseline, and then displays the pattern, based upon that scaling factor.

Listing 3

```
PROCEDURE copy_scale_and_rotate_pattern(sx1, sy1, sx2, sy2: real);

VAR
sx, sy: real        (* Screen coordinates *)
sx_temp: real;
sy_temp: real;
sdx, sdy: real;
spx, spy: integer;  (* Screen pixel coordinates *)
spx_temp: integer;
spy_temp: integer;

px, py: real;       (* Pattern coordinates *)
px_temp: real;
py_temp: real;
pdx, pdy: real;

ppx, ppy: integer;  (* Pattern pixel coordinates *)
ppx_temp: integer;
ppy_temp: integer;

pixel_no: integer;  (* Pixel number *)
srect_width: real;  (* Width  of screen rectangle in pixels *)
srect_length: real; (* Length of screen rectangle in pixels *)
scale_factor: real;

BEGIN
(*****************************************************)
(* Set sx and sy to the screen coordinates of the    *)
(* lower-left corner of the rectangle.               *)
(*****************************************************)
sx := sx1;
sy := sy1;

(*************************************************)
(* Set px and py to the pattern pixel coordinates *)
(* of the lower-left corner of the pattern.      *)
(*************************************************)
px := 0;
py := 0;

(*****************************************************)
(* Calculate the scale factor, and then calculate the   *)
(* horizontal step size and the vertical step size that *)
(* will be used for reading the pattern.  Note: This    *)
(* step size may be a fractional number of pixels.      *)
(*****************************************************)
```

(continued)

Listing 3 *(continued)*

```
srect_width := SQRT(SQR(sx2 - sx1) + SQR(sy2 - sy1));
scale_factor := srect_width/PAT_WIDTH
pdx := 1/scale_factor;
pdy := 1/scale_factor;
srect_length := scale_factor * PAT_LENGTH;

(*****************************************************)
(* Calculate the horizontal step size and the vertical  *)
(* step size that will be used for plotting  the screen *)
(* rectangle.  NOTE: These step sizes may be a          *)
(* fractional number of pixels.                         *)
(*****************************************************)
sdx := (sx2 - sx1)/srect_width;
sdy := (sy2 - sy1)/srect_width;

(*************************************************)
(* Starting with the left-most column, copy each  *)
(* column of pixel values from the pattern into   *)
(* the screen rectangle.                          *)
(*************************************************)
WHILE (TRUNC(sx) <> TRUNC(sx2)) OR
      (TRUNC(sy) <> TRUNC(sy2)) DO
       BEGIN
       (**********************************************)
       (* Starting at the bottom pixel of the column,  *)
       (* copy the pixel values from the pattern into  *)
       (* the screen rectangle.                        *)
       (* NOTE: Since srect_length may be a fractional *)
       (* number of pixels, we must truncate it.       *)
       (**********************************************)
       px_temp := px;
       py_temp := py;
       sx_temp := sx;
       sy_temp := sy;
       FOR pixel_no := 1 to TRUNC(srect_length) DO
              BEGIN
              (*************************************)
              (* Copy a pixel value onto the screen *)
              (*************************************)
              ppx_temp := TRUNC(px_temp);
              ppy_temp := TRUNC(py_temp);
              spx_temp := TRUNC(sx_temp);
              spy_temp := TRUNC(sy_temp);

              read(ppx_temp, ppy_temp, pixel_value);
              plot(spx_temp, spy_temp, pixel_value);
```

(continued)

Listing 3 *(continued)*

```
            (************************************)
            (* Increment the x and y coordinates *)
            (************************************)
            py_temp := py_temp + pdy;
            sx_temp := sx_temp = sdy; (* <- subtle trick: not a
               mistake *)
            sy_temp := sy_temp + sdx; (* <- same here *)

            END; (* FOR pixel_no := ... *)

       (**************************)
       (* Step to the next column *)
       (**************************)
       px := px + pdx;
       sx := sx + sdx;
       sy := sy + sdy;

       END (* WHILE TRUNC(sx) <> ... *)
END; (* PROCEDURE copy_scale_and_rotate_pattern *)
```

This `copy-scale-and-rotate` procedure fully implements the basic algorithm of a pattern transformation engine. It permits the display of 3D motion and rotation simultaneously. Unfortunately, implementing this procedure on a typical general purpose microcomputer results in dismally slow motion. However, you can construct a dedicated high-speed bit-slice transformation engine that executes this algorithm very fast. In fact, with careful design, you can transform and copy 1 pattern pixel every 150 nsecs.

A brief discussion of state machines

The pattern transformation engine is a *state machine*. In its simplest form, a state machine just steps though its microcode memory, which is addressed by an address counter. However, to do anything very useful, a state machine should be able to reload its address counter with a new address, based upon condition codes. This allows you to do conditional jumps and subroutine calls.

Figure 7 shows a simple state machine that's capable of jumping, based upon 4 different condition codes. When the state machine reaches a jump instruction, the ENBL microcode bit enables the multiplexer (MUX). Two bits (S0 and S1) are also routed from the microcode memory to the MUX, where they select 1 of the 4 condition code input bits. The MUX then routes the selected condition code bit to the *Reload* input of the address counter, where it controls whether or not a jump takes place. If the selected condition code level is true, the counter reloads: if the level is false, the counter retains its current address.

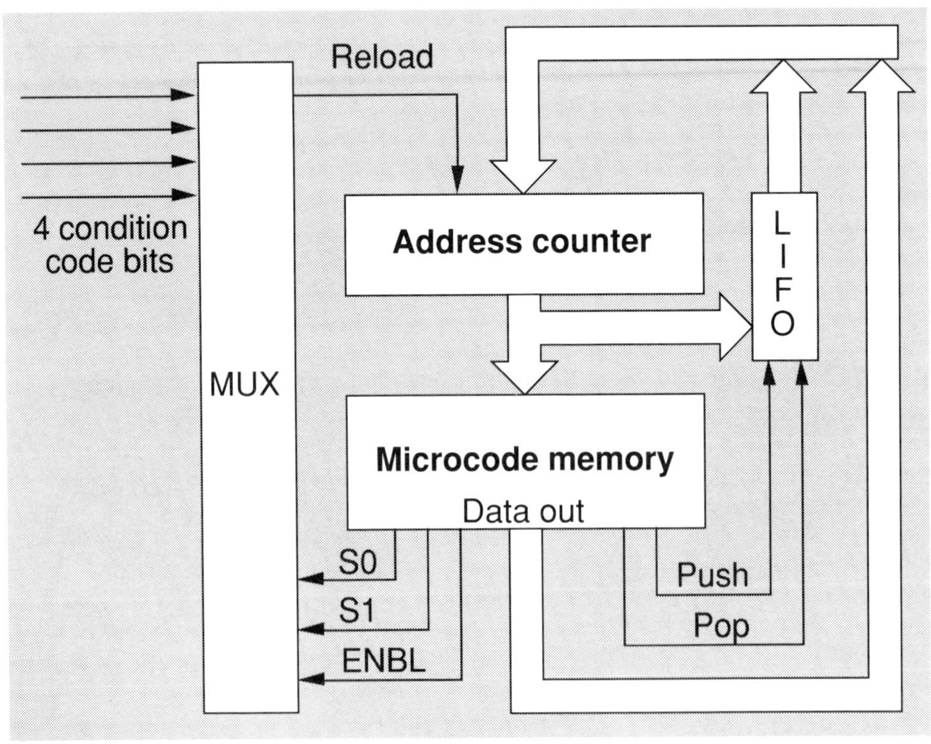

Figure 7

Address bits routed from the microcode memory (or from the *Last In First Out* (LIFO) stack) to the input of the address counter provide the *jump address*. The stack is connected to the output and to the input of the address counter, so it can save (and later restore) addresses. This saving and restoring permits subroutine calls.

This state machine is the central control element of the entire graphics engine. It performs all of the jumps, calls, and returns, and supplies the signal levels that control the rest of the system.

Implementing a pattern transformation engine
The are 4 basic components to a bit-slice pattern transformation engine:

- The pattern scanner
- The color translation logic
- The screen address generator
- The video generator

The pattern scanner
The pattern scanner might consist of a 256 x 256 bit pattern memory, which is addressed by a pair of 8-bit pattern address generators. Each of these 2 address generators can easily be built from an AMD2901 bit-slice processor: one processor calculates the X coordinate values, and the other calculates the Y coordinate values. Each bit-slice processor needs to do 16-bit arithmetic, in order to ensure that the cumulative round-off error doesn't get large enough to distort the screen image. Since only 8 bits are needed to address the pattern memory, there is an assumed binary point in the middle of the 16-bit word.

To display a pattern on the CRT screen, you first load the pattern into the pattern memory, and then load 4 registers and a latch with 5 additional parameters:

- The pattern *step size*
- The pattern *start coordinate*
- The pattern *length* (in pixels)
- The pattern *width* (in pixels)
- The *wrap flag*

The wrap flag determines what will happen when the pattern scanner finishes scanning the last column of the pattern.

If the wrap flag is *set*, the scanner resets itself, and starts scanning at the pattern's start coordinates again. It will then scan the entire pattern again.

If the wrap flag is *reset*, the pattern scanner stops incrementing the pattern (X,Y) coordinates at the last pattern pixel. It will output the intensity value of the last pattern pixel over and over again. This wrap flag can be very useful, as shown in Figure 8.

Each of the pattern scanner's two bit-slice processors performs 3 simple operations: *load*, *step*, and *reset*. Although the 2 processors are driven with a common clock (and run in synchronism with each other) they must each have their own microcode memory, since they usually perform different functions on any given clock cycle.

The Color Translation Logic

All patterns are 2-dimensional 1-bit-deep (black-and-white) images. However, most video terminals are capable of displaying several bits of information per pixel: either as RGB color intensities, or as gray scale. While you could just display black and white images, it's not very difficult to provide a much wider range of colors. All you need is some translation logic to convert the 1-bit pattern bits to multiple-bit video RAM pixels.

Figure 9 shows how this can be done. The level of each output bit from the pattern scanner selects between 2 pairs of 8-bit registers. Each register pair consists of an 8-bit *color register*, and an 8-bit *mask register*.

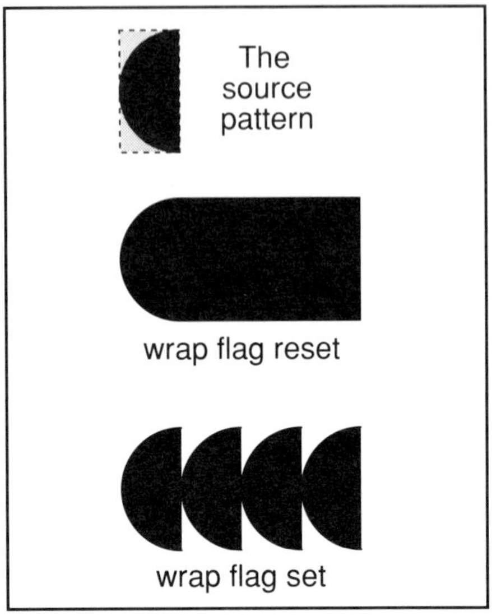

Figure 8

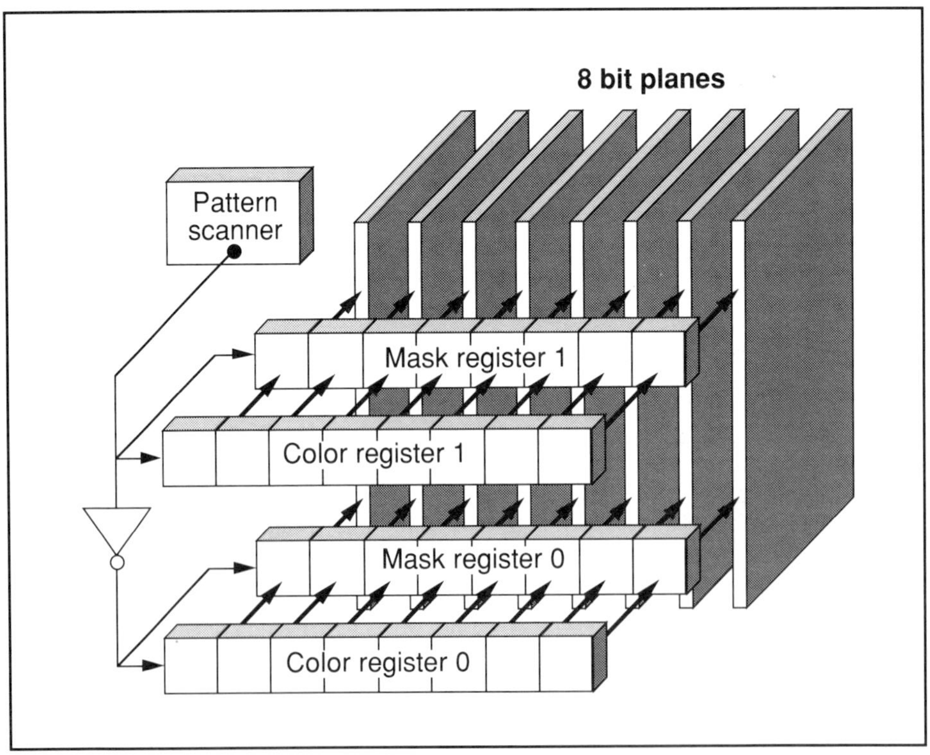

Figure 9

Each bit of the color register corresponds to a single *plane* of the display RAM. Depending upon the type of display terminal, you may be able to display anywhere from 1 to all 8 of these color planes simultaneously. For example, you might use one plane to generate the video bit stream for the *red* input, one plane for the *green* input, and one plane for the *blue* input of an RGB terminal.

Each bit of the *mask* register also corresponds to a single bit plane. It controls whether or not the bit from the color register gets written into video RAM.

To take an example, suppose you've configured your system so that plane 2 generates the red output, plane 1 generates the green output, and plane 0 generates the blue output. You could then configure the 2 register pairs as follows:

color register 0	=	0000 0011	(sky blue)
mask register 0	=	0000 0011	
color register 1	=	0000 0000	
mask register 1	=	0000 0000	

Each time a 0 emerges from the pattern scanner, color register 0 and mask register 0 will be selected. That writes sky blue (green + blue) into the video RAM at the corresponding pixel. Each time a 1 emerges, color register 1 and mask register 1 will be selected. That leaves the video RAM unchanged, since mask register 1 is cleared, to prevent writing to all 8 planes.

This is a very flexible arrangement. It can be used to write simultaneously to any combination of the 8 display planes. It also allows you to copy patterns with *holes* in them into the video RAM, without changing the background seen through the hole. If you set some of the bits in both mask register 0 and mask register 1, you can write one color to screen areas that correspond to the ones of the pattern, and a different color to areas that correspond to the zeros of the pattern.

The Screen Address Generator

While the pattern scanner steps (and reads) through *pattern memory*, the system must also generate the (X, Y) pixel coordinates needed to write into the *video RAM*. That can be accomplished with another pair of 32-bit bit-slice processors. The 32-bit precision of these processors maintains 16 bits of precision after the binary point, so rotations can be done accurately. In addition, the 2 processors are cross-coupled, to allow them to exchange the X and Y coordinates. That's useful when you're drawing circles on the screen.

Each processor has its own microcode memory, allowing it to act independently. Additional hardware (called an *endpoint detector*) can be enabled with a micro-code bit. When enabled, it monitors the screen address during each write operation, and forces the engine to halt when it detects the endpoint coordinates. This is actually an oversimplification. Due to round-off errors, the screen address generator may never generate the precise endpoint coordinates expected by the endpoint generator. The algorithm employed by the endpoint detector checks the X and the Y coordinates independently, during each video RAM access. It stops the screen address generator when both the endpoint X coordinate and the endpoint Y coordinate have been reached, even though they may not have been reached at the same time.

The video RAM

While the pattern is being copied into the video RAM, the system must also maintain a smooth (and very fast) flow of data from the video RAM to the video generator. That's not as easy as it sounds. With both the transformation engine and the video generator competing for access to the video RAM, you have a problem.

The solution is to have several *banks* of interleaved video RAM. If you're going to make interleaving work, you must figure out a way to store the video pixels in the various banks, so that no single bank will be accessed more than once in every 3 clock cycles. (Actually, you can tolerate an occasional access more often than that, as long as the average access rate doesn't exceed 1 in every 3 clock cycles.)

Interleaving works best when the memory locations are accessed in a *predictable* order. Fortunately, the pattern transformation engine typically accesses the screen pixels in a linear fashion. If you can configure the banks of the video RAM so that the pixels along any given line reside in different banks, you can reduce the access rate to any particular bank. Ideally, this interleaving should be effective at all rotation angles.

Suppose you decide to interleave 4 banks of memory: A, B, C, and D. You could assign the screen pixels to the 4 banks as shown in Figure 10. Each box in Figure 10 represents one screen pixel: the memory bank assignments for each pixel are indicated by the letter in the box (A, B, C or D). The pixel arrangement in Figure 10 is actually composed of a repetitive 4 x 4 pixel pattern. (You can see this by comparing the patterns in the 4 quadrants of the pixel array.) This basic 4 x 4 pixel arrangement is repeated over and over, across the entire face of the CRT screen.

The arrows through the array illustrate several line segments that the pattern transformer might draw. Each is labeled with a list of the banks that it encountered as it crossed the array. As you can see, while each segment accesses the banks in a different order, most segments access the banks in a 4-way rotating fashion. In fact, this interleaving arrangement provides an interleave factor of 4 in both horizontal and vertical directions, as well as the 2 primary diagonals. At other angles, the interleaving factor varies, but rarely falls below 2.

Each bank of the interleaved memory is controlled by a state machine. Each state machine can handle up to 3 Mpixels/sec. Given the 4 state machines, the video RAM can be accessed at a rate of 12 Mpixels/sec when the interleaving is optimum. This is a lot faster than the transformation engine can access the video RAM, so there's considerable slack in all but the most unfavorable cases. If these state machines are implemented with PALs, you only need about 20 skinny-DIPs to construct the memory controllers for all 4 banks of 1024 x 1024 video RAM.

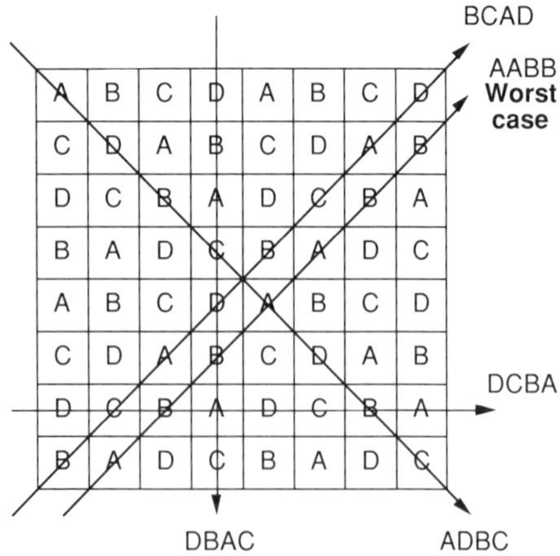

Figure 10

In some cases, the pattern transformation engine may attempt to access a bank before that bank has completed its previous memory cycle. To handle that situation, you need to provide special logic that *stretches* the engine's clock, until the memory cycle completes. Since the expected delay is no more than 150 nsec, this approach makes more sense than to have the state machine enter a polling loop.

The video generator
As mentioned earlier, the screen pixels are arranged in a basic 4 x 4 pattern, which gets repeated over and over, across the face of the CRT screen. Looking back at Figure 10, you can see that all 4 banks of the video RAM must be accessed, in order to get the pixel values needed to generate a single video scan line. This 4-bank access can be done efficiently with a properly designed refresh controller.

If you use 4 video RAM chips (each of which has an internal 256-pixel shift register) you can fetch all 1024 pixels of each scan line in a single memory refresh cycle. The refresh controller gains control of the video RAM at the beginning of each horizontal blanking period and refreshes the memory. It then commands each of the 4 chips to load their 256-pixel internal shift registers.

It's then a simple matter to clock the 1024 pixels out of the 4 shift registers, and into a video DAC, in synchronism with the video clock signal. Brooktree makes a video DAC called the Bt451 which has:

- A high speed (75 MHZ) serial input
- A 256 –> 4096 color look-up table
- Three 4-bit-wide DACs
- A sync combiner

To generate an RGB video signal, all you have to do is feed your digital video data stream (along with a video sync signal and a blanking signal) into the DAC.

But where do you get a good video sync generator? Nobody makes one on a single (or even multiple) chip. The available chips all provide rather crude compromises, which are totally incapable of generating video that approaches broadcast standard sync. Depending upon what type of video display terminal you use, you may also need to generate a special video signal format. In some situations, you may even need to lock your video output onto some external synchronization signal.

To accommodate these requirements (and to provide true RS-170 or RS-343 video sync) you really need a special programmable sync generator. That takes 2 nested state machines; each of which looks much like the microcode address generator. By using this state machine approach, you can easily generate video signals with a resolution of 1/16 μsec on any edge, anywhere in the sync waveform. You can also generate either interlaced or non-interlaced signals, and you can lock these signals onto an external signal.

Feeding the engine
When you design a really fast pattern transformation engine, like the one described above, you find that it's always hungry. To satisfy this hunger, you need to provide some kind of *preprocessor* to feed it. The preprocessor *predigests* the draw list (to get it into

a form that the pattern transforming engine can handle quickly). Then it DMAs the draw list (through an I/O port) to the pattern transformation engine.

The Io pattern transformer

The Io Pattern Transformation Engine (PTE) is shown in Figure 11. The PTE consists of a VMEbus-based 3-board set:

- A *single-board* 68020-based preprocessor
- A *dual-board* microprogrammed engine

The PTE's preprocessor board has 1 Mbyte of dual-ported no-wait-state dynamic RAM. It can pass commands to the engine via the P2 connector, through a dedicated 4K x 16-bit FIFO. Since the preprocessor's software is stored in dynamic RAM, it can also be optimized to suit any given application.

Figure 11

The PTE's address generator is based on the design described in this article. It addresses 2K locations of 96-bit-wide microcode memory, with a cycle time of less than 150 nsecs. The stack permits up to 16 nested subroutine jumps, and there are 12 jump condition codes, which include:

- a carry flag
- a zero flag for both the X and the Y screen addresses
- an endpoint detector
- a loop counter

Since all of the PTE's microcode is stored in RAM, its instruction set can be extended to suit any given application. Having the microcode in RAM also greatly simplifies the debugging of the microcode for these extended instructions. In addition, diagnostic code can be downloaded into the microcode RAM, to test the hardware.

Some special features used on the PTE

In addition to the techniques described in this article, the PTE includes several other features that further enhance its performance. For example, it routes the video RAM plane outputs through a *multiplexer*, on their way to the video DAC. That allows you to double-buffer display images. For example, if you configure the multiplexer to route planes 0-3 to the DAC, you can display the image stored in those planes, while building the next image in planes 4-7. To take a more complex example, you might display 2 planes of text as a constant overlay on top of 2 alternating 3-bit-per-pixel images.

The PTE's 4-way-interleaved video RAM is also designed to permit multiple-pixel writes. This allows you to simultaneously write a single color into all sixteen pixels of a 4 x 4 pixel group. That speeds up area fills and screen clearing. The video RAM chips also permit *page mode* access, which allows you to write to successive pixels in the same *row* of the chip's memory array. This feature can be used to *clear* or *fill* the screen in half the normal time. Using this technique, the screen can be cleared (or area filled) at a rate of 104 Mpixels/sec.

What's a pattern transformation engine good for?
A pattern transformation engine is useful wherever you need fast 2D graphics, as in real-time cockpit instrument simulation for high performance aircraft. (McDonnell Douglas is using a pattern transformation engine for this purpose.) There are also many other, less exotic applications that can benefit from fast graphics.

One example is Computer Aided Design (CAD). While CAD systems generally don't provide real-time animation, they do require a graphics system that draws complex images quickly. The level of human impatience becomes perceptible after about 2 seconds of drawing time, and rises from that point.

Io has developed a PC board layout system which runs on a pattern transformation engine architecture. To give you some idea how quickly this CAD system runs, we have timed it, while it draws the PC layouts for the 3 boards that make up the PTE. Each board is drawn with all pads shown to scale, filled, rounded, and with holes in the middle. All traces are drawn to scale, filled, and radiused at intersections. In other words, the system produces a totally accurate representation in full color on the screen. None of these 3 boards has fewer than 100 chips, and all are 8 layers thick. Yet the system can display each of them (at any scale factor) in under 1/2 second. In other words, it can draw ALL THREE 8-layer boards in less time that it takes you to say "pattern transformation engine".

***Jim Jonas** is vice president of R&D at Io Inc, where he specializes in the design of microcoded systems. He has designed imaging systems based upon a wide variety of input devices, such as CCD cameras and laser scanning microscopes. He also designed the first commercial DNA synthesizer. His hobbies include billiards, and dry French wine.*

***David Forbes** is Chief Engineer at Io, Inc where he is responsible for Io's VMEbus-based products. Dave began his career with microprocessor technology while in high school. In his spare time, Dave enjoys playing bass guitar...in his living room...with his band.*

CHAPTER 21

The VXIbus Hardware Architecture

By David A. Haworth

VXIbus is an extension of the VMEbus open architecture standard. It was developed specifically to support modular plug-in instrumentation allowing you to purchase plug-in instruments from several independent vendors, insert them into a backplane, and operate them as a single system. In this chapter, Dave describes the hardware extensions (the additional signal lines defined by the VXIbus specification) and then describes the protocols used with these lines.

The VXIbus system specification was first published on June 21, 1988. It discusses:

- VMEbus implementation
- Electrical specifications
- Mechanical specifications
- Electromagnetic compatibility
- System power
- VXIbus device operation
- System resources
- VXIbus instruments
- 488-to-VXIbus interfacing
- Command and event formats

It also includes an appendix on backplane design, to help VXIbus designers build backplanes that minimize noise and maximize high frequency performance.

The VXIbus specification allows you to build a wide range of different system configurations. For example, you can build small portable VXIbus systems, consisting of just a few instrument modules, or very large multiple-rack systems, with as many as 255 modules.

Definition of VXIbus terms
Before getting into a detailed discussion of the VXIbus hardware architecture, we first need to define the following terms:

- VXIbus module
- VXIbus device
- VXIbus subsystem

VXIbus modules
The VXIbus module is the basic physical building block from which VXIbus systems are constructed. VXIbus modules come in 4 sizes, called A, B, C,

and D. (See Figure 1) The A and B size modules are similar in size to VMEbus single height and double height boards. However, the C and D size modules are larger Eurocard standard sizes, and have a larger slot-to-slot spacing of 1.2 inches. This 1.2 inch spacing provides enough component height for large analog components and for daughter boards. It also provides enough room for a sheet metal enclosure around each module and a sheet metal chassis shield between module slots.

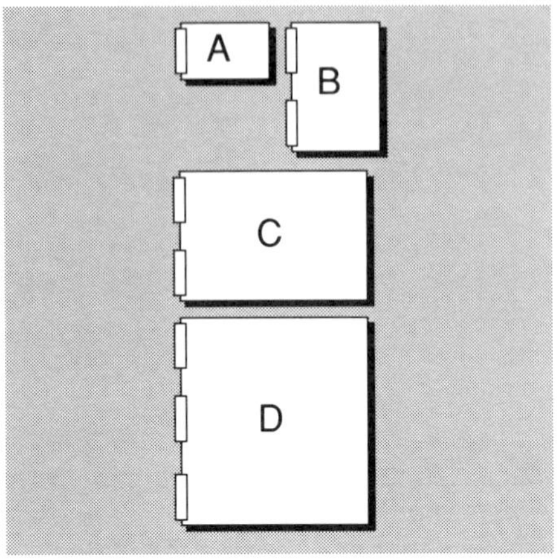

Figure 1

A VXIbus module can be a printed circuit board (similar to most existing VMEbus boards) or it can be housed in a sheet metal enclosure. An enclosed module may contain several printed circuit boards. If an instrument is too large to fit into a single 1.2 inch module, it can be housed in a wider enclosure, which occupies 2 or more slots.

VXIbus devices

While the VXIbus *module* is the *physical* building block for VXIbus systems, the VXIbus *device* is the *logical* building block. A single VXIbus module may contain several VXIbus logical devices. Each VXIbus system can contain up to 255 logical devices. Examples of VXIbus logical devices are:

- MPUs
- Memory
- Disk controllers
- 488-to-VXIbus interfaces
- Analog-to-digital converters
- Digital multimeters
- Counter-timers
- Function Generators
- Logic analyzers
- Scanners

VXIbus subsystems

A *VXIbus subsystem* is housed in a *VXIbus mainframe*, which is the VXIbus equivalent of a VMEbus card rack. A VXIbus mainframe can have a maximum of 13 slots. It houses:

- A slot 0 central timing module
- 1 to 12 instrumentation modules

A *VXIbus system* may consist of several VXIbus subsystems, linked to each other through a VMEbus multicrate interface, or through a local area network.

It's important to understand the distinction between a VXIbus system and a VXIbus subsystem. The instruments in a VXIbus *subsystem* share the same P2 and P3 backplane synchronization signals, and can be tightly synchronized to each other. Instruments in a VXIbus *system* can communicate with each other. However, if they are in different subsystems, they don't share the same backplane synchronization signals. It's very difficult to connect their respective VXIbus synchronization lines (via multicrate interface cables) and still meet the maximum timing delay of 5 nsec, and the maximum skewing of 2 nsec.

The signals on the P1, P2, and P3 connectors

All VXIbus modules connect to the VXIbus backplane through the P1 connector and optionally, through P2, and P3 connectors. Since the VXIbus needs only P1 to operate, only connector P1 is required by the VXIbus specification, regardless of the module size.

P1 provides the 8-bit and 16-bit data transfer capabilities upon which the higher level VXIbus protocols are based. P2 and P3 extends the VXIbus system by adding:

- higher speed 32-bit data transfers
- additional DC voltages
- additional DC power
- local buses, for private high speed module-to-module communication
- system synchronization signals
- a dedicated analog bus
- autoconfiguration capability
- system resources

The VXIbus signals are a superset of the VMEbus P1 and P2 signals. The 2 outer rows of P2 pins are left undefined by the VMEbus specification. VXIbus defines these 2 outer rows of P2 pins, and defines the pins of an additional connector: P3.

The VMEbus signals are grouped into 4 buses:

- Data Transfer Bus
- Arbitration Bus
- Interrupt Bus
- Utility Bus

In a similar fashion, the P2 and P3 VXIbus signals are grouped into 7 buses:

- Clock bus
- Star bus
- Trigger bus
- Local bus
- Analog sum bus
- Module identification bus
- Power distribution bus

The clock bus

The clock bus provides 2 clocking signals and 1 synchronization signal:

- CLK10
- CLK100
- SYNC100

The 10 MHz CLK10 signal is located on P2. The 100 MHz CLK100 signal, and the 100 MHz SYNC100 signal are located on P3. The CLK100 signal is synchronized with the CLK10 signal, so the 2 signals can be used in combination.

Each of these 3 signals are driven by differential drivers, located at the slot 0 end of the backplane. The backplane also provides a buffer for each board slot to minimize the loading of these lines. These built-in buffers ensure that the signal waveforms will not be degraded by excessive module loading, and that variable loading by one module won't cause jitter on the signal, which might affect the timing of some other module using the jittery signal.

When very tight timing control is needed, the CLK100 line is used. The signal transitions on this 100 MHz line are guaranteed to be within 2 nsec of each other along the entire length of the VXIbus backplane.

VXIbus modules can use the SYNC100 signal to gate their incoming CLK100 signal. This allows the slot 0 board to trigger several instruments with a specified CLK100 signal transition. For example, if the slot 0 module asserts the SYNC100 signal before the falling edge of the CLK100 signal, and then negates it before the next falling edge, it can trigger several A/D modules simultaneously. They will all make their measurements on the same CLK100 falling edge.

The VXIbus star bus

The star bus is located on the P3 connector. It's composed of 24 pairs of bidirectional, differential, ECL signal lines. Two pairs (STARX+/STARX- and STARY+/STARY-) connect each of the VXIbus slots to slot 0. (See Figure 2)

The slot 0 board provides a matrix switch, which can cross-connect any pair of differential STAR lines to any other. (It can also cross-connect a single incoming pair to several outgoing pairs, to achieve a broadcast.) Since the maximum delay between the slot 0 module and any other module is 5 nsec, the maximum module-to-module propagation time is 10 nsec.

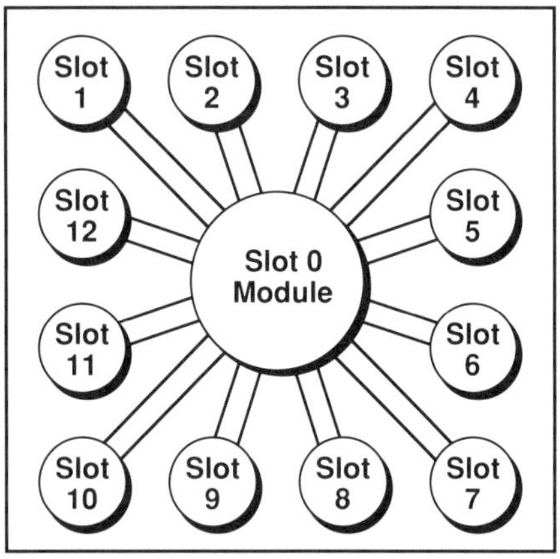

Figure 2

This cross connecting capability allows very high speed signaling between VXIbus modules. These signals can be used for any purpose. For example, one module might sense an incoming strobe signal (from the device under test) and broadcast it over the star bus to several other modules. Those other modules could then use the broadcasted strobe to sample other analog or digital signal lines from the device under test.

The VXIbus trigger bus

The trigger bus has 8 TTL trigger lines and 6 ECL trigger lines. All TTL trigger lines and 2 of the ECL trigger lines are located on P2. The other 4 ECL trigger lines are located on P3.

The VXIbus trigger bus lines are general purpose lines. They can be used either individually or collectively, as a bus. Individual trigger lines can be used for:

- triggering instruments
- clocking

When used collectively, a set of trigger lines can be used as a parallel data transfer bus, while other trigger lines are used to do the handshaking.

The trigger lines are held in an unasserted state by the backplane terminations until they're allocated to a specific VXIbus module by the system software. When a module has been allocated a trigger line, it may then drive that line.

The 8 TTL trigger lines are named TTLTRG0* to TTLTRG7*, and are driven by open-collector drivers. They run the full length of the VXIbus backplane, and have the same terminators as those specified for VMEbus lines. These terminators hold the lines high until some VXIbus module drives them low. Because of their open-collector drivers, these signal lines can only carry clock signals up to 12.5 MHz.

The 6 ECL trigger lines are named ECLTRG0 to ECLTRG5. They can carry clock signals up to 62.5 MHz.

Note: When using the TTL trigger lines to broadcast a clocking signal, you should use the *falling* edge as the synchronizing edge, instead of the rising edge. The rise time on these open-collector signals is slow, and slight variations in the waveform along the backplane can make a significant difference in the time at which various VXIbus modules see their low-to-high transition.

TTL trigger bus protocols

The TTL trigger bus lines are general purpose lines, and manufacturers are permitted to use them as needed. However, the VXIbus specification defines 6 TTL trigger bus protocols:

- Synchronous (SYNC) trigger
- Semisynchronous (SEMI-SYNC) trigger
- Asynchronous (ASYNC) trigger
- Clock transmission
- Data transmission
- Start/stop (STST)

SYNC trigger protocol

The SYNC trigger protocol is used to broadcast a simple trigger pulse to several *acceptor* instruments. It requires no acknowledgment from the acceptor instruments. The maximum trigger repetition rate is 12.5 MHz.

SEMI-SYNC trigger protocol

The SEMI-SYNC protocol is also used to broadcast a trigger pulse, but it also includes a simple acceptor handshake protocol. When the module broadcasting the pulse drives a TTL trigger line low to trigger some operation, each acceptor module holds that same trigger line low, until it's ready for the next operation. Because the trigger line is driven with open-collector drivers, it stays low until the broadcaster and all acceptors release it. When the line goes high again, all acceptors are ready for the next operation.

The SEMI-SYNC protocol can be used to coordinate data transfers over other trigger bus lines. It ensures that both fast and slow acceptor devices can capture the data before the data is removed from the bus. Since the data transfer rate is controlled by the slowest acceptor, no data is lost. When the slowest acceptors on the bus are not involved in a particular data transfer, the transfer rate rises to the limit of the slowest participating acceptor.

ASYNC trigger protocol
The ASYNC trigger protocol allows one module to signal one other module, and receive an acknowledgment. It's done via a pair of TTL trigger lines. The source initiates an operation by asserting the lower-numbered TTL trigger line, and the acceptor then acknowledges on the higher-numbered TTL trigger line.

The ASYNC trigger protocol is especially useful for handshaking between VXIbus subsystems. Although the maximum trigger repetition rate depends upon propagation delay from subsystem to subsystem, it can be as high as 12.5 MHz under ideal conditions.

Clock transmission
A TTL trigger line can be used to broadcast a clock with a frequency up to 12.5 MHz.

Data transmission
Several TTL trigger lines can be grouped together for parallel data transmission. One or two of the TTL trigger lines can be used as strobe or handshake lines, while the rest of the trigger lines are used to transfer the data. If the SYNC or SEMI-SYNC modes are used to transfer data within a subsystem, then 1 trigger line can be reserved for strobing the data. If the ASYNC mode is used to transfer data from one subsystem to another, then 2 trigger lines can be reserved for the strobe and acknowledge.

The data transfers may be synchronized to either the rising edge or the falling edge of the strobe, or to both. For example, you could use 7 TTL trigger lines to route 7-bit data from one module to several others, while using the 8th TTL trigger line for a SYNC strobe. In that case, the maximum broadcast rate would be 87.5 Mbits/sec.

STST protocol
The STST trigger protocol allows you to start and stop a continuous operation in several modules synchronously. One TTL trigger line controls the operation. The slot 0 module might drive the line to its active state to start the operation, and then hold it active until time for the operation to stop. When the line goes active, all participating modules start the operation synchronously on the next CLK10 rising edge. The STST protocol can be used to synchronize the timing of several single-channel instrument modules, allowing them to function as a single time-coordinated multichannel instrument.

ECL trigger lines
The ECL trigger lines run the full length of the VXIbus backplane, and are terminated with 50 ohms to -2 volts at each end. The ECL trigger lines are isolated from other VXIbus lines as much as possible with AC grounds (GND, +5V, -2V, etc) to minimize noise, which causes time jitter. These lines are single-ended ECL, and have a transmission line impedance of 50 ohms. The asserted state is defined as high.

ECL trigger protocols
Like the TTL trigger lines, the ECL trigger lines are general purpose signal lines, and can be used as needed. However, the VXIbus specification describes a set of ECL protocols, similar to the TTL trigger line protocols. These protocols, and their maximum repetition rate are:

- Synchronous trigger: 62.5 MHz
- Semisynchronous trigger: 50 MHz
- Asynchronous trigger: 62.5 MHz
- Clock transmission: 62.5 MHz
- Data transmission: 62.5 MHz

However, the ECL start/stop protocol differs from the TTL start/stop protocol...

Extended start/stop (ESTST) protocol
The VXIbus specification describes an Extended start/stop (ESTST) protocol for use on the ECL trigger lines. This protocol allows higher performance instruments to be synchronized with a 100 MHz clock, as opposed to the 10 MHz clock used for lower performance instruments. The ESTST protocol requires the 100 MHz modules to monitor the TTL trigger line used by the 10 MHz STST protocol, in addition to the ECL trigger line. (This allows 100 MHz instruments to be synchronized to 10 MHz instruments.)

The VXIbus local bus
The local bus is a set of uncommitted signal conductors that provides a private module-to-adjacent-module communication path. (See Figure 3) Except for the end slots (slot 0 and slot 12) each slot has 2 separate local buses: one to the module on its left, and one to the module on its right.

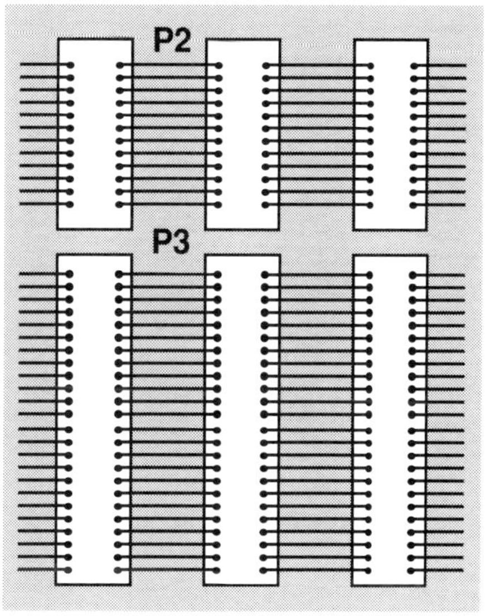

Figure 3

A triple height VXIbus module slot has a total of 72 local bus lines: 36 lines on each side of the module. P2 provides 12 of these lines on each side, and P3 provides the other 24 lines on each side.

Local bus usage
There are 5 classes of local bus usage:

- TTL
- ECL
- ANALOG LOW
- ANALOG MED
- ANALOG HIGH

Figure 4 shows the upper and lower voltage limits permitted for each of these 5 classes.

CLASS	-V LIMIT	+V LIMIT
TTL	-0.50 V	+5.50 V
ECL	-5.46 V	+0.00 V
ANALOG LOW	-5.50 V	+5.50 V
ANALOG MED	-16.0 V	+16.0 V
ANALOG HIGH	-42.0 V	+42.0 V

Figure 4

Why have a local bus?
The local bus provides a versatile method for interconnecting VXIbus modules in a customized way. By using the local bus, manufacturers can avoid the use of front panel ribbon cables between modules. Each local bus is used for a particular purpose, depending on the needs of the particular instrument.

One very useful application for the local bus is pipelined architectures, as shown in Figure 5. Assume the A/D module in slot 6 is converting its front panel analog inputs into a real time digital data stream of 10 bits/word. The data output of the A/D conversion is then routed through the P3 local bus to a digital signal processor (DSP) module in slot 7. The DSP module processes the data (in real time) and outputs the resulting 24-bit floating point values to a memory module in slot 8, again using the P3 local bus. Then a display module in slot 9 gets the floating point values from memory (over the next local bus) and displays them.

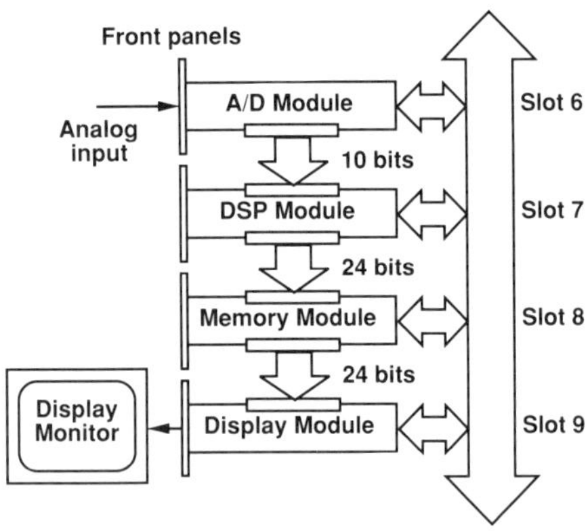

Figure 5

In this example, private module-to-module data paths of different widths are used to allow rapid, concurrent data movement from module to module, without affecting the VMEbus, or any other global VXIbus lines. The result is a measurement system with higher performance than would be possible with the VMEbus alone.

The VXIbus analog SUMBUS
The SUMBUS is a single analog signal that runs the full length of the VXIbus backplane, and is terminated to signal ground through a 50 ohm resistor at each end. Any VXIbus module is permitted to drive this line, using an analog current source driver. The resulting analog voltage on the SUMBUS line is proportional to the sum of the currents injected

into it. Any VXIbus module may monitor the analog voltage level of this line through a high-input-impedance analog amplifier.

The SUMBUS signal line can be used to generate complex waveforms. As shown in Figure 6, four arbitrary waveform generators can be configured and synchronized to output different parts of a waveform to the SUMBUS. The output currents injected by each generator add, producing the final complex waveform. This waveform can then be routed to the front panel of one of the waveform generators, or it can be used by some other VXIbus device. The advantage of this technique is that several, simple, low-cost waveform generators can be used to create very complex waveforms, which otherwise might require a large, expensive waveform generator.

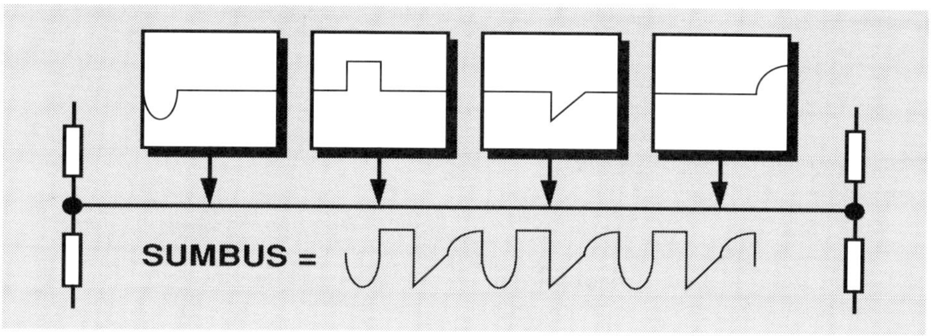

Figure 6

The VXIbus module identification bus
The MODID lines allow the slot 0 module to determine where each logical VXIbus device resides in the VXIbus backplane. These lines run from slot 0 to each of the slots 1 through 12, as shown in Figure 7. A total of 12 MODID lines on P2 connect slot 0 to the other slots in a maximally configured VXIbus subsystem.

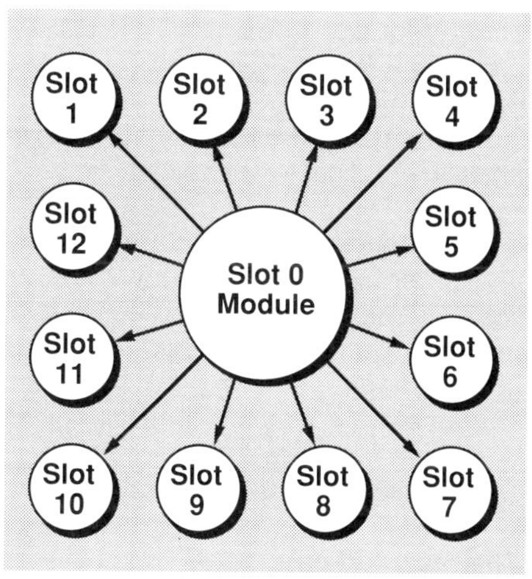

Figure 7

Each VXIbus module is required to provide an 825 ohm resistor between its MODID pin and ground. By determining the resistance between the MODID pin and ground, the slot 0 module can detect the presence of a module in a slot, even if that module has failed. In fact, even if power to the module is lost, the slot 0 module can still detect its presence, as long as the MODID line and the ground return is unbroken.

The slot 0 module determines the physical slot number of each logical device by asserting the MODID line for one slot at a time, and then scanning the A16 map, looking at the MODID bit in each logical device's *configuration register*. The MODID register bit of each logical device indicates the level of the MODID pin of the slot where it resides.

The MODID line can also be used to locate failed VXIbus modules. All that's needed is an LED indicator on the front panel of each VXIbus module that indicates the state of its MODID pin. For example, suppose the slot 0 module determines that there's a module occupying slot 4 but, when it asserts that MODID line, it can't find any configuration register with the MODID bit set. (This indicates that the module in slot 4 is not operating as it should.) The slot 0 module can then simply maintain the MODID line asserted, and alert the user to the failure. The user can then scan the front panel LEDs, to find the failed module.

The VXIbus power distribution bus

The power distribution bus can provide up to 268 watts of DC power to triple height modules. This power is delivered as 7 different regulated DC voltages. Figure 8 shows the voltages and power supplied through each of the VXIbus connectors.

+5V, +12V, and -12V are provided through the VMEbus pins. VMEbus also provides a +5V STANDBY, for devices requiring battery backup. In addition to providing additional +5V, +12V and -12V pins, VXIbus uses P2 to provide +24V and -24V for analog circuits, and -5.2V and -2V for high speed ECL circuits. P3 provides additional power, but no new voltages. While some signal pins of slot 0 differ from slots 1 through 12, the power pins are identical for all VXIbus slots.

These voltages are adequate for most instrumentation. Additional voltages would increase the cost of the VXIbus power supply needlessly. Additional lower voltages (such as the +15V and -15V used for operational amplifiers) can be derived from the +24V, and -24V sources, using onboard regulators. Higher voltages can also be obtained by providing an onboard DC-to-DC converter.

Volts P1	No of pins	Power P1
GND	8	
+ 5 V	3	15 W
+12 V	1	12 W
- 12 V	1	12 W

Volts P2	No of pins	Power P2	Power P1,2
GND	18		
+ 5 V	4	20 W	35 W
+12 V			12 W
- 12 V			12 W
+24 V	1	24 W	24 W
- 24 V	1	24 W	24 W
-5.2 V	5	26 W	26 W
-2 V	2	4 W	4 W

Volts P3	No of pins	Power P3	Power P1,2,3
GND	14		
+ 5 V	5	25 W	60 W
+12 V	1	12 W	24 W
- 12 V	1	12 W	24 W
+24 V	1	24 W	48 W
- 24 V	1	24 W	48 W
-5.2 V	5	26 W	52 W
-2 V	4	8 W	12 W

Figure 8

The reserved pins

The VXIbus has 7 reserved pins. One of these pins is the VMEbus reserved pin in the center row of P2 (RSV1). Two of the additional VXIbus reserved pins (RSV2-RSV3) are on P2, and the remaining 4 reserved pins (RSV4-RSV7) are on P3. The use of these reserved pins will be specified at some future date. Until their use has been specified, VXIbus modules are not permitted to drive or receive signal levels on the reserved pins.

Conclusion

VXIbus has been designed to allow users to purchase instruments from many independent vendors and easily integrate them into a compact, high-performance instrumentation system. It permits the instruments to be tightly synchronized, and its local buses also permit instruments to be configured in a pipelined architecture. This tight time coordination, coupled with an ability to transport data from instrument to instrument very quickly, will permit the use of modular instrumentation in many new applications.

David Haworth is the VXIbus program manager for the Measurement Systems Division at Tektronix. He has worked in the Instrument Group for the last 10 years. Before coming to Tektronix, David taught for 5 years at Purdue University School of Technology.

CHAPTER 22

VXIbus Software Architecture

By Ron Wolfe

From a physical point of view, the VXIbus looks very much like VMEbus. However, VXIbus is more than just a backplane architecture. It includes message-passing and command protocols that permit VXIbus modules to communicate at a high level. These protocols are an extension of the GPIB protocols that have been in use for years. In this chapter, Ron describes how these message-passing protocols work, and how they're tied into the underlying VXIbus hardware architecture. He also describes how VXIbus systems can be supervised by existing GPIB controllers, allowing you to use VXIbus and GPIB technologies together.

The VXIbus registers

The primary difference between VXIbus modules and VMEbus modules is that VXIbus modules have a set of four 16-bit configuration registers, as shown in Figure 1. (Some VXIbus devices also have a set of 6 additional 16-bit registers for message-passing, but this is optional.)

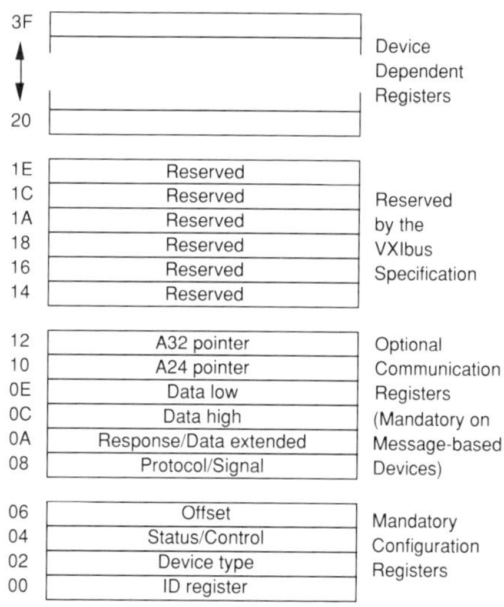

Figure 1

The upper 16 Kbytes of the VMEbus A16 address space is reserved for the registers of VXIbus devices. Each VXIbus device has assigned to it an 8-bit logical address. This logical address may be manually set (via DIP switches) on the device's module, or it may be automatically configured by the system upon power up. The 8-bit logical address specifies where the device's registers appear in the 16 Kbyte address space. This allows up to 256 VXIbus devices in a single VXIbus system. These 8-bit addresses are analogous to the addresses of GPIB devices.

VXIbus modules always have configuration registers

The VXIbus configuration registers are required for all VXIbus devices. They allow the system to identify each VXIbus device, its type, its model, its manufacturer, its address space, and its memory requirements. These configuration registers permit a resource manager device to automatically configure the system upon power-up, including the allocation of the VXIbus memory map.

VXIbus communication modes
To configure and control VXIbus instruments, the VXIbus defines 3 communication modes:

- Register-based
- Message-based word-serial
- Shared memory

Register-based communication
Register-based VXIbus modules are configured and controlled through status and control registers, in a fashion similar to many current VMEbus peripheral boards. While the VXIbus spec clearly defines the functions of the status and control registers used for configuration, it makes no attempt to dictate the function or the number of status and control registers used for communication.

Message-based communication
Besides the mandatory configuration registers, the VXIbus specification also defines an optional set of communication registers, used for message passing. VXIbus modules equipped with these communication registers are called *message-based devices*.

The VXIbus specification prescribes a word-serial protocol, for use when passing messages through these communication registers. The word serial protocol is similar to the GPIB byte-serial protocol, and is based upon the register model of a full duplex serial port. Each message-based device has a 16-bit *data-in register* which other devices can write to, and a 16-bit *data-out register* which other VXIbus devices can read. Handshake bits in the device's *response register* are used to keep things coordinated.

The protocol is called word-serial because the command words and data words are passed through the data-in/data-out registers in a serial fashion.

Shared Memory communication
The VXIbus specification defines commands that allow VXIbus message-based devices to progress to a more sophisticated communication protocol. Using the *shared memory commands*, 2 VXIbus devices can establish a shared buffer in dual-ported VXIbus memory. Once this buffer has been established, one device writes data into the buffer, and then passes a pointer to the other device. This shared memory communication is much faster than the word-serial communication because it eliminates the need to transfer large amounts of data from module to module over the VXIbus.

A VXIbus power-up sequence
The left-most slot of a VXIbus system must always be occupied by a module with VXIbus *slot 0 capabilities*. This slot 0 module provides special system resources such as:

- 10 MHz and 100 MHz backplane clock signals
- MODID system configuration signals
- Synchronization (trigger) signals

The VXIbus *resource manager* is typically implemented in software. It works in combination with the slot 0 device to bring up the system, and to allocate the system resources.

When the system powers up, the resource manager is the only device that's allowed to access the VXIbus.

Identifying the devices
The resource manager's first duty is to identify each device in the system, by searching the VXIbus configuration space (the upper 16 Kbytes of the A16 address space) and locating the configuration registers of all VXIbus devices. Using the MODID lines, it can also assign 8-bit logical addresses to VXIbus devices that support dynamic (switchless) configuration. If all the devices are switchless, it can configure the entire system without any DIP switch setting by the user.

Initiating self-test
After the resource manager identifies each device, it initiates a self-test verification procedure for each device, to confirm that the entire system is operational. The resource manager then reads bits in each device's configuration registers to determine the device's self-test status. If a device fails its internal self-test, the resource manager can use register bits to reset it, or to put the device into a diagnostic state. If the device's health is still in question, the resource manager can even take the device off-line, where it won't interfere with VXIbus activity.

Configuring the address map
Other duties of the resource manager include dynamic placement and management of system memory resources. Each device's configuration registers indicate whether that device has onboard dual-ported memory. If it does, the resource manager finds an unoccupied area of the address map, and writes the base address of that area to the device's offset register. In this fashion, the resource manager dynamically configures the system address map.

Allocating the interrupt lines
The VXIbus specification defines word-serial commands and queries that allow the resource manager to dynamically configure interrupt lines and assign interrupt handler responsibilities to the various devices in the system.

Establishing the chain of command
The resource manager uses word-serial commands to establish a chain-of-command within the system. It does this by *granting servants* to the *commanders* in the system.

The resource manager starts the system running
Once the resource manager has...

- Identified all the devices
- Verified the self-tests
- Configured the address map
- Allocated the interrupt lines
- Established the chain of command

...it initiates system operation by sending the *begin normal operation* word-serial command to all top-level commanders to start the system. Once the system begins normal operation, the resource manager's job is essentially finished. However, it can halt the system, rework the chain-of-command, and then restart the system, if necessary.

The Commander/ Servant hierarchy

The VXIbus hierarchical chain-of-command is an inverted tree, as shown in Figure 2. A *commander* is any device in the hierarchy that has one or more *servant* devices granted to it. A servant is any device under a commander. If the hierarchy has several levels, some devices may be servants to one commander and at the same time, commanders to lower-level servants.

Each VXIbus device can have one and only one commander. A commander has exclusive access to its servant's communication and control registers. Commanders can communicate with message-based servants through the servants' communication registers, using the word-serial protocol. Servants also communicate with their commander through the word-serial protocol. Servants can also communicate asynchronous status and events to their commander through hardware interrupts, or by writing specific messages directly into their commander's signal register.

The VXIbus commander/servant concept provides a mechanism for coordinating the actions of several VXIbus devices. This allows the instruments to collectively perform complex operations, while communicating at the top level through a single commander, via a word-serial command set. For example, servants lower in the hierarchy might perform data acquisition and data processing functions (using shared memory techniques) while synchronizing their operation, using the VXIbus backplane trigger signals.

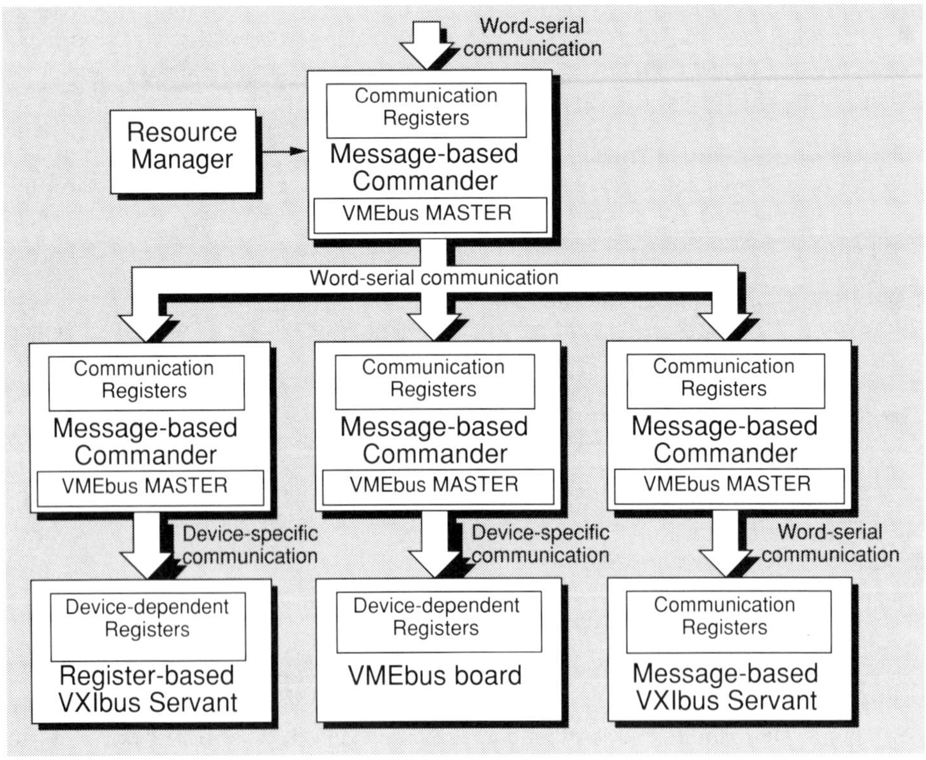

Figure 2

Controlling VXIbus with a GPIB controller

The 8-bit VXIbus logical addresses are similar to the 5-bit GPIB addresses. Also, the VXIbus word-serial communication protocol is very similar to the byte-serial GPIB protocol. Because of this, it's fairly simple to design an interface that translates GPIB commands and addresses into VXIbus commands and addresses. A properly designed GPIB-to-VXIbus interface allows a GPIB controller to control VXIbus instruments transparently: the VXIbus instruments appear to be traditional GPIB instruments.

To ensure that all GPIB commands can be interpreted by VXIbus instruments, the VXIbus specification requires all VXIbus message-based devices to understand a *core set* of word-serial commands. The word-serial commands in this core set allow:

- Data transfer to and from the instrument
- Service requesting by the instrument
- Status polling of the instrument
- Clearing of the instrument
- Triggering of the instrument
- Local (front panel) lockout
- Error reporting

The VXIbus specification describes, in detail, how to map the GPIB commands onto the VXIbus core command set. For example, GPIB commands for...

- Clearing devices (DCL)
- Triggering groups of devices (GET)
- Remote/local control of devices (RL1)
- Service requests (SRQ)
- Serial polling (SPOLL)

...are to be translated directly into equivalent word-serial VXIbus commands.

Additional VXIbus commands

Unlike GPIB instruments, VXIbus instruments can communicate through shared memory, and can tightly synchronize their operations using the VXIbus trigger lines. To support these additional features, the VXIbus command set is an extended version of the GPIB command set. These extensions facilitate:

- Configuring interrupt lines and interrupt handlers
- Constructing a chain-of-command
- Beginning and halting the operation of the system

Using VMEbus boards in VXIbus systems

Devices that don't have configuration registers (such as VMEbus modules) are called *foreign devices*. While foreign devices work fine and, in fact, will be commonly used in VXIbus systems, they can't be recognized and automatically configured by the resource manager. However, the commander/servant concept allows an intelligent message-based VXIbus device to function as a commander, accepting word-serial commands on behalf of a register-based device, and then performing the needed register accesses on the register-based device, to accomplish the commanded operation. Using this method the commander can provide a word-serial interface for VMEbus boards.

The National GPIB-VXI

National Instruments has developed a C-size board, called the GPIB-VXI. (See Figure 3) When plugged into a VXIbus mainframe, it permits a GPIB controller to supervise a VXIbus system by transparently converting the controller's GPIB commands and addresses to VXIbus commands and addresses.

The GPIB-VXI hardware includes:

- a 68070 processor,
- 128 Kbytes of onboard EPROM (expandable 512 Kbytes),
- a private bus for local resources,
- dual-ported VXIbus configuration registers,
- dual-ported VXIbus communication registers that permit word-serial message-passing,
- up to 4 Mbytes of dual-ported RAM for shared memory communication with other VXIbus modules,
- a GPIB interface,
- VXIbus slot 0 capabilities,
- VMEbus slot 1 (system controller) capabilities.

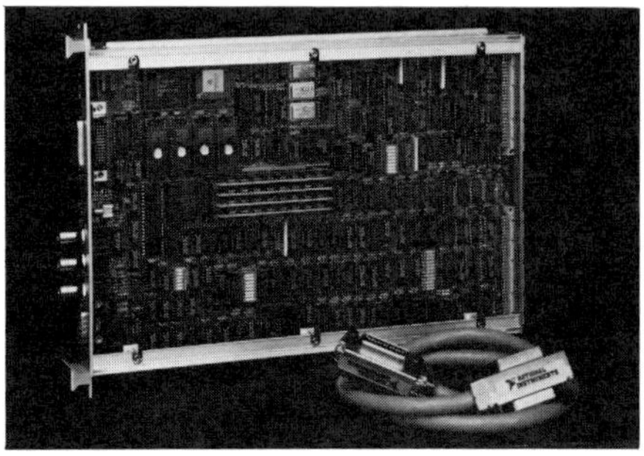

Figure 3

The 128 Kbytes of EPROM contain firmware based on the pSOS real time multitasking operating system.

The slot 0 capabilities permit the GPIB-VXI board to be used as the top-level commander of a VXIbus system, where it performs the VXIbus power-up sequencing and resource management functions, including:

- VXIbus device identification
- Address map configuration
- System self-test management
- Configuration of commander/ servant hierarchies
- Initiation of normal operation
- Configuration and control of the VXIbus trigger lines

The board's slot 0 and system controller capabilities can also be disabled with jumpers, allowing the board to be installed in any slot of a VXIbus system. In that case it can function either as a commander, or as a servant.

The board also allows you to download (or embed) custom software, to create *virtual instruments* that perform complex operations at VXI speeds, yet are controlled with high-level GPIB commands. Once the software is downloaded (or burned into EPROM) these virtual instruments can accept and respond to commands from an external GPIB controller, just as though they were GPIB instruments.

The GPIB-VXI board also permits you to download (or embed) software that provides GPIB-compatible interfaces for register-based VXIbus devices, or for VMEbus boards. With this software, these devices appear to the GPIB controller to be full-function GPIB-compatible instruments, with their own GPIB address.

The onboard hardware and firmware provide a completely transparent interface between a GPIB system and a VXIbus system. This enables VXIbus instruments to be integrated into existing GPIB systems, and to be controlled with existing GPIB controllers and application software. The GPIB-VXI's ability to let users embed their own code inside the mainframe is an important factor. It allows VXI to be used as a platform for developing very sophisticated "super instruments." The user can take advantage of VXI to develop modular, reconfigurable instruments that operate at VXI speeds, yet are controlled using simple high-level GPIB commands. System, integrators can take advantage of VXI's performance, yet hide its complexity from the end user.

In addition to the typical 5-bit *primary addresses* used in most GPIB systems, the GPIB standard also defines 5-bit *secondary addresses*. The GPIB-VXI can be configured to interpret the 5-bit GPIB primary addresses as mainframe addresses (which select one of several VXIbus mainframes attached to the GPIB system) and to map the 5-bit GPIB secondary addresses as 8-bit VXIbus logical device addresses, within that mainframe.

VXIbus programming considerations

Like the GPIB, VXIbus is attractive because it permits you to integrate modular instruments into automated systems. Programming GPIB-based VXIbus systems will require, at the very least, programming skills similar to those for GPIB. However, to take full advantage of VXIbus, you'll need more sophisticated tools than were available for GPIB systems, which were often programmed in easy-to-learn languages such as BASIC. In fact, the acceptance of VXIbus instruments may be limited to high-end applications if the proper software tools are not made available, so that average users can take advantage of VXIbus features.

One problem you will face is that VXIbus instruments don't have the traditional interactive front panel controls and indicators. However, by using graphics on a CRT screen, you can simulate such front panels. This idea is not really new. Software that simulates front panels has been evolving over many years for remote control of GPIB instruments. This software is ideally suited to VXIbus. One example is LabVIEW, which builds virtual instrument hierarchies, complete with animated front panels that represent the remote instruments. (See Figure 4) Another example is LabWindows for IBM PC machines. It uses conventional test-based languages, but enhances them with interactive graphics (including panels with switches) to provide virtual front panels.

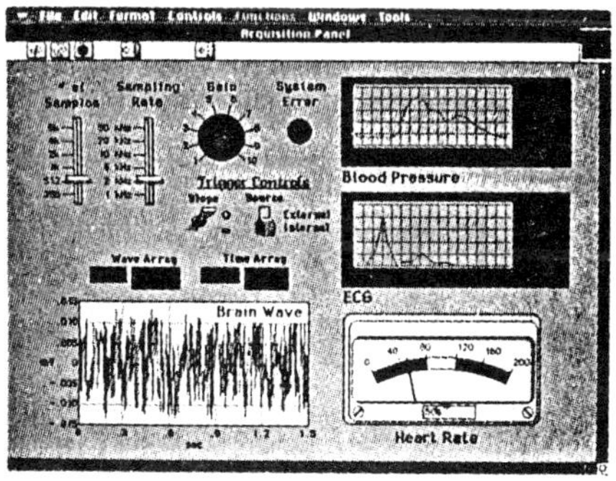

Figure 4

Conclusions

While VXIbus instruments are unlikely to replace all GPIB instruments soon, they will be used in conjunction with the GPIB instruments. Eventually, VXIbus will be as important to the instrumentation industry as GPIB has been. VXIbus allows the design of low-cost, non-intelligent, register-based modules, as well as high-performance, intelligent, message-based modules. Eventually, VXIbus modules will not only span the range of available GPIB instruments, but will also include digital I/O, analog I/O, counters, waveform recorders, processors, communication gateways, memory, and display modules.

Ron Wolfe is the VXIbus product manager at National Instruments. Before assuming his present position, he served 3 years as a hardware design specialist, and 2 years as a product manager for the LabVIEW graphical programming system. He holds a BSEE from the University of Texas.

CHAPTER 23

A Software Foundation For VXIbus

By Glen Myers

VXIbus is truly an extension of the VMEbus standard. It specifies protocols that provide features such as message passing and autoconfiguration. In most VXIbus systems these protocols will be implemented in software. This will make VXIbus systems considerably more complex to implement than earlier GPIB systems: you will need the protocol software, as well as a processor to run it on. In this article, Glen describes the software that you'll need, and suggests that installing an IBM PC-compatible VMEbus board in your system will provide the hardware and software platform upon which to build these protocols. He also points out the benefits that such an approach provides, when it comes to debugging the hardware and software in your VXIbus system.

VXIbus-based systems are typically composed of a collection of intelligent instruments. Just as in GPIB systems, there must be a system controller that runs the test algorithm, sends commands to the various instruments, triggers readings, and records the results. However, VXIbus instruments, unlike their GPIB predecessors, typically have very few front panel controls. At the same time, these newer instruments are, in many cases, more full-featured and more complex than earlier ones. While the VXIbus specification allows the design of non-intelligent VXIbus instruments, most include a processor, in order to take full advantage of the VXIbus features.

GPIB systems are typically controlled by a desk-top PC
During the last few years IBM PC-compatibles have become the primary choice for controlling GPIB-based laboratory and automatic test systems. In addition to the enormous software base that PCs offer, most engineers have used PCs, and feel comfortable with them.

Several companies have recognized this, and have introduced high-level graphical user interface packages specifically for use on GPIB-based systems. Some examples are National Instrument's LabWindows, Wavetek's WaveTest, and Summation's TestWindows.

The use of a desk-top PC to control VXIbus systems is not optimum
The VXIbus designers patterned the VXIbus protocols after the GPIB. That simplifies the job of interfacing VXIbus to GPIB systems. In fact, it's fairly straightforward to connect a PC-based GPIB controller to a VXIbus system. This allows the control of VXIbus instruments in the same way (and with the same controller software) as GPIB

instruments. However, if you want to take full advantage of the VXIbus features, this may not be the best approach.

While VXIbus uses protocols very similar to GPIB, its high-speed backplane allows for much faster command and data transfers. Much of this benefit is lost if the interface between the system controller and the remainder of the system is over the slower GPIB. However, there's more to the problem than just a performance bottleneck. When you study the VXIbus specification, you soon realize that its protocols were designed with the assumption that there would be a processor (and some software) residing inside the VXIbus card rack.

Resource manager software

One required software package is the resource manager. The start-up resource manager is the first program to run after the operating system is booted. For the most part, its functions are dictated by the VXIbus specification. It is responsible for finding and configuring all VXIbus devices at system power-up or reset. Each VXIbus device has an 8-bit logical address, which positions that device's *configuration registers* in the A16 address space. On some instruments these logical addresses are statically defined (via DIP switches). On others, they can be dynamically assigned by the resource manager when the system is powered up. (It does this by using the VXIbus geographic addressing to select the VXIbus module in each slot, and then looking to see whether there are any dynamically configured devices in that slot.)

Unlike VMEbus devices, whose address mapping is static, VXIbus devices specify, in their configuration registers, how much of the global address space they need for local dual-ported RAM, status/control registers, etc. The resource manager determines how much global memory each device needs, and then decides where each device should reside in the overall memory map. Then it writes into a base register on each device, to position that device at the appropriate place in the A24 or A32 address spaces. The resource manager also performs a series of other initialization functions, such as monitoring the self-test status of all devices, and establishing a hierarchy of devices.

Special software is needed to incorporate VMEbus boards

VXIbus was designed to allow the incorporation of VMEbus boards. It's quite likely that many VXIbus systems will incorporate VMEbus memory boards, A/D converter boards, and networking boards. But VMEbus devices don't have the VXIbus configuration registers, so the resource manager won't be able to detect their presence at start-up, and won't be able to figure out what areas of the system memory map they have already claimed. Because of this, the resource manager might configure VXIbus devices to occupy the same address ranges as the VMEbus boards.

One solution to this problem is to have special files that contain the same information about the VMEbus boards that would normally be provided by VXIbus configuration registers. With this information, the resource manager can avoid assigning VXIbus devices to areas already occupied by VMEbus modules, and can provide a consistent logical interface for all modules on the VXIbus. With this consistent interface, the application programmer's job is greatly simplified. The configurator also collects information about the static usage of specific interrupt lines by VMEbus devices, in order to figure out how to allocate usage of those lines.

Shared memory manager software
Another complex piece of software that is specified in detail (but defined as an optional feature) is a shared memory message manager. VXIbus defines a multitude of command words, which are passed over the backplane, using a serial message-passing protocol. These commands are a superset of the GPIB commands, and the specification also describes how to translate GPIB commands into VXIbus commands. In addition to these commands, it defines an optional (considerably more complex) protocol which can be used for moving large amounts of data, using shared-memory queues.

The VXIbus specification implies (but does not detail) additional software
Although the VXIbus specification defines the software environment to a certain degree, much of the needed software is left up to the ingenuity of the manufacturer or system integrator. For example, although the interactions between the resource manager and the VXIbus plug-in devices are defined, the interactions between the resource manager and other higher-level pieces of software are undefined. Once the resource manager has determined the physical makeup of the system (and has configured the devices) higher-level software needs access to information about the system makeup, in order to request system resources. Also, some hardware resources, like the interrupt lines, and the TTL and ECL trigger lines, can't be shared. The specification doesn't explain how higher level software can find out who the lines have been assigned to.

As mentioned earlier, the resource manager must have some means of accessing information about VMEbus boards. However, the VXIbus specification provides no guidance on how this should be done. If it is done with descriptor files, no format for these files is prescribed.

Finally, there is a considerable amount of implied but unspecified software needed for:

- generating and servicing interrupts
- generating and servicing signals
- sending and receiving serial messages
- performing self tests
- interpreting *reset*, *diagnostic*, and *safe* word-serial commands

As an example, because VMEbus (on which VXIbus is based) has a limitation of only 7 hardware interrupt request lines, VXIbus defines an optional alternative mechanism called *signals*, which uses interrupt registers. If a VXIbus device has a signal register (or better, a signal register FIFO) it can be interrupted by simply writing the equivalent of a VMEbus status/ID value into the device's signal register. Ideally, the software at both the sending and receiving end should treat IRQ-line interrupts and signals in the same fashion, so the application programmer doesn't need to know by which method the interrupt is being sent.

Run-time services
Some additional software is needed to provide a variety of services to application programs, after the system has been configured. For example, it would be useful to have software that would provide a communications path between application software and VXIbus devices, while hiding the details of their hardware interfaces. To accomplish this,

you would need to have a set of subroutines that application programs could call whenever they need to:

- send messages to devices
- set up interrupt service routines to respond to signals and interrupts
- copy blocks of memory between the application's data area and the VXIbus global memory
- sense and control the levels of SYSFAIL*, SYSRESET*, BERR*

Development tools

To get a VXIbus system up and running, you'll probably also need some special-purpose development tools. For example, if you have no experience with a certain VXIbus instrument, you may want to be able to control it manually. (This allows you to manually verify the procedure used to control the instrument, before you start writing software.) To operate a VXIbus instrument manually, you need software that allows you to send the instrument a VXIbus message, a signal, or an interrupt, and then watch its response.

You might also want a bus analyzer that continually samples and displays the level of the VXIbus signals, so you can observe the bus activity. By watching such an analyzer, you can get an overall feel for the real-time dynamics of the system. It also allows you to see the cause of a system (or an instrument) hang-up. For example, you can see a stuck interrupt line or a stuck trigger line.

Obviously, if you have to design all of this VXIbus system software, diagnostic hardware, and diagnostic software, you might be a long time getting your VXIbus system running. It would be helpful to have a VXIbus module that you could plug into the bus that would provide you with all of these features.

The EPC-2

The EPC-2, from Radisys, is just such a module. (See Figure 1). It is an IBM PC-compatible VXIbus module that plugs directly into the backplane of a VXIbus mainframe. It includes:

- A 32-bit 80386 processor
- 8 Mbytes of dual-ported DRAM, accessible from VXIbus
- A 40 Mbyte hard disk
- A 3.5" floppy disk drive
- A VGA graphics controller (800x600)
- A GPIB controller
- 2 serial ports
- A parallel port
- A keyboard controller and connector
- A time-of-day clock
- EPROM with PC, SCSI, and VGA BIOSs
- A socket for Intel's new Flash EPROM.

The EPC-2 is packaged as a shielded 2-slot (2.4" wide) module for C-size or D-size VXIbus systems.

The EPC-2 can:

- access all of the VMEbus-defined address spaces
- generate and service all 7 of the VMEbus interrupts
- serve as the slot-1 VMEbus system controller
- serve as a message-based commander
- provide a signal register (with a 2-deep FIFO)
- send and receive 32-bit longword-serial messages
- pass and accept shared memory pointers

It can also be configured as the slot 0 controller, in which case it:

- drives the 10 MHz ECL clock, either from an internal oscillator, or from a front panel external input.
- drives the module identification (MODID) lines, which are used for autoconfiguration.
- drives 8 TTL and 2 ECL trigger lines.

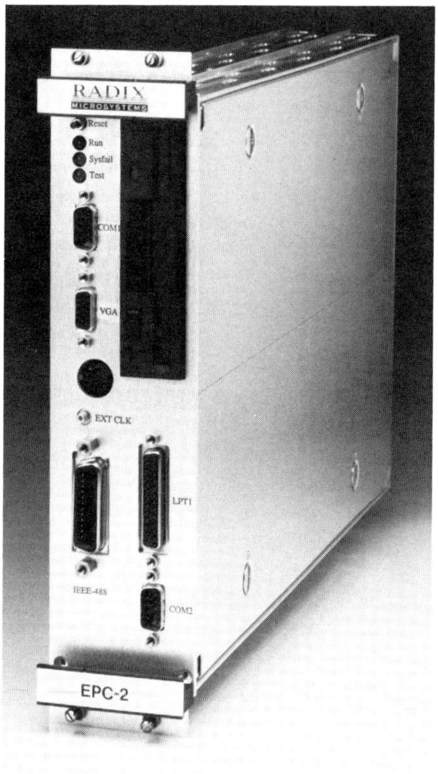

Figure 1

EPC-2 software

EPC-2 comes with a substantial amount of software for developing, debugging, and operating VXIbus systems:

- A start-up resource manager
- A configurator
- An on-line resource manager
- A bus interface manager

In addition to the start-up functions described in the VXIbus specification, the EPC-2's resource manager software performs 2 other functions:

- Naming of devices
- Checking for errors

Naming of devices

Device naming is important because, with dynamic configuration, a device may not always be assigned the same logical address. The device name provides a configuration-independent way for application programs to refer to a device.

The configuration file provides names for instruments in the system. If the EPC-2 resource manager finds no name for a given instrument during system configuration you can have it either assign an arbitrary name, or prompt you for a name.

Checking for errors

The EPC-2's start-up resource manager also performs a thorough check for errors, ambiguities, and undefined states: an important consideration in a system with a new bus containing a lot of devices from a number of manufacturers. The resource manager finishes by displaying the system configuration (and any discovered errors) on the PC screen, and by leaving data structures in memory for the on-line resource manager.

For the display of devices, actual manufacturer and model names are used. This information comes from a file of manufacturer encodings, and model number encodings; the user can update this file as new manufacturers and VXIbus instrument models appear.

The EPC-2 configurator

As mentioned earlier, you can greatly simplify the incorporation of VMEbus boards by providing special files that contain information about the VMEbus boards, which would normally be provided by VXIbus configuration registers. By accessing this information, the resource manager can avoid assigning VXIbus devices to VMEbus memory areas, and can provide a consistent logical interface for all modules on the VXIbus. But how do these files get created?

The EPC-2's *configurator* is a Windows-based program that permits you to create these configuration files by entering information about VMEbus boards that's not otherwise available to the resource manager. Figure 2 shows how this information is entered by the user.

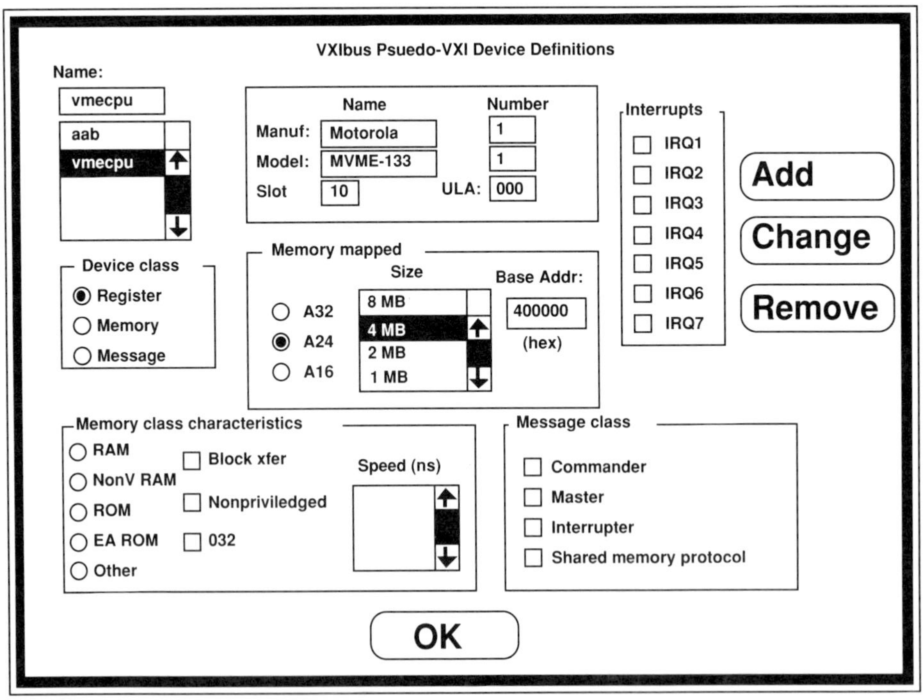

Figure 2

The on-line resource manager

After the resource manager has configured the system at start-up, and normal system operation ensues, the *application software* needs some way to find out what resources are available in the system. For example, what instruments are available? What trigger lines and interrupt lines are available? The EPC-2 software provides this service with a program that resides in the system during normal operation. It's called the *on-line resource manager*. The on-line resource manager consists of a set of routines that the application software can call from either C or BASIC languages. Figure 3 shows how you would call some of these functions from C.

The services provided for VMEbus devices by the configurator (and by the on-line resource manager) provide a uniform application code interface for both VXIbus and VMEbus devices. Also, information about VMEbus board interrupt line usage allows the resource manager to avoid assigning the same lines to VXIbus devices.

The bus interface manager

At the lowest level, the bus interface manager deals with the EPC-2's bus mapping hardware. Mapping hardware is needed because application programs running on the EPC-2's 80386 processor need access to the full range of the VXIbus A16, A24, and A32 address spaces: the sum of which is larger than the 80386's address space. Additionally, when running DOS, these address spaces must be accessible within the 1 Mbyte DOS space.

```
/** Get model number of symbolic device "scannera" **/
i = OLRMGetNumAttr("scannera",0,OLRM_MODEL);

/** Determine if device "wavegen" is a commander **/
if (OLRMGetBoolAttr("wavegen",0,OLRM_CMDR));

/** Get number of devices in the system, and name of 1st device **/
devices = OLRMGetDevCount();
OLRMGetDevName(&devicename,0);

/** Get manufacturer's name of device "wavegen" **/
OLRMGetStringAttr(&manufname,"wavegen",0,OLRM_MANUFACTURER);

/** Try to allocate ECL trigger line 0 **/
i = OLRMAllocate(OLRM_ECLTRG0);

/** Allocate any 2 adjacent TTL trigger lines **/
i = OLRMAllocate(OLRM_TTLTRGANY2);

/** Allocate interrupt line IRQ7 **/
i = OLRMAllocate(OLRM_IRQ7);

/** Give resource i back to the resource manager **/
OLRMDeallocate(i);

/** Send the byte i to the device "scannera" **/
BIMSendMSG("Scannera",0,MSG_byteAVAIL|i);
```

Figure 3

To make this possible, the bus interface manager provides a relocatable 64 Kbyte window that can be positioned anywhere in any of the VXIbus address spaces. Also, because the system might contain devices with a byte ordering opposite to the 80386 (for example, 68000 family MPUs), the mapping hardware allows application software to view VXIbus memory with either Intel or Motorola byte ordering.

The EPC-2's system debugging facilities
The EPC-2 software provides 2 packages that assist you in debugging the system:

- The bus monitor
- The bus probe

These are Windows applications that allow you to watch and interact directly with the VXIbus and with physical devices.

Figure 4 illustrates how the bus monitor window appears to the user on a PC display. The bus monitor samples the levels on the VXIbus lines, and displays the information in its window. This information includes the current state of:

- all interrupt and trigger lines
- the EPC-2 VXIbus registers
- the address-mapping hardware
- the EPC-2 interrupt controller
- the EPC-2 trigger drivers

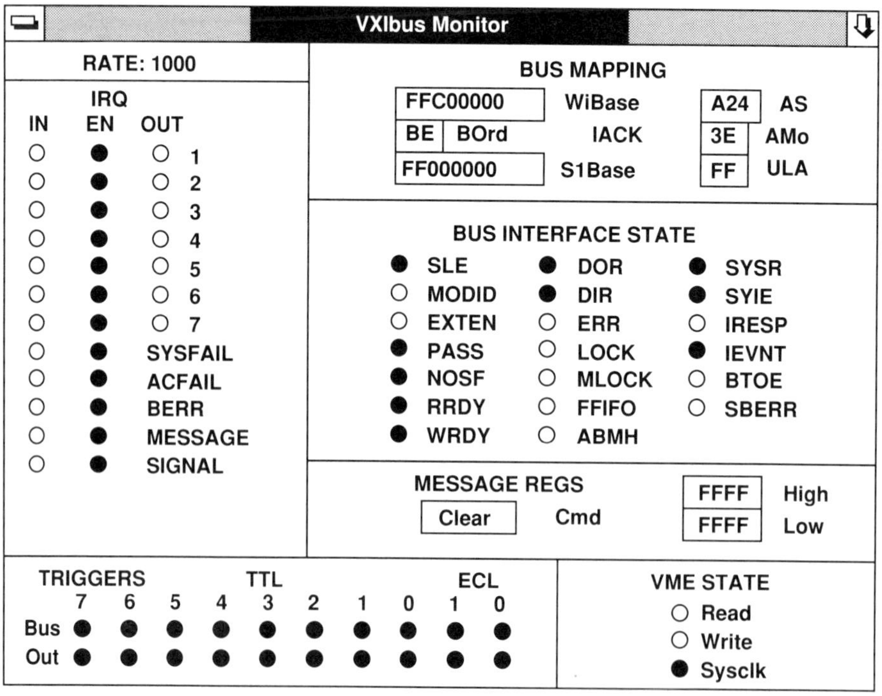

Figure 4

It also indicates the current type of data transfer (read/write) and the last message received.

While the bus monitor is strictly passive, the bus probe is its active counterpart. The bus probe allows you to read or write to memory, assert interrupts and trigger lines, send messages and signals to devices, and reset devices.

The EPC-2's auxiliary processor interface
The VXIbus specification provides for the configuration of multiple-processor systems into hierarchical multi-commander systems. If the commander at the top of the hierarchy is an embedded PC, with a hard disk and a CRT display, its quite likely that you will want to allow other auxiliary processors lower in the hierarchy to access the disk file system, or to display information on the CRT screen.

The EPC-2's Auxiliary Processor Interface (AXPI) software (shown in Figure 5) consists of a server program running as a Windows application on the EPC-2, and an interface routine that can be called by any other processor in the system. The interface routine and the server communicate using VXIbus messages, and shared-memory. Together, they make available to programs running on auxiliary processors the facilities of the PC's operating system. For instance, a program on an auxiliary processor can open, create, read, and write files on the PC's disk drive, as well as create, manipulate, display into, and read from windows on the PC's display.

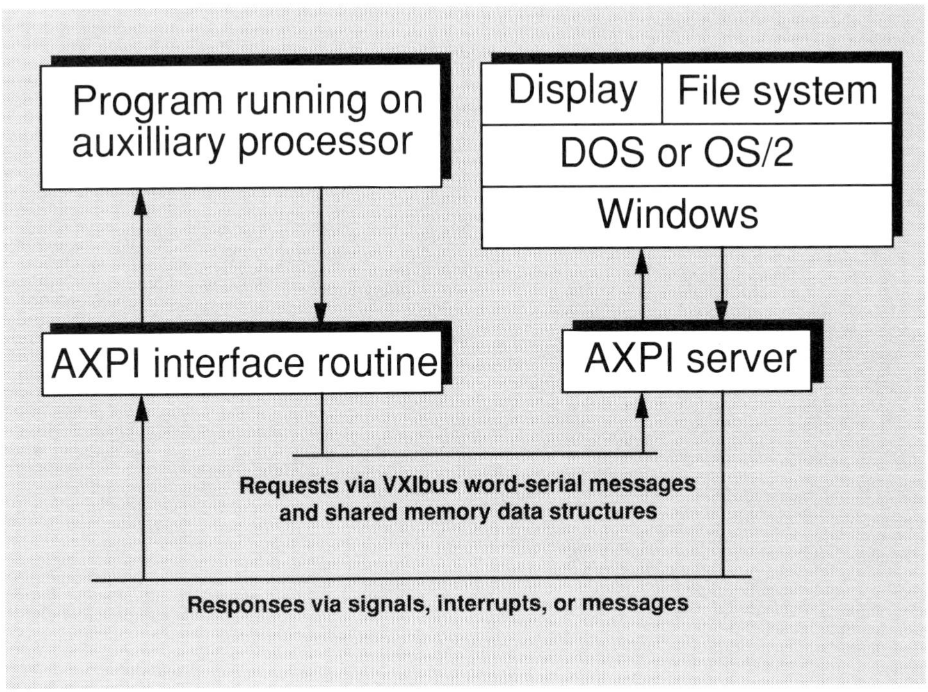

Figure 5

Summary

The VXIbus is going to succeed as a standard for modular instrumentation and test systems. The only question is how quickly? The answer will be determined by the availability of a critical mass of software. By using PC-compatible platforms running Windows/386 or OS/2, you can use existing graphical human-interface and instrument-control packages. By supplementing these software packages with development and diagnostic packages, such as those described in this chapter, you will be able to more quickly take advantage of the new features offered by VXIbus.

Glenford J. Myers is president of Radisys in Beaverton, Oregon. Prior to founding Radisys in 1987, Glen spent 13 years at IBM, and 7 years at Intel, where he worked on the 80386 and 80960 products. He holds BSEE, MS and PhD degrees from Clarkson, Syracuse, and Polytechnic University.

CHAPTER 24

Using VMEbus Boards In VXIbus Systems

By Frank G. Goss

While VXIbus was designed as an upward compatible extension of VMEbus, some VXIbus features could cause you problems when you try to integrate the 2 technologies. In this chapter Frank lists the VXIbus features, describes how they might cause difficulties, and suggests some solutions.

While VMEbus serves processing and control applications well, some additional features are needed to provide for the needs of the instrumentation industry. The VXIbus specification accomplishes this by expanding upon the existing VMEbus specification. In doing so, all of the VMEbus backplane signals and signal drive levels have been kept the same. The VMEbus protocols for the data transfer bus, the arbitration bus, the interrupt bus, and the utility bus have been left unchanged. However, VXIbus specifies some additional features not found on VMEbus:

- usage of the VMEbus *user-defined* pins on P2
- 2 additional plug-in module sizes
- usage of an optional P3 connector
- additional DC power voltages
- electromagnetic compatibility requirements
- an autoconfiguration protocol
- a message-passing protocol
- a command protocol

Usage of the VMEbus user-defined pins on P2
The VMEbus specification does not dictate how the 2 outer rows of pins on the P2 connector are to be used. In VMEbus systems, these 2 rows of pins are typically not bused by the VMEbus backplane. Instead, the P2 signals are simply routed straight through the backplane on wirewrap pins. You can then plug ribbon cables onto the wirewrap pins, to carry I/O signals.

In contrast, VXIbus specifies that some of these P2 pins are to be used for:

- communication between adjacent VXIbus slots
- module identification
- synchronization of VXIbus instruments

Additional sizes of plug-in modules
The VMEbus standard defines 2 board sizes:

- Single height
- Double height

The single height board has a single backplane connector (P1), while the double height board may have either one connector (P1) or 2 connectors (P1 and P2).

The VXIbus standard calls the single height board an A size board, and the double height board a B size board. In addition, it defines 2 additional sizes: C size and D size. The C and D size modules are about twice as deep as the single height and double height VMEbus boards. The C size has 2 backplane connectors (P1 and P2) and the D size has 3 backplane connectors (P1, P2, and P3).

Usage of the optional P3 connector
The VXIbus specification defines an optional P3 connector. The P3 signals provide the same kind of functions as the P2 user-defined pins:

- module-to-module communication
- identification
- synchronization

However, the P3 signals can operate at a much higher clock rate. (100 MHz on P3 vs. 10 MHz on P2.)

Additional DC power sources
The VXIbus specification also provides 4 additional DC power voltages:

- +24 volts
- -24 volts
- -2 volts
- -5.2 volts

The +24V and -24V are useful for analog instrumentation, and the -2V and -5.2V are useful for ECL logic. (ECL signals are used to synchronize VXIbus instruments.)

Electromagnetic compatibility requirements
The VMEbus standard does not address the issue of electromagnetic compatibility (EMC) within a system. VMEbus boards are generally not shielded, and there are no EMC guidelines for the boards. This may not be a problem for digital computer systems but, for sensitive high-performance instrumentation modules, high frequency interference can play havoc with measurements.

To ensure electromagnetic compatibility, VXIbus specifies limits for:

- *Radiated emissions* from the module to the external environment
- *Conducted emissions* from the module to the external environment
- *Module susceptibility* to externally generated emissions

Radiated (RF) emissions are electromagnetic signals that are broadcast into the surrounding environment. *Conducted* emissions are electromagnetic signals that travel along the electrical conductors from the module to the power supply pins on the backplane connector. Module *susceptibility* is a measure of the module's ability to function properly in the presence of external radiated fields.

In addition to the module-to-module EMC requirements, the VXIbus standard requires that both VXIbus modules and VXIbus mainframes conform to an established EMC specification (such as FCC, VDE, or MIL) for radiated emissions. (The particular standard chosen depends upon the intended market of the module.)

Autoconfiguration protocol

The *Module Identification* (MODID) function allows the *resource manager* in slot 0 to locate the devices installed in the VXIbus mainframe. Each VXIbus module slot has a unique signal line, called MODID, which runs to the slot 0 module. Each plug-in module connects an 825 ohm resistor between its MODID pin and ground. By sensing the resistance to ground, the module in slot 0 can determine whether or not a module is installed in each of the other 12 slots.

In addition to being able to verify that a given slot is occupied, the slot 0 module can also use the MODID lines to determine whether the device in that slot is functioning. Each device must have a register bit that reflects the current state of its MODID line. The slot 0 module can simply drive the MODID line for a given slot high, and then read that particular device's register bit. If the device is functioning, the register bit will be true. In this manner the slot 0 board knows that the module occupying that slot is functioning at a minimum level.

Message-passing protocol

The VMEbus specification describes several different types of data transfer bus cycles, which can be used to transfer data between MASTERS and SLAVES. However, it doesn't specify any board-to-board message-passing protocol. The task of creating a message-passing protocol between VMEbus boards is left for the system software designer.

The VXIbus specification goes one step farther, by specifying how the MASTER/SLAVE data transfers are used to pass messages.

Since the VXIbus message-passing protocols are built upon the VMEbus data transfer bus cycles, each VXIbus instrument must support a common set of cycle types. In addition, each VXIbus instrument must be equipped with a block of registers that occupies 64 bytes. This 64-byte block resides within the upper 16 Kbytes of the VMEbus A16 address space. These registers provide a well defined and consistent interface for each VXIbus module, allowing modules from various manufacturers to communicate with each other. These registers can be grouped into 2 categories:

- System configuration registers
- Communication registers

When the VXIbus system is powered up, the resource manager reads the configuration registers to determine whether or not a device has local dual-ported memory. If a device has dual-ported memory, its configuration registers also indicate how much memory. The resource manager, uses this information to allocate space in the system memory map for that memory. By allocating space upon power up, the system ensures that no conflicts arise in memory positioning.

A *sending module* sends a message to a *receiving module* by simply writing the bytes of the message into the communication registers of the receiving module. Alternatively, a sending module can send information via the *memory-based mode,* by writing a *pointer* into the receiver's communication registers. (The receiving module then retrieves the information from global shared memory, using the pointer.)

Command protocol
A VXIbus command set allows the use of high-level commands to control the VXIbus instruments, and to direct data transfers from one instrument to another. These high-level commands are passed as *messages*, and are independent of the internal hardware of each instrument.

Potential problems when using VMEbus boards in a VXIbus system
Because of the features added by VXIbus, you may encounter some difficulties when you install VMEbus boards into VXIbus mainframes. The most common problems will be:

- VMEbus P2 I/O signals may conflict with VXIbus usage
- VMEbus boards may not be deep enough
- VMEbus boards may emit too much radiation
- VMEbus boards don't have configuration registers

VMEbus usage of user-defined P2 connector pins may conflict with VXIbus boards
If a VMEbus board drives the user-defined pins of its P2 connector with signals that are electrically incompatible with the VXIbus definition, component damage could occur. This is the most serious problem that you may encounter when installing VMEbus boards into VXIbus systems. These signal lines must be isolated in some manner.

Fortunately, the physical mounting of the VMEbus boards within a VXIbus mainframe provides a way to isolate these incompatible signals. A small extender board can be installed between the VMEbus board's P1/P2 connectors and the backplane's J1/J2 connectors. The compatible signals are routed straight through this extender board to the VXIbus backplane.

To keep capacitive loading of the bus within allowed limits, the VMEbus specification prohibits the connection of long signal traces to the backplane signal lines. However, since the VXIbus slots are more widely spaced (1.2 inch vs. 0.8 inch) the capacitive loading per slot can be higher. If the signal traces on the extender board are kept short, there should be no loading problems.

VMEbus boards may not be deep enough to engage the backplane in some VXIbus mainframes
The VMEbus double height board is the same height as the VXIbus C size module, so it slides into the card guides of a VXIbus C size mainframe. However, there are 2 problems:

- the depth of the VMEbus double height boards is less than the depth of the C size mainframe
- the upper and lower ends of the VMEbus board's front panel will prevent the board from being pushed in far enough to engage the connectors of the VXIbus backplane.

However, a well designed VXIbus mainframe will have cut-outs along the upper and lower cross members, and will have removable front card guides. The cut-outs allow the VMEbus board's front panel to pass through the front rails, and the removable card guides allow it to be pushed all the way to the back of the mainframe, where it will seat into the backplane properly.

The VMEbus board still must be securely fastened to the mainframe. The wider slot spacing defined by the VXIbus standard allows you to install a sheet metal carrier, which can be secured to the mainframe front rails and also support the front panel of the VMEbus board.

VMEbus boards may emit too much radiation

Installing a VMEbus board into a VXIbus mainframe may present problems, since these boards were not designed to comply with the VXIbus EMC guidelines. It may be necessary to provide sheet metal shields, and additional filtering for the power connections, in order to keep the VMEbus board radiated and conducted emissions within the limits prescribed by the VXIbus standard.

VMEbus boards don't have configuration registers

Current VMEbus boards don't have VXIbus configuration registers. If you install VMEbus boards in your VXIbus system, you'll need to provide the information normally contained in these configuration registers by some other means. One method is to keep a file on disk, with information about each of the VMEbus boards.

Summary

The VXIbus was designed as an extension of VMEbus. The goal was to permit the use of VMEbus boards in VXIbus systems. However, some VXIbus features create problems that must be taken into consideration.

While these problems complicate the integration of VMEbus boards into VXIbus systems, they can be solved. The resulting system gives the user the advantages of both standards. VMEbus boards provide high-speed computation, large memory capacities, and disk storage, while VXIbus modules provide high performance instrumentation. By using VMEbus and VXIbus technology together, you can build very powerful and compact automatic test systems.

Frank Goss is a development engineer for Hewlett-Packard's Loveland Instrumentation Division in Loveland, Colorado. Before transferring to the instrument division in 1983, Frank worked for Hewlett-Packard's New Jersey Division. A native of New Jersey, he obtained his BSEE degree from the New Jersey Institute of Technology. Frank is an avid bicyclist and cross-country skier, and lives with his wife and 2 dogs in Fort Collins.

CHAPTER 25

The SCSI Standard: A Second Generation Of Products

By Bruce Brunette and Mark Stewart

The SCSI (Small Computer System Interface) standard has been widely recognized as a way to provide flexibility in the selection of mass storage peripherals, like Winchester disks. It also allows you to design a single VME-to-SCSI adapter board and to write a single device driver that can then be used to interface to a wide variety of mass storage devices. With the rapid progress in mass storage technology, the SCSI standard also plays an important role in upgrading your VMEbus system. However, many system users have noticed that the performance of systems that use SCSI-based disk drives have not been as good as those that use device-level disk interfaces such as ESDI or SMD. In this chapter Bruce and Mark argue that limitations in the performance of past SCSI-based systems were primarily due to the design of the SCSI interface boards, instead of the SCSI bus itself. According to the authors, disk drive manufacturers have now hired systems designers, and the result is a whole new generation of high performance SCSI drives. However, these new drives demand a new approach to the design of VMEbus-to-SCSI adapter boards. Bruce and Mark discuss features that should be incorporated into 2nd generation VME-to-SCSI boards.

In 1979 Shugart Associates developed a disk interface that they called SASI (Shugart Associates Standard Interface). Their purpose was to provide a standardized interface for their Winchester disk drives. SASI was similar (in some respects) to IBM's multihost, multiperipheral I/O channel architecture, which was designed for use on mainframe computers. However, since SCSI was designed for use on microcomputer systems (which typically have a single hard disk drive) it was much simpler. This simplicity made HOST-to-SASI adapter boards easy and inexpensive to build.

As many manufacturers designed SASI-based disk drives and disk controller boards, it quickly became a de facto industry standard. In 1982, ANSI created a committee (X3T9.2) to develop a formal document that would define a standard based upon SASI.

SASI had been designed to provide a low-cost non-intelligent disk interface. However, with the cost of microprocessors falling rapidly, it seemed likely that future HOST-to-SASI adapter boards (and even disk drives) would include microprocessors. The committee concluded that the SASI definition was inadequate for this, and would need many enhancements, including:

- the addition of multiple host support
- the addition of multiple drive support
- the definition of a *peer-to-peer* protocol, to replace the SASI's *master-slave* protocol

The resulting bus has come to be known as SCSI (Small Computer Systems Interface), and is typically pronounced "skuh-zee". After extensive input and review, ANSI approved SCSI as a standard on June 23, 1986.

The Common Command Set
The SCSI protocols include a long list of commands. Since not all peripherals were likely to support the entire set, this opened the door to incompatibilities. To ensure a basic level of compatibility, an industry group recommended that a subset of the SCSI commands be called the *Common Command Set* (CCS). The ANSI committee accepted this recommendation, and included a list of the CCS commands as an unofficial addendum to the basic SCSI document, recommending the implementation of these commands in *all* SCSI-compatible products.

The CCS consists of 2 sets of commands:

- The REQUIRED commands
- The OPTIONAL commands

The REQUIRED commands must be implemented by all SCSI devices to ensure a basic level of compatibility. The OPTIONAL commands may be implemented at the discretion of the designer.

The CCS REQUIRED commands
The *Inquiry* command requests information about the peripheral. This information includes the type of device (disk, tape, etc) and the version of the SCSI spec to which the peripheral conforms (ISO version, ECMA version, ANSI version).

The *Format unit* command instructs the peripheral to format itself, and to ensure that all data blocks can be accessed. An intelligent VME-to-SCSI adapter may permit the inclusion of a defective block list in this command, for the peripheral to use during the formatting operation. An intelligent SCSI peripheral can also take into account any additional media defects that it finds.

The *Read* command instructs the peripheral to retrieve (and to send back to the VME-to-SCSI adapter) a string of contiguous logical blocks. It specifies a 24-bit beginning block number, and an 8-bit total number of logical blocks in the string.

The *Write* command instructs the peripheral to accept (and store) a string of contiguous logical blocks. It specifies a 24-bit beginning block number, and an 8-bit total number of logical blocks in the string.

The *Read Extended* command instructs the peripheral to retrieve (and to send back to the VME-to-SCSI adapter) a string of contiguous logical blocks. It specifies a 32-bit beginning block number, and a 16-bit total number of logical blocks in the string.

The *Write Extended* command instructs the peripheral to accept (and store) a string of contiguous logical blocks. It specifies a 32-bit beginning block number, and a 16-bit total number of logical blocks in the string.

The *Send diagnostic* command requests the peripheral to perform a diagnostic test upon itself. The peripheral then responds with either a *good* status or a *check condition* status.

The *Test unit ready* command is used to determine whether a peripheral is ready. It doesn't initiate a self test. It just determines the peripheral's current ready status.

The *Read capacity* allows the adapter to find out the block number (and the block *size*, in bytes) of the last logical block of storage on the peripheral. A variation of this command allows the adapter to supply a logical block number, and get back the logical block number of the last block in the same track. (This is helpful when you want to store a frequently accessed data structure on a single track of the disk.)

The *Request sense* command is sent to a peripheral after that peripheral responds to a command with *check condition* status. Upon receiving the *Request sense* command, the peripheral sends further information. A peripheral may return a *check condition* status for several reasons: some of them may simply indicate that the removable media has been changed, while others may indicate something much more serious, such as a non-recoverable ECC error.

The *Reserve unit* command and the *Release unit* command allow a VME-to-SCSI adapter to exclude any other adapters from accessing a peripheral for a period of time. This allows one adapter to make changes to the contents of a peripheral without those changes getting interleaved with changes being initiated by another adapter. If a peripheral is already reserved by one adapter when it receives a *Reserve unit* command from another, it has 2 options:

- It can *reject* the command
- It can *queue* that command, *disconnect*, and then *reconnect* with that adapter when the current reservation is released.

The CCS OPTIONAL commands

The *Reassign blocks* command requests a peripheral to assign a logical block to some other physical block on the physical medium. The adapter typically does this when it detects a defect in the medium currently assigned to the logical block.

The *Mode select* command requests the peripheral controller to return information about the drive that it's controlling. This information includes physical medium parameters (such as bit density, track density, number of sides), as well as the number of blocks on the device, and the block length (in bytes).

The *Mode sense* command is used by the VME-to-SCSI adapter to read back the values it previously passed to a peripheral by a *Mode select* command.

The *Read defect data* command requests the peripheral to supply information about its media defects.

The *Read buffer* and the *Write buffer* commands allow the adapter to transfer data to the peripheral controller's data buffer, and then read the same data back from that buffer, without ever changing the contents of the drive media. This allows the adapter to test the data path to and from the peripheral, as well as the peripheral's buffer.

SCSI can be operated in either of 2 modes:

- synchronous
- asynchronous

After a VME-TO-SCSI adapter broadcasts a SCSI reset command, it sends a *Synchronous Data Transfer Request* message (using the *asynchronous* data transfer mode) to each SCSI peripheral. Each peripheral then responds with its preferred mode of data transfer (asynchronous or synchronous) and (if synchronous) the maximum data rate it can support. (Some SCSI peripherals can handle data rates up to 4 Mbytes per second.)

Once the host has determined a data transfer mode (and a data transfer rate) for each peripheral, it uses that mode and rate in all data transfers between itself and that peripheral, until the SCSI bus is reset again. Since the VME-to-SCSI adapter can interrogate each peripheral independently, a mixture of synchronous and asynchronous SCSI peripherals can be connected to a single SCSI bus.

The SCSI architecture

The SCSI architecture is a very flexible one. In its single-host configuration, it permits a maximum of 7 peripheral controllers on one SCSI bus, as shown in Figure 1. Therefore a fully configured VMEbus system might have 1 VME-to-SCSI adapter and 7 peripheral controllers.

Note: While Figure 1 shows only 1 drive per controller, the SCSI standard allows up to 8 drives per controller, giving a total of 56 drives.

In SCSI systems, a peripheral controller that interfaces several drives to the SCSI bus is typically mounted free-standing, while a peripheral controller that interfaces to a single drive is usually

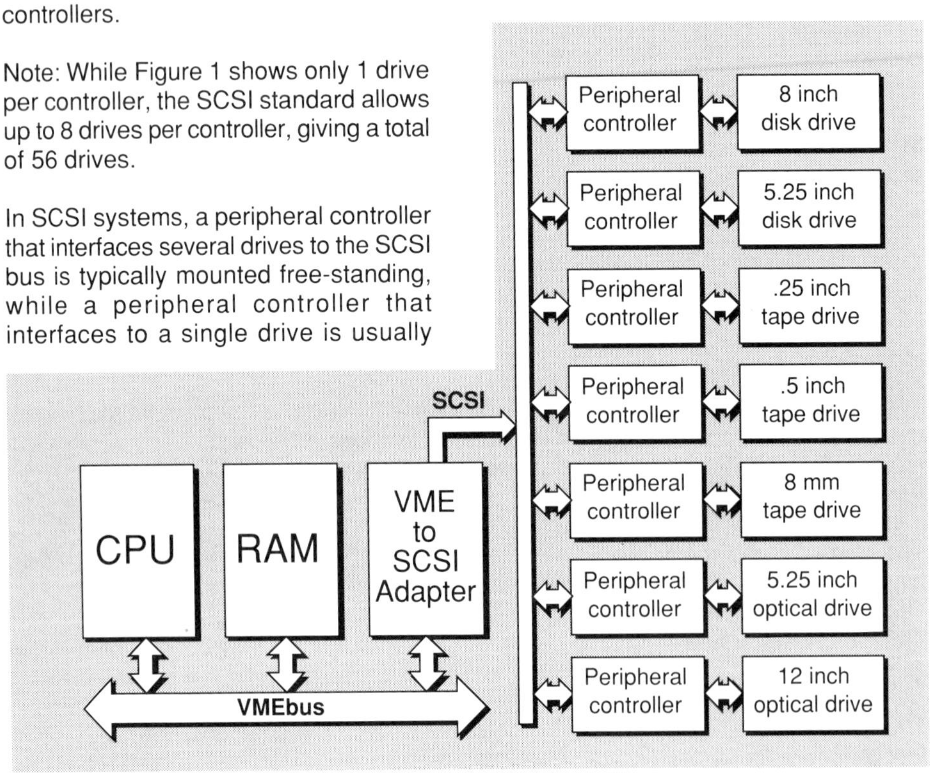

Figure 1

installed *inside* the same enclosure as the drive. (See Figure 2) Newer SCSI-compatible peripherals have the controller built into the drive unit, since this saves the cost of an additional enclosure and power supply. To keep things simple we'll use the terms *peripheral* and *peripheral controller* interchangeably for the remainder of this chapter.

The MPU in the host VMEbus system communicates with a SCSI peripheral through an intelligent VME-to-SCSI adapter board. (This is a plug-in VMEbus board with a SCSI compatible I/O connector.) The operating system running on the VMEbus system doesn't need to know where the data is *physically* located within each peripheral (its physical disk sector number) in order to access that data. In fact, when there are several drives interfaced to the SCSI bus through a single peripheral controller, the host MPU need not even know which of these drives contains the data.

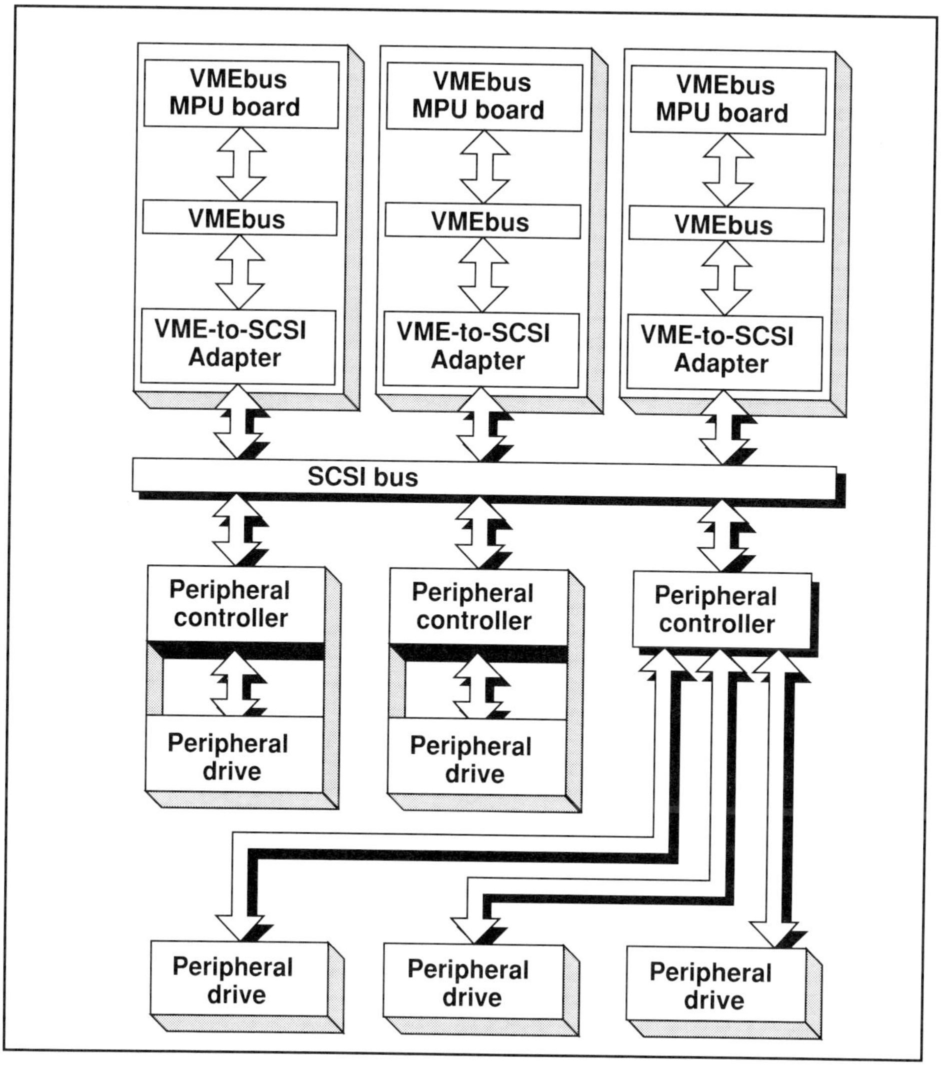

Figure 2

A typical data transfer scenario

To get a feel for how data is transferred over SCSI, let's suppose we have a VMEbus system with an intelligent VME-to-SCSI adapter board. The SCSI port on the adapter board is connected to a single SCSI-compatible disk drive, via a ribbon cable.

The operating system wants to copy a text file (located on the disk) into the VMEbus system's global memory for editing. To do this the operating system calls the VME-to-SCSI adapter *device driver,* and supplies it with:

- the starting logical block number of the file.
- the file length (in blocks).
- the ID number of the SCSI peripheral controller.
- the VMEbus global memory address, where the file is to be copied.

The device driver then builds an *I/O parameter block* that includes VMEbus global addresses, other data parameters, and a *command descriptor block*. The command descriptor block includes:

- the SCSI command code
- the logical unit number (ID) of the SCSI peripheral
- the starting logical block number
- the SCSI command
- the transfer length, in logical blocks

After building this parameter block, the driver passes it to the VME-to-SCSI adapter board's MPU, by passing to it a *pointer* to the global memory location where the parameter block resides. The VME-to-SCSI adapter's MPU then sets up its local DMA controller to copy the I/O parameter block from VMEbus global memory into the adapter's local memory.

When the parameter block has been copied into its local memory, the VME-to-SCSI adapter requests the SCSI bus. When it gains control, it *selects* the appropriate SCSI peripheral controller by sending the *command descriptor block* portion of the parameter block.

If the selected peripheral's access time includes mechanical motion (such as the positioning of a disk drive head) its peripheral controller will typically perform a *logical disconnect* from the SCSI bus at this time. This allows other devices to use the SCSI bus while the disk drive is positioning its head, and reading the data from the disk into a buffer. When the disk drive has finished reading the data, its controller requests the SCSI bus and performs a *logical reconnect* to the VME-to-SCSI adapter.

The VME-to-SCSI adapter then gets the beginning VMEbus address and the number of bytes from its local copy of the parameter block, and writes them into its local VMEbus DMA controller. Then (as the data comes over the SCSI bus from the peripheral) the DMA controller writes it into the appropriate locations in VMEbus global memory. This transfer is typically done in a *burst mode*, to minimize VMEbus usage.

When the SCSI peripheral controller finishes sending the data, it signals completion to the VME-to-SCSI adapter. The VME-to-SCSI adapter's MPU then checks to see if a fault occurred during the data transfer, and writes a final status into the parameter block residing in global VMEbus memory. Then it interrupts the host MPU, using a VMEbus interrupt line.

SCSI's logical addressing simplifies drive selection

Suppose you want to put together a VMEbus file server subsystem, using non-SCSI peripherals. You might select and test an SMD disk controller board and a Pertec tape controller board. Since these controller boards must be configured for the particular model drive you are using, you will need to write one device driver for the disk drive, and another for the tape drive. But, suppose you want to evaluate several SMD disk drives. In order to do that, you would need to write and debug a driver for *each drive* you want to evaluate. If your resources are limited, evaluating more than 1 or 2 drives may be impractical.

However, with SCSI you can design that same file server subsystem with much less time and effort. You simply acquire a VME-to-SCSI adapter board and write (or acquire) a software driver (with separate modules for disk, tape and WORM drives). Then you can easily evaluate the SCSI drives from dozens of different vendors.

Because SCSI driver software uses logical (rather than physical) block addresses, they're much simpler to write. Contrast this with the device driver for an SMD controller board, which must include information about the format of the drive, the spare sector locations, the number of heads and cylinders, etc. Of course, having all of this in the driver allows you to *fine tune* for a specific drive, but this fine tuning will be different for each drive evaluated, and that may require a lot of time and resources.

It's not SCSI's fault

Some people have been disappointed with the performance of SCSI-based systems. However, this disappointing performance is not the fault of the SCSI bus. With a properly designed VME-to-SCSI adapter in the host system, and a properly designed peripheral controller in the disk drive, SCSI can deliver more than adequate performance. For example, suppose you have a UNIX BSD 4.2 system that has a disk drive with the following specifications:

- Average head positioning time = 16 msec
- Rotational time = 16 msec
- Block size = 8 Kbytes
- Data transfer rate = 4 Mbytes per second
- Block transfer time = 2 msec

When you access a block, the average access time will include:

- 16 msec of head positioning time
- 8 msec of rotation time
- 2 msec of data transfer time

This gives a total time of 26 msec per access. If the accesses are random, you would be able to do over 38 accesses per second, giving a throughput of 304 Kbytes per second (38 X 8 = 304). If the SCSI bus were operated at its maximum synchronous data rate of 4 Mbytes per second, it would be able to carry the data transfers of 13 of these disk drives: nearly double the 7 controller limit set by the ANSI standard. If

2-Kbyte blocks were used, the number of drives that could be supported would nearly quadruple.

So, we see that, for a typical UNIX system, one SCSI bus clearly has enough bandwidth. This is true even if the disk drives share the SCSI bus with slower peripherals, such as tape drives. (This assumes that the tape drives buffer their data, and then burst it onto the SCSI bus, rather than transferring the data at the rate that it comes off the tape.) The only exception to this would be parallel head drives, which can transfer data at sustained rates approaching 4 Mbytes per second. For those devices you would need to use a dedicated SCSI bus.

Whose fault is it?
If the SCSI bus is not the limiting factor in most systems, why hasn't the performance of systems been better? The reason is that some important features have been missing from SCSI interface designs.

There are 3 important components in any high-performance SCSI-based VMEbus system:

- a fast peripheral drive
- an efficient SCSI-compatible peripheral controller
- an efficient VME-to-SCSI adapter board

If the design of any one of these components is weak, the total performance may be drastically reduced.

Fast peripheral drives
The 2 key performance features in evaluating disk drives are *average access time*, and the *data transfer rate*. A fast access time ensures that the head gets positioned over the appropriate track of the disk quickly, and a fast data transfer rate ensures that the data will then get moved quickly to (or from) the disk, as the disk spins under the head. You should not assume that, just because a disk drive's SCSI interface moves data at a rate of 1.5 Mbytes per second, the data can be transferred to or from the disk at that rate. If the disk's data transfer rate is only 1.2 Mbytes per second, then the *average* data transfer rate would only be 1.2 Mbytes per second, not 1.5 Mbytes per second.

Efficient SCSI-compatible peripheral controllers
While disk access times and data transfer rates are important, you also need an efficient peripheral controller. SCSI-compatible peripheral controllers (both free-standing controllers and those built into the drives) have increased dramatically in performance. At one time the SASI data transfer rate of 5 Mbits per second was the norm, but SCSI peripheral controllers can now move data at speeds up to the SCSI synchronous limit of 32 Mbits (4 Mbytes) per second.

However, a high data transfer rate is not sufficient (all by itself) to deliver high performance. The peripheral controller board should also be equipped to handle the SCSI protocols efficiently. The 1st generation of built-in peripheral controllers took 2 to 6 msec to process a SCSI command, and their data buffers were small: less than a track. While they provided a data transfer path between the drive electronics and the SCSI bus, they provided little else.

The peripheral controllers built into 2nd generation SCSI drives are greatly improved. They typically provide:

- synchronous data transfer rates up to 4 Mbytes per second
- command processing overhead under 1 msec
- data buffer sizes up to 64 Kbytes.

This faster data transfer rate allows more efficient use of the SCSI bus. The low command overhead often allows immediate access to a disk sector that would otherwise pass under the head before it could be accessed, and would require a complete revolution of the disk. Track buffering allows an entire disk track to be captured (and held) for possible future access. Some examples of SCSI disk drives with high performance peripheral controllers are the CDC Wren V, the Maxtor XT-4000S series, the Hewlett Packard 97536T, and the Micropolis 1580.

The SCSI standard has moved much of the disk controller electronics out of the host systems and into the disk drives. Disk drive manufacturers have realized that they must become subsystem companies. As a result, they have aggressively staffed up with firmware and software engineers. As these new staff members have applied their expertise, new features (such as built-in disk caches) are further increasing SCSI throughput.

Caching bridge controllers are already finding application where response time is more crucial than cost. In a caching bridge controller, the most commonly accessed files are stored in RAM, and are managed by a sophisticated algorithm. This allows access to those files in microseconds, rather than milliseconds.

Efficient VME-to-SCSI adapters

The heart of any SCSI subsystem is the HOST-to-SCSI adapter. A high performance VME-to-SCSI adapter should have:

- A fast VMEbus data transfer rate
- A fast SCSI bus data transfer rate
- A low SCSI command overhead
- An efficient hardware architecture
- Intelligent data handling

In simple terms, a good VME-to-SCSI adapter should be very fast on both ends and very efficient in between.

Fast VMEbus data transfer rates

Most of the 1st generation VME-to-SCSI adapters used off-the-shelf DMA chips like the 68430, which can only achieve data transfer rates of about 2 Mbytes per second. That doesn't even approach the VMEbus's upper limit of 40 Mbytes per second. To really use the VMEbus efficiently you need a *state machine* design. Assuming a VMEbus memory board with a 170 nsec access time, a state machine can provide VMEbus data transfer rates up to about 20 Mbytes per second. That might seem like overkill (since the maximum SCSI data transfer rate is only 4 Mbytes per second) but a faster VMEbus data transfer rate allows the VMEbus data transfers to be done in *bursts*, leaving the bus idle between bursts, for use by other VMEbus masters.

Fast SCSI bus data transfer rates
First generation SCSI interface chips like the NCR 5380 supported only *asynchronous* SCSI data transfers, and moved data at about 1.5 Mbytes per second. However, 2nd generation chips (like the Fujitsu 87030 SCSI protocol chip) also provide *synchronous* data transfer rates, up to 4 Mbytes per second. Synchronous operation is important for 2 reasons:

- Virtually every major disk drive manufacturer has introduced synchronous drives.
- Synchronous operation can reduce the cost of multiple peripheral systems, because you can use a single VME-to-SCSI adapter (instead of 2 or 3). This saves both VMEbus slots and the cost of multiple VME-to-SCSI adapter boards.

One of the benefits of SCSI is that both asynchronous and synchronous peripherals can share the same SCSI bus. In order to take full advantage of this, you need a VME-to-SCSI host adapter that's capable of determining which type of transfer mode each peripheral can handle.

Low SCSI command overhead
You should be cautious about the amount of command processing latency that the VME-to-SCSI adapter (and the SCSI peripheral controller) add to the total subsystem overhead. Some 1st generation VME-to-SCSI adapters took as long as 8 msec to process a SCSI command! When disk drives had average access times of 70 msec or more, this latency was not all that obvious. But, with newer drives now providing 16 msec access times, you really notice the effect of an 8 msec latency on the system's performance.

Efficient Hardware Architecture
VME-to-SCSI adapter board architectures can be classified into 2 groups, based on how they move data between the SCSI bus and the VMEbus:

- The adapter's MPU can copy data from one bus into its onboard RAM buffer, and then copy that data back out to the other bus.
- The adapter can be equipped with a DMA controller that transfers data between the VMEbus and one end of a FIFO, while the adapter simultaneously routes the data between the other end of the FIFO and the SCSI bus.

Since the 1st method uses only the adapter's onboard MPU, it requires very little hardware support. That makes it considerably *cheaper* than the FIFO approach, which requires a FIFO and DMA hardware. However, it also involves a lot of overhead, since it requires 2 block copy operations.

If you use the 2nd method, you must make sure that the FIFO in the adapter is deep enough to keep data moving over the SCSI bus when the VMEbus is temporarily busy. This is especially important when you're using SCSI to interface to disk peripherals that don't have internal track buffers. For example, suppose you request a string of logical blocks from a SCSI-compatible disk. The disk drive moves its head to the appropriate track, and reads the first block into the disk controller's local block buffer. Then the disk controller requests the SCSI bus, and starts to send that block over the SCSI bus, to the VME-to-SCSI adapter board.

Now, suppose some other VMEbus master starts a block transfer cycle, just as the disk controller starts sending its data over the SCSI bus. Data is coming into the adapter's FIFO over the SCSI bus. However, since the adapter can't get control of the VMEbus, the FIFO fills. If the FIFO completely fills, it won't accept any more data. This prevents the disk controller from emptying its block buffer in time to read the next block, as that block rotates under the disk head. The disk will then have to make an entire revolution before that block comes under the head again: a delay of many milliseconds.

The FIFO size is particularly important in systems employing the UNIX operating system, because UNIX periodically copies groups of blocks from system memory to disk, and vice versa. These blocks are typically scattered throughout system memory.

If each of these blocks is copied using a separate SCSI command, there will be a considerable amount of overhead. However, if you provide scatter/gather firmware on the VME-to-SCSI adapter board, you can eliminate this overhead. The scatter/gather firmware accepts a *list* of addresses and byte counts from the UNIX driver (which define the location and size of each block buffer in the group being transferred) and then transfers the contents of each of these block buffers to a sequential logical block on the disk drive. Since a transfer to a sequential string of logical blocks can be accomplished with a single SCSI command, the command overhead is greatly reduced. If the peripheral controller stores sequential logical blocks in contiguous physical blocks on the disk, (which is usually the case) then the entire transfer can often be accomplished without repositioning the disk drive head, and without having to wait for rotational delays between block writes. Once the group of buffers has been saved to a string of contiguous blocks on the disk, they can later be retrieved and restored to system memory quickly.

Of course this technique only works well if the adapter has a deep enough FIFO. If other VMEbus masters are also doing block transfers over the VMEbus, a small FIFO may not be able to continue accepting data from the disk drive while the VMEbus is busy. As a result, the disk controller would not be able to retrieve the entire string of blocks in a single rotation.

So how big should the FIFO be? Suppose you want to copy a group of blocks from system memory to the disk. For each block, the adapter's MPU sets up its local VMEbus DMA controller. Then it waits, while the DMA controller reads the block into the onboard FIFO. When the last byte of the block has been read, the DMA controller interrupts the MPU. The MPU then configures the DMA controller to read the next block into the FIFO.

To maintain the maximum SCSI bus transfer rate, the FIFO must be able to hold enough data to continue emptying to the SCSI bus, while the MPU is preparing the DMA controller to read the next buffer. So how long does that take?

Suppose the MPU takes 150 μsec to respond to the DMA controller's interrupt, and to get it ready to read the next buffer. If the SCSI bus is running at its maximum rate of 4 Mbytes per second, the FIFO will be able to send 600 bytes over the SCSI bus during this 150 μsec. So, to ensure a continuous SCSI data transfer, the FIFO should be at least 600 bytes deep.

These calculations assume that the adapter board's DMA controller will be able to read from global VMEbus memory *immediately*, once it has been configured by the MPU.

However, if some other master is using the VMEbus, the DMA controller won't get immediate access, and the adapter board will need a deeper FIFO.

Intelligent data handling
In addition to providing FIFO hardware, you want to use the adapter's local intelligence to help transfer data into (or out of) the VMEbus system. There are 3 basic ways to do this:

- By doing scatter/gather operations
- By using command queuing
- By using the SCSI disconnect/reconnect feature

We have already talked about the scatter/gather technique.

Command queuing
An adapter with command queuing can accept additional commands, even though it has not yet finished execution of previous commands. In its simplest form, command queuing is where the VME-to-SCSI adapter puts each new incoming command into a queue. Then, when the currently executing command completes, the adapter's MPU selects the next command in the queue for execution.

However, if the adapter is servicing more than one SCSI peripheral, it's more efficient to have it store the incoming commands in several queues: one for each peripheral. Then, each time a command to a peripheral is completed, the adapter can get the next command from that peripheral's command queue, knowing that the peripheral is ready to accept another command.

SCSI disconnect/reconnect
The SCSI disconnect/reconnect feature allows the VME-to-SCSI adapter to do concurrent accesses on several different peripheral devices. For example, if a system has 2 SCSI disk peripherals, the adapter could send a *read* command to the first peripheral. That peripheral could then *logically* disconnect itself from the SCSI bus, copy the data from its disk drive into its local block buffer, and then reconnect, to transfer the data over the SCSI bus. Meanwhile, if the VME-to-SCSI adapter has a command queued for the other disk, it sends that command to that peripheral. The result is overlapped seeks, rather than sequential ones. This provides much higher system performance in multiple-drive systems.

The Introl 300
Figure 3 is a block diagram of the Introl 300 VME-to-SCSI adapter board, which is shown in Figure 4. Since most VMEbus designers have based their systems on the 68000 family (and are familiar with its characteristics) we chose that processor for use in the Introl 300. In addition to its 68000 microprocessor, the Introl 300 uses a 2-Kbyte FIFO to move data quickly between VMEbus memory and the SCSI bus. The adapter's 8-Kbyte local RAM is used only for storing I/O parameter block pointers, which are used by the 68000 to find the parameter blocks in VMEbus global RAM. The board can queue up to 16 I/O parameter blocks, and can issue up to 16 SCSI commands concurrently.

Because the 68000 is not used to do the actual data movement between the VMEbus and the SCSI bus, it's free to schedule SCSI bus activity in an efficient manner. For example, if the processor has sent a command to one SCSI peripheral, and there is another command queued for a 2nd SCSI peripheral, the processor will send the 2nd command while the 1st peripheral is retrieving its data. The resulting overlapped seeks can make a big difference in the overall system performance. The processor also supports scatter/gather operations, and it's able to process any of the commands in the common command set in less than 1 msec.

A programmable state machine is used to DMA data quickly between the adapter's FIFO and VMEbus memory. The DMA data transfers can take place at up to 20 Mbytes per second. The data bus width, and the burst length can be specified for each transfer with appropriate parameters in the I/O parameter block. If the FIFO still has bytes to transfer over VMEbus when the burst length limit is reached, it releases the VMEbus, and then immediately requests it again.

In order to move data over the SCSI bus quickly, the INTROL 300 employs the Fujitsu SCSI protocol chip, and a 65 nsec FIFO. The board supports both an asynchronous data transfer rate of 1.5 Mbytes per second and a synchronous data transfer rate of 4.0 Mbytes per second: the maximum rate possible on the SCSI bus.

To permit its use with both asynchronous and synchronous peripherals, the board determines the mode and the data transfer rate to use with each peripheral when it sends the first command to the peripheral.

The Introl 300 is a good example of a 2nd generation VME-to-SCSI adapter, which can take full advantage of 2nd generation SCSI compatible mass storage devices. The combination of speed, intelligence, and efficiency allows

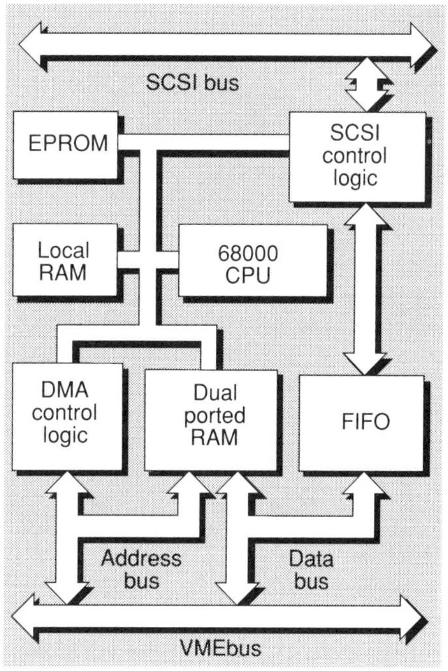

Figure 3

Figure 4

system integrators to build truly high performance systems, while still taking advantage of the software benefits of SCSI.

Bruce Brunette is President of Introl Corporation in St. Paul, Minnesota, which manufactures high performance intelligent peripheral controllers and adapters. Prior to founding Introl, he spent 15 years in engineering, product planning and marketing with Honeywell, Medtronic, and Ciprico. Bruce holds a BSE from the University of Wisconsin at Madison and an MBA from the University of Minnesota.

Mark Stewart is a design engineer for Introl Corporation. Prior to joining Introl, he was a controller design engineer for Control Data Corporation. He was the CDC representative on the IPI ANSI X3T9 committee, and represents Introl on the combined SCSI/IPI committee. Mark has a BS degree in Computer Engineering from Iowa State University.

CHAPTER 26

Designing A VME-To-SCSI Adapter

By Donald Peterson

The Small Computer Systems Interface (SCSI) has grown dramatically in popularity. For example, only a few years ago none of the 5.25 inch Winchester disk drive vendors were selling drives with SCSI interfaces built in, and now most of the models being sold include a SCSI interface. In fact, disk drives with SCSI interfaces are now available from all the major disk drive manufacturers. In this chapter, Don discusses features that should be on a VME-to-SCSI adapter board, to provide the best possible performance.

As a system integrator, you're faced with a wide range of choices when it comes to putting together a disk subsystem. For example, you can purchase a disk drive with a standard *device-level interface*, such as SMD or ESDI, and interface it to your VMEbus system with a VME-to-SMD or VME-to-ESDI disk controller board, as shown in Figure 1.

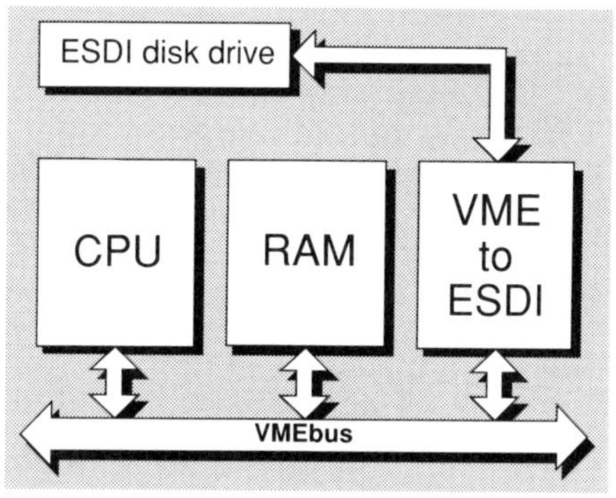

Figure 1

Another alternative is to buy a VME-to-SCSI adapter board. At one time this was not a very appealing approach, since interfacing the SCSI bus to most disk drives was awkward. You typically had to purchase a *bridge controller*, which provided a SCSI interface on one side, and an SMD or ESDI interface on the other, as shown in Figure 2. However, that situation has changed, and today most disk drive manufacturers offer SCSI-compatible drives. With these drives, you can simply daisy chain a cable from the VME-to-SCSI adapter to each of the disk drives in your system, as shown in Figure 3.

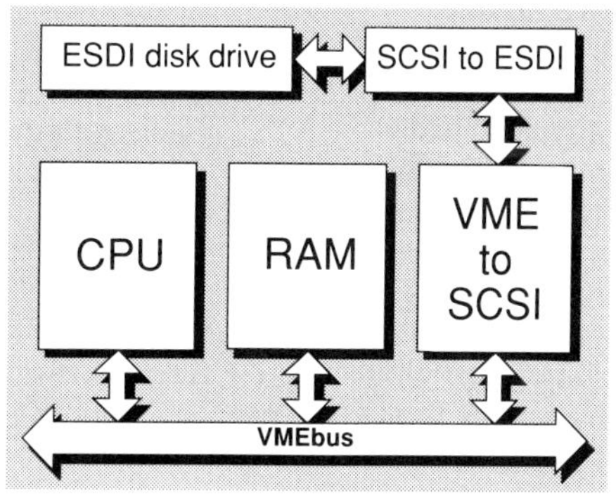

Figure 2

However, although the initial SCSI compatible products claimed conformance to the SCSI specification, they were often incompatible with each other. This was because the SCSI specification allowed for differing interpretations of many of the SCSI bus commands. In an effort to bring order out of chaos a group of companies (led by OMTI, a division of Scientific Micro Systems) formed a committee and selected a subset of the original SCSI commands, which they called the Common Command Set (CCS). Adherence to this standard provides a much higher level of compatibility between products.

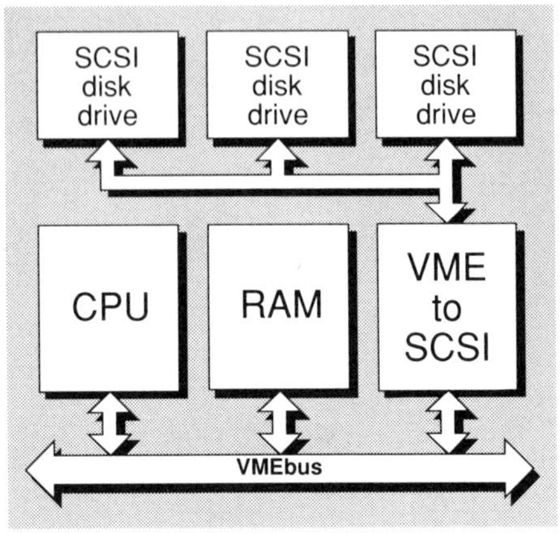

Figure 3

Early SCSI bus users expected to see disk performance comparable to what they had previously seen on *device level* subsystems, such as SMD or ESDI. However, they were disappointed for 2 basic reasons:

- Device-level disk controller vendors had years of hardware and firmware experience behind them, while SCSI vendors had not yet developed that level of expertise.
- The architecture of a SCSI-based subsystem has one additional step in the command flow.

The 1st problem has largely been solved, as SCSI compatible drive manufacturers have gained experience. The 2nd problem has not been so easy to overcome.

Figure 4 shows the flow of control in a typical ESDI-based disk subsystem. Each disk access begins with an operating system call. This activates the software driver for the disk being accessed. The driver builds an I/O parameter block and passes it to the disk controller board, which then translates the logical block number to a physical cylinder, track and sector number.

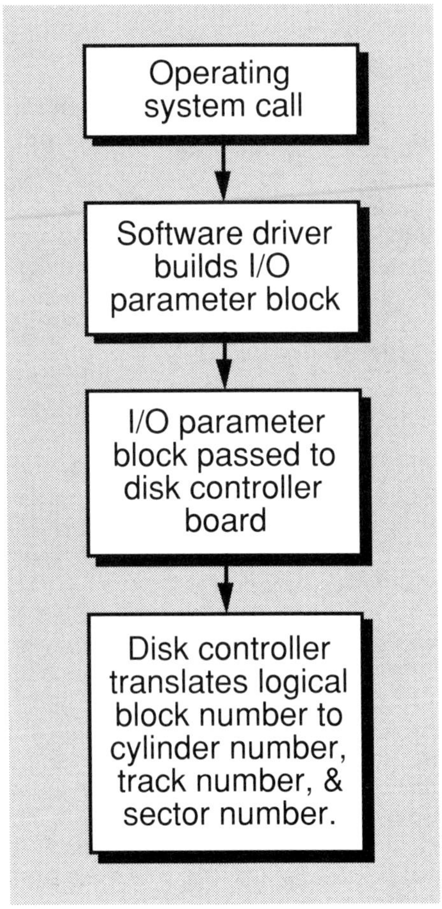

Figure 4

Contrast that with the flow of control in a SCSI-based subsystem, as shown in Figure 5. The operating system call activates the software driver, which builds an I/O parameter block and passes it to the VME-to-SCSI adapter board. The adapter then extracts the SCSI command from the parameter block and sends it over the SCSI bus, where the SCSI compatible disk controller in the drive translates the logical block number to a physical cylinder, track and sector number.

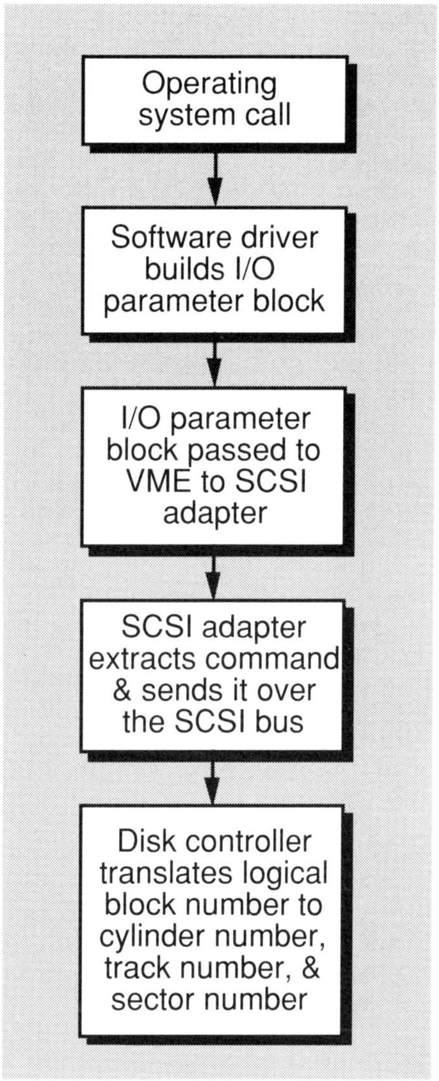

Figure 5

Obviously, there is an extra step in the flow of control in the SCSI-based system. However, with careful design, a SCSI-based system will be able to deliver performance approaching that of ESDI or SMD systems, while still providing much greater flexibility in the selection of disk drives.

There are 6 major factors that affect the performance of a SCSI subsystem:

- The VME-to-SCSI adapter firmware
- The SCSI controller chip used
- The VMEbus interface
- The adapter's buffer architecture
- The operating system's software driver.
- The disk controller that interfaces the disk drive to the SCSI bus

VME-to-SCSI adapter firmware

The adapter firmware should be written to minimize the overhead needed to handle SCSI commands. One technique that works well is to embed the SCSI command block into the I/O parameter block built by the operating system's software driver. The VME-to-SCSI adapter can then simply extract the SCSI command block from the parameter block, and send it directly over the SCSI bus, without any changes. In addition to the speed benefits, this allows the VME-to-SCSI adapter firmware to accept and transmit vendor-unique commands, in addition to the commands of the common command set.

If the adapter firmware supports command queueing, you can also use another technique called *command combining*. The firmware scans and sorts the queued commands, and then combines groups of them into single SCSI bus commands, reducing the SCSI command processing overhead. It also permits scatter/gather operations, where the data from discontiguous memory areas is stored onto the disk (or retrieved from it) with a single SCSI command.

Command sorting also permits the VME-to-SCSI adapter to minimize the amount of head motion needed on each drive to access the required sectors. There are various algorithms you can use to sort the accesses. One popular algorithm is called *C-scan.* It accesses the lowest logical block number, and then the higher block numbers, in ascending order. Assuming that the physical disk blocks are arranged on the disk in the same sequence as the logical blocks (which they usually are) this will access the sectors closest to cylinder zero (near the hub of the disk) and then access the sectors further out toward the edge of the disk. Since the accesses are done in order, random head motion (and the resulting seek latencies) are minimized. When the disk has accessed the sector furthest from the hub, the head is repositioned to the center, and the process begins again.

If the firmware sorts all pending disk access requests by disk drive, it can also improve the performance of a multiple disk subsystem. It does this by sending concurrent commands to several disk drives on the SCSI bus. Note: In order to do this, the firmware must support the SCSI *disconnect/reconnect protocols.*

The SCSI controller chip

Command overhead is critical to the performance of SCSI controller chips. Initial SCSI controller chip designs (such as NCR's 1st generation chip, called the 5380, first sold in 1983) had command overheads in the multiple msec range. However, 2nd generation SCSI chips (such as the Western Digital 33C93A and Emulex ESP) have command overheads under 500 μsecs. The Western Digital chip also includes SCSI drivers and receivers, reducing the component cost and the real estate requirements.

The VMEbus interface

The VMEbus has a theoretical bandwidth of about 40 Mbytes per sec, while the SCSI bus (even in its synchronous mode) has an upper limit of around 5 Mbytes per second. Therefore, you don't want to just directly couple the VMEbus and the SCSI bus. The data would be transferred at too slow a rate over the VMEbus, and would waste much of its available bandwidth.

A better approach is to design the VME-to-SCSI adapter board to accumulate incoming data from the SCSI bus, and then periodically write it to VMEbus memory at a high burst rate. (In the case of data being sent out to the disk, the adapter would read data from the VMEbus in bursts, and then send it out at a slower rate over the SCSI bus.) In order to do this, the adapter must be equipped with either a *cache* or a FIFO (First-In-First-Out) buffer. By using a FIFO, a high performance adapter can transfer bursts of data over the VMEbus at rates over 30 Mbytes per second.

Adapter buffer architecture

The *physical architecture* of a typical SCSI disk subsystem includes 3 components:

- A VME-to-SCSI adapter board in the VMEbus chassis
- A SCSI-compatible disk drive
- A cable from the adapter board to the disk drive

This looks very similar to an ESDI or SMD disk subsystem. For example, in an ESDI subsystem you have:

- A VME-to-ESDI controller board in the VMEbus chassis
- An ESDI-compatible disk drive
- A cable from the controller board to the disk drive

However, the *functional architecture* of the 2 types of subsystems is very different, and should not be confused.

A VME-to-ESDI or VME-to-SMD disk controller board interfaces *directly* to the disk drive's control electronics. In order to do this, the controller must be able to format the drive, handle defects in the disk's media by reassigning sectors, and correct errors as it reads from the disk, based on an ECC code.

A VME-to-SCSI adapter does not handle any of these functions. They are handled by the disk controller circuitry, which is typically mounted *inside* the SCSI compatible disk's enclosure. (In cases where SCSI-to-ESDI or SCSI-to-SMD bridge controllers are used, the controller may be a free-standing board.) The VME-to-SCSI adapter just sends a command over the SCSI bus, requesting the read or write of a *logical block number*. The disk controller in the drive enclosure translates this logical block number into the appropriate cylinder, track and sector numbers, and initiates the control signal sequence needed to position the head over the appropriate track.

When an operating system reads a disk sector, it will quite likely read the next sector some time soon. One technique that improves disk performance is to provide a *multiple sector cache*. Whenever a sector is read from the disk, some of the subsequent sectors on the same track are also read, and stored in the sector cache. Then, if the operating system needs to read the next sector, the contents of that next sector can be made available immediately from the cache, without waiting for the head to position itself over the track again.

Sector caching in the *disk controller* works very well. However sector caching in the VME-to-SCSI *adapter* doesn't work nearly as well. The problem is that the adapter deals only with *logical* block numbers: it doesn't know where one disk track ends and the next begins. As a result, if it reads several subsequent blocks (to fill its local block cache) it may initiate head movements (to an adjacent track) that tie up the disk needlessly for several milliseconds. If another command is received by the adapter while the head movement is taking place, that command won't be processed until the head movement (and the subsequent sector reads) are completed. If the hit rate on the cached blocks is low, this logical block caching might even *decrease* disk performance.

There's another reason that you don't want to try to cache the blocks in the adapter: it requires you to first write each block into the adapter's local memory, and then to copy the block to VMEbus memory. It's much faster to just route an incoming block directly from the SCSI bus, through a FIFO, and onto the VMEbus.

If you use a FIFO in this manner, you must make sure that it's deep enough to continue accepting data from the SCSI bus, even when the VMEbus is temporarily busy. For example, suppose we have a 32-byte-deep FIFO. We first calculate the time required to empty the FIFO over the VMEbus. If we assume that the adapter can transfer data across the VMEbus 32 bits at a time, and that those 32-bit data transfers take place every 400 nsecs, then it would take 3.2 μsecs to empty the 32-byte FIFO.

Next, we calculate the time required to fill the FIFO from the SCSI bus. If we assume that data is coming to the adapter from the SCSI bus at a rate of 5 Mbytes per second, then it would take 6.4 μsecs to fill the FIFO.

So, even with these conservative assumptions, the adapter will be able to keep a 32-byte FIFO from filling, as long as it is given access to the VMEbus at least once every 6.4 μsecs. (You can ensure this by giving the adapter a high bus arbitration priority, as is usually done for most mass storage devices.) In accomplishing the SCSI-to-VME data transfer, the adapter will use only 50% of the VMEbus bandwidth. In general, a 32-byte FIFO provides adequate buffering, because of the high speed of the VMEbus.

The software driver

The operating system's software driver should be designed to minimize overhead, and to take full advantage of the specific features of the VME-to-SCSI adapter, such as scatter/gather DMA transfers.

Writing drivers that work with all SCSI-based peripherals is not as simple as it seems. Disk, tape, and optical SCSI devices each require different driver code, and there are even subtle differences from model to model. Because of this you would be wise to test your device drivers on several different models of SCSI devices.

The disk controller

Any high-performance disk controller should have a fairly large built-in sector cache. This allows the controller to continue reading consecutive sectors after the requested sector has been read. Then, if the next sector access is sequential to the previous one, that sector will already be in the cache.

Some 2nd generation high performance SCSI compatible disks can also *sort* and *combine* sector accesses. This sorting allows the disk to access all of the requested sectors that lie along the same track, without repositioning the head. This feature can be made even more effective if the disk controller is equipped with a feature called *zero latency read.* If a disk controller has zero latency read, and if it's requested to read several sectors scattered along a single track, it can start looking for *any* of the requested sectors, as soon as the head arrives at the track.

Without zero latency read, the disk controller would first look for the *lowest numbered* sector, and then read the additional sectors in ascending order. If the lowest numbered sector just passed the head, none of the sectors would be read until the disk makes one complete revolution. Then, if the highest numbered sector is at the end of the track, the disk would have to make yet another complete revolution, before all the remaining sectors could be read.

With the zero latency read feature, the disk controller can always read all of the needed sectors from the track in only one rotation.

For the most part, 1st generation SCSI compatible disk controllers didn't implement these features. Fortunately, disk drive manufacturers saw the need to upgrade the performance of their SCSI compatible drives. Vendors such as Hewlett-Packard,

Quantum, and Micropolis introduced products that include track buffers, which capture *all* of the sectors of a track in a single revolution of the disk. This allows them to implement read-ahead caching and zero latency reads.

The Rimfire 3510 VME-to-SCSI adapter
Figure 6 is the block diagram of Ciprico's 2nd generation VME-to-SCSI adapter, called the Rimfire 3510. (See also Figure 7) The Rimfire 3510 is controlled by a 10 MHz 80186 microprocessor, which uses a private 16-bit-wide bus to access local RAM and EPROM. The *short burst FIFO* is a custom gate array that provides 32 bytes of buffering between the VMEbus and the SCSI bus. This FIFO has proven to be large enough to decouple the VMEbus and the SCSI bus, without incurring any performance degradation.

Ciprico has also designed another custom gate array called the *Pipelined System Interface* (PSI). It serves as a DMA controller, generating VMEbus addresses. However, it has one special feature that allows it to transfer data much faster than a typical DMA controller: a double-deep *starting address* register and a double-deep *transfer length* register. These double-deep registers allow the onboard 80186 to *preload* the starting address (and the transfer length) of a subsequent DMA operation, while the current

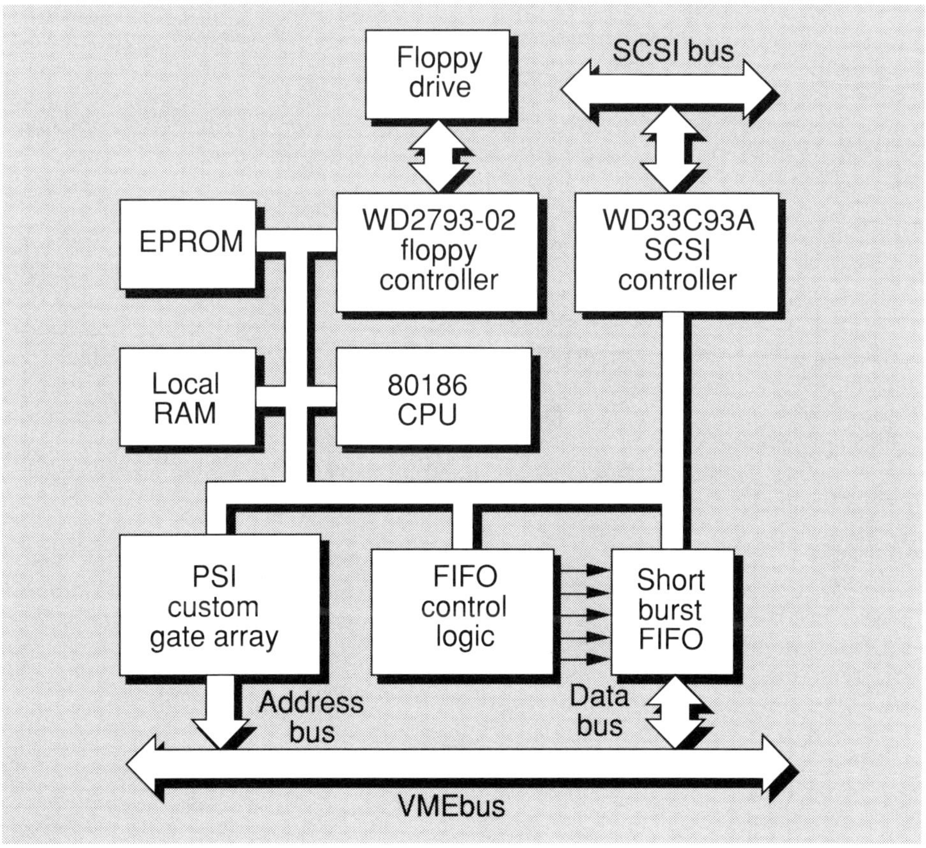

Figure 6

DMA transfer is still in progress. The use of *PSI* and the *short burst FIFO* gate arrays allows a DMA burst rate of greater than 30 Mbytes per second over the VMEbus.

The Western Digital WD33C93A SCSI controller chip is used on the adapter. This 2nd generation intelligent SCSI controller chip (preceded by the WD33C93) operates from a 16 MHz clock, and dramatically reduces SCSI command overhead. The overhead has been reduced from 1.2 milliseconds for the 1st generation chip to less than 500 microseconds.

Figure 7

The SCSI controller chip also includes a 5-byte-deep FIFO. It can support synchronous data rates up to 5 Mbytes per second and asynchronous rates up to 2 Mbytes per second. (The 1st generation WD33C93 only provided 4 Mbytes per second synchronous transfers, and 1.5 Mbytes per second asynchronous transfers.)

The firmware on the Rimfire 3510 does *command sorting* and *command combining*, for both read commands and write commands. This improves performance by minimizing seek and rotational latency delays.

Many users still want to have a floppy disk drive on their system for software distribution and testing. However, there are few (if any) SCSI floppy disk drives available. For this reason, the Rimfire 3510 includes a floppy disk interface, which allows a choice of 3.5-inch or 5.25-inch floppy drives. A portion of the on-board RAM is used as a buffer for the floppy port, so SCSI performance is not degraded by floppy activity. (SCSI transfers are always given first priority by the onboard firmware.) Floppy drives are accessed from the VMEbus just as if they are normal SCSI devices. (The floppy drive is given an unused SCSI ID of FE (hexadecimal). Floppy commands are passed to the adapter board via a parameter block that has a format identical to normal SCSI commands.

Summary

SCSI provides a very flexible method for connecting multiple types of mass storage devices to VMEbus systems. With drive vendors improving the performance of their SCSI compatible drives, the VME-to-SCSI adapter design has become critical. A good adapter should have:

- Minimal overhead in the firmware
- A 2nd generation SCSI controller chip
- A 32-bit, FIFO-buffered VMEbus interface
- An efficient, well-debugged software driver

The resulting SCSI-based mass storage solution will give you the performance and flexibility you need at a reasonable cost.

Donald C. Peterson is the director of marketing at Ciprico, in Plymouth, Minnesota. He has been involved in the marketing of board level products for 12 years. Prior to joining Ciprico in 1986, he worked as product marketing manager at Qualogy in San Jose, California, and held various marketing positions at Intel in Hillsboro, Oregon. He holds a BSEE and MS in business from the University of Wisconsin, Madison.

CHAPTER 27

Building A SCSI Subsystem

By Tom Thawley and Ernest Godsey

The SCSI standard has gained great popularity because it is relatively easy to use. However, the performance of SCSI-based subsystems has sometimes proved to be disappointing. In this chapter Tom and Earnest describe why this has happened, and describe architectural techniques that can greatly improve the performance of your SCSI-based mass storage subsystem.

SCSI has become very popular, especially for use in low-level and mid-level performance applications. Its basic appeal is that it provides a *device-independent* interface to mass storage peripherals. This frees you from having to study the signal timing and electrical specifications of various peripheral types, such as winchester disks, quarter-inch and half-inch tape drives, and optical storage devices. The device-specific details are handled by a peripheral controller, which is either provided by the peripheral manufacturer (as part of the drive unit) or by a supplier that specializes in the design of SCSI-compatible peripheral controllers.

The SCSI bus allows interconnection of up to 8 *devices*, each of which is assigned a *device ID*. When SCSI is used to support a VMEbus system, the VME-to-SCSI adapter board is assigned one of the IDs, leaving the other seven IDs for seven SCSI-compatible peripheral controllers. (See Figure 1)

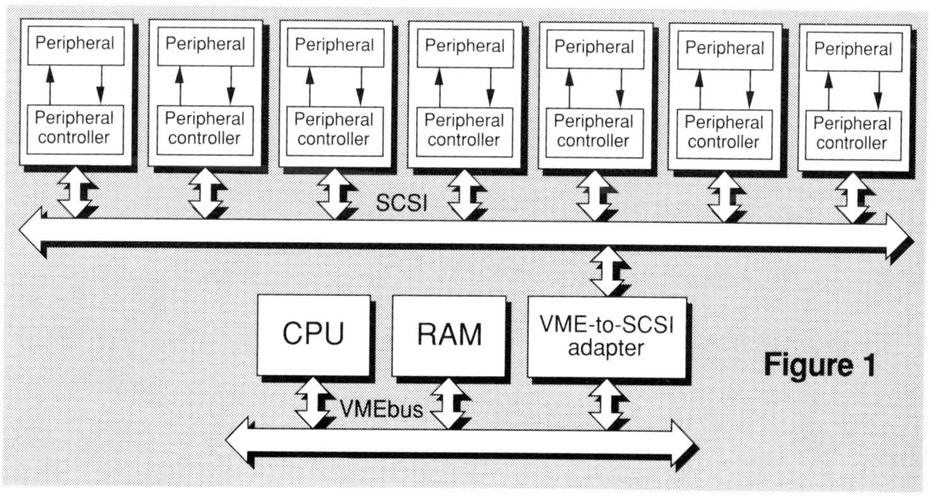

Figure 1

Note: The SCSI spec permits the use of 8 *local unit numbers* (drives) with each SCSI-compatible controller, and 256 *extended ID codes* per local unit. However, very few SCSI peripheral manufacturers are supporting these extensions.

The asynchronous mode

The SCSI bus operates in one of 2 modes: synchronous or asynchronous. In the asynchronous mode, the VME-to-SCSI adapter *selects* a peripheral controller, and then waits for that controller to request a data transfer. Thus, the selected peripheral controller controls the data traffic to (or from) itself, by controlling each data transfer.

The synchronous mode

In the synchronous mode, the VME-to-SCSI adapter *negotiates* with the selected peripheral controller, to agree upon...

- A *clock speed* and an *offset value* for data transfers from the adapter to the peripheral controller.
- A *clock speed* and an *offset value* for data transfers from the peripheral controller to the adapter.

Once this negotiation has been completed, the data transfers commence at the agreed upon clock speeds. For example, if the peripheral controller is sending data to the VME-to-SCSI adapter, it strobes the SCSI request line from its internal clock each time it *sends* a byte over the SCSI bus into the adapter's FIFO. Meanwhile, the adapter strobes the SCSI *acknowledge* line each time it *accepts* a byte from the other end of the FIFO. Since the first byte might not be taken out of the FIFO before several are put into it, the peripheral controller might generate several request strobes before the adapter generates its first acknowledge strobe. The negotiated *offset value* sets a limit on how far the controller's request strobes can get ahead of the adapter's acknowledge strobes before the controller must pause to allow the adapter to catch up.

Asynchronous mode vs. synchronous mode

From a user's point of view, the difference between the 2 data transfer modes is the maximum data rate. The asynchronous mode tops out at 1.5 Mbytes/sec, while the synchronous mode allows data transfers at rates up to 4 Mbytes/sec.

Note: Many SCSI-compatible product suppliers are pushing these specified limits to 2 Mbytes/sec and 5+ Mbytes/sec, respectively.

Asynchronous peripheral controllers can share the same SCSI bus with synchronous ones because SCSI *commands* are always transferred asynchronously. Only the *data transfer mode* gets negotiated.

SCSI products are convenient to use, but not always fast

SCSI products allow you to assemble SCSI-based mass storage subsystems that transfer data reliably. However, the throughput may leave something to be desired, especially when you compare them to ESDI-based or SMD-based disk subsystems.

The reasons for this relatively low performance is two-fold:

- Inadequate peripheral controller designs
- Limitations of the SCSI specification

Many SCSI-compatible peripheral controllers lack performance features that ESDI and SMD controllers have had for years. For example, a *track buffer* in a disk controller allows it to capture and hold all of the sectors of a track. If a disk controller is equipped with a track buffer, the controller can begin reading the sectors of the track into the buffer as soon as the head arrives at the track. This technique (called zero-latency reading) ensures that all of the sectors will be read into the track buffer within a single revolution of the disk.

While zero-latency reading can improve the performance of SCSI-compatible peripherals, the SCSI protocols may still limit performance somewhat. For example, suppose a SCSI command directs the disk controller to get a string of sequential logical sectors that lie on a single track. Although those sectors may lie on the track in ascending order, the head may arrive over the track when the lowest numbered sector has already passed by. In that case, it would make sense to have the controller just send each needed sector over the SCSI bus to the adapter as that sector passes under the disk's read head. However, the SCSI command protocols prevent the controller from doing that because they dictate that the logical blocks be transferred over the SCSI bus in *ascending* order.

In spite of this limitation, SCSI disk controller performance is being improved, and system integrators with higher performance requirements are now considering SCSI for their applications. As specific implementation problems are uncovered (and eliminated) SCSI manufacturers will continue to improve their product performance levels.

First generation SCSI subsystems

First generation SCSI-based subsystems typically employed a free-standing SCSI-compatible peripheral controller card, which interfaces the SCSI bus to some industry standard *device-level* interface. Examples of these devices level interfaces are the:

- ST506 interface for hard disk drives
- ESDI interface for hard disk drives
- SMD for hard disk drives
- QIC-02, or QIC-36 for half-inch or quarter-inch tape drives.

Some peripheral manufacturers package their ESDI or SMD disk drives with a built-in SCSI-to-ESDI, or SCSI-to-SMD peripheral controller. However, the original device-level interface can still be found inside the peripheral enclosure. A SCSI-based mass storage subsystem built with this approach might look like the one shown in Figure 2.

As an alternative many peripheral manufacturers have designed built-in SCSI-compatible peripheral controllers that interface *directly* to the drive electronics, as shown in Figure 3. This eliminates the need for an intermediate device-level interface. However, most of these peripherals are designed for low cost (to appeal to the broadest possible market) and they generally provide less-than-optimum performance.

SCSI system performance can be disappointing

Whether you build your SCSI subsystem based on the approach shown in Figure 1 or Figure 2, its performance can be disappointing. The basic problem is that 1st generation peripheral controllers don't provide a *data buffer* between the peripheral drive and the SCSI bus. Thus, once a peripheral controller is selected by the VME-to-SCSI adapter,

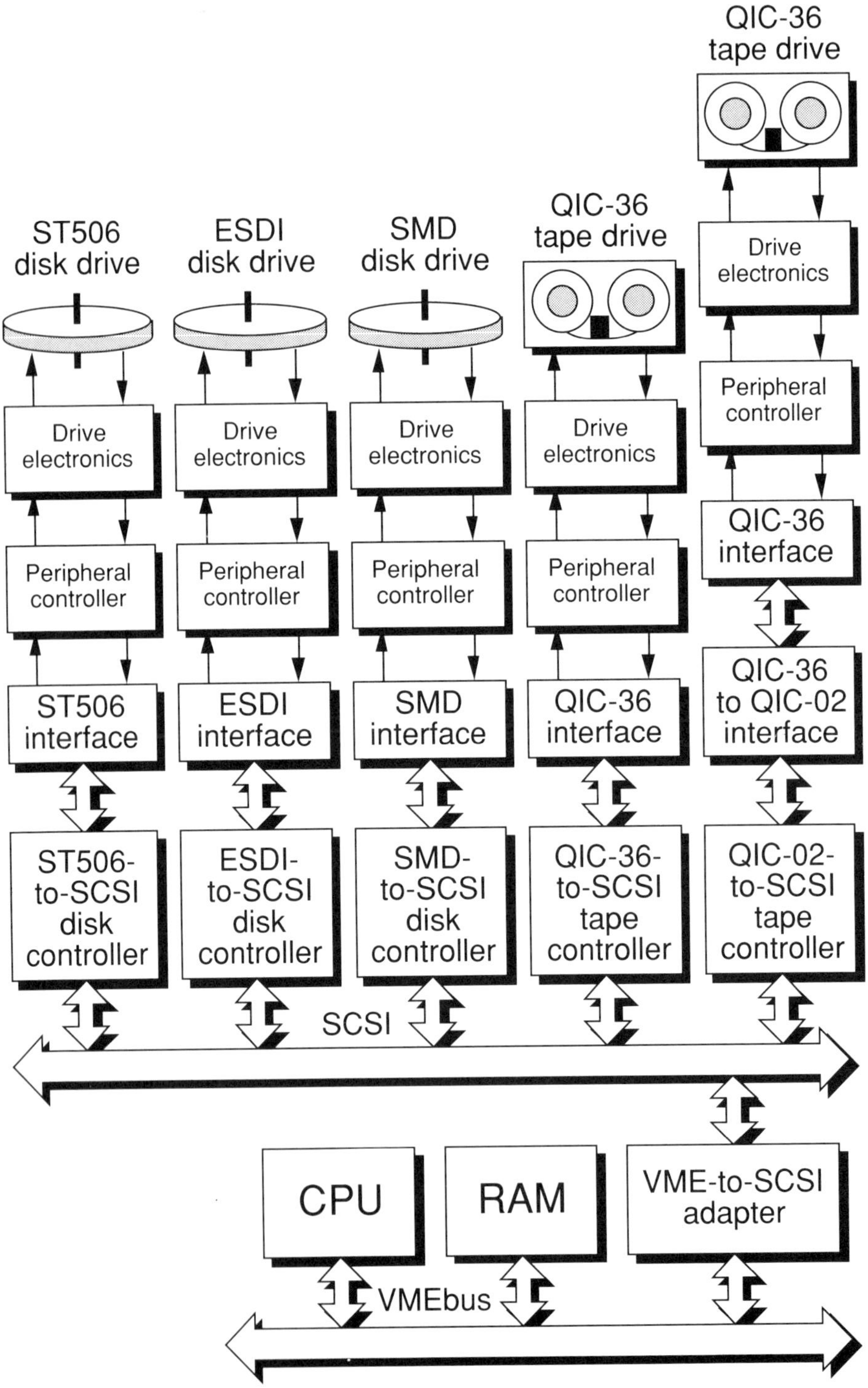
QIC-36
tape drive
ST506
disk drive
ESDI
disk drive
SMD
disk drive
QIC-36
tape drive
Drive
electronics
Drive
electronics
Drive
electronics
Drive
electronics
Drive
electronics
Peripheral
controller
Peripheral
controller
Peripheral
controller
Peripheral
controller
Peripheral
controller
QIC-36
interface
ST506
interface
ESDI
interface
SMD
interface
QIC-36
interface
QIC-36
to QIC-02
interface
ST506-
to-SCSI
disk
controller
ESDI-
to-SCSI
disk
controller
SMD-
to-SCSI
disk
controller
QIC-36-
to-SCSI
tape
controller
QIC-02-
to-SCSI
tape
controller
SCSI
CPU
RAM
VME-to-SCSI
adapter
VMEbus

Figure 2

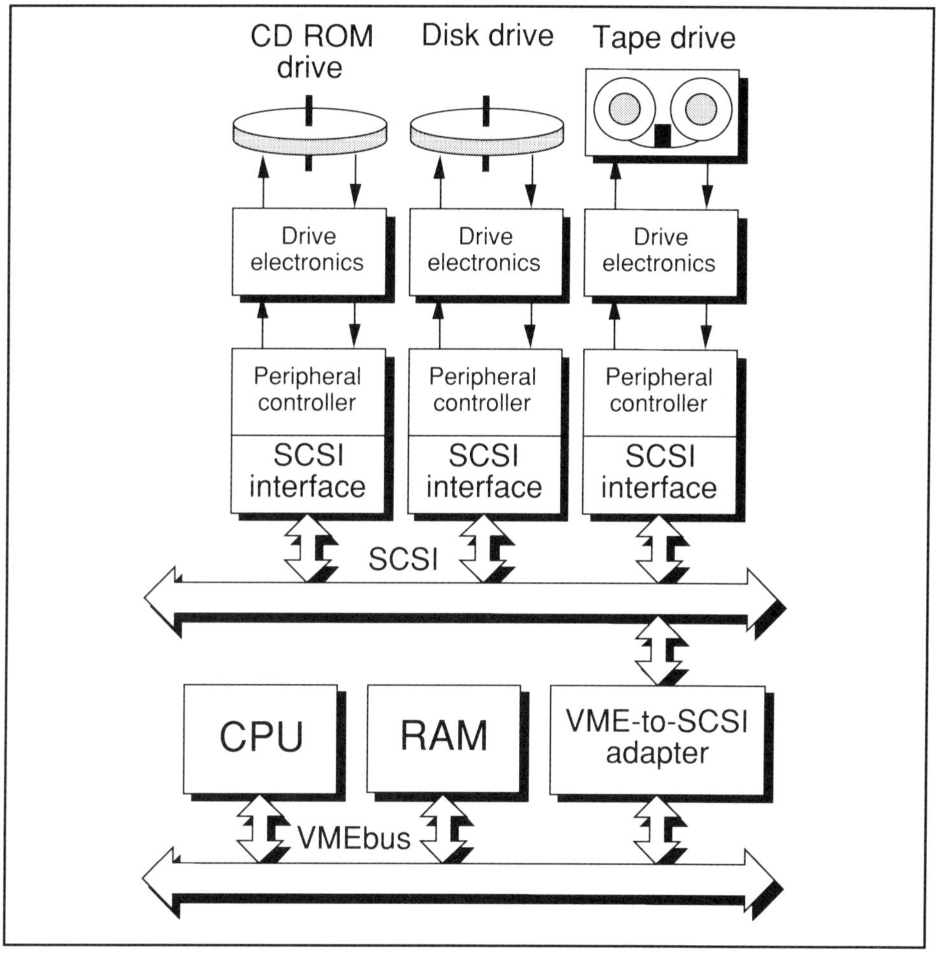

Figure 3

the data transfer takes place across the SCSI bus at the same data rate as that of the selected drive. This is typically *much* slower than the SCSI bus can handle.

This is a problem with both tape drives and disk drives. Tape drives have a *very slow* data transfer rate. Disk drives transfer data much faster. However, there are still pauses in the data transmission whenever the head must be repositioned, to access the next track.

If the peripheral controllers just route the data from the drive directly onto the SCSI bus, they effectively reduce the bandwidth of the bus. For example, suppose you have a disk drive and a tape drive sharing a SCSI bus. The tape drive gets selected, and sends an entire 5 Mbyte file at 250 Kbytes/sec. During this transfer the SCSI bus has an effective bandwidth of 250 Kbytes/sec for 20 seconds. The remaining 95% of the SCSI bandwidth is simply lost during that time, leaving no bandwidth for the disk, and resulting in unsatisfactory disk response time.

The solution: buffering + disconnects
The solution to this problem is to equip each peripheral controller with a buffer that can store the data, as it comes from the drive, and then *burst* this data across the SCSI bus in *packets*. Of course the controller must release the SCSI bus between packet transmissions. This release is accomplished using the SCSI bus *disconnect* protocol.

Once a peripheral controller *disconnects* itself from the SCSI bus, it can perform only *internal* operations. For example, it might read sequential sectors (repositioning the disk head as necessary) until it has read all of the required sectors into its internal buffer. Then the controller *reconnects* itself to the SCSI bus (as soon as the bus is available) and sends the data at the previously negotiated data rate.

Using this disconnect/reconnect protocol, a VME-to-SCSI adapter board can transfer data between itself and one of the SCSI-compatible peripherals, while 6 other (disconnected) peripherals gather data for later transfer. This ability to *overlap* mass storage operations is one of the most important features of the SCSI specification.

Another performance roadblock: command parsing
Some peripheral controllers are *very slow* at parsing the SCSI commands sent to them. If a slow peripheral controller maintains its connection to the SCSI bus while parsing the command, it prevents any traffic over the bus. One way to avoid this is to design the peripheral to disconnect itself immediately, whenever it receives a command. However, if the controller is slow at disconnecting from (and reconnecting to) the bus, this may use up even more bus bandwidth than just holding onto it and parsing the command. In addition, if the peripheral has a data buffer, and if the requested data is already in the buffer, a disconnect/reconnect sequence will just delay its transfer of the data.

The VME-to-SCSI adapter: the 2nd factor in high performance
To work efficiently with a peripheral controller that sends and accepts data in high speed bursts, a VME-to-SCSI adapter should be able to do the same. Unfortunately, many VME-to-SCSI adapters don't have a large enough data buffer to hold multiple data packets. In fact, many have only enough space for a *single* data packet, such as a *sector* of data from a disk drive, or a *block* of data from a tape drive. This means that, before the VME-to-SCSI adapter can accept a block of data from the SCSI bus, the previously received block must be completely written into VMEbus memory. With this single-buffered adapter architecture, all of the improvements made on the peripheral end of the SCSI system are simply wasted, because the VME-to-SCSI adapter becomes the bottleneck.

How large should the adapter's data buffer be? That depends upon how long the adapter might be prevented from using the VMEbus. Suppose data is coming into the adapter from the SCSI bus at a rate of 4 Mbytes/sec, and suppose that the adapter might be prevented from using the VMEbus for up to 1 msec. The adapter would then need to have a 4-Kbyte buffer.

The adapter should fill and empty its buffers quickly
However, you must consider more than just the buffer size. You must also consider how quickly the buffer can be emptied. If your VME-to-SCSI adapter uses a commercial VLSI DMA controller chip to move data from the buffer to VMEbus memory, it may still be a bottleneck. Even the fastest of these chips can move data at only about

3 Mbytes/sec. If you have a high-performance hard disk that sends data over the SCSI bus at 3 Mbytes/sec, even the slightest overhead would cause the adapter's buffer to eventually overflow, regardless of its size.

The data transfer rates between the buffer and the VMEbus memory may also affect the performance of other masters on the VMEbus. For example, if the adapter moves the data at only 10 Mbytes/sec, it may use up much of the available VMEbus bandwidth, substantially impeding other VMEbus traffic whenever SCSI disk transfers are going on. If the adapter's data transfer rates can be raised to 30 Mbytes/sec, most of the VMEbus bandwidth will still be available for use by other masters.

The adapter should be able to queue and sort commands
There is one additional aspect of the adapter's design that you should consider. How does it handle access requests that are directed to different peripherals on the SCSI bus? Ideally, the adapter should keep every possible peripheral continuously accessing data internally. While one peripheral is connected to (and transferring data over) the SCSI bus, the other 6 are disconnected, and are internally accessing data that will be transferred later. In order to do this, the adapter should be able to *queue* incoming commands from the VMEbus, and *sort* those commands, based on the peripheral to which they are directed.

This *command queuing* also permits another form of optimization, called *command grouping*. When the adapter receives separate commands to access 2 sequential logical blocks, it concatenates them into a single command, before transmitting them over the SCSI bus. This eliminates the command overhead of the 2nd command.

Another useful technique (which can be used even when the required blocks are not sequential) is to send the queued commands in ascending order, according to the logical block numbers they access. Since the *physical* blocks on the disk are generally ordered in the same order as the *logical* blocks, this reduces disk head movement, and can profoundly improve disk performance.

If one SCSI bus is good, two are better
Given the fact that the theoretical maximum VMEbus data transfer rate is in excess of 40 Mbytes/sec, and the fact that the SCSI bus theoretical maximum is 4 Mbytes/sec, why not have more than one SCSI bus? For example, if your VME-to-SCSI adapter has 2 SCSI bus interfaces, you can:

- segregate higher-speed disk devices from slower, unbuffered peripherals. This prevents the slower peripherals from blocking access to the disk for prolonged periods of time.
- design a system with 2 high-speed disks. By alternating the storage of a file's blocks between the 2 disks, you can access the file at a rate of 8 Mbytes/sec, instead of 4 Mbytes/sec.

The Interphase V/SCSI 4210
Based upon the principles in this chapter, Interphase designed a high performance VME-to-SCSI adapter called the V/SCSI 4210. It provides 2 SCSI bus interfaces, allowing connection to 14 mass storage peripherals. Both of these SCSI buses can transfer data at rates up to 4 Mbytes/sec simultaneously, giving a total effective data transfer rate of 8 Mbytes/sec.

The V/SCSI 4210 also includes 128 Kbytes of buffering, which is *dynamically* allocated by the onboard 68000. This allocation assigns the available buffer space to the SCSI peripherals dynamically, based on their need at any given time. (This is important when you have several peripherals on the SCSI bus, and the volume of data traffic handled by each varies from time to time.) By optimizing the allocation of the buffers according to the data traffic, the adapter keeps SCSI data transfers moving, and ensures optimum use of the SCSI bus in a way not possible with FIFO-based architectures.

To move data between this buffer and the VMEbus, the adapter uses a state-machine-driven *Buspacket* interface, which allows data to be transferred at rates in excess of 30 Mbytes/sec. This high transfer rate also conserves the VMEbus bandwidth, for use by other VMEbus masters.

The 68000 runs a Simple Multitasking Arbitration in Real Time (SMART) kernel, which enables it to:

- allocate and manage the buffer space
- queue and sort incoming commands
- keep the VMEbus and the 2 SCSI ports active simultaneously.

By doing the command processing and command sorting in parallel with data transfers into (and out of) the onboard buffer, the processor keeps the command processing out of the critical path.

Conclusion

The SCSI bus allows you to put together very high performance mass storage subsystems. However, inadequate peripheral controllers may prevent you from realizing its full potential. By thoroughly understanding where bottlenecks occur, you can more intelligently choose peripherals that will deliver the performance that you need. In addition, you'll find that VME-to-SCSI adapters with features like the ones described in this chapter can compensate for the shortcomings in the available peripherals, and enhance the performance of your SCSI-based system.

Tom Thawley is executive vice president of Interphase Corporation. Before joining Interphase in 1977 as vice president he was the manager of hardware development for Danray, Inc, and a design engineer with Teleswitcher Corporation: both of Dallas. Tom holds a BSEE from the University of Maryland in College Park.

Ernest E. Godsey is director of product marketing at Interphase in Dallas, Texas. He has a BSEE from Texas A&M University and an MBA from the University of Wisconsin. Ernest is a member of Tau Beta Pi, Etta Katta Kappa Nu, and Beta Gamma Sigma honor societies. He has been granted several patents, and has published numerous articles.

CHAPTER 28

SCSI-2: SCSI's High Performance Offspring

By Bill Moren

On June 23rd, 1986 (after more than 4 years of work) the American National Standards Institute approved ANSI X3.131-1986. Slightly over 200 pages in length, the document contains 14 sections and 7 appendices. It defines an interface standard called the Small Computer System Interface (SCSI) bus for connecting peripherals to microcomputers. As the SCSI bus has become widely used in systems of all sizes and performance levels, it has been found to have some performance limitations. In this chapter Bill discusses those limitations, and describes the new version of SCSI (called SCSI-2) which has been designed to allow higher-performance operation.

The Small Computer System Interface (SCSI) bus is a cable bus, used to connect peripherals to microcomputers. Each peripheral can request access to (and gain control of) the SCSI bus, in order to transfer data to (and from) the host system. The bulk of the ANSI standard specifies the bus's physical and logical characteristics, and defines command sets for each of 6 device types:

- Direct-Access (disk)
- Sequential-Access (tape)
- Printers
- Processors
- Write-Once-Read-Many (WORM) optical drive
- Read-Only-Direct-Access optical drive

Most of the commands defined in each command set are used exclusively on that type of device. However, there are some commands that are common to all device types.

SCSI hosts and SCSI targets

The SCSI bus is typically interfaced to a VMEbus-based system through a plug-in board called a VME-to-SCSI adapter, as shown in Figure 1. The VME-to-SCSI adapter provides an intelligent link between the VMEbus and the SCSI bus. Peripheral drives are interfaced to SCSI with a *peripheral controller.*

There are 2 basic types of SCSI devices:

- Initiators
- Targets

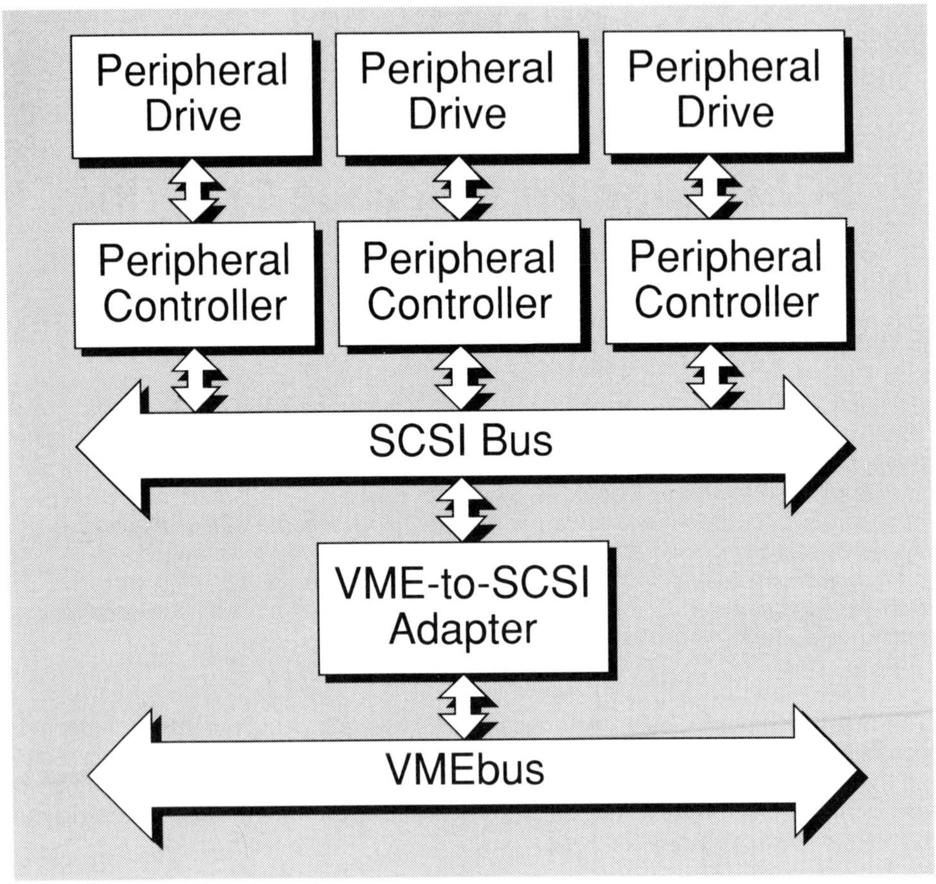

Figure 1

The VME-to-SCSI adapter is typically the *initiator*. It sends commands from the VMEbus system over the SCSI bus, initiating activity in the peripherals. Peripheral controllers are considered *target* devices, since they receive the commands sent by the VME-to-SCSI adapter. Each target can support up to 8 individual drives.

SCSI signal lines and drivers

SCSI provides an 8-bit-wide parallel data transfer bus. The SCSI signals include:

- 9 data signals (including 1 parity signal)
- 9 control signals

Two driver/receiver options are defined:

- Single-ended
- Differential

The *single-ended* drivers allow a maximum cable length of 6 meters, while the *differential* drivers allow a maximum cable length of 25 meters.

Both single-ended and differential configurations use a 50-conductor cable. The signals for the single-ended configuration are shown in Figure 2. Single-ended signal conductors are interleaved with grounded return conductors. The signals for the differential configuration are shown in Figure 3.

1	GND
2	–DB(0)
3	GND
4	–DB(1)
5	GND
6	–DB(2)
7	GND
8	–DB(3)
9	GND
10	–DB(4)
11	GND
12	–DB(5)
13	GND
14	–DB(6)
15	GND
16	–DB(7)
17	GND
18	–DB(P)
19	GND
20	GND
21	GND
22	GND
23	RESERVED
24	RESERVED
25	OPEN
26	TERMPWR
27	RESERVED
28	RESERVED
29	GND
30	GND
31	GND
32	–ATN
33	GND
34	GND
35	GND
36	–BSY
37	GND
38	–ACK
39	GND
40	–RST
41	GND
42	–MSG
43	GND
44	–SEL
45	GND
46	–C/D
47	GND
48	–REQ
49	GND
50	–I/O

Figure 2

1	GND
2	GND
3	+DB(0)
4	–DB(0)
5	+DB(1)
6	–DB(1)
7	+DB(2)
8	–DB(2)
9	+DB(3)
10	–DB(3)
11	+DB(4)
12	–DB(4)
13	+DB(5)
14	–DB(5)
15	+DB(6)
16	–DB(6)
17	+DB(7)
18	–DB(7)
19	+DB(P)
20	–DB(P)
21	DIFFSENS
22	GND
23	RESERVED
24	RESERVED
25	TERMPWR
26	TERMPWR
27	RESERVED
28	RESERVED
29	+ATN
30	–ATN
31	GND
32	GND
33	+BSY
34	–BSY
35	+ACK
36	–ACK
37	+RST
38	–RST
39	+MSG
40	–MSG
41	+SEL
42	–SEL
43	+C/D
44	–C/D
45	+REQ
46	–REQ
47	+I/O
48	–I/O
49	GND
50	GND

Figure 3

The SCSI standard permits the use of either shielded or non-shielded cabling and connectors. For both options, the standard provides mechanical drawings of the connectors.

SCSI bus operation
The SCSI interface is always in 1 of 4 states, called *phases*:

- bus free
- arbitration (optional)
- selection/reselection
- information

The permissible state transitions are depicted in Figure 4.

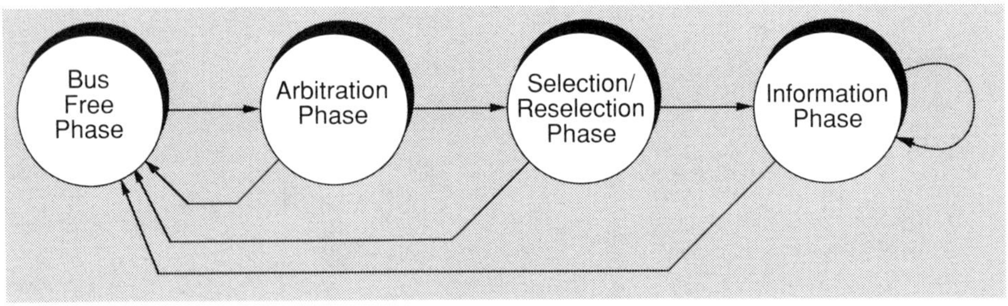

Figure 4

The *information* phase refers collectively to 4 other sub-phases:

- command
- data
- status
- message

During the information phase, information passes between initiators and targets. Each target peripheral controller can accept one command at a time for each of the peripheral drives attached it.

SCSI data transfer modes:
SCSI supports 2 modes of data transfer:

- asynchronous
- synchronous

During *asynchronous* data transfers the receiving device acknowledges receipt of each byte before the next is sent.

The *synchronous* mode allows the sender to send several data bytes over the SCSI bus at a previously established data transfer rate, without waiting for an acknowledge for each byte. To establish the data transfer rate, the initiator uses the asynchronous mode to send a *synchronous data transfer request message* to the target. Within that message is a parameter that specifies a data transfer rate.

Each time it sends a data byte over the SCSI bus synchronously, the sender generates a strobe on the REQ line. If it were sending the data in the asynchronous mode, the sender would then need to wait until the recipient responded with a transition on the ACK line. However, in the synchronous mode, the sender need not wait for the ACK. Instead, it continues to send data bytes at the specified data rate.

As the sender sends bytes to the receiver, the receiver stores them in a FIFO, to await later processing. However, this FIFO has a finite depth, so some provision must be made to avoid FIFO overruns. For this reason the sender is only permitted to send a specified number of bytes before it must stop and wait for an acknowledge on the ACK line (this number is known as the REQ/ACK offset). For example, the initiator specifies the number of bytes a target can send to it, before waiting for an ACK.

The minimum data transfer period permitted by the SCSI specification is 200 nsec. This allows a maximum synchronous data transfer rate of 5 Mbytes/sec. However, the actual synchronous data transfer rates are typically determined by the speed of the SCSI controller chips chosen by the engineer who designs the adapter board. These controller chips are available from:

- Western Digital
- NCR
- Emulex
- National
- Fujitsu

SCSI command descriptor blocks
Commands are sent to peripherals in the form of *command descriptor blocks*. A typical command descriptor block is shown in Figure 5. This block includes:

- an operation code (op code)
- a logical unit number
- a logical block address
- a data transfer length
- a control byte

There are also other types of command descriptor blocks, which permit access to disks with larger storage capacity. (For example, Figure 6 shows a 10-byte block.) In each case, the recipient can

BIT NUMBER

BYTE NUMBER	7	6	5	4	3	2	1	0
0	Operation Code							
1	Logical Unit number			21-bit Logical Block Address				
2								
3								
4	8-bit Transfer Length							
5	Control Byte							

Figure 5

BIT NUMBER

BYTE NUMBER	7	6	5	4	3	2	1	0
0	Operation Code							
1	Logical Unit number			Reserved				
2	32-bit Logical Block Address							
3								
4								
5								
6	Reserved							
7	16-bit Data Transfer Length 16-bit Parameter List Length 16-bit Allocation Length							
8								
9	Control Byte							

Op Code Dependent

Figure 6

deduce the length of the block from the op code in the first byte of the block. The 256 op codes are divided into groups:

- Group 0 = 000x xxxx = 6-byte parameter block
- Group 1 = 001x xxxx = 10-byte parameter block
- Group 2 = 010x xxxx = Reserved
- Group 3 = 011x xxxx = Reserved
- Group 4 = 100x xxxx = Reserved
- Group 5a = 1010 xxxx = Vendor Unique
- Group 5b = 1011 xxxx = Reserved
- Group 6 = 110x xxxx = Vendor Unique
- Group 7 = 111x xxxx = Vendor Unique

SCSI command classes
Within each of these groups, individual op codes are designated as:

- Mandatory
- Extended
- Optional
- Vendor unique
- Reserved

The SCSI standard specifies commands and command descriptor block formats for the *mandatory*, *extended* and *optional* op codes. All SCSI devices must respond properly to *mandatory* commands. If a SCSI device allows system software to configure it in a device-independent fashion, it must do so through the *extended* commands. Devices need not be able to respond to *optional* commands to be considered SCSI compatible.

Vendor unique command codes are available for use at the discretion of an implementer.

Reserved commands are reserved for future standardization.

Figure 7 lists the SCSI commands used to access disk drives. (Disk drives are called *direct access devices* in the SCSI specification.)

The common command set
As peripheral vendors gained experience with SCSI, they found that it didn't go far enough in some respects. For example, important commands, which should have been mandatory were only made optional. Another example was where the SCSI standard left important operations to be implemented with vendor unique commands.

Recognizing the potential for incompatibility, several SCSI disk vendors agreed among themselves to support an informal standard called the *common command set*.

Another problem with SCSI was that the intelligence within each SCSI *peripheral* was sometimes forced to make decisions over which the *host* system should have some influence. For example, when the host system issued a SCSI *format* command to a disk controller on the SCSI bus, it needed to specify how many *spare* sectors to allocate, for later use when media defects were detected. However, the SCSI specification made no provision for passing such a parameter.

Code	Command	SCSI	SCSI-2
00	Test unit ready	O	M
01	Rezero unit	O	O
03	Request sense	M	M
04	Format unit	M	M
07	Reassign blocks	O	O
08	Read	M	M
0A	Write	M	M
0B	Seek	O	O
12	Inquiry	E	M
15	Mode Select	O	O
16	Reserve	O	M
17	Release	O	M
18	Copy	O	O
1A	Mode Sense	O	O
1B	Start/Stop Unit	O	O
1C	Receive diagnostic results	O	O
1D	Send diagnostic	O	M
1E	Prevent/Allow medium removal	O	O
25	Read capacity	E	M
28	Read	E	M
2A	Write	E	M
2B	Seek	O	O
2E	Write and Verify	O	O
2F	Verify	O	O
30	Search data high	O	O
31	Search data equal	O	O
32	Search data low	O	O
33	Set limits	O	O
34	Prefetch	R	O
35	Synchronize cache	R	O
36	Lock/Unlock cache	R	O
37	Read defect data	R	O
39	Compare	O	O
3A	Copy and Verify	O	O
3B	Write buffer	R	O
3C	Read buffer	R	O
3E	Read long	R	O
3F	Write long	R	O
40	Change definition	R	O
41	Write same	R	O
4C	Log select	R	O
4D	Log sense	R	O
55	Mode select	R	O
5A	Mode sense	R	O

M = Mandatory, O = Optional, E = Extended, R = Reserved

Figure 7

To provide for this, the common command set defines a data structure called a *page*. This page (which contains a value for each device-specific parameter stored inside the peripheral) can be passed back and forth between the host and the target.

Even though the common command set was not officially endorsed by ANSI (or by any other recognized standard-setting organization) it was supported by SCSI disk vendors because it ensured compatibility between vendors, and within each vendor's product line.

SCSI has been able to evolve

Many interface standards (once they have been created) remain static until they're eventually obsoleted by a subsequent standard. However, it is possible to evolve an interface standard in a way that permits it to continue serving an industry. That only happens if the industry is involved not only in *developing* the standard, but also in its *modification*, as time passes. The SCSI standard is a good example of this. As the need for higher performance became apparent, peripheral manufacturers evolved SCSI into a new upward-compatible standard, called SCSI-2.

The goals of SCSI-2 are compatibility and connectivity

Many high performance disk users are drawn to SCSI-2 because of its high data transfer rate. However, although SCSI-2 provides high performance, the 3 objectives stated in the scope of the SCSI-2 draft emphasize *compatibility* and *connectivity*:

- "...to provide host computers with device independence within a class of devices."
- "...to provide compatibility with those SCSI-1 devices that support bus parity and that meet conformance level 2 of SCSI-1."
- "...to move device-dependent intelligence out to the SCSI-2 devices."

The last objective emphasizes SCSI's ability to hide the device-specific details of each peripheral drive from the host system. For example, when you interface *directly* to a disk drive, you must understand the physical organization of the disk drive media, including:

- how many heads the disk has.
- how many tracks the disk has.
- how many sectors per track the disk has.
- where the imperfections are in the media.

With a SCSI disk, you don't have to concern yourself with these physical details. A disk just looks like a string of *logical blocks*, starting with block 1, and ending with the size of the disk. The disk appears to have perfect media: no defects require you to remap sectors that lie on blemishes. If the host system reads logical block 100, the SCSI disk controller inside the drive enclosure translates logical block number 100 into the appropriate physical head number, track number, and sector number.

SCSI-2 is upward compatible with SCSI

Even though a considerable amount of functionality has been added to SCSI-2, compatibility with SCSI is still preserved. SCSI and SCSI-2 devices may reside on the same bus and may interact. The physical characteristics of SCSI-2 are the same as SCSI. For example, SCSI-2 uses the same cabling and electrical scheme as SCSI, unless wide data transfers (discussed later) are implemented, in which case a 2nd cable is added. The standard SCSI cable is referred to as the *A-cable* in SCSI-2 nomenclature, and the 2nd cable is called the *B-cable*.

SCSI targets can reject SCSI-2 commands that aren't defined in the SCSI standard. For example, suppose you have a VME-to-SCSI-2 adapter sending commands to a SCSI disk drive. The adapter may attempt to initiate a *wide* data transfer, by sending a message to the disk drive. However, since the SCSI disk drive has no notion of wide data transfers, it doesn't understand the message, and responds by *rejecting* the message, using the SCSI rejection protocol. The SCSI-2 adapter then detects the rejection, and initiates the data transfer using byte-wide SCSI-compatible data transfers.

Note: The actual strategy used by the SCSI-2 host adapter is not prescribed in the SCSI-2 standard. For example, the adapter might first respond to the rejection by initiating a *synchronous* SCSI data transfer, before finally defaulting to an *asynchronous* SCSI data transfer.

Since a SCSI-2 host bus adapter implements the entire mandatory SCSI protocol (and in many cases the entire common command set) it can choose to operate entirely within those protocols. That ensures compatibility with SCSI peripherals.

Since SCSI-2 peripherals respond to all mandatory SCSI commands, they can be directed by SCSI host adapters.

SCSI-2 offers 3 extensions for increased performance
SCSI-2's performance capabilities are the result of 3 extensions:

- *Fast* synchronous data transfers
- *Wide* data transfers
- *Tagged commands*

Fast synchronous data transfers and wide data transfers combine to increase SCSI-2's maximum data transfer rate to 40 Mbytes/sec. The other extension (tagged commands) allows target peripheral controllers to queue multiple commands, and to sequence the execution of those commands for optimum throughput.

Fast synchronous data transfers
The *fast* synchronous data transfer option (fast SCSI) is the best-known SCSI-2 feature. It doubles the 5 MHz maximum SCSI synchronous data transfer rate to 10 MHz. In doing this, the minimum period for synchronous data transfers has been reduced from SCSI's 200 nsec to a minimum of 100 nsec for SCSI-2.

The term *fast SCSI* refers only to the number of *transfers* per second. When fast SCSI transfers are done over a single cable (which permits only 8-bit transfers) data transfers can be done at 10 Mbytes/sec. When 2 cables are used, data can be transferred 32 bits at a time. This permits data transfer rates up to 40 Mbytes/sec.

Fast SCSI is an *optional* mode of operation, and can only be used if both the initiator and the target are equipped to handle it. To initiate a fast SCSI data transfer, the initiator sends a *synchronous data transfer request message* to the target. Embedded in the message are 2 parameters:

- a parameter that specifies the REQ/ACK offset
- a parameter which specifies the data transfer period

The *REQ/ACK offset* tells the target peripheral controller how many REQ pulses it can send during a data transfer before it must wait for an ACK pulse from the adapter.

The *data transfer period* tells the target peripheral controller the data rate to use when sending to the adapter. If the peripheral controller can't send at that rate, it *rejects* the message. Then the adapter sends a new message specifying a lower data rate.

Wide data transfers

The *wide* data transfer mode of operation (often called *wide SCSI*) increases the data path width from 8 bits (SCSI) to either 16 bits or 32 bits (SCSI-2), through the used of a 2nd cable, called the *B cable*. This 2nd cable is a 68-conductor cable that provides 24 additional data bus signals, 3 parity signals, a REQ signal, and an ACK signal. (See Figures 8 and 9) Note: The B cable is only required if wide data transfers are used.

The wider data path is used only for data transfers. Commands and status are still passed using 8-bit transfers. Widening the data path increases data transfer rates by transferring more data per transfer cycle.

Like fast SCSI, wide SCSI is an option. Wide data transfers are initiated with a *wide data transfer request message*. This message includes a parameter that specifies the data transfer width.

SCSI-2 allows commands to be queued in the target

SCSI only permits the VME-to-SCSI adapter to send one command at a time to each peripheral drive. The adapter must then wait for a *status response* from that drive, before sending another command to it.

1	GND
2	GND
3	GND
4	–DB(8)
5	GND
6	–DB(9)
7	GND
8	–DB(10)
9	GND
10	–DB(11)
11	GND
12	–DB(12)
13	GND
14	–DB(13)
15	GND
16	–DB(14)
17	GND
18	–DB(15)
19	GND
20	–DB(P1)
21	GND
22	–ACKB
23	GND
24	GND
25	GND
26	–REQB
27	GND
28	–DB(16)
29	GND
30	–DB(17)
31	GND
32	–DB(18)
33	TERMPWRB
34	TERMPWRB
35	TERMPWRB
36	TERMPWRB
37	GND
38	–DB(19)
39	GND
40	–DB(20)
41	GND
42	–DB(21)
43	GND
44	–DB(22)
45	GND
46	–DB(23)
47	GND
48	–DB(P2)
49	GND
50	–DB(24)
51	GND
52	–DB(25)
53	GND
54	–DB(26)
55	GND
56	–DB(27)
57	GND
58	–DB(28)
59	GND
60	–DB(29)
61	GND
62	–DB(30)
63	GND
64	–DB(31)
65	GND
66	–DB(P3)
67	GND
68	GND

Figure 8

SCSI-2 has *tag messages*, which allow an adapter to send multiple commands to a peripheral drive, without waiting for a status response for previously sent commands. As the SCSI-2 peripheral controller receives these commands, it queues them for later execution.

Tagged commands are commands that are preceded by a tag message. The tag message contains a 1-byte unsigned integer, called the *tag value*. When the target receives a tagged command, it puts the tag value into the queue, along with the command.

There are 2 types of tagged commands:

- *Ordered* tagged commands
- *Unordered* tagged commands

Ordered tagged commands can be used to ensure that commands are executed in a first-come-first-served manner.

Unordered tagged commands are used when the command execution order is not important. The target takes these *unordered* tagged commands, and queues them, just as it would queue *ordered* tagged commands. However, it is then allowed to *sort* the unordered tagged commands, to speed up overall throughput. There are 2 techniques used to improve throughput:

- Command combining
- Seek optimization

Command combining is when the disk controller searches for queued commands that access adjacent sectors, and combines them into a single, multiple-sector disk access. This reduces the number of disk revolutions needed to execute a series of commands.

Seek optimization reduces seek latencies by reordering the sequence of the

1	GND
2	GND
3	+DB(8)
4	–DB(8)
5	+DB(9)
6	–DB(9)
7	+DB(10)
8	–DB(10)
9	+DB(11)
10	–DB(11)
11	+DB(12)
12	–DB(12)
13	+DB(13)
14	–DB(13)
15	+DB(14)
16	–DB(14)
17	+DB(15)
18	–DB(15)
19	+DB(P1)
20	–DB(P1)
21	+ACKB
22	–ACKB
23	GND
24	DIFFSENS
25	+REQB
26	–REQB
27	+DB(16)
28	–DB(16)
29	+DB(17)
30	–DB(17)
31	+DB(18)
32	–DB(18)
33	TERMPWRB
34	TERMPWRB
35	TERMPWRB
36	TERMPWRB
37	+DB(19)
38	–DB(19)
39	+DB(20)
40	–DB(20)
41	+DB(21)
42	–DB(21)
43	+DB(22)
44	–DB(22)
45	+DB(23)
46	–DB(23)
47	+DB(P2)
48	–DB(P2)
49	+DB(24)
50	–DB(24)
51	+DB(25)
52	–DB(25)
53	+DB(26)
54	–DB(26)
55	+DB(27)
56	–DB(27)
57	+DB(28)
58	–DB(28)
59	+DB(29)
60	–DB(29)
61	+DB(30)
62	–DB(30)
63	+DB(31)
64	–DB(31)
65	+DB(P3)
66	–DB(P3)
67	GND
68	GND

Figure 9

queued commands, so that the drive's read heads will never pass over a cylinder for which there is a pending request.

SCSI-2 offers 3 upgrades in the logical interface

Even though most of the fanfare is directed at the *physical* changes that allow higher SCSI-2 data transfer rates, much of the effort put into SCSI-2 involves upgrading the *logical* interface with 3 broad types of enhancements:

- Incorporation of the *common command set*
- Addition of new *logical functions*
- Addition of new *device types*

Incorporation of the common command set

Much of the SCSI common command set has been made mandatory in SCSI-2. An example of this is the use of *pages* to pass parameter values between the host system and SCSI-2 peripherals. For example, several page types have been defined for use with disk drives:

- Caching
- Floppy disks
- Formatting
- Medium types
- Notching
- Partitioning
- Read/Write error recovery
- Rigid disk geometry
- Verify error recovery

SCSI-2 also defines page formats for accessing and modifying the parameters stored in other types of peripherals.

New logical functions

SCSI protocols require a peripheral to return a *status* to the host adapter, upon receiving each command. However, depending upon the peripheral, this might take considerable time. For example, suppose a host system initiates a REWIND operation on one of its SCSI tape drives. The host would not get a *completion status* until the rewind operation is finished.

It would be convenient if the tape drive could initiate the rewind operation, and then return a status to the host, indicating that it had *accepted* the command. Then, at some later time, the drive could notify the host when the rewind was complete.

Unfortunately the SCSI standard has no provision for doing this. However, SCSI-2 defines an *asynchronous event notification* protocol. This protocol allows a peripheral to notify an adapter of some event that has occurred inside the drive. This event need not be the result of a previous command from the host. For example, if someone takes the tape drive off-line to change tapes, the drive could notify the host system that it has been taken off-line.

New device types in SCSI-2
SCSI-2 also defines 3 new device types:

- Scanner
- Medium Changer
- Communications

In addition, SCSI-2 has *redefined* the optical device types previously defined by SCSI...

- write-once-read-many (WORM) direct access
- read-only direct access

...to give 3 new types of optical devices:

- write-once-read-many direct access
- CD-ROM serial access
- read-only or write-many direct access

Each of these SCSI-2 device types have command sets that have been designed to support currently available products.

Migrating from SCSI to SCSI-2
Many current SCSI users want to migrate from SCSI to SCSI-2, as long as that migration doesn't *prevent* them from using existing (proven) SCSI peripherals. To allow for this, the SCSI-2 designers provided a migration path that can be followed in an incremental fashion. (See Figure 10)

Migration step 1: Replace the VME-to-SCSI adapter
A VME-to-SCSI-2 adapter should provide an operational *superset* of the VME-to-SCSI adapter it replaces. To avoid redesign of the existing adapter's software device driver,

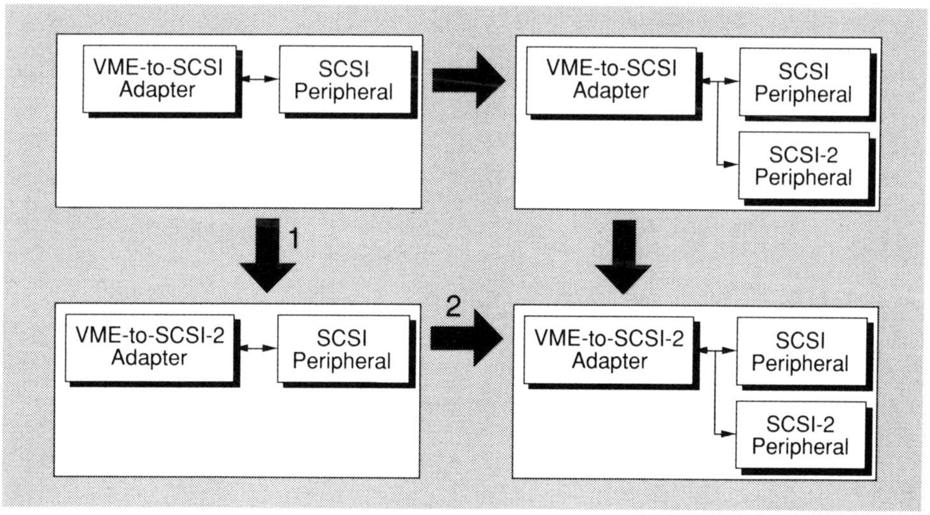

Figure 10

the VME-to-SCSI-2 adapter should also be designed to preserve the existing adapter's register interface. This allows you to substitute the VME-to-SCSI-2 adapter for the VME-to-SCSI adapter, while maintaining the full function of the existing system, with no changes in the SCSI peripherals.

Migration step 2: Add SCSI-2 peripherals
Once the VME-to-SCSI adapter has been replaced with a VME-to-SCSI-2 adapter, SCSI-2 peripheral devices can be added to the system. The SCSI-2 peripherals are simply plugged into the bus. Since the old VME-to-SCSI *software device driver* may not fully use the features of the VME-to-SCSI-2 adapter, you can't exercise all of the SCSI-2 features in these new peripherals. However, you can still test and verify them as SCSI devices. Even though you only use a portion of their full functionality, they can still be useful in the system.

Migration step 3: Update the adapter device drivers
The final step to an incremental SCSI-2 migration is to replace the VME-to-SCSI adapter's *device drivers* with drivers that fully support the extended features of the new VME-to-SCSI-2 adapter. The VME-to-SCSI-2 adapter can then send SCSI-2 messages and commands, when interacting with the peripherals.

Note: If the SCSI-2 system will be using wide data transfers, an additional step is required. You must install a 2nd cable between the VME-to-SCSI-2 adapter and the SCSI-2 peripherals. This cable provides the additional data lines that will be needed to move data 16 bits (or 32 bits) at a time.

An alternative migration path is also shown in Figure 10. SCSI-2 peripherals can be incorporated into the SCSI system before the VME-to-SCSI adapter is upgraded to a VME-to-SCSI-2 adapter. Since the SCSI-2 peripherals recognize all SCSI commands, they can be controlled by the VME-to-SCSI adapter.

PTDs and PDAs: 2 ways to build fast disk systems
There are 2 popular techniques for building high performance disk subsystems:

- Parallel Transfer Disks (PTDs)
- Parallel Disk Arrays (PDAs)

What are PTDs?
PTDs allow very fast data transfer rates by performing parallel data transfers to multiple disk surfaces, which all share a common spindle. The actual data transfer rate is the *product* of the data transfer rate for each surface and the number of surfaces. PTDs are often special versions of standard disks. For example, Fujitsu makes a PTD with an 18 Mbyte/sec data transfer rate. It reads from 6 heads in parallel, at data transfer rates of 3 Mbytes/sec, and routes the data to the host through an SMD-like interface. Its interface differs from SMD in that it has 6 serial data channels, instead of just one. (One channel for each disk head.)

What are PDAs?
A Parallel Drive Array (PDA) is a disk subsystem consisting of several independent disk drives, which are all accessed through a single disk controller. This PDA controller provides fast data transfer rates by using a technique called *striping*. As the controller

writes data to the disks, it divides the data up on a byte-by-byte basis. It then writes the 1st byte to drive 1, the 2nd byte to drive 2, etc.

Data transfers to the various drives in the drive array occur in *parallel*, so the resulting data transfer rate is the product of the number of drives in the array and each individual drive's data transfer rate. Because all of the striping details are handled by the intelligent PDA controller, the host sees the PDA subsystem as a single, fast, large-capacity drive.

Using PTDs and PDAs in microcomputers

PTDs and PDAs have traditionally been used in very expensive, extremely high performance systems, such as supercomputers or special imaging systems. However, extremely fast CISC microprocessors, RISC processors, and multiple-processor VMEbus systems have created a demand for disk I/O performance levels that can only be provided with PTDs and PDAs.

Historically, PTDs and PDAs have been built using proprietary interfaces. However, SCSI-2 now provides a viable interface for PTDs and PDAs. In addition to SCSI-2's performance, it also provides the benefits of the huge base of SCSI compatible disk drives.

The Rimfire 6600

The Rimfire 6600 PDA SCSI-2 disk controller (shown in Figure 11) interfaces 5 ESDI compatible disks to the SCSI-2 bus. (See Figure 12) It contains 6 processors. One is a *supervisory* processor, which controls overall board operation. The other 5 processors each serve as an ESDI controller for one disk drive.

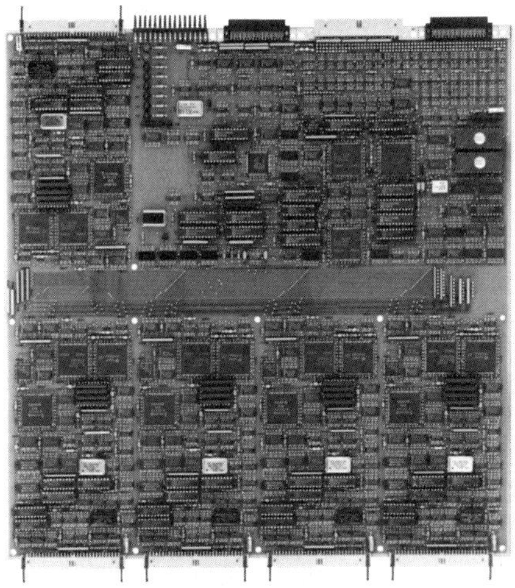

Figure 11

The onboard intelligence provided by these processors allows striping of the data bytes to 4 of the drives, as well as a parity byte to the 5th drive. The parity byte is generated from the data bytes stored on the 4 data drives. (See Figure 13.) Since this striping is hidden from the host computer, the disk array looks (to the host) like a single, fast, large capacity drive.

The actual data transfer rate of this disk array depends on the drives used. With 20 MHz ESDI drives (available from Hitachi and from Hewlett Packard) the

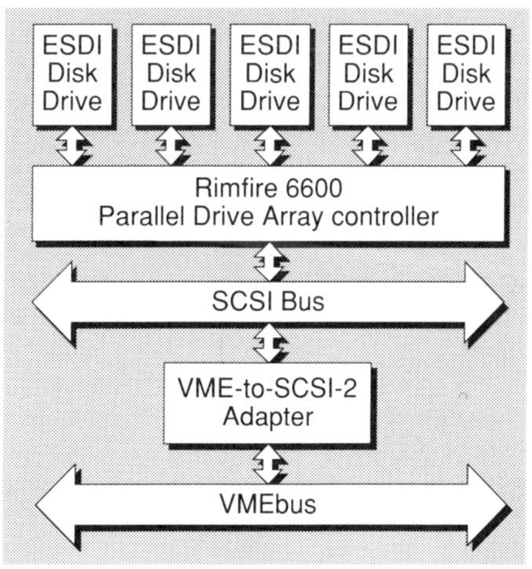

Figure 12

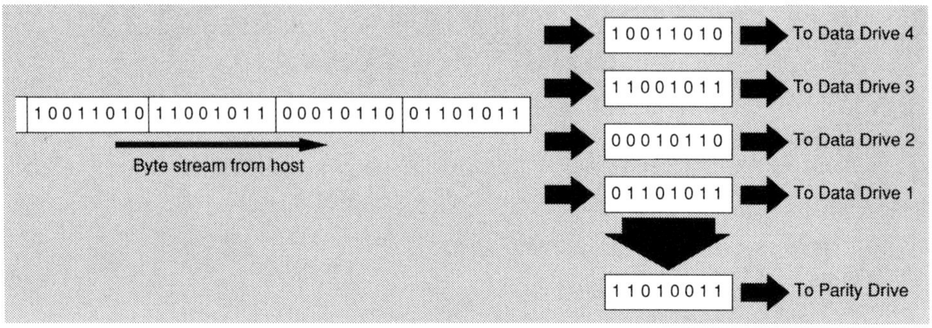

Figure 13

burst data transfer rate is 10 Mbytes/sec, as shown by the following computations:

(20 Mbits/sec)/(8 bits/byte) = 2.5 Mbytes/sec

(2.5 Mbytes/sec) x (4 drives) = 10 Mbytes/sec

The Rimfire 6600 controller board can sustain this 10 Mbyte/sec SCSI-2 data transfer rate indefinitely. However the actual data transfer rate will be lower because, even though the data in each sector is read from the disk at 20 Mbits/sec, the *gaps* between the sectors lower the *average* data transfer rate. In addition, when the head reaches the end of a track, it must be repositioned. This takes time.

In fact, head repositioning times can be quite large. As a result, the average sustained data rate is usually defined to be the rate at which you can read a *track* of data. That depends on how the disk is formatted.

In general, a disk is formatted with a fixed number of bytes per sector, plus a few bytes of overhead, associated with each sector. If you have 512-byte sectors, and 80 bytes of overhead per sector, you will not be able to put as much data on each track as you could with 1024-byte sectors, and the same 80 bytes of overhead per sector.

For example, suppose you have a total unformatted track capacity of 20,000 bytes. If you format it with 512-byte sectors, you might get about 16.5 Kbytes of storage per track. However, if you use 2-Kbyte sectors, you might get 18 Kbytes of storage on the track.

In general, with an optimal sector size, a Rimfire 6600-based PDA can provide sustained rates of about 85% to 90% of the combined burst data transfer rates of the drives.

Surviving a disk failure

The parity drive allows the array to continue operating in the event of a single drive failure. If a single drive failure occurs, the failed drive can be removed, and a replacement drive installed. Then the data from the failed drive is *regenerated* on the replacement drive.

The removal, replacement, and data regeneration process is performed while the Rimfire 6600 continues to execute SCSI-2 commands. The ability to tolerate single drive failures improves the Mean Time Data Availability (MTDA) of the disk subsystem. The MTDA of an array of disk drives predicts the mean time between *simultaneous* failures of two disk drives, and can be calculated as follows...

$$\text{MTDA} = \frac{(\text{MTBF})^2}{(N_d+N_p)(N_d+N_p-1)(\text{MTTR})}$$

Where...

N_d = Number of data drives
N_p = Number of parity drives

For a 5-drive array, with individual drive MTBFs of 40,000 hours, and a Mean Time To Repair (MTTR) for a single drive failure of 72 hours, the calculated MTDA is approximately 1.1 million hours (127 years).

Spindle synchronization
One problem with using a simple parallel drive array is that the various disks may be at different rotational positions when an access is started. Since striping stores data on all the disks, the access cannot be completed until the last drive rotates far enough to bring the needed sector under its read head.

If there were only a single disk drive, the average time would be half a rotation. However, with 5 drives, it's quite likely that at least one drive will need more than half a rotation, so the average time is greater.

To reduce this average time, a technique called *spindle synchronization* is used. One drive serves as a *master*, and the other drives are *slaves*. The master drive provides a *reference* signal, and the other drives synchronize their spindle motion to that reference. Once synchronized, the corresponding sectors on each of the disk drives pass under their respective read heads at the same time. This reduces the average access time to what it would be for a single drive: half a rotation.

The advantages of spindle synchronization were great enough to convince ESDI vendors to *add* the needed signals, as an appendix to the ESDI specification. The Rimfire 6600 controller board uses these signals to tell the various drives in the array who is the master, and who are the slaves. If a drive fails, its dedicated processor (on the 6600 board) detects the failure, and reports it to the supervisory processor. If the supervisory processor detects a failure of the master drive, it selects another drive to be the master drive, allowing the failed drive to be pulled out and replaced.

Note: The information on the parity drive is *not* used to detect *drive* failures. A drive failure is detected by the dedicated processor, based upon CRC errors, etc. The parity information is used, *after* a failed drive has been replaced, to *regenerate the data* .

Cache Memory
To further improve throughput, the Rimfire 6600 controller board has 512-Kbytes of *cache memory*. This cache is physically organized as 128 Kbytes of cache per data

drive. If the controller is commanded to read a block of data, it can *pre-read* an entire track of data from each of the drives into the cache. With 128 Kbytes per drive, the cache can hold 3 to 5 tracks per drive at any given time. Sequential requests (typical of applications requiring PDA technology) can then be satisfied by simply accessing the cache, eliminating the need for additional disk accesses.

The Rimfire 6600's SCSI-2 interface

The Rimfire 6600's SCSI-2 interface supports burst transfers across the SCSI-2 bus at rates up to 20 Mbytes/sec. This is achieved through 32-bit-wide data transfers, and a 5 MHz synchronous data transfer rate. The board also implements tagged commands, allowing it to receive multiple commands from the host. The controller optimizes command execution through command combining and seek optimization.

The Rimfire 3550

Ciprico has also designed a VME-to-SCSI-2 adapter board called the Rimfire 3550. (See Figure 14) This VME-to-SCSI adapter supports both the A cable and the B cable of the SCSI-2 bus, allowing wide (32-bit) data transfers. It also allows the VMEbus system to send tagged commands over SCSI-2. Like the Rimfire 6600, the 3550 supports synchronous SCSI-2 data transfers at 20 Mbytes/sec.

Figure 14

The 3550 board uses a *short-burst FIFO* to buffer the data flow between the SCSI-2 bus and the onboard 2-Kbyte FIFO. (See Figure 15.) This 1st Short-Burst FIFO, and the 2-Kbyte FIFO, accept data from the SCSI-2 bus, while the board waits for permission to use the VMEbus. Once the board gains access to the VMEbus, a 2nd Short-Burst FIFO empties the 2-Kbyte FIFO at VMEbus rates. This 2nd Short-Burst FIFO allows the board to transfer data over the VMEbus in excess of 30 Mbytes/sec.

The adapter also uses a *pipelined system interface* to optimize VMEbus DMA transfers. Its address and counter registers are *pipelined*. This allows the adapter's onboard processor to *preload* the starting address (and the data transfer count) for each DMA transfer while the previous DMA transfer is still in progress.

This pipelined register architecture is especially useful for *scatter/gather* data transfers, often seen in UNIX-based systems. Suppose an application program requests the UNIX operating system to open and read several contiguous sectors from a disk file. UNIX then selects memory buffers from a buffer pool, directing one sector into each buffer. Since the available buffers in the pool aren't usually contiguous in memory, sequential sectors from the disk may not be contiguous in memory. Because of this you need some way to DMA each sector to a different buffer location in memory. The pipelined address and counter registers allow the board to DMA sequential sectors into different areas of memory, without pausing at the end of each DMA transfer to reconfigure the DMA controller.

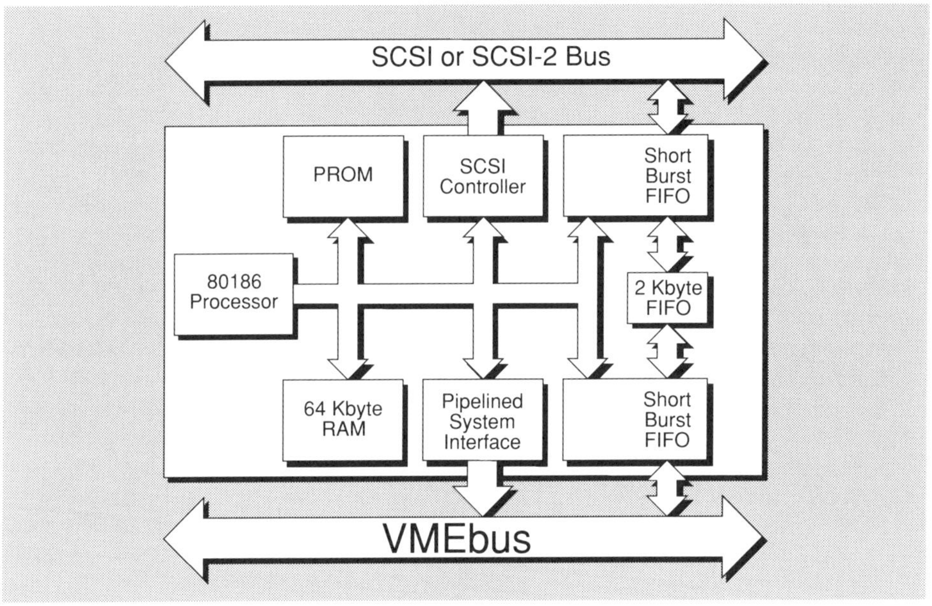

Figure 15

The Rimfire 3550 software interface

The software interface to the Rimfire 3550 VME-to-SCSI-2 adapter consists of parameter blocks, which include SCSI-2 command descriptor blocks (like the ones shown earlier in Figures 5 and 6) plus some additional parameters that are needed to execute the command. For example, in order to transfer data between VMEbus memory and the disk drive, you need to provide a VMEbus address.

A VMEbus processor board builds a *circular queue* of these parameter blocks in VMEbus memory, and then writes a pointer to the 3550 adapter board, to indicate the memory location of the head of the queue. The 3550 board then extracts and executes these parameter blocks, updating the pointer each time.

Since this method of operation is the same as that used on the earlier Rimfire 3510 Series of VME-to-SCSI adapters, the Rimfire 3550 board operates as a VME-to-SCSI adapter when used with 3510 device drivers. If minor coding changes are incorporated into the 3510 drivers, the wide data transfer option, and the tagged command facility of the 3550 can also be used.

Bill Moren has worked for Ciprico since June of 1978, and has served in various engineering and marketing positions prior to his current position as product manager. His current responsibilities include the marketing management of the company's products from concept to obsolescence.

CHAPTER 29

The IPI Standard: The Peripheral Answer To RISC

By James M. Mooney

Because RISC processors are able to execute programs faster than earlier processors, they read and write to disk more often than CISC executing the same software. If the disk subsystem can't keep up with these more frequent disk accesses, the system will become I/O bound, and you won't get the full benefits of RISC technology. Disk drive vendors have responded by marketing faster drives, but it's now clear that the popular disk interfaces are creating bottlenecks by restricting the rapid transfer of data between the host system and the disk drive. To eliminate this bottleneck, the Intelligent Peripheral Interface (IPI) standard has been developed. In this chapter, Jim shows how to design a disk controller board that takes full advantage of the new IPI standard.

The recent introduction of RISC processors has greatly improved the processing performance of microcomputer systems. Single processor RISC workstations and minicomputers provide performance up to 10 or 12 MIPS, while some multiple-processor RISC architectures are exceeding 30 MIPS.

According to David Weaver (See Reference 1), today's RISC machines average about 1.3 to 1.7 clock cycles per instruction.

If we assume...

- 1.5 (clock cycles) / (RISC instruction)
- 3.5 (clock cycles) / (CISC instruction)

...we conclude that we can execute about 2.3 times as many RISC instructions in the same amount of time. However, David also points out that it takes 20 to 40 percent more RISC instructions than CISC instructions to execute the same high-level language program.

If we assume...

- 1.3 (RISC instructions) / (CISC instruction)

...we conclude that a RISC processor would execute a program about 1.8 times faster than a CISC processor with the same clock speed.

Where these calculations are valid, a RISC processor would initiate 1.8 times as many I/O operations as a CISC processor, per unit of time, when executing the same program. So, we see that the introduction of RISC processors alone has almost doubled the demands being placed on disk subsystems.

Disk performance: how it's measured

To quantify disk I/O subsystem performance, 2 different measurements are often used:

- Throughput
- I/Os per second

Throughput (measured in Mbytes/sec) is directly dependent upon how many bits per second pass under the read head of the disk drive. Any design change in the disk drive which increases this data rate is directly reflected in the throughput of the disk.

I/Os per second is a measure of the number of I/O operations the disk subsystem can complete in one second. This measure is of particular importance in transaction processing systems, where the data transfers to (and from) the disk are not large, but may be quite frequent.

Disk subsystems designed to provide a large number of I/O operations per second are not necessarily those with the fastest access times. A good analogy is a high-volume manufacturing plant, which is organized to process the maximum number of units per day, even though the *transit time* of a single manufactured unit through the plant may be several days.

In general, a disk subsystem optimized to process a large number of I/Os per second is well suited to *multitasking* computer systems. That's because, in a multitasking system, the host processor can temporarily set aside any task that's waiting for an I/O operation and resume execution of some other task, whose I/O has completed. As a result, relatively long disk access times don't hurt the system throughput.

Disk performance is not increasing as fast as processor performance

Figure 1 shows that there has been a growing gap between processing performance and disk performance. If we assume that a 1-MIPS processor (1982), running a benchmark program, generates 40 I/Os per second, then an 8-MIPS processor (1989) running the same program would generate 320 I/Os per second.

In 1989 (using existing standard peripherals and controllers) the best possible performance (using SMD-E) was about 80 I/Os per second: a factor of 3 to 4 short of what was needed. Even taking into account the fact that 1989 (RISC) MIPS are based on less powerful instructions, than 1982 (CISC) MIPS, the fact is clear: processor performance is outrunning disk performance.

Graphics demands

Increased processing power has also made it practical to provide animated, interactive graphical user interfaces. However, these new interfaces also demand much more disk I/O performance than earlier, text-only interfaces. For example, if you assume...

- a 30 frame/sec refresh rate

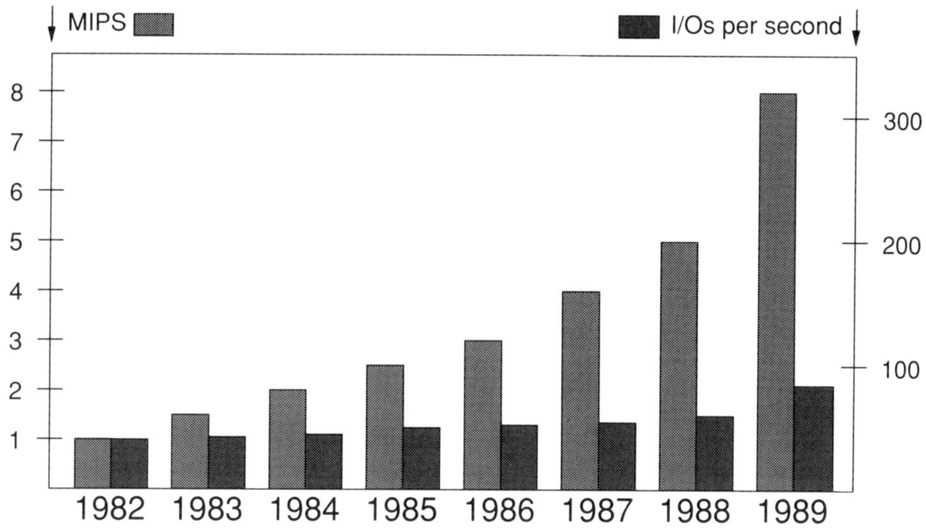

Figure 1

- a 512 x 512 monochrome display
- 8 bits per pixel

...you will need a disk transfer rate of...

(30) x (512) x (512) = 7.8 Mbytes/sec

Things get much worse when you have a high-resolution full-color display. For example, if you assume:

- a 30 frame per second refresh rate
- a 1024 x 1024 color display
- 3 color planes
- 8 bits per color

...you will need a disk transfer rate of...

(30) x (1024) x (1024) x (3) = 94.4 Mbytes/sec

Fortunately, these are *worst case* numbers. In most cases the system doesn't need to read *all* of the displayed pixels from the disk, just to display each new frame. However, data transfer rates can still be extremely high.

Reducing the amount of needed disk I/O
Faced with the inadequate performance of current disk subsystems, engineers are using 3 techniques to reduce the need for disk accesses:

- Adding more system memory
- Using more sophisticated disk caching techniques
- Developing more efficient memory management

While these techniques do reduce the number of disk accesses, they don't solve the problem entirely. Only an entirely new approach to the design of disk subsystems can close the performance gap.

Two popular device-level interface standards
The majority of today's high performance computers (from technical workstations to graphics supercomputers) have disk subsystems based on either the SMD-E or ESDI device-level interface standards.

ESDI provides more compact storage
ESDI drives have 5.25-inch platters, and are used primarily in desk-side workstations, desk-top computers, and in other systems in which packaging size is critical. Although the ESDI disk interface standard permits data transfer rates up to 3 Mbytes/sec, ESDI drives typically transfer data at 15 MHz (1.8 Mbytes/sec). Note: Hewlett Packard offers a 20 MHz (2.5 Mbyte/sec) drive.

ESDI controller boards typically support up to 4 drives, using a single daisy-chain cable, and 4 radial cables. (See Figure 2) The maximum allowed length of each ESDI radial cable is only 3 meters. In small desk-top systems this presents no problem, but in larger systems this cable length dictates where the drives can (and can't) be mounted, with respect to the controller.

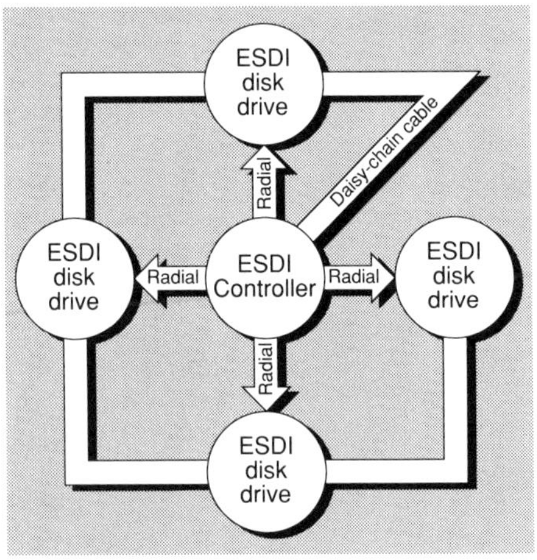

Figure 2

SMD provides higher performance
The original SMD interface allowed data transfers at 1.8 MHz. Later, CDC defined 2 higher speed versions of SMD (2.4 and 3.0 MHz) and called them SMD-E (SMD-Extended). The 3.0 MHz SMD-E interface is the current disk interface of choice for high-performance applications.

Neither SMD or SMD-E has ever been formally adopted as a standard. However, the SMD-E interface is defined by a Control Data Corporation document known as *CDC standard 64712402.*

SMD-E drives are typically used on large workstations, such as compute engines or file servers. The drives typically have 8-inch or 9-inch platters, and transfer data at 3 Mbytes/sec.

Although the SMD-E standard permits a controller to interface to 8 disk drives, most SMD-E controller boards support a maximum of 4 disk drives. The cabling is similar to ESDI: a single daisy-chain cable, and 4 radial cables. However, the radial cables can be up to 50 feet long.

Current disk interfaces have become a bottleneck
Existing disk I/O subsystem standards (such as SMD-E and ESDI) can't provide the kind of performance needed to adequately support the double-digit MIPS performance of the new generation of processors, nor the massive data movements required for today's high-resolution graphics. These old drive interfaces are preventing system designers from getting the full benefit from newer, faster drives.

The primary problem with both SMD-E and ESDI is that they transfer data *serially*, with a maximum data transfer rate of 24 Mbits/sec, or 3 Mbytes/sec. While this 3 Mbyte/sec data transfer rate was adequate for the demands of the last generation of microprocessor-based systems, it's grossly inadequate for the new generation of systems. Moreover, SMD-E and ESDI performance cannot be stretched any further.

IPI: the new high-performance standard
In 1980 a proposal was submitted to an ANSI committee, describing a new high-performance parallel-data-transfer disk interface standard. By 1986, the interface standard was completed and was published as ANSI documents X3.129-1986 and X3.130-1986. It is known as the Intelligent Peripheral Interface (IPI) standard.

IPI is one of the fastest disk interface standards available today, with a maximum data transfer rate of 10 Mbytes/sec. Imprimis offers *2-head parallel* IPI-compatible drives, with data transfer rates of 6 Mbytes/sec. Hitachi offers 4.5 Mbyte/sec drives. Virtually all of the major disk drive manufacturers have introduced IPI drives with data transfer rates above 3 Mbytes/sec.

IPI permits data to be transferred 16 bits at a time over a cable up to 125 meters long. The interface is defined in a layered fashion:

The 1st layer is called IPI-0. It defines a 16-bit *parallel bus* running at 5 MHz.

The 2nd layer is called IPI-1. It defines a *state machine*, and the procedures used to:

- Select a peripheral device
- Deselect a peripheral device
- Poll a peripheral device to determine its status.

The 3rd layer is called IPI-2. It defines an interface that allows the host to access the peripheral at the *physical* level. For example, when accessing a disk drive, the host could specify a cylinder number, a head number, and a sector number within the track.

The 4th layer is called IPI-3. It defines an interface that allows the host to access the peripheral at the *logical* level. For example, when accessing a disk, the host could specify a logical block number, and the disk would calculate the cylinder number, the head number, and the physical sector number needed for the access.

Figure 3 compares the data transfer rates of several available disk controllers to IPI controllers. It shows that:

- IPI more than triples the data transfer rate of the 3 Mbyte/sec SMD-E and ESDI serial interfaces

- IPI doubles the performance of the 5 Mbyte/sec Small Computer System Interface (SCSI)

Cabling an IPI system is easy. An IPI controller can accommodate 8 disk drives, using a single 50-meter daisy-chain cable. This allows you to locate peripheral drives for the convenience of the user, instead of having their locations dictated by cable lengths.

IPI's 10 Mbyte/sec data transfer rate is adequate for most current applications, and its parallel data transfer capability also provides room for further extensions. In fact, the ANSI committee is currently considering methods for enhancing IPI to support 25 Mbyte/sec data transfer rates, with 50 Mbyte/sec and 100 Mbyte/sec options to follow.

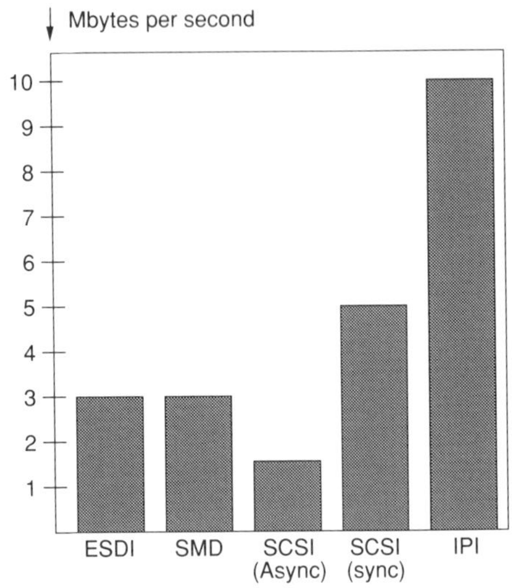

Figure 3

IPI has been mainly used to interface to large 8-inch and 9-inch drives. However most disk drive manufacturers are introducing 5.25-inch IPI drives. This allows you to use the IPI interface in a wide variety of systems, whether performance or compact size is most important.

You can build very large capacity storage subsystems with IPI

Since IPI requires only a single connector (as compared to the multiple connectors needed by SMD-E) there's typically room to provide more than one IPI channel with a single controller board. The only requirement is that there be sufficient board real estate to accommodate a set of signal line drivers and receivers for each connector. (The board's intelligence and data buffering can be shared.) Fortunately, through the use of custom ASICs and surface-mount technology, this is now possible on VMEbus boards.

How to design a high performance IPI disk controller board

To take advantage of the much higher data transfer rates offered by IPI, you need to carefully architect your IPI controller board to move data quickly. This becomes especially important when the VMEbus is heavily used by other masters. To avoid extra disk revolutions, the IPI controller board should be able to transfer data over the VMEbus 2 to 3 times faster than it's coming in over the IPI bus.

Since the data might be transferred over the IPI bus at its full 10 Mbyte/sec rate, your IPI disk controller board should have a VMEbus DMA controller that can transfer data at rates of 20 to 30 Mbytes/sec.

The difference in the data transfer rate of the VMEbus and IPI bus can be handled by equipping the disk controller board with an onboard data buffer: either a cache or a FIFO. (Current cache-based and FIFO-based SMD-E disk controllers are comparable in performance.)

When designing the data buffer, you need to ensure that it can:

- be accessed fast enough (through *both* of its ports simultaneously) to handle the full speed data rate of both the VMEbus DMA controller and the IPI bus.
- hold a *full track* of disk data.
- hold the *parity* received along with the data from the IPI bus.

Note: Parity storage allows the onboard processor to verify that data received from the IPI bus was transferred without any corruption.

You need to include on your disk controller board a *separate* onboard memory array to hold IPI controller commands. This provides 2 advantages:

- It preserves the space in the data buffer, which you need for handling data when traffic becomes congested.
- It allows the processor to parse commands and to setup onboard registers without interfering with data flow between the VMEbus and IPI bus.

In the past, UNIX-based systems have used 512-byte sectors. However, because today's systems are demanding data transfers at a higher rate, the trend is toward 1-Kbyte sectors. Some systems that handle large files even use 2-Kbyte sectors.

With more data being stored in each sector, the 32-bit and 48-bit Error Detection and Correction (EDAC) codes that were once used to detect and correct errors are no longer adequate. Your disk controller board should include hardware to generate 64-bit EDAC codes when data is written to the disk, and to verify the codes when data is read from the disk.

The SV6800 IPI controller board

Figure 4 shows the Xylogics SV6800 VMEbus IPI-2 controller, and Figure 5 shows its architecture. It consists of 6 custom ASICs surrounding a cache buffer. The board has been designed so that it can be expanded from 1 to 2 disk channels.

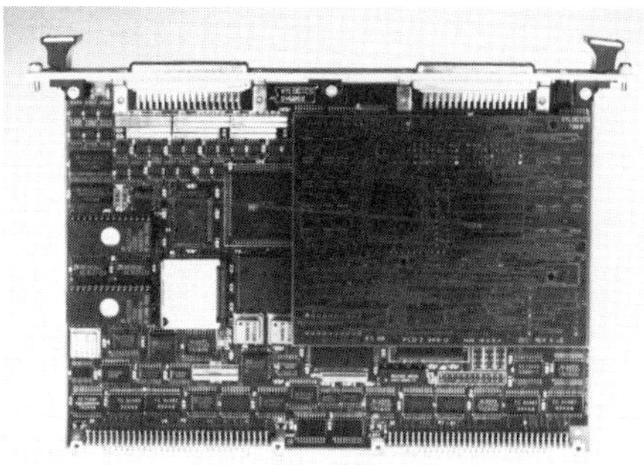

Figure 4

The SV6800 provides quick access to disk sectors through the use of *track buffering*. When an application program on the host processor requests the operating system to read a particular disk sector, the SV6800 reads the *entire track* containing that sector into the data cache. Since the application program is likely to read another sector from the same track, this allows the disk controller to provide the next sector from the cache.

When the host processor is running several programs in a multitasking fashion, consecutive disk reads may not originate from the same program, and will probably not be on the same track. If only one track buffer is available, the buffer will be

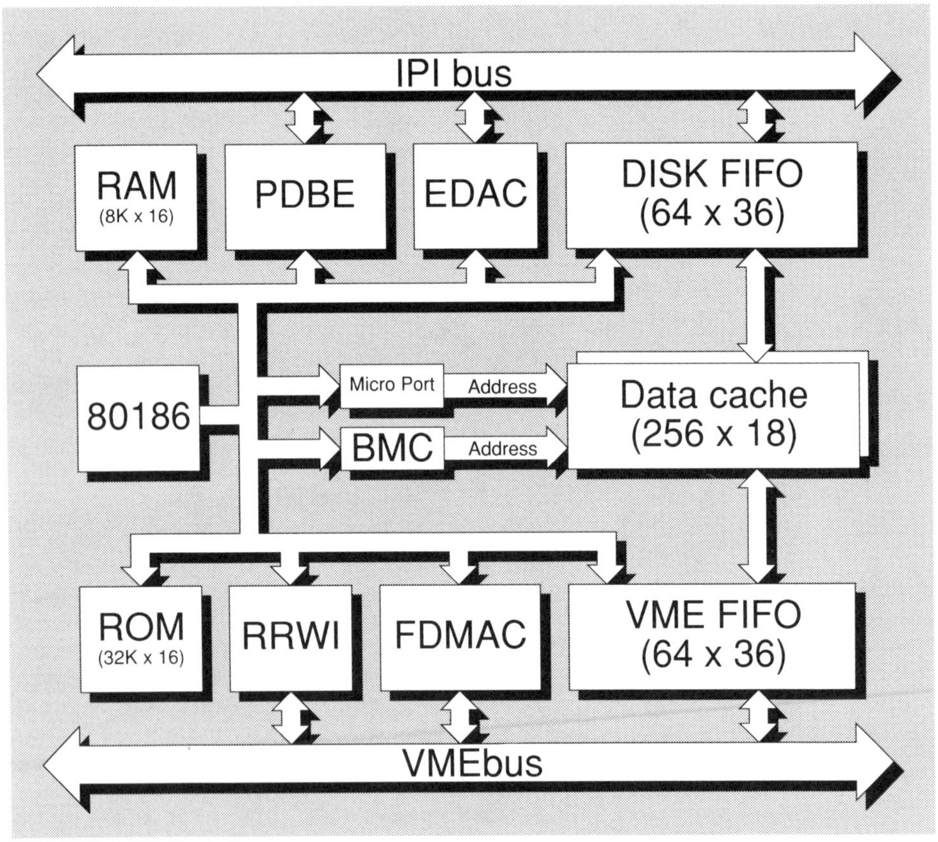

Figure 5

overwritten each time a context switch occurs. If this happens often, most of the advantage of having a track buffer is lost.

For this reason the SV6800 controller board is equipped with a 1 Mbyte (512 Kword) *segmentable cache*. This cache can be subdivided into several track buffers.

Each time a new I/O request comes in, the controller checks to see if the needed sector is already in one of the track buffers. (Since programs tend to access disk data in consecutive sectors, that's quite likely.) If the needed sector is not in any track buffer, the controller overwrites the *least recently used* track buffer with the new track.

The SV6800 VMEbus interface
The VMEbus interface is made up of 3 custom ASICs:

- The Fast DMA Controller (FDMAC)
- The VME FIFO
- The Register Read/Write Interface (RRWI)

The FDMAC chip (working with the VME FIFO) can provide short burst transfers over the VMEbus at rates up to 36 Mbytes/sec, until the VME FIFO empties. Thereafter, it can sustain 30 Mbyte/sec transfers to (or from) any VMEbus memory, as long as the memory responds with a DTACK* in 30 nsecs. Together with some glue logic, this chip can also move data over the VMEbus in block transfer mode.

Data is moved between the VMEbus and the data cache through a bidirectional, high-speed, 36-bit FIFO chip, called the VME FIFO. A VMEbus host processor can command the SV6800 board's processor to configure the VME FIFO chip to transfer data either 16 bits or 32 bits at a time over the VMEbus. The VME FIFO chip provides a buffer between the VMEbus and the data cache, allowing VMEbus data transfers to continue, even when the data cache is momentarily being accessed through one of its other ports.

When designing a new disk controller, it's important to consider the impact it has on *software*. Most of the design effort in systems today is invested in the software. You may find that it's much easier to design your new disk controller to have a similar software-to-hardware interface than to rewrite and debug the system software. This also makes upgrading to the new disk controller a simple matter of removing the old disk controller board, and installing the new one.

For this reason, the Register Read/Write Interface (RRWI) chip was designed to look like the register hardware on previous Xylogics disk controller boards. In fact, it provides the VMEbus host processor with a register interface so similar to these previous boards that you can use the same set of software drivers when upgrading to the SV6800.

The SV6800 IPI disk interface
The IPI disk interface of the SV6800 also consists of 3 ASICs:

- The Parallel Disk Back End (PDBE)
- The Error Detection and Correction chip (EDAC)
- The Disk FIFO

The PDBE chip handles the IPI level 2 protocols. The chip can control up to 8 drives, and has the intelligence needed to read all of the sectors on a single track. The onboard processor simply loads a multiple-sector command into a register on the PDBE chip. The chip then executes the command without further processor assistance, as long as all of the sectors lie on the same cylinder.

Data integrity is provided by the EDAC chip. It generates, appends, and checks:

- a 16-bit Cyclic Redundancy Code (CRC) on each *header field*.
- a 64-bit single-burst Reed-Soloman Error Correction Code (ECC) on each *data field*.

The 16-bit CRC *detects* all single-burst errors of 16 bits or less in the header field.

The 64-bit ECC *detects* all single burst errors of 49 bits or less in the data field, and *corrects* single burst errors up to 17 bits.

Data is moved between the IPI-2 bus and the data cache through the 36-bit-wide DISK FIFO chip (32 bits of data, plus 4 bits of parity). This chip is identical to the VME FIFO chip, and is used for essentially the same purpose. It provides a buffer between the

IPI-2 bus and the data cache, allowing the IPI-2 bus to continue transferring data at the full 10 Mbytes/sec, even when the data cache is being accessed through one of its other ports.

The Buffer Management Chip (BMC) schedules access to the data cache from the following ports:

- Disk 1 FIFO
- Disk 2 FIFO
- VME FIFO
- Micro Port

The BMC is also responsible for refreshing the cache's DRAM. The BMC is designed to support 1-Mbyte, 2-Mbyte, or 4-Mbyte DRAM arrays. It allows interleaved or non-interleaved DRAM access, and 16-bit or 32-bit DRAM accesses.

All accesses to the DRAM are done in *page mode*. The BMC allows a throughput of 80 Mbytes/sec through the cache. However, the VMEbus will limit the *actual* throughput to 40 Mbytes/sec.

The SV6800's onboard intelligence consists of:

- an Intel 80186 processor
- program memory
- a command buffer

The processor does not get involved in moving data. Its 3 primary responsibilities are:

- Decoding commands
- Initiating I/O operations
- Turning on and off the read-ahead feature, as needed to optimize performance

Commands are loaded into the onboard command buffer by a 5-step process:

- The host VMEbus processor writes an *I/O parameter block* into global VMEbus memory.
- The host processor writes a *pointer* into a register of the Register Read/Write Interface chip, to indicate the base address of the parameter block in VMEbus memory.
- The onboard processor uses this pointer to set up the DMA controller.
- The DMA controller copies the parameter block into the onboard data cache.
- The onboard processor copies the parameter block from the data cache into its local command buffer.

The SV7800

The architecture of the Xylogics SV7800 VME IPI-2 controller is identical to the SV6800, except that it provides 2 IPI channels. Both channels can be transferring data simultaneously, providing an aggregate disk data transfer rate twice that of the single-channel IPI-2 disk interface.

The 2nd channel consists of an additional:

- Parallel Disk Back End (PDBE) chip
- Error Detection and Correction (EDAC) chip
- Disk FIFO chip
- Set of IPI signal line drivers and receivers

These chips (along with a 2nd IPI connector) are mounted on a daughter board. Even with this daughter board attached, the SV7800 still only occupies a single VMEbus slot. Connecting a 6 Mbyte/sec IPI-2 drive to each channel of the SV7800 provides an effective throughput of 12 Mbytes/sec. That's 4 times the data transfer rate of today's 3 Mbyte/sec SMD disk drives.

Reference

1. Weaver, David. "SPARC: A scalable processor RISC architecture." *VMEbus Systems,* February 1989 pp 15-22.

James M. Mooney is a senior applications engineer at Xylogics, in Burlington, MA. He earned a Bachelors degree in computer systems from Rochester Institute of Technology. Jim is responsible for technical and marketing support of Xylogics IPI products.

CHAPTER 30

IPI And SCSI-2: Two New Disk Interfacing Standards

By Robert R. Greenfield, Jr. and Ernest E. Godsey

Even after many years of evolution, mechanical disk drives are still the most effective way to store large quantities of information. With each advance in processing technology, these mechanical devices are called upon to store and retrieve data at faster and faster rates. RISC technology is the most recent (and perhaps one of the most celebrated) advance in processing technology. To keep pace, new disk drives have been developed to provide the higher data transfer rates needed for RISC-based systems. However, the older disk interface standards, such as SMD, ESDI, and SCSI have now become bottlenecks: they can't move data between the host system and the disk drives fast enough. To solve this problem, new disk interfacing standards have been developed. In this chapter Bob and Ernie review 2 of the new standards, IPI-2 and SCSI-2, and contrast them with SMD, ESDI, and SCSI.

Do technology advances in microcomputers encourage new applications, or do applications that stretch the limits of current technology, encourage the development of technology advances? You could get into an endless argument over that one, and it's probably pointless to even ask. Whatever the cause, advances are taking place. One such advance, is Reduced Instruction Set Computers (RISC). These new chips promise to more than double the rate at which data is processed.

However, in most systems, faster data processing can only be sustained if you provide faster data storage and retrieval. Speeding up the processing, without also speeding up data transfers to and from the system disk storage, may simply transform a compute-bound system into an I/O-bound system.

For this reason, the introduction of RISC technology is forcing changes in the design of disk drives. However, it's not enough to just increase the data transfer rates of the disk drives. You also need to provide an efficient *interface* between the disk drive and the computer system that it serves. To provide for this, faster disk interfaces (such as IPI and SCSI-2) have been added to those already in wide use.

Today's high performance disk interfaces can be grouped into 2 categories:

- Device-level interfaces
- Generic I/O buses

What is a device-level interface?

A device-level interface is the interface between the raw disk drive and its drive controller. (See Figure 1) Historically, device-level interfaces have simply routed the serial data stream from the disk's read head over a cable to the drive controller. It was the controller's job to interpret this serial data stream. Popular examples of these serial device-level interfaces are ST-506, ESDI and SMD.

Since the data stream from the read-head is routed directly to the controller, device-level interfaces provide the highest possible data transfer rates for a given drive. The data transfer rate between the controller and the disk drive is limited only by the rate at which data can be read from (or written to) the disk surface.

What is a generic I/O bus?

A generic I/O bus moves data between the host system and the disk drive over a *parallel* data bus. (See Figure 2) Because a generic I/O bus provides a consistent physical interface for all peripheral devices, several different device types (such as magnetic disk drives, magnetic tape drives, and optical disk drives) can be supported through a single host adapter.

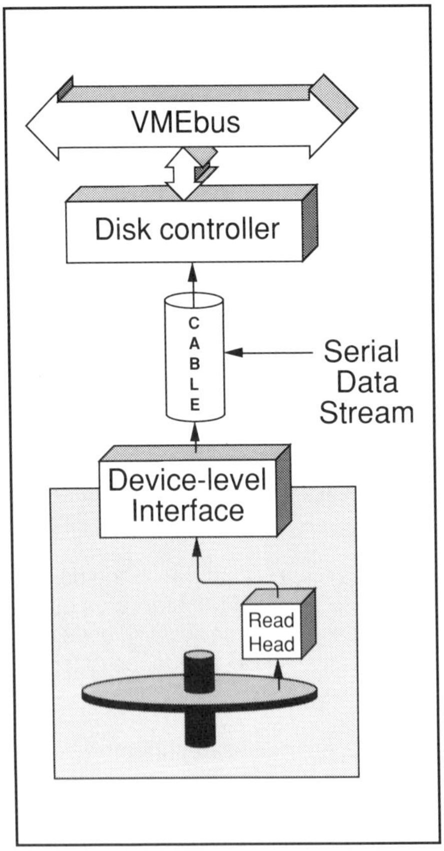

Figure 1

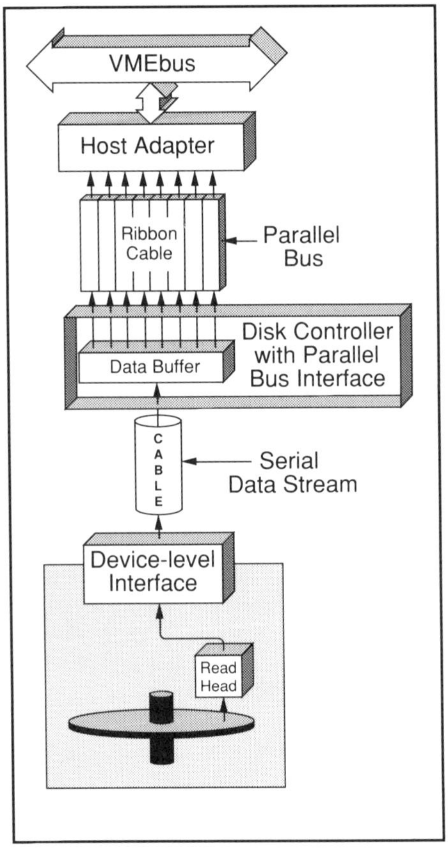

Figure 2

Generic I/O bus interfaces allow the host system to send a *read* command to one drive, and then send another *read* command to a 2nd drive. While the 2nd read command is being sent to the 2nd drive, the 1st drive is positioning its head, and waiting for the required data to rotate under the head.

As the data rotates under the head of the 1st drive, its local controller board captures the data in a hardware buffer. When all the data has been captured, the controller gains access to the bus, and then sends the buffered data back to the host system. By the time the 1st drive's controller empties its data buffer, the 2nd drive's controller has stored data in its buffer, and is ready to send. (See Figure 3) By overlapping the command transfers, the command processing, the disk access, and the data transfers in this way, a generic I/O bus can accomplish more I/O operations in a given amount of time.

Users of *device-level interfaces* expect to be able to:

- transfer data at the maximum possible rate permitted by each drive
- use a single controller to interface to several drives, allowing inexpensive expansion of existing storage capacity

Users of *generic I/O buses* have a slightly different set of needs. They expect to be able to:

- connect a wide variety of device *types* to the bus
- accomplish a large number of *I/O operations per second*, due to the intelligence and the data buffering provided by the controller in each disk drive

Currently popular device-level interfaces
There are 2 device-level interfaces in widespread use today:

- Enhanced Small Device Interface (ESDI)
- Storage Module Devices (SMD)

ESDI is a well established serial device-level interface that's used extensively in workstations, and other high performance desk-top systems. Several features have made it popular:

- Its compact 5.25-inch drive form factor
- Its 2.0 Mbyte/sec data transfer rate
- Drive capacities up to 760 Mbytes

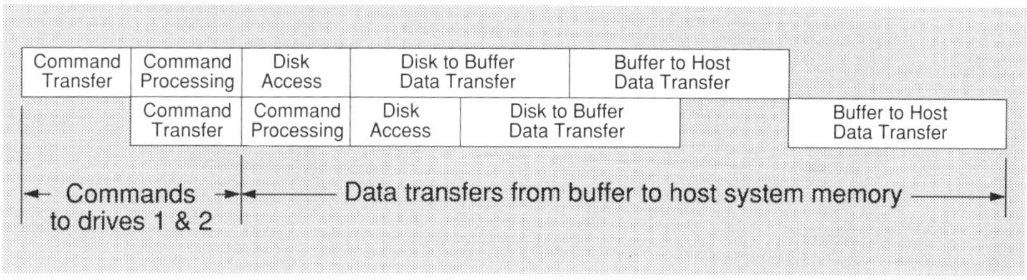

Figure 3

The compact size of ESDI drives also allows you to design the disk drive to be *removable*. Depending upon your application, this might be useful for:

- creating and safeguarding a backup disk
- removing sensitive data from the computer, for storage in a secured vault overnight

SMD drives (Storage Module Devices) have been used for years in networked file servers, data base systems, and other applications that need large storage capacity, and relatively high data transfer rates. The 8-inch, 9-inch, and 14-inch platters in SMD drives make the drive units relatively large, but high-end SMD drives can store up to 1.2 Gbytes of data, and can sustain data transfer rates up to 2.4 Mbyte/sec. SMD is a mature and stable interface that's supported by over 50 device and controller vendors.

Currently popular generic I/O buses

Today there is only 1 widely-used generic I/O bus: SCSI. To interface a device to SCSI, you need to have a device controller board with a SCSI interface. (See Figure 4) In some cases this controller board is mounted separate from the drive. In other cases, the controller resides inside the drive housing, where you can't see it.

As shown in Figure 4, when you use SCSI with your VMEbus system, there are 4 interfaces between the disk drive and the host processor:

- the interface between the drive and the drive controller board
- the interface between the drive controller board and the SCSI bus
- the interface between the SCSI bus and the host adapter board
- the interface between the host adapter board and the VMEbus

SCSI has been in use for years, and has become popular with system designers because of its flexibility. Practically all 5.25-inch and most 8-inch magnetic disk drives can be ordered with (or attached to) a SCSI controller. This provides you with a wide variety of SCSI-compatible disk drives, and permits you to choose the drive that meets your price and performance needs.

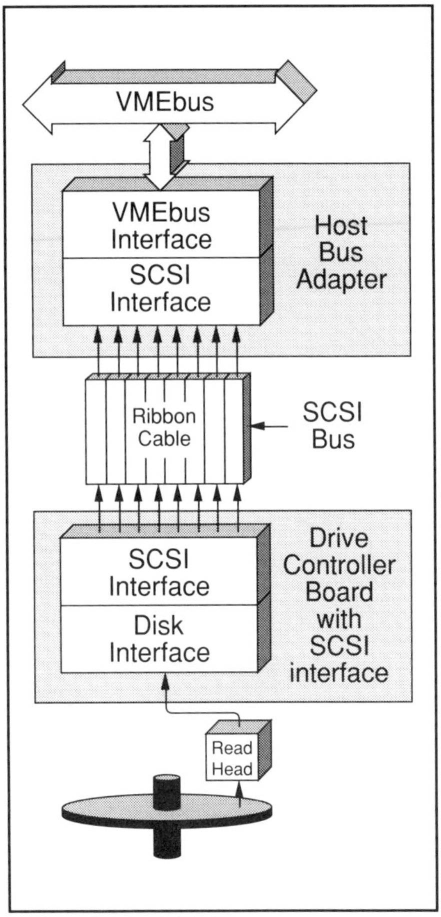

Figure 4

Two ways to measure disk subsystem performance

Disk performance can mean different things to different people. If you're storing high resolution video images on disk you're typically very concerned about how fast you can read those images from (or write those images to) the disk. You can speed up these transfers by storing each image in contiguous sectors. However, the disk head may still need to be repositioned several times: once each time it reaches the end of a track.

On the other hand, if you're designing a transaction processing system, you're probably more concerned about how many disk writes (or reads) can be done per unit of time. Since each transaction involves only a small amount of data, the data transfer rate is not as important as the disk system's ability to queue and handle bursts of traffic, when they occur.

Disk *drive* performance is often specified in terms of data transfer rates. However, when people talk about the data transfer rates of SCSI-compatible drives, they're talking about data transfer rates over the SCSI bus. Since the disk controller (between the SCSI bus and the disk drive) generally provides some data buffering, the SCSI data transfer rate is usually different from the disk drive's data transfer rate at its device-level interface.

ESDI is being enhanced, but it's still serial
In order to meet the demands of new applications, disk drive manufacturers have been stretching the limits of current device-level interface standards. For example, new 5.25-inch ESDI disks have increased data transfer rates from 15 Mbits/sec to 24 Mbits/sec. One disk drive manufacturer is increasing the rotational speed of its drives above 3600 RPM, to further increase data transfer rates. However, there are limits to how far serial interfaces can be stretched. A higher performance interface is needed.

IPI: a new type of device-level bus
In response to this need, a higher performance interface has been developed. It's called the Intelligent Peripheral Interface (IPI) standard. Unlike ESDI and SMD standards (which define only a device-level interface) and unlike SCSI (which defines only an I/O bus) the IPI specification defines *both.* It does this by defining the IPI standard in a *layered* fashion. IPI Levels 0, 1, and 2 (when taken together) define the mechanical, electrical, and signaling details of a device-level interface called IPI-2. This interface is similar (in function) to SMD or ESDI device-level interfaces. IPI Level 3 (IPI-3) defines a set of generic I/O commands, which can be sent over the IPI bus. Level 3 provides a generic I/O bus, similar to SCSI.

IPI Level 0
IPI Level 0 defines the connectors, as well as the signal line drivers and receivers. It permits use of a 50-conductor ribbon cable, to connect the IPI drive to the controller board in the host system. Alternatively, you can use a 50-conductor twisted pair cable, with D-subminiature connectors, or microconnectors.

The electrical specification defines 3 types of line drivers and receivers:

- voltage mode/single ended (limited to 5 meters)
- voltage mode/differential (limited to 15 meters)
- current mode/differential (limited to 75 meters)

IPI Level 1

Unlike earlier *bit-serial* device-level interfaces (ST-506, ESDI and SMD) the IPI interface provides a *parallel* data path between the peripheral drive and the controller board in the host system. (See Figure 5) IPI Level 1 defines how the 24 signal pairs are used to send information over this parallel data path. This parallel data path allows IPI to transfer data much faster than the earlier bit-serial device-level interfaces.

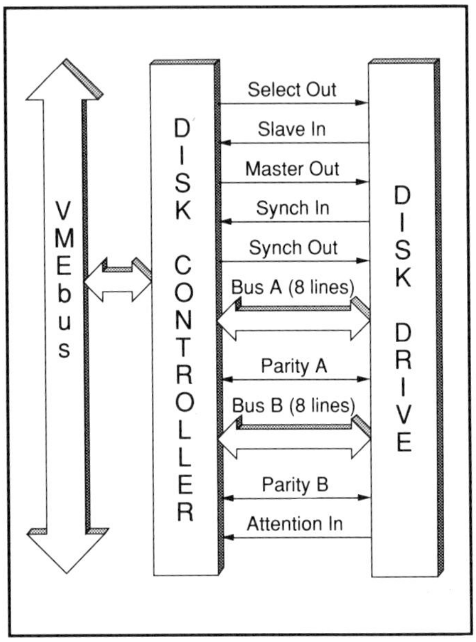

Figure 5

IPI provides 2 independent data paths: Bus A and Bus B. Each of these buses is made up of 9 lines: 8 lines of each bus are used to transfer byte-wide data, while the 9th is used for parity.

These buses are used for 2 purposes:

- to exchange *commands* and *status* between the host and the peripheral
- to exchange *information* between the host and the peripheral

The disk controller in the host system uses Bus A to send 8-bit, Level-1 *commands* to the peripheral. The peripheral uses Bus B to send 8-bit Level-1 *status codes* back to the host system.

During *information* transfers, Bus A and Bus B can operate in either of 2 modes:

- double-octet mode
- single-octet mode

In *double octet mode*, the A Bus and B Bus are used as a single, 16-bit, bidirectional bus. High performance disk drives and other peripherals transfer their information in the double-octet mode.

In *single-octet mode*, the peripheral uses Bus A to send byte-wide information to the host, and the host uses Bus B to send byte-wide information to the peripheral. Thus, by using Bus A and Bus B as unidirectional buses, the single-octet mode effectively reduces the 16-bit-wide data path of the IPI channel to 8-bits.

IPI Level 2

IPI Level 2 is built upon Level 0 and Level 1. Taken together, Levels 0, 1, and 2 provide a device-level interface specification, much like ST-506, ESDI or SMD. (The most important difference is that the IPI device-level interface carries data over a *parallel bus*, instead of a *serial line*.)

IPI Level 3
Level 3 defines a set of generic I/O commands, conceptually similar to SCSI. These commands allow the host system to communicate with different *types* of devices (such as disk drives, tape drives, and network controllers) in a consistent fashion.

Note: The protocols of the IPI specification allow a host system to communicate with level-2 devices and level-3 devices over the same IPI cable.

IPI Level 4
Level 4 of the IPI specification is still under discussion, but the intention is that it will migrate some operating system functions into IPI controllers.

IPI provides fast data transfer rates
IPI disk drives allow you to transfer data very quickly. IPI Level-2 drives with the following specs are available:

- 5.25-inch IPI-2 disk drives, with transfer rates of 4.8 Mbytes/sec, and capacities of 1.2 Gbytes.
- 8-inch IPI-2 disk drives, with transfer rates of 27 Mbytes/sec, and capacities of 3.0 Gbytes.

SCSI-2: a new generic I/O bus
The specification for SCSI-2 has recently been finalized. Whereas SCSI (now called SCSI-1) defines only an 8-bit I/O bus, the SCSI-2 specification allows 8-bit-wide, 16-bit-wide, and 32-bit-wide implementations. SCSI-2 increases the SCSI-1 data transfer rates by using higher clock speeds, and wider data paths. With a single cable, SCSI-2 can transfer data 8 bits at time, at rates up to 10 Mbytes/sec. Adding a 2nd (68-pin) cable widens the data bus to 32 bits, and allows rates up to 40 Mbytes/sec. The dual-cable 16-bit and 32-bit implementations are called *wide SCSI-2*, while single-cable, 8-bit implementations with high clock speeds are called *fast SCSI-2*.

If you're primarily concerned about how many I/O operations you can accomplish per unit of time, SCSI-2 may be a good choice. The intelligence in each SCSI-2 disk drive unit permits overlapped processing of read/write commands from the host system. This allows a SCSI-2-based subsystem to execute a large number of I/O operations per second.

Fast SCSI-2 allows 8-bit data transfers at 10 Mbytes per second. However, today's SCSI-2 peripherals are being designed around ESDI drives, which can only transfer data at about 1.5 Mbytes per second. This allows you to put up to 6 drives on a fast SCSI-2 bus before you saturate the bandwidth.

Why not construct a fast SCSI-2 subsystem from a single 10 Mbytes/sec disk drive, instead of using 6 slower 1.5 Mbyte/sec drives? It would certainly be possible, but such a system doesn't take advantage of SCSI-2's greatest strength. With a single 10 Mbyte/sec drive, you don't benefit from SCSI-2's command overlapping capability.

Interface type	Number of cables needed	Type of devices	Maximum data transfer rate (Mbyte/sec)	Maximum no. of devices
SMD	2	Disk	3	16**
SCSI-1 (8 bits)	1	Tape, Disk & Optical Disk	1.5 (async) 4.0 (sync)	7
Fast SCSI-2 (8 bits)	1	Tape, Disk & Optical Disk	10	7
Wide SCSI-2 (16 bits)	2	Tape, Disk & Optical Disk	10	7
Wide SCSI-2 (32 bits)	2	Tape, Disk & Optical Disk	20	7
Fast + Wide SCSI-2 (16 bits)	2	Tape, Disk & Optical Disk	20	7
Fast + Wide SCSI-2 (32 bits)	2	Tape, Disk & Optical Disk	40	7
ESDI	2	Disk	3	7**
IPI-2	1	Disk	10*	8

* Currently defined in the specification. Higher rates are expected soon.

** Most controllers handle a maximum of 4 peripheral devices

Figure 6

Transfer rates: a comparison

Figure 6 compares the SMD, SCSI-1, SCSI-2, ESDI and IPI interfaces. Figure 7 shows the data transfer rates vs. the disk capacity for ESDI, SMD, and IPI-2. When comparing the performance of SCSI-2 peripherals to IPI-2 peripherals, you can't just compare the quoted data transfer rates. The IPI data transfer rate is dictated by the rate at which data is read from (or written to) the disk surface. The data stream passing through the IPI interface from the disk drive is not stored in a buffer before being sent to the host over the IPI interface. In contrast, the data transfer rate quoted for SCSI peripherals is the rate at which the disk controller's data buffer can be emptied (or filled) over the SCSI-2 bus. Thus the SCSI-2 data transfer rate is a *burst* rate, which typically cannot be sustained over time. In contrast, the IPI data transfer rate is a *continuous* data transfer rate, which can be sustained for an entire disk track.

Choosing a disk interface standard

The driving force behind many new disk drive products is the quest for ever-increasing levels of performance, both in terms of high data transfer rates, and in terms of I/Os

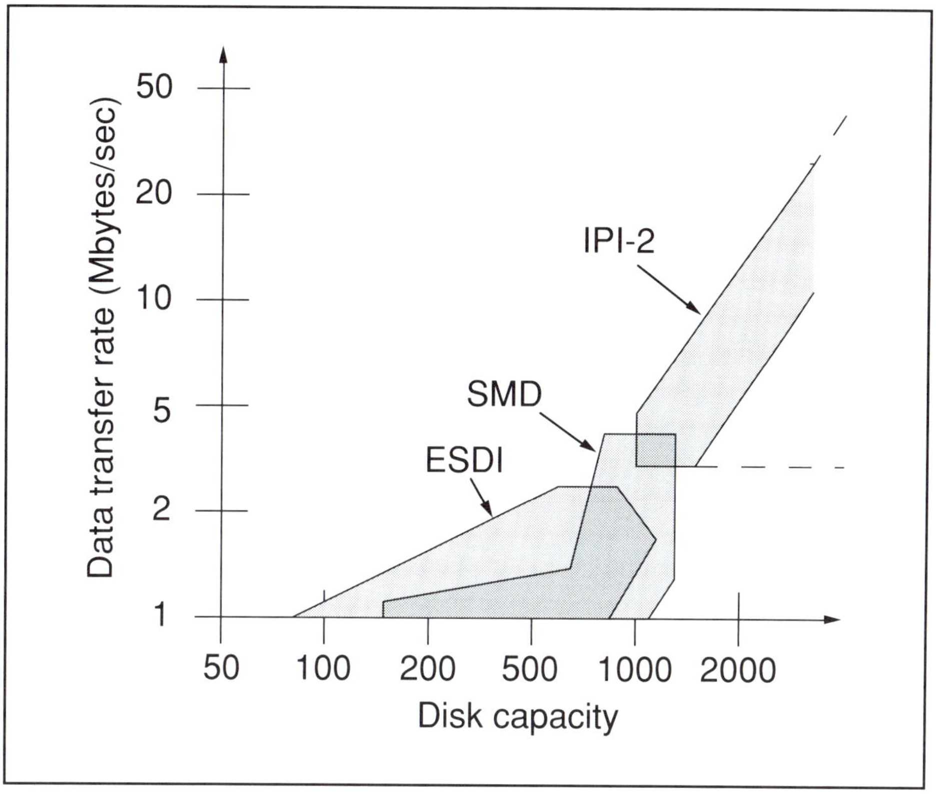

Figure 7

per second. However, raw performance, is not the only criteria that you need to consider when choosing a disk interface for use in your system. You should also consider storage capacity, ease of cabling, and flexibility in the choice of peripherals.

For example, if you're concerned about ease of cabling, you wouldn't use ESDI, SMD, or wide-SCSI-2, since they're all 2-cable solutions. Instead, you would choose fast SCSI-2, or IPI, which use only a single cable.

Each application is different, and no single interface solves all I/O problems. Only by carefully studying the advantages of each in light of your own requirements can you make the correct choice.

A caution

The SCSI-2 specification allows for a lot of design latitude, while maintaining compatibility with SCSI-1. For example, the new higher data transfer rates are *negotiated* between the host and each of the peripherals. This allows data transfer rates as high as 40 Mbytes/sec, while still permitting the use of slower peripherals.

While the SCSI-2 standard allows 40 Mbyte/sec data transfer rates, disk drives don't have to transfer data that fast to be called SCSI-2 compatible. In fact, a SCSI-1 disk drive can be upgraded to SCSI-2 by simply expanding its command set. Thus, if you purchase a low-cost "SCSI-2" disk drive, you may find that it only transfers data at SCSI-1 rates.

Using disk arrays
Neither SCSI-2 nor IPI-2 disk drives currently deliver the full performance promised by their standards. However, if the current drives are too slow for your needs, you can get higher performance by using *several* drives, in a disk array subsystem.

For example, suppose you have an ESDI disk drive that transfers data at a rate of 1.5 Mbytes/sec. By using 4 of these ESDI drives to build a disk array, you can provide an effective 6 Mbytes/sec data transfer rate.

Using striping to speed up access
When using a disk array, you can use a technique called *striping.* Striping is where different portions of a file are stored on different drives, to provide an overall increase in the data transfer rate. For example, the system software might break a file up into sector-sized chunks, and then write consecutive sectors to different drives.

Striping increases the overall data transfer rate. However, since there is no hardware redundancy, if you lose one of the drives you lose the entire file. To guard against this, you need to include one extra drive, as a *parity* drive. Then, if one of the drives fails, you haven't lost any of your data. You can replace the failed drive, and the disk subsystem will then regenerate the information on the failed drive (as a background activity) while the system continues to run.

When using this technique, the old measure of reliability *Mean Time Between Failure* (MTBF) is no longer useful. A new term, *Mean Time Data Availability* (MTDA) describes the average number of hours you can expect between each loss of data, due to the loss of a 2nd drive before the data has been regenerated from the parity data.

Building arrays from ESDI, SMD, SCSI, and IPI drives
The built-in data buffering hardware in SCSI-1 and SCSI-2 drives allows you to build disk arrays by simply connecting several SCSI drives to a single SCSI bus. However, if appropriate data buffering hardware is provided, disk arrays can also be built from ESDI, SMD or IPI-2 drives. In fact, disk arrays being sold today have been implemented using SMD and ESDI, as well as SCSI-1 technology.

Cost: ESDI disk arrays vs IPI-2 drives or SMD drives
Figure 8 compares the performance and the capacity of a 4+1 (4 data drives plus one parity drive) ESDI disk array subsystem to a single IPI-2 disk drive and a single SMD drive. Figure 9 shows the cost of building each of the systems in Figure 8. By the time you add the cost of a 5-drive controller (with data buffering hardware for each ESDI drive) the ESDI-based disk array subsystem turns out to be at least as expensive as a single, high capacity, high speed IPI drive.

As you build larger storage systems (with more and more Gbytes of storage) the cost of an ESDI-based disk array rises more rapidly than the cost of additional IPI-2 and SMD drives, which can be controlled by a single controller board. For example, the Interphase V/IPI4260 controller board (Shown in Figure 10) can control up to 8 IPI

	4 + 1 ESDI Array	IPI	SMD
Average seek time (msec)	16	16	16
Average rotational latency (msec)	8.3	8.3	8.3
Unformatted capacity (Gbytes)	1.5	1.2	1.2
Transfer rate (Mbytes/sec) (See note below)	6.0	4.8	2.4
$ per Mbyte	9.17 +/- 1.17	8.75 +/- .42	6.25 +/- .42

Note: This is the maximum sustainable data rate. The ESDI array may have a higher burst rate (due to buffering in the SCSI controller) but its overall rate will not exceed 6.0 Mbytes/sec.

Figure 8

5 ESDI drives	$ 6,500
Array controller	$ 5,000 (+/- 1000)
SCSI-2 host adapter	$ 2,250 (+/- 750)
Total Cost	$13,750 (+/- 1750)

1 IPI drive	$ 7,000
IPI-2 controller	$ 3,500 (+/- 500)
Total Cost	$10,500 (+/- 500)

1 SMD drive	$ 5,500
SMD controller	$ 2,000 (+/- 500)
Total Cost	$ 7,500 (+/- 500)

Figure 9

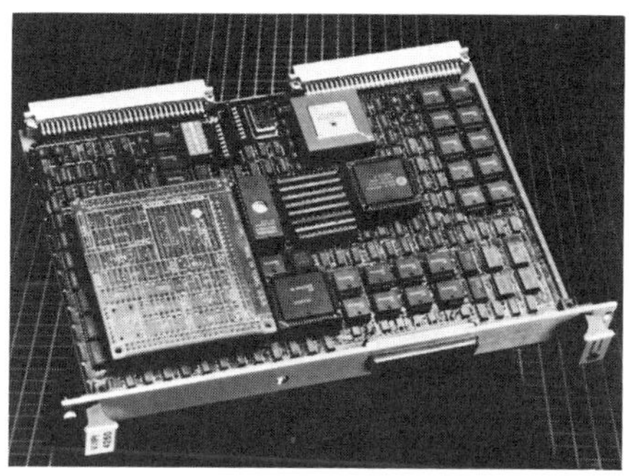

Figure 10

drives. As a result, multiple-Gbyte disk arrays (based on ESDI drives) cost considerably more than an equivalent disk subsystem based upon IPI or SMD drives.

An application demanding very high data transfer rates

High speed graphics and animation systems require very high data transfer rates. Many of today's mini-supercomputers and high-end workstations are dedicated to these types of applications. In fact, high-end workstation vendors have coined the term *graphics supercomputers*, to describe their systems. Whether these systems are used in entertainment, medicine, or seismic exploration, the requirement for high-speed, high-capacity data storage is tremendous.

To take a typical example, a (512 x 512 x 8) graphics screen contains approximately 2.0 Mbytes of data. To simulate movement, the screen must be changed at least 16 times per second, requiring that 32 Mbytes of data be moved to the screen every second. Thus, to produce a 1-minute segment, you need to transfer more than 1.9 Gbytes of data! This 32-Mbyte/sec data transfer rate exceeds even the 27 Mbyte/sec data transfer rate of the fastest 8-inch IPI-2 drives.

Why not provide these higher data transfer rates with an IPI-based disk array? Such a product may eventually be introduced, but (as of today) very few VMEbus users seem to need data transfer rates that high.

Summary

With the introduction of faster RISC processors, the throughput of disk subsystems has become the limiting factor in system performance. For this reason, the type of disk subsystem you use is becoming more critical to system-level performance. Disk arrays are one way to provide increased performance with current disk drive technology.

At the same time, SCSI-2 and IPI-2 products offer you new higher-performance drives for use on disk subsystems. Both of these new standards have strengths, depending upon your application. In order to choose the one that's most appropriate for your own system, you need to understand the underlying philosophy behind each.

Robert R. Greenfield, Jr. is the director of corporate communications at Interphase. He joined Interphase in 1987 from Flexible Computer Corporation, where he was manager of corporate communications. Previously, Bob was manager of communications for the Integrated Systems Division of United Technologies. Bob has a BA in political philosophy, an MA in creative writing, and an MBA from the University of Dallas.

Ernest E. Godsey is director of product marketing at Interphase in Dallas, Texas. He has a BSEE from Texas A&M University and an MBA from the University of Wisconsin. Ernest is a member of Tau Beta Pi, Etta Kappa Nu, and Beta Gamma Sigma honor societies. He also has been granted several patents and has published numerous articles.

CHAPTER 31

An Introduction To Ethernet

By Joe Mannas and Steve Gibson

Why has Ethernet become so popular? In this article, Joe and Steve list the advantages of using Ethernet over other networking approaches. They also explain how Ethernet protocols fit into the OSI model, and they describe the TCP/IP software that's typically used to provide higher-layer protocols in Ethernet systems.

Engineers today can choose from a wide array of networking technologies. *Token ring* (IEEE 802.5) and *token bus* (IEEE 802.4) networks have found favor in real time control applications because they're deterministic: they allow users to calculate the network's worst-case transmission time.

Both token bus and token ring are *broadband* networks, which use 75-Ohm coaxial cabling, similar to that used for cable TV. Broadband systems divide up the coax's total bandwidth into separate *frequency channels.* A modulated carrier signal can be sent over each channel. The channel frequencies and bandwidths are chosen to fit the application. Some of these channels are used to carry audio and video, while other channels carry digital data.

Broadband networks have low bit-error rates. That's because each device that's receiving digital data is only looking at a narrow band of frequencies, and it's not affected by noise outside that band.

However broadband technology is expensive. That's because each node on the network must be equipped with a *modem.* In contrast, the *baseband* technology used by Ethernet is relatively simple and cheap. Ethernet provides a single, high-capacity *digital* channel. Because the digital data is not modulated onto a carrier, modems aren't necessary. All devices on the network share one channel, and take turns sending and receiving bursts (called *packets*) of data.

Ethernet's baseband technology is best at handling light, *bursty* traffic, such as the high speed terminal interaction often seen in office and engineering environments. This type of bursty traffic also occurs in some computer-to-computer interactions, such as file transfers, and disk sharing. However, Ethernet is not as good at handling *sustained,* high-volume traffic, like the traffic often seen in industrial automation applications.

In addition to its low cost, Ethernet has become popular for several other reasons:

- Its 10 Mbit/sec data rate is fast enough for both terminal-to-computer and computer-to-computer applications.
- It's supported by numerous vendors.

- It has been standardized as IEEE 802.3.
- Integrated circuit manufacturers have developed Ethernet controller chips.

Ethernet and the Open Systems Interconnect model

When studying a network, you should consider how its protocols fit into a layered structure called the Open Systems Interconnect (OSI) model (Figure 1). The idea behind layered protocols is that you can change the protocol used in one layer to an alternative protocol, without having to also change the protocols used in the layer above it or below it. The OSI model was defined by the International Standards Organization (ISO), whose U.S. representative is ANSI.

In general, today's standard network protocols are divided into layers that *roughly* conform to the functions described in the OSI model. However, the ISO model is rather fuzzy in defining the *boundaries* between the layers. Because of differing interpretations, one software vendor's implementation of a certain layer won't necessarily work with another vendor's implementation of an adjacent layer. This is especially true of the higher layers, since there has been less time for things to sort themselves out. However, given enough time, the boundaries will become standardized, and you'll be able to choose your favorite protocol for each layer.

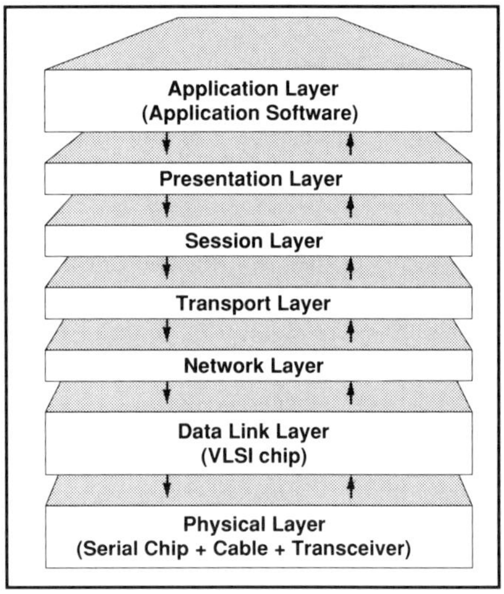

Figure 1

The OSI model is a *connection oriented* model, while Ethernet is a *packet oriented* network. That makes it difficult to draw exact parallels. However, for our purposes, we will say that Ethernet provides most of the functions described in the lowest 2 layers of the OSI model. More specifically, Ethernet specifies:

- The Physical layer
- The interface between the physical layer and the data link layer
- The Data Link layer
- The interface between the data link layer and the network layer

Figure 2 shows how these elements fit together. The Physical layer specifies features such as cabling and data encoding at the *bit* level, while the Data Link layer specifies how data is formatted into *packets*. (A packet is a kind of *electronic envelope*, with a message on the inside, and 2 addresses on the outside: a source address and a destination address).

The Ethernet spec is quite specific about how these lowest 2 layers are to be implemented. In fact, there are VLSI chips available today that implement these 2 layers in hardware. Given these chips, its easy to build Ethernet controller boards that are fully compatible with each other. However, the higher-layer protocols are not discussed in the Ethernet spec. Since these higher-layer protocols are only vaguely defined by the OSI model, system designers must look elsewhere for detailed specification of these layers.

Fortunately, there are 2 standardized high-level protocols that are widely used with Ethernet:

- The Xerox Networking Standard (XNS)
- The Transmission Control Protocol (TCP) and Internet Protocol (IP)

The Xerox Networking Standard (XNS) was the first to be developed. It was used in Xerox's early implementations of Ethernet. In fact, many of today's Ethernet networks still employ XNS.

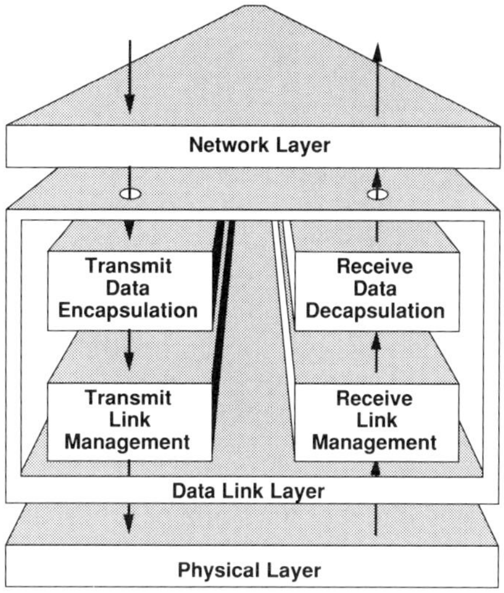

Figure 2

However, the most popular Ethernet protocol today is a combination of Transmission Control Protocol (TCP) and Internet Protocol (IP). TCP/IP protocols have been implemented in a software package that also includes software for some additional protocols, as shown in Figure 3. TCP is roughly equivalent to the OSI's *transport layer* protocols, and IP is roughly equivalent to the OSI's *network layer* protocols.

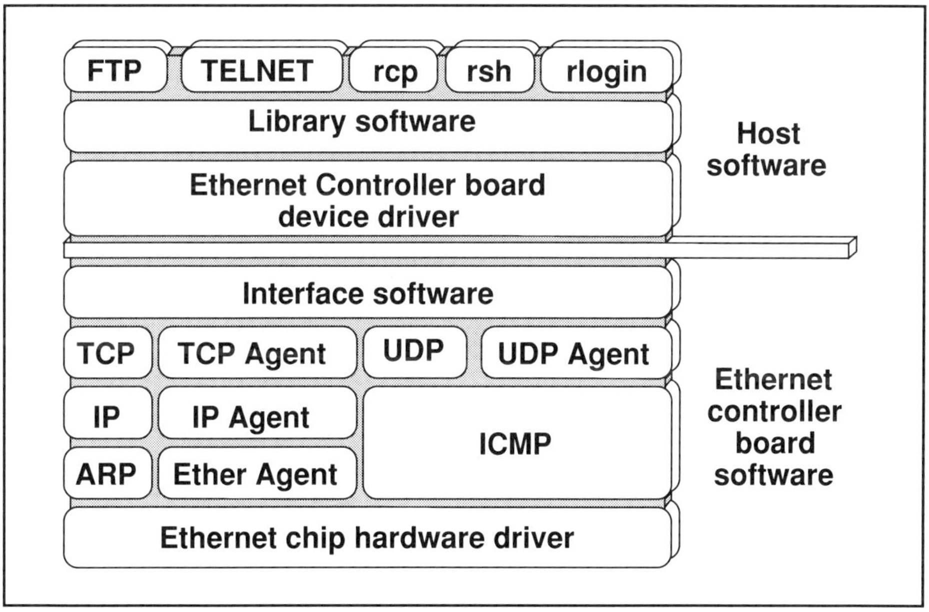

Figure 3

TCP/IP was developed for use on *ARPAnet*, which is a network that originated with the US Department of Defense. ARPAnet was designed to tie together universities, research facilities, and government installations into a massive time sharing system. This system is accessed through remote nodes. People call into the nodes, and the nodes then call each other. ARPAnet has since been renamed the Defense Data Network (DDN).

While the TCP/IP protocols are roughly equivalent to the network layer and the transport layer of the OSI model, they are not exactly equivalent. Largely due to General Motors, there is currently an industry-wide move toward developing a set of carefully defined protocols that conform more closely to the ISO model. The goal is to more clearly define the interfacing between the layers, so you can change the protocol used in one layer without having to change adjacent layers. The hope behind this effort is to create software that will make upgrading to newer protocols straightforward.

Some cheaper versions of Ethernet

In addition to substituting protocols at the higher levels, it's also possible to substitute at the bottom 2 layers. For example, the IEEE 802.3 (Ethernet) standard has been enlarged beyond its original coaxial definition of Ethernet to allow a variation called *thin-wire Ethernet*. Thin-wire Ethernet (widely known as *Cheapernet*) provides less expensive networking than standard Ethernet.

Cheapernet runs at the same 10 Mbit/sec rate as Ethernet, and uses the same physical and data link layer protocols. That allows you to use the same VLSI controller chips. The primary difference between Ethernet and Cheapernet lies in the *cabling*. Ethernet uses 0.4-inch-diameter coaxial cable. This cable is tapped into by a *stinger* that penetrates the outer sheathe of the cable, to make contact with the inner conductor. By contrast, Cheapernet employs 0.25 inch coax, and taps into the coax through standard BNC T-connectors.

Cheapernet coax costs only about 1/6 as much as standard Ethernet coax, and it's much less rigid. That makes it much easier to route through a building. In addition, this flexibility allows a Cheapernet coax to be routed directly to the back of a desktop computer. (Ethernet cables are much too rigid and heavy to permit this.)

Since the Cheapernet coax is routed right up to the Cheapernet controller board, the RF transceiver circuitry can be mounted directly on the controller board. Contrast this with Ethernet coax, where the transceiver circuitry must be mounted onto the cable. Then a secondary cable (containing 4 twisted pairs) is used to connect this transceiver to the Ethernet controller board in the computer. (See Figure 4.)

Like Ethernet, Cheapernet can be constructed from several segments of coax, spliced together by repeaters. Like Ethernet, it can also accommodate up to 1024 nodes. However, Cheapernet users must accept some limitations in exchange for the lower cost:

- The total network span (including repeaters) of Cheapernet cannot exceed 3000 feet (Ethernet allows up to 8000 feet.)
- The end-to-end length of each Cheapernet cable segment cannot exceed 600 feet. (Ethernet allows up to 1600 feet.)
- The number of nodes connected to a single cable segment cannot exceed 30. (Ethernet allows up to 100 nodes.)

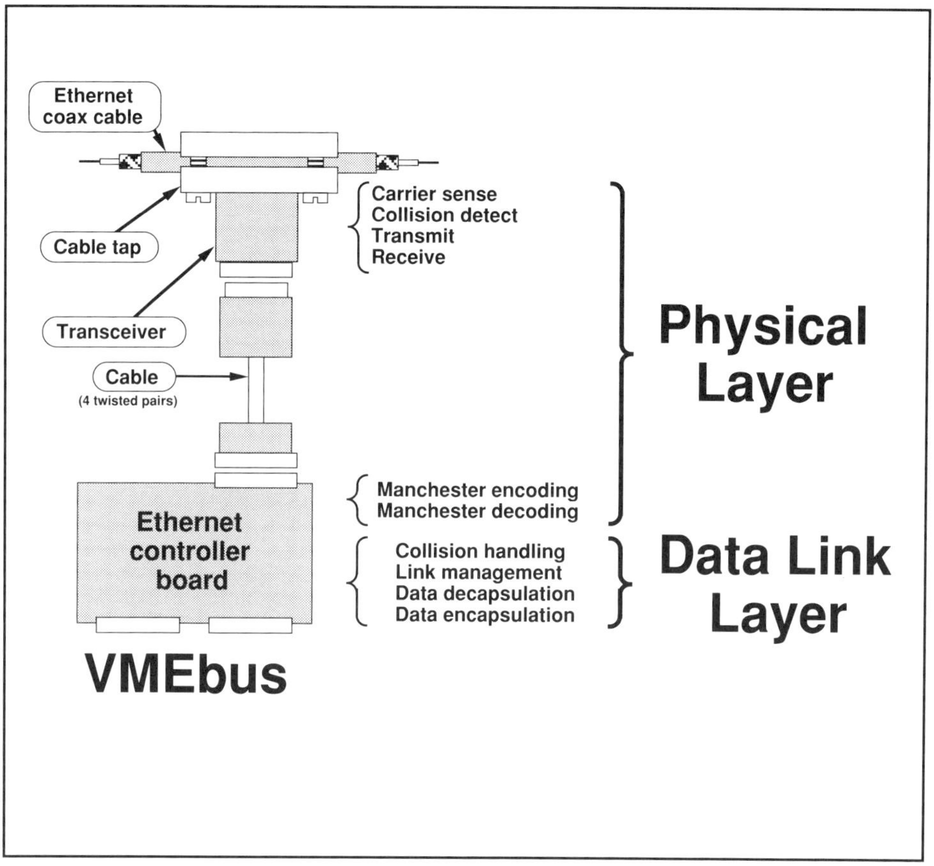

Figure 4

AT&T's *Starlan* is another low-cost version of Ethernet. In fact it's even lower in cost than Cheapernet. It uses a twisted pair cable, instead of coax. While it employs Ethernet protocols, it runs at only 1 Mbit/sec, and can have a maximum cable length of only 500 meters. Starlan is most useful for low-speed terminal interaction.

Ethernet's physical layer implementation

An Ethernet transceiver resides on the coax cable, while a Cheapernet transceiver typically resides on the controller board. However, they both perform the same function. As required by the physical layer of the OSI model, they inject data bits into (and recover data bits from) the network cable.

However, Ethernet's (and Cheapernet's) physical layer also includes a special protocol not discussed in the OSI model, called CSMA/CD. CSMA/CD stands for *Carrier Sense, Multiple Access with Collision Detection*. *Carrier sense* means that each node can detect when another node is transmitting over the network. *Multiple access* means that each node is permitted to send data without getting permission from some centralized control: each node simply verifies that the network isn't busy, and then begins its transmission.

However, if 2 or more nodes start transmitting simultaneously, a *collision* occurs. In order to detect collisions, each transceiver has circuitry that senses how many nodes are sending. That circuitry notifies the controller board when 2 or more nodes are sending simultaneously. The controller board must then respond accordingly.

The transmit sequence works as follows. Any node that wants to transmit must first check for the presence of *signal transitions* on the network. If it detects any signal transitions, then it must wait until the network is quiet. When it detects no signal transitions the node begins transmitting.

Since some other node might begin transmitting at the same time, each node's transceiver monitors the network to detect the presence of 2 or more simultaneous transmissions: an occurrence that indicates a collision. If a collision occurs, each of the colliding devices waits a random amount of time, and then tries to transmit again. This random waiting period makes it unlikely that the same devices will collide a 2nd time.

The transceiver is connected to the Ethernet controller with 4 twisted pairs:

- A transmitted data pair
- A received data pair
- A collision detect pair
- A DC power pair (This pair provides power to the transceiver)

Note: The transceiver just *detects* collisions. It doesn't *do* anything about them, except to notify the controller.

While the transceiver embodies *part* of the physical layer, it is not the *entire* physical layer. For example, the Manchester encoding and decoding of the data bits is performed by the controller. A Manchester encoder combines a serial data stream with a periodic clock signal, as shown in Figure 5.

The resulting signal (which gets sent over the Ethernet link) has either a low-to-high or a high-to-low transition in the middle of each bit cell. Those transitions allow the Manchester decoder to maintain synchronization between its own internal clock and the incoming Manchester encoded signal, using a phase locked loop.

Manchester coding is designed such that the average DC voltage level of the signal is constant. Ethernet's Manchester encoded signal has a 1 volt average DC level. If 2

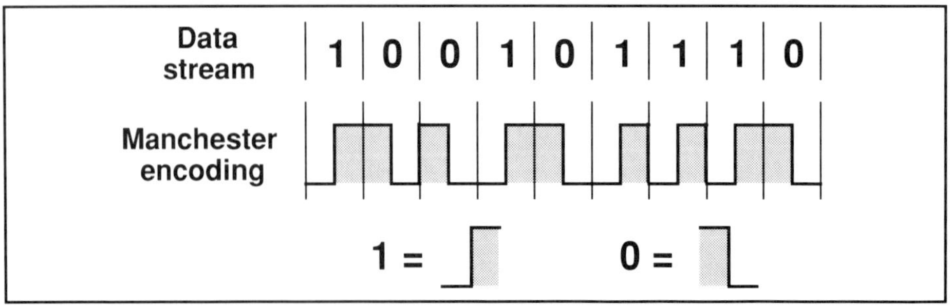

Figure 5

transceivers send signals over Ethernet simultaneously, the result is a signal with an average DC level of 2 volts. Each transceiver detects collisions by monitoring the network's average DC level.

In addition to the encoder/decoder circuitry, each controller board has a 10 MHz oscillator for generating the clock transitions on the outgoing Manchester encoded signal. The board also has line drivers and receivers, which are connected directly to the coax cable.

Some controller boards also include special functions, such as a watchdog timer that prevents the Manchester encoder from spewing a continuous stream of data bits onto the network, in case of a malfunction. There might also be a *squelch* circuit, which prevents collision fragments from being interpreted as messages, or there might be special loop-back circuitry, to allow for self test.

Ethernet's Data Link protocols deal with data packets

The 2nd OSI layer is the Data Link layer: It's the highest layer of protocol specified by the Ethernet standard. The Ethernet's Data Link protocols...

- *encapsulate data* into packets
- *recover data* from packets
- *react to collisions* in the prescribed fashion
- *handle errors* in transmission
- *attach network addresses* to outgoing packets
- *check the network addresses* of incoming packets

Ethernet packets are formatted as shown in Figure 6. The *preamble* permits the receiving node to synchronize its internal clock with the transitions of the incoming signal. After the preamble comes the *destination address*. There are 2 types of destination addresses:

- single node addresses
- multicast addresses

A single node address has a zero in the first bit of its 48-bit destination address field. The remaining 47 bits in the address field specify the *node number* that is to receive the packet. These 47 bits allow individual selection of more than 140 trillion unique Ethernet node addresses.

A *multicast address* allows a node to either broadcast to all nodes, or to

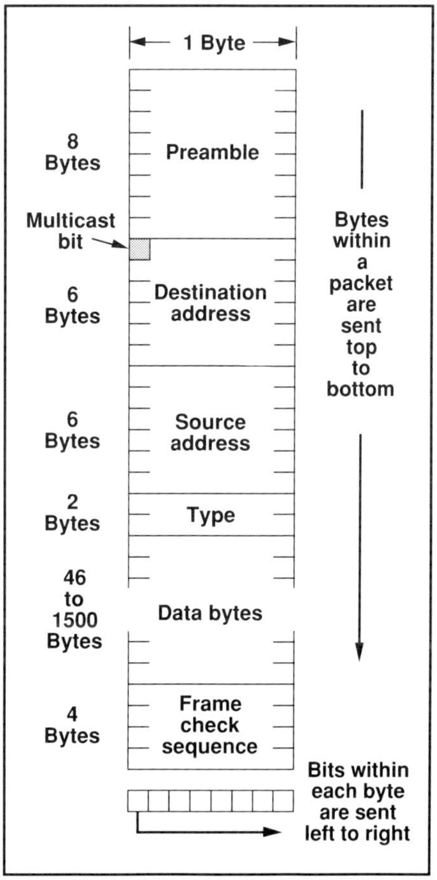

Figure 6

multicast to some specified group of nodes. Multicast addresses always have a one in the first bit of their destination address field. The remaining 47 bits indicate a specific *group number.* If all of these 47 bits are ones, then the packet is being broadcast to all nodes.

Next in the packet is the *source address*, which also consists of 48 bits. After the source address is a *type field* that can be used to specify a packet's data format. This field is used in networks composed of nodes manufactured by several different vendors, or in Ethernets that include gateways to non-Ethernet networks. If the type field is used, its contents are interpreted by higher-level protocols.

After the type field is the *data field.* It must be at least 46 bytes long. This minimum data field length sets a lower limit of 72 bytes on packet size...

Preamble	8 bytes
Header	14 bytes
Data	46 bytes
CRC	4 bytes
	72 bytes

A 4-byte *frame check sequence* follows the data field. It ensures the accuracy of the packet's information. This 4-byte sequence consists of a cyclic redundancy check (CRC) value of all the previous fields in the packet.

The functions of the Data Link layer and Physical layer might seem complex. In fact they're even more complex than the details presented here would indicate. But fortunately, VLSI chips have already been designed to do the job.

TCP/IP protocols provide 2 more layers, for use on top of Ethernet's 2 layers

Above the Data Link layer in the OSI model are the Network and Transport layers. When using Ethernet, you will probably want to implement these layers with the TCP/IP protocols, because these protocols are also used with a wide variety of other networks besides Ethernet. By using TCP/IP, you can use the same software to handle message traffic over a variety of communication links from local area networks to wide area networks, such as the Defense Data Network (DDN).

IP software implements the *Network Layer*, which is the protocol layer that sits on top of Ethernet's Data Link layer. The network layer handles the routing of messages between *computers.* IP accepts messages from (and passes messages to) the TCP layer above it.

TCP establishes a *virtual-wire* connection between a process running on one computer and a process running on another computer. Process A, (running in one system) can use this connection to send messages to process B (running on another system) just as if there was a dedicated I/O port between the 2 systems.

Even though the Ethernet link is typically simulating several virtual-wire connections at the same time, TCP preserves the illusion of a dedicated link. It does this by hiding all of the details involved in the actual sending and receiving of messages, including the detection of packet transmission errors, and retransmission of the packets when necessary.

The TCP/IP software package implements a variety of protocols in addition to TCP and IP. (See Figure 3) The minimum TCP/IP implementation consists of a set of drivers plus software for the following protocols:

- Transmission Control Protocol (TCP)
- Internet Protocol (IP)
- Address Resolution Protocol (ARP)
- User Datagram Protocol (UDP)
- Internetwork Control Message Protocol (ICMP)

IP and TCP provide the network layer and the transport layer, respectively. These 2 layers sit on top of Ethernet's physical layer and data link layer. However, IP's network layer uses 32-bit addresses, while Ethernet's transport layer uses 48-bit addresses. For that reason, another layer must be inserted between them. This layer is known as Address Resolution Protocol (ARP). ARP translates between the 48-bit Ethernet addresses and the 32-bit IP addresses.

UDP is an *alternative transport layer* protocol that may be used in place of TCP. Instead of providing a virtual wire connection between processes, it provides a *mail service* that allows processes to send *addressed packets*. Since UDP doesn't have to maintain several virtual wire connections it requires less overhead processing than does TCP. However, if the sending process needs confirmation that its packet actually reached the destination, UDP must be supplemented with higher-level protocols.

If some controller on the network accepts a message, and then finds that its local host is not able to accept the message, that controller can use ICMP protocols to notify the sender that the message never reached its final destination.

What kind of Ethernet controller do you need?
There are 2 types of Ethernet controller boards:

- Intelligent controller boards
- Dumb controller boards

Intelligent controller boards have an onboard processor to handle the high-level protocol processing. Dumb controllers leave high-level protocol processing to the host MPU. The type of controller you need depends on what kind of system you'll be using it on.

If you use a dumb controller, you may find that the TCP and IP protocol processing consumes more than half of your host MPU's processing capacity. That may be OK on a single-user workstation, especially if the MPU is not being fully used anyway. In fact you might not even notice the slightly slower response times when your workstation sends or receives a message over Ethernet.

However, if your host is a multiuser system, it may generate a lot of Ethernet traffic. Since the MPU may also be heavily loaded by the multiple users it supports, you don't want to add the additional burden of network protocols. An intelligent controller board would be a better choice.

In choosing an Ethernet controller board, you must consider 2 factors:

- How much will your system interact with Ethernet?
- How much of the MPU's time is needed to support system users?

A 3rd factor is cost. Intelligent Ethernet controllers cost about 35% more than dumb controllers. However, if you need most of the system MPU's time to support users, an intelligent controller is usually well worth the additional cost.

How do you interface to an Ethernet controller board?
Whether a controller board is intelligent or dumb, it must provide a well-defined software interface for host software to work with. For dumb controllers the interface is defined by the Ethernet standard. For intelligent controllers the Berkeley UNIX's TCP/IP software defines the interface you will use.

While some controller boards (or software packages) might include additional nonstandard features, you should make sure that they at least provide all of the features specified in the standard.

One example of a useful nonstandard feature is the provision of a *software library* containing routines that allow you to interact directly with the hardware of an intelligent controller board. If you are provided such a library (and if you decide to develop your own customized network layer protocols) you can incorporate those library routines into your software. The library routines save you a lot of coding that would otherwise be necessary just to communicate with the controller board.

There's one important aspect of the interface between host and controller that's not defined by any standard: the amount of local memory that the controller has. Dumb controllers might have only 8 or 16 Kbytes. However, intelligent controllers usually provide at least 128 Kbytes. In an intelligent controller, part of the local memory holds the protocol processing software; the rest is used for I/O buffer space.

As a general rule, the larger the controller's I/O buffer space, the more efficiently it can handle message traffic. That's because whenever its I/O buffer space fills, a controller must refuse any further messages until it processes some of its buffered messages. A larger buffer is less likely to fill before the buffered messages can be processed.

The size of certain critical buffers can have network-wide performance implications. Increasing the amount of buffer space in a heavily used node can speed up the entire network.

An example of an intelligent Ethernet controller board
Most Ethernet controllers support the same standard protocols. However, the way in which various controllers *implement* those protocols can affect the performance of the entire network. For example, having a dedicated MPU on the controller board can provide a great advantage.

CMC's ENP-10 board is shown in Figure 7. It's an intelligent controller board that comes with TCP/IP protocol software (which runs on the board's MPU) and TCP/IP driver and applications software (which runs on the host).

The controller board's network protocol software uses system calls provided by an onboard real time communications executive. These same system calls are also available to any onboard user-generated software. The executive's system call interface allows you to use various higher-layer protocol software packages (such as TCP/IP or XNS) without having to change the lower-level software on the controller board.

Figure 7

The block diagram of an ENP-10 is shown in Figure 8. It includes a 10 MHz 68000 microprocessor, which executes the TCP/IP protocol software, as well as some other overhead software. There is also a high performance version of the ENP-10, with a 10 MHz 68010. It's useful for especially busy nodes.

To handle the lowest 2 layers of protocol (Ethernet's Data Link layer and Physical layer) the ENP-10 has an AMD *Local Area Network Controller for Ethernet* (LANCE) chip set.

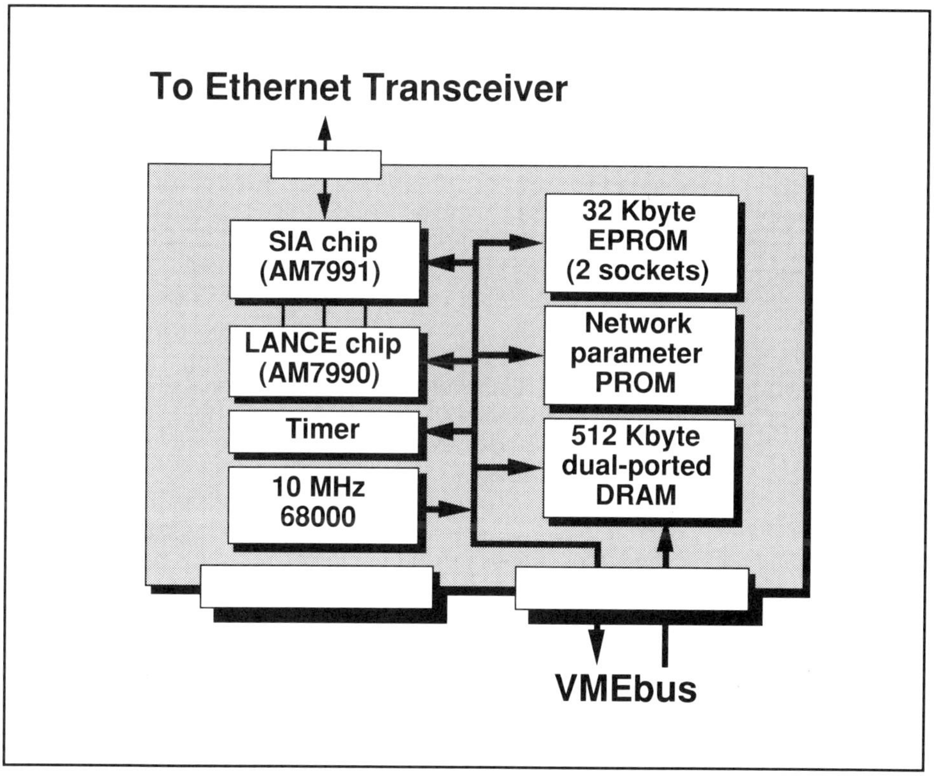

Figure 8

The ENP-10's dual ported RAM can be accessed by either the onboard 68000 or by the VMEbus. The onboard 68000 can also act as a VMEbus master, reading and writing to global VMEbus memory.

In addition to TCP/IP processing, the 68000 has a number of other responsibilities:

- It moves data between onboard dual-ported RAM and global VMEbus RAM.
- It generates VMEbus interrupts.
- It responds to VMEbus interrupts.
- It executes the real time executive.
- It refreshes the dual-ported dynamic RAM.
- It controls various timer functions.
- It runs diagnostics on the controller board.

Each ENP-10 controller board responds to its own world-wide-unique 48-bit Ethernet address. This address is assigned by the Ethernet Address Administration Office at Xerox.

In most systems, you will want to copy the TCP/IP software from a disk into the ENP-10's dual-ported RAM whenever the VMEbus system is powered up. However, the ENP-10 also provides sockets for 2 EPROMs, which provide enough nonvolatile storage to hold the TCP/IP protocol firmware. The EPROM-based approach is most useful on systems that have no local disk (such as industrial robots) but it can also be used where you want to free an additional 40 Kbytes of RAM for use as an I/O buffer.

The basic configuration of the ENP-10 includes 128 Kbytes of dual-ported RAM with parity. This capacity can be increased to 512 Kbytes by changing the RAM chips. The RAM can be accessed by the VMEbus, by the LANCE chip, or by the 68000, in that order of priority. Either the ENP-10 or the host MPU can move data between the ENP-10's local dual-ported RAM and global VMEbus memory.

The software provided with the ENP-10 controller board can be executed by the board's local 68000, or by some other 68000 on the VMEbus. If the ENP-10's 68000 executes the software, then all of the protocols needed for networking (through the 4th layer of the OSI model) are provided by the board.

Of course, it does no good to have a controller board accepting messages from the Ethernet if the host can't get those messages from the controller. To provide for that you need a software *device driver* for the host MPU. These host device drivers are packaged with the ENP-10.

The ENP-10's software package also includes some higher-layer software (in addition to its TCP, IP, UDP, ICMP, and ARP software) which implements the Department of Defense FTP and TELNET protocols. FTP is a file transfer protocol, and TELNET provides virtual terminal connections. (See Figure 3)

The software also includes some other higher-layer software that runs on the host MPU. Included are Berkeley UNIX utilities (See Figure 3) which basically do the same job as FTP and TELNET, but follow UNIX conventions:

- *rcp* (remote copy) provides file transfer services similar to FTP.
- *rsh* (remote shell) allows a process running under UNIX to execute programs on remote machines.
- *rlogin* (remote login) permits a UNIX process to establish connections between itself and some other remote process, which is similar to what TELNET does.

The C-compatible Berkeley *sockets* library is also provided. It allows you to write portable application software which sits on top of (and uses) the TCP/IP software.

Comparing the performance of Ethernet controllers

The throughput of Ethernet controllers is typically specified in bits/sec. However, since many of the bits passed over the Ethernet cable are stripped off as the message rises through the higher layers of protocol, it's important to specify which layer you're talking about when you quote performance figures. Because of the complexities involved, theoretical performance calculations are of limited value. However, you can get a feel for actual throughputs by measuring running networks.

At CMC we put together a small networked system consisting of two 10 MHz Sun 2 workstations. We equipped each workstation with a dumb controller, and dedicated 100% of the Sun's MPU time to servicing network interrupts and running TCP/IP software. We then submitted memory-resident messages to the TCP layer of one station, for transmission over Ethernet to the other station. The messages passed down through all 4 layers of protocol in the sending station, across the Ethernet link, and back up through 4 protocol layers in the receiving station, to form an identical memory-resident message in the 2nd system. Throughputs at the top of the 4 layers of protocol were 700,000 bits/sec.

We then equipped the 2 stations with ENP-10 intelligent controllers, and ran the same test again. The processing load on the Sun's MPU was cut by 50%, and its interrupt load was cut by 33%. At the same time Ethernet throughput rose by almost 50% to 1,000,000 bits/sec.

We suspected that the throughput was being limited by the Sun's MPU, so we allowed the ENP-10 controller to accept messages from Ethernet at its fastest rate, without waiting for the Sun's MPU to accept them. Throughput then rose to 3,100,000 bits/sec.

The ENP-10's MPU was running 2 layers of software: the IP layer and the TCP layer. To gain a little more insight, we configured the controller to run only the IP software layer. Its throughput then rose even higher, to 6,300,000 bits/sec.

In order to test the limits of the ENP-10's hardware, we relieved the 68000 of all protocol processing responsibilities. The throughput went up to nearly 10 Mbits/sec (the theoretical maximum for Ethernet) as long as the ENP-10 didn't have to wait for the Sun's MPU to take messages out of its buffer space.

This test brings up an interesting point. When someone quotes throughput figures, make sure they specify at which layer they are measuring the throughput. And if they're quoting throughput at the IP layer, don't just assume that their controller is also running TCP. It might turn out that the throughput drops drastically when the controller board's MPU is required to also run TCP software.

Summary

Ethernet's complexity can make it seem formidable. But when you study it piece by piece, and organize your thinking according to the OSI model, you see that Ethernet is structured in a straightforward and logical way. In addition, you don't need to understand all of the complex details of the layered protocols just to use Ethernet with your VMEbus system. By using off-the-shelf Ethernet software and controller boards you can build highly efficient networked systems with minimal effort.

Steve Gibson is the systems-level product manager for CMC. Prior to joining CMC as a product marketing engineer in 1985, he worked for ComDesign, a manufacturer of multiplexers and modems. Steve has a BA in economics and a BA in business from Westmont College.

Joe Mannas is CMC's VMEbus and Multibus product manager. He has over 12 years experience in the areas of UNIX software development, technical support, and product marketing. Prior to joining CMC Joe was a UNIX programmer for Bell Labs and for AT&T. He is a graduate of Vassar College, with a BA in Psychology.

CHAPTER 32

How To Design An Efficient Ethernet Controller Board: Just Adding Intelligence Isn't Enough

By Ernest E. Godsey

When you're trying to move messages over Ethernet quickly, you need to consider the entire message path, and remove any bottlenecks. In this chapter, Ernie explains where bottlenecks often occur, and how to architect an Ethernet controller board in a way that eliminates them.

In 1976 R. M. Metcalfe and D. R. Boggs of Xerox wrote a paper entitled *Ethernet: Distributed packet switching for local computer networks.* In it they proposed the use of a collision detection technique to improve the efficiency of *Carrier Sense Multiple Access* (CSMA) protocols. The use of collision detection does improve the efficiency of a CSMA network. However, there are several other variables that also affect efficiency:

- The length of the packets sent over the network
- The length (total propagation delay) of the transmission path
- The number of stations on the network

Since 1976 various authors have published a steady stream of abstract calculations showing how the efficiency of a CSMA/CD network is affected by these variables. One author, Andrew F. Tanenbaum, wrote a definitive text entitled *Computer Networks.* It showed that, for a packet length of 128 bits, on a 1 km Ethernet cable, with 32 stations trying to send, 40% of the Ethernet transmission time results in collision fragments, which end up getting discarded. That means the effective transmission rate for a 10 Mbit/sec Ethernet system of this configuration is only 6 Mbits/sec.

Although this is scarcely better than half of the expected transmission rate, it's still quite good, and would satisfy the needs of many applications. However, in actual practice, the effective transmission rate is much worse. In some cases (where network traffic is extremely heavy) the effective transmission rate is dismal. That's because Ethernet was not designed to handle heavy traffic. But even in cases where traffic is relatively light, Ethernet performance is still often disappointing. The reason? Because the systems *connected to* the Ethernet network are not properly architected.

Networking protocols are structured into *layers*, according to a layered model, known as the Open Systems Interface (OSI) standard. (See Figure 1) When a message is sent by an application program, it first travels downward through the various layers of the protocol, then along the network cable (as a string of bits) and then upward through the protocol layers, to arrive at the destination application program.

Each message must pass through all of the layers in the prescribed order. For that reason, the rate at which messages can be sent from one application program to another is limited by the speed at which the slowest layer processes these messages. If an application program tries to send messages faster than the slowest layer can process them, the messages begin backing up. Eventually, the application program is forced to wait while the slow layer finishes processing its backlog of messages.

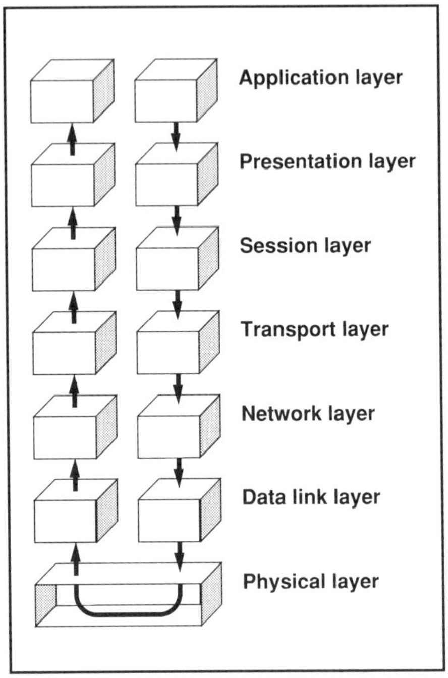

Figure 1

Layered networking protocols can most efficiently be handled in a *pipelined* fashion. In designing any pipelined system, you must be sure that the various processing elements in the pipeline can handle the data flow at the same rate. Thus, in designing an Ethernet system, you must look at how long it takes each layer of protocol to process a message.

The Ethernet standard (IEEE 802.3) prescribes protocols for only the lowest 2 layers: the *physical layer* and the *data link layer.* Since the protocols of these 2 layers are typically implemented in hardware, processing takes place at the same speed that data flows over the Ethernet cable: 10 Mbits/sec.

However, the next 2 OSI layers (the *network layer* and the *transport layer*) are not prescribed by the Ethernet standard. They're typically implemented with software. As a result, these layers are much slower at processing messages than the physical layer and the data link layer. Ironically, *Ethernet* is often blamed for poor performance, when the fault actually lies with slow network layer software and/or slow transport layer software.

So, how do you speed up the network layer and the transport layer? Several VMEbus suppliers have introduced intelligent Ethernet controller boards with onboard MPUs that execute the network and transport layer software.

How does having an onboard MPU help speed things up? Theoretically, this local MPU (unlike the host MPU) can be dedicated *totally* to the execution of the network and transport layer software. That should speed up execution of the software and increase throughput...right? It does...to a degree. But a dedicated MPU (all by itself) may not improve performance much over having the host MPU execute the software unless the board's architecture is well thought out. There are 3 potential problems to consider:

- The processing power of the Ethernet controller board's MPU may not be adequate.
- The processing power of the Ethernet controller board's MPU may not be fully used.
- Message traffic between the Ethernet controller board and the host MPU (over the VMEbus) may interfere needlessly with other VMEbus activities.

The controller board should have a high performance MPU

If you choose a high-performance processor for the controller board it can quickly execute the network layer and transport layer software onboard. That allows it to strip off many of the message bytes before forwarding the message to the host over the VMEbus. This shortens the time needed to forward the message.

If the controller board's MPU is much less powerful than the host's, it may not be very useful. In fact, if the host MPU is lightly loaded with other duties, the quickest approach might be to forward the unprocessed messages to the host, and let the host MPU execute the network layer and transport layer software.

For example, suppose the Ethernet controller board is equipped with an 8 MHz 68000. The extra time needed to move the longer (unprocessed) message over the VMEbus to the 68020 host may be more than compensated for by the 68020's faster execution of the protocol software. Of course, the optimum solution is to equip the controller board with a 68020.

How to get optimum performance from the controller board's MPU

There's no point in choosing a high-performance processor if you design it into an architecture that prevents it from running at full speed. Obviously, you want the processor to be able to fetch its instructions from RAM without wait states. But just providing fast RAM isn't enough. You need to consider how many other devices will be competing with the MPU for access to that RAM. For example if DMA-driven peripheral devices access the RAM for a considerable portion of the time, the effect on MPU performance will be similar to that of slow RAM.

That seems obvious, but many Ethernet controller boards are not designed with this concern in mind. Figure 2 shows a typical Ethernet controller board architecture. It uses the LANCE chip (from AMD or Mostek) which reads data from (and writes data to) shared RAM, using DMA transfers. In this architecture, the LANCE chip is connected directly to the MPU's data bus. The result is predictable. Each time the LANCE chip transmits or receives a packet, it spends several microseconds accessing the shared RAM. During this time, the MPU is forced to stand idle.

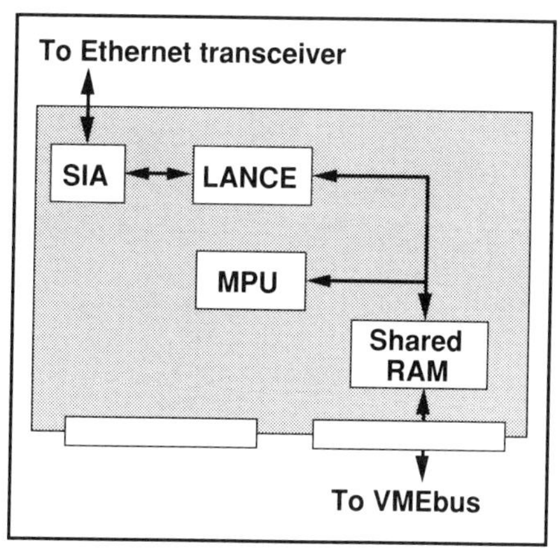

Figure 2

But suppose you recognize this problem, and provide the MPU with its own private instruction memory. The result is an architecture like Figure 3. Since this private memory can't be accessed by any other device, the MPU doesn't have to wait when fetching its instructions.

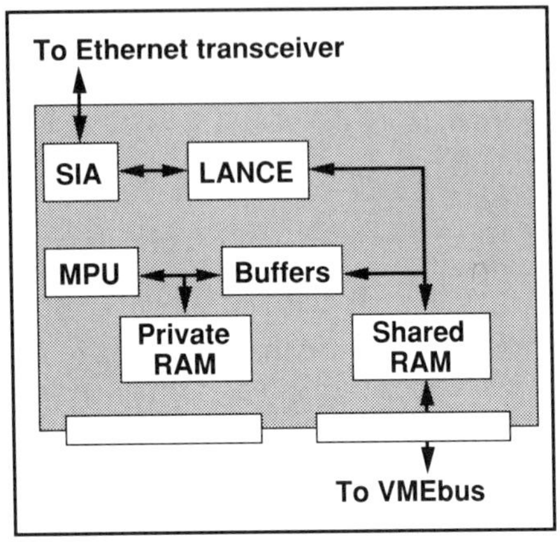

Figure 3

But, sooner or later, the MPU's software must access the message bytes. These message bytes must be kept in shared RAM, since the LANCE chip also needs access to them. But the LANCE chip has very little internal buffering, and is not very tolerant when it comes to sharing memory. If it's not granted access to the memory immediately, it will overrun when receiving a continuous stream of incoming messages.

To avoid overruns, the MPU must *immediately* defer to the LANCE chip whenever the LANCE chip is receiving an Ethernet message. The MPU must then stand idle for the duration of the message transfer. Thus, the MPU can't do any protocol processing concurrently with the reception of new messages.

But, suppose the MPU is permitted to access the shared RAM *occasionally* during a message reception. It could then resume execution, since most of its accesses are to its private instruction RAM.

To allow that, you could redesign the board to include *pipeline registers*, as shown in Figure 4. Whenever the MPU accesses the shared RAM, these pipeline registers prevent overruns by accepting bytes from the LANCE chip. This allows the LANCE chip to go ahead and accept subsequent incoming bytes. The pipeline registers then accumulate these bytes (until the MPU finishes its access) and then write the bytes into the shared RAM.

You need to use the VMEbus efficiently

The speed at which data is transferred over VMEbus depends upon both the master and the slave that are participating in the transfer: the actual VMEbus data transfer time is the *cycle time* of the Master, plus the *access time* of the Slave. Thus, no master design can (by itself) ensure that the VMEbus operates in a high-performance mode. However, the controller board should be designed to transfer data over the VMEbus as fast as each memory board permits.

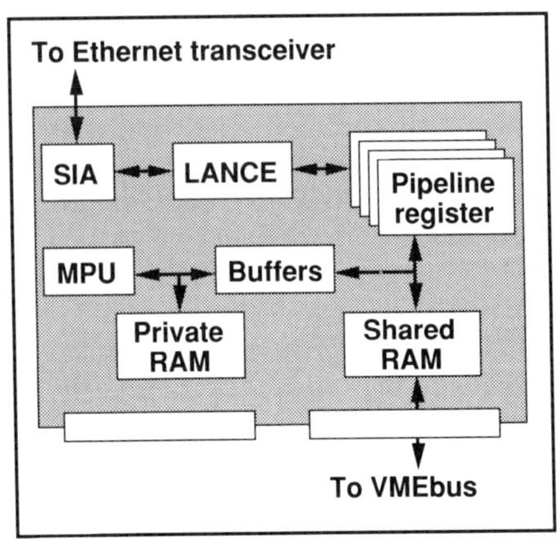

Figure 4

Because all Ethernet message traffic must eventually flow through the VMEbus, it appears that the VMEbus might set an upper limit on the message handling rate of the Ethernet controller board. However, it turns out that this is rarely the case. For example, to maintain an effective transfer rate of 5 Mbits/sec over the Ethernet, the controller board must move only 625 Kbytes of data over the bus each second. That's hardly straining the 40 Mbyte/sec upper limit that VMEbus provides. In fact, even if the Ethernet controller board transfers 1 byte at a time (and takes a full microsecond for each transfer) it would only be using the VMEbus 62.5% of the time. However, if you look at this from another point of view, you realize that only 37.5% of the available VMEbus bandwidth is available for other masters on the bus. In most applications that's simply not acceptable. The solution is obvious. The controller board should move data over the VMEbus several bytes at a time.

An Ethernet controller board that's designed to move data 16 bits at a time would be better. Given the rate of message traffic on most of today's Ethernet systems, that would be entirely adequate. That's because the rate at which the controller receives incoming messages from the Ethernet is limited by the speed of the *sender's* protocol software.

However, as high-performance controller board architectures (like the one shown in Figure 4) become more popular, the average message sending rate will rise, and your controller board will need to handle incoming message traffic more efficiently. Otherwise it will use up too much VMEbus time. That can be done by transferring data 32 bits at a time.

A properly designed 32-bit Ethernet controller board should have no problem keeping its VMEbus usage within reasonable limits. Even with the theoretical maximum Ethernet transfer rate of 10 Mbits/sec, a 32-bit controller board only needs to move 1.25 Mbytes/sec over VMEbus. If a 32-bit controller board can achieve a VMEbus data transfer rate of 30 Mbyte/sec, it will only use the VMEbus 4.17% of the time.

An example of a high performance Ethernet controller board

Interphase offers a 68020-based (16 MHz) intelligent Ethernet controller board called the V/Ethernet 4207 Eagle, which is architected to give maximum performance. (See Figure 5) All of the board's internal data paths are 32 bits wide. The MPU's private RAM consists of 128 Kbytes of zero-wait-state static RAM.

The 32-bit data bus that connects the 68020 to its private memory is partitioned off from the rest of the controller board's 32-bit local data bus. Data movement between the LANCE chip and the 512 Kbytes of shared RAM doesn't interfere with the 68020's instruction fetches from its private RAM. Incoming and outgoing Ethernet messages are stored in the 512-Kbyte shared RAM. Pipeline registers move data into and out of this shared RAM whenever the 68020 and the VMEbus stop accessing it.

The shared RAM can be accessed as a slave from VMEbus, with an access time of 300 nsec. The board can also act as a master. By using Interphase's *Buspacket* interface, the controller board can move data between its local shared RAM and other global VMEbus RAM at rates in excess of 30 Mbytes/sec. (This assumes 30 nsec VMEbus memory access times.)

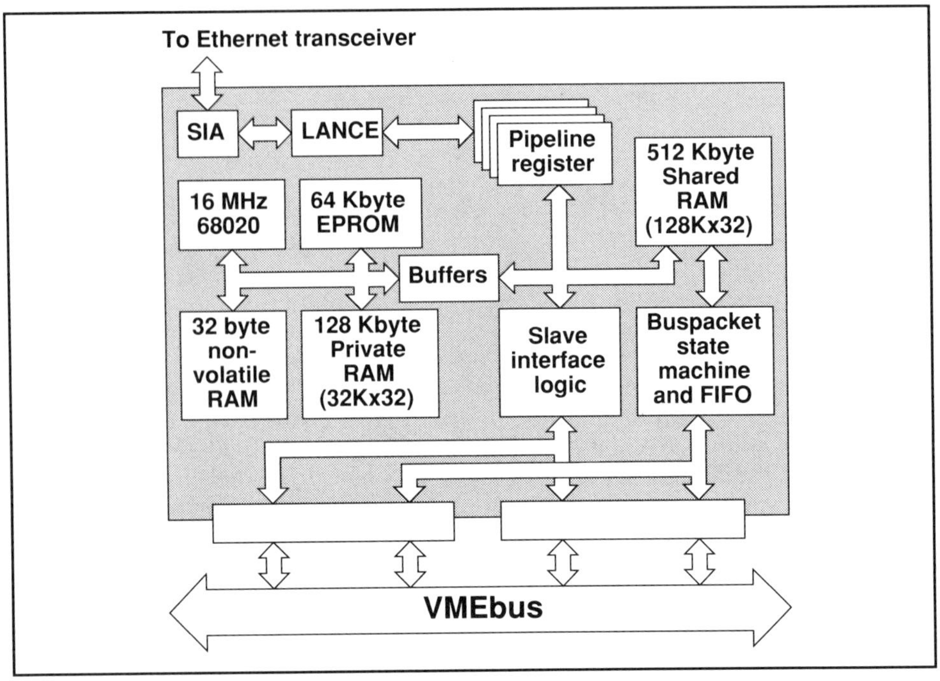

Figure 5

Protocol software can either be downloaded to the local shared RAM from the host (over the VMEbus) or it can be kept in EPROM on the board. (In the latter case, the 68020 would typically copy this EPROM firmware into its private RAM for faster execution.)

Summary

Although Ethernet is sometimes accused of being very slow, the problem is often that Ethernet's capabilities just aren't being fully utilized. That's because traditional intelligent Ethernet controller board architectures prevent their local MPUs from executing higher-layer protocol software efficiently.

However, by designing controller boards properly, the protocol software can be executed much faster. In addition, if the controller board's VMEbus interface is designed properly, its load on the VMEbus can be minimized, improving the performance of the entire VMEbus system as well. The end result is higher data transfer rates over both the VMEbus and the Ethernet local area network.

Ernest E. Godsey is director of product marketing at Interphase in Dallas Texas. He has a BSEE from Texas A&M University and an MBA from the University of Wisconsin. Ernest is a member of Tau Beta Pi, Etta Katta Kappa Nu, and Beta Gamma Sigma honor societies. He also has been granted several patents, and has published numerous articles.

CHAPTER 33

How To Use VMEbus And Ethernet To Build Real-Time Distributed Systems

By Jerry Fiddler and Leslie Kirby

With the widespread availability of off-the-shelf backplane and local area network hardware, it's fairly easy to assemble a multiple-processor system from off-the-shelf components. However, it's not so easy to create the software you need to make the processors in such a system work together. In this chapter Jerry and Leslie describe how you can use existing TCP/IP software to provide the communication links you need between the various MPUs in a multiprocessing system.

With today's off-the-shelf microprocessor technology you can assemble very low cost real-time multiprocessing systems. However, to be useful, such a system must provide some way for the various processors to *communicate* with each other. Different applications demand different communication characteristics. For example, in applications where you need to move a lot of data quickly, a backplane might be the most appropriate communication path. In other applications, where you are more concerned about communicating over relatively long distances, a local area network would be more appropriate.

However, communication involves more than just providing a *hardware path* for the message flow. There's still the job of creating the *software* that initiates and regulates the message traffic. What's the solution to that problem? The one being chosen by many engineers is the Berkeley UNIX networking standard TCP/IP. Among other things, this software package includes routines that implement Internet Protocol, from which it derives the last portion of its name. Internet protocol provides for transparent routing of messages between MPUs. This protocol is largely independent of the type of hardware link that's used to actually carry the messages between the MPUs. As a result, it can be used where the message traffic must be routed over different types of physical media, such as a backplane and a coaxial cable.

How does TCP/IP work?

TCP/IP can be thought of as a common communication protocol that allows you to interconnect several MPUs to function as a single distributed processing system. Operating systems that support the TCP/IP protocols provide special system calls that allow application programs to:

- access files on remote MPUs
- login to operating systems running on remote MPUs
- execute subroutines on remote MPUs

It has become popular to structure networking protocols into a hierarchy of layers. The bottom layer consists of the network hardware. In many cases this hardware sends the messages by a serial communications link, such as an Ethernet coaxial cable. However, the messages can just as easily be sent over a parallel communications link, such as VMEbus backplane.

On top of the hardware layer lies the *Internet* family of software modules. Internet software implements 3 main protocols:

- Internet Protocol (IP)
- User Datagram Protocol (UDP)
- Transmission Control Protocol (TCP)

What is Internet Protocol?
Internet Protocol (IP) is used to send messages from MPU to MPU. Each MPU on a Local Area Network (LAN) has a unique *Internet address*. The sending IP software accepts a message from its own local MPU (along with an Internet address), combines the message with the address (to form a *packet*) and sends the packet to the MPU indicated by the Internet address.

In some cases, the 2 MPUs might be on different LANs. In order to deliver the message, there must then be at least one MPU which is interfaced to *both* LANs. That MPU then acts as a *gateway*, accepting the packet from one LAN, and sending it over the other LAN.

When a message is too long to send over a local area network as a single packet, the sending IP software breaks it up into several packets. Then it sends each of these packets to the destination IP software separately. As these packets are sent, the destination IP software *reassembles* them into the original message.

To ensure that none of the packets gets garbled during transmission, the sending IP software calculates (and incorporates) a *checksum* into each packet, as it sends it. Upon arrival of each packet, the destination IP software verifies this checksum. If it finds the packet to be corrupted, it signals the sending IP software to retransmit it.

The IP protocol layer uses the network hardware to send packets from MPU to MPU. (See Figure 1) If there's only one application program running on each of the communicating MPUs, each application program could simply call the subroutines within the IP software package. However, there will usually be *several* application programs running on each MPU. (When several application programs run on a single microprocessor, they're often called *processes*.) These processes will need to *share* the single communication link provided by the network hardware. To accomplish this, they need some kind of *multiplexing* software. This multiplexing can be provided with an intermediate software layer, between the processes and the IP layer.

There are 2 popular methods for multiplexing a communication link. One of these methods is modeled after a *postal system* and the other is modeled after a *telephone system*. These 2 models are embodied in 2 software packages that work with the IP software. The 1st is called User Datagram Protocol (UDP) and the 2nd is called Transmission Control Protocol (TCP). (See Figure 1)

What does User Datagram Protocol do?
To draw an analogy, if we view each MPU as a kind of post office, then the IP layer serves as a *delivery service* between post offices. However, since post offices typically serve many subscribers, we also need *postmen*. The postmen accept letters from subscribers and carry them to the local post office. After those letters are sent to the destination post office, postmen deliver each letter to the appropriate subscriber.

User Datagram Protocol (UDP) is the postman that accepts letters (datagrams) from subscribers (processes) and delivers them to the post office. After the IP layer (the delivery service) transports these datagrams to the destination post office, UDP then delivers them to the appropriate processes.

The UDP layers sits on top of the IP layer. (In other words, the UDP software calls the subroutines in the IP software.) UDP associates a unique *port number* (a kind of *street address*) with each communicating process. When some process wants to send a datagram, UDP accepts the datagram, (along with a destination port number) and then calls the subroutines in the IP layer to send the datagram to the appropriate MPU. When the datagram arrives at the destination MPU, the destination UDP software checks the port number attached to the datagram, and then directs the datagram to the appropriate local process.

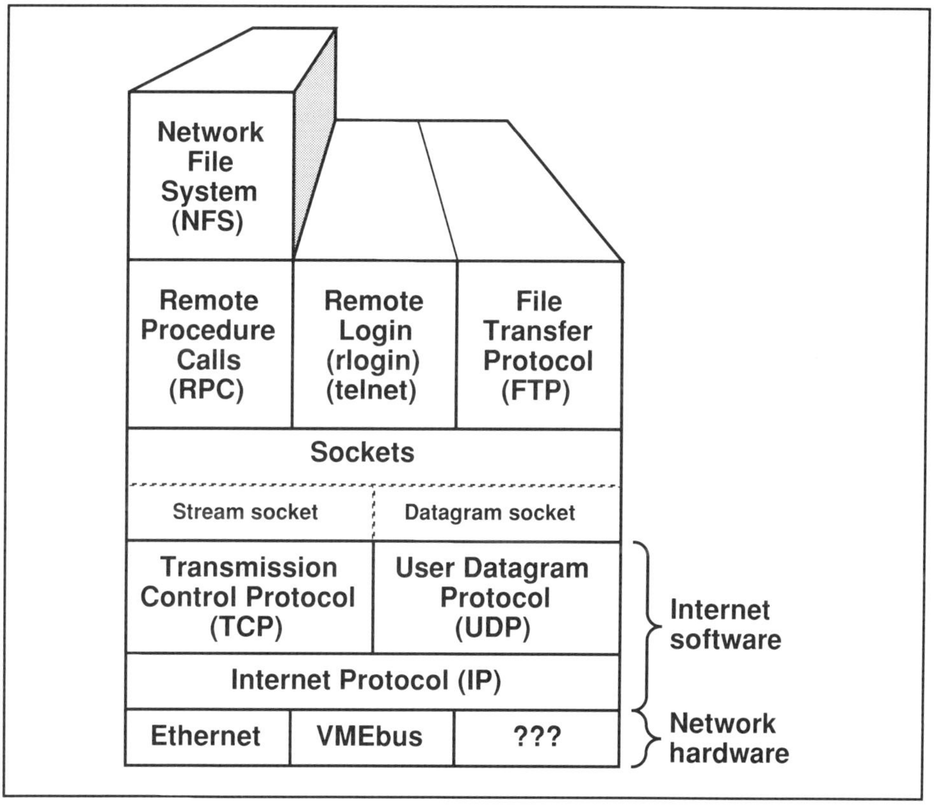

Figure 1

While UDP is very useful, it has some limitations:

- It provides no way for the receiving process to acknowledge receipt of the datagram. Because of this, the sending process doesn't know when the datagram has actually been received.
- Since the IP layer may have to delay final delivery of a particular datagram until it can retransmit some garbled packet, there's no guarantee that the datagrams *sent* first, will also *arrive* first at the destination.
- In *rare* cases, the destination process may actually receive 2 duplicate messages. (This might occur when UDP detects that a gateway has failed while sending a message, and then sends the message a 2nd time through an alternate gateway.)

What does Transmission Control Protocol do?
UDP is good for sending *information* from one process to another. However, if you want to coordinate the *activities* of 2 processes, you might prefer to have the 2 processes communicate in a more *interactive* fashion, as if they were talking over a telephone line. That's what Transmission Control Protocol (TCP) is for.

TCP (like UDP) sits on top of the IP layer, and calls subroutines in the IP layer. However, the kind of communication that TCP supports is quite different from UDP. TCP is a *connection based* protocol. Before 2 processes can communicate, they must first establish a bidirectional *connection*. That's done through the use of *port numbers* (a kind of *phone number*) which TCP assigns to each process. Once the connection has been established, the 2 processes can send *byte streams* to each other.

Byte stream communication is quite different from datagram communication. Either process can send bytes whenever it chooses, and can stop sending whenever it chooses. Any communication conventions are a private matter between the 2 connected processes: TCP simply provides a bidirectional connection between them.

We mentioned earlier that UDP does not include any formal mechanism that allows the destination process to indicate receipt of a datagram. However, TCP verifies successful transmission of each byte string before it sends any subsequent byte strings. If a particular byte string does not arrive at the destination, all subsequent byte strings must wait until that byte string has been retransmitted successfully. Because of this continuous verification process, the bytes in the byte stream are always received at the destination in the same order in which they were sent.

TCP also provides *flow control*. Flow control allows the destination to temporarily block any further byte transmissions until it has finished processing the bytes it has already received.

What's a socket?
We've now talked about how UDP and TCP provide for communication between processes. But, how does a process actually *use* these protocols? It does so by creating a *socket*, which is a software interface to a network. (See Figure 1)

There are 2 popular types of sockets:

- Stream sockets
- Datagram sockets

Stream sockets

Stream sockets use TCP/IP for interprocess communication. When each process creates its own *stream socket*, that socket is assigned a unique *port number*. (The stream socket is analogous to a telephone, and the port number is analogous to a telephone number.) Any process with a stream socket can request a connection to any other stream socket. In other words, one process can dial up another process by providing a *network address* (area code) and a *port number* (phone number). The process being dialed then *accepts* the connection (answers the phone).

Once the dialed process accepts the connection, the 2 processes can communicate. This connection works just as if there is a dedicated physical circuit between the 2 processes. However, this circuit is actually only a *virtual circuit*, simulated by the TCP layer, which actually multiplexes several connections over a single local area network link. This virtual circuit provides bidirectional, reliable, error-free, socket-to-socket communication.

Datagram sockets

Each process can also create a *datagram socket*, in much the same way it creates a *stream socket*. When a process creates a datagram socket, a unique *port number* is assigned to it. (The datagram socket is analogous to a mailbox, and the port number is analogous to a street address.) Any process with a datagram socket can send datagrams to any other datagram socket by specifying a network address (a post office) and a port number (street address). There is no need to establish a connection before sending each datagram.

There is nothing to prevent a process from creating several sockets. In fact, a process could even create several stream sockets and several datagram sockets.

It might appear from this discussion that any communication that goes through a socket goes over a communication link, such as a local area network or a VMEbus backplane. However, that's not necessarily the case. One process can use a socket to send a message to another process that's running on the same MPU. In fact, a process can even send a message to itself!

Higher-level network protocols and utilities

It seems that sockets provide just about all the communication capability that any application programmer might want. However, the TCP/IP software package also includes additional layers of very useful higher-layer protocols. (See Figure 1)

For example, suppose you want to allow a process running on one MPU to login (as a user) to an operating system running on some other MPU. Or suppose you want your process to get a copy of a file that's in the file system of some other MPU. Better yet, why not allow your process to access the remote file, just as if were a *local* file? Maybe you would even like to be able to *call a subroutine* whose code resides within the address space of some other MPU!

That may sound like a big order, but there are utilities in the TCP/IP software package that can do all of this. For example, the `rlogin` software utility allows a UNIX process to login to some other operating system over a local area network. The `Telnet` utility is similar, but runs under non-UNIX operating systems, and allows processes to do the same thing. Remote login is especially useful when you're creating and compiling code in one system, and then downloading the code into another system for debugging.

FTP (File Transfer Protocol) is a software utility that allows a process to transfer files between systems, over a network. It also allows a process to read the directories of a remote file system.

RPC (Remote Procedure Calls) is a utility that allows a program running under one MPU to call a subroutine running under another MPU, over the network. It's so transparent that the calling program and called subroutine need not even be aware that they run on different MPUs. This is often a less cumbersome way to communicate across the network than creating a connection, or exchanging datagrams.

NFS (Network File System) unifies all of the file systems in a network into one **gigantic** file system, where any process running on any MPU can access all remote files, just as if they were all local.

How is network addressing done?

Some distributed systems provide multiple communication links. For example, you might interconnect several VMEbus card racks with an Ethernet local area network. Each VMEbus backplane allows the MPUs within that VMEbus card rack to communicate with each other, while the Ethernet link allows the MPUs in each VMEbus card rack to communicate with those in other card racks.

To link each VMEbus backplane to the Ethernet, we could use the MPU on an intelligent Ethernet controller board to execute IP-level software. The board then acts as a *gateway*. When a process running inside one VMEbus card rack sends a message to a process running in some other card rack, the gateway routes the message from its own VMEbus backplane over the Ethernet. Then the gateway in the destination card rack receives the message, and routes it over its own local VMEbus backplane. The function of both gateways is transparent to the sending and receiving processes.

To ensure that each message gets routed to the right MPU, IP assigns a *network address* to each MPU. TCP/IP and UDP/IP each extend this network address with a *port number.* This port number is assigned to a particular *socket.* This socket is associated with one (and only one) process running on that MPU.

The network address is written in a standard dot notation: 4 integers separated by periods. For example, MPU number 4 on network 90 would be indicated by 90.0.0.4. In a similar fashion, MPU number 5, on the same network would be 90.0.0.5. An MPU on network 93 might be indicated by 93.0.0.4.

Some MPUs are interfaced to more than one network. In our previous example, the MPU on the intelligent Ethernet controller board was interfaced to both the Ethernet and to the VMEbus. If the VMEbus backplane is known as *network 91* and the Ethernet network as *network 90*, the Ethernet controller's MPU would have 2 network addresses. For example, if its MPU number is 4, its VMEbus network address would be 91.0.0.4, and its Ethernet network address would be 90.0.0.4.

TCP/IP is only part of the answer
When you build a distributed system, it makes sense to use TCP/IP software. However, this software only provides you with a way to send messages between MPUs. It doesn't provide you with an *operating system* for each MPU. Having a good operating system is important when it comes to debugging the application software in your distributed system.

Suppose you're developing a distributed, real-time automation system for a manufacturing plant. This system will consist of several real-time VMEbus-based systems connected to a supervisory system (and to each other) via an Ethernet link. Each real-time system consists of a VMEbus card rack containing several MPUs.

You'll want to provide each MPU with an operating system. UNIX will do the job nicely for the supervisory system, but it's not very useful for supporting real-time activities. On the other hand, real-time operating systems that provide prompt, predictable response times, often have rather limited software development tools. Rather than trying to create a single operating system that does it all, we could build a distributed system that uses both kinds of operating systems.

But how do you get these different operating systems to communicate with each other? That's where TCP/IP enters the picture. For the supervisory system you need to find a version of UNIX that supports TCP/IP. Then you need UNIX software drivers for the particular Ethernet controller board you plan to use. For the MPU boards in the real-time VMEbus card racks you need to find a real-time operating system that supports a full implementation of TCP/IP, with drivers for both an Ethernet controller board and for VMEbus.

Once you have that, you can build a distributed system consisting of a supervisory MPU (running UNIX) and other MPUs, each running a real-time operating system. The various MPUs can use TCP/IP to communicate with each other over the VMEbus backplanes, and over the Ethernet link. Application-level processes running on the supervisory UNIX system could send commands to processes running on the real-time operating systems.

To develop your application software efficiently, you want to be able to:

- use modern editors, compilers, source-level debuggers, and source code control utilities to develop your application software on the supervisory system.
- download files (over the Ethernet link) from the supervisory system (on which you have developed them) to the various MPUs in the VMEbus card racks.
- login to the operating system of the real-time MPU, in order to debug the software that you just downloaded into it.

UNIX provides all of the software development tools that you need, and the FTP utility in the TCP/IP software package allows you to download files. The `rlogin` utility package provides protocols for logging into the remote real-time operating system.

However, downloading the files and logging into the remote operating system is only the first step. You'll also need *debugging utilities* that are designed specifically for debugging distributed systems. Traditional approaches (which use ROM-based debuggers) just don't get the job done. Things happen too fast in a distributed system to allow debugging with that kind of tool.

What you need is a properly integrated debugging system that allows you to download object code from the supervisory system to your real-time system (from anywhere in the plant) and then login to the real-time operating system, without having to physically move from place to place. Then (when you uncover a bug) you want to login to the supervisory UNIX system, to gain immediate access to its high-quality tools, such as its editors and compilers.

After you have the whole distributed plant automation system debugged and running, you want to allow a UNIX operator at the supervisory system to login to any of the real-time systems, determine its status, and possibly adjust operating parameters. When no operator is present, you want the UNIX system to automatically keep track of inventory, and to respond to alarms sent to it by the real-time systems.

VxWorks: a real-time operating system with TCP/IP support

All of this sounds pretty good, but first you need a real-time operating system that fully supports TCP/IP: an operating system like VxWorks. VxWorks is a real-time multitasking operating system which fully supports TCP/IP. It supports:

- Sockets for interprocess communication across a network.
- Remote File Access over a network.
- Remote Login to other operating systems across a network.
- Remote Procedure Invocation, which allows a process under VxWorks to call a subroutine on another MPU over a network, without even being aware that the subroutine resides on another MPU.

Since all processes running under VxWorks communicate with each other using sockets, you can easily restructure the system hardware, without having to modify the application software. For example, if you need more computing power, you can add extra MPUs by inserting additional MPU boards into the VMEbus backplane, or even by connecting an additional VMEbus card rack to the Ethernet link. Since the application software is insulated from the particular hardware configuration (by the use of sockets) it's very tolerant of changes in the hardware.

This flexibility can be very useful during development. For example, suppose you have a team of programmers developing a VMEbus system. That system includes one supervisory MPU board thats runs UNIX, and several other MPU boards that run VxWorks. You know that it's risky to develop the software directly in the VMEbus card rack, since a bug in one programmer's application code might crash the entire UNIX system. However, you need to make available the UNIX development tools while the development is going on.

The solution is to move the UNIX MPU board into a 2nd VMEbus card rack, and then connect the 2 card racks together with an Ethernet link. That keeps them loosely coupled, while maintaining the advantages of a tightly coupled system. Then, after the bugs have been removed from the application software, the UNIX MPU board can be reinstalled in the original VMEbus card rack for final checkout, without changing any software.

All of this sounds great, but will it really work?

An actual application

The University of California and the California Institute of Technology have built a huge multiple-mirror telescope on Mauna Kea in Hawaii. The effective mirror size is 10 meters, making it the largest optical telescope in the world. A 10-meter monolithic mirror is impossible to grind using current technology. Even if it could be ground, it would be impossible to transport, prohibitively heavy to support and aim, and would deform under its own weight. The solution is to use 36 two-meter hexagonal mirrors, each weighing approximately 880 lb. Each hexagonal mirror is positioned by a set of three actuators (with 4 nanometer resolutions) which are repositioned twice per second.

Obviously, this telescope includes a lot of electronics. Because of its physical size, a distributed control system is needed. There's one system that controls the dome, a 2nd that points the telescope, a 3rd that controls the main mirrors, and a 4th that controls the secondary mirror. These systems are all interconnected with Ethernet. In our discussion here, we'll discuss only the main mirror control system.

The main mirror control system is shown in Figure 2. It consists of a VAX and a VMEbus card rack containing several MPU boards. The VAX and VMEbus card rack are connected to each other (as well as to all of the other systems) by an Ethernet link. Also attached to the Ethernet are some VMEbus-based Sun workstations. The workstations are used only for development of the real-time mirror control software, and will be removed from the final system.

The VMEbus card rack contains 12 MPU boards, each running VxWorks:

- Three 68020 MPUs (Heurikon HK68/V2F boards)
- Nine 68010 MPUs (Ironics IV-1624 boards)

The VAX is the supervisory processor. It issues high-level commands, and provides information to the telescope control MPU. That information includes the telescope's current position.

The 68020 telescope control MPU is the local supervisory processor in the VMEbus subsystem. It calculates the needed positional corrections for the 36 hexagonal mirrors, and issues commands to the nine 68010 mirror control MPUs.

The 68020 communications MPU serves as a gateway between the Ethernet and the VMEbus backplane. While this job could be performed by one of the other MPUs in the VMEbus card rack, the addition of a dedicated communication MPU reduces the processing load on the other MPUs.

The 68020 monitoring MPU is a watchdog for the entire system. Its function is optional, and it will likely be removed from the final system.

Each of the nine 68010 mirror control MPUs controls 4 hexagonal mirrors. These MPUs take orders from (and report back to) the telescope control MPU.

There are 2 separate networks in this distributed control system. The first is the Ethernet (network addresses 90.0.0.X). The other is the VMEbus backplane (network addresses 91.0.0.X). One of the MPUs (the Communications MPU) is on both networks: and has 2 Internet addresses. It acts as the gateway between the 2 networks.

Because TCP/IP handles the routing of messages between these 2 networks transparently, each of the MPUs on the VMEbus backplane has access to all of the resources on the Ethernet (including those in the Sun Workstations and in the VAX) even though none of these MPUs (except the communications MPU) is actually on the Ethernet.

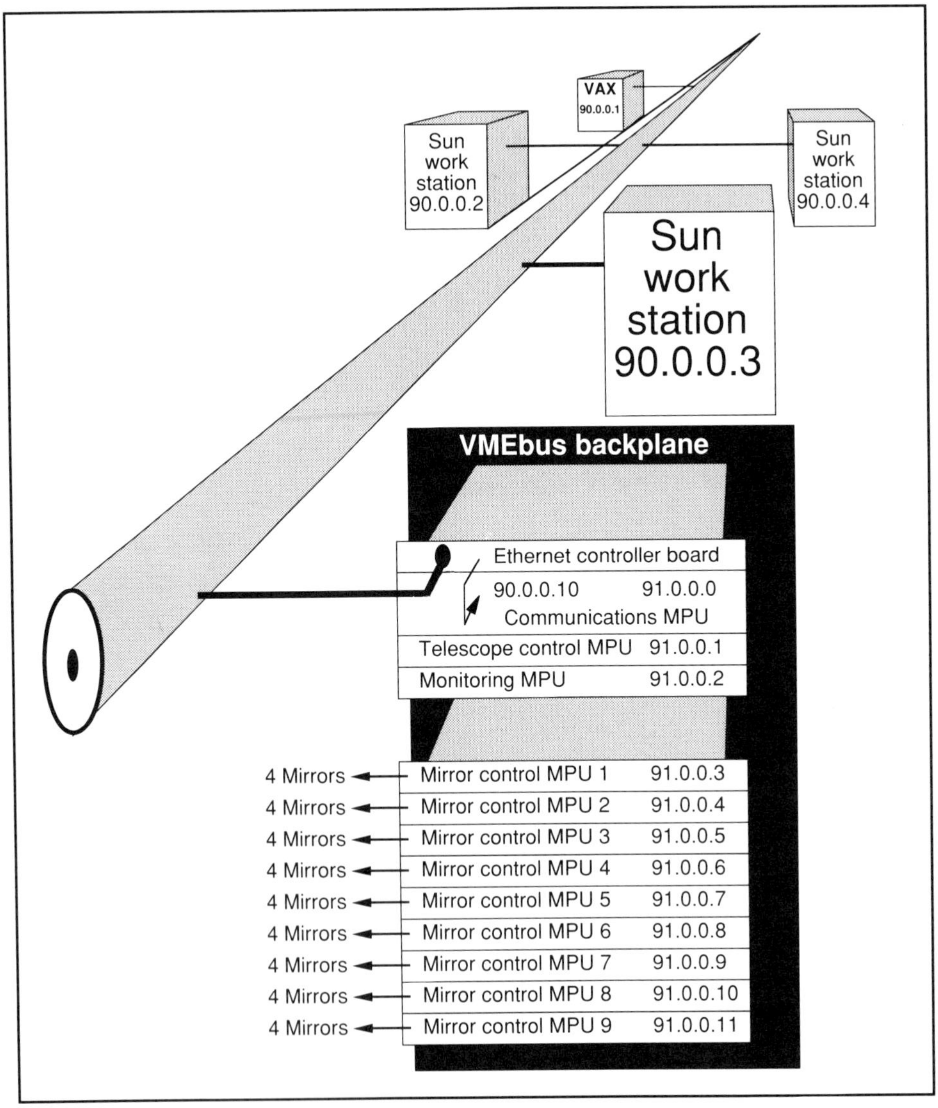

Figure 2

When a UNIX process on the VAX communicates with a process on one of the backplane MPUs (which is running under VxWorks) the communications MPU routes the messages between the 2 networks in a transparent fashion. Likewise, any MPU on the VMEbus can download code from the VAX, use the VAX as a remote file server, or even execute remote procedure calls on the VAX.

Software development for the mirror control system begins on the UNIX-based Sun workstation. Using UNIX software development and management tools (such as text editors, compilers, assemblers, the `make` utility, etc) programmers can design and implement the application software in C.

The application software is then compiled in the usual way, with the UNIX C compiler. Although the resulting compiled code typically uses a library of subroutines provided with VxWorks, it need not be linked to those library subroutines before it's downloaded. Instead, as VxWorks downloads the compiled application code modules into its local memory, it reads their symbol tables. Then it downloads (and links) the appropriate library subroutines. After all the code has been downloaded into the VxWorks system, the programmer can login to VxWorks, and use its shell program to interactively invoke and test individual application subroutines, or even complete processes.

VxWorks remembers the symbol tables from the downloaded object modules, allowing the programmer to use symbolic names for data and subroutines. During the debugging process, the programmer can examine variables, call subroutines, spawn tasks, disassemble code in memory, set breakpoints, and obtain subroutine call tracebacks, all using the original symbolic names. VxWorks also traps program errors detected by the hardware (such as illegal memory references or illegal instructions) and reports them to the programmer.

In addition, VxWorks allows the programmer to use the source language debuggers on the Sun workstation to debug downloaded software. The debugger itself runs on the workstation, but communicates with a *debug server* on the VxWorks system, over Ethernet.

After development and debugging have been completed, the programmer removes all VxWorks debugging facilities. That permits the system to run with minimal resources. At that point, the application software code can be linked with VxWorks, and put into ROM.

This distributed telescope control system provides several advantages:

- It provides enormous computing power at a low cost.
- Its architecture is very flexible. If more processing power is ever needed for the mirrors, more MPUs can be added, or higher-speed MPU's can be used, with no change in the software.
- It allows for close interactive monitoring of the system's performance. For example, each MPU in the mirror control system can be easily monitored and controlled directly from the development system at any time.
- The system is largely hardware independent. Any of the MPUs (or even the workstations) can be replaced with products from alternate vendors with little or no impact on the software.

Summary
A properly designed distributed multiprocessing system provides interprocess communication that hides the details of the underlying hardware. In doing so, it allows you to partition your real-time application software into logical subprograms, without concerning yourself with the hardware architecture of the system. Functions such as data collection, data analysis, and machine control can then be assigned to several different MPUs. If later experience indicates that this initial partitioning is not optimal, you can change it, without having to modify your application software.

Jerry Fiddler is the founder and president of Wind River Systems. He has a BA in music, and an MS in computer science. His leisure activities include volleyball, guitar playing, music composition, airplane flying, and avoiding harmonic convergences.

Leslie Kirby is an engineer at Wind River Systems. She has a BS in computer science from UC Berkeley. In addition to her work, Leslie enjoys drumming, dancing and sharpshooting with her water pistol.

CHAPTER 34

How To Download Files Into Your VMEbus System *FAST*

By Mike Fedde, Ed Theobald, Roger Larson, and Dick Eue

In this chapter Mike, Ed, Roger and Dick talk about their own experiences using hardware emulators and RS-232 links. The problems they encountered with RS-232 downloads from their VAX led them to develop an Ethernet protocol for downloading object files. They describe how that protocol works, and they compare it to using an RS232 link. The C source code that they provide for implementing the protocol should be useful, even if you download your files from some other type of host system.

Like many other VMEbus system developers, you may decide to develop your application software on a large, multiuser operating system, and then download the compiled object code into your VMEbus system for debugging. There are good reasons to do this:

- Multiuser systems usually provide better facilities for team programming.
- Multiuser systems are typically backed up at regular intervals. This provides insurance against any catastrophic failures.
- Multiuser systems typically provide some level of system security. This prevents unauthorized people from accessing critical files.

With the tools available on a multiuser system, you can typically develop your software in a very efficient manner. Before very long, you finish the initial stage of debugging, and you're ready to download the object code into your VMEbus system. However, at that point your pace of development slows down considerably.

The reason for this is that downloading large files over a simple RS232 link takes a lot of time. For example, suppose you want to download a 256-Kbyte file at 9600 baud...

256 X (1024 bytes) = 262,144 bytes

(9600 baud) X (1 bit/sec/baud) = 9600 bits/sec

(1 start bit/byte) + (8 data bits/byte) + (1 stop bit/byte) = 10 bits/byte

(9600 bits/sec)/(10 bits/byte) = 960 bytes/sec

(262,144 bytes)/(960 bytes/sec) = 273 sec = 4.55 min

A few minutes of download time might not seem too bad. However, when you're forced to endure it over and over it can become unbearable. To make the situation even worse, many multiuser operating systems handle RS232 serial I/O in a very interrupt-intensive manner. After sending each character, they wait until the I/O port interrupts, before sending the next. If there are a lot of users (and/or jobs) running on the system, the interrupt latency may even get long enough to prevent a 9600 baud serial link from carrying its full capacity of 960 characters/sec. This can greatly slow down serial transmissions. For example, when our own VMS-based system is heavily loaded, it sometimes takes more than an hour just to download a 256 Kbyte file!

The target system
In designing any download protocol, you must consider the target system configuration. There are 2 popular configurations...

- A target system equipped with an in-circuit emulator
- A target system equipped with debugging ROMs

Figure 1 shows a target system equipped with an in-circuit emulator. The emulator has a cable with a plug, which is inserted into the system's CPU chip socket in place of the microprocessor chip. Inside the emulator is a microprocessor with an additional address map that can be *overlayed* on top of the target system's address map.

When the overlay is inactive, the emulator's processor sees only the target system's address map, and thus emulates the normal operation of the target system. When the overlay is activated, the emulator's processor sees its own ROM/RAM overlayed on top of the target's address map. By executing this overlayed ROM/RAM code, the emulator can accept and execute debug commands from a CRT terminal. For example, the user might command the emulator to:

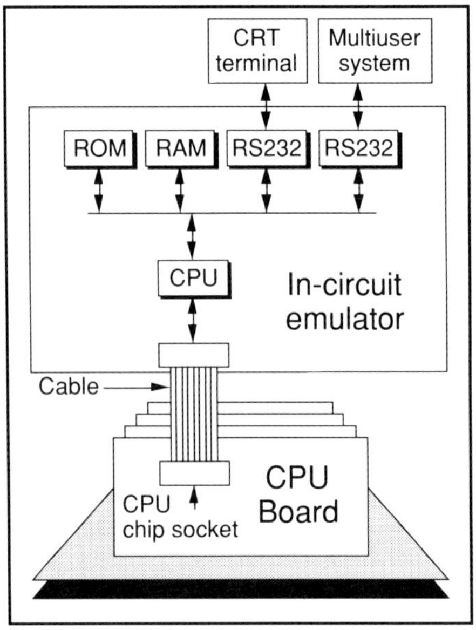

Figure 1

- display or change the contents of a certain target memory location
- download a file through its 2nd serial port, and store that file in the target system's memory for later execution

At the conclusion of a download, the emulator can be configured to either request further commands from the user, or to automatically start executing the downloaded program.

In-circuit emulators are useful, but expensive. Debugging ROMs provide a much cheaper way to debug a target system. By including some extra ROM sockets (and a pair of RS232 serial ports) in the target system, you can include some debugging code that does essentially the same job as the code in the emulator's overlay ROMs. If you're willing to accept certain compromises, this approach eliminates the need for an in-circuit emulator. The debugging ROMs perform the functions of the emulator: accepting commands from the user through one RS232 serial port, and downloading files through the other port.

We initially chose to use an in-circuit emulator, partly because it supported symbolic debugging. However, the extra time required to create and download the symbol table made symbolic debugging impractical. (We could debug faster by studying source code listings or by inserting PRINT statements into the source code and recompiling.) We also encountered some hardware timing problems which we eventually traced to the hardware limitations of the in-circuit emulator.

Since our target VMEbus system already included some rudimentary debugging code (which permitted us to set breakpoints and to start execution at a given address) we decided to enlarge it to support downloads. The target system also included a CMC Ethernet controller board and (since several members of the design team had experience with Ethernet) we decided to use the Ethernet port on our VAX to improve the speed of our downloads.

Block vs. character I/O devices
Operating systems typically support 2 types of I/O devices:

- Block-oriented devices
- Character-oriented devices

When outputting to a *character-oriented* device, the operating system copies one character (byte) into the I/O controller board's output register, and then moves on to some other activity, while the controller actually sends the character. When the controller board finishes sending the character, it generates an *interrupt.* The operating system then responds to this interrupt by copying the next character into the controller board's output register.

This process repeats, with the controller board generating an interrupt each time it finishes sending a character. As you can imagine, these interrupts place a heavy burden on the operating system. If the operating system is already heavily loaded with other interrupts its interrupt latency goes up, and the controller ends up waiting for new characters to send. As a result, the character transmission rate slows way down.

A *block-oriented* I/O device (like the VAX's Ethernet controller) is much more efficient for our purposes. When outputting through a block-oriented device, the operating system passes a *block* of characters to the I/O controller board. The controller board then sends the entire block before generating an interrupt.

By using the Ethernet link, we were able to achieve some startling results. We took a file that previously required 45 minutes to download over the RS232 link and sent it over the Ethernet link. It took only 30 seconds! With further refinements (including the incorporation of checksums) we brought the transmission time down to only a few seconds, while maintaining 100% reliability.

The download software

The download software consists of 3 parts:

- A format converter program, which runs on the VAX host system
- A packet sender program, which runs on the VAX host system
- A packet receiver program, which runs on the VMEbus target system

The *format converter* program reads a Motorola S-Record object code file and generates a *network-layer packet file*, called RAM_ETH. (See Figure 2) It accomplishes this conversion in 2 steps:

- It verifies that all memory addresses of the downloaded object code are within the address range of the target system's memory.
- It *packetizes* the object code, and puts the resulting network-layer packets into the network-layer packet file.

This packetizing is done by repeatedly:

- reading a portion of the object code from the S-Record file into a 1400-byte buffer
- prepending a transport-layer header, a network-layer header, and a packet length, to form a *network-layer data packet*

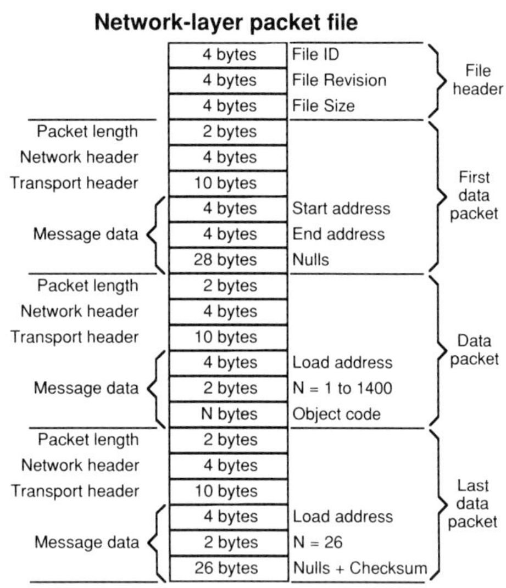

Figure 2

When this process has been completed, the network-layer packet file contains:

- A first data packet
- One or more data packets
- A last data packet

The file begins with a *first data packet*, which will be sent to begin the download sequence. This packet specifies the starting and ending addresses in the target's memory where the downloaded object code should be stored.

The file then contains one or more *data packets*. Each of these packets may contain up to 1400 bytes of the actual downloaded object code, and each specifies the memory address in the target system where its downloaded code is to be stored.

The file ends with a *last data packet*, which is sent to complete the download sequence. This packet contains a checksum of all bytes previously sent during the download.

The *packet sender* program reads the entire network-layer packet file into host RAM, and then sends the packets (one at a time) over the Ethernet link. As shown in

Figure 2, each network-layer data packet in the file includes a *packet length*, a *network header*, and a *transport header*. The packet sender program strips off the packet length, and then prepends a data-link header. The resulting *data-link-layer packet* (which gets sent over the Ethernet link) looks like Figure 3.

In addition to the 3 types of data packets, there are 4 other types of data-link-layer packets:

- Connect packets
- Request packets
- Confirm packets
- Error packets

These 4 types of packets are used to initiate, and to coordinate the download process. (See Figure 4)

When a user at the target system requests a download, the target system sends a *connect packet*, to establish an Ethernet *connection* between itself and the download server. The host system then responds by sending a connect packet back to the target system. Once this exchange has taken place, a connection is established.

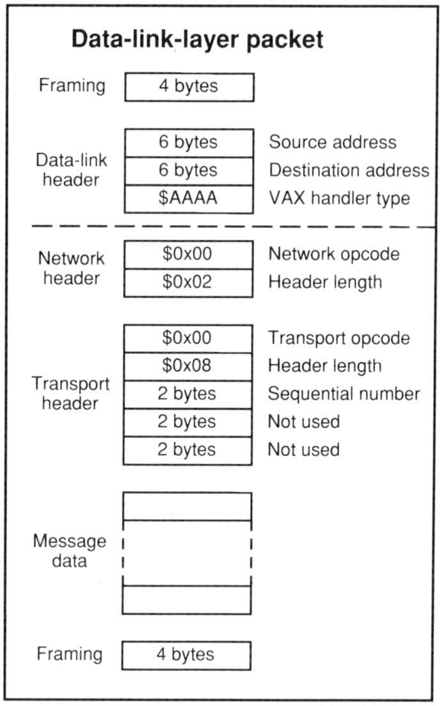

Figure 3

After a connection has been established, the target system sends a *request packet* to initiate a download. The request packet includes the target system's Ethernet address and the name of the file it wants the host to download.

Each time the host downloads a data packet, the target system acknowledges receipt of that data packet with a *confirm packet*, whose sequence number matches that of the data packet it just received. The target system only acknowledges receipt of the data packet if that packet's sequence number follows immediately after the sequence number of the previously accepted data packet.

If the host system finds itself unable to complete the download, it sends an *error packet*, to prematurely terminate the download sequence. When the target system receives this error packet, it aborts the download, and then prints the message contained in the error packet on the target system's console, for the user to read.

If a data packet gets lost, the target system will eventually time out and send an *error packet*, containing the sequence number of the last valid data packet it received. This causes the host to start sending packets again, beginning with the next packet in the sequence. Likewise, if the target system detects an out-of-sequence data packet, it

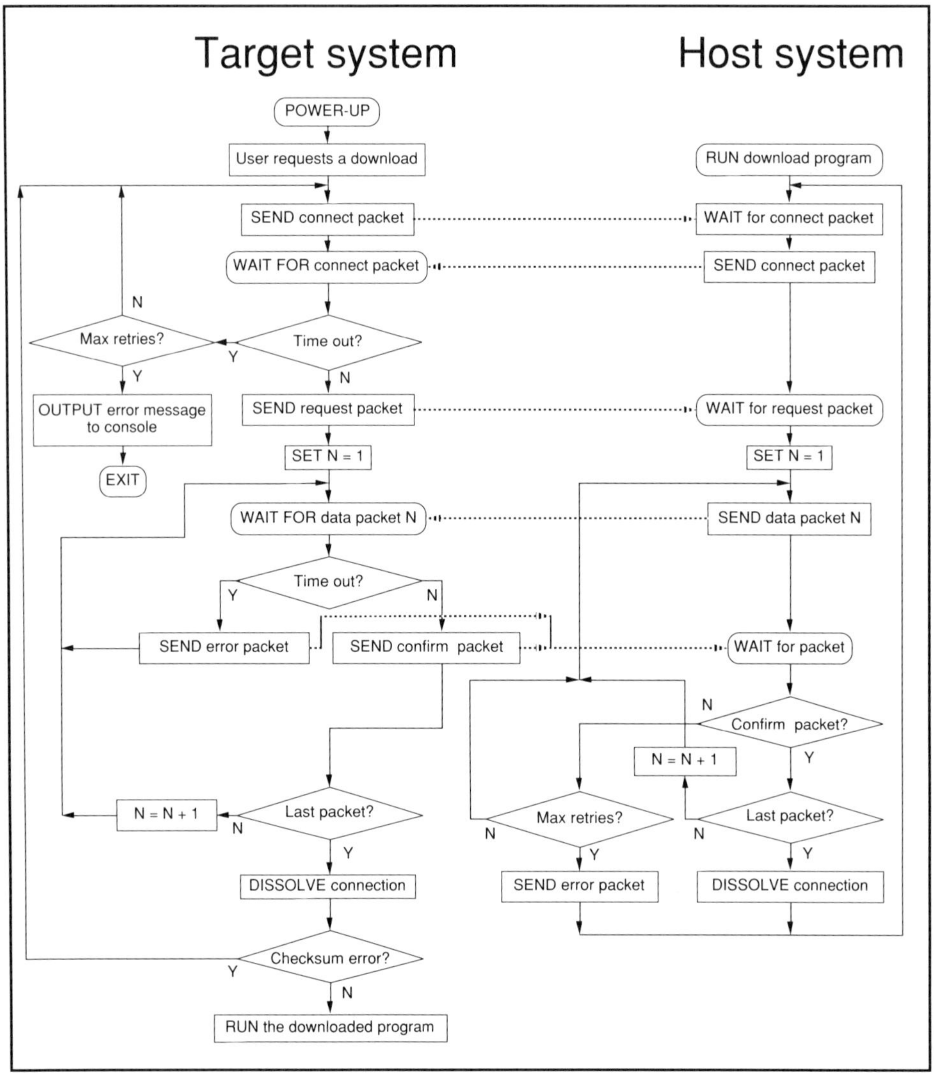

Figure 4

sends an error packet with the sequence number of the last packet it received. The use of sequence numbers assures that the target system receives all of the packets.

Note: All 7 types of data-link-layer packets are identical to the format shown in Figure 3. They differ only in the length and content of their data fields. Figure 5 shows the data format for each type of packet.

After the target system acknowledges the last packet, both the host system and the target system dissolve the connection. The target system then verifies the checksum. If the checksum is valid (and it almost always is) the target system runs the downloaded program. If the checksum is not valid, the target system initiates another download.

In some cases the host may not be able to complete the download. For example, it might not be able to find a network-layer packet file with the specified pathname. When that happens, it terminates the download sequence by sending an error packet (with an ASCII message) to the target system. The target system then aborts the download and displays the ASCII message on its system console.

Packet type	Content of data field
Connect	36-byte ASCII message
Request	80-byte ASCII path & filename of download file
First data	4-byte start address + 4-byte end address + 28 null bytes
Data	4-byte load address + 2 byte size + 1 to 1400 object code bytes
Last data	4-byte load address + 2-byte size + 26 null bytes (incl. checksum)
Confirm	36 null bytes
Error	36-byte ASCII message

Figure 5

Implementing the packet protocol

In our case, the host system is a DEC VAX running the VMS operating system. DEC provides 2 different drivers for its Ethernet controller boards:

- a driver for its Unibus board (used on the VAX)
- a driver for its Q-bus board (used on the Micro-VAX)

Before these drivers can be used to send packets over Ethernet, you must make an ASSIGN system call. This assigns the Ethernet driver to the channel number where the Ethernet controller board is installed. We do this by calling the function `init_eth()`, as shown in Listing 1.

The function `init_eth()` first assigns the Ethernet driver to a channel number. Then it makes a `QIOW` system call to configure the driver (and the controller board) for the mode of operation that you need. `QIOW` expects 12 parameters, which are passed as a structure called `setenparm`. Listing 1 shows the parameter values that `init_eth()` uses to call `QIOW`. (The `QIOW` system call is discussed in the *System Services* section of the VAX/VMS manuals, so we won't go into detail here on the meaning of each parameter.)

When `QIOW` is called, execution of the calling process (which is currently executing the function `init_eth()` is temporarily suspended, while the system call is executed. The system call first checks to make sure that it has been passed the proper number of parameters, that the parameters are within the prescribed range, etc. If there is anything wrong with the parameters, the `QIOW` system call doesn't configure the Ethernet driver. Instead, it just returns an error indication. Odd error values, greater than 1, are warning numbers. Even values (including 0) are error numbers.

The IF statement that calls `QIOW` (see Listing 1) checks the returned status value. If it is anything but 1, `init_eth()` returns a value of 2 to its caller, indicating that the controller has not yet been configured.

Assuming that all parameters are valid, the system call configures the driver and initializes the Ethernet controller board. Then it returns a status word value of 1. The

IF statement that called `QIOW` tests this status word, finds that all of the parameters were OK, and execution of `init_eth()` continues with the next IF statement. That statement checks the I/O completion status block (`iosb`) to see whether its first entry contains a value of 1 (`SS$_NORMAL`). If it contains any other value, then something went wrong: the Ethernet driver and/or the Ethernet controller board is not ready. (Note: The meaning of each parameter in the `iosb` is discussed in DEC's device driver documentation.) However, if all goes well, `init_eth()` returns with a value of 1, and the Ethernet interface is ready to send and receive packets.

Once the Ethernet driver and controller board are ready, you can send packets with the `QIO` system call. Like `QIOW`, the `QIO` system call expects to get 12 parameters. The first 6 are identical to those passed to `QIOW`. The last 6 are specific to the Ethernet driver.

Listing 2 shows part of the code for a function `eth_xmt()` that can be called to send a packet over the Ethernet link. Similar code is used to provide a function `eth_rcv()` for receiving a packet. (The host must be able to read incoming packets, so it can receive the confirm packets and the error packets from the target system during the download.)

The `QIO` system call, provided by the VMS operating system, does not have a time-out feature. However, we were able to achieve a time-out as follows:

- Call the `SETIMR` system call, to start a timeout.
- Call the `QIO` system call, to send the packet.
- Call the `WFLOR` system call, wait for an incoming packet or for the timer to time out.
- Call the `READEF` system call, to determine whether we returned from `WFLOR` because of an incoming packet, or because the timer timed out.

Note: The `SETIMR`, `WFLOR`, and `READEF` system calls are discussed in the VMS System Services manual.

Our VMEbus target system includes a 68010 processor board and a CMC Ethernet controller board, which uses a LANCE Ethernet chip. When the target system is powered up, (or when it is reset) it waits 5 seconds for operator input. If it detects no input, then it initiates a download by calling the function `miz()`, shown in Listing 3. (The function `miz()` gets the Ethernet address of the download server (the VAX) and the path name for the packet file from its local, nonvolatile RAM.

The function `miz()` calls the function `connected()` (See Listing 4) to establish an Ethernet connection between the target system and the download server.

Listing 5 shows the main program `main()`, which runs on the VAX to provide the download server. It calls the function `read_file()` (See Listing 6) to read the packet file into RAM. Each time it sends a packet to the target, it calls the function `get_packet()` (See Listing 7) to get the confirm packet or the error packet from the target system. Note: `main()` also calls some other low-level functions, which are not included here because of space limitations. However, the comments in `main()` explain what they do.

This Ethernet download technique has proven to be very reliable (much more reliable than RS232) and a big time saver. While it takes some effort to implement the protocols, the listings in this article should provide much of what you need. We have found it well worth the effort required to implement it: our software developers can now use the VAX as a central software development system without wasting a lot of time on downloading. Because the software provided here is written in C, you should have little difficulty in using it for downloading from other software development systems, such as PCs or workstations.

Mike Fedde is the project leader for display technology and hardware for image storage and retrieval systems at 3M. He has a BSEE from South Dakota State University and an MSEE from the University of Minnesota. He holds 3 patents in computer architecture and data compression.

Ed Theobald has been employed as a technician at 3M since 1981. He is currently completing a BS degree in Physics at the University of Minnesota.

Roger Larson is a 1983 graduate of the University of Minnesota, with a BS in computer science. While working at Molecular Genetics in Minneapolis, he published 2 papers on algorithms and analysis software for genetic engineering. As a software engineer for Data General, he received the *Contribution of the Quarter* award from the Information Systems Division in 1985. Currently he is working on image storage and retrieval systems for the Image and Information Technologies Sector at 3M Corporation.

Dick Eue is a software consultant with Broadway and Seymour and has 23 years experience in computer hardware and software. He is presently working on image storage and retrieval systems at 3M. Dick specializes in UNIX and VAX VMS systems.

Listing 1

```
/******************** FUNCTION init_eth ***********************/
/* This function accepts a pointer to a VAX channel number.    */
/* It calls the VMS system calls ASSIGN and QIOW to assign and */
/* initialize an Ethernet driver to that channel.  Then it     */
/* returns an integer status value, indicating whether the     */
/* Ethernet interface is ready or not.                         */
/***************************************************************/

int init_eth(eth_chn)
unsigned short *eth_chn;

{
$DESCRIPTOR(eth_dev_nam_vax,"XEA0");    /* Ethernet VAX device name       */
$DESCRIPTOR(eth_dev_nam_micro,"XQA0"); /* Ethernet micro-VAX device name */

/***************************************************************************/
/* Declare a structure to hold the parameters for initializing the Ethernet */
/* driver and controller board.  Then declare a structure that will hold a  */
/* pointer to, and the size of, the parameter structure.                    */
/***************************************************************************/

SETENPARM setenparm;
struct dsc$descriptor_s setparmdsc;

/***************************************************************/
/* Declare a block of 4 integers that will be used to hold     */
/* the status of the Ethernet channel when it gets initialized */
/***************************************************************/

unsigned short eth_iosb[4];

/*************************************************************/
/* Since we don't know whether we are running this code on   */
/* a VAX or a micro-VAX, we first try to assign the VAX      */
/* Ethernet driver to the channel.  If that doesn't work,    */
/* we try to assign the micro-VAX Ethernet driver to the     */
/* channel.  If neither works, then we return from this      */
/* function with a value of 0.                               */
/*************************************************************/

if ((SYS$ASSIGN(&eth_dev_nam_vax,eth_chn,0,0) & 0x01) != 1)
   {
   if ((SYS$ASSIGN(&eth_dev_nam_micro,eth_chn,0,0) & 0x01) != 1)
      {
      return(0);
      }
   }
```

(continued)

Listing 1 *(continued)*

```
/*****************************************************************/
/* If we get down to this point, the driver has been assigned to the  */
/* channel. Now we need to call the QIOW system call with 12 Ethernet */
/* set-up parameters                                                  */
/*****************************************************************/
setenparm.bscod  = NMA$C_CPCLI_BUS;
setenparm.bufsiz = RCVBUFLEN;       /* actual buffer size       */
setenparm.nbcod  = NMA$C_PCLI_BFN;
setenparm.numbuf = 7;               /* actual number of buffers */
setenparm.psbcod = NMA$C_PCLI_PAD;
setenparm.psb    = NMA$C_STATE_OFF; /* pad short buffers        */
setenparm.ptcod  = NMA$C_PCLI_PTY;
setenparm.pt     = 0xabcd;          /* arbitrary protocol type  */
swap_word(&setenparm.pt);           /* swap the protocol type   */
setenparm.pmcod  = NMA$C_PCLI_PRM;
setenparm.pm     = NMA$C_STATE_OFF; /* promiscuous mode         */
setenparm.dccod  = NMA$C_PCLI_DCH;
setenparm.dc     = NMA$C_STATE_OFF; /* data chaining            */
setenparm.crccod = NMA$C_PCLI_CRC;
setenparm.crc    = NMA$C_STATE_ON;  /* generate crc on transmit */

/*****************************************************************/
/* Set the parameter descriptor structure to point to the base address */
/* of the parameter structure, and to indicate its length in bytes.    */
/*****************************************************************/

setparmdsc.dsc$w_length  = sizeof(SETENPARM);
setparmdsc.dsc$a_pointer = &setenparm;

/*********************************************************************/
/* Make a system call to QIOW to set up the driver and the Ethernet       */
/* controller board. Test to see if the call was successful. If so, return */
/* a value of 1. If not, return an error value.                           */
/*********************************************************************/

if ((SYS$QIOW(0, eth_chn, IO$_SETMODE | IO$M_CTRL | IO$M_STARTUP, eth_iosb,
   0, 0, 0, &setparmdsc, 0, 0, 0, 0) & 0x01) != 1)
   {
   return(2);    /* bad parameters */
   }
if (eth_iosb[0] != SS$_NORMAL)
   {
   return(4);    /* Something went wrong in QIOW */
   }
return(1);       /* All went well */

} /* End of init_eth() */
```

Listing 2

```
/************************ PART OF eth_xmt() **************************/
/* This is the last portion of a function that gets called to send a   */
/* packet over the Ethernet link.  (For brevity we have left out the   */
/* portion that copies a data packet from the packet file into the     */
/* ethernet output buffer (eth_out_buf).                               */
/*                                                                     */
/* The following code attempts to send the contents of the buffer by   */
/* calling the QIO system call.  It skips the first 2 bytes of the     */
/* buffer, which contains the data link header (the packet length).    */
/* If QIO returns any status value other than 1, then this code logs   */
/* the status value in a file and returns a zero to the caller to      */
/* indicate that the packet transmission was unsuccessful.  If QIO     */
/* returns a 1, then all went well, and the function returns a 1.      */
/***********************************************************************/
if (((status = SYS$QIO(0, eth_chn, IO$_WRITEVBLK, eth_out_iosb, write_eth_ast,
    0, &eth_out_buf->data[sizeof(DAT_LNK_HDR) + 1], PKTBUFSIZ, 0, 0,
    &dest_addr, 0 )) & 0x01) != 1)
   {
   sprintf(buffer, "write_eth_ast >>> QIO status: %d", status);

   /********************************/
   /* Log the error to a log file */
   /********************************/
   write_error(buffer);

   /**********************************/
   /* Inform the caller of the error */
   /**********************************/
   return(0);
   }
else return(1); /* all went well */
```

Listing 3

```
/*************************** FUNCTION miz() ****************************/
/* This function is called to download a packet file over the Ethernet. */
/* It calls the function connected() to establish a connection between  */
/* the VMEbus target system and the VAX.  Then it gets the filename of  */
/* the packet file to be downloaded out of nonvolatile RAM and sends    */
/* a request packet with that filename to initiate the download.        */
/*                                                                      */
/* CALLING PARAMETERS: none                                             */
/*                                                                      */
/* RETURNS: TRUE  = The download was successful                         */
/*          FALSE = The download was unsuccessful                       */
/*                                                                      */
/* ASSUMPTIONS:                                                         */
/* Nonvolatile RAM is available.                                        */
/************************************************************************/
```

(continued)

Listing 3 *(continued)*

```
GLOBAL VOID miz()
{
TEXT filename[80],
     ret_error,
     *load_file_addr,
     *nvr_begin_ptr,
     *nvr_download_ptr;

BOOL file_error,
     addr_error;

USHORT try_count,
       type;

/*****************/
/* Initialization */
/*****************/
file_name_p = &filename[0];
nvr_download_ptr = (TEXT *) nvr_download;
load_file_addr = (TEXT *) &nvr_download->load_file[0];
nvr_begin_ptr = (TEXT *) NVR_BEGIN;
addr_error = FALSE;
init_d7();                        /* initialize the D7 registor to 1   */
clear_mem(ram_base, RAM_SIZE);    /* Clear the VMEbus target's memory  */
copy_buffer(nvr_begin_ptr,        /* Copy the download values from the */
            nvr_download_ptr,     /* nonvolatile memory into RAM.      */
            sizeof(LOAD_NODES));

do  /* while not check sum error */
   {
   putfmt("\n********** Load GNIB Software via the Network **********\n");

   init_lance(); /* Initialize the lance chip     */
   setup();      /* Setup the Lance ring pointers */

   /*************************************************************/
   /* If the VMEbus target system is not connected to the download */
   /* server, then return from this function with a boolean value  */
   /* of FALSE, indicating that the download was unsuccessful.     */
   /*************************************************************/
   if (!connected(nvr_download->host_addr))
      {
      addr_error = TRUE;

      putfmt("Host not found at address: (%+04hi %+08hi)\n",
             nvr_download->host_addr.lna_global,
             nvr_download->host_addr.lna_local);

      return(FALSE);          /* Bad download address */

      } /* End if (! connected) */
```

(continued)

Listing 3 *(continued)*

```
/*******************************************************/
/* Copy the download filename from the nonvolatile RAM into */
/* a 16-byte buffer called "filename"                       */
/*******************************************************/
copy_buffer(load_file_addr, filename, 16);

file_error = FALSE;

/****************************************************************/
/* If the string found in "filename" is not a valid filename,      */
/* then output it to the target system's console, and prompt       */
/* the user for a new filename, until a valid filename is entered. */
/****************************************************************/
while (!valid_filename(filename))
   {
   putfmt("GNIB network load filename (%p): ", filename);
   get_string(filename,50);
   putfmt("\n");
   }

/*****************************************************************/
/* If the filename is not NULL then...                              */
/* 1. Copy it to the nonvolatile RAM.                               */
/* 2. Recompute and store the new checksum value for the LOAD_NODES */
/*    portion of the nonvolatile RAM.                               */
/* 3. Copy the updated download data into the nonvolatile RAM.      */
/*****************************************************************/
if (filename[0] != '\0')
   {
   copy_buffer(filename, load_file_addr, 16);
   nvr_download->nvr_checksum =
      nvr_chksum(nvr_download_ptr, (sizeof(LOAD_NODES) - 1));
   copy_buffer(nvr_download_ptr, nvr_begin_ptr, sizeof(LOAD_NODES));
   }

putfmt("Sending request for file: %p\n", filename);

/*****************************************/
/* Set the packet sequence number to zero */
/*****************************************/
last_seq_num = 0;

/************************/
/* Make a Request packet */
/************************/
make_packet (REQUEST_TYPE);
```

(continued)

Listing 3 *(continued)*

```
do  /* while (type != RANGE_TYPE) */
   {
   /********************************************/
   /* Put the Request packet in the outgoing queue */
   /********************************************/
   queue_packet ();

   /*****************************************************/
   /* Set the number of tries to 1000, and call the function  */
   /* transceive_packet.  transceive_packet sends the queued  */
   /* Request packet to the VAX and then waits for the VAX to */
   /* send the "First data packet".  (The First data packet   */
   /* contains the target memory address range where the      */
   /* downloaded file is to be stored.)                       */
   /*****************************************************/
   try_count = 1000;
   if (!tranceive_packet (&try_count)) {return(FALSE);}
   if (try_count == 0)
      {
      putfmt ("\nGnib Error: Request timed out\n");
      return(FALSE);
      }

   /*****************************************************/
   /* If the VAX responded to the target's Request packet with   */
   /* an Error packet (containing an error message) then display */
   /* the error message to the target system's console, and      */
   /* return a boolean value of FALSE to the caller, to indicate */
   /* that the download was unsuccessful.                        */
   /*****************************************************/
   type = *(recv_buffer_ptr + REC_PKT_TYP_OFFSET);
   if (type == ERROR_TYPE)
      {
      save_err_mess();
      putfmt("%p\n", err_mess_buf + 1);
      return(FALSE);
      }

   } while (type != RANGE_TYPE);

   /*************************************************/
   /* Acknowledge the receipt of the previous packet, and */
   /* input another packet.                               */
   /*************************************************/
   rec_data(&ret_error);
```

(continued)

Listing 3 *(continued)*

```
    /****************************************************/
    /* Get the status value out of the First data packet   */
    /* and compare it with all of the known error numbers. */
    /* If it matches a known error number (or if it's an   */
    /* unrecognized error number) set the file_error flag. */
    /****************************************************/
    if ((ret_error == ERR_BAD_FILENAME) ||
        (ret_error == ERR_PACKET_SIZE) ||
        (ret_error == ERR_BAD_REV) ||
        (ret_error == ERR_EMPTY_FILE) ||
        (ret_error == ERR_BAD_ADDR) ||
        (ret_error == ERR_BAD_ID))
       {
       file_error = TRUE;
       }
    else if ((ret_error == ERR_NO_ERROR) ||
             (ret_error == ERR_STATUS_MESS))
            {
            file_error = FALSE;
            }
         else
            {
                putfmt("Unrecognized Error code\n");
                file_error = TRUE;
            }

  /**********************************************************************/
  /* If the file_error flag is set, then return to the caller with a        */
  /* boolean value of FALSE, to indicate that the download was unsucessful. */
  /**********************************************************************/
     if (file_error) {return(FALSE);}

  /************************************************************/
  /* If there was no error, then output a message to the console  */
  /* indicating that the download has been completed.             */
  /************************************************************/
     if (ret_error == ERR_NO_ERROR) {putfmt("\n*** Load complete ! \n");}
  }

  /****************************************************/
  /* If the checksum does not work, then hang up here,  */
  /* until a timeout occurs.                            */
  /****************************************************/
  while (sum_checker() == FALSE);

  /*****************************************************************/
  /* If the checksum works, then return to the caller with a boolean   */
  /* value of TRUE to indicate that the download was successful.       */
  /*****************************************************************/
  return(TRUE);

} /* END of miz() */
```

Listing 4

```
/*************************** FUNCTION connected() **************************/
/* This function runs on a VMEbus target system and establishes a connection */
/* between the target system and the VAX download server by calling the      */
/* function transceive_packet().It sends the Connect packet up to 3 times,   */
/* in an attempt to get the download server to respond with its own connect  */
/* packet. If it fails all 3 times, it gives up and returns a value of FALSE */
/* to the caller.                                                            */
/*****************************************************************************/
LOCAL BOOL connected(dest_addr)
   LNA dest_addr;
{
SHORT try_count;
/*****************************************/
/* Set the packet sequence number to zero */
/*****************************************/
 last_seq_num = 0;

 putfmt("\nTrying to connect to host using:\n");
 putfmt("   Global address: %+04hi\n", nvr_download->host_addr.lna_global);
 putfmt("   Local address: %+08hi\n", nvr_download->host_addr.lna_local);

 vax_type = 0xAAAA;

/************************/
/* Make a Connect packet */
/************************/
 make_packet(CONNECT_TYPE);

/*********************************************/
/* Put the Connect packet in the outgoing queue */
/*********************************************/
queue_packet();

/***************************************************************************/
/* Call the function transceive_packet.  transceive_packet sends the queued   */
/* connect  packet to the VAX and then waits for the VAX to respond with its */
/* own connect packet If the VAX responds to the target's Connect packet with */
/* its own Connect packet then transcieve_packet will return a TRUE value.    */
/* If the number of tries was not exhausted, then return a boolean value of   */
/* TRUE to the caller, to indicate that a connection has been established.    */
/***************************************************************************/
try_count = 3;
if (tranceive_packet(&try_count) && try_count)
   {
   putfmt("*** Connected ***\n");
   return(TRUE);
   }
```

(continued)

Listing 4 *(continued)*

```
/****************************************************************************/
/* If the VAX responds to the target's Connect packet with an Error packet    */
/* (containing an error message) then display the error message to the target */
/* system's console, and return a boolean value of FALSE to the caller, to    */
/* indicate that the connection was not made.                                 */
/****************************************************************************/
if (error_code & ERROR_FLAG_MASK)
   {
   putfmt("%p\n", err_mess_buf + 1);
   return(FALSE);
   }

/****************************************************************************/
/* If the VAX responds with neither a Connect packet nor an Error packet then */
/* display a message on the target system's console, and return a boolean    */
/* value of FALSE to the caller to indicate that the connection was not made. */
/****************************************************************************/
putfmt("GNIB Network Error: host/server not found\n");
return(FALSE);
}
```

Listing 5

```
/************************* FUNCTION main () **************************/
/* This program runs on a VAX to provide an Ethernet download server. */
/* It responds to 2 different packet types: Connect packets and       */
/* Request packets.                                                   */
/*                                                                    */
/* Connect packets are sent to the VAX by the target system to        */
/* establish an Ethernet connection.  (These Connect packets contain  */
/* the Ethernet address of the target system.)                        */
/*                                                                    */
/* Request packets are sent to the VAX by the target system to        */
/* initiate the download of a file.  (These Request packets contain   */
/* the filename of the file to be downloaded.)                        */
/*                                                                    */
/* The VAX can establish only one connection at a time.  If it        */
/* receives a Connect packet while the download of a file is in       */
/* progress, it will break the existing connection and establish the  */
/* new connection.                                                    */
/*                                                                    */
/* This program makes no attempt to log errors which occur during the */
/* download.  It simply sends an Error packet to the target system    */
/* with an error number and an ASCII error message.                   */
/*********************************************************************/
```

(continued)

Listing 5 *(continued)*

```
main()
{
/********************************************************/
/* Declare 2 character arrays: one for use as a buffer for */
/* incoming packets, and the other for outgoing packets.   */
/********************************************************/
char inbuffer[PACKET_SIZE];
char outbuffer[TOTAL_PACKET_SIZE];

/**********************************************************************/
/* Declare an 80-character array to hold the pathname of the packet file, */
/* and a 6-byte array to hold the 48-bit Ethernet address of the target.  */
/**********************************************************************/
char     type, filename[80], cmc_ether_add[6];

/*********************************/
/* Declare miscelaneous variables */
/*********************************/
short int i,k,
          j, *iptr, *optr,
          count, seq_number = 0,
          p_count,
          ret_error = ERR_NO_ERROR,
          ep_size;

long int vme_beg_addr,
         vme_end_addr;

int fend; /* Number of packet file bytes not yet sent */
long int buffer_size,
         chan;

/************************************************************************/
/* Declare a variable to hold the target memory address to which the current */
/* byte of the packet file will be downloaded.                              */
/************************************************************************/
long int destin_add;

BOOL     use_ether_add;

/***************************************************/
/* Initialize the buffer pointer and the base address */
/* of the disk read buffer to NULL values.          */
/***************************************************/
buffer_ptr = beg_buf_ptr = 0; /* NULL */

/********************************************************/
/* Build the data structures needed to read the packet file */
/********************************************************/
initialize(filename);
```

(continued)

Listing 5 *(continued)*

```
/****************************************************************/
/* Initialize the Ethernet driver and Ethernet controller board. */
/****************************************************************/
eth_init();

/************************************************************************/
/* The following do-while loop waits for a Request packet from the  target  */
/* system, and then downloads the file named in that request packet.        */
/************************************************************************/
do                          /* REPEAT FOREVER */
   {
   /***************************************************************/
   /* Read a request packet from Ethernet into the input buffer. */
   /***************************************************************/
   use_ether_add = FALSE;
   if (NOT get_packet(inbuffer, use_ether_add, &ret_error, &seq_number))
      {
      goto handle_error;
      }

   /************************************************************************/
   /* Get the Ethernet address of the target system from the request packet */
   /************************************************************************/
   cmc_addr.global = header_address.source_global;
   cmc_addr.local = header_address.source_local;

   /********************************************************************/
   /* Set the following flag to ensure that the target ethernet address */
   /* received in all subsequent data packets will be verified against  */
   /* the address received in the request packet.                       */
   /********************************************************************/
   use_ether_add = TRUE;

   /***************************************************************/
   /* Get the packet type from the request packet, and verify it. */
   /***************************************************************/
   type = inbuffer[BUF_TYPE_INDEX];
   if (type != REQUEST_TYPE)
      {
      ret_error = ERR_WRONG_PACK_TYPE;
      goto handle_error;
      }
   else
      /************************************************************/
      /* If the packet is a request packet, then get the pathname */
      /* of the packet file to be downloaded out of the packet.   */
      /************************************************************/
      {
      filename[0] = '\0';
```

(continued)

Listing 5 *(continued)*

```
restart:
      if ((inbuffer[BUF_TYPE_INDEX] == REQUEST_TYPE) &&
          (strcmp(&inbuffer[FILENAME_OFFSET], &filename[0]) != 0))
         {
         /***************************************************************/
         /* If the current filename does not match the filename in the      */
         /* request packet, check to see if some area of memory has         */
         /* already been allocated as a disk read buffer.  If so, free it.  */
         /***************************************************************/
         if (beg_buf_ptr != NULL)
            {
            free(beg_buf_ptr);
            beg_buf_ptr = NULL;
          }
      /*******************************************************************/
      /* Copy, from the Request packet, the pathname of the packet file to   */
      /* be downloaded into the character array filename, and append a null. */
      /*******************************************************************/
      for (i = 0, j = FILENAME_OFFSET; inbuffer[j] != '\0' && i < 81; ++i, ++j)
         {
         filename[i] = inbuffer[j];
         }
      filename[i] = '\0';

      /*******************************************/
      /* Set up FAB with requested packet file name */
      /*******************************************/
      fab.fab$l_fna = filename;
      fab.fab$b_fns = strlen(filename);

      /**************************************/
      /* Open the packet file for an RMS read */
      /**************************************/
      rms_status = sys$open(&fab);
      if (rms_status != RMS$_NORMAL)
         {
         ret_error = ERR_BAD_FILENAME;
         goto handle_error;
         }
      /*****************************************************************/
      /* Get the number of blocks and the block size of the packet file. */
      /* Then calculate the size of the file (in bytes) and allocate a   */
      /* memory buffer large enough to hold the entire file.             */
      /*****************************************************************/
      chan = fab.fab$l_stv;
      buffer_size = fab.fab$l_alq * fab.fab$w_bls;
      beg_buf_ptr = (unsigned char *) malloc(buffer_size);
      if (beg_buf_ptr == NULL)
         {
         ret_error = ERR_BAD_FILENAME;
         goto handle_error;
         }
```

(continued)

Listing 5 *(continued)*

```
      /*******************************************/
      /* Read the packet file into the read buffer. */
      /* Then get the revision number and check it. */
      /*******************************************/
      buffer_ptr = beg_buf_ptr;
      if (read_file(chan,buffer_size) != ERR_NO_ERROR)
         {
         goto handle_error;
         }

      if (NOT rev_ok(buffer_ptr, &ret_error))
         {
         goto handle_error;
         }
   } /* End if (...strcmp(&inbuffer[FILENAME], &filename[0]) */

/*********************************************************/
/* Set the buffer pointer to point to the first byte in the  */
/* disk read buffer after the file header.  (In other words, */
/* set it to point to the first byte of the first packet.)   */
/*********************************************************/
buffer_ptr = beg_buf_ptr;
buffer_ptr += sizeof(file_size) + sizeof(header);

/*********************************************************************/
/* Copy the first packet from the disk read buffer to the output buffer */
/*********************************************************************/
ep_size = TOTAL_PACKET_SIZE;
fend = (packet(outbuffer, &ep_size, &ret_error, &seq_number));
if (ret_error != ERR_NO_ERROR)
   {
   goto handle_error;
   }

/*********************************************************************/
/* The following do-while loop sends the packet that's in the output    */
/* buffer. Then it copies the next packet from the read buffer into the */
/* output buffer. It repeats this process over and over until all the   */
/* packets have been sent.                                              */
/*********************************************************************/
do /* until end of packet file */
   {
   count = 0;
```

(continued)

Listing 5 *(continued)*

```
/*******************************************************************/
/* Send the data packet in the output buffer to the target system    */
/* over and over until the target system sends an Acknowledge packet */
/* with the same sequence number, or until it sends an Error packet. */
/*******************************************************************/
do
   {
   count++;
   /**************************************/
   /* Send a data packet to target system */
   /**************************************/
   eth_xmt(buffer_ptr, ep_size, &header_address.source_global);

   /*******************************************************************/
   /* Call get_packet to receive the acknowledge packet from the target */
   /* system.  get_packet inputs a packet from the Ethernet and then    */
   /* checks the packet's source address to make sure that it came from */
   /* the target system.                                                */
   /*******************************************************************/
   if (NOT get_packet(inbuffer, use_ether_add, &ret_error, &seq_number))
      {
      goto handle_error;
      }

   /******************************************************************/
   /* If the packet received from the target was neither an          */
   /* acknowledge packet, nor an Error packet, check to see if       */
   /* it's a Request packet.  If it is, then restart the download. */
   /* Otherwise return an error indication to the caller, and        */
   /* terminate without completion of the download.                  */
   /******************************************************************/
   if (wrong_type(inbuffer[BUF_TYPE_INDEX]))
      {
      if (inbuffer[BUF_TYPE_INDEX] == REQUEST_TYPE)
         {
         ret_error = ERR_REQUEST_PACKET;
         }
      else
         {
         ret_error = ERR_WRONG_PACK_TYPE;
         }
      goto handle_error;
      }
   iptr = &inbuffer[IN_SEQ_INDEX];
   optr = buffer_ptr + OUT_SEQ_INDEX;
   }
```

(continued)

Listing 5 *(continued)*

```
/*************************************************************/
/* If we get to this point, the packet received from the target */
/* was either an Acknowledge packet, or an Error packet.  If it */
/* is an Acknowledge packet, then it will have a matching       */
/* sequence number, and we will drop out of the loop.  If it is */
/* an Error packet, then it will have the sequence number of    */
/* the previous packet and we'll go around the loop once again. */
/*************************************************************/
while(*iptr != *optr);

/**********************************************************/
/* When we exit the loop above, the data packet has been    */
/* successfully sent and acknowledged by the target system. */
/**********************************************************/

/*********************************************************/
/* The packet has been successfully sent, so increment the */
/* disk read buffer pointer, so that it points to the      */
/* beginning of the next packet in the read buffer.        */
/*********************************************************/
buffer_ptr += ep_size;

/*************************************************************/
/* If we did not yet reach the end of the disk read buffer the  */
/* last time we copied a packet into the output buffer, copy    */
/* the next packet from the read buffer into the output buffer, */
/* and go back to the top of the loop to send it.               */
/*************************************************************/
if (!fend)
   {
   ep_size = PACKET_SIZE;
   fend = (packet(outbuffer, &ep_size, &ret_error, &seq_number));
   if (ret_error != ERR_NO_ERROR)
      {
      goto handle_error;
      }
   }
 } while(!fend);

/*************************************************************/
/* If we have reached the end of the read buffer, we drop out   */
/* of the do-while loop above.  We have finished sending the    */
/* packet file, so we no longer need the disk read buffer, and  */
/* we free it.  Then we set the buffer pointer to a NULL value, */
/* to indicate that no disk read buffer is allocated.           */
/*************************************************************/
free(beg_buf_ptr);
beg_buf_ptr = NULL;

}  /* End of else part of...if(type != REQUEST_TYPE)
```

(continued)

Listing 5 *(continued)*

```
      if (ret_error != ERR_NO_ERROR)
         {
handle_error:
         ep_size = PACKET_SIZE;
         switch(ret_error)
            {
            case ERR_REQUEST_PACKET:
               goto restart;
               break;

            case ERR_PACKET_SIZE :
               error_packet(ret_error, $err_bad_pack_size, outbuffer,
                            &ep_size, &seq_number);
               break;

            case ERR_BAD_FILENAME :
               $err_bad_filename[13] = '\0';
               strcat($err_bad_filename,filename);
               strcat($err_bad_filename, "\n\0");
               error_packet(ret_error, $err_bad_filename, outbuffer,
                            &ep_size, &seq_number);
               break;

            case ERR_SERVICE_TIMEOUT :
               error_packet(ret_error, $err_service_timeout, outbuffer,
                            &ep_size, &seq_number);
               break;

            case ERR_WRONG_PACK_TYPE :
               error_packet(ret_error, $err_wrong_pack_type, outbuffer,
                            &ep_size, &seq_number);
               break;

            case ERR_BAD_REV :
               error_packet(ret_error, $err_bad_rev, outbuffer,
                            &ep_size, &seq_number);
               break;

            case ERR_EMPTY_FILE :
               error_packet(ret_error, $err_empty_file, outbuffer,
                            &ep_size, &seq_number);
               break;

            case ERR_BAD_ID :
               error_packet(ret_error, $err_bad_id, outbuffer,
                            &ep_size, &seq_number);
               break;

            } /* End of switch */
```

(continued)

Listing 5 *(continued)*

```
      /*******************************************************************/
      /* Send the error packet to target and loop to start a new download */
      /*******************************************************************/
      eth_xmt(outbuffer, ep_size, &header_address.source_global);
      ret_error = ERR_NO_ERROR;

      } /* End of if (ret_error != ERR_NO_ERROR) */
   } while ( TRUE ); /* REPEAT FOREVER */
} /* End of main() */
```

Listing 6

```
/*********************** FUNCTION read_file() ****************************/
/* This function accepts the I/O channel number of the disk, and the size */
/* of the disk read buffer.  It then reads the packet file from the disk  */
/* into the buffer.                                                       */
/*************************************************************************/
int read_file(chan,buffer_size)
   long int chan; 1
   long int buffer_size;

{
long int status;
short int iosb[4];

/****************************************************************/
/* Make a system call to QIOW to read the packet file into the    */
/* buffer in memory.  Then test to see if the call was successful.*/
/* If so, return a value of 1.  If not, then STOP.                */
/****************************************************************/
if (((status = SYS$QIOW(0,chan,IO$_READVBLK,
      iosb,0,0, buffer_ptr,buffer_size,1,0,0,0)) &  0x01) != 1)
   {
   LIB$STOP(status);
   }

if(((iosb[0] &  0x01) != 1) && (iosb[0] != SS$_ENDOFFILE) )
   {
   LIB$STOP(iosb[0]);
   }

if (((status = SYS$DASSGN(chan)) &  0x01) != 1)
   {
   LIB$STOP(status);
   }

return(ERR_NO_ERROR); /* Everything is OK */
} /* End of FUNCTION read_file() */
```

Listing 7

```
/***************************** FUNCTION get_packet() *************************/
/* This function reads a packet in from the Ethernet. If use_eth_add is TRUE, */
/* then it checks the packet's source address to make sure that the packet    */
/* came from the target system.  However, if the packet is a Connect packet,  */
/* this function sends a Connect packet back to the target system that sent   */
/* it, thus dropping the current connection in favor of the new one. Then it  */
/* reads packets from the new target system.                                  */
/******************************************************************************/
BOOL get_packet(inbuffer, use_ether_add, ret_error, seq_num)
   char inbuffer[PACKET_SIZE];
   BOOL use_ether_add;
   short int *ret_error, *seq_num;

{
short int p_count,
          I,j,
          type,
          ep_size;

char outbuffer[PACKET_SIZE],
     tmp_ether_add[6];

BOOL valid_pkt;

*ret_error = ERR_NO_ERROR;
p_count = 0;

/*************************************************************/
/* The following do-while loop inputs packets from the Ethernet */
/* until we get one from the target                          */
/*************************************************************/
do
   {
   /*************************************************************/
   /* If we have received 9 packets, and none have been from the */
   /* target then give up and return a boolean value of FALSE.   */
   /*************************************************************/
   if (p_count > SERVICE_RETRY_COUNT)
      {
      *ret_error = ERR_SERVICE_TIMEOUT;
      return(FALSE);
      }

   /*******************************/
   /* Input a packet from Ethernet */
   /*******************************/
   eth_rcv(inbuffer, PACKET_SIZE);
   valid_pkt = TRUE;
   type = inbuffer[BUF_TYPE_INDEX];
```

(continued)

Listing 7 *(continued)*

```
   /******************************************************************/
   /* If the packet was a Connect packet, then send a Connect packet */
   /* back to target system.  Then set valid_pkt FALSE, so we will   */
   /* input another packet from the new target system.               */
   /******************************************************************/
   if (type == CONNECT_TYPE)
      {
      connect_packet(outbuffer, &ep_size, seq_num);
      eth_xmt(outbuffer, ep_size, &header_address.source_global);
      valid_pkt = FALSE;
      }

   /**********************************************************************/
   /* If the packet wasn't a Connect packet, and if we are supposed      */
   /* to verify that the packet came from the target system, then do so. */
   /**********************************************************************/
   else
   if (use_ether_add)
      {
      if ((header_address.source_global == cmc_addr.global) &&
          (header_address.source_local == cmc_addr.local))
         {
         valid_pkt = TRUE;
         }
      else
         {
         valid_pkt = FALSE;
         }
      p_count++;
      }
   } while(!valid_pkt);
return(TRUE);
} /* End of FUNCTION get_packet() */
```

CHAPTER 35

Image Processing: A Tutorial

By Robert I. Birenbaum and David R. King

Several factors have contributed to the rapid growth of the image processing market. Recent advances in semiconductor technology (such as higher density dynamic RAMs, digital signal processor chips, and gate arrays) have made high-performance image processing hardware affordable. In addition, open architectures like the VMEbus allow users to incorporate this image processing hardware into standardized systems. Great strides have also been made in image processing software. Standard software packages for applications in the natural sciences, medicine, machine vision, and publishing are making image processing technology available to more users than ever before. In this chapter Bob and Dave discuss the basic principles of image processing, and describe how you can build pipelined image processing systems on VMEbus.

The term *image processing* is used to describe activities ranging from document scanning to the visual abilities of automated robots that install windshields into automobiles. However, for our purposes here, we'll define image processing to be the capture and manipulation of *real-world* images.

Images have traditionally been captured with either a standard vidicon camera, or a solid-state camera. However, high resolution cameras and other specialized input devices are now being used in a variety of applications. For example:

- Infrared cameras capture images that depict the amount of radiated heat.
- X-ray detectors look *through* objects (such as printed-circuit boards) to find hidden flaws.
- Scanning electron microscopes greatly magnify small objects for analysis.

Regardless of the type of device used to capture the image, the output signal usually conforms to some well-established video standard. The most common display format in use today is known as the RS-170 standard in the U.S. and Japan, and as the CCIR standard in Europe and many other parts of the world.

This standard was developed by the Electronic Industries Association for the commercial broadcast industry in 1957. Because it was developed for use with *analog* video equipment, it is less than ideal for the *digital* video equipment used today. However, the image processing industry has still adopted RS-170 as a de facto standard, simply because of the widespread availability of low-cost video equipment that conforms to it.

The RS-170 standard specifies *timing pulses* as part of the analog video signal. These pulses allow the display device to synchronize its scanning to the incoming analog signal. The transmitted RS-170 image is sent at a rate of 30 *frames* of video information per second. Each frame contains 2 *fields*: an *odd field* and an *even field.* The odd field contains the odd numbered scan lines of the image, and the even field contains the even numbered scan lines.

Figure 1 shows how the scan lines of these 2 fields are interleaved on a screen, to produce the frame. One field is drawn first, from the top of the screen to the bottom. Then the other field is drawn, with its lines *interlaced* with the first. This process of alternately drawing the even and the odd fields repeats continuously, which is why RS-170 is described as an *interlaced video standard.*

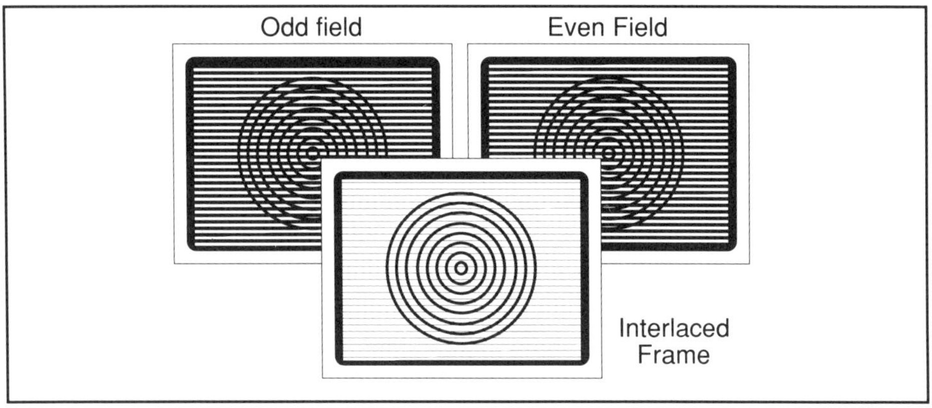

Figure 1

Each RS-170 field consists of 262.5 scan lines. (The top scan line of each field starts in the middle.) That gives 525 horizontal scan lines per frame. Each scan line consists of an analog video waveform representing the brightness of the scan line as the electron beam is scanned across the screen. To ensure that the scan lines are properly aligned in the reassembled image, a *horizontal sync pulse* follows each scan line, as shown in Figure 2.

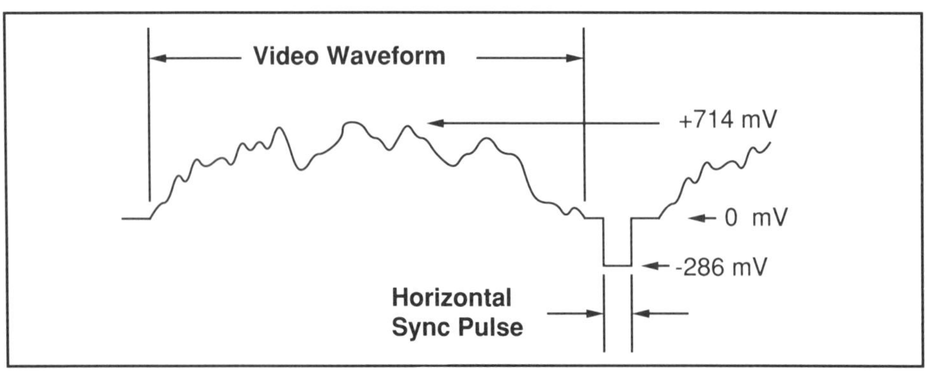

Figure 2

The video waveform varies between 0 volts (black) and +714 millivolts (white). The horizontal sync pulse at the end of each scan line drops to -286 millivolts. A pause after the sync pulse allows the electron beam in the monitor to reposition itself at the left side of the screen, to begin the next scan line.

At the end of each field there is a *vertical sync pulse.* When a video monitor detects this vertical sync pulse it repositions the electron beam back at the top, left-hand corner of the screen. To prevent the electron beam from illuminating the screen as it sweeps upward, the beam is turned off. This is called *vertical blanking.*

Most image processors that accept an RS-170 video signal sample only 480 of the 525 scan lines, at a rate that gives either 640 or 512 pixels per scan line. The other 45 horizontal scan lines represent the *vertical blanking interval.*

A sample rate of 640 pixels per line produces square pixels on the screen: their height equals their width. That's useful, since you don't need to do extra computations to scale the vertical distances with respect to the horizontal distances: a line 30 pixels long is the same length whether it's vertical or horizontal.

A sample rate of 512 pixels does not produce square pixels. However, it simplifies the computation of pixel addresses. That's because the address of the memory location for each pixel differs from the address of the pixels above it and below it only in the most significant bits.

Note: Despite the fact that an RS-170 video frame is represented by 2 fields, those 2 fields are stored in memory as one image, with the pixels for the even and the odd scan lines interleaved, as they appear on the screen.

As image processing hardware becomes more efficient, higher resolution displays, with formats such as VGA, SuperVGA, 8514 and even 1024 x 1024 are now being used. These newer formats provide much better image quality than the RS-170 standard.

Image processor architectures

Today's image processing systems come in many different forms. The simplest systems consist of a personal computer (or a workstation) in which a *frame grabber* board has been installed. The frame grabber receives the incoming video image, and the host computer does the actual image processing with software. However, since the image processing algorithms are implemented in software, the system may be too slow for some applications.

Image processing hardware can speed things up considerably. This specialized hardware can be provided by plugging boards into a backplane-based architecture, such as VMEbus. A host *computer* on the bus controls the operation of the image processing boards, and monitors their operation. Figure 3 shows the basic elements of such an image processing system.

Capturing the image

The image processor includes a *digitizer* that samples the incoming video signal with an analog-to-digital (A/D) converter, typically at 100 nsec intervals. Each sample represents a single picture element, or *pixel.* The precision of the A/D converter determines the number of bits stored for each pixel. For example, if an 8-bit A/D

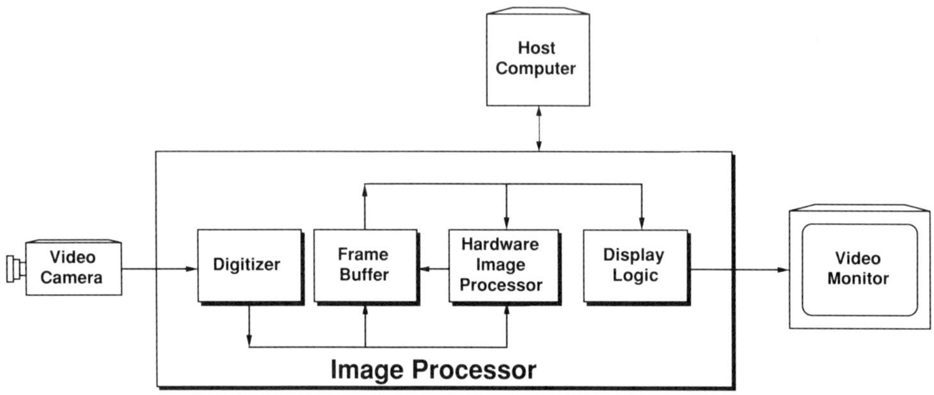

Figure 3

converter is used, each pixel is stored as an 8-bit value, representing 1 of 256 possible gray-scale levels.

Many newer image processors are also equipped to accept video input from non-standard video input devices. The input device provides some kind of timing signals that allow a *variable-scan digitizer* to acquire the image. These timing signals typically include horizontal and vertical blanking signals, and a pixel clock.

Some image processors can also accept *digital* image input. This eliminates the need for an A/D converter. The digital input port is often designed to accept high-accuracy 12-bit (or even 16-bit) pixel values.

As the image is digitized (or acquired through a digital input port) it is stored in a special-purpose memory called a *frame buffer*. The locations within this frame buffer memory are accessed by specifying a row and a column. The frame buffer often includes special addressing hardware, to provide for panning, scrolling and zooming.

Some image processing systems have more than one frame buffer. This can be very useful when you want to process an image quickly, since you can use one frame buffer to hold the original image and another frame buffer to store the processed image.

Other systems have large frame buffers that allow storage of high-resolution images. For example systems that receive images from orbiting satellites have very high resolutions: sometimes as high as 4,096 by 4,096 pixels. (Note: Present CRT technology doesn't allow the display of such a large image on a single screen). A frame buffer large enough to hold the entire 4096 x 4096 image speeds interpretation, since it eliminates the need for disk accesses.

Processing the image

Once an image has been stored in a frame buffer, it's available for processing. A special-purpose *hardware image processor* is often used to:

- read the pixels out of the frame buffer
- perform operations on the pixels
- return the pixels to the same (or to some other) frame buffer

Displaying the image
Display logic converts the processed image back into an analog signal that can drive a video monitor. This conversion takes place in real time, at a rate of 30 frames (60 fields) per second for RS-170 displays, and at a higher rate for high-resolution and non-interlaced displays.

The display logic can be designed to display images in black and white, color, or pseudocolor. Pseudocolor images are created by assigning different colors to the various shades of gray in a black-and-white image. Since the human eye can distinguish many more colors than shades of gray, pseudocolor images are much easier to interpret than black-and-white images.

Image processing algorithms
Many different algorithms are used to process images. Some algorithms produce an image that is easier for humans to interpret. For example, an algorithm might be used to improve the contrast of an image. Other algorithms are used to make it easier for a computer to interpret the image. For example, an algorithm that enhances the edges of an object in an image may help a robot figure out how to pick the object up.

Image processing operations can be categorized into 4 classes:

- Point operations
- Frame operations
- Group operations
- Statistical operations

Point operations
Point operations work on each pixel individually, mapping the pixel's intensity to some new value. This can be used to enhance contrast. For example, if an image has pixel intensities ranging from 0 through 127 (when values up to 255 are accommodated by the system), each pixel value can be multiplied by 2, thereby increasing the contrast of the displayed image.

Figure 4 shows the mapping function for another point operation called *binarization*. When an image is binarized, each pixel value is changed to either black or white, depending upon whether or not its value is within a specified grayscale range. In many applications the captured grayscale image is a high-contrast silhouette. Binarization greatly reduces the amount of memory required to store such an image. The reduced image is also much simpler to process.

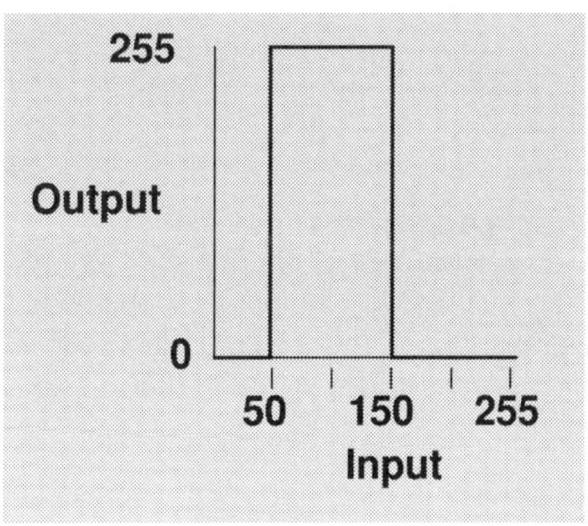

Figure 4

Using look-up tables to do point operations

Look-up tables (LUTs) are high-speed static RAMs that are used to perform point operations. A serial data stream of pixel values is passed, one-by-one, to the LUT. Each pixel value acts as an address to the LUT's RAM. The output of the LUT is the data stored at the specified RAM address. This data value becomes the new pixel value in the serial output data stream from the LUT. (Before the LUT can be used, the host computer must load the LUT's RAM locations with values that produce the desired translation.)

LUTs in the display logic can be used to transform a monochrome image to a pseudocolor image. Each pixel in the monochrome pixel stream is used to simultaneously address the RAM of 3 separate LUTs. One of these LUTs generates a stream of output pixels for the red channel, one for the green channel, and one for the blue channel. These LUTs can be configured to produce a distinct RGB color for each gray-scale intensity in the original image.

Most image processors contain at least 2 sets of LUTs: an input set and an output set. (See Figure 5) Input LUTs are between the A/D converter and the frame buffer, and are used to translate pixel values prior to storage in the frame buffer. Output LUTs are between the frame buffer and the D/A converter, and they translate pixel values prior to display on a monitor. (The pseudocolor example given above is one application of output LUTs.)

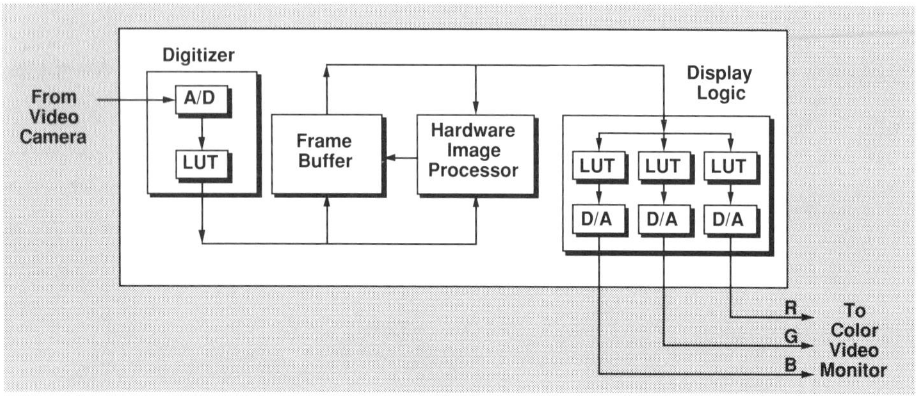

Figure 5

Frame operations

Frame operations use the pixel values from 2 or more frames to create a composite frame, typically on a pixel-by-pixel basis. In *image averaging*, the corresponding pixel values in 2 or more frames (sometimes several hundred) are averaged, to generate one frame. Image averaging can *dramatically* reduce the amount of random noise in an image.

Figures 6 and 7 show a typical application of image averaging. Figure 6 is an image of a silicon wafer captured by a scanning electron microscope. The image is "noisy" due to the high gain of the electronics within the microscope. By capturing several frames, and then averaging them, you can greatly increase the clarity of the image, as shown in Figure 7.

There are several of kinds of frame processes. *Image subtraction* highlights the *differences* between one frame and another. *Slow-pass filtering* eliminates rapidly moving objects from an image, leaving only the background. *Logical operations* can be used to cut out a portion of one image and paste it into another.

In many frame operations, the intermediate results need to have more precision than the final image. For example, during image averaging, the sum of two 8-bit pixel values may give a 9-bit result. If the system cannot store and operate on the larger values, it will introduce errors into the image as it processes it. To avoid this, the frame buffer should be able to hold more than 8 bits per pixel, and any image processing hardware should be designed to do the necessary operations on these larger values. In fact, to ensure satisfactory results when image averaging with a large number of frames, the frame buffer and the image processing hardware should be able to handle at least 16 bits of precision.

Group operations

Group operations (also known as *neighborhood operations*) operate on localized regions within a single image, such as a 5 by 5 neighborhood of pixels. *Convolutions* and *morphological operations* fall into this category.

Neighborhood processes are done by creating a special pattern of weighting coefficients (called a kernel) which is then centered over each pixel in the original image, to compute the value of a pixel in the new image.

For example, Figure 8 shows a 3 by 3 kernel of weighting coefficients. This kernel is centered over each pixel of the original image. Then that pixel (and each of its 8 closest neighbors) is multiplied by the coefficients contained

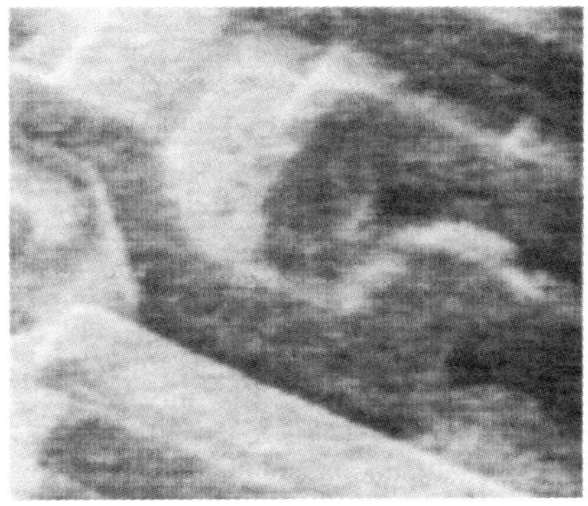

Figure 6

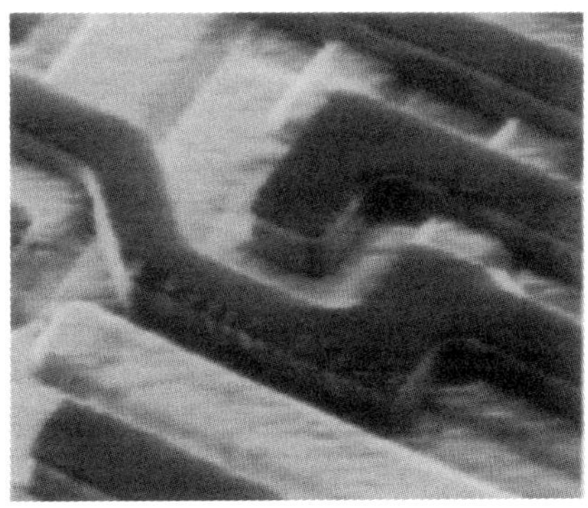

Figure 7

-1	-1	-1
-1	8	-1
-1	-1	-1

Figure 8

in the corresponding locations of the kernel. The 9 products are then summed, scaled and stored as one pixel of a new (processed) image. (See Figure 9.) By sequentially centering the kernel over every pixel in the original image, and by storing the resulting scaled sum of products in a 2nd frame buffer, a new (transformed) image is generated.

By changing the weights in the kernel, you can enhance or blur details, or you can emphasize edges. For example, you might create a *median filter* that calculates the median gray level of 9 pixels, to remove spike-type noise from the image.

Another example is where you create a kernel for *erosion* of an object, which clips off the outermost pixels, making it smaller. You could also create a pattern to *dilate* an object, which places additional pixels around the object, making it larger. By first eroding and then dilating objects you can remove unwanted burrs that develop as a result of other processing operations. Erosion and dilation fall into the type of neighborhood processing known as *morphology*.

The pattern shown in Figure 8 is an example of a *Laplacian filter*, which emphasizes the *edges* within an image. Since the 9 weighting coefficients in the kernel sum to 0, regions of the image with a constant pixel intensity will produce zero-valued pixels in the resulting image. However, in regions of the image where there is a *gradient* in the pixel intensities, a non-zero result will occur. The stronger the gradient, the more positive the result, giving maximum emphasis to edges. By using other patterns of weighting coefficients, you can emphasize only horizontal or vertical edges. Figure 10 is an image of a magnesium wheel that was run through a Laplacian filter, which highlights the edges.

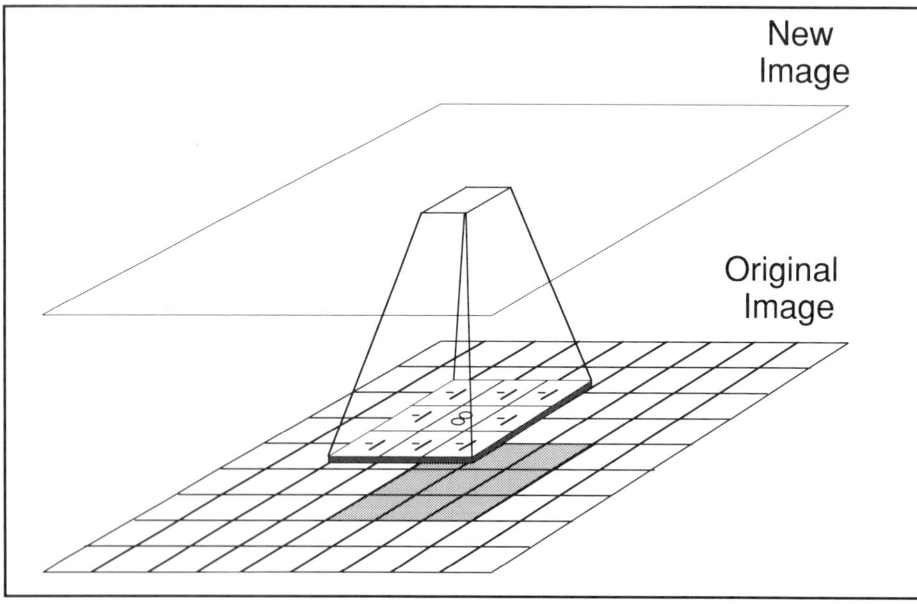

Figure 9

Statistical operations

Statistical operations don't modify the image. Rather, they produce relevant statistics that can be used when doing subsequent image processing operations. For example, using statistical information you can plot an *intensity histogram*, that shows the number of pixels with each gray level, as in Figure 11. These kinds of statistics are helpful when performing operations such as contrast enhancement.

Figure 10

Automatic inspection uses another statistical process called *feature extraction*. The goal of feature extraction is to detect flaws in the object being inspected. One way to do feature extraction is to designate a particular gray level as a feature. Then, when the object is scanned, the (x,y) coordinates of each pixel with that gray-level are retained.

What kind of system do you need?

Image processing is still an emerging technology that's changing very rapidly. While image processing *operations* are easily understood, the *sequence* in which you want to apply these operations depends heavily on the image itself, and the desired result. Often, you need to experiment to find out how to best process the image. In view of this, you should take into consideration the following factors when putting together an image processing system:

- Throughput
- Flexibility
- Camera interface
- Display characteristics
- Software
- Host computer
- Compatible products
- Upgrading possibilities

Throughput

At what rate must the image be processed? For example, if an image processor cannot process the image fast enough, an automated handler may not be able to keep up with parts moving down an assembly line.

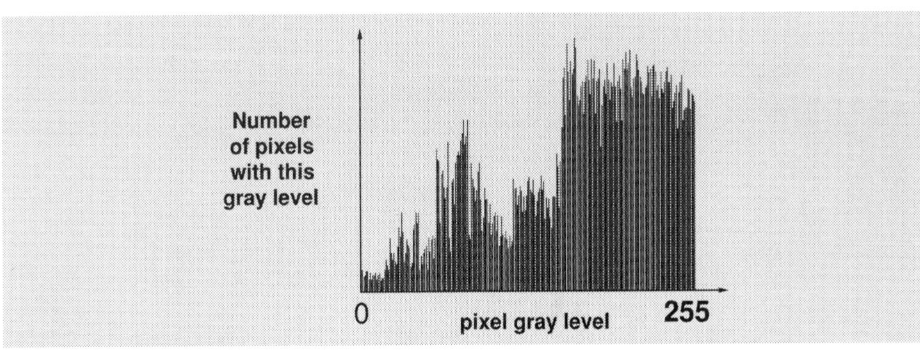

Figure 11

Flexibility
How much flexibility do you need? Will you be processing images captured from a controlled environment? Can you alter the environment to make the captured image easier to process?

Camera interface
Will you be using a standard RS-170 camera, or will you need an image processor with a more flexible variable-scan interface? (High-resolution cameras are available, but are typically more expensive than their RS-170-compatible counterparts.) Will your camera have a digital output, eliminating the need for a digitizer in the system?

Display characteristics
What is the resolution of the display? Will you be displaying monochrome, pseudocolor, or true-color images? In general, the higher the resolution of the display, the more costly the monitor will be.

Software
How much time can you afford to spend designing custom software? You may be able to save considerable time by *acquiring* the low-level software you need. For example, a complete set of software driver routines will allow you to focus on your application software, saving you development costs.

Host computer:
What host computer do you want to use with the image processor? You may want to use a familiar one. Perhaps the *cost* of the host computer is a factor.

Compatible products
The availability of compatible hardware and software products will speed up the integration of the image processing system. Cameras, monitors and low-level software that have been prequalified to work with the image processing hardware can cut months off the development of your system software.

Upgrading possibilities:
What if you decide that you need more image processing power than you originally anticipated? (This often turns out to be the case.) Can you get compatible image processing hardware that allows you to speed things up, without scrapping all of your previously developed software?

Using a pipelined architecture
Image processing consists largely of doing a finite sequence of operations repeatedly on each pixel of 1 or 2 input images. This kind of processing can be done well with a *pipelined* architecture, like the one shown in Figure 12. With a pipelined architecture,

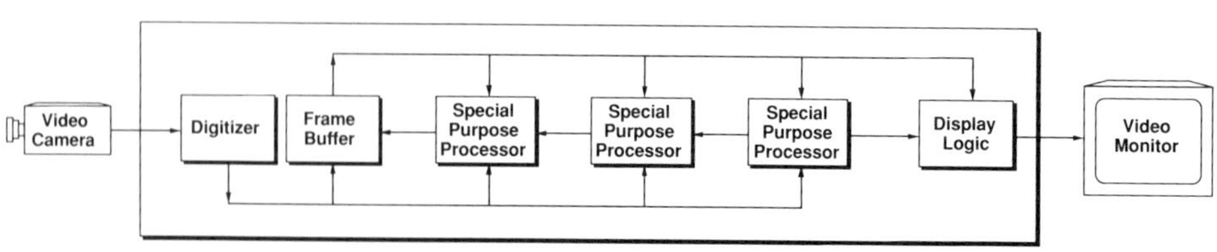

Figure 12

the image data is routed through a series of special-purpose processor modules that are each designed to do one (or a few) operations efficiently. If the system is flexible, additional special-purpose processor modules can be inserted into the pipeline. In that way, when new custom semiconductor components become available, you can easily incorporate new special-purpose processor modules into your system.

Many pipelined architectures use the same frame buffer for both the source and the destination of an image processing operation. This can be done efficiently if the frame buffer is designed to permit a read operation and a write operation to the same location in the same memory cycle.

However, if there are several stages in the pipeline, the pixel value entering the pipeline will not emerge in processed form until some time later. That means the processed pixel will get written back to the frame buffer in a *different* location from where it was read. As a result, the new processed image will be stored in the frame buffer *offset* from its original position.

This offsetting of the image is known as *spin*. To compensate for spin, the host computer can change the contents of the frame buffer's pan and scroll registers each time a frame goes through the pipeline. (As shown in Figure 13, the pan and scroll registers specify the location in the frame buffer which corresponds to the upper left-hand corner of the image.)

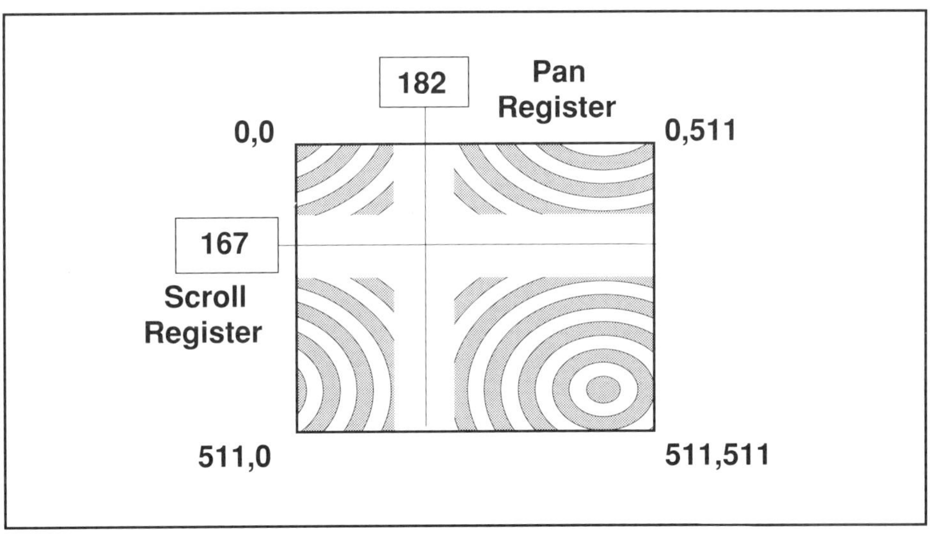

Figure 13

Spin compensation puts a considerable processing burden on the host computer, unless you incorporate *hardware spin compensation.* Hardware spin compensation adjusts the position of the image, based on the delay in the processing pipeline, so the host computer never needs to get involved. Newer memory architectures have independent read and write ports that accept a different address for a read operation and for a write operation. This eliminates the spin problem completely.

In many cases you don't need to process an entire image. It may be sufficient to process only a small portion of it, called the *area-of-interest* (AOI). An AOI is a user-defined rectangular portion of the image that's read out of the frame buffer, processed, and then returned to the frame buffer. Since only a small portion of the image is processed, performance is improved.

The Series 150

The Series 150 VMEbus board line from Imaging Technology is an example of a modular, pipelined image processing architecture. It allows users to select only those functions required by the application, and it is supported by an extensive library of image processing subroutines.

The *ADI-150* analog/digital board accepts video signals from a standard RS-170 camera. It contains an 8-bit A/D converter that provides 256 levels of gray, and three D/A converters for output to an RGB color monitor. The A/D converter has a look-up table on its output, and each D/A converter has a look-up table: one for red, one for green, and one for blue.

The *VSI-150* variable-scan board provides a flexible interface for both RS-170 and for nonstandard video sources, including those providing digital image data. The VSI-150 can also display pseudocolor images, and has an onboard frame buffer that allows it to acquire new images while previously-acquired images are being processed.

The *DP-150* display processor board is a dedicated, high-resolution display board. It provides advanced graphics hardware support and generates 12-bit-per-pixel images, with display resolutions up to 1024 x 1024 pixels.

The *FB-150* frame buffer contains four 512 x 512 x 8-bit frame buffers for storing up to 4 medium-resolution images. Two of these frame buffers can be used together, to hold intermediate processing results with up to 16 bits of accuracy. Another version of the FB-150 has expanded memory to allow storage of 1024 x 1024 high-resolution images.

The *ALU-150* pipeline processor performs most of the commonly used image processing operations in real time, including image averaging, subtraction, logical operations, minimum/maximum calculation, and conditional processing.

The *IPA-150* image processing accelerator is a programmable, 32-bit floating-point processor. It provides highly accurate processing, and allows users to implement operations (such as fast Fourier transforms) that don't lend themselves well to a pipelined architecture.

The *RTMP-150* real-time modular processor is a configurable pipeline processor for performing highly specialized image processing operations. Four plug-on computational modules allow the board to perform convolutions with 1 or 2 simultaneous kernels, morphology, median filters, binary correlation, and in-line 16-bit look-up table operations.

The *HF-150* histogram/feature extractor performs real-time intensity histograming and feature extraction, based on pixel values.

The *ITEX Software* is a supporting library of hundreds of subroutines that control all operations of the Series 150 boards. ITEX includes an interactive interpreter that eliminates the need for repetitive compilation of application software, saving time during the development process.

These series 150 boards can be plugged into a VMEbus system in any order, to form a customized pipeline. The number and the order of the processing modules in the pipeline determines the processing capabilities of the resulting pipelined image processing system.

Figure 14 shows one arrangement. The Analog/Digital interface at the left digitizes an incoming RS-170 signal, and routes the resulting stream of 8-bit pixel values to the Vid*eo data input bus.*

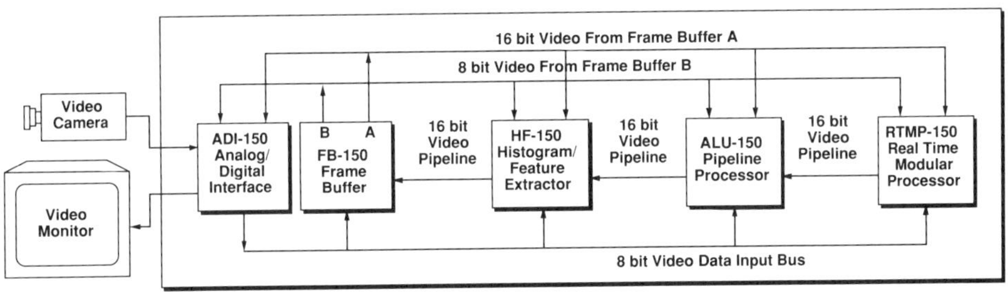

Figure 14

The frame buffer board stores these pixel values in one of its internal frame buffers. (Frame buffer A is 16 bits wide and frame buffer B is 8 bits wide.)

Pixels in frame buffer A can be output over the 16-bit bus labeled "16 bit video data from frame buffer A." Pixels in frame buffer B can be output over the 8-bit bus labeled "8 bit video data from frame buffer B." These pixels can then be used by any of the 3 boards shown in the right half of Figure 14, or by the display logic in the Analog/Digital interface board at the left in Figure 14.

If the real-time modular processor at the right in Figure 14 uses these pixels to generate a transformed image, the pixels of that new image pass through the 16-bit video pipeline to the pipeline processor for further processing. The pipeline processor then passes the resulting output image to the Histogram/Feature Extractor, which passes its output image to the frame buffer, for storage.

Since these 3 special purpose image processor boards compute simultaneously, each board is able to process one pixel for each pipeline clock cycle. This effectively cuts the total processing time by about 66%.

Summary
By installing image processing systems into existing standard open systems, like VMEbus, users can capture, process, and display real-world images economically. This low-cost approach results in wider usage of image processing equipment. Wider usage, in turn, allows engineers to improve image processing techniques. Thus, open architectures, like VMEbus, are accelerating the advance of image processing.

Robert I. Birenbaum is vice president of marketing at Imaging Technology. He has a BSEE and an MS in computer science from Union College, and an MBA from Boston University. Bob's previous experience includes key technical contributions to the development of the MicroVAX chip set, while at Digital Equipment Corporation, and various marketing positions at Imaging Technology. He has often acted as an industry spokesperson on the issues and challenges facing the emerging image processing industry.

David R. King is a hardware project manager at Imaging Technology, where he has managed the design of several successful state-of-the-art products. He has a BSEE from Southwestern Massachusetts University, and an MSEE from Northeastern University. Dave's previous experience includes the design of signal processing hardware at Raytheon. He spends his spare time oil painting, skiing, and playing basketball.

CHAPTER 36

Using Image Processing Techniques In Machine Vision Systems

By Bruce MacDougall

Recent advances have produced a new generation of board-level image processing systems that rival minicomputer-based and mainframe-based imaging systems that were considered state of the art only a few years ago. The reduced cost (and the reduced size) of these board-level imaging systems now allows you to use machine vision techniques in applications where earlier image processing systems would have been too expensive. In this chapter, Bruce describes image processing techniques and architectures, and shows how you can use board-level products to build VMEbus-based machine vision systems.

With the recent introduction of powerful image processing boards, you can now build useful and cost effective VMEbus-based machine vision systems. However, you first need to understand the image processing techniques and algorithms used in machine vision applications. Where system cost is an important factor, you also need to know how to trade off hardware and software alternatives.

What is machine vision?

Machine vision is the use of digital imaging techniques to inspect a workpiece, a product, or a tool, or to otherwise provide a visual interface to a control system and, ultimately, to a controlled machine. The controlled machine typically consists of a robot, a numerically controlled machine tool, or a conveyor subsystem in a manufacturing process.

The process of machine vision involves extracting relevant information from an image, and then making this information available to other system resources, in a form they can use. This information may include:

- the *position* of the object
- the *size* of the object
- the *orientation* of the object
- the presence or absence of a feature, or a set of features

Steps in the machine vision process

The machine vision process consists of 4 steps:

- Image acquisition and digitization
- Image preprocessing
- Segmentation, and object labeling
- Extraction of object characteristics (features)

Image acquisition and digitization

Before data can be extracted from an image, it must be represented in a digital format. This is usually accomplished by digitizing the analog video signal received from a camera, to form a 2D array of pixel intensities. A typical pixel array might have 512 rows and 512 columns. If each pixel is represented as a single byte, the entire image occupies 256 Kbytes.

In many applications, the video camera is equipped with special optics, such as a microscope. In other applications specialized devices (such as scanning electron microscopes) generate video signals directly.

Image preprocessing

If the lighting conditions are ideal, and if the resulting image has good contrast, preprocessing may not be needed. However, when conditions are not ideal, preprocessing is usually needed to allow extraction of useful information from a poor image. The algorithm used to do this depends upon the specific application.

Image processing algorithms

Image processing algorithms are typically constructed with 4 basic types of image processing operations:

- single point operations
- multiple point operations
- neighborhood operations
- statistical operations

However, there is no standard method for selecting *which* operations to use, or for choosing the *order* in which these operations are to be applied to the incoming video images. Those decisions must be based on the system engineer's experience. This makes the overall system performance highly dependent upon the experience of the engineer. Since small performance differences in today's machine vision systems can make the difference between success or failure, the engineer's experience may determine the overall success or failure of the entire venture.

Single point operation

A single point operation produces a single output pixel value for each input pixel value. (See Figure 1) Single point processes are typically implemented with a hardware look-up table that specifies an output intensity for each possible input intensity. This look-up table is usually implemented with a static RAM. Each incoming pixel is used as a RAM address. The data read from that RAM address becomes the outgoing pixel. Look-up tables are fast enough to transform incoming pixel streams into output streams at video rates.

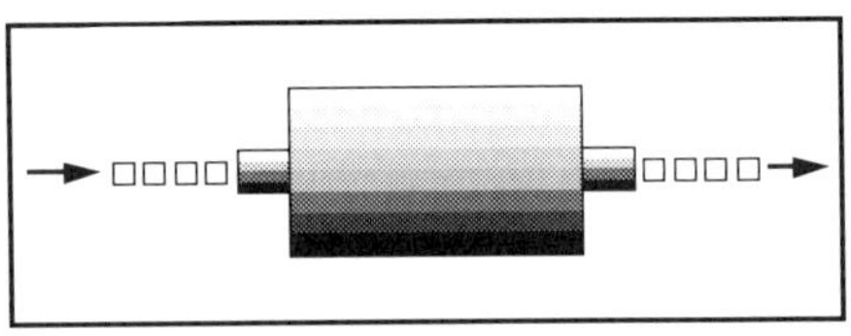

Figure 1

The look-up table RAM locations are loaded with appropriate intensity values *before* the image processing begins. Figures 2 through 5 illustrate several useful point

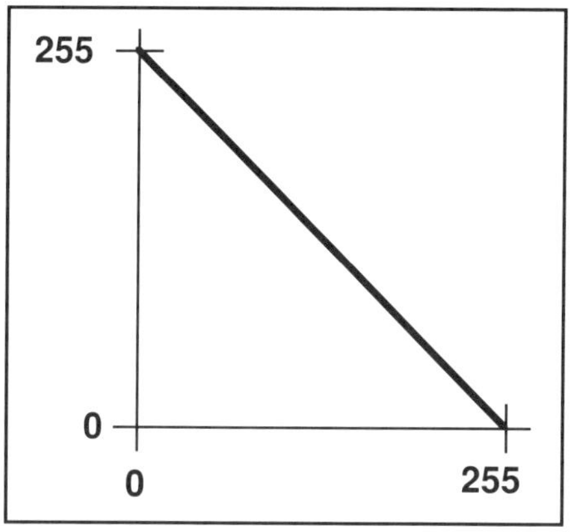

Figure 2

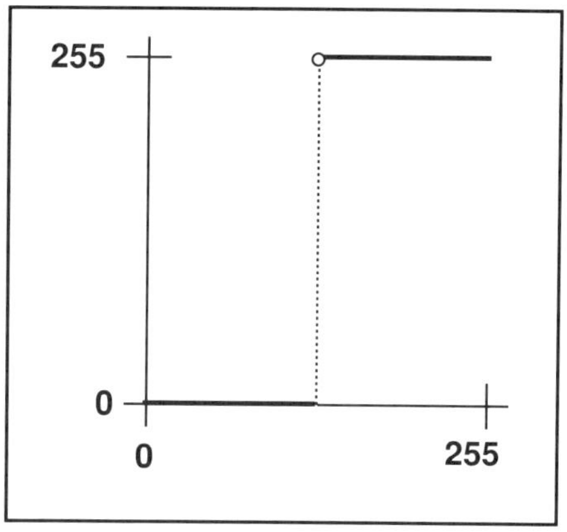

Figure 3

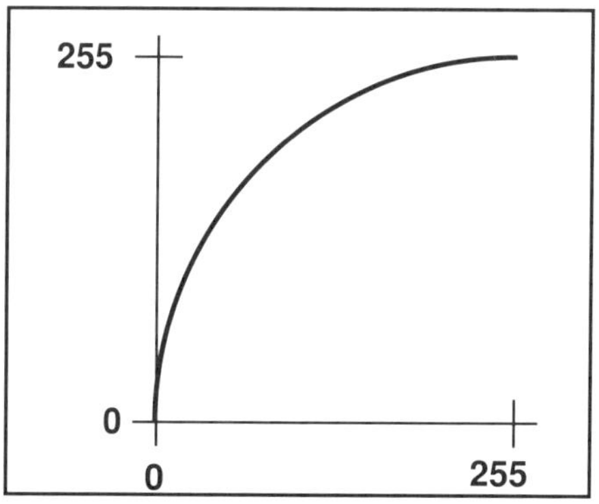

Figure 4

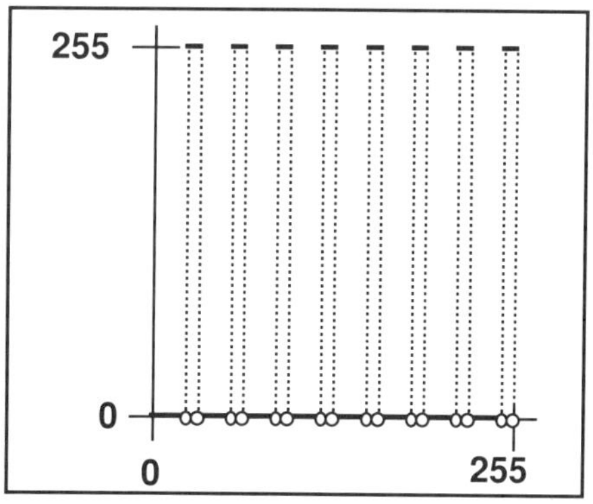

Figure 5

transformations functions. The input pixel intensity is represented by the horizontal axis, and the output intensity is represented by the vertical axis.

One type of point process, known as *thresholding*, transforms all pixel intensities at or below a certain threshold intensity to 0 (black) and all those above the threshold to 255 (white). (See Figure 3) Thresholding is used in situations where there is good contrast between the object of interest and the background. Once thresholded, an image may be represented in a binary format, where a 1 represents each object pixel, and a 0 represents each background pixel. If the original image consists of 8-bit pixels, this reduces the amount of image data to 1/8th of the original.

Multiple point operations

Multiple point operations can be used to combine 2 images, pixel by pixel. The pixel streams from 2 frame buffers are routed into an arithmetic or a logical operation. That operation then generates a stream of output pixels, which is stored in a 3rd frame buffer, as part of a *combined image*.

This method is often used to inspect assemblies on a production line. The incoming image of the assembly is subtracted from a *reference image*, which is held in a frame buffer. Any differences show up as a non-zero area in the resulting output image. This technique is useful for inspecting printed circuit boards, prior to wave soldering, since differences indicate missing components.

Neighborhood operations

Neighborhood operations compute each output pixel intensity, based upon the pixel intensities in a *neighborhood* around each input pixel. Since each output pixel intensity is calculated from several input pixels, neighborhood operations require much more computation than point processes.

Some machine vision systems speed up neighborhood processing with special hardware. In fact, very high performance systems dedicate a processor to each of the pixels in the input neighborhood. This increases the throughput to a level approaching that of point processes.

Convolution is the best known form of neighborhood processing. A 2D array of coefficients (known as a *kernel*) is swept across the image matrix, generating an output pixel for each input neighborhood. The simplest (and most commonly used) kernel consists of 9 coefficients, as shown in Figure 6.

To perform a convolution, each pixel in a neighborhood is multiplied by the corresponding coefficient in the kernel. Then the products are summed. After scaling, this sum becomes a pixel intensity for the new image.

This process is computationally intensive: a 3 x 3 kernel convolution involves 9 multiplications and 8 additions per pixel. When applied to a 512 x 512 matrix, there's a grand total of 4.5 million arithmetic operations! Unless special purpose hardware is available, this process takes several seconds. Larger kernel convolutions require even more computations.

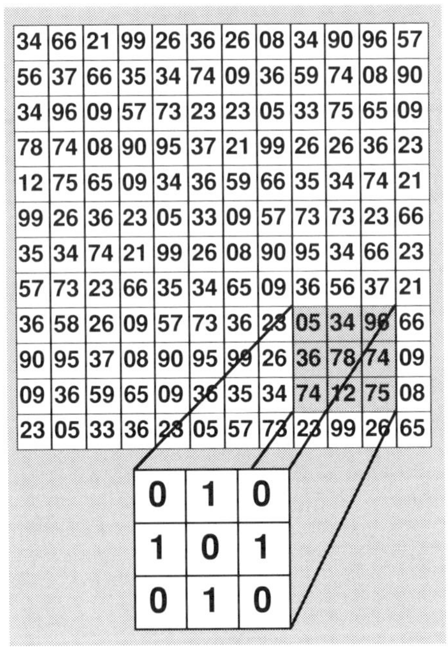

Figure 6

Statistical operations

Statistical operations are used to extract information about how often a pixel intensity (or possibly a feature) occurs in an image. The resulting statistics are typically represented as histograms.

Intensity histograms

Intensity histograms contain the most basic statistical information that can be derived from an image. They chart the number of image pixels with each discrete intensity level. (See Figure 7) The horizontal axis represents the intensity level, with 0 (black) on the

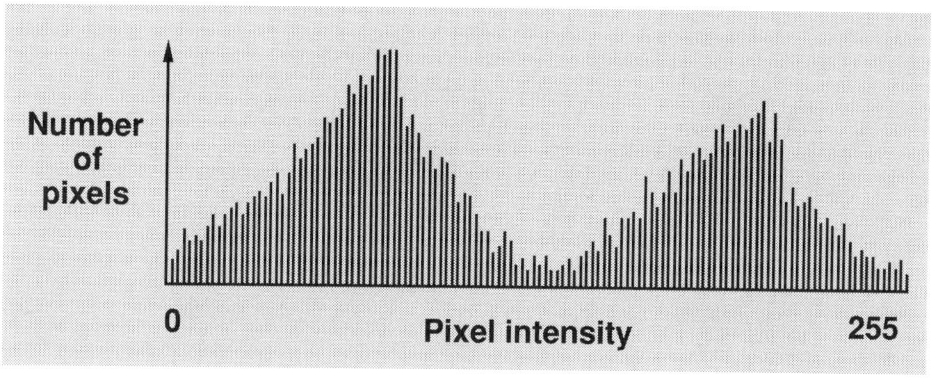

Figure 7

extreme left and 255 (white) on the extreme right. The vertical axis indicates how many pixels of each intensity there are in the image. A histogram like the one shown in Figure 7 might be obtained from an image containing a large number of relatively dark (low intensity) pixels and a lesser number of light (high intensity) pixels. This is typical of a light colored object on a dark (but not black) background.

As objects are moved into (or out of) the video camera's field of view, intensity histograms can change radically. However, the total area under the histogram curve is always the same. That's because the area under the histogram's curve depends only on the total number of pixels in the image, which is constant.

Profile histograms

Profile histograms are sometimes called intensity profiles. They chart the sum of all pixel intensities along a given axis. For example, an X-axis profile histogram is generated by summing the pixel intensities in each column of pixels, and graphing these against the column number. Y-axis profile histograms can be generated in a similar manner, by summing the rows of pixels.

Figure 8 shows an image, along with its X-axis and Y-axis profile histograms. Profile histograms can also be generated at an angle from the X and Y axis. 45-degree oblique profiles are often used.

To determine the number, the size, and the position of objects in an image, you first threshold the image. Then you use the resulting binary image to generate profile histograms. If the image contains only a single object, you can use the X-axis and Y-axis profile histograms to determine the X and Y coordinates of the object's center of mass. If the object has an irregular shape, you can calculate moments around the center of mass, and possibly use them to determine the object's orientation. You can also determine the area of the object by simply finding the area under either of the profiles.

Image Segmentation

Image segmentation is the process of separating image areas occupied by potentially interesting objects from areas making up the image background. Once this has been done, it's easier to distinguish between objects of interest and the background.

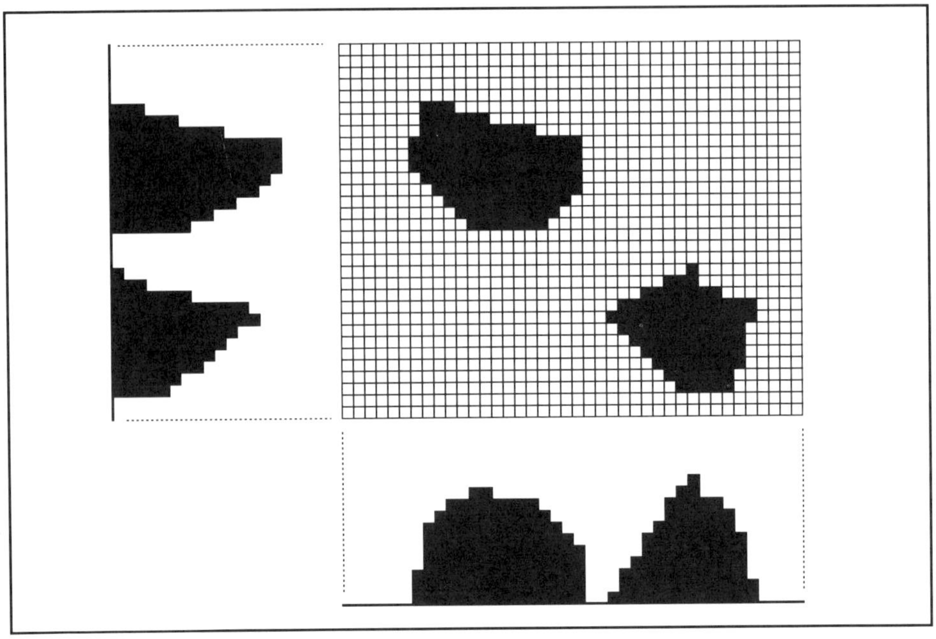

Figure 8

In some cases segmentation can be achieved by simply thresholding the image. Given a bright object on a dark background, a well chosen threshold will segment the image into background and foreground areas. If uneven lighting prevents the use of thresholding, intensity profiles can often be used to define rectangular areas of interest, which bound the image objects.

When an image contains several objects of interest it's helpful to label each object. One way to do this is to assign unique intensity values to the pixels of each object. Later, this allows you to determine which object a pixel belongs to. For example, a thresholded image could be labeled by assigning ones to each pixel in the 1st object, twos to each pixel in the 2nd object, and so on. This segmenting and labeling greatly simplifies the feature extraction process.

Feature extraction

Up to this point we've been talking about methods for processing the images to remove unwanted information (such as the background) and to enhance portions of the image that are likely to be important. These image processing and image enhancement techniques have been found to be useful whether the resulting image is to be interpreted by machines or by humans.

However, it could be argued that none of this image processing is really machine vision. A human who views the processed image on a video screen, might say "It's a tea cup", but an image processing system draws no such conclusions. Vision is a process by which the image is *interpreted* rather than just *processed*.

Image preprocessing and image segmentation are steps that facilitate the final step: *feature extraction*. For example, if the image contains a lot of high frequency visual noise, you might use convolution to remove that noise. Then you might use thresholding to segment the image into object and background. These steps produce an image from which *features* can more easily be extracted.

So, what is a feature? A feature is some geographically-independent numerical information about the object such as its:

- Area
- Center of mass
- Length
- Breadth
- Perimeter
- Orientation

Feature extraction is the process by which this geographically independent numerical information is derived from the image. These numerical values are the final product of the machine vision process.

In most cases, a single feature (by itself) is not sufficient to identify the object. However, when that feature appears together with several other features, it allows the object to be identified.

Morphological processing
An object's *morphology* is a set of characteristic features that (taken together) identify the object. To facilitate feature extraction, machine vision systems often use a technique called *morphological processing*.

Morphological processing is accomplished by doing logical operations on pixel neighborhoods. The goal of morphological processing is to remove from the image things that cannot possibly be features, while preserving those things which might be features. If done properly, this morphological processing makes the real features easier to identify.

A system doesn't need to know anything about what features are being sought in order to convolve an image. (For example, a system can convolve an image to enhance edges without knowing the shape or size of the objects in the image.) However, to do morphological processing (which removes non-features from the image) the system must first know something about WHAT features are being sought.

This knowledge is provided to the vision system in the form of *structuring elements*. A structuring element is a 2D array of boolean values with a specific size and shape. For example, a simple structuring element might define a neighborhood consisting of a pixel and its 8 immediate neighbors. Larger and differently shaped structuring elements are also possible.

One of the most basic characteristics of a feature is its size. Using this fact, a vision system can remove non-features from an image by simply removing anything smaller than the smallest possible feature. The size of the smallest possible feature can be specified with a special structuring element. Given this structuring element, the system can do morphological processing to remove anything smaller.

Morphological processing can be done on either binary (thresholded) images or on gray-scale images. There are 4 popular morphological operations that are used on thresholded images:

- Erosion
- Dilation
- Opening
- Closing

Erosion
An *eroded image* is generated by:

- laying the structuring element over each pixel of a thresholded input image, to define a pixel neighborhood.
- generating a corresponding output pixel intensity, by logically ANDing all pixels within this neighborhood.

Unless all the pixels within the neighborhood are 1 (white), this results in a 0 (black) pixel.

Applying this algorithm to all pixels in a thresholded image will *erode* (eliminate) object features smaller than the structuring element, as shown in Figure 9. In this example, the irregularly shaped object is eroded by the circular structuring element. The resulting eroded object shows that features smaller than the structuring element are removed. In this case, the narrow projection toward the right is removed.

Dilation
Dilation acts in a complementary fashion to erosion. In a dilated image, the output pixels are the logical OR of all pixels within the input pixel's neighborhood. Therefore, any input pixel with at least one white pixel in its neighborhood generates a white output pixel. Figure 10 shows how an object is made larger by the dilation.

Opening
Performing an erosion followed by a dilation, with the same structuring element, is called *opening*. The erosion removes objects and features which are smaller than the structuring element, but it also decreases the size of the surviving objects. The subsequent dilation restores the surviving objects to their original size. Opening is useful for removing unwanted bridges between objects, which may result from shadowing.

Closing
Closing is accomplished by applying a dilation followed by an erosion. (These are the same operations used for opening, except in reverse order.) Closing joins objects which are closer together than the size of the structuring element.

Higher-order openings and closings
A 2nd order opening can be accomplished by eroding twice, and then dilating twice. Other higher-order openings and closings can be accomplished in a similar fashion.

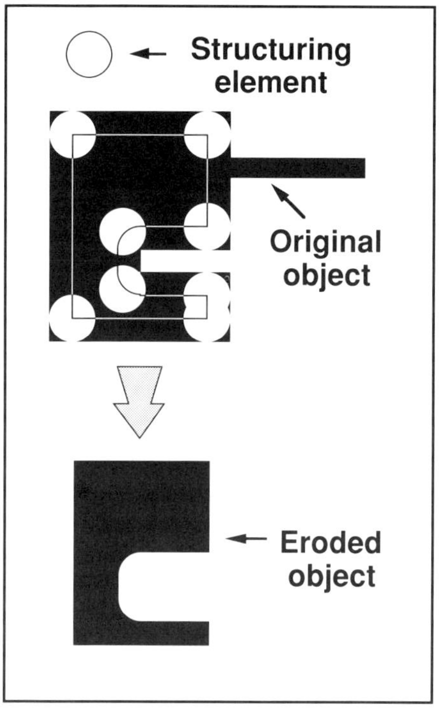

Figure 9

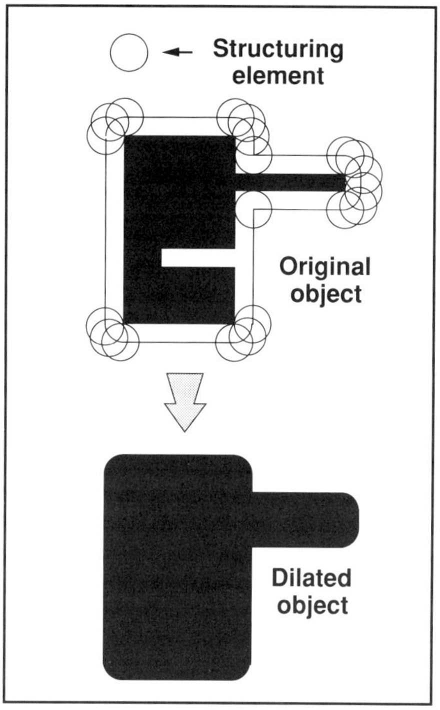

Figure 10

Erosion can also be done on gray-scale images. In that case, the output pixel intensity is the *minimum* pixel intensity within the input pixel's neighborhood. Likewise, gray-scale dilation outputs the *maximum* pixel intensity within the neighborhood of each input pixel.

Skeletonizing
There is one other popular type of morphological processing. However, unlike the previous techniques, this method *radically* transforms the shape of the object from its original 2D shape into a *skeleton*. This skeleton consists of a set of connected line segments. (See Figure 11) The shape of the original object, and the transformation rules, determine the shape of the skeleton.

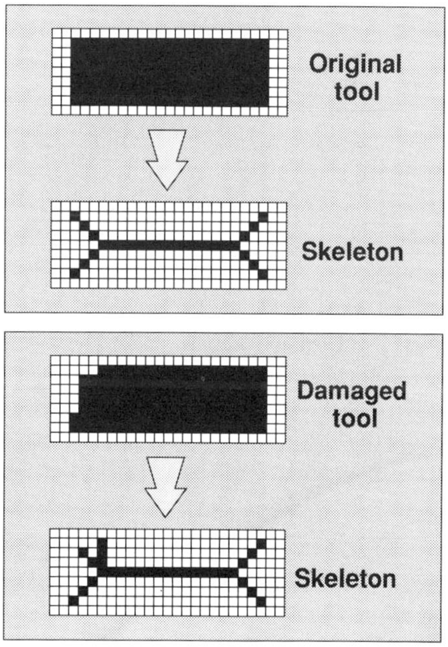

Figure 11

The transformation rules are expressed as a set of *neighborhood templates*, which are applied to the pixels of the original image. Several different template sets are popular: each set generates a different skeletal representation. Figure 12 shows the template set used to generate *L-type* skeletons. These skeletons are generated by the algorithm shown in Listing 1.

For vision applications image processing algorithms must be implemented in hardware

Image processing algorithms may be implemented in software on virtually any computer. However, achieving the performance levels required in most machine vision applications requires the use of specially designed hardware.

To support single point operations (which convert a single input pixel into a single output pixel) you need only a simple look-up table, consisting of RAM. However, it takes a lot more hardware to implement multiple point operations, such as convolution.

For less demanding applications, you may be able to do multiple point operations fast enough with closely interacting hardware and software. But, for the more demanding applications, you will need to implement the operations directly in hardware, with software doing only the initialization.

The elements of an image processing system

The most important elements of an image processing system are:

- the input device
- the look-up tables
- the frame buffers
- the image data ALUs
- the output device

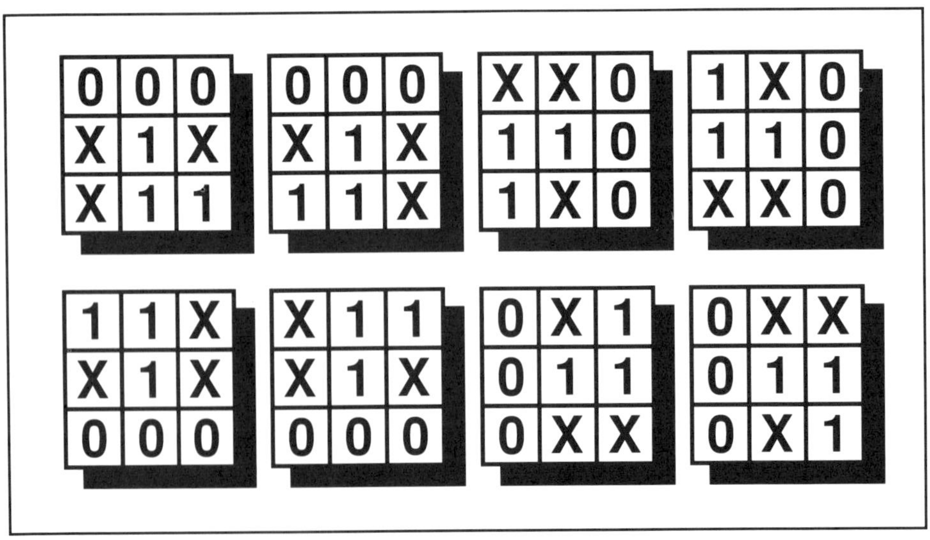

Figure 12

Listing 1

```
input_buffer  := input_frame_buffer;
output_buffer := temp_buf_1;

REPEAT
  no_of_pixels_changed := 0;

  FOR template := 1 TO 8 DO
    BEGIN
    FOR Y := 1 TO pixels_per_column DO
      BEGIN
      FOR X := 1 TO pixels_per_row DO
        BEGIN
          IF match(template, input_pixel_neighborhood(X,Y))
            THEN BEGIN
                 output_buffer[X,Y] := black;
                 no_of_pixels_changed :=
                      no_of_pixels_changed + 1;
                 END
            ELSE output_buffer[X,Y] := input_buffer[X,Y];
        END; (* FOR X := 1... *)
      END; (* FOR Y := 1... *)
    END; (* FOR template := 1... *)
  IF (output_buffer = temp_buf_1)
    THEN BEGIN
         output_buffer := temp_buf_2;
         input_buffer  := temp_buf_1;
         END;
    ELSE BEGIN
         output_buffer := temp_buf_1;
         input_buffer  := temp_buf_2;
         END;
UNTIL no_of_pixels_changed = 0;
skeleton_buffer := input_buffer;
```

The input device
The input analog-to-digital conversion device provides an interface between the input video signal and the internal digital data paths. An analog-to-digital conversion rate of 10 MHz allows real time digitization of a standard (RS-170) video signal.

The look-up tables
Incoming digitized video data is usually routed through a look-up table, which is loaded with appropriate intensity values before the image processing begins. This look-up table is fast enough to transform incoming pixel streams at video rates. (Outgoing video data is also often routed through a look-up table, prior to being converted back into an analog RS-170 video signal.)

The frame buffer
The pixel values from the look-up table are stored in a frame-buffer. Typical image processing systems have at least 1 Mbyte of frame buffer RAM. VMEbus imaging systems usually have dual-ported or triple-ported frame buffer RAMs. Dual-ported frame buffers provide input and output data ports. Triple-ported frame buffers also allow a host processor to access the image pixels. Some frame buffers are designed to permit reconfiguration, such as (512 X 512 X 16 bits) versus (1024 X 1024 X 8 bits).

The image data ALU
The image data ALU performs functional transformations on image data, at video rates, using both incoming images and previously stored images as operands. The ALU output may be stored in a frame buffer, or it may be fed directly to a video display device.

The output device
The output device may either generate an analog RS-170 video signal, or it may output a higher frequency signal, to drive a higher-resolution CRT display.

Processing color images
Color image processing can be done by first decoding the composite NTSC color video input signals into their red, green, and blue primary constituents. Each of these 3 primaries is then treated as a separate image. Some color image processing systems provide 3 parallel data paths, as well as multiple frame buffers and look-up tables. Other architectures use 3 separate image processing systems, operated in parallel.

The structure of a practical image processing system
An idealized image processing system consists of a pipelined architecture, with each step of the pipeline being either a look-up table or a processing element. However, the hardware cost of building such a system usually makes it impractical. If you can accept a lower level of performance, you can reduce hardware costs by *recirculating* the video data through a single ALU, as is done in the system shown in Figure 13. It consists of:

- an input look-up table (LUT)
- a single pixel ALU
- a frame buffer
- an output look-up table.

The result is an imperfect pipeline, which can do all of the operations that a pipeline can do, but not concurrently. For example, this system can't digitize and store an

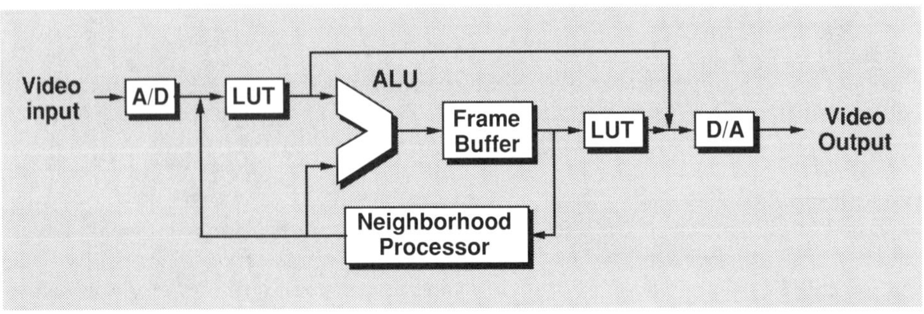

Figure 13

incoming video image into one frame buffer, while simultaneously displaying the contents of another frame buffer.

Types of image processing systems
There are 3 types of image processing systems:

- Special purpose systems
- Multiple-board pipeline systems
- Single-board general purpose systems

Special purpose image processors
A special purpose image processor is a system designed specifically to execute a predefined algorithm. It typically performs that particular algorithm very efficiently, but it may be very inefficient when performing other algorithms. If you decide to change the algorithm, you may have to scrap the entire system, and then start over with a new design.

Multiple-board pipeline image processors
Pipeline imaging systems allow you to build custom architectures by arranging boards in the required sequence. Data movement from one stage of the pipeline to the next takes place over a byte-wide pixel bus. Unfortunately, the maximum performance of a pipelined system is limited by the speed of its pixel bus. So, as technology advances, you may not be able to take full advantage of faster processors.

Single-board general purpose image processors
A single-board general purpose image processor is cheaper than a modular system, and it offers a level of flexibility not found in special purpose systems. However, you must choose the board carefully: if its performance proves inadequate, you'll have to abandon it, in favor of a faster board. On the other hand, unlike the multiple-board image processor systems, single-board systems aren't limited by the speed of a pixel bus, so higher performance boards can always be substituted to take advantage of improvements in image processing technology.

A machine vision application
Numerically controlled machine tools are capable of producing accurately machined parts completely automatically, by following digitally coded machining instructions. However, before the machining process begins, a blank must be loaded onto the table, and an origin established. Since the physical coordinates of the origin depend upon the position of the blank, this usually requires the involvement of a human operator.

Once the machining process begins, the machine's control system can monitor the rotation and the torque of the tool head, but the operator is still be needed to visually inspect the work as it progresses, and to check for tool head damage or wear.

A machine vision subsystem would allow the machine tool's control system to sense the presence of (and the position of) the workpiece, and to visually determine the condition of the tool head automatically. Figure 14 shows a hypothetical scene. The video camera's field of view encompasses both the tool and the workpiece, and both appear as dark areas against a light background.

We'll assume that the home position of the tool is known in advance. This is a fair assumption, given the nature of the machine. However, the position and fine orientation of the workpiece is not known.

We want the vision system to:

- inspect the tool for gross imperfections, such as would occur if the tool were broken.
- gauge the tool width, in order to compensate for wear.
- gauge the workpiece position, in order to set the machining origin.

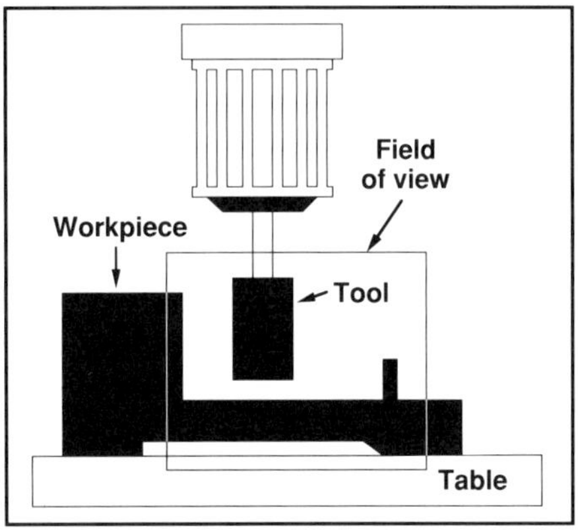

Figure 14

The tool can be inspected in several axes by rotating it and repeating the measurement. However, the workpiece can't be rotated, so we need a 2nd camera to determine its orientation. After processing all of these images, we want the vision system to output:

- a PASS/FAIL indication of tool integrity
- a measure of tool wear
- an (X,Y) position for the workpiece, with respect to the tool's home position.

Given a properly lit, well focused video image, the machine vision system begins by digitizing the image and storing the pixels in a frame buffer. Each digitized video frame contains 256 Kbytes of data, organized as 512 rows and 512 columns. Dark areas (such as the tool and workpiece) appear as low intensity pixels, while light areas are represented by high intensity pixels.

Preprocessing enhances the image without affecting the fundamental characteristics of the objects or the image features. It is accomplished in 3 steps:

- A look-up table thresholds each pixel value, to produce a binary image. Since the intensity histogram doesn't change very much from one image to the next, the system can use the histogram from the previous image to build the look-up table. By using the previous histogram, the system can transform the image pixels in real time, as the image is being digitized.
- A morphological *opening* is performed on the thresholded image, to eliminate all objects smaller than the chosen structuring element.
- A profile histogram is done on a row or a column known to contain pixels representing the workpiece. This allows the system to determine whether or not the workpiece is present.

After this preprocessing, objects of interest appear as black (0) on a white (1) background. This initial segmentation (into foreground and background) facilitates object labeling.

Feature extraction

The first step in the vision process is to move the black image of the tool out of the frame buffer and into a 2nd frame buffer. This is accomplished by logically ANDing a *possible tool area template* with the digitized image, and storing the resulting binary pixel values in a new frame buffer.

Hopefully the tool image and the workpiece image are now separated, and reside in different frame buffers. As a final verification, a D-type morphological skeleton is generated on both frame buffers. The D-Type skeleton reduces each object to a single pixel. An intensity histogram then reveals the total number of objects in the scene. In this case, this result should yield one in each buffer.

The features of both the tool and the workpiece may now be extracted. Performing an L-type skeleton (illustrated earlier) on the tool, and counting the number of triple points reveals any differences in shape resulting from a damaged or broken tool. This morphological technique is independent of size. The number of triple points is the same for similarly shaped objects, even if the size of their image is different due to varying distances from the camera. However, even a small nick in the tool will result in a radically different skeleton.

Tool wear is gauged by measuring the overall tool area with an intensity histogram. The vertical and horizontal dimensions are measured using profile histograms.

The workpiece orientation and dimensions are measured in a similar manner. The parameters thus recovered are used to control the machine tool bed, in order to correctly position the workpiece.

The Matrox MVP-VME

The above example illustrates how you might use an image processor as the front end for a machine vision system. Minicomputer-based and mainframe-based image processors have been available for many years, but only recently (with the appearance of high performance buses like VMEbus) have reasonably priced real-time image processors and image processing software become available. These products now permit you to build customized machine vision systems for use in industrial control applications.

Matrox has designed an entire image processing system on a single VMEbus board. It is supported by software that can be called upon by the user's high-level application code to execute any of the image processing algorithms described in this article, including the morphological operations. The Matrox MVP-VME, with its optional NP module, provides:

- up to 2 Mbytes of frame buffer RAM
- 3 input look-up tables
- 3 output look-up tables
- an NTSC decoder, for color decoding
- hardware support, for neighborhood processing
- an onboard 68000 processor, for command management

The optional NP module accelerates neighborhood operations by doing 3 X 3 convolutions on a 512 X 512 image in 100 msec. Software support consists of

256 Kbytes of resident code, executed by the 68000 processor. By executing this resident code, the 68000 can read and execute high-level commands from lists written into the board's 512 Kbytes of onboard RAM. User code can also be loaded into the local RAM, for execution by the MPU. Also available is Visilog, a user-friendly morphological development system, developed by the French firm Noesis.

Summary

Recent price-performance shifts in the digital imaging field have made the use of machine vision techniques viable, even for common industrial control applications. Vision systems have also been introduced into such diverse applications as railway track inspection and building security. This expansion of applications has occurred, in part, due to the availability of low-cost VMEbus-compatible image processing boards, which have been designed for integration into VMEbus systems.

Bruce C. MacDougall is a financial analyst with the Canadian Broadcasting Corporation, where he's involved with technological and strategic planning. Prior to joining the Canadian Broadcasting Corporation, Bruce gained experience with imaging and graphics at Matrox Electronic Systems. He holds an MBA from Queen's University, Kingston, and a BSE from McGill University in Montreal. Bruce spends his spare time in Kingston, Ontario, perusing antique markets and attending folk festivals.

CHAPTER 37

Using Pipelined Architectures To Design Real-Time Machine Vision Systems

By Robert I. Birenbaum

Machine vision systems used on fast-moving automated manufacturing lines must be able to analyze images quickly. In fact, it's hard to build systems that can do the required computations fast enough. In this chapter, Bob describes a machine vision technique called "template matching" and describes a pipelined architecture that can do the required calculations in real time.

Suppose you want to design a computerized system for installation alongside a conveyer belt, where it will inspect manufactured components as they pass by. This system must:

- detect the presence of the component.
- verify that the component is oriented properly.
- verify that the component is the proper size and shape.

Figure 1 shows how you could position a video camera over the conveyer belt to capture an image of the component, as it passes beneath. (We'll assume that the color of the component contrasts well with the color of the conveyer belt.) The RS-170 video output of the camera is then routed to a *video digitizer*. The digitizer generates a 512 X 480 array of 8-bit pixels, each representing a gray-scale value. (0 = black, 255 = white)

Figure 1

Once the image has been captured, digitized, and stored in RAM, an MPU compares it (pixel-by-pixel) to a *template*, which is also stored in RAM. This template is an array of pixel values, typically much smaller than the captured 512 X 480 image.

The MPU first compares the template to the cluster of pixels in the upper left-hand corner of the captured image, as shown in Figure 2. Obviously, if the gray-scale intensity of every pixel in the cluster matches the corresponding template pixel perfectly, an object has been detected. The MPU could record this by writing a value of 1 into the upper-left element of a 512 X 480 array of binary values, as shown in Figure 2. (If any of the pixels in the cluster fails to match the template, the MPU could write a value of 0 into that location.)

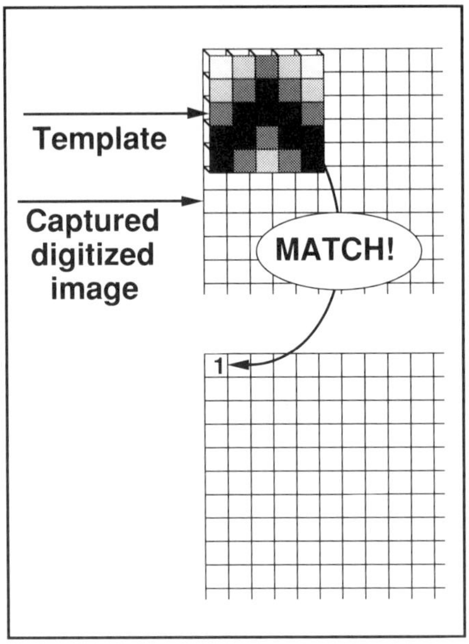

Figure 2

Since the object might be anywhere on the conveyer belt, the MPU must then shift the template one pixel to the right, and repeat this comparison, as shown in Figure 3. This shifting and comparing is repeated for each pixel of the captured image, even if the MPU detects an object at some point in the image. (There may be more than one object in the image.)

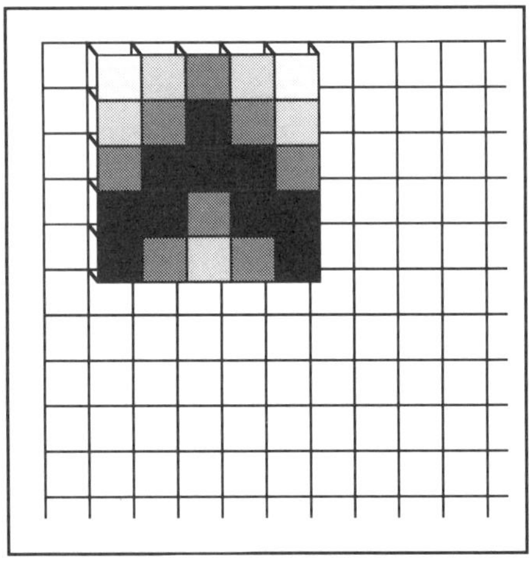

Figure 3

By the time the MPU finishes scanning the entire image, it has created a 512 X 480 array of binary values, as shown in Figure 4, where each binary value corresponds to a location of the component within the captured video image.

Note: When the template is compared to the last few pixels in each row (and to the pixels in the last few rows of the image) there won't be enough image pixels to compare to every template pixel. Since we want the resulting binary array to be the same size as the original image (for reasons that will become clear later) the best solution is to simply set the binary values that correspond to those positions to 0.

Thresholding provides some tolerance for lighting changes

Humans are very good at compensating for variable lighting. In fact, our eyes and brain adjust so readily and so automatically that we don't even notice small lighting changes. For example, we have no difficulty recognizing an object at different times of day, as the sun's intensity or its angle changes. If a light bulb's intensity dims with age, we're usually not even aware of the change. We're also able to interpret the reflected glare from the surface of an object, even when it changes with a repositioning of the light source. However, all of these lighting effects influence the digitized pixel values of a video image, and the template matching method described above completely fails, even with small lighting variations.

A technique called *thresholding* can be used to make the system more tolerant to lighting variations. It's especially useful when the camera is looking at a silhouette of an object against a contrasting background.

1	0	0	0	0	0	0	0	0	0
0	0	0	0	0	0	0	0	0	0
0	0	0	0	0	0	0	1	0	0
0	0	0	0	0	0	0	0	0	0
0	0	0	0	0	0	0	0	0	0
0	0	0	0	0	0	0	0	0	0
0	0	0	0	0	0	0	0	0	0
0	0	0	0	0	0	0	0	0	0
0	0	0	1	0	0	0	0	0	0
0	0	0	0	0	0	0	0	0	0

Figure 4

Thresholding is used to modify the captured image *before* it's compared to the template. Assume we have a digitized image with 8-bit pixels. To modify the image, the MPU must be given a *threshold intensity value* from 0 to 255. It then scans the pixels in the captured image, comparing each pixel to the threshold value. If the pixel value is greater than the threshold value, the MPU raises its value to 255 (white). If the pixel value is less than or equal to the threshold value, the MPU lowers it value to zero (black). The resulting pixel array is called a *binary image*.

Once this process has been completed, each pixel in the resulting image will have 1 of 2 values (0 or 255). If the contrast between the object and the background is good, you can change the lighting intensity that's used to capture the image considerably before it changes the thresholded binary image.

Thresholding can introduce spurious pixel values

While thresholding is a very useful technique, it can cause problems. For example, suppose the edge of a silhouetted object passes through the *middle* of some of the pixels in the captured image, as shown in Figure 5A. The digitized intensity of those pixels will be between the background (black) and the object (white). Depending upon

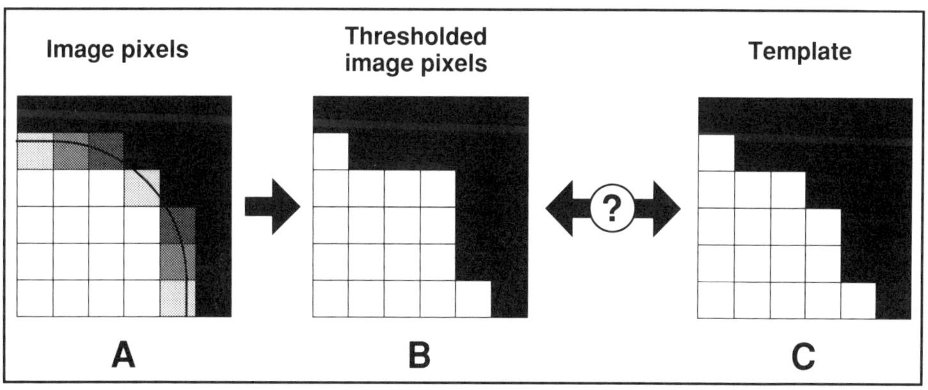

Figure 5

the precise value chosen for the threshold, each of these digitized pixel values might either be increased to white, or decreased to black. As shown in Figures 5B and 5C, when the thresholded cluster of pixels is compared to the template, a single pixel might cause a mismatch, and prevent the system from recognizing the object at that location.

A similar problem arises when there is *low contrast* between the object and the background. When contrast is low some of the object's pixel intensities may drop below the threshold value (possibly due to shadowing) and might be classified as background pixels. At the same time, reflections from the background surface might raise the intensity of some background pixels to the point where they're classified as part of the object.

Correlation coefficients allow some pixel variations

Obviously, we need some way to detect thresholded pixel patterns that are similar (but possibly not identical to) the template. Otherwise a single spurious pixel intensity may prevent us from recognizing the presence of the object.

In the previous example we thresholded the input image, and then compared the template to each cluster of pixels in the thresholded image. The result of each comparison was a binary answer to the question "Is the component located here?" But suppose we allow for an analog answer. For example, suppose we allow for...

- NO
- EXTREMELY UNLIKELY
- VERY UNLIKELY
- UNLIKELY
- HARD TO SAY
- LIKELY
- VERY LIKELY
- EXTREMELY LIKELY
- YES

To do this, you need to use an array of *integers*, instead of the array of binary values shown in Figure 4. These integers are called *correlation coefficients*, and they indicate how close the cluster of thresholded pixels matches the binary template. When using correlation coefficients, a single spurious pixel value in the captured image won't change the answer from YES to NO. Instead, it changes the answer from YES to EXTREMELY LIKELY.

The simplest way to generate this correlation coefficient is to:

- threshold the image
- compare the binary template (pixel by pixel) to each cluster of pixels in the binary image
- increment a counter each time a match is found between an image pixel and a template pixel

A perfect match with a 3 X 3 template will result in a count of 9 (YES). If the match is close, but not perfect, the correlation coefficients may be 7 or 8. (Figure 6 shows what an array of correlation coefficients might look like near an object.)

Once these correlation coefficients have been determined, the MPU can:

- determine whether an object similar to the template lies anywhere within the image, by comparing each of the coefficients to a threshold value.
- determine the exact locations of objects in the image, by analyzing all of the coefficients that exceed the threshold value.

Shadows can cause problems
Up to this point we've talked about an idealized situation: one in which we're trying to locate an object based upon its silhouette. Unfortunately, it's not usually that simple. In many cases, shadows can make things very difficult. For example, suppose you use a point light source to illuminate the object, as shown in Figure 7. Because the light source is located to one side of the object, the opposite side of the object (facing away from the light) may be poorly illuminated.

Filling shadows may cause uneven lighting
Figure 8 shows how you can position a 2nd light source on the other side of the object, to provide *fill light*. This 2nd light illuminates the portion of the object that was previously in the shadow.

Unfortunately, it also produces *uneven* lighting. Most of the pixels of the conveyer belt are illuminated by both light sources and some may be brighter than partially shaded pixels of the object. This uneven lighting prevents you from using a simple thresholding method, since a threshold low enough to select the entire object might also select a portion of the background.

Convolution can be used to handle uneven lighting
One solution to this problem is to look for *abrupt* changes of intensity. To do this, you need a kind of *high-pass filter*, which emphasizes abrupt intensity

4	4	4	4	5	4	5	5
5	5	5	6	5	5	5	5
4	5	6	8	7	6	4	5
4	4	4	6	5	4	5	4
5	5	4	4	4	3	5	4
4	5	4	5	5	4	5	5

Figure 6

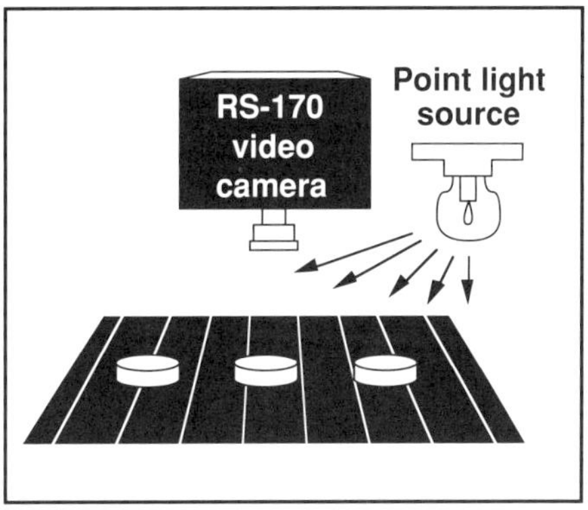

Figure 7

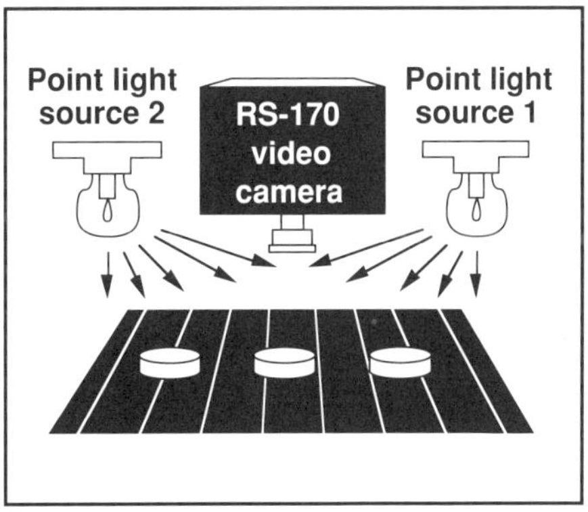

Figure 8

transitions between neighboring pixels. This high-pass filtering is usually done by performing a *2D convolution.*

Convolution can be visualized as shown in Figure 9, where the original digitized video image is used as input to the convolver. Another *convolved image* emerges as output. The pixels of the *convolved image* are brightest where the intensity changes are the most abrupt in the original image.

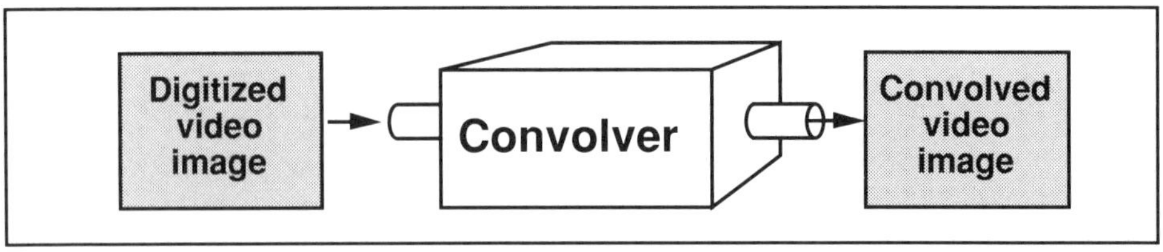

Figure 9

How is convolution done? To convolve an image, you first define a 2D array of *weighting coefficients* (called a kernel) as shown in Figure 10. The size of the kernel array is arbitrary. In general, a larger kernel allows detection of more subtle intensity transitions, but it also increases the number of calculations.

-2	-2	-2
-2	16	-2
-2	-2	-2

Figure 10

The weighting coefficients are used as *multipliers*, to generate a new image from the original. Figure 11 shows how this is done with the 3 X 3 kernel of Figure 10. The multiplications and additions are performed as shown, and the total is then *normalized*, by dividing it by the number of elements in the kernel: in this case 9. Each normalized result becomes a pixel value in the new convolved image.

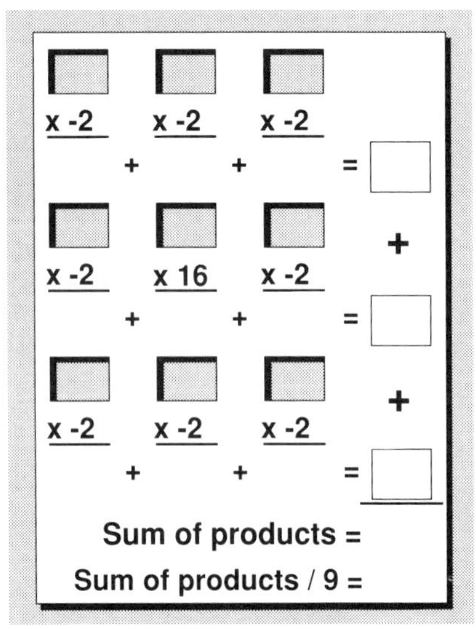

Figure 11

The weighting coefficients in the kernel can be selected to either emphasize or de-emphasize particular pixel patterns. The coefficients chosen in Figure 10 emphasize *abrupt intensity transitions.*

Once a convolved image emerges from the convolver, what do you do with it? You treat it just like any other image: you threshold it, and then compare it to a binary template. However, since the convolved image looks very different from the original, you can't just compare the convolved image to a template made from an image of the ideal object. Instead, you must produce a *convolved template* by convolving an image of the idealized object, and then thresholding it to produce a binary template.

Convolution won't always get the job done

As shown above, high-pass filters can be used to emphasize abrupt intensity transitions in an image. However, in doing so, high-pass filters make the signal-to-noise ratio much worse, especially if there's a lot of high frequency noise in the original image.

Suppose you capture an image of an object with a rough, irregular surface. The image will have lots of small dark spots and bright spots. The dark spots appear wherever there's a small depression. The bright spots appear at the higher elevations. Assuming that the details of the texture are unimportant, these dark and bright spots are a form of *visual noise*. If this visual noise is present near the edge of the object, high pass filtering may enhance this noise to the point where it's impossible to trace the edge of the object accurately.

Erosion eliminates visual noise at object boundaries

Suppose you need to eliminate visual noise from an image of a white object in front of a black background. You could do this by passing the image through a *minimum filter*, as shown in Figure 12.

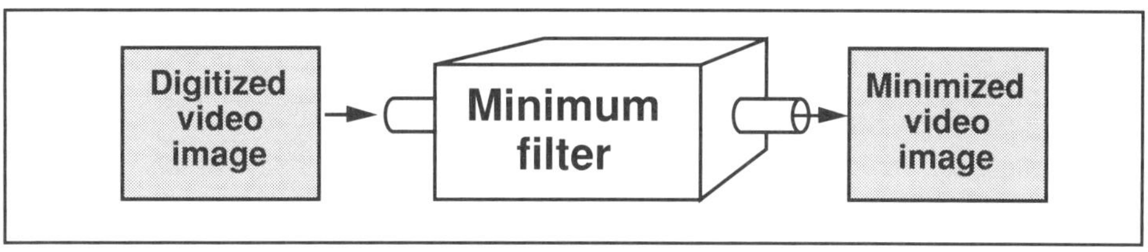

Figure 12

For each input pixel, the minimum filter looks at all of the pixels within a prescribed neighborhood, and outputs the lowest (blackest) pixel value it finds within that neighborhood. As a result, any input pixel with a background pixel in its neighborhood will generate an output pixel of background intensity.

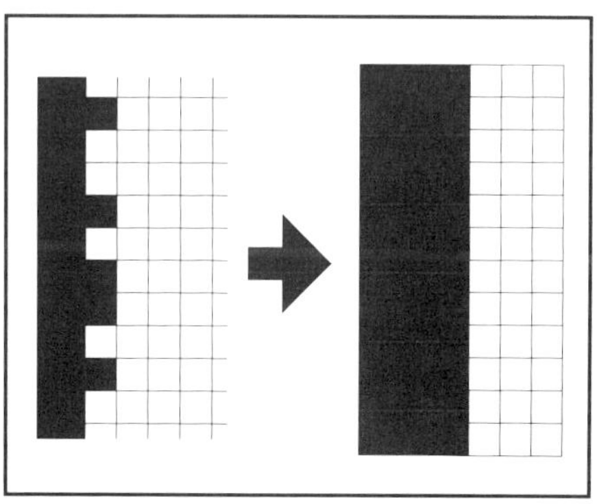

Figure 13

Figure 13 shows what happens when you pass an image of a white object on a black background through a minimum filter with a 3 X 3 neighborhood. The filter effectively reduces the size of the object image by one pixel and, in the process, clips off the spurious pixels that resulted from visual noise. This clipping process in called *erosion*.

Using erosion to compensate for uneven lighting

Earlier we studied how fill light can be used to illuminate shaded areas, and found that this fill light results in uneven lighting. This prevents you from using a simple thresholding method, since a threshold high enough to select the entire object might also select a portion of the background.

But suppose you capture an image of just the background, with no object in front of it. Then suppose you capture another image that includes both the background and an object. You could then compensate for the uneven lighting by *subtracting* the background intensity level (pixel by pixel) from the image that contains the object. The background intensity of this *difference image* would be much more uniform, allowing you to use thresholding to distinguish between the object and the background.

While this approach may work, it requires you to capture 2 images instead of just one. If you have objects traveling on a conveyer belt, you would have to leave empty sections on the belt, just so you could sample the background lighting level. To avoid this, you can use a technique called *top-hat filtering*, which builds an artificial background.

Figure 14 shows how a top-hat filter is constructed. An image containing the object is routed through a minimum filter. For each input pixel, this filter finds the blackest pixel in its neighborhood, and outputs that pixel. However, the neighborhood size is chosen to be large enough to ensure that every pixel of the object has at least one background pixel in its neighborhood. As a result, the object does not appear in the output image. (Each of the white object pixels is replaced by a dark background pixel.) In place of the object is an artificially generated background, whose intensity matches the real background at the boundaries of the object.

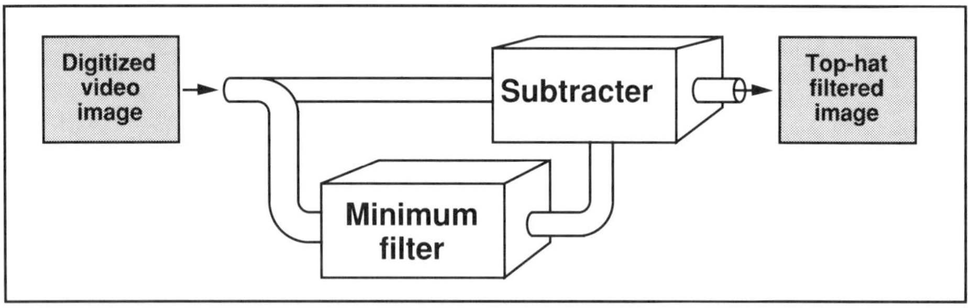

Figure 14

After generating this artificial background, the top-hat filter subtracts it (pixel-by-pixel) from the original image, to generate a *difference image*. Since the pixel intensity of the artificial background is determined by the pixel intensity of the surrounding real background, it compensates for the uneven lighting in the original image. This allows you to threshold the difference image, and distinguish between the object and the background.

A simple vision system
Figure 15 shows a simple vision system. It consists of:

- an RS-170 video camera
- a video digitizer
- a frame buffer
- a histogram generator
- an MPU
- a template, stored in RAM

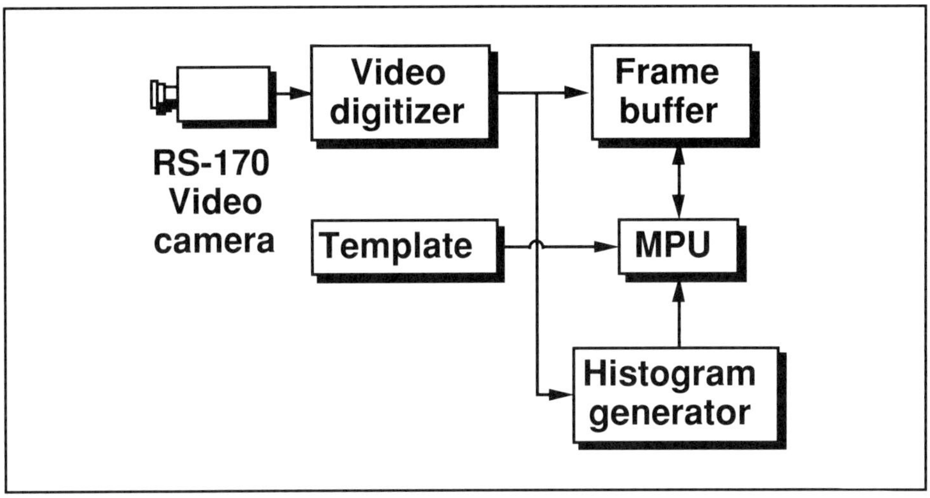

Figure 15

The RS-170 video camera produces 30 analog frames per second. The video digitizer samples the video waveform at a rate of 10 Mpixels/second, to produce a 512 X 480 pixel image. As it digitizes each pixel, it writes the pixel value into the frame buffer.

As the digitizer writes each pixel into the frame buffer, the histogram generator keeps a tally of the number of pixels at each intensity level. If lighting conditions are good (and if there is a good contrast between the object of interest and the background) the histogram will look like Figure 16. When the digitizer writes the last pixel into the frame buffer, the histogram is complete, and the MPU runs a simple algorithm to select a threshold value. Then it:

- scans through the frame buffer, using the threshold value to generate a binary image.
- scans through the frame buffer again, comparing the binary image to the template.

Real-time implementations demand specialized hardware
The simple system described above gets the job done, but it's not very fast. In order to generate the binary image, the MPU must access the memory twice for each pixel location in the frame buffer:

- The 1st access reads the pixel value.
- The 2nd access writes the thresholded binary value.

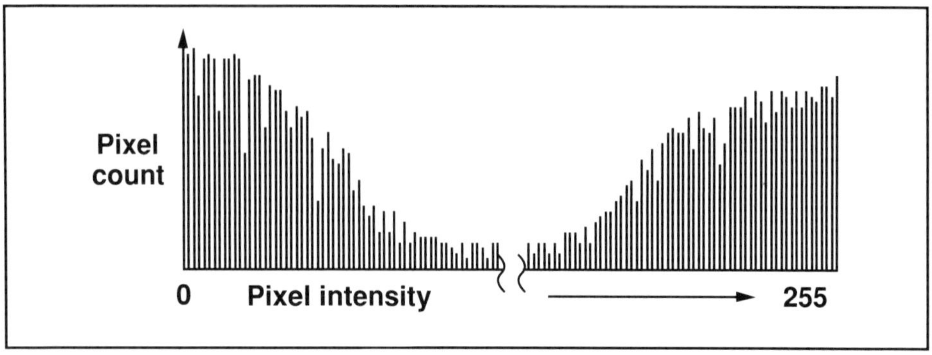

Figure 16

In addition, the comparison with the template doesn't begin until the binary image has been completely generated.

The process of thresholding and template matching is so straightforward that it's easy to underestimate how long it takes to do the job. The difficulty is not in the complexity of the calculations, but in the *number* of calculations. For example, when comparing a 16 X 16 template to a 512 X 480 image, you must do 256 pixel comparisons (and counter increment operations) for *each* of the 245,760 pixels in the 512 X 480 image. If you want to do this fast enough to keep up with a manufacturing line, you'll need some kind of specialized hardware.

You can speed things up considerably by using a *binary correlator*, like the one shown in Figure 17. When the system is initialized, the MPU writes a binary template into the correlator. Then (as described earlier) while the image pixels are digitized and written into the frame buffer, a histogram is computed. As before, the MPU uses the histogram to calculate a suitable threshold value.

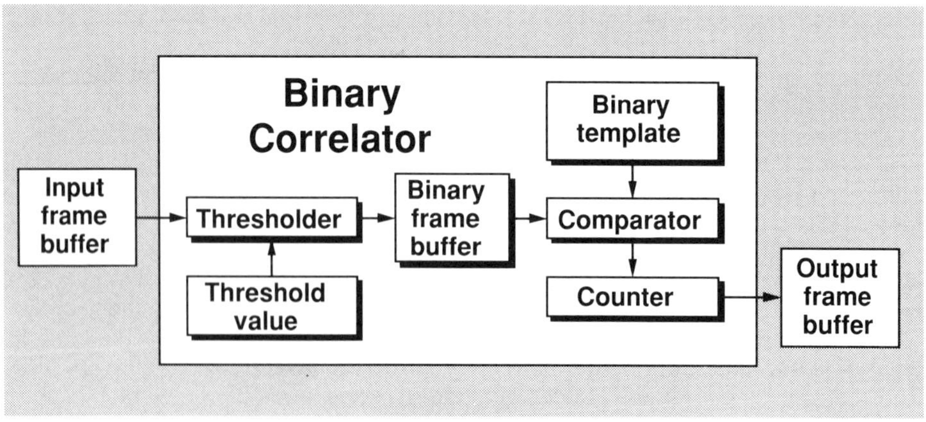

Figure 17

However, instead of comparing this threshold value to each pixel (to create a binary image) the MPU simply writes the threshold value into a register in the correlator. When the correlator hardware receives this 8-bit threshold value, it:

- *reads* each pixel from the frame buffer
- *compares* that pixel value to the threshold value, to calculate a binary value
- *stores* the binary value in its own internal binary frame buffer

When it has stored enough binary pixels to do the first comparison with the template, it begins the template comparisons. (The reading of the image pixels and the generation of the rest of the binary pixels continues, while the template comparison is going on.)

As the correlator compares the template pixels to those in its internal binary frame buffer, it increments a counter for each pixel match. When it finishes comparing all of the template pixels, it outputs the count, as a correlation coefficient. By this time, the thresholder has generated enough additional binary pixels to allow the comparator to compare the template with the adjacent cluster of binary pixels.

The correlation coefficient range depends upon the size of the template. For example, a 16 X 16 template will produce correlation coefficients from 0 to 256, while a 32 X 32 template will produce coefficients from 0 to 1024.

The binary correlator's output string of correlation coefficients forms a 512 X 480 array, like the one in Figure 6. This array is stored in a 2nd frame buffer. The MPU then analyzes these coefficients to determine where the peak correlation coefficients lie.

Convolution takes a lot of calculation

As in the case of template matching, convolution involves a lot of calculation. For a 3 X 3 kernel there will be 9 multiplications and 8 additions for *each* of the 245,760 pixels in the 512 X 480 pixel image. If the video camera is generating 30 images per second, there will be...

$$(9 + 8)\ (245{,}760)\ (30) = 125{,}337{,}600$$

...operations per second. If the kernel size is increased to 8 X 8 (a more useful size) there will be...

$$(64 + 63)\ (245{,}760)\ (30) = 936{,}345{,}600$$

...operations per second. Again, to do this kind of calculating in real time, you obviously need specialized hardware.

The Imaging Technology Series 150

Imaging Technology's Series 150 product line consists of 14 double-height VMEbus-compatible boards, which can be selected and configured for vision applications. Four of the most widely used boards are the:

- ADI-150 video digitizer
- FB-150 triple frame buffer
- RTMP-150 real time modular processor
- HF-150 histogram generator and feature extractor

Each of these boards has a VMEbus interface that allows a VMEbus master to configure, control, and determine the status of the board.

Figure 18 is an oversimplified block diagram of the ADI-150 board, which is the video input/output device for series 150 systems. It consists of 2 parts:

- a video digitizer that samples an analog RS-170 video signal, and routes the resulting 8-bit pixel values to the VDI bus.
- a video controller that accepts 8-bit pixel values from either the VDA bus or the VDB bus, and generates an analog RS-170 output signal to a video monitor.

As the ADI-150 digitizes an incoming video signal, it typically routes the 8-bit pixels over the VDI bus to an FB-150 frame buffer board, which is shown in Figure 19. The FB-150 contains a 16-bit frame buffer and two 8-bit frame buffers (A and B).

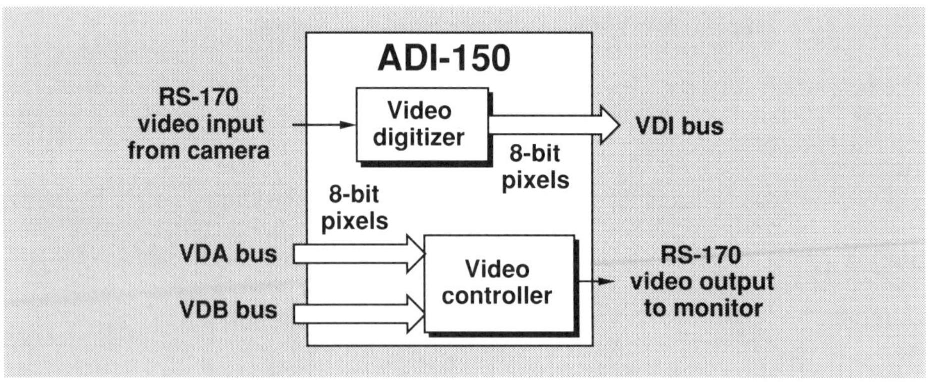

Figure 18

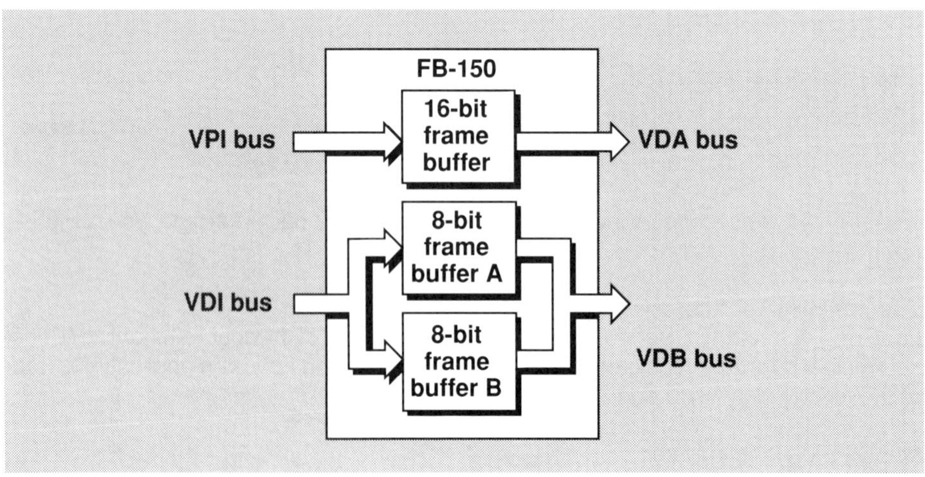

Figure 19

If a video frame has been stored in the FB-150's frame buffer A, its pixel values can then be sent over the VDB bus to some other board. In a similar fashion, the pixel values in frame buffer B can also be sent over the VDB bus.

The 2 remaining boards listed above are *pipeline processors*, which look like Figure 20. These boards accept pixel values from:

- the video digitizer board, over the VDI bus
- frame buffer A or frame buffer B, over the VDB bus
- the 16-bit frame buffer, over the VDA bus
- a board earlier in the video pipeline, via the VPI port

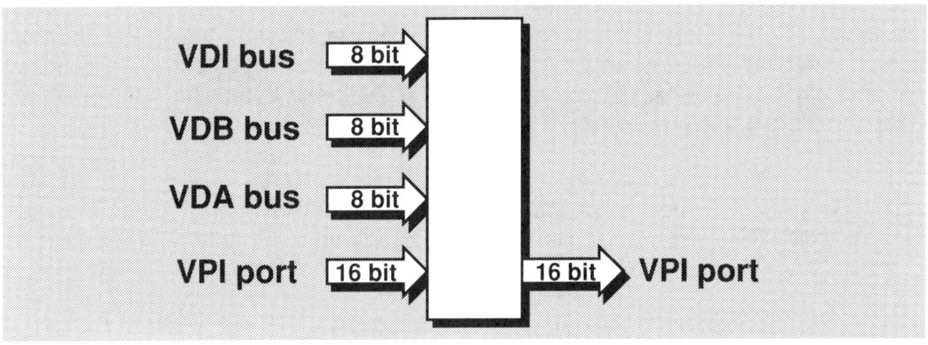

Figure 20

One of these 2 boards (the RTMP-150) is a general purpose pipeline processor board, which can be customized by the installation of special plug-on modules. For example, you can plug-on a real-time convolver module, or a binary correlator module.

Figure 21 shows how you can build a customized video processing pipeline by inserting these processor boards into a VMEbus chassis. The ADI-150 board digitizes an analog RS-170 signal from a video camera. The resulting digitized pixels are sent over the Video Data Input (VDI) bus to the RTMP-150 board, which is equipped with a convolver module. When it has received and stored enough pixels in its internal frame buffer, the RTMP-150 begins convolving the image. As it convolves the image, it outputs the convolved pixels to an HF-150 histogram generator board through its VPI port. The histogram generator generates a pixel count for each intensity, and passes along the pixel values to the FB-150 frame buffer board for storage.

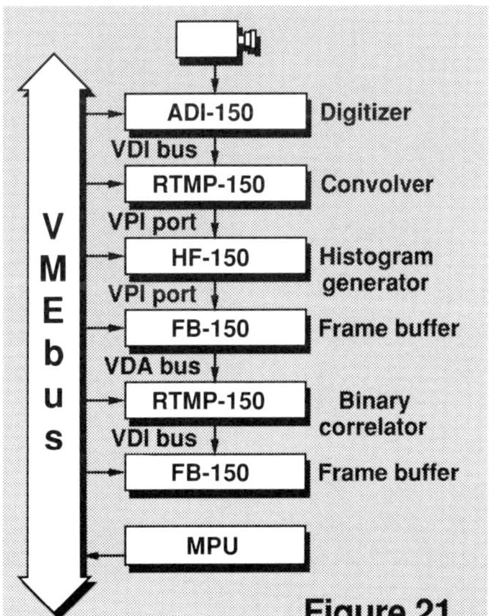

Figure 21

When the RTMP-150 finishes convolving the image, the histogram generator board (HF-150) contains a complete histogram. The MPU then analyzes this histogram and writes an appropriate threshold value into the second RTMP-150 board, which is equipped with a binary correlator module. That RTMP-150 then generates a binary image, compares the binary image (pixel-by-pixel) to a template (up to 32 X 32), and outputs an array of correlation coefficients to another frame buffer board.

Summary
Real time machine vision applications demand very high speed calculation. Because the image pixels come into the system in a serial fashion, and because the mathematical operations can (for the most part) be done on them serially, a pipelined architecture works very well. This pipelined architecture can be constructed from customized VMEbus boards that each do one operation (or a few operations) very efficiently. The VMEbus P2 connector allows interconnection of the various boards in a pipelined fashion. By using these pipelined architectures you can build systems capable of processing images orders of magnitude faster than with a conventional general purpose microprocessor.

Robert I. Birenbaum is vice president of marketing at Imaging Technology. He has a BSEE and an MS in computer science from Union College, and an MBA from Boston University. Bob's previous experience includes key technical contributions to the development of the MicroVAX chip set, while at Digital Equipment Corporation, and various marketing positions at Imaging Technology. He has often acted as an industry spokesperson on the issues and challenges facing the emerging image processing industry.

CHAPTER 38

Motion Control With Stepper Motors

By William A. Gowan

In many machine control applications, a computer is called upon to move an object from one place to another in a precise manner. For example, the machining of a piece of metal can only be done accurately if the motion of the piece being machined (or possibly the machine tool's head) can be accurately controlled. In this chapter Bill describes how you can use stepper motors to accomplish precise motion control at a reasonable cost.

While other options exist, motion control is typically accomplished using some kind of electric motor. Of the various types of motors available, steppers are very popular. There are several reasons for this:

- Steppers are relatively inexpensive.
- The circuitry required to drive steppers is simpler and less expensive than for other motor types.
- Steppers require very little maintenance, since they have no brushes to service.
- Steppers provide accurate and reproducible displacements. This eliminates the need for feedback devices, such as encoders.
- Steppers provide a very high *holding torque*, once they have been moved to the specified position. This is especially important when you're driving time variant loads.
- Steppers are essentially digital devices, rather than analog. This simplifies the software used to control them. For example, there is typically no need to resort to Laplace or Fourier transform algorithms.

There are 3 kinds of stepper motors:

- Permanent magnet stepper motors
- Variable reluctance stepper motors
- Hybrid stepper motors

Permanent magnet stepper motors
Figure 1 shows a permanent magnet (PM) stepper. Its rotor is a permanent magnet, eliminating the need for armature brushes. Its stator consists of 2 pairs of windings: A and B.

In Figure 1, windings A1 and A2 have been energized, to generate a pair of North/South poles, and windings B1 and B2 have been energized to generate a 2nd pair of North/South poles. The combination of these 2 magnetic fields causes the permanent magnet rotor to align itself as shown in Figure 1.

To turn the rotor clockwise 90 degrees you would de-energize windings A (as shown in Figure 2) and then re-energize A1 and A2 in the opposite direction (as shown in Figure 3). An entire rotational sequence is shown in Figure 4. Because this sequence resembles a person *walking*, we say that the rotor is being "walked" through a rotation.

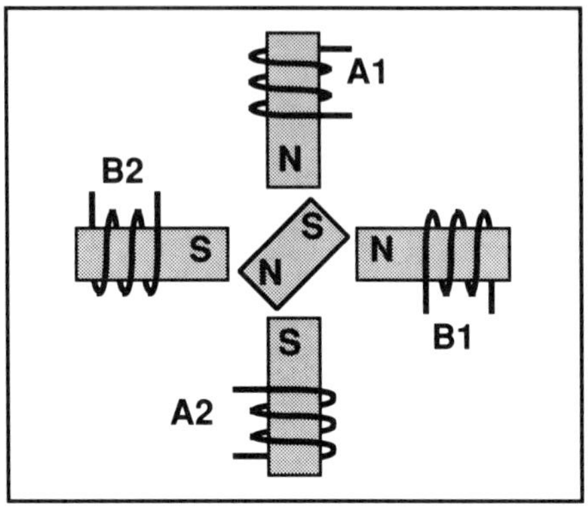

Figure 1

PM steppers employ a rotor that's magnetized perpendicular to its axis of rotation. The stator is composed of an even number of windings (usually 4) which are always energized in pairs. Permanent magnet steppers are simple and inexpensive to manufacture, but they offer only very coarse rotation angles. The stepper in Figures 1 to 3 steps in increments of 45 or 90 degrees, depending upon how it's driven.

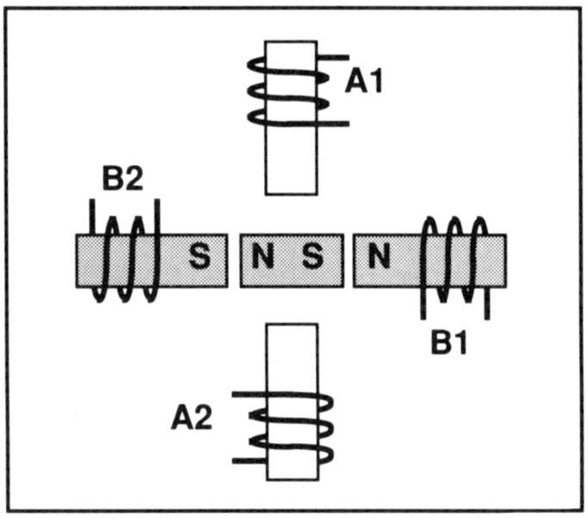

Figure 2

Variable reluctance stepper motors
Figure 5 shows a Variable reluctance (VR) stepper. Like PM steppers, VR steppers have several sets of windings that can be energized to walk the rotor through its rotation. (The VR stepper in Figure 5 has 3 sets of windings: A, B and C. However, unlike the windings in a PM stepper, the windings in a VR stepper each energize *several* pairs of poles, around the periphery of stator. Each winding in Figure 5 energizes 4 poles of the stator.

Another difference between the windings in a PM stepper and those of a VR stepper is that the VR stepper's windings all generate the *same* magnetic polarity. Thus, in Figure 5, when we energize winding A, every 4th pole in the stator is a South pole. The *magnetic circuit* is closed through the rotor's shaft, as shown in Figure 6.

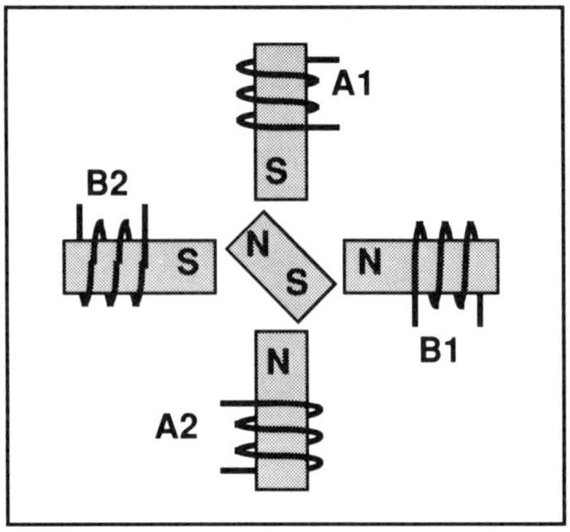

Figure 3

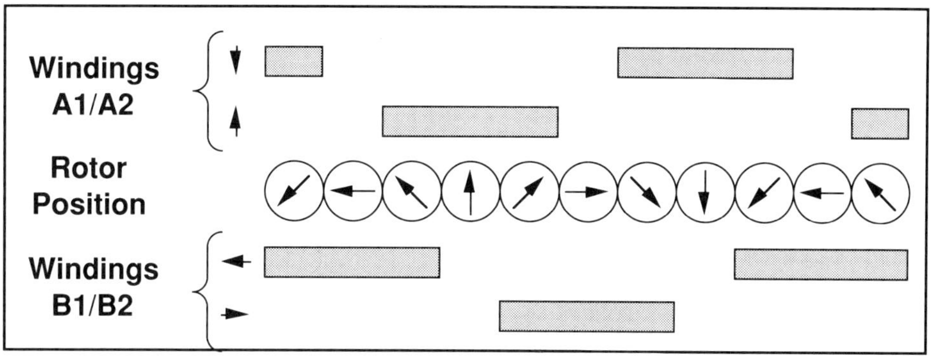

Figure 4

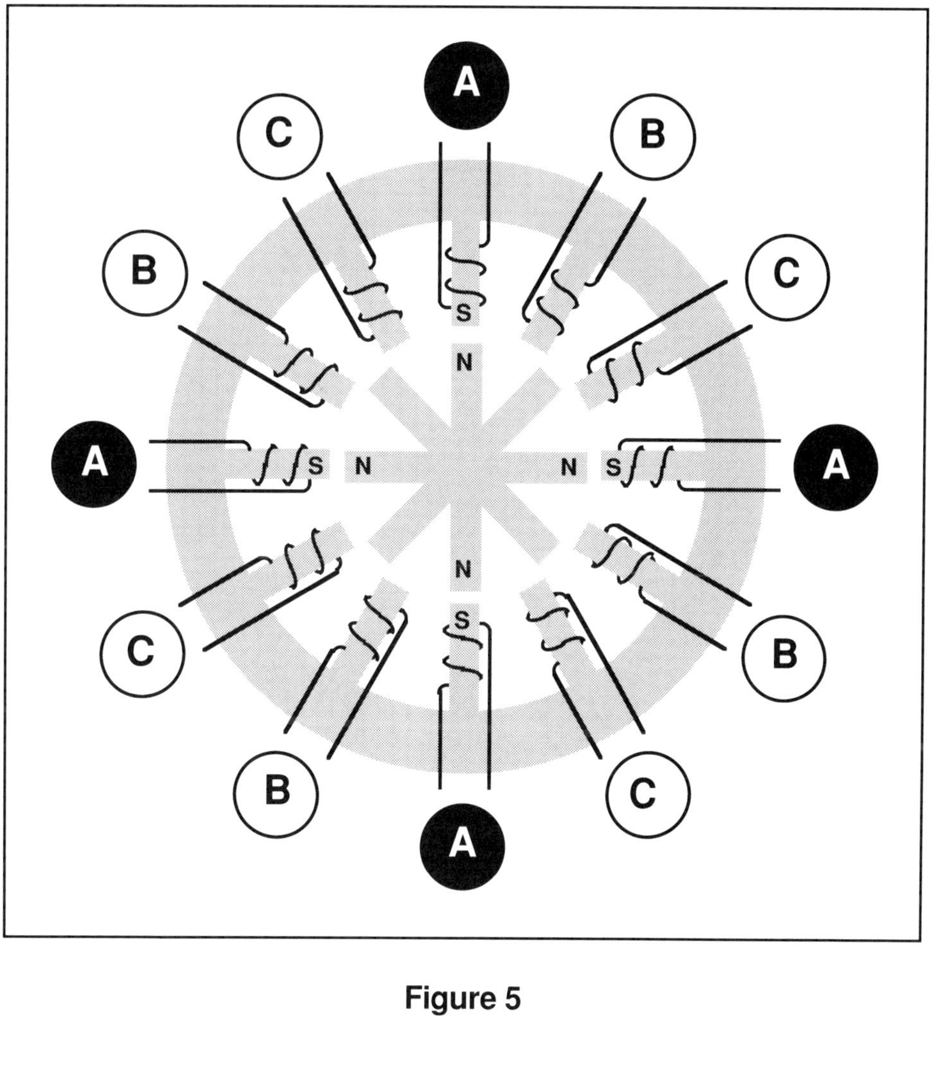

Figure 5

In order to turn the rotor in Figure 5, we first energize winding B. (See Figure 7) That causes the rotor to turn 7.5 degrees counter-clockwise. Then we de-energize winding A, and the rotor turns another 7.5 degrees counter-clockwise, to the position shown in Figure 8. Thus, by energizing the various windings in sequence, we can walk the rotor around its entire rotation. Note that each step results in two 7.5 degree rotations, for a total of 15 degrees. If more resolution is needed, a VR stepper with more poles can be used. VR steppers are available with enough poles to provide 5 degree increments of rotation.

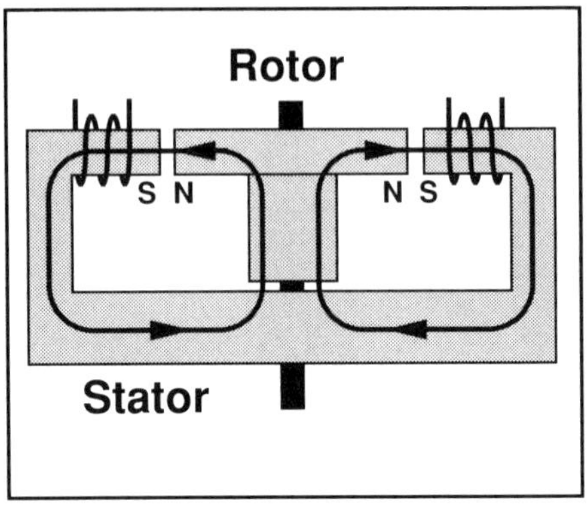

Figure 6

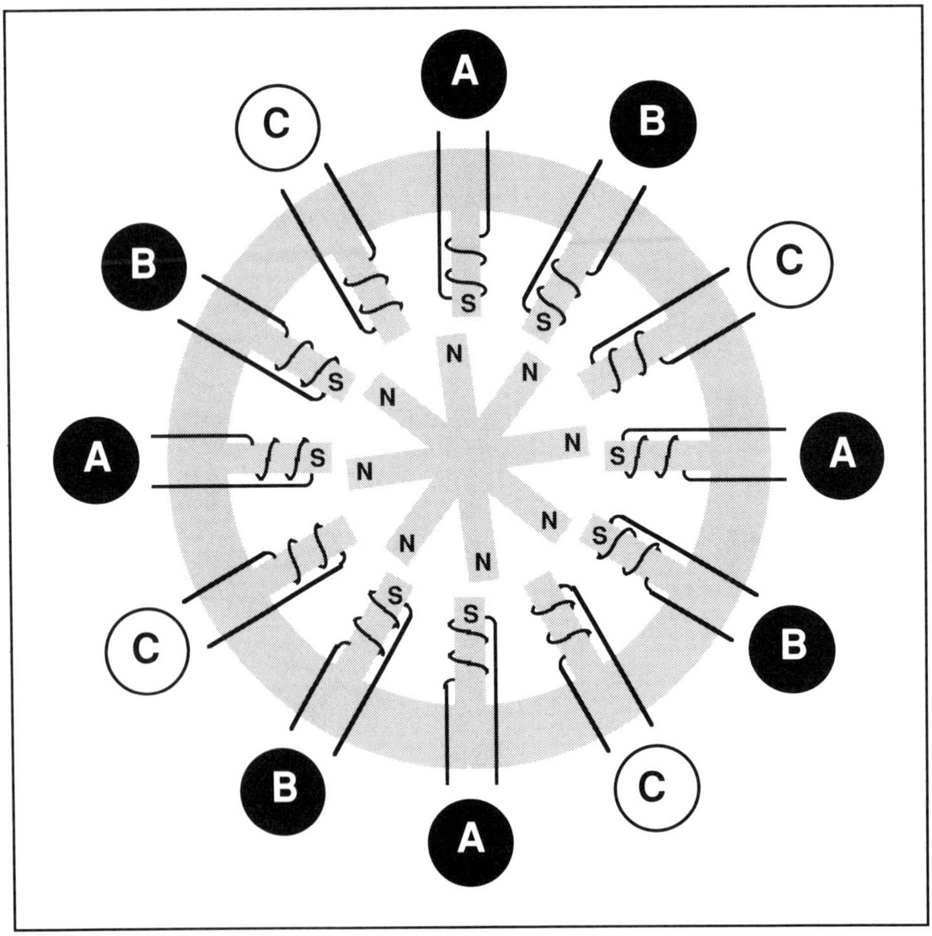

Figure 7

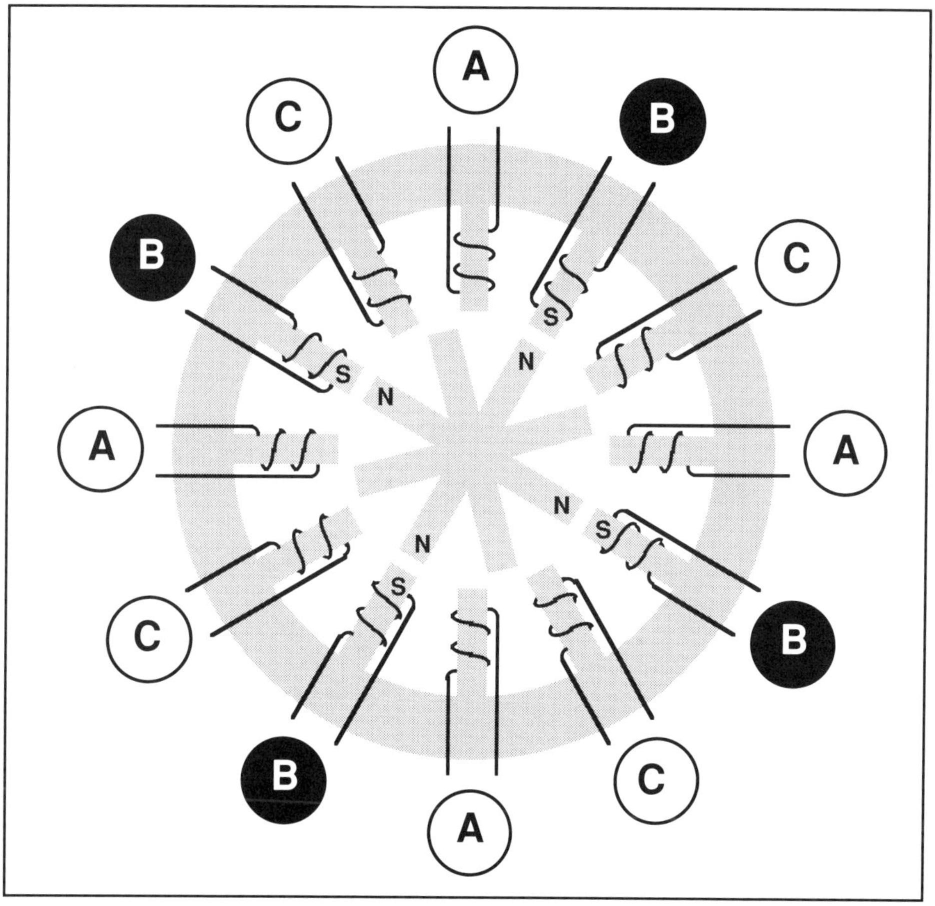

Figure 8

Hybrid stepper motors

By far, the most common type of stepper motor in use today is the *hybrid* stepper. It combines the best features of the PM and VR steppers. The hybrid stepper has a permanent magnet rotor that's actually a kind of double rotor. A cross section of the rotor and stator is shown in Figure 9. The stator is a *cylinder* with two rings of poles: one at each end. The permanent magnet rotor is a spindle that's inserted into the cylindrical stator. The stator closes the magnetic circuit between the poles at the 2 ends of the rotor.

Most hybrid steppers employ a special technique to provide small rotational steps. Figure 5 previously showed how a *rotor* with more than 2 poles can provide rotational steps smaller than would otherwise be possible, given a fixed number of stator poles. (In that example, we stepped 15 degrees per step, despite the fact that the stator poles were 30 degrees apart.)

If you try to design a rotor with a large number of poles, you'll find that there isn't enough room to fit the stator windings around the periphery. Hybrid steppers solve this problem by machining many *notches* into the stator poles, as shown in Figure 10. These notches create several *minipoles* for each stator winding.

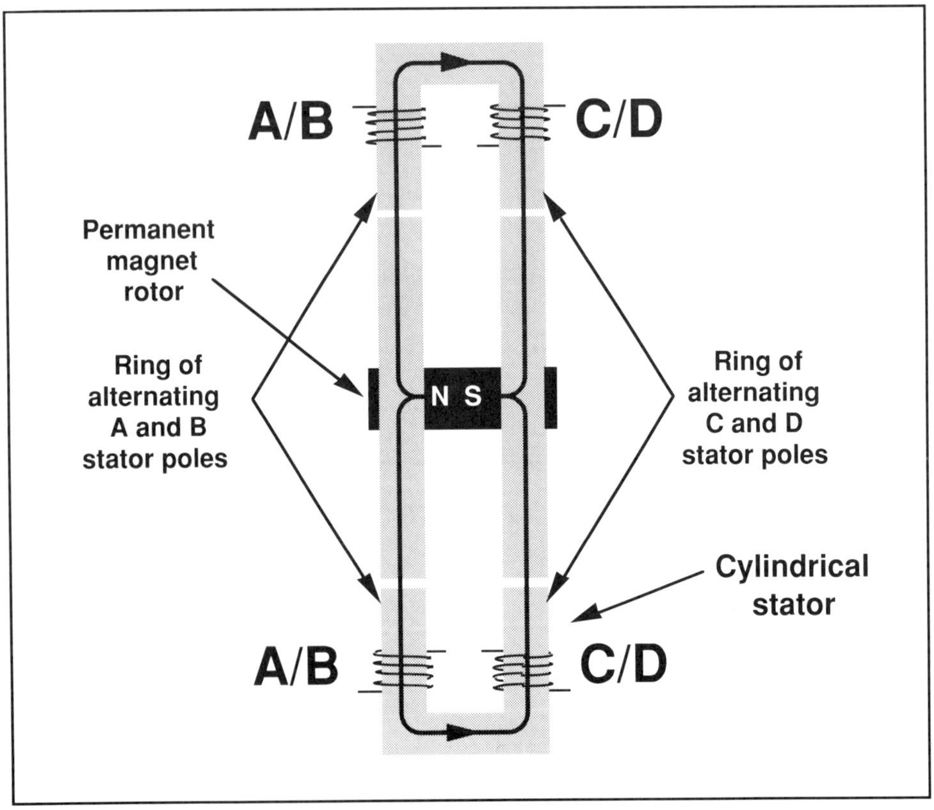

Figure 9

The minipoles of the even-numbered stator poles are shifted 1/2 step from those on the odd-numbered stator poles. This has the effect of *interleaving* the minipoles, just as if they were arranged in an alternating fashion around the periphery of the stator. As the rotor is walked, its poles align first with the minipoles on one stator pole and then with the minipoles on a different stator pole.

This technique allows very small incremental rotations. Some hybrid steppers offer step angles as small as .72 degrees. However, the most common step angle is 1.8 degrees, which provides 200 steps per revolution.

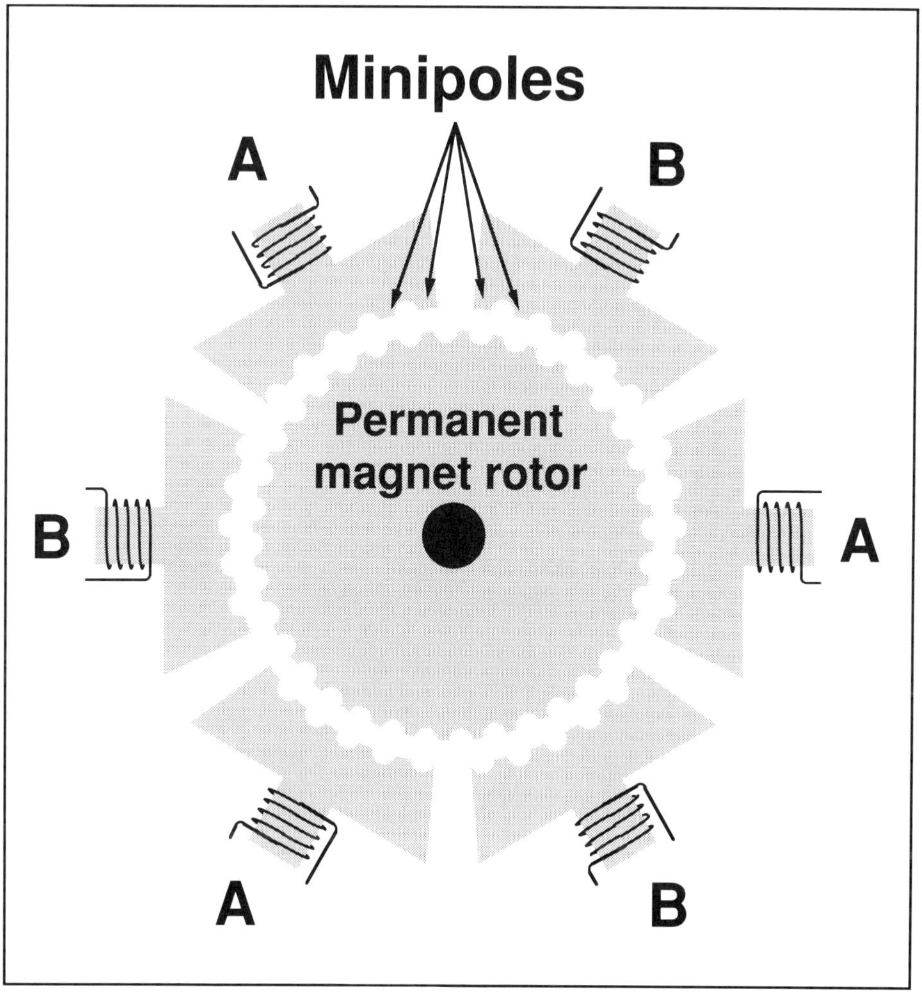

Figure 10

Unipolar versus bipolar motors

Most hybrid steppers are *unipolar* motors that have 4 windings, wired as shown in Figure 11. These are called *8-lead* unipolar motors. However, some unipolar steppers are wired as shown in Figure 12: a *6-lead* unipolar. Since you only need to drive current through a unipolar stepper's windings in one direction, you can use a simpler motor driver circuit.

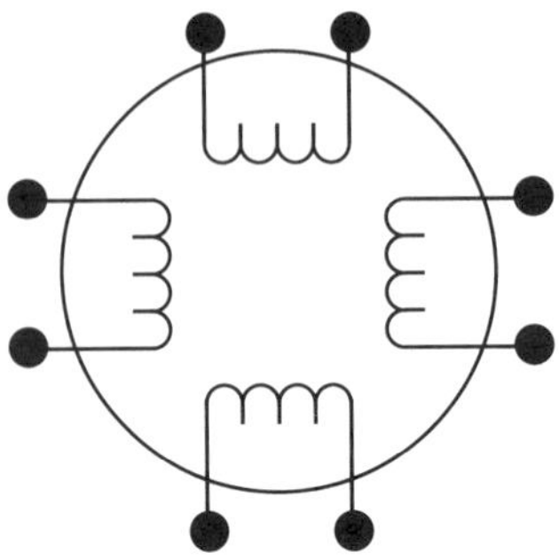

4 winding, 8 lead

Figure 11

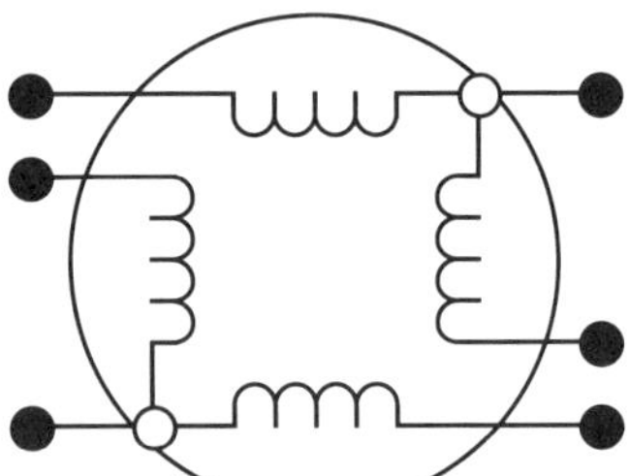

4 winding, 6 lead

Figure 12

Bipolar hybrid steppers are also available. Bipolar hybrid steppers have 2 windings, as shown in Figure 13. They work just like the permanent magnet stepper shown earlier in Figures 1, 2, and 3: the direction of the current through the windings is reversed to step the rotor.

Note: Unipolar 8-lead and 6-lead motors can also be used as bipolar steppers, simply by driving 2 of their windings in series. The 2 series windings look, electrically, like one of the windings in a bipolar motor.

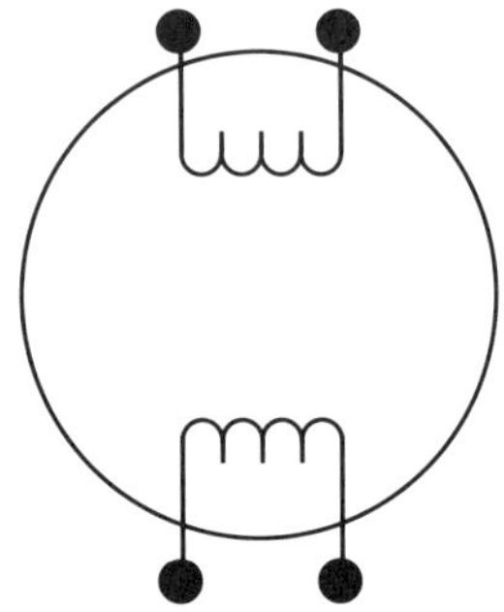

2 winding, 4 lead

Figure 13

Microstepping allows much finer resolutions

With the simple on/off approach described above, the resolution of movement is limited by the number of poles in the stepper motor. However, an *analog* approach permits better resolution. For example, you can position the rotor at several different points between 2 poles by simultaneously driving opposing windings with differing currents. This method is called *microstepping*.

Microstepping allows for very refined positioning. Movements of 1/125th of a step are possible. With 200 steps this allows 25,000 microsteps per revolution!

What does a stepper-based motion control system look like?

Figure 14 shows a typical computer-driven motion control system that uses a stepper motor. It consists of 6 basic components:

- A computer
- A controller (sometimes called an indexer)
- A driver
- A DC power supply
- A stepper motor
- A mechanical linkage, which drives the load

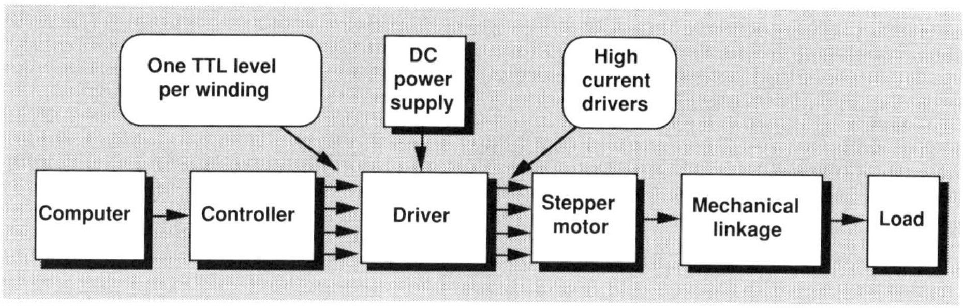

Figure 14

When the computer writes into the controller's registers, the controller changes the levels of its 4 TTL output signals. Each of these 4 output signals corresponds to one of the stepper's 4 windings. Upon detecting an active TTL level, the driver energizes the corresponding winding, causing the rotor to rotate.

Figure 15 shows another type of system. It's similar to the one in Figure 14, except that it uses a *translator* instead of a driver. The controller in this system has only 2 TTL output signals: a *direction* signal and a *step* signal. The level of the direction signal indicates the direction of the stepper motor rotation, and the step signal indicates when the step is to take place.

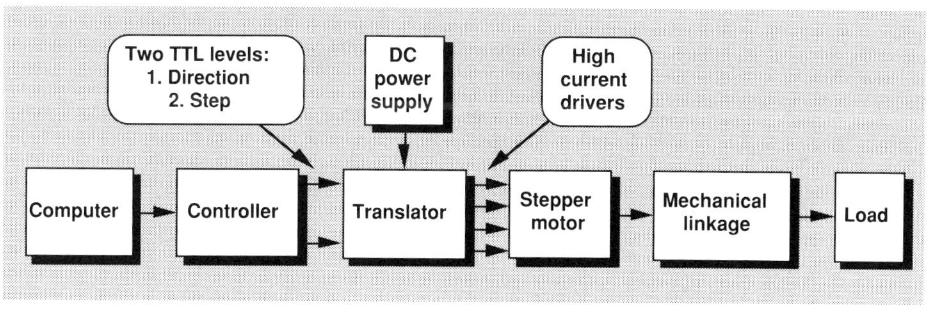

Figure 15

Designing a stepper-based motion controller

In designing a motion controller, you start by evaluating the load, and then work backwards. This process involves 4 steps:

- choosing the mechanical linkage
- choosing the stepper motor
- choosing the motor driver
- choosing the motor controller

Choosing the mechanical linkage

The first element to consider is the mechanical linkage. You may need to include some kind of special linkage between the stepper motor and the load. For example, you may need linear motion, instead of the stepper's rotary motion, or you may need more torque than a stepper can provide. There are several popular mechanical linkages:

- *Leadscrews* convert rotary motion into linear motion. They're often used where long linear motions are needed, such as on X-Y tables.
- *Gear drives* can be used to increase the torque or the speed of the stepper, much like an electrical transformer is used to step up the voltage or the current of an AC signal.
- *Tangential drives* (such as conveyer belts or pulleys) can also be used to convert rotary to linear motions.

Choosing a stepper motor

Once you have chosen a mechanical linkage, you must then decide what type of stepper motor to use. In choosing the stepper, you should consider the following factors:

- How much static friction must be overcome?
- How quickly do you need to accelerate the load?
- How fast must the motor run?
- How much holding torque do you need?
- How quickly must the load be moved?

Static friction

The total static friction (*stiction*) is the sum of the stiction of:

- the stepper
- the mechanical linkage
- the load

This stiction dictates the minimum *starting torque* that the stepper must provide, just to get the load moving.

Load acceleration

Assuming that the motor has adequate torque to overcome stiction, the maximum load acceleration will depend upon:

- The rotor's inertia
- The mechanical linkage's inertia
- The load's inertia
- The starting torque of the motor

Since the inertia of the rotor, the mechanical linkage, and the load are typically hard to change, the only option you may have open to you is the choice of the motor.

Motor speed

You will typically need to move the load at some prescribed velocity. That velocity (along with the mechanical linkage) will dictate how fast the motor must be able to run. You must choose a motor with adequate running speed.

Holding torque

Once the load has been moved into position, you will often need to provide a specified holding torque to keep it there. For example, when a stepper positions an object for machining, the cutting tool might generate a force on the machined object, which manifests itself as a torque on the motor shaft. The stepper must have enough *holding torque* to maintain the machined object in its proper position.

Selecting the motor driver

Once you have selected a stepper motor, you must then select a driver (or translator) to drive the windings of the stepper motor.

Figure 16 shows how a unipolar motor driver energizes the stepper's windings. S1 to S4 represent open collector drive transistors, which are turned on to energize each winding. Figure 17 shows the sequence in which these transistors are turned on. Notice that, in each static position, the rotor is held in position by 2 energized windings. In other words it sits midway between 2 of the stator's poles. Rotation in the opposite direction can be accomplished by reversing the sequence shown in Figure 17.

Note: The degrees of rotation that result from one pass through the chart in Figure 17 depends upon how many poles the stepper has.

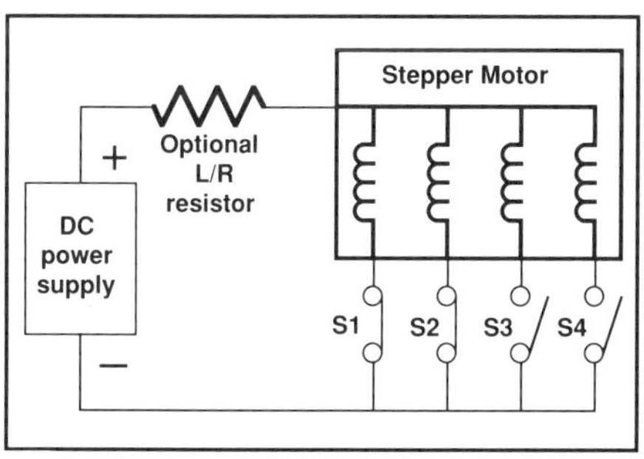

Figure 16

Step	S1	S2	S3	S4
1	ON			ON
2	ON	ON		
3		ON	ON	
4			ON	ON
1	ON			ON

Figure 17

A bipolar stepper is driven in a different fashion, as shown in Figure 18. Figure 19 shows the sequence. This driver can drive current through the windings in either direction. The reason for putting a bipolar driver on both ends of the windings (instead of just connecting one end of each winding to ground) is that it only requires 1 DC power supply.

Because the rotor of a bipolar stepper is a permanent magnet, it can be simultaneously attracted by one pole, and repelled by another. For this reason a bipolar stepper has a lot of torque at low speeds.

When selecting a driver/translator, it's important to choose the proper type of output stage. There are 3 types:

- L/R
- Bi-level
- Chopper

Stepper motor manufacturers typically provide their customers with graphs that show the torque of their steppers, as a function of speed. However, the actual torque also depends upon what kind of circuit is *driving* the stepper.

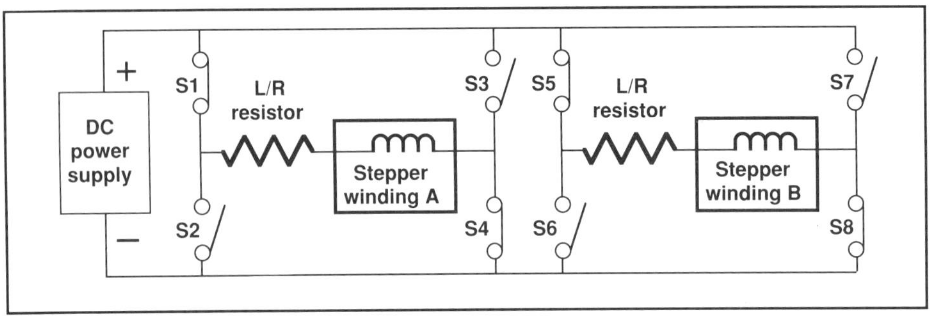

Figure 18

Step	S1	S2	S3	S4	S5	S6	S7	S8
1	ON			ON	ON			ON
2		ON	ON		ON			ON
3		ON	ON			ON	ON	
4	ON			ON		ON	ON	
1	ON			ON	ON			ON

Figure 19

For example, the torque/speed curve of a stepper driven by an L/R driver is different from the curve when it's driven by a chopper driver. In view of this, some manufacturers provide two curves: an L/R curve, and a chopper curve.

L/R motor drivers

L/R drivers are the most widely used. In an L/R driver, a current limiting resistor is inserted in series with the stepper's windings, as shown in Figure 18.

Each winding in a stepper has a certain inductance and a certain resistance. The winding resistance (and the power supply voltage) determines the DC current level through the winding. Given a fixed drive voltage, a stepper motor designer can increase the holding torque of a stepper by either *decreasing* the winding resistance, or by *increasing* the number of turns on each pole of the stator. The latter approach is generally better, since it doesn't increase the steady state power dissipation appreciably.

However, as you add turns to the motor's windings, you also increase the winding *inductance*. With a large inductance it takes longer for the winding current to rise to its steady state DC level. When the motor is running very fast, the winding current doesn't have time to reach its steady state level before it's turned back off. Since motor torque is proportional to the winding current, the torque falls off at higher speeds, and the motor runs slow under load.

You can speed up a stepper by increasing the DC power supply voltage used to drive the windings. For a given inductance, this causes the winding current to rise more quickly. However, the winding's *steady state* power dissipation is proportional to the *square* of the drive voltage. That produces an unwanted side effect: when you stop stepping the rotor, the motor's windings burn out.

Using a limiting resistor

To prevent the motor from burning out, you can insert a current limiting resistor in series with each of the motor's windings. When you first turn on a winding, the current through the limiting resistor is zero. Since the drop across the resistor is zero, the full power supply voltage is applied across the winding.

Since the $\partial i/\partial t$ of the current through the inductive winding is proportional to the voltage across it, this causes the current to rise rapidly. As the current increases, the drop across the resistor also increases, slowing the rise of the current. Eventually, the current levels off at a level that provides adequate holding torque, without burning out the motor.

A typical series resistance value is 5 times the rated winding resistance. With this resistance inserted, the DC drive voltage can be increased by a factor of 6, without risking burnout of the winding.

With a typical stepper, this technique doubles (or possibly even triples) the speed at which the torque of the stepper falls below a given value. If this is not enough, the resistance could be increased to 8 times that of windings (with a corresponding increase in the DC drive voltage). That can increase the speed by a factor of six.

Figure 20 shows the torque/speed curves for a typical stepper with different series resistor values. Other steppers may have different curves, but their curves will generally be of the same shape as those in Figure 20.

L/R driving techniques are useful for motor speeds up to 2500 steps per second. However, if you try to apply this technique to obtain higher speeds, you run into problems. Each time you double the series resistance (and the power supply voltage) you double the power dissipation of the resistor. That calls for a high-wattage resistor, and a much more expensive power supply. It also means that almost all of the power being used by the system is simply converted to heat by the series resistor, without doing any useful work.

Bi-level motor drivers

A bi-level driver provides a more elegant solution. It applies an initial pulse of high voltage to get the winding current up quickly. Once the winding current reaches a level adequate to provide the needed torque, the bi-level driver switches off the high voltage, and maintains the current with a lower voltage source. This eliminates the need for series resistors.

Bi-level drivers are more expensive than L/R drivers because they must have 2 power supplies. For that reason they aren't the best choice for low-speed stepper operation. However, at speeds above 2500 steps per second, they provide adequate torque without the inefficiencies of L/R drivers.

Chopper motor drivers

Ideally, you would like to have a motor that provides a perfectly flat torque curve. In other words, you want the motor to deliver the same torque, no matter what speed you run it. Since the torque of the motor is proportional to its winding current, you would

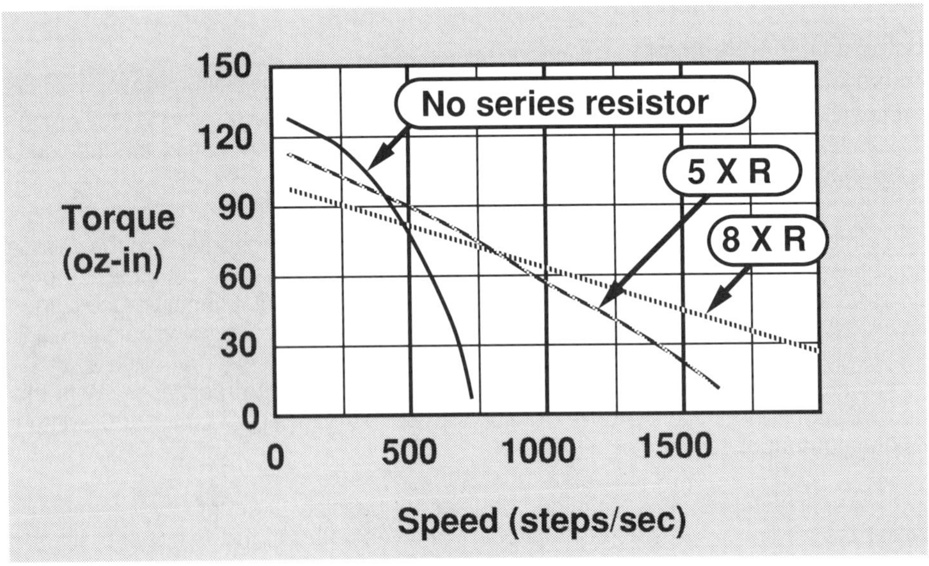

Figure 20

need to drive the motor from a constant current source. That's the idea behind chopper drivers.

A chopper driver drives the stepper with a high-voltage, pulsed DC waveform. When the driver first applies this high voltage to the inductive winding of the stepper, the winding's impedance is high, and so the current through the winding is low. However, when the current through the winding rises to a specified level, the driver begins to switch the DC off and on, generating a pulsed waveform. The duty cycle of this waveform determines the average current through the winding, providing the desired torque.

Chopper drivers have historically been more expensive than bi-level drivers. However, since they offer excellent torque at very high speeds, they're often the best choice.

Figure 21 shows the speed versus torque curves for a typical stepper motor, when driven by L/R, bi-level, and chopper drivers. Unlike L/R curves (which tend to have the same general shape, regardless of the stepper motor used) bi-level and chopper curves vary widely from one model of stepper to the next. Some curves may look quite different from the ones shown here.

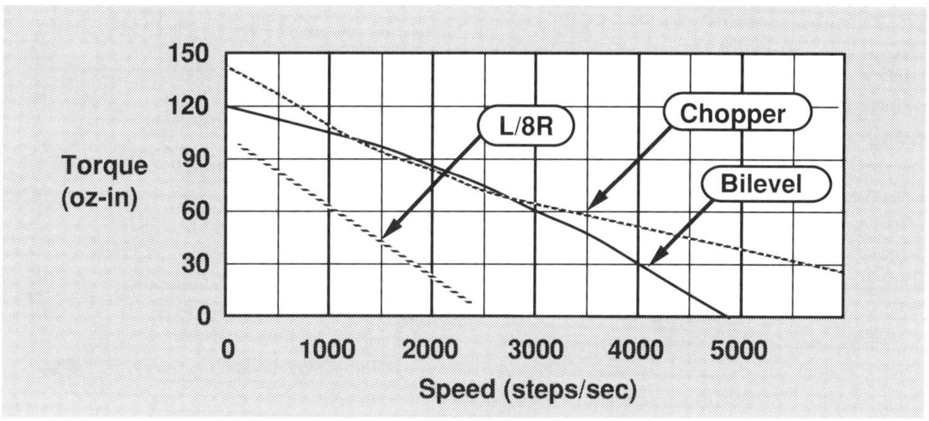

Figure 21

Selecting the motor controller

Once you have chosen a driver (or translator) for the stepper, you'll need to select a controller. Some applications only require a very simple stand-alone controller. It's preprogrammed with thumbwheel switches, and it simply steps through a repetitive series of motions. However, other applications may require an *intelligent* controller, which can accept motion commands from a computer in real time. Intelligent controllers are useful in control systems that are expected to tailor their motions to a changing environment.

Intelligent controllers come in 2 basic forms:

- those that receive commands through an I/O port: often an RS-232 port.
- those that plug into a backplane.

Because the latter are more tightly coupled to the computer, they're typically more flexible. They're also less expensive, since no additional enclosure is needed to house them. For these reasons, Matrix Corporation offers plug-in controllers.

Matrix motor controllers
Matrix offers 2 basic models of intelligent controllers.

- The MS-SSC, which is a single-channel stepper controller
- The MS-DSC, which is a dual-channel stepper controller.

Both boards are single-height intelligent VMEbus slaves.

When the VMEbus system is powered up, these boards first execute an internal initialization sequence. Then, the main MPU board in the VMEbus system tells them what type of stepper they'll be driving. After configuring themselves for the motor type, they're ready to act upon any commands that the MPU board writes into their on-board registers.

The commands supported by the SSC and DSC boards are shown in Figure 22. Once the MPU board writes a command into the controller's registers, execution of the command is automatic. This allows the MPU board to work on other tasks. The stepper controller boards can be programmed to generate a VMEbus interrupt when they finish executing the command.

Comand list

1. **Single step**
2. **Move at constant speed**
3. **Move at constant speed until limit switch**
4. **Move at full speed until limit switch**
5. **Move to base point**
6. **Accelerate/Decelerate**
7. **Decelerate and stop**
8. **Emergency stop**

Figure 22

In addition to normal single stepping, the MS-SSC and MS-DSC allow the stepper to be moved in half-step increments. In either case the maximum step rate is limited only by the motor: the boards can drive motors at rates exceeding 10,000 steps/sec.

Some applications use limit switches. One common way to do this is to have the MPU board monitor the status of a limit switch, and issue start and stop commands to the controller board when necessary. However, the MS-SSC and MS-DSC controllers are intelligent, and can monitor limit switches, without support from the MPU board.

Each stepper channel can be equipped with up to 5 limit switches. The *base point* limit switch defines a *home position*. Two limit switches can be positioned to control motion in the clockwise direction. In that case, the 1st switch causes the controller to begin a programmed deceleration, and the 2nd switch serves as an emergency stop. The other 2 limit switches can be positioned to control counter-clockwise rotation, in a similar fashion.

The MS-SSC and MS-DSC provide 4 TTL output signals, which can be used to control either 3-phase or 4-phase steppers. These output signals are ribbon-cable compatible with Matrix Corporation's unipolar driver board (USD) and bipolar driver board (BSD). The USD is an L/R driver that sinks up to 8 amps per phase, and can switch up to 70 volts. The BSD is a bipolar L/R driver that provides up to 5 amps per phase, at up to 50 volts.

The MS-SSC and the MS-DSC also provide step/direction TTL output signals. These outputs are useful when using drivers other that Matrix's USD or BSD boards, such as:

- a special driver to drive a 5-phase stepper
- a chopper driver for high-speed applications
- a microstepped driver for accurate positioning.

Figure 23 shows a system for controlling one stepper of an X-Y table. The MS-DSC plugs directly into the VMEbus chassis. A 26-pin ribbon cable connects channel 0 of the MS-DSC to the unipolar motor driver (USD). The USD has a terminal block that allows insertion of L/R resistors in series with the windings of the unipolar stepper.

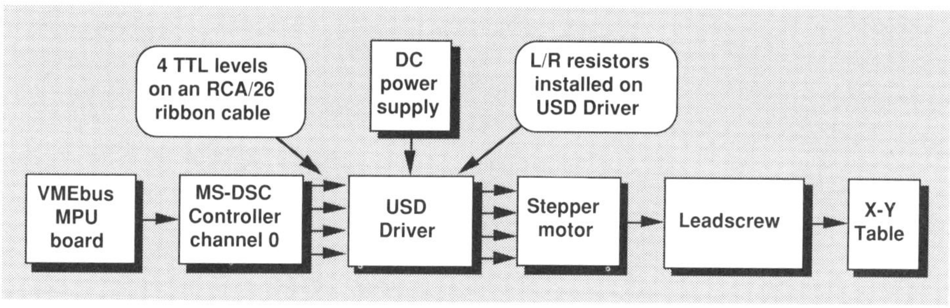

Figure 23

The resulting modular VMEbus system is compact, powerful, and inexpensive. The USD's 8-amp current drive allows the use of almost any stepper motor. The intelligence and programmability of the MS-DSC controller allows it to be programmed to operate the stepper in virtually any ramp-up/run/ramp-down pattern. At the same time, this system is considerably less expensive per motor than most packaged motion control systems.

Bill Gowan is engineering manager for Matrix Corporation. He has a BSEE from North Carolina State University, and is a licensed professional engineer. Bill has been involved in development of new board-level products for STD bus and VMEbus since 1983. His hobbies include handball and backpacking.

CHAPTER 39

Motion Control With Servo Motors

By Jacob Tal

In the past, motion control systems have been designed from scratch by servo experts. However, with the availability of sophisticated off-the-shelf components, the construction of motion control systems has been simplified, and can now be done by non-specialists. In this article, Jacob describes how you can use off-the-shelf components to design your own customized motion control system.

Motion control systems can be classified into 2 groups:

- Open loop systems
- Closed loop systems.

Open loop systems typically use stepper motors to provide the required motion. Since the stepper motor is rotated in an incremental fashion, there's no need to *sense* its rotational position.

Open loop systems provide 2 advantages:

- They're simple
- They're inexpensive

Closed loop systems, on the other hand, use *position sensors* to continuously monitor the position of the *actuators*. This feedback allows the system to continuously correct the position of the load, as necessary, regardless of any external disturbances.

Closed loop systems provide 2 advantages:

- They provide tighter control of the positioning of the load, especially in the presence of varying loads and disturbances.
- They provide greater design flexibility

Closed loop motion control systems may use various actuators including:

- Electrical actuators
- Hydraulic actuators
- Pneumatic actuators

Although the actuators may vary, the basic principles of closed loop operation remain the same. In this article I will focus on the use of electric motors.

The system elements of a closed loop system
Figure 1 shows a closed loop motion control system. The *amplifier* (often called the *driver*) accepts a ±10 volt analog signal from the *motion controller* and amplifies it to provide a proportional *current*. That current is then used to drive the *servo motor*.

Note: The instantaneous motor drive current determines the *torque* that the motor provides. In typical applications the peak drive current may sometimes go as high as 10 Amps.

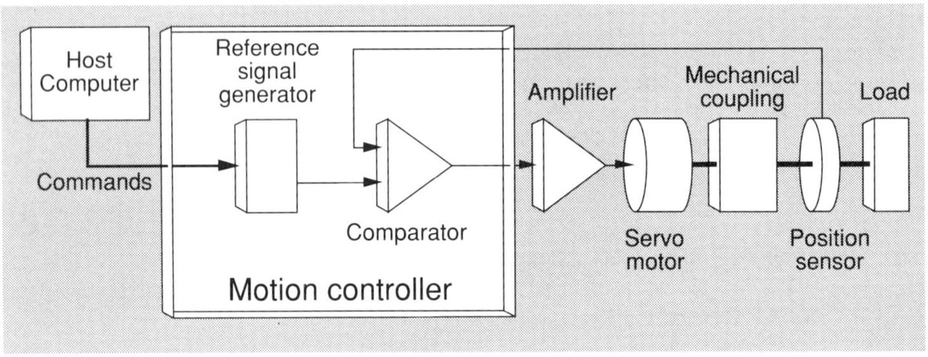

Figure 1

If the *load* is coupled directly to the motor shaft, the *position sensor* can also be attached directly to the same shaft. However, if a mechanical coupling is used, it may absorb some of the shaft's rotation under high torque. In that case you may need to couple the sensor directly to the load, in order to get an accurate measure of the load's position. In either case, the analog positional feedback signal from the sensor is routed back to the motion controller.

As shown in Figure 1, the simplest type of motion controller includes an analog *comparator* that compares the level of the position sensor's feedback signal to the analog *reference* signal. This comparator produces an analog output signal equal to the *difference* between these 2 analog inputs. (This difference is called the *position error.*) This analog error signal is then amplified, and used to drive the motor in the direction that reduces the positional error.

If the reference signal is constant, the motor's movement will eventually reduce the comparator's output to essentially zero. However, if the reference signal is constantly *changing* (as it often is in a motion control system) the comparator's output will also be constantly changing, driving the motor with the analog signal needed to make its position track the waveform on the reference input.

However (due to inertial effects) the motor won't be able to respond instantaneously to rapid changes in the reference signal. As a result, transitory position errors will occur. The magnitude of these errors (and the speed with which they are reduced) depends upon the *dynamic* characteristics of the system. These dynamic characteristics include:

- the propagation delay through the comparator.
- the propagation delay through the amplifier.
- the mechanical inertia of the motor and the load.

Since the error signal produced by the comparator is routed through the amplifier (to drive the motor) the *gain* of the amplifier determines how *vigorously* the control system corrects any positional errors. The value of this gain is important. If it's too low, the motor may not have enough torque to overcome the *friction* in the load, and in the mechanical linkages. On the other hand, if the amplifier gain is too high, the system will *overreact* to position errors, driving the motor so vigorously that it *overshoots*, due to inertial effects. Whenever this overshoot occurs, the motor must then be driven in the *reverse* direction, to restore it to the desired position.

You can prevent this inertial overshoot by reducing the gain. However, if the load has a lot of inertia, you'll need to reduce the gain to a very low value. This will cause the load to respond sluggishly to sudden changes in the *reference* signal.

To improve the system response, you can insert a *filter* between the comparator and the amplifier, as shown in Figure 2. This filter includes 2 elements:

- A programmable gain *amplifier*
- A programmable gain *differentiator*

This filter can be represented by the following equation:

$$\text{output} = K_1\,(\text{input}) + K3\ \partial/\partial t(\text{input})$$

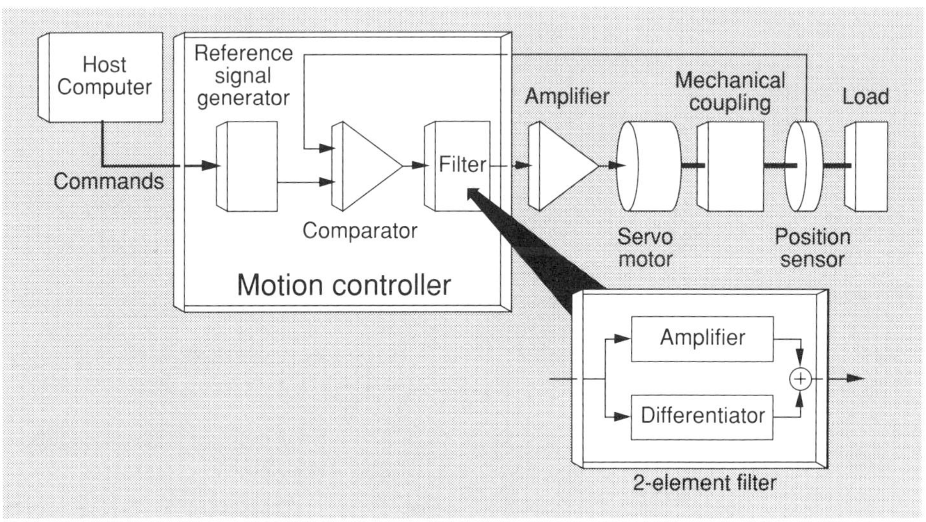

Figure 2

This filter accepts the analog error signal from the comparator and passes it along to the amplifier. However, in doing so, it also adds the *time derivative* of the error signal. The filter has the effect of *accentuating* any *changes* in the error signal. A *rapid change* at the filter input produces a *large change* at the filter output.

Figure 3 shows ideal input and output waveforms for such a filter. Note the "spike" which occurs at the output of the filter whenever a rapid transition occurs on its input. This spike injects a momentary surge of current through the motor, producing a high torque, and causing it to accelerate rapidly.

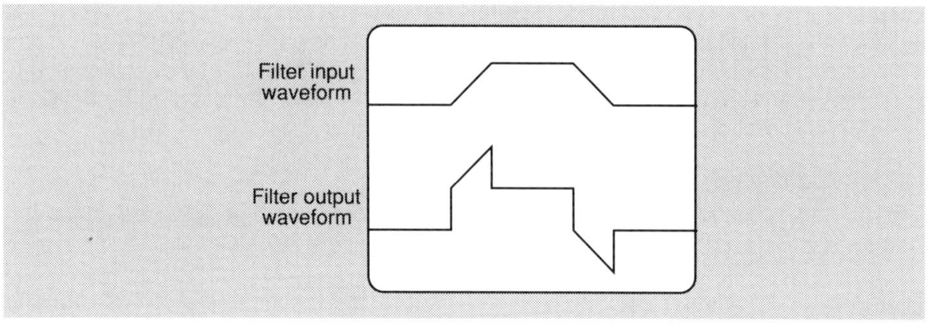

Figure 3

While the above filter can improve system *response times*, by making it more responsive to sudden changes in the reference signal, it can also create problems. If the position of the load is suddenly changed by some *externally induced* mechanical disturbance, the comparator will produce a sudden change in its output error signal. This sudden change in the error signal (induced by the *external disturbance*) is indistinguishable from a sudden change induced by a change in the *reference* signal.

Upon receiving this sudden input change, the differential filter produces a current surge, to restore the motor to its original position. However, if the external mechanical disturbance is *cyclic*, it may quickly restore the motor to its original position. The resulting current surge then serves only to drive the motor *beyond* its original position. This produces a positional error in the *opposite* direction.

If the cycle time of the external cyclic disturbance is about the same as the loop delay time, the high-gain amplification effect (from the differentiator in the filter) can produce a *self-sustained oscillation.*

One way to suppress this oscillation, is to increase the contribution of the differential element of the filter by increasing K_3. This results in a steeper response to position error. This steeper response effectively shortens the response delay, and helps stabilize the system.

Although such a filter produces a fast and stable response, it doesn't guarantee accurate *positioning* of the load. This is because, while the motor will drive the load *close* to the desired position, it will stop as soon as its torque (produced by the falling position error) drops to the level of the friction. As a result, the motor might stop anywhere within a given range, known as the *dead band.*

To eliminate this positional error, you need to introduce a 3rd filter element: the integrator. (See Figure 4) If the motor is stuck at a position a slight distance from the desired position (due to friction) the small (but continuous) position error to the integrator causes its output to increase continuously, until the torque of the motor finally rises to the point where it overcomes the friction, and the motor moves the load. As a result, the integrator assures that the motor ends up exactly at the correct position.

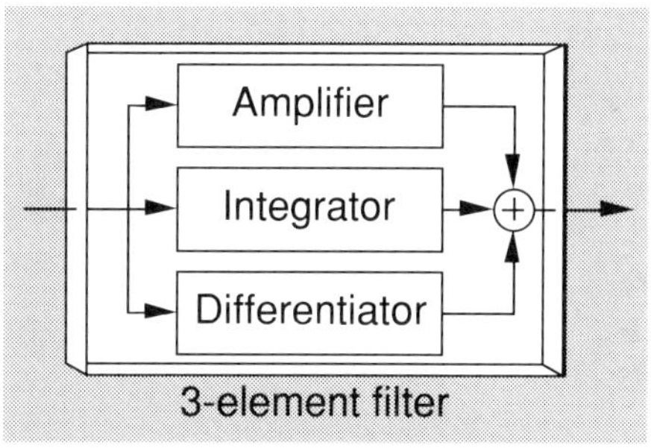

Figure 4

A 3-element filter can be described by the following equation:

$$\text{output} = K1\ (\text{input}) + K2 \int(\text{input}) + K3\ \partial/\partial t(\text{input})$$

...which has 3 terms...

- A Proportional term
- A Integral term
- A Differential term

...and is often called a *PID filter*. The combination of the proportional and differential terms provide *stability* and *fast response*, while the integral term provides *accurate positioning*.

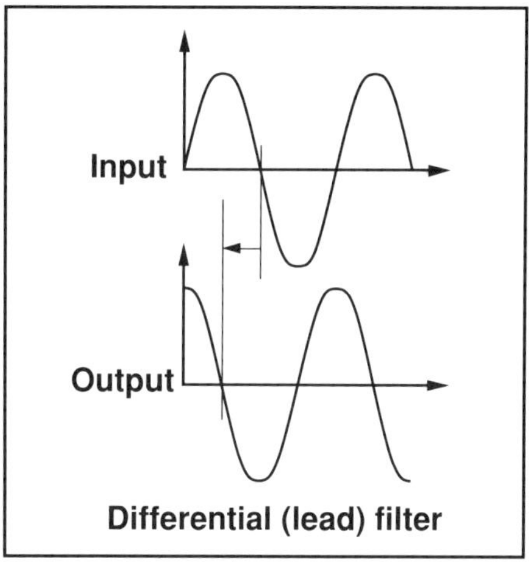

Figure 5

Another way to describe this filter is to describe its response to *sinusoidal* inputs. Figure 5 shows the output produced by the *differential* filter element when a sinusoidal waveform is presented to its input. Its output is also sinusoidal, but the waveform *leads* the input by 90 degrees. For this reason the differential filter is sometimes called a *lead filter*. Figure 6 shows the output of the *integral* filter when a sinusoidal waveform is presented to its input. Because its output waveform lags behind the input by 90 degrees, it is called a *lag filter*.

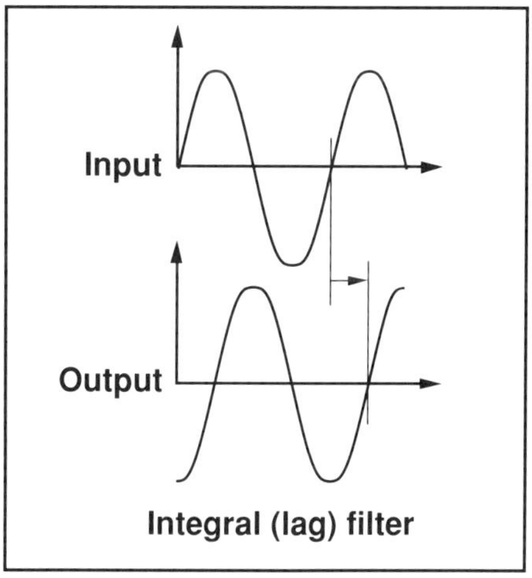

Figure 6

When proportional, integral, and differential filter elements are combined (to form a PID filter) they are often collectively called a *lead-lag* filter.

Designing a motion control system

The design procedure for motion control systems includes 3 steps:

- Selection of the system elements
- System integration and tuning
- Motion programming

Step 1: Selection of the system elements

Because motion control hardware is designed in a modular fashion, you can purchase a servo motor, an amplifier and a motion controller from 3 different vendors, and then integrate them into a working system. However, you need to know how to select components that are appropriate for your particular application.

Motor selection

There are 2 basic types of servo motors:

- Brush motors
- Brushless motors

Unless you have some special application, you'll probably find the relatively low-cost brush motors adequate. However, you may need to use brushless motors in applications where they must run:

- in a vacuum.
- in a clean-room.
- in an explosive environment.
- at high speeds (above 4000 RPM).

Once you choose the motor *type*, you must then choose a motor *size*. There are 2 main parameters to consider here:

- the maximum required *torque*
- the maximum required *speed*

The maximum torque and speed requirements depend upon the characteristics of the load, and the mechanical linkage, which might include gearing.

For example, suppose you need to rotate a load with a moment of inertia of 50 oz-in-sec^2 half a revolution in one second. This is a rather slow motion, so it makes sense to gear down the motor. (This gearing will allow you to use a smaller motor, with less torque.)

At a given ambient temperature (and under a given load) a servo motor may have a *maximum duty cycle*. If you fail to take this duty cycle into account, you may overheat the motor. In this application we will assume that the motor spends 1.00 second rotating the load, and then 0.35 seconds idle.

Given a list of servo motors of varying sizes (torques and speeds) you want to select the smallest motor that will do the job without overheating. Once you have done that, you'll need to figure out the *gear ratio* needed to best match the torque/speed characteristics of the motor to the load. Finally, you'll need to compute how much motor drive current and drive voltage the amplifier needs to provide.

This can all get rather complicated. Fortunately most servo motor suppliers provide their customers with computer-based selection software. This software helps you analyze your motion requirements, and select the most suitable motor.

One example of such a program is the Motion Component Selector (MCS) program from Galil. This program prompts you for important parameter values and then produces a "report card", indicating the suitability of any particular Galil motor for that particular application. In the above example you would enter the load inertia, the speed and the duty cycle of the motion. The program then scans its internal list of motors, and selects the ones that are most suitable. You then select one of these motors, and the program computes the optimum gear ratio (to match the motor's torque and speed to the load) and specifies how much drive current and drive voltage are needed. Figure 7 shows the screen display produced for the application described above.

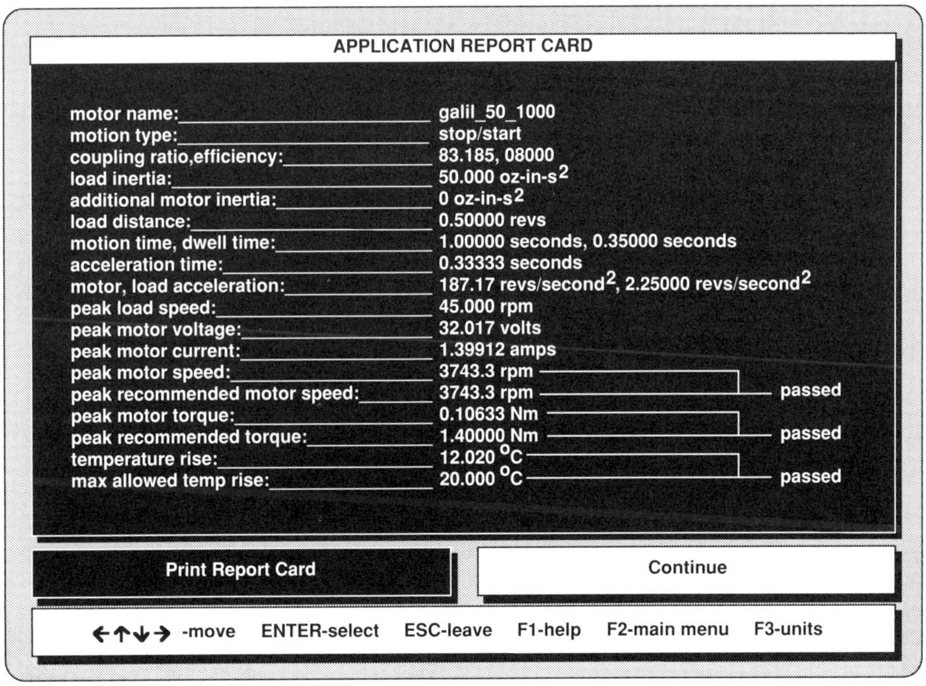

Figure 7

Amplifier selection

Your choice of amplifiers should be based on the current and voltage drive requirements of the motor you have chosen. Specifically, you should determine:

- the maximum DC *current* that the amplifier can source and sink
- the maximum DC *voltage* that the amplifier can deliver.

The *torque* provided by a servo motor is proportional to the *drive current* flowing through it. The proportionality constant is called Kt. Thus, the current required to provide torque T_g is:

(1) $I = T_g/K_t$

For an ideal servo motor, the rotational speed of the motor (S) is proportional to the voltage across its windings. However, real servo motors have a non-zero winding resistance (R) which must be taken into account when computing motor speed. Thus, the voltage required to drive the motor at a speed S is:

(2) $V = (Kt \times S) + (I \times R)$

For example, suppose you choose a servo motor with the following specifications:

Tg = 50 oz-in

Kt = 10 oz-in/Amp

S = 3000 RPM

R = 2 Ω

You first convert these parameters to a consistent set of units, such as the MKS units...

Tg = 0.355 Newton meters

Kt = 0.071 Newton meters/Amp

S = 314 radians/second

R = 2 Ω

Then, using Equations (1) and (2)...

I = (0.355)/(.071) = 5 Amps

V = (0.071 x 314) + (5 x 2) = 32.3 Volts

You can then use these computed current and voltage drive requirements to choose a servo amplifier.

Note: It's a good idea to add a 25% margin to these computed current and voltage values.

Selecting a sensor
Your sensor choice will depend upon the type of mechanical motion you are trying to control. If the motion is rotational, you will probably want to sense it with a rotational position sensor. This sensor can be attached directly to the motor shaft.

There are 3 types of rotational sensors:

- Incremental encoders
- Absolute encoders
- Resolvers

Incremental encoders are optically-based devices. As the encoder's shaft rotates, a transparent disk inside it rotates between a photosensor and a light source. Lines inscribed on the disk generate pulses as the shaft turns. Incremental encoders are the most commonly used rotational sensors, since they're simple and inexpensive.

Absolute encoders are rarely used anymore because they're relatively expensive. They're used only when very high precision is required; 16 to 24 bits per revolution.

Resolvers are magnetically-based devices. Although they're more complex to use, they're becoming more popular because they provide *absolute* position information, and because they're relatively *rugged*, allowing them to be used in harsh environments. Since a resolver is essentially an analog device, its resolution depends upon the electronic circuitry used to decode its position signal. Typical resolutions range from 12 to 16 bits per revolution.

If the servo motor is being used to control *linear* motion in your system, you'll probably want to use a sensor designed to measure linear displacement. There are 2 common types of linear position sensors:

- Incremental linear sensors
- Magnetic linear sensors

Incremental linear sensors work on the same principles as the incremental encoder, except that they use a rectangular glass plate, instead of a glass disk. They provide precise position sensing, but they're not as rugged as magnetic linear sensors.

Magnetic linear sensors work on the same principles as resolvers. They're more rugged than incremental linear sensors, but they're also less precise.

Regardless of whether you use a rotational sensor or a linear sensor, you should consider both its *resolution* and its *accuracy*. A sensor's *resolution* specifies the minimum *movement* it can detect. Its *accuracy* specifies how precisely it can indicate the absolute linear or rotational *position*.

It's good practice to select a sensor with a resolution about 4 times better than the positional accuracy you need to maintain. This allows the motion controller to detect small position errors (and respond to them) before they exceed the allowed positioning tolerances.

Note: Using sensors with a higher resolution than this offers no perceptible benefits, and may even cause system instability. This is because the digital counter that's driven by the digital output of a high resolution sensor produces a larger count change, for a given angular rotation. This wider swing effectively increases the gain of the system and (unless it can be compensated for elsewhere in the control loop) may lead to oscillations.

Controller selection

Motion controllers have now become sufficiently standardized to allow you to purchase them from any of numerous sources. In choosing your motion controller, you should take into account 4 important factors:

- The host interface requirements
- The number of motors that must be controlled
- The type of coordination between the motors
- The velocity profiles

The host interface that you need depends upon what type of computer you plan to use to control the motor. The 3 most common host interfaces are the PC-bus, the VMEbus and an RS232 serial port.

Step 2: System integration and tuning

Once you have selected the servo motor, the amplifier, the sensor, and the motion controller, you'll need to couple them together, to form the complete motion control system. Theoretically, if you know the inertia and friction of the load (and of the mechanical linkages) you can compute optimum values for the PID filter's K1, K_2, and K_3. The resulting system would provide optimum *accuracy* and *stability*, as well as a *fast response*.

Unfortunately you rarely know the system characteristics precisely enough to use this theoretical approach. As a result, you will need to *tune* the system for optimum performance by adjusting the 3 constants (K_1, K_2, and K_3) until the system exhibits acceptable accuracy, responsiveness and stability. Note: There are several methods for accomplishing this. References [1] and [2] describe some of them.

There are also several computer programs that perform this tuning *automatically*, by measuring the system's response to a *step function* on the reference signal. One example is the Servo Design Kit software package from Galil, which tunes the PID filter and then tests the resulting system response. Figure 8 shows a typical step response of a system.

Once an acceptable set of PID filter parameters has been chosen, the control system is ready for programming.

Step 3: Motion programming

In addition to closing the control loop (to ensure that the load tracks the input reference signal) a motion controller should be able to accept and execute a variety of *motion commands* from the host computer. This allows you to specify the desired motion trajectory as a string of commands. Ideally, these commands would be represented by ASCII mnemonics, such as BG to begin a programmed motion and ST to stop the

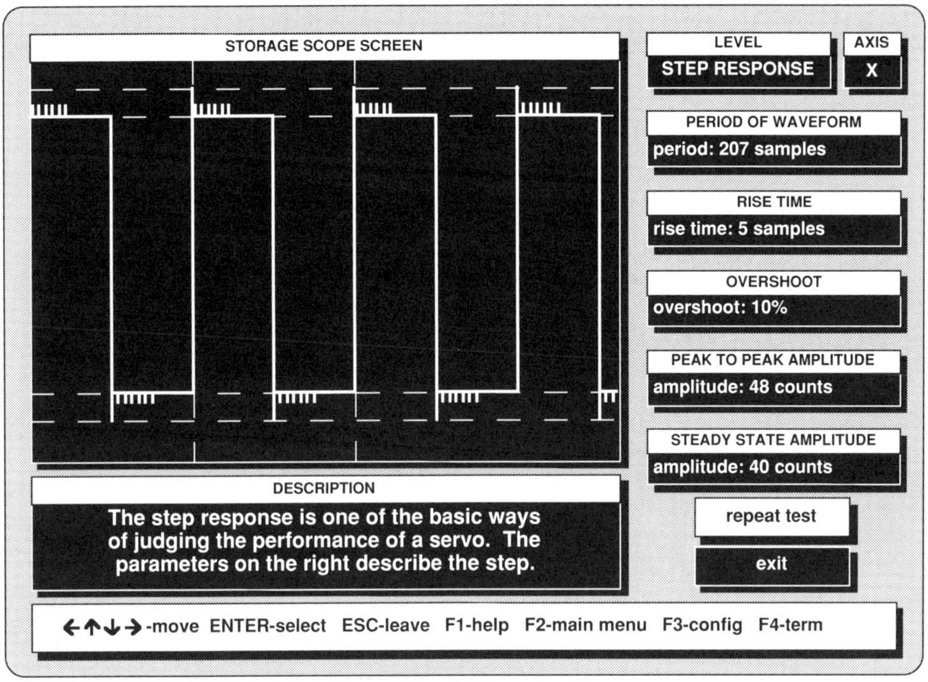

Figure 8

motion. If the motion controller can control several motors, BGX might begin the motion of motor X, BGY the motion of motor Y, and BGZ the motion of motor Z.

For example, the Galil DMC-530 motion controller board (shown in Figure 9) controls up to 3 servo motors, and accepts commands from its host, via dual-ported VMEbus registers. The commands may be sent to the controller board either in binary or in ASCII form. The following discussion provides some examples.

Example 1: One-dimensional motion
Suppose you want to move a fragile load a specified distance along a linear track. This move requires exactly 10 revolutions of the servo motor. If the encoder resolution is 4000 counts per turn, you'll need to move a total distance of 40,000 counts.

Figure 9

The initial velocity of the load is zero. It will first be accelerated and then decelerated, with its velocity falling to zero at the final position. Since the fragile load will only tolerate an acceleration of 1000 revolutions per second^2 (4,000,000 counts per second^2) you'll need to specify that as the maximum. To keep the velocity from becoming too high at the midpoint of the movement, you will also need to specify a maximum velocity. A velocity of 50 revolutions per second (200,000 counts per second) is appropriate.

The resulting velocity profile is trapezoidal, as shown in Figure 10. The ASCII commands required to accomplish this motion are shown in Figure 11. These commands can be sent to the board in either binary or ASCII form. While the binary form is faster, the ASCII form is easier to understand. To keep things simple here, I have shown the ASCII form.

Example 2: Two-dimensional motion
Two-dimensional motion requires that the motion controller coordinate the motion of 2 servo motors. For example, suppose an (X,Y) table holds a work-piece onto which glue is being dispensed. The objective is to follow the path shown on Figure 12, at a uniform surface speed of 10,000 counts per second.

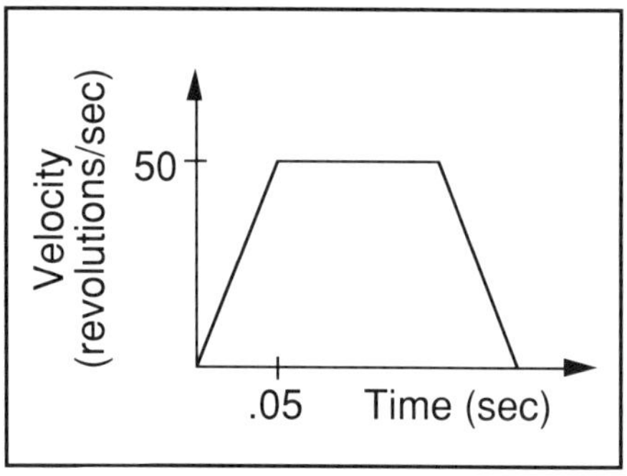

Figure 10

The motion controller accepts special instructions that define:

- Straight line segments (vectors)
- Circular arcs

Each line segment is defined with a VP instruction, which defines a segment that starts at the current position, and ends at the (X,Y) coordinates given with the command.

Each arc, on the other hand, is specified by a CR instruction, which specifies the *radius*, the *starting angle* and the *traversed angle*.

The velocity and acceleration are specified by a VS command and a VA command, respectively. The complete program is shown in Figure 13.

Instruction	Interpretation
PR 40000	Distance
SP 200000	Slew Speed
AC 4000000	Acceleration
BGX	Start Motion

Figure 11

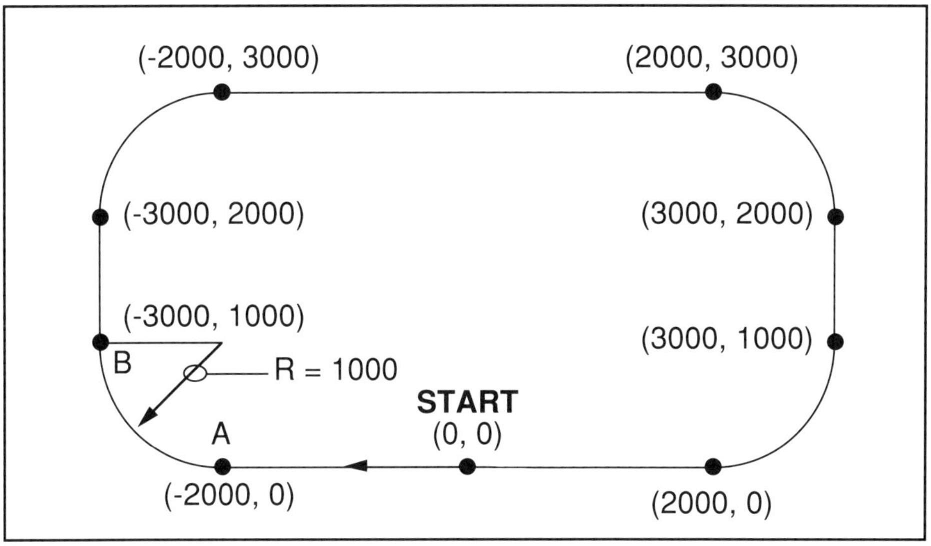

Figure 12

Instruction	Interpretation
#C	Label
VP -2000,0	
CR 1000,270,-90	
VP -3000,2000	
CR 1000,180,-90	Instructions
VP 2000,3000	describing
CR 1000,90,-90	the segments
VP 3000,1000	of the path
CR 1000,0,-90	
VP 0,0	
VS 5000	Vector speed
VA 20000	Vector acceleration
BGS	Start the motion
EN	End program

Figure 13

Example 3: Backlash compensation

If the mechanical linkage in a closed-loop system includes a significant amount of "play", the system may oscillate, as it repeatedly reverses the direction of the motor in its attempts to correct for overshoots. One example is where a leadscrew is used to convert the rotational motion of the motor to linear motion. If the backlash of the leadscrew is large (compared to the positioning tolerance) the system will have difficulty precisely positioning the load.

The solution to this problem is to use 2 position sensors:

- A *rotational* position sensor on the motor shaft
- A *linear* position sensor on the leadscrew

The *rotational* sensor is used to close the control loop in a tight fashion, without any backlash. That prevents oscillations. It allows the closed loop system to position the load within the limits of the leadscrew backlash. The *linear* sensor is then used to compute the amount of correction needed, and repeated *small* motions are initiated, until the load is precisely positioned. Figure 14 shows the program used to accomplish this.

Instruction	Interpretation
#DUAL	Label
CF,1	Configure for dual loop
DE 0	Set starting position to zero
PR 50000	Distance
SP 50000	Speed
AC 200000	Acceleration
BGX	Start X motion
#LOOP	Label
AMX	Wait for completion of motion
ER=50000-_DEX_TEX	Compute position error
AB=@ABS[ER]	Find absolute value
JP#EXIT,AB<3	Exit if error is small
PR ER	Perform correction
BGX	Start motion
JP #LOOP	Repeat the process
#EXIT	Exit
EN	End program

Figure 14

Example 4: Adaptive programming
Simple motion controllers are programmed to step though a fixed sequence of movements, as specified by a list of ASCII commands. However, more sophisticated motion controllers can be programmed with *symbolic variables*. Those variables can be re-evaluated during program execution. This permits the design of *adaptive* motion control systems, which adjust their behavior, based upon sensory input from the environment around them.

To take a simple example, suppose you have access to a rotating shaft that's running at a variable speed. (We'll call this shaft the "target".) You want to build a system that controls a motor (which we'll call the "follower") in such a way that its rotational position tracks the position of the target.

You could build such a system with a 2-channel motion controller board, which includes a channel X and a channel Y. The rotational position of the target is monitored by a position sensor, whose output is routed into channel Y, as shown in Figure 15. The follower motor is driven (and monitored) through channel X.

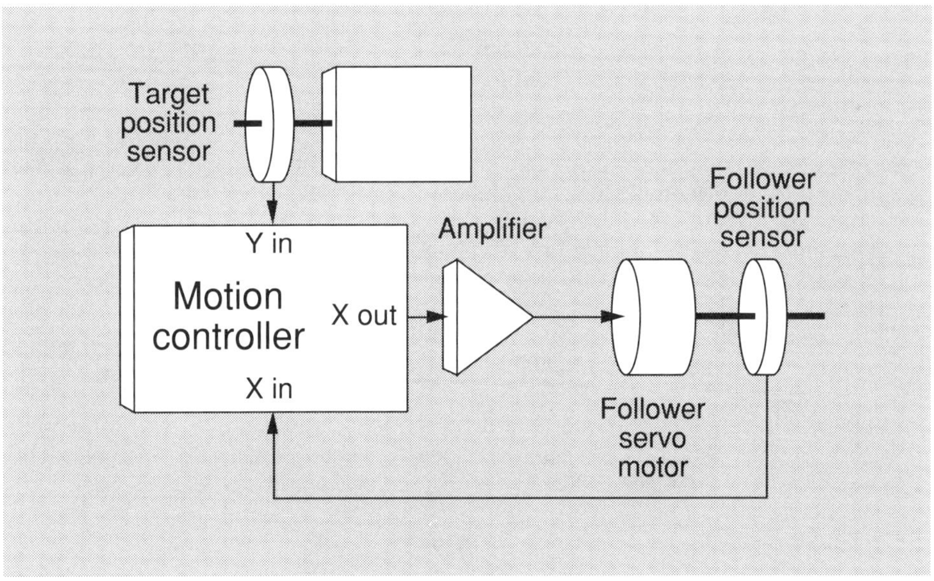

Figure 15

To synchronize the motion of the follower with the target, you would repeatedly execute a 3-step loop:

- Determine the difference in the rotational positions of the target and follower, as indicated by their respective encoders.
- Compute a speed proportional to this difference.
- Change the speed of the follower motor to the computed value.

The program that does this is shown in Figure 16.

Instruction	Interpretation
#F	Program name
DP 0,0	Define current (X,Y) position as zero
JG 0	Set initial X speed to zero
BGX	Start X
#L	Label
V1=PY-PX	Find the position difference
V2=V1*10	Compute the speed
JG V2	Modify the speed
JP #L	Repeat the process

Figure 16

Conclusion

The design of motion control systems has traditionally been a very specialized art, involving a lot of computation, as well as a certain amount of "tinkering" with PID filter parameters. However, the technology has now progressed to the point where you can use software tools to select off-the-shelf system elements. Additional system tools greatly simplify the job of tuning your system for optimum accuracy, responsiveness, and stability. This simplification has allowed much wider use of motion control techniques. Servo systems are now being used for such diverse tasks as cutting materials with a rotating knife, positioning semiconductor wafers under a microscope, and synchronizing label applicators at assembly-line speeds.

In many of the new applications (including the label applicator above) the motion controller must synchronize its motion with external events. To allow for synchronization, a new generation of intelligent *adaptive* motion controllers has been developed.

References

[1] *DC Motors, Speed Controls, Servo Systems.* Engineering Handbook by Electro-Craft Corporation.
[2] Control Systems Design, V. W. Eveleigh. McGraw Hill, 1972.
[3] Motion Control Applications, J. Tal. Published by Galil Motion Control, 575 Maude Ct., Sunnyvale, CA 94086.

Jacob Tal has been involved with motion control systems for over 2 decades. During that time he has taught at the University of Utah, and has worked as a consultant. Eight years ago Jacob co-founded Galil Motion Control. His research interests include advanced applications of motion controllers.

APPENDIX A

VMEbus Manufacturers

VMEbus Product Directory

Processor boards

Motorola-based processor boards

68000/68010

8.0 MHz
ELTEC ELEKTRONIK
FORCE COMPUTERS
MOTOROLA
PEP MODULAR COMPUTERS
PHILIPS EXPORT
RADSTONE TECHNOLOGY
ROTEC
SYSTEMFORSCHUNG
THEMIS COMPUTER
TL INDUSTRIES

10 MHz
BVM LIMITED
COMPCONTROL B.V.
DMS DORSCH MICROSYSTEMS
ELTEC ELEKTRONIK
EMBEDDIX EUROPE
FORCE COMPUTERS
GENERAL MICRO SYSTEM
HEURIKON
HTEC LIMITED
IDEATION
IMP
KONTRON ELEKTRONIK
MATRIX
MICRO-LINK PRODUCTS
MICROSYS
MIZAR/INTEGRATED SOLUTIONS
MORNING STAR TECHNOLOGIES
MOTOROLA
OMNIBYTE
PERFORMANCE TECHNOLOGIES
PHILIPS EXPORT
RADSTONE TECHNOLOGY
SYSTEMFORSCHUNG
TL INDUSTRIES
XYCOM

12.0 MHz
ELTEC ELEKTRONIK
IDEATION
IRONICS INCORPORATED
KONTRON ELEKTRONIK
KWS COMPUTERSYSTEME
MEN MIKRO ELEKTRONIK
PEP MODULAR COMPUTERS
ROTEC
SYSTEMFORSCHUNG

12.5 MHz
BVM LIMITED
CELOGIC
CREATIVE ELECTRONIC SYSTEMS
EKF-ELEKTRONIK-MESSTECHNIK
EMBEDDIX EUROPE
FORCE COMPUTERS
GENERAL MICRO SYSTEM
HEURIKON
HTEC LIMITED
KONTRON ELEKTRONIK
MATRIX
MICROSYS
MIZAR/INTEGRATED SOLUTIONS
MOTOROLA
OETTLE + REICHLER
OMNIBYTE
PERFORMANCE TECHNOLOGIES
TECNINT
THEMIS COMPUTER

16.0 MHz
COMPCONTROL B.V.
ELTEC ELEKTRONIK
GENERAL MICRO SYSTEM
MEN MIKRO ELEKTRONIK
MICROSYS
OETTLE + REICHLER
PERFORMANCE TECHNOLOGIES
ROTEC
SYSTEMFORSCHUNG
TECNINT
WESPERCORP

16.7 MHz
FORCE COMPUTERS

20.0 MHz
MEN MIKRO ELEKTRONIK

68HC000

10.0 MHz
ELTEC ELEKTRONIK
IKON

12.0 MHz
MEN MIKRO ELEKTRONIK

12.5 MHz
CELOGIC
DYNATEM
GreenSpring Computers
PEP MODULAR COMPUTERS

16.0 MHz
DYNATEM
MEN MIKRO ELEKTRONIK

16.7 MHz
PEP MODULAR COMPUTERS

20.0 MHz
MEN MIKRO ELEKTRONIK

68020

12.0 MHz
DMS DORSCH MICROSYSTEMS
Io, INCORPORATED
PHOENIX DIGITAL
SYNERGY MICROSYSTEMS
TADPOLE TECHNOLOGY

12.5 MHz
FORCE COMPUTERS
IDEATION
IRONICS INCORPORATED
Io, INCORPORATED
KONTRON ELEKTRONIK
MICROSYS
MOTOROLA
OETTLE + REICHLER
PEP MODULAR COMPUTERS
PERFORMANCE TECHNOLOGIES
PHILIPS EXPORT
PROTEUS Gesellschaft
RADSTONE TECHNOLOGY
TADPOLE TECHNOLOGY
XYCOM

16.0 MHz
AITECH
BVM LIMITED
CREATIVE ELECTRONIC SYSTEMS
ELTEC ELEKTRONIK
FORCE COMPUTERS
GENERAL MICRO SYSTEM
GreenSpring Computers
HTEC LIMITED
IMP
IRONICS INCORPORATED
Io, INCORPORATED
KONTRON ELEKTRONIK
MATRIX
MEN MIKRO ELEKTRONIK
MICRO-LINK PRODUCTS
MICROSYS
OETTLE + REICHLER
OMNIBYTE
PEP MODULAR COMPUTERS
SBE
SYNERGY MICROSYSTEMS
SYSTEMFORSCHUNG
TADPOLE TECHNOLOGY
TECNINT
THEMIS COMPUTER

16.7 MHz
DY 4 SYSTEMS
EKF-ELEKTRONIK-MESSTECHNIK
EMBEDDIX EUROPE
FORCE COMPUTERS
IDEATION
MATRIX
MIZAR/INTEGRATED SOLUTIONS
MOTOROLA
PEP MODULAR COMPUTERS
PERFORMANCE TECHNOLOGIES
RADSTONE TECHNOLOGY
XYCOM

20.0 MHz
AEROFLEX LABORATORIES
CELOGIC
CETIA
CYCLONE MICROSYSTEMS
DUUAL COMPUTER
DY 4 SYSTEMS
DYNATEM
ELTEC ELEKTRONIK
EMBEDDIX EUROPE
ENCORE COMPUTER
FORCE COMPUTERS
GENERAL MICRO SYSTEM
IDEATION
IMP
IRONICS INCORPORATED
Io, INCORPORATED
MEN MIKRO ELEKTRONIK
MICROSYS
MIZAR/INTEGRATED SOLUTIONS
MORNING STAR TECHNOLOGIES
MOTOROLA
PEP MODULAR COMPUTERS
PERFORMANCE TECHNOLOGIES
RADSTONE TECHNOLOGY
SBE
SYNERGY MICROSYSTEMS
TADPOLE TECHNOLOGY
TECNINT
THEMIS COMPUTER

24.0 MHz
CREATIVE ELECTRONIC SYSTEMS
FORCE COMPUTERS

25.0 MHz
COMPUTADORES MODULARES
DMS DORSCH MICROSYSTEMS
DY 4 SYSTEMS
ELTEC ELEKTRONIK
EMBEDDIX EUROPE
FORCE COMPUTERS
GENERAL MICRO SYSTEM
HEURIKON
IDEATION
IMP
KONTRON ELEKTRONIK
MATRIX
MEN MIKRO ELEKTRONIK
MICROSYS
MIZAR/INTEGRATED SOLUTIONS
MOTOROLA
OETTLE + REICHLER
OMNIBYTE
PEP MODULAR COMPUTERS

PERFORMANCE TECHNOLOGIES
PHOENIX DIGITAL
PROTEUS Gesellschaft
SBE
SYNERGY MICROSYSTEMS
TECNINT
UNISYS/CONVERGENT

33.0 MHz
DYNATEM
EKF-ELEKTRONIK-MESSTECHNIK
FORCE COMPUTERS
GENERAL MICRO SYSTEM
Io, INCORPORATED
MATRIX
MICROSYS
OETTLE + REICHLER
SYNERGY MICROSYSTEMS
WESPERCORP

68030

16.0 MHz
AITECH
COMPCONTROL B.V.
GENERAL MICRO SYSTEM
IMP
Io, INCORPORATED
MEN MIKRO ELEKTRONIK
MICROSYS
PHOENIX DIGITAL
TADPOLE TECHNOLOGY
TECNINT

16.7 MHz
MOTOROLA
PHILIPS EXPORT
PROTEUS Gesellschaft
RADSTONE TECHNOLOGY

20.0 MHz
COMPCONTROL B.V.
DY 4 SYSTEMS
ELTEC ELEKTRONIK
FORCE COMPUTERS
GENERAL MICRO SYSTEM
GreenSpring Computers
IMP
MEN MIKRO ELEKTRONIK
MICROPROCESS systemes
MICROSYS
MOTOROLA
PHILIPS EXPORT
SBE
SYNERGY MICROSYSTEMS
THEMIS COMPUTER

25.0 MHz
COMPUTADORES MODULARES
DUUAL COMPUTER
DY 4 SYSTEMS
EKF-ELEKTRONIK-MESSTECHNIK
ELTEC ELEKTRONIK
EMBEDDIX EUROPE
FORCE COMPUTERS
GENERAL MICRO SYSTEM
HCL AMERICA
IMP
IRONICS INCORPORATED
KWS COMPUTERSYSTEME
LYNX REAL-TIME SYSTEMS
MATRIX
MEN MIKRO ELEKTRONIK
MICROSYS
MIZAR/INTEGRATED SOLUTIONS
MOTOROLA
PEP MODULAR COMPUTERS
PERFORMANCE TECHNOLOGIES
PHILIPS EXPORT
PHOENIX DIGITAL
PROTEUS Gesellschaft
SBE
SYNERGY MICROSYSTEMS
THEMIS COMPUTER

30.0 MHz
FORCE COMPUTERS
KONTRON ELEKTRONIK

32.0 MHz
MOTOROLA
RADSTONE TECHNOLOGY

33.0 MHz
BVM LIMITED
CETIA
CREATIVE ELECTRONIC SYSTEMS
CYCLONE MICROSYSTEMS
GENERAL MICRO SYSTEM
HEURIKON
IRONICS INCORPORATED
KONTRON ELEKTRONIK
MEN MIKRO ELEKTRONIK
MOTOROLA
RADSTONE TECHNOLOGY
SYNERGY MICROSYSTEMS
TADPOLE TECHNOLOGY

50.0 MHz
ELTEC ELEKTRONIK
EMBEDDIX EUROPE
GENERAL MICRO SYSTEM
HEURIKON
IMP
MIZAR/INTEGRATED SOLUTIONS
MOTOROLA
PROTEUS Gesellschaft
RADSTONE TECHNOLOGY
SYNERGY MICROSYSTEMS

68040

20.0 MHz
RADSTONE TECHNOLOGY

25.0 MHz
COMPCONTROL B.V.
DY 4 SYSTEMS
FORCE COMPUTERS

MOTOROLA
OETTLE + REICHLER
PEP MODULAR COMPUTERS
SYSTEMFORSCHUNG
TADPOLE TECHNOLOGY
THEMIS COMPUTER

33.0 MHz
CETIA
FORCE COMPUTERS
GENERAL MICRO SYSTEM
MATRIX
OETTLE + REICHLER
TADPOLE TECHNOLOGY
THEMIS COMPUTER

40.0 MHz
EMBEDDIX EUROPE
MIZAR/INTEGRATED SOLUTIONS
RADSTONE TECHNOLOGY

50.0 MHz
COMPCONTROL B.V.
HEURIKON
OETTLE + REICHLER

88100

20.0 MHz
ELTEC ELEKTRONIK
IMP
MOTOROLA
TADPOLE TECHNOLOGY

25.0 MHz
AITECH
ELTEC ELEKTRONIK
FORCE COMPUTERS
TADPOLE TECHNOLOGY

33.0 MHz
CETIA
MOTOROLA
TADPOLE TECHNOLOGY

Other processor boards

80186
GE FANUC AUTOMATION

80286
HTEC LIMITED

80386
DMS DORSCH MICROSYSTEMS
DYNATECH COMPUTER SYSTEMS
FORCE COMPUTERS

68070
KWS COMPUTERSYSTEME
MICROPROCESS systemes
PROTEUS Gesellschaft
pro_VME INDUSTRIELLE RECHNERTECHNIK

80960CA
CYCLONE MICROSYSTEMS
HEURIKON
TADPOLE TECHNOLOGY

AM29000
IRONICS INCORPORATED
YARC SYSTEMS

Clipper
DU PONT PIXEL SYSTEMS

F9450 (Fairchild)
TASCO ELECTRONIC SERVICES

i860
TAURUS TECHNOLOGIES

J11
CREATIVE ELECTRONIC SYSTEMS
HTEC LIMITED

LSI Logic CMOS CPU
TASCO ELECTRONIC SERVICES

MDC281 (McDonnell Douglas)
TASCO ELECTRONIC SERVICES

MIPS R3000
CREATIVE ELECTRONIC SYSTEMS
HTEC LIMITED
LOCKHEED SANDERS
OMNIBYTE

NEC V30
HTEC LIMITED

NS32016
JANZ COMPUTER AG

NS32332
JANZ COMPUTER AG

PACE1750A
TASCO ELECTRONIC SERVICES

rtVAX300
AEON SYSTEMS
LOGICAL DESIGN GROUP
TEWS DATENTECHNIK

RTX-2000 (Harris)
PERITEK
SILICON COMPOSERS
VME Inc.

RTX-2001 (Harris)
VME Inc.

SPARC
EMBEDDIX EUROPE
KONTRON ELEKTRONIK
MIZAR/INTEGRATED SOLUTIONS
SUN MICROSYSTEMS

WE32200
IDEATION

DSP processors

C30
SKY COMPUTERS

DSP56XXX
OETTLE + REICHLER

DSP32C
SYMINEX S.A.

DSP96002 (Motorola)
SPECTRUM SIGNAL PROCESSING

MC56001
CREATIVE ELECTRONIC SYSTEMS
Data Cell
LOUGHBOROUGH SOUND IMAGES
SPECTRUM SIGNAL PROCESSING

TMS32XXX
ATLANTA SIGNAL PROCESSORS
PENTEK
PENTLAND SYSTEMS
STRUCK

TS68931
THEMIS COMPUTER

Other DSP processors
DATACUBE
DSP SYSTEMS
IMAGING TECHNOLOGY
INTERACTIVE CIRCUITS AND SYSTEMS

Multiple processor boards

320C30 x 3 + 68030
PC/M

320C30 x 3
PENTEK

68020 x 4
PC/M

68030 x 6
PC/M

88100
MOTOROLA
TADPOLE TECHNOLOGY

DSP56001 x 4
SIGNAL PROCESSING INNOVATIONS

Transputer
INCAA COMPUTERS BV
PARACOM
PROTEUS Gesellschaft

Special-purpose processors

Floating point vector processors
CSP
Impact Technologies
MERCURY COMPUTER SYSTEMS

Integer array processors
Data Cell

Floating point array processors
ANALOGIC/CDA
ARRAY MICROSYSTEMS
CELOGIC
CSP
DSP SYSTEMS
DY 4 SYSTEMS
SKY COMPUTERS

Floating point accellerators
DU PONT PIXEL SYSTEMS
DYNATEM
Io, INCORPORATED

Integer accellerators
SKY COMPUTERS

Application accellerator
SKY COMPUTERS

FFT processor
Data Cell

Serial protocol processor
SYSTECH

Multiprocessing synchronizer module

IBM PC-compatible VME products

Processor
AITECH
AJIDA TECHNOLOGIES
DYNATECH COMPUTER SYSTEMS
ELTEC ELEKTRONIK
INTERGON
LOGICAL DESIGN GROUP
NATIONAL INSTRUMENTS
NISSHO ELECTRONICS
OETTLE + REICHLER
PHILIPS EXPORT
RADISYS
XYCOM

EGA
HTEC LIMITED
XYCOM

VGA
PHILIPS EXPORT
RADISYS

AT-to-VME adapter board
HTEC LIMITED
INTERGON
RADISYS
XYCOM

Winchester + floppy

Winchester
INTERGON

Floppy
INTERGON
LOGICAL DESIGN GROUP

Flash disk
INTERGON
PROTEUS Gesellschaft
RADISYS

SCSI host adapter
RADISYS

SCSI + floppy
RADISYS

GPIB

RADISYS

Modem
RADISYS

2 RS232 ports
RADISYS

2 RS422 ports
RADISYS

Ethernet controller
RADISYS

Atari-compatible VME products

Atari-ST-compatible
pro_VME INDUSTRIELLE RECHNERTECHNIK

Memory

DRAM

512 Kbyte
BVM LIMITED
COMPCONTROL B.V.
EMBEDDIX EUROPE
IRONICS INCORPORATED
KONTRON ELEKTRONIK
MIZAR/INTEGRATED SOLUTIONS
PEP MODULAR COMPUTERS
RADSTONE TECHNOLOGY
SYSTEMFORSCHUNG

1 Mbyte
BVM LIMITED
GENERAL MICRO SYSTEM
GreenSpring Computers
MATRIX
MICRO MEMORY
MICRO-LINK PRODUCTS
MOTOROLA
OETTLE + REICHLER
PEP MODULAR COMPUTERS
PERFORMANCE TECHNOLOGIES

2 Mbyte
AITECH
BVM LIMITED
CLEARPOINT RESEARCH
COMPCONTROL B.V.
EMBEDDIX EUROPE
FORCE COMPUTERS
GreenSpring Computers
IDEATION
IRONICS INCORPORATED
KONTRON ELEKTRONIK
KWS COMPUTERSYSTEME
LOGICAL DESIGN GROUP
MICRO MEMORY
MICROSYS
MIZAR/INTEGRATED SOLUTIONS
MOTOROLA
OETTLE + REICHLER
PERFORMANCE TECHNOLOGIES
PHILIPS EXPORT
RADSTONE TECHNOLOGY
SYSTEMFORSCHUNG
THEMIS COMPUTER

4 Mbyte
CHRISLIN INDUSTRIES
CLEARPOINT RESEARCH
COMPCONTROL B.V.
EMBEDDIX EUROPE
ENCORE COMPUTER
FORCE COMPUTERS
GENERAL MICRO SYSTEM
GreenSpring Computers
HCL AMERICA
HTEC LIMITED
IDEATION
KONTRON ELEKTRONIK
LOGICAL DESIGN GROUP
MICRO MEMORY
MIZAR/INTEGRATED SOLUTIONS
MOTOROLA
OETTLE + REICHLER
PERFORMANCE TECHNOLOGIES
PHILIPS EXPORT

8 Mbyte
CHRISLIN INDUSTRIES
CLEARPOINT RESEARCH
COMPCONTROL B.V.
ELTEC ELEKTRONIK
EMBEDDIX EUROPE
FORCE COMPUTERS
GreenSpring Computers
HCL AMERICA
IDEATION
KWS COMPUTERSYSTEME
MICRO MEMORY
MICROSYS
MIZAR/INTEGRATED SOLUTIONS
MOTOROLA
OETTLE + REICHLER
PERFORMANCE TECHNOLOGIES
PHILIPS EXPORT
RADSTONE TECHNOLOGY
THEMIS COMPUTER

12 Mbyte
FORCE COMPUTERS
MICRO MEMORY

16 Mbyte
BVM LIMITED
CHRISLIN INDUSTRIES
CLEARPOINT RESEARCH
EMBEDDIX EUROPE
FORCE COMPUTERS
GENERAL MICRO SYSTEM
GreenSpring Computers
HTEC LIMITED
MICRO MEMORY
MICRO-LINK PRODUCTS
MIZAR/INTEGRATED SOLUTIONS
MOTOROLA

OETTLE + REICHLER
PERFORMANCE TECHNOLOGIES
RADSTONE TECHNOLOGY

32 Mbyte
CHRISLIN INDUSTRIES
CLEARPOINT RESEARCH
FORCE COMPUTERS
MICRO MEMORY
MOTOROLA
PERFORMANCE TECHNOLOGIES

48 Mbyte
CLEARPOINT RESEARCH

64 Mbyte
CHRISLIN INDUSTRIES
CLEARPOINT RESEARCH
MICRO MEMORY
MOTOROLA

128 Mbyte
MICRO MEMORY

DRAM with EDAC

256 Kbyte
RADSTONE TECHNOLOGY

512 Kbyte
RADSTONE TECHNOLOGY

1 Mbyte
CLEARPOINT RESEARCH
RADSTONE TECHNOLOGY

2 Mbyte
CLEARPOINT RESEARCH
RADSTONE TECHNOLOGY

4 Mbyte
CLEARPOINT RESEARCH
DY 4 SYSTEMS
MICRO MEMORY
MOTOROLA

8 Mbyte
CLEARPOINT RESEARCH
MICRO MEMORY
MOTOROLA

12 Mbyte
MICRO MEMORY

16 Mbyte
CLEARPOINT RESEARCH
MICRO MEMORY
MOTOROLA

Dual ported VME/VSB DRAM

1 Mbyte
MICRO MEMORY

2 Mbyte
CREATIVE ELECTRONIC SYSTEMS
MICRO MEMORY
NISSHO ELECTRONICS
SBE

3 Mbyte
MICRO MEMORY

4 Mbyte
AEROFLEX LABORATORIES
CHRISLIN INDUSTRIES
CREATIVE ELECTRONIC SYSTEMS
DUUAL COMPUTER
MICRO MEMORY
MOTOROLA
NISSHO ELECTRONICS
RADSTONE TECHNOLOGY
TECNINT

8 Mbyte
CHRISLIN INDUSTRIES
CREATIVE ELECTRONIC SYSTEMS
MICRO MEMORY
NISSHO ELECTRONICS
RADSTONE TECHNOLOGY
TECNINT

12 Mbyte
MICRO MEMORY

16 Mbyte
CHRISLIN INDUSTRIES
COMPCONTROL B.V.
CREATIVE ELECTRONIC SYSTEMS
DUUAL COMPUTER
MICRO MEMORY
NISSHO ELECTRONICS
TECNINT

32 Mbyte
CHRISLIN INDUSTRIES
COMPCONTROL B.V.
MICRO MEMORY
TECNINT

64 Mbyte
CHRISLIN INDUSTRIES
COMPCONTROL B.V.
GENERAL MICRO SYSTEM
MICRO MEMORY

128 Mbyte
GENERAL MICRO SYSTEM

Dual ported VME/VSB DRAM with EDAC

2 Mbyte
CLEARPOINT RESEARCH

4 Mbyte
CLEARPOINT RESEARCH
MICRO MEMORY

8 Mbyte
CLEARPOINT RESEARCH
MICRO MEMORY

12 Mbyte
MICRO MEMORY

16 Mbyte
CLEARPOINT RESEARCH
MICRO MEMORY

Dual ported VME/VSB SRAM

256 Kbyte

ANALOGIC/CDA
NISSHO ELECTRONICS

512 Kbyte
NISSHO ELECTRONICS

1 Mbyte
NISSHO ELECTRONICS

Dual ported VME/VMX DRAM

1 Mbyte
MICRO MEMORY
PERFORMANCE TECHNOLOGIES

2 Mbyte
MICRO MEMORY
PERFORMANCE TECHNOLOGIES

4 Mbyte
PERFORMANCE TECHNOLOGIES

8 Mbyte
PERFORMANCE TECHNOLOGIES

Dual ported VME/PMEbus (Proteus)

4 Mbyte
PROTEUS Gesellschaft

8 Mbyte
PROTEUS Gesellschaft

DRAM (nonvolatile)

4 Mbyte
GreenSpring Computers

SRAM

32-pin sockets
AITECH

192 Kbyte
IRONICS INCORPORATED

256 Kbyte
NISSHO ELECTRONICS
PHILIPS EXPORT

512 Kbyte
ELTEC ELEKTRONIK
NISSHO ELECTRONICS
PHILIPS EXPORT

1 Mbyte
CREATIVE ELECTRONIC SYSTEMS
FORCE COMPUTERS
KONTRON ELEKTRONIK
NISSHO ELECTRONICS
PHILIPS EXPORT

2 Mbytes
MICROSYS

4 Mbytes
TASCO ELECTRONIC SERVICES

6 Mbytes
PC/M

16 Mbytes
CSP
TASCO ELECTRONIC SERVICES

ROM/RAMEPROM

28-pin sockets
AITECH
BVM LIMITED
CELOGIC
COMPCONTROL B.V.
COMPUTADORES MODULARES
DY 4 SYSTEMS
EKF-ELEKTRONIK-MESSTECHNIK
EMBEDDIX EUROPE
GENERAL MICRO SYSTEM
GreenSpring Computers
HTEC LIMITED
IRONICS INCORPORATED
JANZ COMPUTER AG
KONTRON ELEKTRONIK
KWS COMPUTERSYSTEME
MATRIX
MICRO MEMORY
MICROSYS
MIZAR/INTEGRATED SOLUTIONS
MOTOROLA
OETTLE + REICHLER
PHILIPS EXPORT
RADSTONE TECHNOLOGY
SYSTEMFORSCHUNG
THEMIS COMPUTER

32-pin sockets
AEROFLEX LABORATORIES
DY 4 SYSTEMS
EKF-ELEKTRONIK-MESSTECHNIK
ELTEC ELEKTRONIK
ENCORE COMPUTER
FORCE COMPUTERS
GD CALIFORNIA
GENERAL MICRO SYSTEM
IDEATION
LOGICAL DESIGN GROUP
MICROSYS
OETTLE + REICHLER
PEP MODULAR COMPUTERS
PHILIPS EXPORT
RADSTONE TECHNOLOGY
ROTEC
TECNINT
XYCOM

32-pin sockets + 40-pin sockets
ELTEC ELEKTRONIK
ESD - ELECTRONIC SYSTEM DESIGN

CMOS SRAM (Nonvolatile)

32 Kbyte
ANALYTEK, LTD.
MIZAR/INTEGRATED SOLUTIONS

64 Kbyte
MICRO MEMORY
OVATION SYSTEMS
THEMIS COMPUTER

128 Kbyte
GreenSpring Computers
MICRO MEMORY
OVATION SYSTEMS
RADSTONE TECHNOLOGY

256 Kbyte
JANZ COMPUTER AG
MICRO MEMORY
MOTOROLA
OVATION SYSTEMS
THEMIS COMPUTER

512 Kbyte
FORCE COMPUTERS
MICRO MEMORY
MOTOROLA
OETTLE + REICHLER
OVATION SYSTEMS
RADSTONE TECHNOLOGY
THEMIS COMPUTER

1 Mbyte
BVM LIMITED
HTEC LIMITED
MICRO MEMORY
MOTOROLA
OETTLE + REICHLER
OVATION SYSTEMS

2 Mbyte
DYNATEM
FORCE COMPUTERS
MICRO MEMORY
OETTLE + REICHLER

3 Mbyte
MICRO MEMORY

4 Mbyte
CELOGIC
MATRIX
MICRO MEMORY
OETTLE + REICHLER
THEMIS COMPUTER

6 Mbyte
MICRO MEMORY

8 Mbyte
MICRO MEMORY
OETTLE + REICHLER

28-pin sockets
EMBEDDIX EUROPE
MATRIX

CMOS Battery back-up unit
PEP MODULAR COMPUTERS

Mass memory subsystems

512 Mbyte
ZITEL

2 Gbyte
ZITEL

Hardened VME products

Full MIL-SPEC
AEROFLEX LABORATORIES
AITECH
BICC-VERO ELECTRONICS
DY 4 SYSTEMS
ENCORE COMPUTER
Impact Technologies
LOCKHEED SANDERS
MALCO MICRODOT
MOTOROLA
RADSTONE TECHNOLOGY
RIBCO
RUGGED DIGITAL SYSTEMS

Ruggedized
AITECH
ALPHI TECHNOLOGY
CODAR TECHNOLOGY
COMPUTADORES MODULARES
EG&G BIRTCHER
LOCKHEED SANDERS
MATRIX
MOTOROLA
OMNIBYTE
RADSTONE TECHNOLOGY
THEMIS COMPUTER

Mass storage

Plug-in VME disk modules

1 winchester
PHILIPS EXPORT
RADSTONE TECHNOLOGY

2 winchesters
RADSTONE TECHNOLOGY

1 floppy
RADSTONE TECHNOLOGY

2 floppies
RADSTONE TECHNOLOGY

Plug-in VME disk modules with controllers

1 floppy
MATRIX
PHILIPS EXPORT
RADSTONE TECHNOLOGY

1 floppy + 1 tape + SCSI
COMPCONTROL B.V.

2 floppies
COMPUTADORES MODULARES
RADSTONE TECHNOLOGY

2 floppies +SCSI
ACT-VME
COMPCONTROL B.V.

1 winchester
RADISYS

RADSTONE TECHNOLOGY

1 winchester + 1 floppy

ACT-VME
FORCE COMPUTERS
PHILIPS EXPORT

1 winchester + SCSI

PHILIPS EXPORT

1 winchester + 1 floppy + SCSI

COMPCONTROL B.V.

1 winchester + 1 tape + SCSI

COMPCONTROL B.V.

1 winchester + 1 optical + SCSI

COMPCONTROL B.V.

2 winchesters + SCSI

COMPCONTROL B.V.

2 cart tape drives + SCSI

COMPCONTROL B.V.

Winchester controller board

2 winchesters + 2 floppies

PHILIPS EXPORT

2 winchesters + 3 floppies

BVM LIMITED

2 winchesters + 4 floppies

MOTOROLA
OETTLE + REICHLER
PEP MODULAR COMPUTERS
PHILIPS EXPORT
XYCOM

3 winchesters + 4 floppies

FORCE COMPUTERS
RADSTONE TECHNOLOGY

4 winchesters + 4 floppies

EMBEDDIX EUROPE
KONTRON ELEKTRONIK
MIZAR/INTEGRATED SOLUTIONS

Floppy controller board

2 floppies

OETTLE + REICHLER
ROTEC
SYSTEMFORSCHUNG

3 floppies

CLEARPOINT RESEARCH
PEP MODULAR COMPUTERS

4 floppies

CELOGIC
ELTEC ELEKTRONIK
EMBEDDIX EUROPE
KONTRON ELEKTRONIK
MATRIX
MICROSYS
MIZAR/INTEGRATED SOLUTIONS
THEMIS COMPUTER

8 floppies

BVM LIMITED

Other mass storage controllers

SASI controller

EMBEDDIX EUROPE
MIZAR/INTEGRATED SOLUTIONS
MOTOROLA
RADSTONE TECHNOLOGY
SYSTEMFORSCHUNG

SCSI host adapter

AITECH
ALPHI TECHNOLOGY
BVM LIMITED
CELOGIC
CETIA
CIPRICO
COMPCONTROL B.V.
DY 4 SYSTEMS
EKF-ELEKTRONIK-MESSTECHNIK
EMBEDDIX EUROPE
FORCE COMPUTERS
GreenSpring Computers
HCL AMERICA
IDEATION
IMP
INTERPHASE
INTROL
Io, INCORPORATED
JANZ COMPUTER AG
KONTRON ELEKTRONIK
KWS COMPUTERSYSTEME
LOGICAL DESIGN GROUP
MATRIX
MEN MIKRO ELEKTRONIK
MICRO-LINK PRODUCTS
MICROPROCESS systemes
MIZAR/INTEGRATED SOLUTIONS
MOTOROLA
NATIONAL INSTRUMENTS
PEP MODULAR COMPUTERS
PERFORMANCE TECHNOLOGIES
PHILIPS EXPORT
RADSTONE TECHNOLOGY
SBE
SYSTEMFORSCHUNG
VME Inc.
XYCOM
XYLOGICS

SCSI-2 host adapter

CIPRICO
MEN MIKRO ELEKTRONIK

SCSI + floppy

DMS DORSCH MICROSYSTEMS
DUUAL COMPUTER
DYNATEM
ESD - ELECTRONIC SYSTEM DESIGN
GENERAL MICRO SYSTEM
IMP
KONTRON ELEKTRONIK

MATRIX
MICROSYS
MOTOROLA
OETTLE + REICHLER
TECNINT
THEMIS COMPUTER

SCSI + GPIB
GENERAL MICRO SYSTEM

SMD-E
CIPRICO
DUUAL COMPUTER
INTERPHASE
IRONICS INCORPORATED
STOLLMANN
UNISYS/CONVERGENT
XYLOGICS

ESDI
CIPRICO
DUUAL COMPUTER
INTERPHASE
XYLOGICS

IPI-2 controller
XYLOGICS

Half-inch tape controllers
CIPRICO
DUUAL COMPUTER
HCL AMERICA
INTERPHASE
MOTOROLA
STOLLMANN
UNISYS/CONVERGENT

Streamer tape controllers (QIC-02)
PEP MODULAR COMPUTERS
RADSTONE TECHNOLOGY
STOLLMANN

Bubble memory
BUBBL-TEC
MemTech
RADSTONE TECHNOLOGY

Parallel disk controller
IBIS SYSTEMS
PRIMARY RATE

Floppy-to-SCSI adapter
IDEATION
PHILIPS EXPORT

Solid state mass storage

Solid state disk controller
DATARAM

SCSI solid state disk drive
COMPCONTROL B.V.
CONTROLEX

Solid state disk drive
PHILIPS EXPORT

Plug-in cartridge memory
BVM LIMITED
CELOGIC

EEPROM
HTEC LIMITED

LAN controllers

ARCNET
COMENDEC
COMPCONTROL B.V.
PROTEUS Gesellschaft

Ethernet/Cheapernet (IEEE 802.3)
BVM LIMITED
CETIA
CMC
DR. ING. SEUFERT
DY 4 SYSTEMS
DYNATEM
EKF-ELEKTRONIK-MESSTECHNIK
ELTEC ELEKTRONIK
EMBEDDIX EUROPE
FORCE COMPUTERS
GENERAL MICRO SYSTEM
GreenSpring Computers
HCL AMERICA
IDEATION
IMP
INTERPHASE
IRONICS INCORPORATED
Io, INCORPORATED
JANZ COMPUTER AG
KONTRON ELEKTRONIK
MATRIX
MEN MIKRO ELEKTRONIK
MICROPROCESS systemes
MICROSYS
MIZAR/INTEGRATED SOLUTIONS
MOTOROLA
OETTLE + REICHLER
PEP MODULAR COMPUTERS
PHILIPS EXPORT
PSITECH
RADISYS
RADSTONE TECHNOLOGY
SBE
STOLLMANN
STRUCK
SYSTEMFORSCHUNG
TADPOLE TECHNOLOGY
TECNINT
THEMIS COMPUTER
UNISYS/CONVERGENT

FDDI
CMC
FORMATION
INTERPHASE
JANZ COMPUTER AG
STOLLMANN

MAP/TOP (IEEE 802.4)
FAIRCHILD DATA
MOTOROLA

Token ring (IEEE 802.5)
COMPCONTROL B.V.
ELTEC ELEKTRONIK
FORMATION
INTERPHASE
OMNIBYTE
SBE

Token bus
ELTEC ELEKTRONIK

Bitbus
ELTEC ELEKTRONIK

Telecommunications

European 2.048 MHz CEPT
NEWBRIDGE MICROSYSTEMS

Modem
CELOGIC

ISDN
CELOGIC
CMC
PRIMARY RATE
STOLLMANN

T1
NEWBRIDGE MICROSYSTEMS
PRIMARY RATE
SBE

E1
SBE

Serial controllers

Asynchronous serial controllers

RS232 only

2 port
KONTRON ELEKTRONIK
METACOMP
MICROPROCESS systemes
MICROSYS
MIZAR/INTEGRATED SOLUTIONS
MOTOROLA

4 port
BVM LIMITED
CELOGIC
COMPCONTROL B.V.
ENCORE COMPUTER
HTEC LIMITED
IDEATION
IRONICS INCORPORATED
JANZ COMPUTER AG
MEN MIKRO ELEKTRONIK
METACOMP
MICRO-LINK PRODUCTS
MICROSYS
PEP MODULAR COMPUTERS
PHILIPS EXPORT
SYSTEMFORSCHUNG
XYCOM

6 port
ELTEC ELEKTRONIK
KONTRON ELEKTRONIK
MICROSYS
MOTOROLA
PROTEUS Gesellschaft

7 port
DMS DORSCH MICROSYSTEMS

8 port
BVM LIMITED
CELOGIC
COMPCONTROL B.V.
DUUAL COMPUTER
FORCE COMPUTERS
GENERAL MICRO SYSTEM
HCL AMERICA
IDEATION
IRONICS INCORPORATED
Io, INCORPORATED
JANZ COMPUTER AG
MICROPROCESS systemes
MOTOROLA
PERFORMANCE TECHNOLOGIES
RADSTONE TECHNOLOGY
SBE
THEMIS COMPUTER
XYCOM

12 port
Io, INCORPORATED

16 port
ALPHI TECHNOLOGY
ARIEL SYSTEMS
CO-SYSTEMS INC.
ESD - ELECTRONIC SYSTEM DESIGN
HCL AMERICA
IDEATION
IMP
Io, INCORPORATED
MIZAR/INTEGRATED SOLUTIONS
PERFORMANCE TECHNOLOGIES
RADSTONE TECHNOLOGY
SIMPACT ASSOCIATES
TADPOLE TECHNOLOGY
XYLOGICS

RS232/RS422

1 port
KWS COMPUTERSYSTEME
MEN MIKRO ELEKTRONIK

2 port
CELOGIC
ESD - ELECTRONIC SYSTEM DESIGN
GreenSpring Computers
MATRIX
PEP MODULAR COMPUTERS

3 port
ROTEC

4 port
CETIA

DMS DORSCH MICROSYSTEMS
DY 4 SYSTEMS
EKF-ELEKTRONIK-MESSTECHNIK
EMBEDDIX EUROPE
GreenSpring Computers
Io, INCORPORATED
KONTRON ELEKTRONIK
KWS COMPUTERSYSTEME
MATRIX
MICRO-LINK PRODUCTS
MICROSYS
MIZAR/INTEGRATED SOLUTIONS
OETTLE + REICHLER
PHILIPS EXPORT
THEMIS COMPUTER
UNISYS/CONVERGENT
VME MICROSYSTEMS INTERNATIONAL
XYCOM

6 port
FORCE COMPUTERS
MOTOROLA
RADSTONE TECHNOLOGY
SYSTEMFORSCHUNG

8 port
ALPHI TECHNOLOGY
CENTRAL DATA
DY 4 SYSTEMS
EKF-ELEKTRONIK-MESSTECHNIK
EMBEDDIX EUROPE
FORCE COMPUTERS
GreenSpring Computers
KONTRON ELEKTRONIK
MIZAR/INTEGRATED SOLUTIONS
OETTLE + REICHLER
RADSTONE TECHNOLOGY
XYCOM

12 port
AITECH
MICROSYS
TECNINT

16 port
DYNATEM
EKF-ELEKTRONIK-MESSTECHNIK
EMBEDDIX EUROPE
ESD - ELECTRONIC SYSTEM DESIGN

RS485

4 port
XYCOM

8 port
FORMATION

TTL

8 port
XYCOM

SDLC/HDLC

1 port (isolated)
DMS DORSCH MICROSYSTEMS

2 port
MATRIX
THEMIS COMPUTER

4 port
BVM LIMITED
EMBEDDIX EUROPE
MATRIX
MIZAR/INTEGRATED SOLUTIONS
PHILIPS EXPORT
SBE

Synchronous serial controllers

X.25
CELOGIC
CETIA
COMPCONTROL B.V.
FORMATION
IMP
METACOMP
MIZAR/INTEGRATED SOLUTIONS
MOTOROLA
RADSTONE TECHNOLOGY
SBE
STOLLMANN
THEMIS COMPUTER

X.21 (SNA)
PEP MODULAR COMPUTERS
STOLLMANN

Other serial controllers

Serial concentrator subsytems
ARIEL SYSTEMS
BICC-VERO ELECTRONICS
METACOMP
PERFORMANCE TECHNOLOGIES
SYSTECH

FERA
TECNINT

Fiber optic
AUGMENT SYSTEMS
CONCISE TECHNOLOGY
CREATIVE ELECTRONIC SYSTEMS
KWS COMPUTERSYSTEME
PEP MODULAR COMPUTERS
PHOENIX DIGITAL
RKB
STRUCK
SYSTRAN
VME MICROSYSTEMS INTERNATIONAL

Peripheral I/O controllers

Parallel Centronics I/O
BVM LIMITED
CETIA
ELTEC ELEKTRONIK
EMBEDDIX EUROPE
IDEATION
IKON
LOGICAL DESIGN GROUP

MIZAR/INTEGRATED SOLUTIONS
MOTOROLA
PEP MODULAR COMPUTERS

Serial + Centronics
COMPUTADORES MODULARES
DUUAL COMPUTER
KONTRON ELEKTRONIK
MOTOROLA

Counter/timer/real-time clock
ACT-VME
BVM LIMITED
COMPCONTROL B.V.
KONTRON ELEKTRONIK
THEMIS COMPUTER

Kitchen sink peripheral I/O board
ESD - ELECTRONIC SYSTEM DESIGN
GreenSpring Computers
KWS COMPUTERSYSTEME
LOGICAL DESIGN GROUP
MATRIX
MIZAR/INTEGRATED SOLUTIONS
PC/M
PEP MODULAR COMPUTERS

Audio I/O
COMPUSONICS
CREATIVE STRATEGIES PTY.
Data Cell
MICROVOICE
PENTEK
SIGNAL PROCESSING INNOVATIONS
VIGRA
WESPERCORP

Versatec plotter
IKON

Special cables

RS232 cable
THEMIS COMPUTER

GPIB cable
THEMIS COMPUTER

Peripheral I/O transition cables
COMPCONTROL B.V.
ESD - ELECTRONIC SYSTEM DESIGN
MOTOROLA
TL INDUSTRIES
XYCOM

VME-to-?

32-series HSD interface (Encore/Gould)
VME MICROSYSTEMS INTERNATIONAL

ADCCP (Air Force Systems Integration Std)
STANFORD TELECOM

Allen Bradley 1770 Industrial I/O
MATRIX

Allen Bradley Data Highway
SUTHERLAND-SCHULTZ

Allen Bradley Remote I/O
SUTHERLAND-SCHULTZ

ARINC 429
MSS GROUP

BI bus (VAX)
AEON SYSTEMS
CREATIVE ELECTRONIC SYSTEMS
VME MICROSYSTEMS INTERNATIONAL

Bitbus
XYCOM

C-Bus (DEC)
IRTV mbH

CAMAC crate controller
BERGOZ
BI RA SYSTEMS
CREATIVE ELECTRONIC SYSTEMS
STRUCK

DR11-W DMA interface (DEC)
IKON
MDB SYSTEMS
VME MICROSYSTEMS INTERNATIONAL

FASTBUS
STRUCK

Floating Point Systems mini-supercomputer
ADVANCED PROCESSING LABORATORIES

GPBS-bus (MIOS I/O Subsystem)
PHILIPS EXPORT

GPIB Talker/Listener
BVM LIMITED
CELOGIC
EKF-ELEKTRONIK-MESSTECHNIK
ELTEC ELEKTRONIK
Io, INCORPORATED
KWS COMPUTERSYSTEME
MIZAR/INTEGRATED SOLUTIONS
PHILIPS EXPORT
SYSTEMFORSCHUNG

GPIB Talker/Listener/Controller
BVM LIMITED
COMPCONTROL B.V.
DMS DORSCH MICROSYSTEMS
EMBEDDIX EUROPE
GreenSpring Computers
HTEC LIMITED
IDEATION
IRONICS INCORPORATED
KONTRON ELEKTRONIK
MEN MIKRO ELEKTRONIK
MICROSYS
MIZAR/INTEGRATED SOLUTIONS
MOTOROLA
NATIONAL INSTRUMENTS
OETTLE + REICHLER
RADSTONE TECHNOLOGY
SIGNAL PROCESSING INNOVATIONS
THEMIS COMPUTER

HP-9000 (Hewlett-Packard)
HEWLETT-PACKARD - WORKSTATION GROUP

Hyperchannel
IKON

I/O channel (Motorola)
MOTOROLA
PEP MODULAR COMPUTERS

IBM 360/370 I/O channel
ANDREW/KMW SYSTEMS
DATA/WARE DEVELOPMENT
Dr. Materna
FORMATION
STOLLMANN

IBM RISC/System 6000
NATIONAL INSTRUMENTS

Interbus-S (Phoenix Contact)
PHOENIX CONTACT

L-Bus (Data General)
INDOCOMP SYSTEMS

Metrabus (Keithley-Metrabyte)
KEITHLEY METRABYTE

Microchannel
AMTRONICS
BEECO INCORPORATED
BIT 3 COMPUTER
NATIONAL INSTRUMENTS

Mil-Std 1553B
AEROFLEX LABORATORIES
AITECH
BALLARD TECHNOLOGY
COMPUTADORES MODULARES
DIGITAL TECHNOLOGY
DY 4 SYSTEMS
ENCORE COMPUTER
GENERAL MICRO SYSTEM
ILC DATA DEVICE
MOTOROLA
PEP MODULAR COMPUTERS
RADSTONE TECHNOLOGY
SBS ENGINEERING
SCI TECHNOLOGY

Modicon 984 800 series I/O network
SUTHERLAND-SCHULTZ

Multibus I
BIT 3 COMPUTER
PERFORMANCE TECHNOLOGIES

Multibus II
AEON SYSTEMS

MXIbus (National Instruments)
NATIONAL INSTRUMENTS

NTDS (Mil-Std 1397B)
ANTARES GROUP
ENCORE COMPUTER
G.E.T. ENGINEERING
GET ENGINEERING
ROCKWELL INTERNATIONAL
SABTECH INDUSTRIES

NUbus (Mac II)
AUGMENT SYSTEMS
BERGOZ
BIT 3 COMPUTER
CREATIVE ELECTRONIC SYSTEMS
LEXTEL
SIEMENS-ALBIS SA
TMC SrL
pro_VME INDUSTRIELLE RECHNERTECHNIK

PC-bus
BIT 3 COMPUTER
CREATIVE ELECTRONIC SYSTEMS
NATIONAL INSTRUMENTS
PENTEK
STRUCK
ULTRALINK INTERNATIONAL
XYCOM
pro_VME INDUSTRIELLE RECHNERTECHNIK

PC20 PLC line
PHILIPS EXPORT

Q-Bus
AEON SYSTEMS
BI RA SYSTEMS
BIT 3 COMPUTER
CREATIVE ELECTRONIC SYSTEMS
LOGICAL DESIGN GROUP
PERFORMANCE TECHNOLOGIES
STRUCK
TEWS DATENTECHNIK

S-bus (Sun SPARCstation)
BIT 3 COMPUTER
NATIONAL INSTRUMENTS
PERFORMANCE TECHNOLOGIES

Series 300 workstation (Hewlett-Packard)
HEWLETT-PACKARD - WORKSTATION GROUP

STANAG-Std 4156
G.E.T. ENGINEERING
ROCKWELL INTERNATIONAL

Transputer processing array
PARACOM

Unibus
BI RA SYSTEMS
JORWAY

VERSAbus
HVE ENGINEERING

VMEbus
AEON SYSTEMS
AKEBIA
BIT 3 COMPUTER
CREATIVE ELECTRONIC SYSTEMS
GENERAL MICRO SYSTEM
HVE ENGINEERING
IRONICS INCORPORATED
LEXTEL
MUPAC
PERFORMANCE TECHNOLOGIES

STRUCK
SYSTRAN
VME MICROSYSTEMS INTERNATIONAL
pro_VME INDUSTRIELLE RECHNERTECHNIK

VSB-to-?

CAMAC crate controller
STRUCK

FASTBUS
STRUCK

VSB
STRUCK

Sun 3 and Sun 4 Compatible products

6U height

Board set
AEROFLEX LABORATORIES

9U height

Turn-key ystems

Development system + real-time subsystem
ADVANCED PROCESSING LABORATORIES

Disk controllers

SMD-E disk controllers
CIPRICO
INTERPHASE

SCSI disk controllers
INTERPHASE
NATIONAL INSTRUMENTS

SCSI-2 disk controllers
INTERPHASE

ESDI disk controllers
INTERPHASE

IPI-2 disk controllers
XYLOGICS

Serial controllers

Serial controllers
SBE

RS232 serial controllers
SIMPACT ASSOCIATES

X.25 communications controller
SBE

FDDI controller
XYLOGICS

Other Sun compatible boards

Data acquisition board
INTERACTIVE CIRCUITS AND SYSTEMS

GPIB talker/listener/controller
NATIONAL INSTRUMENTS

Transputer board
PARACOM

Video boards

CRT graphics controllers
PARALLAX GRAPHICS

Video frame buffer
SUN MICROSYSTEMS

Image processor
ANALOGIC/CDA
GEMS OF CAMBRIDGE
IMAGING TECHNOLOGY

Mechanical/prototyping products

EMI shield
HYBRICON

Front panel
HYBRICON

Ejectors
HYBRICON

Mechanical adapters
BUSTRONIC
ELMA ELECTRONIC
HYBRICON

SUN board accessories
BUSTRONIC
PHILIPS EXPORT

NULL boards (Daisy-chain jumpers only)
BUSTRONIC

Extender boards
DAWN VME PRODUCTS

Prototyping board with S-bus interface

Backplanes
BUSTRONIC
DAWN VME PRODUCTS
ELMA ELECTRONIC
HYBRICON
INTERLOGIC INDUSTRIES
MUPAC

Card cage
BUSTRONIC
HYBRICON

Card cage + backplane + power supply
MUPAC

Card cage + backplane + power supply + enclosure
BUSTRONIC
ELMA ELECTRONIC

Card cage + enclosure
ELMA ELECTRONIC
Io, INCORPORATED

Industrial/process I/O

General purpose controllers

Industrial I/O subsystems
ACROMAG
ALLEN-BRADLEY CO.

VME MICROSYSTEMS INTERNATIONAL

Machine control systems
AITECH
VME MICROSYSTEMS INTERNATIONAL

PLCs
ALLEN-BRADLEY CO.
GE FANUC AUTOMATION

Motion control

Servo motor drivers
CREONICS
DELTA COMPUTER SYSTEMS
DMS DORSCH MICROSYSTEMS
DYNATEM
EKF-ELEKTRONIK-MESSTECHNIK
ELTEC ELEKTRONIK
GALIL MOTION CONTROL
GreenSpring Computers
HYDAC ELECTRONIC
INDUSTRIAL INDEXING SYSTEMS
LYNX REAL-TIME SYSTEMS
MATRIX
MICROSYS
OETTLE + REICHLER
OREGON MICRO SYSTEMS
PEP MODULAR COMPUTERS
PHILIPS EXPORT
TECNINT
TEWS DATENTECHNIK
THEMIS COMPUTER
WESPERCORP
WHEDCO INCORPORATED

Unipolar motor drivers
MATRIX

Stepper motor drivers
MATRIX
PEP MODULAR COMPUTERS
WHEDCO INCORPORATED

Synchro/Resolver
ASTROSYSTEMS
COMPUTADORES MODULARES
DMS DORSCH MICROSYSTEMS
MIDI INGENIERIE
TRANSMAGNETICS
VME MICROSYSTEMS INTERNATIONAL

Absolute encoder
COMPCONTROL B.V.
ROTEC
TECNINT
WHEDCO INCORPORATED

Incremental encoder
ROTEC

Incremental analog tracker
ROTEC

Low speed A/D (<1 MHz)

10-bit A/D
DSP SYSTEMS

12-bit A/D
ACROMAG
ANGLIA TECHNOLOGY LIMITED
BVM LIMITED
CELOGIC
COMPUTADORES MODULARES
CREATIVE ELECTRONIC SYSTEMS
DATEL
DMS DORSCH MICROSYSTEMS
Data Cell
ELTEC ELEKTRONIK
ESD - ELECTRONIC SYSTEM DESIGN
GreenSpring Computers
INCAA COMPUTERS BV
IRONICS INCORPORATED
JANZ COMPUTER AG
KEITHLEY METRABYTE
KONTRON ELEKTRONIK
MATRIX
MEN MIKRO ELEKTRONIK
MICROPROCESS systemes
MICROSYS
MIDI INGENIERIE
MIZAR/INTEGRATED SOLUTIONS
OETTLE + REICHLER
PENTLAND SYSTEMS
PEP MODULAR COMPUTERS
PHILIPS EXPORT
RADSTONE TECHNOLOGY
ROTEC
TECNINT
THEMIS COMPUTER
VME MICROSYSTEMS INTERNATIONAL
XYCOM

12-bit + sign A/D
GreenSpring Computers
PENTEK

14-bit A/D
ACROMAG
DATEL

16-bit A/D
ANALOGIC
DATEL
DMS DORSCH MICROSYSTEMS
Data Cell
Data Translation, Inc.
ESD - ELECTRONIC SYSTEM DESIGN
INCAA COMPUTERS BV
IRONICS INCORPORATED
JANZ COMPUTER AG
MEN MIKRO ELEKTRONIK
PENTEK
PENTLAND SYSTEMS
SIGNAL PROCESSING INNOVATIONS
STRUCK
VME MICROSYSTEMS INTERNATIONAL

17-bit A/D

INCAA COMPUTERS BV

High speed A/D (>1 MHz)

8-bit A/D
STRUCK

12-bit A/D
DSP SYSTEMS
Data Cell
EMBEDDIX EUROPE
JANZ COMPUTER AG
PENTEK
PENTLAND SYSTEMS
TECNINT

16-bit A/D
DSP SYSTEMS

D/A

8-bit D/A
DMS DORSCH MICROSYSTEMS
GreenSpring Computers

12-bit D/A
ACROMAG
ANGLIA TECHNOLOGY LIMITED
BVM LIMITED
CELOGIC
COMPUTADORES MODULARES
DATEL
Data Cell
Data Translation, Inc.
EKF-ELEKTRONIK-MESSTECHNIK
ELTEC ELEKTRONIK
EMBEDDIX EUROPE
ESD - ELECTRONIC SYSTEM DESIGN
HYTEC ELECTRONICS
IRONICS INCORPORATED
JANZ COMPUTER AG
KONTRON ELEKTRONIK
MATRIX
MEN MIKRO ELEKTRONIK
MICROSYS
MIDI INGENIERIE
MIZAR/INTEGRATED SOLUTIONS
OETTLE + REICHLER
PENTLAND SYSTEMS
PEP MODULAR COMPUTERS
PHILIPS EXPORT
RADSTONE TECHNOLOGY
ROTEC
SIGNAL PROCESSING INNOVATIONS
VME MICROSYSTEMS INTERNATIONAL
XYCOM

12-bit D/A(current output)
ELTEC ELEKTRONIK
VME MICROSYSTEMS INTERNATIONAL

14-bit D/A
DMS DORSCH MICROSYSTEMS
OETTLE + REICHLER

16-bit D/A
DATEL
GreenSpring Computers
INCAA COMPUTERS BV
JANZ COMPUTER AG
MEN MIKRO ELEKTRONIK
OETTLE + REICHLER
VME MICROSYSTEMS INTERNATIONAL

16-bit D/A (current output)
ACROMAG

18-bit D/A
Data Cell
HYTEC ELECTRONICS

D/A + A/D

10-bit A/D + 10-bit D/A
HYDAC ELECTRONIC

12-bit A/D + 12-bit D/A
ACROMAG
COMPCONTROL B.V.
DATEL
DY 4 SYSTEMS
DYNATEM
Data Translation, Inc.
IRONICS INCORPORATED
KWS COMPUTERSYSTEME
MATRIX
PENTLAND SYSTEMS
PEP MODULAR COMPUTERS
PHILIPS EXPORT
SYSTEMFORSCHUNG
THEMIS COMPUTER
TRANSMAGNETICS
VME MICROSYSTEMS INTERNATIONAL
XYCOM

12-bit A/D + 12-bit D/A + 30 V/48 mA output
ACROMAG

14-bit A/D + 12-bit D/A
Data Translation, Inc.

16-bit A/D + 12-bit D/A
DATEL
Data Translation, Inc.
VME MICROSYSTEMS INTERNATIONAL
XYCOM

Analog accessories

Thermocouple input: 12-bit + sign A/D
DATEL
ESD - ELECTRONIC SYSTEM DESIGN
ROTEC
TECNINT
VME MICROSYSTEMS INTERNATIONAL

RTD: 12-bit + sign
DATEL
VME MICROSYSTEMS INTERNATIONAL

RTD Multiplexors
DATEL

LVDT Demod/Mux
VME MICROSYSTEMS INTERNATIONAL

Strain gauge multiplexors
DATEL

Voltage-to-current converter
ESD - ELECTRONIC SYSTEM DESIGN

Voltage-to-frequency converter
VME MICROSYSTEMS INTERNATIONAL

Analog isolation amplifier
VME MICROSYSTEMS INTERNATIONAL

Analog signal conditioners
Data Cell
IRONICS INCORPORATED
KEITHLEY METRABYTE
PENTLAND SYSTEMS
SIGNAL PROCESSING INNOVATIONS
VME MICROSYSTEMS INTERNATIONAL
XYCOM

Misc. analog I/O accessories
ACROMAG

Analog multiplexors

With sample-and-hold
ANALOGIC
DATEL
Data Translation, Inc.
KEITHLEY METRABYTE
VME MICROSYSTEMS INTERNATIONAL

Without sample-and-hold
ACROMAG
ANALOGIC
ANALOGIC/CDA
DATEL
KEITHLEY METRABYTE
MEN MIKRO ELEKTRONIK
VME MICROSYSTEMS INTERNATIONAL

Thermocouple
ACROMAG
DATEL

Data acquisition products

Data acquisition board
ANALOGIC
ANALYTEK, LTD.
Data Translation, Inc.
EKF-ELEKTRONIK-MESSTECHNIK
HYTEC ELECTRONICS
INTERACTIVE CIRCUITS AND SYSTEMS
VME MICROSYSTEMS INTERNATIONAL

Data acquisition memory
DSP SYSTEMS

Data acquisition controller
CREATIVE ELECTRONIC SYSTEMS
DATACUBE
Data Cell

12-bit A/D for data acq. controller
BVM LIMITED
DATEL
PENTEK

16-bit A/D for data acq. controller
DATEL

Analog instruments

Frequency counter
EIP Microwave

Frequency synthesizer
PENTEK
SCITEQ

Voltage reference (AC/DC)
SIGNAL PROCESSING INNOVATIONS

Digital Output

5 V/48 mA
ACROMAG
AEROFLEX LABORATORIES

30 V/100 mA
ACROMAG
MEN MIKRO ELEKTRONIK

30 V/0.5 A
EKF-ELEKTRONIK-MESSTECHNIK

35 V/0.5 A
VME MICROSYSTEMS INTERNATIONAL

1.5W @ 0.5A
MICRO-LINK PRODUCTS

24 V/0.5 A
PEP MODULAR COMPUTERS

48 V/600 mA
VME MICROSYSTEMS INTERNATIONAL

60 V/2.0 A
MICROPROCESS systemes

Triac > 110 VAC
MICROPROCESS systemes
VME MICROSYSTEMS INTERNATIONAL

Opto-isolated
CELOGIC
COMPCONTROL B.V.
COMPUTADORES MODULARES
DMS DORSCH MICROSYSTEMS
ESD - ELECTRONIC SYSTEM DESIGN
GreenSpring Computers
KONTRON ELEKTRONIK
MICROSYS
OETTLE + REICHLER
PEP MODULAR COMPUTERS
PHILIPS EXPORT
ROTEC
TEWS DATENTECHNIK
VME MICROSYSTEMS INTERNATIONAL
XYCOM

Relay
CELOGIC
DYNATEM
JANZ COMPUTER AG
KONTRON ELEKTRONIK
MEN MIKRO ELEKTRONIK
MICROPROCESS systemes
MICROSYS
OETTLE + REICHLER
PENTLAND SYSTEMS

PHILIPS EXPORT
THEMIS COMPUTER
VME MICROSYSTEMS INTERNATIONAL
XYCOM

Relay driver
VME MICROSYSTEMS INTERNATIONAL

RS422/485 drivers
VME MICROSYSTEMS INTERNATIONAL

Digital Input

Opto-isolated
CELOGIC
COMPCONTROL B.V.
COMPUTADORES MODULARES
DMS DORSCH MICROSYSTEMS
DYNATEM
ESD - ELECTRONIC SYSTEM DESIGN
GreenSpring Computers
JANZ COMPUTER AG
KONTRON ELEKTRONIK
MICROPROCESS systemes
MICROSYS
OETTLE + REICHLER
PENTLAND SYSTEMS
PEP MODULAR COMPUTERS
PHILIPS EXPORT
ROTEC
TECNINT
TEWS DATENTECHNIK
THEMIS COMPUTER
VME MICROSYSTEMS INTERNATIONAL
XYCOM

TTL
ELTEC ELEKTRONIK
MEN MIKRO ELEKTRONIK
PEP MODULAR COMPUTERS
VME MICROSYSTEMS INTERNATIONAL

> 5 V levels
VME MICROSYSTEMS INTERNATIONAL

Contact debounce
VME MICROSYSTEMS INTERNATIONAL
XYCOM

Digital Input + Output

TTL
ACROMAG
CELOGIC
COMPCONTROL B.V.
DYNATEM
Data Translation, Inc.
EKF-ELEKTRONIK-MESSTECHNIK
ELTEC ELEKTRONIK
GreenSpring Computers
IRONICS INCORPORATED
Io, INCORPORATED
JANZ COMPUTER AG
KEITHLEY METRABYTE
KONTRON ELEKTRONIK
MATRIX
MEN MIKRO ELEKTRONIK
MICRO-LINK PRODUCTS
MICRODIMENSIONS
MICROSYS
MOTOROLA
OETTLE + REICHLER
PENTEK
PENTLAND SYSTEMS
PROTEUS Gesellschaft
RADSTONE TECHNOLOGY
SYSTEMFORSCHUNG
THEMIS COMPUTER
VME MICROSYSTEMS INTERNATIONAL
XYCOM

TTL input + 48 V/50 mA output
ELTEC ELEKTRONIK

TTL input + 24 V/100 mA output
HYDAC ELECTRONIC

TTL: Opto 22 compatble
DATEL
GENERAL MICRO SYSTEM
GreenSpring Computers
IRONICS INCORPORATED
LOGICAL DESIGN GROUP
MATRIX
MIDI INGENIERIE
XYCOM

TTL + RS232
EMBEDDIX EUROPE
MICRODIMENSIONS

> 5 V input + 35 V/600 mA output
VME MICROSYSTEMS INTERNATIONAL

Opto-isolated
DMS DORSCH MICROSYSTEMS
DY 4 SYSTEMS
EKF-ELEKTRONIK-MESSTECHNIK
ELTEC ELEKTRONIK
ESD - ELECTRONIC SYSTEM DESIGN
FORCE COMPUTERS
HYDAC ELECTRONIC
JANZ COMPUTER AG
KWS COMPUTERSYSTEME
LOGICAL DESIGN GROUP
MICRODIMENSIONS
PEP MODULAR COMPUTERS
PHILIPS EXPORT
ROTEC
SYSTEMFORSCHUNG
TECNINT
THEMIS COMPUTER

Opto-isolated input + relay output
RADSTONE TECHNOLOGY

RS485 drivers/recievers
FORMATION
GENERAL MICRO SYSTEM

Digital I/O accessories
ACROMAG

Misc. industrial I/O products

Digital + serial
MATRIX

Digital signal level converters
STRUCK

Counter/timer
ANALYTEK, LTD.
EKF-ELEKTRONIK-MESSTECHNIK
ELTEC ELEKTRONIK
EMBEDDIX EUROPE
GreenSpring Computers
KEITHLEY METRABYTE
KONTRON ELEKTRONIK
MEN MIKRO ELEKTRONIK
MICROPROCESS systemes
MIZAR/INTEGRATED SOLUTIONS
MOTOROLA
OETTLE + REICHLER
PENTLAND SYSTEMS
PEP MODULAR COMPUTERS
SYSTEMFORSCHUNG
VME MICROSYSTEMS INTERNATIONAL
XYCOM

Ribbon-to-terminal-strip adapters
ACROMAG
DATEL
ESD - ELECTRONIC SYSTEM DESIGN
KEITHLEY METRABYTE

IRIG precision timing generator
KINEMETRICS/TrueTime
KSI

Precision time code generator
BANCOMM

Kitchen sink industrial I/O board
HTEC LIMITED
KONTRON ELEKTRONIK
MEN MIKRO ELEKTRONIK
THEMIS COMPUTER
XYCOM

Mezzanine products

Base boards/Mezzanine boards

2701-5XXX (Ideation)
IDEATION

2701-8XXX (Ideation)
IDEATION

4282 (Pentek)
PENTEK

4283 (Pentek)
PENTEK

APEX (Radstone)
Data Cell
RADSTONE TECHNOLOGY

Base board for mezzanine I/O boards
CELOGIC
COMPCONTROL B.V.
DMS DORSCH MICROSYSTEMS
DUUAL COMPUTER
ESD - ELECTRONIC SYSTEM DESIGN
FORCE COMPUTERS
GreenSpring Computers
IDEATION
JANZ COMPUTER AG
KONTRON ELEKTRONIK
MEN MIKRO ELEKTRONIK
METACOMP
MICRODIMENSIONS
MICROPROCESS systemes
MICROSYS
OETTLE + REICHLER
OMNIBYTE
PENTEK
PEP MODULAR COMPUTERS
RADSTONE TECHNOLOGY
SIGNAL PROCESSING INNOVATIONS
SYSTEMFORSCHUNG
TEWS DATENTECHNIK
THEMIS COMPUTER

CPU (Microsys)
MICROSYS

CPU2 (KWS Computer Systeme)
KWS COMPUTERSYSTEME

DCPU (Dynatem)
DYNATEM

DYANA (Signal Processing Inovations)
SIGNAL PROCESSING INNOVATIONS

E PAK (Performance Technologies)
PERFORMANCE TECHNOLOGIES

E32DE CPU-30 (Microprocess)
MICROPROCESS systemes

E32DE DM-00 (Microprocess)
MICROPROCESS systemes

E32DE IDM-00 (Microprocess)
MICROPROCESS systemes

EMX 7126 (Mizar)
MIZAR/INTEGRATED SOLUTIONS

EMX 7136 (Mizar)
MIZAR/INTEGRATED SOLUTIONS

EXMbus (Philips/Radisys)
HTEC LIMITED
INTERGON
LOGICAL DESIGN GROUP
PHILIPS EXPORT
RADISYS

I/O PAK (Performance Technologies)
PERFORMANCE TECHNOLOGIES

Industry Pack (Greenspring computers/SBE)
GreenSpring Computers
SBE

IV-3272 (Ironics)
IRONICS INCORPORATED

IV-9001 (Ironics)
IRONICS INCORPORATED

KJUPIO-X (Kontron/Mizar)
KONTRON ELEKTRONIK

KN030 (Kontron)
KONTRON ELEKTRONIK

M-modules (Men Mikro)
MEN MIKRO ELEKTRONIK

Maxbus
DATACUBE

MEX (Microsys)
MICROSYS

MODULbus (Janz Computers)
JANZ COMPUTER AG

MPVME100X (Systemforchung)
SYSTEMFORSCHUNG

MPVME102X (Systemforchung)
SYSTEMFORSCHUNG

MPVME1500 (Systemforchung)
SYSTEMFORSCHUNG

MVE 01/02 (Microsys)
MICROSYS

MXbus (Kontron/Embeddix)
EMBEDDIX EUROPE
KONTRON ELEKTRONIK
MIZAR/INTEGRATED SOLUTIONS

MZ 7122/24/30 (Mizar)
MIZAR/INTEGRATED SOLUTIONS

MZ71XX (Mizar/Kontron/Integrated Solutions)
KONTRON ELEKTRONIK
MIZAR/INTEGRATED SOLUTIONS

PB-XXX (Pep Modular)
PEP MODULAR COMPUTERS

PCM-100 (Tews Datentechnik)
TEWS DATENTECHNIK

PEP (Pep Modular)
PEP MODULAR COMPUTERS

PEX (Radstone)

PG2050 (Philips)
PHILIPS EXPORT

SAMbus (General Micro System)
GENERAL MICRO SYSTEM

TSVME 10X/12X (Themis)
THEMIS COMPUTER

TSVME 12X (Themis)
THEMIS COMPUTER

UIC 01 (Microsys)
MICROSYS

V68/3X0 (Io, Inc)
Io, INCORPORATED

V68/3X0 (Io, Inc)
Io, INCORPORATED

V68/32 (Io, Inc)
Io, INCORPORATED

VCOM-33 (SBE)
SBE

VME9100D (Logical Design Group)
LOGICAL DESIGN GROUP

VME9100D (Logical Design Group)
LOGICAL DESIGN GROUP

VMEcomm20 (Metacomp)
METACOMP

VMEM-S1 (Pep Modular)
PEP MODULAR COMPUTERS

XVME-42X (Xycom)
XYCOM

XVME-682
XYCOM

XVME-683 (Xycom)
XYCOM

Mezzanine prototyping boards

(PB-XXX (Pep Modular)
PEP MODULAR COMPUTERS

DCPU (Dynatem)
DYNATEM

FLEXI
CELOGIC

IBAM (Eltec)
ELTEC ELEKTRONIK

MAXbus (Datacube)
DATACUBE

MODULObus
CELOGIC

MXbus
EMBEDDIX EUROPE
KONTRON ELEKTRONIK

MXbus (Kontron/Mizar)
MIZAR/INTEGRATED SOLUTIONS

SAMbus (General Micro System)
GENERAL MICRO SYSTEM

VIBus (Eltec)
ELTEC ELEKTRONIK

Mezzanine documentation

MODULObus specification
CELOGIC

Video

CRT graphics controllers

512 x 458 x 8
IDEATION

512 x 512 x 1
MICROGRAPHICS

512 x 512 x 3
CELOGIC

512 x 512 x 4
GRAPHIC STRATEGIES

512 x 512 x 4

512 x 512 x 8
CELOGIC
MATROX ELECTRONIC SYSTEMS

640 x 400 x 4
KWS COMPUTERSYSTEME

640 x 400 x 8
EKF-ELEKTRONIK-MESSTECHNIK

640 x 480 x 4
DMS DORSCH MICROSYSTEMS
GRAPHIC STRATEGIES
OETTLE + REICHLER
PSITECH

640 x 480 x 8
TECNINT

640 x 512 x 4
MICROGRAPHICS

680 x 480 x 8
ELTEC ELEKTRONIK
IDEATION
MATROX ELECTRONIC SYSTEMS

680 x 480 x 16
ELTEC ELEKTRONIK

768 x 512 x 8
PROTEUS Gesellschaft

800 x 400 x 1
OETTLE + REICHLER

800 x 600 x 8
IDEATION
TECNINT

1024 x 512 x 4
CREATIVE ELECTRONIC SYSTEMS
THEMIS COMPUTER

1024 x 768 x 4
DMS DORSCH MICROSYSTEMS
ESD - ELECTRONIC SYSTEM DESIGN
MICROPROCESS systemes
OETTLE + REICHLER
PROTEUS Gesellschaft

1024 x 768 x 8
BVM LIMITED
IDEATION
MEN MIKRO ELEKTRONIK

1024 x 780 x 1
ELTEC ELEKTRONIK

1024 x 800 x 9
COMPUTADORES MODULARES

1024 x 1024 x 1
UNIVISION TECHNOLOGIES
VIGRA

1024 x 1024 x 4
GRAPHIC STRATEGIES
GreenSpring Computers
IRONICS INCORPORATED
MICRO-LINK PRODUCTS
PEP MODULAR COMPUTERS
PSITECH
SYSTEMFORSCHUNG

1024 x 1024 x 8
GRAPHIC STRATEGIES
IDEATION
KWS COMPUTERSYSTEME
PARALLAX GRAPHICS
PERITEK
RADSTONE TECHNOLOGY
TECNINT
THEMIS COMPUTER
UNIVISION TECHNOLOGIES

1024 x 1024 x 10
UNIVISION TECHNOLOGIES

1024 x 1024 x 12
IMAGING TECHNOLOGY
UNIVISION TECHNOLOGIES

1024 x 1024 x 24
CREATIVE ELECTRONIC SYSTEMS
PERITEK
UNIVISION TECHNOLOGIES

1024 x 1024 x 28
PSITECH

1150 x 910 x 1
UNIVISION TECHNOLOGIES

1150 x 910 x 10
UNIVISION TECHNOLOGIES

1150 x 910 x 24
UNIVISION TECHNOLOGIES

1160 x 870 x 4
FORCE COMPUTERS

1280 x 1024 x 1
MIZAR/INTEGRATED SOLUTIONS
UNIVISION TECHNOLOGIES

1280 x 1024 x 1
EMBEDDIX EUROPE

1280 x 1024 x 4
AKEBIA
CETIA
COMPCONTROL B.V.
DY 4 SYSTEMS
ELTEC ELEKTRONIK

1280 x 1024 x 8
GRAPHIC STRATEGIES
JUPITER SYSTEMS
RADSTONE TECHNOLOGY
VIGRA

1280 x 1024 x 8
AKEBIA
CETIA
COMPCONTROL B.V.

DATACUBE
EKF-ELEKTRONIK-MESSTECHNIK
ELTEC ELEKTRONIK
MATROX ELECTRONIC SYSTEMS
MICROSYS
PROTEUS Gesellschaft
TADPOLE TECHNOLOGY

1280 x 1024 x 10
UNIVISION TECHNOLOGIES

1280 x 1024 x 12
PSITECH

1280 x 1024 x 24
DATACUBE
PERITEK
UNIVISION TECHNOLOGIES

1600 x 1200 x 1
PROTEUS Gesellschaft

1600 x 1200 x 8
CETIA

1664 x 1280 x 16
DYNATECH COMPUTER SYSTEMS

2048 x 1024 x 4
ELTEC ELEKTRONIK
MOTOROLA
PHILIPS EXPORT

2048 x 1024 x 5
COMPUTADORES MODULARES

2048 x 1024 x 8
ELTEC ELEKTRONIK
Io, INCORPORATED
MOTOROLA

2048 x 2048 x 12
DATEL
UNIVISION TECHNOLOGIES

1 Mbyte: 8 bits/pixel
DMS DORSCH MICROSYSTEMS

2 Mbyte: 8 bits/pixel
HTEC LIMITED

4 Mbyte: 8 bits/pixel
DR. ING. SEUFERT
HEURIKON

8 Mbyte: 32 bits/pixel
ELTEC ELEKTRONIK

8 Mbyte: 16 bits/pixel
ELTEC ELEKTRONIK

Frame buffers

512 x 512 x 8
DATACUBE
IMAGING TECHNOLOGY

512 x 512 x 8
DATACUBE
RECOGNITION TECHNOLOGY

1024 x 768 x 8
DU PONT PIXEL SYSTEMS

2048 x 1024 x 8
DATACUBE

512 Kbytes
CELOGIC

2 Mbytes
DATACUBE
ELTEC ELEKTRONIK

4 Mbytes
ELTEC ELEKTRONIK
IMAGING TECHNOLOGY

8 Mbytes
ELTEC ELEKTRONIK

40 Mbytes
DU PONT PIXEL SYSTEMS

Video digitizers

12-bit video A/D + D/A
AKEBIA

16-bit video A/D
IMAGING TECHNOLOGY

Display drivers

Character-oriented display controller
CELOGIC
COMPUTADORES MODULARES
IRONICS INCORPORATED
KSI

LCD display controller
KWS COMPUTERSYSTEME
MICROSYS
OETTLE + REICHLER

Electroluminescent display controller
OETTLE + REICHLER

Plasma display controller
OETTLE + REICHLER

CRT display terminal
THEMIS COMPUTER

Graphics + keyboard + mouse

Graphics accessories
AKEBIA
ELTEC ELEKTRONIK

Cable kit

Specialized video products

Video signal regenerator
SOUTHERN RESEARCH TECHNOLOGIES

Frame grabbers
AUGMENT SYSTEMS
BVM LIMITED
CELOGIC
COGNEX
DU PONT PIXEL SYSTEMS
Data Translation, Inc.
ELTEC ELEKTRONIK
OVATION SYSTEMS
PSITECH

Frame grabber + CRT display controller
DATACUBE
IMAGING TECHNOLOGY
MATROX ELECTRONIC SYSTEMS
RECOGNITION TECHNOLOGY

Video image resampling processor
ELTEC ELEKTRONIK
RGB SPECTRUM

Digital mixer
AKEBIA

Image processors/machine vision
COGNEX
DATACUBE
DMS DORSCH MICROSYSTEMS
DU PONT PIXEL SYSTEMS
Data Translation, Inc.
ELTEC ELEKTRONIK
IMAGING TECHNOLOGY
INTERACTIVE CIRCUITS AND SYSTEMS
THEMIS COMPUTER

Real-time video overlay
DATACUBE
DU PONT PIXEL SYSTEMS
DYNATEM
INSTRUMENTATION TECHNOLOGY SYSTEMS
RGB SPECTRUM
SOUTHERN RESEARCH TECHNOLOGIES

Turn-key systems

Desktop/rackmount systems

SPARC/SunOS
ADVANCED PROCESSING LABORATORIES
EMBEDDIX EUROPE
MATRIX
MIZAR/INTEGRATED SOLUTIONS

SPARC/VXWorks
EMBEDDIX EUROPE
MIZAR/INTEGRATED SOLUTIONS

SPARC/VADS/works
MIZAR/INTEGRATED SOLUTIONS

MS-DOS
COMPUTADORES MODULARES

680X0/Unix
DUUAL COMPUTER
ELTEC ELEKTRONIK
EMBEDDIX EUROPE
FORCE COMPUTERS
IRONICS INCORPORATED
Impact Technologies
JANZ COMPUTER AG
MERIT TECHNOLOGY
MIZAR/INTEGRATED SOLUTIONS
MOTOROLA
THEMIS COMPUTER

680X0/Uniplus
IMP

690X0/VADS/works
EMBEDDIX EUROPE
MIZAR/INTEGRATED SOLUTIONS

680X0/VXWorks
EMBEDDIX EUROPE
MIZAR/INTEGRATED SOLUTIONS

680X0/PDOS
FORCE COMPUTERS
MICRO-LINK PRODUCTS
PEP MODULAR COMPUTERS
PERFORMANCE TECHNOLOGIES
ZOLTECH

680X0/OS9
IDEATION
JANZ COMPUTER AG
KWS COMPUTERSYSTEME
MIZAR/INTEGRATED SOLUTIONS
PEP MODULAR COMPUTERS
THEMIS COMPUTER
XYCOM

680X0/SunOS
MIZAR/INTEGRATED SOLUTIONS

88100/Unix
JANZ COMPUTER AG

WE3220X/Unix
IDEATION
JANZ COMPUTER AG

NSXXX/Unix
JANZ COMPUTER AG

RSX-11M
CREATIVE ELECTRONIC SYSTEMS

80X86/MS-DOS
XYCOM

OS/9
BVM LIMITED
COMPCONTROL B.V.
EMBEDDIX EUROPE
HTEC LIMITED

Unix
ELTEC ELEKTRONIK
HEURIKON
MATRIX

PSOS
ENCORE COMPUTER

ADA
ENCORE COMPUTER

No operating system
GreenSpring Computers
IRONICS INCORPORATED
MEN MIKRO ELEKTRONIK

Special purpose systems

Image processing systems
CELOGIC

Transaction processing systems
AITECH

Development systems
AITECH
DYNATEM
Data Cell
ELTEC ELEKTRONIK

Systems without a chassis

Plug-in self-contained system modules
KWS COMPUTERSYSTEME

System board set (680X0/OS9)
THEMIS COMPUTER

Board accessories

DIN 41612 connectors
AMERACE
BICC-VERO ELECTRONICS
ELECTRONIC PRECISION TECHNOLOGY
PANDUIT
ROBINSON NUGENT
SCHROFF
SOURIAU

Ejectors
BUSTRONIC
DAWN VME PRODUCTS
EG&G BIRTCHER
SCANBE

Front panels, handles & hardware
DAWN VME PRODUCTS
EG&G BIRTCHER
ELMA ELECTRONIC
PHILLIPS COMPONENTS
SCANBE

Card rack accessories

Card guides
EG&G BIRTCHER
RITTAL

Divider kits
AUGAT
BICC-VERO ELECTRONICS
ELECTRONIC SOLUTIONS
PHILLIPS COMPONENTS
RITTAL

Cable retainers
PHILIPS EXPORT

Null boards (Daisy-chain jumpers only)
ACT-VME
BICC-VERO ELECTRONICS
BUSTRONIC
COMPCONTROL B.V.
FORMATION
SCHROFF

System reset board
SCHROFF

Power fail monitor board
COMPCONTROL B.V.

Backplane products

Backplanes

VME backplanes
ALPHAPAC
AUGAT
BICC-VERO
BICC-VERO ELECTRONICS
BUSTRONIC
CELOGIC
DATACON dataCon
DATACON dataCon
DAWN VME PRODUCTS
DEAN MICROSYSTEMS
DUUAL COMPUTER
EKF-ELEKTRONIK-MESSTECHNIK
ELECTRONIC SOLUTIONS
ELMA ELECTRONIC
ELTEC ELEKTRONIK
FORCE COMPUTERS
GRAHAM ELECTRONICS MANUFACTURING
HYBRICON
INTERLOGIC INDUSTRIES
MALCO MICRODOT
MATRIX
MUPAC
OETTLE + REICHLER
PEP MODULAR COMPUTERS
PHILIPS EXPORT
RADSTONE TECHNOLOGY
RIBCO
ROTEC
SCANBE
SCHROFF
STANDARD LOGIC
SYSTEM-KONTAKT
TECNINT
XIXIN

NULL backplanes (for P2 airflow control)
BUSTRONIC
DAWN VME PRODUCTS

VSB backplanes
ALPHAPAC
AUGAT
BUSTRONIC
DAWN VME PRODUCTS
HYBRICON
MALCO MICRODOT
MUPAC
RIBCO
SCHROFF
TECNINT

VMXbus backplanes
BUSTRONIC
DAWN VME PRODUCTS
HYBRICON
INTERLOGIC INDUSTRIES
PERFORMANCE TECHNOLOGIES
SCHROFF

AMXbus P2 backplanes (VMIC)
VME MICROSYSTEMS INTERNATIONAL

SRTbus P2 backplanes (VMIC)
VME MICROSYSTEMS INTERNATIONAL

I/O channel backplanes
BICC-VERO ELECTRONICS

Backplane accessories

VMEbus terminators
ALPHAPAC
DAWN VME PRODUCTS
DEAN MICROSYSTEMS
MALCO MICRODOT
RIBCO
SCHROFF
XIXIN

VSB terminators
MALCO MICRODOT
RIBCO

J2 connector shrouds
DAWN VME PRODUCTS
EG&G BIRTCHER

End-to-end backplane splicers
DAWN VME PRODUCTS
HYBRICON
SCHROFF

System controllers

Basic functions
COMPCONTROL B.V.
DUUAL COMPUTER
IDEATION
IRONICS INCORPORATED
KWS COMPUTERSYSTEME
MOTOROLA
PERFORMANCE TECHNOLOGIES
PHILIPS EXPORT
ZOLTECH

Basic functions + system resources
CREATIVE ELECTRONIC SYSTEMS
DY 4 SYSTEMS
FORCE COMPUTERS
IRONICS INCORPORATED
JANZ COMPUTER AG
MOTOROLA
PHILIPS EXPORT
PROTEUS Gesellschaft
XYCOM

System packaging

Fan units
ACT-VME
AUGAT
INTERLOGIC INDUSTRIES
MATRIX
PHILIPS EXPORT
RITTAL
XIXIN

Card cages
AUGAT
BICC-VERO ELECTRONICS
BUSTRONIC
DATACON dataCon
DAWN VME PRODUCTS
EG&G BIRTCHER
ELECTRO-SPACE FABRICATORS, INC
ELECTRONIC SOLUTIONS
FORCE COMPUTERS
INTERLOGIC INDUSTRIES
MEN MIKRO ELEKTRONIK
MUPAC
RITTAL
SCANBE
SCHROFF
STANDARD LOGIC
XIXIN

Enclosures
ELMA ELECTRONIC
MICRO-LINK PRODUCTS
MUPAC
PHILIPS EXPORT
RITTAL
SCHROFF
TRACEWELL ENCLOSURES
XIXIN
ZOLTECH

VMEbus plug-in disk drive enclosures
ELMA ELECTRONIC
ELTEC ELEKTRONIK
INTERLOGIC INDUSTRIES
SCANBE

Powered disk drive enclosures
DYNA FIVE
PHILIPS EXPORT

Card cage + backplane
HYBRICON
IRONICS INCORPORATED
MATRIX
MICRO-LINK PRODUCTS
PHILIPS EXPORT
UNISYS/CONVERGENT

Card cage + backplane + power supply
ELECTRONIC SOLUTIONS
ESD - ELECTRONIC SYSTEM DESIGN
Impact Technologies
MUPAC
PEP MODULAR COMPUTERS
PHILIPS EXPORT
RKB

Card cage + backplane + enclosure
ELECTRONIC SOLUTIONS
ELMA ELECTRONIC
FORCE COMPUTERS
GRAHAM ELECTRONICS MANUFACTURING

Card cage + backplane + power supply + enclosure
AEROFLEX LABORATORIES
AITECH
AUGAT
BICC-VERO
BICC-VERO ELECTRONICS
BIT 3 COMPUTER
BUSTRONIC
CODAR TECHNOLOGY
DATACUBE
DAWN VME PRODUCTS
DMS DORSCH MICROSYSTEMS
DY 4 SYSTEMS
DYNA FIVE
ELECTRONIC SOLUTIONS
ELMA ELECTRONIC
ESD - ELECTRONIC SYSTEM DESIGN
FORCE COMPUTERS
FORMATION
GRAHAM ELECTRONICS MANUFACTURING
HTEC LIMITED
HYBRICON
INCAA COMPUTERS BV
IRONICS INCORPORATED
MATRIX
MOTOROLA
PHILIPS EXPORT
RADISYS
RADSTONE TECHNOLOGY
RUGGED DIGITAL SYSTEMS
SCHROFF
STRUCK
TEXAS INSTRUMENTS
TRACEWELL ENCLOSURES
XIXIN
ZOLTECH

Card cage + backplane + disk subsystem + power supply
PROTEUS Gesellschaft

Card cage + processor + system controller
VME MICROSYSTEMS INTERNATIONAL

Miscellaneous Eurorack hardware
RITTAL

Miscellaneous enclosure hardware
RITTAL

Power supplies

AC-to-DC
AUGAT
BICC-VERO ELECTRONICS
DAWN VME PRODUCTS
EKF-ELEKTRONIK-MESSTECHNIK
ELMA ELECTRONIC
FORCE COMPUTERS
INTERLOGIC INDUSTRIES
PDS TECHNOLOGIES
PHILIPS EXPORT
RADSTONE TECHNOLOGY
VME MICROSYSTEMS INTERNATIONAL
XIXIN

DC-to-DC
AITECH
PDS TECHNOLOGIES

+5V STDBY backup unit
DY 4 SYSTEMS
RADSTONE TECHNOLOGY

VME prototyping aids

Prototype board without VMEbus interface
AUGAT
BICC-VERO ELECTRONICS
BVM LIMITED
DATACON dataCon
DOUGLAS ELECTRONICS
ELECTRONIC PRECISION TECHNOLOGY
ELECTRONIC SOLUTIONS
EMBEDDIX EUROPE
HYBRICON
INTERLOGIC INDUSTRIES
JANZ COMPUTER AG
LOGICAL DESIGN GROUP
MICRO-LINK PRODUCTS
MIZAR/INTEGRATED SOLUTIONS
MUPAC
SCANBE
SCHROFF
STANDARD LOGIC

Prototype board with VMEbus interface
BICC-VERO ELECTRONICS
CELOGIC
DAWN VME PRODUCTS
ELTEC ELEKTRONIK
KONTRON ELEKTRONIK
LOGICAL DESIGN GROUP
PEP MODULAR COMPUTERS
RADSTONE TECHNOLOGY
SIGNAL PROCESSING INNOVATIONS
XYCOM

EPROM Programmer
DMS DORSCH MICROSYSTEMS
EKF-ELEKTRONIK-MESSTECHNIK

VMEbus interface chips
DY 4 SYSTEMS
NEWBRIDGE MICROSYSTEMS
PERFORMANCE TECHNOLOGIES
PLX TECHNOLOGY
SBE
SCHWEBER ELECTRONICS

Debugging accessories

VME extender boards
AUGAT
BUSTRONIC

DAWN VME PRODUCTS
DEAN MICROSYSTEMS
DOUGLAS ELECTRONICS
ELECTRONIC PRECISION TECHNOLOGY
ELECTRONIC SOLUTIONS
ELMA ELECTRONIC
HYBRICON
INTERLOGIC INDUSTRIES
NATIONAL INSTRUMENTS
RADSTONE TECHNOLOGY
SCANBE
SCHROFF
STANDARD LOGIC
XYCOM

VME load boards
ACT-VME
DAWN VME PRODUCTS

NTDS bus analyzers
ANTARES GROUP

VMEbus analyzers
CONCISE TECHNOLOGY
CREATIVE ELECTRONIC SYSTEMS
CYCLONE MICROSYSTEMS
MUPAC
NISSHO ELECTRONICS
ULTRAVIEW
VMETRO
pro_VME INDUSTRIELLE RECHNERTECHNIK

VSB analyzers
MUPAC
VMETRO

VMEbus stimulators
CONCISE TECHNOLOGY
VMETRO

Power supervisor
CONCISE TECHNOLOGY

Scope probe bus access board
DEAN MICROSYSTEMS

Program analyzers/optimizers
ELTEC ELEKTRONIK
VMETRO

Some special features

CMOS (100%)
ALPHI TECHNOLOGY
COMPUTADORES MODULARES
DY 4 SYSTEMS
DYNATEM
WESPERCORP

VSB boards
ELTEC ELEKTRONIK
EMBEDDIX EUROPE
GENERAL MICRO SYSTEM
IDEATION
INTERLOGIC INDUSTRIES
IRONICS INCORPORATED
Io, INCORPORATED
KONTRON ELEKTRONIK
MICROSYS
MIZAR/INTEGRATED SOLUTIONS
RADSTONE TECHNOLOGY
TECNINT
THEMIS COMPUTER

VMXbus boards
FORCE COMPUTERS

HMP boards (HCL America)
HCL AMERICA

I/O channel (Motorola)
MOTOROLA

VME64
PERFORMANCE TECHNOLOGIES

VMEbus Manufacturers' Addresses
USA And Canada

ABB PROCESS AUTOMATION
Telephone: 414-785-3248

ACROMAG
Telephone: 313-624-1541
Fax: 313-624-9234

ACT-VME
Telephone: 215-322-9220
Fax: 215-322-9227

ADVANCED PROCESSING LABORATORIES
Telephone: 619-546-8626
Fax: 619-546-0278

AEON SYSTEMS
Telephone: 505-828-9120
Fax: 505-828-9115

AEROFLEX LABORATORIES
Telephone: 516-694-6700
Fax: 516-694-6771

AITECH
Telephone: 408-720-9400
Fax: 408-746-3630

AJIDA TECHNOLOGIES
Telephone: 707-545-7777
Fax: 707-575-3210

ALLEN-BRADLEY CO.
Telephone: 216-646-5000
Fax: 216-646-3525

ALPHAPAC
Telephone: 603-641-6699
Fax: 603-622-1773

ALPHI TECHNOLOGY
Telephone: 602-838-2428
Fax: 602-838-4477

AMERACE
Telephone: 201-992-8400

AMTRONICS
Telephone: 504-831-0969
Fax: 504-831-0691

ANALOGIC
Telephone: 800-451-1816
Fax: 508-532-6097

ANALOGIC/CDA
Telephone: 800-237-1011
Fax: 508-977-9220

ANALYTEK, LTD.
Telephone: 408-745-1114
Fax: 408-745-1894

ANDREW/KMW SYSTEMS
Telephone: 512-338-3000
Fax: 512-338-3199

ANTARES GROUP
Telephone: 619-223-4311
Fax: 619-223-4759

APPLIED CONCEPTS
Telephone: 619-453-0090
Fax: 619-453-6715

ARIEL SYSTEMS
Telephone: 619-598-7222
Fax: 619-598-7227

ARRAY MICROSYSTEMS
Telephone: 719-540-3888
Fax: 719-540-3970

ASTROSYSTEMS
Telephone: 516-328-1600
Fax: 516-328-1658

ATLANTA SIGNAL PROCESSORS
Telephone: 404-892-7265
Fax: 404-892-2512

AUGAT
Telephone: 800-999-9VME
Fax: 508-699-6717

AUGMENT SYSTEMS
Telephone: 617-271-0230
Fax: 617-275-4461

BALLARD TECHNOLOGY
Telephone: 206-782-8704
Fax: 206-782-8174

BANCOMM
Telephone: 408-578-4161
Fax: 408-578-4165

BEECO INCORPORATED
Telephone: 317-547-1717
Fax: 317-547-1729

BI RA SYSTEMS
Telephone: 505-881-8887
Fax: 505-888-0651

BICC-VERO ELECTRONICS
Telephone: 203-288-8001
Fax: 203-287-0062

BIT 3 COMPUTER
Telephone: 612-881-6955
Fax: 612-881-9674

BUBBL-TEC
Telephone: 415-829-8700
Fax: 415-829-9796

BURR BROWN
Telephone: 800-548-6132
Fax: 602-741-3895

BUSTRONIC
Telephone: 415-490-7388
Fax: 415-490-1853

C & C TECHNOLOGY
Telephone: 708-231-0015
Fax: 708-231-0345

CADENT TECHNOLOGY
Telephone: 714-952-7803

CENTRAL DATA
Telephone: 800-482-0315
Fax: 217-359-6904

CHRISLIN INDUSTRIES
Telephone: 800-468-0736
Fax: 818-991-3490

CIPRICO
Telephone: 612-559-2034

CLEARPOINT RESEARCH
Telephone: 508-435-2000
Fax: 508-435-7504

CODAR TECHNOLOGY
Telephone: 303-776-0472
Fax: 303-776-1806

COGNEX
Telephone: 617-449-6030
Fax: 617-449-4013

COLORADO DATA SYSTEMS
Telephone: 303-762-1640
Fax: 303-781-0253

COMENDEC LTD./ DALLASTONE
Telephone: 603-647-8168
Fax: 603-624-2466

COMMUNICATION MACHINERY CORP. (CMC)
Telephone: 805-968-4262
Fax: 800-CMC-8023

COMPCONTROL
Telephone: 408-356-3817
Fax: 408-356-1755

COMPUSONICS
Telephone: 415-494-1184
Fax: 415-494-3040

CONTROLEX
Telephone: 818-780-8877

CREONICS OPERATION
Telephone: 603-448-6300
Fax: 603-448-5659

CSP, INCORPORATED
Telephone: 617-272-6020
Fax: 508-663-0150

CSPI
Telephone: 617-272-6020

CYCLONE MICROSYSTEMS
Telephone: 203-786-5536

CYCLONE MICROSYSTEMS
Telephone: 203-786-5536

CoSYSTEMS
Telephone: 408-748-2190
Fax: 408-988-0785

DATA TRANSLATION
Telephone: 508-481-3700
Fax: 508-481-8620

DATA/WARE DEVELOPMENT
Telephone: 619-453-7660
Fax: 619-453-2794

dataCon
Telephone: 215-443-8400
Fax: 215-672-9867

DATACUBE
Telephone: 508-535-6644
Fax: 508-535-5643

DATARAM
Telephone: 609-799-0071
Fax: 609-799-6734

DATEL
Telephone: 508-339-3000
Fax: 508-339-6356

DATRON WAVETEK
Telephone: 619-450-0071
Fax: 619-277-6221

DAWN VME PRODUCTS
Telephone: 415-657-4444

DELTA COMPUTER SYSTEMS
Telephone: 206-254-8688
Fax: 206-254-5435

DIGITAL TECHNOLOGY
Telephone: 513-443-0412
Fax: 513-226-0511

DOUGLAS ELECTRONICS
Telephone: 415-483-8770

DSP SYSTEMS
Telephone: 714-630-1330
Fax: 714-630-6320

DUUAL COMPUTER
Telephone: 415-785-8890
Fax: 415-785-5854

DY 4 SYSTEMS
Telephone: 613-596-9911
Fax: 613-596-0574

DYNA FIVE
Telephone: 714-525-8795
Fax: 714-525-9310

DYNATECH COMPUTER SYSTEMS
Telephone: 415-964-7400
Fax: 415-969-3350

DYNATEM
Telephone: 714-855-3235
Fax: 714-770-3481

EG & G BIRTCHER
Telephone: 818-442-2572
Fax: 818-442-8127

EIP MICROWAVE
Telephone: 408-945-1477
Fax: 408-945-0977

ELECTRO-SPACE FABRICATORS
Telephone: 215-682-7181

ELECTRONIC PRECISION TECHNOLOGY
Telephone: 617-661-2454
Fax: 617-661-2989

ELECTRONIC SOLUTIONS
Telephone: 800-854-7086

ELMA ELECTRONIC
Telephone: 415-656-3400

ELTEC
Telephone: 818-449-1558

EMERGE SYSTEMS
Telephone: 407-723-0444
Fax: 407-723-0463

ENCORE COMPUTER
Telephone: 301-499-4700
Fax: 301-441-9087

EYRING
Telephone: 801-YES-PDOS
Fax: 801-374-8339

FAIRCHILD DATA
Telephone: 602-949-1155

FORCE COMPUTERS
Telephone: 408-370-6300
Fax: 408-374-1146

FORMATION
Telephone: 800-257-0452
Fax: 609-234-8543

G.E.T. ENGINEERING
Telephone: 619-443-8295
Fax: 619-443-8613

GALIL MOTION CONTROL
Telephone: 415-964-6494

GD CALIFORNIA
Telephone: 415-847-9660
Fax: 415-847-9663

GE FANUC AUTOMATION
Telephone: 804-978-5000
Fax: 804-978-5205

GENERAL MICRO SYSTEM
Telephone: 714-625-5475
Fax: 714-621-4400

GRAHAM ELECTRONICS MANUFACTURING (GEM)
Telephone: 800-628-4095
Fax: 619-747-9326

GRAPHIC STRATEGIES
Telephone: 800-336-2733
Fax: 408-744-0235

GREENSPRING COMPUTERS
Telephone: 415-327-1200
Fax: 415-327-3808

HCL AMERICA
Telephone: 408-733-0480
Fax: 408-733-0482

HEURIKON
Telephone: 608-831-0900
Fax: 608-831-4249

HEWLETT-PACKARD
Telephone: 303-226-3800

HVE ENGINEERING
Telephone: 800-TRANSL8
Fax: 408-376-2299

HYBRICON
Telephone: 508-772-5422
Fax: 508-772-2963

IBIS SYSTEMS
Telephone: 818-706-2505

ICL DATA DEVICE
Telephone: 516-567-5600

ICS ELECTRONICS
Telephone: 408-263-5500
Fax: 408-263-5896

IDEATION
Telephone: 714-879-7712
Fax: 213-454-6191

IKON
Telephone: 206-728-6465
Fax: 206-728-1633

IMAGING TECHNOLOGY
Telephone: 617-938-8444

IMP
Telephone: 408-429-1338
Fax: 408-429-9342

IMPACT TECHNOLOGIES
Telephone: 408-988-4980
Fax: 408-988-5049

IMPACT TECHNOLOGIES
Telephone: 408-988-4980
Fax: 408-988-5049

INDOCOMP SYSTEMS
Telephone: 313-666-9715
Fax: 313-666-1001

INDUSTRIAL INDEXING SYSTEMS
Telephone: 716-924-9181
Fax: 716-924-2169

INFORMATION MACHINES
Telephone: 201-766-9539
Fax: 201-754-3918

INSTRUMENTATION TECHNOLOGY SYSTEMS
Telephone: 818-886-2034
Fax: 818-886-7573

INTECOLOR
Telephone: 404-623-9145
Fax: 404-623-9163

INTERACTIVE CIRCUITS AND SYSTEMS, LTD.
Telephone: 800-267-9794
Fax: 613-749-9461

INTERGON
Telephone: 800-68VMEAT
Fax: 713-974-6059

INTERLOGIC INDUSTRIES
Telephone: 516-420-8111
Fax: 516-420-8007

INTERPHASE
Telephone: 214-919-9000
Fax: 214-919-9200

INTROL
Telephone: 612-631-7600
Fax: 612-631-7802

IO TECH
Telephone: 216-439-4091
Fax: 216-439-4093

IRONICS INCORPORATED
Telephone: 607-277-4060
Fax: 607-272-5787

Io, INCORPORATED
Telephone: 602-792-0969
Fax: 602-798-1489

JORWAY
Telephone: 516-997-8120

JUPITER SYSTEMS
Telephone: 415-523-9000
Fax: 415-523-9151

KEITHLEY METRABYTE
Telephone: 508-880-3000
Fax: 508-880-0179

KINEMETRICS
Telephone: 707-528-1230
Fax: 707-527-6640

KSI (Division of Odetics)
Telephone: 915-589-9769
Fax: 915-593-2080

LEX ELECTRONICS
Telephone: 800-447-5270

LEXTEL
Telephone: 617-245-5017

LOCKHEED SANDERS
Telephone: 603-885-6022
Fax: 603-885-7264

LOGICAL DESIGN GROUP
Telephone: 919-851-1101
Fax: 919-851-2844

LYNX REAL-TIME SYSTEMS
Telephone: 408-370-2233

MALCO MICRODOT
Telephone: 215-699-5373
Fax: 215-699-5536

MATRIX
Telephone: 919-231-8000
Fax: 919-231-8001

MATROX ELECTRONIC SYSTEMS LTD.
Telephone: 514-685-2630
Fax: 514-685-2853

MDB SYSTEMS
Telephone: 714-998-6900

MEM TECH
Telephone: 408-970-8900
Fax: 408-986-0656

MERCURY COMPUTER SYSTEMS
Telephone: 508-458-3100
Fax: 508-458-9580

MERIT TECHNOLOGY
Telephone: 214-248-2502

METACOMP
Telephone: 619-578-1800
Fax: 619-578-7965

MICRO MEMORY
Telephone: 818-998-0070
Fax: 818-998-4459

MICRO-LINK PRODUCTS
Telephone: 800-428-6155
Fax: 317-848-2254

MICRODIMENSIONS
Telephone: 800-423-7252
Fax: 216-974-1648

MICROENERGY
Telephone: 708-653-5900

MICROGRAPHICS
Telephone: 414-478-2889
Fax: 414-478-3689

MICROVOICE
Telephone: 714-859-1091
Fax: 714-951-7405

MICROWARE SYSTEMS
Telephone: 515-224-1929
Fax: 515-224-1352

MIZAR / INTEGRATED SOLUTIONS
Telephone: 214-446-2664
Fax: 214-242-5997

MORNING STAR TECHNOLOGIES
Telephone: 800-558-7827
Fax: 614-459-5054

MOTOROLA
Telephone: 800-624-8999

MULTIPROCESSOR TOOLSMITHS
Telephone: 613-727-8707

MUPAC
Telephone: 800-92-MUPAC
Fax: 508-588-0498

NATIONAL INSTRUMENTS
Telephone: 800-IEEE-488

NEWBRIDGE MICROSYSTEMS
Telephone: 800-267-7231
Fax: 613-592-1320

NISSHO ELECTRONICS (U.S.A.)
Telephone: 714-261-8811

ODETICS
Telephone: 714-774-5000
Fax: 714-774-9432

OMNIBYTE
Telephone: 800-638-5022

OREGON MICRO SYSTEMS

OREGON SOFTWARE
Telephone: 503-245-2202
Fax: 503-245-8449

PANDUIT
Telephone: 708-532-1800
Fax: 708-887-0230

PARACOM
Telephone: 708-293-9500
Fax: 708-884-9065

PARALLAX GRAPICS
Telephone: 408-727-2220
Fax: 408-980-5139

PC/M
Telephone: 415-829-8700
Fax: 415-829-9796

PDS TECHNOLOGIES
Telephone: 203-925-0123

PENTEK
Telephone: 201-767-7100
Fax: 201-767-1890

PEP MODULAR COMPUTERS
Telephone: 800-228-1737
Fax: 412-279-6860

PERFORMANCE SEMICONDUCTOR
Telephone: 408-734-9000
Fax: 408-734-0258

PERFORMANCE TECHNOLOGIES
Telephone: 716-586-6727
Fax: 716-586-6707

PERITEK
Telephone: 415-531-6500
Fax: 415-530-8563

PHILLIPS COMPONENTS
Telephone: 714-855-4263
Fax: 714-583-9337

PHOENIX CONTACT
Telephone: 717-944-1300
Fax: 717-944-1625

PHOENIX DIGITAL
Telephone: 602-483-7393
Fax: 602-483-7391

PLX TECHNOLOGY
Telephone: 800-759-3753
Fax: 415-960-0479

POLARIS COMMUNICATIONS
Telephone: 503-643-1533
Fax: 503-643-6988

PRIMARY RATE INCORPORATED
Telephone: 603-898-1800
Fax: 603-898-11999

PSITECH
Telephone: 714-964-7818
Fax: 714-968-7884

RACAL-DANA INSTRUMENTS
Telephone: 714-859-8999
Fax: 714-859-2505

RADISYS
Telephone: 800-950-0044
Fax: 503-690-1228

RADSTONE TECHNOLOGY
Telephone: 201-391-2700
Fax: 201-391-2899

READY SYSTEMS
Telephone: 800-228-1249
Fax: 408-736-3400

RECOGNITION TECHNOLOGY
Telephone: 508-366-4800

RGB SPECTRUM
Telephone: 415-848-0180
Fax: 415-848-0971

RHODE ISLAND BACKPANEL COMPANY (RIBCO)
Telephone: 800-338-4096
Fax: 401-849-4948

RITTAL
Telephone: 800-4PS-4000
Fax: 513-390-5599

RKB
Telephone: 301-843-5925

ROBINSON NUGENT
Telephone: 812-945-0211
Fax: 812-945-0804

ROCKWELL INTERNATIONAL
Telephone: 714-762-1476

RUGGED DIGITAL SYSTEMS
Telephone: 415-966-1770
Fax: 415-966-8110

SABTECH INDUSTRIES
Telephone: 714-970-5311
Fax: 714-970-5370

SBE
Telephone: 800-347-COMM
Fax: 415-680-1427

SBS ENGINEERING
Telephone: 800-437-0007
Fax: 505-345-9429

SCANBE
Telephone: 800-227-0557
Fax: 818-444-3953

SCHROFF
Telephone: 401-732-3770
Fax: 401-738-7988

SCI TECHNOLOGY
Telephone: 205-882-4569
Fax: 205-882-4652

SCITEQ ELECTRONICS
Telephone: 619-292-0500
Fax: 619-292-9120

SILICON COMPOSERS
Telephone: 415-322-8763
Fax: 415-322-8778

SIMPACT ASSOCIATES
Telephone: 619-565-1865
Fax: 619-565-4112

SKY COMPUTERS
Telephone: 508-250-1920
Fax: 508-250-0036

SOFTWARE COMPONENTS GROUP
Telephone: 408-437-0700
Fax: 408-437-0711

SOURIAU
Telephone: 805-257-4830
Fax: 805-257-2383

SOUTHERN RESEARCH TECHNOLOGIES
Telephone: 205-581-2900
Fax: 205-581-2903

SPECTRUM SIGNAL PROCESSING
Telephone: 604-438-7266

STANDARD LOGIC
Telephone: 714-779-2897
Fax: 714-779-3499

STANFORD TELECOM
Telephone: 408-748-1010
Fax: 408-980-5756

STORAGE CONCEPTS
Telephone: 714-852-8511
Fax: 714-557-5064

SUN MICROSYSTEMS
Telephone: 415-960-1300

SYNERGY MICROSYSTEMS
Telephone: 619-753-2191
Fax: 619-753-0903

SYSTECH
Telephone: 619-453-8970
Fax: 619-453-0238

SYSTEM STREGIES
Telephone: 212-279-8400

SYSTRAN
Telephone: 513-252-5601
Fax: 513-258-2729

TADPOLE TECHNOLOGY
Telephone: 800-232-6656
Fax: 512-338-4462

TALON INSTRUMENTS
Telephone: 714-599-0690

TASCO ELECTRONIC SERVICES
Telephone: 714-635-0550
Fax: 714-535-3458

TAURUS TECHNOLOGIES
Telephone: 800-486-V860
Fax: 804-873-2102

TEKTRONIX
Telephone: 206-253-5525
Fax: 206-253-6075

TEXAS INSTRUMENTS
Telephone: 415-829-6600
Fax: 415-829-3119

THEMIS COMPUTER
Telephone: 415-734-0870
Fax: 415-734-0873

THOMSON-CSF/CETIA
Telephone: 415-325-6170
Fax: 415-325-6226

TL INDUSTRIES
Telephone: 419-666-8144
Fax: 419-666-6534

TRACEWELL ENCLOSURES
Telephone: 800-848-4525
Fax: 614-846-4450

TRANSMAGNETICS
Telephone: 516-293-3100
Fax: 516-293-3793

ULTRALINK INTERNATIONAL
Telephone: 800-852-6273
Fax: 702-782-2128

ULTRASCIENCE
Telephone: 708-256-0080
Fax: 708-256-0097

ULTRAVIEW
Telephone: 415-657-9501
Fax: 415-657-0927

UNISYS/CONVERGENT
Telephone: 408-435-3482

UNIVERSAL TEST EQUIPMENT
Telephone: 714-956-5337
Fax: 714-956-7250

UNIVISION TECHNOLOGIES
Telephone: 617-221-6700
Fax: 617-221-6777

UniFLEX COMPUTING, LIMITED
Telephone: 800-486-1000
Fax: 919-490-2903

VIGRA
Telephone: 619-483-1197
Fax: 619-483-7531

VISTA CONTROL SYSTEMS
Telephone: 505-662-2484
Fax: 505-662-3956

VME Inc.
Telephone: 408-946-3833
Fax: 408-946-0351

VME MICROSYSTEMS INTERNATIONAL
Telephone: 800-322-3616
Fax: 205-882-0859

VMETRO
Telephone: 713-266-6430
Fax: 713-266-6919

WANDEL & GOLTERMANN
Telephone: 919-941-5730
Fax: 919-941-5751

WAVETEK SAN DIEGO
Telephone: 619-279-2200
Fax: 619-565-9558

WESPERCORP
Telephone: 714-261-0606
Fax: 714-261-0894

WESPERCORP
Telephone: 714-261-0606
Fax: 714-261-0894

WHEDCO INCORPORATED
Telephone: 313-665-5473

WIND RIVER SYSTEMS
Telephone: 415-428-2623
Fax: 415-428-0540

XTAR
Telephone: 619-271-4440

XYCOM
Telephone: 800-367-7300
Fax: 313-429-1010

XYLOGICS
Telephone: 617-272-8140
Fax: 617-273-5392

YARC SYSTEMS
Telephone: 818-889-4388
Fax: 818-889-2658

ZITEL
Telephone: 800-622-5020
Fax: 408-943-0353

ZOLTECH
Telephone: 818-780-1800
Fax: 818-780-1978

Europe

AKEBIA LTD. (U.K.)
Telephone: 01-546 4908

ANGLIA TECHNOLOGY LIMITED (England)
Telephone: 0603-789432

BERGOZ (France)
Telephone: 33-50.41.00.89
Fax: 33-50.41.01.99

BVM LIMITED (England)
Telephone: 0703 270770
Fax: 0489 783589

CELOGIC S.A. (France)
Telephone: (+33)67.52.52.02
Fax: (+33)67.52.59.04

COMENDEC LTD. (U.K.)
Telephone: 021-359-0998
Fax: 021-359-0433

CREATIVE ELECTRONIC SYSTEMS S.A. (Switzerland)
Telephone: (41.22)792.57.45
Fax: (41.22)792.57.48

CREATIVE STRATEGIES PTY. LTD. (Australia)
Telephone: 61 2 427 5526
Fax: 61 2 427 4450

DATA CELL LTD. (U.K.)
Telephone: 734-333666
Fax: 734-332827

DEAN MICROSYSTEMS LTD. (England)
Telephone: 07357-5155
Fax: 07357-2165

DMS DORSCH MICROSYSTEMS GmbH (Germany)
Telephone: (49)04632-1411
Fax: (49)04632-1422

DR. MATERNA GmbH (Germany)
Telephone: (+49)231 559900
Fax: (+49)231 5599100

DU PONT PIXEL SYSTEMS LTD. (England)
Telephone: 44-01-541 1944
Fax: 44-01-546-5335

EKF-ELEKTRONIK-MESSTECHNIK GMBH (Germany)
Telephone: 02381-12630
Fax: 02381-15067

EMBEDDIX EUROPE (U.K.)
Telephone: 44-296-435214
Fax: 44-296-25861

ESD-ELECTRONIC SYSTEM DESIGN (Germany)
Telephone: 0511/3563380
Fax: 0511/3563100

EUROPEL SYSTEMS (ELECTRONICS) LTD. (England)
Telephone: (+44)635 31074
Fax: (+44)635 31070

FERRANTI COMPUTER SYSTEMS (U.K.)
Telephone: 06333-71111
Fax: 06333-61177

FRANKE SOFTWARE (Germany)
Telephone: 0241-51 21 70

GEMS OF CAMBRIDGE LTD. (U.K.)
Telephone: 44 223 214958
Fax: 44 223 215361

HTEC LIMITED (England)
Telephone: (+703)581555

HYDAC ELECTRONIC GmbH (Germany)
Telephone: 0 68 97 509-01
Fax: 0 68 97 509-404

IRTV mbH (Germany)
Telephone: (+49)40 298014-0
Fax: (+49)40 298014-37

JANZ COMPUTER AG (Germany)
Telephone: 49-5251-5197
Fax: 49-5251-55941

KDM-INGENIEURBURO (Germany)
Telephone: 0 60 63 45 22
Fax: 0049 60 63 46 64

KONTRON MESSTECHNIK (Germany)
Telephone: 8165-77611
Fax: 8165-77512

KWS COMPUTERSYSTEME GmbH (Germany)
Telephone: 07243-78068
Fax: 07243-79196

LOUGHBOROUGH SOUND IMAGES LTD. (England)
Telephone: 0509-231843
Fax: 0509-262433

MEN MIKRO ELEKTRONIK GMBH (Germany)
Telephone: 0911-332755
Fax: 0911-396686

METRONIX (Germany)
Telephone: 0531-377077

MICROPROCESS systems (France)
Telephone: (1)47 68 80 80
Fax: (1)47 88 97 85

MICROSYS ELECTRONICS GMBH (Germany)
Telephone: 089/ 63801-0

MICROSYSTEM SERVICES LTD. (U.K.)
Telephone: 0494-441661

MIDI INGENIERIE (France)
Telephone: 61 80 17 61
Fax: 61 54 76 11

OETTLE+REICHLER DATENTECHNIK GmbH (Germany)
Telephone: 0821-157094
Fax: 0821-30257

OVATION SYSTEMS LTD (U.K.)
Telephone: 0844 279638
Fax: 0844 279071

PARACOM GmbH (Germany)
Telephone: +49(241) 166000
Fax: +49(241) 1660050

PHILIPS EXPORT B.V. (The Netherlands)
Telephone: +31 40 786167

pro_VME INDUSTRIELLE RECHNERTECHNIK (Germany)
Telephone: 0049/6223/72029
Fax: 0049/6223/71873

PROTEUS Gesellschaft fur Datentechnik mbH (Germany)
Telephone: 49 721 693015
Fax: 49 721 174263

ROTEC GmbH (Germany)
Telephone: 07222/81336
Fax: 07222/81330

SIEMENS-ALBIS SA (Switzerland)
Telephone: +41 21 631 31 11
Fax: +41 21 634 09 90

SIGINAL PROCESSING INNOVATIONS (Belgium)
Telephone: 32 2-641-2844

STOLLMANN GMBH (Germany)
Telephone: 40/389003-0
Fax: 40/5905997

STRUCK (Germany)
Telephone: 04109/55-0
Fax: 04109/5533

SWAROVSKI ELECTRONIC (Austria)
Telephone: 5224 22450
Fax: 5224 2335

SYSTEM-KONTAKT (Germany)
Telephone: 0 71 36 831-0
Fax: 07136-831122

SYSTEMFORSCHUNG (Germany)
Telephone: 49 228 229029

TECNINT (Italy)
Telephone: +39458620183
Fax: +39458620645

TEWS DATENTECHNIK GmbH (Germany)
Telephone: 04101/42637
Fax: 04101/45753

WORDSWORTH TECHNOLOGY LTD. (England)
Telephone: 0 959 63208
Fax: 0 959 64426

XIXIN LIMITED (U.K.)
Telephone: 0582 400340/ 402637

APPENDIX B

VXIbus Manufacturers

VXIbus Product Directory

VXI ACCESSORIES

Cables
HEWLETT-PACKARD
L-COM

DC-to-DC Power Supplies
HEWLETT-PACKARD

Daisy Chain Jumper Board
DAWN VME PRODUCTS

Extender Boards
AUGAT
BICC-VERO ELECTRONICS
INTERLOGIC INDUSTRIES
MUPAC
STRUCK
STANDARD LOGIC

Front Panels
DAWN VME PRODUCTS

Injector/Ejector Handles
BUSTRONIC
ELMA ELECTRONIC

Local Area Networks
EVIRA

Mainframe Extenders
NATIONAL INSTRUMENTS

Mass Storage
HEWLETT-PACKARD

Mechanical Adapters
DAWN VME PRODUCTS
HEWLETT-PACKARD
ICS ELECTRONICS
TEKTRONIX/CDS

Prototyping Modules
AUGAT
BICC-VERO ELECTRONICS
DELTA TECHNOLOGIES
HEWLETT-PACKARD
ICS ELECTRONICS
MUPAC
RACAL-DANA INSTRUMENTS
STRUCK
STANDARD LOGIC
TEKTRONIX/CDS

SUMBUS Protection Modules
BICC-VERO ELECTRONICS

Serial Interfaces
HEWLETT-PACKARD
ICS ELECTRONICS

Shield Boxes For Modules
BUSTRONIC
ELMA ELECTRONIC
HEWLETT-PACKARD
ICS ELECTRONICS
MUPAC

Test Interface Adapters
HEWLETT-PACKARD
TTI TESTRON
VIRGINIA PANEL

VXIbus Analyzers
HEWLETT-PACKARD
ULTRAVIEW
VMETRO

VXI CONTROLLERS

Embedded VXI Controllers
HEWLETT-PACKARD
LOGICAL DESIGN GROUP
NATIONAL INSTRUMENTS
RACAL-DANA INSTRUMENTS
RADISYS
SPECTRUM SIGNAL PROCESSING
TEKTRONIX/CDS

External VXI Computers with MXI
NATIONAL INSTRUMENTS

MAC-to-VXI
STRUCK

Slot 0 Modules
HEWLETT-PACKARD
NH RESEARCH
NATIONAL INSTRUMENTS
RACAL-DANA INSTRUMENTS
TEKTRONIX/CDS
WESTEST ENGINEERING

VXI ENCLOSURES

Backplanes
AUGAT
BICC-VERO ELECTRONICS
BUSTRONIC
DAWN VME PRODUCTS
ELMA ELECTRONIC
INTERLOGIC INDUSTRIES
MUPAC
SCHROFF INC.

Mainframes-Card Racks-Enclosures
AUGAT
BICC-VERO ELECTRONICS
BUSTRONIC

DAWN VME PRODUCTS
ELMA ELECTRONIC
HEWLETT-PACKARD
INTERLOGIC INDUSTRIES
KNURR USA
MUPAC
NATIONAL INSTRUMENTS
RACAL-DANA INSTRUMENTS
STRUCK
SCHROFF INC.
TEKTRONIX/CDS
TRACEWELL ENCLOSURES

VXI INSTRUMENTS

A/D, DA Converters
ANALOGIC
GIORDANO ASSOCIATES
HEWLETT-PACKARD
NATIONAL INSTRUMENTS
TASCO ELECTRONICS
TEKTRONIX/CDS
TRANSMAGNETICS

AC Voltmeters
NH RESEARCH
TEKTRONIX/CDS

ARINC 429 Testers
TASCO ELECTRONICS
TEKTRONIX/CDS

Amplifiers
CAL-AV LABS
HEWLETT-PACKARD

Attenuators
CAL-AV LABS

Counters
EIP MICROWAVE
HEWLETT-PACKARD
RACAL-DANA INSTRUMENTS
TEKTRONIX/CDS

Data Acquisition
ANALOGIC
TEKTRONIX/CDS

Digital I/O
C&H ENGINEERING
HEWLETT-PACKARD
RACAL-DANA INSTRUMENTS
SERENDIPITY
TALON INSTRUMENTS
TEKTRONIX/CDS
WESTEST ENGINEERING

Digital Multimeters
HEWLETT-PACKARD
RACAL-DANA INSTRUMENTS
TEKTRONIX/CDS

Digital Oscilloscopes
HEWLETT-PACKARD

FET Multiplexers/Switches
NSD

Frequency Response Analyzers
SCHLUMBERGER TECHNOLOGIES

Frequency Standards
RACAL-DANA INSTRUMENTS

LCR Meters
VERETEST

Logic Analyzers
HEWLETT-PACKARD

MATE-STD-IAC
TEKTRONIX/CDS

MIL-STD-1553
DIGITAL TECHNOLOGY
EXCALIBUR SYSTEMS
ILC DATA DEVICE
NSD
TEKTRONIX/CDS

Microwave Equipment
HEWLETT-PACKARD
TEKTRONIX/CDS

Pattern/Word Generators
HEWLETT-PACKARD
KIKUSUI INTERNATIONAL
RACAL-DANA INSTRUMENTS
TEKTRONIX/CDS

Phase Angle Generators
TRANSMAGNETICS

Phase Angle Voltmeters
HEWLETT-PACKARD
NORTH ATLANTIC

Power Meters
HEWLETT-PACKARD

Power Supplies
DELTRON
ELGAR
INTERLOGIC INDUSTRIES
KIKUSUI INTERNATIONAL
TRANSMAGNETICS
VERETEST
WESTCOR

Programmable Attenuators
QUARTZLOCK INSTRUMENTS

Programmable Loads
NH RESEARCH

Programmable Resistors
STRUCK
TEKTRONIX/CDS

Pulse Generators
AAI

RF Multiplexers/Switches
HEWLETT-PACKARD
NH RESEARCH
NSD
QUARTZLOCK INSTRUMENTS
TEKTRONIX/CDS

Relay Multiplexers/Switches
A.D. DATA SYSTEMS
C&H ENGINEERING
HEWLETT-PACKARD
NH RESEARCH
RACAL-DANA INSTRUMENTS
TASCO ELECTRONICS
TEKTRONIX/CDS

Sample And Hold
ANALOGIC

Signal Generators
HEWLETT-PACKARD

Sweep Generators
ELMA ELECTRONIC

Synchro-to-digital/Digital-to-synchro
ILC DATA DEVICE
NORTH ATLANTIC
TEKTRONIX/CDS
TRANSMAGNETICS

Threshold Detector
KIKUSUI INTERNATIONAL

Timers
NH RESEARCH
RACAL-DANA INSTRUMENTS

Waveform Digitizers
HEWLETT-PACKARD
TEKTRONIX/CDS

Waveform Generators
WAVETEK

Word Generators with D/A Output
GIORDANO ASSOCIATES INC.
TASCO ELECTRONICS

VXI SOFTWARE

Software
ADONIS
ALYSYS
ANALOGIC
AUTOMATED REASONING
HEWLETT-PACKARD
MICROWARE SYSTEMS
NATIONAL INSTRUMENTS
RADISYS
TESTNIQUES INC.
TYX
TEKTRONIX/CDS
VXI SOLUTIONS
WAVETEK

VXI SYSTEMS

Development Systems
AUGAT
ELMA ELECTRONIC
KIKUSUI INTERNATIONAL

Integrated Test Systems
APPLIED DIAGNOSTICS
ASCOR
BLAKELL ATE LTD.
GIORDANO ASSOCIATES
HEWLETT-PACKARD
ICS ELECTRONICS
KIKUSUI INTERNATIONAL
NH RESEARCH
TYX
TALON INSTRUMENTS

Starter Systems
HEWLETT-PACKARD
RADISYS
TEKTRONIX/CDS

System Integration Services
C&H ENGINEERING
HEWLETT-PACKARD
PROTEUS
RO-MAR

VXI-TO-???

VXI-to-GPIB
ICS ELECTRONICS

VXI-to-MXI
HEWLETT-PACKARD
TRANSMAGNETICS

VXI-to-T1
PRIMARY RATE

VXIbus Manufacturers' Addresses

A.D. Data Systems Corp.
770 Basket Rd
Webster, NY 14580-0478
Tele: (716) 265-1600
Fax: (716) 265-1689

AAI Corp.
Mail Stop 208/106 P.O. Box 126
Hunt Valley, MD 21031
Tele: (301) 527-2515

Adonis Corp.
12310 NE 8th St
Bellevue, WA 98005
Tele: (206) 451-4742

Alsys, Inc.
67 S Bedford St
Burlington, MA 01803-5152
Tele: (617) 270-0030
Fax: (617) 270-6882

Analogic Corp.
360 Audobon Rd
Wakefield, MA 01180
Tele: (800) 446-8936
Fax: (508) 531-1266

Applied Diagnostics, Inc.
121 Industry Dr
Pittsburgh, PA 15275
Tele: (412) 787-3002
Fax: (412) 787-9836

Ascor, Inc.
47790 Westinghouse Dr
Fremont, CA 94539
Tele: (415) 490-8819
Fax: (415) 490-8493

Augat, Inc.
452 John Dietsch Blvd
Attleboro Falls, MA 02763
Tele: (508) 699-9800
Fax: (508) 699-6717

Automated Reasoning Corp.
25 Davids Drive
Hauppauge, NY 11788
Tele: (516) 231-7777
Fax: (516) 231-7075

BICC-VERO Electronics, Inc.
1000 Sherman Ave
Hamden, CT 06514
Tele: (800) 242-2863
Fax: (203) 287-0062

Blakell ATE LTD.
Woodbridge Industrial Park
Wimborne/Doset BH21 6SP
United Kingdom
Tele: (44) 202 823381
Fax: (44) 202 823388

BUSTRONIC CORP.
5680 Stewart Ave
Fremont, CA (415) 490-7388
Tele: (415) 490-7388
Fax: (415) 490-1853

C&H Engineering, Inc.
8705 Shoal Creek, STE 107
Austin, TX 78758
Tele: (512) 467-7444
Fax: (512) 834-9165

CAL-AV LABS, INC.
515B Westchester Dr
Campbell, CA 95008
Tele: (408) 371-0666
Fax: (408) 371-0672

Dawn VME Products
47073 Warm Springs Blvd
Fremont, CA 94539
Tele: (415) 657-4444
Fax: (415) 657-3274

Delta Technolgies, Inc.
978 Salem St
Aurora, CO 80011
Tele: (719) 6159
Fax: (206) 254-5435

Deltron, Inc.
290 Wissahickon Ave
North Wales, PA 19454
Tele: (215) 699-9261
Fax: (215) 699-2310

Digital Technology, Inc.
2300 Edwin C. Moses Blvd
Dayton, OH 45408
Tele: (513) 443-0412
Fax: (513) 226-0511

EIP Microwave, Inc.
1589 Centre Pointe Dr
Milpitas, CA 95035
Tele: (408) 945-1477
Fax: (408) 945-0977

Elgar Corp.
9250 Brown Deer Rd
San Diego, CA 92121
Tele: (619) 450-0085
Fax: (619) 458-0267

ELMA Electronic, Inc.
41440 Christy St
Fremont, CA 94538
Tele: (415) 656-3400
Fax: (415) 656-3400

EVIRA S.A.
Z.I. Des Playes
83500 LaSeyne Sur Mer France
Tele: 33 94 302500

Excalibur Systems, Inc.
P.O. Box 6839
Fresh Meadows, NY 11365
Tele: (718) 357-3500
Fax: (718) 357-3544

Giordano Associates, Inc.
703 60th St Court E STE A
Bradenton, FL 34208
Tele: (813) 746-8753
Fax: (201) 202-9416

Hewlett Packard
P.O. Box 301
Loveland, CO 80539
Tele: (303) 679-3207
Fax: (303) 679-5953

ICS Electronics Corp.
744 S. Hillview Dr
Milpitas, CA 95035-5545
Tele: (408) 263-5500
Fax: (408) 263-3896

ILC Data Device Corp.
105 Wilbur Pl
Bohemia, NY 11716
Tele: (516) 567-5600
Fax: (516) 567-7358

Interlogic Industries
85 Marcus Dr
Melville, NY 11747-4294
Tele: (516) 420-8111
Fax: (516) 420-8007

Kikusui International Corp.
19601 Mariner Ave
Torrance, CA 90503
Tele: (800) 545-8784
Fax: (213) 371-4662

Knurr USA, Inc.
1890 N. Voyager Ave
Simi Valley, CA 93063
Tele: (805) 526-7733

L-com, Inc.
1755 Osgood St
N Andover, MA 01845
Tele: (508) 682-6936
Fax: (508) 689-9484

Logical Design Group, Inc.
6301 Chapel Hill Rd
Raleigh, NY 27607
Tele: (919) 851-1101
Fax: (919) 851-2844

Microware
1900 NW 114 St
Des Moines, IA 50322
Tele: (515) 224-1929
Fax: (515) 224-1352

Mupac Corp.
10 Mupac Dr
Brockton, MA 02401
Tele: (508) 588-6110
Fax: (508) 588-0498

N.H. Research, Inc.
16601 Hale Ave
Irvine, CA 92714
Tele: (714) 474-3900
Fax: (714) 474-7062

National Instruments Corp.
6504 Bridge Point Pkwy
Austin, TX 78730-5039
Tele: (800) 433-3488
Fax: (512) 794-8411

North Atlantic Inc.
60 Plant Ave
Hauppauge, NY 11788
Tele: (516) 582-6500
Fax: (516) 582-8079

NSD Corp.
Box 2888
Winnetka, CA 91036

Primary Rate Incorporated
15 Ermer Rd, STE 210
Salem, NH 03079
Tele: (603) 898-1800
Fax: (603) 898-1199

Proteus Corp.
5827 S Rapp Str
LIttleton, CO 80120
Tele: (303) 795-5944
Fax: (303) 795-7225

Quartzlock Instruments
Gothic/Plymouth Rd
Totnes/Devon TQ9 5LH
England
Tele: (0)803 862062
Fax: (0) 803 867962

Racal-Dana Instruments, Inc.
4 Goodyear St
Irvine, CA 92718-2002
Tele: (714) 859-8999
Fax: (714) 859-2505

RadiSys Corp.
19545 NW Von Neuman Dr
Beaverton , OR 97006
Tele: (503) 690-1229
Fax: (503) 690-1228

RO-MAR, LTD.
2268 Executive Cir
Colorado Springs, CO 80908
Tele: (719) 540-0600
Fax: (719) 576-9105

Schlumberger Technologies
829 Middlesex Turnpike
Billerica, MA 01821
Tele: (508) 671-9700
Fax: (508) 671-9704

Schroff Inc.
170 Commerce Dr
Warwick, RI 02886
Tele: (401) 732-3770
Fax: (401) 738-7988

Serendipity
5030 Allan Rd
Bethesda, MD 20816
Tele: (301) 229-5999
Fax: (301) 229-6259

Spectrum Signal Processing Inc.
1500 W Park Dr
Westborough, MA 01581
Tele: (508)366-7355
Fax: (508) 898-2772

Standard Logic, Inc.
2841 E La Palma Ave
Anaheim, CA 92806
Tele: (714) 632-9292
Fax: (714) 632-9699

STRUCK
Bakerbarg 6
D 2000 Tangstedt
W. Germany
Tele: 4 0955 9
Fax: 041095533

Talon Instruments
150 E Arrow Highway
San Dimas, CA 91773
Tele: (714) 599-0690
Fax: (714) 599-6529

TASCO Electronics, Inc.
1950 W Corporate Way
Anaheim, CA 92801
Tele: (714) 635-0550
Fax: (714) 535-3458

Tektronix/CDS
P.O. Box 3500, M/S C1-949
Vancouver, WA 98668
Tele: (800) TKE-WIDE
Fax: (303) 781-0253

TESTniques Inc.
P.O. Box 27401
Minneapolis, MN 55427
Tele: (612) 533-4107

Tracewell Enclosures, Inc.
567 Enterprise Dr
Westerville, OH 43081
Tele: (800) 848-4525
Fax: (614) 846-4450

Transmagnetics, Inc.
210 Adams Boulevard
Farmingdale, NY 11735
Tele: (516) 293-3100
Fax: (516) 293-3793

TTI Testron, Inc.
41 Century Dr
Woonsocket, RI 02895
Tele: (401) 766-9100
Fax: (401) 766-9017

TYX Corp.
1851 Alexander Bell Dr, STE 200
Reston, VA 22091-4384
Tele: (703) 264-1080
Fax: (703) 264-1090

Ultraview Corp.
475 Yampa Way
Fremont, CA 94539
Tele: (415) 657-9501
Fax: (415) 657-0927

VereTest Inc.
45 E Main St
Southborough, MA 01772
Tele: (508) 485-5522
Fax: (508) 485-8771

Virginia Panel Corp.
1400 New Hope Rd P.O. Box 1407
Waynesboro, VA 22980
Tele: (703) 949-8376
Fax: (703) 942-2856

VMETRO, Inc.
2500 Wilcreest #550
Houston, TX 77042
Tele: (713) 266-6430
Fax: (713) 266-6919

VXI Solutions, Inc.
1412 Tower Square, STE 1
Ventura, CA 93003
Tele: (805) 654-1318
Fax: (895) 644-8128

Wavetek, Inc.
9045 Balboa Ave
San Diego, CA 92123
Tele: (619) 279-2200
Fax: (619) 450-0325

Westcor Corp.
485-100 Alberto Way
Los Gatos, CA 95032
Tele: (408) 395-7050
Fax: (408) 395-1518

WesTest Engineering Corp.
1470 N Main
Bountiful, UT 84010
Tele: (801) 298-7100

The VMEbus Specification

Revision C.1

THE HISTORY OF THE VMEbus

The architectural concepts of VMEbus date back to Motorola's development of the second generation 68000 microprocessor in the late 1970's. A team of engineers at Motorola Microsystems led by Jack Kister, designed a 68000 development system called the EXORmacs. The backplane of the EXORmacs was called VERSAbus. While coordinating the efforts of his team, Jack wrote a 41 page bus description of VERSAbus which was published in November of 1979. The first EXORmacs was shipped in January 1980.

Shortly after that, John Black was assigned the job of developing a new line of board products called VERSAmodules. Upon review of the existing VERSAbus specification, he concluded that it was insufficient to guarantee compatibility between independently designed boards. As a result, he and Jack Kister began refining the documentation.

As VERSAmodule boards were designed, inadequacies in the existing documentation were uncovered and corrected. In December of 1980, Motorola published a much expanded 188 page version of the VERSAbus specification. One month later that company shipped its first VERSAbus-based single board computer, called the VERSAmodule 1. During the next six months, as John continued refining the specification, Motorola published 3 more revisions and in July of 1981 it published the fifth and last revision, which consisted of 240 pages.

During this same time Motorola's European Microsystems group in Munich, Germany proposed the development of a VERSAbus-like product line based on the Eurocard board size. To demonstrate the concept, Max Loesel and Sven Rau developed three prototype boards: a 68000 CPU board, a dynamic RAM board, and a static RAM board. They named the bus VERSAbus-E.

Shortly after this, Motorola licensed both Mostek (Carrolton, Texas) and N.V. Philips/Signetics (Sunnyvale, California) to second source the 68000 and co-develop support chips for the 68000 family. Since both of these companies planned to manufacture board products, Motorola invited them to discuss the possibility of jointly supporting a bus. As a result of that meeting, all three companies agreed to support VERSAbus-E, which they renamed VMEbus.

John Black of Motorola, Craig Mackenna of Mostek, and Cecil Kaplinsky of Signetics, acting as technical representatives for their respective companies, took the fifth revision of the VERSAbus specification and developed the first VMEbus specification, Revision A. In October 1981, the three companies announced their joint support of VMEbus at the "System 81" show in Munich, Germany and placed the VMEbus specification in the public domain.

In August of 1982 the VMEbus manufacturer's group published a second version of the specification, called Revision B. Revision B refined the specifications for the signal line drivers and also brought the mechanical specifications more into line with the developing IEC 297-3 Eurocard standard.

During 1982 the French delegation to the International Electrotechnical Commission (IEC) proposed Revision B as an international standard. The IEC established a subcommittee under its 47b committee to start formal standardization of VMEbus, calling it the IEC 821 BUS. Mira Pauker of Philips was given editorial responsibility.

In March of 1983 the IEEE Microprocessor Standards Committee voted to establish a committee of their own to standardize VMEbus, which the IEEE later gave the name IEEE P1014. Wayne Fischer was appointed chairman. John Black agreed to serve as chairman of the P1014 technical subcommittee.

The VMEbus Manufacturers Group, the IEC, and the IEEE distributed copies of Revision B for comment. John Black and Shlomo Pri-Tal incorporated changes from all three sources into a common document. The resulting document, called Revision C.1, was first published in 1985.

ACKNOWLEDGMENTS

The development of the Revision C.1 was a team effort which involved experts from the US, the UK, West Germany, France and many other nations. It is truly an internationally developed standard.

However there are several individuals whose contribution is worthy of special recognition. John Black, now of Micrology pbt, took a brief bus specification (VERSAbus) and expanded it to enough detail that boards could be designed independently and still be compatible with one another. John also served as chairman of the IEEE P1014 technical subcommittee during the standardization of VMEbus. Craig Mackenna, now of Signetics, worked with John, starting with the first revision of the VMEbus specification and following through to the completion of Revision C.1. Without John's and Craig's efforts the VMEbus would not be what it is today.

Max Loesel and Sven Rau proposed and demonstrated the feasability of applying the Eurocard mechanical standard to the VERSAbus. Eike Waltz, of Schroff, deserves special recognition for his extensive contribution to the mechanical chapter which was incorporated into Revision C. Wayne Fischer, of Force Computers, serving as chairman of the IEEE P1014 Standard Committee was responsible for guiding that committee during its first three years, and Shlomo Prital served as chairman during the remainder of the project. Mira Pauker, of Philips, served as president of the IEC 821 BUS editorial committee. The IEC's timely standardization of VMEbus can largely be attributed to Mira's persistence and determination.

Also worthy of recognition are Paul Borril, of University College London, who contributed to the electrical specifications and Ken Smith, of BICC VERO, who contributed to the mechanical specifications.

TABLE OF CONTENTS

CHAPTER 1 INTRODUCTION TO THE VMEbus SPECIFICATION ....Page

CHAPTER 2 DATA TRANSFER BUS

CHAPTER 5 UTILITIES

CHAPTER 6 ELECTRICAL SPECIFICATIONS

LIST OF ILLUSTRATIONS

LIST OF TABLES

CHAPTER 1

INTRODUCTION TO THE VMEbus SPECIFICATION

1.1 VMEbus SPECIFICATION OBJECTIVES

The VMEbus specification defines an interfacing system used to interconnect data processing, data storage, and peripheral control devices in a closely coupled hardware configuration. The system has been conceived with the following objectives:

a. To allow communication between devices on the VMEbus without disturbing the internal activities of other devices interfaced to the VMEbus.

b. To specify the electrical and mechanical system characteristics required to design devices that will reliably and unambiguously communicate with other devices interfaced to the VMEbus.

c. To specify protocols that precisely define the interaction between the VMEbus and devices interfaced to it.

d. To provide terminology and definitions that describe system protocol.

e. To allow a broad range of design latitude so that the designer can optimize cost and/or performance without affecting system compatibility.

f. To provide a system where performance is primarily device limited, rather than system interface limited.

1.2 VMEbus INTERFACE SYSTEM ELEMENTS

1.2.1 Basic Definitions

The VMEbus structure can be described from two points of view: its mechanical structure and its functional structure. The mechanical specification describes the physical dimensions of subracks, backplanes, front panels, plug-in boards, etc. The VMEbus functional specification describes how the bus works, what functional modules are involved in each transaction, and the rules which govern their behavior. This section provides informal definitions for some basic terms used to describe both the mechanical and the functional structure of the VMEbus.

1.2.1.1 Terms Used To Describe The VMEbus Mechanical Structure

VMEbus BACKPLANE — A printed circuit (PC) board with 96 pin connectors and signal paths that bus the connector pins. Some VMEbus systems have a single PC board, called the J1 backplane. It provides the signal paths needed for basic operation. Other VMEbus systems also have an optional second PC board, called a J2 backplane. It provides the additional 96 pin connectors and signal paths needed for wider data and address transfers. Still others have a single PC board that provides the signal conductors and connectors of both the J1 and J2 backplanes.

BOARD — A printed circuit (PC) board, its collection of electronic components, and either one or two 96 pin connectors that can be plugged into VMEbus backplane connectors.

SLOT — A position where a board can be inserted into a VMEbus backplane. If the VMEbus system has both a J1 and a J2 backplane (or a combination J1/J2 backplane) each slot provides a pair of 96 pin connectors. If the system has only a J1 backplane, then each slot provides a single 96 pin connector.

SUBRACK — A rigid framework that provides mechanical support for boards inserted into the backplane, ensuring that the connectors mate properly and that adjacent boards do not contact each other. It also guides the cooling airflow through the system, and ensures that inserted boards do not disengage themselves from the backplane due to vibration or shock.

1.2.1.2 Terms Used To Describe The VMEbus Functional Structure

Figure 1-1 shows a simplified block diagram of the functional structure, including the VMEbus signal lines, backplane interface logic, and functional modules.

BACKPLANE INTERFACE LOGIC — Special interface logic that takes into account the characteristics of the backplane: its signal line impedance, propagation time, termination values, etc. The VMEbus specification prescribes certain rules for the design of this logic based on the maximum length of the backplane and its maximum number of board slots.

FUNCTIONAL MODULE — A collection of electronic circuitry that resides on one VMEbus board and works together to accomplish a task.

DATA TRANSFER BUS — One of the four buses provided by the VMEbus backplane. The Data Transfer Bus allows MASTERS to direct the transfer of binary data between themselves and SLAVES. (Data Transfer Bus is often abbreviated DTB.)

DATA TRANSFER BUS CYCLE — A sequence of level transitions on the signal lines of the DTB that result in the transfer of an address or an address and data between a MASTER and a SLAVE. The Data Transfer Bus cycle is divided into two portions, the address broadcast and then zero or more data transfers. There are 34 types of Data Transfer Bus cycles. They are defined later in this chapter.

MASTER — A functional module that initiates DTB cycles in order to transfer data between itself and a SLAVE module.

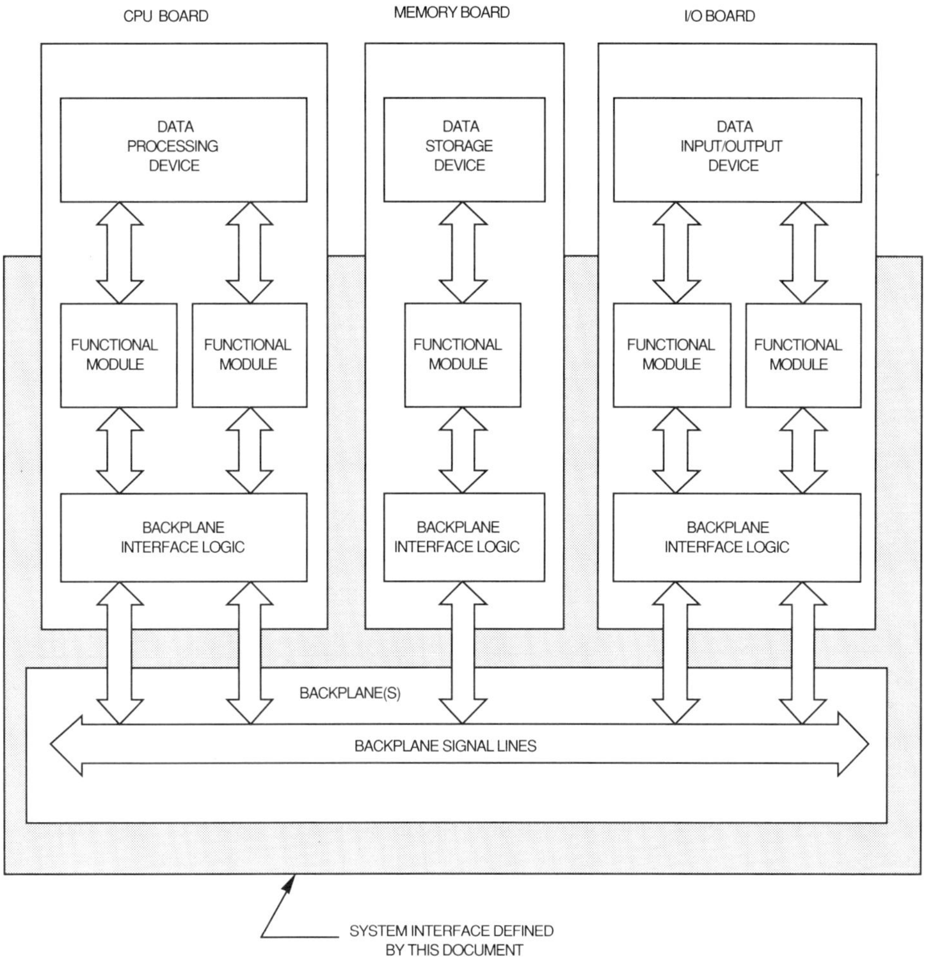

Figure 1-1. System Elements Defined By This Document

SLAVE — A functional module that detects DTB cycles initiated by a MASTER and, when those cycles specify its participation, transfers data between itself and the MASTER.

LOCATION MONITOR — A functional module that monitors data transfers over the DTB in order to detect accesses to the locations it has been assigned to watch. When an access to one of these assigned locations occurs, the LOCATION MONITOR generates an on-board signal.

BUS TIMER — A functional module that measures how long each data transfer takes on the DTB and terminates the DTB cycle if a transfer takes too long. Without this module, if the MASTER tries to transfer data to or from a non-existent SLAVE location it might wait forever. The BUS TIMER prevents this by terminating the cycle.

PRIORITY INTERRUPT BUS — One of the four buses provided by the VMEbus backplane. The Priority Interrupt Bus allows INTERRUPTER modules to send interrupt requests to INTERRUPT HANDLERS.

INTERRUPTER — A functional module that generates an interrupt request on the Priority Interrupt Bus and then provides STATUS/ID information when the INTERRUPT HANDLER requests it.

INTERRUPT HANDLER — A functional module that detects interrupt requests generated by INTERRUPTERS and responds to those requests by asking for STATUS/ID information.

DAISY-CHAIN — A special type of VMEbus signal line that is used to propagate a signal level from board to board, starting with the first slot and ending with the last slot. There are four bus grant daisy-chains and one interrupt acknowledge daisy-chain on the VMEbus.

IACK DAISY-CHAIN DRIVER — A functional module which activates the interrupt acknowledge daisy-chain whenever an INTERRUPT HANDLER acknowledges an interrupt request. This daisy-chain ensures that only one INTERRUPTER will respond with its STATUS/ID when more than one has generated an interrupt request.

ARBITRATION BUS — One of the four buses provided by the VMEbus backplane. This bus allows an ARBITER module and several REQUESTER modules to coordinate use of the DTB.

REQUESTER — A functional module that resides on the same board as a MASTER or INTERRUPT HANDLER and requests use of the DTB whenever its MASTER or INTERRUPT HANDLER needs it.

ARBITER — A functional module that accepts bus requests from REQUESTER modules and grants control of the DTB to one REQUESTER at a time.

UTILITY BUS — One of the four buses provided by the VMEbus backplane. This bus includes signals that provide periodic timing and coordinate the power-up and power-down of VMEbus systems.

SYSTEM CLOCK DRIVER — A functional module that provides a 16 MHz timing signal on the Utility Bus.

SERIAL CLOCK DRIVER — A functional module that provides a periodic timing signal that synchronizes operation of the VMSbus. (Although the VMEbus specification defines a SERIAL CLOCK DRIVER for use with the VMSbus, and although it reserves two backplane signal lines for use by that bus, the VMSbus protocol is completely independent of the VMEbus). Timing specifications for the SERIAL CLOCK DRIVER are given in Appendix C.

POWER MONITOR MODULE — A functional module that monitors the status of the primary power source to the VMEbus system, and signals when that power has strayed outside the limits required for reliable system operation. Since most systems are powered by an AC source, the POWER MONITOR is typically designed to detect drop-out or brown-out conditions on AC lines.

SYSTEM CONTROLLER BOARD — A board which resides in slot 1 of a VMEbus backplane and has a SYSTEM CLOCK DRIVER, an ARBITER, an IACK DAISY-CHAIN DRIVER, and a BUS TIMER. Some also have a SERIAL CLOCK DRIVER, a POWER MONITOR or both.

1.2.1.3 Types Of Cycles On The VMEbus

READ CYCLE — A DTB cycle used to transfer 1, 2, 3, or 4 bytes from a SLAVE to a MASTER. The cycle begins when the MASTER broadcasts an address and an address modifier. Each SLAVE captures the address modifier and address, and checks to see if it is to respond to the cycle. If so, it retrieves the data from its internal storage, places it on the data bus and acknowledges the transfer. The MASTER then terminates the cycle.

WRITE CYCLE — A DTB cycle used to transfer 1, 2, 3, or 4 bytes from a MASTER to a SLAVE. The cycle begins when the MASTER broadcasts an address and address modifier and places data on the DTB. Each SLAVE captures the address and address modifier and checks to see if it is to respond to the cycle. If so, it stores the data and then acknowledges the transfer. The MASTER then terminates the cycle.

BLOCK READ CYCLE -- A DTB cycle used to transfer a block of 1 to 256 bytes from a SLAVE to a MASTER. This transfer is done using a string of 1, 2, or 4 byte data transfers. Once the block transfer is started, the MASTER does not release the DTB until all of the bytes have been transferred. It differs from a string of read cycles in that the MASTER broadcasts only one address and address modifier (at the beginning of the cycle). Then the SLAVE increments this address on each transfer so that the data for the next transfer is retrieved from the next higher location.

BLOCK WRITE CYCLE — A DTB cycle used to transfer a block of 1 to 256 bytes from a MASTER to a SLAVE. The block write cycle is very similar to the block read cycle. It uses a string of 1, 2, or 4 byte data transfers and the MASTER does not release the DTB until all of the bytes have been transferred. It differs from a string of write cycles in that the MASTER broadcasts only one address and address modifier (at the beginning of the cycle). Then the SLAVE increments this address on each transfer so that the data from the next transfer is stored in the next higher location.

READ-MODIFY-WRITE CYCLE — A DTB cycle that is used to both read from, and write to, a SLAVE location without permitting any other MASTER to access that location. This cycle is most useful in multiprocessing systems where certain memory locations are used to provide semaphore functions.

ADDRESS-ONLY CYCLE — A DTB cycle that consists of an address broadcast, but no data transfer. SLAVES do not acknowledge ADDRESS-ONLY cycles and MASTERS terminate the cycle without waiting for an acknowledgment.

INTERRUPT ACKNOWLEDGE CYCLE — A DTB cycle, initiated by an INTERRUPT HANDLER, that reads a STATUS/ID from an INTERRUPTER. An INTERRUPT HANDLER generates this cycle whenever it detects an interrupt request from an INTERRUPTER and it has control of the DTB.

1.2.2 Basic VMEbus Structure

The VMEbus interface system consists of backplane interface logic, four groups of signal lines called "buses", and a collection of "functional modules" which can be configured as required. The functional modules communicate with each other using the backplane signal lines.

The "functional modules" defined in this document are used as vehicles for discussion of the bus protocol, and need not be considered a constraint to logic design. For example, the designer might choose to design logic which interacts with the VMEbus in the manner described, but uses different on-board signals, or monitors other VMEbus signals. VMEbus boards might be designed to include any combination of the functional modules defined by this document.

The VMEbus functional structure can be divided into four categories. Each consists of a bus and its associated functional modules which work together to perform specific duties. Figure 1-2 shows the VMEbus functional modules and buses. Each category is briefly summarized below.

Data Transfer — Devices transfer data over the Data Transfer Bus (DTB), which contains data and address pathways and associated control signals. Functional modules called MASTERS, SLAVES, INTERRUPTERS, and INTERRUPT HANDLERS use the DTB to transfer data between each other. Two other modules, called a BUS TIMER and an IACK DAISY-CHAIN DRIVER also assist them in this process.

DTB Arbitration — Since a VMEbus system can be configured with more than one MASTER or INTERRUPT HANDLER, a means is provided to transfer control of the DTB between them in an orderly manner and to guarantee that only one controls the DTB at a given time. The Arbitration Bus modules (REQUESTERS and ARBITER) coordinate the control transfer.

Priority Interrupt — The priority interrupt capability of the VMEbus provides a means by which devices can request service from an INTERRUPT HANDLER. These interrupt requests can be prioritized into a maximum of seven levels. INTERRUPTERS and INTERRUPT HANDLERS use the Priority Interrupt Bus signal lines.

Utilities — Periodic clocks, initialization, and failure detection are provided by the Utility Bus. It includes two clock lines, a system reset line, a system fail line, an AC fail line, and a serial data line.

1.3 VMEbus SPECIFICATION DIAGRAMS

As aids to defining or describing VMEbus operation, several types of diagrams are used, including:

a. Timing diagrams that show the timing relationships between signal transitions. The times involved will have minimum and/or maximum limits associated with them. Some of the times specified on these diagrams specify the behavior of the backplane interface logic, while other times specify the interlocked behavior of the functional modules.

b. Sequence diagrams that are similar to a timing diagram but show only the interlocked timing relationships of the functional modules. This diagram is intended to show a sequence of events, rather than to specify the times involved. For example, a sequence diagram might indicate that module A cannot generate signal transition B until it detects module C's generation of signal transition D.

c. Flow diagrams that show a stream of events as they would occur during a VMEbus operation. The events are stated in words and result from interaction of two or more functional modules. The flow diagram describes VMEbus operations in a sequential manner and, at the same time, shows interaction of the functional modules.

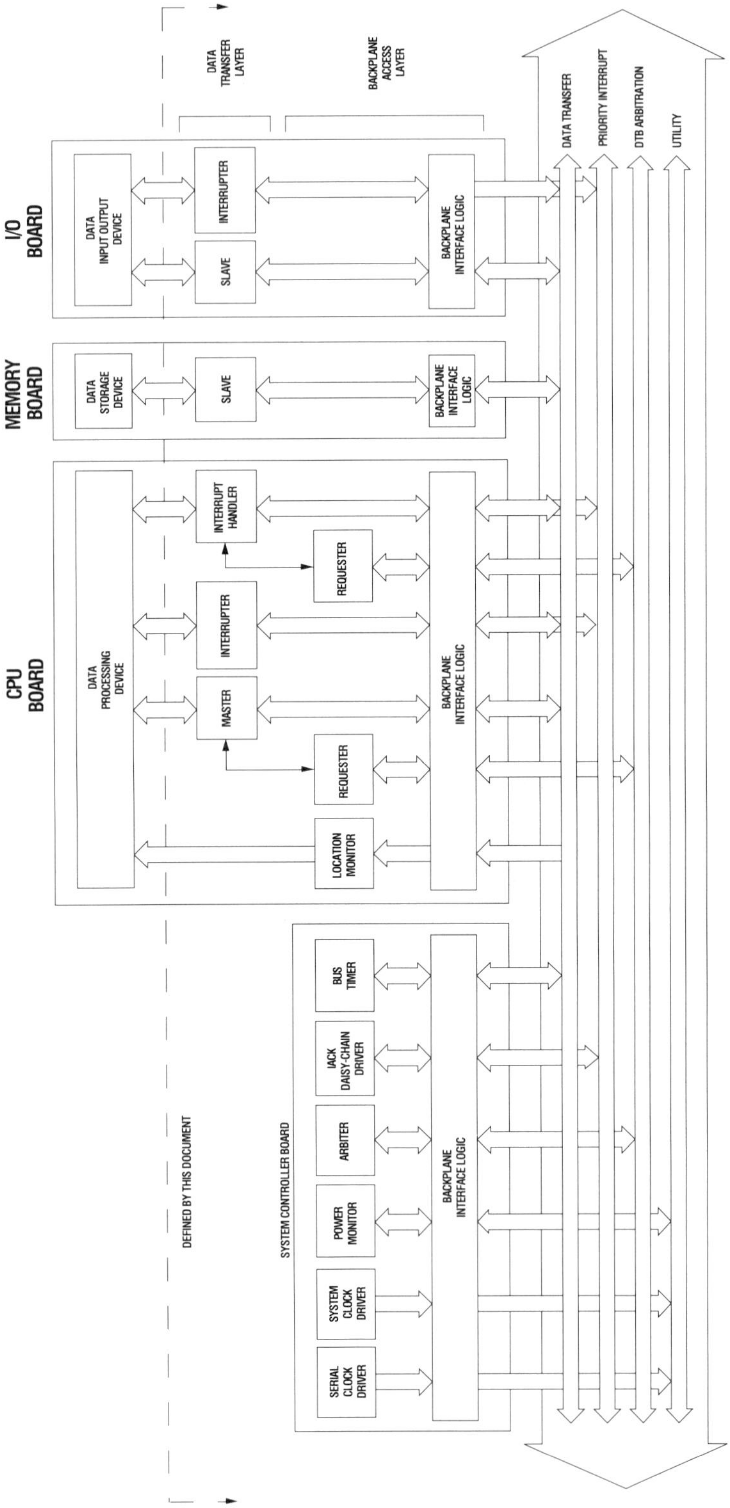

Figure 1-2. Functional Modules And Buses Defined By This Document

1.4 SPECIFICATION TERMINOLOGY

To avoid confusion, and to make very clear what the requirements for compliance are, many of the paragraphs in this document are labeled with keywords that indicate the type of information they contain. The keywords are listed below:

RULE
RECOMMENDATION
SUGGESTION
PERMISSION
OBSERVATION

Any text not labeled with one of these keywords describes the VMEbus structure or operation. It is written in either a descriptive or a narrative style. These keywords are used as follows:

RULE chapter.number:
Rules form the basic framework of the VMEbus specification. They are sometimes expressed in text form and sometimes in the form of figures, tables, or drawings. All VMEbus rules **MUST** be followed to ensure compatibility between VMEbus designs. Rules are characterized by an imperative style. The upper-case words **MUST** and **MUST NOT** are reserved exclusively for stating rules in this document and are not used for any other purpose.

RECOMMENDATION chapter.number:
Wherever a recommendation appears, designers would be wise to take the advice given. Doing otherwise might result in some awkward problems or poor performance. While VMEbus has been designed to support high performance systems, it is possible to design a VMEbus system that complies with all the rules, but has abysmal performance. In many cases, a designer needs a certain level of experience with VMEbus in order to design boards that deliver top performance. Recommendations found in this document are based on this kind of experience, and are provided to designers to speed their traversal of the learning curve.

SUGGESTION chapter.number:
In the VMEbus specification, a suggestion contains advice which is helpful but not vital. The reader is encouraged to consider the advice before discarding it. Some design decisions that need to be made in designing VMEbus boards are difficult until experience has been gained with the VMEbus. Suggestions are included to help a designer who has not yet gained this experience. Some suggestions have to do with designing boards that can be easily reconfigured for compatibility with other boards, or with designing the board to make the job of system debugging easier.

PERMISSION chapter.number:
In some cases a VMEbus rule does not specifically prohibit a certain design approach, but the reader might be left wondering whether that approach might violate the spirit of the rule, or whether it might lead to some subtle problem. Permissions reassure the reader that a certain approach is acceptable, and will cause no problems. The upper-

case word **MAY** is reserved exclusively for stating permissions in this document and is not used for any other purpose.

OBSERVATION chapter.number:
Observations do not offer any specific advice. They usually follow naturally from what has just been discussed. They spell out the implications of certain VMEbus rules and bring attention to things that might otherwise be overlooked. They also give the rationale behind certain rules, so that the reader understands why the rule must be followed.

1.4.1 Signal Line States

The VMEbus specification describes its protocol in terms of levels and transitions on bus lines.

A signal line is always assumed to be in one of two levels or in transition between these levels. Whenever the term "high" is used, it refers to a high TTL voltage level. The term "low" refers to a low TTL voltage level. A signal line is "in transition" when its voltage is moving between these levels. (See Chapter 6 for voltage thresholds used on the VMEbus.)

There are two possible transitions which can appear on a signal line, and these are called "edges". A rising edge is the time during which a signal level makes its transition from a low level to a high level. The falling edge is the time during which a signal level makes its transition from a high level to a low level.

Some bus specifications prescribe maximum or minimum rise and fall times for these edges. The problem with doing this is that board designers have very little control over these times. If the backplane is heavily loaded, the rise and fall times will be long. If it is lightly loaded, these times might be short. Even if designers know what the maximum and minimum loading will be, they still need to spend time in the lab, experimenting to find out which drivers will provide the needed rise and fall times.

In fact, rise and fall times are the result of a complex set of interactions involving the signal line impedances of the backplane, its terminations, the source impedance of the drivers, and the capacitive loading of the signal line. In order to trade off all of these factors the board designer would have to study transmission line theory, as well as certain specific parameters of drivers and receivers which are not normally found in most manufacturers' data sheets.

Recognizing all of this, the VMEbus standard doesn't specify rise and fall times. Instead, it specifies the electrical characteristics for drivers and receivers and suggests how to design the backplane. It also tells designers how the worst case bus loading will affect the propagation delay of these drivers so that they can ensure that the VMEbus timing is met before building a board. If VMEbus designers follow these propagation delay guidelines, their boards will operate reliably with other VMEbus compatible boards under worst case conditions.

1.4.2 Use Of The Asterisk (*)

To help define usage, signal mnemonics have an asterisk suffix where required:

a. An asterisk (*) following the signal name of signals which are level significant denotes that the signal is true or valid when the signal is low.

b. An asterisk (*) following the signal name of signals which are edge significant denotes that the actions initiated by that signal occur on a high to low transition.

OBSERVATION 1.1:
The asterisk is inappropriate for the asynchronously running clock lines SYSCLK and SERCLK. There is no fixed phase relationship between these clock lines and other VMEbus timing.

1.5 PROTOCOL SPECIFICATION

There are two layers of VMEbus protocol. The lowest VMEbus layer, called the backplane access layer, is composed of the backplane interface logic, the Utility Bus modules, and the Arbitration Bus modules. The VMEbus data transfer layer, is composed of the Data Transfer Bus and Priority Interrupt Bus modules. Figure 1-2 shows this layering.

OBSERVATION 1.2:
The signal lines used by the data transfer layer modules form a special class because they are driven by different modules at different times. They are driven with line drivers that can be turned on and off at each board, based upon signals generated in the backplane access layer. It is very important that their turn-on and turn-off times be carefully controlled to prevent two drivers from attempting to drive the same signal line to different levels. Special timing diagram notation is used in this document to specify their turn-on and turn-off times. It is shown in Figure 1-3.

There are two basic kinds of protocol used on the VMEbus: closed loop protocols and open loop protocols. Closed loop protocols use interlocked bus signals while open loop protocols use broadcast bus signals.

1.5.1 Interlocked Bus Signals

An interlocked bus signal is sent from a specific module to another specific module. The signal is acknowledged by the receiving module. An interlocked relationship exists between the two modules until the signal is acknowledged.

For example, an INTERRUPTER can send an interrupt request which is answered later with an interrupt acknowledge signal (no time limit is prescribed by the VMEbus specification). The INTERRUPTER doesn't remove the interrupt request until the INTERRUPT HANDLER acknowledges it.

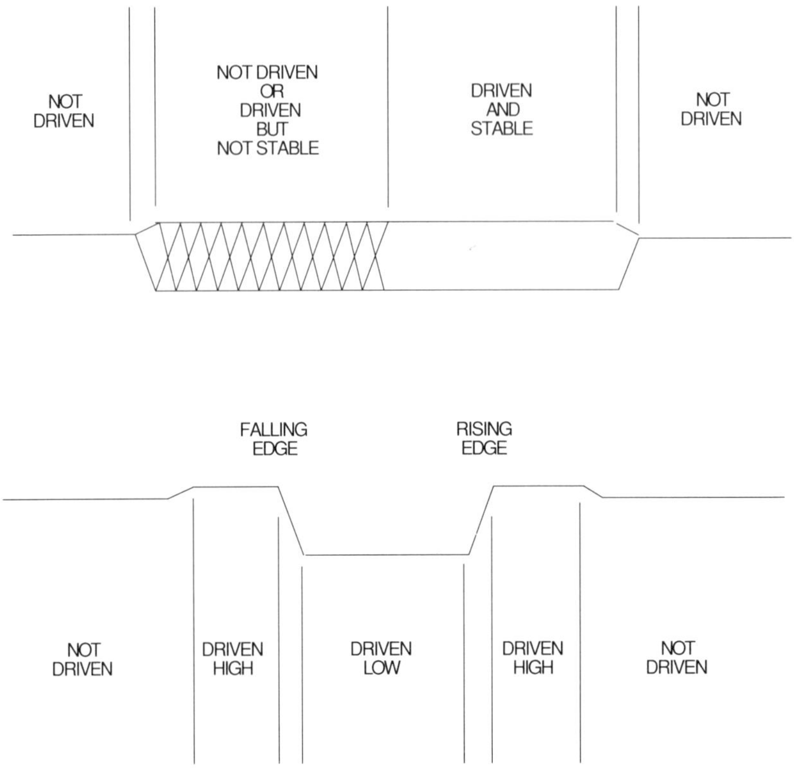

Figure 1-3. Signal Timing Notation

Interlocked bus signals coordinate internal functions of the VMEbus system, as opposed to interacting with external stimuli. Each interlocked signal has a source module and a destination module within the VMEbus system.

The address strobe and data strobes are especially important interlocking signals. They are interlocked with the data transfer acknowledge and bus error signals and coordinate the transfer of addresses and data which are the basis for all information flow between modules in the data transfer layer.

1.5.2 Broadcast Bus Signal

A module generates a broadcast signal in response to an event. There is no protocol for acknowledging a broadcast signal. Instead, the broadcast is maintained for a minimum specified time, long enough to assure that all appropriate modules detect the signal. Broadcast signals might be activated at any time, irrespective of any other activity taking place on the bus. They are each sent over a dedicated signal line. Some examples are the system reset and AC failure lines. These signal lines are not sent to any specific module, but announce special conditions to all modules.

1.6 SYSTEM EXAMPLES AND EXPLANATIONS

A protocol specification describes, in detail, the behavior of the various functional modules. It discusses how a module responds to a signal without saying where the signal came from. Because of this, a protocol specification does not give the reader a complete picture of what is going on over the bus. To help the reader, the VMEbus specification provides examples of typical VMEbus operations. Each example shows one possible sequence of events: other sequences are also possible. In providing these examples, there is the danger that readers will assume that the sequence shown in the example is the only legal one. To help readers avoid this trap, all examples are given in a narrative style, using the present tense. This is in contrast to the imperative style used when giving rules for compliance with the VMEbus specification.

CHAPTER 2

VMEbus DATA TRANSFER BUS

2.1 INTRODUCTION

The VMEbus includes a high speed asynchronous parallel Data Transfer Bus (DTB). Figure 2-1 shows a typical VMEbus system, including all of the DTB functional module types. MASTERS use the DTB to select storage locations provided by SLAVES, and to transfer data to or from those locations. Some MASTERS and SLAVES use all of the DTB lines, while others use only a subset.

LOCATION MONITORS monitor data transfers between MASTERS and SLAVES. When an access is done to one of the byte location(s) that it monitors, a LOCATION MONITOR generates an on-board signal. For example, it might signal its on-board processor by means of an interrupt request. In such a configuration, if processor board A writes into a location of the global VMEbus memory that is monitored by processor B's LOCATION MONITOR, processor B will be interrupted.

After a MASTER initiates a data transfer cycle it waits for the responding SLAVE to respond before finishing the cycle. The asynchronous definition of the VMEbus allows a SLAVE to take as long as it needs to respond. If a SLAVE fails to respond because of some malfunction, or if the MASTER accidentally addresses a location where there is no SLAVE, the BUS TIMER intervenes, allowing the cycle to be terminated.

2.2 DATA TRANSFER BUS LINES

The Data Transfer Bus lines can be grouped into three categories:

Addressing lines	Data lines	Control lines
A01-A31	D00-D31	AS*
AM0-AM5		DS0*
DS0*		DS1*
DS1*		BERR*
LWORD*		DTACK*
		WRITE*

OBSERVATION 2.1:
The two data strobes (DS0* and DS1*) serve a dual function:
a. The levels of these two data strobe lines are used to select which byte(s) are accessed.
b. The edges of the data strobes are also used as timing signals which coordinate the transfer of the data between the MASTER and SLAVE.

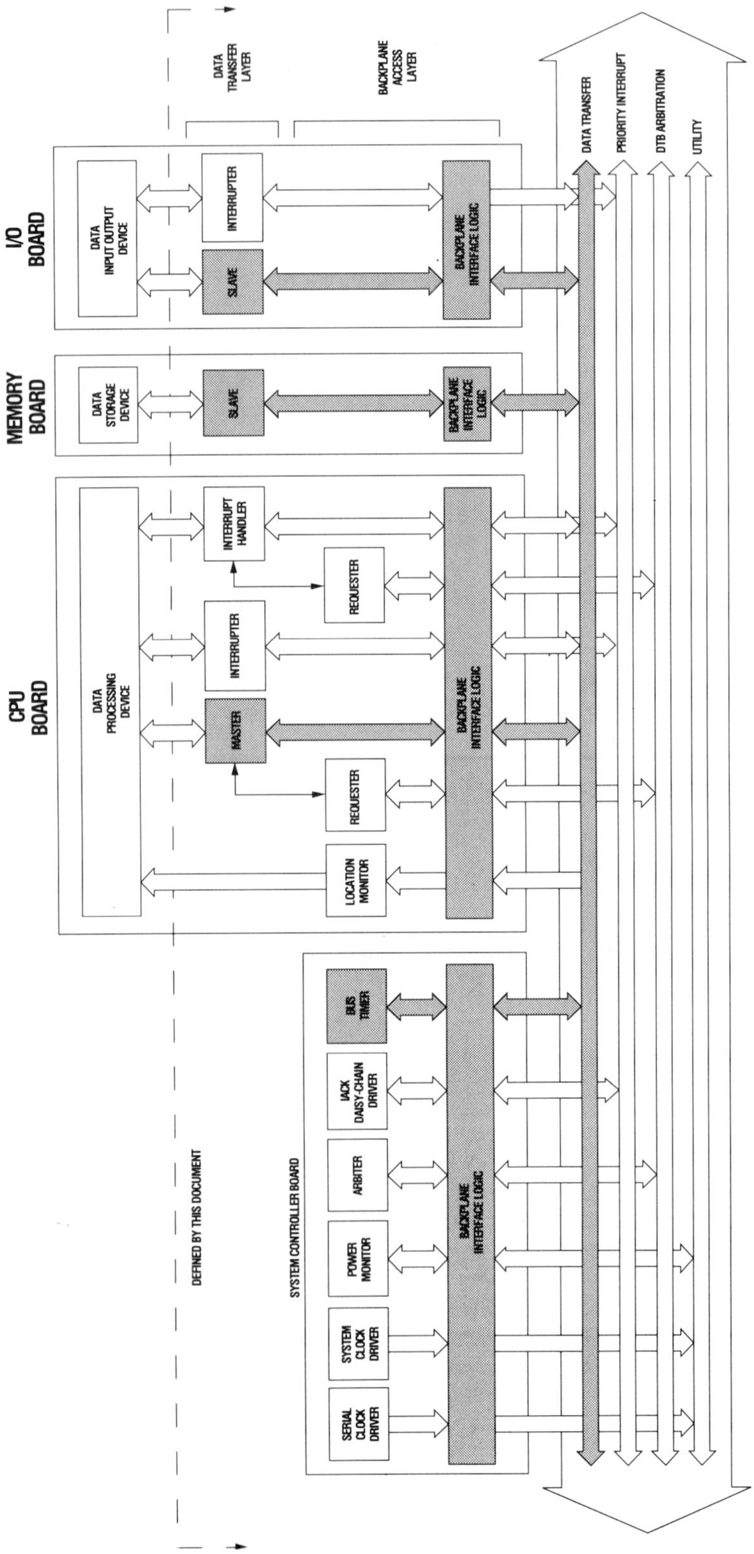

Figure 2-1. Data Transfer Bus functional block diagram

2.2.1 Addressing Lines

The smallest addressable unit of storage is the byte location. Each byte location is assigned a unique binary address. Each byte location can be assigned to one of four categories, according to the least significant two bits of its address. (See Table 2-1)

Table 2-1. The Four Categories of Byte Locations

Category	Byte address
BYTE(0)	XXXXXX.....XXXXXX00
BYTE(1)	XXXXXX.....XXXXXX01
BYTE(2)	XXXXXX.....XXXXXX10
BYTE(3)	XXXXXX.....XXXXXX11

A set of byte locations, whose addresses differ only in the two least significant bits, is called a 4-byte group or a BYTE(0-3) group. Some, or all, of the bytes in a 4-byte group can be accessed simultaneously by a single DTB cycle.

MASTERS use address lines A02-A31 to select which 4-byte group will be accessed. Four additional lines (DS1*, DS0*, A01, and LWORD*) are then used to select which byte location(s) within the 4-byte group are accessed during the data transfer. Using these four lines, a MASTER can access 1, 2, 3, or 4 byte locations simultaneously, depending upon the type of cycle. The 34 possible cycle types with the corresponding levels of these four lines are listed in Table 2-2.

OBSERVATION 2.2:
In cycles where both data strobes are driven low, one data strobe might go low slightly after the other. In this case the signal levels indicated in Table 2-2 are the final levels.

OBSERVATION 2.3:
Given the 4 signal line levels shown in Table 2-2, there are 16 possible combinations of levels. Of these 16, there are two illegal combinations that are not used (see Table in RULE 2-1).

RULE 2.1:
MASTERS **MUST NOT** generate DTB cycles where the final levels of DS0*, DS1*, A01, and LWORD* are either of the following illegal combinations:

DS1*	DS0*	A01	LWORD*
high	low	high	low
low	high	high	low

PERMISSION 2.1:
MASTERS that generate BYTE(1-2) READ or BYTE(1-2) WRITE cycles **MAY** generate either of the two combinations described in RULE 2.1 briefly as transition states (i.e. while one data strobe has fallen, but the other has not).

OBSERVATION 2.4:
Whenever a MASTER drives LWORD* low and A01 high it drives both data strobe low. (Any other combination is illegal.) VMEbus board designers can take advantage of this to simplify the logic on SLAVES.

PERMISSION 2.2:
To simplify the required logic, SLAVES which respond to BYTE(1-2) READ and BYTE(1-2) WRITE cycles **MAY** be designed without logic to distinguish between these cycles and the two illegal cycles described in RULE 2.1.

Table 2-2. Signal Levels Used To Select Which Byte Location(s) Are Accessed During A Data Transfer

Type of cycle	DS1*	DS0*	A01	LWORD*
ADDRESS-ONLY	high	high	← Note 1 → See Pg 20	
Single even byte transfers				
BYTE(0) READ or WRITE	low	high	low	high
BYTE(2) READ or WRITE	low	high	high	high
Single odd byte transfers				
BYTE(1) READ or WRITE	high	low	low	high
BYTE(3) READ or WRITE	high	low	high	high
Double byte transfers				
BYTE(0-1) READ or WRITE	low	low	low	high
BYTE(2-3) READ or WRITE	low	low	high	high
Quad byte transfers				
BYTE(0-3) READ or WRITE	low	low	low	low

(Continued on Page 19)

Table 2-2. (Continued)

Type of cycle	DS1*	DS0*	A01	LWORD*
Single byte block transfers				
SINGLE BYTE BLOCK READ or WRITE	← Note 2 → See Pg 20			high
Double byte block transfers				
DOUBLE BYTE BLOCK READ or WRITE	low	low	Note 3 See Pg 20	high
Quad byte block transfers				
QUAD BYTE BLOCK READ or WRITE	low	low	low	low
Single byte RMW transfers				
BYTE(0) READ-MODIFY-WRITE	low	high	low	high
BYTE(1) READ-MODIFY-WRITE	high	low	low	high
BYTE(2) READ-MODIFY-WRITE	low	high	high	high
BYTE(3) READ-MODIFY-WRITE	high	low	high	high
Double byte RMW transfers				
BYTE(0-1) READ-MODIFY-WRITE	low	low	low	high
BYTE(2-3) READ-MODIFY-WRITE	low	low	high	high
Quad byte RMW transfers				
BYTE(0-3) READ-MODIFY-WRITE	low	low	low	low
Unaligned transfers				
BYTE(0-2) READ or WRITE	low	high	low	low
BYTE(1-3) READ or WRITE	high	low	low	low
BYTE(1-2) READ or WRITE	low	low	high	low

Table 2-2. (Notes)

Notes:

1. During ADDRESS-ONLY cycles, both data strobes are maintained high, but the A01 and LWORD* lines might be either high or low.

2. During single byte block transfers, the two data strobes are alternately driven low. Either data strobe might be driven low on the first transfer. If the first accessed byte location is BYTE(0) or BYTE(2), then DS1* is driven low first. If the first accessed byte location is BYTE(1) or BYTE(3), then DS0* is driven low first. A01 is valid only on the first data transfer (i.e. until the SLAVE drives DTACK* or BERR* low the first time) and might be either high or low depending upon which byte the single byte block transfer begins with. If the first byte location is BYTE(0) or BYTE(1), then A01 is low. If the first byte location is BYTE(2) or BYTE(3), then A01 is high.

An example of a Single byte block transfer cycle which starts with BYTE(2) is given below:

		DS1*	DS0*	A01	LWORD*
First data transfer	BYTE (2)	low	high	high	high
↓	BYTE (3)	high	low	X	X
	BYTE (0)	low	high	X	X
	BYTE (1)	high	low	X	X
Last data transfer	BYTE (2)	low	high	X	X

X = high or low

3. During a Double byte block transfer, the two data strobes are both driven low on each data transfer. A01 is valid only on the first data transfer (i.e. until the SLAVE drives DTACK* or BERR* low the first time) and might be either high or low depending upon what double byte group the double byte block transfer begins with. If the first double byte group is BYTE(0-1), then A01 is low. If the first double byte group is BYTE(2-3), then A01 is high.

An example of a Double byte block transfer cycle which starts with BYTE(2-3) is given below:

		DS1*	DS0*	A01	LWORD*
First data transfer	BYTE (2-3)	low	low	high	high
↓	BYTE (0-1)	low	low	X	X
	BYTE (2-3)	low	low	X	X
Last data transfer	BYTE (0-1)	low	low	X	X

X = high or low

2.2.2 Address Modifier Lines

There are 6 address modifier lines. They allow the MASTER to pass additional binary information to the SLAVE during data transfers. Table 2-3 lists all of the 64 possible address modifier (AM) codes and classifies each into one of three categories:

Defined
Reserved
User defined

The defined address modifier codes can be further classified into three categories:

Short addressing AM codes indicate that address lines A02-A15 are being used to select a BYTE(0-3) group.

Standard addressing AM codes indicate that address lines A02-A23 are being used to select a BYTE(0-3) group.

Extended addressing AM codes indicate that address lines A02-A31 are being used to select a BYTE(0-3) group.

RULE 2.2:
Except for the User defined codes, the codes defined in Table 2-3 **MUST NOT** be used for purposes other than those specified.

RULE 2.3:
VMEbus SLAVE boards **MUST NOT** respond to reserved address modifier codes.

OBSERVATION 2.5:
Reserved address modifier codes are for future enhancements. If SLAVE boards respond to these codes, incompatibilities might result at some future date, when usage of these codes is defined.

PERMISSION 2.3:
User defined codes **MAY** be used for any purpose which board vendors or board users deem appropriate (page switching, memory protection, MASTER or task identification, privileged access to resources, etc.).

Table 2-3. Address Modifier Codes

HEX CODE	ADDRESS MODIFIER 5	4	3	2	1	0	FUNCTION
3F	H	H	H	H	H	H	Standard Supervisory Block Transfer
3E	H	H	H	H	H	L	Standard Supervisory Program Access
3D	H	H	H	H	L	H	Standard Supervisory Data Access
3C	H	H	H	H	L	L	Reserved
3B	H	H	H	L	H	H	Standard Non-Privileged Block Transfer
3A	H	H	H	L	H	L	Standard Non-Privileged Program Access
39	H	H	H	L	L	H	Standard Non-Privileged Data Access
38	H	H	H	L	L	L	Reserved
37	H	H	L	H	H	H	Reserved
36	H	H	L	H	H	L	Reserved
35	H	H	L	H	L	H	Reserved
34	H	H	L	H	L	L	Reserved
33	H	H	L	L	H	H	Reserved
32	H	H	L	L	H	L	Reserved
31	H	H	L	L	L	H	Reserved
30	H	H	L	L	L	L	Reserved
2F	H	L	H	H	H	H	Reserved
2E	H	L	H	H	H	L	Reserved
2D	H	L	H	H	L	H	Short Supervisory Access
2C	H	L	H	H	L	L	Reserved
2B	H	L	H	L	H	H	Reserved
2A	H	L	H	L	H	L	Reserved
29	H	L	H	L	L	H	Short Non-Privileged Access
28	H	L	H	L	L	L	Reserved
27	H	L	L	H	H	H	Reserved
26	H	L	L	H	H	L	Reserved
25	H	L	L	H	L	H	Reserved
24	H	L	L	H	L	L	Reserved
23	H	L	L	L	H	H	Reserved
22	H	L	L	L	H	L	Reserved
21	H	L	L	L	L	H	Reserved
20	H	L	L	L	L	L	Reserved

L = low signal level H = high signal level

(Continued on Page 23)

Table 2-3. (Continued)

HEX CODE	ADDRESS MODIFIER 5	4	3	2	1	0	FUNCTION
1F	L	H	H	H	H	H	User defined
1E	L	H	H	H	H	L	User defined
1D	L	H	H	H	L	H	User defined
1C	L	H	H	H	L	L	User defined
1B	L	H	H	L	H	H	User defined
1A	L	H	H	L	H	L	User defined
19	L	H	H	L	L	H	User defined
18	L	H	H	L	L	L	User defined
17	L	H	L	H	H	H	User defined
16	L	H	L	H	H	L	User defined
15	L	H	L	H	L	H	User defined
14	L	H	L	H	L	L	User defined
13	L	H	L	L	H	H	User defined
12	L	H	L	L	H	L	User defined
11	L	H	L	L	L	H	User defined
10	L	H	L	L	L	L	User defined
0F	L	L	H	H	H	H	Extended Supervisory Block Transfer
0E	L	L	H	H	H	L	Extended Supervisory Program Access
0D	L	L	H	H	L	H	Extended Supervisory Data Access
0C	L	L	H	H	L	L	Reserved
0B	L	L	H	L	H	H	Extended Non-Privileged Block Transfer
0A	L	L	H	L	H	L	Extended Non-Privileged Program Access
09	L	L	H	L	L	H	Extended Non-Privileged Data Access
08	L	L	H	L	L	L	Reserved
07	L	L	L	H	H	H	Reserved
06	L	L	L	H	H	L	Reserved
05	L	L	L	H	L	H	Reserved
04	L	L	L	H	L	L	Reserved
03	L	L	L	L	H	H	Reserved
02	L	L	L	L	H	L	Reserved
01	L	L	L	L	L	H	Reserved
00	L	L	L	L	L	L	Reserved

L = low signal level H = high signal level

RECOMMENDATION 2.1:
To allow VMEbus users to tailor the use of the user defined address modifier codes to their own needs, decode them in a flexible way on SLAVE boards. Users can then configure the board to give any decoding required by their system.

OBSERVATION 2.6:
Socketed PROMS and FPLAS provide a flexible way for decoding the address modifier codes.

SUGGESTION 2.1:
Where SLAVES are manufactured with a programmed device (e.g. a PROM or a FPLA) installed in the socket, program the device so that the SLAVE responds to the following AM codes:

A16 SLAVES with D08(O), D08(EO), D16, or D32	capability:	29, 2D
A24 SLAVES with D08(O), D08(EO), D16, or D32	capability:	39, 3A, 3D, and 3E
A32 SLAVES with D08(O), D08(EO), D16, or D32	capability:	09, 0A, 0D, and 0E
A24 SLAVES with BLT	capability:	3B, 3F
A32 SLAVES with BLT	capability:	0B, 0F

The mnemonics A16, A24, and A32 are defined in Table 2-9. The mnemonics D08(O), D08(EO), D16, D32, and BLT are defined in Tables 2-10, and 2-11.

2.2.3 Data Lines

VMEbus systems can be built with a backplane configuration that provides either 16 data lines (D00-D15), or 32 data lines (D00-D31). Backplane configurations that provide 16 data lines allow a MASTER to access only two byte locations simultaneously, while those with 32 data lines allow up to four byte locations to be accessed. When the MASTER has selected 1, 2, 3, or 4 byte locations, using the method described in Section 2.2.1, it can transfer binary data between itself and those locations over the data bus. Table 2-4 shows how the data lines are used to move data during each of the 34 cycle types.

PERMISSION 2.4:
The data sender (MASTER for a write cycle; SLAVE for a read cycle) may drive data lines which are not used to transfer data.

Table 2-4. Use Of The Data Lines To Move Data During Each Of The 34 Cycle Types

During the following types of cycles...	the data lines are used to transfer data as shown below: D24-D31	D16-D23	D08-D15	D00-D07
ADDRESS-ONLY	←	no bytes transferred		→
Single even byte transfers				
BYTE(0) READ or WRITE			BYTE(0)	
BYTE(2) READ or WRITE			BYTE(2)	
Single odd byte transfers				
BYTE(1) READ or WRITE				BYTE(1)
BYTE(3) READ or WRITE				BYTE(3)
Double byte transfers				
BYTE(0-1) READ or WRITE			BYTE(0)	BYTE(1)
BYTE(2-3) READ or WRITE			BYTE(2)	BYTE(3)
Quad byte transfers				
BYTE(0-3) READ or WRITE	BYTE(0)	BYTE(1)	BYTE(2)	BYTE(3)
Single byte block transfers				
SINGLE BYTE BLOCK READ or WRITE			← Note 1 → See Pg 27	
Double byte block transfers				
DOUBLE BYTE BLOCK READ or WRITE			← Note 2 → See Pg 27	

(Continued on Page 26)

Table 2-4. (Continued)

During the following types of cycles...	the data lines are used to transfer data as shown below: D24-D31	D16-D23	D08-D15	D00-D07
Quad byte block transfers				
QUAD BYTE BLOCK READ or WRITE	BYTE(0)	BYTE(1)	BYTE(2)	BYTE(3)
Single byte RMW transfers				
BYTE(0) READ-MODIFY-WRITE			BYTE(0)	
BYTE(1) READ-MODIFY-WRITE				BYTE(1)
BYTE(2) READ-MODIFY-WRITE			BYTE(2)	
BYTE(3) READ-MODIFY-WRITE				BYTE(3)
Double byte RMW transfers				
BYTE(0-1) READ-MODIFY-WRITE			BYTE(0)	BYTE(1)
BYTE(2-3) READ-MODIFY-WRITE			BYTE(2)	BYTE(3)
Quad byte RMW transfers				
BYTE(0-3) READ-MODIFY-WRITE	BYTE(0)	BYTE(1)	BYTE(2)	BYTE(3)
Unaligned transfers				
BYTE(0-2) READ or WRITE	BYTE(0)	BYTE(1)	BYTE(2)	
BYTE(1-3) READ or WRITE		BYTE(1)	BYTE(2)	BYTE(3)
BYTE(1-2) READ or WRITE		BYTE(1)	BYTE(2)	

Table 2-4. (Notes)

Notes:

1. During Single byte block transfers, data is transferred 8 bits at a time over D00-D07 or D08-D15. One example of this is given below:

	D24-D31	D16-D23	D08-D15	D00-D07
First data transfer				BYTE(1)
↓			BYTE(2)	
				BYTE(3)
			BYTE(0)	
				BYTE(1)
			BYTE(2)	
Last data transfer				BYTE(3)

2. During a Double byte block transfer, data is transferred 16 bits at a time over D00-D15. One example of this is given below:

	D24-D31	D16-D23	D08-D15	D00-D07
First data transfer			BYTE(2)	BYTE(3)
↓			BYTE(0)	BYTE(1)
			BYTE(2)	BYTE(3)
			BYTE(0)	BYTE(1)
			BYTE(2)	BYTE(3)
Last data transfer			BYTE(0)	BYTE(1)

2.2.4 Data Transfer Bus Control Lines

The following signal lines are used to control the movement of data over the data transfer lines:

AS*	Address Strobe
DS0*	Data Strobe Zero
DS1*	Data Strobe One
BERR*	Bus Error
DTACK*	Data Transfer Acknowledge
WRITE*	Read/Write

2.2.4.1 AS*

A falling edge on this line informs all SLAVE modules that the address is stable and can be captured.

2.2.4.2 DS0* And DS1*

In addition to their function in selecting byte locations for data transfer, as described in Section 2.2.1, the data strobes also serve additional functions. On write cycles, the first data strobe falling edge indicates when the MASTER has placed valid data on the data bus. On read cycles, the first rising edge tells the SLAVE when it can remove valid data from the data bus.

OBSERVATION 2.7:
VMEbus MASTERS are not permitted to drive either of the data strobes low before driving AS* low. However, due to the fact that AS* might be more heavily loaded on the backplane than the data strobes, SLAVES and LOCATION MONITORS might detect a falling edge on a data strobe, before they detect the falling edge on AS*.

PERMISSION 2.5:
VMEbus SLAVES and LOCATION MONITORS **MAY** be designed to capture the address when they detect a falling edge on a data strobe instead of on the falling edge of AS*.

OBSERVATION 2.8:
VMEbus SLAVES and LOCATION MONITORS that capture the address on the falling edge of the data strobe(s) need not monitor AS*.

OBSERVATION 2.9:
In order to take full advantage of address pipelining as described in section 2.4.2, or to perform block read and write cycles, a SLAVE should capture the address on the falling edge of AS*.

2.2.4. 3 DTACK*

The SLAVE drives DTACK* low to indicate that it has successfully received the data on a write cycle. On a read cycle, the SLAVE drives DTACK* low to indicate that it has placed data on the data bus.

2.2.4.4 BERR*

BERR* is driven low by the SLAVE or by the BUS TIMER to indicate to the MASTER that the data transfer was unsuccessful. For example, if a MASTER tries to write to a location which contains Read-Only Memory, the responding SLAVE might drive BERR* low. If the MASTER tries to access a location that is not provided by any SLAVE, the BUS TIMER would drive BERR* low after waiting a specified period.

OBSERVATION 2.10:
The BERR* line is a convenience which is useful when debugging VMEbus systems. It also allows system failures to be detected quickly during normal operation. Not all VMEbus systems will need this capability.

PERMISSION 2.6:
VMEbus SLAVES **MAY** be designed without a driver for BERR*.

SUGGESTION 2.2:
Design SLAVES to respond with a falling edge on BERR* in the following situations:

- **a.** When a D08(O), D08(EO), or D16 SLAVE is requested to do a quad byte cycle.
- **b.** When a D08(O) or D08(EO) SLAVE is requested to do a double byte cycle.
- **c.** When a non-UAT SLAVE is requested to do an unaligned transfer. (i.e. A triple byte transfer or a double byte BYTE(1-2) transfer.)
- **d.** When a SLAVE detects an uncorrectable error in the data it retrieves from its internal storage during a read cycle.

The mnemonics D08(O), D08(EO), D16, and UAT are defined in Tables 2-10 and 2-15.

RULE 2.4:
D08(O), D08(EO), and D16 SLAVES **MUST NOT** respond with a falling edge on DTACK* during a quad byte cycle if they do not have quad byte capability.

RULE 2.5:
D08(O) and D08(EO) SLAVES **MUST NOT** respond with a falling edge on DTACK* during a double byte cycle if it does not have double byte capability.

RULE 2.6:
A SLAVE **MUST NOT** respond with a falling edge on DTACK* during an unaligned transfer cycle, if it does not have UAT capability.

2.2.4.5 WRITE*

WRITE* is a level significant signal line strobed by the falling edge of the first data strobe. It is used by the MASTER to indicate the direction of data transfer operations. When WRITE* is driven low, the data transfer direction is from the MASTER to the SLAVE. When WRITE* is driven high, the data transfer direction is from the SLAVE to the MASTER.

2.3 DTB MODULES - BASIC DESCRIPTION

In addition to the ADDRESS-ONLY cycle, the DTB protocol defines 33 different cycle types that can be used to transfer data. Each of these 34 cycle types can be used in any of three addressing modes: short (16-bit), standard (24-bit), and extended (32-bit). The capabilities of MASTERS, SLAVES and LOCATION MONITORS are described by a list of mnemonics that show what cycle types they can generate, accept, or monitor respectively. (This will be described in more detail later in this chapter.)

Sections 2.3.1 through 2.3.4 provide block diagrams for the four types of DTB functional modules: MASTER, SLAVE, LOCATION MONITOR, and BUS TIMER.

RULE 2.7:
Output signal lines shown with solid lines in Figures 2-2 through 2-5 **MUST** be driven by the module, unless it would always drive them high.

OBSERVATION 2.11:
IF an output signal line is not driven,
THEN terminators on the backplane ensure that it is high.

RULE 2.8:
Input signal lines shown with solid lines in Figures 2-2 through 2-5 **MUST** be monitored and responded to in the appropriate fashion.

OBSERVATION 2.12:
RULES and PERMISSIONS for driving and monitoring signal lines shown with dotted lines in Figures 2-2 through 2-5 are given in Tables 2-5, 2-6, and 2-8.

2.3.1 MASTER

The block diagram of the MASTER is shown in Figure 2-2. The dotted lines in the diagram show signals whose use varies among the various types of MASTERS. Table 2-5 specifies how the various types of MASTERS use these lines. Further information about how the various types of MASTERS drive the address lines, the data lines, LWORD*, DS0*, and DS1* is given in Tables 2-19, 2-20, 2-21.

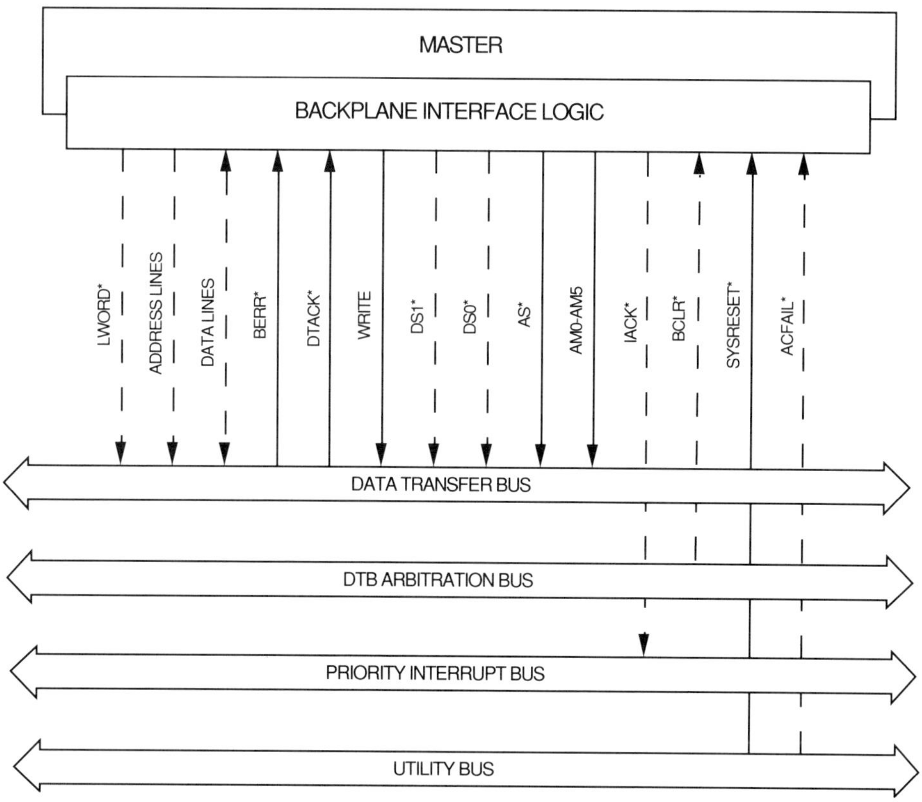

Figure 2-2. Block Diagram: MASTER

Table 2-5. RULES And PERMISSIONS That Specify The Use Of The Dotted Lines By The Various Types Of MASTERS

Type of MASTER	Use of dotted lines
D08(EO)	**MUST** drive DS0* and DS1*, but not both low on the same data transfer. **MUST** monitor and drive D00-D15. **MUST NOT** drive IACK* low. **MAY** or **MAY** not drive LWORD*. **MAY** or **MAY** not drive or monitor D16-D31.
D16	**MUST** drive DS0* and DS1*. **MUST** monitor and drive D00-D15. **MUST NOT** drive IACK* low. **MAY** or **MAY** not drive LWORD*. **MAY** or **MAY** not drive or monitor D16-D31.
D32	**MUST** drive DS0*, DS1*, and LWORD*. **MUST** monitor and drive D00-D31. **MUST NOT** drive IACK* low.
A16	**MUST** drive A01-A15. **MAY** or **MAY** not drive A16-A31
A24	**MUST** drive A01-A23. **MAY** or **MAY** not drive A24-A31
A32	**MUST** drive A01-A31.
ALL	**MAY** OR **MAY** not monitor BCLR*, or ACFAIL* (see Chapters 3 and 5).

Notes:
1. The mnemonics D08(EO), D16, and D32 are defined in Table 2-10.
2. The mnemonics A16, A24, and A32 are defined in Table 2-9.

2.3.2 SLAVE

The block diagram of the SLAVE is shown in Figure 2-3. The dotted lines in the diagram show signals whose use varies among the various types of SLAVES. Table 2-6 shows how the various types of SLAVES use these lines. Further information about how the various types of SLAVES drive the data lines is given in Table 2-22.

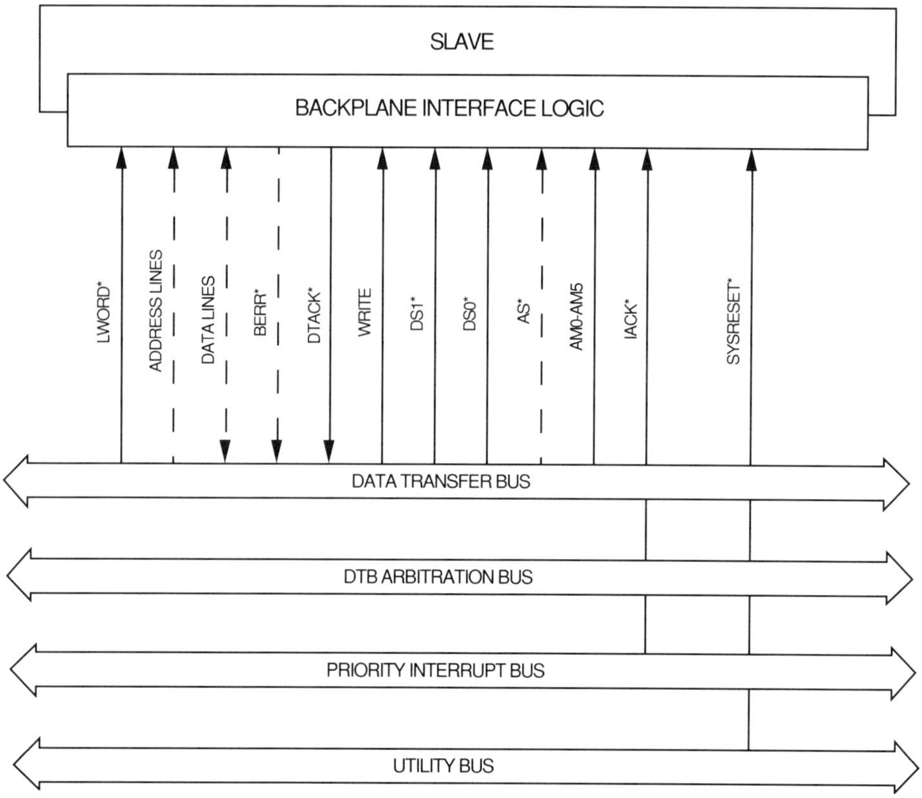

Figure 2-3. Block Diagram: SLAVE

Table 2-6. RULES And PERMISSIONS That Specify The Use Of The Dotted Lines By The Various Types Of SLAVES

Type of SLAVE	Use of dotted lines
D08(O)	**MUST** monitor and drive D00-D07. **MAY** or **MAY** not monitor AS*. **MAY** or **MAY** not monitor or drive D08-D31.
D08(EO)	**MUST** monitor and drive D00-D15. **MAY** or **MAY** not monitor AS*. **MAY** or **MAY** not monitor or drive D16-D31.
D16	**MUST** monitor and drive D00-D15. **MAY** or **MAY** not monitor AS*. **MAY** or **MAY** not monitor or drive D16-D31.
D32	**MUST** monitor and drive D00-D31. **MAY** or MAY not monitor AS*.
A16	**MUST** monitor A01-A15. **MAY** or **MAY** not monitor A16-A31.
A24	**MUST** monitor A01-A23. **MAY** or **MAY** not monitor A24-A31.
A32	**MUST** monitor A01-A31.
ALL	**MAY** or **MAY** not drive BERR*.

Notes:
1. The mnemonics D08(O), D08(EO), D16, and D32 are defined in Table 2-10.
2. The mnemonics A16, A24, and A32 are defined in Table 2-9.

2.3.3 BUS TIMER

The block diagram of the BUS TIMER is shown in Figure 2-4. BUS TIMERS can be designed to drive BERR* low after various periods of time. Table 2-7 shows how the BTO( ) mnemonic is used to describe the various types of BUS TIMERS.

OBSERVATION 2.13:
The dotted DTACK* and BERR* lines shown in Figure 2-4 allow implemention of a BUS TIMER in one of two ways:

- **a.** To drive BERR* low when the first data strobe stays low for longer than the bus time-out period, regardless of the levels on the DTACK* and BERR* lines.
- **b.** To drive BERR* low when the first data strobe stays low for longer than the bus time-out period, but only if both DTACK* and BERR* are high at the point of time-out.

Table 2-7. Use Of The BTO( ) Mnemonic To Specify The Time-Out Period Of BUS TIMERS

The Following Mnemonic	When Applied to a	Means that it
BTO(x)	BUS TIMER	drives BERR* low if the first data strobe stays low for longer than x microseconds.

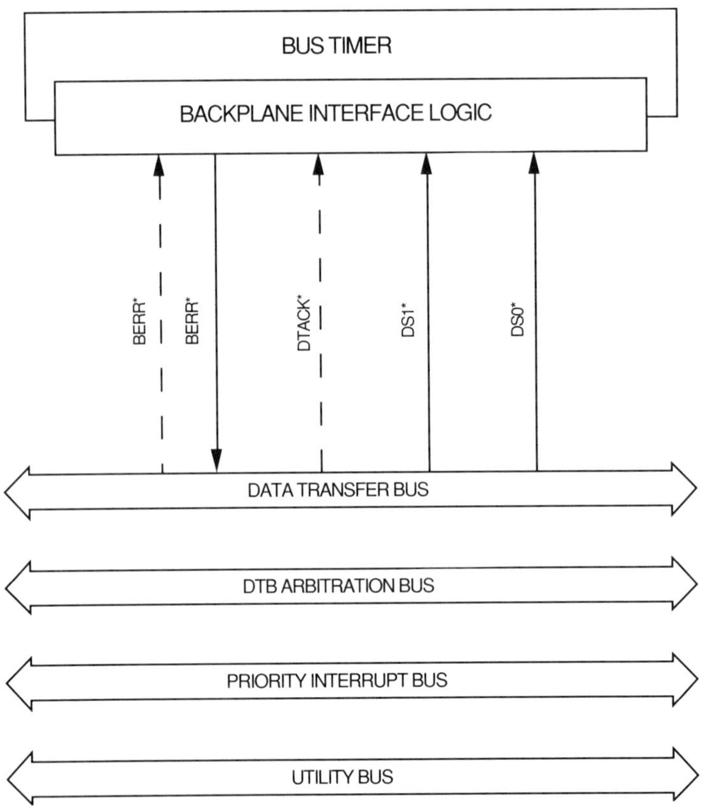

Figure 2-4. Block Diagram: BUS TIMER

2.3.4 LOCATION MONITOR

The block diagram of the LOCATION MONITOR is shown in Figure 2-5. The dotted lines in the diagram show signals whose use varies among the various types of LOCATION MONITORS. Table 2-8 shows how the various types of LOCATION MONITORS use these lines.

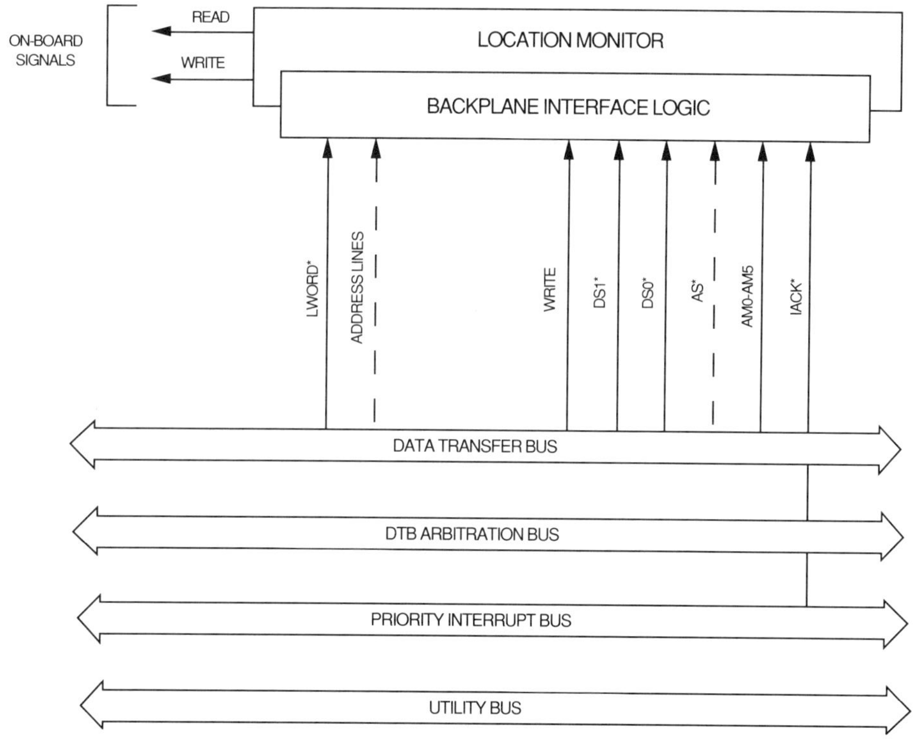

Figure 2-5. Block Diagram: LOCATION MONITOR

Table 2-8. RULES And PERMISSIONS That Specify The Use Of The Dotted Lines By The Various Types Of LOCATION MONITORS

Type of LOCATION MONITOR	Use of dotted lines
A16	**MUST** monitor A01-A15. **MAY** or **MAY** not monitor A16-A31.
A24	**MUST** monitor A01-A23. **MAY** or **MAY** not monitor A24-A31.
A32	**MUST** monitor A01-A31.
ALL	**MAY** or **MAY** not monitor AS*.

Note: The mnemonics A16, A24, and A32 are defined in Table 2-9.

2.3.5 Addressing Modes

MASTERS broadcast an address over the DTB at the beginning of each cycle. This broadcasted address might be a 16, 24 or 32-bit address, depending upon the capabilities of the MASTER broadcasting it. The 16 bit addresses are "short addresses", the 24 bit addresses are "standard addresses", and the 32 bit addresses are "extended addresses".

Table 2-9 shows the various mnemonics used to describe the addressing capabilities, and how each is used to describe MASTERS, SLAVES, and LOCATION MONITORS.

Table 2-9. Mnemonics That Specify Addressing Capabilities

The Following Mnemonic	When Applied to a	Means that it
A16	MASTER	can generate cycles with short (16 bit) addresses.
	SLAVE	can accept cycles with short (16 bit) addresses.
	LOCATION MONITOR	can monitor cycles with short (16 bit) addresses.
A24	MASTER	can generate cycles with standard (24 bit) addresses.
	SLAVE	can accept cycles with standard (24 bit) addresses.
	LOCATION MONITOR	can monitor cycles with standard (24 bit) addresses.
A32	MASTER	can generate cycles with extended (32 bit) addresses.
	SLAVE	can accept cycles with extended (32 bit) addresses.
	LOCATION MONITOR	can monitor cycles with extended (32 bit) addresses.

The MASTER broadcasts an Address Modifier (AM) code along with each address to tell SLAVES whether the address is short, standard, or extended.

Short addresses are generated by A16 MASTERS, and accepted by A16 SLAVES. Standard addresses are generated by A24 MASTERS, and accepted by A24 SLAVES. Extended addresses are generated by A32 MASTERS, and accepted by A32 SLAVES.

Short addressing is intended primarily for addressing I/O devices. It allows A16 SLAVES to be designed with less logic, since they do not have to decode as many address lines. While I/O boards can be designed to decode standard addresses and extended addresses, short addressing usually makes this unnecessary.

Standard and extended addressing modes are intended primarily for addressing memory, although there is no rule against designing I/O boards that also respond to these addressing modes. Standard and Extended addressing modes allow much larger addressing ranges.

RULE 2.9:
SLAVE boards **MUST** decode all of the address modifier lines.

OBSERVATION 2.14:
RULE 2.9 allows a SLAVE to differentiate short addresses, standard addresses, and extended addresses.

OBSERVATION 2.15:
In addition to the three modes of addressing described here, there is a fourth mode which is used on interrupt acknowledge cycles (see chapter 4). These interrupt acknowledge cycles can be distinguished from data transfer cycles by the fact that the IACK* signal line is low instead of high.

RULE 2.10:
Whenever a MASTER broadcasts an address over the address bus, it **MUST** ensure that IACK* is high.

PERMISSION 2.7:
A MASTER **MAY** either drive IACK* high during the address broadcast or it **MAY** leave IACK* undriven. (The bus terminators will then hold it high.)

RULE 2.11:
SLAVES **MUST NOT** respond to DTB cycles when IACK* is low.

RECOMMENDATION 2.2:
Since many systems will include a mixture of A16, A24, and A32 SLAVES, include A16 and A24 capability on 32-bit CPU cards in addition to the A32 capability, and include A16 capability on CPU cards with A24 capability.

2.3.6 Basic Data Transfer Capabilities

There are four basic data transfer capabilities associated with the DTB: D08(EO) (Even and Odd byte), D08(O) (Odd byte only), D16, and D32. These capabilities allow flexibility when interfacing different types of processors and peripherals to the bus.

Eight-bit processors can be interfaced to the bus as D08(EO) MASTERS. Sixteen-bit processors can be interfaced to the bus as D16 MASTERS. The D16 SLAVE module is useful for interfacing 16 bit memory devices or 16 bit I/O SLAVES to the DTB.

Many existing peripheral chips have registers that are only 8 bits wide. While these chips often have several of these registers, they cannot provide the contents of two registers simultaneously when a D16 MASTER attempts to access two adjacent locations with a double byte read cycle. These 8-bit peripheral ICs can be interfaced to the DTB as a D08(O) SLAVE. D08(O) SLAVES provide only BYTE(1) or BYTE(3) locations and respond only to single byte accesses. (Since single byte accesses to these odd byte locations always take place over D00-D07, this simplifies the D08(O) SLAVE'S interface logic.)

RECOMMENDATION 2.3:
Since most 16-bit microprocessors can also access memory 8 bits at a time, include D08(EO) MASTER capability on 16-bit CPU boards in addition to the D16 capability. In addition to allowing 8 bit data transfers to and from memory, this also allows them to access D08(O) SLAVES.

RECOMMENDATION 2.4:
Since most 32-bit microprocessors can also transfer data to and from memory 8 and 16 bits at a time, include D08(EO) and D16 capability on 32-bit CPU boards in addition to D32 capability. The D08(EO) capability not only allows 8-bit data transfers to and from memory, but it also allows the MASTER to access D08(O) SLAVES.

OBSERVATION 2.16:
It might seem logical to define "even byte only" SLAVES which respond to the even byte memory locations adjacent to the D08(O) SLAVES. But, this cannot be done because there is only one data transfer acknowledge line. If a MASTER were to select both an even byte and an odd byte location simultaneously, by doing a double byte transfer, both SLAVES would drive the same data acknowledge (DTACK*) line and the MASTER would not know whether both boards had acknowledged the access.

RECOMMENDATION 2.5:
Since many systems will include MASTERS with various data transfer capabilities, include D08(EO) and D16 capability on 32-bit SLAVE (memory) boards in addition to the D32 capability, and include D08(EO) capability on 16-bit SLAVE boards in addition to the D16 capability.

OBSERVATION 2.17:
Since D08(O) SLAVES respond only to odd byte addresses, they cannot provide contiguous memory. D08(O) SLAVES are useful only for I/O, status, or control registers, while D08(EO), D16 and D32 SLAVES are also useful for memory.

Table 2-10 shows the various mnemonics used to describe the basic data transfer capabilities, and how each is used to describe MASTERS, SLAVES, and LOCATION MONITORS.

Table 2-10. Mnemonics That Specify Basic Data Transfer Capabilities

The Following Mnemonic	When Applied to a	Means that it
D08(EO)	MASTER	can generate the following cycles:
	SLAVE	can accept the following cycles:
	LOCATION MONITOR	can monitor the following cycles:
		Single byte read cycles: BYTE(0) READ BYTE(1) READ BYTE(2) READ BYTE(3) READ
		Single byte write cycles: BYTE(0) WRITE BYTE(1) WRITE BYTE(2) WRITE BYTE(3) WRITE
D08(O)	SLAVE	can accept the following cycles:
		Single byte read cycles: BYTE(1) READ BYTE(3) READ
		Single byte write cycles: BYTE(1) WRITE BYTE(3) WRITE

(Continued on Page 43)

Table 2-10. (Continued)

The Following Mnemonic	When Applied to a	Means that it
D16	MASTER	can generate the following cycles:
	SLAVE	can accept the following cycles:
	LOCATION MONITOR	can monitor the following cycles: Double byte read cycles: BYTE(0-1) READ BYTE(2-3) READ Double byte write cycles: BYTE(0-1) WRITE BYTE(2-3) WRITE
D32	MASTER	can generate the following cycles:
	SLAVE	can accept the following cycles:
	LOCATION MONITOR	can monitor the following cycles: Quad byte read cycle: BYTE(0-3) READ Quad byte write cycle: BYTE(0-3) WRITE

Note: (EO) is Even and Odd, (O) is Odd only.

2.3.7 Block Transfer Capabilities

MASTERS often access several memory locations in ascending order. When this is the case, block transfer cycles are very useful. They allow the MASTER to provide a single address, and then access data in that location and those at higher addresses, without providing additional addresses.

When a MASTER initiates a block transfer cycle, the responding SLAVE latches the address into an on-board address counter. The MASTER, upon completing the first data transfer, (i.e. driving data strobes high) does not allow the address strobe to go high. Instead, it repeatedly drives the data strobe(s) low in response to data transfer acknowledgments from the SLAVE, and transfers data to or from sequential memory locations in ascending order.

To access the next location(s), the SLAVE increments an on-board counter that generates the address for each transition of the data strobe(s).

OBSERVATION 2.18:
Block transfer cycles of indefinite length are not allowed, because this complicates the design of memory boards. Specifically, all block transfer SLAVES (the one that responds, as well as those that do not) would need to latch the initial address and then increment the address counter on each bus transfer. All SLAVES would then have to decode the incremented address to see if the block transfer has crossed a board boundary into their address range. While this is certainly possible, such address decoding typically limits access times of the SLAVE. To simplify the design of these SLAVES, and to permit faster access times, RULE 2.12 has been formulated.

RULE 2.12:
Block transfer cycles **MUST NOT** cross any 256 byte boundary.

OBSERVATION 2.19:
RULE 2.12 limits the maximum length of block transfers to 256 bytes. However, knowing that only A01 through A07 will change during the course of the block transfer simplifies the design of block transfer SLAVES. The upper address lines only have to be decoded once, at the beginning of the block transfer cycle, allowing much faster access times on all subsequent data transfers.

OBSERVATION 2.20:
In some cases it might be necessary to transfer a large block of data which crosses one or more 256-byte boundaries. In such a case, if the hardware on the board which does the block transfer is designed to recognize the arrival at a 256-byte boundary, it can momentarily drive AS* high and then initiate another block transfer without the intervention of system software.

The block read cycle is very similar to a string of read cycles. Likewise, the block write cycle is very similar to a string of write cycles. The difference is that only the initial address is broadcast by the MASTER and the address strobe is held low during all of the data transfers.

OBSERVATION 2.21:
Control of the DTB cannot be transferred during block transfers because the address strobe is held low through all of the data transfers, and control of the DTB can only be transferred while the address strobe is high.

Table 2-11 lists the various mnemonics used to describe block transfer capabilities and how each is used to describe MASTERS, SLAVES, and LOCATION MONITORS.

Table 2-11. Mnemonics That Specify Block Transfer Capabilities

The Following Mnemonic	When Applied to a	Means that it
BLT	D08(EO) MASTER	can generate the following cycles:
	D08(EO) SLAVE	can accept the following cycles:
	D08(EO) LOCATION MONITOR	can monitor the following cycles: Block read cycles: SINGLE BYTE BLOCK READ Block write cycles: SINGLE BYTE BLOCK WRITE
	D16 MASTER	can generate the following cycles:
	D16 SLAVE	can accept the following cycles:
	D16 LOCATION MONITOR	can monitor the following cycles: Block read cycles: DOUBLE BYTE BLOCK READ Block write cycles: DOUBLE BYTE BLOCK WRITE
	D32 MASTER	can generate the following cycles:
	D32 SLAVE	can accept the following cycles:
	D32 LOCATION MONITOR	can monitor the following cycles: Block read cycles: QUAD BYTE BLOCK READ Block write cycles: QUAD BYTE BLOCK WRITE

2.3.8 Read-Modify-Write Capability

In multiprocessor systems which share resources such as memory and I/O, a method is needed to allocate these resources. One very important goal of this allocation algorithm is to ensure that a resource being used by one task cannot be used by another at the same time. The problem is best described by an example:

Two processors in a distributed processing system share a common resource (e.g., a printer). Only one processor can use the resource at a time. The resource is allocated by a bit in memory — i.e., if the bit is set, the resource is busy; if it is cleared, the resource is available. To gain use of the resource, processor A reads the bit and tests it to determine whether it is cleared. If the bit is cleared, processor A sets the bit to lock out processor B. This operation takes two data transfers: a read to test the bit, and a write to set the bit. However, a difficulty might arise if the bus is given to processor B between these two transfers. Processor B might then also find the bit cleared and assume the resource is available. Both processors will then set the bit in the next available cycle and attempt to use the resource.

This conflict is avoided by defining a read-modify-write cycle which prevents transferring control of the DTB between the read portion and the write portion of the cycle. This cycle is very similar to a read cycle immediately followed by a write cycle. The difference is that the address strobe is held low during both transfers. This ensures that, unlike a read cycle followed by a write cycle, control of the DTB cannot be transferred during a read-modify-write cycle, as this is only possible while the address strobe is high.

Table 2-12 lists the various mnemonics used to describe read-modify-write capabilities and how each is used to describe MASTERS, SLAVES, and LOCATION MONITORS.

Table 2-12. Mnemonics That Specify Read-Modify-Write Capabilities

The Following Mnemonic	When Applied to a	Means that it
RMW	D08(EO) MASTER	can generate the following cycles:
	D08(EO) SLAVE	can accept the following cycles:
	D08(EO) LOCATION MONITOR	can monitor the following cycles: Single byte read-modify-write cycles: BYTE(0) READ-MODIFY-WRITE BYTE(1) READ-MODIFY-WRITE BYTE(2) READ-MODIFY-WRITE BYTE(3) READ-MODIFY-WRITE
	D08(O) SLAVE	can accept the following cycles: Single byte read-modify-write cycles: BYTE(1) READ-MODIFY-WRITE BYTE(3) READ-MODIFY-WRITE
	D16 MASTER	can generate the following cycles:
	D16 SLAVE	can accept the following cycles:
	D16 LOCATION MONITOR	can monitor the following cycles: Double byte read-modify-write cycles: BYTE(0-1) READ-MODIFY-WRITE BYTE(2-3) READ-MODIFY-WRITE
	D32 MASTER	can generate the following cycles:
	D32 SLAVE	can accept the following cycles:
	D32 LOCATION MONITOR	can monitor the following cycles: Quad byte read-modify-write cycles: BYTE(0-3) READ-MODIFY-WRITE

2.3.9 Unaligned Transfer Capability

Some 32-bit microprocessors store and retrieve data in an unaligned fashion. For example, a 32-bit value might be stored in four different ways, as shown in Figure 2-6.

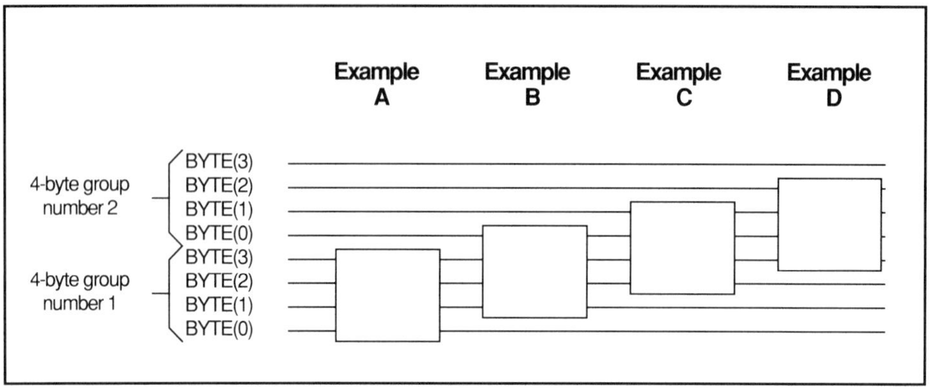

Figure 2-6. Four Ways That 32 Bits Of Data Might Be Stored In Memory

The MASTER can transfer the 32 bits of data using several different sequences of DTB cycles. For example, it can transfer the data one byte at a time, using four single byte data transfers. However, a MASTER can accomplish the transfer much quicker by using one of the cycle sequences shown in Table 2-13.

OBSERVATION 2.22:
The sequences shown in Table 2-13 would be typical of a MASTER that accesses the byte locations in ascending order. The VMEbus protocol does not require this.

As shown in Table 2-13, each of these 32-bit transfers can be accomplished with a combination of single byte and double byte transfers. However, examples B and D require three bus cycles when done this way. Because of this, the DTB protocol also includes two triple byte transfer cycles. When used in combination with a single byte cycle, these triple byte cycles allow data to be stored as shown in examples B and D using only two bus cycles.

Some 32-bit microprocessors also store and retrieve data 16 bits at a time, in an unaligned fashion, as shown in Figure 2-7.

Table 2-13. Transferring 32 Bits Of Data Using Multiple Byte Transfer Cycles

Example	Cycle sequences used to accomplish the transfer	Data bus lines used	Byte locations accessed (See Figure 2-6)
A	Quad byte transfer	D00-D31	Grp 1, BYTE(0-3)
B	Single byte transfer	D00-D07	Grp 1, BYTE(1)
	Double byte transfer	D00-D15	Grp 1, BYTE(2-3)
	Single byte transfer	D08-D15	Grp 2, BYTE(0)
	or		
	Triple byte transfer	D00-D23	Grp 1, BYTE(1-3)
	Single byte transfer	D08-D15	Grp 2, BYTE(0)
C	Double byte transfer	D00-D15	Grp 1, BYTE(2-3)
	Double byte transfer	D00-D15	Grp 2, BYTE(0-1)
D	Single byte transfer	D00-D07	Grp 1, BYTE(3)
	Double byte transfer	D00-D15	Grp 2, BYTE(0-1)
	Single byte transfer	D08-D15	Grp 2, BYTE(2)
	or		
	Single byte transfer	D00-D07	Grp 1, BYTE(3)
	Triple byte transfer	D08-D31	Grp 2, BYTE(0-2)

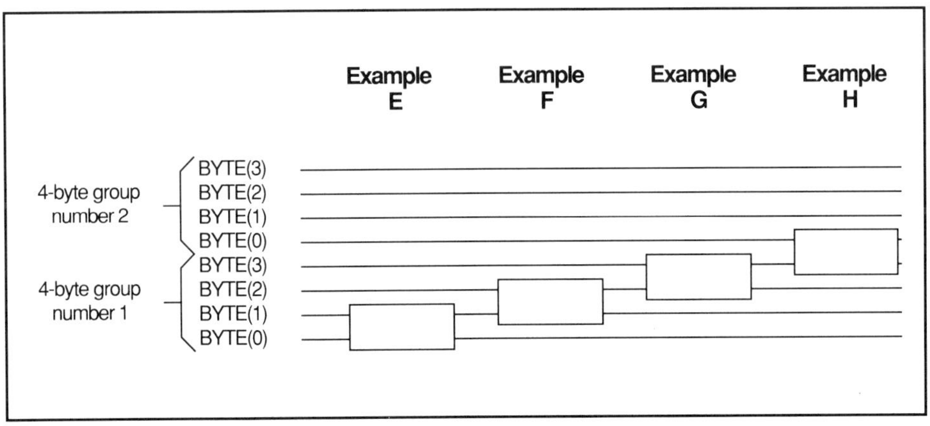

Figure 2-7. Four Ways That 16 Bits Of Data Might Be Stored In Memory

The MASTER can transfer the 16 bits of data using several different sequences of DTB cycles as listed in Table 2-14.

OBSERVATION 2.23:
The sequences listed in Table 2-14 would be typical of a MASTER that accesses the byte locations in ascending order. The VMEbus protocol does not require this.

As shown in Figure 2-13, the 16-bit transfer in example F can be accomplished with two single byte transfers. However, this requires two bus cycles. Because of this, the DTB protocol also includes a double byte transfer cycle which allows data to be stored as shown in example F using only one bus cycle.

OBSERVATION 2.24:
Since unaligned transfers make use of all 32 data lines, only D32 MASTERS and SLAVES can do unaligned transfers.

Table 2-15 lists how the unaligned transfer (UAT) mnemonic is used to describe MASTERS, SLAVES, and LOCATION MONITORS.

Table 2-14. Transferring 16 Bits Of Data Using Multiple Byte Transfer Cycles

Example	Cycle sequences used to accomplish the transfer	Data bus lines used	Byte locations accessed (See Figure 2-7)
E	Double byte transfer	D00-D15	Grp 1, BYTE(0-1)
F	Single byte transfer Single byte transfer or Double byte transfer	D00-D07 D08-D15 D08-D23	Grp 1, BYTE(1) Grp 1, BYTE(2) Grp 1, BYTE(1-2)
G	Double byte transfer	D00-D15	Grp 1, BYTE(2-3)
H	Single byte transfer Single byte transfer	D00-D07 D08-D15	Grp 1, BYTE(3) Grp 2, BYTE(0)

Table 2-15. Mnemonic That Specifies Unaligned Transfer Capability

The Following Mnemonic	When Applied to a	Means that it
UAT	D32 MASTER	can generate the following cycles:
	D32 SLAVE	can accept the following cycles:
	D32 LOCATION MONITOR	can monitor the following cycles:
		Triple byte read cycles: BYTE(0-2) READ BYTE(1-3) READ
		Triple byte write cycles: BYTE(0-2) WRITE BYTE(1-3) WRITE
		Double byte read cycle: BYTE(1-2) READ
		Double byte write cycle: BYTE(1-2) WRITE

2.3.10 ADDRESS-ONLY Capability

The ADDRESS-ONLY cycle is the only cycle on the DTB that is not used to transfer data. It begins as a typical DTB cycle, with the address, address modifier code, IACK* and LWORD* lines becoming valid and the address strobe falling after a set-up time. However, the data strobes are never driven low. After holding the various lines strobed by the address strobe stable for a prescribed minimum period, the MASTER finishes the cycle without waiting for DTACK* or BERR* to go low. (The ADDRESS-ONLY cycle is also the only type of DTB cycle that does not require a response in order to complete.)

Table 2-16 lists how the ADDRESS-ONLY mnemonic (ADO) is used to describe MASTERS and SLAVES.

OBSERVATION 2.25:
ADDRESS-ONLY cycles can be used to enhance board performance by allowing a CPU board to broadcast an address before it has determined whether or not that address selects a SLAVE on the VMEbus. Broadcasting the address in this fashion allows VMEbus SLAVES to decode the address concurrently with the CPU board.

RECOMMENDATION 2.6:
Since MASTERS might generate ADDRESS-ONLY cycles, design SLAVES to include ADO capability.

Table 2-16. Mnemonic That Specifies ADDRESS-ONLY Capability

The Following Mnemonic	When Applied to a	Means that it
ADO	MASTER	can generate ADDRESS-ONLY cycles
	SLAVE	can tolerate ADDRESS-ONLY cycles without loss or alteration of stored data.

2.3.11 Interaction Between DTB Functional Modules

Data transfers take place between MASTERS and SLAVES. The MASTER is the module controlling the transfer. The SLAVE which recognizes the address as its own is the responding SLAVE, and all other SLAVES are non-responding SLAVES.

After initiating a data transfer cycle, the MASTER waits for a response from the responding SLAVE. When the MASTER detects the response from the SLAVE it drives the data strobes and address strobe high, terminating the cycle. The SLAVE responds by releasing its response line.

OBSERVATION 2.26:
Although the address and data timing are largely independent, there are two exceptions. First, the MASTER waits until it has driven AS* low before driving either of the data strobes low. Second, the SLAVE acknowledges both the address strobe and the data strobes with either DTACK* or BERR*.

RULE 2.13:
IF a SLAVE responds to a data transfer cycle,
THEN it **MUST** either drive DTACK* low or it **MUST** drive BERR* low, but not both.

OBSERVATION 2.27:
Because of possible bus skew, due to different loading of the address strobe and the data strobes, the falling edge of the data strobes might be detected by the SLAVE slightly before the address strobe falling edge.

OBSERVATION 2.28:
The WRITE* line is high/low to identify a read/write cycle before the first data strobe is driven low, and remains stable until both data strobes are high.

RULE 2.14:
Before driving the data bus, the MASTER **MUST** ensure that the previous responding SLAVE has stopped driving the data bus. It does so by verifying that DTACK* and BERR* are both high before it drives the data strobe(s) to low on any cycle, and before it drives any of the data lines during a write cycle.

RULE 2.15:
At the end of a read cycle the responding SLAVE **MUST** release the data bus before allowing DTACK* to go high.

RULE 2.16:
When the MASTER reads data from the SLAVE, the SLAVE **MUST** maintain valid data on the data bus until the MASTER returns the first data strobe to high.

SUGGESTION 2.3:
For optimum performance, design MASTERS so that they drive the data strobes high as soon as possible after DTACK* or BERR* goes low. Also, design SLAVES so that they release the data bus and DTACK* as soon as possible after detecting the data strobes high. This allows the maximum data transfer rate on the bus.

OBSERVATION 2.29:
Addressing information on the bus might change soon after a module drives DTACK* or BERR* low, and before the MASTER drives the data strobes high.

A third type of module, called the LOCATION MONITOR, monitors the data transfer and generates either or both of two on-board signals whenever an access is done to a byte location that it monitors. If the access is a write cycle, then the on-board WRITE signal is generated. If the access is a read cycle, then the on-board READ signal is generated. If a read-modify-write cycle is performed, then both on-board signals are generated.

If the cycle takes too long, a fourth module, called a BUS TIMER intervenes by driving BERR* low, completing the data transfer handshake, and allowing the bus to resume operation.

RULE 2.17:
There is a strict interlock between the rising and falling edges of the data strobes and DTACK*/BERR*. Once a MASTER has driven its data strobe(s) low it **MUST NOT** drive its data strobe(s) high and finish a transfer without first receiving a data transfer acknowledge or a bus error response.

OBSERVATION 2.30:
A board containing a processor, which needs to direct data transfers between itself and other VMEbus boards, would contain a MASTER module. If the same board also contained memory accessible from the VMEbus, it would also contain a SLAVE module. A floating point processor or intelligent peripheral controller might receive commands through a SLAVE interface from a general purpose processor board. It then might act as a MASTER to access global VMEbus memory to execute the command it has been given.

2.4 TYPICAL OPERATION

MASTERS initiate data transfers over the DTB. The addressed SLAVE then acknowledges the transfer. After receiving the data transfer acknowledge, the MASTER terminates the data transfer cycle. The asynchronous nature of the DTB allows the SLAVE to control the time taken for the transfer.

Before doing any data transfers, a MASTER has to be granted exclusive control of the DTB. This ensures that multiple MASTERS will not try to use the DTB at the same time. The MASTER gains control of the DTB using the modules and signal lines of the Arbitration Bus. (Section 2.5 explains this further) The following discussion presumes that the MASTER has already been granted and assumed control of the DTB.

2.4.1 Typical Data Transfer Cycles

Figure 2-8 shows a typical single byte read cycle. To start the transfer, the MASTER drives the addressing lines with the desired address and address modifier code. Since this example is a BYTE(1) READ CYCLE, the MASTER drives LWORD* high and A01 low. Since it is not doing an interrupt acknowledge cycle, it does not drive IACK* low. The MASTER then waits for a specified set-up time before driving AS* low, to allow the address lines and the address modifier lines to stabilize before the SLAVES sample them.

Each SLAVE determines whether it should respond by examining the levels on the address lines, the address modifier lines, and IACK*. While this is happening, the MASTER drives WRITE* high to indicate a read operation. The MASTER then verifies that DTACK* and BERR* are high to ensure that the SLAVE from the previous cycle is no longer driving the data bus. If this is the case, the MASTER then drives DS0* low, while keeping DS1* high.

The responding SLAVE then determines which 4-byte group and which byte of that group is to be accessed, and starts the transfer. After it has retrieved the data from its internal storage and placed it on data bus lines D00-D07, the SLAVE signals the MASTER by driving DTACK* low. The SLAVE then holds DTACK* low and maintains the data valid for as long as the MASTER holds DS0* low.

When the MASTER receives DTACK* driven to low, it captures the data on D00-D07, releases the address lines, and drives DS0* and AS* to high. The SLAVE responds by releasing D00-D07 and releasing DTACK* to high.

OBSERVATION 2.31:
The MASTER in Figure 2-8 releases all of the DTB lines at the end of the data transfer. This is not required unless the MASTER'S REQUESTER released BBSY* during the data transfer as described in Section 2.5.

The cycle flow for double byte and quad byte data transfer cycles are very similar to the single byte cycle. Flow diagrams for these cycles are shown in Figures 2-9 and 2-10.

MASTER

ADDRESS THE SLAVE

Present address
Present address modifier
Drive LWORD* high
Drive IACK* high
Drive AS* to low

SPECIFY DATA DIRECTION

Drive WRITE* high

SPECIFY DATA WIDTH

Wait until DTACK* high and
BERR* high (indicates that
previous SLAVE is no longer
driving data bus)
Drive DS0* to low and DS1* to high

SLAVE

PROCESS ADDRESS

Receive address
Receive address modifier
Receive LWORD* high
Receive IACK* high
Receive AS* low
If address is valid for this SLAVE
Then select on-board device

FETCH DATA

Receive WRITE* high
Read data from selected device
Receive DS1* high
Receive DS0* low
Present data on lines D00-D07

(Continued on sheet 2)

Figure 2-8. An Example Of A Single Byte Read Cycle (Sheet 1 of 2)

RESPOND TO MASTER

Drive DTACK* to low

ACQUIRE DATA

Receive data on lines D00-D07
Receive DTACK* low

TERMINATE CYCLE

If last cycle then
  Release address lines
  Release address modifier lines
  Release LWORD*
  Release IACK*
Endif
Drive DS0* to high
Drive AS* to high

END TERMINATION

If last cycle then
  Release DS0* and DS1*
  Release AS*
Else go to ADDRESS THE SLAVE
Endif

END RESPONSE TO MASTER

Receive AS* and DS0* high
Release D00-D07

ACKNOWLEDGE TERMINATION

Release DTACK*

Figure 2-8. An Example Of A Single Byte Read Cycle (Sheet 2 of 2)

MASTER

ADDRESS THE SLAVE

Present address
Present address modifier
Drive LWORD* high
Drive IACK* high
Drive AS* to low

SPECIFY DATA DIRECTION

Drive WRITE* low

SPECIFY DATA WIDTH

Wait until DTACK* high and BERR* high (indicates that previous SLAVE is no longer driving data bus)
Place data on D00-D15
Drive DS0* and DS1* to low

SLAVE

PROCESS ADDRESS

Receive address
Receive address modifier
Receive LWORD* high
Receive IACK* high
Receive AS* low
If address is valid for this SLAVE then select on-board device

STORE DATA

Receive WRITE* low
Receive DS1* low
Receive DS0* low
Capture data from lines D00-D15
Write data into selected device

RESPOND TO MASTER

Drive DTACK* to low

(Continued on sheet 2)

Figure 2-9. An Example Of A Double Byte Write Cycle (Sheet 1 of 2)

TERMINATE CYCLE

Receive DTACK* low
If last cycle then
  Release address lines
  Release address modifier lines
  Release data lines
  Release LWORD*
  Release IACK*
Endif
Drive DS0* and DS1* to high
Drive AS* to high

END TERMINATION

If last cycle then
  Release DS0* and DS1*
  Release AS*
Else go to ADDRESS THE SLAVE
Endif

ACKNOWLEDGE TERMINATION

Receive AS*, DS0*, and DS1* high
Release DTACK*

Figure 2-9. An Example Of A Double Byte Write Cycle (Sheet 2 of 2)

MASTER

ADDRESS THE SLAVE

Present address
Present address modifier
Drive LWORD* low
Drive IACK* high
Drive AS* to low

SPECIFY DATA DIRECTION

Drive WRITE* low

SPECIFY DATA WIDTH

Wait until DTACK* high and BERR* high (indicates that previous SLAVE is no longer driving data bus)
Place data on D00-D31
Drive DS0* and DS1* to low

SLAVE

PROCESS ADDRESS

Receive address
Receive address modifier
Receive LWORD* low
Receive IACK* high
Receive AS* low
If address is valid for this SLAVE
Then select on-board device

STORE DATA

Receive WRITE* low
Receive DS1* low
Receive DS0* low
Latch data from lines D00-D31
Write data into selected device

RESPOND TO MASTER

Drive DTACK* to low

(Continued on sheet 2)

Figure 2-10. An Example Of A Quad Byte Write Cycle (Sheet 1 of 2)

TERMINATE CYCLE

Receive DTACK* low
If last cycle then
    Release address lines
    Release address modifier lines
    Release data lines
    Release LWORD*
    Release IACK*
Endif
Drive DS0* and DS1* to high
Drive AS* to high

END TERMINATION

If last cycle then
    Release DS0* and DS1*
    Release AS*
Else go to ADDRESS THE SLAVE
Endif

ACKNOWLEDGE TERMINATION

Receive AS*, DS0*, and DS1* high
Release DTACK*

Figure 2-10. An Example Of A Quad Byte Write Cycle (Sheet 2 of 2)

2.4.2 Address Pipelining

The VMEbus strobes the address and data with separate strobe signals. This allows a MASTER to broadcast the address for the next cycle while the data transfer for the previous cycle is still in progress. This is called "address pipelining".

PERMISSION 2.8:
As soon as a SLAVE drives DTACK* or BERR* low, the MASTER **MAY** change the address, and, after driving AS* high for a minimum time, drive AS* low again.

For example, when a SLAVE drives DTACK* low on a read cycle, the MASTER can place a new address on the address bus while it is reading the data from the bus. This amounts to an overlapping of one cycle with the next one, and permits greater speeds on the VMEbus.

RULE 2.18:
Since address pipelining can occur, SLAVES **MUST NOT** be designed on the assumption that they will never encounter pipelined cycles.

OBSERVATION 2.32:
The responding SLAVE might recognize its address and respond very quickly on the DTACK* or BERR* lines. Since the MASTER is permitted to remove the address after the responding SLAVE drives DTACK* or BERR* low, non-responding SLAVES might not be able to decode the addressing information before the MASTER removes it from the bus.

SUGGESTION 2.4:
Design SLAVES to capture the addressing information on the falling edge of AS*.

OBSERVATION 2.33:
Because a MASTER might broadcast a new address while a previous cycle is finishing, the designer needs to ensure that the assertion of the second address strobe does not invalidate the first address if it is still needed by on-board logic to maintain the data on the bus.

PERMISSION 2.9:
MASTERS **MAY** be designed without the ability for address pipelining. (e.g. They **MAY** wait until the responding SLAVE releases DTACK* or BERR* before driving AS* low for the next cycle.)

OBSERVATION 2.34:
A MASTER might drive AS* low for a new cycle before it drives data strobes high from the previous cycle. Because of this, there might be a period when the address strobe for the new cycle, as well as at least one data strobe from the previous cycle coincide during the cycle overlap.

SUGGESTION 2.5:
To assure reliable operation, design SLAVES to initiate data transfers to and from the bus on the falling **edge** of the data strobes, instead of a simultaneous low **level** on the address strobe and data strobes.

2.5 DATA TRANSFER BUS ACQUISITION

RULE 2.19:
Before transferring any data on the DTB, a MASTER **MUST** get permission to use it.

Several MASTERS might want to use the DTB at the same time. The process which determines which MASTER can use the DTB is called arbitration and is discussed in Chapter 3. Because arbitration is closely tied to the operation of the DTB, it is briefly described here.

Figure 2-11 provides two examples that show possible sequences when a MASTER (called MASTER A) finishes using the DTB and allows arbitration to take place.

In Example 1, MASTER A, partway into its last transfer, indicates that it no longer needs the DTB. It does this by having its REQUESTER release the bus busy (BBSY*) signal line. Since MASTER A gives this early notice that the DTB will soon be available, the arbitration is done during this last data transfer. The arbitration is completed and MASTER B is granted permission to use the DTB before MASTER A has finished its cycle, but it waits until MASTER A releases AS*. (This assures that MASTER B will not start driving the DTB before MASTER A has finished with its last data transfer.)

In Example 2, MASTER A waits until after its last transfer (i.e., after AS* is released) before releasing BBSY*. In this case the DTB is idle while the arbitration is done. MASTER B is then granted the bus and, since AS* is already high, it begins using the DTB immediately.

RULE 2.20:
Once a MASTER'S REQUESTER releases BBSY* to high, the MASTER **MUST NOT** drive AS* from high to low (i.e., it **MUST NOT** begin a new cycle) until its REQUESTER receives a new bus grant.

2.6 DTB TIMING RULES AND OBSERVATIONS

This section describes the timing RULES and OBSERVATIONS that govern the behavior of MASTERS and SLAVES. This timing information is in the form of figures and tables:

Table 2-17 lists a timing table and timing diagrams that specify MASTER, SLAVE, and LOCATION MONITOR operation.

Table 2-18 defines the various mnemonics that are used in Tables 2-19 through 2-21.

Tables 2-19 through 2-21 specify the use of the DTB signals.

Tables 2-22 through 2-27 specify the timing parameters of the DTB. (The reference numbers used in Tables 2-24 through 2-27 correspond to the timing parameter numbers in Tables 2-22 and 2-23.)

Figures 2-12 through 2-15 specify the timing RULES and OBSERVATIONS during address broadcasting time.

Figures 2-16 through 2-21 specify the timing RULES and OBSERVATIONS for MASTERS, SLAVES, and LOCATION MONITORS during data transfer time.

Figures 2-22 through 2-24 specify the timing RULES and OBSERVATIONS for MASTERS and SLAVES between DTB cycles.

Figure 2-25 is the timing diagram for MASTER, SLAVE, and BUS TIMER during a timed-out cycle.

Figure 2-26 shows the timing during the mastership transfer of the DTB.

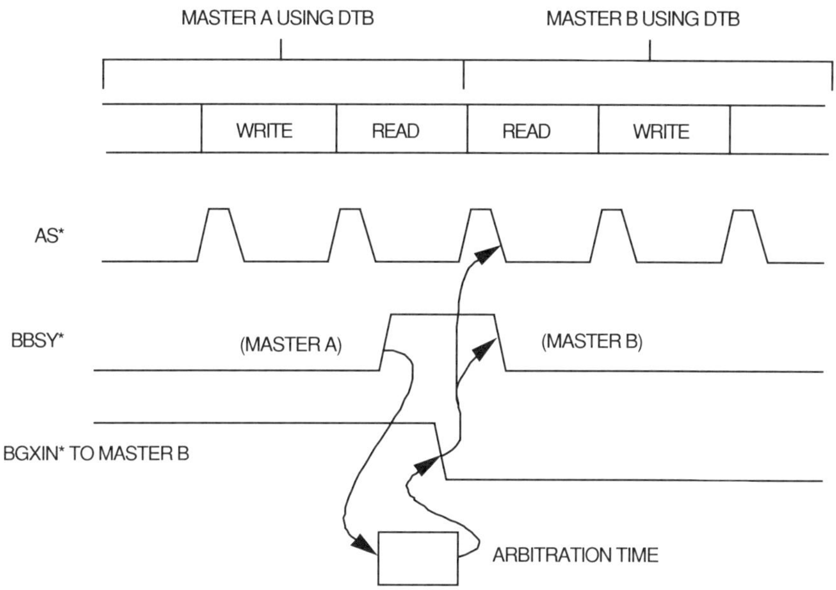

Example 1 — Arbitration DURING The Last Data Transfer

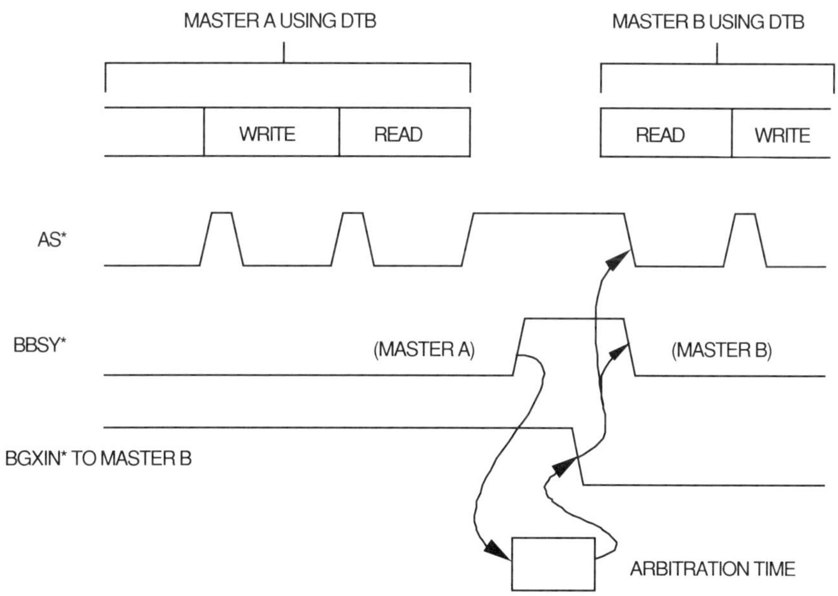

Example 2 — Arbitration AFTER The Last Data Transfer

Figure 2-11. Data Transfer Bus MASTER Exchange Sequence

In order to meet the specified timing RULES, board designers need to take into account the worst case propagation delays of the bus drivers and receivers used on their VMEbus boards. The propagation delay of the drivers depends on their output loads and manufacturers specifications do not always give enough information to calculate the propagation delays under various loads. To help the VMEbus board designer, some suggestions are offered in Chapter 6.

The OBSERVATIONS specify the timing of incoming lines signal transitions. These times can be relied upon as long as the backplane loading RULES in Chapter 6 are not violated. The RULES for the bus terminators in Chapter 6 guarantee that the timing parameters for signal lines that are released after they have been driven, are met.

Typically, for each timing RULE there is a corresponding OBSERVATION. However, the time that is guaranteed in the OBSERVATION might differ from the time specified by the RULE. For example, a careful inspection of the timing diagrams shows that the MASTER is required to provide 35 nanoseconds of address and data set-up time, but the SLAVE is only guaranteed 10 nanoseconds. This is because the address and data bus drivers are not always able to drive the backplane's signal lines completely through the threshold region from low to high until the transition propagates to the end of the backplane and is reflected back. The falling edge of the address and data strobes, however, typically cross the 0.8-volt threshold without waiting for a reflection. The resulting set-up time at the SLAVE is the MASTER'S set-up time less two bus propagation times.

A special notation has been used to describe the data strobe timing. The two data strobes (DS0* and DS1*) will not always make their transitions simultaneously. For purposes of these timing diagrams, DSA* represents the first data strobe to make its transition (whether that is DS0* or DS1*). The broken line shown while the data strobes are stable is to indicate that the first data strobe to make a falling transition might not be the first to make its rising transition - i.e., DSA* can represent DS0* on its falling edge and DS1* on its rising edge.

Table 2-17. Timing Diagrams That Define MASTER, SLAVE, And LOCATION MONITOR Operation (See Table 2-22 For Timing Values)

Mnemonic	Type of cycles	Address Broadcast Timing Diag Figure(s)	Data Transfer Timing Diag Figure
ADO	ADDRESS-ONLY	2-12	N/A
D08(EO)	Single even byte transfer		
	BYTE(0) READ	2-12 & 2-13	2-16
	BYTE(2) READ	2-12 & 2-13	2-16
	BYTE(0) WRITE	2-12 & 2-13	2-18
	BYTE(2) WRITE	2-12 & 2-13	2-18
D08(EO)	Single odd byte transfers		
or	BYTE(1) READ	2-12 & 2-13	2-16
	BYTE(3) READ	2-12 & 2-13	2-16
D08(O)			
	BYTE(1) WRITE	2-12 & 2-13	2-18
	BYTE(3) WRITE	2-12 & 2-13	2-18
D16	Double byte transfers		
	BYTE(0-1) READ	2-12 & 2-13	2-17
	BYTE(2-3) READ	2-12 & 2-13	2-17
	BYTE(0-1) WRITE	2-12 & 2-13	2-19
	BYTE(2-3) WRITE	2-12 & 2-13	2-19
D32	Quad byte transfers		
	BYTE(0-3) READ	2-12 & 2-13	2-17
	BYTE(0-3) WRITE	2-12 & 2-13	2-19
D08(EO):BLT	Single byte block transfers		
	SINGLE BYTE BLOCK READ	2-12 & 2-14	2-16
	SINGLE BYTE BLOCK WRITE	2-12 & 2-14	2-18

(Continued on Page 66)

Table 2-17. (Continued)

Mnemonic	Type of cycles	Address Broadcast Timing Diag Figure(s)	Data Transfer Timing Diag Figure
D16:BLT	Double byte block transfers		
	DOUBLE BYTE BLOCK READ	2-12 & 2-14	2-17
	DOUBLE BYTE BLOCK WRITE	2-12 & 2-14	2-19
D32:BLT	Quad byte block transfers		
	QUAD BYTE BLOCK READ	2-12 & 2-14	2-17
	QUAD BYTE BLOCK WRITE	2-12 & 2-14	2-19
D08(EO):RMW	Single byte RMW transfers		
	BYTE(0) READ-MODIFY-WRITE	2-12 & 2-15	2-20
	BYTE(1) READ-MODIFY-WRITE	2-12 & 2-15	2-20
	BYTE(2) READ-MODIFY-WRITE	2-12 & 2-15	2-20
	BYTE(3) READ-MODIFY-WRITE	2-12 & 2-15	2-20
D16:RMW	Double byte RMW transfers		
	BYTE(0-1) READ-MODIFY-WRITE	2-12 & 2-15	2-21
	BYTE(2-3) READ-MODIFY-WRITE	2-12 & 2-15	2-21
D32:RMW	Quad byte RMW transfers		
	BYTE(0-3) READ-MODIFY-WRITE	2-12 & 2-15	2-21
D32:UAT	Unaligned transfers		
	BYTE(0-2) READ	2-12 & 2-13	2-16
	BYTE(1-3) READ	2-12 & 2-13	2-16
	BYTE(1-2) READ	2-12 & 2-13	2-17
	BYTE(0-2) WRITE	2-12 & 2-13	2-18
	BYTE(1-3) WRITE	2-12 & 2-13	2-18
	BYTE(1-2) WRITE	2-12 & 2-13	2-19

Tables 2-19, 2-20, and 2-21 show how the various signal lines of the DTB are used to broadcast addresses and to transfer data. These Tables are referenced by the various timing diagrams that follow. In order to keep these Tables compact, mnemonics are used to describe when and how the various lines are driven. These mnemonics are defined in Table 2-18.

Table 2-18. Definitions Of Mnemonics Used In Tables 2-19, 2-20, And 2-21

Mnemonic	Description	Comments
DVBM	DRIVEN VALID BY MASTER	RULE 2.21: The MASTER **MUST** drive DVBM lines to a valid level
DLBM	DRIVEN LOW BY MASTER	RULE 2.22: The MASTER **MUST** drive DLBM lines to a low level.
DHBM	DRIVEN HIGH BY MASTER	RULE 2.23: The MASTER **MUST** drive DHBM lines to a high level.
dhbm?	DRIVEN HIGH BY MASTER?	PERMISSION 2.10: The MASTER **MAY** drive dhbm? lines high. RULE 2.24: The MASTER **MUST NOT** drive dhbm? lines low.
dxbm?	DRIVEN BY MASTER?	PERMISSION 2.11: The MASTER **MAY** drive dxbm? lines, or it **MAY** leave these lines undriven. (When dxbm? lines are driven, they carry no valid information.)

(Continued on Page 68)

Table 2-18. (Continued)

Mnemonic	Description	Comments
DVBS	DRIVEN VALID BY SLAVE	RULE 2.25: The SLAVE **MUST** drive DVBS lines to a valid level.
dxbs?	DRIVEN BY SLAVE?	PERMISSION 2.12: The SLAVE **MAY** drive dxbs? lines, or it **MAY** leave these lines undriven. (When dxbs? lines are driven, they carry no valid information.)
DVBB	DRIVEN VALID BY BOTH SLAVE AND MASTER	RULE 2.26: During the "read" portion of a read-modify-write cycle, the SLAVE **MUST** drive DVBB lines with valid data. During the "write" portion of a read-modify-write cycle, the MASTER **MUST** drive DVBB lines with valid data.
dxbb?	DRIVEN BY BOTH SLAVE AND MASTER?	PERMISSION 2.13: During the "read" portion of a read-modify-write cycle, the SLAVE **MAY** drive dxbb? lines, or it **MAY** leave them undriven. During the "write" portion of a read-modify-write cycle, the MASTER **MAY** drive dxbb? lines, or it **MAY** leave them undriven. (When dxbb? lines are driven, they carry no valid information.)

Table 2-19. Use Of Addressing Lines To Select A 4-Byte Group

Mnemonic	Addressing mode	A02-A15 See Note	A16-A23	A24-A31	IACK*
A16	SHORT	DVBM	dxbm?	dxbm?	dhbm?
A24	STANDARD	DVBM	DVBM	dxbm?	dhbm?
A32	EXTENDED	DVBM	DVBM	DVBM	dhbm?

Note: A01 is used in conjunction with LWORD* and the two data strobes to select which of the four bytes within the 4-byte group is accessed. (See Table 2-20).

Table 2-20. Use Of The DS1*, DS0*, A01, And LWORD* Lines To Select The Byte(s) Within A 4-Byte Group.

Mnemonic	Type of cycles	DS1*	DS0*	A01	LWORD*
ADO	ADDRESS-ONLY	dhbm?	dhbm?	dxbm?	dxbm?
D08(EO)	Single even byte transfers				
	BYTE(0) READ or WRITE	DLBM	dhbm?	DLBM	dhbm?
	BYTE(2) READ or WRITE	DLBM	dhbm?	DHBM	dhbm?
D08(EO)	Single odd byte transfers				
	BYTE(1) READ or WRITE	dhbm?	DLBM	DLBM	dhbm?
	BYTE(3) READ or WRITE	dhbm?	DLBM	DHBM	dhbm?
D16	Double byte transfers				
	BYTE(0-1) READ or WRITE	DLBM	DLBM	DLBM	dhbm?
	BYTE(2-3) READ or WRITE	DLBM	DLBM	DHBM	dhbm?

(Continued on Page 70)

Table 2-20. (Continued)

Mnemonic	Type of cycles	DS1*	DS0*	A01	LWORD*
D32	Quad byte transfers				
	BYTE(0-3) READ or WRITE	DLBM	DLBM	DLBM	DLBM
D08(EO) :BLT	Single byte block transfers				
	SINGLE BYTE BLOCK READ or WRITE	← NOTE 1 → See Pg 71			dhbm?
D16:BLT	Double byte block transfers				
	DOUBLE BYTE BLOCK READ or WRITE	DLBM	DLBM	NOTE 2 See Pg 71	dhbm?
D32:BLT	Quad byte block transfers				
	QUAD BYTE BLOCK READ or WRITE	DLBM	DLBM	DLBM	DLBM
D08(EO) :RMW	Single byte RMW transfers				
	BYTE(0) READ-MODIFY-WRITE	DLBM	dhbm?	DLBM	dhbm?
	BYTE(1) READ-MODIFY-WRITE	dhbm?	DLBM	DLBM	dhbm?
	BYTE(2) READ-MODIFY-WRITE	DLBM	dhbm?	DHBM	dhbm?
	BYTE(3) READ-MODIFY-WRITE	dhbm?	DLBM	DHBM	dhbm?
D16:RMW	Double byte RMW transfers				
	BYTE(0-1) READ-MODIFY-WRITE	DLBM	DLBM	DLBM	dhbm?
	BYTE(2-3) READ-MODIFY-WRITE	DLBM	DLBM	DHBM	dhbm?
D32:RMW	Quad byte RMW transfers				
	BYTE(0-3) READ-MODIFY-WRITE	DLBM	DLBM	DLBM	DLBM

(Continued on Page 71)

Table 2-20. (Continued)

Mnemonic	Type of cycles	DS1*	DS0*	A01	LWORD*
D32:UAT	Unaligned transfers				
	BYTE(0-2) READ or WRITE	DLBM	dhbm?	DLBM	DLBM
	BYTE(1-3) READ or WRITE	dhbm?	DLBM	DLBM	DLBM
	BYTE(1-2) READ or WRITE	DLBM	DLBM	DHBM	DLBM

Notes:

1. During Single byte block transfers, the two data strobes are alternately driven low. Either data strobe might be driven low on the first transfer. If the first accessed byte location is BYTE(0) or BYTE(2), then the MASTER drives DS1* to low first. If the first accessed byte location is BYTE(1) or BYTE(3), then it drives DS0* to low first. A01 is valid only on the first data transfer (i.e. until the SLAVE drives DTACK* or BERR* low the first time) and might be either high or low depending upon which byte the single byte block transfer begins with. If the first byte location is BYTE(0) or BYTE(1), then the MASTER drives A01 to low. If the first byte location is BYTE(2) or BYTE(3), then the MASTER drives A01 to high.

An example of the use of DS0*, DS1*, A01, and LWORD* during a Single byte block transfer cycle which starts with BYTE(2) is given below:

	DS1*	DS0*	A01	LWORD*
First data transfer BYTE(2)	DLBM	DHBM	DHBM	dhbm?
↓ BYTE(3)	DHBM	DLBM	dxbm?	dxbm?
BYTE(0)	DLBM	DHBM	dxbm?	dxbm?
BYTE(1)	DHBM	DLBM	dxbm?	dxbm?
Last data transfer BYTE(2)	DLBM	DHBM	dxbm?	dxbm?

2. During a Double byte block transfer, A01 is valid only on the first data transfer (i.e. until the SLAVE drives DTACK* or BERR* low the first time) and is driven either high or low depending upon what double byte group the double byte block transfer begins with. If the first double byte group is BYTE(0-1), then the MASTER drives A01 to low. If the first double byte group is BYTE(2-3), then the MASTER drives A01 to high.

Table 2-21. Use Of The Data Bus Lines To Transfer Data

Mnemonic	Type of cycles	D24-D31	D16-D23	D08-D15	D00-D07
ADO	ADDRESS-ONLY	dxbm?	dxbm?	dxbm?	dxbm?
D08(EO)	Single even byte transfers				
	BYTE(0) READ	dxbs?	dxbs?	DVBS	dxbs?
	BYTE(2) READ	dxbs?	dxbs?	DVBS	dxbs?
	BYTE(0) WRITE	dxbm?	dxbm?	DVBM	dxbm?
	BYTE(2) WRITE	dxbm?	dxbm?	DVBM	dxbm?
D08(EO) or D08(O) (SLAVE only)	Single odd byte transfers				
	BYTE(1) READ	dxbs?	dxbs?	dxbs?	DVBS
	BYTE(3) READ	dxbs?	dxbs?	dxbs?	DVBS
	BYTE(1) WRITE	dxbm?	dxbm?	dxbm?	DVBM
	BYTE(3) WRITE	dxbm?	dxbm?	dxbm?	DVBM
D16	Double byte transfers				
	BYTE(0-1) READ	dxbs?	dxbs?	DVBS	DVBS
	BYTE(2-3) READ	dxbs?	dxbs?	DVBS	DVBS
	BYTE(0-1) WRITE	dxbm?	dxbm?	DVBM	DVBM
	BYTE(2-3) WRITE	dxbm?	dxbm?	DVBM	DVBM
D32	Quad byte transfers				
	BYTE(0-3) READ	DVBS	DVBS	DVBS	DVBS
	BYTE(0-3) WRITE	DVBM	DVBM	DVBM	DVBM

(Continued on Page 73)

Table 2-21. (Continued)

Mnemonic	Type of cycles	D24-D31	D16-D23	D08-D15	D00-D07
D08(EO):BLT	Single byte block transfers				
	SINGLE BYTE BLOCK READ	dxbs?	dxbs?	← Note 1 → (See Pg 74)	
	SINGLE BYTE BLOCK WRITE	dxbm?	dxbm?	← Note 1 →	
D16:BLT	Double byte block transfers				
	DOUBLE BYTE BLOCK READ	dxbs?	dxbs?	DVBS	DVBS
	DOUBLE BYTE BLOCK WRITE	dxbm?	dxbm?	DVBM	DVBM
D32:BLT	Quad byte block transfers				
	QUAD BYTE BLOCK READ	DVBS	DVBS	DVBS	DVBS
	QUAD BYTE BLOCK WRITE	DVBM	DVBM	DVBM	DVBM
D08(EO):RMW	Single byte RMW transfers				
	BYTE(0) READ-MODIFY-WRITE	dxbb?	dxbb?	DVBB	dxbb?
	BYTE(1) READ-MODIFY-WRITE	dxbb?	dxbb?	dxbb?	DVBB
	BYTE(2) READ-MODIFY-WRITE	dxbb?	dxbb?	DVBB	dxbb?
	BYTE(3) READ-MODIFY-WRITE	dxbb?	dxbb?	dxbb?	DVBB
D16:RMW	Double byte RMW transfers				
	BYTE(0-1) READ-MODIFY-WRITE	dxbb?	dxbb?	DVBB	DVBB
	BYTE(2-3) READ-MODIFY-WRITE	dxbb?	dxbb?	DVBB	DVBB
D32:RMW	Quad byte RMW transfers				
	BYTE(0-3) READ-MODIFY-WRITE	DVBB	DVBB	DVBB	DVBB

(Continued on Page 74)

Table 2-21. (Continued)

Mnemonic	Type of cycles	D24-D31	D16-D23	D08-D15	D00-D07
D32:UAT	Unaligned transfers				
	BYTE(0-2) READ	DVBS	DVBS	DVBS	dxbs?
	BYTE(1-3) READ	dxbs?	DVBS	DVBS	DVBS
	BYTE(1-2) READ	dxbs?	DVBS	DVBS	dxbs?
	BYTE(0-2) WRITE	DVBM	DVBM	DVBM	dxbm?
	BYTE(1-3) WRITE	dxbm?	DVBM	DVBM	DVBM
	BYTE(1-2) WRITE	dxbm?	DVBM	DVBM	dxbm?

Note:
1. During Single byte block transfers, data is transferred 8 bits at a time over D00-D07 or D08-D15. A single byte block read example is given below:

	D08-D15	D00-D07
First data transfer	DVBS	dxbs?
↓	dxbs?	DVBS
	DVBS	dxbs?
	dxbs?	DVBS
	DVBS	dxbs?
	dxbs?	DVBS
Last data transfer	DVBS	dxbs?

Table 2-22. MASTER, SLAVE, And LOCATION MONITOR Timing Parameters

PARAMETER NUMBER	MASTER See also Table 2-24		SLAVE See also Table 2-25		LOCATION MONITOR See also Table 2-26	
	MIN	MAX	MIN	MAX	MIN	MAX
1	0					
2	0					
3	60					
4	35		10		10	
5	40		30		30 Note 3	
6	0		0			
7	0		0			
8	35		10			
9	0		0			
10	0		-10 Note 4		-10 Note 4	
11	40		30		30	
12	35		10		10	
13		10		20		20
14	0		0			
15	0		0			
16	0		0			
17	40		30		30	
18	0		0			
19	40		30		30	
20	0		0			
21	0		0			
22	0		0			
23	10		0		0	
24A	0					
24B	0					
25		25				
26	0		0			
27	-25		0			
28	30	2T	30			
29	0		0			
30	0		0			
31	0		0			
32			10		10	
33			30		30	

NOTES:

1. All times are in nsec.
2. T = the time-out value in microseconds.
3. Correct value. (Previous version showed 3 nsec.)
4. Correct value. (Previous version showed 10 nsec.)

Table 2-23. BUS TIMER Timing Parameters (See Also Table 2-27)

PARAMETER NUMBER	MASTER MIN	MASTER MAX
28	T	2T
30	0	

NOTE: T = the time-out value in microseconds.

Table 2-24. MASTER, Timing RULES And OBSERVATIONS

Note: The numbers correspond to the timing parameters specified in Table 2-22 on page 75.

1.	**RULE 2.27:** When taking control of the DTB, the MASTER **MUST NOT** drive any of IACK*, AM0-AM5, A01-A31, LWORD*, D00-D31, WRITE*, DS0*, DS1*, or AS* until after the previous MASTER allows AS* to rise above the low level. **OBSERVATION 2.35:** Chapter 3 describes how a MASTER'S REQUESTER is granted use of the DTB.
2.	**RULE 2.28:** When taking control of the DTB, the MASTER **MUST NOT** drive any of IACK*, AM0-AM5, A01-A31, LWORD*, D00-D31, WRITE*, DS0*, DS1*, or AS* until after its REQUESTER is granted the bus. **OBSERVATION 2.36:** Chapter 3 describes how a MASTER'S REQUESTER is granted use of the DTB.
3.	**RULE 2.29:** When taking control of the DTB, the MASTER **MUST NOT** drive AS* low until this time after the previous MASTER allows AS* to rise above the low level. **OBSERVATION 2.37:** RULE 2.29 ensures that timing parameter 5 for SLAVES is guaranteed when there is an interchange of the DTB mastership.

(Continued on Page 77)

Table 2-24. (Continued)

4.	**RULE 2.30:** The MASTER **MUST NOT** drive AS* low until IACK* has been high, and the required lines among A01-A31, AM0-AM5, and LWORD* have been valid for this minimum time. **OBSERVATION 2.38:** Tables 2-19 and 2-20 specify the lines among A01-A31, AM0-AM5, and LWORD* that a MASTER is required to drive.
5.	**RULE 2.31:** When using the DTB for two consecutive cycles, the MASTER **MUST NOT** drive AS* low until it has been high for this minimum time.
6.	**RULE 2.32:** After a read cycle, the MASTER **MUST NOT** drive any of D00-D31 until both DTACK* and BERR* are high.
7.	**RULE 2.33:** During read cycles, the MASTER **MUST NOT** drive DSA* low until it has released all of D00-D31.
8.	**RULE 2.34:** During write cycles, the MASTER **MUST NOT** drive DSA* low until the required lines among D00-D31 have been valid for this minimum time. **OBSERVATION 2.39:** Table 2-21 specifies the lines among D00-D31 that a MASTER is required to drive.
9.	**RULE 2.35:** The MASTER **MUST NOT** drive DSA* low until both DTACK* and BERR* are high.
10.	**RULE 2.36:** The MASTER **MUST NOT** drive DSA* low until it has driven AS* low.
11.	**RULE 2.37:** The MASTER **MUST NOT** drive DSA* low until DS0* and DS1* have both been simultaneously high for this minimum time.
12.	**RULE 2.38:** The MASTER **MUST NOT** drive DSA* low until WRITE* has been valid for this minimum time.

(Continued on Page 78)

Table 2-24. (Continued)

13.	**RULE 2.39:** During cycles where the MASTER drives both data strobes low, it **MUST** drive DSB* low within this maximum time after it drives DSA* low. **OBSERVATION 2.40:** Timing parameter 13 does not apply to transfers where only one data strobe is driven low.
14.	**RULE 2.40:** During all data transfer cycles, except read-modify-write cycles, the MASTER **MUST** hold the address valid and maintain the appropriate level on LWORD* until it detects the first falling edge on DTACK* or BERR*. **OBSERVATION 2.41:** During all data transfer cycles, except block transfer and read-modify-write cycles, there will be only one falling edge on DTACK* or BERR*.
15.	**RULE 2.41:** During read-modify-write cycles, the MASTER **MUST** hold the address valid and maintain the appropriate level on LWORD* until it detects the second falling edge on DTACK* or BERR*.
16.	**RULE 2.42:** During all data transfer cycles, the MASTER **MUST** maintain a valid address modifier code, and ensure that IACK* stays high until it detects the last falling edge on DTACK* or BERR*. **OBSERVATION 2.42:** During all data transfer cycles, except block transfer and read-modify-write cycles, there will be only one falling edge on DTACK* or BERR*.
17.	**RULE 2.43:** The MASTER **MUST NOT** change the levels on IACK*, A01-A31, AM0-AM5, and LWORD* for this minimum time after it drives AS* low.
18.	**RULE 2.44:** During all data transfer cycles, the MASTER **MUST** hold AS* low until it detects the last falling edge on DTACK* or BERR*.
19.	**RULE 2.45:** The MASTER **MUST** hold AS* low for this minimum time.
20.	**RULE 2.46:** Once a MASTER has driven DSA* low, it **MUST** maintain that line low until it detects DTACK* or BERR* low.

(Continued on Page 79)

Table 2-24. (Continued)

21.	**RULE 2.47:** Once a MASTER has driven DSB* low, it **MUST** maintain that line low until it detects DTACK* or BERR* low.
22.	**RULE 2.48:** During write cycles, once the MASTER has driven DSA* low, it **MUST NOT** change any of D00-D31 until it detects DTACK* or BERR* low.
23.	**RULE 2.49:** Once a MASTER has driven DSA* low, it **MUST NOT** change the level on the WRITE* line, until this minimum time after both data strobes are high.
24A.	**RULE 2.50:** **IF** the MASTER drives or releases AS* to high after its REQUESTER releases BBSY*, **THEN** it **MUST** release IACK*, AM0-AM5, A01-A31, LWORD*, D00-D31, WRITE*, DS0*, and DS1* before allowing AS* to rise above the low level. **OBSERVATION 2.43:** Chapter 3 describes how a MASTER'S REQUESTER releases the BBSY* line.
24B.	**RULE 2.51:** **IF** the MASTER drives or releases AS* to high before its REQUESTER releases BBSY*, **THEN** it **MUST** release as*, IACK*, AM0-AM5, A01-A31, LWORD*, D00-D31, WRITE*, DS0*, and DS1* before allowing its REQUESTER to release BBSY*. **OBSERVATION 2.44:** Chapter 3 describes how a MASTER'S REQUESTER releases the BBSY* line.
25.	**RULE 2.52:** **IF** the MASTER drives or releases AS* to high after its REQUESTER releases BBSY*, **THEN** it **MUST** release AS* within this time after allowing AS* to rise above the low level. **OBSERVATION 2.45:** Chapter 3 describes how a MASTER'S REQUESTER releases the BBSY* line.

(Continued on Page 80)

Table 2-24. (Continued)

26.	**OBSERVATION 2.46:** Timing parameter 26 guarantees that during read cycles, the data bus will not be driven until the MASTER drives DSA* low.
27.	**OBSERVATION 2.47:** During read cycles, the MASTER is guaranteed that the data bus will be valid within 27 nsec after DTACK* goes low. This time does not apply to cycles where the SLAVE drives BERR* low instead of DTACK*.
28.	**OBSERVATION 2.48:** The MASTER is guaranteed that neither DTACK* nor BERR* will go low until this minimum time after it drives DSA* low. The BUS TIMER guarantees the MASTER that if DTACK* has not gone low after its time-out period has elapsed, and within twice its time-out period, then the BUS TIMER will drive BERR* low.
29.	**OBSERVATION 2.49:** During read cycles, the MASTER is guaranteed that the data bus will remain valid until it drives DSA* high.
30.	**OBSERVATION 2.50:** Timing parameter 30 guarantees that neither DTACK* nor BERR* will go high until the MASTER drives both DS0* and DS1* high.
31.	**OBSERVATION 2.51:** During read cycles, the MASTER is guaranteed that the data bus has been released by the time DTACK* and BERR* are high.

Table 2-25. SLAVE, Timing RULES And OBSERVATIONS

Note: The numbers correspond to the timing parameters specified in Table 2-22 on page 75.

4.	**OBSERVATION 2.52:** All SLAVES are guaranteed that IACK*, A01-A31, AM0-AM5, and LWORD* have been valid for this minimum time when they detect a falling edge on AS*.
5.	**OBSERVATION 2.53:** All SLAVES are guaranteed this minimum high time on AS* between DTB cycles.
6.	**OBSERVATION 2.54:** During read cycles, the responding SLAVE is guaranteed that none of D00-D31 will be driven by any other module until the responding SLAVE releases DTACK* and BERR* to high.
7.	**OBSERVATION 2.55:** During read cycles, the responding SLAVE is guaranteed that the data bus will be released by all other modules by the time DSA* goes low.
8.	**OBSERVATION 2.56:** During write cycles, the responding SLAVE is guaranteed that the data bus has been valid for this minimum time when it detects a falling edge on DSA*.
9.	**OBSERVATION 2.57:** The responding SLAVE is guaranteed that neither DS0* nor DS1* will go low until DTACK* and BERR* from the previous cycle have gone high.
10.	**OBSERVATION 2.58:** Due to bus skew, SLAVES on the DTB might detect a falling edge on DSA* before detecting the falling edge on AS*. However, SLAVES are guaranteed that a falling edge on DSA* will not precede the falling edge on AS* by more than this time.
11.	**OBSERVATION 2.59:** SLAVES are guaranteed this minimum time during which both data strobes are simultaneously high between consecutive data transfers.
12.	**OBSERVATION 2.60:** SLAVES are guaranteed that WRITE* has been valid for this minimum time before a falling edge on DSA*.
13.	**OBSERVATION 2.61:** **IF** both data strobes are going to be driven low, **THEN** the responding SLAVE is guaranteed that DSB* will go low within this maximum time after DSA* has gone low.

(Continued on Page 82)

Table 2-25. (Continued)

14.	**OBSERVATION 2.62:** During all data transfer cycles except read-modify-write cycles, the responding SLAVE is guaranteed that the address and LWORD* remain valid until it drives DTACK* or BERR* low for the first time, provided that it does so within the bus time-out period.
15.	**OBSERVATION 2.63:** During all read-modify-write cycles, the responding SLAVE is guaranteed that the address and LWORD* remain valid until it drives DTACK* or BERR* low for the second time, provided that it does so within the bus time-out period.
16.	**OBSERVATION 2.64:** The responding SLAVE is guaranteed that IACK* and AM0-AM5 remain valid until it drives DTACK* or BERR* low for the last time, provided that it does so within the bus timeout period.
17.	**OBSERVATION 2.65:** SLAVES are guaranteed that IACK*, A01-A31, AM0-AM5, and LWORD* will remain valid for this minimum time after the falling edge of AS*. During ADDRESS-ONLY cycles this time is guaranteed by the MASTER. During all other cycle types, this time is guaranteed by timing parameters 10, 14, 16, and 28.
18.	**OBSERVATION 2.66:** The responding SLAVE is guaranteed that AS* will remain low until it drives DTACK* or BERR* low, provided that it does so within the bus timeout period.
19.	**OBSERVATION 2.67:** SLAVES are guaranteed that the AS* will remain low for this minimum time.
20.	**OBSERVATION 2.68:** The responding SLAVE is guaranteed that once DSA* goes low, it will remain low until it drives DTACK* or BERR* low, provided that the SLAVE does so within the bus time-out period.
21.	**OBSERVATION 2.69:** The responding SLAVE is guaranteed that once DSB* goes low, it will remain low until it drives DTACK* or BERR* low, provided that the SLAVE does so within the bus time-out period.
22.	**OBSERVATION 2.70:** During write cycles, the responding SLAVE is guaranteed that the data bus will remain valid until it drives DTACK* or BERR* low, provided that it does so within the bus time-out period.

(Continued on Page 83)

Table 2-25. (Continued)

23.	**OBSERVATION 2.71:** The responding SLAVE is guaranteed that the WRITE* line remains valid until both data strobes are high.
26.	**RULE 2.53:** During read cycles, the responding SLAVE **MUST NOT** drive the data bus until DSA* goes low.
27.	**RULE 2.54:** During read cycles, the responding SLAVE **MUST NOT** drive DTACK* before it drives the data lines with valid data. **OBSERVATION 2.72:** RULE 2.54 does not apply to cycles where the responding SLAVE drives BERR* low instead of DTACK*.
28.	**RULE 2.55:** The responding SLAVE **MUST** wait this minimum time after DSA* goes low before driving DTACK* or BERR* low.
29.	**RULE 2.56:** During read cycles, once the responding SLAVE has driven DTACK* low, it **MUST NOT** change D00-D31 until DSA* goes high.
30.	**RULE 2.57:** Once the responding SLAVE has driven DTACK* or BERR* low, it **MUST NOT** release it until the SLAVE detects both DS0* and DS1* high.
31.	**RULE 2.58:** During read cycles, the responding SLAVE **MUST** release all of D00-D31 before releasing DTACK* or BERR* to high.
32.	**OBSERVATION 2.73:** SLAVES are guaranteed that IACK*, LWORD*, A00-A31, and AM0-AM5 have been valid for this minimum time when they detect a falling edge on DSA*. This time is derived from timing parameters 4 and 10.
33.	**OBSERVATION 2.74:** During data transfer cycles, SLAVES are guaranteed that DS0* and/or DS1* will remain low for at least this minimum time. This time is derived from timing parameter 28, where the responding SLAVE is required to wait a minimum time before driving BERR* or DTACK* to low.

Table 2-26. LOCATION MONITOR, Timing Observations

Note: The numbers correspond to the timing parameters specified in Table 2-22 on Page 75.

4.	**OBSERVATION 2.75:** The LOCATION MONITOR is guaranteed that IACK*, A01-A31, AM0-AM5, and LWORD* have been valid for this minimum time when it detects a falling edge on AS*.
5.	**OBSERVATION 2.76:** The LOCATION MONITOR is guaranteed this minimum high time on AS* between DTB cycles.
10.	**OBSERVATION 2.77:** Due to bus skew, LOCATION MONITORS on the DTB might detect a falling edge on DSA* before detecting the falling edge on AS*. However, the LOCATION MONITOR is guaranteed that the falling edge on DSA* will not precede the falling edge on AS* by more than this time.
11.	**OBSERVATION 2.78:** The LOCATION MONITOR is guaranteed this minimum time during which both data strobes are simultaneously high between consecutive data transfers.
12.	**OBSERVATION 2.79:** The LOCATION MONITOR is guaranteed that WRITE* has been valid for this minimum time when it detects a falling edge on DSA*.
13.	**OBSERVATION 2.80:** **IF** both data strobes are going to be driven low, **THEN** the LOCATION MONITOR is guaranteed that DSB* will go low within this maximum time after DSA* has gone low.
17.	**OBSERVATION 2.81:** The LOCATION MONITOR is guaranteed that IACK*, A01-A31, AM0-AM5, and LWORD* will remain valid for this minimum time after the falling edge of AS*. During ADDRESS-ONLY cycles this time is guaranteed by the MASTER. During all other cycle types, this time is derived from timing parameters 10, 14, 16, and 28.
19.	**OBSERVATION 2.82:** The LOCATION MONITOR is guaranteed that the AS* will remain low for this minimum time.
23.	**OBSERVATION 2.83:** The LOCATION MONITOR is guaranteed that the WRITE* line remains valid until both data strobes go high.

(Continued on Page 85)

Table 2-26. (Continued)

32.	**OBSERVATION 2.84:** The LOCATION MONITOR is guaranteed that IACK*, LWORD*, A01-A31, and AM0-AM5 have been valid for this minimum time when it detects a falling edge on DSA*.
33.	**OBSERVATION 2.85:** During data transfer cycles, the LOCATION MONITOR is guaranteed that DS0* and/or DS1* will remain low for at least this minimum time. This time is derived from timing parameter 28, where the responding SLAVE is required to wait a minimum time before driving BERR* or DTACK* to low.

Table 2-27. BUS TIMER, Timing RULES

Note: The numbers correspond to the timing parameters specified in Table 2-23 on page 76.

28.	**RULE 2.59:** The BUS TIMER **MUST** wait at least its time-out time, but no longer than twice its timeout time, after the first data strobe goes low before driving BERR* low.
30.	**RULE 2.60:** Once it has driven BERR* low, the BUS TIMER **MUST NOT** release BERR* until it detects both DS0* and DS1* high.

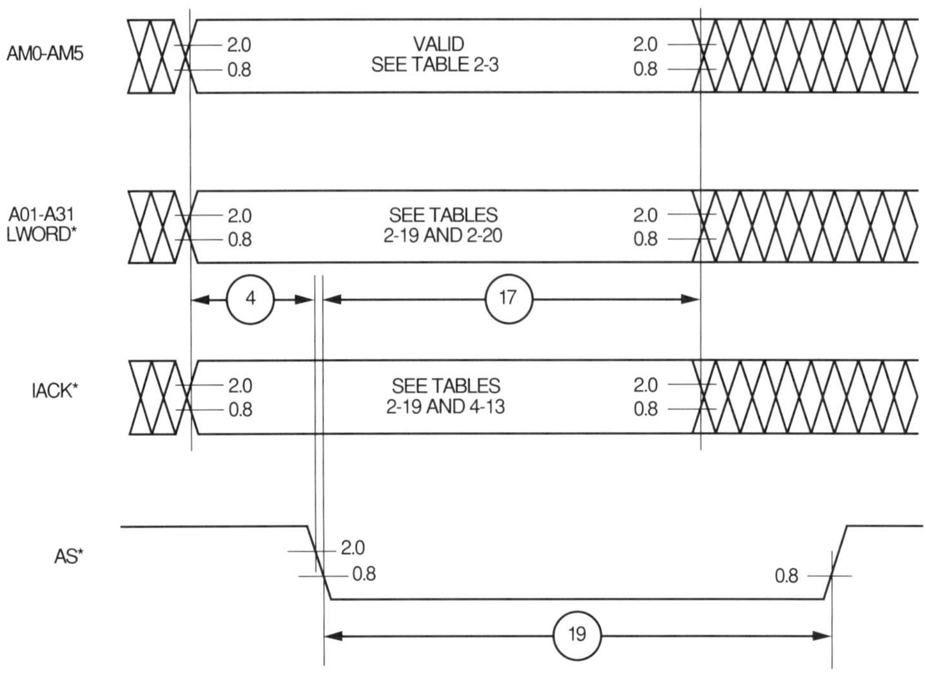

Figure 2-12. Address Broadcast Timing

ALL CYCLES

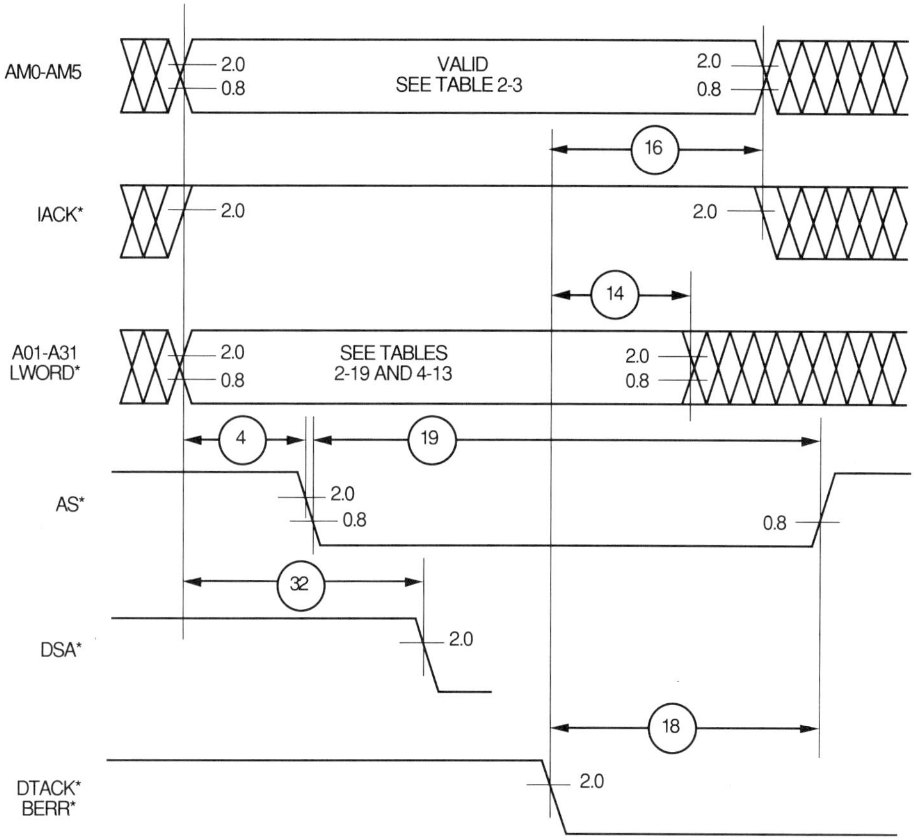

Figure 2-13. MASTER, Responding SLAVE, And LOCATION MONITOR Address Broadcast Timing

Single even byte transfers
Single odd byte transfers
Double byte transfers
Quad byte transfers
Unaligned transfers

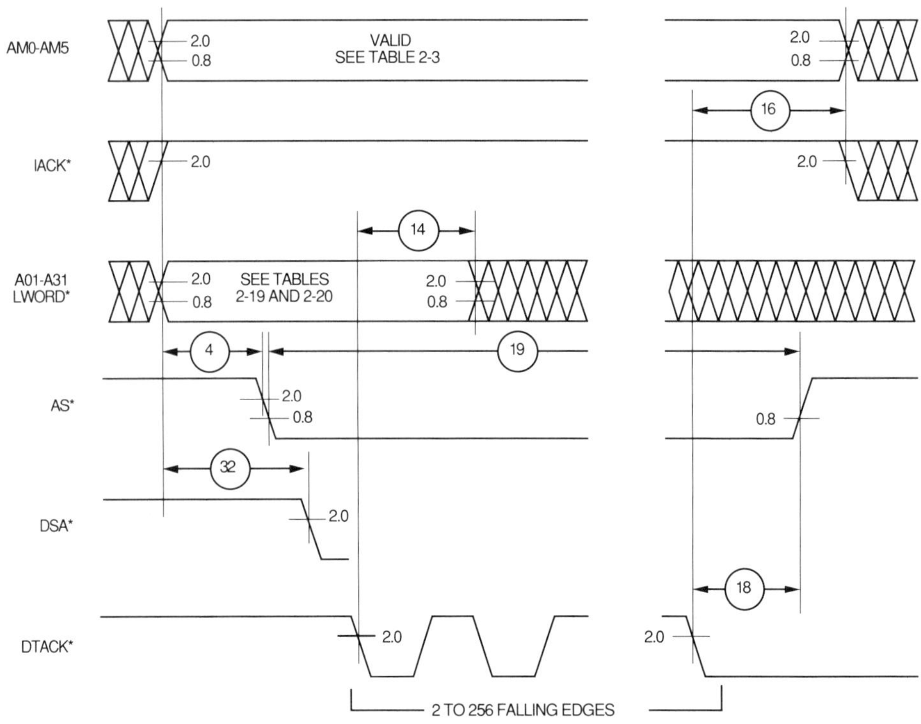

Figure 2-14. MASTER, SLAVE, And LOCATION MONITOR
Address Broadcast Timing

Single byte block transfers
Double byte block transfers
Quad byte block transfers

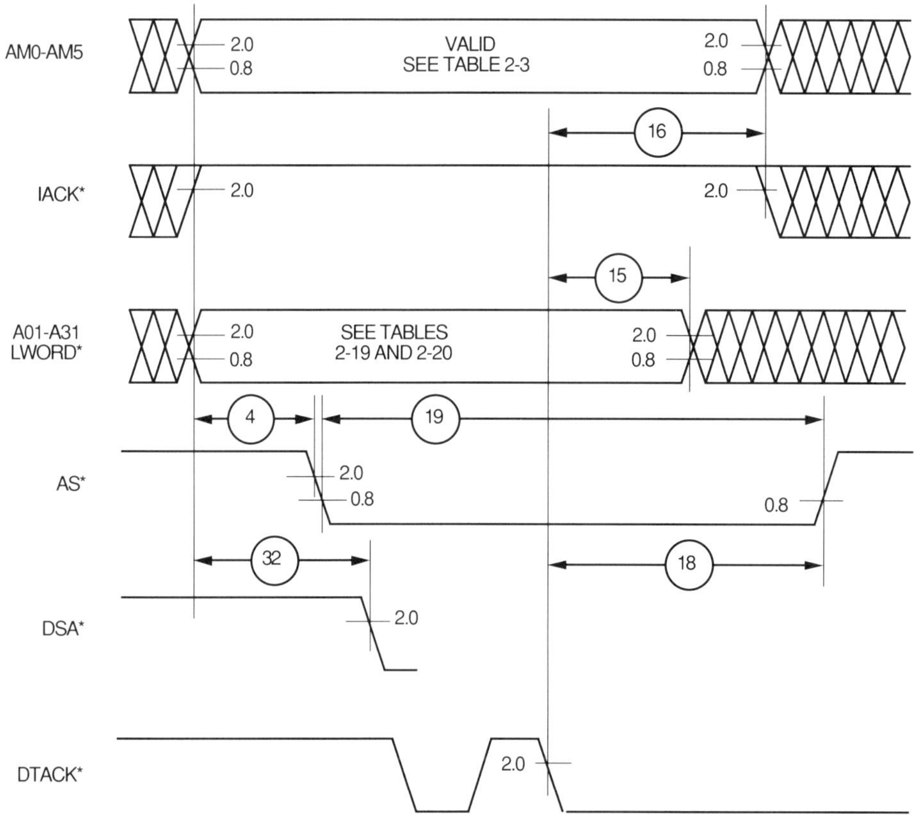

Figure 2-15. MASTER, SLAVE, And LOCATION MONITOR
Address Broadcast Timing

Single byte RMW cycles
Double byte RMW cycles
Quad byte RMW cycles

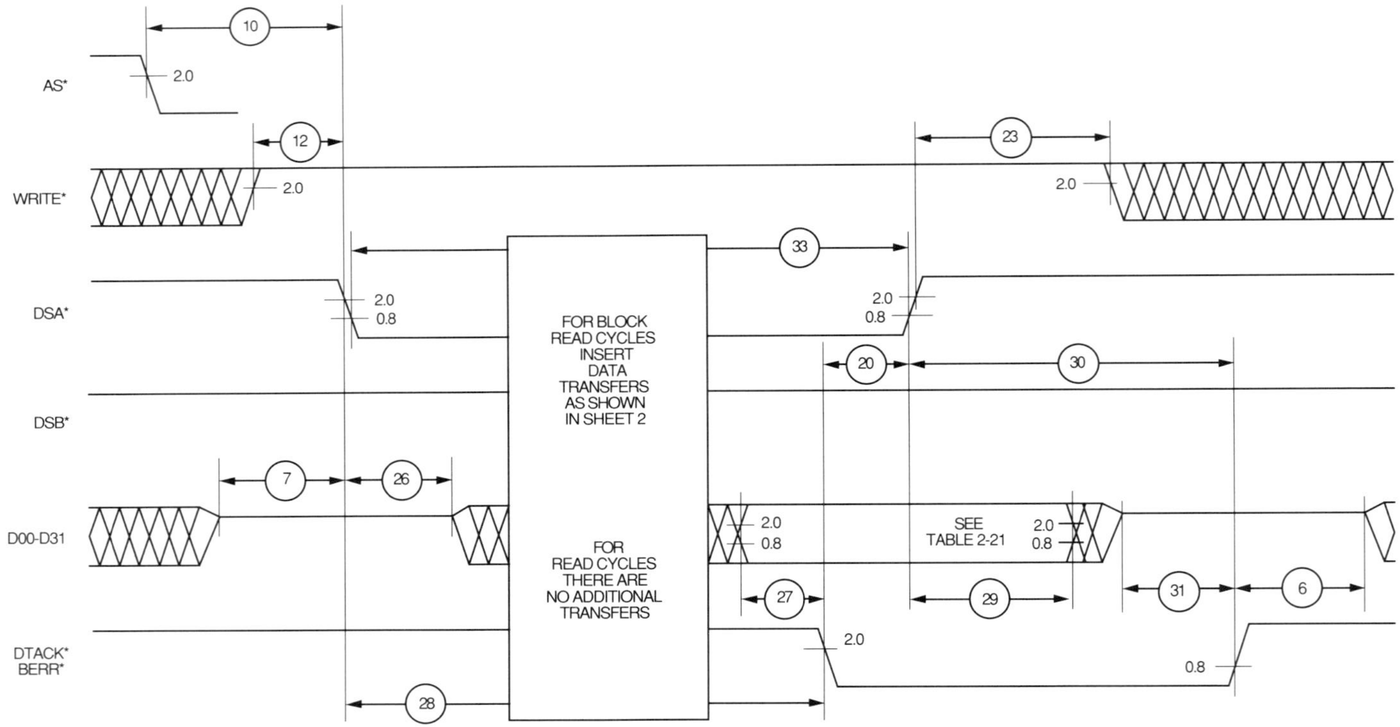

Figure 2-16. MASTER, SLAVE, And LOCATION MONITOR (Sheet 1 of 2)
Data Transfer Timing

Byte(2) Read	Byte(1) Read
Byte(0) Read	Byte(3) Read
Byte(0-2) Read	Byte(1-3) Read
Single Byte Block Read	

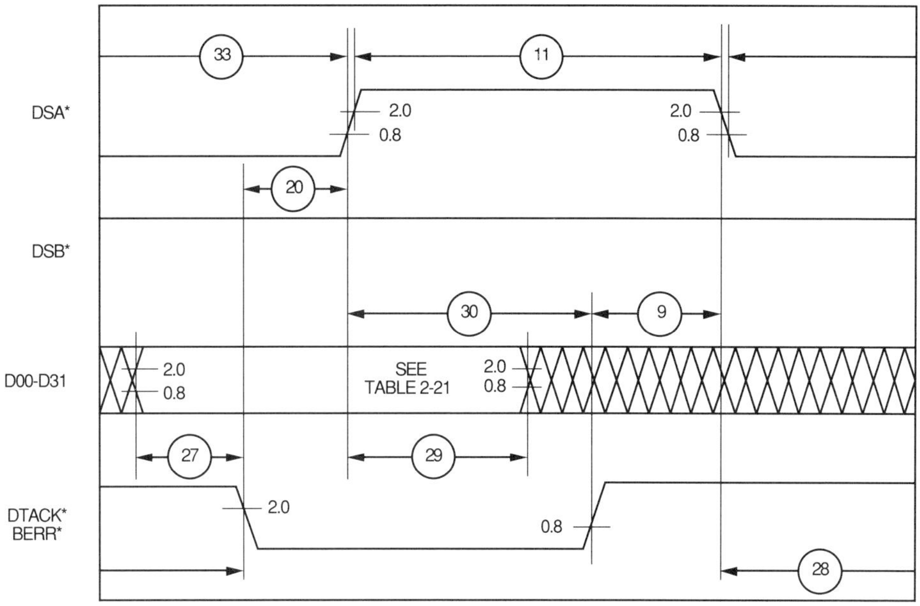

Figure 2-16. MASTER, SLAVE, And LOCATION MONITOR (Sheet 2 of 2)
Data Transfer Timing

Byte(0) Read	Byte(1) Read
Byte(2) Read	Byte(3) Read
Byte(0-2) Read	Byte(1-3) Read
Singe Byte Block Read	

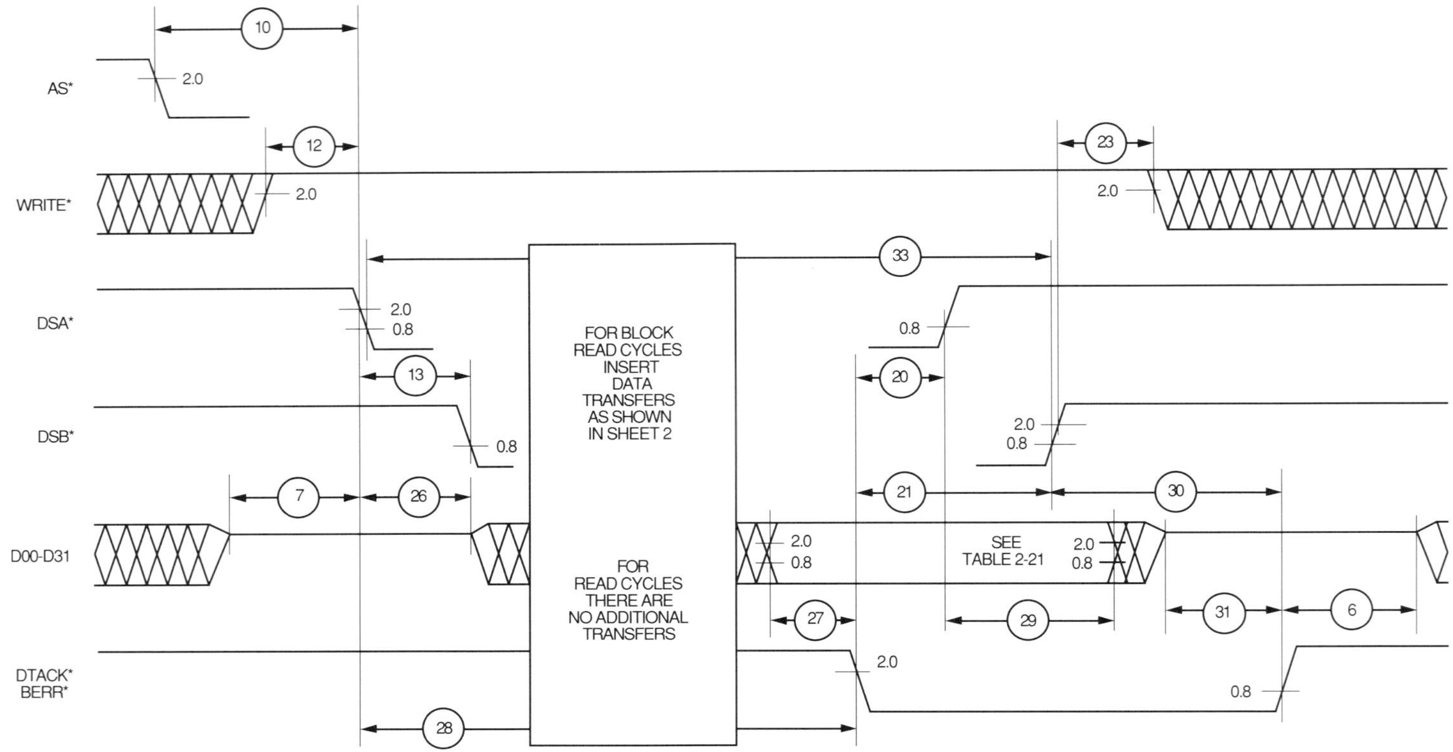

Figure 2-17. MASTER, SLAVE, And LOCATION MONITOR (Sheet 1 of 2)
Data Transfer Timing

Byte(0-1) Read	Byte(2-3) Read
Byte(0-3) Read	Byte(1-2) Read
Double Byte Block Read	Quad Byte Block Read

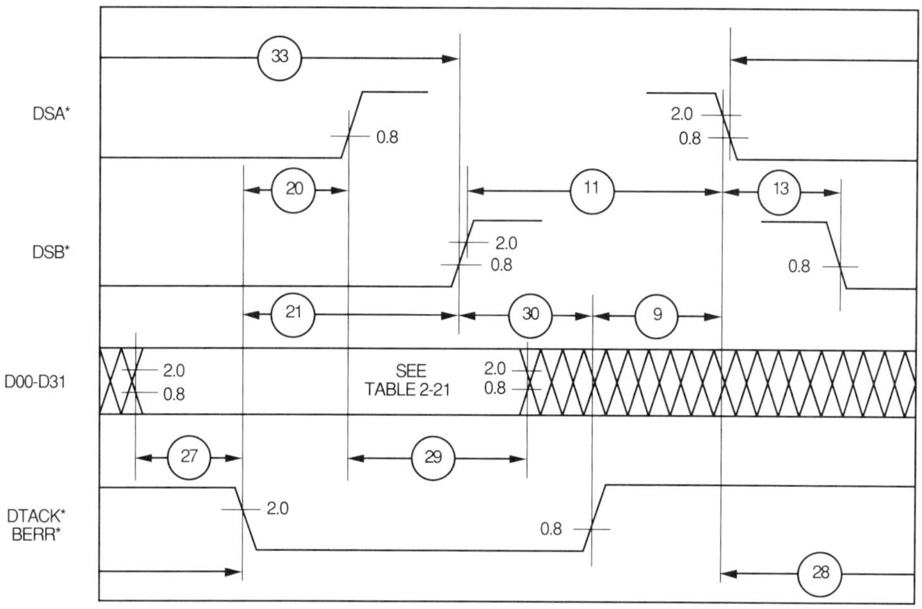

Figure 2-17. MASTER, SLAVE, And LOCATION MONITOR (Sheet 2 of 2)
Data Transfer Timing

Byte(0-1) Read	Byte(2-3) Read
Byte(0-3) Read	Byte(1-2) Read
Double Byte Block Read	Quad Byte Block Read

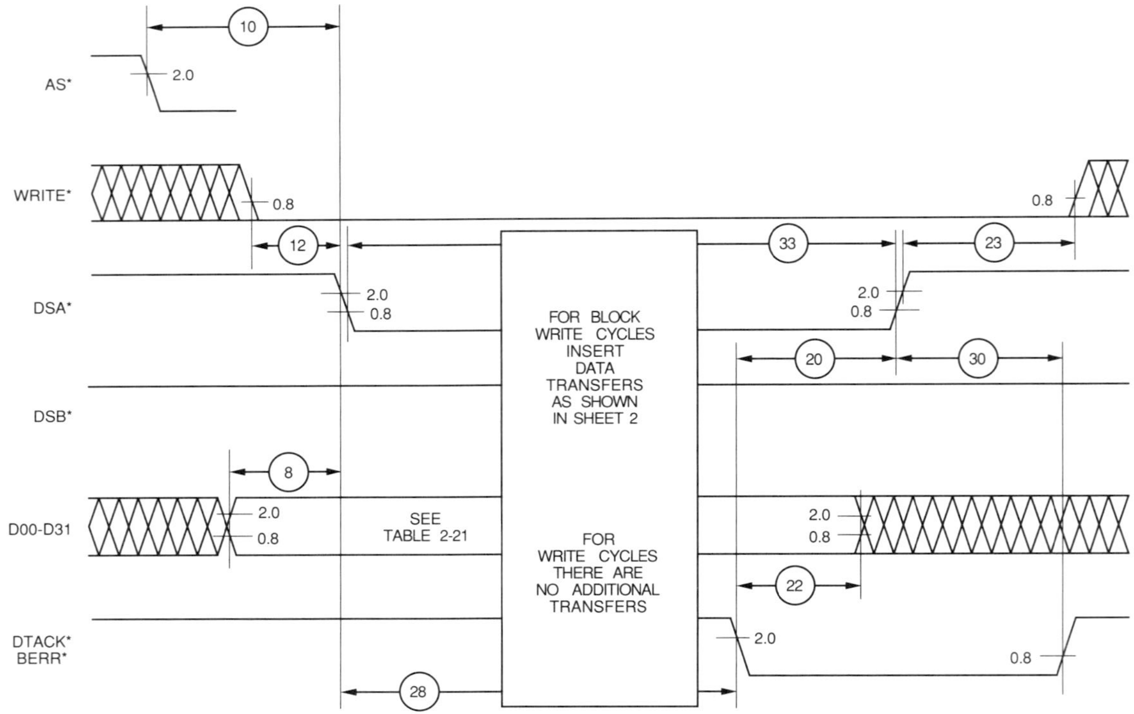

Figure 2-18. MASTER, SLAVE, And LOCATION MONITOR (Sheet 1 Of 2)
Data Transfer Timing

Byte(0) Write	Byte(1) Write
Byte(2) Write	Byte(3) Write
Byte(0-2) Write	Byte(1-3) Write
Single Byte Block Write	

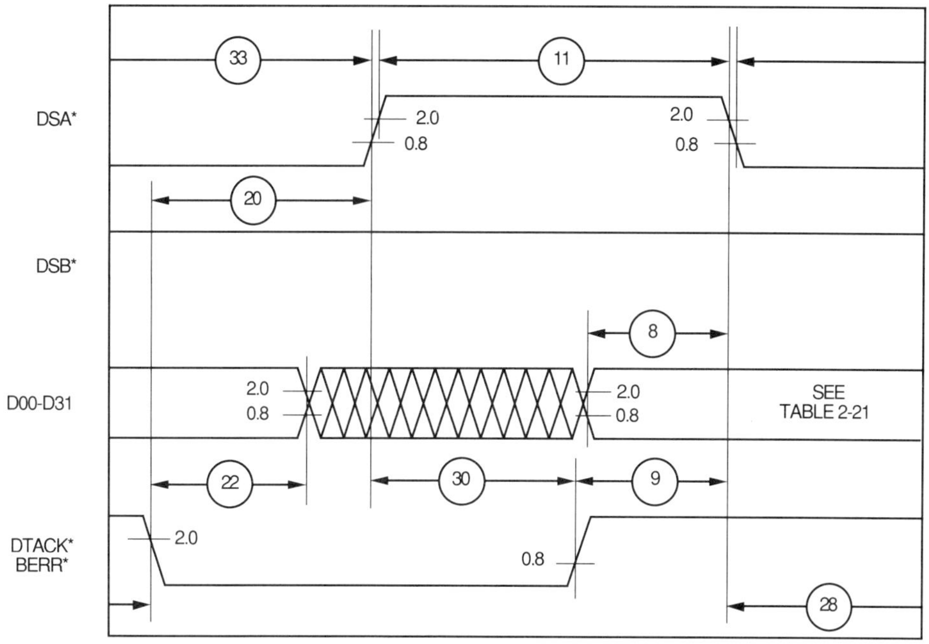

Figure 2-18. MASTER, SLAVE, And LOCATION MONITOR (Sheet 2 of 2) Data Transfer Timing

Byte(0) Write
Byte(2) Write
Byte(0-2) Write
Single Byte Block Write

Byte(1) Write
Byte(3) Write
Byte(1-3) Write

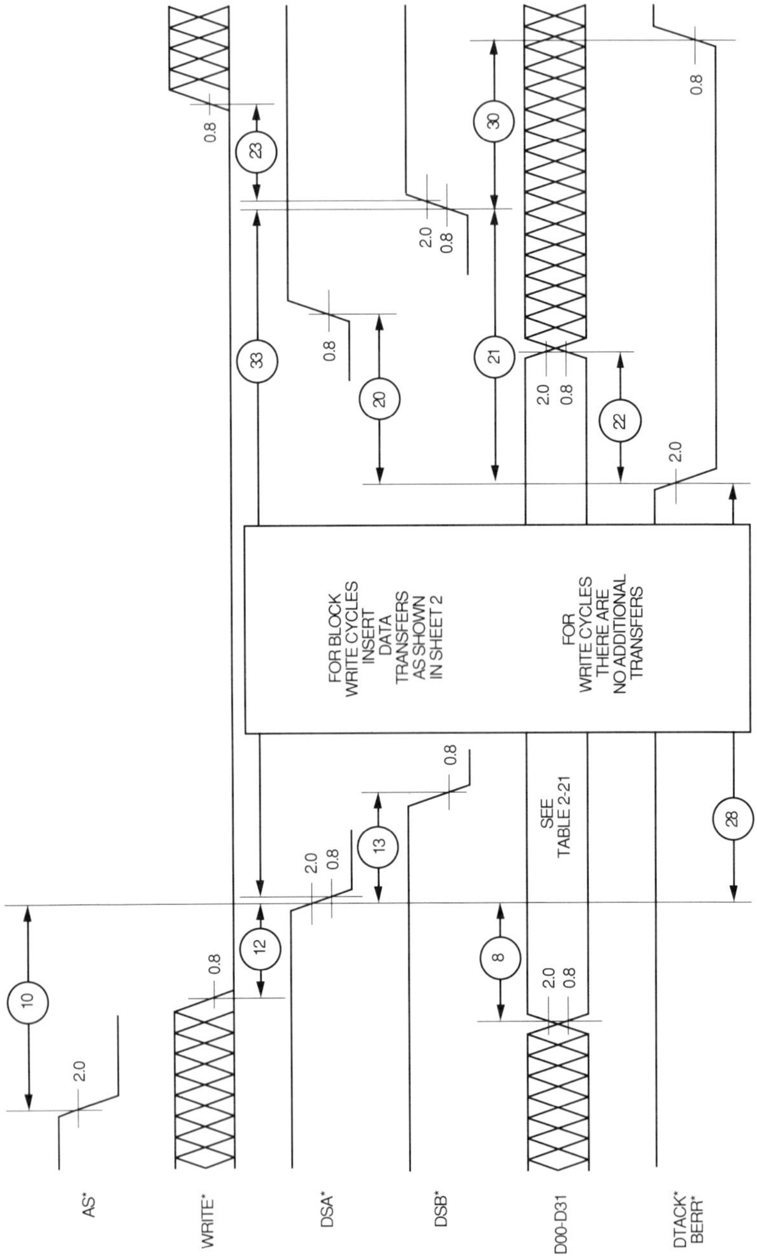

Figure 2-19. MASTER, SLAVE, and LOCATION MONITOR (Sheet 1 of 2)
Data Transfer Timing

Byte(0-1) Write	Byte(2-3) Write
Byte(0-3) Write	Byte(1-2) Write
Double Byte Block Write	Quad Byte Block Write

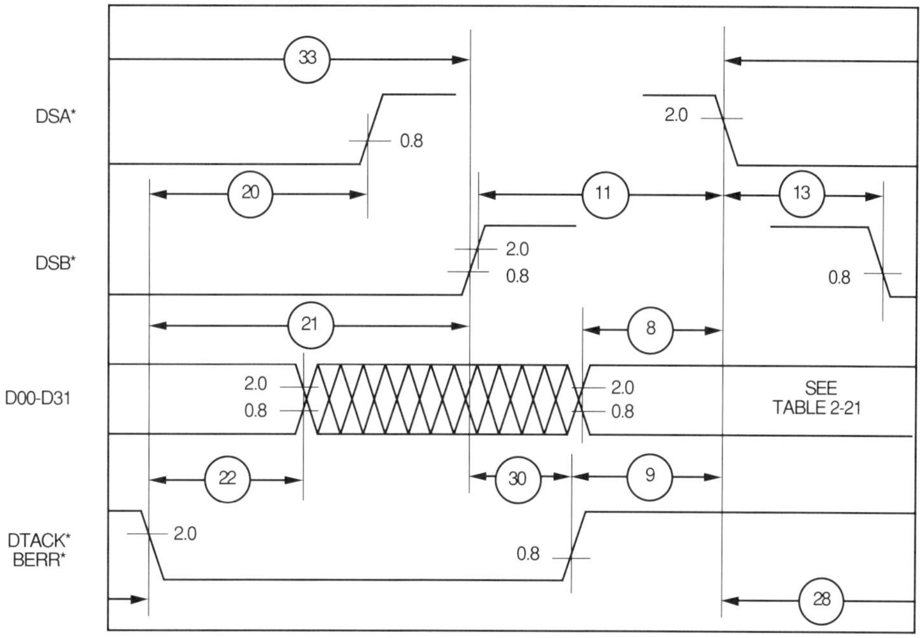

Figure 2-19. MASTER, SLAVE, and LOCATION MONITOR (Sheet 2 of 2)
Data Transfer Timing

Byte(0-1) Write	Byte(2-3) Write
Byte(0-3) Write	Byte(1-2) Write
Double Byte Block Write	Quad Byte Block Write

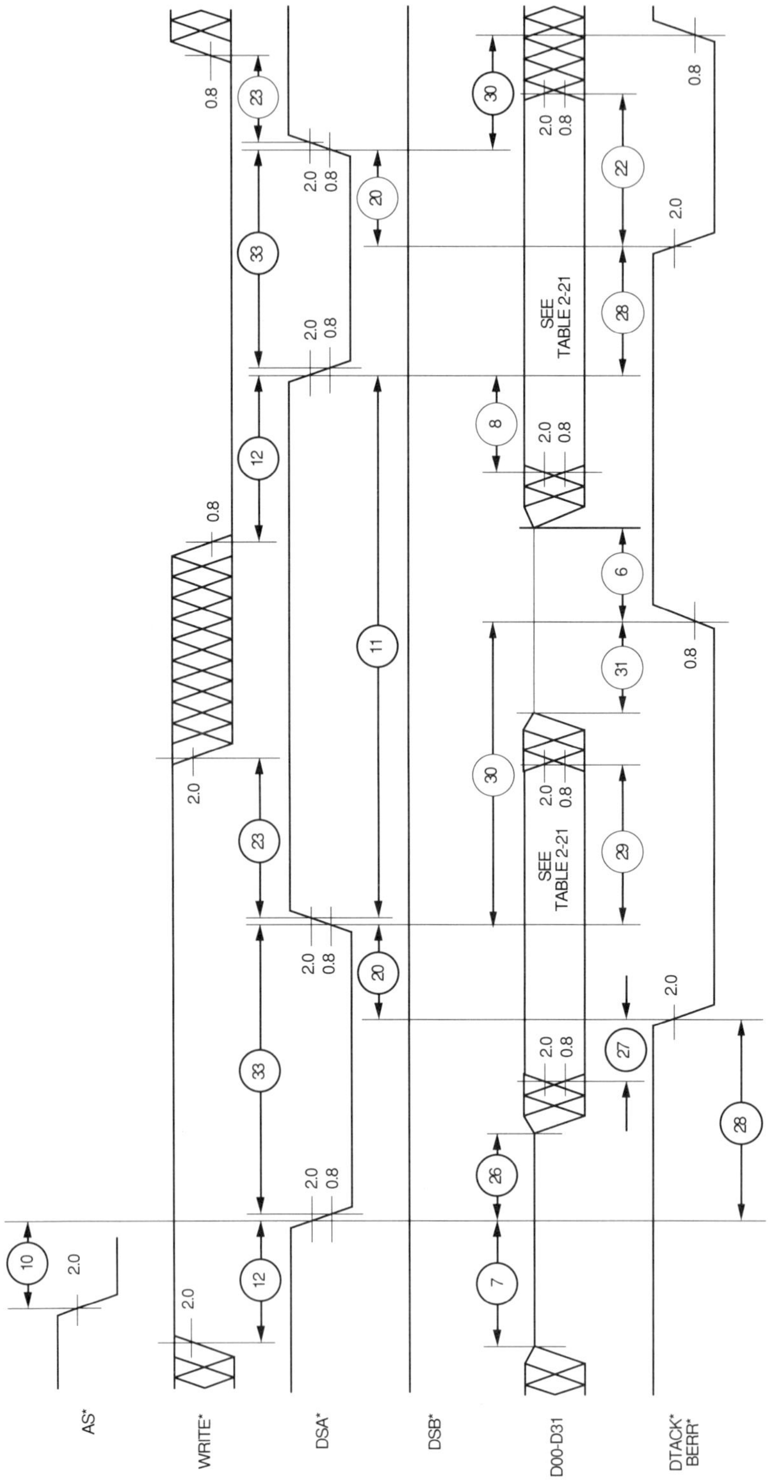

Figure 2-20. MASTER, SLAVE, And LOCATION MONITOR
Data Transfer Timing

Single byte RMW cycles

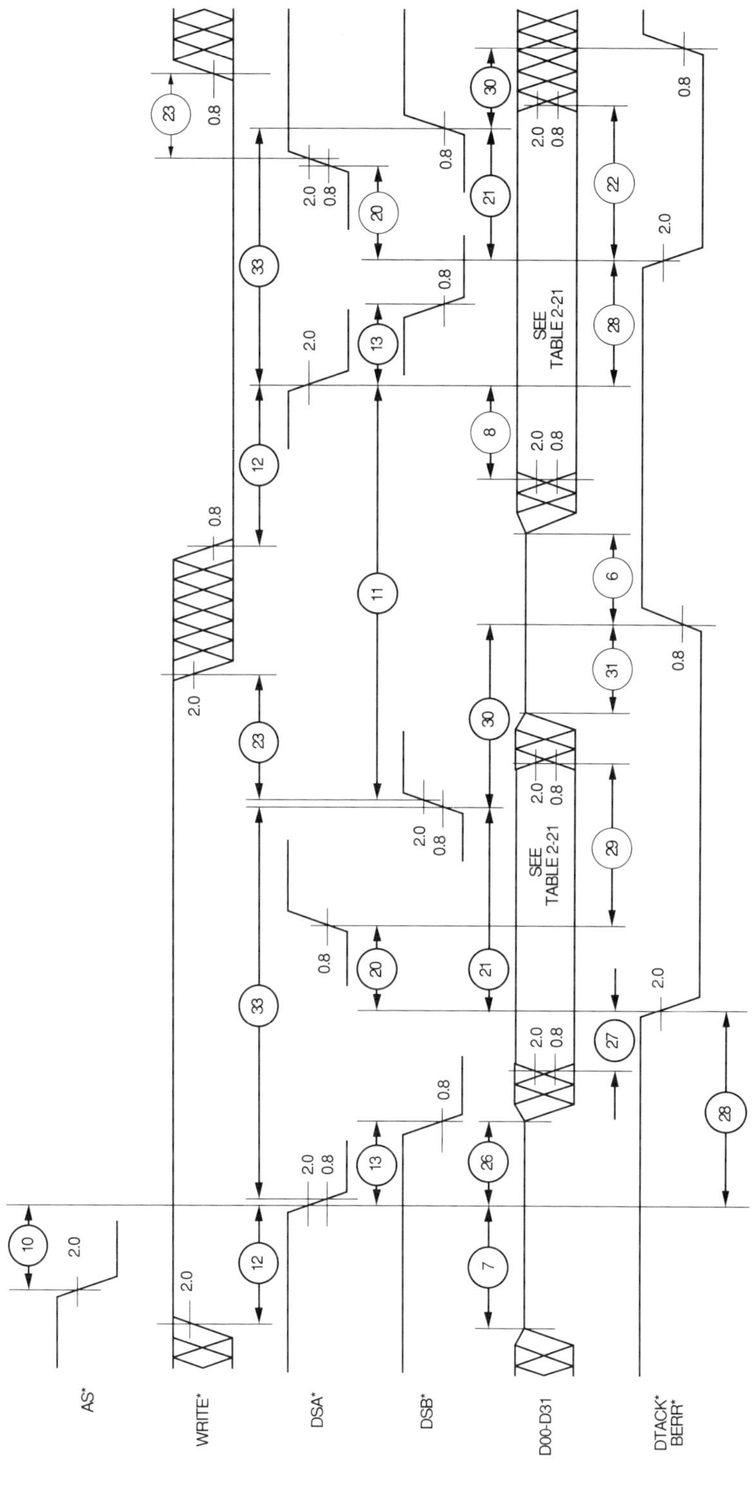

Figure 2-21. MASTER, SLAVE, And LOCATION MONITOR
Data Transfer Timing

Double Byte RMW cycles
Quad Byte RMW cycles

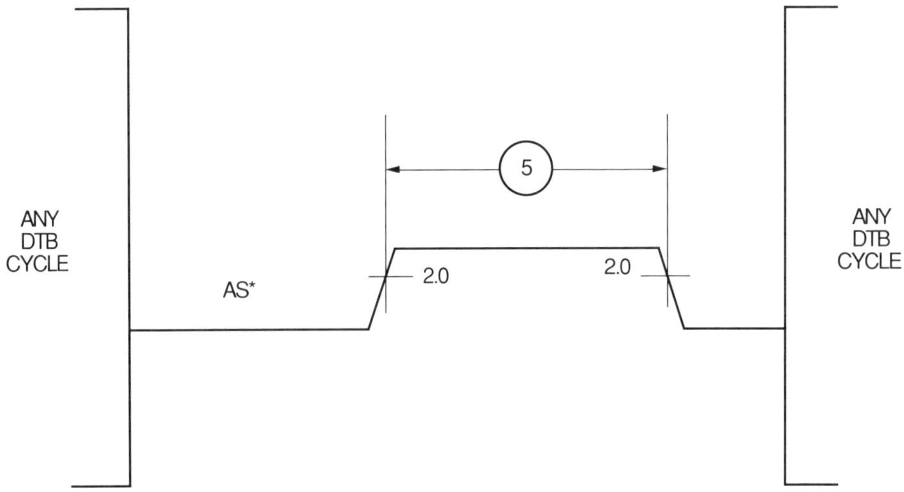

Figure 2-22. Address Strobe Intercycle Timing

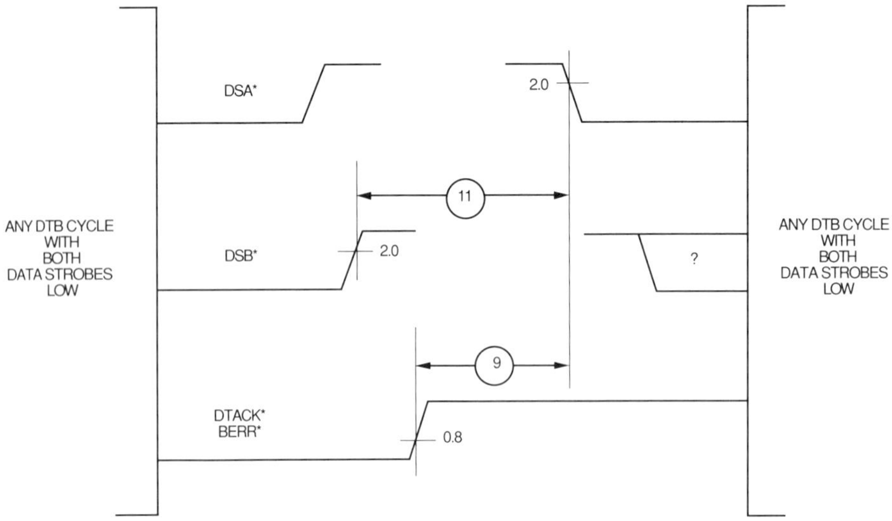

Figure 2-23. Data Strobe Intercycle Timing
A cycle where both data strobes go low followed by a cycle where one or both data strobes go low

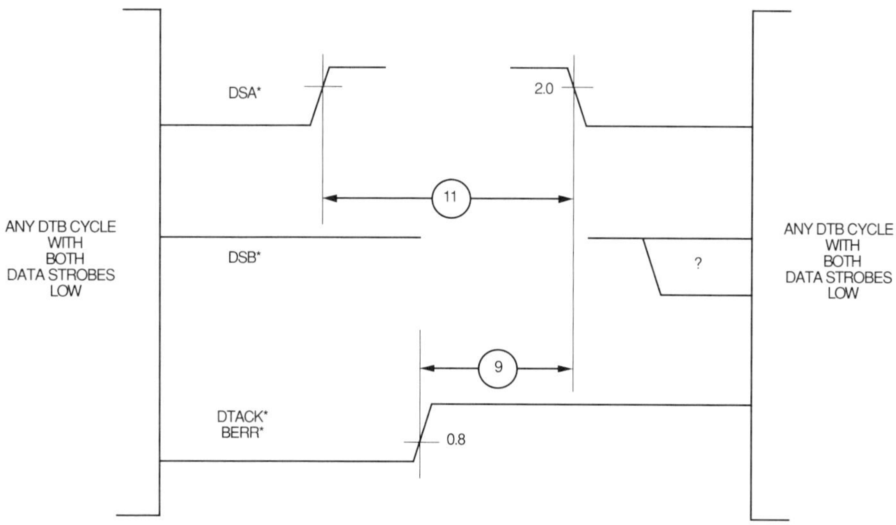

Figure 2-24. Data Strobe Intercycle Timing
A cycle where one data strobe goes low followed by a cycle where one or both data strobes go low

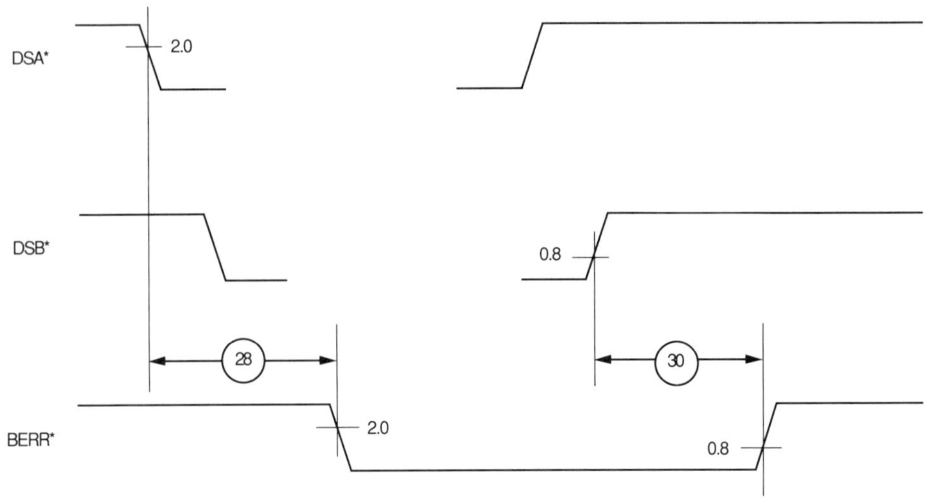

Figure 2-25. MASTER, SLAVE, And BUS TIMER
Data Transfer Timing

Timed-Out Cycle

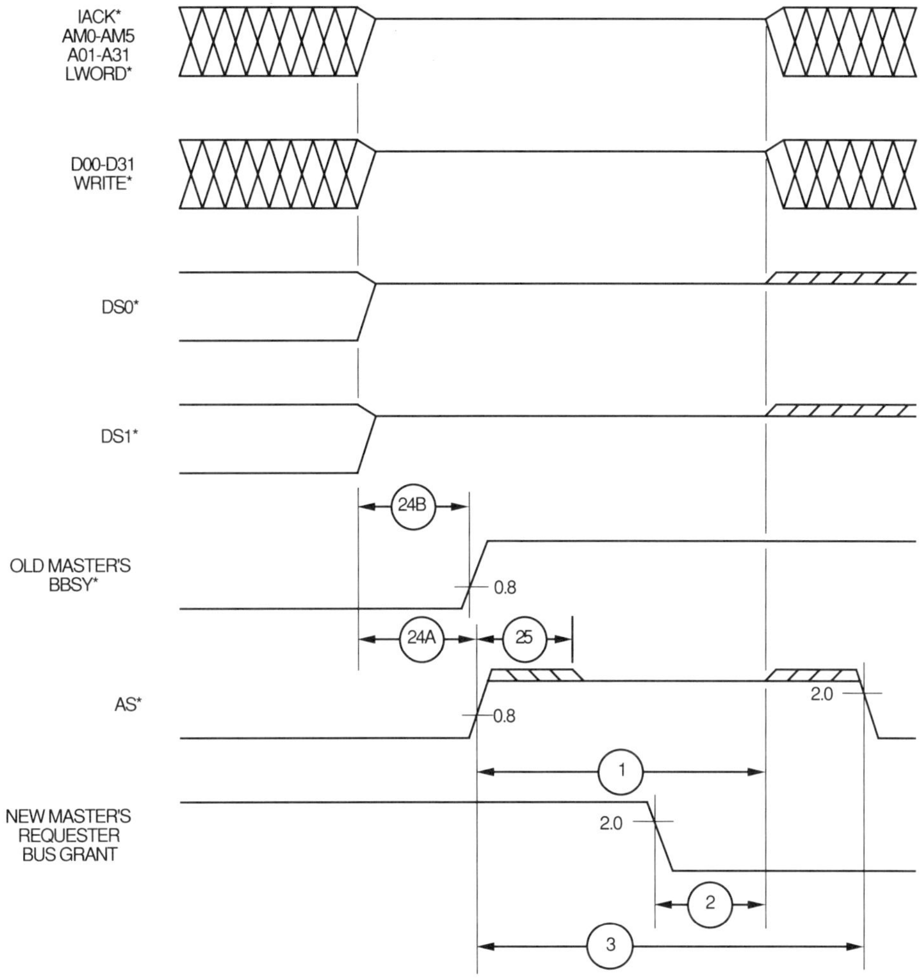

Figure 2-26. MASTER DTB Control Transfer Timing

CHAPTER 3

DATA TRANSFER BUS ARBITRATION

3.1 BUS ARBITRATION PHILOSOPHY

As microprocessor costs decrease, it is becoming more cost effective to design systems with multiple processors sharing global resources.

The most fundamental of these global resources is the Data Transfer Bus through which all other global resources are accessed. Therefore, any system that supports multiprocessing needs to provide an efficient allocation method for the Data Transfer Bus. Because speed of allocation is vital, a hardware allocation scheme is the only practical alternative. The VMEbus meets this need with its Arbitration subsystem. (See Figure 3-1).

The VMEbus arbitration subsystem:

a. Prevents simultaneous use of the bus by two MASTERS.
b. Schedules requests from multiple MASTERS for optimum bus use.

3.1.1 Types Of Arbitration

When several boards request use of the DTB simultaneously, the arbitration subsystem detects these requests and grants the bus to one board at a time. The decision of which board is granted the bus first depends upon what scheduling algorithm is used. Many algorithms are possible. The VMEbus describes three: prioritized, round-robin, and single level.

Prioritized arbitration assigns the bus according to a fixed priority scheme where each of four bus request lines has a priority from highest (BR3*) to lowest (BR0*).

Round-robin arbitration assigns the bus on a rotating priority basis. When the bus is granted to the REQUESTER on bus request line BR(n)*, then the highest priority for the next arbitration is assigned to bus request line BR(n-1)*.

Single level arbitration only accepts requests on BR3*, and relies on BR3*'s bus grant daisy-chain to arbitrate the requests.

PERMISSION 3.1:
Scheduling algorithms other than priority, round-robin, or single level **MAY** be used. For example, an ARBITER'S algorithm might give highest priority to BR3*, but grant the bus to BR0* through BR2* on a round-robin basis.

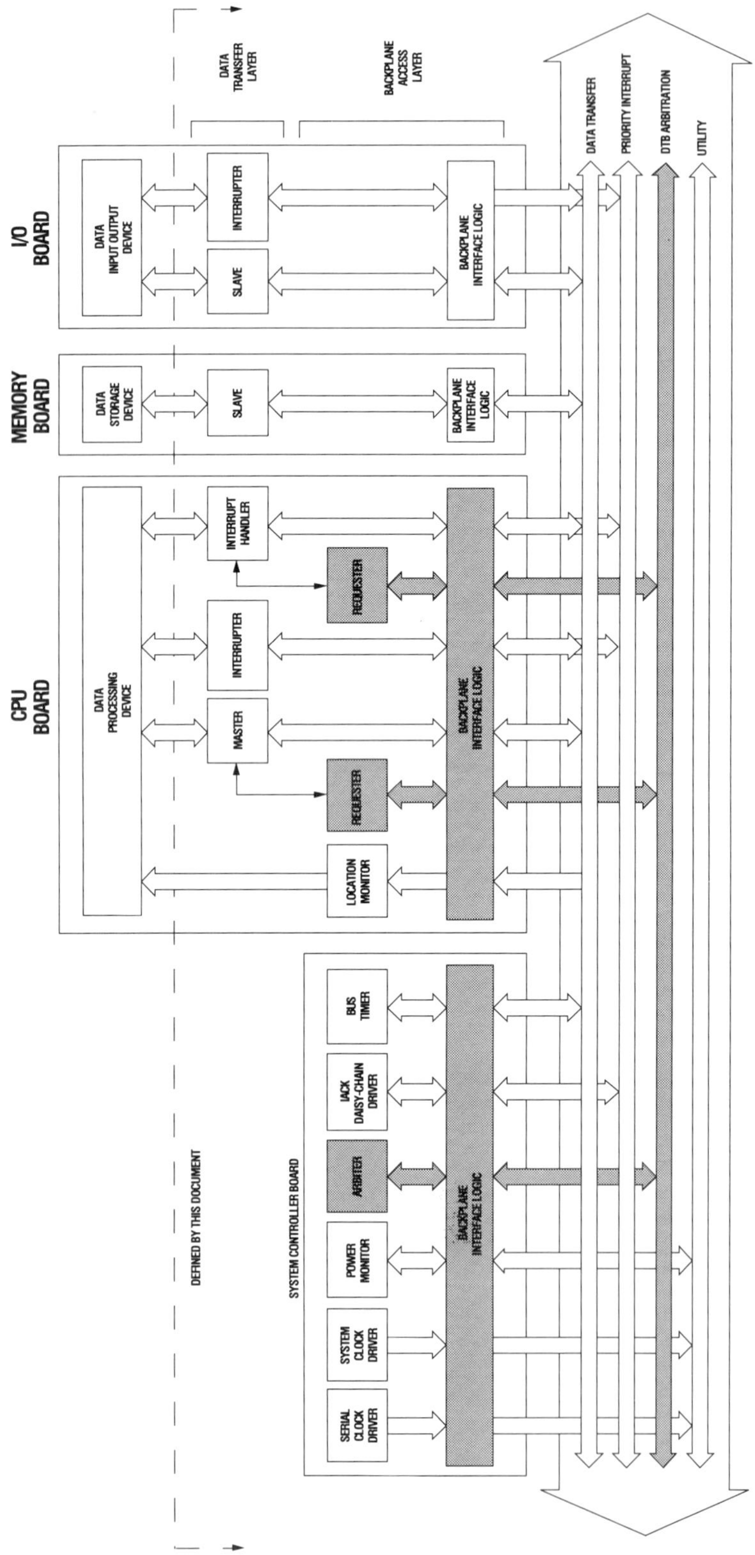

Figure 3-1. Arbitration Bus Functional Block Diagram

3.2 ARBITRATION BUS LINES

The Arbitration Bus consists of six bused VMEbus lines and four daisy-chained lines. These daisy-chained lines require special signal names. The signals entering each board are called "Bus Grant IN" lines (BGxIN*), while the signals leaving each board are called "Bus Grant OUT" lines (BGxOUT*). The lines which leave slot n as BGxOUT* enter slot n+1 as BGxIN*. This is illustrated in Figure 3-2.

OBSERVATION 3.1:
In all descriptions in this chapter, the terms BRx*, BGxIN*, and BGxOUT* are used to describe the bus request and bus grant lines, where x takes on any value from zero to three.

In the VMEbus arbitration system, a REQUESTER module drives the following lines:

1 bus request line	(one of BR0* through BR3*)
1 bus grant out line	(one of BG0OUT* through BG3OUT*)
1 bus busy line	(BBSY*)

RULE 3.1
IF a VMEbus board does not generate bus requests on some bus request levels,
THEN it **MUST** propagate the daisy-chain signals for those levels from its BGxIN* lines to its respective BGxOUT* lines.

PERMISSION 3.2:
The propagation for the unused lines of the bus grant daisy-chain can be done using jumpers or active logic. The latter approach allows selection of the request level under software control, while the former results in faster propagation through the daisy-chain.

Three types of ARBITERS are described in the VMEbus standard. They are:

PRIoritized	(PRI)
Round-Robin-Select	(RRS)
SinGLe level	(SGL)

The operation of these three types of ARBITERS is described in Section 3.3.

A PRI ARBITER drives the following:

1 bus clear line	(BCLR*)
4 bus grant lines	(Slot 1 BG0IN* through BG3IN*) if the ARBITER'S board also has a REQUESTER.
	(Slot 1 BG0OUT* through BG3OUT*) if the ARBITER'S board does not have a REQUESTER.

An RRS ARBITER drives the four Slot 1 BGxIN* or BGxOUT* lines and, optionally, the BCLR* line.

A SGL ARBITER drives only BG3IN* or BG3OUT* at slot 1.

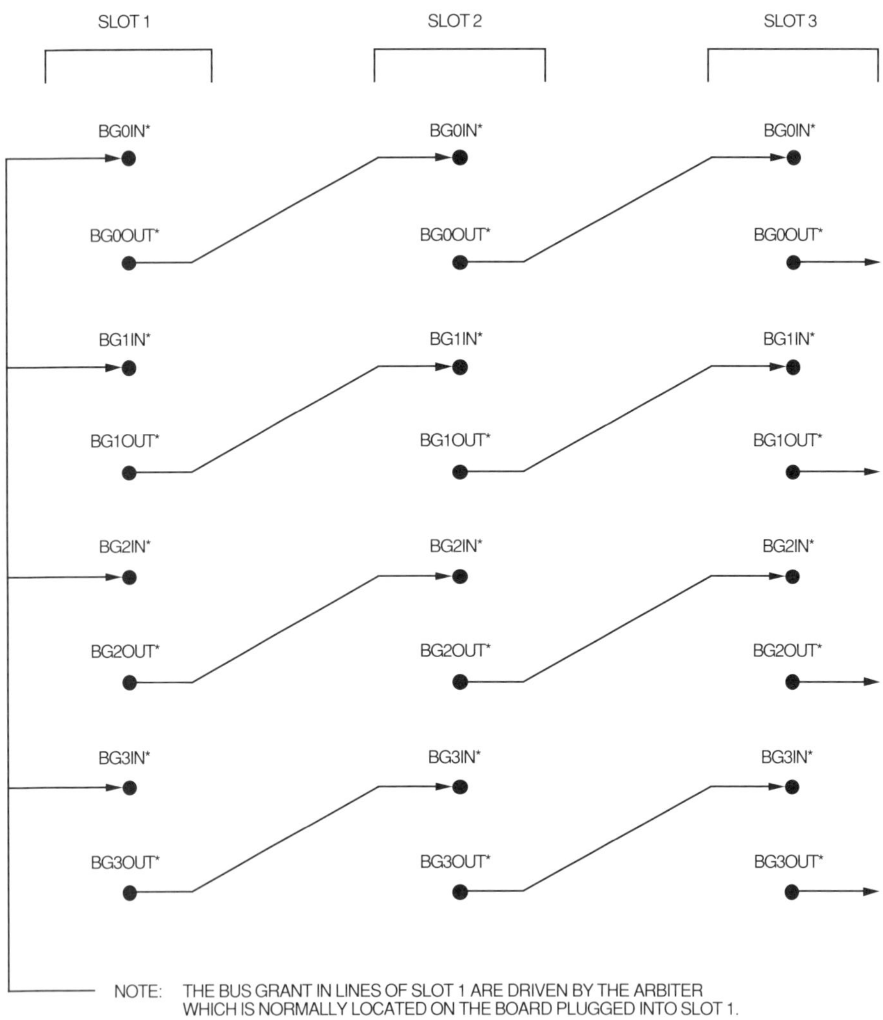

Figure 3-2. Illustration Of The Daisy-Chained Bus Grant Lines

Two additional lines are connected with the arbitration system during power-up and power-down sequencing: SYSRESET* and ACFAIL*. While their impact on the arbitration system is included in this chapter, these lines will be discussed further in Chapter 5.

3.2.1 Bus Request And Bus Grant Lines

The bus request lines are used by each REQUESTER to request use of the DTB. The bus grant lines allow the ARBITER to award use of the bus. It does this by driving a bus grant daisy-chain line low. This low level propagates down the daisy-chain, typically passing through several boards in the process. If a board never uses a particular request/grant level, the signal is passed through that board. Where the board uses a request/grant level x, the corresponding signal BGxIN* is gated on board. If its on-board REQUESTER is currently requesting the DTB on that level, it does not pass the low level on to its BGxOUT*. Otherwise, it passes on the low level.

RULE 3.2:
IF a VMEbus backplane slot is not occupied by a board, and if there are boards farther down the daisy-chain.
THEN jumpers **MUST** be installed at the empty slot to pass through the daisy-chain signal.

OBSERVATION 3.2:
The backplane mechanical specification in Chapter 7 describes a provision for the installation of jumpers at each slot.

RULE 3.3:
The ARBITER **MUST** always be located in slot 1.

3.2.2 Bus Busy Line (BBSY*)

Once a REQUESTER has been granted control of the Data Transfer Bus via the bus grant daisy-chain, it drives BBSY* low. It then has control of the DTB until it releases BBSY*, allowing the ARBITER to grant the DTB to some other REQUESTER.

3.2.3 Bus Clear Line (BCLR*)

The PRI ARBITER drives BCLR* low to inform the MASTER, currently in control of the DTB, when a higher priority request is pending. The current MASTER does not have to relinquish the bus within any prescribed time limit. It can continue transferring data until it reaches an appropriate stopping point, and then allow its on-board REQUESTER to release BBSY*.

PERMISSION 3.3:
Although RRS ARBITERS are not required to drive the BCLR* line they **MAY** do so.

SUGGESTION 3.1:
IF an RRS ARBITER drives the BCLR* line low.
THEN design it to do so whenever there is a request pending on any of the non-granted bus request lines

3.3 FUNCTIONAL MODULES

The arbitration subsystem is composed of several modules:

a One ARBITER
b. One or more REQUESTERS

Figures 3-3 and 3-4 provide block diagrams for the two types of Arbitration Bus modules.

RULE 3.4:
Output signal lines shown with solid lines in Figure 3-3 and 3-4 **MUST** be driven by the module, unless it would always drive them high.

RULE 3.5:
Input signal lines shown with solid lines in Figures 3-3 and 3-4 **MUST** be monitored and responded to in the appropriate fashion.

OBSERVATION 3.3:
RULES and PERMISSIONS for driving and monitoring the signal lines shown with dotted lines in Figures 3-3 and 3-4 are given in Tables 3-1 and 3-2.

OBSERVATION 3.4:
IF an output signal line is not driven.
THEN terminators on the backplane ensure that it is high.

OBSERVATION 3.5:
Although SYSRESET* and ACFAIL* are not specified as part of the Arbitration Bus, they are important here because MASTERS, which are paired with REQUESTERS, respond to these signal lines. (SYSRESET* and ACFAIL* are driven by the POWER MONITOR module which is discussed in Chapter 5.)

3.3.1 ARBITER

The ARBITER is a functional module that decides which REQUESTER should be granted control of the DTB when several request it simultaneously. There are many possible algorithms that could be used to make this decision. Three types of ARBITERS are described in this specification: a prioritized (PRI) ARBITER, a round-robin (RRS) ARBITER, and a single level (SGL) ARBITER.

An ARBITER responds to incoming bus requests and grants the DTB to the appropriate REQUESTER with one of the bus grant lines.

When the ARBITER detects BBSY* high, and after it detects one or more bus requests, it issues a bus grant, corresponding to the highest priority bus request.

When the REQUESTER receives the bus grant, it drives BBSY* low and signals to its on-board MASTER or INTERRUPT HANDLER that it has been granted the DTB. After its on-board MASTER or INTERRUPT HANDLER finishes using the DTB, the REQUESTER releases BBSY*. The resulting rising edge of BBSY* enables the ARBITER to issue another bus grant, based upon the levels of the bus request lines at that time.

In addition to the arbitration provided by the ARBITER, a secondary level of arbitration is provided by the bus grant daisy-chains. Because of these daisy-chains, REQUESTERS sharing a common request line are prioritized by slot position. The REQUESTER closest to slot 1 has the highest priority.

SGL ARBITERS respond only to bus requests on BR3* and depend on the BG3IN*/BG3OUT* daisy-chain to do the arbitration.

The PRI ARBITER prioritizes the four bus request lines, from BR0* (the lowest) to BR3* (the highest), and responds with BG0IN* through BG3IN*, as appropriate. A PRI ARBITER also informs any MASTER, currently in control of the bus, when a higher level request is pending by driving BCLR* low.

To visualize an RRS ARBITER, consider a mechanical switch being driven by a stepping motor. Each position on the switch connects a bus request line to its corresponding bus grant line. When the bus is busy, the switch is stopped on the current level. Upon release of the bus, the switch steps one position lower (i.e., from BR(n)* to BR(n-1)*) and tests for a request. It continues this scanning operation until a request is found, sending a bus grant over the appropriate line.

PERMISSION 3.4:
An ARBITER **MAY** be designed with a built-in time-out feature that causes it to withdraw a bus grant if BBSY* is not driven low by a REQUESTER within a prescribed time.

OBSERVATION 3.6:
The time used by the ARBITER allowed by PERMISSION 3.4 needs to be longer than the longest possible bus grant daisy-chain propagation delay time, plus the time the slowest REQUESTER takes to generate BBSY*.

RULE 3.6:
Except for a time-out situation where no REQUESTER responds, once the ARBITER grants the bus to a REQUESTER it **MUST NOT** generate a new bus grant before that REQUESTER generates a rising edge on the BBSY* line. (The REQUESTER generates a rising edge by driving BBSY* low and then releasing it.)

OBSERVATION 3.7:
IF an ARBITER uses a "snapshot" of the request lines taken prior to the rising edge of BBSY*.
THEN it might grant the bus to a REQUESTER that has since removed its request.

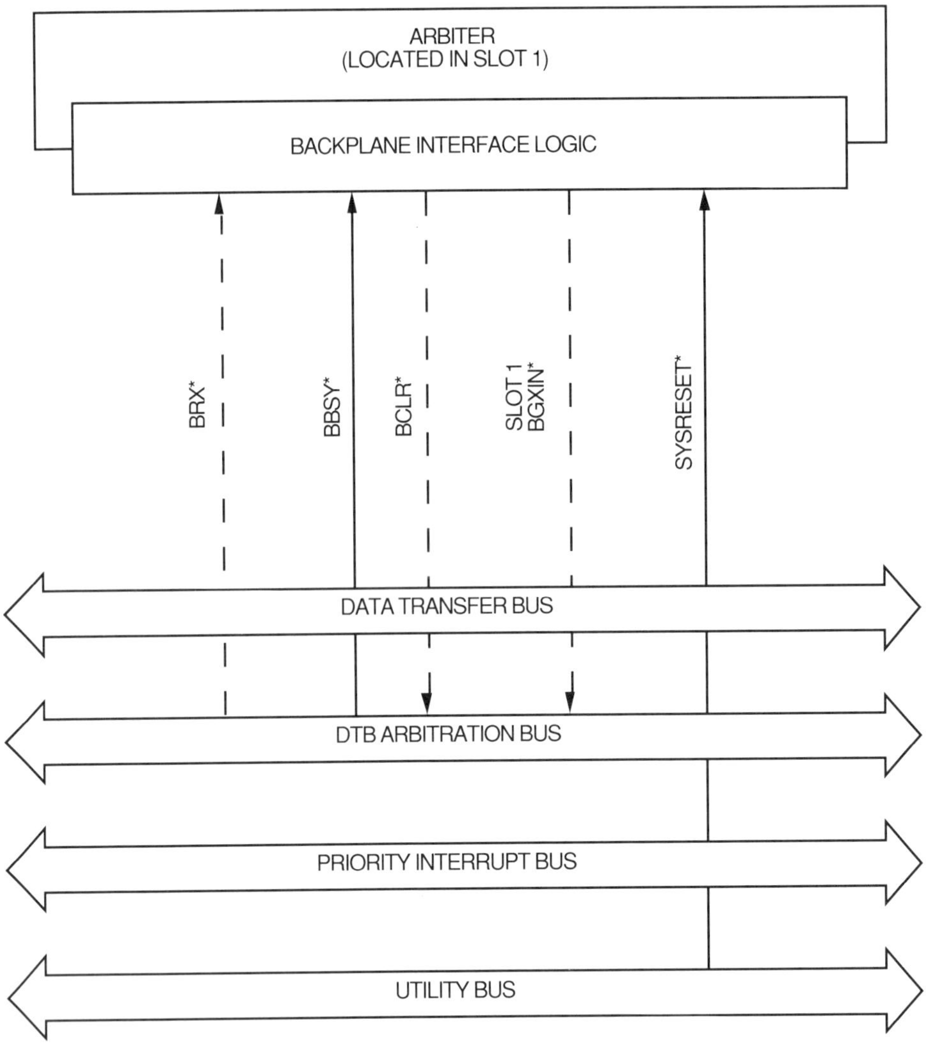

Note: The RULES and PERMISSIONS for monitoring and driving the dotted lines are given in Table 3-1.

Figure 3-3. Block Diagram: ARBITER

Table 3-1. RULES and PERMISSIONS That Specify The Use Of The Dotted Lines By The Various Types Of ARBITERS Defined By This Document

Type of ARBITER	Use of dotted line
SGL	**MUST** drive slot 1 BG3IN*or BG3OUT* (See Note). **MUST** ensure slot 1 BG01N* - BG2IN* are high. **MUST** monitor BR3*. **MAY** or **MAY** not drive BCLR*, or slot 1 BG0IN*-BG2IN*. **MAY** or **MAY** not monitor BR0*-BR2*.
RRS	**MUST** drive slot 1 BG0IN*-BG3IN* or BG0OUT*-BG3OUT* (See Note). **MUST** monitor BR0*-BR3*. **MAY** or **MAY** not drive BCLR*.
PRI	**MUST** drive slot 1 BCLR* and BG0IN*-BG3IN*, or BG0OUT*-BG3OUT* (See Note). **MUST** monitor BR0*-BR3*.

Note: The ARBITER drives BGXIN* if a REQUESTER resides on its board, and drives BGXOUT* if no REQUESTER resides on its board.

3.3.2 REQUESTER

Each REQUESTER in the system:

a. Monitors the DEVICE WANTS BUS signal from its on-board MASTER or INTERRUPT HANDLER and generates a bus request when the DTB is needed.

b. If it detects a low level on its BGxIN* line, and its on-board MASTER or INTERRUPT HANDLER does not need the DTB, then it passes on that low level to its BGxOUT*

c. If it detects a low level on its BGxIN* line, and its on-board MASTER or INTERRUPT HANDLER needs the DTB, it generates an on-board DEVICE GRANTED BUS signal to indicate the DTB is available, and drives the BBSY* signal low.

Two types of REQUESTERS are described in this specification: a Release When Done (RWD) REQUESTER and a Release On Request (ROR) REQUESTER.

The RWD REQUESTER releases BBSY* when the MASTER or INTERRUPT HANDLER drives the on-board DEVICE WANTS BUS signal false.

The ROR REQUESTER does not release BBSY* when its on-board DEVICE WANTS BUS signal goes false unless some other REQUESTER on the bus drives one of the bus request lines low. It monitors the four bus request lines and releases BBSY* only if another bus request is pending. ROR REQUESTERS reduce the number of arbitrations initiated by a MASTER which is generating a large percentage of the bus traffic. See Figure 3-4.

Assuming that the REQUESTER'S DEVICE WANTS BUS input is true, when it receives a bus grant it does 3 things:

a. It drives BBSY* to low.
b. It releases its low on BRx.
c. It drives the DEVICE GRANTED BUS on-board signal true, allowing the MASTER or INTERRUPT HANDLER to initiate bus transfers.

These events might occur in any order. It is even possible, although meaningless, that the MASTER or INTERRUPT HANDLER might not use the bus in response to this particular grant.

However, the following RULES apply:

RULE 3.7:
The REQUESTER **MUST** drive BBSY* to low for at least 90 nanoseconds.

RULE 3.8:
The REQUESTER **MUST** release BUS REQUEST to high.

RULE 3.9:
The REQUESTER **MUST** maintain BBSY* low for at least 30 nanoseconds after it releases its bus request.

OBSERVATION 3.8:
The 30 nanosecond delay between the bus request's rising edge and the BBSY* rising edge ensures that the ARBITER does not mistakenly interpret the old bus request as a new one and issue another grant.

RULE 3.10:
The REQUESTER **MUST** hold BBSY* low until its bus grant goes high.

OBSERVATION 3.9:
RULE 3.10 ensures that the BBSY* transition to low has been seen by the ARBITER and that all segments of the bus grant daisy-chain have returned to high, in preparation for the next arbitration.

PERMISSION 3.5:
IF a REQUESTER has a bus request pending and, if it sees some other REQUESTER drive BBSY* low
THEN it **MAY** withdraw its request by releasing the bus request line.

RULE 3.11:

IF a REQUESTER withdraws a bus request without having first been granted the bus,

THEN it **MUST** wait to do so until BBSY* goes low and it **MUST** do so within 50 nanoseconds after BBSY* goes low.

SUGGESTION 3.2:

Design REQUESTERS so that they pass on the bus grant daisy-chain as fast as possible after receipt of a bus grant. This will improve system performance.

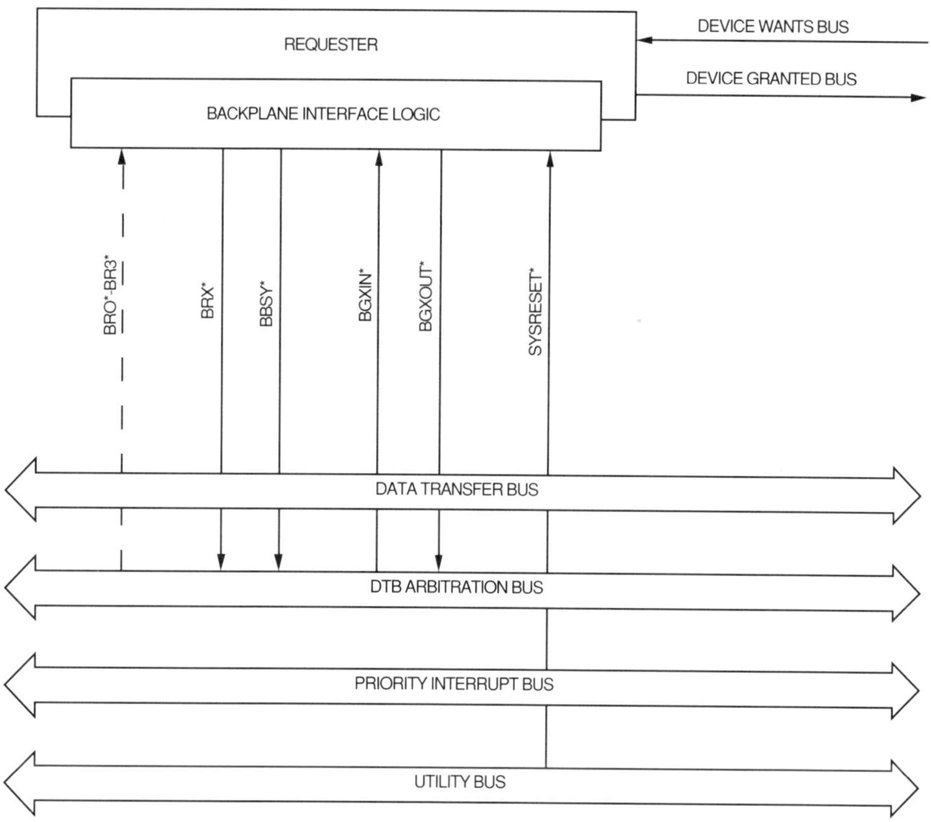

Note: The RULES and PERMISSIONS for monitoring the dotted lines are given in Table 3-2.

Figure 3-4. Block Diagram: REQUESTER

Table 3-2. RULES And PERMISSIONS That Specify The Use Of The Dotted Lines By The Various Types Of REQUESTERS Defined By This Document

Type of REQUESTER	Use of dotted line
RWD	**MAY** or **MAY NOT** monitor BR0*-BR3*.
ROR	**MUST** monitor BR0*-BR3*.

3.3.3 Data Transfer Bus MASTER

3.3.3.1 Release Of The DTB

The bus arbitration protocol determines how and when the DTB is granted to the various MASTERS and INTERRUPT HANDLERS in the system. It does not, however, control when MASTERS and INTERRUPT HANDLERS release the DTB.

MASTERS and INTERRUPT HANDLERS use several criteria in deciding when to release the DTB. INTERRUPT HANDLERS give up the bus after their interrupt acknowledge cycle. MASTERS give up the bus when they finish their data transfers.

Some MASTERS also monitor the ACFAIL* and BCLR* VMEbus signals. Both of these signals inform the MASTER that the DTB is needed for some higher priority activity. In the case of BCLR*, the MASTER'S design determines how long it takes to release the bus. For example, a MASTER on a disk controller board might not be able to relinquish the bus during a disk sector transfer without loss of data, so it might keep the bus until the sector transfer is finished. ACFAIL* informs the MASTER that an AC power loss has been detected, and whatever problems the MASTER will face in surrendering the bus are insignificant compared to the needs of the total system.

RECOMMENDATION 3.1:
Design MASTERS to release the DTB within 200 microseconds after ACFAIL* goes low, except to participate in the ensuing power failure activities.

OBSERVATION 3.10:
The 200 microsecond time specified in RECOMMENDATION 3.1 is intended to provide enough time for an orderly shut-down of the system.

Whatever criteria are used to decide when to release the DTB, arbitration is done before some other MASTER or INTERRUPT HANDLER begins using it. This arbitration takes place either during the last data transfer or after that transfer, depending on how much notice the MASTER or INTERRUPT HANDLER gives to its on-board REQUESTER.

PERMISSION 3.6:
MASTERS and INTERRUPT HANDLERS **MAY** release the DTB either during or after their last data transfer.

For example, if the MASTER notifies its on-board REQUESTER that it no longer wants the bus during its last data transfer, the REQUESTER releases BBSY* and arbitration takes place during the last transfer. But if the MASTER waits until the last transfer has completed before signaling its on-board REQUESTER, the DTB remains idle while the arbitration is done. (This was illustrated in Section 2.5.1)

Chapters 2 and 4 contain RULES that pertain to the release of the DTB.

SUGGESTION 3.3:
Design block transfer MASTER boards so that they signal their REQUESTER to release BBSY* during the last data strobe of the block transfer. If it is released at the beginning of the block transfer, high priority bus requests initiated during the block transfer might not be taken into account by the ARBITER until the next arbitration cycle.

3.3.3.2 Acquisition Of The DTB

To ensure that no DTB line is ever driven to opposite states by two MASTERS or INTERRUPT HANDLERS, when these modules take control of the DTB they are constrained by certain rule.

RULE 3.12:
When a MASTER or INTERRUPT HANDLER is given control of the DTB by its on-board REQUESTER, it **MUST** wait until it detects AS* high before turning on its DTB drivers.

OBSERVATION 3.11:
IF the previous MASTER or INTERRUPT HANDLER releases the bus DURING its last data transfer,
THEN RULE 3.12 ensures that the data transfer will be finished before the new MASTER or INTERRUPT HANDLER starts using the DTB. (If the previous MASTER or INTERRUPT HANDLER waited until the data transfer was finished before releasing the bus, AS* will already be high.)

3.3.3.3 Other Information

RECOMMENDATION 3.2:
To allow for prompt servicing of interrupt requests and for optimum use of the DTB, design MASTERS so that they release the DTB as soon as possible after they detect BCLR* low.

PERMISSION 3.7:
A MASTER or INTERRUPT HANDLER **MAY** have more than one REQUESTER, where each REQUESTER generates bus requests on a different bus request line.

OBSERVATION 3.12:
Where a MASTER or INTERRUPT HANDLER has two or more REQUESTERS, it can do high priority data transfers using one REQUESTER and low priority transfers using another REQUESTER.

3.4 TYPICAL OPERATION

3.4.1 Arbitration Of Two Different Levels Of Bus Request

Figures 3-5 and 3-6 illustrate the sequence of events that takes place when two REQUESTERS send simultaneous bus requests to a PRI ARBITER on different bus request lines. When the sequence begins, each of the REQUESTERS drives its respective bus request line low (REQUESTER A drives BR1* and REQUESTER B drives BR2*). The ARBITER detects BR1* and BR2* low simultaneously, and it drives BG2IN* low to its own slot (slot 1). That BG2IN* signal is monitored by REQUESTER B (also in slot 1). When REQUESTER B detects BG2IN* low, it responds by driving BBSY* low. REQUESTER B then releases the BR2* line and informs its own MASTER (MASTER B) that the DTB is available. When BBSY* goes low, the ARBITER drives BG2IN* of slot 1 high.

When MASTER B completes its data transfer(s), and signals that fact by driving DEVICE WANTS BUS false, REQUESTER B releases BBSY*, provided that its BG2IN* has been received high and 30 nanoseconds have elapsed since it released BR2*.

The ARBITER interprets the release of BBSY* as a signal to arbitrate any current bus requests. Since BR1* is the only bus request being driven low, the ARBITER grants the DTB to REQUESTER A by driving BG1IN* low. REQUESTER A responds by driving BBSY* low. When MASTER A completes its data transfer(s), and signals that fact by driving DEVICE WANTS BUS false, REQUESTER A releases BBSY*, provided that its BG1IN* has been received high and 30 nanoseconds have elapsed since it released BR1*.

In this example, since no bus request lines are low when REQUESTER A releases BBSY*, the ARBITER waits until it detects a bus request.

OBSERVATION 3.13:
The description illustrated in Figures 3-5 and 3-6 would hold for both PRI and RRS ARBITERS, unless we consider an RRS ARBITER where the last active request was level BR2*. In this case, the ARBITER would process the BR1* request first and then proceed to the BR2* request.

OBSERVATION 3.14:
BBSY* and the bus grants are fully interlocked as shown in Figure 3-6:

1) The ARBITER does not drive the bus grant high until it detects BBSY* low.
2) The REQUESTER does not release BBSY* to high until it detects the bus grant high.
3) The ARBITER does not drive the next bus grant low until it detects BBSY* high.
4) The next REQUESTER does not drive BBSY* low until it detects the bus grant low.

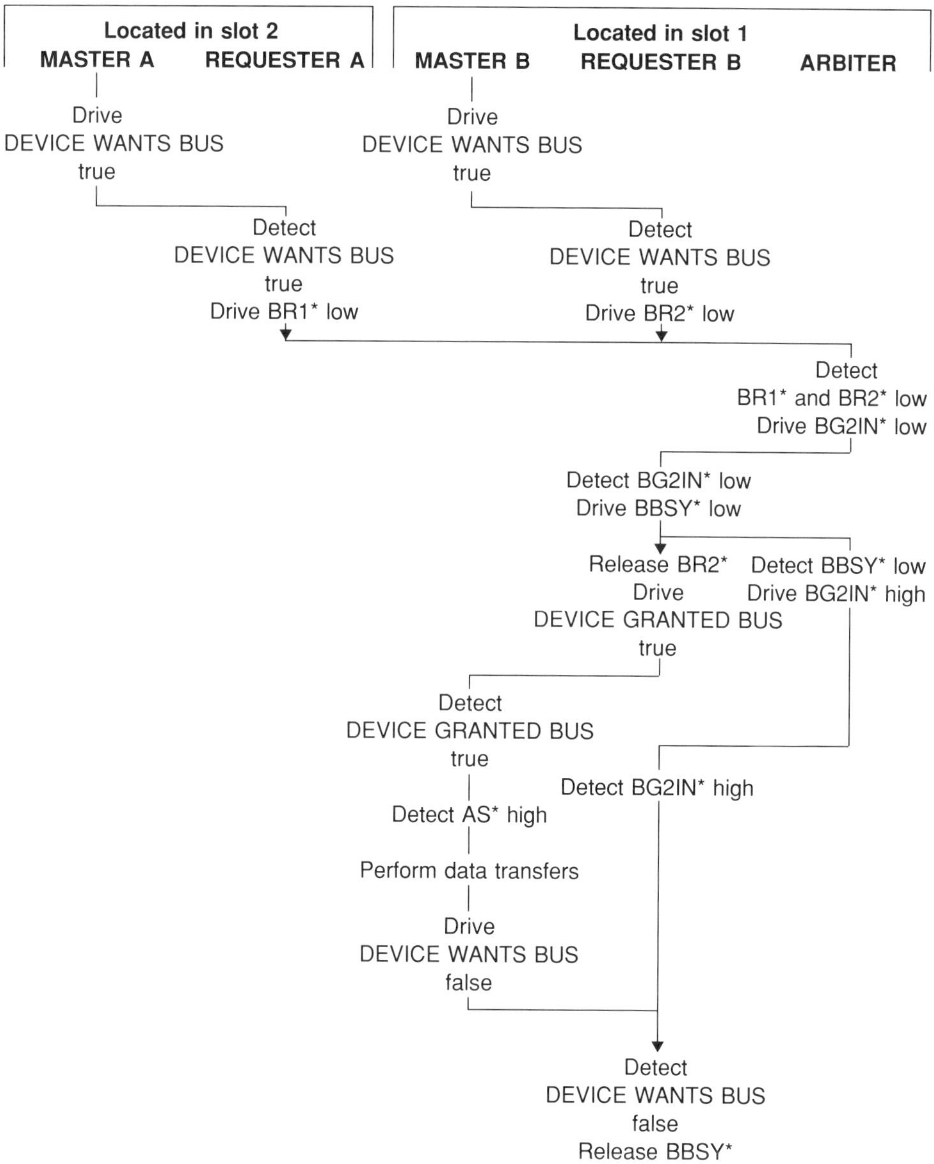

(Continued on sheet 2)

Note: DEVICE WANTS BUS and DEVICE GRANTED BUS are on-board signals between the MASTER and its REQUESTER. (See Figure 3-4)

Figure 3-5. Arbitration Flow Diagram (Sheet 1 of 2)
Two REQUESTERS, Two Request Levels

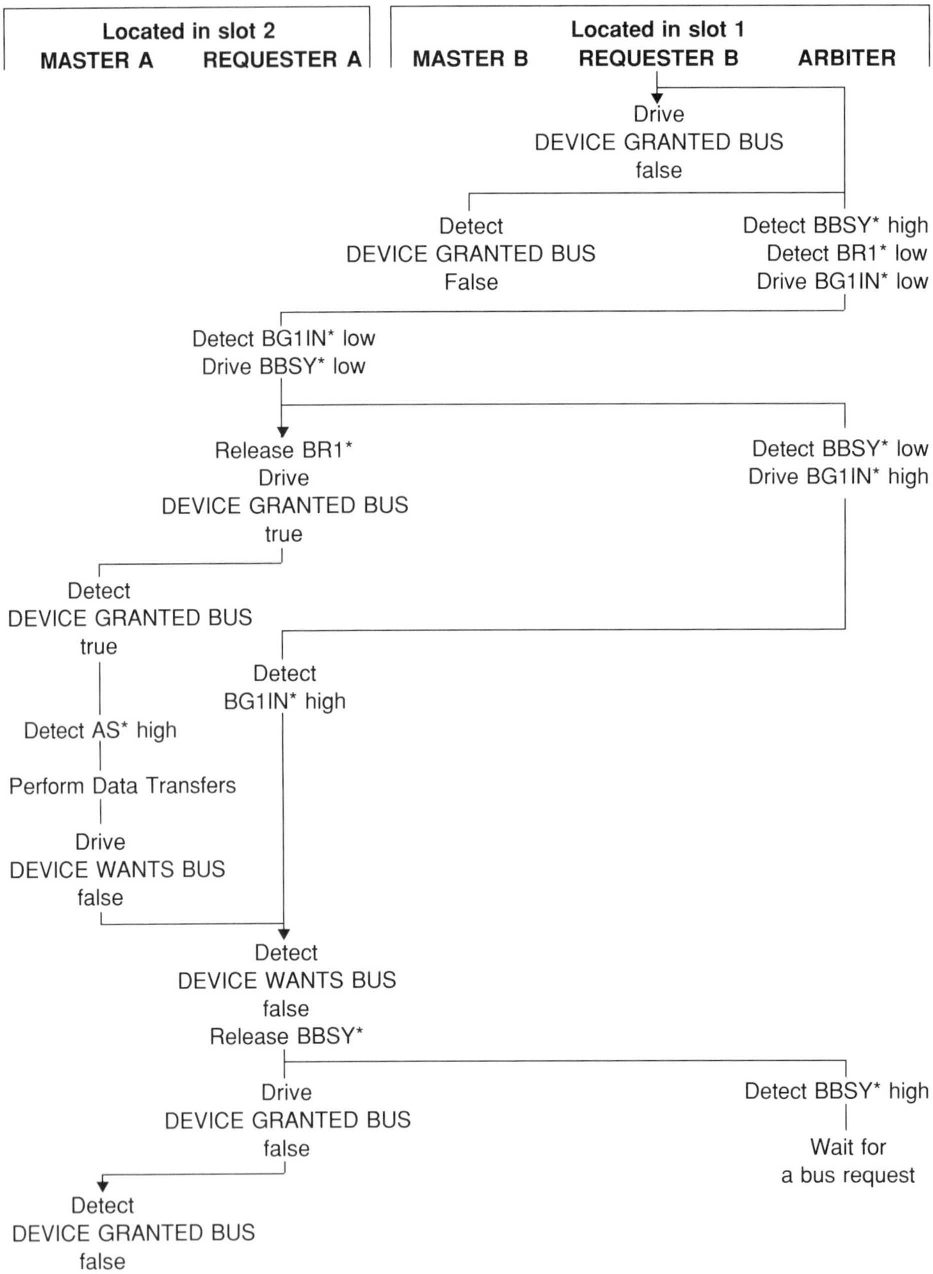

Figure 3-5. Arbitration Flow Diagram (Sheet 2 of 2)
Two REQUESTERS, Two Request Levels

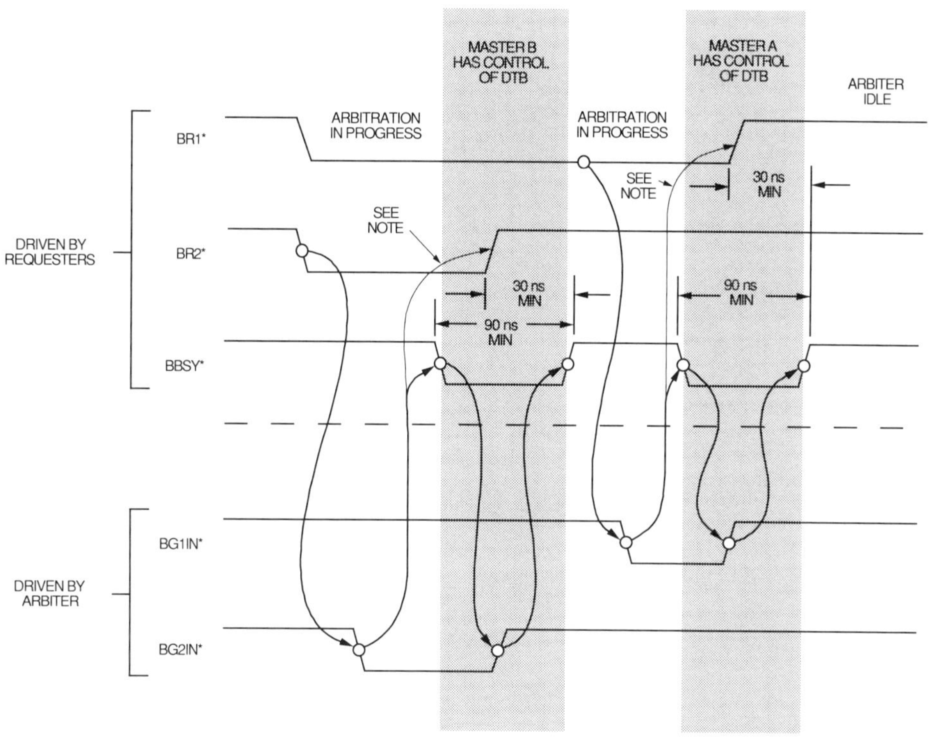

Note:

In this example each REQUESTER maintains its bus request line low until it is granted the DTB. In some cases a REQUESTER might release its bus request line without receiving a bus grant (see Section 3.3.2).

Figure 3-6. Arbitration Sequence Diagram
Two REQUESTERS, Two Request Levels

3.4.2 Arbitration Of Two Bus Requests On The Same Bus Request Line

Figures 3-7 and 3-8 illustrate the sequence of events which take place when an ROR REQUESTER and an RWD REQUESTER send simultaneous requests to a PRI ARBITER on a common bus request line. In this example, the ARBITER and RWD REQUESTER are located on the system controller board in slot 1, with the ROR REQUESTER located in slot 2. When the sequence begins, both of the REQUESTERS drive BR1* low simultaneously. The ARBITER then drives BG1IN* low to its own slot (slot 1). That BG1IN* signal is monitored by REQUESTER A (also in slot 1). When REQUESTER A detects BG1IN* low, it responds by driving BBSY* low. REQUESTER A then releases BR1* and informs MASTER A that the DTB is available.

OBSERVATION 3.15:
Even though REQUESTER A releases BR1*, REQUESTER B continues to drive it low (see Figures 3-7 and 3-8).

After detecting BBSY* low, the ARBITER drives BG1IN* high. When MASTER A has completed its data transfer(s), it drives DEVICE WANTS BUS false. When REQUESTER A detects this, and when the 30 nanoseconds delay since the release of BR1* has been satisfied, REQUESTER A releases BBSY*.

The ARBITER interprets the release of BBSY* as a signal to arbitrate any current bus requests. Since the BR1* line is still low, the ARBITER drives BG1IN* low again. When REQUESTER A detects BG1IN* low, it drives its BG1OUT* low because it does not need the DTB. REQUESTER B then detects the low on its BG1IN* and responds by driving BBSY* low. When the ARBITER detects the low on BBSY*, it drives BG1IN* high, which causes REQUESTER A to drive its BG1OUT* high.

Some time later, when MASTER B has finished its data transfers, it drives DEVICE WANTS BUS false, indicating that it has finished using the DTB.

Since REQUESTER B is an ROR REQUESTER, it does not release BBSY*, but keeps it driven low. In the event that MASTER B needs to use the DTB again, no arbitration will be required. In this example, however, REQUESTER A drives BR1* low, indicating a need to use the DTB, and REQUESTER B (which is monitoring the bus request lines) releases the BBSY* line. The ARBITER then grants the DTB to REQUESTER A.

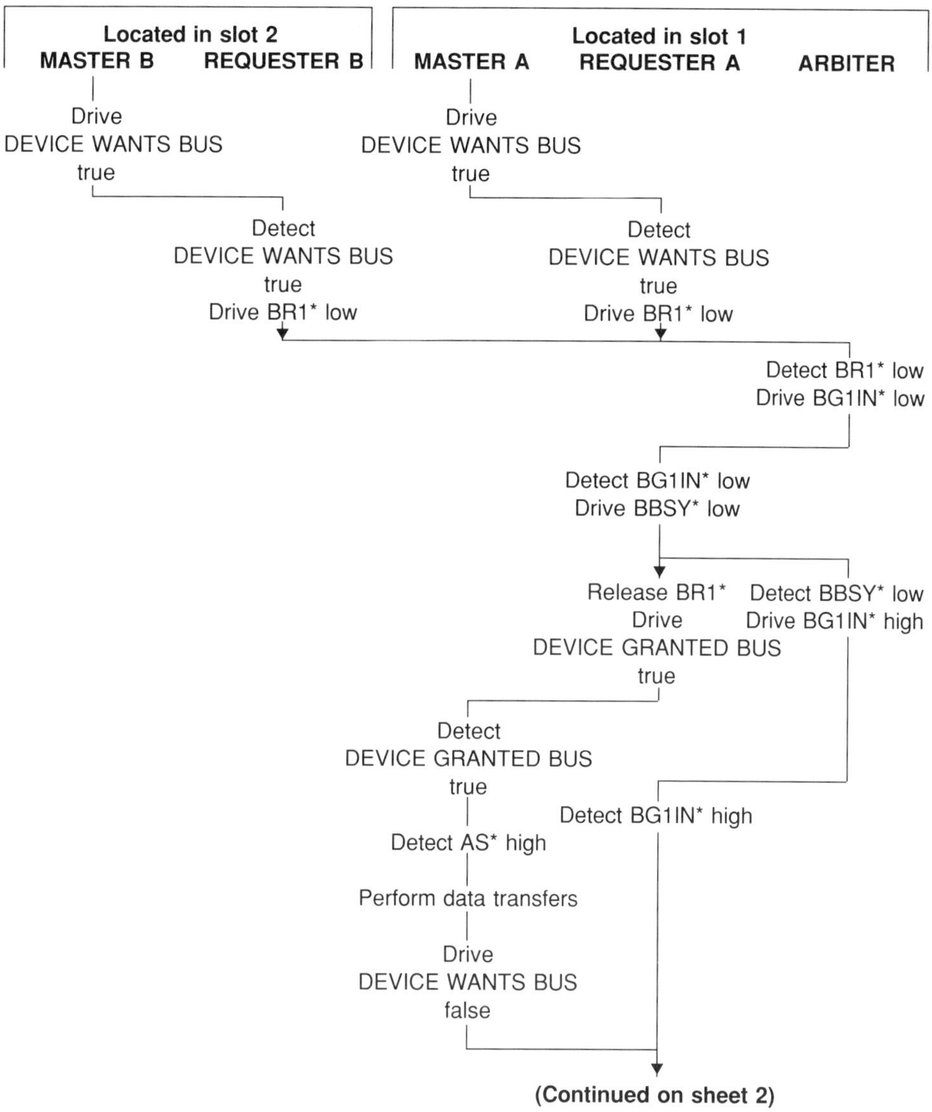

Note: DEVICE WANTS BUS and DEVICE GRANTED BUS are on-board signals between the MASTER and its REQUESTER. (See Figure 3-4)

Figure 3-7. Arbitration Flow Diagram (Sheet 1 of 3)
Two REQUESTERS, Same Request Level

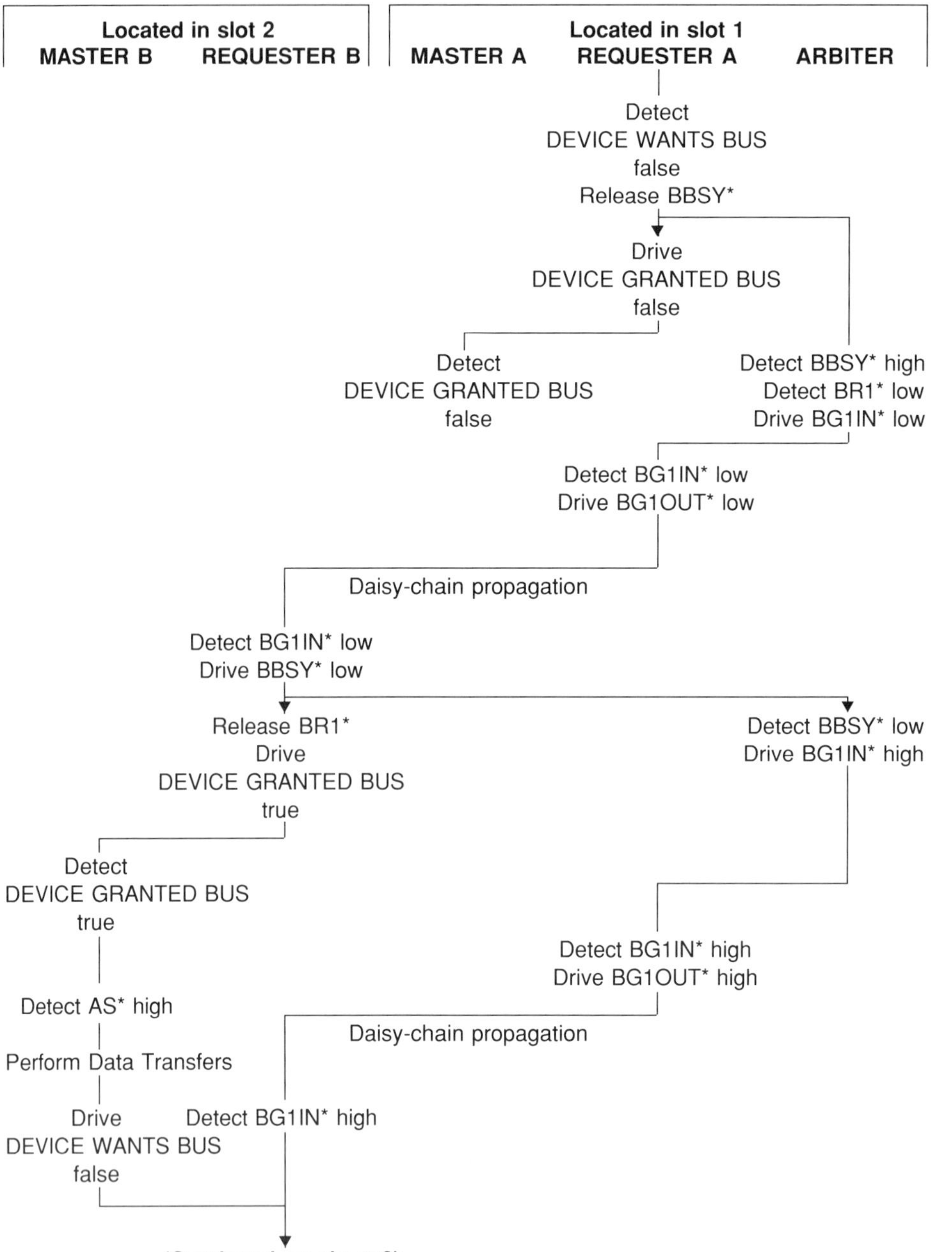

(Continued on sheet 3)

Figure 3-7. Arbitration Flow Diagram (Sheet 2 of 3)
Two REQUESTERS, Same Request Level

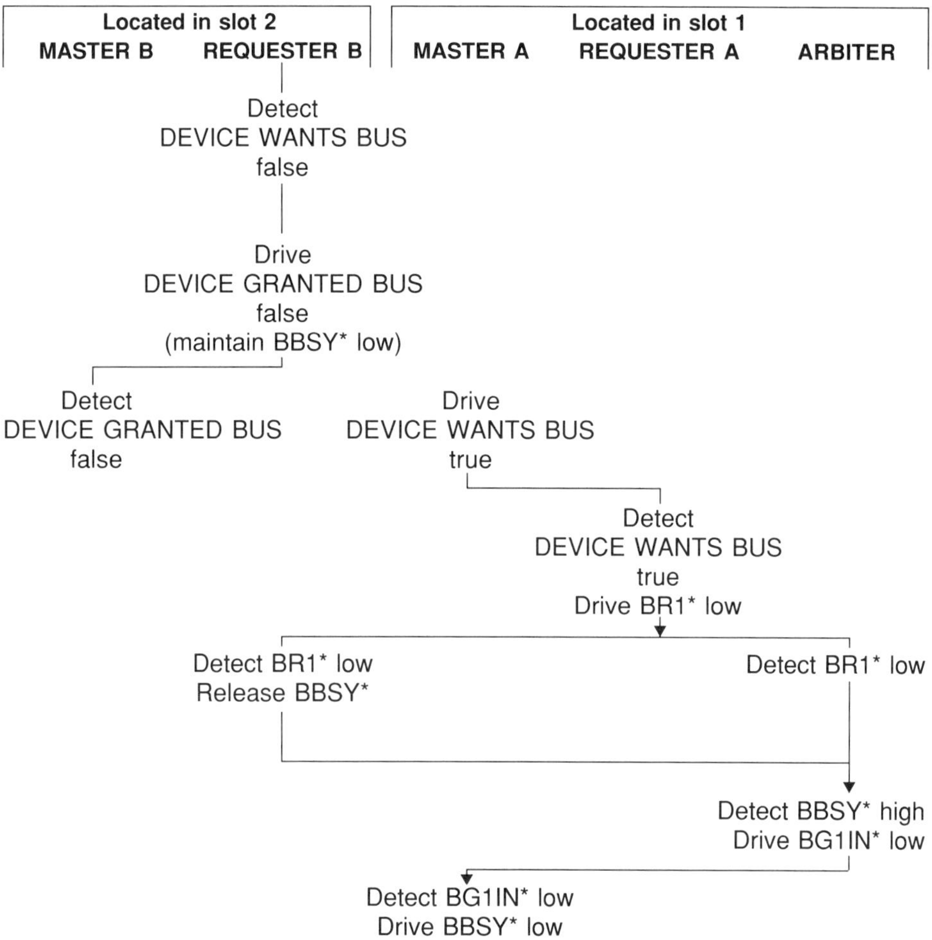

Figure 3-7. Arbitration Flow Diagram (Sheet 3 of 3)
Two REQUESTERS, Same Request Level

3.5 RACE CONDITIONS BETWEEN MASTER REQUESTS AND ARBITER GRANTS

Suppose that there are two REQUESTERS: REQUESTER A and REQUESTER B, that share a common bus request line. REQUESTER B, which is farther down the daisy-chain, requests the bus and the ARBITER drives the corresponding bus grant line low. This bus grant arrives at REQUESTER A just as MASTER A signals that it wants the bus. If REQUESTER A is improperly designed, this situation might cause it to momentarily drive its BGxOUT* line low and then high again resulting in a low-going transient.

RULE 3.13:
REQUESTERS **MUST** be designed to ensure that no momentary low-going transients are generated on their BGxOUT* out line.

OBSERVATION 3.16:
If the REQUESTER is designed such that it latches the state of the on-board DEVICE WANTS BUS line upon the falling edge of its bus grant in line, and if that signal is in transition when the falling edge occurs, the outputs of the latch will sometimes oscillate, or remain in the threshold region between the high and low levels, for a short time. Because of this, the VMEbus specification does not set a time limit for the REQUESTER to pass along the bus grant. It only prohibits the REQUESTER from generating a low-going transient on its BGxOUT* line which might be interpreted as a bus grant by a REQUESTER further down the daisy-chain.

PERMISSION 3.8:
IF a REQUESTER detects that its on-board MASTER needs the bus between the time that it receives a bus grant intended for another REQUESTER and the time it would pass that bus grant on.
THEN it **MAY** treat the bus grant as its own. In this case the other REQUESTER will maintain its bus request until another bus grant is issued.

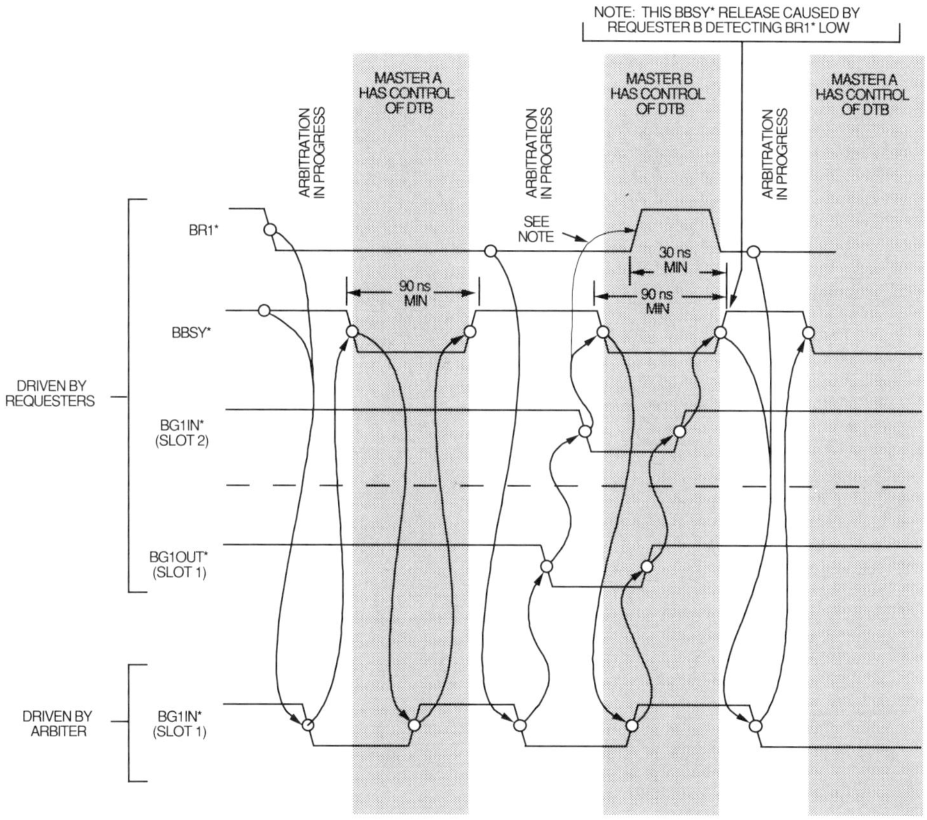

Note: In this example each REQUESTER maintains its bus request line low until it is granted the DTB. In some cases a REQUESTER might release its bus request line without receiving a bus grant (see Section 3.3.2).

Figure 3-8. Arbitration Sequence Diagram
Two REQUESTERS, Same Request Level

CHAPTER 4

PRIORITY INTERRUPT BUS

4.1 INTRODUCTION

The VMEbus includes a Priority Interrupt Bus which provides the signal lines needed to generate and service interrupts. Figure 4-1 shows a typical VMEbus system. INTERRUPTERS use the Priority Interrupt Bus to send interrupt requests to INTERRUPT HANDLERS which respond to these requests.

Any system which has interrupt capability includes software routines that are called interrupt service routines, and are invoked by the interrupts. Interrupt subsystems can be classified into two groups:

a. Single handler systems, which have only one INTERRUPT HANDLER that receives and services all bus interrupts.

b. Distributed systems, which have two or more INTERRUPT HANDLERS that receive and service bus interrupts.

4.1.1 Single Handler Systems

In a single handler system, all interrupts are received by one INTERRUPT HANDLER, and all interrupt service routines are executed by one processor. Figure 4-2 shows the interrupt structure of a single handler system. This type of architecture is well suited to machine or process control applications, where a supervisory processor coordinates the activities of dedicated processors. The dedicated processors are typically interfaced to the machine or the process being controlled.

The supervisory processor is the destination for all bus interrupts, servicing them in a prioritized manner. The dedicated processors are not required to service interrupts from the bus, and can give primary attention to controlling a machine or process.

4.1.2 Distributed Systems

Figure 4-3 shows the interrupt structure of a distributed system. This system includes two or more INTERRUPT HANDLERS, each servicing only a subset of the bus interrupts. In a typical implementation, each of the INTERRUPT HANDLERS resides on a different processor board. This type of architecture is well suited to distributed computing applications, where multiple, co-equal processors execute the application software. As each of the co-equal processors executes part of the system software, it might need to communicate with the other processors. In the distributed system, each processor services only those interrupts directed to it, establishing dedicated communication paths among all processors.

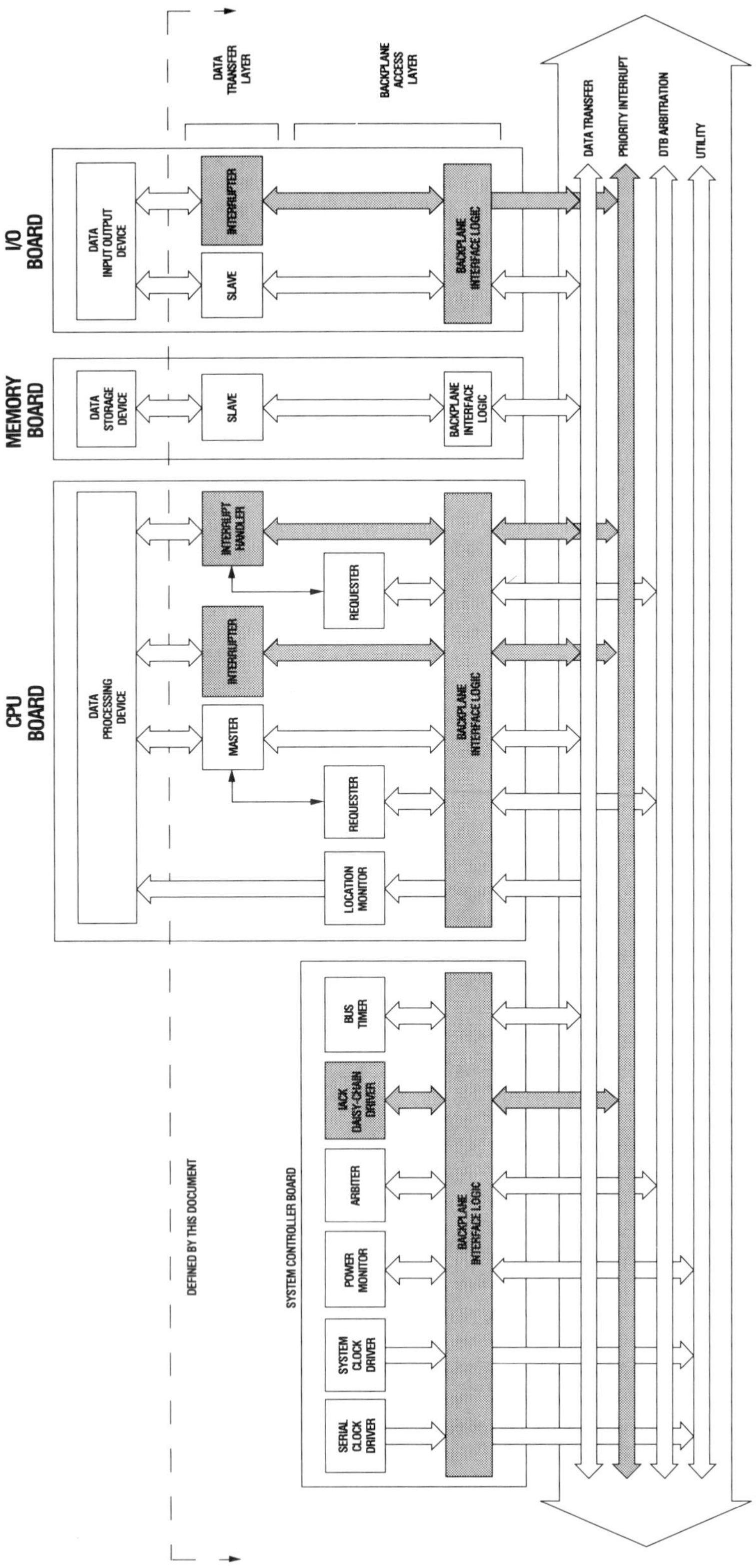

Figure 4-1. Priority Interrupt Bus Functional Block Diagram

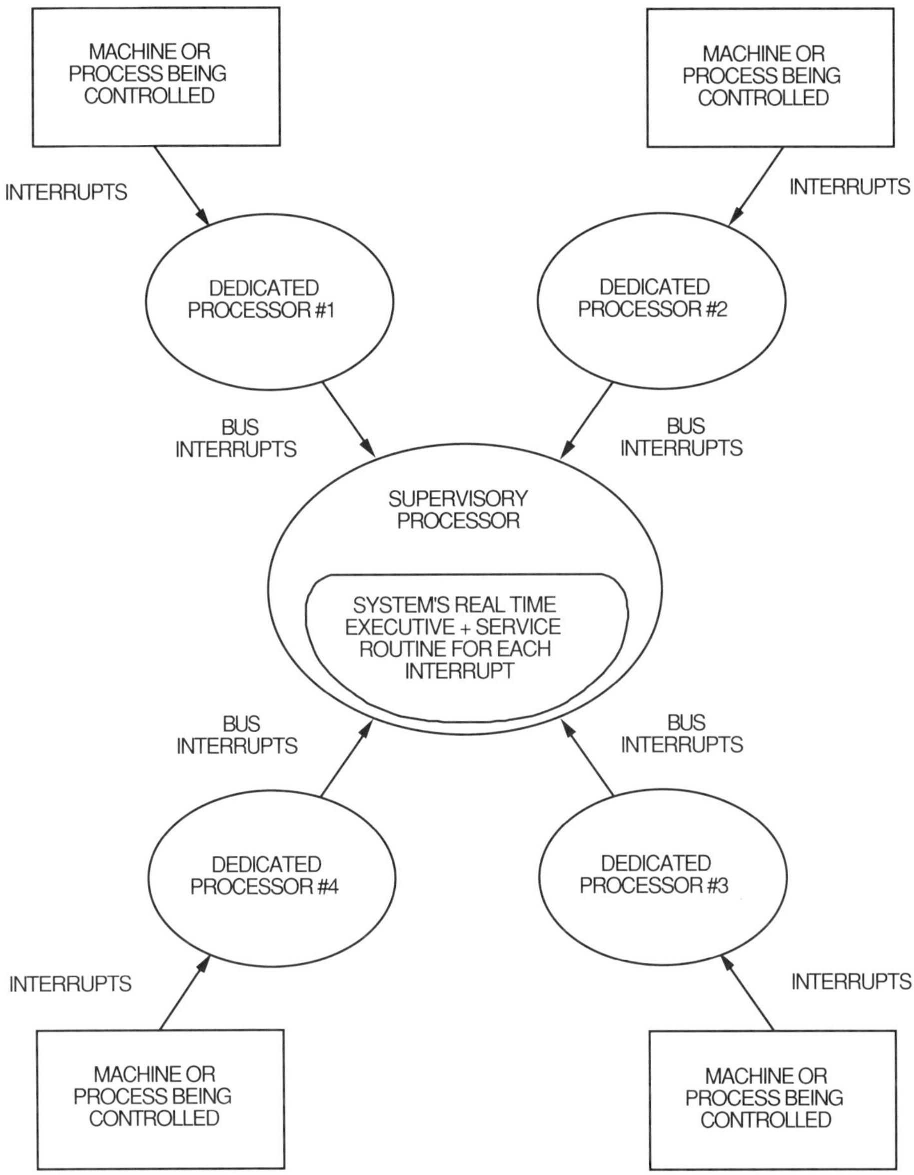

Figure 4-2. Interrupt Subsystem Structure: Single Handler System

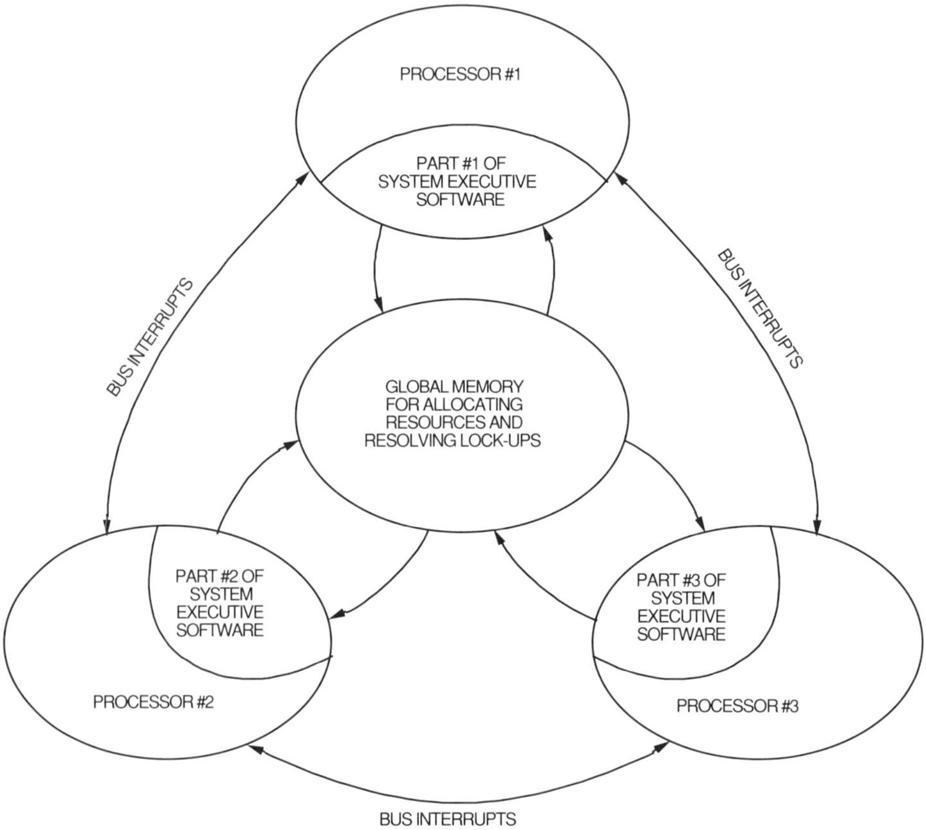

Figure 4-3. Interrupt Subsystem Structure: Distributed System

4.2 PRIORITY INTERRUPT BUS LINES

The Data Transfer Bus, the Arbitration Bus, and the Priority Interrupt Bus are all used in the process of generating and handling bus interrupts.

The following discussion of the Priority Interrupt Bus assumes that the reader understands the operation of the Data Transfer Bus described in Chapter 2, and the Arbitration Bus described in Chapter 3.

The Priority Interrupt Bus consists of seven interrupt request signal lines, one interrupt acknowledge line, and one interrupt acknowledge daisy-chain:

IRQ1*	Interrupt Request 1
IRQ2*	Interrupt Request 2
IRQ3*	Interrupt Request 3
IRQ4*	Interrupt Request 4
IRQ5*	Interrupt Request 5
IRQ6*	Interrupt Request 6
IRQ7*	Interrupt Request 7
IACK*	Interrupt Acknowledge
IACKIN*/IACKOUT*	Interrupt Acknowledge Daisy-Chain

4.2.1 Interrupt Request Lines

INTERRUPTERS request interrupts by driving an interrupt request line low. In a single handler system, these interrupt request lines are prioritized, with IRQ7* having the highest priority.

4.2.2 Interrupt Acknowledge Line

The IACK* line runs the full length of the backplane and is connected to the IACKIN* pin of slot 1 (see Figure 4-4). When driven low, the IACKIN* pin causes the IACK DAISY-CHAIN DRIVER, located in slot 1, to propagate a falling edge down the interrupt acknowledge daisy-chain.

4.2.3 Interrupt Acknowledge Daisy-Chain - IACKIN*/IACKOUT*

Each of the seven interrupt request lines can be shared by two or more INTERRUPTER modules. The interrupt acknowledge daisy-chain assures that only one INTERRUPTER responds to the interrupt acknowledge cycle. This daisy-chain line passes through each board on the VMEbus. Each INTERRUPTER that is driving an interrupt request line low waits for a falling edge to arrive at its IACKIN* daisy-chain input. Only upon receiving this falling edge does an INTERRUPTER respond to an interrupt acknowledge cycle. It does not pass the falling edge on down the daisy-chain, preventing other INTERRUPTERS from responding to the interrupt acknowledge cycle.

RULE 4.1:
IF a VMEbus backplane slot is not occupied by a board, and if there are boards farther down the interrupt acknowledge daisy-chain,
THEN jumpers **MUST** be installed at the empty slot to pass through the daisy-chain signal.

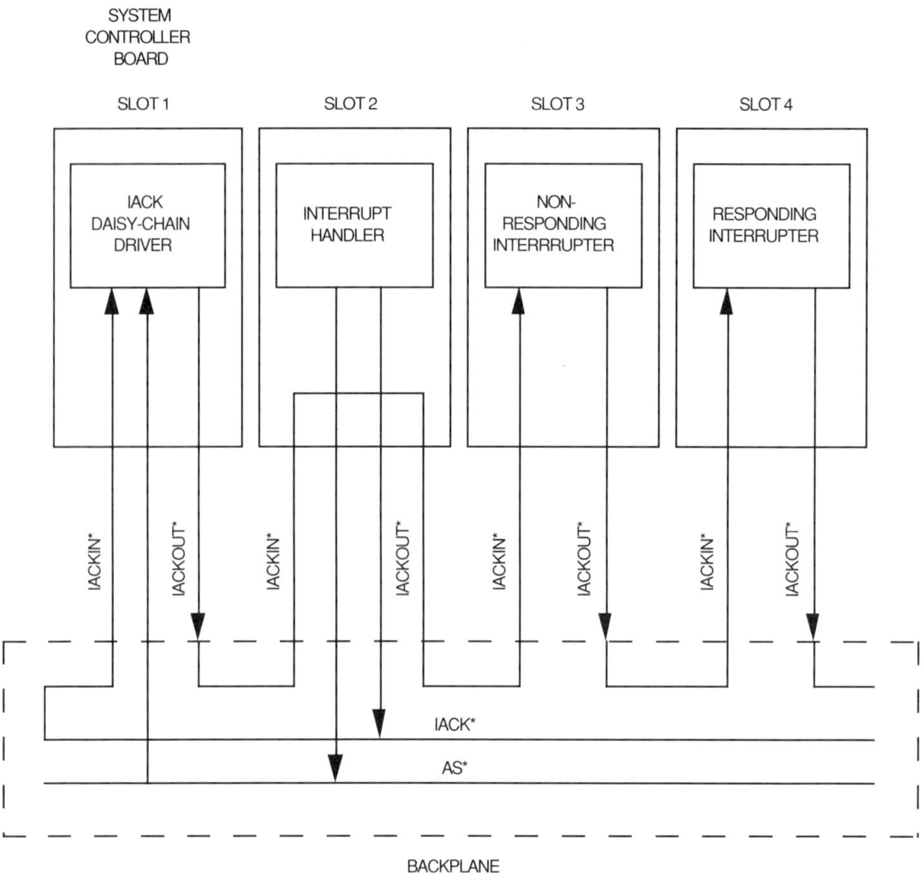

Figure 4-4. IACKIN*/IACKOUT* DAISY-CHAIN

4.3 PRIORITY INTERRUPT BUS MODULES - BASIC DESCRIPTION

There are three types of functional modules associated with the Priority Interrupt Bus: INTERRUPTERS, INTERRUPT HANDLERS, and IACK DAISY-CHAIN DRIVERS. The capabilities of INTERRUPT HANDLERS and INTERRUPTERS are described by a list of mnemonics that show what interrupt acknowledge cycle types they can generate and accept, respectively.

Sections 4.3.1 through 4.3.3 provide block diagrams for the three types of Priority Interrupt Bus modules: INTERRUPT HANDLER, INTERRUPTER, and IACK DAISY-CHAIN DRIVER.

RULE 4.2:
Output signal lines shown with solid lines in Figures 4-5 through 4-7 **MUST** be driven by the module, unless it would always drive them high.

OBSERVATION 4.1:
IF an output line is not driven,
THEN terminators on the backplane ensure that it is high.

RULE 4.3:
Input signal lines shown with solid lines in Figures 4-5 through 4-7 **MUST** be monitored and responded to in the appropriate fashion.

OBSERVATION 4.2:
RULES and PERMISSIONS for driving and monitoring signal lines shown with dotted lines in Figures 4-5 through 4-7, are given in Tables 4-1 and 4-2.

4.3.1 INTERRUPT HANDLERS

The INTERRUPT HANDLER is used to accomplish several tasks:

a. It prioritizes the incoming interrupt requests within its assigned group of interrupt request lines (highest of IRQ1*-IRQ7*).

b. It uses its on-board REQUESTER to request the DTB and, when granted use of the DTB, initiates an interrupt acknowledge cycle, reading a STATUS/ID from the INTERRUPTER being acknowledged.

c. It initiates the appropriate interrupt servicing sequence, based on the information received in the STATUS/ID.

OBSERVATION 4.3:
The VMEbus specification does not dictate what will happen during the interrupt servicing sequence. Servicing of the interrupt might or might not involve use of the VMEbus.

The INTERRUPT HANDLER uses the DTB to read a STATUS/ID from the INTERRUPTER. In this respect, the INTERRUPT HANDLER acts like a MASTER and the INTERRUPTER acts like a SLAVE. However, there are four important differences. The INTERRUPT HANDLER:

a. always drives IACK* low.
b. is not required to drive the address modifier lines.
c. only uses the lowest three address lines (A01-A03).
d. never drives the data bus.

The INTERRUPT HANDLER always drives IACK* low when it accesses the bus. The MASTER either drives it high or does not drive it at all.

The INTERRUPT HANDLER does not have to drive the address modifier lines with a valid code, and it only drives the lowest three address lines (A01-A03) with valid information. The levels of these three address lines indicate which of the seven interrupt request lines is being acknowledged, as shown in Table 4-7. A MASTER drives 15, 23, or 31 address lines (depending on the addressing mode) with the address of the SLAVE being accessed, and provides an address modifier code on the address modifier lines.

The INTERRUPT HANDLER does not drive the data lines (i.e., it does not "write" to the INTERRUPTER) and does not have to drive the WRITE* line. A MASTER uses the data lines to a SLAVE bidirectionally and, during normal use, drives WRITE* low or high as required.

A block diagram of the INTERRUPT HANDLER is shown in Figure 4-5.

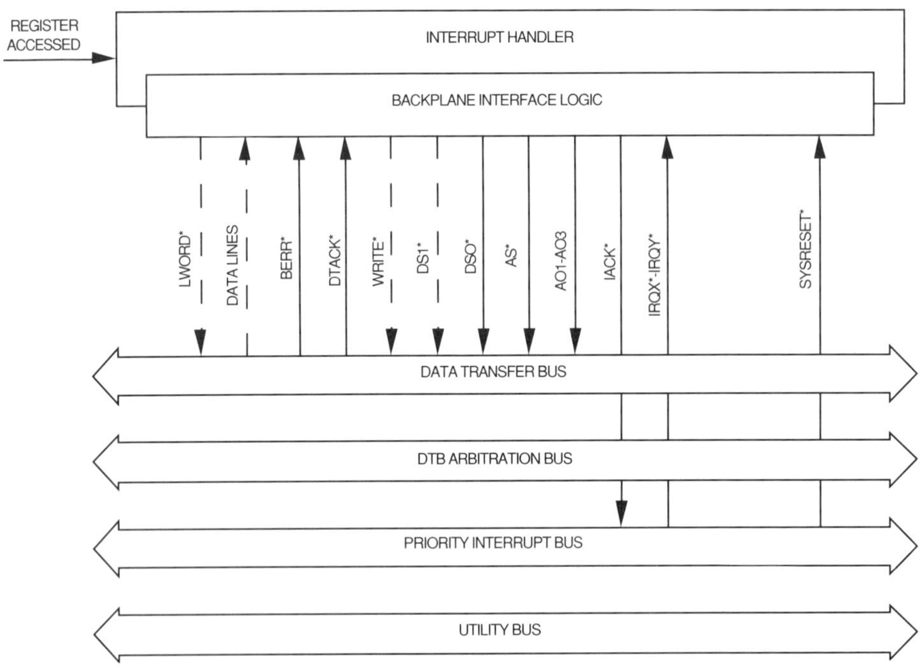

Note:

The RULES and PERMISSIONS for monitoring and driving the dotted lines are given in Table 4-1.

Figure 4-5. Block Diagram: INTERRUPT HANDLER

Table 4-1. RULES And PERMISSIONS That Specify The Use Of The Dotted Lines In Figure 4-5 By The Various Types Of INTERRUPT HANDLERS

Type of INTERRUPT HANDLER	Use of dotted lines
D08(O)	**MUST** monitor D00-D07. **MAY** or **MAY** not drive LWORD* and DS1*. **MAY** or **MAY** not monitor D08-D31.
D16	**MUST** drive DS1*. **MUST** monitor D00-D15. **MAY** or **MAY** not drive LWORD*. **MAY** or **MAY** not monitor D16-D31.
D32	**MUST** drive DS1* and LWORD*. **MUST** monitor D00-D31.
ALL	**MUST NOT** drive WRITE* low.

Note:
The mnemonics D08(O), D16, and D32 are defined in Table 4-5.

4.3.2 INTERRUPTER

The INTERRUPTER functions as follows:

a. It requests an interrupt from the INTERRUPT HANDLER which monitors its interrupt request line.

b. **IF** it receives a falling edge on the interrupt acknowledge daisy-chain input,
THEN
  IF it is requesting an interrupt **AND** the levels on the three valid address lines correspond to the interrupt request line it is using, **AND** the width of the requested STATUS/ID is either equal to, or greater than the size it can supply,
  THEN it supplies a STATUS/ID,
  ELSE it passes the falling edge down the interrupt acknowledge daisy-chain.

Each INTERRUPTER module drives only one interrupt request line. The VMEbus specification describes a board that generates interrupt requests on several interrupt lines as having several INTERRUPTER modules.

PERMISSION 4.1:
Since the INTERRUPTER is just a conceptual model, logic on a VMEbus board **MAY** be shared between several INTERRUPTER modules.

The INTERRUPTER uses one of seven lines to request an interrupt. It then monitors the lowest three lines of the address bus (A01-A03), the IACKIN*/IACKOUT* daisy-chain, and optionally IACK*, to determine when its interrupt is being acknowledged. When acknowledged, it places its STATUS/ID on the data bus and signals the INTERRUPT HANDLER that the STATUS/ID is valid by driving DTACK* low.

There are five primary differences in the use of the DTB by the INTERRUPTER and the SLAVE. The INTERRUPTER:

- **a.** only responds when its IACKIN* is low.
- **b.** does not have to monitor the address modifier lines.
- **c.** only monitors the lowest three address lines.
- **d.** does not monitor the WRITE* line.
- **e.** is permitted to respond with data of a different size than that requested.

The SLAVE monitors AS*, and interprets a falling edge on AS* as the signal that a valid bus cycle is in progress. It then proceeds to decode the appropriate number of address lines (15, 23, or 31), and the address modifier lines, and based on this information it determines whether it was addressed. However, the SLAVE responds only if IACK* is high.

The INTERRUPTER, on the other hand, interprets the falling edge on its IACKIN* line as a signal that it can respond to the interrupt acknowledge cycle in progress. It decodes only the lowest three address lines (A01-A03), ignoring the address modifier lines.

The INTERRUPTER does not need to monitor WRITE*, since it is never written to. SLAVES need to monitor WRITE* so that they can distinguish read cycles from write cycles.

The INTERRUPTER places a STATUS/ID on the bus, and responds with DTACK*, even if the LWORD*, DS1*, and DS0* lines call for a STATUS/ID whose width is greater than the INTERRUPTER is able to provide. For example, the INTERRUPT HANDLER might drive LWORD* and both data strobes low, indicating that it will read 32-bits of STATUS/ID from D00-D31, but a D08(O) INTERRUPTER would still respond with its 8-bit STATUS/ID on D00-D07. In contrast, when a SLAVE cannot provide the requested data width, it either responds with BERR* or does not respond at all, typically resulting in a bus time-out.

OBSERVATION 4.4:
When an INTERRUPTER places a STATUS/ID on the data bus, any undriven data lines are read by the INTERRUPT HANDLER as high because of the bus terminators. For example, if a D16 INTERRUPT HANDLER initiates a double byte interrupt acknowledge cycle, a D08(O) INTERRUPTER would place an 8-bit STATUS/ID on D00-D07. The upper 8 bits, read by the INTERRUPT HANDLER from D08-D15, are read as ones (high), since they are not driven by the D08(O) INTERRUPTER.

RULE 4.4:
Before responding to an interrupt acknowledge cycle, the INTERRUPTER:

1. **MUST** have an interrupt request pending.
2. The level of that request **MUST** match the level indicated on A01-A03.
3. The width of the requested STATUS/ID **MUST** be equal to or greater than the size it can respond with.
4. It **MUST** have received an incoming falling edge on its IACKIN* daisy-chain input.

IF any of these four conditions are not met.
THEN the INTERRUPTER **MUST NOT** respond to the interrupt acknowledge cycle.
IF condition 4 is met, but either 1, 2, or 3 is not.
THEN the INTERRUPTER **MUST** pass the falling edge of IACKIN* to the next INTERRUPTER module in the daisy-chain.

A block diagram of the INTERRUPTER is shown in Figure 4-6.

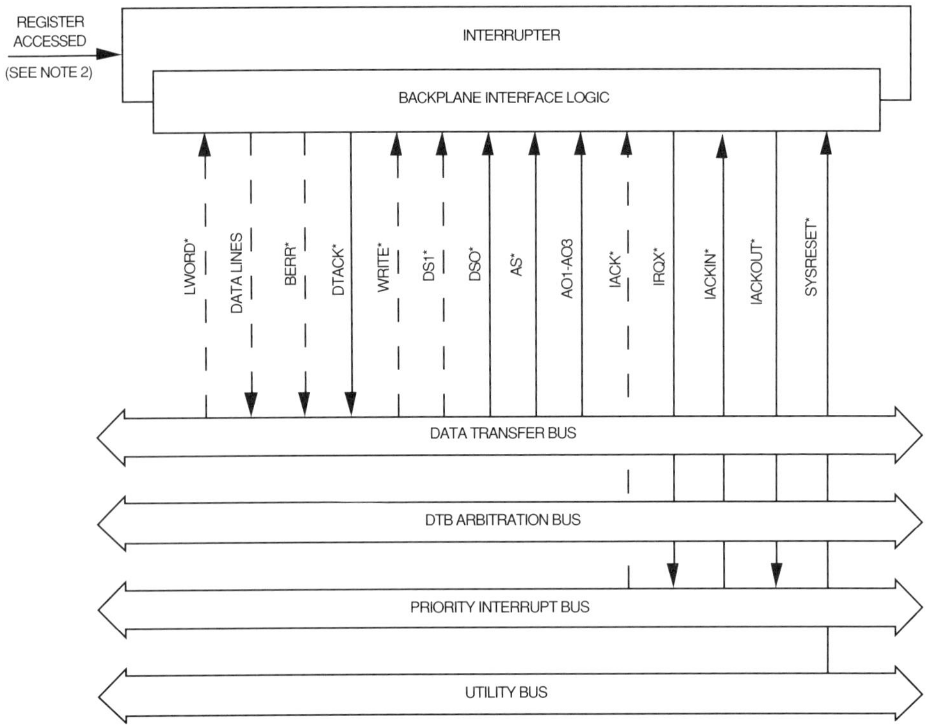

Notes:

1. The RULES and PERMISSIONS for driving and monitoring the dotted lines are given in Table 4-2.
2. This input signal is present on RORA INTERRUPTERS only.

Figure 4-6. Block diagram: INTERRUPTER

Table 4-2. RULES And PERMISSIONS That Specify The Use Of The Dotted Lines In Figure 4-6 By The Various Types Of INTERRUPTERS

Type of INTERRUPTER	Use of dotted lines
D08(O)	**MUST** drive D00-D07. **MUST NOT** drive D08-D31 low. has no reason to monitor LWORD* or DS1*.
D16	**MUST** monitor DS1*. **MUST** drive D00-D15. **MUST NOT** drive D16-D31 low. has no reason to monitor LWORD*.
D32	**MUST** monitor DS1* and LWORD*. **MUST** drive D00-D31.
ALL	**MAY** or **MAY** not monitor WRITE* and IACK*. **MAY** or **MAY** not drive BERR*.

Note: The mnemonics D08(O), D16, and D32 are defined in Table 4-5.

4.3.3 IACK DAISY-CHAIN DRIVER

The IACK DAISY-CHAIN DRIVER is another module that interacts with INTERRUPT HANDLERS and INTERRUPTERS to coordinate the servicing of interrupts. It generates a falling edge on the interrupt acknowledge daisy-chain each time an INTERRUPT HANDLER initiates an interrupt acknowledge cycle.

A block diagram of the IACK DAISY-CHAIN DRIVER is given in Figure 4-7.

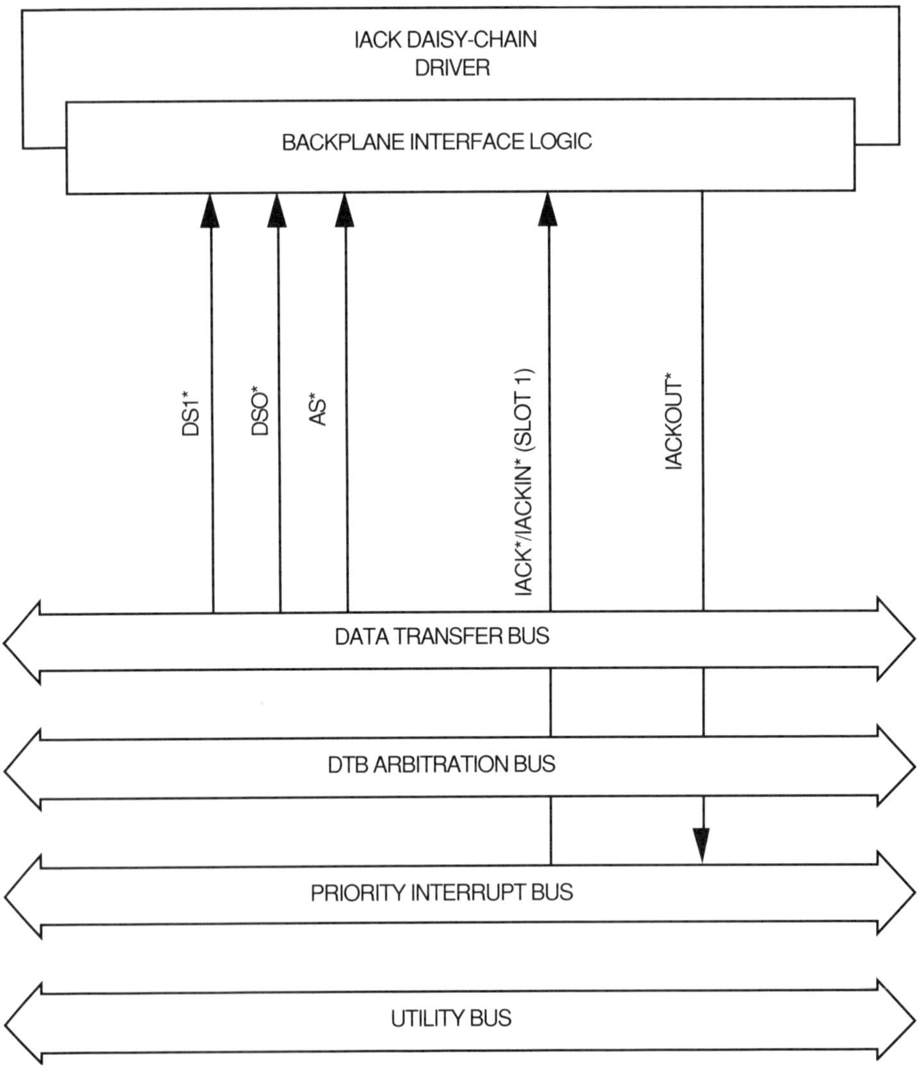

Figure 4-7. Block Diagram: IACK DAISY-CHAIN DRIVER

4.3.4 Interrupt Request Handling Capabilities

INTERRUPT HANDLERS can be designed to handle interrupt requests received on one to seven interrupt request lines. Table 4-3 shows how the IH( ) mnemonic is used to describe the interrupt handling capabilities of INTERRUPT HANDLERS.

Table 4-3. Use Of The IH( ) Mnemonic To Specify Interrupt Request Handling Capabilities

The Following Mnemonic	When Applied to an	Means that it
IH(x-y)	INTERRUPT HANDLER	can generate interrupt acknowledge cycles in response to interrupt requests on lines IRQx* through IRQy*.
IH(x)	INTERRUPT HANDLER	can generate interrupt acknowledge cycles in response to interrupt requests on line IRQx*.

4.3.5 Interrupt Request Generation Capabilities

INTERRUPTERS can be designed to generate an interrupt request on any of the seven interrupt request lines. Table 4-4 shows how the I( ) mnemonic is used to describe the interrupt request generation capabilities of INTERRUPTERS.

Table 4-4. Use Of The I( ) Mnemonic To Specify Interrupt Request Generation Capabilities

The Following Mnemonic	When Applied to an	Means that it
I(x)	INTERRUPTER	can generate interrupt requests on line IRQx*

4.3.6 STATUS/ID Transfer Capabilities

There are three STATUS/ID transfer capabilities: D08(O), D16 and D32. Table 4-5 shows how these mnemonics are used to describe the interrupt handling capabilities of INTERRUPT HANDLERS and INTERRUPTERS.

Table 4-5. Mnemonics That Specify STATUS/ID Transfer Capabilities

The Following Mnemonic	When Applied to an	Means that it
D08(O)	INTERRUPTER	responds to 8-bit, 16-bit, and 32-bit interrupt acknowledge cycles by providing an 8-bit STATUS/ID on D00-D07.
	INTERRUPT HANDLER	generates 8-bit interrupt acknowledge cycles in response to the requests on the interrupt request line(s) and reads an 8-bit STATUS/ID from D00-D07.
D16	INTERRUPTER	responds to 16-bit and 32-bit interrupt acknowledge cycles by providing a 16-bit STATUS/ID on D00-D15
	INTERRUPT HANDLER	generates 16-bit interrupt acknowledge cycles in response to the requests on the interrupt request line(s) and reads a 16-bit STATUS/ID from D00-D15.
D32	INTERRUPTER	responds to 32-bit interrupt acknowledge cycles by providing a 32 bit STATUS/ID on D00-D31.
	INTERRUPT HANDLER	generates 32-bit interrupt acknowledge cycles in response to the requests on the interrupt request line(s) and reads a 32-bit STATUS/ID from D00-D31.

4.3.7 Interrupt Request Release Capabilities

Many widely used peripheral ICs generate interrupt requests. Unfortunately, there is no standard method for indicating to these ICs when it is time for them to remove their interrupt request from the bus. Three methods are used:

a. When the relevant processor senses an interrupt request from a peripheral device, it enters an interrupt service routine, and READS a status register in the device. The peripheral device interprets this read cycle on its status register as a signal to remove its interrupt request.

b. When the relevant processor senses an interrupt request from a peripheral device, it enters an interrupt service routine, and WRITES to a control register in the device. The peripheral device interprets this write cycle to its control register as a signal to remove its interrupt request.

c. When the relevant processor senses an interrupt request from a peripheral device, it reads a STATUS/ID from the device. The peripheral device interprets this read cycle as a signal to remove its interrupt request.

The VMEbus specification calls INTERRUPTERS that use methods a and b Release On Register Access (RORA) INTERRUPTERS, and those that use method c Release On AcKnowledge (ROAK) INTERRUPTERS. Figure 4-8 shows how an ROAK INTERRUPTER releases its interrupt request line when the INTERRUPT HANDLER reads its STATUS/ID and how an RORA INTERRUPTER releases its interrupt request upon an access to a control or status register.

OBSERVATION 4.5:
The SLAVE that provided the access to the INTERRUPTER'S control or status register is typically on the same board as the INTERRUPTER, and it generates an on-board signal to the INTERRUPTER when it has completed the register access.

RULE 4.5:
An RORA INTERRUPTER **MUST NOT** release its interrupt request line before it detects a falling edge on DSA* during the register access cycle. It **MUST** release the interrupt request line within 2 microseconds after the last data strobe goes high at the end of the register access cycle.

RULE 4.6:
An ROAK INTERRUPTER **MUST NOT** release its interrupt request line before it detects a falling edge on DSA* during the interrupt acknowledge cycle which acknowledges its interrupt, and it **MUST** release its interrupt request line within 500 nanoseconds after the last data strobe goes high at the end of the STATUS/ID read cycle.

RULE 4.7:
Both RORA and ROAK INTERRUPTERS **MUST** provide a STATUS/ID during the interrupt acknowledge cycle that was initiated in response to their interrupt request.

RULE 4.8:
After an INTERRUPT HANDLER initiates an interrupt acknowledge cycle, and reads the STATUS/ID from an RORA INTERRUPTER, it **MUST** ignore the low level on the interrupt request line for 2 microseconds after its on-board signal REGISTER ACCESSED goes true.

OBSERVATION 4.6:
RULE 4.8 prevents the INTERRUPT HANDLER from misinterpreting the low level on that line as a new interrupt request.

OBSERVATION 4.7:
The MASTER that accesses the INTERRUPTER'S control or status register is typically on the same board as the INTERRUPT HANDLER, and it generates an on-board signal to the INTERRUPT HANDLER when it has completed the register access.

PERMISSION 4.2:
IF a procedure is established to allow the MASTER to signal the INTERRUPT HANDLER that an access to the INTERRUPTER'S control or status registers has taken place.
THEN the MASTER and INTERRUPT HANDLER **MAY** reside on different boards.

Table 4-6 shows how the RORA and ROAK mnemonics are used to describe INTERRUPTERS.

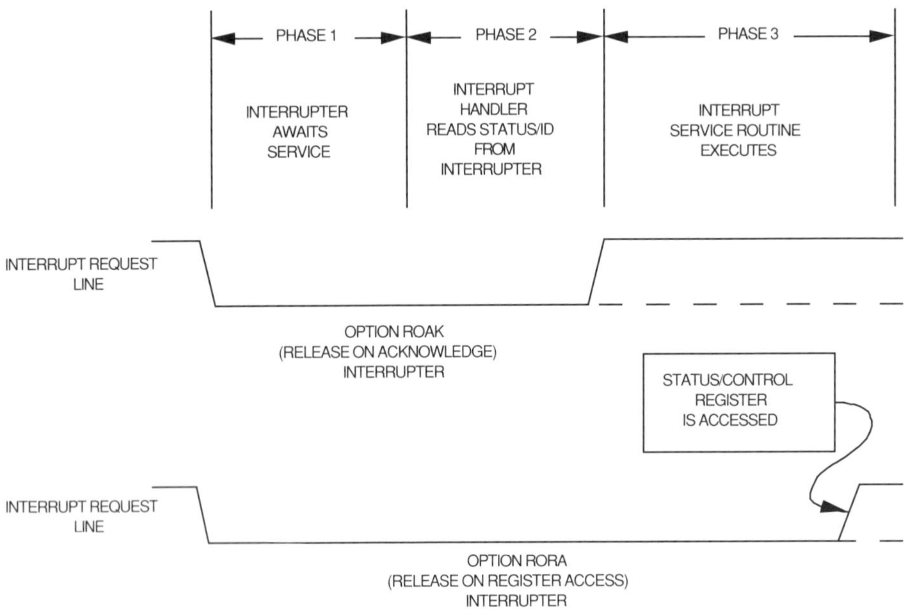

Figure 4-8. Release Of Interrupt Request Lines By ROAK And RORA INTERRUPTERS

Table 4-6. MnemonicsThat Specify Interrupt Request Release Capabilities

The Following Mnemonic	When Applied to a	Means that it
RORA	INTERRUPTER	releases its interrupt request line when some MASTER accesses an on-board status or control register.
ROAK	INTERRUPTER	releases its interrupt request line when its STATUS/ID is read during an interrupt acknowledge cycle.

4.3.8 Interaction Between Priority Interrupt Bus Modules

In the following discussions, several on-board signals are defined to describe the interaction between the INTERRUPTER and INTERRUPT HANDLER modules and other on-board logic. These signals are only intended to illustrate the information which is passed to and from the modules, rather than to define their designs.

PERMISSION 4.3:
VMEbus boards **MAY** be designed with on-board signals that differ from those used in the following discussions.

Figure 4-4 shows how the IACKIN*/IACKOUT* daisy-chain is routed through a typical configuration of boards on the VMEbus.

The IACK* line runs the full length of the backplane and can be driven by any INTERRUPT HANDLER that has control of the DTB. The backplane connects IACK* to the IACKIN* pin of slot 1. The IACK DAISY-CHAIN DRIVER resides in slot 1 and monitors the level of slot 1's IACKIN* line.

When an INTERRUPT HANDLER drives IACK* (and slot 1's IACKIN*) low, and then drives DSA* low, the IACK DAISY-CHAIN DRIVER generates a falling edge on its IACKOUT* pin. This pin is connected to the IACKIN* pin of slot 2. A jumper on the board in slot 2 routes the falling edge on the IACKIN* pin to the IACKOUT* pin, and through the backplane to the IACKIN* pin of the board in slot 3. The INTERRUPTER in slot 3 does not have a pending interrupt request, so it passes on the falling edge to its IACKOUT* pin. The INTERRUPTER in slot 4 then detects the falling edge on its IACKIN* line and responds by placing its STATUS/ID on the data bus, and then driving DTACK* low.

PERMISSION 4.4:
An INTERRUPTER **MAY** reside on the system controller board, installed in slot 1, along with the IACK DAISY-CHAIN DRIVER. Figure 4-9 shows how the two modules would be connected.

PERMISSION 4.5:
More than one INTERRUPTER **MAY** reside on a board. Figure 4-10 shows how this might be done.

OBSERVATION 4.8:
In some cases, board designers might not know whether or not the board they are designing will be installed in slot 1, or in some other slot of a VMEbus system.

RECOMMENDATION 4.1:
IF a board includes both an IACK DAISY-CHAIN DRIVER and an INTERRUPTER, and might or might not be installed in slot 1.
THEN design it as shown in Figure 4-9.

PERMISSION 4.6:
Several boards containing IACK DAISY-CHAIN DRIVERS **MAY** be installed in a VMEbus system.

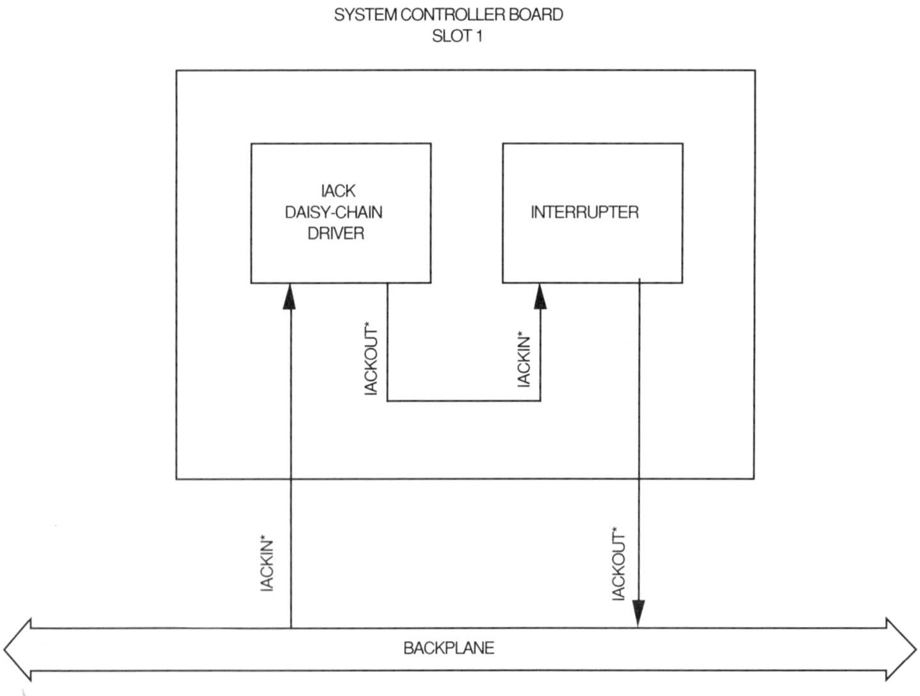

Figure 4-9. An IACK DAISY-CHAIN DRIVER And An INTERRUPTER On The Same Board

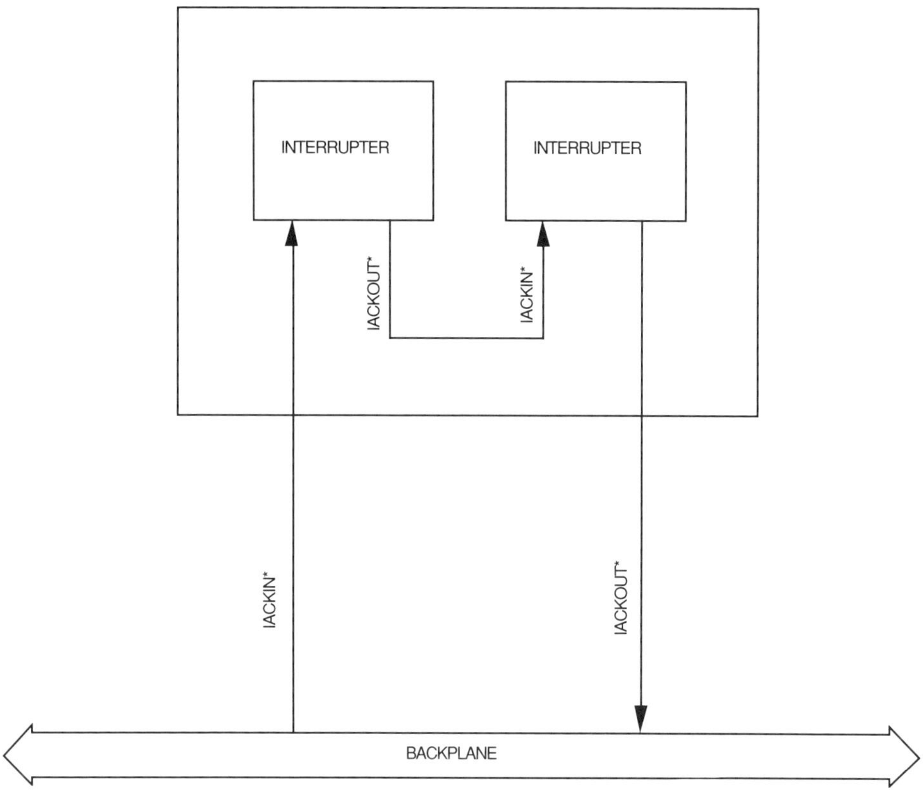

Figure 4-10. Two INTERRUPTERS On The Same Board

4.4 TYPICAL OPERATION

A typical interrupt sequence can be divided into three phases:

Phase 1: The interrupt request phase.
Phase 2: The interrupt acknowledge phase.
Phase 3: The interrupt servicing phase.

Figure 4-11 illustrates the timing relationships between the three phases.

Phase 1 starts when an INTERRUPTER drives an interrupt request line low and ends when the INTERRUPT HANDLER gains control of the DTB. During phase 2 the INTERRUPT HANDLER uses the DTB to read the INTERRUPTER'S STATUS/ID. During phase 3 an interrupt servicing routine is executed. (This might or might not involve data transfers on the VMEbus.)

The protocol for the interrupt subsystem describes the module interaction required during phase 1 and phase 2. Any data transfers which take place during phase 3 will follow the Data Transfer Bus protocol described in Chapter 2.

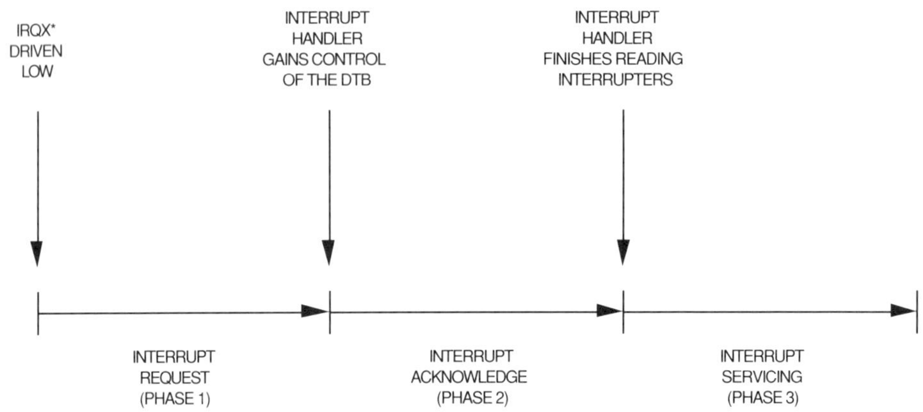

Figure 4-11. The Three Phases Of An Interrupt Sequence

4.4.1 Single Handler Interrupt Operation

In single handler interrupt systems, the seven interrupt request lines are all monitored by a single INTERRUPT HANDLER. The interrupt request lines are prioritized such that IRQ7* has the highest priority, and IRQ1* has the lowest priority. When the INTERRUPT HANDLER detects simultaneous requests on two interrupt request lines, it acknowledges the highest priority request first.

4.4.2 Distributed Interrupt Operation

Distributed interrupt systems contain from two to seven INTERRUPT HANDLERS. For purposes of the following discussion, distributed interrupt systems will be considered in two groups:

a. distributed interrupt systems with seven INTERRUPT HANDLERS,

b. distributed interrupt systems with two to six INTERRUPT HANDLERS.

4.4.2.1 Distributed Interrupt Systems With Seven INTERRUPT HANDLERS

In distributed interrupt systems with seven INTERRUPT HANDLERS, each of the interrupt request lines is monitored by a separate INTERRUPT HANDLER. Each INTERRUPT HANDLER gains control of the DTB before it reads the STATUS/ID from INTERRUPTERS driving its interrupt request line.

OBSERVATION 4.9:
There is no specified relationship between the interrupt request line that an INTERRUPT HANDLER services and the bus request line used by its on-board REQUESTER. For example, an INTERRUPT HANDLER that services IRQ7* might have a REQUESTER that uses BR0*, and an INTERRUPT HANDLER that services IRQ1* might have a REQUESTER that uses BR3*. It is clear from this that there is no implied interrupt priority between lines serviced by different INTERRUPT HANDLERS.

Figure 4-12 illustrates a distributed interrupt system where INTERRUPT HANDLER A monitors IRQ2* and has an on-board REQUESTER which requests the DTB on BR2*. INTERRUPT HANDLER B monitors IRQ5* and has an on-board REQUESTER which requests the DTB on BR3*. Two INTERRUPTERS simultaneously drive IRQ2* and IRQ5* low, and the two INTERRUPT HANDLERS cause their on-board REQUESTERS to drive BR2* and BR3* low simultaneously. In this example, priority arbitration is used and, since both bus requests go low together, the ARBITER first grants control of the DTB to INTERRUPT HANDLER B's REQUESTER, and INTERRUPT HANDLER A waits until B has finished using the DTB.

OBSERVATION 4.10:
If round-robin arbitration is used, either of the INTERRUPT HANDLERS described in Figure 4-12 might be granted the bus first.

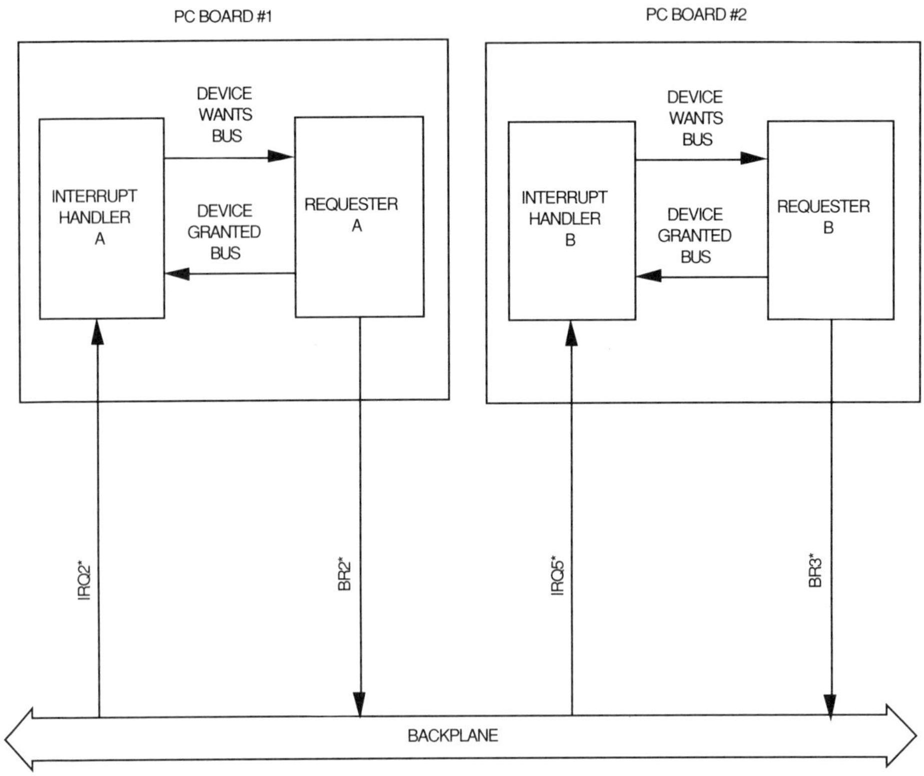

Figure 4-12. Two INTERRUPT HANDLERS, Each Monitoring One Interrupt Request Line

4.4.2.2 Distributed Interrupt Systems with Two to Six INTERRUPT HANDLERS.

It is also possible to configure a distributed interrupt system in which two or more of the interrupt request lines are monitored by a single INTERRUPT HANDLER. Figure 4-13 illustrates a system configured with two INTERRUPT HANDLERS in which INTERRUPT HANDLER A monitors IRQ1*-IRQ4*, and INTERRUPT HANDLER B monitors IRQ5*-IRQ7*. In this case, the IRQ1*-IRQ4* lines are prioritized; IRQ4* = highest priority for INTERRUPT HANDLER A, and the IRQ5*-IRQ7* lines are prioritized; IRQ7* = highest priority for INTERRUPT HANDLER B. The DTB arbitration still determines which INTERRUPT HANDLER is granted the use of the DTB first.

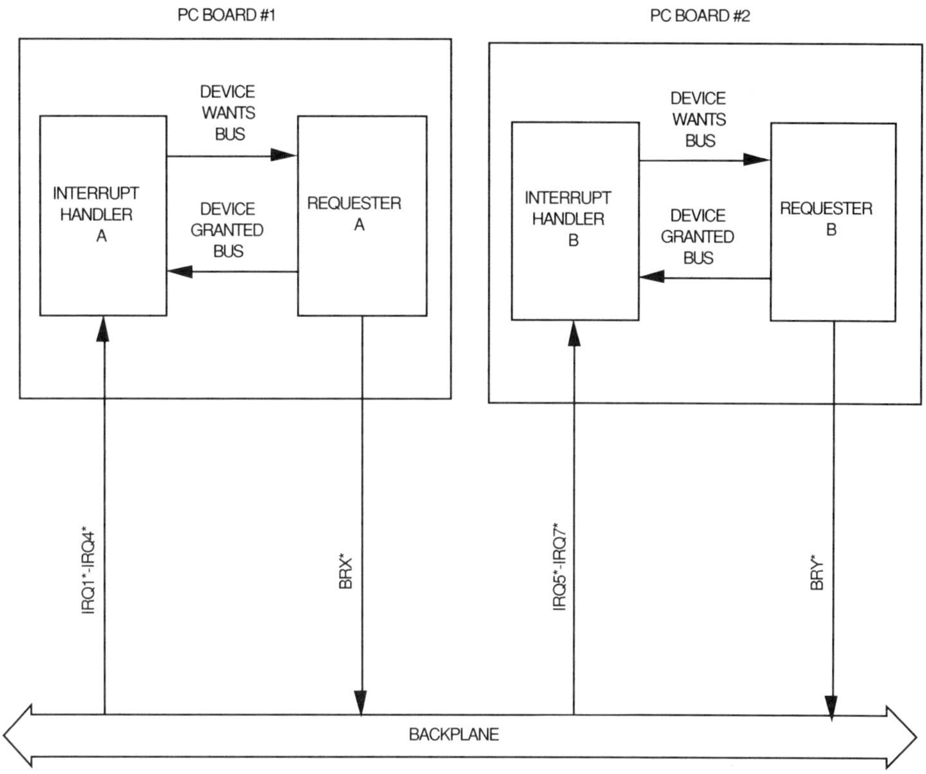

Figure 4-13. Two INTERRUPT HANDLERS, Each Monitoring Several Interrupt Request Lines

4.4.3 Example: Typical Single Handler Interrupt System Operation

Figure 4-14 illustrates the operation of a single handler interrupt system in which one INTERRUPT HANDLER monitors and prioritizes all seven interrupt lines. At the top of the diagram, a MASTER is using the DTB to move data within the system at a bus request level of 2. An INTERRUPTER in slot 3 requests an interrupt by driving IRQ4* low. When the INTERRUPT HANDLER detects the low level on IRQ4* it sends a signal to its on-board REQUESTER, indicating that it needs the bus. This REQUESTER then drives BR3* low. Upon detecting the bus request, the ARBITER drives BCLR* low, indicating that a higher priority REQUESTER is waiting for the DTB. (This example assumes a PRI ARBITER). When MASTER A detects the low level on BCLR*, it stops moving data and allows its requester to relinquish control of the DTB, and release BBSY*.

OBSERVATION 4.11:
The active MASTER is not required to relinquish the DTB within any specified time, but a prompt response to the BCLR* line allows the interrupt to be serviced quicker.

When the ARBITER detects BBSY* high, it grants the DTB to REQUESTER B, which informs its INTERRUPT HANDLER that the DTB is available (see Figures 2-26 and 2-27). The INTERRUPT HANDLER then puts a 3-bit code on the lower three address lines, indicating that it is acknowledging the interrupt request on the IRQ4* line (see Table 4-7). At the same time, it drives IACK* low, indicating that it is acknowledging an interrupt, and drives AS* low. The low level on IACK* is connected, by a signal trace in the backplane, to the IACKIN* pin of slot 1 and causes the IACK DAISY-CHAIN DRIVER to generate a falling edge down the IACKIN*/IACKOUT* daisy-chain.

When the INTERRUPTER detects a falling edge on its incoming daisy-chain line (IACKIN*) it checks the lower three address bits to see if they match the interrupt request line which it is driving low. Since the 3-bit code matches the line on which it is making its interrupt request, when the INTERRUPTER detects the data strobe(s) low, it places its STATUS/ID on the data bus and drives the DTACK* line low. When the INTERRUPT HANDLER detects the low DTACK*, it reads the STATUS/ID and activates the appropriate interrupt service routine.

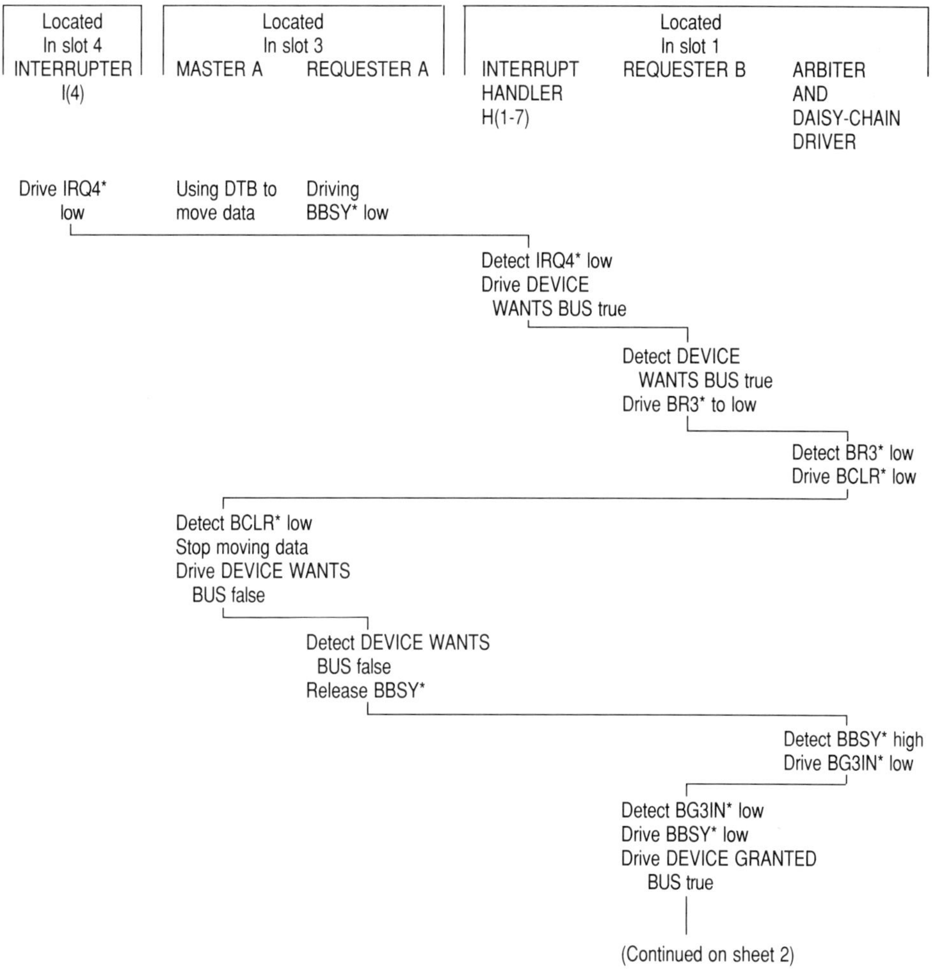

Figure 4-14. Typical Single Handler Interrupt System Operation Flow Diagram (Sheet 1 of 2)

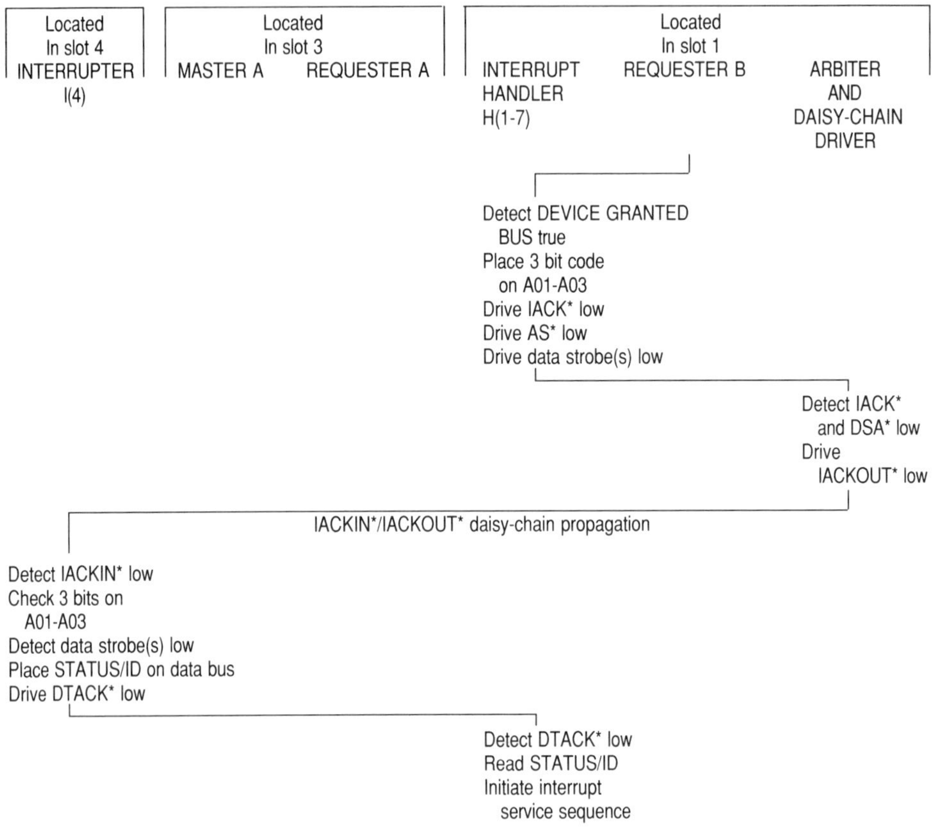

Figure 4-14. Typical Single Handler Interrupt System Operation Flow Diagram (Sheet 2 of 2)

Table 4-7. 3-Bit Interrupt Acknowledge Code

Interrupt line being acknowledged	Use of the address bus to broadcast the 3-bit interrupt acknowledge code		
	A03	A02	A01
IRQ1*	L	L	H
IRQ2*	L	H	L
IRQ3*	L	H	H
IRQ4*	H	L	L
IRQ5*	H	L	H
IRQ6*	H	H	L
IRQ7*	H	H	H

H = High level
L = Low level

4.4.4 Example: Prioritization of Two Interrupts in a Distributed Interrupt System

Figure 4-15 illustrates the operation of a distributed interrupt system with two INTERRUPT HANDLERS. INTERRUPT HANDLER A monitors IRQ1*-IRQ4*, while INTERRUPT HANDLER B monitors IRQ5*-IRQ7*. INTERRUPT HANDLER A treats IRQ4* as its highest priority interrupt, while INTERRUPT HANDLER B treats IRQ7* as its highest priority interrupt. At the top of the diagram, INTERRUPTER C drives IRQ3* low, and INTERRUPTER D drives IRQ6* low. Both INTERRUPT HANDLERS detect their respective interrupt request lines low, and both simultaneously indicate to their on-board REQUESTER that they need the DTB. Both the REQUESTERS drive BR3* low. Upon detecting BR3* low, the DTB ARBITER drives BG3IN* low on slot 1. This low signal is passed down the BG3IN*/BG3OUT* daisy-chain until it is detected by the REQUESTER B in slot 4. This REQUESTER then signals its on-board INTERRUPT HANDLER B that the DTB is available. INTERRUPT HANDLER B then reads the STATUS/ID from INTERRUPTER D.

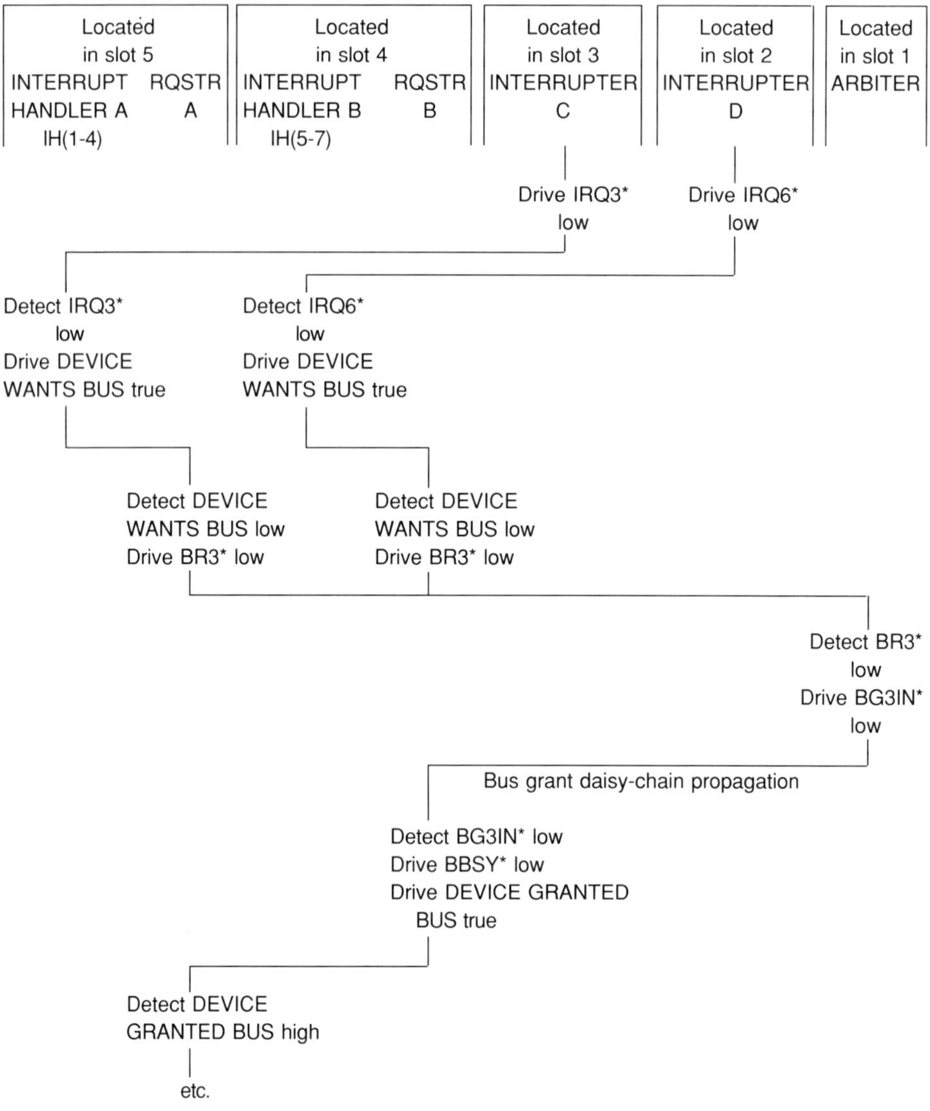

Figure 4-15. Typical Distributed Interrupt System With Two INTERRUPTHANDLERS, Flow Diagram

4.5 PRIORITY INTERRUPT BUS TIMING RULES AND OBSERVATIONS

This section describes the timing RULES and OBSERVATIONS that govern the behavior of INTERRUPT HANDLERS, INTERRUPTERS, and IACK DAISY-CHAIN DRIVERS during the selection of the responding INTERRUPTER (i.e. the INTERRUPTER that is to provide its STATUS/ID in response to the interrupt acknowledge cycle). This timing information is in the form of figures and tables.

The interrupt acknowledge cycle begins with the selection of the responding INTERRUPTER. This is called the INTERRUPTER selection portion of the cycle. Once an INTERRUPTER responds, the INTERRUPT HANDLER reads the STATUS/ID from it. This is called the STATUS/ID transfer portion of the cycle.

When the INTERRUPT HANDLER initiates an interrupt acknowledge cycle, there might be INTERRUPTERS between it and the INTERRUPTER being acknowledged that either:

- **a.** do not have an interrupt pending,
- **b.** have an interrupt pending on a different interrupt request line than the one being acknowledged.

Although these INTERRUPTERS do not respond with a STATUS/ID, they do participate in the interrupt acknowledge cycle by passing the falling edge from their IACKIN* line on to their IACKOUT* line. For this reason, these INTERRUPTERS are called participating INTERRUPTERS.

The first INTERRUPTER in the daisy-chain that has an interrupt pending on the interrupt request line that is being acknowledged responds with a STATUS/ID. For this reason it is called the responding INTERRUPTER. All other INTERRUPTERS are called non-participating INTERRUPTERS.

Table 4-8 lists timing Tables and timing diagrams that specify INTERRUPT HANDLER and INTERRUPTER operation.

Table 4-9 lists timing Tables and timing diagrams that specify IACK DAISY-CHAIN DRIVER operation.

Table 4-10 lists timing Tables and timing diagrams that specify participating INTERRUPTER operation.

Table 4-11 lists timing Tables and timing diagrams that specify responding INTERRUPTER operation.

Table 4-12 defines the mnemonics that are used in **Tables 4-13 through 4-15**.

Tables 4-13 through 4-15 specify the use of bus signal lines by the Priority Interrupt Bus functional modules.

Tables 4-16 through 4-19 specify the timing parameters for the Priority Interrupt Bus functional modules. (The reference numbers used in Tables 4-17 through 4-19 correspond to the timing parameter numbers in Table 4-16.)

Figures 4-16 through 4-23 are timing diagrams that specify the timing during interrupt acknowledge cycles.

Figure 4-24 specifies additional inter-cycle timing for the IACKIN*/IACKOUT* daisy-chain.

All of the RULES and OBSERVATIONS associated with the figures listed below also apply to INTERRUPT HANDLERS, INTERRUPTERS, and IACK DAISY-CHAIN DRIVERS.

Figures 2-22 through 2-24 in Chapter 2 specify the timing for the address and data strobes between data transfer cycles.

Figure 2-25 specifies the timing of a timed-out cycle.

Figures 2-26 and 2-27 specify the timing during mastership transfer.

In order to meet the specified timing, board designers have to take into account the worst case propagation delays of the bus drivers and receivers used on their VMEbus boards. The propagation delay of the drivers depends on their output loads. However, manufacturer specifications do not always give enough information to calculate the propagation delays under various loads. To help the VMEbus board designer, some suggestions are offered in Chapter 6.

The OBSERVATIONS specify the timing of signal transitions on incoming lines. These times can be relied upon as long as the backplane loading RULES in Chapter 6 are not violated. The RULES for the bus terminators in Chapter 6 guarantee that the timing parameters for signal lines that are released after they have been driven, are met.

Typically, for each timing RULE there is a corresponding timing OBSERVATION. However, the time that is guaranteed in the OBSERVATION might differ from the time specified by the RULE. For example, a careful inspection of the timing diagrams shows that the INTERRUPT HANDLER is required to provide 35 nanoseconds of address set-up time, but the INTERRUPTER is only guaranteed 10 nanoseconds. This is because the address drivers are not always able to drive the backplane's signal lines completely through the low to high threshold region, until the transition propagates to the end of the backplane and is reflected back. The falling edge of the address strobe, however, typically crosses the 0.8-volt threshold without waiting for a reflection. Therefore, the resulting set-up time at the INTERRUPTER is the INTERRUPT HANDLER'S set-up time less two bus propagation times.

A special notation has been used to describe the data strobe timing. The two data strobes (DS0* and DS1*) do not always make their transitions simultaneously. For purposes of these timing diagrams, DSA* represents the first data strobe to make its transition (whether that is DS0* or DS1*). DSB* represents the second data strobe to make its transition (whether that is DS1* or DS0*). The broken line shown while the data strobes are stable indicates that the first data strobe to make a falling transition might not be the first to make its rising transition -- i.e., DSA* might represent DS0* on its falling edge and DS1* on its rising edge.

Table 4-8. Timing Diagrams That Define INTERRUPT HANDLER And INTERRUPTER Operation

Mnemonic	Type of cycle	INTERRUPTER Selection Timing Diag Figures	STATUS/ID Transfer Timing Diag Figure
D08(O)	SINGLE BYTE STATUS/ID READ	2-12 & 4-16	4-20
D16	DOUBLE BYTE STATUS/ID READ	2-12 & 4-16	4-21
D32	QUAD BYTE STATUS/ID READ	2-12 & 4-16	4-21

Note: See Tables 4-16 and 4-17 for timing parameters

Table 4-9. Timing Diagrams That Define IACK DAISY-CHAIN DRIVER Operation

Type of cycle	INTERRUPTER Selection Timing Diagram
SINGLE BYTESTATUS/ID READ	Figure 4-17
DOUBLE BYTE STATUS/ID READ	Figure 4-17
QUAD BYTE STATUS/ID READ	Figure 4-17

Note: See Tables 4-16 and 4-19 for timing values

Table 4-10. Timing Diagrams That Define Participating INTERRUPTER Operation

Type of cycle	INTERRUPTER Selection Timing Diagram
SINGLE BYTE STATUS/ID READ	Figure 4-18
DOUBLE BYTE STATUS/ID READ	Figure 4-18
QUAD BYTE STATUS/ID READ	Figure 4-18

Note: See Tables 4-16 and 4-18 for timing values

Table 4-11. Timing Diagrams That Define Responding INTERRUPTER Operation

Mnemonic	Type of cycle	INTERRUPTER selection Timing Diag	STATUS/ID Transfer Timing Diag
D08(O)	SINGLE BYTE STATUS/ID READ	Figure 4-19	Figure 4-22
D16	DOUBLE BYTE STATUS/ID READ	Figure 4-19	Figure 4-23
D32	QUAD BYTE STATUS/ID READ	Figure 4-19	Figure 4-23

Notes: See Tables 4-16 and 4-18 for timing parameters

Table 4-12. Definitions Of Mnemonics Used In Tables 4-13, 4-14, And 4-15

Mnemonic	Description	Comments
DVBIH	DRIVEN VALID BY INTERRUPT HANDLER	RULE 4.9: The INTERRUPT HANDLER **MUST** drive DVBIH lines to a valid level.
DLBIH	DRIVEN LOW BY INTERRUPT HANDLER	RULE 4.10: The INTERRUPT HANDLER **MUST** drive DLBIH lines to a low level.
DHBIH	DRIVEN HIGH BY INTERRUPT HANDLER	RULE 4.11: The INTERRUPT HANDLER **MUST** drive DHBIH lines to a high level.
dhbih?	DRIVEN HIGH BY INTERRUPT HANDLER?	PERMISSION 4.7: The INTERRUPT HANDLER **MAY** drive dhbih? lines high. RULE 4.12: The INTERRUPT HANDLER **MUST NOT** drive dhbih? lines low during the cycle.
dxbih?	DRIVEN BY INTERRUPT HANDLER?	PERMISSION 4.8: The INTERRUPT HANDLER **MAY** drive dxbih? lines during the cycle, or it **MAY** leave dxbih? lines undriven. (When the line is driven it carries no valid information.)
DVBI	DRIVEN VALID BY INTERRUPTER	RULE 4.13: The INTERRUPTER **MUST** drive DVBI lines a to valid level.
dhbi?	DRIVEN BY INTERRUPTER?	PERMISSION 4.9: The INTERRUPTER **MAY** drive dhbi? lines high. RULE 4.14: The INTERRUPTER **MUST NOT** drive dhbi? lines low.

(Continued on Page 164)

Table 4-12. (Continued)

Mnemonic	Description	Comments
dxbi?	DRIVEN BY INTERRUPTER?	PERMISSION 4.10: The INTERRUPTER **MAY** drive dxbi? lines during the cycle, or it **MAY** leave the line undriven. (When the line is driven, it carries no valid information.)

Table 4-13. Use Of Addressing Lines During Interrupt Acknowledge Cycles

Interrupt line being acknowledged	A03	A02	A01	IACK*
IRQ1*	DLBIH	DLBIH	DHBIH	DLBIH
IRQ2*	DLBIH	DHBIH	DLBIH	DLBIH
IRQ3*	DLBIH	DHBIH	DHBIH	DLBIH
IRQ4*	DHBIH	DLBIH	DLBIH	DLBIH
IRQ5*	DHBIH	DLBIH	DHBIH	DLBIH
IRQ6*	DHBIH	DHBIH	DLBIH	DLBIH
IRQ7*	DHBIH	DHBIH	DHBIH	DLBIH

Table 4-14. Use Of The DS1*, DS0*, LWORD*, And WRITE* Lines During Interrupt Acknowledge Cycles

Mnemonic	Type of cycles	DS1*	DS0*	LWORD*	WRITE*
D08(O)	Single byte interrupt acknowledge	dhbih?	DLBIH	dhbih?	dhbih?
D16	Double byte interrupt acknowledge	DLBIH	DLBIH	dhbih?	dhbih?
D32	Quad byte interrupt acknowledge	DLBIH	DLBIH	DLBIH	dhbih?

Table 4-15. Use of the data bus lines to transfer the STATUS/ID

Mnemonic	Type of cycles	D24-D31	D16-D23	D08-D15	D00-D07
D08(O)	Single, double, and quad byte interrupt acknowledge	dhbi?	dhbi?	dhbi?	DVBI
D16	Double and quad byte interrupt acknowledge	dhbi?	dhbi?	DVBI	DVBI
D32	Quad byte interrupt acknowledge	DVBI	DVBI	DVBI	DVBI

Table 4-16. INTERRUPT HANDLER, INTERRUPTER, And IACK DAISY-CHAIN DRIVER Timing Parameters

PARAMETER NUMBER	INTERRUPT HANDLER See Table 4-17		INTERRUPTER See Table 4-18		IACK DAISY-CHAIN DRIVER See Table 4-19	
	MIN	MAX	MIN	MAX	MIN	MAX
1	0					
2	0					
3	60					
4	35		10			
5	40		30		30	
6			0			
7			0			
9	0		0			
10	0		-10 Note 3			
11	40		30			
12	35		10			
13		10		20		
14	0		0			
16	0		0			
18	0		0			
19	40		30		30	
20	0		0			

(Continued on Page 166)

Table 4-16. (Continued)

PARAMETER NUMBER	INTERRUPT HANDLER See Table 4-17 MIN	MAX	INTERRUPTER See table 4-18 MIN	MAX	IACK DAISY-CHAIN DRIVER See Table 4-19 MIN	MAX
21	0		0			
23	10		0			
24A	0					
24B	0					
25		25				
26	0		0			
27	-25		0			
28	30	2T	30			
29	0		0			
30	0		0			
31	0		0			
32			10		10	
34			40		40	
35				30	0 Note 3	30
36			0			
37			0			
38A			0			
38B			0			
39				40		
40			30		30	
41			0			
42					30	
43			30			

NOTES:

1. All times are in nanoseconds.
2. T = the time-out value.
3. This value was not in the previous edition.

Table 4-17. INTERRUPT HANDLER, Timing RULES And OBSERVATIONS

Note: The numbers correspond to the timing parameters specified in Table 4-16 on pages165 and 166.

1.	**RULE 4.15:** When taking control of the VMEbus, the INTERRUPT HANDLER **MUST NOT** drive any of IACK*, A01-A03, LWORD*, WRITE*, DS0*, DS1* or AS* until after the previous MASTER or INTERRUPT HANDLER allows AS* to rise above the low level. **OBSERVATION 4.12:** Chapter 3 describes how an INTERRUPT HANDLER'S REQUESTER is granted use of the DTB.
2.	**RULE 4.16:** When taking control of the DTB, the INTERRUPT HANDLER **MUST NOT** drive any of IACK*, A01-A03, LWORD*, WRITE*, DS0*, DS1*, or AS* until after it receives DEVICE GRANTED BUS true. **OBSERVATION 4.13:** Chapter 3 describes how an INTERRUPT HANDLER'S REQUESTER is granted use of the DTB.
3.	**RULE 4.17:** When taking control of the DTB, the INTERRUPT HANDLER **MUST NOT** drive AS* low until this time after the previous MASTER or INTERRUPT HANDLER allows AS* to rise above the low level. **OBSERVATION 4.14:** RULE 4.17 ensures that timing parameter 5 for INTERRUPTERS and SLAVES is guaranteed when there is an interchange of the DTB mastership.
4.	**RULE 4.18:** The INTERRUPT HANDLER **MUST NOT** drive AS* low until IACK* has been low, and LWORD* and A01-A03 have been valid for this minimum time.
5.	**RULE 4.19:** When using the DTB for two consecutive cycles, the INTERRUPT HANDLER **MUST NOT** drive AS* low until it has been high for this minimum time.
9.	**RULE 4.20:** The INTERRUPT HANDLER **MUST NOT** drive DSA* low until both DTACK* and BERR* are high.
10.	**RULE 4.21:** The INTERRUPT HANDLER **MUST NOT** drive DSA* low before it has driven AS* low.

(Continued on Page 168)

Table 4-17. (Continued)

11.	**RULE 4.22:** The INTERRUPT HANDLER **MUST NOT** drive DSA* low until DS0* and DS1* have been simultaneously high for this minimum time.
12.	**RULE 4.23:** The INTERRUPT HANDLER **MUST NOT** drive DSA* low until WRITE* has been high for this minimum time.
13.	**RULE 4.24:** During double byte or quad byte STATUS/ID read cycles, the INTERRUPT HANDLER **MUST** drive DSB* low within this maximum time after it drives DSA* low. **OBSERVATION 4.15:** Timing parameter 13 does not apply to single byte STATUS/ID reads.
14.	**RULE 4.25:** During all interrupt acknowledge cycles, the INTERRUPT HANDLER **MUST** hold the 3 bit interrupt acknowledge code valid, and maintain the appropriate level on LWORD* until it detects a falling edge on DTACK* or BERR*.
16.	**RULE 4.26:** During all interrupt acknowledge cycles, the INTERRUPT HANDLER **MUST** maintain IACK* low until it detects a falling edge on DTACK* or BERR*.
18.	**RULE 4.27:** The INTERRUPT HANDLER **MUST** hold AS* low until it detects DTACK* or BERR* low.
19.	**RULE 4.28:** The INTERRUPT HANDLER **MUST** hold AS* low for this minimum time.
20.	**RULE 4.29:** Once an INTERRUPT HANDLER has driven DSA* low, it **MUST** maintain it low until it detects DTACK* or BERR* low.
21.	**RULE 4.30:** Once an INTERRUPT HANDLER has driven DSB* low, it **MUST** maintain it low until it detects DTACK* or BERR* low.
23.	**RULE 4.31:** Once an INTERRUPT HANDLER has driven DSA* low, it **MUST** maintain a high on the WRITE* line, until this minimum time after both data strobes are high.

(Continued on Page 169)

Table 4-17. (Continued)

24A.	**RULE 4.32:** **IF** the INTERRUPT HANDLER drives or releases AS* to high after its REQUESTER releases BBSY*, **THEN** it **MUST** release IACK*, A01-A03, LWORD*, WRITE*, DS0*, and DS1* before allowing AS* to rise above the low level.
24B.	**RULE 4.33:** **IF** the INTERRUPT HANDLER drives or releases AS* to high before its REQUESTER releases BBSY*, **THEN** it **MUST** release IACK*, A01-A03, LWORD*, WRITE*, DS0*, and DS1* before changing its DEVICE WANTS BUS signal from true to false.
25.	**RULE 4.34:** **IF** the INTERRUPT HANDLER drives or releases AS* to high after its REQUESTER releases BBSY*, **THEN** it **MUST** release AS* within this time after allowing it to rise above the low level.
26.	**OBSERVATION 4.16:** Timing parameter 26 guarantees that the data bus will not be driven until the INTERRUPT HANDLER drives DSA* low.
27.	**OBSERVATION 4.17:** The INTERRUPT HANDLER is guaranteed that the data bus will be valid within this time after DTACK* goes low. This time does not apply to cycles where the INTERRUPTER drives BERR* low instead of DTACK*.
28.	**OBSERVATION 4.18:** The INTERRUPT HANDLER is guaranteed that neither DTACK* nor BERR* will go low until this minimum time after it drives DSA* low. The BUS TIMER guarantees the INTERRUPT HANDLER that if DTACK* has not gone low after its time-out period has elapsed, and within twice its time-out period, then the BUS TIMER will drive BERR* low.
29.	**OBSERVATION 4.19:** The INTERRUPT HANDLER is guaranteed that the data bus remains valid until it drives DSA* high.
30.	**OBSERVATION 4.20:** This guarantees that neither DTACK* nor BERR* goes high until the INTERRUPT HANDLER drives both DS0* and DS1* high.
31.	**OBSERVATION 4.21:** The INTERRUPT HANDLER is guaranteed that the data bus has been released by the time DTACK* and BERR* are high.

(Continued on Page 170)

Table 4-18. INTERRUPTER, Timing RULES And OBSERVATIONS

Note: The numbers correspond to the timing parameters specified in Table 4-16.

4.	**OBSERVATION 4.22:** INTERRUPTERS are guaranteed that IACK*, LWORD*, and A01-A03 have been valid for this minimum time when they detect a falling edge on AS*.
5.	**OBSERVATION 4.23:** All INTERRUPTERS are guaranteed this minimum high time on AS* between DTB cycles.
6.	**OBSERVATION 4.24:** The responding INTERRUPTER is guaranteed that none of D00-D31 will be driven by any other module until the responding INTERRUPTER releases DTACK* and BERR* to high.
7.	**OBSERVATION 4.25:** The responding INTERRUPTER is guaranteed that the data bus will be released by all other modules by the time DSA* goes low.
9.	**OBSERVATION 4.26:** The responding INTERRUPTER is guaranteed that neither DS0* nor DS1* will go low until DTACK* and BERR* from the previous cycle have gone high.
11.	**OBSERVATION 4.27:** INTERRUPTERS are guaranteed this minimum time during which both data strobes are simultaneously high between cycles.
12.	**OBSERVATION 4.28:** INTERRUPTERS are guaranteed that WRITE* has been high for this minimum time when they detect a falling edge on DSA*.
13.	**OBSERVATION 4.29:** **IF** both data strobes are going to be driven low, **THEN** the responding INTERRUPTER is guaranteed that DSB* will go low within this maximum time after DSA*. And therefore: **IF** the DSB* does not go low within this maximum time, **THEN** the responding INTERRUPTER assumes that it is to respond with a single byte STATUS/ID.

(Continued on Page 171)

Table 4-18. (Continued)

14.	**OBSERVATION 4.30:** The responding INTERRUPTER is guaranteed that LWORD* and A01-A03 will remain valid until it drives DTACK* or BERR* low, provided it does so within the bus time-out period.
16.	**OBSERVATION 4.31:** The responding INTERRUPTER is guaranteed that IACK* will remain low until it drives DTACK* or BERR* low, provided it does so within the bus time-out period.
18.	**OBSERVATION 4.32:** The responding INTERRUPTER is guaranteed that AS* will remain low until it drives DTACK* or BERR* low, provided that it does so within the bus time-out period.
19.	**OBSERVATION 4.33:** INTERRUPTERS are guaranteed that the AS* will remain low for this minimum time.
20.	**OBSERVATION 4.34:** The responding INTERRUPTER is guaranteed that once DSA* goes low, it will remain low until the INTERRUPTER has driven DTACK* or BERR* low, provided that the INTERRUPTER does so within the bus time-out period.
21.	**OBSERVATION 4.35:** The responding INTERRUPTER is guaranteed that once DSB* goes low, it will remain low until the INTERRUPTER has driven DTACK* or BERR* low, provided that the INTERRUPTER does so within the bus time-out period.
23.	**OBSERVATION 4.36:** INTERRUPTERS are guaranteed that the WRITE* line remains high until both data strobes are high.
26.	**RULE 4.35:** The INTERRUPTER **MUST NOT** drive the data bus until DSA* goes low.
27.	**RULE 4.36:** The responding INTERRUPTER **MUST NOT** drive DTACK* low before it drives the data lines with a valid STATUS/ID. **OBSERVATION 4.37:** This time does not apply to cycles where the INTERRUPTER drives BERR* low instead of DTACK*.

(Continued on Page 172)

Table 4-18. (Continued)

28.	**RULE 4.37:** The responding INTERRUPTER **MUST** wait this minimum time after DSA* goes low before driving DTACK* or BERR* low.
29.	**RULE 4.38:** Once the responding INTERRUPTER has driven DTACK* low, it **MUST NOT** change D00-D31 until DSA* goes high.
30.	**RULE 4.39:** Once the responding INTERRUPTER has driven DTACK* or BERR* to low, it **MUST NOT** release it until it detects both DS0* and DS1* high.
31.	**RULE 4.40:** The responding INTERRUPTER **MUST** release all of D00-D31 before releasing DTACK* and BERR* to high.
32.	**OBSERVATION 4.38:** The responding INTERRUPTER is guaranteed that IACK*, LWORD*, and A01-A03 have been valid for this minimum time when it detects a falling edge on DSA*. This time is derived from timing parameters 4 and 10.
34.	**OBSERVATION 4.39:** The INTERRUPTER is guaranteed that AS* has been low for this minimum time, when it detects a falling edge on IACKIN*.
35.	**RULE 4.41:** A participating INTERRUPTER **MUST** drive its IACKOUT* high within this maximum time after the rising edge on AS*.
36.	**RULE 4.42:** The INTERRUPTER **MUST NOT** drive the data bus until its IACKIN* line goes low.
37.	**RULE 4.43:** **IF** a participating INTERRUPTER drives the data bus, **THEN** it **MUST** release it before driving its IACKOUT* line low.
38A.	**RULE 4.44:** A participating INTERRUPTER **MUST NOT** drive its IACKOUT* line low, until it detects its IACKIN* line low.
38B.	**RULE 4.45:** The responding INTERRUPTER **MUST NOT** drive its DTACK* line low, until it detects its IACKIN* line low.

(Continued on Page 173)

Table 4-18. (Continued)

39.	**OBSERVATION 4.40:** This time guarantees that each INTERRUPTER'S IACKIN* line will go high within this time after the rising edge on AS*. This time is derived from timing parameter 35, where the IACK DAISY-CHAIN DRIVER and participating INTERRUPTERS are required to drive their IACKOUT* line high within a maximum time.
40.	**OBSERVATION 4.41:** All INTERRUPTERS are guaranteed that their IACKIN* line will stay high for this minimum time between consecutive DTB cycles.
41.	**OBSERVATION 4.42:** This time guarantees that A01-A03 and LWORD* remain valid until this time after the participating INTERRUPTER drives its IACKOUT* low, provided it does so within the bus time-out period.
43.	**OBSERVATION 4.43:** This time guarantees that AS* remains low for this minimum time after the participating INTERRUPTER drives its IACKOUT* low, provided it does so within the bus time-out period.

Table 4-19. IACK DAISY-CHAIN DRIVER, Timing RULES And OBSERVATIONS

Note: The numbers correspond to the timing parameters specified in Table 4-16.

OBSERVATION 4.44:
Since the backplane connects IACK* to the slot 1 IACKIN*, these two signals are equivalent. And therefore, all RULES and OBSERVATIONS that apply to one, are applicable to the other as well.

5.	**OBSERVATION 4.45:** The IACK DAISY-CHAIN DRIVER is guaranteed this minimum high time on AS* between DTB cycles.
19.	**OBSERVATION 4.46:** The IACK DAISY-CHAIN DRIVER is guaranteed that the AS* will remain low for this minimum time. This time is derived from timing parameters 8, 16, and 27 of the INTERRUPTER.
32.	**OBSERVATION 4.47:** The IACK DAISY-CHAIN DRIVER is guaranteed that IACK* (and the slot 1 IACKIN*) has been valid for this minimum time when it detects a falling edge on DSA*.
34.	**RULE 4.46:** **IF** the IACKIN* line is low when the IACK DAISY-CHAIN DRIVER detects a falling edge on DSA*, **THEN** it **MUST** drive its IACKOUT* line low, but it **MUST NOT** do so until this time after the falling edge on DSA*. **OBSERVATION 4.48:** The IACK DAISY-CHAIN DRIVER does not drive its IACKOUT* line low every time DSA* goes low. It only does so when the IACK* line is low as well, indicating that an interrupt acknowledge cycle is in progress.
35.	**RULE 4.47:** **IF** the IACK DAISY-CHAIN DRIVER drives its IACKOUT* line low, **THEN** it **MUST** drive its IACKOUT* high within this time after the rising edge of AS*.
40.	**RULE 4.48:** The IACK DAISY-CHAIN DRIVER **MUST NOT** drive IACKOUT* low until it has been high for this minimum time.
42.	**OBSERVATION 4.49:** **IF** the IACK DAISY-CHAIN DRIVER drives its IACKOUT* line low within the bus time-out period, **THEN** this time guarantees that IACK* (and the slot 1 IACKIN*) remains valid for this minimum time.

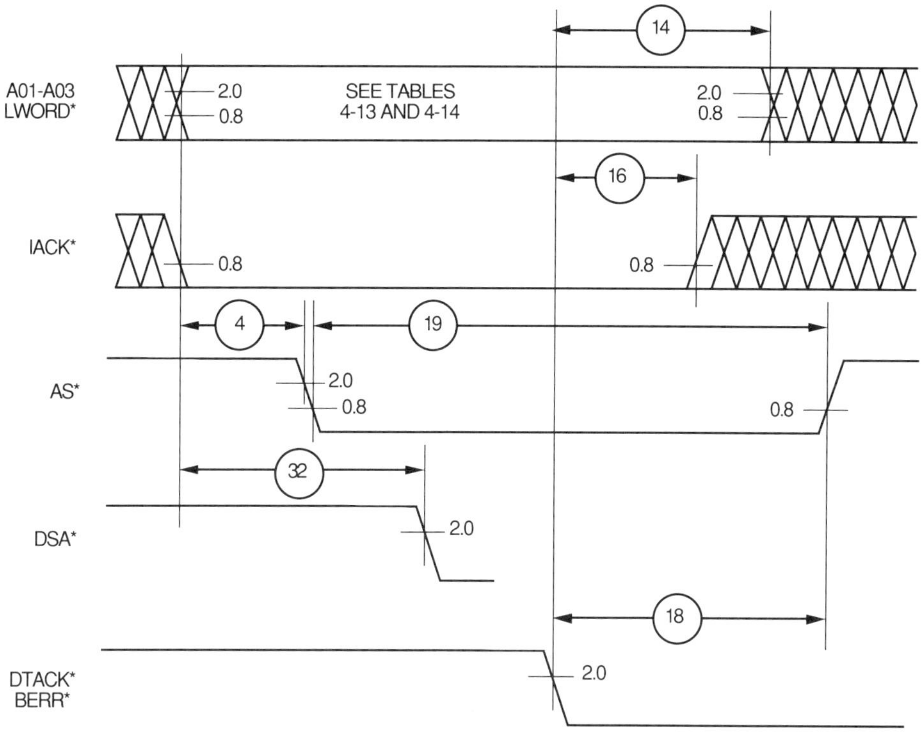

Figure 4-16. INTERRUPT HANDLER And INTERRUPTER
INTERRUPTER Selection Timing
SINGLE, DOUBLE And QUAD BYTE INTERRUPT ACKNOWLEDGE CYCLES

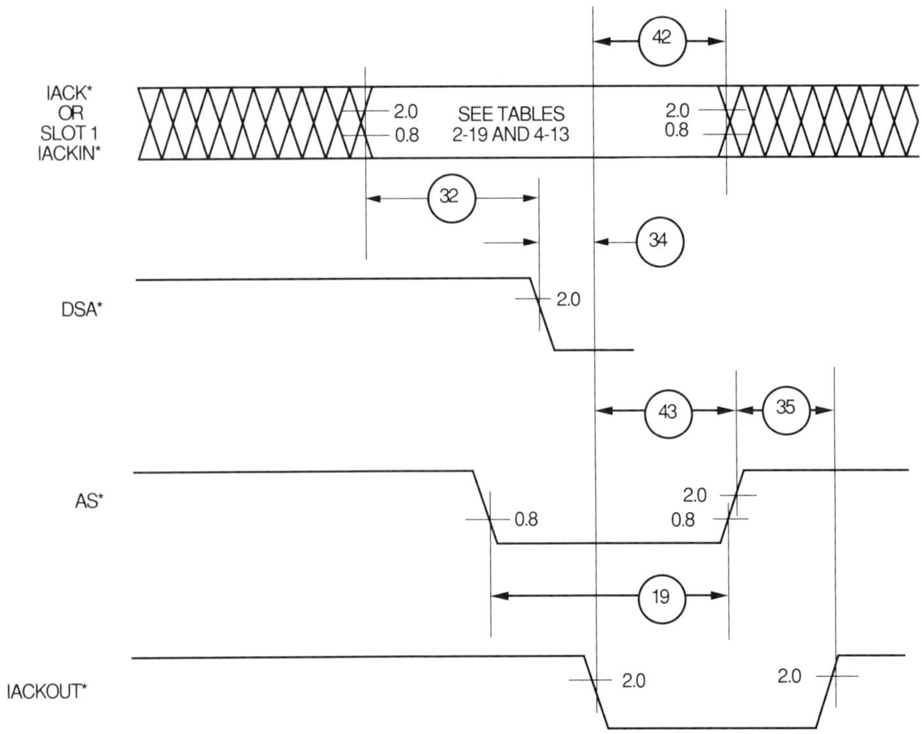

Figure 4-17. IACK DAISY-CHAIN DRIVER
INTERRUPTER Selection Timing
SINGLE, DOUBLE, And QUAD BYTE INTERRUPT ACKNOWLEDGE CYCLES

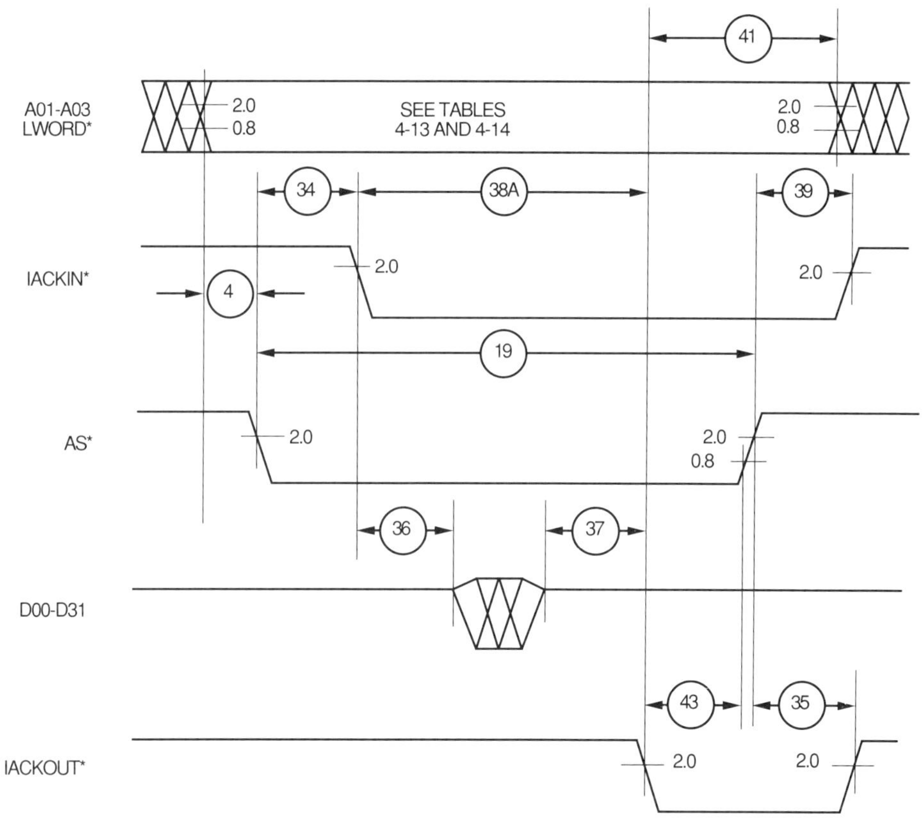

Figure 4-18. Participating INTERRUPTER
INTERRUPTER Selection Timing
SINGLE, DOUBLE, And QUAD BYTE INTERRUPT ACKNOWLEDGE CYCLES

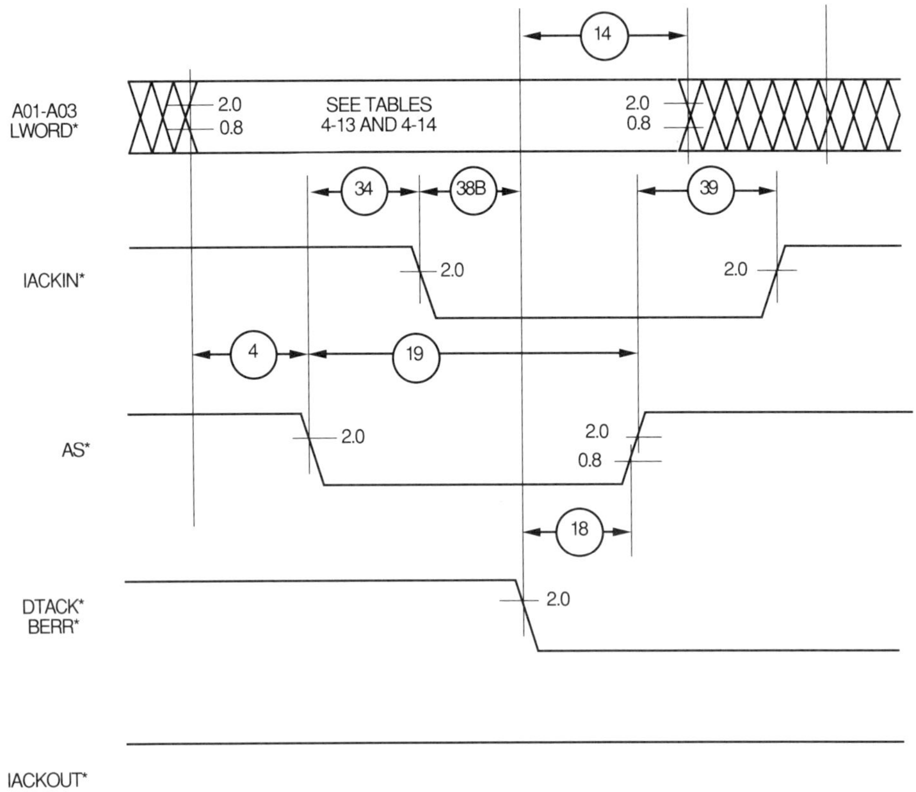

Figure 4-19. Responding INTERRUPTER
INTERRUPTER Selection Timing
SINGLE, DOUBLE, And QUAD BYTE INTERRUPT ACKNOWLEDGE CYCLES

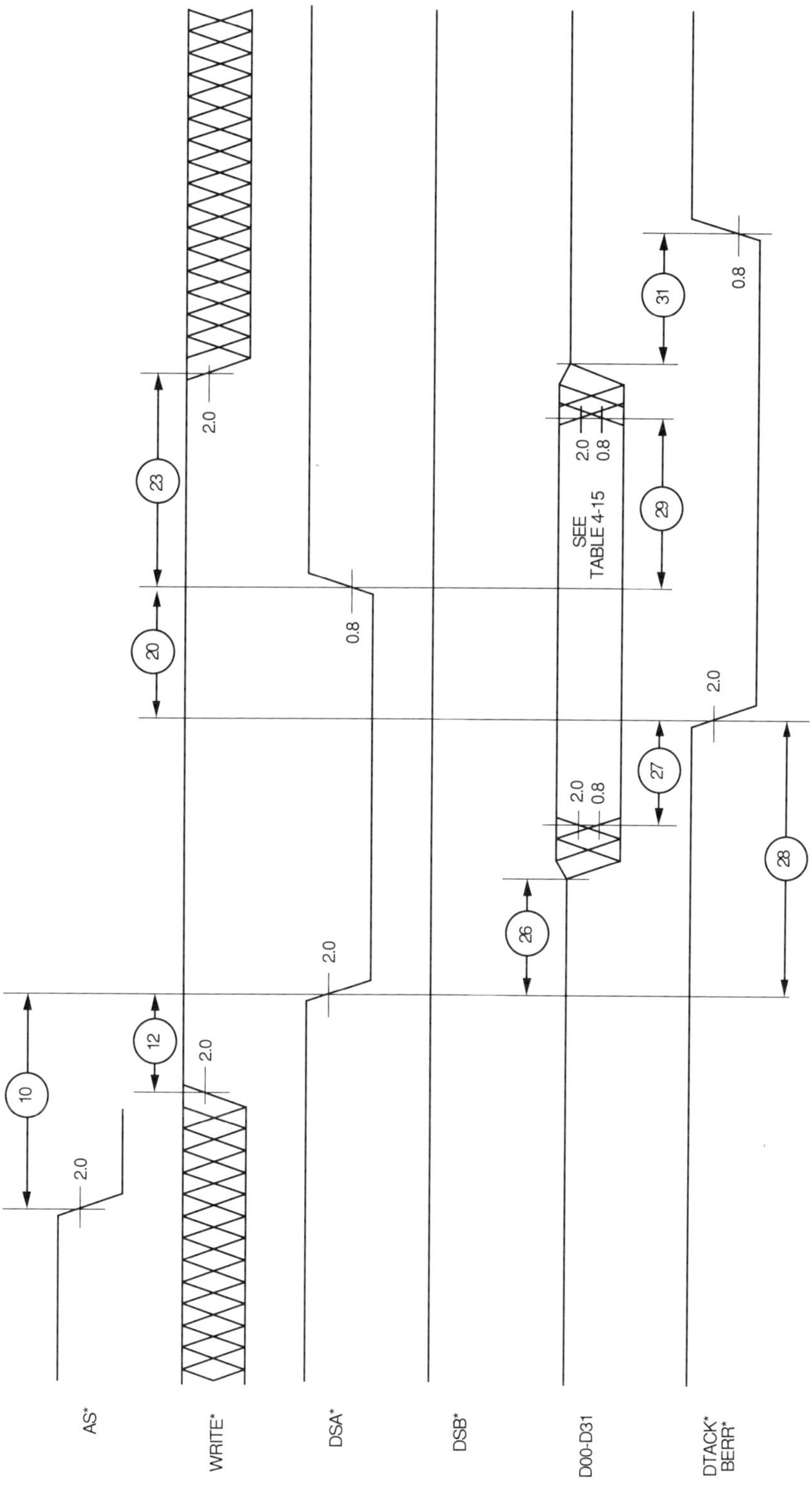

Figure 4-20. INTERRUPT HANDLER
STATUS/ID Transfer Timing
SINGLE BYTE INTERRUPT ACKNOWLEDGE CYCLE

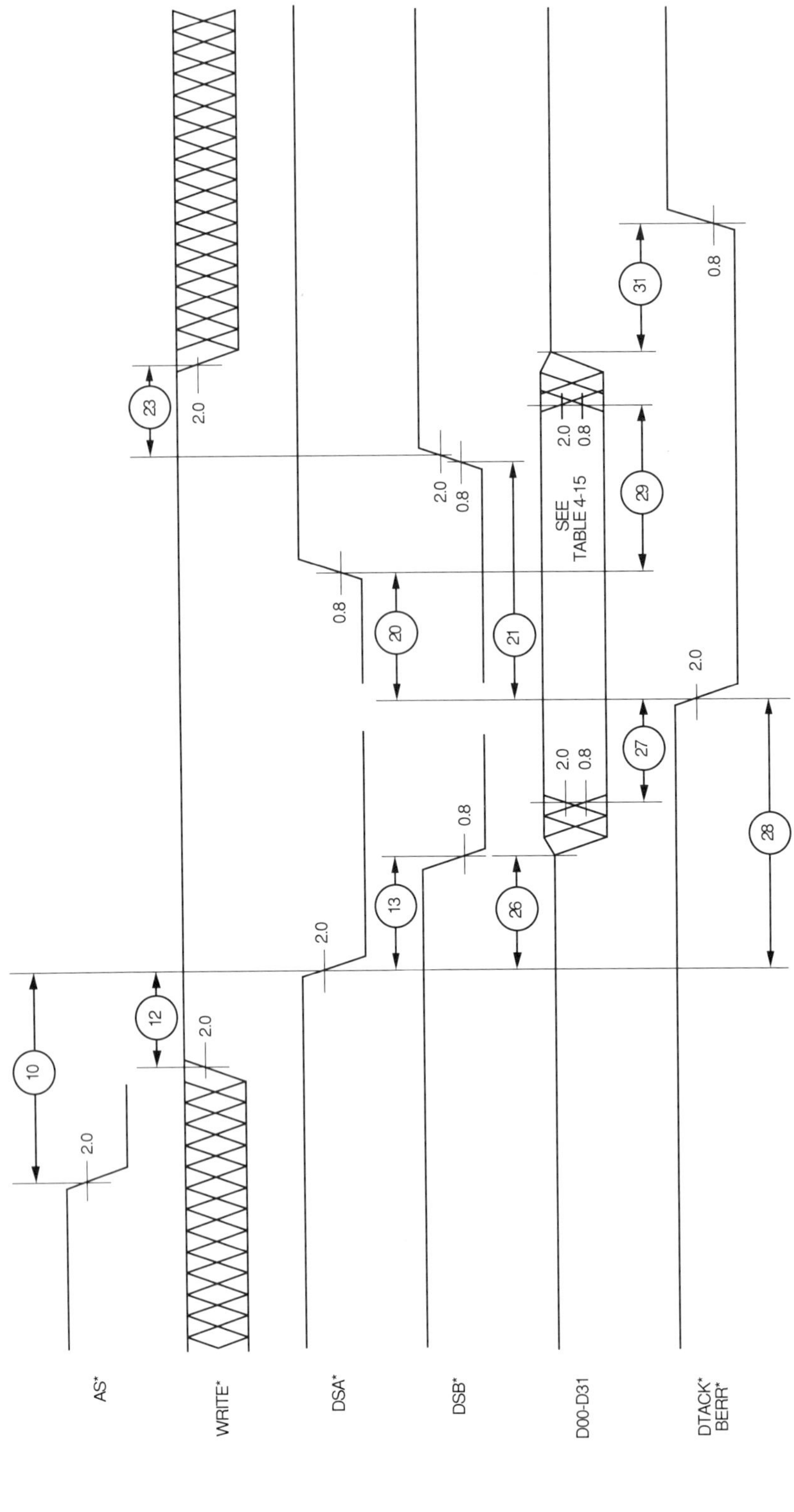

Figure 4-21. INTERRUPT HANDLER
STATUS/ID Transfer Timing
DOUBLE BYTE INTERRUPT ACKNOWLEDGE CYCLE
QUAD BYTE INTERRUPT ACKNOWLEDGE CYCLE

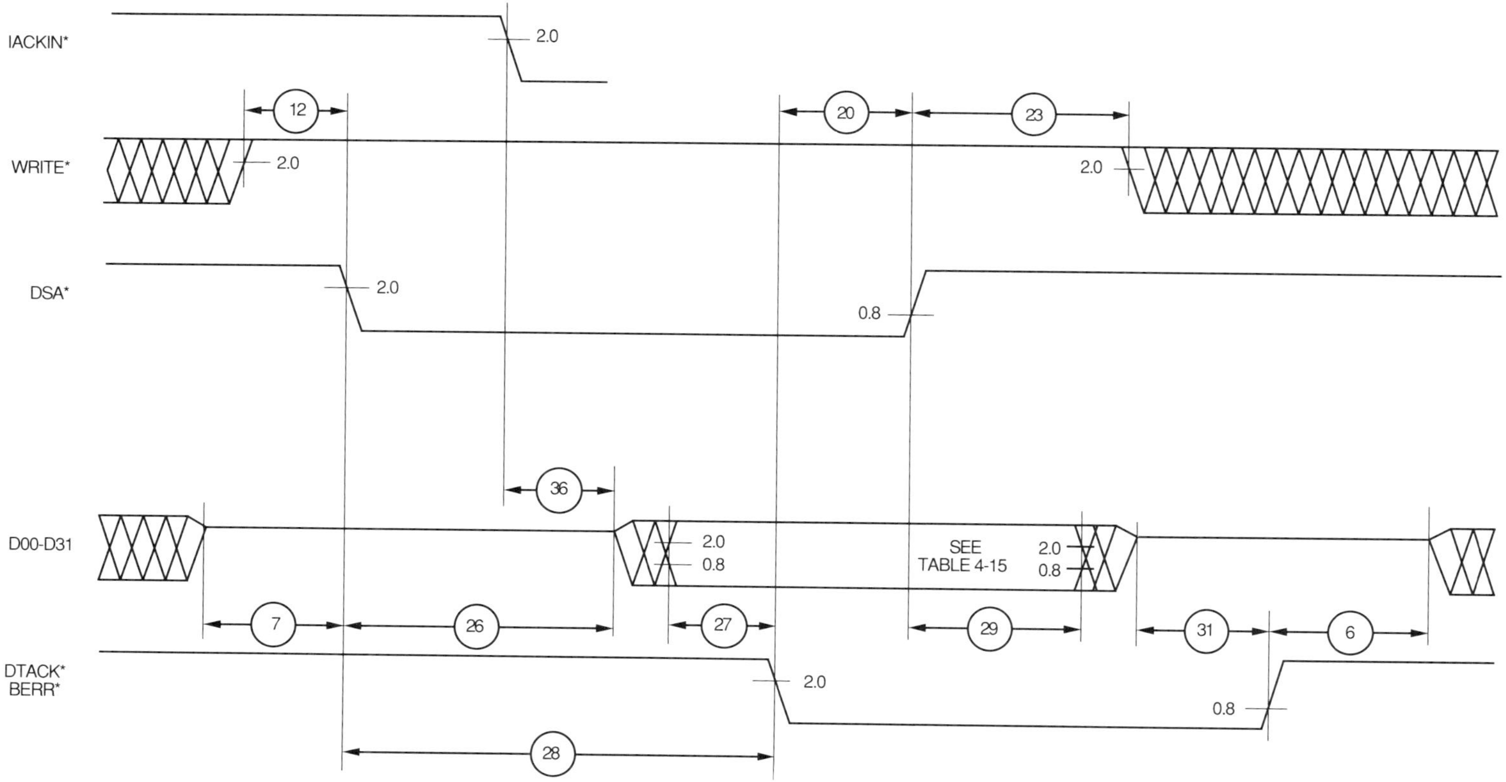

Figure 4-22. Responding INTERRUPTER
STATUS/ID Transfer Timing
SINGLE BYTE INTERRUPT ACKNOWLEDGE CYCLE

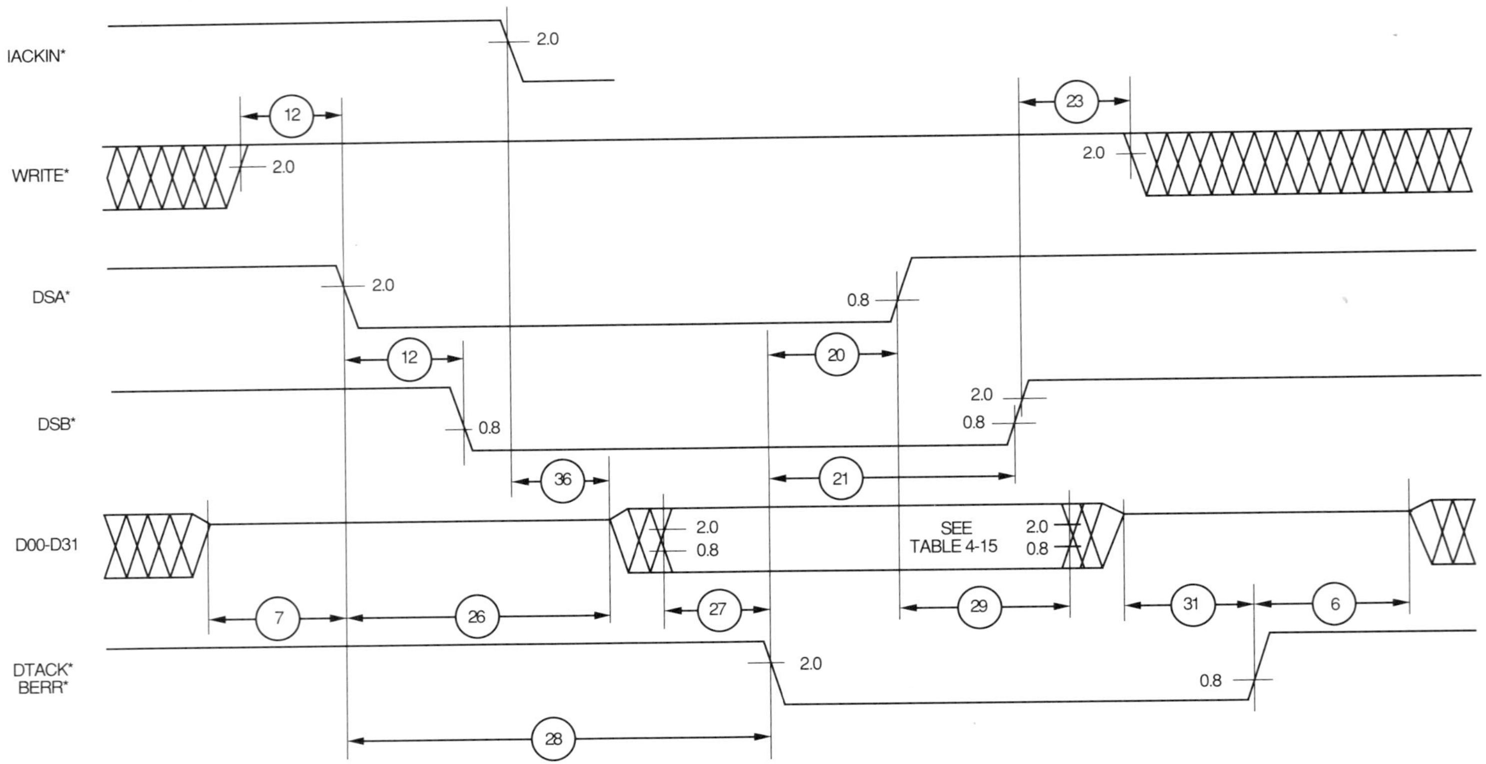

Figure 4-23. Responding INTERRUPTER
STATUS/ID Transfer Timing
DOUBLE BYTE INTERRUPT ACKNOWLEDGE CYCLE
QUAD BYTE INTERRUPT ACKNOWLEDGE CYCLE

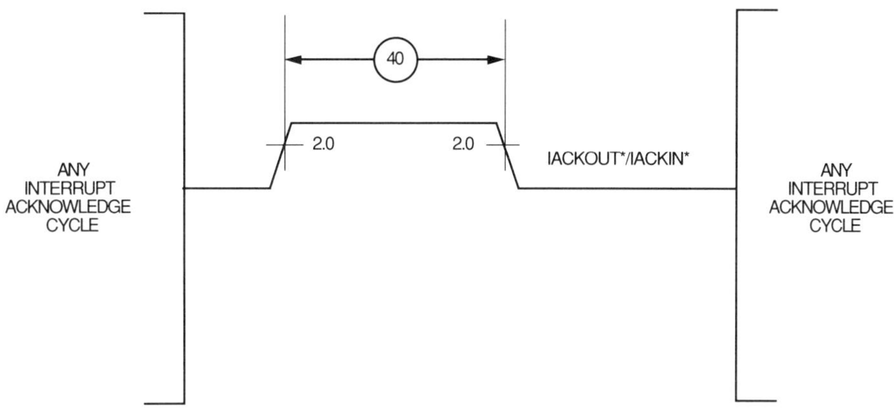

Figure 4-24. IACK DAISY-CHAIN DRIVER, Responding INTERRUPTER, And Participating INTERRUPTER IACK Daisy-Chain Inter-Cycle Timing

CHAPTER 5

VMEbus UTILITY BUS

5.1 INTRODUCTION

This chapter identifies and defines the signal lines and modules which provide utility functions for the VMEbus. The Utility Bus supplies periodic timing, initialization and diagnostic capability for the VMEbus (See Figure 5-1).

5.2 UTILITY BUS SIGNAL LINES

The Utility Bus signal lines are listed below:

System Clock	(SYSCLK)
Serial Clock	(SERCLK)
Serial Data	(SERDAT*)
AC Fail	(ACFAIL*)
System Reset	(SYSRESET*)
System Failure	(SYSFAIL*)

5.3 UTILITY BUS MODULES

5.3.1 The SYSTEM CLOCK DRIVER

The system clock is an independent, non-gated, fixed frequency, 16 MHz, 50 percent (nominal) duty cycle signal. The SYSCLK driver is located on the system controller in board slot one (see Chapter 1). It provides a known time base that is useful for counting off time delays. Figure 5-2 shows the SYSTEM CLOCK DRIVER timing diagram.

OBSERVATION 5.1:
SYSCLK has no fixed phase relationships with other VMEbus timing.

5.3.2 The SERIAL CLOCK DRIVER

The SERIAL CLOCK DRIVER provides a fixed frequency, special waveform signal used by VMSbus modules that reside on VMEbus boards. Its waveform is specified in the VMSbus specification. For the convenience of designers, the timing parameters in effect at the time that this document has been published are provided in Appendix C.

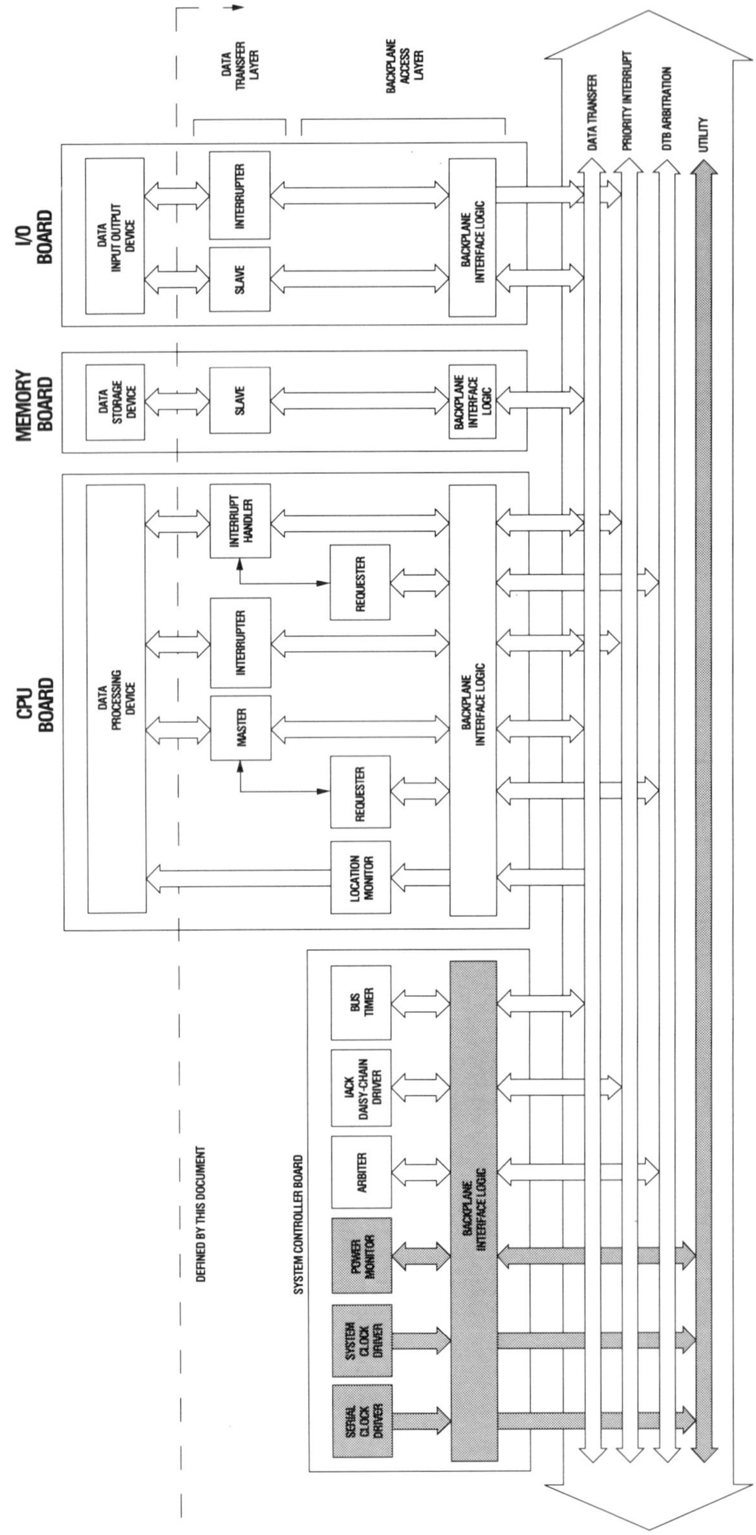

Figure 5-1. Utility Bus Block Diagram

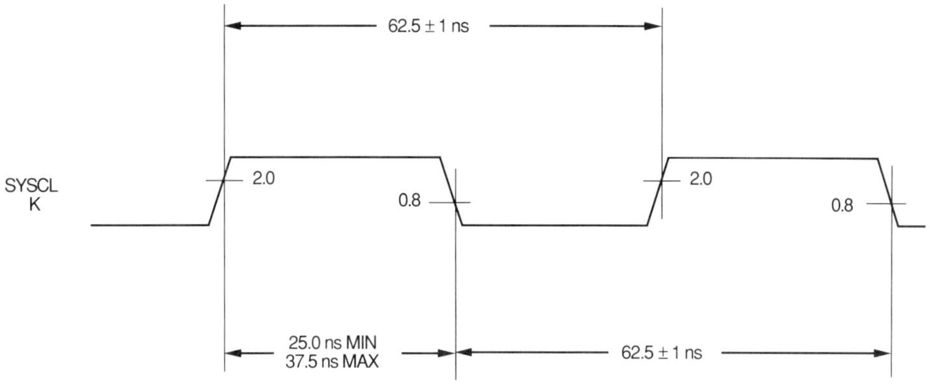

Figure 5-2. SYSTEM CLOCK DRIVER Timing Diagram

5.3.3 The POWER MONITOR

Figure 5-3 is the block diagram for the POWER MONITOR module. This module detects power failures and signals the VMEbus system in time to effect orderly shutdown. When power is then reapplied to the system the POWER MONITOR ensures that all other VMEbus modules are initialized.

The POWER MONITOR might also monitor a manually operated push button and initialize the VMEbus system whenever that button is depressed by the operator.

The ACFAIL* and SYSRESET* transitions, and the point at which the system DC voltages violate specification, have certain timing relationships. These relationships are shown in Figures 5-4 and 5-5.

PERMISSION 5.1:
VMEbus systems **MAY** be built with or without a POWER MONITOR module.

RULE 5.1:
POWER MONITORS **MUST** comply with the timing given in Figures 5-4 and 5-5.

PERMISSION 5.2:
The SYSRESET* line **MAY** be driven low by any VMEbus board to initialize the system from a manual push-button. Where a board drives SYSRESET*, but does not drive ACFAIL*, the timing in Figure 5-4 and Figure 5-5 does not apply.

RULE 5.2:
Whenever any board drives SYSRESET* low, it **MUST** hold SYSRESET* low for a minimum period of 200 milliseconds.

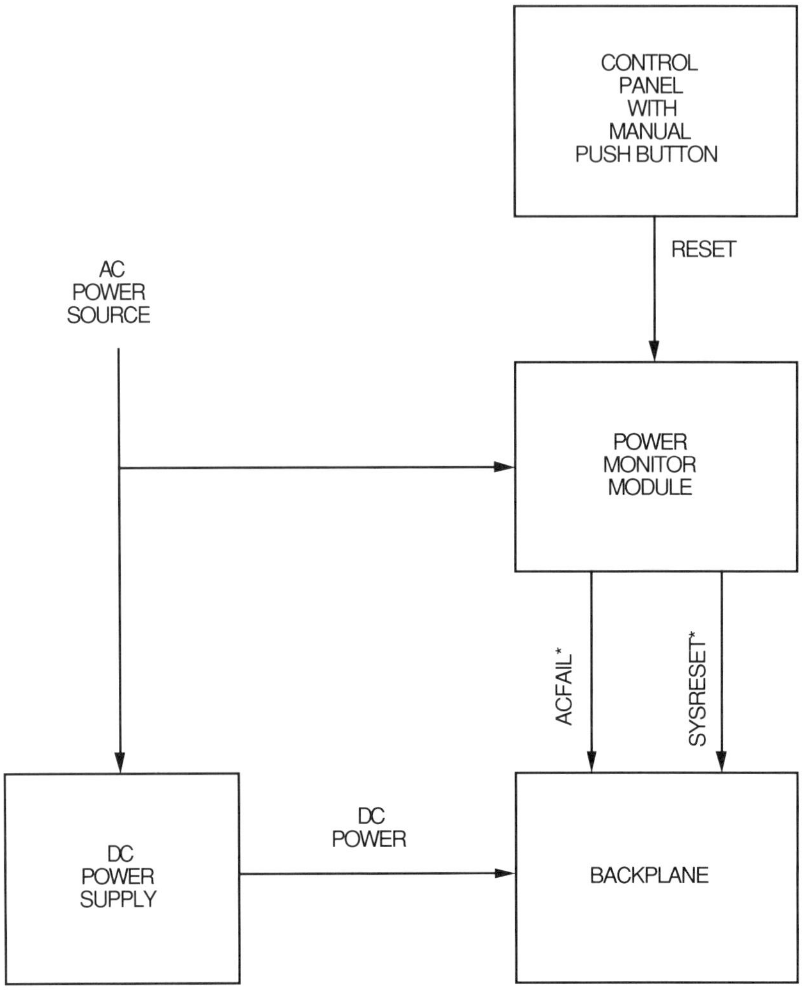

Figure 5-3. Block Diagram Of POWER MONITOR Module

5.4. SYSTEM INITIALIZATION AND DIAGNOSTICS

The VMEbus provides protocols which allow the system to be shut down and powered up in an orderly manner. Two signal lines are used in the power-down and power-up sequence: ACFAIL* and SYSRESET*. Another signal line is used in the power-up sequence: SYSFAIL*.

The following specifies the behavior of VMEbus functional modules during the power-down sequence:

RECOMMENDATION 5.1:
Design MASTERS so that they do not request the bus for any purpose except power-fail activity, after ACFAIL* has been low for 200 microseconds.

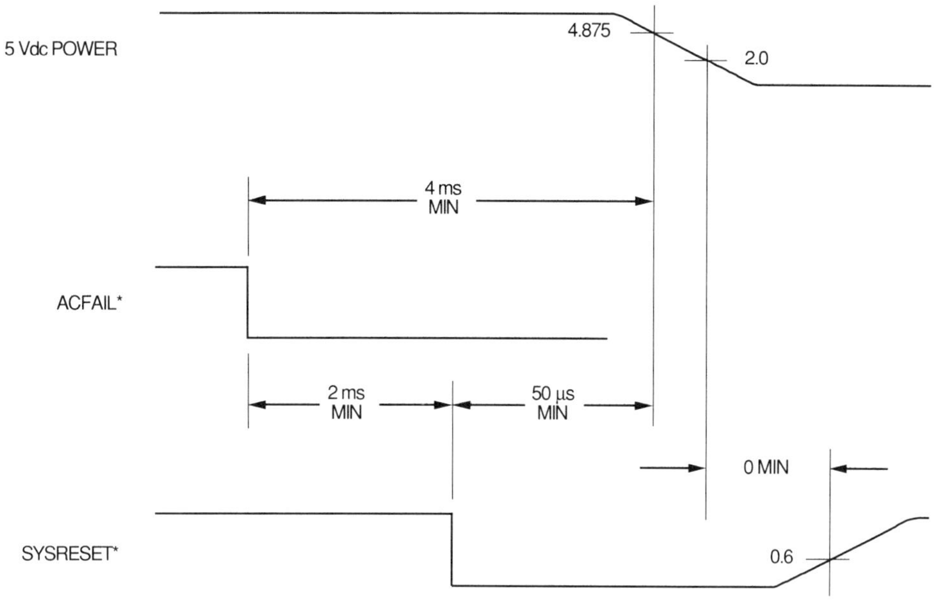

Figure 5-4. POWER MONITOR Power Failure Timing

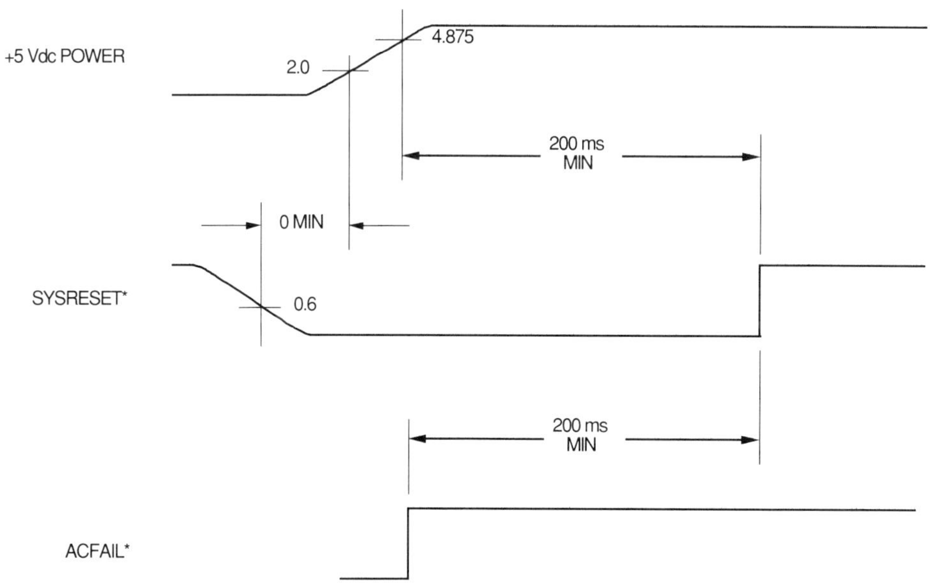

Figure 5-5. POWER MONITOR System Restart Timing

RECOMMENDATION 5.2:

IF MASTERS and INTERRUPT HANDLERS have a bus request pending prior to detecting ACFAIL* low,

THEN they are to limit their subsequent non-power-fail activity to 200 microseconds.

OBSERVATION 5.2:

Bus accesses required to save and restore system data to VMEbus memory depend upon the application, and are not specified here. (The operating system has to assure that data saved during the shut-down process is restored prior to system operation.) In the case of a multiprocessing system, this might require some processor to processor communication.

System Reset (SYSRESET*) is an open-collector line driven by a POWER MONITOR module, or by any board in response to a push-button switch closure.

OBSERVATION 5.3:

Special circuitry is needed where push-button reset switches are used, to ensure that switch bounce does not cause the board to violate the 200 milliseconds minimum SYSRESET* low time.

RULE 5.3:

The SYSTEM CLOCK driver **MUST** continue to provide the specified SYSCLK waveform regardless of the state of the SYSRESET* line.

PERMISSION 5.4:

When SYSRESET* goes low, any board that requires more than 200 milliseconds to complete its initialization **MAY** turn on its SYSRESET* driver to maintain SYSRESET* low for the required period.

RULE 5.4:

IF the +5 Vdc power source is within its specified range when SYSRESET* goes low,

THEN functional modules **MUST** satisfy the timing RULES given in Table 5-1 within the specified time after SYSRESET* goes low.

RULE 5.5:

IF SYSRESET* is low when the +5 Vdc enters its specified range,

THEN functional modules **MUST** satisfy the timing RULES given in Table 5-1 and then MUST refrain from driving specified lines until SYSRESET* goes high.

RULE 5.6:

After satisfying the RULES in Table 5-1, functional modules **MUST NOT** change the state of their drivers until SYSRESET* goes high, unless the +5 Vdc power source exits its specified range.

RULE 5.7:

IF the +5 Vdc power source is within the specified range when SYSRESET* goes low, and a MASTER or INTERRUPT HANDLER is driving AS*, DS0*, or DS1* low.

THEN it **MUST** maintain these strobes low long enough to satisfy the minimum low times given in Chapters 2 and 4.

Table 5-1. Module Drive During Power-Up And Power-Down Sequences

MODULE	MUST REFRAIN FROM DRIVING	AFTER SYSRESET HAS BEEN LOW FOR
MASTERS and INTERRUPT HANDLERS	AS*, DS0*, or DS1* from high to low	5 μsec
MASTERS and INTERRUPT HANDLERS	IACK*, LWORD*, AS*, DS0*, DS1*, AM0-AM5, A01-A31, WRITE*, or D00-D31	20 μsec
SLAVES and INTERRUPTERS	D00-D31, DTACK*, or BERR*	30 μsec
INTERRUPTERS	IRQ1*-IRQ7*	30 μsec
BUS TIMER	BERR*	30 μsec
ARBITER	BG0IN*-BG3IN* from high to low	5 μsec
ARBITER	BG0IN*-BG3IN* low	30 μsec
REQUESTERS	BBSY*	30 μsec

SYSFAIL* is an open collector line that is held low when the system is powered-up and remains low until system self-tests are complete (see Figure 5-6). The following applies:

SUGGESTION 5-1:
On intelligent MASTER boards, include a locally accessible control register bit that is initialized to drive SYSFAIL* low when power is first applied to the board. This permits the board's local intelligence to do a local self-test and release SYSFAIL* only if the self-test passes.

SUGGESTION 5-2:
Design non-intelligent boards with a globally accessible control register bit that is initialized to drive SYSFAIL* low. This allows a MASTER on the VMEbus to run a test on the non-intelligent board, and then write to the global control register bit, releasing the board's SYSFAIL* driver.

SUGGESTION 5-3:
Where a SYSFAIL* control register bit is included on VMEbus boards, provide a status LED on the board's front panel to indicate the status of the control register bit. Then, if a system failure is indicated by the SYSFAIL* signal line, a visual inspection will help determine which board has failed.

RULE 5.8:
IF SYSFAIL* control register bits are included on VMEbus boards,
THEN they **MUST** drive SYSFAIL* low within 50 milliseconds after SYSRESET* goes low, as shown in Figure 5-6.

PERMISSION 5.4:
A VMEbus board **MAY** also drive SYSFAIL* low at any time during normal operation to indicate that it has detected some kind of failure.

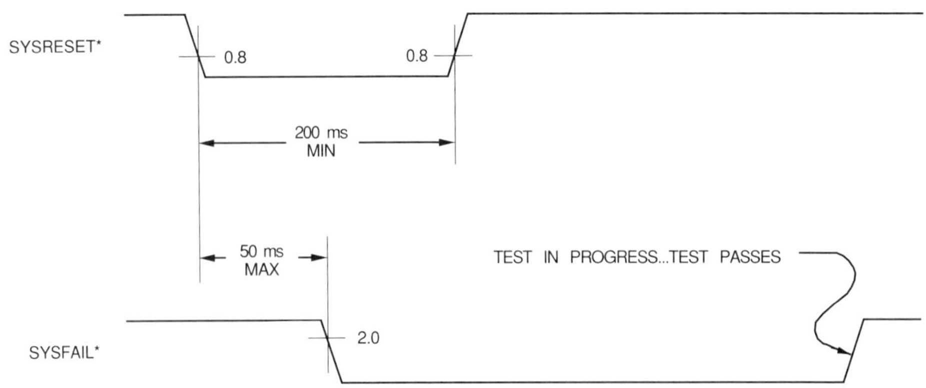

Figure 5-6. SYSRESET* And SYSFAIL* Timing Diagram

5.5 POWER PINS

Figure 5-7 gives the current rating for the VMEbus power pins at various temperatures.

OBSERVATION 5.4:
Some connector pins have a slightly higher contact resistance than others when plugged into the backplane. This produces unbalanced current flow in pins which are paralleled. Suppose that two pins are paralleled and are carrying a total of two amperes of current. If the contact resistance on one is 1 milliohm and the other is 2 milliohms, then one pin will be carrying only 0.67 amperes while the other carries 1.33 amperes.

RULE 5.9:
VMEbus connector pins **MUST** be capable of carrying the currents shown by the solid line in Figure 5-7.

OBSERVATION 5.5:
If one or more power pins fail completely, all of the load current flows through the remaining pins. For example, if half of the pins fail, the remaining pins carry twice the normal current. Depending upon the load current, this might cause damage to these remaining good pins.

SUGGESTION 5-4:
When designing a VMEbus board with a high current load, divide the board's area into zones which are each powered by a separate power grid. Don't connect these grids to each other on the VMEbus board. Instead, connect each to its own VMEbus power pin.

OBSERVATION 5.6:
IF a double height VMEbus board which draws more power than its P1 connector can provide is plugged into a subrack which contains only a J1 backplane,
THEN its P1 power pins will overheat, and might be damaged.

5.6 RESERVED LINE

RULE 5.10:
The VMEbus specification labels one signal line as RESERVED. The RESERVED line is terminated and bused. This line is set aside for future use and **MUST NOT** be used in any VMEbus board designs.

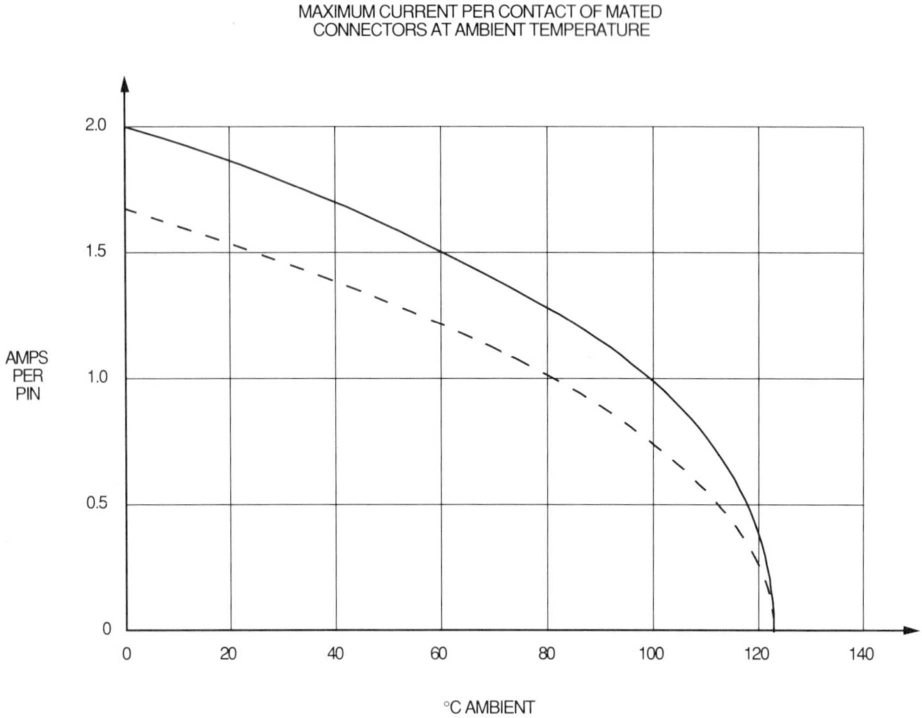

Notes:

1. The dotted line shows the upper limit for current that can be safely drawn per +5 Vdc power pin where two or more pins are connected to a common on-board power grid.

2. The solid line shows the upper limit for current that can be safely drawn per +5 Vdc power pin where each pin is connected to a separate power grid.

Figure 5-7. Current Rating For Power Pins

CHAPTER 6

ELECTRICAL SPECIFICATIONS

6.1 INTRODUCTION

The transmission of data between VMEbus boards such as processors, memories and I/O devices takes place over one or two backplanes, depending on the design. The rules in this chapter ensure proper timing, minimal noise and minimal cross talk problems on the backplane signal lines. The design of VMEbus backplanes is governed by the following RULES:

RULE 6.1:
VMEbus backplanes **MUST NOT** have any signal conductors longer than 500 mm (19.68 inches).

RULE 6.2:
VMEbus backplanes **MUST NOT** have more than 21 slots.

RULE 6.3:
For those lines requiring termination (see Section 6.7) the backplane **MUST** provide some means for terminating them at both ends of the signal line.

RULE 6.4:
The backplane **MUST** provide power conductors for distribution of +5V, +5 STDBY, +12V, and -12V to all of the power pins specified in Section 7.6.

RULE 6.5:
The backplane **MUST** provide ground connections to all of the ground pins specified in Section 7.6.

PERMISSION 6.1:
VMEbus signal lines are normally driven by bipolar drivers, but any technology which complies with this specification **MAY** be used.

6.2 POWER DISTRIBUTION

Power in a VMEbus system is distributed on the backplane(s) as regulated direct current (DC) voltages. The available voltages are:

+5 Vdc — This is the main power source for most VMEbus systems. Most of the system circuitry, including TTL logic, MOS microprocessors, and memories, requires this voltage.

±12 Vdc — These are often used for powering RS232C drivers. They are also sometimes used for powering MOS and analog devices. In some cases -5 Vdc bias voltage or -5.2 Vdc ECL voltages are also derived from the -12Vdc source using on-board regulators. These supplies normally don't supply as much power to the VMEbus system as the +5Vdc source does.

+5 Vdc STDBY — This is used to sustain memory, time-of-day clocks, etc, when the +5Vdc power is lost.

6.2.1 DC Voltage Specifications

Table 6-1 summarizes the DC voltage specifications. The listed specifications are the maximum allowed variance as measured at the connector pins of any card plugged into the backplane.

RECOMMENDATION 6.1:
Design and connect backplanes so that the power supply sense point is located somewhere near the center of the backplane, and as close as possible to the point where power is introduced into the backplane.

OBSERVATION 6.1:
Placing the power supply sense point near the power input point prevents boards near the power input point from receiving too high a voltage.

Table 6-1. Bus Voltage Specification

MNEMONIC	DESCRIPTION	ALLOWED VARIATION (see OBSERVATION 6.2)	RIPPLE/NOISE BELOW 10 MHz (Peak-to-Peak)
+5V	+5 Vdc	+0.25V / -0.125V	50 mV
+12V	+12 Vdc power	+0.60V / -0.36V	50 mV
-12V	-12 Vdc power	-0.60V / +0.36V	50 mV
+5V STDBY	+5 Vdc standby	+0.25V / -0.125V	50 mV
GND	Ground	REFERENCE	

OBSERVATION 6.2:
The non-symmetric variation given in Table 6-1 ensures that the DC power remains within the tolerance required by most ICs despite the typical voltage drops that occur in the power distribution network.

OBSERVATION 6.3:
The power consumed by some systems fluctuates over a wide range during normal system operation. For example, dynamic memory refreshing might cause significant fluctuations if large amounts of memory are refreshed at one time. In this case the response time of the voltage distribution system becomes important.

RECOMMENDATION 6.2:
Use bypass capacitors on VMEbus boards to minimize the effects of power transients.

6.2.2 Pin And Socket Connector Electrical Ratings

RULE 6.6:
The 96 pin connector used by the VMEbus **MUST** provide the following:

Voltage rating:	≥	100 volts DC, isolation pin to pin
Contact resistance:	≤	50 milliohms, at rated current
Insulation resistance:	≥	100 Megohms, pin to pin

6.3 ELECTRICAL SIGNAL CHARACTERISTICS

RULE 6.7:
VMEbus boards **MUST NOT** drive any backplane signal line to a higher steady-state voltage than the highest voltage on any of its +5V power pins, or to a lower steady-state voltage than the lowest voltage on any of its GND pins.

RULE 6.8:
VMEbus boards **MUST** use drivers and receivers that meet the following characteristics:

Steady-state driver	low output level	≤ 0.6 V
Steady-state receiver	low input level	≤ 0.8 V
Steady-state driver	high output level	≥ 2.4 V
Steady-state receiver	high input level	≧ 2.0 V

Figure 6-1 gives a simple graphic representation of these levels.

VMEbus boards drive the backplane lines with three-state, open collector, and totem-pole drivers. Section 6.4 specifies the drive and loading requirements for the various signal lines. Section 6.7 provides a summary, showing which types of drivers are used to drive each signal line.

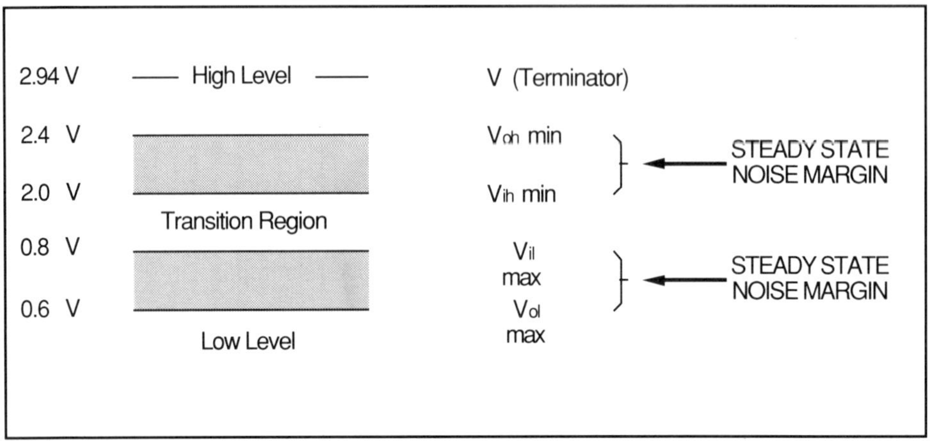

Figure 6-1. VMEbus Signal Levels

RULE 6.9:
When making voltage threshold measurements on a VMEbus board to verify compliance with timing specifications, the ground reference **MUST** be taken from the board's ground pin nearest the signal pin being measured, and the signal voltage **MUST** be measured on the board's connector pin.

6.4 BUS DRIVING AND RECEIVING REQUIREMENTS

This section defines the driver and receiver specifications for all VMEbus signal lines. Table 6-2 lists all of the signals and shows which of the following sub-sections discusses it.

6.4.1 Bus Driver Definitions

Totem-pole, three-state, and open-collector drivers are defined as follows:

Totem-pole — an active driver in both states which sinks current in the low state and sources current in the high state. Totem-pole drivers are used on signals having only a single driver per line (e.g., daisy-chain lines).

Three-state — similar to a totem-pole driver except that it can go to a high impedance state (drivers turned off) in addition to the low and high logic states. Three-state drivers are used for lines that can be driven by several devices at different points on the bus (e.g., address or data lines). Only one of these drivers can be active at any one time.

Open-collector — sinks current in the low state but sources no significant current in the high state. Terminating resistors on the backplane ensure that the signal line voltage rises to a high level whenever it is not driven low. Open-collector drivers are used for signal lines which can be driven by several devices simultaneously (e.g., interrupt and bus request lines).

Table 6-2. Bus Driving And Receiving Requirements

SIGNAL NAME	SUB-SECTION 6.4.2.x
A01-A31	2
ACFAIL*	5
AM0-AM5	2
AS*	1
BBSY*	5
BCLR*	3
BERR*	5
BG0OUT*-BG3OUT*	4
BR0*-BR3*	5
D00-D31	2
DS0*	1
DS1*	1
DTACK*	5
IACK*	2, 5
IACKOUT*	4
IRQ1*-IRQ7*	5
LWORD*	2
SERCLK	3
SYSCLK	3
SYSFAIL*	5
SYSRESET*	5
WRITE*	2

6.4.2 Driving And Loading RULES For All VMEbus Lines

RULE 6.10:
All VMEbus boards **MUST** provide clamping on each VMEbus signal line that they monitor to prevent negative excursions below -1.5 V.

OBSERVATION 6.4:
Standard 74LSxxx and 74Fxxx devices have internal clamping diodes on their inputs that will satisfy the clamping requirement specified in RULE 6.10.

RULE 6.11:
VMEbus receivers **MUST** guarantee detection of a high logic level above a threshold of 2.0 volts, as shown in Figure 6-1.

RULE 6.12:
VMEbus receivers **MUST** guarantee detection of a low logic level below a threshold of 0.8 volts, as shown in Figure 6-1.

PERMISSION 6.2:
A three-state driver **MAY** be used as a totem-pole driver if its output is permanently enabled.

6.4.2.1 Driving And Loading RULES For High Current Three-State Lines

(AS*, DS0*, DS1*)

RULE 6.13:
IF a VMEbus board drives AS*, DS0*, or DS1*,
THEN its drivers for these lines **MUST** meet the following specifications:

Low state sink current:	IOL	≥	64 mA
Low state voltage:	VOL	≤	0.6V at IOL = 64 mA
High state source current:	IOH	≥	3 mA
High state voltage:	VOH	≥	2.4V at IOH = 3 mA
Minimum source current with board pin grounded:	IOS	≥	50 mA at 0V
Maximum source current with board pin grounded:	IOS	≤	225 mA at 0V

RULE 6.14:
When drivers are turned off, VMEbus boards **MUST** limit their loading of AS*, DS0*, and DS1* to the following values:

Current sourced by board at 0.6 V, including leakage current:	IOZL+IIL	≤	450 uA
Current sunk by board at 2.4 V, including leakage current:	IOZH+IIH	≤	100 uA
Total capacitive load on signal, including signal trace:	CT	≤	20 pF

OBSERVATION 6.5:
The source and sink currents listed in RULES 6.13 and 6.14 include both driver and receiver currents sourced and sunk on the board.

SUGGESTION 6.1:
Use 74S241 or 74F241/244 devices to drive the lines AS*, DS0*, and DS1*.
Use 74LS240, 74LS241, or 74LS244 devices to receive the lines AS*, DS0*, and DS1*.

6.4.2.2 Driving And Loading RULES For Standard Three-State Lines

(A01-A31, D00-D31, AM0-AM5, IACK*, LWORD*, WRITE*)

RULE 6.15:
IF a VMEbus board drives the lines A01-A31, D00-D31, AM0-AM5, IACK*, LWORD*, or WRITE*,
THEN its drivers for these lines **MUST** meet the following specifications:

Low state Sink current:	IOL	≥	48 mA
Low state voltage:	VOL	≤	0.6V at IOL = 48 mA
High state source current:	IOH	≥	3 mA
High state voltage:	VOH	≥	2.4V at IOH = 3 mA
Minimum source current with board pin grounded:	IOS	≥	50 mA at 0V
Maximum source current with board pin grounded:	IOS	≤	225 mA at 0V

RULE 6.16:
When drivers are turned off, VMEbus boards **MUST** limit their loading of the lines A01-A31, D00-D31, AM0-AM5, IACK*, LWORD*, and WRITE* to the following values:

Current sourced by board at 0.6 V, including leakage current:	IOZL+IIL	≤	700 uA
Current sunk by board at 2.4 V, including leakage current:	IOZH+IIH	≤	150 uA
Total capacitive load on signal, including signal trace:	CT	≤	20 pF

OBSERVATION 6.6:
The source and sink currents specified in RULES 6.15 and 6.16 include both driver and receiver currents sourced and sunk on the board.

SUGGESTION 6.2:
Use 74ALS645-1, 74F244, 74AS573, or 74AS580 devices to drive the lines A01-A31, D00-D31, AM0-AM5, IACK*, LWORD*, and WRITE*.

Use 74LS240, 74LS241, or 74LS244 devices to receive the lines A01-A31, D00-D31, AM0-AM5, IACK*, LWORD*, and WRITE*.

Use 74ALS645-1, 74ALS245A-1, 74ALS646-1, or 74ALS648-1 devices to transceive the lines A01-A31, D00-D31, AM0-AM5, IACK*, LWORD*, and WRITE*.

6.4.2.3 Driving And Loading RULES For High Current Totem-Pole Lines

(SERCLK, SYSCLK, BCLR*)

RULE 6.17:
VMEbus systems **MUST** have no more than one board driving each of the lines SERCLK, SYSCLK, or BCLR*. Its drivers for these lines **MUST** meet the following specifications:

Low state sink current:	IOL	≥	64 mA
Low state voltage:	VOL	≤	0.6V at IOL = 64 mA
High state source current:	IOH	≥	3 mA
High state voltage:	VOH	≥	2.4V at IOH = 3 mA
Minimum source current with board pin grounded	IOS	≥	50 mA at 0 V
Maximum source current with board pin grounded	IOS	≤	255 mA at 0V

RULE 6.18:
All VMEbus boards **MUST** limit their loading of the lines SERCLK, SYSCLK, and BCLR* to the following values:

Current sourced by board at 0.6 V, including leakage current:	IOZL+IIL	≤	600 uA
Current sunk by board at 2.4 V, including leakage current:	IOZH+IIH	≤	50 uA
Total capacitive load on signal, including signal trace, for system controllers (which have drivers).	CT	≤	20 pF
Total capacitive load on signal, including signal trace, for other boards (which have no drivers).	CT	≤	12 pF

OBSERVATION 6.7:
The source and sink currents specified in RULES 6.17 and 6.18 include both driver and receiver currents sourced and sunk on the board.

SUGGESTION 6.3:
Use 74S241 or 74F241/244 devices to drive the lines SERCLK, SYSCLK, and BCLR*.
Use 74LS240, 74LS241 or 74LS244 devices to receive the lines SERCLK, SYSCLK, and BCLR*.

6.4.2.4 Driving And Loading RULES For Standard Totem-Pole Lines.

(BG0OUT*-BG3OUT*/BG0IN*-BG3IN*, IACKOUT*/IACKIN*)

RULE 6.19:
IF a VMEbus board drives the lines BG0OUT*-BG3OUT*/BG0IN*-BG3IN*, or IACKOUT*/IACKIN*,
THEN its drivers for these lines **MUST** meet the following specifications:

Low state sink current:	IOL	$\geq$	8 mA
Low state voltage:	VOL	$\leq$	0.6V at IOL = 8 mA
High state source current:	IOH	$\geq$	400 uA
High state voltage:	VOH	$\geq$	2.7V at IOH = 400 uA

RULE 6.20:
All VMEbus boards **MUST** limit their loading of each of the lines BG0OUT*-BG3OUT*/BG0IN*-BG3IN*, and IACKOUT*/IACKIN* to the following values:

Current sourced by board at 0.6 V, including leakage current:	IOZL+IIL	$\leq$	600 uA
Current sunk by board at 2.4 V, including leakage current:	IOZH+IIH	$\leq$	50 uA
Total capacitive load on signal, including signal trace:	CT	$\leq$	20 pF

OBSERVATION 6.8:
The source and sink currents specified in RULES 6.19 and 6.20 include both driver and receiver currents sourced and sunk on the board.

SUGGESTION 6.4:
Use any standard device that meets the specifications above to drive the lines BG0OUT*-BG3OUT*/BG0IN*-BG3IN*, and IACKOUT*/IACKIN*.

Use 74LS240, 74LS241, or 74LS244 devices to receive the lines BG0OUT*-BG3OUT*/BG0IN*-BG3IN*, and IACKOUT*/IACKIN*.

6.4.2.5 Driving And Loading RULES For Open-Collector Lines

(BR0*-BR3*, BBSY*, IRQ1*-IRQ7*, DTACK*, BERR*, SYSFAIL*, SYSRESET*, ACFAIL*, IACK* and SERDAT*)

RULE 6.21:

IF a VMEbus board drives the lines BR0*-BR3*, BBSY*, IRQ1*-IRQ7*, DTACK*, BERR*, SYSFAIL*, SYSRESET*, ACFAIL*, IACK* or SERDAT*,

THEN its drivers for these lines **MUST** meet the following specifications:

Low state sink current:	IOL	$\geq$	48 mA
Low state voltage:	VOL	$\leq$	0.6V at IOL = 48 mA

RULE 6.22:

All VMEbus boards **MUST** limit their loading of the lines BR0*-BR3*, BBSY*, IRQ1*-IRQ7*, DTACK*, BERR*, SYSFAIL*, SYSRESET*, ACFAIL*, IACK* and SERDAT* to the following values:

Current sourced by board at 0.6 V, including leakage current:	IOZL+IIL	$\leq$	400 uA (DTACK* and BERR*)
		$\leq$	600 uA (all others)
Current sunk by board at 2.4 V, including leakage current:	IOZH+IIH	$\leq$	50 uA
Total capacitive load on signal, including signal trace:	CT	$\leq$	20 pF

OBSERVATION 6.9:

The sink current specified in RULES 6.21 and 6.22 includes both driver and receiver currents sourced and sunk on the board.

SUGGESTION 6.5:

Use 74S38 devices to drive the lines BR0*-BR3*, BBSY*, IRQ1*-IRQ7*, DTACK*, BERR*, SYSFAIL*, ACFAIL*, IACK* and SERDAT*.

Use 74LS240, 74LS241, or 74LS244 devices to receive the lines BR0*-BR3*, BBSY*, IRQ1*-IRQ7*, DTACK*, BERR*, SYSFAIL*, SYSRESET*, ACFAIL*, IACK* and SERDAT*.

SUGGESTION 6.6:

Since most TTL drivers do not work reliably when the +5 Vdc power source is out of its specified range, drive SYSRESET* on a POWER MONITOR module with a driver built from a discrete high-gain small signal transistor.

6.5 BACKPLANE SIGNAL LINE INTERCONNECTIONS

The VMEbus is a high performance interface system. Its design takes into account transmission line effects on the backplane. The address and data set-up times specified in Chapters 2 and 4 take into account the fact that most drivers available today do not reliably drive backplane signal lines from the low to the high level until there is a reflection from the end of the bus. Although these reflections serve a useful purpose, they cannot be excessive or ringing will result. The following paragraphs specify the backplane characteristics that achieve the desired result.

6.5.1 Termination Networks

RULE 6.23:
Termination networks **MUST** be used on each end of all VMEbus signal lines except the daisy-chain lines.

OBSERVATION 6.10:
The terminations in the VMEbus serve four purposes:

- They reduce reflections from the ends of the backplanes.
- They provide a high state pull-up for open-collector drivers.
- They restore the signal lines to the high level when three-state devices are disabled.
- They provide a standing current for the driver sink transistor to switch off, causing the signal line to rise more swiftly on positive transitions.

The Thevenin equivalent of the termination is shown in Figure 6-2. The voltage divider also shown provides this termination value.

OBSERVATION 6.11:
IF a maximum tolerance of ±5% is maintained on the resistor values and source voltage used in the resistor network shown in Figure 6-2.
THEN the circuit shown will meet the tolerances shown for the Thevenin equivalent.

OBSERVATION 6.12:
The resistor network shown in Figure 6-2 presents its Thevenin equivalent impedance only when its +5V source is adequately decoupled to ground by a bypass capacitor.

RECOMMENDATION 6.3:
Provide a bypass capacitor with a value in the range of 0.01 to 0.1 uF as close as possible to the Vcc pin of each resistor termination package.

PERMISSION 6.3:
Any resistor network and voltage source **MAY** be used to provide the termination, as long as they provide the Thevenin equivalent shown in Figure 6-2.

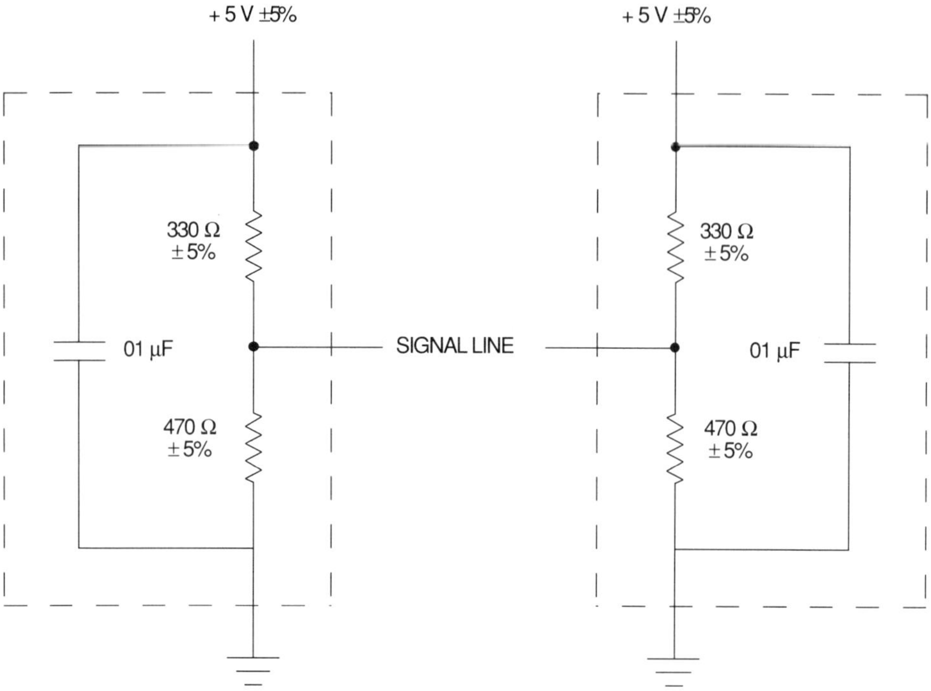

RESISTOR NETWORKS THAT PROVIDE
THE REQUIRED TERMINATIONS

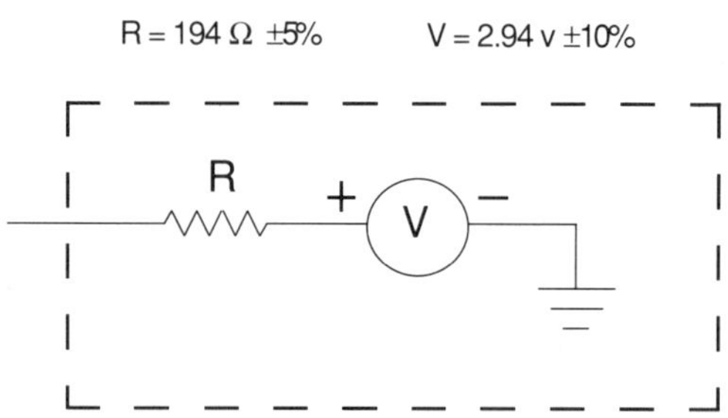

THEVINEN EQUIVALENT
FOR EACH TERMINATOR

Figure 6-2. Standard Bus Termination

6.5.2 Characteristic Impedance

Each signal line in the backplane has an associated characteristic impedance Zo. This characteristic impedance is important because discontinuities in Zo (due to capacitive effect and loads on the bus) and mismatches between Zo and the terminations can cause distortions of signal waveforms.

Figure 6-3 shows a microstrip signal line cross section which is the normal configuration for a multilayer backplane signal line. The Zo is a function of the width and thickness of the line, the thickness of the dielectric, and its relative dielectric constant. Figure 6-4 shows characteristic impedance versus microstrip line width for common thickness of fiberglass-epoxy board. More information on Microstrip lines can be found in the MECL System Design Handbook, Motorola, 1983.

The terminations on the VMEbus signal lines reduce distortion of their signal waveforms. Although a perfect impedance match (which totally eliminates distortions due to reflections) is not maintained between the termination networks and the signal lines, it is important not to allow too great a mismatch, as might be the case if a signal line's Zo value is too low.

RECOMMENDATION 6.4:
When designing a VMEbus backplane, choose a signal line width and board thickness that gives a Zo (as calculated from Figure 6-4) as close as possible to 100 ohms.

The actual characteristic impedance of a backplane signal line is called the effective characteristic impedance (Zo'), and will be lower than Zo, due to the capacitance of plated through holes and connector pins. This additional capacitance makes Zo' go below 100 ohms. Although plated-through holes are necessary to accommodate connectors, other holes should be kept to a minimum.

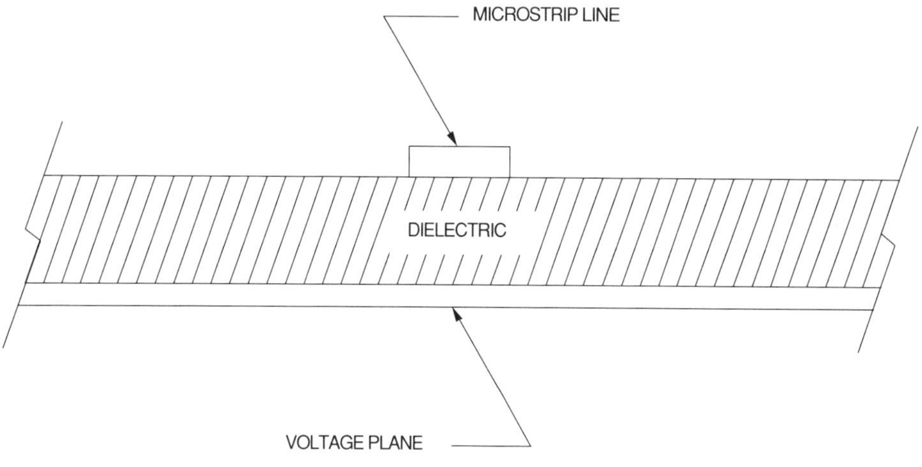

Figure 6-3. Backplane Microstrip Signal Line Cross Section

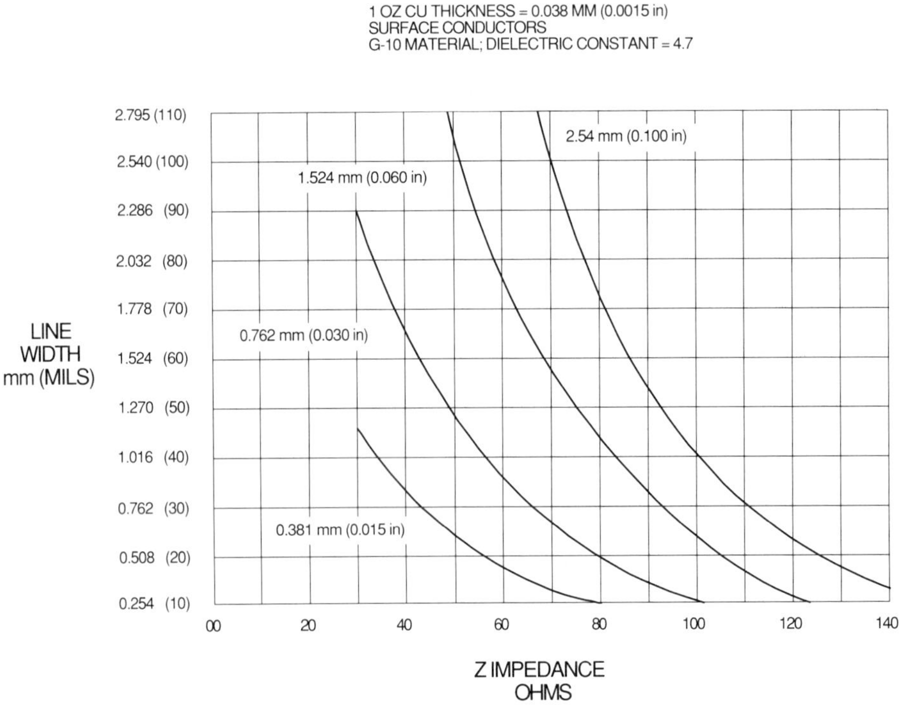

Figure 6-4. Zo Versus Line Width

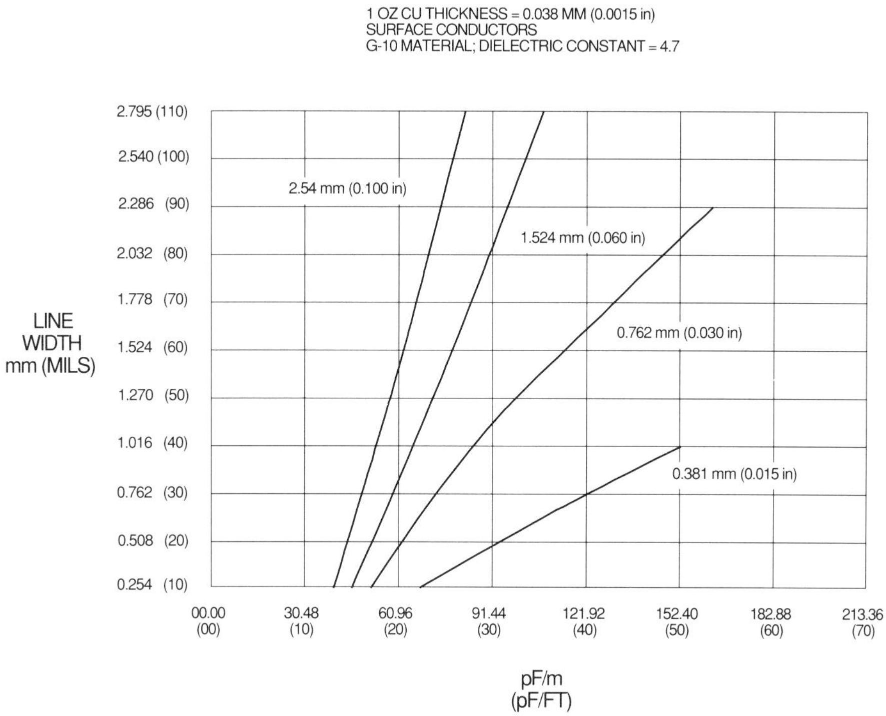

Figure 6-5. Co Versus Line Width

The backplane signal line impedance (without any boards plugged into the backplane) can be calculated with the following equation:

$$Zo' = \frac{Zo}{\sqrt{1 + Cd/Co}}$$

where:

Zo	=	The impedance of the microstrip line, ignoring the loading effects of plug-in pc boards, connectors, and plated-through holes. (See Figure 6-4)
Cd	=	The distributed capacitance, per unit of distance, of the plated-through holes, and backplane connectors.
Co	=	The intrinsic line capacitance, per unit of distance, of the microstrip line, ignoring the loading effects of plug-in pc boards, connectors, and plated-through holes. (See Figure 6-5)
Zo'	=	The backplane signal line impedance, including the loading effects of connectors, and plated-through holes but excluding the loading effects of plug-in pc boards.

OBSERVATION 6.13:
Typical Zo' values for a VMEbus backplane, with no boards inserted, range from 50 to 60 ohms. If this impedance is 50 ohms, or higher, it will provide satisfactory operation.

6.5.3 Additional Information

RULE 6.24:
Circuit traces from the 96 pin connectors to the on-board circuitry **MUST NOT** have a length of greater than 50.8 mm (2 inches).

OBSERVATION 6.14:
IF the trace from the 96 pin connector to on-board circuitry branches,
THEN the length of each branch is added to get the total length specified in RULE 6.24.

RULE 6.25:
There **MUST NOT** be more than one driver driving the SYSCLK and SERCLK lines.

RULE 6.26:
IF the SYSTEM CLOCK DRIVER and the SERIAL CLOCK DRIVER modules are provided,
THEN they **MUST** be installed in slot 1 of the backplane.

OBSERVATION 6.15:
Locating the SYSTEM CLOCK DRIVER and the SERIAL CLOCK DRIVER on the board in slot 1 minimizes the distortion of their waveforms from the reflections off the ends of the backplane.

SUGGESTION 6.7:
If actual capacitance loading values cannot be obtained from manufacturer specifications sheets, the following values can be used to estimate the total capacitive loading of a VMEbus board:

The typical capacitance of a receiver	3 - 5 pf
The typical capacitance of a driver	10 - 12 pf
The typical capacitance of a transceiver	15 - 18 pf
The typical capacitance of a 50.8 mm (2 inch) PC trace	2 - 3 pf

OBSERVATION 6.16:
Circuit traces which run parallel to each other, such as in a backplane, sometimes induce signal transitions in each other. This phenomenon is commonly known as cross talk. When designing VMEbus backplanes, the spacings of lines and their position relative to ground and power planes have a large effect on the amount of cross talk observed.

SUGGESTION 6.8:
Propagation delays through bus drivers depend on how heavily they are loaded and VMEbus signal lines typically represent heavy loads. This has to be taken into account when calculating worst case timing. If the manufacturer's data sheet for the driver gives a propagation delay for a 300 pf load, use that to do the worst case calculations. If the only propagation delay values are for a 30 pf load, add 10 ns to the propagation delay and 15 ns to the turn-on delay.

6.6 USER DEFINED SIGNALS

RECOMMENDATION 6.5:
If a board has a 96 pin connector in its P2 location, do not allow any of the P2 pins to be driven to a voltage greater than ±15V. This reduces the likelihood of serious damage to the VMEbus system in the event that a signal trace from one of these pins is accidently shorted to some other signal line.

6.7 SIGNAL LINE DRIVERS AND TERMINATIONS

This section summarizes the types of drivers which have to be used for each of the signal lines on the VMEbus.

In order to simplify Table 6-3, an abbreviated notation is used to describe the various types of drivers. The notations used are shown below:

Totem-pole (high current)	-	TP HC
Totem-pole (standard)	-	TP STD
Three-state (high current)	-	3 HC
Three-state (standard)	-	3 STD
Open-collector	-	OC

For detailed specifications, see Section 6.4.

Table 6-3. Bus Driver Summary

SIGNAL MNEMONIC	SIGNAL NAME	DRIVER TYPE	BUSED AND TERMINATED?
A01-A31 (31 lines)	ADDRESS BUS	3 STD	YES
ACFAIL*	AC POWER FAILURE	OC	YES
AM0-AM5 (6 lines)	ADDRESS MODIFIER	3 STD	YES
AS*	ADDRESS STROBE	3 HC	YES
BBSY*	BUS BUSY	OC	YES
BCLR*	BUS CLEAR	TP HC	YES
BERR*	BUS ERROR	OC	YES
BG0IN*-BG3IN* BG0OUT*-BG3OUT* (Daisy-chain)	BUS GRANT DAISY-CHAIN	TP STD	NO
BR0*-BR3* (4 lines)	BUS REQUEST	OC	YES

(Continued on Page 213)

Table 6-3. (Continued)

SIGNAL MNEMONIC	SIGNAL NAME	DRIVER TYPE	BUSED AND TERMINATED?
D00-D31 (32 lines)	DATA BUS	3 STD	YES
DS0*-DS1* (2 lines)	DATA STROBES	3 HC	YES
DTACK*	DATA TRANSFER ACKNOWLEDGE	OC	YES
IACK*	INTERRUPT ACKNOWLEDGE	3 STD or OC	YES
IACKIN*/IACKOUT* (Daisy-chain)	INTERRUPT ACKNOWLEDGE DAISY-CHAIN	TP STD	NO
IRQ1*-IRQ7* (7 lines)	INTERRUPT REQUEST	OC	YES
LWORD*	LONGWORD	3 STD	YES
RESERVED	RESERVED	——	YES
SERCLK	SERIAL CLOCK	TP HC	YES
SERDAT*	SERIAL DATA	OC	YES
SYSCLK	SYSTEM CLOCK	TP HC	YES
SYSFAIL*	SYSTEM FAILURE	OC	YES
SYSRESET*	SYSTEM RESET	OC	YES
WRITE*	WRITE	3 STD	YES

CHAPTER 7

MECHANICAL SPECIFICATIONS

7.1 INTRODUCTION

Information is provided in this chapter to ensure that the VMEbus board assemblies, backplanes, subracks, and associated mechanical accessories are dimensionally compatible.

The mechanical dimensions given in this chapter conform to IEC publications 297-1, 297-3, 297-3A and 603-2. The electrical characteristics for VMEbus connectors, as specified in Chapters 5 and 6 supersede publication 603-2 where they differ.

IEC publication 603-2 describes a family of connector types which are identified by labels of the form:

603-2-IEC-xxxxxx-xxx

All of the P1/J1 and P2/J2 connectors used on VMEbus boards and backplanes are members of this family. In this chapter, the label 603-2-IEC-xxxxxx-xxx is used when referring to all of these connector types as a group. The label 603-2-IEC-C096Mx-xxx is used when referring to the 96-pin male connector types within this family which are used on VMEbus boards. 603-2-IEC-C096Fx-xxx is used when referring to the 96-pin female connector types which are used on VMEbus backplanes.

Figure 7-1 is a front view of a 19-inch width subrack which shows how single height and double height VMEbus boards can be mixed in a single subrack. Boards are inserted into the subrack from the front, in a vertical plane with the component face of the board on the right.

PERMISSION 7.1:
A VMEbus system **MAY** be composed of single height boards, double height boards, or a mixture of both.

RULE 7.1:
Single height VMEbus subracks **MUST** have a single "J1" backplane.

RULE 7.2:
Double height VMEbus subracks **MUST** have

1) a "J1" backplane mounted in the upper portion of the subrack,

OR

2) a "J1" and a "J2" backplane, with the J1 backplane mounted in the upper portion and the J2 backplane mounted in the lower portion.

OR

3) a double height backplane which provides both J1 and J2 connectors.

RULE 7.3:
VMEbus backplanes **MUST NOT** have more than 21 slots.

PERMISSION 7.2:
When using backplanes of fewer than 21 slots, the VMEbus subrack **MAY** be less than the standard 482.6 mm (19 inches) rack size.

RULE 7.4:
Except for the rack width, which varies depending on the number of slots it supports, all subrack dimensions **MUST** agree with those given in this chapter to ensure mechanical compatibility between the boards and the subrack.

7.2 VMEbus BOARDS

RECOMMENDATION 7.1:
Make VMEbus boards 1.6 ±0.2 mm (0.063 ±0.008 inch) thick.

OBSERVATION 7.1:
The thickness of VMEbus boards is important because the subrack's guide rails are designed to accommodate boards of this thickness. Thicker boards might not fit into the guides and thinner boards might not be guided properly into the backplane's J1 and J2 connectors.

OBSERVATION 7.2:
The dimensioning of the 603-2-IEC-xxxxxx-xxx connectors provides a certain distance between the connector's mounting face and the center line of each of the connector's pins. This ensures that the P1 and P2 connector pin centers will align properly with the J1 and J2 connectors of the backplane.

PERMISSION 7.3:
VMEbus boards **MAY** be designed with board thicknesses greater than 1.6 mm (0.063 inch) if:

1. the thickness of the top and bottom edges of the board, which fit into the guide rails, is reduced to 1.6 mm (0.63 inch) for distance of 2.5 mm (0.098 inch) from the top and bottom edges of the board (See Figure 7-2 and Figure 7-3) and,
2. the mounting surface provided by the board for the IEC 602-3 connector(s) is 4.07 mm (0.160 inch) from the interboard separation plane. (See Figure 7-5)

Two board sizes are defined as standard VMEbus boards: single height and double height (see Figure 7-2 and Figure 7-3).

7.2.1 Single Height Boards

OBSERVATION 7.3:
A single height VMEbus board is 100mm (3.937 inch) high and 160 mm (6.299 inch) deep with an area of approximately 160.0 sq. cm (24.8 sq. inch).

RULE 7.5:
All single height boards **MUST** be designed according to the dimensions given in Figure 7-2.

RULE 7.6:
The hole pattern for the 96 pin 603-2-IEC-CO96Mx-xxx P1 connector **MUST** be as shown in Figure 7-2.

SUGGESTION 7.1:
Use the PC layout grid shown in Figure 7-2.

OBSERVATION 7.4:
Modular front panel hardware is available from several manufacturers which, when mounted on the grid shown in Figure 7-2, properly aligns with the front panel grid shown in Figure 7-7.

PERMISSION 7.4:
Components, other than the 603-2-IEC-CO96Mx-xxx connector, **MAY** be placed in such a way that they do not align with the grid pattern.

7.2.2 Double Height Boards

OBSERVATION 7.5:
A double height VMEbus board is 233.35 mm (9.187 inch) high and 160 mm (6.299 inch) deep with an area of approximately 373.4 sq. cm (57.9 sq. inch).

RULE 7.7:
All double height boards **MUST** be designed according to the dimensions given in Figure 7-3.

RULE 7.8:
The hole pattern for the 96 pin 603-2-IEC-CO96Mx-xxx P1 connector **MUST** be as shown in Figure 7-3.

RULE 7.9:
IF a 96 pin 603-2-IEC-CO96Mx-xxx connector is used for P2,
THEN its hole pattern **MUST** be as shown in Figure 7-3.

OBSERVATION 7.6:
As in the case of the single height boards above, the grid shown in Figure 7-3 properly aligns modular front panel components with the front panel grid shown in Figure 7-8.

PERMISSION 7.5:
Components, other than the 603-2-IEC-xxxxxx-xxx connectors, **MAY** be placed in such a way that they do not align with the grid pattern.

OBSERVATION 7.7:
There is a discontinuity of 1.27 mm (0.05 inch) between the 2.54 mm (0.10 inch) grid patterns for the upper and lower half of the board.

7.2.3 Board Connectors

The single height board has only one connector on its back edge. It is called the P1 connector. A double height board has either one or two connectors on its back edge. If it has one connector, that connector is called the P1 connector and is located on the upper half of the back edge. If it has two connectors the upper one is called the P1 connector and the lower one is called the P2 connector.

RULE 7.10:
The P1 and P2 connectors of all VMEbus boards **MUST** meet or exceed the mechanical specifications of a 603-2-IEC-C096Mx-xxx class 2 connector, and **MUST** be mounted as shown in Figure 7-4.

OBSERVATION 7.8:
603-2-IEC class 2 connectors have a minimum mechanical endurance of 400 insertion/extraction cycles.

OBSERVATION 7.9:
The symmetry symbol in the box below each board outline in Figure 7-4 sets an upper limit on how much the connector's center line can be tilted with respect to the board's lower edge. This limit is described in the following rule.

RULE 7.11:
The perpendicular distance (d1) from the board's lower edge to the point A in Figure 7-4 **MUST NOT** differ from its perpendicular distance (d2) to point B by more than 0.3 mm (0.012 inch).

RULE 7.12:
IF a VMEbus board is designed to use the center row of P2 for address or data bus expansion, or the board requires more power than the P1 connector can provide,
THEN a 96 pin 603-2-IEC-CO96Mx-xxx P2 connector **MUST** be provided and it **MUST** be mounted as shown in Figure 7-4.

PERMISSION 7.6:
Where neither address nor data bus expansion is required, and where the board does not require more power than the P1 connector can provide, any 603-2-IEC-xxxxxx-xxx connector **MAY** be used for P2 on a double height VMEbus board or the board **MAY** be designed without a P2 connector.

PERMISSION 7.7:
On double height boards, the two outside rows of pins on the P2 connector **MAY** be used to provide user defined connections (See Section 7.6.2).

PERMISSION 7.8:
I/O cables **MAY** be connected to the front edge of a VMEbus board. No connectors are prescribed for these cable connections.

SUGGESTION 7.2:
Where possible, avoid the use of cable connections to the front edge of VMEbus boards. This makes it much easier to install and remove boards from the subrack during maintenance.

7.2.4 Board Assemblies

The board assembly typically consists of a PC board, as defined above with either one or two 603-2-IEC-xxxxxx-xxx connectors affixed to the board's back edge, electronic components, and an optional front panel with handles. For more detail on the front panels see Section 7.3.

RULE 7.13:
Solder filets, tracking, and components on VMEbus boards **MUST NOT** be closer than 2.5mm (0.098 inch) from the top and bottom edges of the board to guarantee clearance between them and the board guides. Figure 7-2 and Figure 7-3 show these dimensions.

Figure 7-5 shows a cross sectional view of a PC board, its front panel, its connector and the backplane. The dimensions given are nominal values and are based upon the dimensions given in the other drawings in this chapter.

7.2.5 Board Widths

PERMISSION 7.9:
VMEbus boards **MAY** be designed to occupy more than one slot.

VMEbus boards designed to occupy a single slot of the subrack are called single width boards.

7.2.6 VMEbus Board Warpage, Lead Length and Component Height.

During the manufacturing process boards sometimes become warped.

RULE 7.14:
The sum of warpage and component lead length **MUST** be less than or equal to 2.47mm (0.097 inch) from where the solder side of an ideal (unwarped) board would be, and the sum of component height and warpage (in the other direction) **MUST** be less than or equal to 13.71mm (0.54 inch) plus an integral multiple (N) of 20.32mm (0.8 inch), from where the component side of an ideal (unwarped) board would be. (Where N = the number of slots that the board occupies minus 1.)

OBSERVATION 7.10:
During insertion into the subrack, the component leads of a board might contact the right edge of the front panel to its left. For this reason, the board should be inserted carefully to avoid bending the component leads.

SUGGESTION 7.3:
Where possible, trim the component lead lengths to 1.52 mm (0.06 inch). This makes installation and removal of boards more convenient. Since the maximum allowable warpage is affected by the length of these component leads, trimming them allows boards with larger warpage to meet the specifications.

OBSERVATION 7.11:
The dimensions given in SUGGESTION 7.3 ensure that there will be a clearance of at least 2.54 mm (0.1 inch) between components of each board and the component leads of the board to its right. This space allows adequate air flow and prevents additional warpage and vibration from causing interboard contact.

RULE 7.15:
All VMEbus boards **MUST** be measured, after they are assembled, to ensure that the combination of board warpage, component lead length, and component height do not exceed the specified limits when the board is inserted into a subrack. In order to properly make these measurements, the board **MUST** be placed in a subrack (or a similar test fixture) while the measurements are taken.

Figure 7-6 shows a board (single or double height) in a subrack, and shows how the combination of board warpage, component lead length, and component height are to be measured. The interboard separation planes defined in that Figure 7-6 provide the reference from which the measurements are made.

OBSERVATION 7.12:
A special test fixture, which simulates a subrack, is helpful in speeding up the measurements shown in Figure 7-6.

7.3 FRONT PANELS

This section provides the mechanical specifications for the single and double height board front panels and associated hardware.

PERMISSION 7.10:
VMEbus boards **MAY** be manufactured with or without front panels.

RECOMMENDATION 7.2:
Use front panels and associated hardware to prevent VMEbus boards from vibrating out of the subrack, and to guide airflow through the subrack.

RULE 7.16:
IF front panels are used,
THEN screws **MUST** be provided on those panels to secure the top and bottom of the panels to the subrack, and their screw threads **MUST** be M2.5 X 0.45 pitch (see Figure 7-7).

Figure 7-7 shows a single height, single width front panel. Figure 7-8 shows a double height, single width front panel. The grid format on the rear face of these front panels are aligned with the board grids in Figure 7-2 and Figure 7-3 respectively.

SUGGESTION 7.4:
Install front panel components such as Light Emitting Diodes (LEDs) and switches so that their centers are aligned with a front panel grid point.

7.3.1 Handles

PERMISSION 7.11:
The VMEbus board designer **MAY** design VMEbus board front panels with or without handles.

RECOMMENDATION 7.3:
Provide handles to make VMEbus boards easier to remove from the subrack.

OBSERVATION 7.13:
The handles available from various manufacturers vary somewhat in overall shape.

SUGGESTION 7.5:
Choose handles whose depth and height conform to the dimensions shown in Figure 7-7 and Figure 7-8.

RECOMMENDATION 7.4:
When mounting handles on VMEbus board front panels, choose one or more of the locations shown in Figure 7-7, Figure 7-8, Figure 7-11 and Figure 7-12.

PERMISSION 7.12:
On single height boards, handles **MAY** be installed in any of the following combinations

1. Top only
2. Bottom only
3. Top and bottom

PERMISSION 7.13:
On double height boards handles **MAY** be installed in any of the following combinations:

1. Top only
2. Middle only
3. Bottom only
4. Top and middle
5. Bottom and middle
6. Top and bottom

OBSERVATION 7.14:
When double and single height boards share the same subrack, handles in the center of the double height front panels align with the handles of single height boards, forming an unbroken line and giving a more consistent appearance.

OBSERVATION 7.15
Removal of double height boards that have both a P1 and a 96-pin P2 connector requires up to 180 N (40.5 lbf) of extraction force. Placing handles at the top and bottom of double height boards makes removal of these boards easiest.

7.3.2 Front Panel Mounting

RECOMMENDATION 7.5:
IF front panels are used,
THEN keep the reserved areas shown in Figure 7-9 and Figure 7-10 free of components to allow for installation of front panel mounting brackets. Locate the holes used to mount these brackets as shown in Figure 7-2 and Figure 7-3.

RECOMMENDATION 7.6:
IF front panels are used on double height boards,
THEN provide at least one center mounting bracket at one of the two locations shown in Figure 7-10.

RULE 7.17:
IF front panels are used,
THEN the test dimension shown in Figure 7-9 and Figure 7-10, from the rear face of the front panel to the rear face of the connector, **MUST** be maintained.

OBSERVATION 7.16:
The test dimensions from the rear face of the front panel to the front face of the backplane guarantee that the P1 and P2 connectors will be fully engaged and that the front panel fasteners will be able to secure the board into the subrack.

7.3.3 Front Panel Dimensions

All dimensioning of the front panel is done from a datum point 0.15 mm (0.006 inch) to the left of its top left corner as viewed from the front.

RULE 7.18:
Single width front panels **MUST** be designed to the dimensions shown in Figure 7-7 and Figure 7-8.

RECOMMENDATION 7.7:
Make front panels 2.5 mm (0.098 inch) nominal thickness.

OBSERVATION 7.17:
The 20.02 mm (0.788 inch) width of the single slot front panels is 0.30 mm (0.012 inch) narrower than the 20.32 mm (0.8 inch) slot spacing. This prevents mechanical, interference between adjacent front panels due to tolerances in the board assemblies and subracks.

RULE 7.19:
IF a board occupies more than one slot and has a front panel,
THEN the width of that front panel **MUST** be 20.02 mm (0.788 inch) plus an integral multiple (N-1) of 20.32mm (0.8 inch), where N is the number of slots that the board occupies.

RULE 7.20:
Single slot front panels **MUST** be equipped with one fastener at the top and another at the bottom, located as shown in Figure 7-7 and Figure 7-8.

7.3.4 Filler Panels

Filler panels are sometimes used where the backplane positioning leaves a gap on the left or right end of the subrack's front panel or where there are empty slots. These filler panels require no mounting brackets because they are not attached to printed circuit boards. They are secured to the subrack by screws or quarter-turn fasteners on their top and bottom ends like the front panels of VMEbus boards.

RULE 7.21:
Filler panels **MUST** be designed to conform to the dimensions given in Figure 7-11 and Figure 7-12.

RULE 7.22:
Single slot filler panels **MUST** be equipped with one fastener at the top and another at the bottom, located as shown in Figure 7-11 and Figure 7-12.

RECOMMENDATION 7.8:
Use filler panels on VMEbus based systems to maintain proper air flow within the subrack and to improve the appearance of the assembled VMEbus system.

SUGGESTION 7.6:
Equip filler panels with handles. This will give a more consistent appearance when they are installed in the same subrack with VMEbus boards that have front panels and handles.

SUGGESTION 7.7:
Provide additional mounting holes on filler panels wider than 101.60 mm (4.0 inch) to provide better attachment to the subrack.

7.3.5 Board Ejectors/Injectors

OBSERVATION 7.18:
Several vendors offer different types of board ejectors/injectors that make insertion and removal of VMEbus boards easier. Unfortunately, ejectors/injectors designed for use with one vendor's products do not always work with other vendor's products. Because no agreement has been reached between the various manufacturers, the VMEbus specification does not specify any as the preferred type.

OBSERVATION 7.19:
The insertion force for a single 603-2-IEC-CO96Mx-xxx connector can be as high as 90 N (20.23 lbf).

PERMISSION 7.14:
VMEbus boards **MAY** be equipped with any type of ejectors, injectors, or retainers as long as they do not make the products they are used on incompatible with boards or subracks designed to the VMEbus specifications.

7.4 BACKPLANES

The primary backplane is designated as the J1 backplane. In some cases this is the only backplane in the VMEbus system. When a double height subrack is used, this backplane is mounted in the upper portion of that subrack. When the expanded VMEbus is used, a second backplane, designated the J2 backplane, is installed below the J1 backplane in the lower portion of the subrack. This J2 backplane buses only the center row (row b) of the P2 connector pins, allowing the two outer rows (rows a and c) to be used to implement the VMXbus or for any other user defined functions.

Board slots are designated 1, 2, 3, ... 21, with the slot numbering starting at the left end of the subrack, as viewed from the front. The daisy-chain propagation starts with slot 1 and goes to 21.

RULE 7.23:
J1 VMEbus backplanes **MUST** bus all signals in all slots except for the daisy-chained signals. (See Section 7.6.1)

RULE 7.24:
When a J2 backplane is used to provide for 32 bit wide address and data transfers, it **MUST** bus all pins of the center row (row b) of those slots for which it has connectors. (See Section 7.6.2)

RULE 7.25:
The 96 pin 603-2-IEC-CO96Fx-xxx connectors **MUST** be used on all J1 VMEbus backplanes and on all J2 expansion backplanes.

RULE 7.26:
All VMEbus J1 backplanes **MUST** have some provision for jumpering the interrupt acknowledge and bus grant daisy-chains when boards are not plugged into a slot.

SUGGESTION 7.8:
To provide for the jumpering of the J1 backplane daisy-chains, use 603-2-IEC-CO96Fx-xxx connectors that have wire-wrap pins.

SUGGESTION 7.9:
IF J1 connectors without wire-wrap pins are used,
THEN place all of the jumper pins for daisy-chains next to the J1 connector that they are jumpering.

PERMISSION 7.15:
As long as all other requirements are met, and proper spacing is maintained between the J1 and J2 connectors (See Figure 7-13), the J1 and J2 backplanes **MAY** be designed as a single PC board.

SUGGESTION 7.10:
Use 603-2-IEC-CO96Fx-xxx connectors with wire-wrap pins for the connectors on J2 backplanes. This allows attachment of ribbon cables and secondary backplanes to those pins.

7.4.1 Backplane Dimensional Requirements

Figure 7-13 depicts a 21 slot backplane. Figure 7-14 shows where optional threaded studs can be provided to connect DC power cables to the backplane.

PERMISSION 7.16:
VMEbus backplanes **MAY** be designed with up to 21 slots.

RECOMMENDATION 7.9:
When designing a backplane with fewer than 21 slots, make the width

$$(N \times 20.32\text{mm})-1.44\text{mm}, \quad +0/-0.3\text{mm}$$

where N is the number of slots. This allows two backplanes to be installed next to each other without wasting a slot position in the subrack.

RECOMMENDATION 7.10:
Do not design backplanes with widths greater than 425.28mm (16.743 inch). Backplanes longer than this might not fit into widely available subracks.

PERMISSION 7.17:
VMEbus backplanes **MAY** be designed with or without threaded studs for power cable connections to the backplane. (See Figure 7-13 and Figure 7-14).

RULE 7.27:
Except for the width dimension which varies with the number of slots, and the optional threaded studs, VMEbus backplanes **MUST** be designed to the dimensions given in Figure 7-13 and Figure 7-14.

OBSERVATION 7.20:
The backplane dimensions shown in Figure 7-14 repeat every 20.32 mm (0.8 inch).

7.4.2 Signal Line Termination Networks

RULE 7.28:
VMEbus backplanes **MUST** either have built-in terminations or they **MUST** provide some way to connect plug-on terminator boards for terminating all of the signal lines indicated in Section 6.7.

PERMISSION 7.18:
The termination circuitry **MAY** either be built onto the ends of the backplane or it **MAY** be provided by separate modules that plug onto either the front or the back of the backplane at the two end slots.

RULE 7.29:
Backplane signal trace lengths, including any plug-on terminator boards, **MUST NOT** exceed 508 mm (20.0 inches).

OBSERVATION 7.21:
The entire board width of a 21 slot VMEbus backplane is required to accommodate the 603-2-IEC-xxxxxx-xxx connectors. Separate plug-on terminator modules are typically needed for such a backplane.

7.5 ASSEMBLY OF VMEbus SUBRACKS

This section shows how VMEbus subracks are assembled. All horizontal dimensions are from the left hand edge of the front opening of the subrack.

7.5.1 Subracks And Slot Widths

Figure 7-15 shows a typical double height 21-slot subrack.

PERMISSION 7.19:
Double height subracks **MAY** be used to house double height boards or they **MAY** be sub-divided by the installation of a bracket, into two single height sections, one above the other.

OBSERVATION 7.22:
The bracket allowed in PERMISSION 7.19 provides two board guides: the lower guide for the board above it and the upper guide for the board below it.

SUGGESTION 7.11:
Where possible, install the left-most (slot 1) board guides so that their center line is 3.27mm (0.129 inches) from the left end of the rack opening. This provides sufficient clearance for the component leads of the board installed in that slot, and doesn't waste any horizontal space. (If this board guide is installed farther to the right there will not be room for 21 slots.)

7.5.2 Subrack Dimensions

RULE 7.30:
All double height VMEbus subracks **MUST** meet all of the dimensional requirements given in Figure 7-15, except for the width dimension, which varies according to how many slots are in the subrack.

RULE 7.31:
All single height VMEbus subracks **MUST** meet all of the dimensional requirements given in Figure 7-15, except for the width dimension, which varies according to how many slots are in the subrack, and the vertical distance between the lower and upper board guides, which is 100.2 mm, +0.4/-0.0 mm (3.94 inches +0.015/-0.0 inches) instead of 233.55 mm (9.19 inches).

OBSERVATION 7.23:
The dimension from the front panel mounting surface to the front face of the backplane is particularly critical, since it guarantees correct connector engagement.

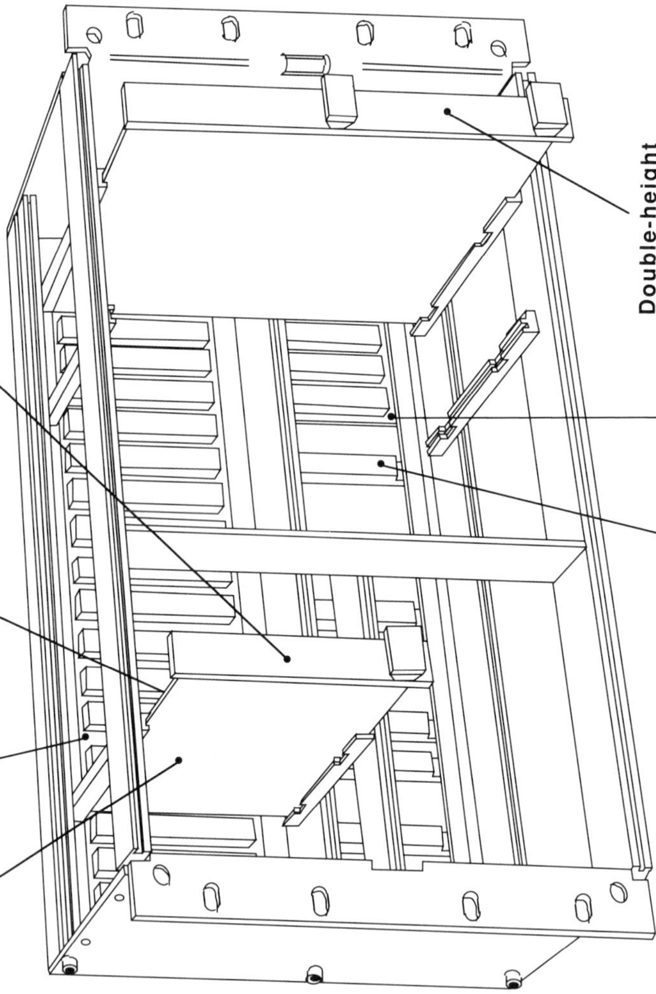

Figure 7-1. Subrack With Mixed Board Sizes

NOTES:

1. All dimensions are shown in millimeters. Inch dimensions are shown in parentheses.
2. This grid is provided to help the board designer to align components with the front panel grid.
3. **RULE 7.32:** Boards **MUST** be 1.6 mm ±0.2 mm (0.063 inch ±0.008 inch) thick in the card guide area.
4. This dimension was not specified in the previous edition.

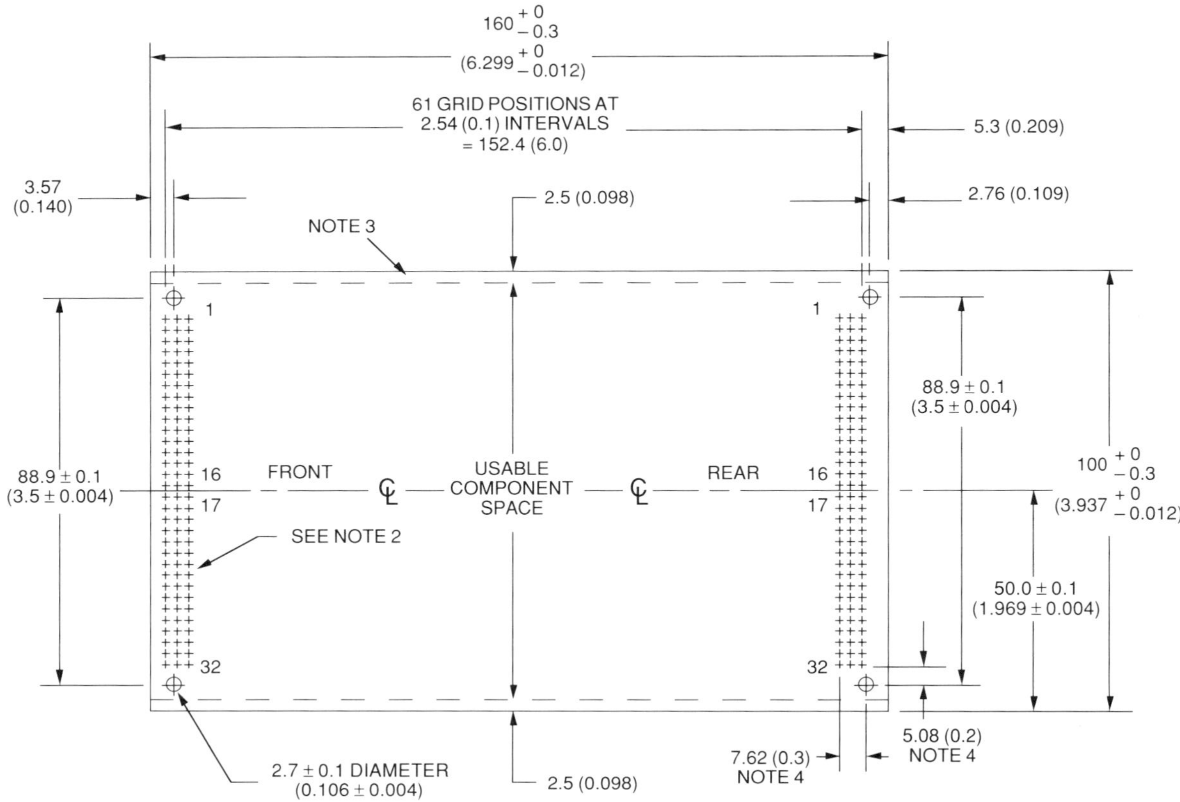

Figure 7-2. Single Height Board: Basic Dimensions

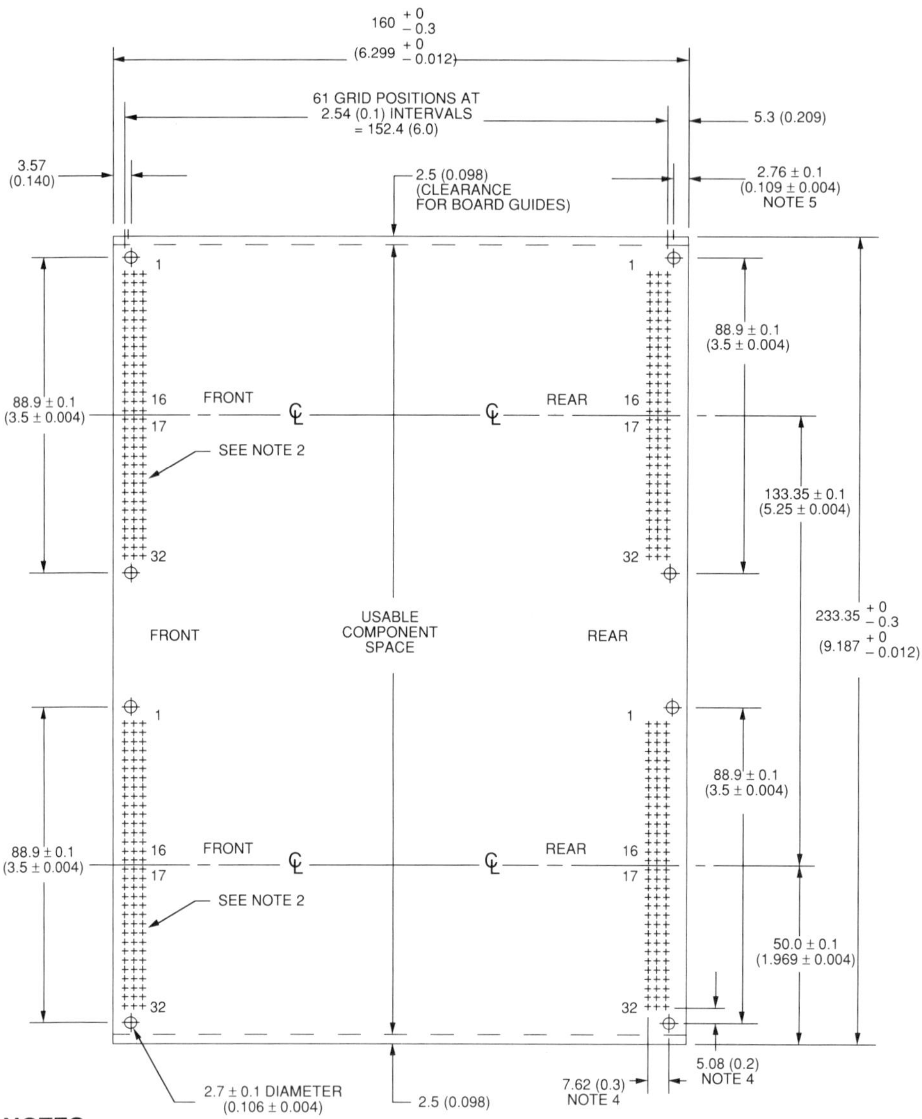

NOTES:

1. All dimensions are shown in millimeters. Inch dimensions are shown in parentheses.
2. This grid is provided to help the board designer to align components with the front panel grid.
3. **RULE 7.33:**
   Boards **MUST** be 1.6 mm ±0.2 mm (0.063 inch ±0.008 inch) thick in the card guide area.
4. This dimension was not specified in the previous edition.
5. This tolerance was not specified in the previous edition.

Figure 7-3. Double Height Board: Basic Dimensions

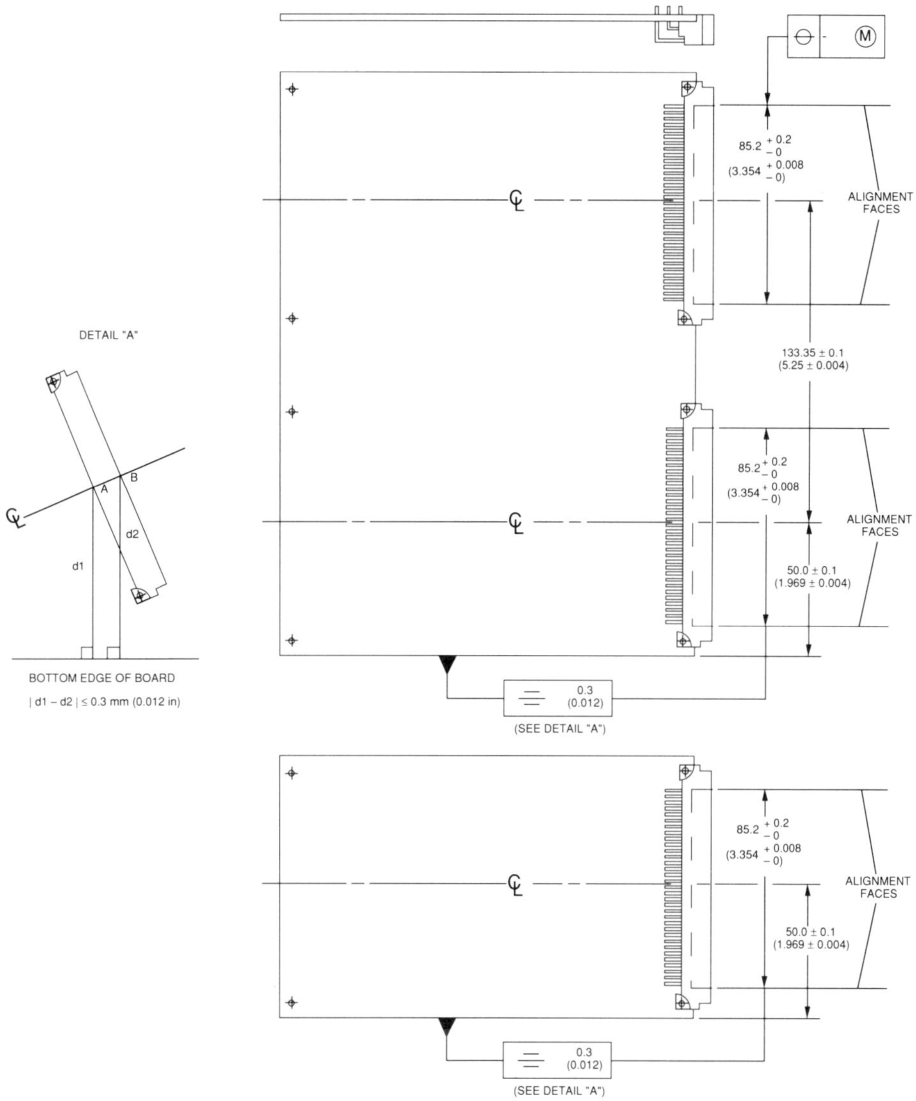

NOTE: All dimensions are shown in millimeters. Inch dimensions are shown in parentheses.

Figure 7-4. Connector Position On Single And Double Height Boards

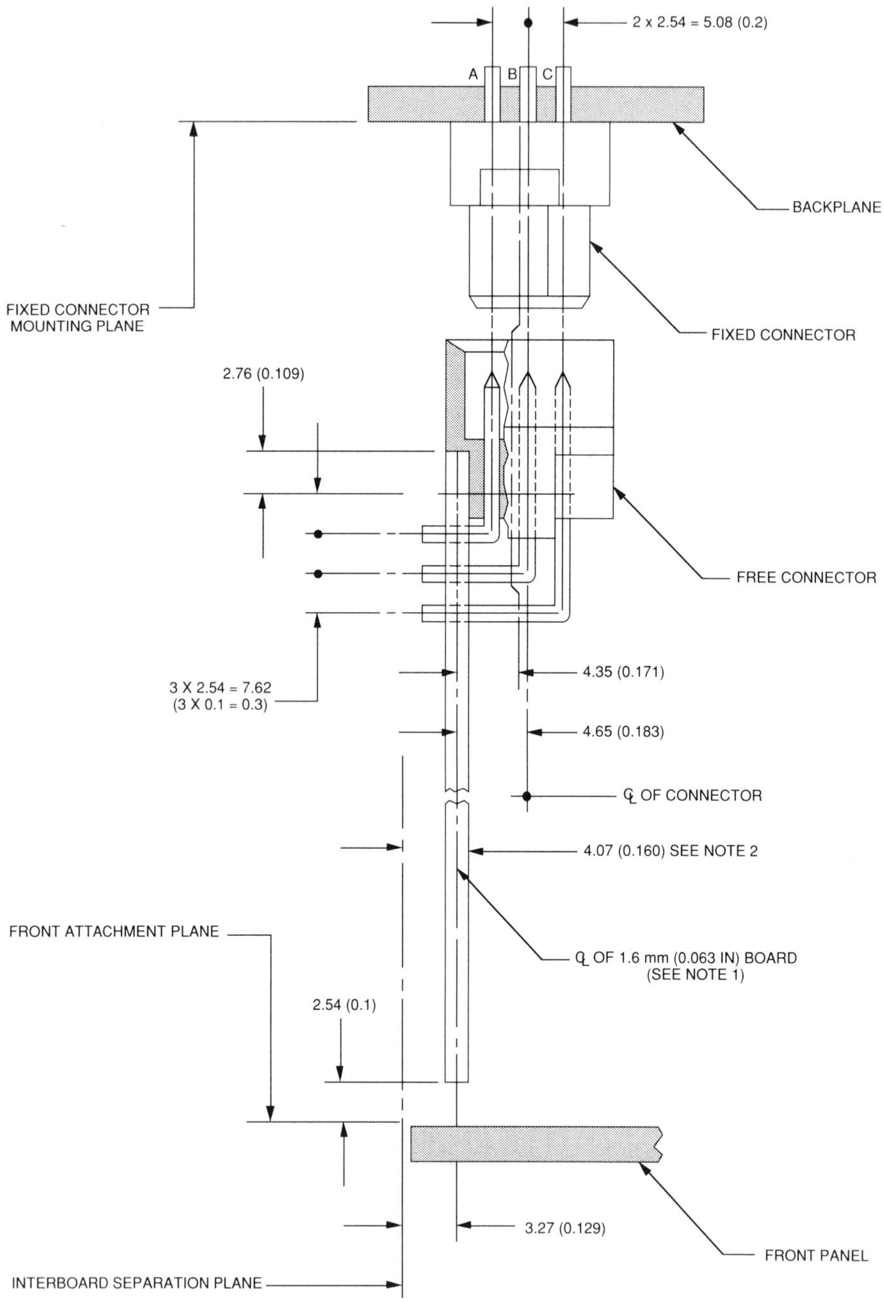

NOTES:

1. For additional information regarding the allowed thickness of boards, see Section 7.2.
2. This 4.07 mm (0.160 in) dimension is the same regardless of the thickness of the board.

Figure 7-5. Cross Sectional View Of Board, Connector, Backplane, And Front Panel

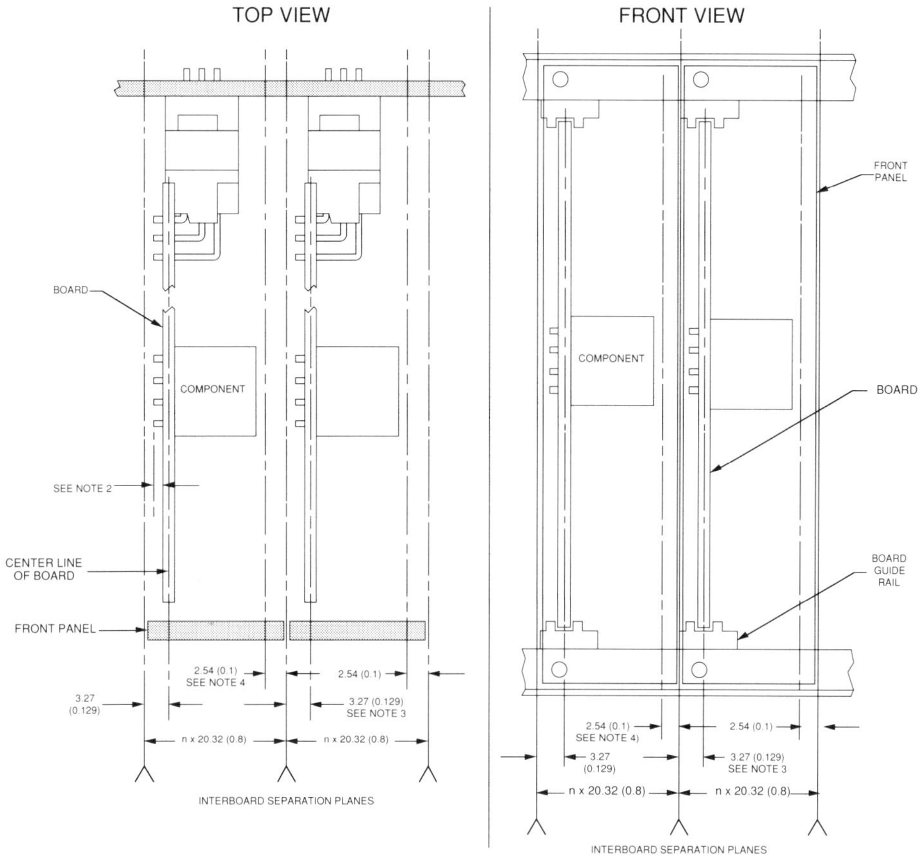

NOTES:

1. All dimensions are shown in millimeters. Inch dimensions are shown in parentheses.
2. **SUGGESTION 7.12:**
   Trim component lead lengths to no more than 1.52 millimeters (0.06 inch).
3. **RULE 7.34:**
   Component leads and components mounted on the solder side of the board **MUST NOT** protrude through the interboard separation plane after it has been completely seated in the backplane.
4. **RULE 7.35:**
   Components mounted on the component side of the board **MUST NOT** be any closer than 2.54 millimeters (0.1 inch) to the interboard separation plane after it has been completely seated in the backplane.

Figure 7-6. Component Height, Lead Length And Board Warpage

NOTES:

1. All dimensions are shown in millimeters. Inch dimensions are shown in parentheses.
2. Dimensions given for height and depth of handles are suggestions only.
3. **RECOMMENDATION 7.11:** Locate the mounting hole 7.62 mm (0.3) in from the interboard separation plane.
   Permission 7.20: The mounting hole **MAY** be located 12.7 mm (0.5 in) from the interboard separation plane.

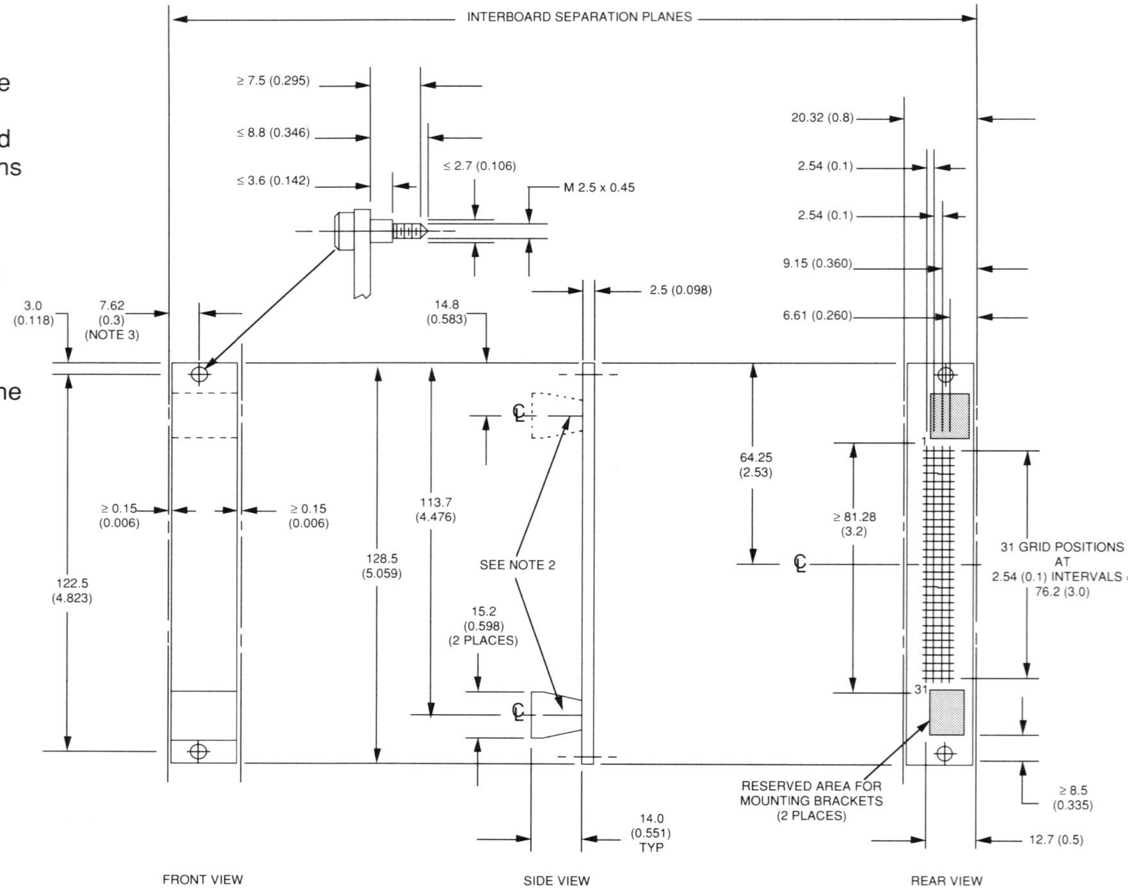

Figure 7-7. Single Height, Single Width Front Panel

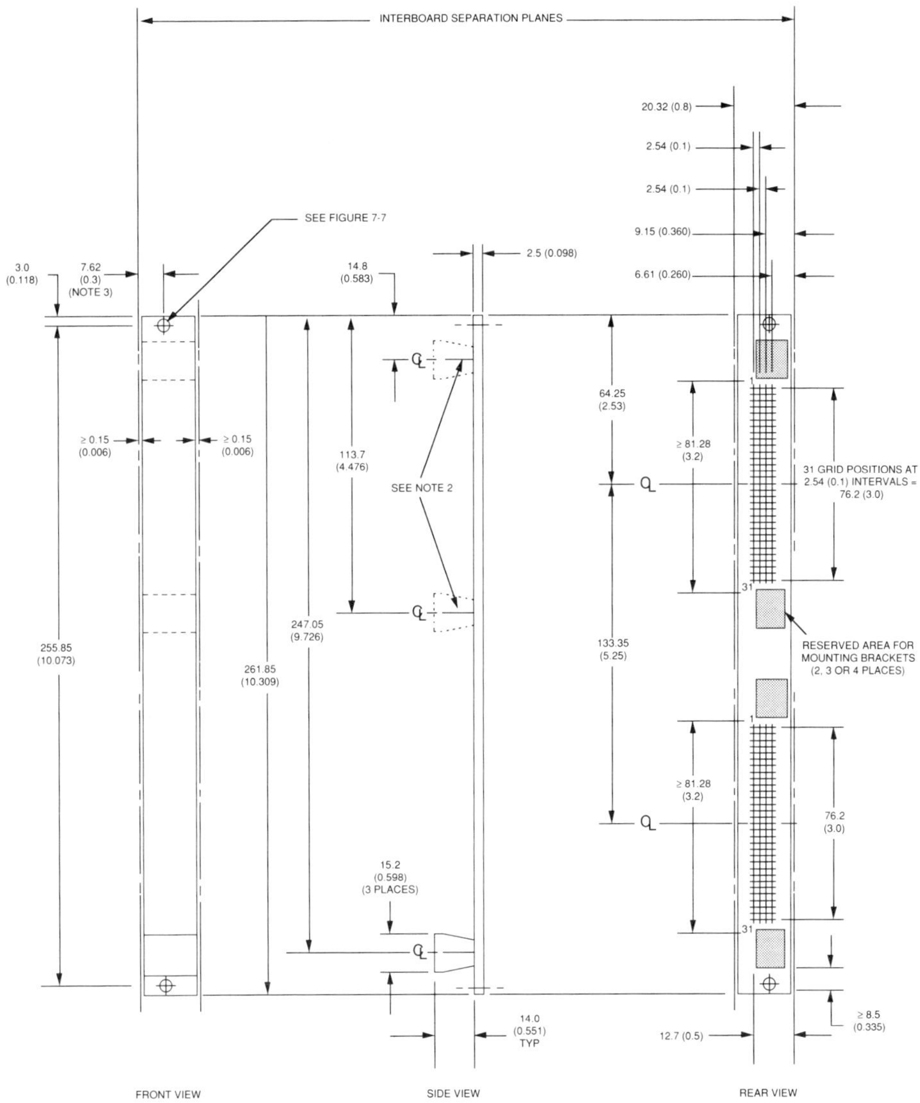

NOTES:

1. All dimensions are shown in millimeters. Inch dimensions are shown in parentheses.
2. Dimensions given for height and depth of handles are suggestions only.
3. **RECOMMENDATION 7.12:**
   Locate the mounting hole 7.62 mm (0.3 in) from the interboard separation plane.
   PERMISSION 7.21:
   The mounting hole **MAY** be located 12.7 mm (0.5 in) from the interboard separation plane.

Figure 7-8. Double Height, Single Width Front Panel

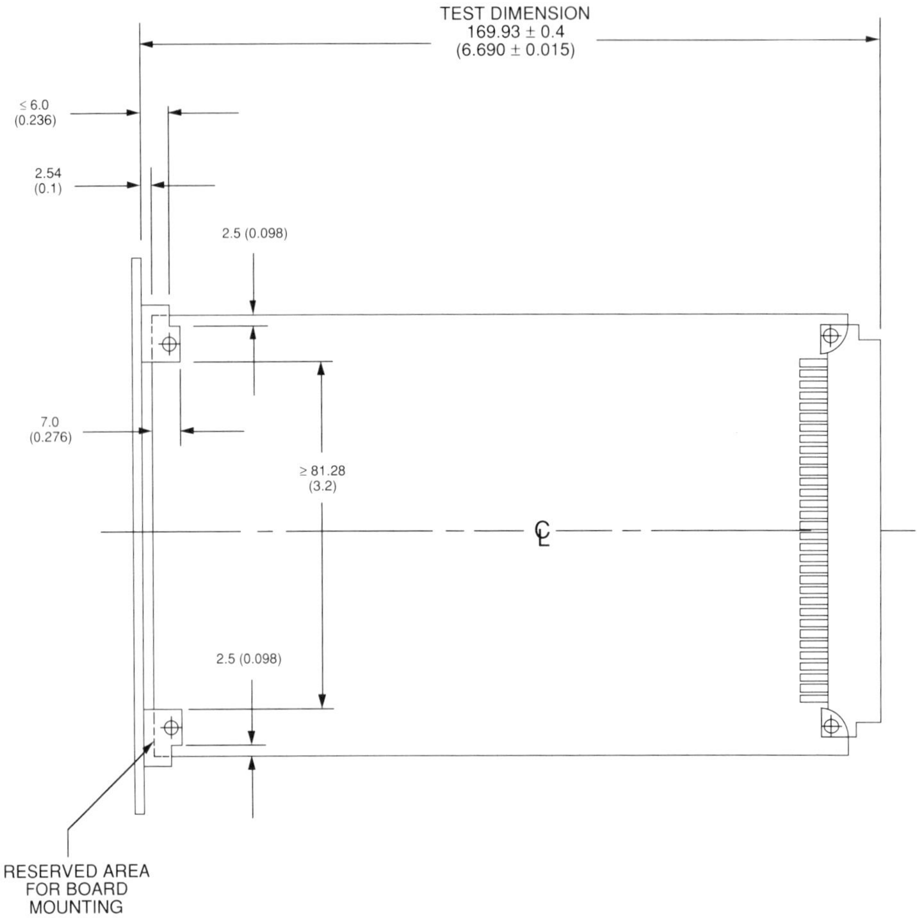

NOTE:

1. All dimensions are shown in millimeters. Inch dimensions are shown in parentheses.

Figure 7-9. Front Panel Mounting Brackets And Dimension Of Single Height Boards

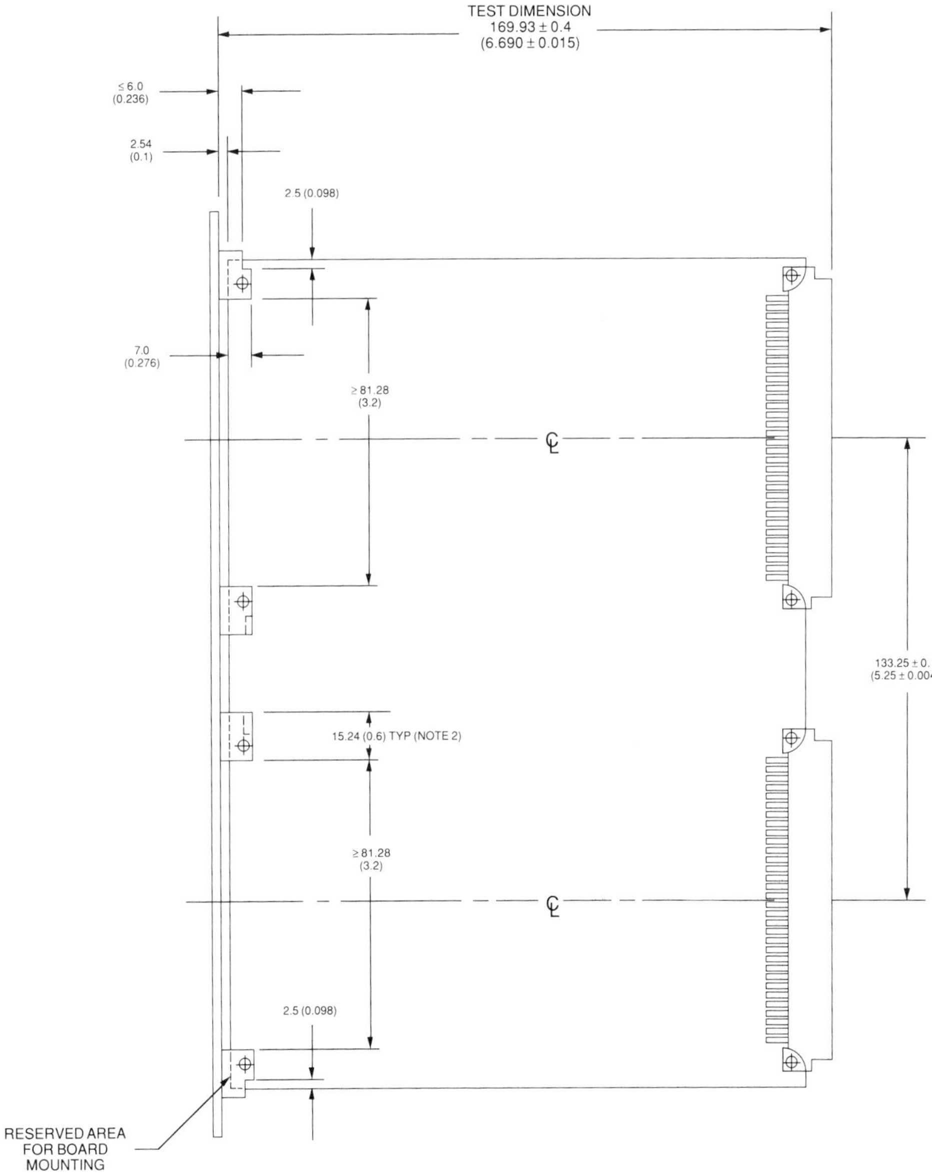

NOTE:

1. All dimensions are shown in millimeters. Inch dimensions are shown in parentheses.
2. Previous edition showed 15.11 (0.6) TYP

Figure 7-10. Front Panel Mounting Brackets And Dimension Of Double Height Boards

NOTES:

1. All dimensions are shown in millimeters. Inch dimensions are shown in parentheses.
2. **RECOMMENDATION 7.13:** Where a panel is more than 50.8 millimeters (2.0 inch) wide use at least 4 mounting holes; two at the top and two at the bottom.
3. Dimensions given for height and depth of handles are suggestions only.
4. Corrected to read "N+1" (Previous edition read "N")

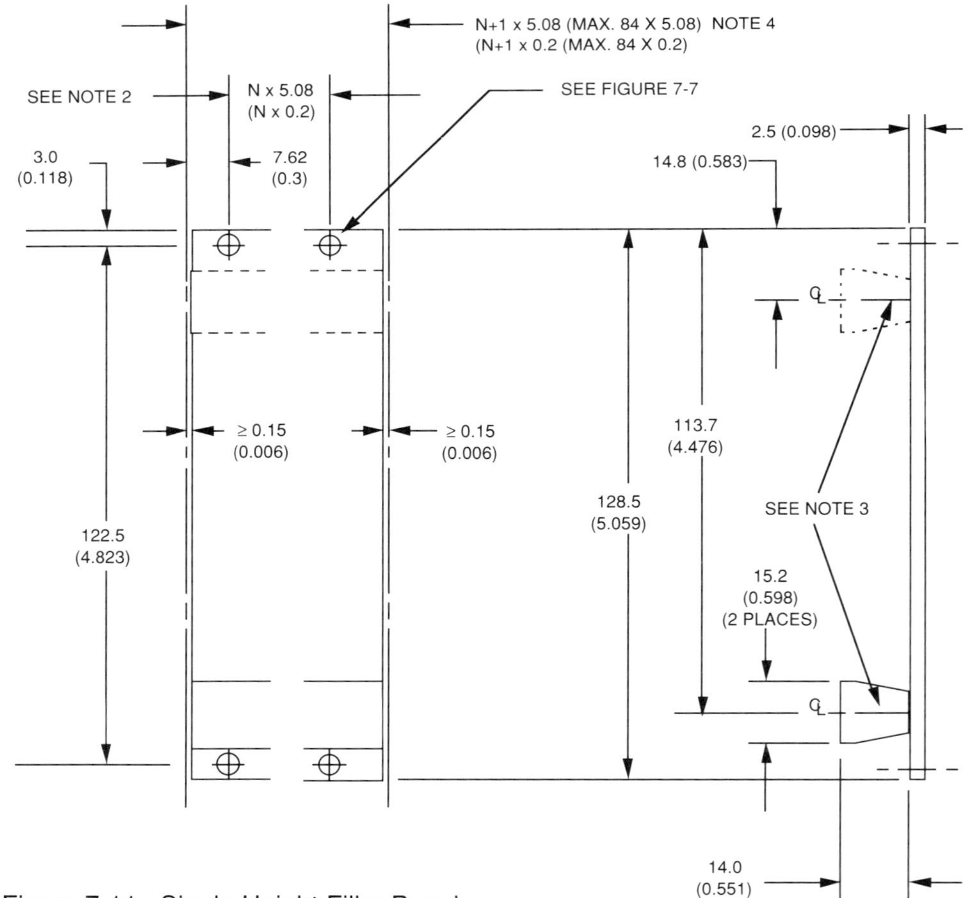

Figure 7-11. Single Height Filler Panel

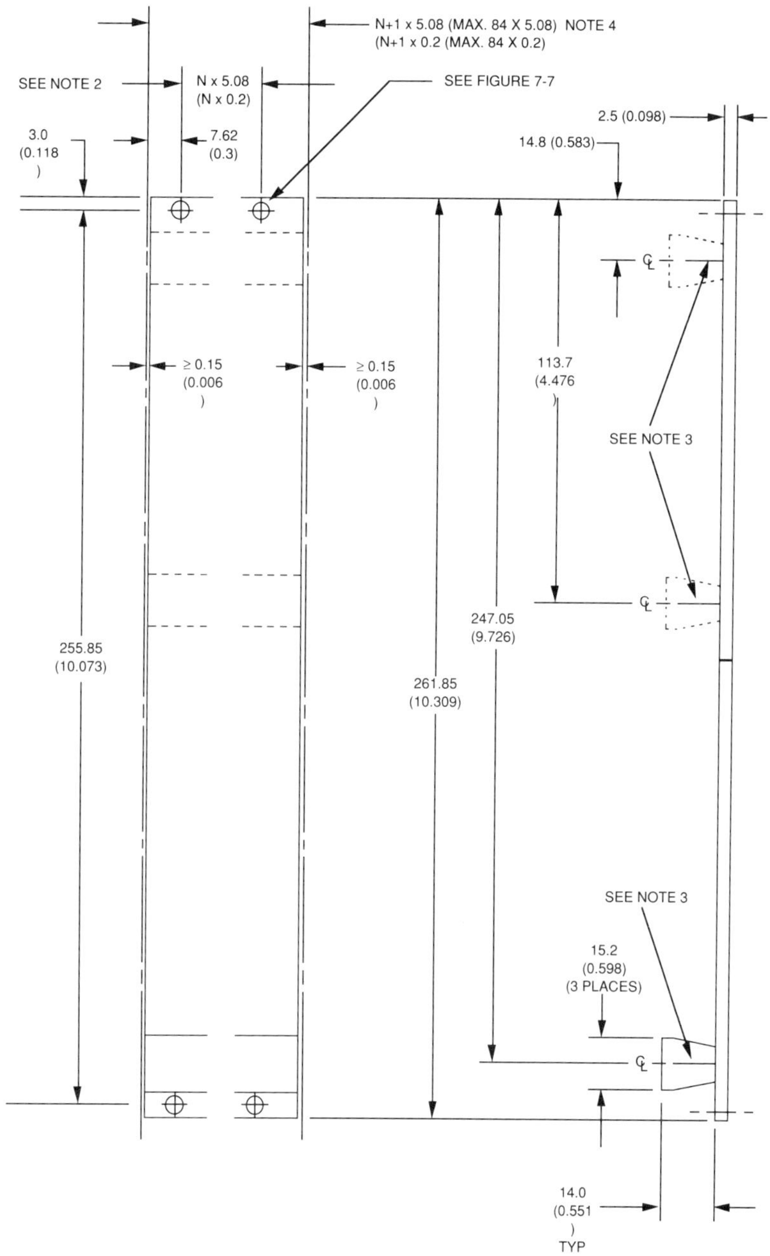

NOTES:

1. All dimensions are shown in millimeters. Inch dimensions are shown in parentheses.
2. **RECOMMENDATION 7.14:**
   Where a panel is more than 50.8 millimeters (2.0 inch) wide use at least 4 mounting holes; two at the top and two at the bottom.
3. Dimensions given for height and depth of handles are suggestions only.
4. Corrected to read "N+1" (Previous edition read "N")

Figure 7-12. Double Height Filler Panel

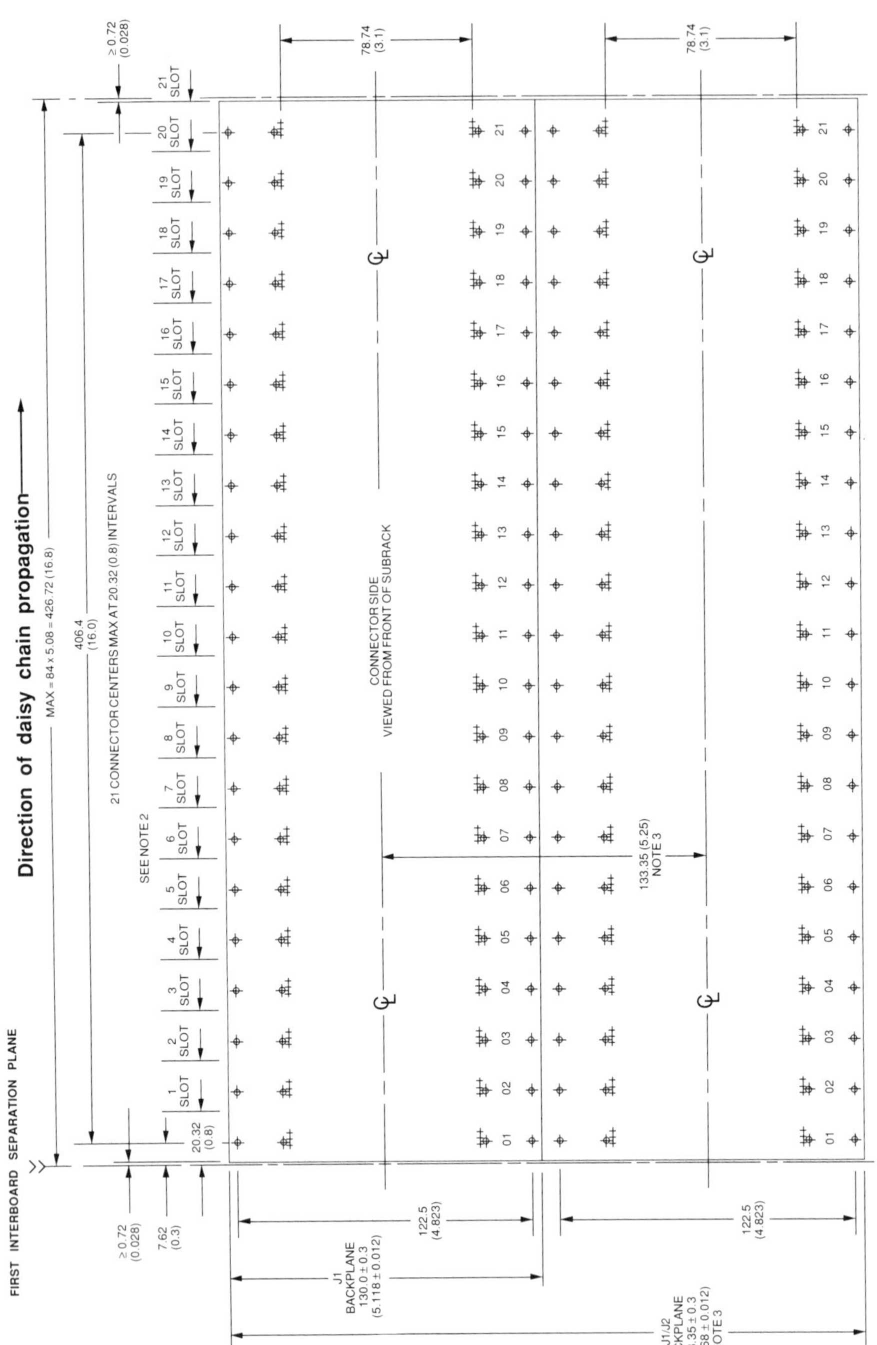

NOTES:

1. All dimensions are shown in millimeters. Inch dimensions are shown in parentheses.
2. Backplane width varies depending on the number of slots.
3. This dimension was not specified in the previous edition.

Figure 7-13. Backplane Overall Dimensions

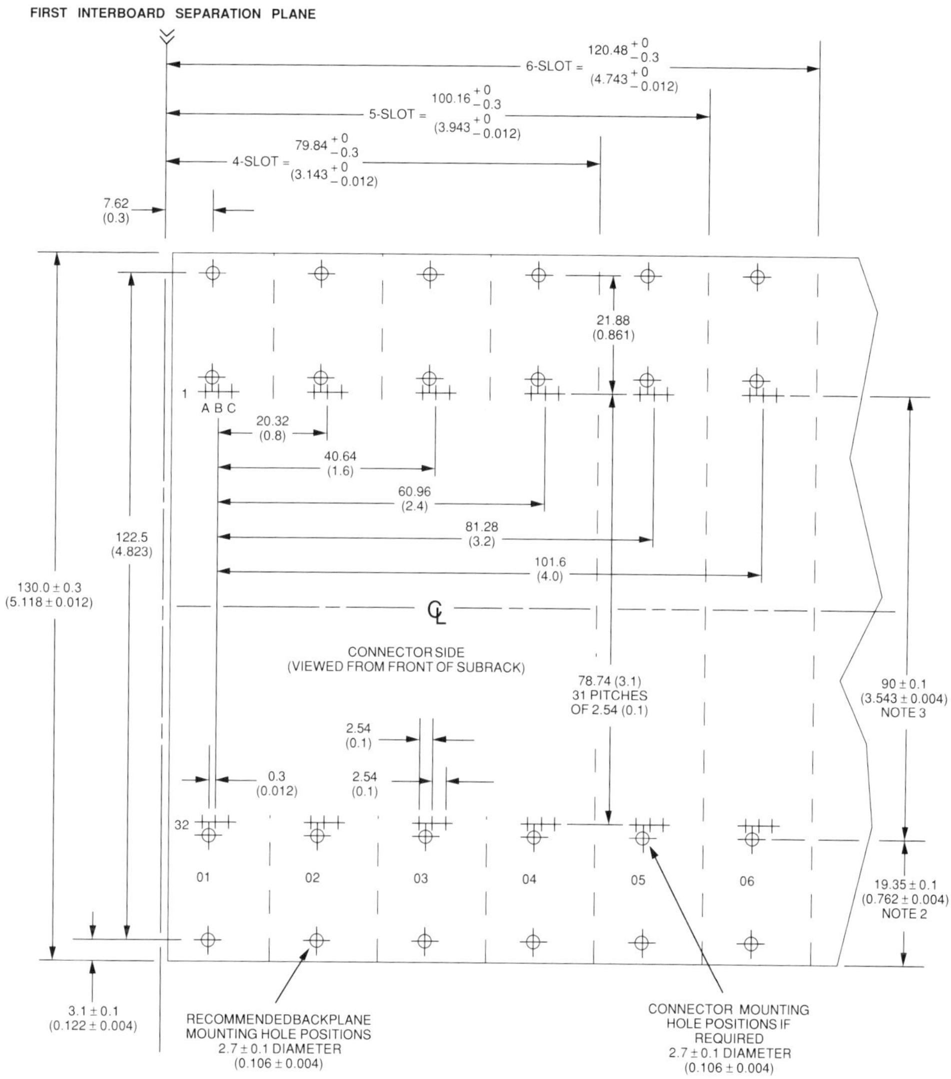

NOTES:

1. All dimensions are shown in millimeters. Inch dimensions are shown in parentheses.
2. This dimension was not specified in the previous edition.
3. This tolerance was not specified in the previous edition.

Figure 7-14. Backplane Detailed Dimensions

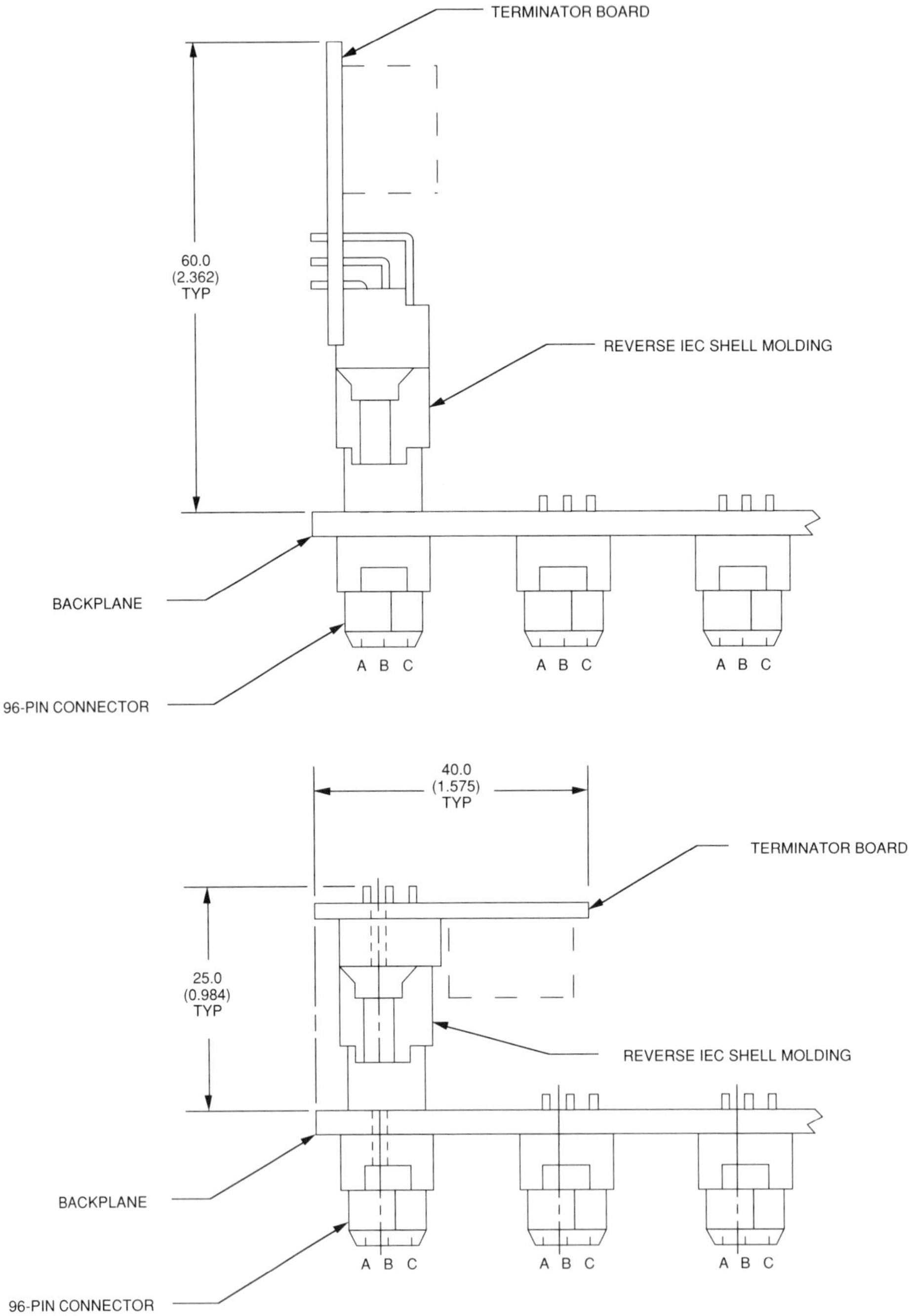

Figure 7-15. "Off-Board Type" Backplane Termination (Viewed From Top Of Backplane)

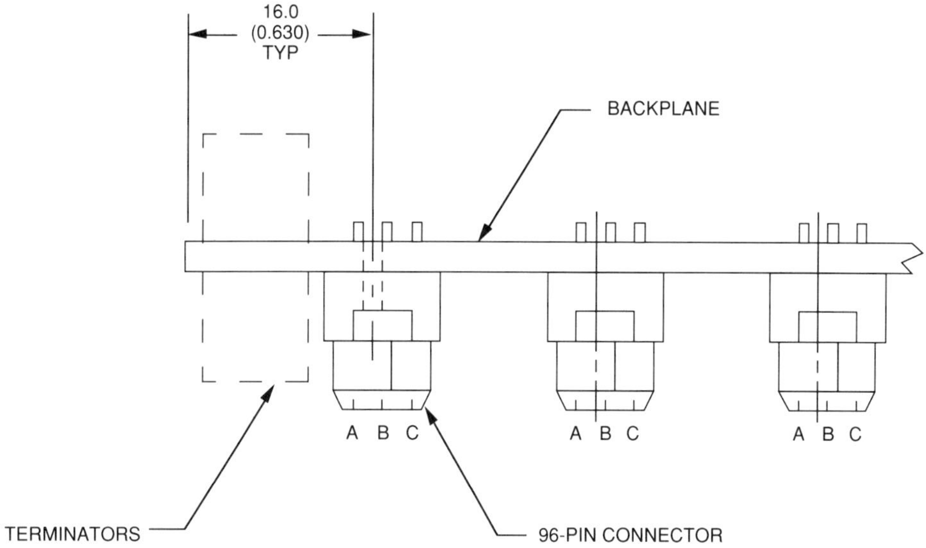

Figure 7-16. "On-Board Type" Backplane Termination
(Viewed From Top Of Backplane)

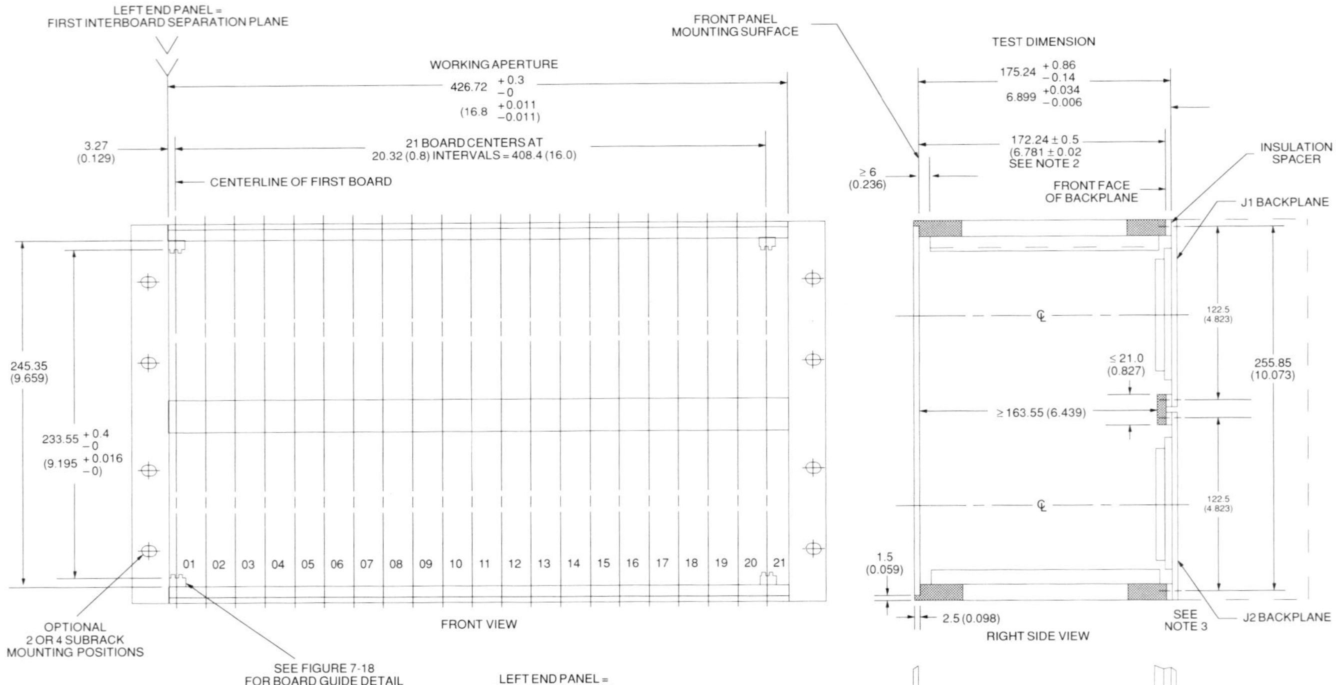

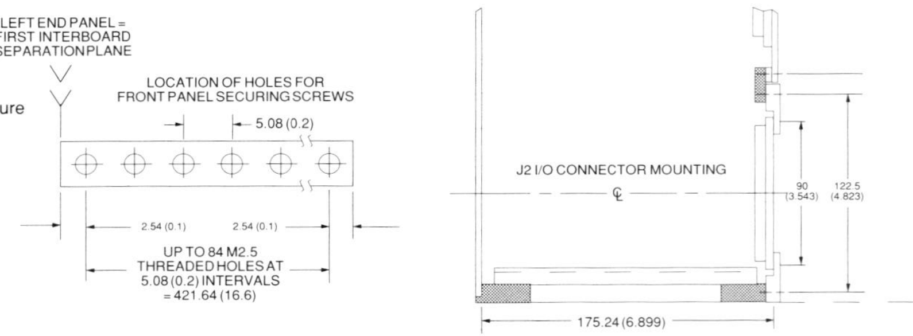

NOTES:

1. All dimensions are shown in millimeters. Inch dimensions are shown in parentheses.
2. **RULE 7.36:**
   The thickness of the insulation spacer (shown in the right side view) **MUST** be chosen to assure that the face of the backplane is the correct distance from the front panel mounting surface.
   OBSERVATION 7.24:
   The thickness of the spacer specified in RULE 7.36 might vary from one vendor to another.
3. **PERMISSION 7.22:**
   Side plates MAY be extended to the rear as required.
4. For additional information, see IEC 297-1 and IEC 297-3.

Figure 7-17. 21 Slot Subrack

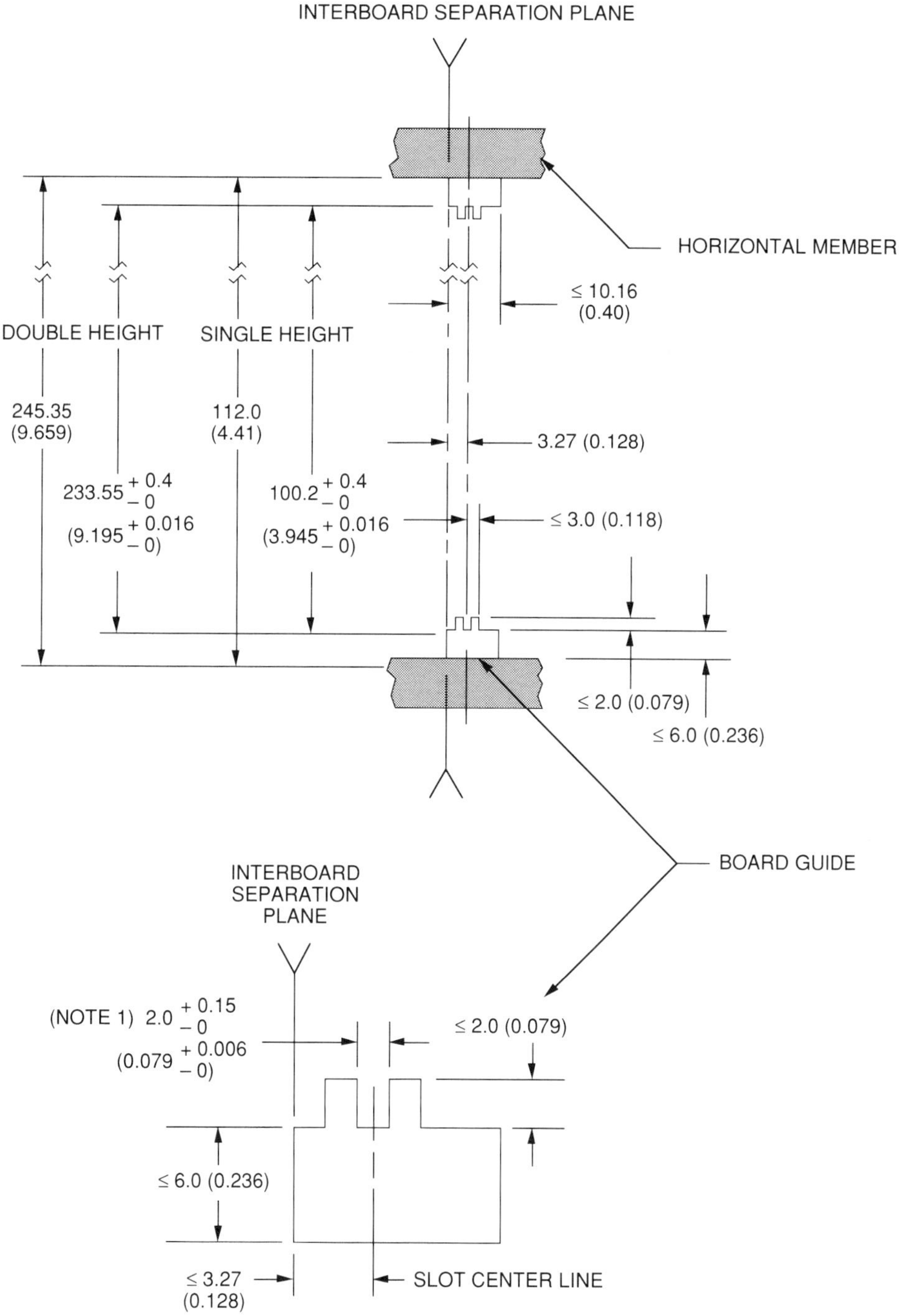

NOTE:

1. This dimension was not specified in the previous edition.

Figure 7-18. Board Guide Detail

7.6 VMEbus BACKPLANE CONNECTORS AND VMEbus BOARD CONNECTORS

7.6.1 Pin Assignments For The J1/P1 Connector

Table 7-1 provides signal names for the J1/P1 connector pins. (The connector consists of three rows of pins labeled rows a, b, and c.)

Table 7-1. J1/P1 Pin Assignments

PIN NUMBER	ROW a SIGNAL MNEMONIC	ROW b SIGNAL MNEMONIC	ROW c SIGNAL MNEMONIC
1	D00	BBSY*	D08
2	D01	BCLR*	D09
3	D02	ACFAIL*	D10
4	D03	BG0IN*	D11
5	D04	BG0OUT*	D12
6	D05	BG1IN*	D13
7	D06	BG1OUT*	D14
8	D07	BG2IN*	D15
9	GND	BG2OUT*	GND
10	SYSCLK	BG3IN*	SYSFAIL*
11	GND	BG3OUT*	BERR*
12	DS1*	BR0*	SYSRESET*
13	DS0*	BR1*	LWORD*
14	WRITE*	BR2*	AM5
15	GND	BR3*	A23
16	DTACK*	AM0	A22
17	GND	AM1	A21
18	AS*	AM2	A20
19	GND	AM3	A19
20	IACK*	GND	A18
21	IACKIN*	SERCLK (1)	A17
22	IACKOUT*	SERDAT*(1)	A16
23	AM4	GND	A15
24	A07	IRQ7*	A14
25	A06	IRQ6*	A13
26	A05	IRQ5*	A12
27	A04	IRQ4*	A11
28	A03	IRQ3*	A10
29	A02	IRQ2*	A09
30	A01	IRQ1*	A08
31	-12V	+5VSTDBY	+12V
32	+5V	+5V	+5V

Note: (1) See Appendix C for further information on the use of these signals.

7.6.2 Pin Assignments For The J2/P2 Connector

Table 7-2 provides signal names for the J2/P2 connector pins. (The connector consists of three rows of pins labeled rows a, b, and c.)

Table 7-2. J2/P2 Pin Assignments

PIN NUMBER	ROW a SIGNAL MNEMONIC	ROW b SIGNAL MNEMONIC	ROW c SIGNAL MNEMONIC
1	User Defined	+5V	User Defined
2	User Defined	GND	User Defined
3	User Defined	RESERVED	User Defined
4	User Defined	A24	User Defined
5	User Defined	A25	User Defined
6	User Defined	A26	User Defined
7	User Defined	A27	User Defined
8	User Defined	A28	User Defined
9	User Defined	A29	User Defined
10	User Defined	A30	User Defined
11	User Defined	A31	User Defined
12	User Defined	GND	User Defined
13	User Defined	+5V	User Defined
14	User Defined	D16	User Defined
15	User Defined	D17	User Defined
16	User Defined	D18	User Defined
17	User Defined	D19	User Defined
18	User Defined	D20	User Defined
19	User Defined	D21	User Defined
20	User Defined	D22	User Defined
21	User Defined	D23	User Defined
22	User Defined	GND	User Defined
23	User Defined	D24	User Defined
24	User Defined	D25	User Defined
25	User Defined	D26	User Defined
26	User Defined	D27	User Defined
27	User Defined	D28	User Defined
28	User Defined	D29	User Defined
29	User Defined	D30	User Defined
30	User Defined	D31	User Defined
31	User Defined	GND	User Defined
32	User Defined	+5V	User Defined

APPENDIX A

GLOSSARY OF VMEbus TERMS

A16
A type of module that provides or decodes an address on address lines A01 through A15.

A24
A type of module that provides or decodes an address on address lines A01 through A23.

A32
A type of module that provides or decodes an address on address lines A01 through A31.

ARBITRATION
The process of assigning control of the DTB to a REQUESTER.

ADDRESS-ONLY CYCLE
A DTB cycle that consists of an address broadcast, but no data transfer. SLAVES do not acknowledge ADDRESS-ONLY cycles and MASTERS terminate the cycle without waiting for an acknowledgment.

ARBITER
A functional module that accepts bus requests from REQUESTER modules and grants control of the DTB to one REQUESTER at a time.

ARBITRATION BUS
One of the four buses provided by the VMEbus backplane. This bus allows an ARBITER module and several REQUESTER modules to coordinate use of the DTB.

ARBITRATION CYCLE
An ARBITRATION CYCLE begins when the ARBITER senses a bus request. The ARBITER grants the bus to a REQUESTER, which signals that the DTB is busy. The REQUESTER terminates the cycle by releasing the bus busy signal which causes the ARBITER to sample the bus requests again.

BACKPLANE (VMEbus)
A printed circuit (PC) board with 96 pin connectors and signal paths that bus the connector pins. Some VMEbus systems have a single PC board, called the J1 backplane. It provides the signal paths needed for basic operation. Other VMEbus systems also have an optional second PC board, called a J2 backplane. It provides the additional 96 pin connectors and signal paths needed for wider data and address transfers. Still others have a single PC board that provides the signal conductors and connectors of both the J1 and J2 backplanes.

BACKPLANE INTERFACE LOGIC

Special interface logic that takes into account the characteristics of the backplane: its signal line impedance, propagation time, termination values, etc. The VMEbus specification prescribes certain rules for the design of this logic based on the maximum length of the backplane and its maximum number of board slots.

BLOCK READ CYCLE

A DTB cycle used to transfer a block of 1 to 256 bytes from a SLAVE to a MASTER. This transfer is done using a string of 1, 2, or 4 byte data transfers. Once the block transfer is started, the MASTER does not release the DTB until all of the bytes have been transferred. It differs from a string of read cycles in that the MASTER broadcasts only one address and address modifier (at the beginning of the cycle). Then the SLAVE increments this address on each transfer so that the data for the next cycle is retrieved from the next higher location.

BLOCK WRITE CYCLE

A DTB cycle used to transfer a block of 1 to 256 bytes from a MASTER to a SLAVE. The block write cycle is very similar to the block read cycle. It uses a string of 1, 2, or 4 byte data transfers and the MASTER does not release the DTB until all of the bytes have been transferred. It differs from a string of write cycles in that the MASTER broadcasts only one address and address modifier (at the beginning of the cycle). Then the SLAVE increments this address on each transfer so that the next transfer is stored in the next higher location.

BOARD

A printed circuit (PC) board, its collection of electronic components, and either one or two 96 pin connectors that can be plugged into VMEbus backplane connectors.

BUS TIMER

A functional module that measures how long each data transfer takes on the DTB and terminates the DTB cycle if a transfer takes too long. Without this module, if the MASTER tries to transfer data to or from a non-existent SLAVE location it could wait forever for a SLAVE to respond. The BUS TIMER prevents this by terminating the cycle.

D08(O)

A SLAVE that sends and receives data 8 bits at a time over D00-D07,

OR

an INTERRUPT HANDLER that receives 8 bit STATUS/IDs over D00-D07,

OR

an INTERRUPTER that sends 8 bit STATUS/IDs over D00-D07.

D08(EO)

A MASTER that sends or receives data 8 bits at a time over either D00-D07 or D08-D15,

OR

a SLAVE that sends and receives data 8 bits at a time over either D00-D07 or D08-D15.

D16

A MASTER that sends and receives data 16 bits at a time over D00-D15,

OR

a SLAVE that sends and receives data 16 bits at a time over D00-D15,

OR

an INTERRUPT HANDLER that receives 16 bit STATUS/IDs over D00-D15,

OR

an INTERRUPTER that sends 16 bit STATUS/IDs over D00-D15.

D32

A MASTER that sends and receives data 32 bits at a time over D00-D31,

OR

a SLAVE that sends and receives data 32 bits at a time over D00-D31,

OR

an INTERRUPT HANDLER that receives 32 bit STATUS/IDs over D00-D31,

OR

an INTERRUPTER that sends 32 bit STATUS/IDs over D00-D31.

DAISY-CHAIN

A special type of VMEbus signal line that is used to propagate a signal level from board to board, starting with the first slot and ending with the last slot. There are four bus grant daisy-chains and one interrupt acknowledge daisy-chain on the VMEbus backplane.

DATA TRANSFER BUS

One of the four buses provided by the VMEbus backplane. The DATA TRANSFER BUS allows MASTERS to direct the transfer of binary data between themselves and SLAVES. (DATA TRANSFER BUS is often abbreviated DTB.)

DATA TRANSFER BUS CYCLE

A sequence of level transitions on the signal lines of the DTB that result in the transfer of an address or an address and data between a MASTER and a SLAVE. There are 34 types of data transfer bus cycles.

DTB

An acronym for DATA TRANSFER BUS.

FUNCTIONAL MODULE

A collection of electronic circuitry that resides on one VMEbus board and works together to accomplish a task.

IACK DAISY-CHAIN DRIVER

A functional module which activates the interrupt acknowledge daisy-chain whenever an INTERRUPT HANDLER acknowledges an interrupt request. This daisy-chain ensures that only one INTERRUPTER will respond with its STATUS/ID when more than one has generated an interrupt request on the same level.

INTERRUPT ACKNOWLEDGE CYCLE

A DTB cycle, initiated by an INTERRUPT HANDLER, that reads a "STATUS/ID" from an INTERRUPTER. An INTERRUPT HANDLER generates this cycle when it detects an interrupt request from an INTERRUPTER and it has control of the DTB.

PRIORITY INTERRUPT BUS

One of the four buses provided by the VMEbus backplane.The PRIORITY INTERRUPT BUS allows INTERRUPTER modules to send interrupt requests to INTERRUPT HANDLER modules, and INTERRUPT HANDLER modules to acknowledge these interrupt requests.

INTERRUPTER

A functional module that generates an interrupt request on the INTERRUPT BUS and then provides STATUS/ID information when the INTERRUPT HANDLER requests it.

INTERRUPT HANDLER

A functional module that detects interrupt requests generated by INTERRUPTERS and responds to those requests by asking for STATUS/ID information.

LOCATION MONITOR

A functional module that monitors data transfers over the DTB in order to detect accesses to the locations it has been assigned to watch. When an access occurs to one of these assigned locations, the LOCATION MONITOR generates an on-board signal.

MASTER

A functional module that initiates DTB cycles in order to transfer data between itself and a SLAVE module.

POWER MONITOR MODULE

A functional module that monitors the status of the primary power source to the VMEbus system and signals when the power has strayed outside the limits required for reliable system operation. Since most systems are powered by an AC source, the power monitor is typically designed to detect drop-out or brown-out conditions on AC lines.

READ CYCLE

A DTB cycle used to transfer 1, 2, 3, or 4 bytes from a SLAVE to a MASTER. The cycle begins when the MASTER broadcasts an address and an address modifier. Each SLAVE captures this address and address modifier, and checks to see if it is to respond to the cycle. If so, it retrieves the data from its internal storage, places it on the data bus, and acknowledges the transfer. Then the MASTER terminates the cycle.

READ-MODIFY-WRITE CYCLE

A DTB cycle that is used to first read from, and then write to, a SLAVE location without permitting any other MASTER to access that location during that cycle. This cycle is most useful in multiprocessing systems where certain memory locations are used to control access to certain systems resources. (i.e., implementation of semaphores)

REQUESTER

A functional module that resides on the same board as a MASTER or INTERRUPT HANDLER and requests use of the DTB whenever its MASTER or INTERRUPT HANDLER needs it.

SERIAL CLOCK DRIVER

A functional module that provides a periodic timing signal that synchronizes operation of the VMSbus. (Although the VMEbus specification defines a SERIAL CLOCK DRIVER for use with the VMSbus, and although it reserves two backplane signal lines for use by that bus, the VMSbus protocol is completely independent of the VMEbus). A specification for the timing of the signal generated by the SERIAL CLOCK DRIVER is given in Appendix C.

SLAVE

A functional module that detects DTB cycles initiated by a MASTER and, when those cycles specify its participation, transfers data between itself and the MASTER.

SLOT

A position where a board can be inserted into a VMEbus backplane. If the VMEbus system has both a J1 and a J2 backplane (or a combination J1/J2 backplane) each slot provides a pair of 96 pin connectors. If the system has only a J1 backplane, then each slot provides a single 96 pin connector.

SUBRACK

A rigid framework that provides mechanical support for boards inserted into the backplane, ensuring that the connectors mate properly and that adjacent boards do not contact each other. It also guides the cooling airflow through the system, and ensures that inserted boards do not disengage themselves from the backplane due to vibration or shock.

SYSTEM CLOCK DRIVER

A functional module that provides a 16 MHz timing signal on the UTILITY BUS.

SYSTEM CONTROLLER BOARD

A board which resides in slot 1 of a VMEbus backplane and has a SYSTEM CLOCK DRIVER, a DTB ARBITER, an IACK DAISY CHAIN DRIVER, and a BUS TIMER. Some also have a SERIAL CLOCK DRIVER, a POWER MONITOR or both.

UAT

A MASTER that sends or receives data in an unaligned fashion,

OR

a SLAVE that sends and receives data in an unaligned fashion.

UTILITY BUS

One of the four buses provided by the VMEbus backplane. This bus includes signals that provide periodic timing and coordinate the power-up and power-down of VMEbus systems.

WRITE CYCLE

A DTB cycle used to transfer 1, 2, 3, or 4 bytes from a MASTER to a SLAVE. The cycle begins when the MASTER broadcasts an address and address modifier and places data on the DTB. Each SLAVE captures this address and address modifier, and checks to see if it is to respond to the cycle. If so, it stores the data and then acknowledges the transfer. The MASTER then terminates the cycle.

APPENDIX B

VMEbus CONNECTOR/PIN DESCRIPTION

INTRODUCTION

This appendix describes the VMEbus signal lines. The following table identifies the VMEbus signals by signal mnemonic, and describes the signal characteristics.

VMEbus Signal Identification

SIGNAL MNEMONIC	SIGNAL NAME AND DESCRIPTION
A01-A15	ADDRESS bus (bits 1-15) - Three-state driven address lines that are used to broadcast a short, standard, or extended address.
A16-A23	ADDRESS bus (bits 16-23) - Three-state driven address lines that are used in conjunction with A01-A15 to broadcast a standard or extended address.
A24-A31	ADDRESS bus (bits 24-31) - Three-state driven address lines that are used in conjunction with A01-A23 to broadcast an extended address.
ACFAIL*	AC FAILURE - An open-collector driven signal which indicates that the AC input to the power supply is no longer being provided or that the required AC input voltage levels are not being met.
AM0-AM5	ADDRESS MODIFIER (bits 0-5) - Three-state driven lines that are used to broadcast information such as address size, cycle type, and/or MASTER identification.
AS*	ADDRESS STROBE - A three-state driven signal that indicates when a valid address has been placed on the address bus.
BBSY*	BUS BUSY - An open-collector driven signal driven low by the current MASTER to indicate that it is using the bus. When the MASTER releases this line, the resultant rising edge causes the ARBITER to sample the bus grant lines and grant the bus to the highest priority requester.

SIGNAL MNEMONIC	SIGNAL NAME AND DESCRIPTION
BCLR*	BUS CLEAR - A totem-pole driven signal, generated by an ARBITER to indicate when there is a higher priority request for the bus. This signal requests the current MASTER to release the DTB .
BERR*	BUS ERROR - An open-collector driven signal generated by a SLAVE or BUS TIMER. This signal indicates to the MASTER that the data transfer was not completed.
BG0IN*-BG3IN*	BUS GRANT (0-3) IN - Totem-pole driven signals generated by the ARBITER and REQUESTERS. "Bus grant in" and "bus grant out" signals form bus grant daisy chains. The "bus grant in" signal indicates, to the board receiving it, that it may use the DTB.
BG0OUT*-BG3OUT*	BUS GRANT (0-3) OUT - Totem-pole driven signals generated by REQUESTERS. The bus grant out signal indicates to the next board in the daisy-chain that it may use the DTB.
BR0*-BR3*	BUS REQUEST (0-3) - Open-collector driven signals generated by REQUESTERS. A low level on one of these lines indicates that some MASTER needs to use the DTB.
D00-D31	DATA BUS - Three-state driven bidirectional data lines used to transfer data between MASTERS and SLAVES.
DS0*, DS1*	DATA STROBE ZERO, ONE - Three-state driven signals used in conjunction with LWORD* and A01 to indicate how many data bytes are being transferred (1, 2, 3, or 4). During a write cycle, the falling edge of the first data strobe indicates that valid data is available on the data bus. On a read cycle, the rising edge of the first data strobe indicates that data has been accepted from the data bus.
DTACK*	DATA TRANSFER ACKNOWLEDGE - An open-collector driven signal generated by a SLAVE. The falling edge of this signal indicates that valid data is available on the data bus during a read cycle, or that data has been accepted from the data bus during a write cycle. The rising edge indicates when the SLAVE has released the data bus at the end of a READ CYCLE.
GND	The DC voltage reference for the VMEbus system.

SIGNAL MNEMONIC	SIGNAL NAME AND DESCRIPTION
IACK*	INTERRUPT ACKNOWLEDGE - An open-collector or three-state driven signal used by an INTERRUPT HANDLER acknowledging an interrupt request. It is routed, via a backplane signal trace, to the IACKIN* pin of slot 1, where it is monitored by the IACK DAISY-CHAIN DRIVER.
IACKIN*	INTERRUPT ACKNOWLEDGE IN - A totem-pole driven signal. The IACKIN* and IACKOUT* signals form a daisy-chain. The IACKIN* signal indicates to the VMEbus board receiving it that it is allowed to respond to the INTERRUPT ACKNOWLEDGE CYCLE that is in progress.
IACKOUT*	INTERRUPT ACKNOWLEDGE OUT - A totem-pole driven signal. The IACKIN* and IACKOUT* signals form a daisy-chain. The IACKOUT* signal is sent by a board to indicate to the next board in the daisy-chain that it is allowed to respond to the INTERRUPT ACKNOWLEDGE CYCLE that is in progress.
IRQ1*-IRQ7*	INTERRUPT REQUEST (1-7) - Open-collector driven signals, generated by an INTERRUPTER, which carry interrupt requests. When several lines are monitored by a single INTERRUPT HANDLER the highest numbered line is given the highest priority.
LWORD*	LONGWORD - A three-state driven signal used in conjunction with DS0*, DS1*, and A01 to select which byte location(s) within the 4 byte group are accessed during the data transfer.
RESERVED	RESERVED - A signal line reserved for future VMEbus enhancements.
SERCLK	SERIAL CLOCK - A totem-pole driven signal which is used to synchronize the data transmission on the VMSbus.
SERDAT*	SERIAL DATA - An open-collector driven signal which is used for VMSbus data transmission.
SYSCLK	SYSTEM CLOCK - A totem-pole driven signal which provides a constant 16-MHz clock signal that is independent of any other bus timing.
SYSFAIL*	SYSTEM FAIL - An open-collector driven signal that indicates that a failure has occurred in the system. This signal may be generated by any board on the VMEbus.

SIGNAL MNEMONIC	SIGNAL NAME AND DESCRIPTION
SYSRESET*	SYSTEM RESET - An open-collector driven signal which, when low, causes the system to be reset.
WRITE*	WRITE - A three-state driven signal generated by the MASTER to indicate whether the data transfer cycle is a read or a write. A high level indicates a read operation; a low level indicates a write operation.
+5V STDBY	+5 Vdc STANDBY - This line supplies +5 Vdc to devices requiring battery backup.
+5V	+5 Vdc Power - Used by system logic circuits.
+12V	+12 Vdc Power - Used by system logic circuits.
-12V	-12 Vdc Power - Used by system logic circuits.

APPENDIX C

USE OF THE SERCLK AND SERDAT* LINES

Two signal lines on the VMEbus backplane (SERCLK and SERDAT*) are designated for use by the VMSbus and provide a serial communication link between boards. The protocol used on the VMSbus is outside the scope of this document. Since system controller board designers will want to include circuitry on their boards to drive SERCLK, this appendix provides the necessary information.

SERCLK, like SYSCLK has no fixed timing relationship with any VMEbus signal (except SERDAT* which carries data bits that are synchronized to SERCLK).

The drivers and receivers for SERCLK and SERDAT* are specified in Chapter 6.

Figure C-1 and Table C-1 show the required timing parameters for the SERCLK signal line. A SERCLK waveform with these timing values can be derived from a 32 Mhz clock source. The timing values in Table C-1 are for use when the SERCLK and SERDAT* lines are not extended beyond the VMEbus backplane. If these signals are extended to carry intersystem information, each of the timing values given in Table C-1 should be multiplied by a common scaling factor, greater than 1, to allow for the increased SERCLK and SERDAT* propagation times.

RECOMMENDATION C.1:
When designing the SERIAL CLOCK DRIVER module, take into account the fact that the SERCLK line driver's propagation delays for the rising and falling edges will likely be different. This difference is emphasized when the SERCLK line is heavily loaded. When calculating the driver's propagation delays, use the delays specified on the manufacturer's data sheet for a 300 pf capacitive load. If the only propagation delays given are for a 30 pf load, add 10 nSec to each of the propagation delays.

SUGGESTION C.1:
Design the SERIAL CLOCK DRIVER so that it can be jumpered to work from various stages of a binary counter that is driven by a 32 MHz clock source. This allows the selection of 32 MHz, 16 MHz, 8 MHz etc., as the base frequency for the SERIAL CLOCK DRIVER and makes it easy to select a frequency appropriate for the length of the SERCLK and SERDAT* lines.

OBSERVATION C.1:
If a 32 Mhz clock source is used to generate the SERCLK waveform, it can also be used to generate the VMEbus's 16 MHz SYSCLK signal.

SUGGESTION C.2:
To allow multiple boards that include SERIAL CLOCK DRIVERS to be installed in the same backplane, design them with a jumper that disconnects the SERIAL CLOCK DRIVER from the SERCLK line.

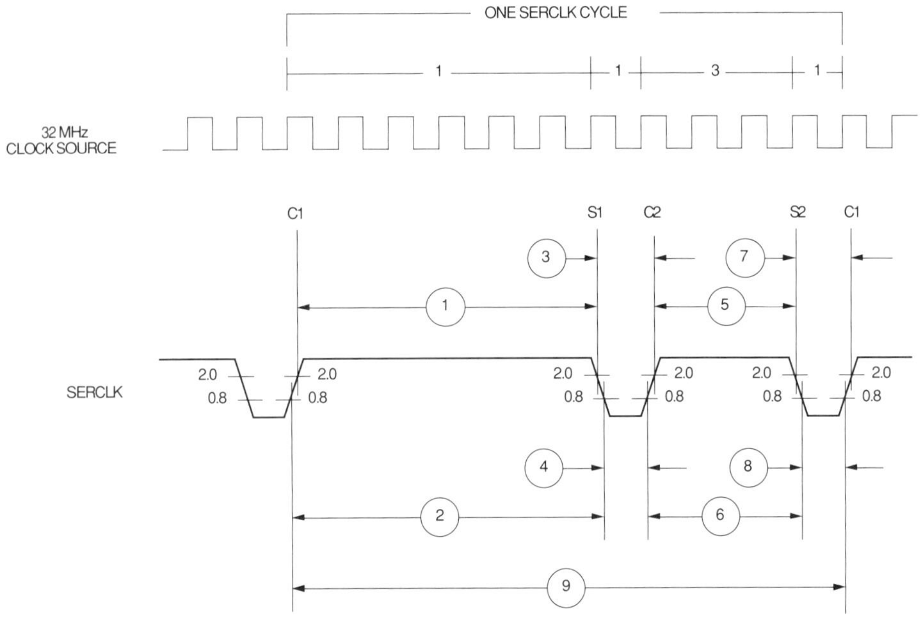

Note:
See Table C-1 on the following page for timing values.

Figure C-1. SERCLK Timing Diagram

Table C-1. SERCLK Timing Values

PARAMETER NUMBER	MIN	MAX
1	167	
2		194
3		51
4	25	
5	74	
6		100
7		51
8	25	
9	340	347

Note: All timing values are in nanoseconds.

The VXIbus Specification

Revision 1.3

VMEbus Extensions for Instrumentation System Specification Revision 1.3 is authored by the VXIbus Consortium, Inc. and its sponsor members:

A/S Bruel & Kjaer
Colorado Data Systems, Inc.
John Fluke Mfg. Co., Inc.
GenRad, Inc.
Hewlett-Packard Co.
Keithley Instruments, Inc.
National Instruments, Corp.
Racal-Dana Instruments, Inc.
Tektronix, Inc.
Wavetek, Inc.

NOTICE

TABLE OF CONTENTS

LIST OF FIGURES

LIST OF TABLES

SECTION A

Introduction To VXIbus Specifications

This document discusses the VXIbus family of VMEbus compatible modular instrumentation. It is intended to be used by designers interested in generating compatible components for the system.

The architectural concepts of the VMEbus date back to Motorola's development of the 68000 microprocessor in the late 1970's. In late 1979 Motorola published a brief description of a 68000-oriented bus known as VERSAbus. Several revisions followed, the last being in July 1981.

At the same time, a new printed circuit board standard (IEC 297-3) was being developed, known as the "Eurocard" standard. In October 1981, Motorola, Mostek, and Signetics announced their agreement to support a line of cards based on the VERSAbus with Eurocard module dimensions, which was renamed VMEbus. There have been three releases of the VMEbus specifications, with the most recent, Revision C.1, published in April 1985. The VMEbus was recently standardized as IEEE 1014.

The marketplace has demonstrated the open system nature of VMEbus. There are thousands of VMEbus cards available from a multitude of vendors. These cards are primarily computer oriented cards, though there are some data acquisition cards available which are primarily aimed at industrial processing and control.

There have been numerous requests for instrumentation modules for VMEbus, particularly from the United States Department of Defense. Reduction of the size of ATE equipment is a major factor driving the Defense Department toward VMEbus, though the high bandwidth of the VMEbus backplane is also considered a benefit, particularly for digital test and digital signal processing applications. The United States Air Force, Navy, and Army have programs addressing this issue, known as MATE, CASS, and IFTE respectively.

The most serious impediment to VMEbus based instrumentation is the lack of standards beyond that of VMEbus. In the spring of 1987, technical representatives from Colorado Data Systems, Hewlett Packard, Racal Dana, Tektronix, and Wavetek formed an ad hoc committee to engineer the additional standards necessary for an open architecture instrumentation bus based on VMEbus, the Eurocard standard, and other instrumentation standards such as IEEE 488.2. In July of 1987 they announced their agreement to support a common architecture of VMEbus Modular Instrumentation, named the VXIbus.

A.1 MANUFACTURER ID NUMBERS

The VXIbus specification provides a mechanism for identifying the manufacturer of every VXIbus device. This consists of a 12 bit manufacturer ID number (0 - 4095) which may be read over the VMEbus (See Section C.2.1.1.2, "Configuration Registers"). The list of ID numbers is maintained by the VXIbus Consortium, and is available to the public upon request. Each VXIbus device manufacturer has exactly one Manufacturer ID number. Numbers are assigned to manufacturers in decreasing order beginning with number 4095. To obtain an ID number, submit the following application:

Application for Obtaining a VXIbus Manufacturer's Identification Number

Company Name: ______________________________ Date: ________

Company Contact: ______________________________

Title or Position: ______________________________

Address : ______________________________

____

Phone Number : ______________________________

Mail to the VXIbus Consortium:

VXIbus Consortium Inc.
P.O. Box 1736
Vancouver, WA 98668 U.S.A.

A.2 VXIbus OVERVIEW

A.2.1 Introduction

The goal of the VXIbus is to define a technically sound modular instrument standard based on the VMEbus that is open to all manufacturers and is compatible with present industry standards.

VXIbus is an acronym for VMEbus extensions for instrumentation. The VXIbus specification details the technical requirements of VXIbus compatible components, such as mainframes, backplanes, power supplies, and modules. Before studying the VXIbus architecture, one should become familiar with the VMEbus and its specification.

A.2.2 VMEbus Background

The VMEbus is an open system architecture primarily focused at computer systems, though there presently is a limited offering of instrumentation. VMEbus modules are approximately six inches deep and come in two heights; about four inches and nine inches. The VXIbus specification refers to these as the A and B sizes respectively. The precise dimensions are specified by the Eurocard standard, which describes a family of printed circuit boards and their associated DIN connector locations. VMEbus modules are designed for 0.8 inch slot-to-slot spacing. The A size board has a single 96-pin connector known as P1, while the B size may include a P1 and P2 connector. Each of these DIN connectors consists of three rows of 32 pins apiece, on 0.1-inch centers. Typically, these boards are positioned vertically in a frame with the P1 connector closest to the top. Neither the VMEbus nor the VXIbus mandates a physical orientation, since orientation is only an implementation issue not needed for compatibility. Many VMEbus systems are designed to accept boards horizontally.

The VMEbus specification allows a maximum of 21 modules. VMEbus makes no particular provision for an extension chassis or frame-to-frame communication. Multiple frame systems can be created by electrically buffering the VMEbus (at the loss of some bandwidth between cages) or by using standard data communication links that disguise the underlying VMEbus architecture. There are no EMC (electromagnetic compatibility) requirements dictated by VMEbus, either conducted or radiated, nor are there power dissipation limits or chassis cooling requirements. VMEbus has left these issues to the system integrator, while VXIbus addresses these issues more rigorously.

Although electrically and logically similar to the 68000 microprocessor architecture, the VMEbus interface has been specified broadly enough that it is not dependent on any particular processor, and many processors are already supported on VMEbus, including the 80386. Many of the simpler VMEbus boards do not have processors at all.

A minimum VMEbus system requires only the P1 connector. All handshaking, arbitration, and interrupt support exists on P1, with P2 used to expand the system to 32 bits of address and data (A32 and D32). P1 supports 16-bit and 24-bit addressing (A16 and A24), as well as 8 and 16 bit data paths (D08 and D16). The extra lines needed for A32 and D32 are contained on the center row of P2, while the outer rows are user defined. These undefined pins are typically used for interface connections, such as allowing a module to drive a chassis-mounted connector, access an internal disk drive, or provide for module-to-module communication. VSB (VMEbus Subsystem Bus) is a standard "subsystem bus" that has defined P2 as an additional communication path for up to six modules. Multiple VSBs may exist within any one VMEbus system. This is important to note, because VXIbus defines a subsystem of up to 13 modules and, like VSB, multiple VXIbus subsystems may exist within any one VXIbus system.

A.2.3 The VXIbus Extensions

VXIbus retains P1 and the center row of P2 exactly as defined by VMEbus. This includes the 5 volt and ±12 volt power pins on P1, and the additional 5 volt pins on P2. VXIbus includes the A and B card sizes, and these modules remain totally VMEbus compatible. However, VXIbus has made substantial additions to the VMEbus specification oriented towards instrumentation that can best be described as an electromechanical superset and a logical subset.

A.2.3.1 VXIbus Modules

VXIbus has added two Eurocard module sizes of about 13 inch depth, referred to as the C and D sizes. These modules are 9 and 14 inches high respectively, and are placed on 1.2-inch centers. The C Eurocard is the same height as the VMEbus B size board, and may sport both the P1 and P2 connectors. The D size module is a triple-height Eurocard that may include a P3 connector, in addition to P1 and P2. The 1.2-inch module width allows implementation of high-density instrumentation modules, while allowing enough space for shielding both sides of a module and inserting an optional chassis shield. It also has the added benefit of allowing a high degree of compatibility with the shorter and narrower A and B sizes by allowing them to be mounted on full-length board carriers or adapters. These carriers/adapters may also shield the sides of standard VMEbus cards, giving them a high degree of electromagnetic compatibility with VXIbus systems.

A.2.3.2 VXIbus Subsystems

A VXIbus system may have up to 256 devices, including one or more VXIbus subsystems. A VXIbus subsystem consists of a central timing module (referred to as a Slot 0 module) and up to 12 additional instrument modules. P2 and P3 are completely defined in a VXIbus subsystem. These 13 modules conveniently fill a standard 19-inch cabinet when mounted vertically on 1.2 inch centers. Many VXIbus systems will consist only of a single frame with these 13 modules. A common configuration will load the Slot 0 module with system resources such as the VXIbus-mandated timing generation, the VMEbus-required system controller functions, and a data communication port such as IEEE 488 or RS-232. Slot 0 may also include optional instrumentation. The other positions are general-purpose slots for the user to mix and match modules. A single VXIbus subsystem may have less than 12 additional slots, but may not have more. Any combination of VXIbus subsystems may exist within a VXIbus system. For instance, one VXIbus system may consist of a frame with one Slot 0 and 12 VXIbus modules extended to another frame that has a Slot 0 module adjacent to three instrument slots, another Slot 0 module with five instrument slots, and four standard VMEbus slots of undefined P2.

A.2.3.2.1 P2 Connector Definition

As mentioned previously, a VXIbus subsystem defines all P2 and P3 pins. The VXIbus P2 adds a 10 MHz ECL clock, ECL and analog supply voltages, ECL trigger lines, TTL trigger lines, an analog summing bus, a module identification line, and a daisy chain structure, known as the local bus. The trigger lines serve primarily as resources for signaling between instruments in a VXIbus subsystem, while the local bus lines are preferred for use within a multiple-module instrument set (installed in adjacent slots). The daisy chain local bus use is left to the module manufacturer to define, and several classes of electrical signals are permitted. Allowed signals are TTL, ECL, low voltage analog, and analog up to 42 volts. A keying mechanism on the faceplate of each module indicates its local bus class, and prevents incompatible classes from accidentally being placed adjacently, and potentially causing a destructive condition. Typical uses of the local bus include creating an internal analog bus or a chain of serial digital signal processors. There are a total of 24 local bus pins on P2, (12 lines in and 12 lines out for each slot) thus creating a 12-line bus that may or may not be passed on to adjacent slots.

A.2.3.2.2 P3 Connector Definition

The VXIbus P3 connector adds many of the same resource types as described for P2, but is aimed at higher performance instrumentation. Included on P3 is a 100 MHz clock and sync signal, additional power pins of the same supply voltages, more ECL trigger lines, and 24 additional lines (48 pins) of daisy chain local bus. Also defined on P3 is a "star" trigger system, where precision ECL trigger signals are routed through Slot 0, which acts as a cross point switch. This allows very precisely matched trigger timing between modules, regardless of module position.

A.2.3.3 VXIbus System Architecture

The VXIbus device protocols define how modules are granted non-conflicting portions of the VMEbus address space. A device is typically a single module, but this is not required. Several devices may exist on a single module, and a single device may consist of multiple modules. 256 devices may exist in any one VXIbus system, and are referred to by logical device addresses ranging from 0 to 255. A VXIbus system configuration space is defined in the upper 16K of the 64K A16 address space. Each device is granted a total of 64 bytes in this space, which is sufficient for many of the simpler devices. Devices requiring additional address space have their address requirements readable in a defined register in the A16 address space. A "resource manager" reads this value shortly after power-on, and then assigns the requested memory space by writing the module's new VMEbus address into the device's offset register. This method positions a device's additional memory space in the A24 (16 Mbyte) or A32 (4 Gbyte) address space. If present day VMEbus cards are used in a system, the resource manager must position the VXIbus devices around the space taken by the standard VMEbus cards.

Higher-level communication protocols are defined, to allow sharing of interface modules and other devices by multiple manufacturers.

A.2.3.4 VXIbus Specification Structure

In structure, the VXIbus specification parallels much of the VMEbus C.1 specification. Some VXIbus specification sections have the same titles as the VMEbus chapters. Other sections are devoted to topics not addressed by the VMEbus specification. The VXIbus document contains more than just the rules needed for VXIbus compliance; included are design tips and suggestions added to help the designer create compatible components. The same format of RULES, RECOMMENDATIONS, SUGGESTIONS, and OBSERVATIONS as VMEbus is used. As experience is gained creating VXIbus devices and using the specification, appropriate updates to the specification will occur.

A.3 DOCUMENT STRUCTURE

This document is organized in sections, with each section discussing a particular independent level of the implementation. In the event where a section parallels another standard document (*such as VMEbus*), subsections will have corresponding numbers. Several appendices, containing clarifications and extensions, follow the body of the specification.

A.4 SPECIFICATION OBJECTIVES

This document defines a set of RULES and RECOMMENDATIONS for constructing a component which will interface to the VXIbus family. The specifications cover the complete spectrum from basic hardware issues (such as card dimensions) to recommendations for communications protocols. This specification has the following objectives:

1. To allow communication among devices in a unambiguous fashion.

2. To allow for physical size reduction of standard *rack-&-stack* instrumentation systems.

3. To allow for software cost reduction in test system integration by the use of common interfaces to analogous capabilities.

4. To provide higher system throughput for test systems, through the use of higher- bandwidth channels for inter-device communication, and the use of new protocols specifically designed to enhance throughput.

5. To provide test equipment which could be used in military IAC (Instrument on a Card) systems.

6. To provide the ability to implement new functionality in test systems, through the use of *virtual instruments*.

7. To define how to implement *multi-module instruments* within the framework of this standard.

A.5 DEFINITION of TERMS

Throughout this document you will see the following headings on paragraphs. These headings identify the contents of the paragraph:

RULE:
Rules **must** be followed to ensure compatibility for cards in the system. A rule is characterized by the use of the words **SHALL** and **SHALL NOT**. These words are not used for any purpose other than stating rules.

RECOMMENDATION:
Recommendations consist of advice to implementors which will affect the usability of the final device. Discussions of particular hardware to enhance throughput would fall under a recommendation. These should be followed to avoid problems, and to obtain optimum performance.

SUGGESTION:
A suggestion contains advice which is helpful but not vital. The reader is encouraged to consider the advice before discarding it. Suggestions are included to help the novice designer with areas of design that can be problematic.

PERMISSION:
Permissions are included to clarify the areas of the specification that are not specifically prohibited. Permissions reassure the reader that a certain approach is acceptable, and will cause no problems. The word **MAY** is reserved for indicating permissions.

OBSERVATION:
Observations spell out implications of rules and bring attention to things that might otherwise be overlooked. They also give the rationale behind certain rules, so that the reader understands why the rule must be followed.

Any text which appears without a heading should be considered as description of the standard.

SECTION B

VXIbus Implementation Of VMEbus Specs

This document describes additions to and recommendations for the VMEbus specifications.[1] This section is organized in a parallel fashion to the VMEbus specification document. Sections B.1 - B.7 are additions to the corresponding VMEbus Revision C.1 Chapters 1-7, and Section B.8 is new material not contained in the VMEbus specifications.

B.1 INTRODUCTION

The intent of the VXIbus implementation of the VMEbus specifications is to produce an electromechanical superset and logical subset of the VMEbus specifications, in order to establish compatibility for all VXIbus systems. It is expected that all boards designed to the VMEbus specification will be compatible with this implementation both mechanically and electrically. However, a subset of VMEbus protocols (relating to data transfer, arbitration, and interrupts) has been specified, and may limit the utility of some standard VMEbus products in a VXIbus system. Mechanical extensions are defined to allow for a superset of board sizes.

B.2 DATA TRANSFER BUS

B.2.1 Address Modifiers

RECOMMENDATION B.2.1:
It is recommended that the address modifiers listed in the VMEbus specification as USER DEFINED not be used.

B.2.2 DTACK* Operation

RULE B.2.1:
A SLAVE **SHALL** pull DTACK* or BERR* low within 20 μsec after a data strobe goes low.

RECOMMENDATION B.2.2:
A SLAVE should pull DTACK* low within 1 μsec after a data strobe goes low.

RULE B.2.2:
A SLAVE **SHALL** release DTACK* and BERR* within 5 μsec after the data strobe(s) goes high.

[1] See References at the end of Appendix IV (page VXI-246)

RECOMMENDATION B.2.3:
A SLAVE should release DTACK* and BERR* within 0.5 μsec after the data strobe(s) goes high.

B.2.3 Bus Timer Operation

RULE B.2.3:
The BUS TIMER **SHALL** be BTO(≥100).

OBSERVATION B.2.1:
The BTO(100) capability minimum is to allow for chassis-to-chassis communication. There is a potential for each chassis to take as much as 20 μsec to respond to a data transfer.

B.3 DATA TRANSFER BUS ARBITRATION

RULE B.3.1:
The BUS GRANT daisy-chain lines **SHALL** be used or passed by all single boards and board assemblies.

RECOMMENDATION B.3.1:
The arbiter should implement priority-based arbitration between the 4 Bus Request/Grant levels.

B.4 PRIORITY INTERRUPT

RULE B.4.1:
The INTERRUPT ACKNOWLEDGE daisy-chain lines **SHALL** be used (or passed) by all single boards and board assemblies.

B.5 UTILITIES

B.5.1 Power Monitor

RULE B.5.1:
Every VXIbus mainframe **SHALL** provide a VMEbus POWER MONITOR.

RECOMMENDATION B.5.1:
A power monitor should drive ACFAIL* low a minimum of 8 msec before driving SYSRESET* low, and a minimum of 10 msec before +5 Vdc power drops below 4.875 volts, when a power failure is detected. See Figure 5-4 of the VMEbus Specification, Rev. C.1.

B.5.2 Power Pins

See Section B.8, "EMC and System Power", for power pin specifications.

B.6 ELECTRICAL SPECIFICATIONS

RULE B.6.1:
All conductors on P1, P2, and P3 **SHALL** meet the requirements of Section 6.1 of the VMEbus Rev C.1 Electrical Specification.

B.6.1 The P1 Connector

The P1 connector shall remain as defined by the VMEbus specification (Rev C.1).

B.6.2 The VXIbus Subsystem P2 Connector

The VXIbus subsystem bus defines the outer rows (user defined pins) for instrumentation use. The center row retains the VMEbus specification. Alternate definitions of the outer two rows of P2 are permitted in VXIbus systems (not VXIbus subsystems) with a "Segmented Backplane" architecture, where other VMEbus subsystem buses, such as VSB, may exist. A VXIbus subsystem bus consists of a system resource module defined as a Slot 0 module with up to 12 adjacent modules in increasing slot numbers. The terms *VXIbus* and *VXIbus subsystem* bus are used interchangeably in this section of the VXIbus specification.

The VXIbus P2 connector delivers resources to modules particularly oriented toward instrumentation. On P2 it adds:

- -5.2V, -2V, ±24V, and additional +5V power.
- 10 MHz differential clock.
- 2 parallel ECL trigger lines.
- 8 parallel TTL trigger lines.
- 1 module identification pin.
- 12 lines of manufacturer-defined local bus lines that connect to adjacent modules.
- A 50 Ω terminated analog summing bus.

The Slot 0 module serves as a system resource, and has a modified P2 functionality for central handling of the module identification pins. The Slot 0 module, while providing these common resources, may also contain other devices and instrumentation.

PIN NUMBER	ROW a SIGNAL MNEMONIC	ROW b SIGNAL MNEMONIC	ROW c SIGNAL MNEMONIC	PIN NUMBER
1	ECLTRG0	+5V	CLK10+	1
2	–2V	GND	CLK10–	2
3	ECLTRG1	RSV1	GND	3
4	GND	A24	–5.2V	4
5	LBUSA00	A25	LBUSC00	5
6	LBUSA01	A26	LBUSC01	6
7	–5.2V	A27	GND	7
8	LBUSA02	A28	LBUSC02	8
9	LBUSA03	A29	LBUSC03	9
10	GND	A30	GND	10
11	LBUSA04	A31	LBUSC04	11
12	LBUSA05	GND	LBUSC05	12
13	–5.2V	+5V	–2V	13
14	LBUSA06	D16	LBUSC06	14
15	LBUSA07	D17	LBUSC07	15
16	GND	D18	GND	16
17	LBUSA08	D19	LBUSC08	17
18	LBUSA09	D20	LBUSC09	18
19	–5.2V	D21	–5.2V	19
20	LBUSA10	D22	LBUSC10	20
21	LBUSA11	D23	LBUSC11	21
22	GND	GND	GND	22
23	TTLTRG0*	D24	TTLTRG1*	23
24	TTLTRG2*	D25	TTLTRG3*	24
25	+5V	D26	GND	25
26	TTLTRG4*	D27	TTLTRG5*	26
27	TTLTRG6*	D28	TTLTRG7*	27
28	GND	D29	GND	28
29	RSV2	D30	RSV3	29
30	MODID	D31	GND	30
31	GND	GND	+24V	31
32	SUMBUS	+5V	–24V	32

TABLE B.1. P2 Pin Definitions: Slots 1-12

B.6.2.1 CLK10

CLK10 is a 10 MHz system clock. It is sourced from Slot 0 and distributed to Slots 1-12 on P2. The Slot 0 output is differential ECL. It is buffered on the backplane and distributed to each module slot as a single source, single destination differential ECL signal. The CLK10 is individually buffered on the backplane for each slot position to provide a high level of inter-module isolation and relax the loading rules for modules. See Figure B.1.

PIN NUMBER	ROW a SIGNAL MNEMONIC	ROW b SIGNAL MNEMONIC	ROW c SIGNAL MNEMONIC	PIN NUMBER
1	ECLTRG0	+5V	CLK10+	1
2	–2V	GND	CLK10–	2
3	ECLTRG1	RSV1	GND	3
4	GND	A24	–5.2V	4
5	MODID12	A25	LBUSC00	5
6	MODID11	A26	LBUSC01	6
7	–5.2V	A27	GND	7
8	MODID10	A28	LBUSC02	8
9	MODID09	A29	LBUSC03	9
10	GND	A30	GND	10
11	MODID08	A31	LBUSC04	11
12	MODID07	GND	LBUSC05	12
13	–5.2V	+5V	–2V	13
14	MODID06	D16	LBUSC06	14
15	MODID05	D17	LBUSC07	15
16	GND	D18	GND	16
17	MODID04	D19	LBUSC08	17
18	MODID03	D20	LBUSC09	18
19	–5.2V	D21	–5.2V	19
20	MODID02	D22	LBUSC10	20
21	MODID01	D23	LBUSC11	21
22	GND	GND	GND	22
23	TTLTRG0*	D24	TTLTRG1*	23
24	TTLTRG2*	D25	TTLTRG3*	23
25	+5V	D26	GND	25
26	TTLTRG4*	D27	TTLTRG5*	26
27	TTLTRG6*	D28	TTLTRG7*	27
28	GND	D29	GND	28
29	RSV2	D30	RSV3	29
30	MODID00	D31	GND	30
31	GND	GND	+24V	31
32	SUMBUS	+5V	–24V	32

TABLE B.2. P2 Pin Definitions: Slot 0

RULE B.6.2:
The CLK10 frequency (sourced from Slot 0) **SHALL** be 10 MHz. Its accuracy **SHALL** be equal to or better than ±100 ppm (.01%) over its specified operating temperature and time.

RECOMMENDATION B.6.1:
A Slot 0 module should allow CLK10 to be derived from an external frequency source.

OBSERVATION B.6.1:
An external frequency source would permit the use of an accurate frequency reference, such as a Rubidium Standard, and would facilitate synchronization of multiple VXIbus mainframes.

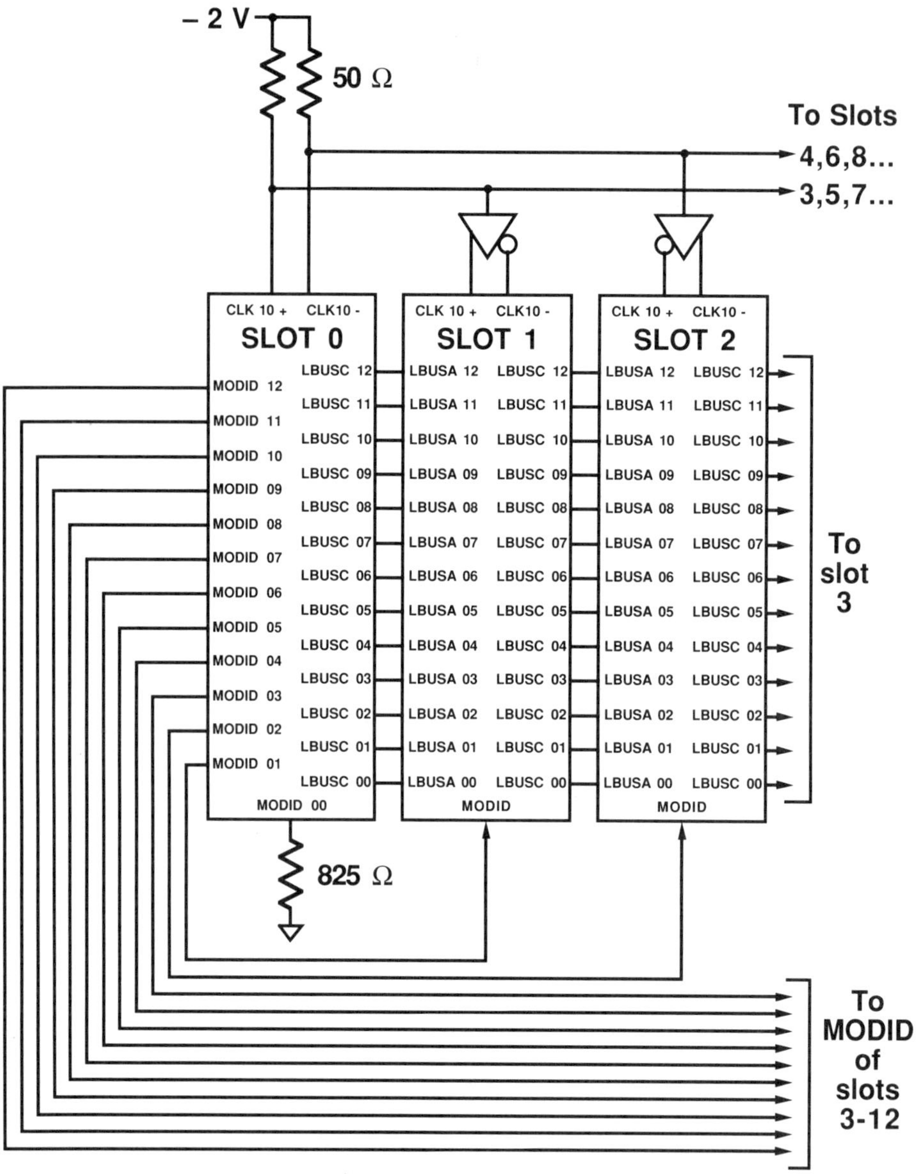

Figure B.1. CLK10, MODID, and LBUS backplane signal routing

RULE B.6.3:
The CLK10 duty cycle **SHALL** be 50% ±5%, when measured at the 50% transition level.

RULE B.6.4:
IF CLK10 is switched between different clock sources,
THEN the minimum pulse width (high or low) **SHALL NOT** be less than 30 nsec or more than 10 µsec during switching. The minimum time between two successive transitions of the same polarity **SHALL NOT** be less than 80 nsec.

RULE B.6.5:
Each slot's CLK10 **SHALL** be driven by a unique backplane buffer output.

RULE B.6.6:
CLK10 **SHALL** be distributed differentially from Slot 0.

RULE B.6.7:
The backplane CLK10 distribution traces **SHALL** be designed for 50 Ω.

RULE B.6.8:
IF a module accesses the CLK10 signals,
THEN it **SHALL** provide 50 Ω termination on CLK10+ and CLK10-, with no more than two equivalent ECL loads.

RULE B.6.9:
The absolute delay of CLK10 from Slot 0 to any module **SHALL NOT** exceed 8 nsec.

OBSERVATION B.6.2:
Either single-ended input buffers or differential input buffers can be used to buffer and fan out CLK10 on the backplane. Typical components are the 10H101 and 10H116 buffers.

B.6.2.2 MODID Lines

The MODID lines allow a logical device to be identified with a particular physical location or slot. These lines are sourced from the VXIbus Slot 0 module to the Slot 1 and higher modules. There is one line to each module, located on P2 pin A30. A total of 12 MODID lines will connect between Slot 0 and the other modules in a maximally configured VXIbus subsystem. In addition to these 12 lines, Slot 0 has its own MODID line (MODID00). The purpose of the MODID lines is to:

1. Detect the presence of a module in a slot, even if that module has failed.
2. Identify the geographical location (slot number) of a particular device.
3. Indicate by lamp (or other means) the actual physical location of the module.

Slot 0 detects the presence of a module by the module pulling down its MODID line with a fixed resistor to ground. This allows any module, even if failed and unpowered, to be detected.

The slot number of a device is identified by the Slot 0 module asserting a particular MODID line, and polling the MODID bit in each module's A16 autoconfiguration space to detect the selected device.

A visual indicator, such as a lamp, may be placed adjacent to a slot (or even on the module) to illuminate when that particular MODID line is driven true. This may be used to quickly identify the location of any module, including failed modules. Refer to Figure B.2 for rules for loading and driving the MODID lines.

RULE B.6.10:
Each Slot 0 MODIDxx receiver **SHALL** comply with the VMEbus loading rules for High Current 3 State lines (VMEbus Revision C.1, Rule 6.14).

RULE B.6.11:
Each Slot 0 MODIDxx driver **SHALL** comply with the VMEbus driving rules for Standard 3 State lines (VMEbus Revision C.1, Rule 6.15).

RULE B.6.12:
Each Slot 0 device **SHALL** provide a 16.9 KΩ pull up resistor from each MODID line to +5 volts.

RULE B.6.13:
IF a non-Slot 0 module accesses the MODID pin,
THEN it **SHALL** provide an 825 Ω pull down resistor to ground, in parallel with no more than 100 μA of leakage current on its MODID line.

OBSERVATION B.6.3:
16.9 KΩ and 825 Ω are standard 1% metal film resistor values.

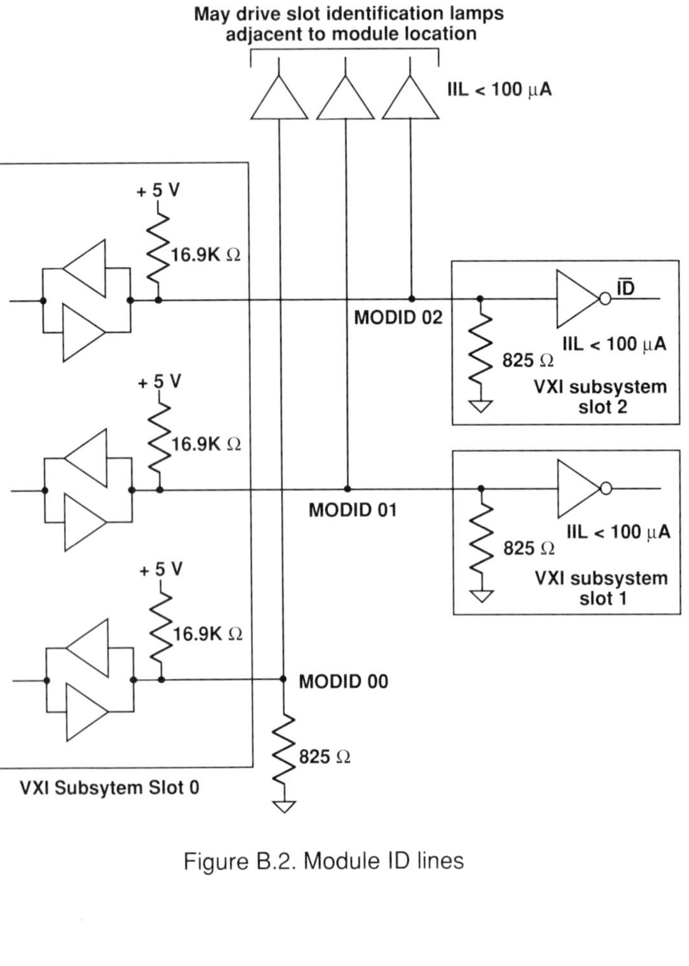

Figure B.2. Module ID lines

RULE B.6.14:
A VXIbus subsystem backplane **SHALL** provide an 825 Ω pull down resistor to ground on the MODID00 line.

RULE B.6.15:
The backplane **SHALL NOT** sink or source more than 100 μA of leakage current to any MODID line.

PERMISSION B.6.1:
A VXIbus subsystem Slot 0 module **MAY** drive its MODID line true to illuminate its own module location lamp.

OBSERVATION B.6.4:
The proper placement of a Slot 0 module in a Slot 0 position may be verified by a read of its MODID00 state when all Slot 0 MODID drivers are disabled. A low state indicates that the module is in a Slot 0 position. A high state indicates that it is in another position.

B.6.2.3 TTLTRG0-7* (TTL Trigger Lines)

The TTLTRG* lines are open collector TTL lines used for intermodule communication. Any module, including Slot 0, may drive these lines and receive information on these lines. They are general-purpose lines that may be used for trigger, handshake, clock, or logic state transmission. They are in a non-asserted (high) state until allocated by the user under program control. Some standard allocation procedures are defined. Standard protocols, synchronous (SYNC), semi-synchronous (SEMI-SYNC), asynchronous (ASYNC), and start/stop (STST) are defined. To compensate for the larger rise time caused by the passive pullup termination, these protocols specify separate timing requirements for trigger sources and trigger acceptors. When used for logic state transmission, setup and hold times with respect to a clock edge are defined. Other protocols may be defined by a manufacturer.

OBSERVATION B.6.5:
The VMEbus Specification Revision C.1 recommends a 100 Ω system impedance for open collector lines. Although the TTLTRG* lines will be terminated and meet the drive and load requirements for the VMEbus, the characteristic impedance of the traces on the backplane may be closer to 50 Ω. This allows ECL traces to share the same backplane layer, and may keep backplanes at a manufacturable number of planes and thickness for component insertion. This should not significantly affect performance of the TTLTRG* lines.

RULE B.6.16:
The backplane **SHALL** terminate all TTLTRG* lines as specified by the VMEbus specification, Rev. C.1, Figure 6-2, Standard Bus Termination.

RULE B.6.17:
A module's interface to any TTLTRG* signal line **SHALL** conform to the driving and loading rules for open collector lines (VMEbus Specification Revision C.1, Section 6.4.2.5).

RULE B.6.18:
All TTLTRG* line drivers **SHALL** be unasserted within one second after SYSRESET* becomes unasserted.

RULE B.6.19:
A module **SHALL** allocate any TTLTRG* lines to specific functions in groups of one, two, or four.

RULE B.6.20:
IF a module uses single line TTLTRG* groups,
THEN that module **SHALL** be user-programmable to connect to any TTLTRG* line.

RULE B.6.21:
IF a module uses two-line TTLTRG* groups,
THEN that module **SHALL** be user programmable to connect all defined TTLTRG* group pairs. The defined TTLTRG* group pairs are lines 0 and 1, lines 2 and 3, lines 4 and 5, and lines 6 and 7.

RULE B.6.22:
IF a module uses four-line TTLTRG* groups,
THEN that module **SHALL** be user-programmable to connect to all defined TTLTRG* four line groups. The defined TTLTRG* four line group are lines 0, 1, 2, and 3; and lines 4, 5, 6, and 7.

B.6.2.3.1 Standard TTLTRG* Protocols

B.6.2.3.1.1 TTLTRG* Synchronous (SYNC) Trigger Protocol

The TTLTRG* SYNC trigger protocol is a single-line broadcast trigger that does not require an acknowledge from any acceptors.

RULE B.6.23:
A TTLTRG* SYNC trigger source **SHALL** assert the trigger for a minimum time of T1 and **SHALL NOT** reassert the trigger for a minimum time of T2, as shown in Figure B.3.

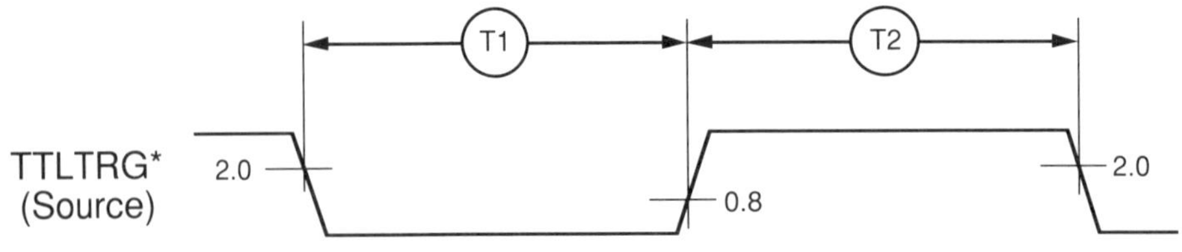

Parameter Label	Min	Max	Description
T1	30 nsec		Minimum source assertion time
T2	50 nsec		Minimum source non-assertion time

Figure B.3. TTLTRG* Synchronous (SYNC) Trigger Protocol

RULE B.6.24:
A TTLTRG* SYNC trigger acceptor **SHALL** accept any trigger asserted for ≥10 nsec, following a ≥10 nsec deassertion.

B.6.2.3.1.2 TTLTRG* Semi-Synchronous(SEMI-SYNC) Trigger Protocol

The TTLTRG* SEMI-SYNC trigger protocol is a single-line broadcast, multiple-acceptor handshake protocol. A single source asserts the trigger for a minimum time of T1, as shown in Figure B.4. If the trigger needs to be acknowledged, the acceptor(s) also asserts the trigger within a time of T3, and unasserts the trigger when it is ready to proceed with the next operation. The source recognizes acknowledge completion when the source and all acceptors have unasserted the trigger line. The source may re-assert the trigger after a minimum time of T2, following the completion of the previous acknowledge.

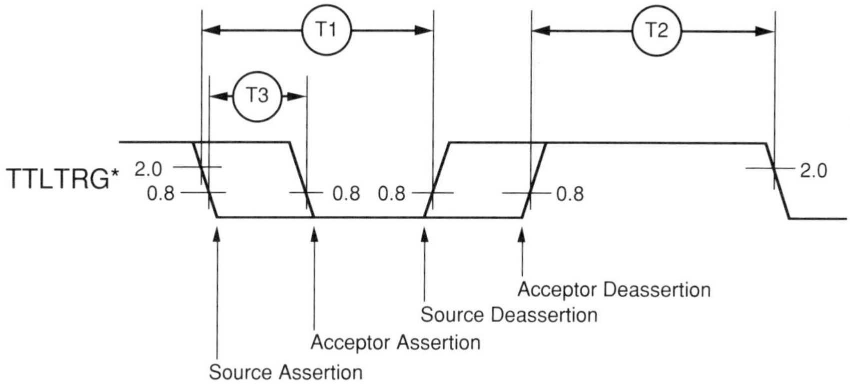

Parameter Label	Min	Max	Description
T1	50 nsec		Min. Source Assertion Time
T2	50 nsec		Min. Unassertion Time
T3	0 nsec	40 nsec	Max. Acceptor Response Time

Figure B.4. TTLTRG* Semi-Synchronous (SEMI-SYNC) Trigger Protocol

RULE B.6.25:
A TTLTRG* SEMI-SYNC trigger source **SHALL** meet the timing requirements for parameters T1 and T2 shown in Figure B.4.

RULE B.6.26:
A TTLTRG* SEMI-SYNC trigger acceptor **SHALL** meet the timing requirements for parameter T3 shown in Figure B.4.

RULE B.6.27:
A TTLTRG* SEMI-SYNC trigger acceptor **SHALL** accept any trigger asserted for ≥10 nsec, following a ≥10 nsec deassertion.

B.6.2.3.1.3 TTLTRG* Asynchronous (ASYNC) Trigger Protocol

The TTLTRG* ASYNC trigger protocol is a two-line, single-source, single-acceptor protocol. The source initiates an operation by asserting the lower-numbered line of the allocated TTLTRG* line pair, while the acceptor acknowledges by asserting the higher-numbered line of the TTLTRG* line pair. This is a useful trigger mode for handshaking between VXIbus modules and external instruments or between VXIbus mainframes.

RULE B.6.28:
A module implementing the TTLTRG* ASYNC trigger protocol **SHALL** source only on the lower-numbered line of a TTLTRG* line pair.

RULE B.6.29:
A module implementing the TTLTRG* ASYNC trigger protocol **SHALL** acknowledge only on the higher-numbered line of a TTLTRG* line pair.

RULE B.6.30:
IF a module implements the TTLTRG* ASYNC trigger protocol
THEN it **SHALL** meet the timing requirements shown in Figure B.5.

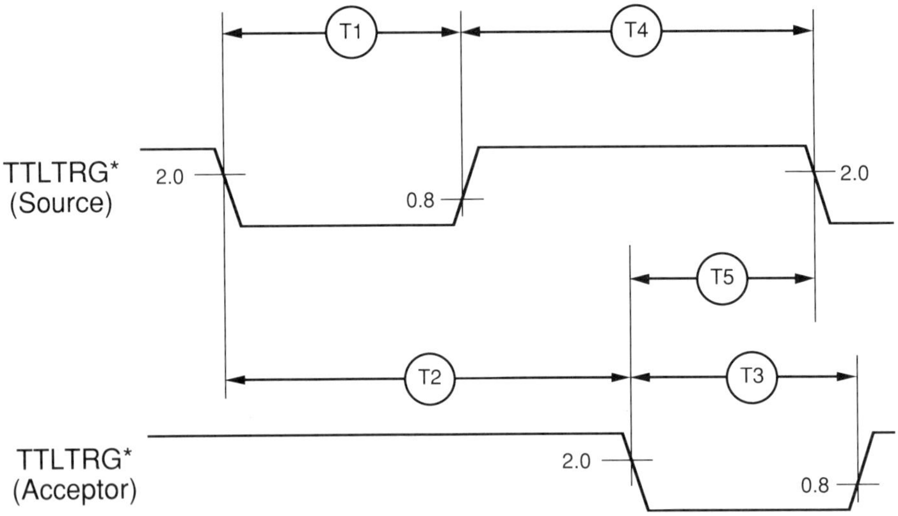

Parameter Label	Min	Max	Description
T1	30 nsec		Minimum Source assertion time
T2	0 nsec		Minimum Acceptor response time
T3	30 nsec		Minimum Acceptor assertion time
T4	50 nsec		Minimum Source non-assertion time
T5	0 nsec		Minimum Acceptor Acknowledge to source re-assertion time

Figure B.5. TTLTRG* Asynchronous (ASYNC) Trigger Protocol

RULE B.6.31:
A TTLTRG* ASYNC trigger source or acceptor **SHALL** accept any trigger asserted for ≥10 nsec, following a ≥10 nsec deassertion.

B.6.2.3.1.4 TTLTRG* Clock Transmission

A TTLTRG* line may be used for a clock signal transmission.

RULE B.6.32:
IF A TTLTRG* line is used for clock transmission,
THEN it **SHALL** meet the timing requirements described for the TTLTRG* SYNC trigger protocol.

OBSERVATION B.6.6:
In a fully loaded VXIbus subsystem, the TTLTRG* rise time will approach 40 nsec. For this reason, the falling edge is used for triggering.

B.6.2.3.1.5 TTLTRG* Data Transmission

The TTLTRG* lines can also be used for data transmission, where one of the TTLTRG* lines is used as a clock. Data may be synchronized to either the rising or falling edge of the clock, or both.

RULE B.6.33:
IF a source uses TTLTRG* lines for data transmission on a falling clock edge,
THEN the source **SHALL** meet the timing requirements shown in Figure B.6.

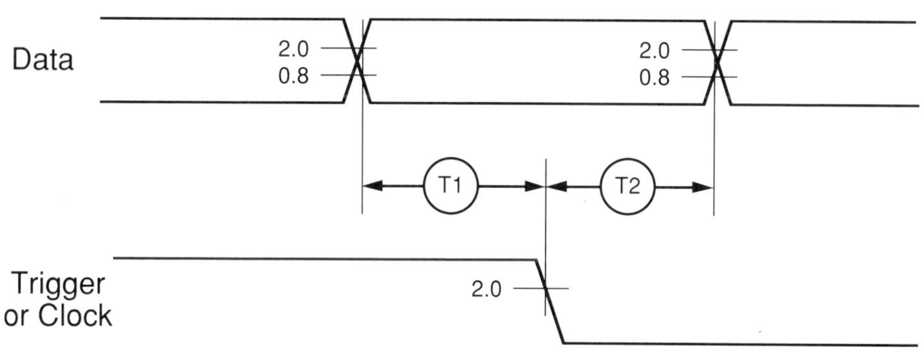

Parameter Label	Min	Max	Description
T1 T2	40 nsec 10 nsec		Data setup time Data hold time

Figure B.6. TTLTRG* Data Transmission on falling clock edge

RULE B.6.34:
IF a source uses TTLTRG* lines for data transmission on a rising clock edge,
THEN the source **SHALL** meet the timing requirements shown in Figure B.7.

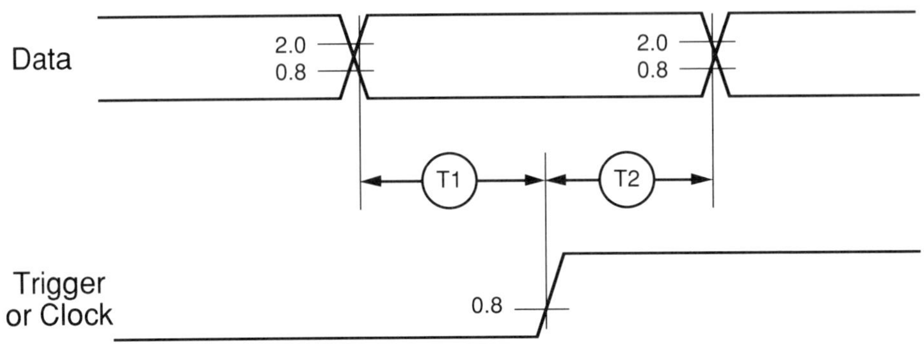

Parameter Label	Min	Max	Description
T1	40 nsec		Data setup time
T2	40 nsec		Data hold time

Figure B.7. TTLTRG* Data Transmission on rising clock edge

RULE B.6.35:
A TTLTRG* data acceptor **SHALL** accept any data having data setup and hold times of ≥7 nsec each.

B.6.2.3.1.6 Start/Stop (STST) Protocol

The Start/Stop protocol (STST) provides a method for starting and stopping module clusters synchronously. One TTLTRG* line is driven from the Slot 0 module, and its state signifies START or STOP operation. All participating modules respond to this line synchronously at the next CLK10 rising edge. Slot 0 has the responsibility to synchronize the START/STOP signal with CLK10 so that setup and hold times are guaranteed for all acceptors. This may require an arbiter in Slot 0 to synchronize external events.

RULE B.6.36:
IF a module implements STST protocol,
THEN the START/STOP line **SHALL** be user-programmable to connect to any TTLTRG* line.

RULE B.6.37:
IF a Slot 0 module implements the STST protocol,
THEN it **SHALL** meet the timing requirements shown in Figure B.8.

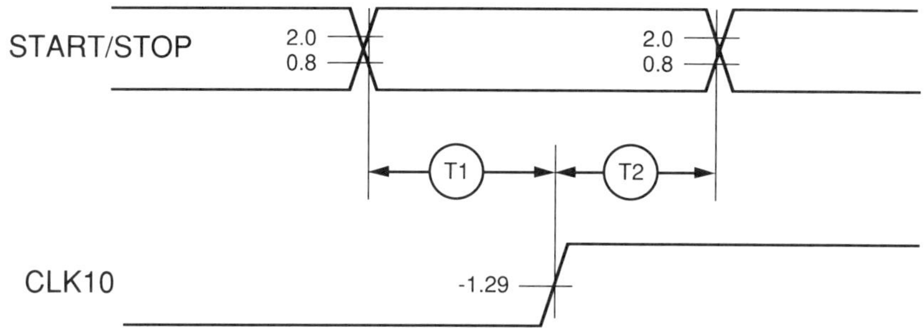

Parameter Label	Min	Max	Description
T1 T2	50 nsec 15 nsec		START/STOP setup time START/STOP hold time

Figure B.8. TTLTRG* START/STOP timing

RULE B.6.38:
A TTLTRG* STST acceptor **SHALL** accept any STST command having data setup and hold times of ≥ 7 nsec each.

RULE B.6.39:
When implementing the STST protocol, START **SHALL** be represented by a low (asserted) state on the START/STOP TTLTRG* line, and STOP **SHALL** be represented by a high (unasserted) state on the START/STOP TTLTRG* line.

B.6.2.3.1.7 External Trigger Buffering

OBSERVATION B.6.7:
The TTLTRG* loading and driving rules require that these lines be buffered when extended outside a mainframe.

RECOMMENDATION B.6.2:
When extending a trigger operation external to a VXIbus chassis, the buffer module should source and accept signals as described in the Standard External Trigger Buffer below.

B.6.2.3.1.7.1 Standard External Trigger Buffer

To maximize the compatibility between multiple VXIbus mainframes or a VXIbus mainframe and external instruments, a standard buffering scheme for TTLTRG* lines is recommended. A source has an output series impedance of 50 Ω. Its open-circuit low-level output is ≤ 0.4 V, and its open-circuit high-level output is ≥ 4.2 V. A 50 Ω cable is recommended for connection from the source module to other VXIbus mainframes or instruments. For highest performance, terminate the cable with 50 Ω at the furthest point from the source. The acceptor's threshold level is centered at

1.5 V, with a loading impedance of ≥ 5 KΩ. The TTLTRG* sense remains uninverted (Asserted state is low). See Figure B.9.

OBSERVATION B.6.8:
The Standard External Trigger Buffer requires only simple buffering to implement the SYNC and ASYNC trigger protocols.

OBSERVATION B.6.9:
The Standard External Trigger Buffer does not provide for acceptors to assert low on the same line the source is driving. This is an issue only when extending the SEMI-SYNC trigger protocol. For this protocol, it is necessary to translate to a multiline protocol (such as the ASYNC protocol) external to the chassis in which the source and each acceptor have unique connections on the face plate of the buffer module. The acknowledge will be asserted by the buffer module within the time T3, as required (See Figure B.4). The trigger line will be unasserted only after all acceptors have acknowledged the trigger.

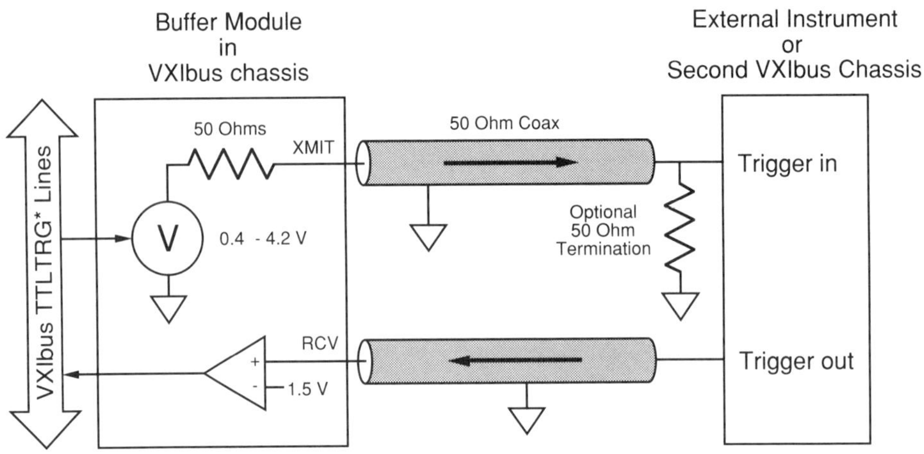

Figure B.9. External Trigger Buffering

B.6.2.3.1.8 Trigger Protocol Example

Two contrasting examples of using trigger protocols follow. The first involves a DMM and a set of relay multiplexer cards. The second involves a test sequencer and a set of modules simultaneously testing the DUT.

In the first example, a DMM and a set of relay multiplexers are configured as a multiple channel data acquisition system. The TTLTRG* lines are used to coordinate the DMM readings with the relay sequencing. The DMM is configured as a SEMI-SYNC source, and the relay cards as SEMI-SYNC acceptors. After each reading, the DMM momentarily asserts TTLTRG* to request a different input channel. It then waits for the switching operation (which may involve several relay modules) to complete. Each affected relay module immediately asserts TTLTRG* and keeps it asserted until it has completed its change of state. The DMM waits until all relay modules have deasserted TTLTRG*, and then initiates its next reading.

In the second example, a test sequencer module paces other modules through a test suite. A different sequence of tests and actions is downloaded to each module. TTLTRG* controls the execution of each list. Using the SEMI-SYNC protocol, the test sequencer initiates each step of the test by momentarily asserting TTLTRG*. Each of the other modules then asserts TTLTRG* and executes the next item in its test/action list, asserting TTLTRG* until done. After all the modules have deasserted TTLTRG*, the test sequencer initiates the next step. Exceptions, such as errors or test failures, which may require interruption of the test sequence, are reported through some other channel, such as VMEbus interrupts.

The trigger protocols may be used in combination. For example, a series of DMM readings is initiated by a test sequencer as one step in a stimulus/response test sequence paced by TTLTRG1*. The DMM, while asserting TTLTRG1*, then executes its measurement sequence, coordinating with its relay modules via TTLTRG2*. Upon completion of its measurement sequence, the DMM deasserts TTLTRG1*, allowing the test to proceed.

The trigger protocols may be used to generate any action, as long as the timing and allocation rules are followed.

B.6.2.4 ECLTRG0-1 (ECL Trigger Lines)

The ECLTRG lines are intended to be an intermodule timing resource. These two lines are bused the length of the VXIbus subsystem backplane. Any module (including the Slot 0 module) may drive these lines and receive information from these lines. These lines are single-ended ECL, and have a system impedance of 50 Ω. The asserted state is defined as logical high.

RULE B.6.40:
Each ECLTRG line **SHALL** be terminated with 50 Ω to -2 V at each end of the backplane. This requires that any driver be able to drive 25 Ω.

RULE B.6.41:
The maximum ECLTRG trace length from the P2 DIN connector pad on any module **SHALL NOT** exceed 1.0 inch.

RECOMMENDATION B.6.3:
The transmission line impedance connecting the P2 DIN connector to the ECLTRG receiver/driver circuitry should be 50 Ω.

RULE B.6.42:
All modules that connect to the ECLTRG lines **SHALL** use forward-biased and reverse-biased Schottky diodes from the signal lines to the transition bias voltage V_{BB} (V_{BB} = -1.29 V). Place the diodes as close to the receiver/driver as possible. Figure B.10 shows a typical ECL interface.

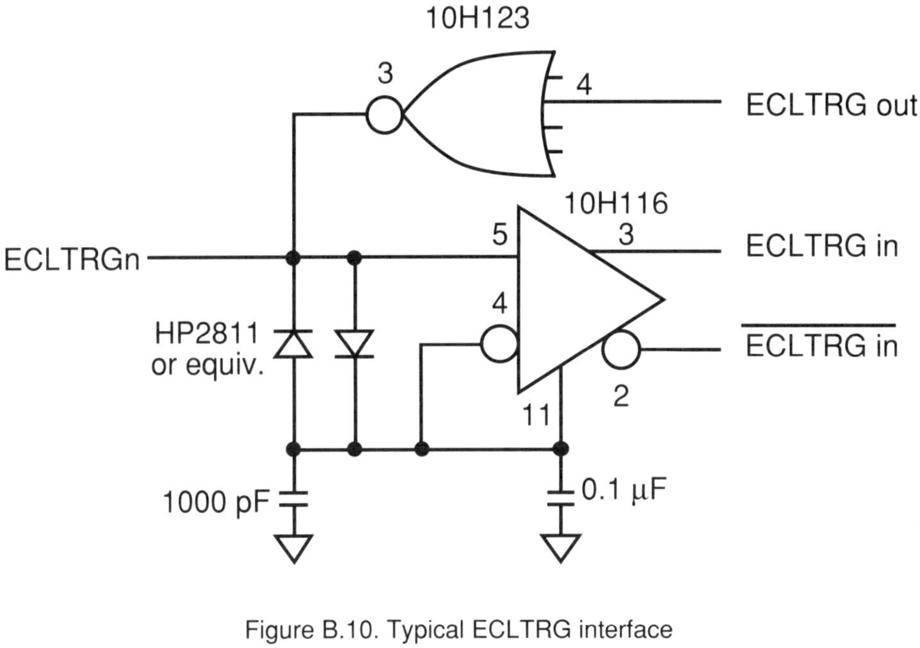

Figure B.10. Typical ECLTRG interface

OBSERVATION B.6.10:
Some 10KH ECL integrated circuits provide this bias voltage as an output. If such a device is being used to receive the signal from the bus, then this output should be used for diode clamps. Otherwise, a bias voltage generation circuit with a low equivalent impedance should be used.

OBSERVATION B.6.11:
If a module has an emitter-coupled output to the bus and this output is low, the emitter junction of the driver appears to the output as a diode to V_{BB}. When the bus is being driven from a different module, the emitter junction on this module will serve to clamp the ringing on the falling edge of the signal. Therefore, when using a module interface which includes an emitter coupled output, only one external diode to V_{BB} is necessary to clamp the ringing on the rising edge.

RULE B.6.43:
IF a module only receives from the ECLTRG lines, and does not drive these lines
THEN it **SHALL** have either the Schottky diodes specified for an ECLTRG driver, or a 150 Ω resistor in series with the transmission line connecting the ECL trigger lines on the P2 DIN connector to the line receiver.

RULE B.6.44:
The VBB power supply used to bias the ECLTRG Schottky clamping diodes **SHALL** be adequate to ensure proper operation.

RECOMMENDATION B.6.4:
Use two capacitors in parallel as shown in Figure B.10. One, a high frequency, low ESR 1000 pF capacitor, should be connected as close to the clamp diodes as possible. The reactance of this capacitor should remain capacitive up to at least 2 GHz. The placement of the other (a 0.1 μF capacitor) is not as critical.

RULE B.6.45:
Each module **SHALL NOT** have more than one equivalent driver/receiver load on any ECLTRG line.

RULE B.6.46:
IF a module drives the ECLTRG bus
THEN it **SHALL** use ECL drivers, which provide a LOW output level which is less than the -2V termination supply, e.g. 10H123.

OBSERVATION B.6.12:
Using drivers which produce a high impedance LOW level for ECLTRG lines eliminates the glitching which occurs when ECL devices are WIRE-OR'ed.

RULE B.6.47:
IF a module connects to an ECLTRG line
THEN it **SHALL** unassert that line within one second after SYSRESET* becomes unasserted.

RULE B.6.48:
ECLTRG lines **SHALL** be allocated in groups of one or two, and meet the same allocation rules as defined for the TTLTRG* lines.

B.6.2.4.1 Standard ECLTRG Protocols

The ECLTRG lines have a set of defined protocols corresponding to the TTLTRG* protocols. These protocols are defined in the following sections.

B.6.2.4.1.1 ECLTRG Synchronous (SYNC) Trigger Protocol

The ECLTRG SYNC trigger protocol is a single-line-broadcast trigger that does not require an acknowledge from any acceptors.

RULE B.6.49:
A ECLTRG SYNC trigger source **SHALL** assert the trigger for a minimum time of T1 and **SHALL NOT** reassert the trigger for a minimum time of T2 as shown in Figure B.11.

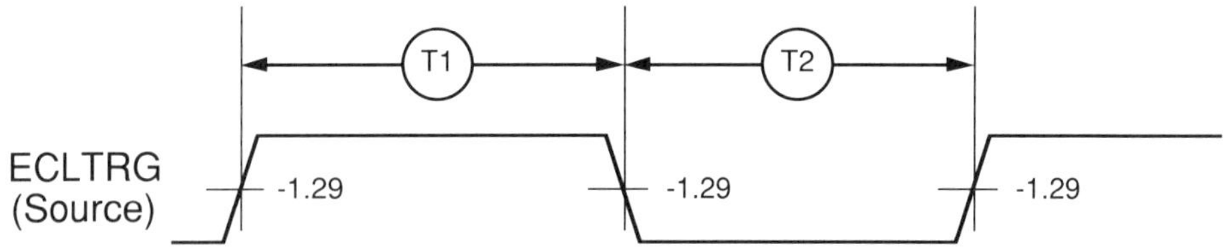

Parameter Label	Min	Max	Description
T1	8 nsec		Minimum source assertion time
T2	8 nsec		Minimum source non-assertion time

Figure B.11. ECLTRG Synchronous (SYNC) Trigger Protocol

RULE B.6.50:
An ECLTRG SYNC trigger acceptor **SHALL** accept any trigger asserted for ≥ 6 nsec, following a ≥ 6 nsec deassertion.

B.6.2.4.1.2 ECLTRG Semi-Synchronous (SEMI-SYNC) Trigger Protocol

The ECLTRG SEMI-SYNC trigger protocol is a single-line-broadcast, multiple-acceptor handshake protocol. A single source asserts the trigger for a minimum time of T1, as shown in Figure B.12. If the trigger needs to be acknowledged, the acceptor(s) also asserts the trigger within a time of T3, and unasserts the trigger when it is ready to proceed with the next operation. The source recognizes acknowledge completion when the source and all acceptors have unasserted the trigger line. The source may re-assert the trigger after a minimum time of T2, following the completion of the previous acknowledge.

RULE B.6.51:
An ECLTRG SEMI-SYNC trigger source **SHALL** meet the timing requirements for parameters T1 and T2 shown in Figure B.12.

RULE B.6.52:
An ECLTRG SEMI-SYNC trigger acceptor **SHALL** meet the timing requirements for parameter T3 shown in Figure B.12.

RULE B.6.53:
An ECLTRG SEMI-SYNC trigger acceptor **SHALL** accept any trigger asserted for ≥ 13 nsec, following a ≥ 3 nsec deassertion.

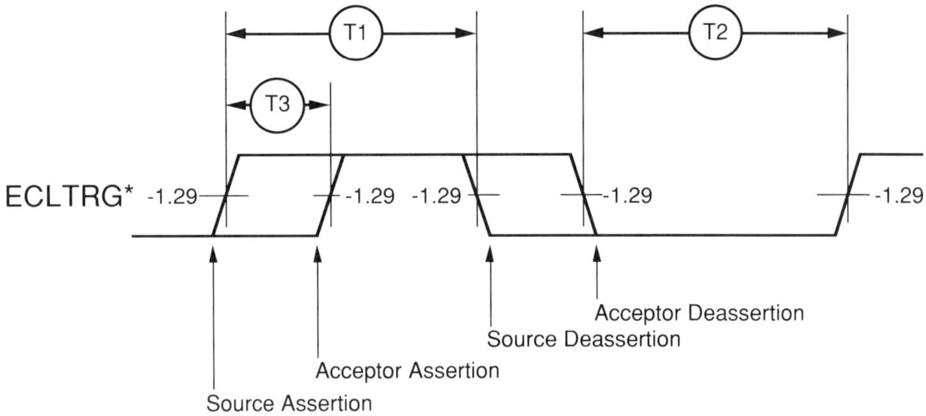

Parameter Label	Min	Max	Description
T1	20 nsec		Min. Source Assertion Time
T2	5 nsec		Min. Unassertion Time
T3	0 nsec	15 nsec	Max. Acceptor Response Time

Figure B.12. ECLTRG Semi-Synchronous (SEMI-SYNC) Trigger Protocol

B.6.2.4.1.3 ECLTRG Asynchronous (ASYNC) Trigger Protocol

The ECLTRG ASYNC trigger protocol is a two-line, single-source, single-acceptor protocol. The source initiates an operation by asserting the lower-numbered line of the allocated ECLTRG line pair, while the acceptor acknowledges by asserting the higher-numbered line of the ECLTRG line pair. This is a useful trigger mode for handshaking between VXIbus modules and external instruments.

RULE B.6.54:
A module implementing the ECLTRG ASYNC trigger protocol **SHALL** source only on the lower-numbered line of an ECLTRG line pair.

RULE B.6.55:
A module implementing the ECLTRG ASYNC trigger protocol **SHALL** acknowledge only on the higher-numbered line of an ECLTRG line pair.

RULE B.6.56:
IF a module implements the ECLTRG ASYNC trigger protocol
THEN it **SHALL** meet the timing requirements shown in Figure B.13.

RULE B.6.57:
An ECLTRG ASYNC trigger source or acceptor **SHALL** accept any trigger asserted for $\geq$ 6 nsec, following a $\geq$ 6 nsec deassertion.

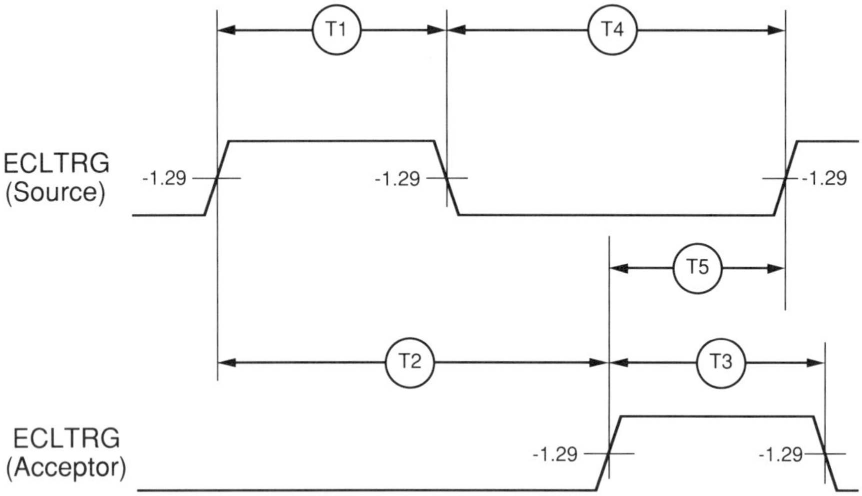

Parameter Label	Min	Max	Description
T1	8 nsec		Minimum Source assertion time
T2	0 nsec		Minimum Acceptor response time
T3	8 nsec		Minimum Acceptor assertion time
T4	8 nsec		Minimum Source non-assertion time
T5	0 nsec		Minimum Acceptor Acknowledge to source re-assertion time

Figure B.13. ECLTRG Asynchronous (ASYNC) Trigger Protocol

B.6.2.4.1.4 ECLTRG Clock Transmission

An ECLTRG line may be used for a clock signal transmission.

RULE B.6.58:
IF an ECLTRG line is used for clock transmission,
THEN it **SHALL** meet the timing requirements described for the ECLTRG SYNC trigger protocol.

B.6.2.4.1.5 ECLTRG Data Transmission

The ECLTRG lines can also be used for data transmission, where one of the ECLTRG lines is used as a clock. Data may be synchronized to either the rising or falling edge of the clock, or both.

RULE B.6.59:
IF a source uses ECLTRG lines for data transmission,
THEN the source **SHALL** meet the timing requirements shown in Figure B.14.

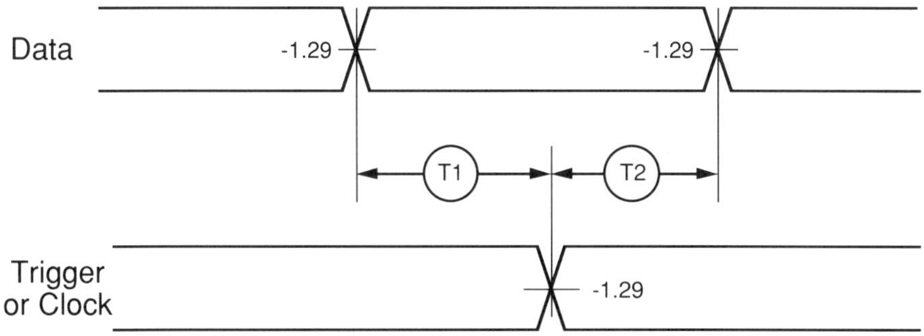

Parameter Label	Min	Max	Description
T1	8 nsec		Data setup time
T2	4 nsec		Data hold time

Figure B.14. ECLTRG Data Transmission

RULE B.6.60:
An ECLTRG data acceptor **SHALL** accept any data with a setup time of ≥ 6 nsec and a hold time of ≥ 2 nsec .

B.6.2.4.1.6 Extended Start/Stop (ESTST) Protocol

Extended START/STOP (ESTST) protocol is the natural extension of the START/STOP (STST) protocol to D size modules. Because the modules may contain P3 (and use CLK100 as their instrument clock) three additional signals have been added to the STST protocol (CLK100, SYNC100, and an ECLTRG line). The ECLTRG line determines whether an ESTST operation is initiated, while the same TTLTRG line used to transmit the START/STOP value for STST devices also determines the START or STOP operation for ESTST devices.

In the ESTST protocol, a Slot 0 SYNC100 driver (See Sec. B.6.3.2, SYNC100) selects a particular CLK100 clock edge to initiate the START or STOP command. An ECLTRG line indicates whether the SYNC100 pulse is related to the ESTST protocol. This line is known as the ESTST qualifier line. Since the CLK10 and CLK100 signals are synchronized, the Slot 0 module can always qualify a CLK100 edge having a fixed phase relationship to CLK10. This fixed-phase relationship, over a series of ESTST operations, guarantees fixed and repeatable delays between modules paced by CLK10 (STST) and those paced by CLK100 (ESTST). The TTLTRG* line used to indicate START/STOP for the STST protocol also determines START or STOP for the ESTST protocol.

RULE B.6.61:
IF a module implements the ESTST protocol,
THEN the module **SHALL** be user programmable, to select any ECLTRG line as the ESTST qualifier line.

RULE B.6.62:
ESTST qualifier true **SHALL** be represented by a high state of the selected ECLTRG ESTST qualifier line, and ESTST qualifier false **SHALL** be represented by a low state.

RULE B.6.63:
A Slot 0 module implementing the ESTST protocol **SHALL** meet the timing requirements of Figure B.15.

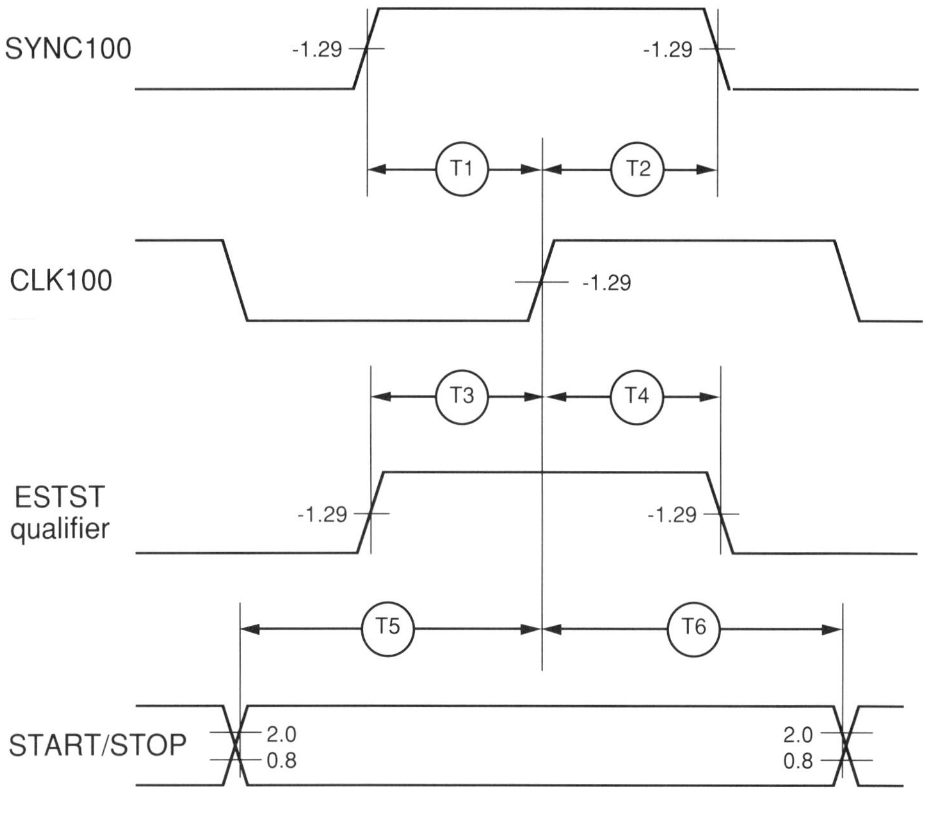

Parameter Label	Min	Max	Description
T1	4.0 nsec	8.0 nsec	SYNC100 setup time
T2	3.5 nsec	8.0 nsec	SYNC100 hold time
T3	13.0 nsec		ESTST Qualifier setup time
T4	11.0 nsec		ESTST Qualifier hold time
T5	50.0 nsec		START/STOP setup time
T6	15.0 nsec		START/STOP hold time

Figure B.15. CLK100, SYNC100, and ECLTRG START/STOP Timing

RULE B.6.64:
A Slot 0 module implementing the ESTST protocol **SHALL** meet the timing requirements for the STST protocol.

RULE B.6.65:
An ESTST acceptor **SHALL** accept SYNC100 setup times ≥ 2 nsec and hold times ≥ 0.5 nsec, ESTST qualifier setup times ≥ 4 nsec and hold times ≥ 2 nsec, and STST setup times ≥ 7 nsec and hold times ≥ 7 nsec, with respect to the CLK100 edge.

B.6.2.4.1.7 External Trigger Buffering

RECOMMENDATION B.6.5:
Since the ECLTRG asserted sense is opposite the TTLTRG* asserted sense, the ECLTRG signal should be inverted by the Standard External Trigger Buffer. This allows compatibility of external ECLTRG and TTLTRG* signals.

OBSERVATION B.6.13:
By meeting the loading and driving rules for ECLTRG lines, a module is prohibited from extending these lines external to the chassis, without buffering these lines. For example, to properly buffer the SEMI-SYNC trigger protocol, the acknowledge must be asserted by the buffer card within the time T3, as required in Figure B.12. The trigger line will be unasserted by the buffer card when the external acceptor has acknowledged the trigger. See Section B.6.2.3.1.7.1, "Standard External Trigger Buffer."

B.6.2.5 SUMBUS

The SUMBUS is an analog summing node that is bused the length of the VXIbus subsystem backplane. It is intended that any module be able to drive or receive this line. Each module can drive this line using an analog current source driver. Each module may receive from this bus through a high impedance receiver, such as a high impedance analog buffer amplifier.

RULE B.6.66:
The SUMBUS **SHALL** be terminated to signal ground through 50 ±1 Ω, at each end of the backplane.

RULE B.6.67:
A receiver on the SUMBUS **SHALL** have an equivalent input impedance of ≥ 10 KΩ, in parallel with ≤ 3 pF.

RULE B.6.68:
A SUMBUS output current source **SHALL** have an equivalent output impedance of ≥ 10 KΩ in parallel with ≤ 4 pF.

RULE B.6.69:
A SUMBUS output current source **SHALL** have ± 0.8 V minimum compliance.

RULE B.6.70:
A SUMBUS output current source **SHALL NOT** source more than 40 mA.

RULE B.6.71:
The backplane **SHALL** supply a SUMBUS protection network that clamps the SUMBUS voltage within the range ± 3.0 V for all SUMBUS input currents in the range of ± 520 mA.

RULE B.6.72:
The SUMBUS protection network **SHALL NOT** source or sink more than ± 1.0 μA, nor load with greater than 30 pF, for SUMBUS voltages in the range ± 1.0 V.

OBSERVATION B.6.14:
Low leakage diodes might be required to meet the SUMBUS leakage requirement.

SUGGESTION B.6.1:
Because all modules might drive the SUMBUS at the same time, a module designer should consider the total dynamic range of a SUMBUS output current source, and be able to adjust the maximum output current accordingly. This will keep the signal within the compliance region.

RULE B.6.73:
SUMBUS output current sources and receivers **SHALL NOT** be located more than 1.0 inch from the P2 DIN SUMBUS pad.

RULE B.6.74:
When a module's SUMBUS module output current source is disabled, the module **SHALL NOT** source more than ±1.0 μA into the SUMBUS.

B.6.2.6 LOCAL BUS

The Local Bus is a daisy chained bus. It is defined by adjacently installed modules. The backplane connects from the LBUSC pins of Slot N to the LBUSA pins of Slot N+1. (See Figure B.1.) No Local Bus lines are extended beyond Slot 12. Several classes of signal levels are allowed and may be transmitted via this bus. To ensure that modules which produce incompatible levels are not installed in adjacent slots, a keying mechanism has been provided. The Local Bus key provides support for TTL, ECL and analog communication. Each module indicates (with a mechanical key on its faceplate, which classes of signals it may source or accept nondestructively on each side of its Local Bus. The particular key position for each of the signal classes (as well as sensor-only devices) are specified in the Mechanical Specifications (Section B.7). Sensor-only devices are modules that can non-destructively accept signals of a number of classes, but do not source any signals of their own.

RULE B.6.75:
Modules **SHALL NOT** drive more than ±42 V onto any LBUS line.

RULE B.6.76:
Modules **SHALL NOT** drive more than 500 mA DC current into any LBUS line.

NUMBER	CLASS	-LIMIT	+LIMIT	DRIVE LIMIT
1	TTL	-0.5 V	+5.5 V	200 mA
2	ECL	-5.46 V	0.0 V	50 mA
3	ANALOG LOW	-5.5 V	+5.5 V	50 Ω Drive
4	ANALOG MED	-16.0 V	+16.0 V	500 mA
5	ANALOG HIGH	-42.0 V	+42.0 V	500 mA
6	RESERVED			

TABLE B.3. Logical Classes of LBUS Signals

RULE B.6.77:
The backplane resistance between any LBUS line and any adjacent pin **SHALL** be greater than 10.0 MΩ.

RULE B.6.78:
The di/dt on any LBUS line **SHALL NOT** exceed ±100 mA/nsec.

RECOMMENDATION B.6.6:
Multi-module instrument systems which use the Local Bus for intra-system communications should be built with the anchor and expander protocol. An anchor module has no connections on LBUSA lines. It may drive or receive LBUSC lines. An expander module may drive or receive LBUSA or LBUSC lines. Expanders should be placed in adjacent slots (of higher number) to the anchor. This protocol always guarantees compatibility between adjacent module groups without empty slots.

RULE B.6.79:
Any module that accesses the LBUS lines **SHALL** indicate its class on its local bus keys.

RULE B.6.80:
A module **SHALL NOT** exceed the voltage specified for its class under any loading conditions.

OBSERVATION B.6.15:
If a module provides a direct connection (an electrical short) between LBUSA lines and LBUSC lines, the keying mechanism should indicate that the module will source on LBUSA and LBUSC any class it accepts on either LBUSC or LBUSA respectively.

RULE B.6.81:
A module **SHALL NOT** indicate on its key the RESERVED signal class.

OBSERVATION B.6.16:
The keying mechanism cannot prevent a destructive condition in all situations. It is good practice for module manufacturers to protect their modules from destruction by any signals within the class that the module accepts.

RULE B.6.82:
The Slot 0 P2 LBUSA key **SHALL** indicate TTL class.

RULE B.6.83:
The Slot 0 P3 LBUSA and LBUSC keys **SHALL** indicate ECL class.

RULE B.6.84:
The mainframe (if it supports VXIbus Subsystem P2) **SHALL** provide a Slot 0 P2 LBUSC key which indicates TTL class.

RULE B.6.85:
The mainframe (if it supports VXIbus Subsystem P3) **SHALL** provide a Slot 0 P3 LBUSC key which indicates ECL class.

PERMISSION B.6.2:
A device which non-destructively senses signals up to ±16 volts, but does not source any signals (or provide direct connection between LBUSA and LBUSC) **MAY** provide a single local bus key tab for ±16 V sensor-only operation.

PERMISSION B.6.3:
A device which non-destructively senses all classes of signals up to ±42 volts, but does not source any signals (or provide direct connection between LBUSA and LBUSC) **MAY** provide no local bus key tabs for ±42 V sensor-only operation. Such a device can not cause a destructive condition.

OBSERVATION B.6.17:
A device that does not access the local bus does not provide any local bus keys.

OBSERVATION B.6.18:
Size A, B, and C modules do not provide P3 local bus keys.

B.6.2.7 RSV2-3 (Reserved)

These pins are reserved.

RULE B.6.86:
Modules and backplanes **SHALL NOT** make any connections to RSV pins.

B.6.3 The VXIbus Subsystem P3 Connector

For higher performance instrumentation, the VXIbus also defines the P3 connector. As with P2, alternate pin definitions of P3 may coexist in VXIbus systems, but any one VXIbus subsystem must exist in contiguous slots. Again, Slot 0 plays a unique role in delivering system resources on P3, such as high-speed clocking and trigger routing. In particular, VXIbus P3 adds:

- Additional +5V, -5.2V, -2V, ±24V, and ±12V power.
- A 100 MHz differential clock, which is synchronous with the P2 10 MHz clock.
- A synchronizing signal for 100 MHz clock edge selection.

- 4 additional ECL trigger lines, for a total of 6.
- 24 additional local bus lines, for a total of 36.
- "Star Trigger" lines for precision module-to-module timing.

B.6.3.1 CLK100

CLK100 is a 100 MHz system clock. It is sourced from Slot 0, and distributed to slots 1-12 on P3. The Slot 0 CLK100 output is differential ECL. It is buffered on the backplane and distributed to each module slot as a single-source, single-destination, differential ECL signal, in a manner similar to CLK10. The distribution delays are matched to provide a tight timing relationship between modules. The backplane buffering provides a high level of inter-module isolation.

RULE B.6.87:
The CLK100 frequency sourced from Slot 0 **SHALL** be 100 MHz. Its accuracy **SHALL** be equal to or better than ±100 ppm (.01%) over its specified operating temperature and time.

OBSERVATION B.6.19:
A Slot 0 module may generate CLK100 internally, or it may derive CLK100 from an external frequency source.

RULE B.6.88:
CLK100 duty cycle **SHALL** be 50% ± 20%, when measured at the 50% transition level.

RULE B.6.89:
IF CLK100 is switched between different clock sources,
THEN the minimum clock width, high or low, **SHALL NOT** be less than 2.5 nsec, or more than 1.0 μsec, during switching. The minimum time between two successive transitions of the same polarity **SHALL NOT** be less than 9.5 nsec.

RULE B.6.90:
CLK100 **SHALL** be synchronous to CLK10.

OBSERVATION B.6.20:
The absolute timing relationship between CLK10 and CLK100 is not specified. This relationship may be produced by deriving CLK10 from CLK100, or by deriving CLK100 from CLK10.

RULE B.6.91:
The CLK100 and SYNC100 backplane distribution network **SHALL NOT** insert more than 2.0 nsec timing skew between CLK100 and SYNC100 at any slot.

OBSERVATION B.6.21:
If a 10H101 distribution buffer is used for CLK100 or SYNC100, the trace lengths should be matched within 1 nsec: approximately 15 cm (6 inches).

PIN NUMBER	ROWa SIGNAL MNEMONIC	ROWb SIGNAL MNEMONIC	ROWc SIGNAL MNEMONIC	PIN NUMBER
1	ECLTRG2	+24V	+12V	1
2	GND	–24V	–12V	2
3	ECLTRG3	GND	RSV4	3
4	–2V	RSV5	+5V	4
5	ECLTRG4	–5.2V	RSV6	5
6	GND	RSV7	GND	6
7	ECLTRG5	+5V	–5.2V	7
8	–2V	GND	GND	8
9	LBUSA12	+5V	LBUSC12	9
10	LBUSA13	LBUSC15	LBUSC13	10
11	LBUSA14	LBUSA15	LBUSC14	11
12	LBUSA16	GND	LBUSC16	12
13	LBUSA17	LBUSC19	LBUSC17	13
14	LBUSA18	LBUSA19	LBUSC18	14
15	LBUSA20	+5V	LBUSC20	15
16	LBUSA21	LBUSC23	LBUSC21	16
17	LBUSA22	LBUSA23	LBUSC22	17
18	LBUSA24	2V	LBUSC24	18
19	LBUSA25	LBUSC27	LBUSC25	19
20	LBUSA26	LBUSA27	LBUSC26	20
21	LBUSA28	GND	LBUSC28	21
22	LBUSA29	LBUSC31	LBUSC29	22
23	LBUSA30	LBUSA31	LBUSC30	23
24	LBUSA32	+5V	LBUSC32	24
25	LBUSA33	LBUSC35	LBUSC33	25
26	LBUSA34	LBUSA35	LBUSC34	26
27	GND	GND	GND	27
28	STARX+	–5.2V	STARY+	28
29	STARX–	GND	STARY–	29
30	GND	–5.2V	-5.2V	30
31	CLK100+	-2V	SYNC100+	31
32	CLK100–	GND	SYNC100–	32

TABLE B.4. P3 Pin Definitions.

RULE B.6.92:
Each slot CLK100 **SHALL** be driven by a unique backplane buffer output.

RULE B.6.93:
CLK100 **SHALL** be distributed differentially from Slot 0.

RULE B.6.94:
The backplane CLK100 distribution traces **SHALL** be designed for 50 Ω.

PIN NUMBER	ROWa SIGNAL MNEMONIC	ROWb SIGNAL MNEMONIC	ROWc SIGNAL MNEMONIC	PIN NUMBER
1	ECLTRG2	+24V	+12V	1
2	GND	–24V	–12V	2
3	ECLTRG3	GND	RSV4	3
4	–2V	RSV5	+5V	4
5	ECLTRG4	–5.2V	RSV6	5
6	GND	RSV7	GND	6
7	ECLTRG5	+5V	–5.2V	7
8	–2V	GND	GND	8
9	STARY12+	+5V	STARX01+	9
10	STARY12–	STARY01–	STARX01–	10
11	STARX12+	STARX12–	STARY01+	11
12	STARY11+	GND	STARX02+	12
13	STARY11–	STARY02–	STARX02–	13
14	STARX11+	STARX11–	STARY02+	14
15	STARY10+	+5V	STARX03+	15
16	STARY10–	STARY03–	STARX03–	16
17	STARX10+	STARX10–	STARY03+	17
18	STARY09+	–2V	STARX04+	18
19	STARY09–	STARY04–	STARX04–	19
20	STARX09+	STARX09–	STARY04+	20
21	STARY08+	GND	STARX05+	21
22	STARY08–	STARY05–	STARX05–	22
23	STARX08+	STARX08–	STARY05+	23
24	STARY07+	+5V	STARX06+	24
25	STARY07–	STARY06–	STARX06–	25
26	STARX07+	STARX07–	STARY06+	26
27	GND	GND	GND	27
28	STARX+	–5.2V	STARY+	28
29	STARX–	GND	STARY–	29
30	GND	–5.2V	–5.2V	30
31	CLK100+	–2V	SYNC100+	31
32	CLK100–	GND	SYNC100–	32

TABLE B.5. P3 Slot 0 Pin Definitions.

RULE B.6.95:
The CLK100 backplane distribution network **SHALL NOT** insert more than 2.0 nsec timing skew between CLK100 signals at any two slots.

RULE B.6.96:
The absolute delay of CLK100 from Slot 0 to any module **SHALL NOT** exceed 8 nsec.

RULE B.6.97:
IF a module accesses the CLK100 signals
THEN it **SHALL** provide 50 Ω termination on CLK100+ and CLK100–, with no more than two equivalent ECL loads.

RECOMMENDATION B.6.7:
Use a 10H101 to buffer and fan out CLK100.

B.6.3.2 SYNC100

The synchronization signal SYNC100 is used to synchronize multiple devices, with respect to a given rising edge of CLK100. This provides very tight time coordination between modules. The SYNC100 signal is distributed from Slot 0 to slots 1-12, with individual backplane buffers for each slot. The SYNC100 signal can be compared to the GPIB Group Execute Trigger command in function, but with greatly improved time coordination. As the Slot 0 module drives the SYNC100 and CLK100 signals, it may also be programmed to drive one or more ECLTRG lines, in order to qualify the SYNC100 action. An example of this protocol is the ESTST protocol, described under ECLTRG Trigger Protocols.

A Slot 0 module implementing the SYNC100 function will provide an arbiter to synchronize external events to CLK100, and to meet the guaranteed setup and hold times for the SYNC100 signal. This guarantees that all affected modules will trigger on the same CLK100 clock edge. SYNC100 is a nominally 10 nsec pulse, and may be initiated by any type of external or internal event: an external BNC trigger, a software register write, the state of any of the TTLTRG*, ECLTRG, LBUS or STAR lines, or any other signal synchronized by the Slot 0 arbiter. Refer to the timing diagram on Figure B.15.

OBSERVATION B.6.22:
SYNC100 delay matching relative to CLK100 is called out under CLK100 matching rules.

RECOMMENDATION B.6.8:
Use a 10H101 to buffer and fan out SYNC100.

RULE B.6.98:
Slot 0 modules **SHALL** provide ≥ 4.0 nsec of setup time from SYNC100 to the rising edge of CLK100.

RULE B.6.99:
Slot 0 modules **SHALL** provide ≥ 3.5 nsec of hold time from the rising edge of CLK100 to SYNC100.

RECOMMENDATION B.6.9:
Slot 0 modules that provide SYNC100 support should be designed such that the SYNC100 recycle rate is ≤ 50 nsec .

RULE B.6.100:
Each module slot SYNC100 **SHALL** be driven by a unique backplane buffer output.

RULE B.6.101:
SYNC100 **SHALL** be distributed differentially from Slot 0.

RULE B.6.102:
The backplane SYNC100 distribution traces **SHALL** be designed for 50 Ω.

RULE B.6.103:
The SYNC100 backplane distribution network **SHALL NOT** insert more than 2.0 nsec timing skew between SYNC100 signals at any two slots.

RULE B.6.104:
IF a module accesses the SYNC100 signals
THEN it **SHALL** provide 50 Ω termination on SYNC100+ and SYNC100–, with no more than two equivalent ECL loads.

B.6.3.3 STARX and STARY

STARX and STARY provide inter-module asynchronous communication. Two STAR lines are connected between each module slot and Slot 0. Slot 0 may provide a cross matrix switch which can programmably route signals between any two STARX or STARY lines. It may also broadcast a signal received on one STAR line to a group of STAR lines. The STAR lines are bidirectional, providing additional flexibility.

RULE B.6.105:
The STARX and STARY backplane distribution network **SHALL NOT** insert more than 2.0 nsec timing skew between any two STAR signals.

RULE B.6.106:
STARX and STARY **SHALL** be driven differentially.

RULE B.6.107:
The backplane STAR distribution traces **SHALL** be designed for 50 Ω.

RULE B.6.108:
IF a module accesses the STAR signals
THEN it **SHALL** provide a 50 Ω termination on STAR+ and STAR–, with no more than a single equivalent ECL driver load and a single equivalent ECL receiver load.

RULE B.6.109:
Modules that can both drive and receive STARX or STARY **SHALL** use bidirectional ECL drivers, e.g. 10H123. When the module is programmed to receive, the STAR+ and STAR– drivers **SHALL** be forced into the high impedance state (Vol ≤ -2.0 V)

RULE B.6.110:
A D-size Slot 0 module **SHALL** provide a default differential signal to all STARX and STARY lines that are not being driven by another module.

PERMISSION B.6.4:
If a Slot 0 module does not support STAR bus routing, it may meet the above rule by connecting a 50 Ω resistor between STAR+ lines and ground.

RULE B.6.111:
The absolute delay of any STAR line between Slot 0 and any module **SHALL NOT** exceed 5 nsec .

RULE B.6.112:
The backplane **SHALL NOT** connect to the STARX+, STARX-, STARY+, and STARY- pins of the Slot 0 backplane connector.

PERMISSION B.6.5:
The Slot 0 module **MAY** access the STARX+, STARX-, STARY+, and STARY- pins.

OBSERVATION B.6.23:
Allowing all modules (including Slot 0) to access the STAR lines allows construction of modules that may operate in any slot position, including Slot 0.

B.6.3.4 ECLTRG2-5

These lines are functionally the same as the ECLTRG0-1 lines on P2 (described in Section B.6.2.4 *ECLTRG0-1*) and meet the same rules.

B.6.3.5 LBUS12-35

These lines are functionally the same as the LBUS00-11 lines on P2 (described in Section B.6.2.6 *Local Bus*) and meet the same rules.

B.6.3.6 RSV4-7

These lines are functionally the same as the RSV2-3 lines on P2 (described in Section B.6.2.7 *RSV2-3*) and meet the same rules.

B.6.4 Backplane

The backplane is the common signal and power distribution network for all modules. As such, it must be very well designed, so as not to introduce noise into the system.

OBSERVATION B.6.24:
Noise should be minimized on the ECLTRG Lines, Clock lines, and SUMBUS lines. It is desirable to maintain time jitter to $\leq$ 25 psec. This requires that noise be limited to $\pm$ 9 mV, assuming an ECL slew rate of 2.75 V/nsec .

RULE B.6.113:
All backplane ECL transmission paths **SHALL** be designed for 50 Ω nominal characteristic impedance.

OBSERVATION B.6.25:
When TTL lines share a layer of the backplane with 50 Ω ECL transmission lines, the TTL signal characteristic impedance may also be near 50 Ω. This should provide satisfactory operation (See the VMEbus Specification, Rev. C.1, Observation 6.13).

SUGGESTION B.6.2:
See the MECL System Design Handbook, Motorola, for more backplane and micro-strip design information.

B.6.4.1 ECL Signal Levels

The signal levels present on the backplane ECL lines are summarized in Table B.6.

Symbol	Parameter	Minimum	Typical	Maximum	Units
V_{OH}	Output logic level "1" (HIGH)	-0.980	-0.9	-0.735	Volts
V_{OL}	Output logic level "0" (LOW)	-1.950	-1.75	-1.630	Volts
V_{BB}	Logic Switching Threshold		-1.29		Volts

Table B.6. Backplane ECL signal levels

B.7 MECHANICAL SPECIFICATIONS

B.7.1 Introduction

Information is provided in this subsection to ensure that VXIbus modules, backplanes, mainframes, and associated mechanical accessories are dimensionally compatible.

The mechanical dimensions given in this subsection are based on ANSI/IEEE 1101 and IEC publications 297-3 and 603-2. The electrical characteristics for VXIbus connectors, as specified here, supersede publication 603-2 where they differ.

IEC publication 603-2 describes a family of connector types which are identified by labels of the form:

603-2-IEC-xxxxxx-xxx

All of the P1/J1, P2/J2 and P3/J3 connectors used on VXIbus modules and backplanes are members of this family. In this chapter, the label 603-2-IEC-xxxxxx-xxx is used when referring to all of these connector types as a group. The label 603-2-IEC-C096Mx-xxx is used when referring to the 96-pin male connector types within this family which are used on the VXIbus modules. 603-2-IEC-C096Fx-xxx is used when referring to the 96-pin female connector types which are used on VXIbus backplanes.

Figure B.16 is a front view of a typical VXIbus mainframe. In the mainframe shown, modules are inserted into the mainframe from the front, in a vertical plane, with the component face of the module board on the right.

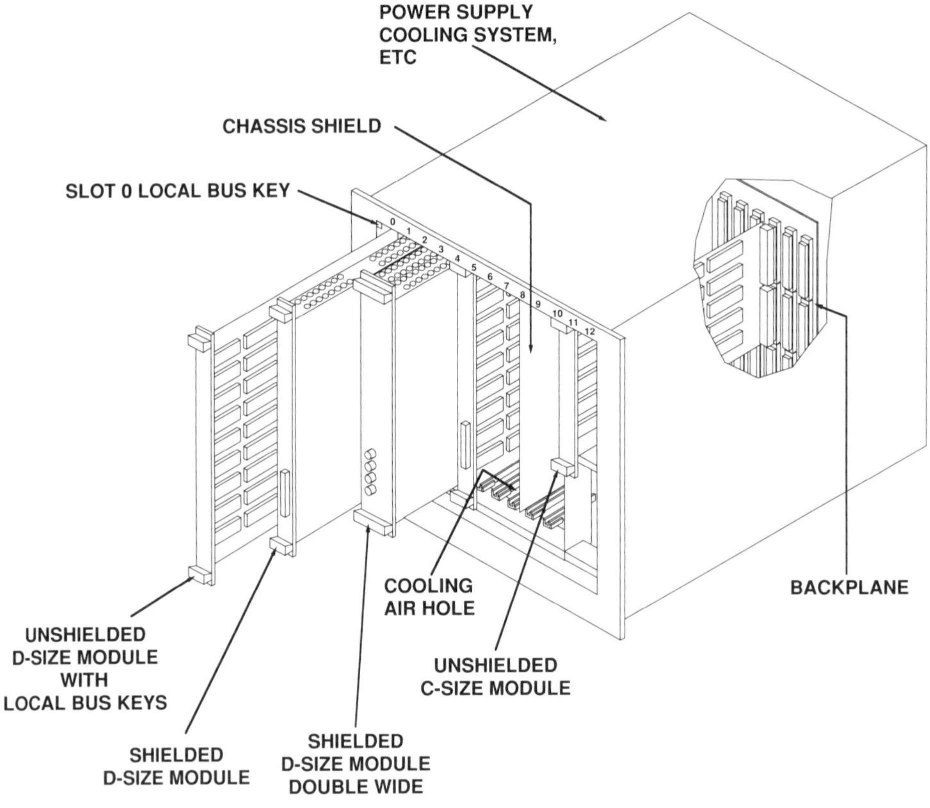

Figure B.16. Typical D-size mainframe

B.7.2 Module Specifications

B.7.2.1 VXIbus Boards and Modules

RULE B.7.1:
VXIbus modules **SHALL** fit into one of four size classifications:

1. Single height modules *typically* containing a board with dimensions 100 mm (3.937 in) high x 160 mm (6.299 in) deep will be referred to as "A" Size modules.

2. Double height modules *typically* containing a board with dimensions 233.35 mm (9.187 in) high x 160 mm (6.299 in) deep will be referred to as "B" Size modules.

3. Double height modules *typically* containing a board with dimensions 233.35 mm (9.187 in) high x 340 mm (13.386 in) deep will be referred to as "C" Size modules.

4. Triple height modules *typically* containing a board with dimensions 366.7 mm (14.437 in) high x 340 mm (13.386 in) deep will be referred to as "D" Size modules.

RULE B.7.2:
A size and B size boards and board assemblies **SHALL** be designed according to the dimensions given in the VMEbus specification. [1]

OBSERVATION B.7.1:
A size and B size boards and board assemblies are designed for a mainframe with 20.32 mm (0.8 in) spacing between slots.

RULE B.7.3:
C size and D size modules **SHALL** be contained within the envelope specified in Figures B.19, B.20, and B.23.

RULE B.7.4:
C size and D size modules **SHALL** be designed for a mainframe with 30.48 mm (1.2 in) spacing between slots.

OBSERVATION B.7.2:
Typical C size and D size modules will contain a board as shown in Figures B.17 and B.18.

RULE B.7.5:
Modules **SHALL** provide the board edge feature (as specified in Figure B.23) to allow proper insertion into a mainframe. This feature **SHALL** be 1.6 $\pm$ 0.2 mm (0.063 $\pm$ 0.008 in) thick.

RECOMMENDATION B.7.1:
Design C size and D size modules around a board as specified in Figures B.17 and B.18. This will ensure proper position of connectors, front panel, card edge, etc.

[1] See Appendix IV (page VXI-246)

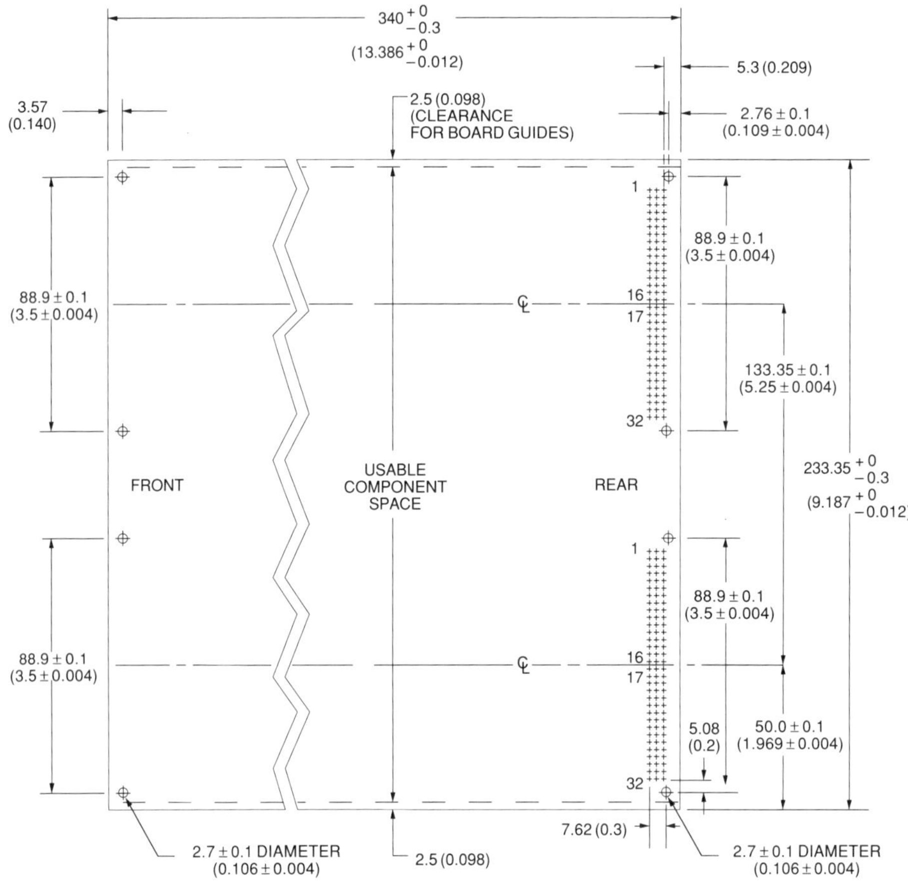

NOTES:
1. All dimensions are shown in millimeters. Inch dimensions are shown in parentheses.
2. Front panel mounting holes are suggestions only.
3. OBSERVATION: P1 is required for VXI devices. P2 is optional.
4. OBSERVATION: Board outline may be reduced to accomodate shields.

Figure B.17. C Size board

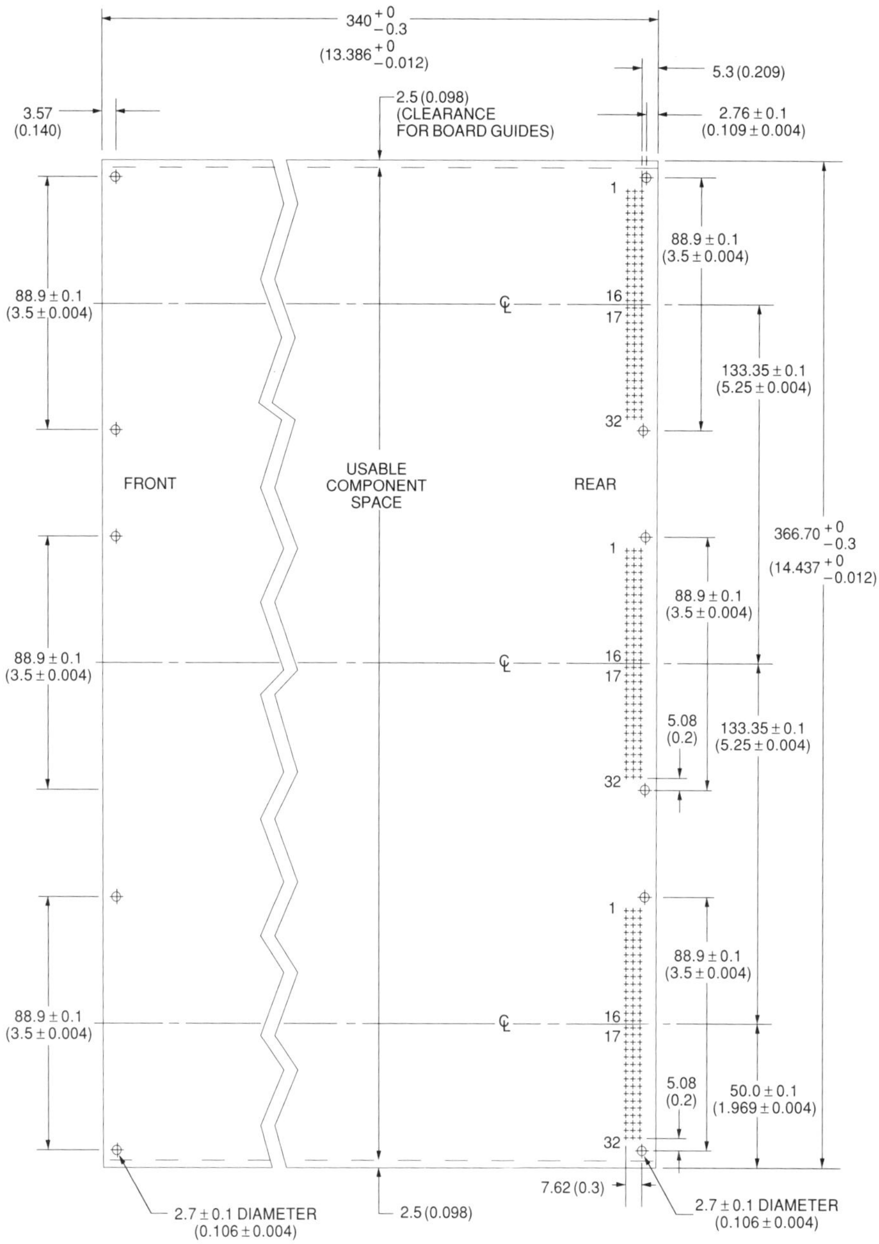

NOTES:

1. All dimensions are shown in millimeters. Inch dimensions are shown in parentheses.
2. Front panel mounting holes are suggestions only.
3. OBSERVATION: P1 is required for VXI devices. P2 and P3 are optional.
4. OBSERVATION: Board outline may be reduced to accomodate shields.

Figure B.18. D Size board

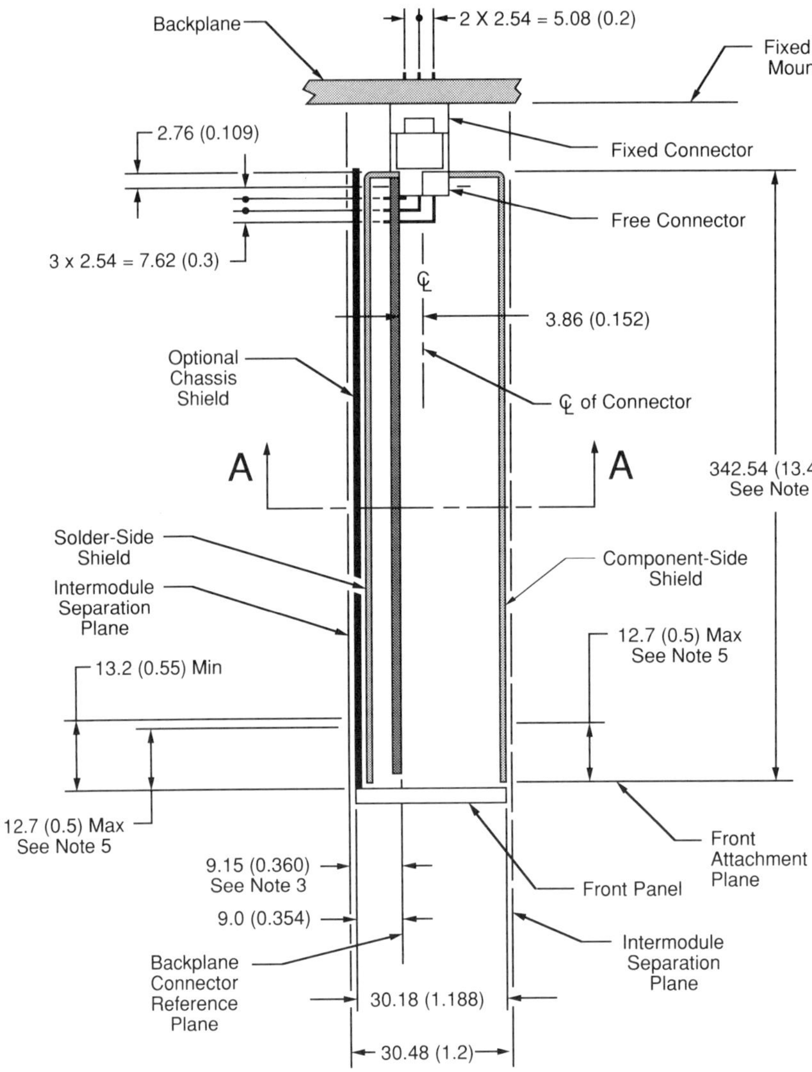

NOTES:
1. All dimensions are shown in millimeters. Inch dimensions are shown in parentheses.
2. For additional information regarding the allowed thickness of boards, refer to the section "Module Width".
3. This 9.15 mm (0.360 inch) dimension is the same regardless of the thickness of the board.
4. Refer to the section 'Component Height, Lead Length, Shield Height and Warpage.
5. This area may be used for electrical contact with compatible adjacent modules. Refer to the section 'Module Shielding'.
6. View A-A. See Figure 'Module Envelope, Front View' and Figure 'Module Edge Guide Feature'.
7. Refer to the section 'Backplanes' rule on maximum component height, and the section 'Module Shielding' observation on box type plug-in units.

Figure B.19. Module envelope, top view

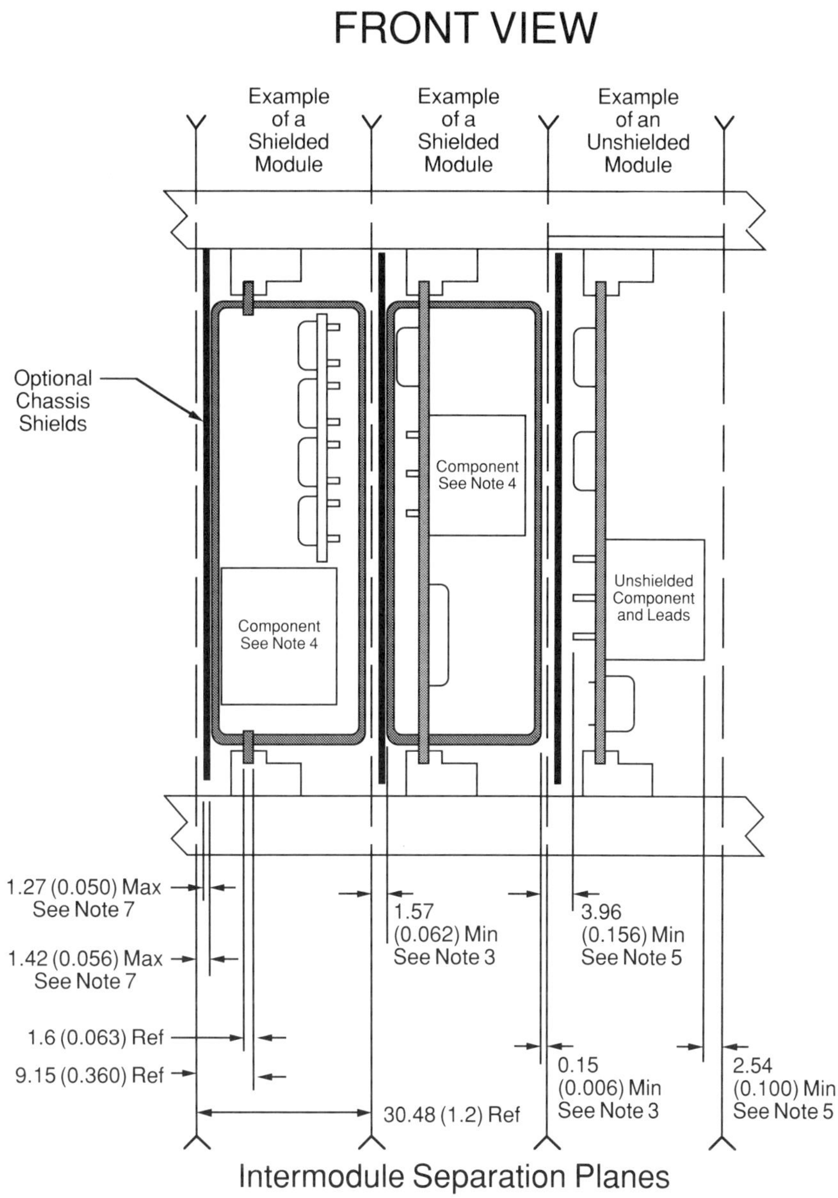

NOTES:
1. All dimensions are shown in millimeters. Inch dimensions are shown in parentheses.
2. Refer to the section, 'Component Height, Lead Length, Shield Height and Warpage'.
3. Minimum shield clearance refers to module when installed and includes the effects of warpage'. RECOMMENDATION: Nominal clearance should be at least 0.5 (0.020).
4. No restrictions are placed on components inside a shielded module.
5. Applies only to unshielded components and leads.
6. OBSERVATION: Extreme care must be taken with unshielded modules. Safety, handling, voltage, warpage, etc. should be considered.
7. RULE: IF provided, THEN the chassis shield MUST conform to these dimensions.

Figure B.20. Module envelope, front view

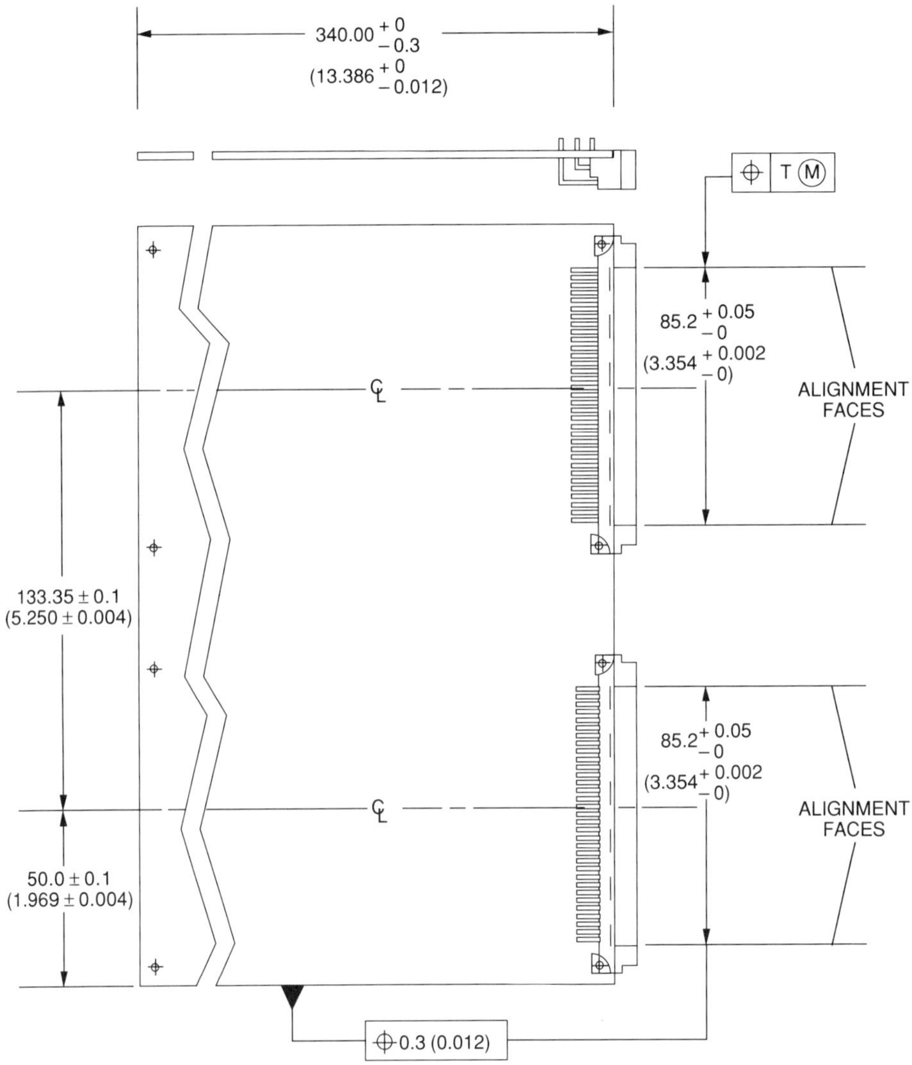

NOTE:
1. All dimensions are shown in millimeters. Inch dimensions are shown in parentheses.

Figure B.21. C size board assembly connector positions

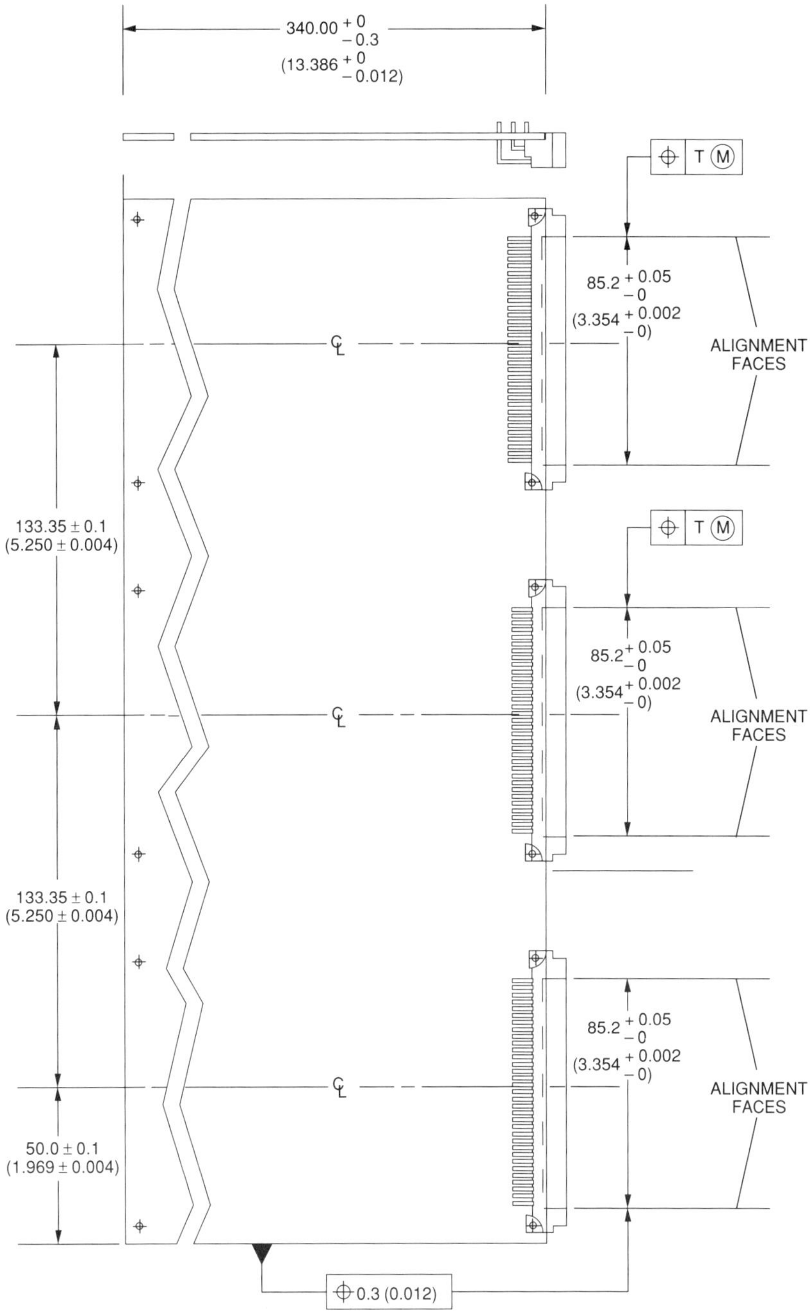

NOTE:
1. All dimensions are shown in millimeters. Inch dimensions are shown in parentheses.

Figure B.22. D size board assembly connector positions

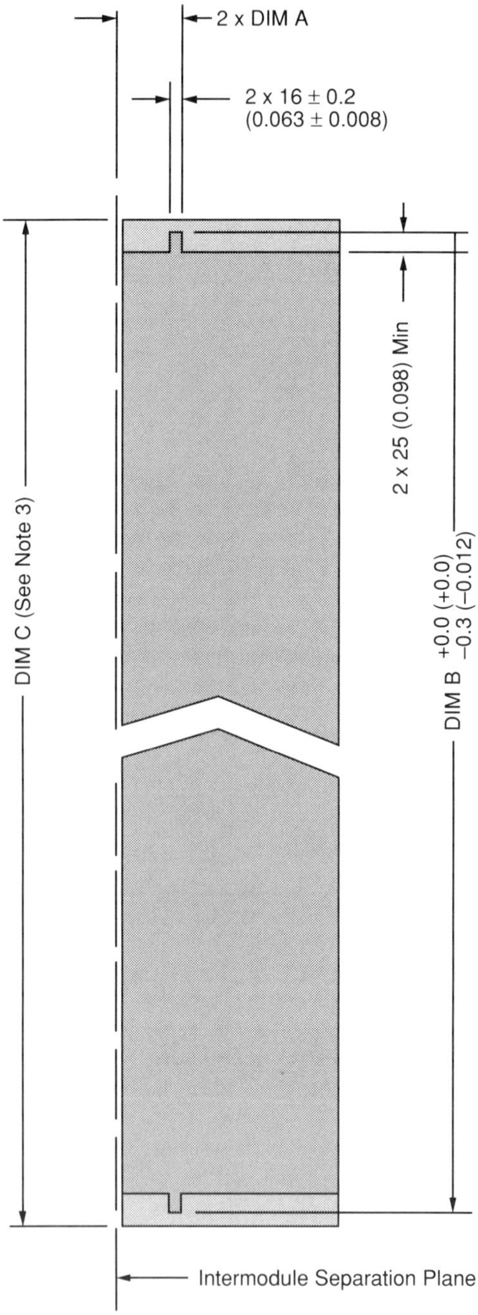

Module Size	DIM A	DIM B	DIM C
A	4.07 (0.160)	100.0 (3.927)	105.3 (4.146)
B	4.07 (0.160)	233.35 (9.187)	238.65 (9.396)
C	9.15 (0.360)	233.35 (9.187)	238.65 (9.396)
D	9.15 (0.360)	366.70 (14.437)	372.0 (14.646)

NOTES:
1. All dimensions are shown in millimeters. Inch dimensions are shown in parentheses.
2. Refer to the Figure 'Module Envelope. Top View'. View A-A
3. Restriction imposed by injector surfaces. Refer to the section 'Module Shielding', and to the figure 'Module Injection Surface'.

Figure B.23. Module edge guide feature

B.7.2.1.1 Board Connectors

A size modules have only one backplane connector, called the P1 connector. B size and C size modules have either one or two backplane connectors, called the P1 and P2 connectors. D size modules have one, two, or three connectors, called the P1, P2, and P3 connectors. In all modules, the P1 connector is mandatory, P2 and P3 are optional.

RULE B.7.6:
P1, P2, and P3 connectors **SHALL** meet or exceed the mechanical specifications of a 603-2-IEC-C096Mx-xxx class 2 connector.

OBSERVATION B.7.3:
603-2-IEC class 2 connectors have a minimum mechanical endurance of 400 insertion/extraction cycles.

RULE B.7.7:
The mounting location of the P1, P2, and P3 connectors **SHALL** be consistent with the hole patterns specified in Figures B.17 and B.18.

RULE B.7.8:
The P1, P2, and P3 connectors **SHALL** be positioned as shown in Figures B.21 and B.22.

B.7.2.1.2 Board Assemblies

The board assembly typically consists of a PC board with connectors (as defined above) and electronic components.

RULE B.7.9:
Solder fillets, traces, shielding, and components on VXIbus boards **SHALL NOT** be closer than 2.5 mm (0.098 inch) from the top and bottom edges of the board to guarantee clearance between them and the board guides. Figures B.17 and B.18 show these dimensions for C size and D size modules.

RULE B.7.10:
IF a module contains components sensitive to orientation (e.g. mercury switches),
THEN that module's specifications and labels **SHALL** clearly state its limitations.

RECOMMENDATION B.7.2:
An orientation-sensitive module should be designed to be installed vertically with its component side to the right, or horizontally with its component side up.

B.7.2.1.3 Module Width

PERMISSION B.7.1:
C size or D size modules that require more than the standard spacing **MAY** be designed to occupy more than one slot, adding an integer multiple of 30.48 mm (1.2 in) to the module width.

OBSERVATION B.7.4:
Component height, lead length, shield positions, etc of a typical module are shown in Figure B.20.

PERMISSION B.7.2:
Module shields (as shown in Figure B.19 and Figure B.20) **MAY** be omitted (or their positions changed)but must remain within the defined envelope.

RECOMMENDATION B.7.3:
Include module shields to improve electromagnetic compatibility (EMC), ease of handling, etc.

B.7.2.1.4 Board Warpage, Lead Length, Shield Height, and Component Height

RULE B.7.11:
All A size and B size board assemblies **SHALL** meet the VMEbus specification 7.2.6.[1]

RULE B.7.12:
For shielded C size and D size modules, the outer surface of the solder-side shield, including warpage, **SHALL** be greater than or equal to 1.57 mm (0.062 inch) from the intermodule separation plane when installed. The outer surface of the component-side shield (including warpage) **SHALL** be greater than or equal to 0.15mm (0.006 inch) from the intermodule separation plane, when installed. Refer to Figure B.20.

RECOMMENDATION B.7.4:
To account for warpage, the nominal solder-side shield should be greater than 2.07mm (0.082 inch) from the intermodule separation plane. The nominal component-side shield should be greater than 0.65mm (0.026 inch) from the intermodule separation plane. Refer to Figure B.20.

RULE B.7.13:
For unshielded C and D size modules, leads and solder-side components, including warpage, **SHALL** be greater than or equal to 3.96mm (0.156 inch) from the intermodule separation plane when installed. Components (including warpage) **SHALL** be greater than or equal to 2.54mm (0.100 inch) from the intermodule separation plane, when installed. Refer to Figure B.20.

B.7.2.2 Front Panel

This section provides the mechanical specifications for the double-height and triple-height module front panels and associated hardware. Figure B.24 shows a

[1] See Appendix IV (page VXI-246)

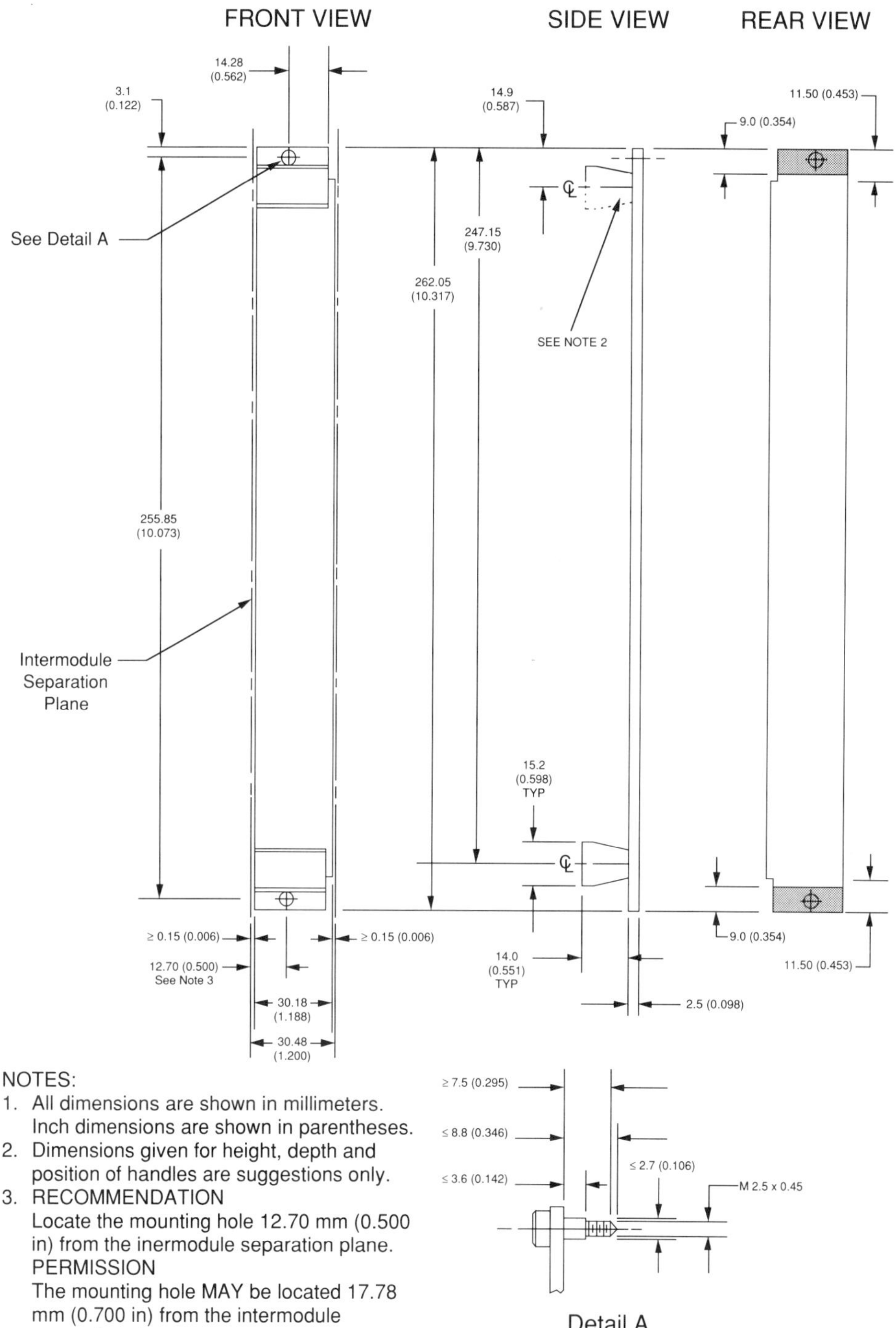

NOTES:
1. All dimensions are shown in millimeters. Inch dimensions are shown in parentheses.
2. Dimensions given for height, depth and position of handles are suggestions only.
3. RECOMMENDATION
   Locate the mounting hole 12.70 mm (0.500 in) from the inermodule separation plane.
   PERMISSION
   The mounting hole MAY be located 17.78 mm (0.700 in) from the intermodule separation plane.

Figure B.24. C size Front panel details

double-height, single-width (C size) front panel. Figure B.25 shows a triple-height, single-width (D size) front panel.

RULE B.7.14:
All modules **SHALL** include a front panel.

RULE B.7.15:
Each module, regardless of size, **SHALL** have a front panel conforming to the size of the mainframe it is used in.

PERMISSION B.7.3:
Additional hardware **MAY** be used to fill a front panel gap when inserting a small module into a larger mainframe.

RULE B.7.16:
Front panel screws **SHALL** be provided to secure the top and bottom of the panel to the mainframe, and their screw threads **SHALL** be M2.5 x 0.45 pitch. See Figure B.24.

RULE B.7.17:
Front panels **SHALL** make contact with chassis ground in the darkened area surrounding the retention screws in REAR VIEW of Figure B.24 and Figure B.25 This surface **SHALL** be conductive, smooth, and free of all protrusions.

B.7.2.2.1 Front Panel Mounting

RULE B.7.18:
All modules **SHALL** comply with the test dimension shown in Figure B.28 and Figure B.29. This dimension is from the rear face of the front panel to the rear face of the backplane connector.

B.7.2.2.2 Front Panel Dimensions

RULE B.7.19:
Single-width C size and D size front panels **SHALL** be designed to the dimensions shown in Figure B.24 and Figure B.25 respectively.

RULE B.7.20:
IF a C size or D size module occupies more than one slot
THEN the width of the front panel **SHALL** be 30.18 mm (1.188 inch) plus an integral multiple (N) of 30.48 mm (1.200 inch), where N is the number of slots that the module occupies, minus one.

RULE B.7.21:
IF local bus lockout keys are required on a module,
THEN the front panel **SHALL** be notched as shown in the FRONT VIEW and REAR VIEW of Figure B.24 and Figure B.25.

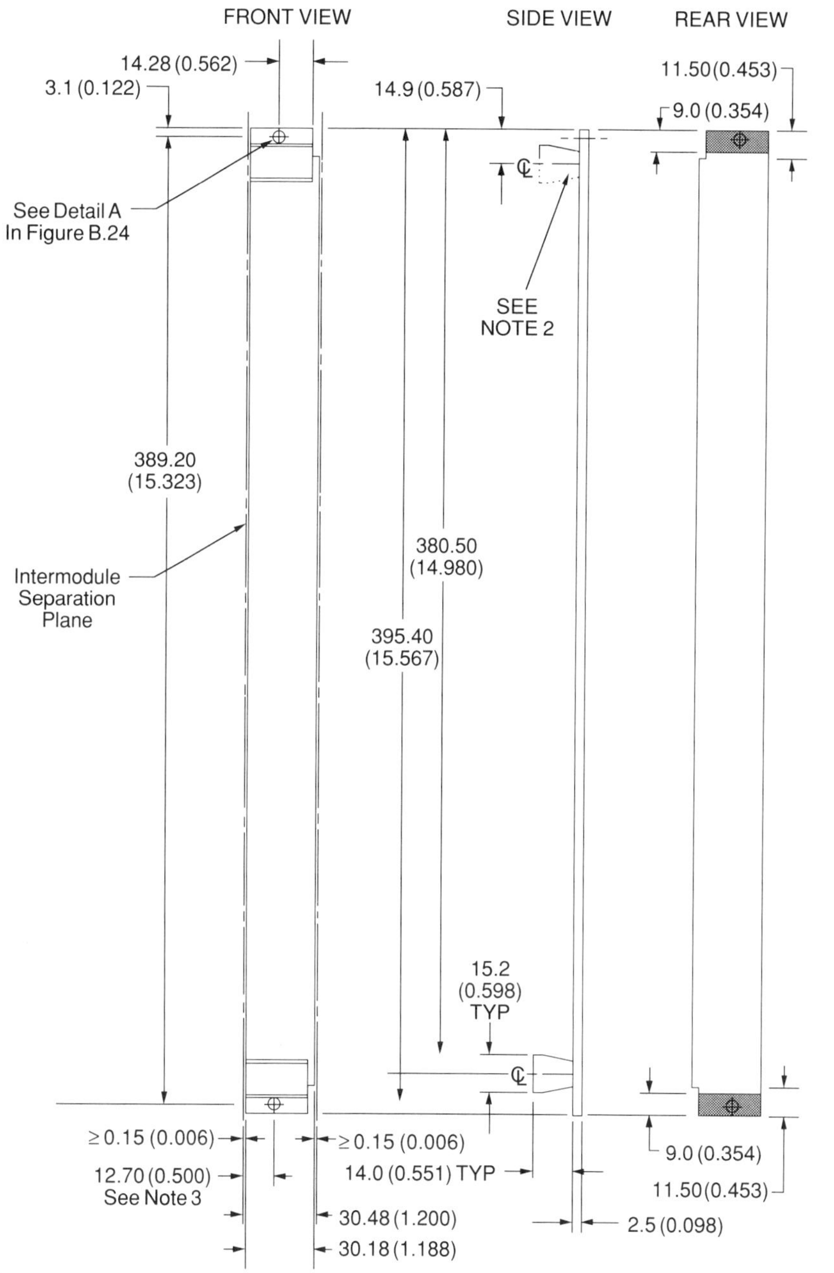

NOTES:
1. All dimensions are shown in millimeters. Inch dimensions are shown in parentheses.
2. Dimensions given for height, depth and position of handles are suggestions only.
3. RECOMMENDATION
   Locate the mounting hole 12.70 mm (0.500 in) from the intermodule separation plane.
   PERMISSION
   The mounting hole MAY be located 17.78 mm (0.700 in) from the intermodule separation plane.

Figure B.25. D size Front panel details

SUGGESTION B.7.1:
Include the lockout key notch on all front panels. This will simplify removal of a module when the adjacent module is keyed.

B.7.2.2.3 Filler Panels

RULE B.7.22:
Filler panels **SHALL** be used in any empty slots.

OBSERVATION B.7.5:
Filler panels are often required by safety regulations, and may be necessary for cooling and EMC performance.

RULE B.7.23:
C size and D size filler panels **SHALL** conform to dimensions given in Figure B.26 and Figure B.27 respectively.

RULE B.7.24:
A size and B size filler panels **SHALL** conform to VMEbus specification subsection 7.3.4.[1]

B.7.2.2.4 Board Injectors/Ejectors

Insertion and extraction forces of the module, relative to the mainframe, may be very high. These forces are greatest for D size modules that use all three backplane connectors.

RECOMMENDATION B.7.5:
Modules using three or more of the 96 pin 603-2-IEC-C096Mx-xxx connectors should provide an injector/ejector system.

OBSERVATION B.7.6:
The insertion force for three 96 pin 603-2-IEC-C096Mx-xxx connectors can be as high as 270 N (60.69 lbf).

PERMISSION B.7.4:
An injector and/or ejector system **MAY** be provided on any module size.

RULE B.7.25:
Modules equipped with ejectors or injectors **SHALL NOT** interfere with mainframes designed to the VXIbus or VMEbus specification. See Figure B.39 and Figure B.40.

OBSERVATION B.7.7:
Refer to Section B.7.3.3, (Injection/Ejection), for more information.

[1] See Appendix IV (page VXI-246)

B.7.2.2.5 Front Panel Interface

RULE B.7.26:
External wiring to a module, if any, **SHALL** exit the module only through the front panel area.

OBSERVATION B.7.8:
External wiring is defined as any connectors, cables, etc. that carry signals or interfaces out of the mainframe in which the module is installed. The above rule does not exclude module-to-module wiring.

RULE B.7.27:
Connector and front panel interfaces **SHALL NOT** leave significant gaps in the front panel, or disrupt proper airflow within the system.

OBSERVATION B.7.9:
Although the terms "significant gaps" and "proper air flow" can be rather broadly interpreted, it is intended that a manufacturer be very cautious about the front panel components chosen.

B.7.2.2.6 Front Panel Handles

RULE B.7.28:
Handles **SHALL NOT** interfere with the local bus lockout keys of adjacent modules.

SUGGESTION B.7.2:
Handle width and position should comply with the dimensions shown in Figure B.24 and Figure B.25. This will avoid interference with local bus lockout keys.

OBSERVATION B.7.10:
Injection and ejection constraints may require the use of a handle with dimensions different than those in Figure B.24 and Figure B.25.

B.7.2.3 Module Shielding

Shields are often included in modules to improve EMC performance, ease of handling, etc. Refer to Figure B.19 and Figure B.20.

PERMISSION B.7.5:
Module shields **MAY** be tied to chassis ground or circuit ground.

RULE B.7.29:
C size and D size modules **SHALL** provide room for the optional mainframe chassis shield, as shown in Figure B.19 and Figure B.20.

OBSERVATION B.7.11:
The envelope (as defined in Figure B.19) ensures that module shields do not contact adjacent modules.

OBSERVATION B.7.12:
The use of box-type plug-in units (as specified by IEC 297-3 and ANSI/IEEE 1101) might be limited. Maximum module length, injector surfaces, and non-removable card guides might be incompatible. Refer to Figure B.19, Figure B.23, and Figure B.40.

PERMISSION B.7.6:
A grounding shield **MAY** be provided around each 96 pin connector as shown in Figure B.36. This will be referred to as the connector shield.

OBSERVATION B.7.13:
Connector shield contacts might or might not be removable. However, removable contacts may be advisable for some modules. The module manufacturer should evaluate the applicability of removable contacts for each module.

PERMISSION B.7.7:
A module **MAY** make electrical contact with a compatible adjacent module at any point in the area near the front panel, as specified in Figure B.19. This contact **MAY** be implemented by means of gasketing, clips, etc.

RULE B.7.30:
IF gasketing is installed for contact with a compatible adjacent module,
THEN it **SHALL** be installed on the solder side only, and it **SHALL** be compressible to fit within the intermodule separation plane.

RECOMMENDATION B.7.6:
Gasketing installed for contact with a compatible adjacent module should be removable. This will allow installation adjacent to unshielded or otherwise incompatible modules.

RECOMMENDATION B.7.7:
The areas specified in Figure B.19 for contact with adjacent modules should be chassis ground and conductive.

B.7.2.4 Module Cooling

RULE B.7.31:
Cooling air **SHALL** flow as shown in Figure B.31, Figure B.32, and Figure B.41. Airflow direction is from P3 to P1.

RULE B.7.32:
Cooling requirements **SHALL** be established for all modules and included in product specifications. The specification **SHALL** include the maximum ambient operating temperature and the operating point. (Example: 55 °C ambient maximum operating temperature. For 10°C internal rise, 0.5 mm H_2O @ 1.5 liters/sec.)

RECOMMENDATION B.7.8:
Label the module with the cooling specification. This will facilitate system integration. See Figure B.44.

RECOMMENDATION B.7.9:
Establish a "resistance" pressure drop versus airflow graph, and include it in the product specifications. This is useful for more detailed system integration. Example:

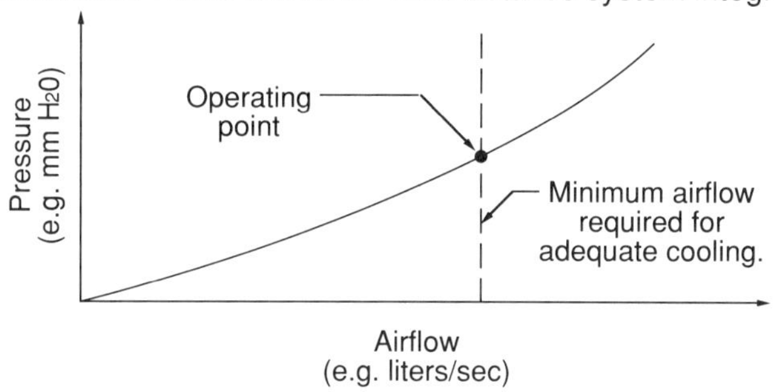

OBSERVATION B.7.14:
An example of simplified test set-up and procedures for establishing module cooling requirements is shown in Figure B.42.

OBSERVATION B.7.15:
Great care must be taken with the design of modules that require cooling air to flow in parallel paths. Examples are modules with multiple boards and modules with components on the solder side of the main board. In particular, the power dissipation of solder side components may need to be limited because of restrictions in the cooling airflow. The module designer is responsible for assuring that resistance to air flow is consistent with the cooling requirements for each parallel path.

B.7.2.5 Module Power

RECOMMENDATION B.7.10:
Attach an abbreviated version of the power supply voltage and current requirements to the module. This will facilitate system integration.

B.7.2.6 Module Keying

Lockout keys are provided to serve as an impediment to local bus incompatibilities.

RULE B.7.33:
IF modules access a local bus,
THEN lockout keys **SHALL** be provided, consistent with the particular bus accessed. The lockout keys **SHALL** conform to the specifications in Figure B.30.

RULE B.7.34:
IF lockout keys are required on a module,
THEN they **SHALL** be installed by the manufacturer, consistent with the definition of the product.

OBSERVATION B.7.16:
Lockout keys are not intended to provide failsafe protection. Accidental damage may still occur if incompatible modules are forced next to each other on the local bus.

B.7.2.7 Module Environmental

RECOMMENDATION B.7.11:
Test all modules for temperature, humidity, vibration, shock, etc., according to the procedures described in IEC Publication 68. Include the results in product specifications, along with module mass.

OBSERVATION B.7.17:
It is a system integrator's responsibility to select modules and mainframes appropriate for the application's environmental requirements.

B.7.3 Mainframe Specifications

RULE B.7.35:
VXIbus mainframes **SHALL** fit into one of four size classifications:

1. A mainframe that provides the J1 backplane, and only allows the insertion of an A size module, will be referred to as an "A" size mainframe, and will conform to the VMEbus specifications[1] for subracks.

2. A mainframe that allows the insertion of a B size module, maximum, will be referred to as a "B" size mainframe, and will conform to the VMEbus specifications[1] for subracks. The J1 backplane is mandatory. J2 is optional.

3. A mainframe that allows the insertion of a C size module, maximum, will be referred to as a "C"-size mainframe. The J1 backplane is mandatory. J2 is optional.

4. A mainframe that allows the insertion of a D size module, maximum, will be referred to as a "D" size" mainframe. The J1 backplane is mandatory. J2 and J3 are optional.

RECOMMENDATION B.7.12:
Allowances should be made for the insertion of A size modules into B size mainframes, the insertion of A size and B size modules into C size mainframes, and the insertion of A size, B size, and C size modules into D size mainframes.

OBSERVATION B.7.18:
Additional hardware and removable card guides may be required to allow the insertion of smaller modules into a larger mainframe.

RULE B.7.36:
IF the insertion of smaller modules is allowed into a larger mainframe,
THEN the mainframe manufacturer **SHALL** specify a procedure and/or a means for the insertion of the smaller modules.

PERMISSION B.7.8:
Mechanical and/or electrical means may be used for the adaptation of a smaller module to a larger mainframe.

[1] See Appendix IV (page VXI-246)

RULE B.7.37:
C size and D size mainframes **SHALL** be designed for 30.48 mm (1.2 in) slot spacing, as shown in Figure B.31 and Figure B.32.

RULE B.7.38:
A size and B size mainframes **SHALL** be designed to 20.32 mm (0.8 inch) spacing.

RULE B.7.39:
C size and D size mainframes **SHALL** conform to the dimensions and tolerances shown in Figure B.31 and Figure B.32. This ensures that connectors mate properly and that modules will fit in the mainframe without interference.

RULE B.7.40:
With the mainframe oriented with the modules vertical, P1 at the top, and the component side of module boards to the right, Slot 0 **SHALL** be the left-most slot of any VXIbus subsystem.

OBSERVATION B.7.19:
This rule does not mandate mainframe orientation, it only establishes the position of Slot 0, relative to mainframe orientation.

SUGGESTION B.7.3:
IF vertical orientation of the modules is desired,
THEN Slot 0 should be at the left.
IF horizontal orientation of the modules is desired,
THEN Slot 0 should be at the bottom.

RULE B.7.41:
Card guides **SHALL** be made of non-conductive material, or be isolated from chassis ground.

OBSERVATION B.7.20:
Removable card guides may permit the use of box style plug-in units, as specified by IEC 297-3 and ANSI/IEEE 1101.

B.7.3.1 Backplanes

RULE B.7.42:
Each backplane **SHALL** be a single monolithic board, within any slot position.

RULE B.7.43:
Backplane connectors for C size and D size mainframes **SHALL** be positioned as shown in Figures B.33 and B.34.

PERMISSION B.7.9:
Backplanes **MAY** deviate from Figure B.33 and Figure B.34 in structure and in mounting methods.

OBSERVATION B.7.21:
Typical C size and D size backplanes are shown in Figure B.33 and Figure B.34 for reference. Adherence to these drawings will ease the task of mounting and aligning the backplane within the mainframe.

RULE B.7.44:
IF components other than connectors are placed on the component side of a C size or D size backplane,
THEN they **SHALL** be placed outside the shaded areas specified in Figure B.35, and they **SHALL** be less than 12.2 mm (0.480 inch) in height.

RECOMMENDATION B.7.13:
Mainframes should allow shielding around each 96-pin backplane connector. This shield typically provides a connection between a ground ring on the backplane and a module. See the references to the "connector shield" in Section B.7.2.3, "Module Shielding". Refer also to Rule B.7.44 above, Figure B.35, and Figure B.36.

RULE B.7.45:
Connector shields **SHALL** comply with Figure B.36.

RULE B.7.46:
IF conductive backplane traces are exposed within the ground ring area defined by Figure B.35,
THEN the potential of those traces **SHALL** be circuit ground.

RULE B.7.47:
Components **SHALL NOT** be located near backplane connectors, as specified in Figure B.35.

RULE B.7.48:
The 96 pin 603-2-IEC-C096Fx-xxx connectors **SHALL** be used on all J1, J2, and J3 positions.

RULE B.7.49:
The VXIbus J1 bus **SHALL** have some provision for jumpering the interrupt acknowledge and bus grant daisy-chains when boards are not plugged into a slot.

B.7.3.2 Grounding

OBSERVATION B.7.22:
Mainframes may connect circuit ground to chassis ground.

RULE B.7.50:
All mainframes **SHALL** provide chassis ground to the module front panels in the area of the front panel mounting screws. The chassis ground contact area is shown in Figure B.24 for C size modules, and in Figure B.25 for D size modules.

B.7.3.3 Injection/Ejection

RECOMMENDATION B.7.14:
Mainframes should provide a surface for module extraction, as shown in Figure B.39.

RECOMMENDATION B.7.15:
Mainframes should provide a surface for module injection, as shown in Figure B.40.

RULE B.7.51:
Mainframes **SHALL NOT** interfere with injector/ejector systems conforming to Figures B.39 and B.40.

B.7.3.4 Mainframe Shielding

PERMISSION B.7.10:
Mainframes **MAY** allow the installation of a chassis potential shield, as shown in Figure B.19, and in Figure B.20.

RULE B.7.52:
IF mainframe chassis shields are provided,
THEN they **SHALL** be removable.

RULE B.7.53:
IF the shield potential of a module is not chassis ground, or the module is unshielded,
THEN any adjacent chassis shield **SHALL** be insulated.

PERMISSION B.7.11:
IF the shield potential of a module is defined as chassis ground,
THEN electrical contact between the module and adjacent chassis shields **MAY** be allowed.

RECOMMENDATION B.7.16:
IF a mainframe has provision for chassis shields,
THEN both insulated and non-insulated shields should be available.

B.7.3.5 Mainframe Cooling

RULE B.7.54:
IF a mainframe provides cooling air,
THEN it **SHALL** supply the air to each slot in the minimum area specified in Figure B.41.

RULE B.7.55:
IF a mainframe provides cooling air,
THEN that mainframe **SHALL** provide room for the exhausted cooling air to exit each slot in the minimum area specified in Figure B.41.

PERMISSION B.7.12:
A mainframe **MAY** provide for additional cooling air entry and exhaust in areas outside those specified in Figure B.41.

RULE B.7.56:
Cooling specifications **SHALL** be established for all mainframes and included in the product specifications. These specifications **SHALL** include a plot of "worst slot" pressure drop versus flow rate through the slot. See Figure B.44.

RECOMMENDATION B.7.17:
Label the mainframe with a copy of the "worst slot" graph. This will facilitate system integration. See Figure B.44.

OBSERVATION B.7.23:
"Worst slot" refers to worst case airflow through a slot, versus the pressure drop imposed by a module in the slot. In most cases, this means identifying the slot with the least airflow, and testing it with all other slots empty.

OBSERVATION B.7.24:
Airflow can often be increased by blocking empty slots. However, this technique is not valid when testing the "worst slot", since the existence of empty slots cannot be assumed.

SUGGESTION B.7.4:
Establish plots of pressure drop versus airflow for each slot in the mainframe, and include them in the product specifications.

OBSERVATION B.7.25:
Airflow plots of each slot (and the entire mainframe) are useful for more detailed system integration.

OBSERVATION B.7.26:
The cooling specifications required by VXI are sufficient for most system integration. Some systems may require more detailed analysis and specification. Examples are variable or high-power systems, proprietary systems, borderline cases, etc.

OBSERVATION B.7.27:
An example of simplified test set-up, and procedures for establishing mainframe cooling specifications, is shown in Figure B.43.

B.7.3.6 Mainframe Power

RECOMMENDATION B.7.18:
Label the mainframe with an abbreviated version of the power supply specifications. This label should specify the available current for each of the supply voltages, and state whether circuit ground is connected to chassis ground. This will facilitate system integration.

B.7.3.7 Keying

RULE B.7.57:
Mainframes **SHALL** provide local bus lockout keys for Slot 0. These keys **SHALL** allow only TTL-compatible modules on P2 and only ECL-compatible modules on P3. See Figure B.30.

OBSERVATION B.7.28:
Mainframe lockout keys typically consist of a tab or recess to the left of Slot 0. See Figure B.16.

B.7.3.8 Mainframe Environmental

RECOMMENDATION B.7.19:
Test mainframes for temperature, humidity, vibration, shock, etc., according to the procedures described in IEC Publication 68. Include the results in product specifications.

SUGGESTION B.7.5:
Include specifications for the vibration and shock environment experienced by modules installed in a mainframe during testing of the mainframe. This should include a variety of module weights and slot positions.

OBSERVATION B.7.29:
It is a system integrator's responsibility to select modules and mainframes appropriate for the application's environmental requirements.

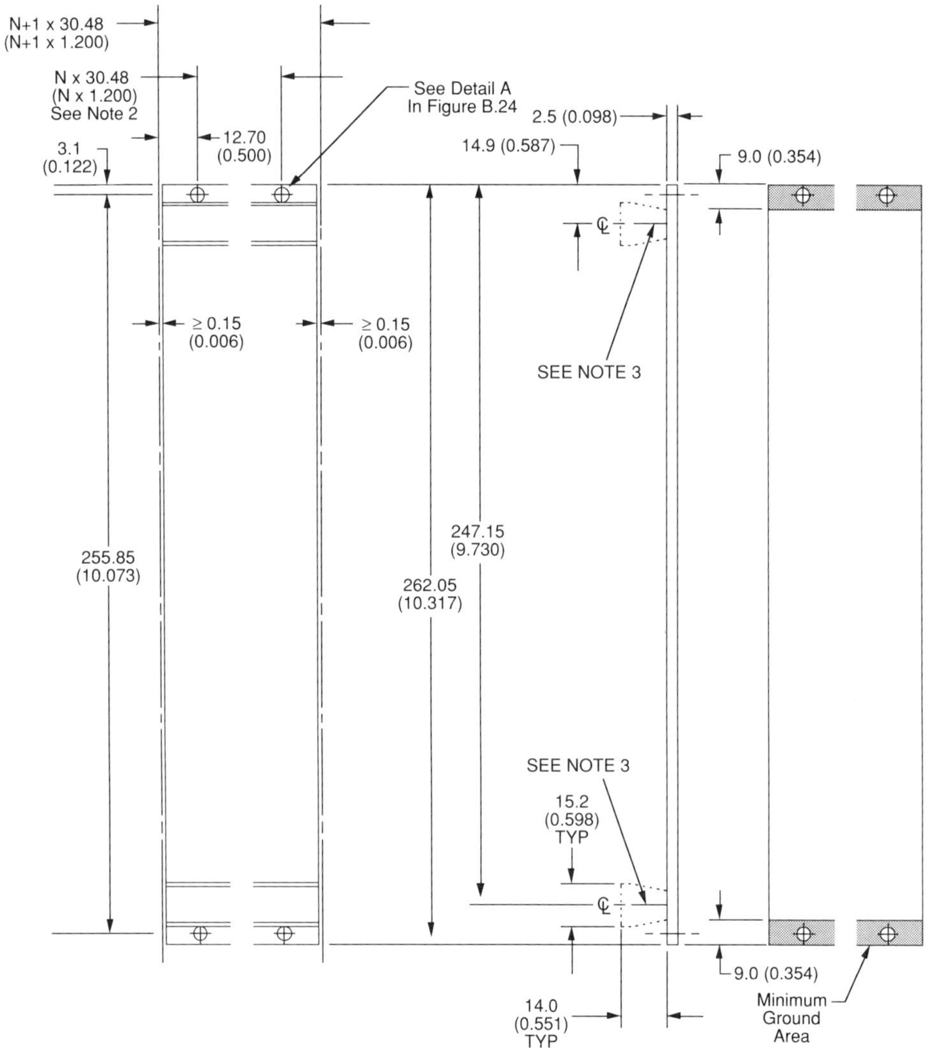

NOTES:

1. All dimensions are shown in millimeters. Inch dimensions are shown in parentheses.
2. RECOMMENDATION
   Where a panel is more than 30.48 mm (1.200 in) wide use at least 4 mounting holes: two at the top and two at the bottom.
3. SUGGESTION
   IF handles are used on filler panels.
   THEN they should be positioned as shown in Front Panel Detail.

Figure B.26. C size filler panel

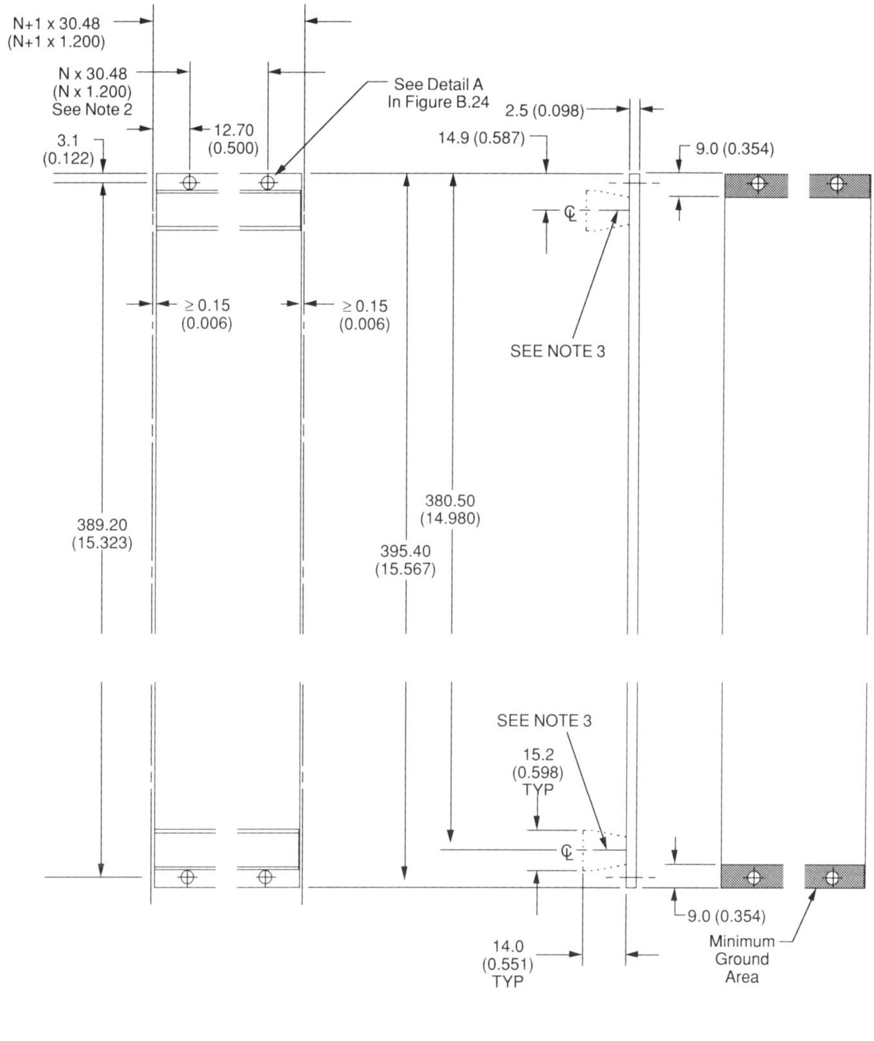

NOTES:
1. All dimensions are shown in millimeters. Inch dimensions are shown in parentheses.
2. RECOMMENDATION
   Where a panel is more than 30.48 mm (1.200 in) wide use at least 4 mounting holes: two at the top and two at the bottom.
3. SUGGESTION
   IF handles are used on filler panels.
   THEN they should be positioned as shown in Front Panel Detail.

Figure B.27. D size filler panel

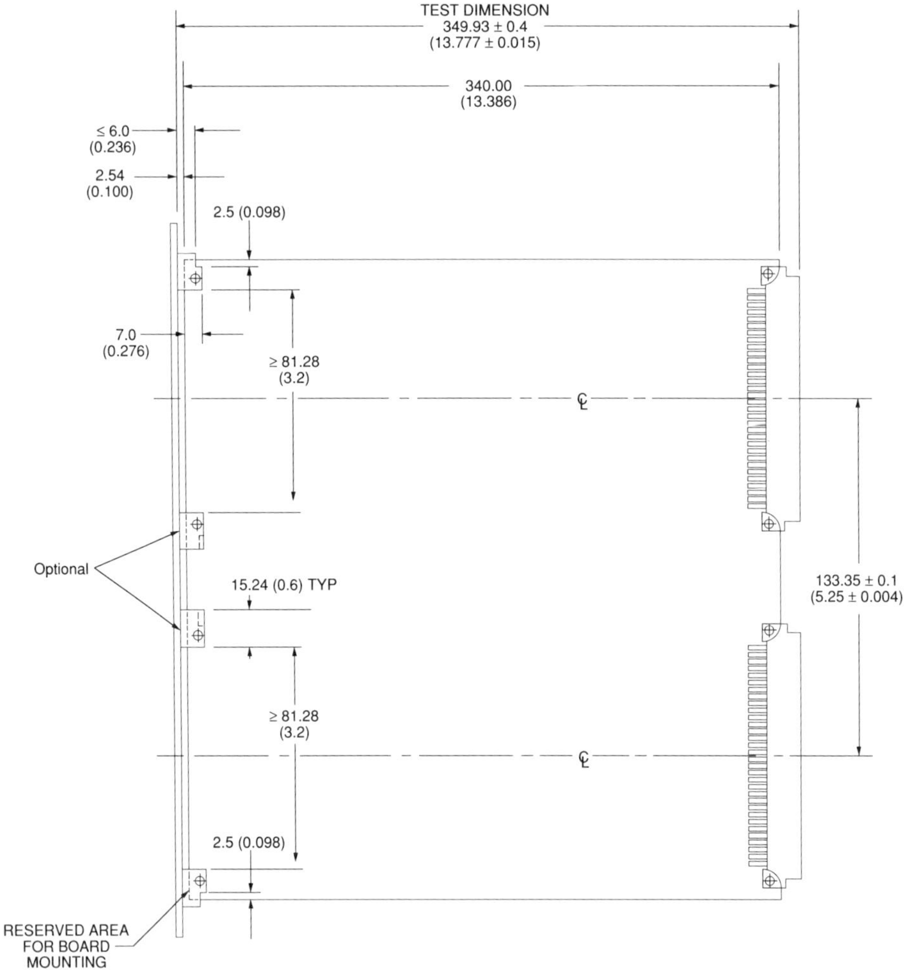

NOTES:
1. All dimensions are shown in millimeters. Inch dimensions are shown in parentheses.
2. Dimensions for frontpanel mounting brackets are suggestions only. Mounting methods may vary.

Figure B.28. Typical C size front panel mounting and dimensions

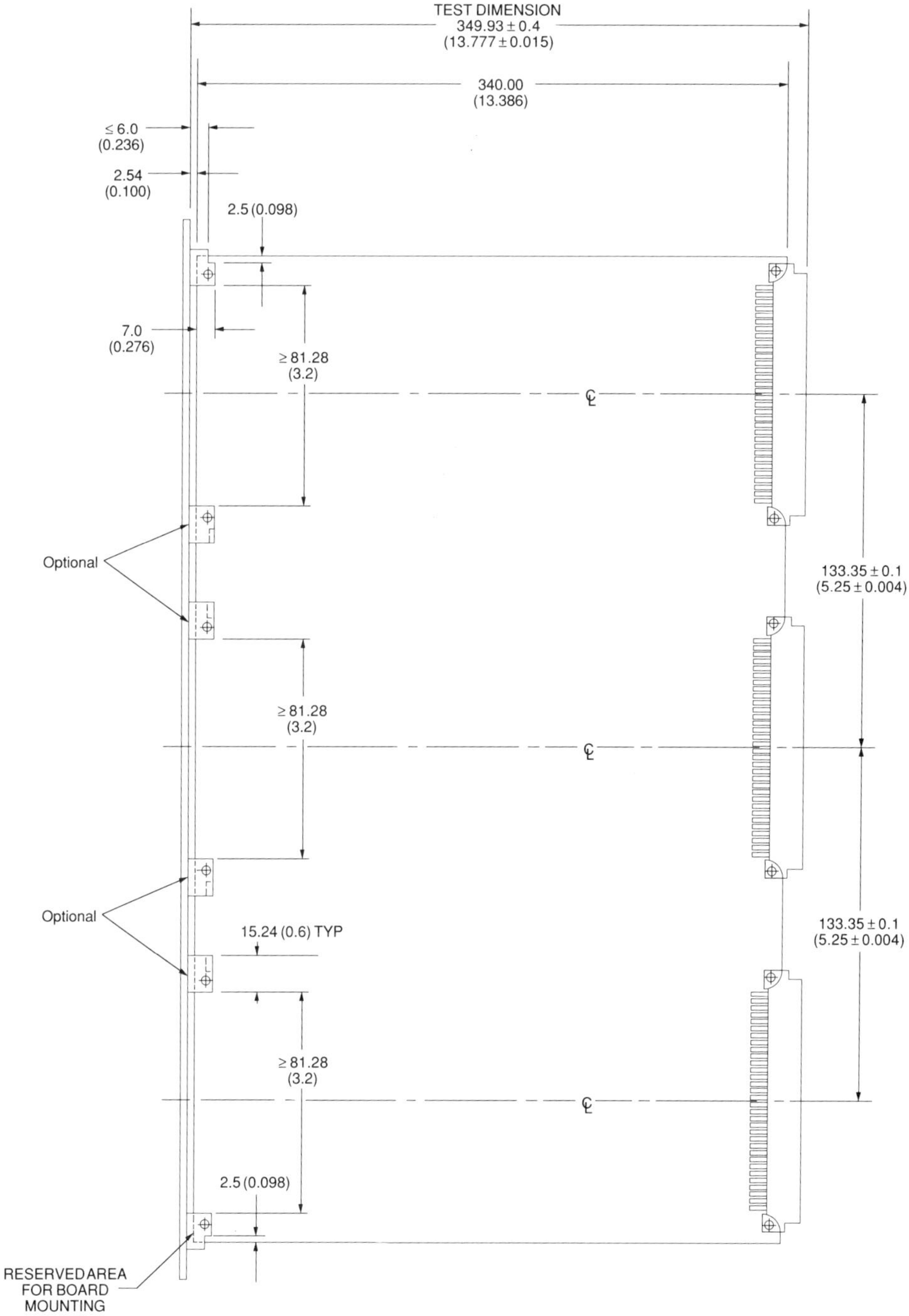

NOTES:
1. All dimensions are shown in millimeters. Inch dimensions are shown in parentheses.
2. Dimensions for frontpanel mounting brackets are suggestions only. Mounting methods may vary.

Figure B.29. Typical D size front panel mounting and dimensions

NOTES:
1. All dimensions are shown in millimeters. Inch dimensions are shown in parentheses.
2. The example shown is keyed for TTL on P2 and Analog Low on P3.

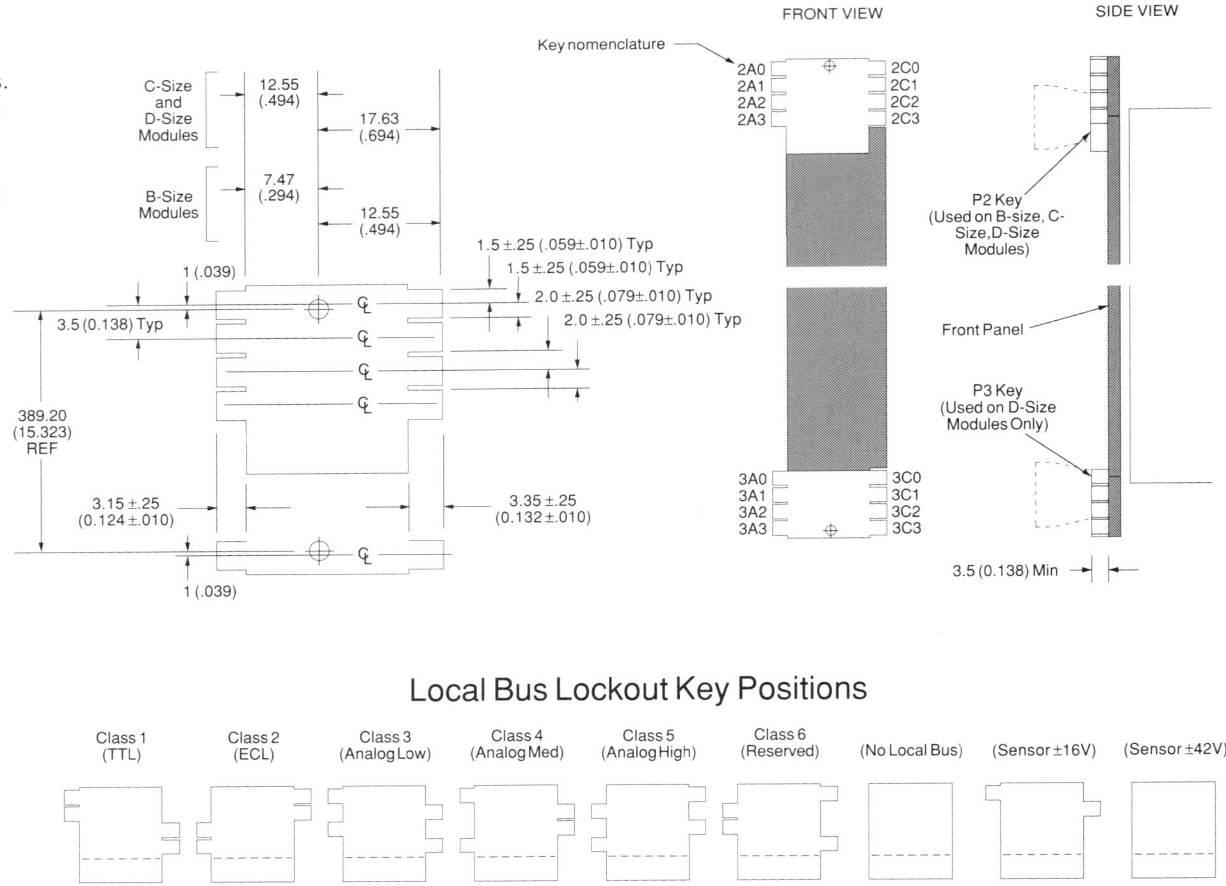

Figure B.30. Local bus lockout key details

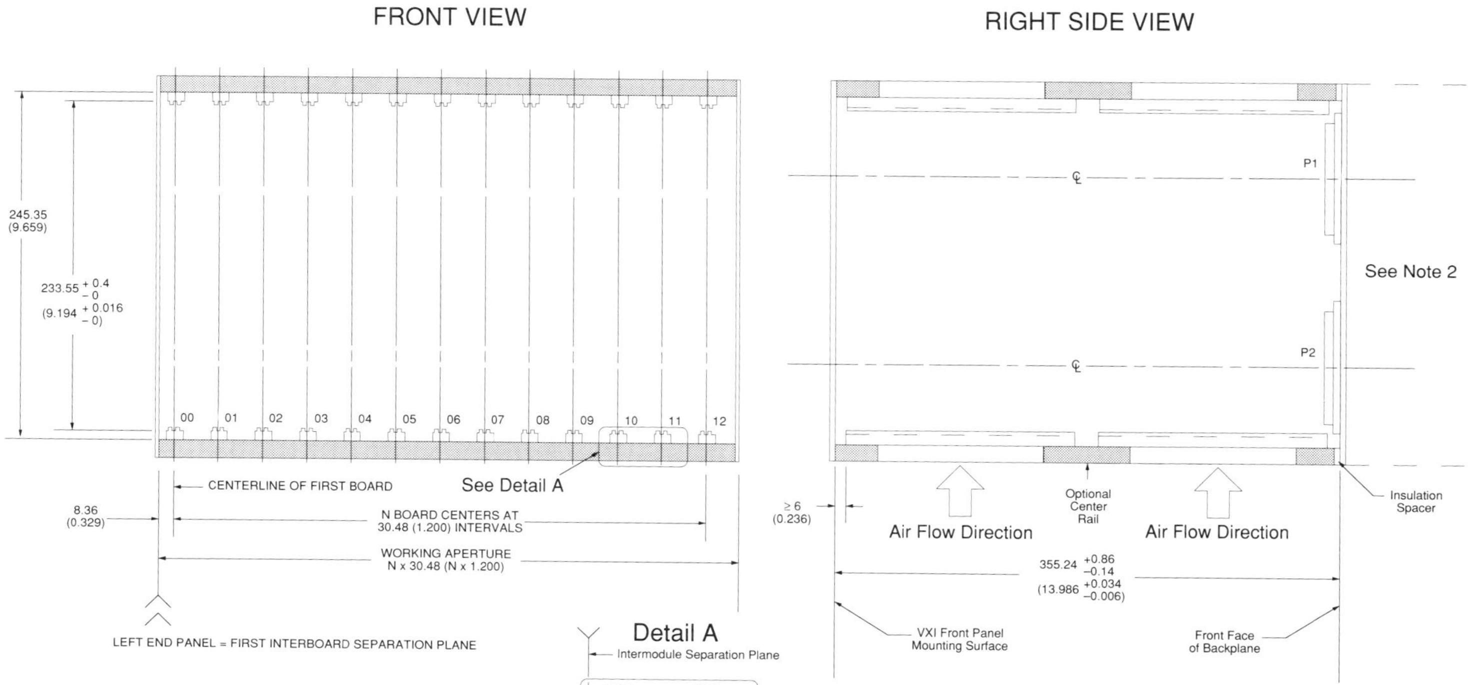

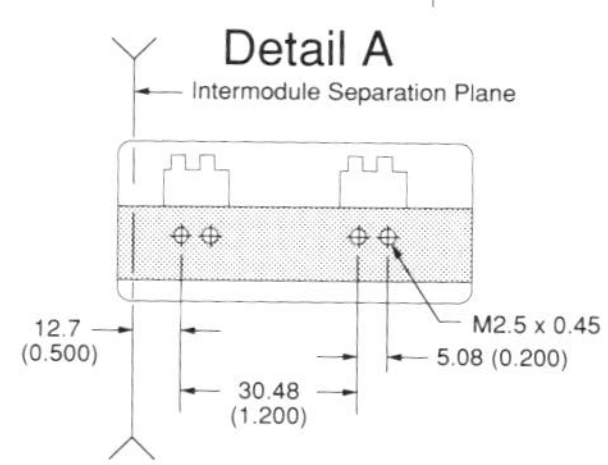

NOTES:
1. All dimensions are shown in millimeters. Inch dimensions are shown in parentheses.
2. PERMISSION
   Side plates MAY be extended to the rear as required.

Figure B.31. C size mainframe drawing

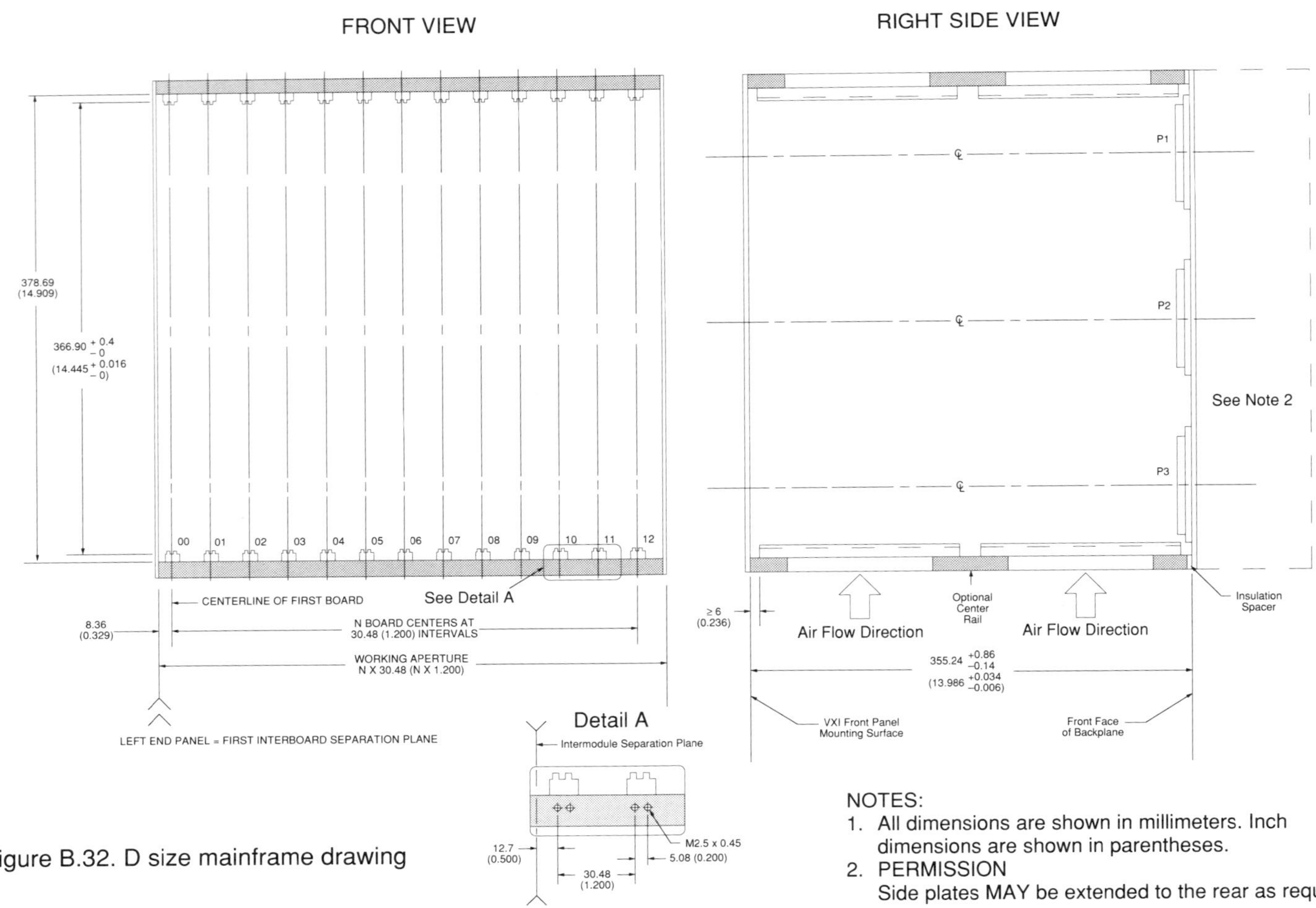

Figure B.32. D size mainframe drawing

NOTES:
1. All dimensions are shown in millimeters. Inch dimensions are shown in parentheses.
2. PERMISSION
   Side plates MAY be extended to the rear as required.

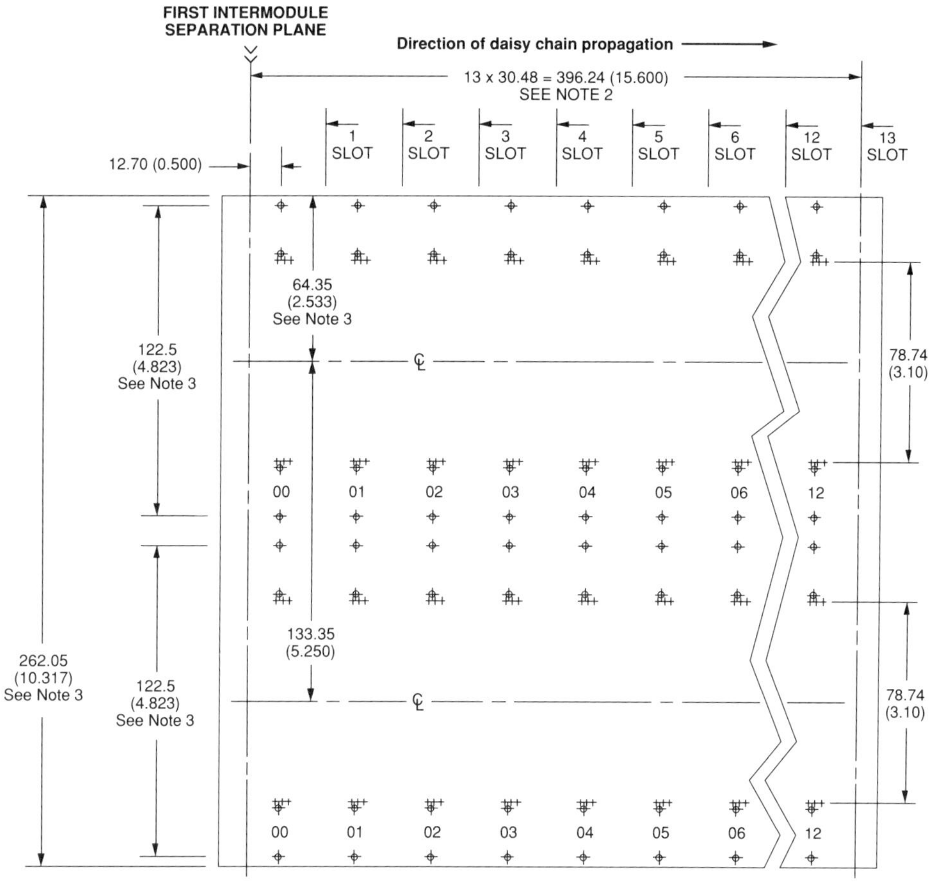

Connector side, as viewed from the front of the mainframe

NOTES:
1. All dimensions are shown in millimeters. Inch dimensions are shown in parentheses.
2. Backplane width varies depending on the number of slots.
3. These dimensions are shown as suggestions only. (See Backplane subsection)

Figure B.33. C size backplane drawing

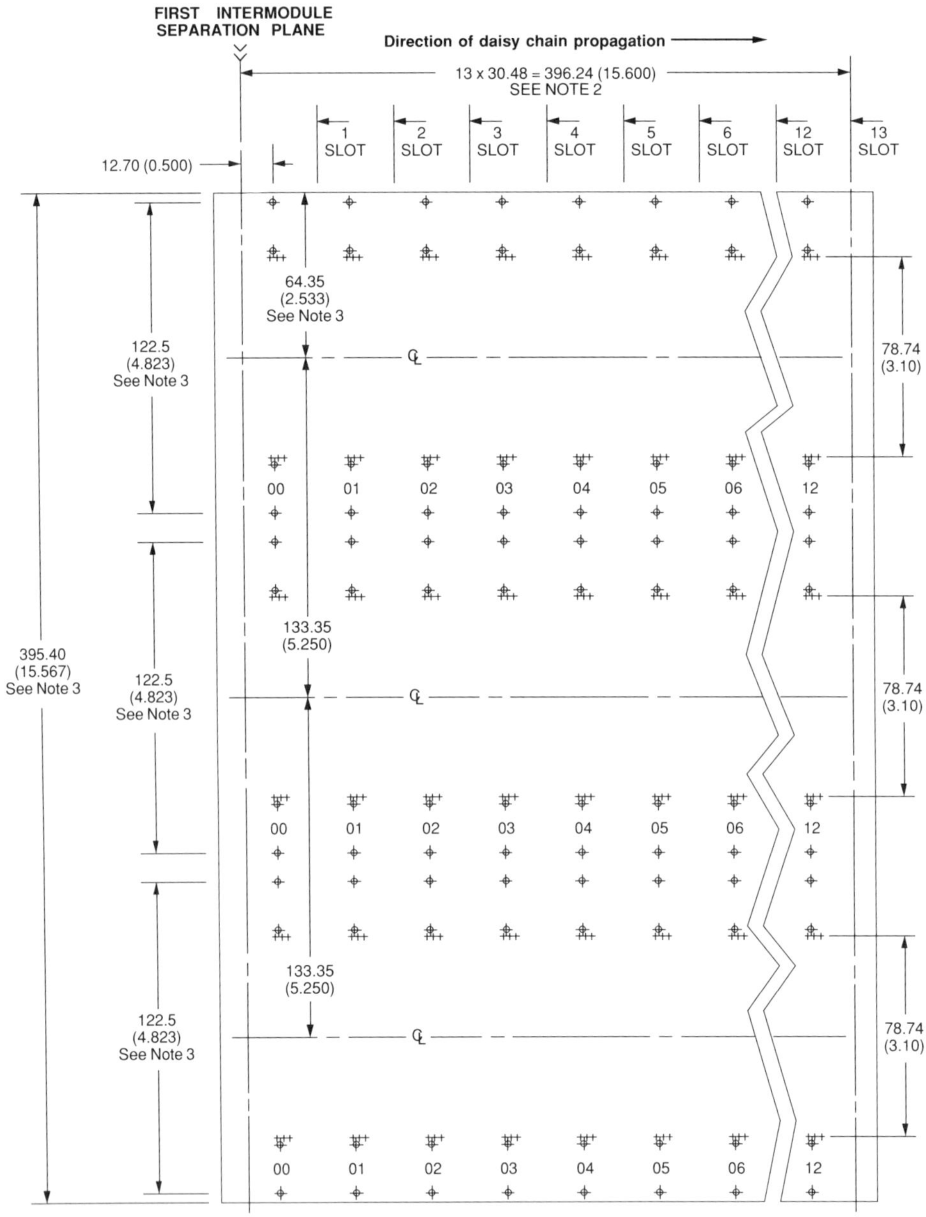

Connector side, as viewed from the front of the mainframe

NOTES:

1. All dimensions are shown in millimeters. Inch dimensions are shown in parentheses.
2. Backplane width varies depending on the number of slots.
3. These dimensions are shown as suggestions only. (See Backplane subsection)

Figure B.34. D size backplane drawing

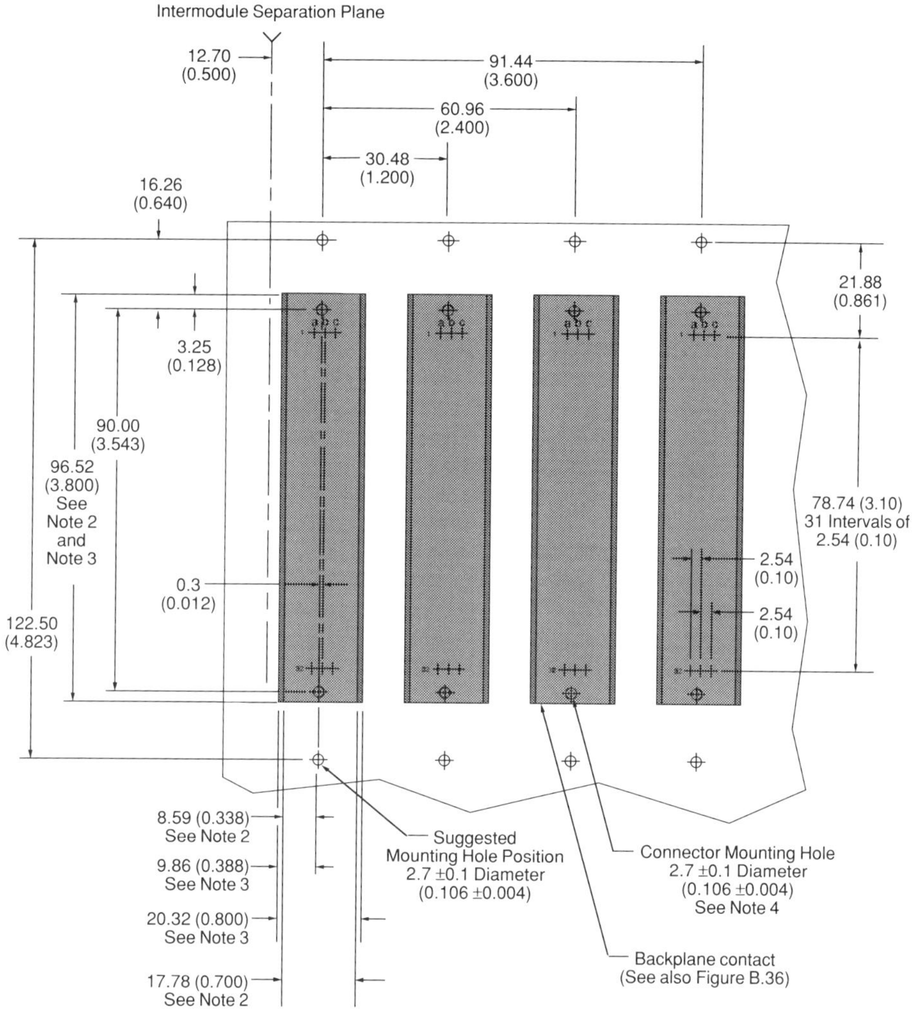

NOTES:
1. All dimensions are shown in millimeters. Inch dimensions are shown in parentheses.
2. Minimum area of ground plane for Module Shield grounding ring around all: J1, J2 and J3 connectors. (See Mainframe Shielding subsection)
3. No components are allowed in this area except 96-pin connector.
4. Connector mounting hole may be required by connector shields, etc.

Figure B.35. Detailed backplane dimensions

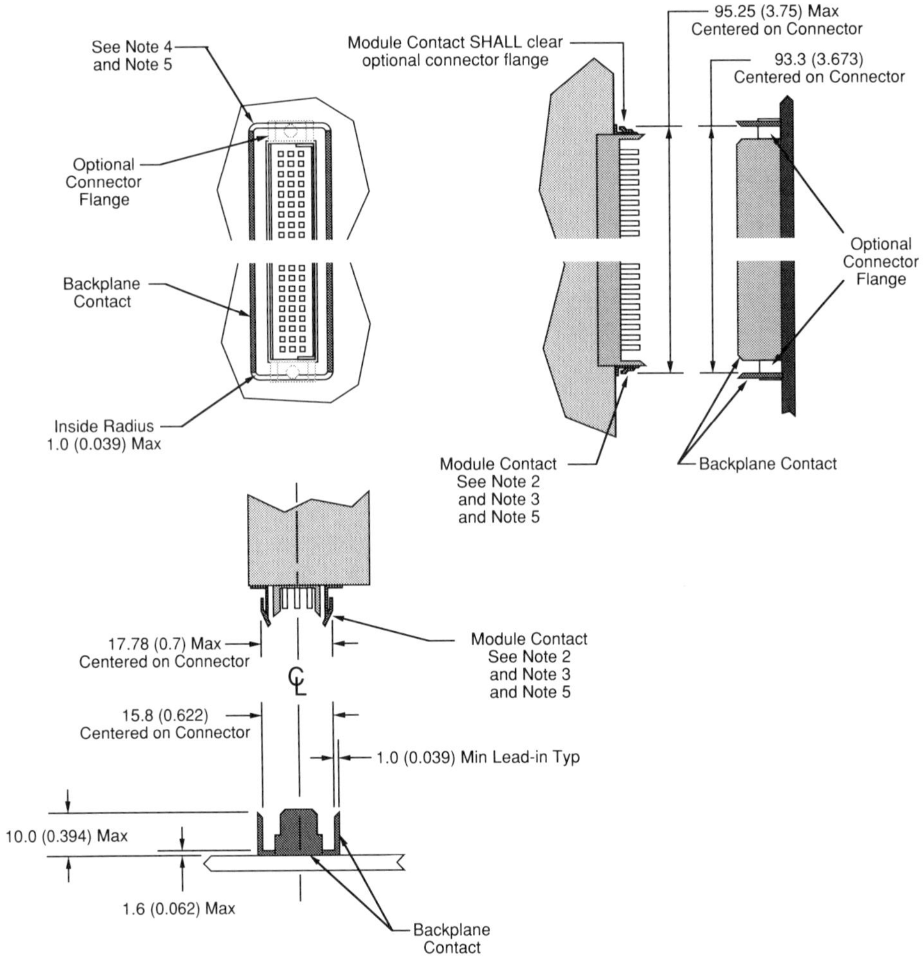

NOTES:

1. All dimensions are shown in millimeters. Inch dimensions are shown in parentheses.
2. RULE: Module contacts SHALL be compatible with the backplane contact dimensions shown.
3. RULE: Module contact design SHALL NOT impede mating of the backplane connectors. Sufficient spring contact flexure and lead-in is required.
4. PERMISSION: Module contacts and backplane contacts MAY or MAY not extend around the end of the backplane connector as shown.
5. RULE: All module contacts SHALL be compatible with backplane contacts that extend around the end of the backplane connector.
6. PERMISSION: Module contacts and backplane contacts MAY or MAY not be removable.
7. OBSERVATION: Removable contacts allow the user to eliminate the connector shield. This may be useful for some applications. Examples are modules that connect circuit ground directly to chassis ground. Module manufacturers should evaluate the applicability of removable contacts for each module
8. SUGGESTION: Mount backplane contacts to the backplane connector mounting hole.

Figure B.36. Backplane connector shield

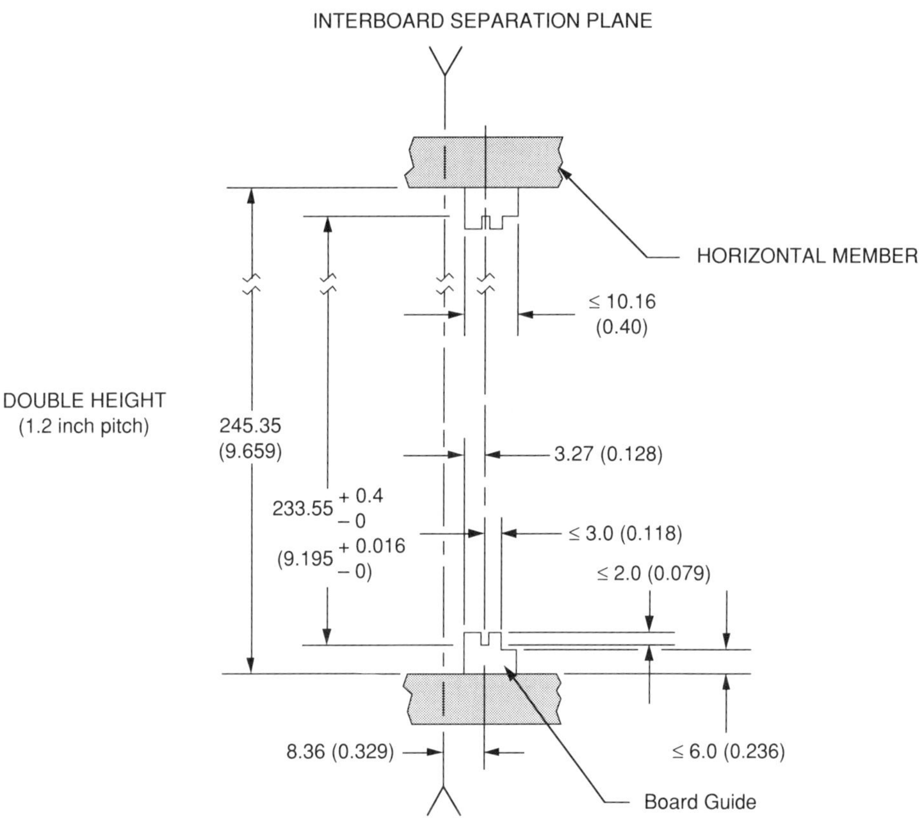

Figure B.37. C size board guide detail

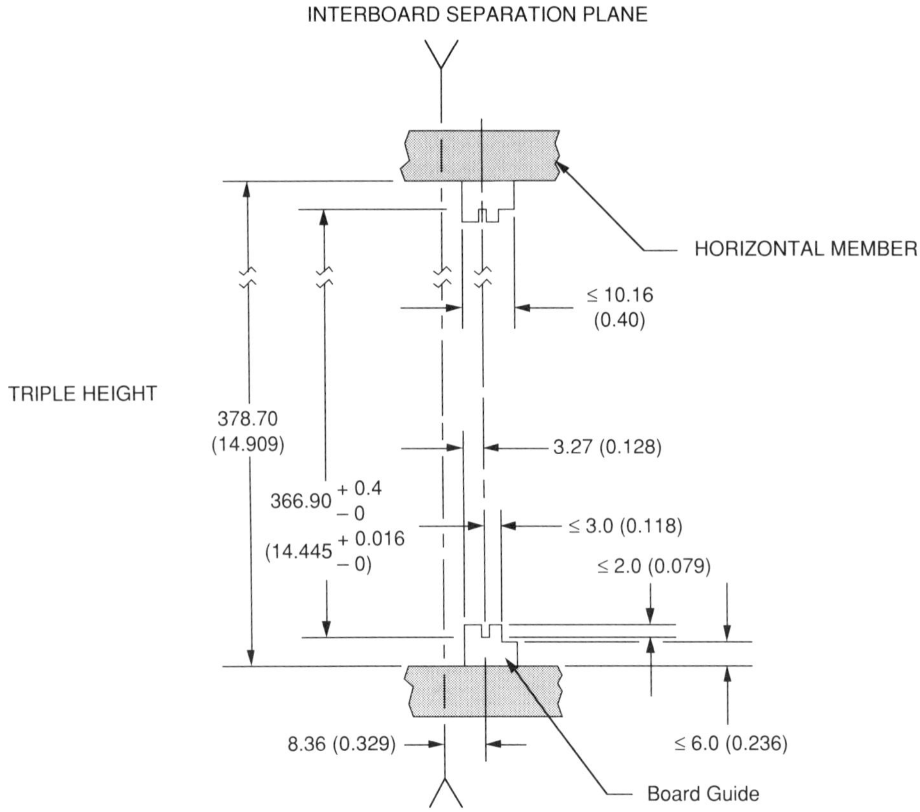

Figure B.38. D size board guide detail

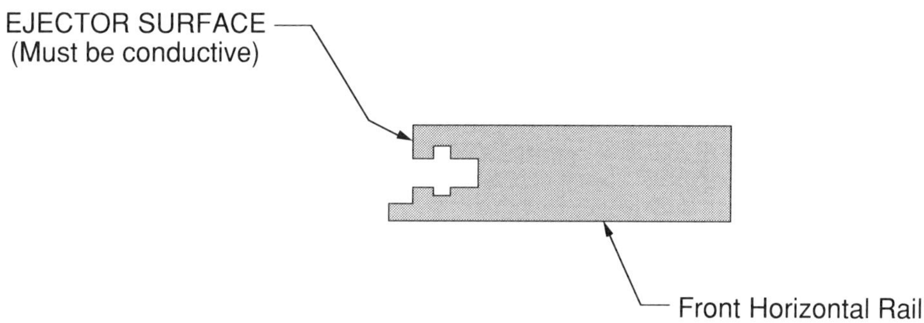

Figure B.39. Module extraction surface

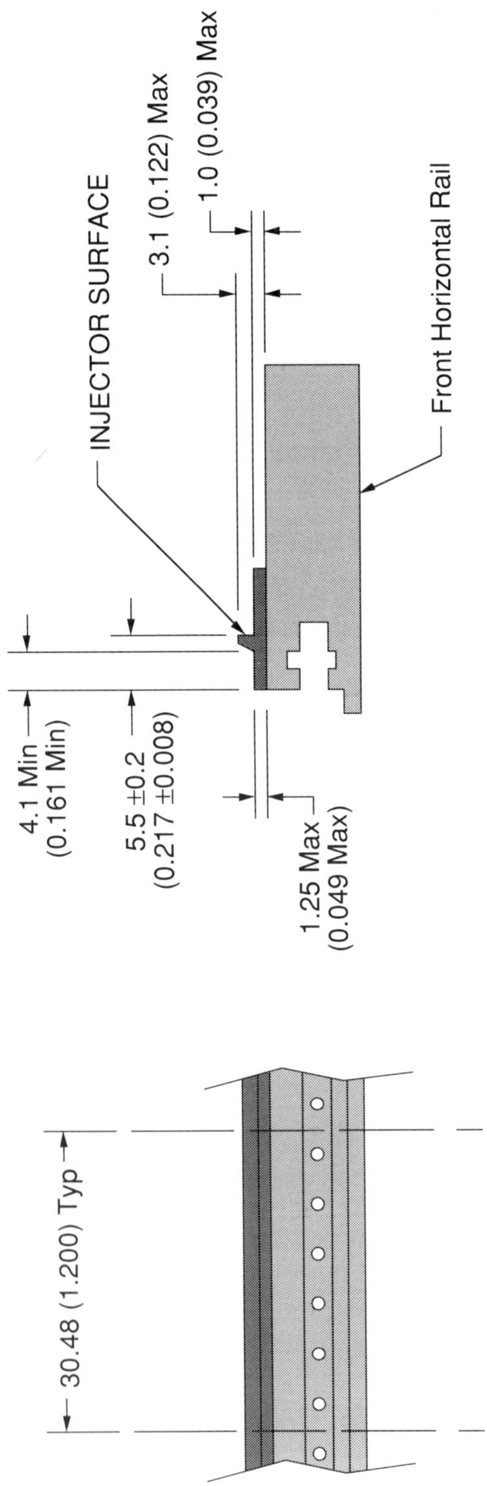

Figure B.40. Module injection surface

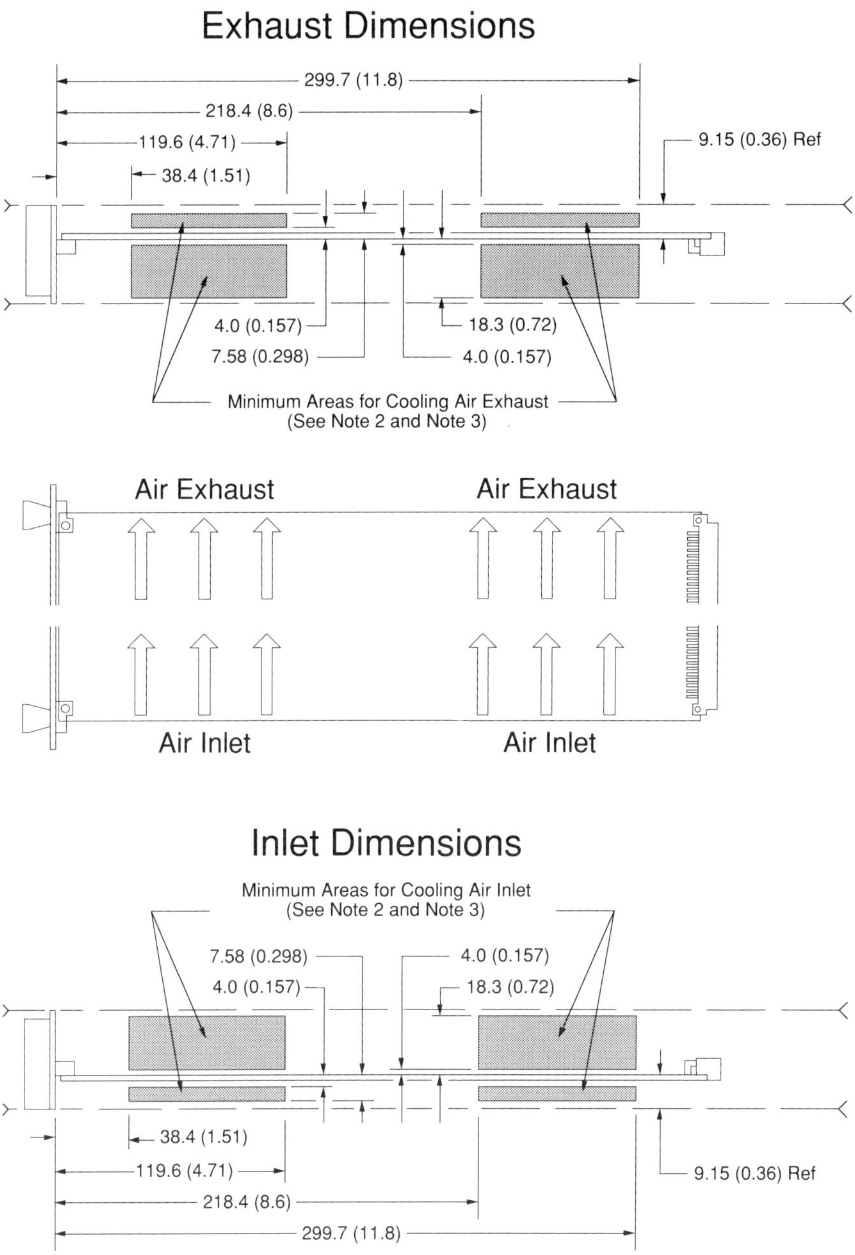

NOTES:
1. All dimensions are shown in millimeters. Inch dimensions are shown in parentheses.
2. Minimum areas apply to mainframes only. Module designers should use them as a guideline. For example, many modules may not require cooling air on the solder side.
3. Mimimum areas may or may not be completely open. They may consist of screens, filters, hold patterns, etc.

Figure B.41. C size and D size module inlet and exhaust areas

PROCEDURE

1. Build a test fixture as shown. The inlet and exhaust areas of the fixture should match those specified in Figure B.41
2. Turn fan on low.
3. Allow system to reach steady-state.
4. Measure P1 and P2. P = P1 - P2
5. Determine airflow (G) by sampling several points in the duct with a hot-wire anemometer.*
6. Plot point (G,P).
7. Increase the fan speed and repeat steps 3-7 until the relationship between P and G is sufficiently defined.
8. Establish the mimium airflow requirements of the module based on power consumption, reliability needs, and the intended application.
9. The operating point is defined as the intersection between the minimum airflow required and P versus G. See the subsection on module cooling.

*Note: Great care is required to reduce the effects of boundary layers, turbulence, flow direction, etc. For more detailed information on airflow testing, refer to published standards such as ANSI/AMCA 210-85, etc.

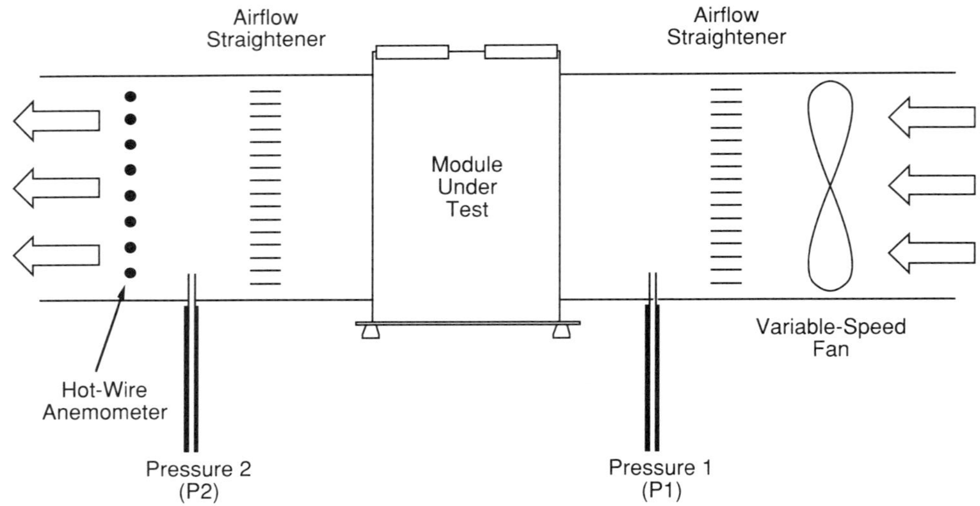

Figure B.42. Simplified test for module cooling specifications

PROCEDURE

1. Build a test fixture as shown, same size as a module. The inlet and exhaust areas of the text fixture should not restrict the flow of air within the module envelope. See Figure B.19.
2. With all other slots empty, place the fixture in a single slot. Place filler panels in all the empty slots.
3. Insert the variable restrictor with least resistance to airflow. Turn on the mainframe.
4. Allow system to reach steady-state.
5. Measure P1 and P2. P = P1 - P2
6. Determine airflow (G) by sampling several points in the fixture with a hot-wire anemometer.*
7. Plot point (G,P).
8. Change the airflow restrictor, slightly increasing the resistance to airlow. Repeat steps 4-8 until the relationship between P and G is sufficiently defined.
9. Repeat steps 2-8 for other slots until the 'worst slot' is identified.
10. It may be necessary to specify more than one curve (P versus G) if the mainframe has features such as variable speed fans, large airflow differences between slots, etc.

*Note: Great care is required to reduce the effects of boundary layers, turbulence, flow direction, etc. For more detailed information on airflow testing, refer to published standards such as ANSI/AMCA 210-85, etc.

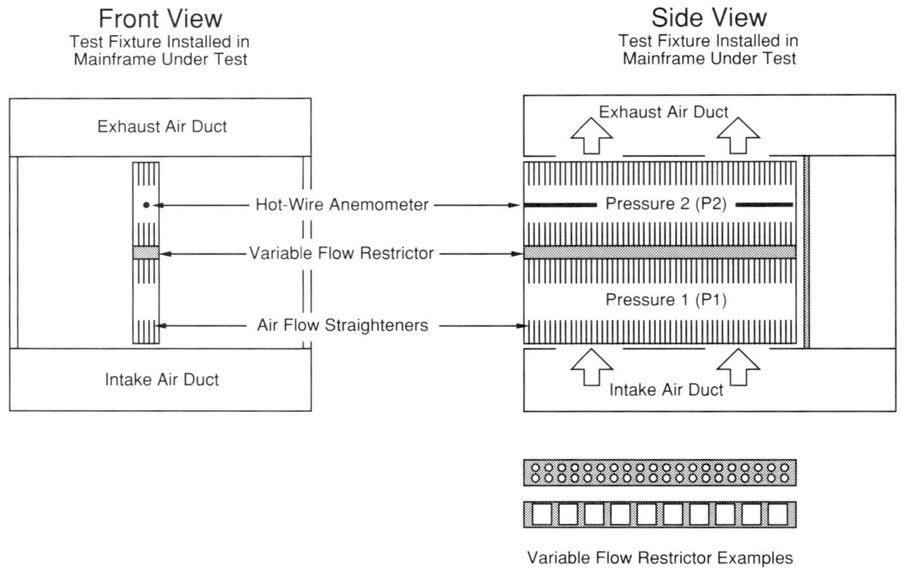

Figure B.43. Simplified test for mainframe cooling specifications

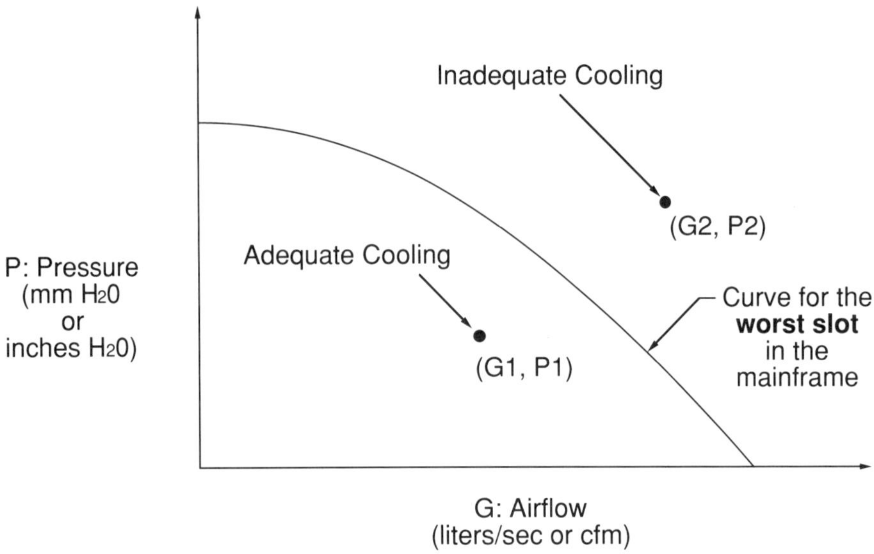

IF the module operating point (G1,P1) falls BELOW the worst slot curve,
THEN the module WILL receive adequare cooling.

IF the module oeprating point (G2,P2) falls ABOVE the worst slot curve,
THEN the module WILL NOT receive adequate cooling.

Note:

IF the module operating point (G,P) falls on or near the worst slot curve.
THEN more detailed analysis may be rquired to determine whether cooling is adequate.

Figure B.44. Example of pressure versus airflow for system integration

B.8 EMC and SYSTEM POWER

B.8.1 Introduction

The VXIbus backplane is governed by the following rules.

RULE B.8.1:
VXIbus subsystems **SHALL** provide power conductors for distribution of +5V, +5 STDBY, +12V, -12V, +24V, -24V, -5.2V, and -2V to all of the power pins specified in the pin assignments.

B.8.2 Power Distribution

Power in a VXIbus system is distributed on the backplane as regulated direct current (DC) voltages. The available voltages are:

+5 Vdc This is the main power source for most VXIbus systems. All systems require this voltage for bus communications.

±12 Vdc These are often used for powering analog devices, disc drives, and communication interfaces.

±24 Vdc These are used for powering analog signal sources, and to derive other voltage levels (e.g. ±15 Vdc) as needed, using on-board regulators.

–5.2 Vdc This is used for ECL devices.

–2 Vdc This is used for termination of ECL loads.

+5 Vdc STDBY This is used to sustain memory, clocks, etc, when the +5 Vdc power is lost.

B.8.3 Power Pins

RULE B.8.2:
The current flow per pin on any connector **SHALL** comply with Figure 5-7 in the VMEbus specification, based on temperature rise considerations in the target system.

OBSERVATION B.8.1:
A 1-Amp-per-pin limit allows the VMEbus-rated connector to operate at 55°C ambient.

B.8.4 DC Voltage Specifications

RULE B.8.3:
Any voltage furnished by the mainframe power supply **SHALL** comply with the maximum allowed variation and maximum allowed DC Load Ripple/Noise in Table B.7, up to the mainframe's maximum rated current (I_{MP}).

VOLTAGE	DESCRIPTION	ALLOWED VARIATION		DC LOAD RIPPLE/NOISE	INDUCED RIPPLE/NOISE
+5V	+5 Vdc	+0.25V	−0.125V	50 mV	50 mV
+12V	+12 Vdc	+0.60V	−0.36V	50 mV	50 mV
−12V	−12 Vdc	−0.60V	+0.36V	50 mV	50 mV
+24V	+24 Vdc	+1.20V	−0.72V	150 mV	150 mV
−24V	−24 Vdc	−1.20V	+0.72V	150 mV	150 mV
−5.2V	−5.2 Vdc	−0.26V	+0.156V	50 mV	50 mV
−2V	−2 Vdc	−0.10V	+0.10V	50 mV	50 mV
+5V STDBY	+5 Vdc standby	+0.25V	−0.125V	50 mV	50 mV

Table B.7. Power Supply Voltage Specifications

DC Load Ripple/Noise is the maximum Periodic And Random Deviations (PARD) generated by the power supply under all DC load conditions up to the Mainframe Peak Current, measured as the peak-to-peak voltage in the DC to 10 MHz bandwidth.

Induced Ripple/Noise is the additional peak-to-peak ripple that may exist on the backplane power supply pins, due to injected current from operating modules. This is the V_{MAX} parameter in Figure B.46 and in Figure B.48.

Mainframe Peak Current (I_{MP}) is the rated DC current of a particular voltage supplied by the mainframe.

Mainframe Dynamic Current (I_{MD}) is the rated dynamic current capacity of a particular voltage supplied by the mainframe for the frequencies from 2 Hz to 1 GHz.

RULE B.8.4:
Power supplies *source* power to the modules and **SHALL NOT** be used to *sink* power, other than transient current from protection circuits.

OBSERVATION B.8.2:
The non-symmetric variation given in Table B.7 ensures that the DC power remains within the tolerance required by most ICs, despite the typical voltage drops that occur in the power distribution network. Placing the power supply sense point near the power input point prevents boards near the power input point from receiving too high a voltage.

RULE B.8.5:
All voltages which are supplied by a mainframe **SHALL** tolerate peak-to-peak current variations of I_{MP} from DC to 2 Hz without generating peak-to-peak voltage variations in the DC to 2 Hz frequency range exceeding the Induced Ripple/Noise limit of Table B.7.

RULE B.8.6:
All voltages which are supplied by a mainframe **SHALL** tolerate peak-to-peak current variations, as specified by the Maximum Rated Mainframe Dynamic Output Current (I_{MD}) in Figure B.45, without generating peak-to-peak voltage variations

exceeding the limits specified in Figure B.46. For the test method see subsection B.8.7.1 "Induced Ripple Noise Test of Mainframes". I_{MD} is the rated dynamic current of the mainframe supply. V_{MAX} is the peak-to-peak Induced Ripple/Noise specified in Table B.7.

OBSERVATION B.8.3:
The rated Mainframe Dynamic Current (I_{MD}) is chosen by the manufacturer to ensure compliance with the above Induced Ripple/Noise specification.

B.8.5 Power Management

The power compatibility of mainframes and modules is accomplished by documenting the *mainframe* output power (both peak and dynamic) and *module* input power (both peak and dynamic). Mainframe power supplies will identify all voltage and current capabilities which are present in the mainframe. Modules will identify all voltage and input current requirements for proper operation. This documentation of the power requirements will allow the system integrator to verify compatibility between mainframe output power capability and module supply requirements.

Successful integration of a VXIbus instrument system depends on the power compatibility of mainframes and modules. The individual dynamic currents of modules will add together to form complex dynamic currents that the mainframe must tolerate in order to maintain its ripple specifications. The VXIbus rules establish a minimum level of mainframe and module documentation that the manufacturer must comply with.

Peak Module Current (I_{Pm}) is the maximum rated instantaneous current (measured in the DC to 10 MHz bandwidth) which is drawn by a module for a particular supply voltage during normal operation.

Dynamic Module Current (I_{Dm}) is the maximum rated dynamic current required by the module. I_{Dm} is chosen by the manufacturer so that the module meets the conducted emission test from 2 Hz to 1 GHz.

RULE B.8.7:
Mainframe Peak Current (I_{MP}) and Mainframe Dynamic Current (I_{MD}) **SHALL** be documented by the mainframe manufacturer for every voltage, even if not supported.

RECOMMENDATION B.8.1:
Mainframes should be marked with current ratings of all voltages listed, even if some of those voltages are not supplied. Example: +5V STDBY = 0.0 A Peak, 0.0 A Dynamic (if not supported).

RULE B.8.8:
Peak Module Current (I_{Pm}) and Dynamic Module Current (I_{Dm}) **SHALL** be documented by the module manufacturer for all power supply voltages used.

RECOMMENDATION B.8.2:
Modules should be marked with input current specifications for all voltages used.

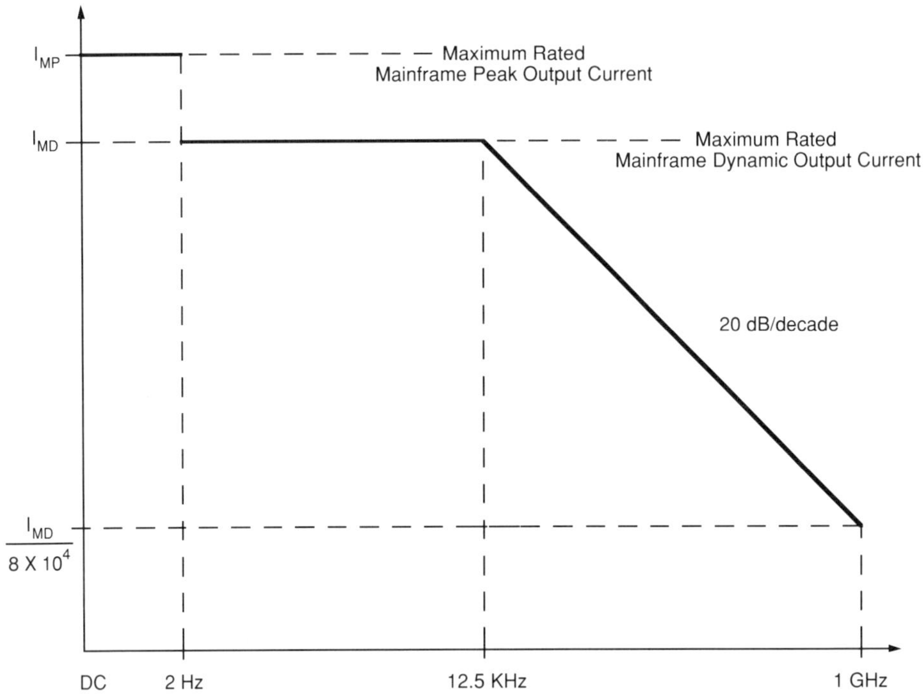

Figure B.45. Mainframe Load Current

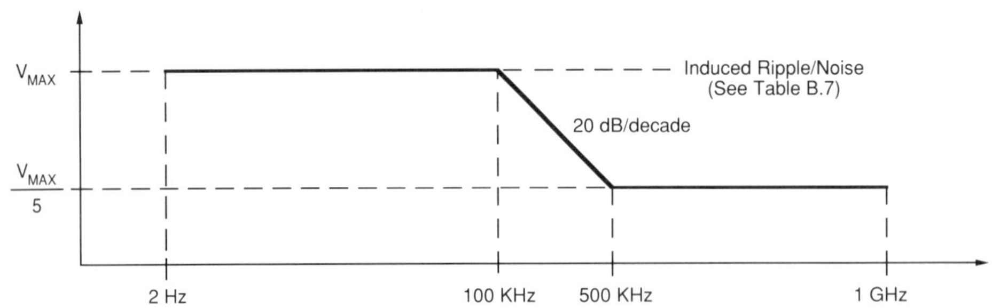

Figure B.46. Mainframe Induced Voltage Limits

OBSERVATION B.8.4:
The module peak current is measured over a 10 MHz bandwidth to ensure that contributions from all potentially significant frequency components are included. For proper operation, the peak current capacity of the mainframe (I_{MP}) is required to exceed the sum of the Peak Module Currents (I_{Pm}) for each supply used. If the dynamic current capacity of the mainframe (I_{MD}) exceeds the sum of the dynamic current requirements of the modules (I_{Dm}), then proper power supply ripple performance is also guaranteed. However, if the dynamic currents of the modules exceed that of the mainframe, the specified ripple limits might be exceeded, and more detailed specifications are required from the mainframe and module manufacturer(s).

The final determination of compatibility may be application specific. The mainframe manufacturer may provide a curve that shows the mainframe dynamic current versus frequency. The module manufacturer may provide dynamic module input current versus frequency for particular applications. For example, a pulse generator that can drive 10 volts into 50 ohms could have curves for 50 ohms, 75 ohms, etc. The module manufacturer may also provide voltage ripple specifications for the module if it is more tolerant than the VXIbus specification defined in Table B.7.

OBSERVATION B.8.5:
The required current specifications of mainframes and modules reflect the products in an operational state, after power up. Mainframes and modules might act differently during the power-up process.

RECOMMENDATION B.8.3:
Mainframes should handle large inrush currents during power up gracefully, without shutting down.

RECOMMENDATION B.8.4:
Design a module to present a reasonable load to the mainframe power supplies during the power up period. The DC current required by the module, when it is powered by any DC voltage between 0 and V_{supply}, should not exceed the current drawn from an equivalent resistive load of V_{supply}/I_{Pm}. Inrush current, due to a module's supply capacitance while the power supply voltage is increasing, is allowed to exceed this value. Limit the module's input capacitance to 200 μF or less on each supply voltage used. Design modules that have large turn on currents (such as may be required for ovens and disk drives) to comply with the module's peak and dynamic current specifications. These restrictions allow a VXIbus system to behave gracefully during the power-up period.

B.8.6 Electromagnetic Compatibility (EMC) of Modules

B.8.6.1 Conducted Emissions

RULE B.8.9:
The maximum instantaneous current drawn by a module **SHALL NOT** exceed the Peak Module Current (I_{Pm}) specified by its manufacturer for any power supply voltage used.

RULE B.8.10:
Each module **SHALL** have less conducted emissions than the levels specified by the Maximum Rated Dynamic Input Current in Figure B.47 on all power supply voltages used. For the test method see B.8.7.2. I_{Dm} is the Dynamic Module Current of the module, as specified by its manufacturer.

RULE B.8.11:
IF a module's I_{Dm} or I_{Pm} is dependent on a particular configuration or set of configurations,
THEN the module manufacturer **SHALL** document the configuration assumptions that allow the module to meet the conducted emission specifications.

B.8.6.2 Conducted Susceptibility

RULE B.8.12:
Each module **SHALL** tolerate conducted susceptibility levels specified in Figure B.48 on all the power supply voltages used. For the test method see B.8.7.3. V_{MAX} is the induced ripple/noise value specified in Table B.7.

RECOMMENDATION B.8.5:
Module Electrostatic Discharge (ESD) return path to chassis ground should be through the faceplate. The faceplate chassis ground is the primary return path. The energy discharged to any backplane signal pin should be at a low enough level that it does not disrupt the normal operation of the mainframe or the module e.g. change of state of data lines, etc.

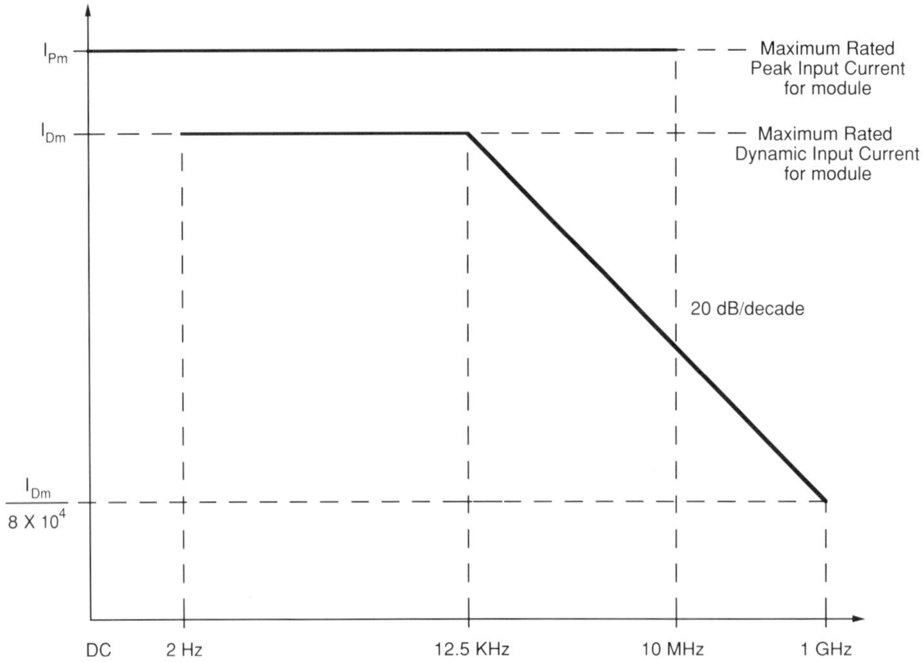

Figure B.47. Module Conducted Emissions

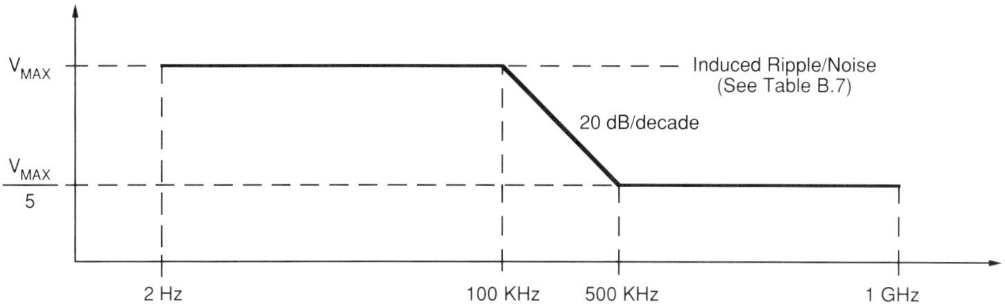

Figure B.48. Module Susceptibility Level

B.8.6.3 Radiated Emissions

Mainframes and modules will be specified to conform to chosen standards appropriate to the intended market, e.g. FCC, VDE, or MIL Spec. for far-field radiated emissions. Each module should not contribute more than its fair share of the total allowable radiation. For example in a 13-module mainframe, no module should contribute more than 1/13 of the allowable radiation if its radiation is coherent. Module-to-module interference is a close-field radiation problem. The term close-field is used to indicate those fields within a few centimeters of the surface of the modules.

RULE B.8.13:
The module manufacturer **SHALL** specify under what conditions the mainframe containing the module will meet the chosen radiated emission requirements.

RECOMMENDATION B.8.6:
Twelve identical single-slot modules, plus the Slot 0 module, should stay within the chosen limits of the mainframe regulatory requirement. This is a coherency test.

OBSERVATION B.8.6:
If one module is tested alone, the radiation per slot should be 1/13 of the allowable far-field specification, if the radiation is coherent. Module radiation allowed is proportional to the number of slots occupied.

OBSERVATION B.8.7:
If the module complies with the coherency test described in the above recommendation, then the module manufacturer can just claim compliance to the regulatory radiation requirement, with no additional documentation. If the module does not comply with the coherency test, then the burden is on the module manufacturer to document under what configurations the mainframe containing the module will meet the regulatory radiation requirements.

RULE B.8.14:
The close-field magnetic emissions of A size and B size modules (when measured on the intermodule separation plane identified by the shaded area of Figure B.53) **SHALL NOT** exceed the limit shown in Figure B.49. For the test method see subsection B.8.7.4 (Close-Field Magnetic Emissions Test).

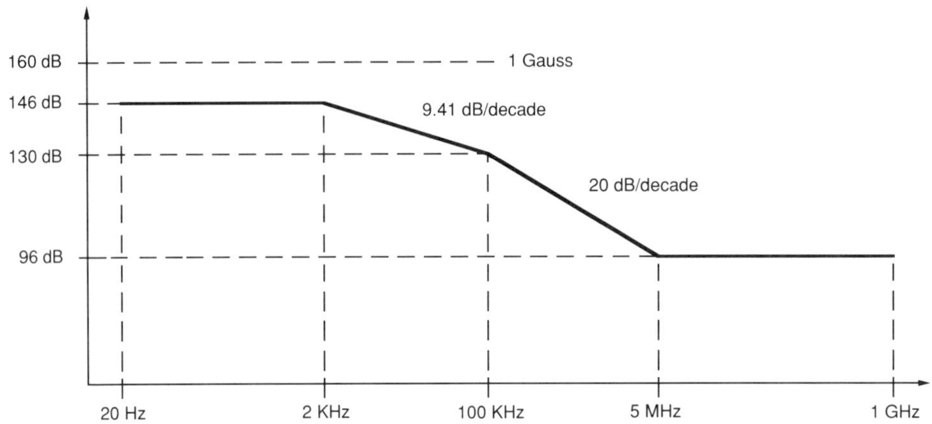

Figure B.49. A size and B size maximum close-field emissions (dB above 1 picotesla)

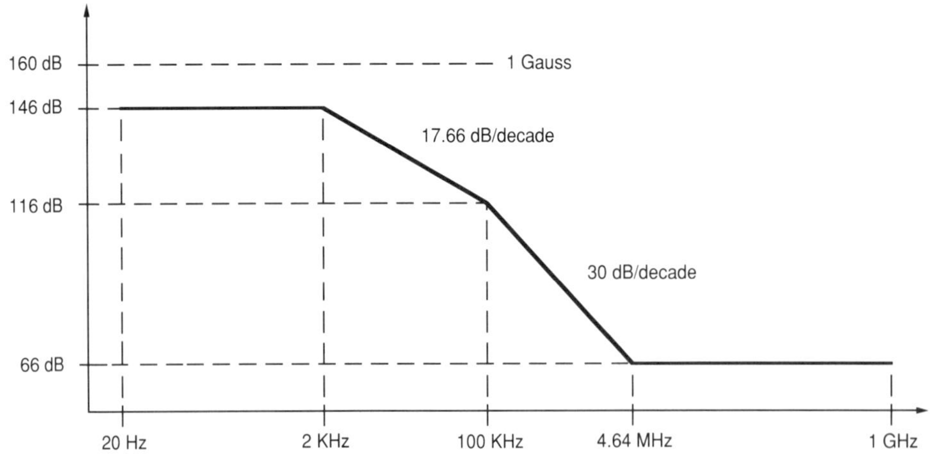

Figure B.50. C size and D size maximum close-field emissions (dB above 1 picotesla)

RULE B.8.15:
The close-field magnetic emissions of C size and D size modules (when measured on the intermodule separation plane identified by the shaded area of Figure B.54) **SHALL NOT** exceed the limit shown in Figure B.50. For the test method see subsection B.8.7.4, "Close-Field Magnetic Emissions Test".

RULE B.8.16:
IF an A size or a B size module is inserted in a C size or a D size mainframe,
THEN the close-field magnetic emissions (when measured on the C size or D size intermodule separation plane identified by the shaded area of Figure B.55) **SHALL NOT** exceed the limit shown in Figure B.49. For the test method see subsection B.8.7.4, "Close-Field Magnetic Emissions Test".

RULE B.8.17:
IF an A size or a B size module is inserted in a C size or a D size mainframe,
THEN the close-field magnetic emissions (when measured on the C size or D size intermodule separation plane identified by the shaded area of Figure B.56) **SHALL NOT** exceed the limit as shown in Figure B.50. For the test method see subsection B.8.7.4, "Close-Field Magnetic Emissions Test".

PERMISSION B.8.1:
A size and B size modules **MAY** use the shield of an adapter to meet the close-field magnetic emissions limit when they are installed into a C size or a D size mainframe.

RECOMMENDATION B.8.7:
The A size and B size module manufacturer should document the available adapter(s) or procedure(s) that allow a module to meet the close-field magnetic compatibility limits, when the module is installed into a C size or a D size mainframe.

OBSERVATION B.8.8:
The integration of A size VXI, B size VXI, or standard VME modules into C size or D size mainframes will require some forethought and planning. The system integrator should carefully consider the adapter choice, as well as the module placement, to preserve electromagnetic compatibility.

OBSERVATION B.8.9:
For modules without covers or shields, the surface for close-field testing is defined as the intermodule separation plane. The close-field magnetic emissions and susceptibility of the backplane are not specified.

RECOMMENDATION B.8.8:
Place all significant sources of radiated emissions (such as switching power converters or crystal oscillators) within an area 15 cm (6 inches) or less from the backplane connectors. Place all sources of radiated emissions as close as possible to the backplane connectors, and as far as possible from the board guides.

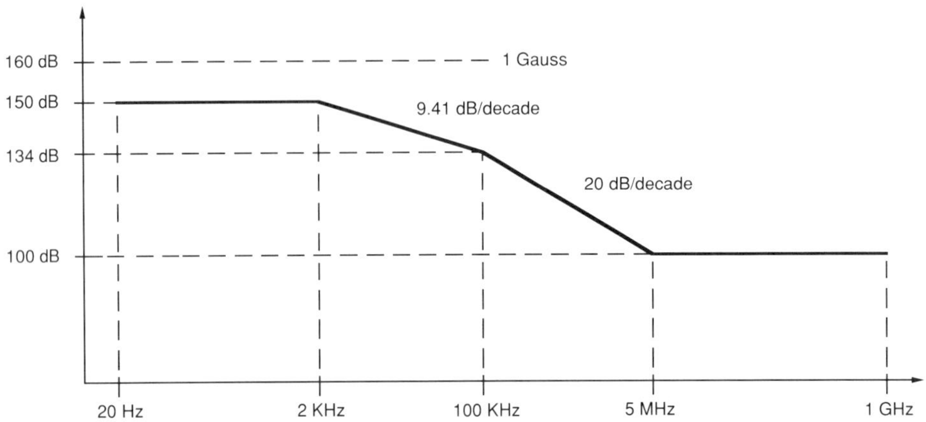

Figure B.51. A size and B size minimum close-field susceptibility (dB above 1 picotesla)

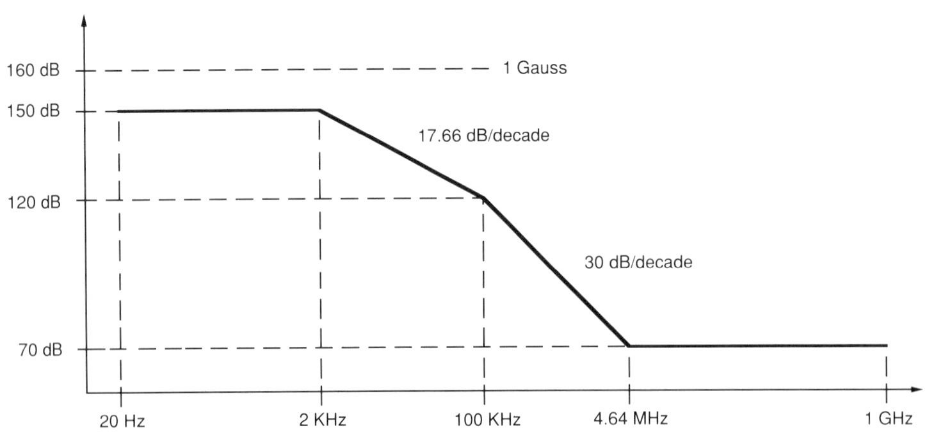

Figure B.52. C size and D size minimum close-field susceptibility (dB above 1 picotesla)

OBSERVATION B.8.10:
There is no close-field EMC specification for backplanes. Module EMC requirements start 38 mm (1.5 inches) from the backplane surface. A connector shield like the one described in the Mechanical Specification might be needed to comply with the close-field magnetic emission specification.

RECOMMENDATION B.8.9:
All C size and D size backplanes should provide a conductive surface in the grounding area, as described in the Mechanical Specification.

OBSERVATION B.8.11:
There are no specified limits for broadband or non-repetitive close-field emissions.

RECOMMENDATION B.8.10:
Place all significant sources of broadband emissions as close as possible to the backplane connectors, and as far as possible from the board guides.

B.8.6.4 Radiated Susceptibility

Module-to-module interference is a close-field susceptibility problem. The term close-field is used to indicate those fields within a few centimeters of the surface of the modules. Close-field magnetic emission limits and susceptibility levels should be met by using good engineering design practices.

RULE B.8.18:
The close-field magnetic susceptibility of A size and B size modules (when measured on the intermodule separation plane identified by the shaded area of Figure B.53) **SHALL** operate as specified at the level shown in Figure B.51. For the test method see subsection B.8.7.5, "Close-Field Magnetic Susceptibility Test".

RULE B.8.19:
The close-field magnetic susceptibility of C size and D size modules (when measured on the intermodule separation plane identified by the shaded area of Figure B.55) **SHALL** operate as specified at the level shown in Figure B.51. For the test method see subsection B.8.7.5, "Close-Field Magnetic Susceptibility Test".

RULE B.8.20:
The close-field magnetic susceptibility of C size and D size modules (when measured on the intermodule separation plane identified by the shaded area of Figure B.56) **SHALL** operate as specified at the level shown in Figure B.52. For the test method see subsection B.8.7.5, "Close-Field Magnetic Susceptibility Test".

RECOMMENDATION B.8.11:
Place all circuits susceptible to radiated emissions (such as front-end input circuits) in the portion of a module greater than 15 cm (6 inches) from the backplane connectors. Place these circuits as close as possible to the front face plate, and as far as possible from the board guides.

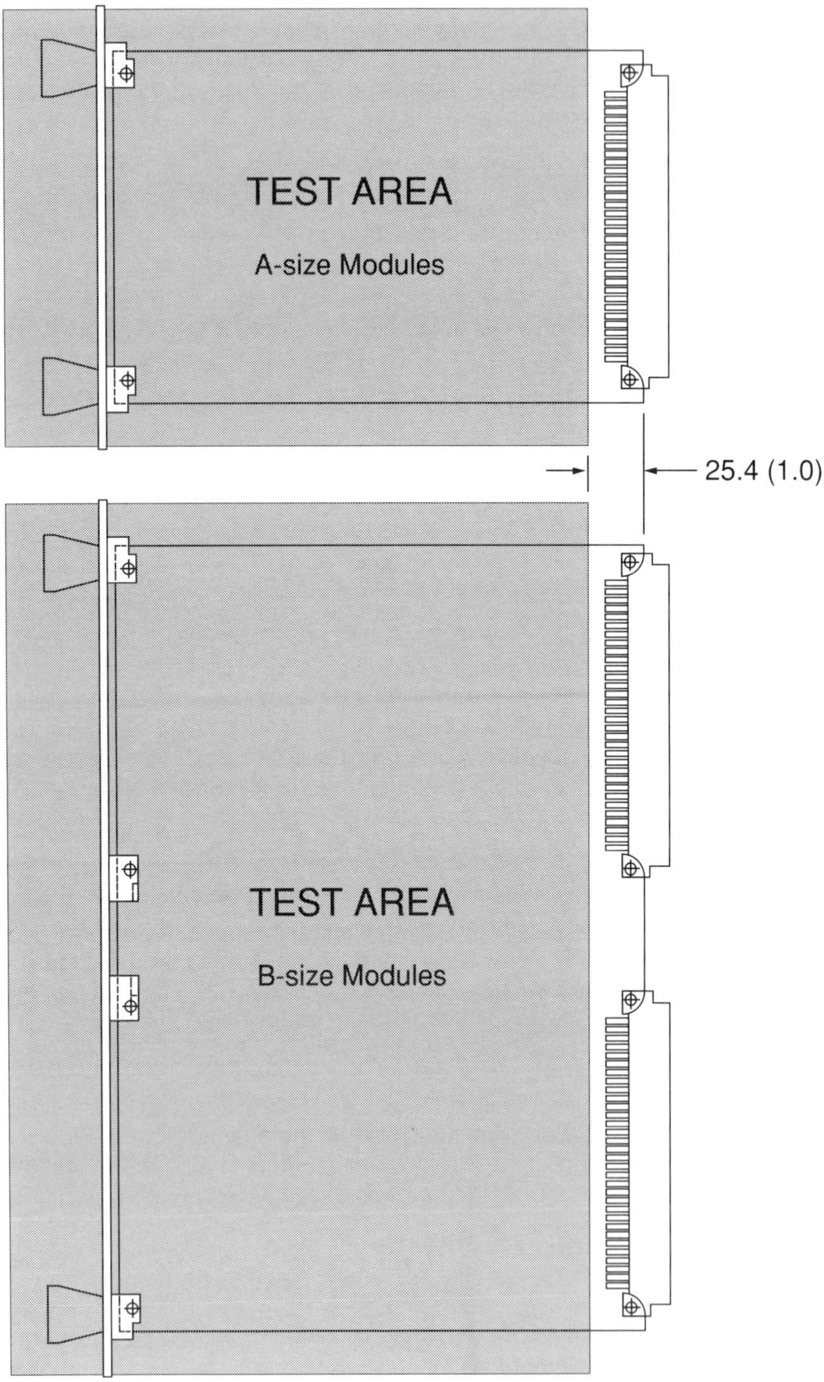

Figure B.53. A size and B size EMC Test Areas

OBSERVATION B.8.12:
There are no specified limits for broadband or non-repetitive close-field susceptibility.

RECOMMENDATION B.8.12:
Place all circuits susceptible to broadband or non-repetitive emissions as close as possible to the front face plate, and as far as possible from the board guides.

RECOMMENDATION B.8.13:
Electric close-fields should be evaluated on all unshielded modules.

OBSERVATION B.8.13:
Electric field emissions are easily contained with Faraday shields. Unshielded modules should be checked for excessive electric field emissions, particularly at areas where high voltage swings are present. A length of coax with one or two inches of outer conductor stripped back at the tip is a useful close-field probe for making relative electric field measurements. A shield around the generating module, around the susceptible module, or between the modules, will usually eliminate any electric field interference problems.

B.8.7 Suggested Test Methods

The test methods described in this section are provided as suggestions to the manufacturers of VXIbus mainframes and modules.

In the following sections it is assumed that all measurements (which are made using a spectrum analyzer) are for single-frequency stationary signals. Therefore the bandwidth and sweep rate must be chosen to accurately measure only one frequency component at a time. For modules or mainframes which may generate non-stationary signals, (for example, swept signals or transients) the specified techniques cannot be used, but the manufacturers should point out in their documentation what limitations, if any, may arise because of these signals.

At higher frequencies, a relatively narrow bandwidth may be necessary to obtain sufficient sensitivity.

Note that all *induced* and *conducted* tests are based on peak-to-peak measurements, while the *radiated* tests are based on RMS measurements.

B.8.7.1 Induced Ripple/Noise Test Of Mainframes

The Induced Ripple/Noise test requires applying a load of the magnitude and frequencies shown by the Maximum Rated Mainframe Dynamic Output Current in Figure B.45, while monitoring the power supply voltage to guarantee that the induced voltage ripple does not exceed that of Figure B.46.

Set the DC load on the power supply to be half of the rated current (1/2 I_{MP}). For the low frequency tests, use a time domain measurement method by driving an active load with a low frequency oscillator to the levels specified, while measuring the induced ripple with an oscilloscope or AC voltmeter. Monitor the voltage noise present

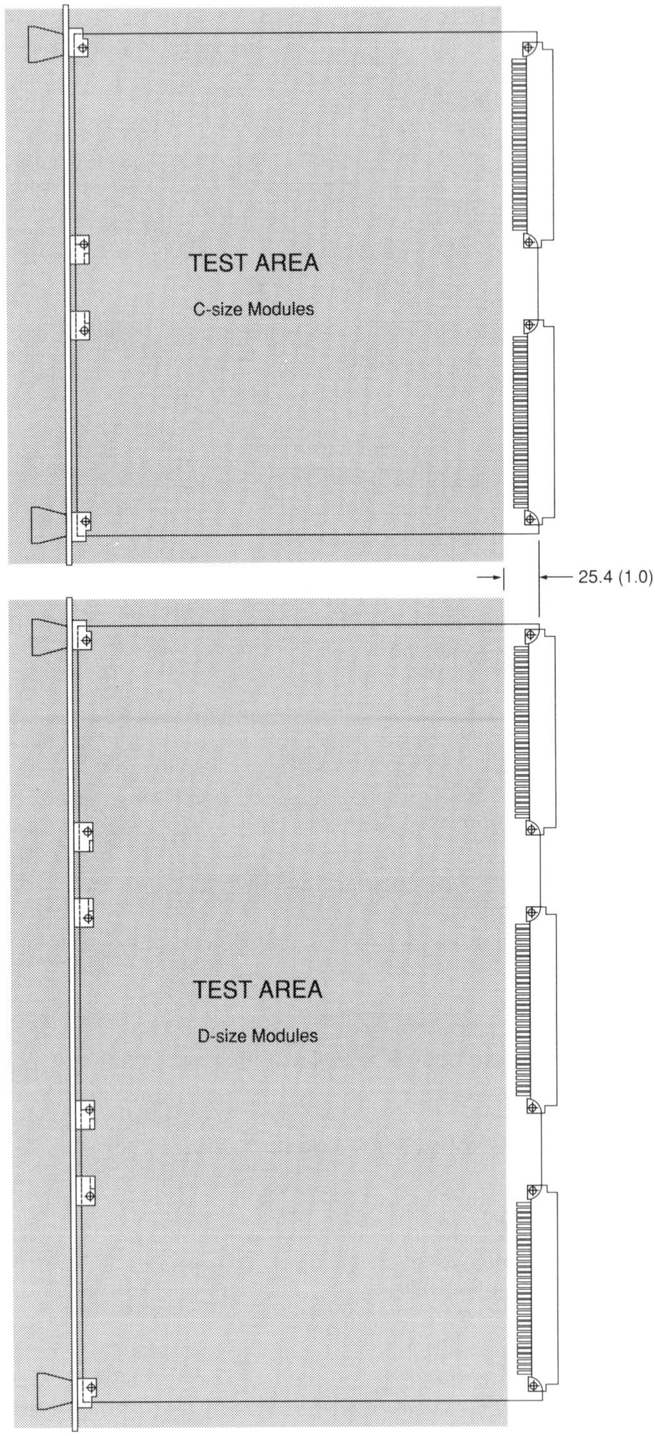

Figure B.54. C size and D size EMC Test Areas

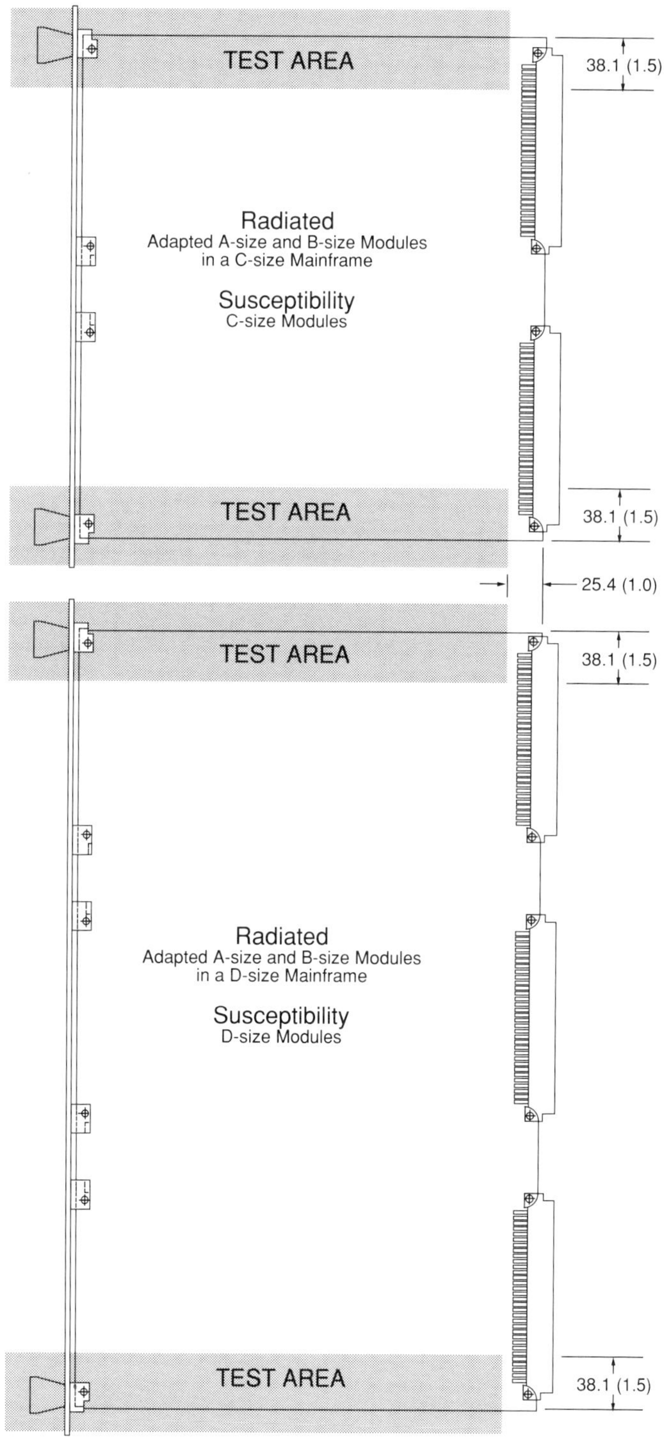

Figure B.55. Adapted A size, B size, C size, and D size EMC Test Areas

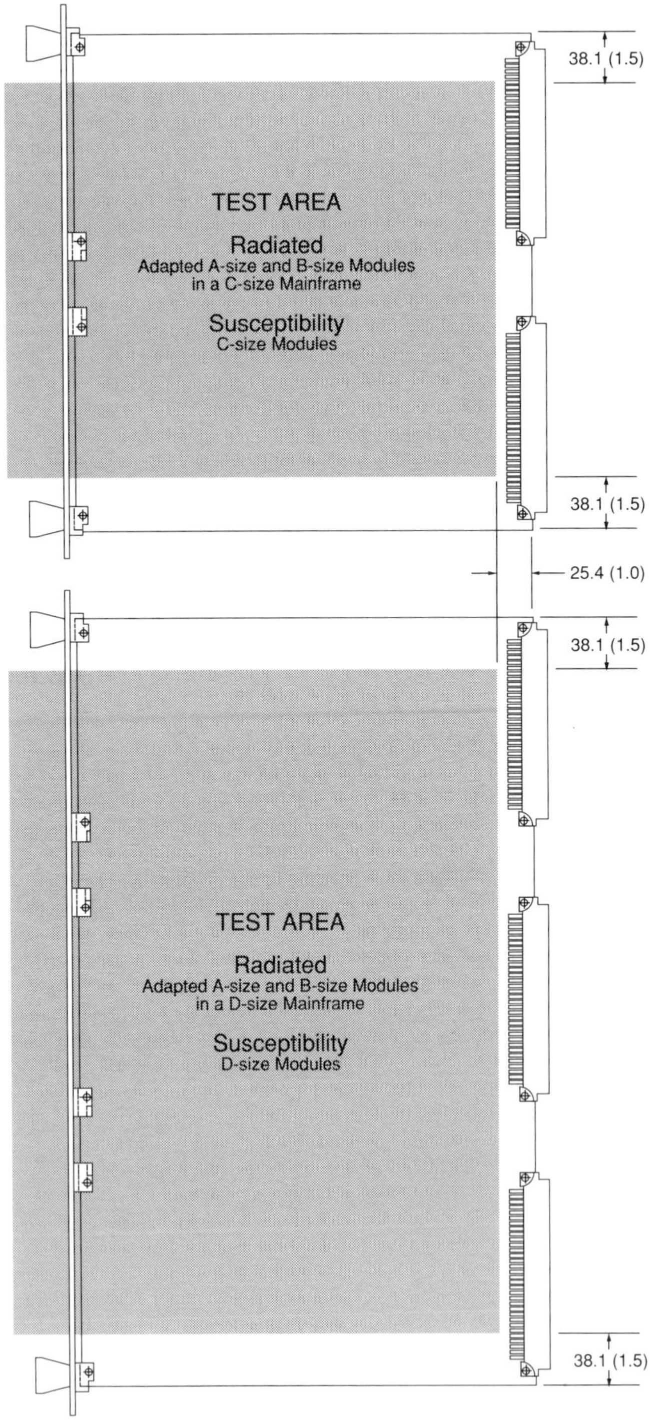

Figure B.56. Adapted A size, B size, C size, and D size EMC Test Areas

at an adjacent slot. For the higher-frequency conducted-emission tests, drive a load from a tracking oscillator while monitoring the induced voltage at an adjacent slot with a spectrum analyzer.

B.8.7.2 Conducted Emissions Test Of Modules

The testing of a module's dynamic current should exclude all effects of the mainframe. Because the mainframe provides current for backplane terminations, clocks, and other devices requiring dynamic current, voltage ripple may exist on power pins at various frequencies. This power supply ripple voltage may cause an apparent dynamic current flow into the module that is not caused by any current requirements of the module, and should not be included as part of the module's dynamic current. By supplying voltages to the module under test from independent power supplies, the mainframe ripple performance will be isolated from the module dynamic current requirements.

The module being tested should be operating at its worst case current-versus-frequency requirements.

A current sensor, in series with each supply of the module, is used to measure the AC portion of the power supply load. Check with an oscilloscope that the module meets the low frequency (up to 12.5 KHz) conducted emission requirements. A spectrum analyzer is used to measure the higher-frequency content of the current. The module is exercised to find the highest readings. Appropriate current sensors may include resistors, current probes, and Hall effect devices.

An alternative to a current sensor for high frequency conducted emissions (>50 MHz) is to place an appropriate air core inductor (4.7 μH) in series with the backplane power output pin and the card power input pin. Then measure the voltage with an AC-coupled spectrum analyzer at the card power input pin and at the card ground pin, using a 50 Ω cable, terminated at the spectrum analyzer. The current is then the measured voltage, divided by 50.

This test should be conducted using a backplane having an impedance of less than $0.2 \times V_S/I_S$ in parallel with a ≥ 0.1 μF bypass capacitor at the measured slot, where V_S is the Module Susceptibility voltage level (Figure B.48) and I_S is the Module Conducted current level (Figure B.47).

B.8.7.3 Conducted Susceptibility Test Of Modules

Induce a voltage between each backplane power supply and module at the maximum voltage shown in Figure B.46, as measured from the module supply pin to ground. Search for potential frequencies where performance is degraded, due to the presence of injected noise. For low frequency testing, use a series resistor driven by an active load to induce the specified voltage level. For higher frequencies use current probes appropriate to the frequency range being tested. Be sure the current probe is rated for the DC current being supplied to the module.

B.8.7.4 Close-Field Magnetic EmissionsTest

The equipment used to measure the close-field magnetic emissions of a module consists of close-field magnetic probe and a spectrum analyzer that operates over the frequency range specified in Figure B.49 or Figure B.50. To cover the frequency range, different probes and spectrum analyzers may be required. It is required that the close-field probes have calibrated antenna factors, and that the magnetic loop be within 5 mm of the end of the probe. With the close-field magnetic probe connected to the spectrum analyzer, move the probe tip over the entire module surface to search for potential radiating spots such as seams and holes. At all areas where magnetic radiation is detected, the probe should be oriented to maximize the probe output, as displayed on the spectrum analyzer. Using the antenna factors of the close-field probe, calculate the magnetic flux density for each frequency of radiation detected. Flux density levels may not exceed the levels of the appropriate figure.

B.8.7.5 Close-Field Magnetic Susceptibility Test

The equipment used to test the susceptibility of a module consists of a close-field magnetic probe connected to a signal generator that operates over the frequency range specified in Figure B.51 or Figure B.52. To cover the frequency range, different probes and signal generators may be required. It is required that the close-field probes have calibrated antenna factors, and that the magnetic loop be within 5 mm of the end of the probe. Connect the close-field magnetic probe to the signal generator. For each frequency tested, set the signal generator to produce a magnetic flux density at the limit given by the appropriate figure for that frequency. Move the probe tip over the entire module surface to search for potential spots where the performance of the module is degraded, due to the presence of a magnetic field. The probe tip should be within 1 cm of the surface of the module.

B.8.7.6 Examples Of Close Field Magnetic Probes

Examples of close-field magnetic probes that meet the above requirements are:

Magnetek Part Number 1846 for frequencies from 20 Hz to 10 kHz

Hewlett-Packard Model 11941 for frequencies from 10 kHz to 30 MHz

Hewlett-Packard Model 11940 for frequencies from 30 MHz to 1 GHz

An EMCO Model 7405 Near-Field Probe Set may also be used to complement the above probes.

An amplifier may be needed with any probe, to compensate for large probe antenna factors.

SECTION C

System Architecture

C.1 VXIbus SYSTEM ARCHITECTURE OVERVIEW

The VXIbus architecture allows a wide range of instruments, interface cards, or computers from different manufacturers to coexist as modules in the same cardcage. The range of the applications being addressed by different cardcages in the VXIbus family is very broad, and hence the architecture is intentionally left open to accommodate this flexibility. VXIbus does not define a specific system hierarchy or topology. Nor does it specify the type of microprocessor, the type of operating system or the type of interface to the host computer. What VXIbus defines is a foundation (or platform) upon which the above decisions can be made, while ensuring compatibility between different manufacturers. Figure C.1. shows *some* of the possible system configurations, and may help illustrate the flexibility that is allowed by the VXIbus standard.

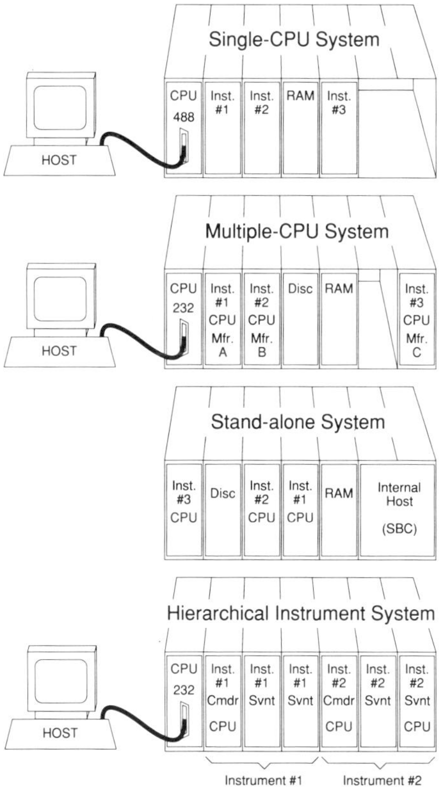

Figure C.1. Typical System Configurations

As shown by the examples in Figure C.1, there are many topologies possible within the VXIbus architecture. These include single CPU systems that centralize the intelligence for all their instrument modules, multiple CPU systems with distributed intelligence, and stand-alone mainframes that contain a host computer running user applications. Each of these topologies has different communication needs, which are met by a layered set of communication protocols illustrated in Figure C.2.

In order to support automatic system and memory configuration, the VXIbus standard specifies a minimum amount of capability on each device, regardless of its function. A VXIbus device (see Section C.2, "Device Operation"), being the lowest element of a hierarchy, has a set of "Configuration Registers", all totally accessible from P1 on VMEbus. This set of registers allows the system to identify the device, its class, model and manufacturer, address space (A16, A24, A32), and memory requirements. VXIbus devices with only this minimum level of capability are called **Register Based Devices**. As an example, the single CPU system shown in Figure C.1 may implement all of its instruments using only Register Based Devices with device-specific protocols describing how the CPU communicates with each of those instruments. VXIbus also defines Memory Devices, (See Section C.2.3, "Memory Devices") that can be identified as RAM, ROM, or other memory types, and can be configured in contiguous blocks of memory, based on speed or memory type.

If a higher level of communication ability between modules is desired in the system, VXIbus defines a class of devices called **Message Based Devices**. In addition to the configuration registers mentioned above, Message Based Devices are required to have communication registers that are accessible to other modules in the system. Each device in the system can then use specific communication protocols such as VXIbus's Word

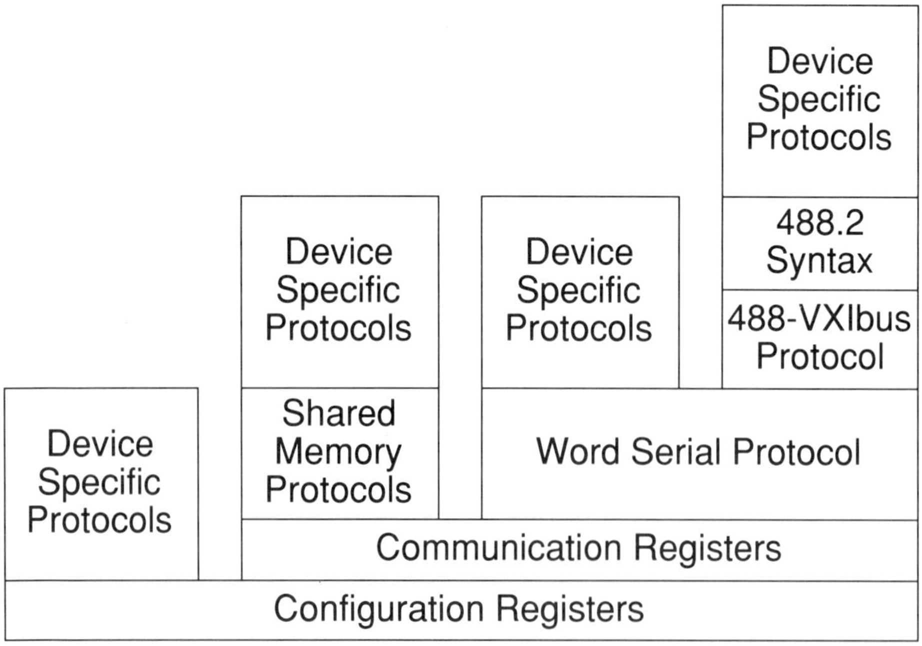

Figure C.2. VXIbus Communication Layers

Serial Protocol (See Section C.3.3.1, "Serial Protocols") to communicate with other devices. The multi-CPU system shown in Figure C.1 is a good example of this type of system, where each instrument is a Message Based Device, and is therefore capable of receiving instructions from a host or a common host interface. Note that by agreeing on a class of communication protocols, such as Word Serial, different manufacturers can now build any of those instruments and be assured of compatibility with the rest of the system. Higher level communication protocols (referred to as Shared Memory Protocols in Figure C.2) can later be defined using these platforms.

Communication between VXIbus devices is based on a hierarchical device relationship involving COMMANDERS and SERVANTS. For many systems (such as the single CPU system shown in Figure C.1) only the first level of this hierarchy exists. In that example, the CPU and host interface device is a Commander which controls the 3 Instrument Servants. A Commander is able to initiate communication with its Servants based on their capabilities. If the Servants are Message Based Devices, such communication can take place using **Commands**, which are described by the Word Serial Protocols. Communication with Register Based Servants is device dependent, and may vary between systems. The Commander/Servant hierarchy can be nested, since a Message Based Device may be a Commander, as well as a Servant to the next level above it. The hierarchical instrument system of Figure C.1 is an example of such a system. Instrument #1 and #2 are shown as multi-device instruments that have two Servants each, while the instruments themselves may be Servants to the host interface.

A common element that instruments may desire to share among them is the interface to the host computer, as shown in the multi-CPU system example in Figure C.1 The 488-VXIbus interface device (see Section D.2, "488-VXIbus Interface") is an example of such a device. VXIbus defines this as a function-specific Message Based Device, with communication protocols that support IEEE-488 functions. This device, if provided, may be shared by the instruments needing to be accessed by the host. Other host or peripheral interfaces such as RS-232, LAN, or Human Interfaces can be designed in a similar fashion.

Regardless of the topology of the VXIbus cardcage, (See Figure C.1) VMEbus's multiple-bus-master architecture demands a minimal amount of system-related functions that are performed by a central **Resource Manager** (see Section C.4.1, "Resource Manager") The following are some of the functions a Resource Manager might need to perform, depending on the topology of the system.

* *Configuring the Address map*. To configure the A24, A32 address maps and assign blocks of addresses to the devices requiring them.

* *Establishing the System Hierarchy.* In a multi-CPU system, the potential for more than one processor to try and control another device exists. The Commander/Servant hierarchies established by the Resource Manager prevent this type of conflict.

* *Allocating Shared System Resources.* A central Resource Manager (e.g. operating system) is needed in systems that share blocks of memory for communication, or share other hardware resources, such as trigger buses.

* *Performing System Self Test and Diagnostics.*

* *Initializing all Commanders in the system.*

C.2 DEVICE OPERATION

C.2.1 Device Overview

Devices are the simplest logical component in the VXIbus system. Normally a device will consist of one VXIbus module. However, multi-board devices and multi-device boards are permitted. There is a unique Logical Address associated with each device in the system. Examples of devices are computers, multimeters, multiplexers, oscillators, operator interfaces, and counters. VXIbus devices may be categorized by their supported protocols into four classes, as illustrated in Figure C.3.

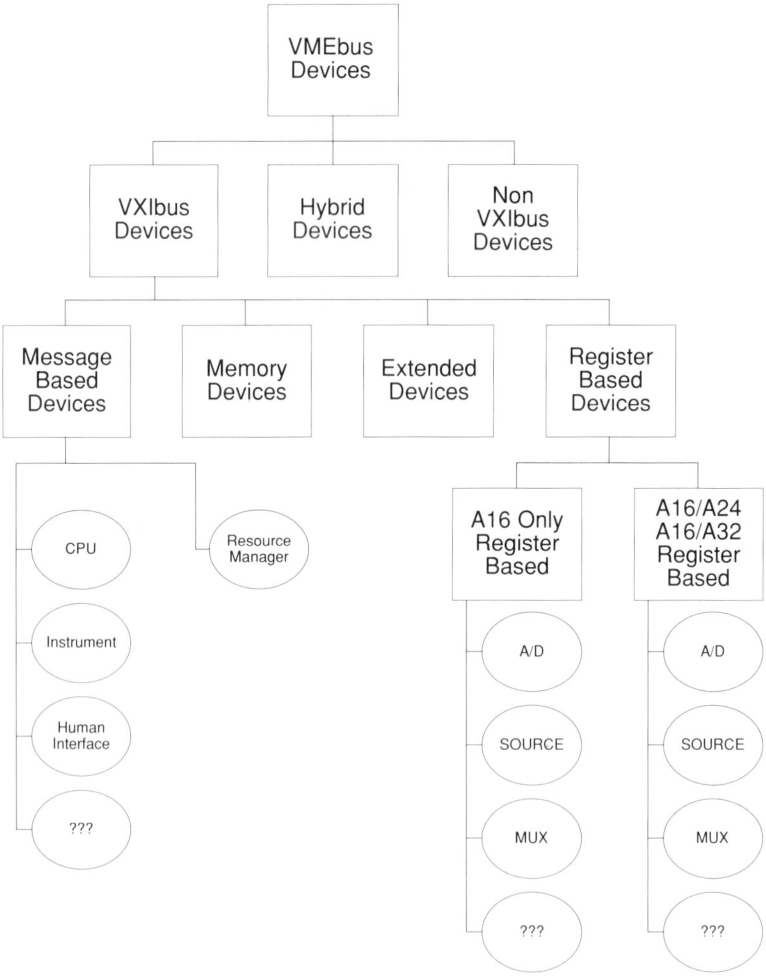

Figure C.3. Device Classifications

* *Message Based Devices* support the VXIbus configuration and communication protocols. This category includes only devices with Commander and/or command-based Servant elements. Examples of Message Based Devices are any devices with local intelligence that require a level of communication capability such as DMM's, Spectrum Analyzers, Display Controllers, 488-VXIbus interface devices, Switch controllers, etc.

* *Register Based Devices* support VXIbus register maps, but no VXIbus communication protocols. This category is comprised of devices having Register Based Servant elements. Typically Register Based Devices are simple inexpensive devices, such as simple switches, digital I/O cards, simple serial interface cards and, in general, any card that requires little or no local intelligence. More sophisticated devices could also be implemented as Register Based Devices that rely on Message Based Devices for their instructions.

* *Memory Devices* have configuration registers, and contain certain attributes of a Memory Device such as type and access time, etc. but have no other VXIbus defined registers or protocols. Bubble memory, RAM and ROM cards are in this category.

* *Extended Devices* are special-purpose VXIbus devices that have configuration registers so that they can be identified by the system. This category of devices will allow for definition of future device classes to support further levels of compatibility.

Hybrid Devices are VMEbus-compatible devices that know about VXIbus devices, and have the ability to communicate with them (or make use of them) but don't themselves comply with VXIbus device requirements. Existing VMEbus boards with software to make use of VXIbus devices are an example of this class of device.

Non-VXIbus devices are VMEbus devices that do not comply with, or use, any of the VXIbus requirements.

C.2.1.1 Device Slave Capabilites

All VXIbus devices have a common set of VMEbus slave attributes. These include standard memory maps, common addressing modes, and standard register definitions.

C.2.1.1.1 Device Addressing

All VXIbus devices have registers located within 64-byte blocks in the A16 address space. The base address of a device's registers is determined by the device's unique Logical Address. The Logical Address selector is an eight-bit selector which has 256 unique settings. The Logical Address is the value indicated by this selector, and corresponds to bits 6 - 13 of the device register base address. Bits 14 and 15 of the base address are both 1, and the base address is

$$(V \times 64) + 49152$$

where V is the device's Logical Address.

If a device requires fewer than 64 bytes of address space, all of it may reside within the defined 64-byte block. Otherwise, it has additional registers in the A24 or A32 address space. The base address of a device's A24 or A32 registers is programmed via an Offset register, in the configuration portion of its A16 registers.

RULE C.2.1:
Every VXIbus device **SHALL** have a non-volatile means of selecting Logical Addresses and implement the defined A16 addressing scheme.

RULE C.2.2:
IF a VXIbus device has A24 or A32 registers,
THEN the base address of such registers **SHALL** be programmable via its Offset register in its A16 configuration area.

RULE C.2.3:
All VXIbus slaves **SHALL** implement A16 addressing modes, and have configuration registers available in the A16 address space.

PERMISSION C.2.1:
VXIbus slaves **MAY** also implement A24 or A32 addressing modes.

RULE C.2.4:
IF a VXIbus slave implements A24 addressing modes
THEN it **SHALL NOT** implement A32 addressing modes.

OBSERVATION C.2.1:
IF a VXIbus slave implements A32 addressing modes
THEN it cannot implement A24 addressing modes.

RULE C.2.5:
VXIbus slaves **SHALL** implement D16.

RECOMMENDATION C.2.1:
VXIbus slaves should implement D08 (EO).

PERMISSION C.2.2:
VXIbus slaves **MAY** implement D32.

IF	THEN					
	A16	A24	A32	D08(EO)	D16	D32
SLAVE A16	•	MAY	MAY	REC.	SHALL	MAY
SLAVE A24	SHALL	•	N.A.	REC.	SHALL	MAY
SLAVE A32	SHALL	N.A.	•	REC.	SHALL	MAY

N.A.- Not Allowed
REC. - Recommended

C.2.1.1.2 Configuration Registers

The following diagram contains the register map for the Configuration registers of a VXIbus device. These registers are located in the A16 address space by the Logical Address selector, as described previously.

$3F_{16}$	DEVICE DEPENDENT REGISTERS
$1F_{16}$	DEVICE CLASS DEPENDENT REGISTERS
06_{16}	Offset Register
04_{16}	Status/Control Register
02_{16}	Device Type
00_{16}	ID/Logical Address Register

The fields of the above table are defined as follows:

ID Register: A read of this 16-bit register provides information about the device's configuration. Its contents are defined in the following table.

Bit #	15 - 14	13 - 12	11 - 0
Contents	Device Class	Address Space	Manufacturer ID

- *Device Class*: This field indicates the classification of the VXIbus device according to the following table.

Value	Class
00	Memory
01	Extended
10	Message Based
11	Register Based

- *Address Space*: This field indicates the addressing mode(s) of the device's operational registers, according to the following table:

Value	Mode
00	A16/A24
01	A16/A32
10	RESERVED
11	A16 Only

- *Manufacturer ID*: This number uniquely identifies the manufacturer of the device. The list of ID numbers is maintained by the VXIbus Consortium. Each VXIbus device manufacturer has exactly one Manufacturer ID number. Numbers are assigned to manufacturers in decreasing order, beginning with number 4095. See Section A.1, "Manufacturer ID Numbers" for information on obtaining a manufacturer ID number.

OBSERVATION C.2.2:
ID number 3840 ($F00_{16}$) is reserved for user-customized devices, and any other unique special-function device.

Logical Address: This 16-bit write register is defined by the optional Dynamic Configuration protocol. See Section F, "Dynamic Configuration".

Device Type: This 16-bit register contains a device-dependent type identifier. The read contents of this register are defined below:

Bit #	15 - 12	11 - 0
Contents	Required Memory	Model Code

- *Required memory* [Only required for A16/A24 and A16/A32 devices] These 4 bits contain a number *m*, which is between 0 and 15. The required memory usage is defined as

$$256^{a} \times 2^{(23-m)}$$

where *a* is the value of the *Address Space* field in the ID register. This equation gives the amount of A24 or A32 memory space (in bytes) resident on the device. This also means that the device must decode addresses to this resolution.

OBSERVATION C.2.3:
The above algorithm for determining memory requirements allows devices with minimal memory requirements to let the *Required Memory* bits float to a high level, and minimizes hardware requirements for low-cost devices.

In the case of an A16-only device, these four bits are the upper bits of the Model Code.

OBSERVATION C.2.4:
The minimum size of an A24 device's memory is 256 bytes. The maximum size of the memory map is 1/2 of the complete A24 VMEbus address space.

OBSERVATION C.2.5:
The minimum size of an A32 device's memory is 65536 bytes (64 Kbytes). The maximum size of the memory map is 1/2 of the complete A32 VMEbus address space.

SUGGESTION C.2.1:
It would be a good idea to limit the memory space of any device to 1/4 of the available address range for the device. This would facilitate having more than one device in each possible address space.

- *Model Code:* This field contains a unique card identifier, which is defined by the manufacturer. In the case of an A16-only device, this field occupies all 16 bits of the Device Type register.

OBSERVATION C.2.6:
Model codes 0-255 ($0\text{-}FF_{16}$) are reserved for Slot 0 devices. See Rules C.4.16 and C.4.17 in Section C.4.3, "VXIbus Subsystem Slot 0". A write to the Device Type register location has no effect. Its use is reserved for VXIbus definition.

Status Register: A *read* of this register provides information about the Device's status, as defined in the following table.

Bit #	15	14	13 - 4	3	2	1 - 0
Contents	A24/A32 Active	MODID*	Device Dependent	Ready	Passed	Device Dependent

- *A24/A32 Active:* This bit is only required for A16/A24 or A16/A32 devices. A one in this field indicates that its A24 or A32 registers can be accessed. This bit reflects the state of the Control register's A24/A32 *Enable* bit.

RULE C.2.6:
The *A24/A32 Active* bit in an A16/A24 and A16/A32 device **SHALL** be cleared (set to zero) whenever SYSRESET* is asserted, and changed only by a write of the A24/A32 *Enable* bit in the device's Control register.

OBSERVATION C.2.7:
In an A16-only device the *A24/A32 Active* bit is device dependent.

- *MODID*:* A one in this field indicates that the device is not selected via the P2 MODID line. A zero indicates that the device is selected by a high state on the P2 MODID line.

RULE C.2.7:
IF a device does not implement MODID capability (P1 only device)
THEN the device's MODID* bit **SHALL** always be set to one.

RULE C.2.8:
IF A device implements MODID capability,
THEN the device's *MODID** bit **SHALL** always be set or reset within 1 μsec of the module's MODID line being set or reset.

- *Ready:* This field serves two purposes:

1. After the power-on initialization sequence (see Section C.2.1.2, "Device Initialization and Diagnostics") a one (in this field in combination with a zero in the *Passed* field) indicates that the device has failed register initialization.

2. After the power-on initialization sequence, a one in this field (in combination with a one in the *Passed* field) indicates that the device is ready to execute its full set of operational commands.

OBSERVATION C.2.8:
The *Ready* bit indicates that a device is ready to accept its full set of operational commands. This "readiness" is determined in a device-dependent manner. A standard definition of readiness for Message Based Devices is given in Section C.2.4.5, "Message Based Device Configuration".

- *Passed:* A zero in this field indicates that the device is either executing or has failed its self test. A one indicates that the self test has successfully completed.

RULE C.2.9:
IF a device does not implement a power-on self test,
THEN its *Passed* bit **SHALL** always be set to one.

PERMISSION C.2.3:
The device-dependent bits of the Status register **MAY** be defined to indicate the state of the corresponding bits of the Control register.

Control Register: A write to this register causes specific actions to be executed by the device.

These actions are described in the following table.

Bit #	15	14 - 2	1	0
Contents	A24/A32 Enable	Device Dependent	Sysfail Inhibit	Reset

- *A24/A32 Enable:* A one in this field enables access to the device's A24 or A32 VMEbus registers. A zero disables such access.

- *Device dependent:* The bits in this field are available to the device designer for device-specific and/or application-specific control.

OBSERVATION C.2.9:
A Resource Manager which lacks device-specific knowledge of a device will always write a one to each of the device-dependent bits of a device's Control register whenever it writes to that Control register to change any of the VXIbus defined bits.

RECOMMENDATION C.2.2:
The device-dependent bits of the Control register should be implemented so that writing ones to these bits will not disturb the initialized state of the device. This will permit a Resource Manager or Commander (which lacks device-specific knowledge) to write the VXIbus-defined bits without causing some undesired activity within the device.

- Sysfail Inhibit: A one in this field disables the device from driving the SYSFAIL* line.

- Reset: A one in this field forces the device into a reset state.

RULE C.2.10:
IF a device's Commander writes a one to the device's *Reset* bit,
THEN the Commander must not write a zero to that device's *Reset* bit within the following 100 µsec.

See Section C.2.1.2, "Device Initialization and Diagnostics", for more information on the operation of the above bits.

Offset Register: This register is needed only for A16/A24 and A16/A32 devices. This 16-bit read/write register defines the base address of this device's A24 or A32 operational registers. The m+1 most significant bits of the Offset register are the values of the m+1 most significant bits of the device's A24 or A32 register addresses, where m is the value of the *Required Memory* field of the device's Device Type register. The (15-m) least significant bits of the Offset register are meaningless. Thus, the Offset register bits 15 through (15-m) map to the address lines A_{23} through A_{23-m} for A24 registers, or to lines A_{31} through A_{31-m} for A32 registers.

OBSERVATION C.2.10:
In the case of an A16-only device, the Offset register's location is a device-dependent operational register.

OBSERVATION C.2.11:
A read of the Offset register always returns the address offset most recently written to the Offset register.

OBSERVATION C.2.12:
The Dynamic Configuration protocol defines an additional use of the Offset register. See Section F, "Dynamic Configuration".

C.2.1.1.3 Device Class Dependent Registers

The definition of the VXIbus Device Class Dependent Registers varies with the class of the device, i.e Message Based, Register Based, etc..

C.2.1.1.4 Device-dependent Registers

VXIbus Device-dependent Registers are user definable.

C.2.1.1.5 Address Modifiers

RULE C.2.11:
VXIbus device A16 registers **SHALL** respond to address modifiers 29_{16} and $2D_{16}$ only.

RULE C.2.12:
VXIbus device A24 registers **SHALL** respond to address modifiers $3D_{16}$ and $3E_{16}$.

PERMISSION C.2.4:
VXIbus device A24 registers **MAY** respond to any of address modifiers 39_{16}, $3A_{16}$, $3B_{16}$, and $3F_{16}$.

RULE C.2.13:
VXIbus device A24 registers **SHALL NOT** respond to any address modifier other than 39_{16}, $3A_{16}$, $3B_{16}$, $3D_{16}$, $3E_{16}$, or $3F_{16}$.

RULE C.2.14:
VXIbus device A32 registers **SHALL** respond to address modifiers $0D_{16}$ and $0E_{16}$.

PERMISSION C.2.5:
VXIbus device A32 registers **MAY** respond to any of address modifiers 09_{16}, $0A_{16}$, $0B_{16}$, or $0F_{16}$.

RULE C.2.15:
VXIbus device A32 registers **SHALL NOT** respond to any address modifier other than 09_{16}, $0A_{16}$, $0B_{16}$, $0D_{16}$, $0E_{16}$, or $0F_{16}$.

OBSERVATION C.2.13:
The preceding rules prevent any overlapping of the defined VMEbus address spaces. They do not prohibit the use of VMEbus user-defined address modifiers within a VXIbus system.

C.2.1.2 Device Initialization And Diagnostics

The VMEbus provides a minimal self-test reporting facility (SYSFAIL*). VXIbus devices implement an extended reporting strategy that addresses the ambiguities of the VMEbus specification. In addition, the VXIbus initialization process provides for a VXIbus device to initialize its own configuration registers.

C.2.1.2.1 Overview

The VXIbus initialization sequence provides the following services:

- Assertion of SYSFAIL* at power-on to indicate a self-test condition.
- Status register indication of self-test status.
- Optional faceplate LED indication of self-test status, labeled Failed.
- Time limit for completion of a normal self-test.
- Control register bits for disabling SYSFAIL* driver and resetting the device.
- A waiting period for initialization of the device's configuration registers.

The initialization sequence is implemented with an internal state machine, the *Passed* and *Ready* bits of the Status register, the *Reset* and *Sysfail Inhibit* bits of the Control register, a VMEbus SYSFAIL* driver, and the Failed LED.

A typical power-on initialization sequence proceeds as follows.

1. At power-on SYSRESET* is asserted, forcing the device into a reset condition. At this time the *Passed*, *Reset*, *Ready*, and *Sysfail Inhibit* bits are all cleared to zero. The device asserts SYSFAIL* and lights its Failed LED (if present).

2. When SYSRESET* is de-asserted, the device begins its self test. This self test initializes the device's capability to execute VXIbus communication.

3. If the device fails its self test, it leaves its *Passed* and *Ready* bits cleared, SYSFAIL* asserted, and its Failed LED lighted. After 4.9 seconds, these conditions indicate a self-test failure.

4. When the device successfully completes its self test, it sets its *Passed* bit to one, de-asserts SYSFAIL*, and turns off its Failed LED (if present). It is then ready to commence VXIbus communications.

5. If the device requires further initialization (as in the case of a Message Based Device) its *Ready* bit remains cleared. When it is ready to accept its normal operational commands, it sets its *Ready* bit to one.

6. A VMEbus Master may turn off a failed device's SYSFAIL* driver by setting its *Sysfail Inhibit* bit to one. It may also force the device into a reset condition by setting the device's *Reset* bit.

7. When the *Reset* bit is cleared, the device begins another self-test sequence.

A device may initialize its configuration registers any time within the 4.9 seconds after the de-assertion of SYSRESET*, typically in the following sequence.

1. At power-on, SYSRESET* is asserted, forcing the device into a reset condition and disabling all VMEbus access to its configuration registers. The device asserts SYSFAIL* and lights its Failed LED (if present).

2. When SYSRESET* is de-asserted, the device begins initialization of its configuration registers. At this point VMEbus access remains disabled, SYSFAIL* remains asserted, and the Failed LED is lighted (if present).

3. After initializing the *Passed*, *Reset*, and *Sysfail Inhibit* bits to zero and initializing the *Ready* bit to one, the device may enable VMEbus access to its configuration registers.

4. If the device is unable to initialize the *Passed*, *Reset*, *Sysfail Inhibit*, and *Ready* bits, its configuration registers will remain inaccessible to the VMEbus.

5. After initializing all of its configuration registers, the device enables VMEbus access to its registers (if it hasn't already), clears the *Ready* bit, and begins its self test.

6. If the device fails to completely initialize its configuration registers, it enables VMEbus access to its registers (if it hasn't already), leaves its *Passed* bit cleared, its *Ready* bit set, SYSFAIL* asserted, and its Failed LED lighted (if present). After 4.9 seconds, these conditions indicate an initialization error.

7. When the device successfully completes both its configuration register initialization and its self test, it sets its *Passed* bit to one, de-asserts SYSFAIL*, and turns off its Failed LED (if present). It is then ready to commence VXIbus communications.

C.2.1.2.2 Self Test Operation

The initialization sequence is characterized by eight states, as illustrated in Figure C.4.

- *Hard Reset:* This is the state of the device when the SYSRESET* line is true. While in this state the device is inactive and its Status and Control registers may be initialized. The SYSFAIL* line is driven low, and the failed LED is lighted.

- *Config Reg Init:* During this optional state, the device configures its own configuration registers. This may include setting its Logical Address and its ID and Device Type register values, as well as initializing its Status, Control, and Offset registers. Initially, a device's configuration registers may be inaccessible via the VMEbus. In this state, the device continues to assert SYSFAIL* and its optional Failed LED is lighted. The CONFIF REG INIT state, and any additional self test, must complete within 4.9 seconds of the SYSRESET* de-assertion or the clearing of the *Reset* bit. If the configuration register initialization fails before the device enables VMEbus access to its configuration registers, the device remains in the CONFIG REG INIT state until the next Hard Reset. Before register access is enabled, the *Passed* bit is set to zero, and the *Ready* bit is set to one. The ID and Device Type registers may be initialized either before or after register access is enabled.

- *Self Test:* In the SELF TEST state, the device tests and initializes the functionality necessary for VXIbus communications. This test must complete within 4.9 seconds of the SYSRESET* de-assertion or the clearing of the *Reset* bit. If the self test fails, the device enters the FAILED state. If it passes, it enters the PASSED state. While in the SELF TEST state, the device's VMEbus interface is active. However the device is incapable of responding to any commands other than *Reset* and *Sysfail Inhibit*. While in the SELF TEST state, the *Passed* bit remains zero, the *Ready* bit is also cleared to zero, and SYSFAIL* remains asserted.

- *Init Failed:* This state is entered if the configuration register initialization fails after the device has enabled access of its configuration registers. While in the INIT FAILED state, SYSFAIL* is asserted, the *Passed* bit is set to zero, and the *Ready* bit is set to one. This state may be exited by either a Hard Reset or a Soft Reset.

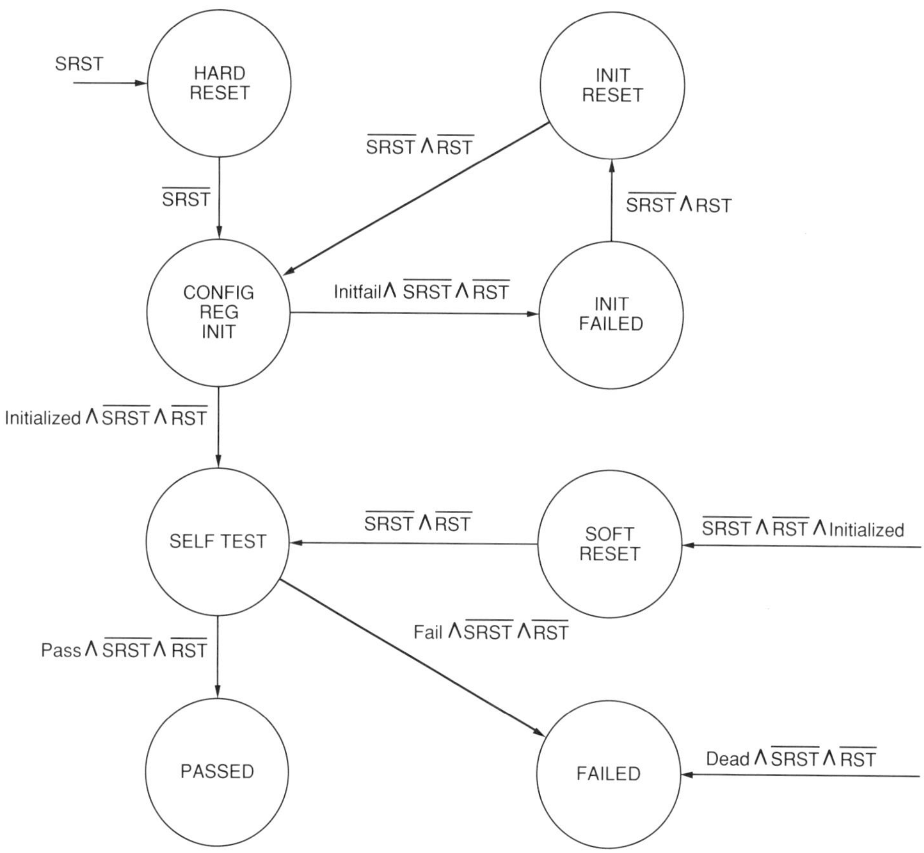

Figure C.4. Self Test State Diagram

OBSERVATION C.2.14:
The output indications are identical for the CONFIG REG INIT and INIT FAILED states. Therefore a completely failed device that is incapable of even beginning to initialize its configuration registers will correctly indicate an INIT FAILED status.

- *Failed:* This state is entered if the self test fails. While in the FAILED state, the *Passed* bit and the *Ready* bit remain zero, and SYSFAIL* remains asserted. This state may be exited by either a Hard Reset or a Soft Reset.

OBSERVATION C.2.15:
The output indications are identical for the SOFT RESET, SELF TEST, and FAILED states. Therefore a completely failed device that is incapable of even beginning a self test will correctly indicate a failed status.

- *Passed*: This state is entered when the self test has successfully terminated. The SYSFAIL* line is de-asserted, and the *Passed* bit of the Status register is set to one. While in the PASSED state, the device is capable of fully supporting VXIbus communications.

OBSERVATION C.2.16:
The PASSED state can be further subdivided in a device-dependent manner. For Message Based Devices three sub-states are defined in Section C.2.4.4, "Message Based Device Operation".

- *Soft Reset:* This state is entered when the *Reset* bit of the Control register is set to one. While in this state, a device is inactive, interrupts which are pending are unasserted, and all pending bus requests are removed. While in this state the device's VMEbus slave interface is active, however the device is incapable of responding to any commands other than *Reset* and *Sysfail Inhibit*. While in this state the Status register must accurately reflect the state of the device.

OBSERVATION C.2.17:
The VMEbus Specification (Section 3.3.2, "Requester") defines the protocols for proper removal of bus requests.

- *Init Reset:* This state is similar to the SOFT RESET state, except that the device's configuration registers have not been initialized (and therefore are meaningless).

The mnemonics used in the self test state diagram are listed in the following table.

SELF TEST Mnemonics			
Mnemonic		**Name**	**Type**
SRST	=	SYSRESET* asserted (low)	VMEbus line
RST	=	Reset (1=true)	Control register bit
Fail	=	Self Test Failed	Internal signal
Pass	=	Self Test Passed	Internal signal
Dead	=	Device Failure	Internal signal
Initialized	=	Config Regs Initialized	Internal signal
Initfail	=	Init Config Regs Failed	Internal signal

OBSERVATION C.2.18:
The *Fail*, *Pass*, and *Dead* signals are generated by the Initialization process as sequencing signals to itself.

OBSERVATION C.2.19:
The *Dead* signal is generated by the device whenever it detects a catastrophic failure during normal operation.

The states of the initialization sequence produce the Status register indications given in the following table.

SELF TEST Status Bits		
State	**Status Register Bits** **Passed**	**Ready**
HARD RESET	(0)	(0/1)
CONFIG REG INIT	(0)	(1)
INIT FAILED	0	1
SELF TEST	0	0
FAILED	0	0
PASSED	1	0/1
SOFT RESET	0	0
INIT RESET	0	1

OBSERVATION C.2.20:
The Status register is inaccessible during the HARD RESET state. However, in many cases its bits will be initialized by SYSRESET* to indicate the device's state (CONFIG REG INIT or SELF TEST) immediately following HARD RESET.

OBSERVATION C.2.21:
The device's configuration registers may be inaccessible throughout the CONFIG REG INIT state. The Status register indications shown are meaningful only after VMEbus access has been enabled.

The status of the SYSFAIL* driver and the optional failed LED may be derived from the states of the *passed* status register bit and the *Sysfail Inhibit* control register bit, as described in the following table.

SELF TEST Indicators			
Register Bits		**Indicators**	
Passed Status Bit	*Sysfail Inhibit Control Bit*	*SYSFAIL* Driver*	*FAILED LED*
0	0	LOW	ON
0	1	HIGH	ON
1	X	HIGH	OFF

The VMEbus SYSRESET* line and the Control register interact to force the device into particular states, as described in the following table.

SYSRESET* and Control Register Interactions			
SYSRESET* Line	Reset Bit	Current State	Effect
LOW	X	X	HARD RESET state
↑	X	HARD RESET	Begin CONFIG REG INIT (if implemented) or SELF TEST
HIGH	1	SELF TEST, PASSED, or FAILED	SOFT RESET State
HIGH	1	INIT FAILED	INIT RESET State
HIGH	↓	SOFT RESET	Begin SELF TEST
HIGH	↓	INIT RESET	Begin CONFIG REG INIT

RULE C.2.16:
All devices implementing power-on self tests **SHALL** implement the HARD RESET, SELF TEST, PASSED, FAILED, and SOFT RESET states of the self test function.

RULE C.2.17:
Any device not implementing the initialization sequence **SHALL NOT** assert SYSFAIL* or clear its *Passed* bit to zero at any time.

RULE C.2.18:
A device **SHALL NOT** implement a CONFIG REG INIT and SELF TEST combination of more than 4.9 seconds total duration.

RULE C.2.19:
A device's ID and Device Type registers **SHALL** be valid before it enters the SELF TEST state.

PERMISSION C.2.6:
A device **MAY** execute self test operations while in the CONFIG REG INIT state, provided that its *Ready* bit is set to one, until its configuration registers have been initialized.

RULE C.2.20:
A device's VXIbus interface **SHALL** be fully operational before entering the PASSED state.

PERMISSION C.2.7:
A device **MAY** enter the FAILED state any time it detects a catastrophic failure during normal operation.

OBSERVATION C.2.22:
The *Sysfail Inhibit* bit of the Control register is used to disable the device's SYSFAIL* driver.

OBSERVATION C.2.23:
The Failed LED is optional. Other LED's such as a SYSFAIL* inhibit indicator may also be optionally implemented.

OBSERVATION C.2.24:
Once a device has enabled access to its configuration registers, they will remain accessible until the next Hard Reset.

RULE C.2.21:
Once a device enters either the INIT FAILED or SELF TEST state, it **SHALL NOT** change its Logical Address, except as provided for by Dynamic Configuration (see Section F, "Dynamic Configuration"), until the next Hard Reset.

OBSERVATION C.2.25:
A Soft Reset will not affect a device's Logical Address, configuration register accessibility, Offset register, or *A24/A32 Active* bit.

RULE C.2.22:
After either a Hard Reset or a Soft Reset, a device **SHALL** remain deactivated until activated by its Commander.

RULE C.2.23:
A deactivated device **SHALL NOT** assert any VMEbus IRQ lines, or initiate any data transfer cycles.

OBSERVATION C.2.26:
A Commander activates a device in a device-dependent manner. A standard method of activating Message Based Devices is given in Section C.2.4.5, "Message Based Device Configuration".

RULE C.2.24:
A VXIbus device, other than a Resource Manager (see Section C.4.1, "Resource Manager"), **SHALL NOT** autonomously activate itself.

C.2.1.3 Priority Interrupts

Use of priority interrupts is optional for VXIbus devices. However all VXIbus devices recognize a standard format for the STATUS/ID word that is provided in response to interrupt acknowledge cycles.

OBSERVATION C.2.27:
Priority interrupts provide a mechanism for routing information between a device with an Interrupter and a device with an Interrupt Handler (on the same IRQ line). The Interrupt Handler might not be the intended recipient of the information. If an Interrupt Handler receives an interrupt which it recognizes to be intended for another device, it may forward the information carried by the interrupt to the intended recipient, via either a Signal or another interrupt.

C.2.1.3.1 Interrupters

RULE C.2.25:
VXIbus devices with Interrupter capability **SHALL** allow the user to configure which Interrupt Request line (IRQ1*-IRQ7*) is used.

RECOMMENDATION C.2.3:
VXIbus Interrupters should implement Release On Acknowledge (ROAK).

OBSERVATION C.2.28:
Message Based Devices that are Interrupters are required to implement ROAK. See Section C.2.4, "Message Based Devices".

RULE C.2.26:
Interrupters which implement D16 STATUS/ID Transfer capabilities **SHALL** also implement D08(O) STATUS/ID Transfer capabilities.

RULE C.2.27:
Interrupters which implement D32 STATUS/ID Transfer capabilities **SHALL** also implement D16 STATUS/ID Transfer capabilities and D08(O) STATUS/ID Transfer capabilities.

RULE C.2.28:
VXIbus devices **SHALL** implement combined D08(O), D16, and D32 interrupt response modes such that the size of the STATUS/ID word is selected according to the type of interrupt acknowledge cycle detected, according to the following table:

STATUS/ID Word Size			
Cycle Type	D08	Combined Mode D08/16	D08/16/32
8 Bit	8	8	8
16 Bit	8	16	16
32 Bit	8	16	32

OBSERVATION C.2.29:
The VMEbus specification requires that D08(O) Interrupters respond to 8-bit, 16-bit, and 32-bit interrupt acknowledge cycles with an 8-bit STATUS/ID word. D16 Interrupters respond to 16-bit and 32-bit cycles with a 16-bit word.

OBSERVATION C.2.30:
The preceding three rules guarantee that (no matter what width STATUS/ID read cycle the Interrupt Handler generates) the Interrupter will respond with the appropriate width of STATUS/ID.

OBSERVATION C.2.31:
Many Interrupt Handlers generate only D08(O) STATUS/ID read cycles. The preceding three rules guarantee that Interrupters which respond to D16 and D32 STATUS/ID cycles will also be able to respond to Interrupt Handlers which request D08(O) responses.

RULE C.2.29:
All VXIbus Interrupters **SHALL** provide STATUS/ID responses to interrupt acknowledge cycles according to the format defined in the following table.

Bit #	31 - 16	15 - 8	7 - 0
Contents	D32 Extension	Cause/ Status	Logical Address

The fields of the STATUS/ID word are defined below:

- *D32 Extension:* These optional bits may be provided by a D32 Interrupter, in order to give additional information about its status or the cause of the interrupt. The value $FFFF_{16}$ is reserved to mean *No Extension Given.* Otherwise the format of this field is unspecified.

OBSERVATION C.2.32:
If a D08 (or D16) Interrupt Handler reads a 32-bit STATUS/ID word, the upper 24 (or 16) bits of data will be lost.

- *Cause/Status:* These bits give the cause of the interrupt and/or reflect the Interrupter's Status register. The value FF_{16} is reserved to mean *No Cause/Status Given.*

OBSERVATION C.2.33:
If a D08 Interrupt Handler reads a 16-bit STATUS/ID word, the upper 8 bits of data will be lost.

- *Logical Address:* This field contains the Logical Address of the Interrupter device.

RECOMMENDATION C.2.4:
A VXIbus Interrupter should provide a means of masking VMEbus interrupts.

C.2.1.3.2 Interrupt Handlers

RULE C.2.30:
VXIbus devices with Interrupt Handler capability **SHALL** allow the user to configure which Interrupt Request line (IRQ1*-IRQ7*) is monitored.

RECOMMENDATION C.2.5:
VXIbus Interrupt Handlers should implement D16 or D32 interrupt acknowledge cycles.

SUGGESTION C.2.2:
For compatibility with Non-VXIbus Interrupters, design VXIbus Interrupt Handlers to include D08(O) interrupt acknowledge cycle generation capability.

C.2.1.4 VMEbus Master Capabilities

As discussed in the specific device class sections, Master capability is optional for Register Based and Extended VXIbus Devices, prohibited for Memory Devices, and required for some Message Based Devices. The capabilities are intended to ensure compatibility with all VXIbus slaves, as well as with many non-VXIbus slaves.

C.2.1.4.1 Bus Requesters

VXIbus devices implement a fair requester protocol, which requires that a level *n* requester (after receiving a bus grant via BG*n*IN*) may not request the bus again until at least 30 nsec after the next low-to-high transition of Bus Request line BR*n**. This forces a Round Robin type of bus management on a single Bus Request/Grant level. This protocol is also known as Request on No Request (RONR).

RULE C.2.31:
All VXIbus devices with Master capability **SHALL** implement the Fair Requester protocol.

OBSERVATION C.2.34:
The above rule provides a more egalitarian environment than the standard VMEbus bus request protocols. This matches instrumentation requirements better than the default prioritization of devices provided by the daisy chains.

RULE C.2.32:
VXIbus devices with Master capability **SHALL** use Bus Request/Grant level 3 as a default.

RECOMMENDATION C.2.6:
VXIbus devices with Master capability should allow the user to configure which Bus Request/Grant level is used.

PERMISSION C.2.8:
VXIbus devices with Master capability **MAY** provide the capability of selecting another type of Requester.

OBSERVATION C.2.35:
The above permission allows designers of high performance boards with real time constraints to implement bus locking. Not implementing a Fair Requester can cause system performance difficulties.

RECOMMENDATION C.2.7:
IF a device has the Fair Requester disabled
THEN it should be moved to another Bus Request/Grant level.

RECOMMENDATION C.2.8:
A VMEbus Master should implement cycle-by-cycle arbitration. A Master should not hold the bus while doing computation. In the case of a read from a device, a subsequent operation, and then write to another device, the Master should allow arbitration of the VMEbus after the initial read.

PERMISSION C.2.9:
A VMEbus Master **MAY** maintain control of the data transfer bus for the time required to complete a Block Transfer Cycle.

OBSERVATION C.2.36:
The VMEbus Specification allows for the possibility of a Master holding the data transfer bus (DTB) indefinitely.

C.2.1.5 Terminating Operation

A functioning VXIbus device terminates its operation in an orderly and consistent manner. This allows for putting devices off-line without adversely affecting other operating devices. It also allows for efficient reconfiguration of the VXIbus hierarchy. Four different methods are available to a Commander or a Resource Manager for terminating the operations of a device.

- *Hard Reset:* When the SYSRESET* line goes true, all devices immediately terminate their operation. Hard Reset is used only when it is necessary to promptly terminate the operation of all devices in a VXIbus system. A Hard Reset will result in the system being configured by the Resource Manager.

RECOMMENDATION C.2.9:
Asserting SYSRESET* during operation is a drastic measure, and should only be initiated to emulate a power-up cycle.

- *Soft Reset:* A Commander may terminate the operation of one of its Servants by writing a one to the *Reset* bit in the Servant's Control register.

RULE C.2.33:
A Commander **SHALL NOT** Soft Reset any device not in its Servant list.

PERMISSION C.2.10:
The Resource Manager **MAY** Soft Reset a Top Level device.

- *VXIbus commands:* A Commander may terminate the operation of a Message Based Servant by sending it one of a number of low-level VXIbus commands. Refer to Section C.2.4.7, "Terminating Operation", for a detailed description of these commands.

RULE C.2.34:
Only a device's Commander may terminate the operation of that device by sending it a VXIbus device termination command.

- *Device-dependent* termination: A Commander may terminate the operation of a non Message Based Device in a device-dependent manner, known to that Commander. The Servant ends up in a well-defined state (refer to Section C.2.2.4, "Terminating Operation").

RULE C.2.35:
A Commander **SHALL NOT** use a device-dependent method to terminate the operation of a Message Based Device.

RULE C.2.36:
When a device is terminated by a Soft Reset, a VXIbus command, or in a device-dependent manner, it **SHALL** retain the contents of its ID register, Device Type register, Offset register, Logical Address, and the state of its A24/A32 enable bit.

C.2.2 Register Based Devices

A Register Based Device is generally a slave-only device. Communication with one of these devices is usually accomplished via reads and writes of its registers. However, Register Based Devices may also use interrupts. One example of this class of device is a relay multiplexor device, which gets switched by writes to specific device registers. Register Based Devices are the simplest VXIbus devices, and are suitable for low-cost implementations.

C.2.2.1 Data Transfer Capabilities

OBSERVATION C.2.37:
The VXIbus protocols provide no standard mechanism for identifying Register Based Devices with Master capability, or for the regulation of the address space accessed by such devices. It is recognized that limited Master capability may, in some cases, be important for Register Based Servants. However, a Register Based Device with Master capability could make errant writes into unprotected address spaces unless such accesses are very closely guarded (such as in the case of a DMA capable disc interface). Care should be exercised to prevent the autonomous operation of a Register Based Device outside of the limits defined by its Commander.

C.2.2.2 Priority Interrupts

Register Based Devices are not required to use VMEbus interrupts. Some Register Based Devices may use interrupts for status reporting.

RECOMMENDATION C.2.10:
IF a Register Based Device has a status register bit to indicate a busy or not ready state, AND that bit can normally be asserted for periods longer than fifty microseconds,
THEN the device should have interrupt capability that may be enabled to be asserted at the end of the busy/not ready period.

OBSERVATION C.2.38:
The preceding recommendation is intended to minimize backplane traffic due to excessive polling of Status registers.

OBSERVATION C.2.39:
A Register Based Interrupter responds to interrupt acknowledge cycles with a STATUS/ID word conforming to the standard VXIbus format defined in Section C.2.1.3.1 "Interrupters".

The standard VXIbus STATUS/ID word format is illustrated in the following table.

Bit #	31 - 16	15 - 8	7 - 0
Contents	D32 Extension	Cause/ Status	Logical Address

OBSERVATION C.2.40:
The Cause/Status and D32 Extension fields of the STATUS/ID word are optional. Their bit definitions are entirely device dependent for Register Based Devices.

C.2.2.3 Register Based Device Registers

Register Based Devices have both VXIbus standard configuration registers and device-dependent operational registers. The location and quantities of these registers are determined from the device's Logical Address, and other information stored by its configuration registers (See Section C.2.1.1.2, "Configuration Registers"). The *Address Space* field of the ID register defines the location of the operational registers as follows.

A16 Only: The operational registers occupy addresses 06_{16} - $3F_{16}$ relative to the device's A16 base address.

A16/A24: The operational registers occupy addresses 08_{16} - $3F_{16}$ relative to the device's A16 base address and a block of A24 addresses located by the device's Offset register.

A16/A32: The operational registers occupy addresses 08_{16} - $3F_{16}$ relative to the device's A16 base address and a block of A32 addresses located by the device's Offset register.

OBSERVATION C.2.41:
All Register Based Device operational registers have device-dependent functionality.

C.2.2.4 Terminating Operation

Three different methods are available for terminating the operation of a Register Based Device. These are Hard Reset, Soft Reset and Device-dependent Termination. The orderly termination of device operation allows for deactivating a device without adversly affecting other operating devices. It also allows for efficient reconfiguration of the VXIbus hierarchy.

OBSERVATION C.2.42:
When a VXIbus Register Based Device is deactivated it may not assert any VMEbus IRQ lines or intiate any data transfers.

C.2.2.4.1 Hard Reset

Upon receipt of a Hard Reset (assertion of SYSRESET*), a Register Based Device enters, as does any other device class, the HARD RESET state. From that state a device continues its initialization until it ends in one of the possible states, as defined in Section C.2.1.2, "Device Initialization and Diagnostics". Hard Reset is used only when it is necessary to promptly terminate operation of all devices in a VXIbus system. A Hard Reset always affects the entire VXIbus system, and will result in the system being configured by the Resource Manager.

C.2.2.4.2 Soft Reset

A Commander may terminate the operation of a Register Based Servant by setting the Reset bit in the Servant's Control register to one. When the Commander clears the Reset bit, a Register Based Device enters (as does any other device class) the SELF TEST state (refer to Section C.2.1.2, "Device Initialization and Diagnostics"). If successful, the device will then enter the PASSED state. At this point, the device will be in a benign state, with no links to any other devices.

OBSERVATION C.2.43:
A Soft Reset is the only non-device-dependent method available to a Commander for terminating the operation of a Register Based Servant.

OBSERVATION C.2.44:
A Register Based Device that has been Soft Reset may be inconsistent with the remainder of the system. It is important that such a device not be activated until all inconsistencies have been resolved.

RULE C.2.37:
Before a Commander reactivates a device (after having Soft Reset that device) the Commander **SHALL** guarantee that the device is in a state consistent with the rest of the system.

OBSERVATION C.2.45:
A Register Based Device which has been Soft Reset is activated by its Commander in a device-dependent manner.

RULE C.2.38:
A Register Based Device which has been Soft Reset **SHALL** retain its assigned Logical Address if it has been dynamically configured (refer to Section F, "Dynamic Configuration"). It **SHALL** also retain any A24/A32 memory assignment.

C.2.2.4.3 Device-dependent Termination

Soft Resetting a device will cause it to go through its self test. It may be desirable for a Commander to simply deactivate a Register Based Device, without having to go through self test.

PERMISSION C.2.11:
A Register Based Device **MAY** have a device-dependent method other than Soft Reset that allows the device's Commander to terminate the device's operation.

OBSERVATION C.2.46:
A Register Based Device which has been deactivated in a device-dependent manner is also reactivated by its Commander in a device-dependent manner.

RULE C.2.39:
Before a Commander reactivates a device (after having terminated its operation in a device-dependent manner) the Commander **SHALL** guarantee that the device is in a state consistent with the rest of the system.

C.2.3 Memory Devices

VXIbus Memory Devices provide either permanent or temporary data storage in the form of blocks of ROM, RAM, etc. in the VMEbus A24 or A32 address space. The addresses of these memory blocks are configured via the VXIbus configuration registers.

C.2.3.1 Memory Device Registers

A Memory Device has both A16 configuration registers and A24 or A32 operational registers. In addition it has an ATTRIBUTE REGISTER, which indicates the nature of its operational memory registers.

C.2.3.1.1 Memory Device Attribute Register

The attribute register is a read-only register which lists important characteristics of the Memory Device. The following diagram contains the configuration register map for a Memory Device.

$3F_{16}$	DEVICE DEPENDENT REGISTERS
08_{16}	Attribute Register
00_{16}	CONFIGURATION REGISTERS

The Attribute Register has the following format.

Attribute Register:

Bit #	15 - 14	13	12	11	10 - 8	7	6 - 4	3 - 0
Contents	Memory Type	N/S	BT*	N_P*	Access Speed	D32*	RESERVED	Device Dependent

- **Memory Type:** These two bits contain information about the type of memory contained in this device.

BIT # 15	14	Meaning
1	1	RAM
0	1	ROM
1	0	OTHER
0	0	RESERVED

- **N/S** A one in this position indicates that the device is accessible in both Non-Privileged and Supervisory modes. A zero in this position indicates that the device is accessible only in Supervisory mode.

- **BT*** A zero in this position indicates that the device has Block Transfer Capability.

- **N_P*** The meaning of this field is dependent on the *memory type* field. If the memory type is RAM (11), a zero in this field indicates that the RAM is non-volatile. If the memory type is ROM (01), a zero in this field indicates that the device is electrically programmable.

- **Access Speed** The value in this field indicates the memory access times, as listed in the following table.

BIT #			Access Time (t)
10	9	8	
0	0	0	0 nsec $\leq t < 50$ nsec
0	0	1	50 nsec $\leq t < 100$ nsec
0	1	0	100 nsec $\leq t < 200$ nsec
0	1	1	200 nsec $\leq t < 400$ nsec
1	0	0	400 nsec $\leq t < 800$ nsec
1	0	1	800 nsec $\leq t < 1600$ nsec
1	1	0	1600 nsec $\leq t$
1	1	1	Device dependent

OBSERVATION C.2.47:
The access time is the maximum delay from the assertion DS0* or DS1* until the MEMORY device asserts DTACK*.

- **D32*** A zero in this position indicates that the device has D32 capability, in addition to its D16 and/or D08 capability.

- **RESERVED** Reserved for future definition, and must default to all ones (7_{16}).

- **Device dependent** These bits may be defined by the device's manufacturer. A write to the Attributes Register location has no effect. Its use is reserved for VXIbus definition.

C.2.3.1.2 Memory Device Operational Registers

A Memory Device's operational registers are its functional memory locations. The location and quantities of these registers are determined from information stored in the device's configuration registers (See Section C.2.1.1.2, "Configuration Registers"). The *Address Space* field of the ID register defines the location of the operational registers as follows.

A16 Only: This configuration is undefined for Memory devices.

A16/A24: The operational registers occupy a block of A24 addresses located by the device's Offset register.

A16/A32: The operational registers occupy a block of A32 addresses located by the device's Offset register.

PERMISSION C.2.12:
A Memory Device's A16 device-dependent registers **MAY** be used for device-specific purposes.

C.2.4 Message Based Devices

Message Based Devices support the VXIbus configuration and communication protocols. All Message Based Devices are capable of at least a minimum set of standard communication protocols. Examples of Message Based Devices are any sophisticated devices with local intelligence that require a level of communication capability such as digital multimeters, spectrum analyzers, display controllers, 488-VXIbus interface devices, switch controllers, etc.

C.2.4.1 Data Transfer Capabilities

Message Based Devices use a standard set of registers to implement the VXIbus messaging protocols. These registers operate in the A16 address space and require D16 capability.

RULE C.2.40:
All Message Based Commanders **SHALL** have A16 and D16 Master capability.

RECOMMENDATION C.2.11:
All Message Based Devices should have A16 and D16 Master capability.

OBSERVATION C.2.48:
Servant-only devices with A16 Master capability are able to send status messages to their associated Commanders efficiently, without the use of priority interrupts.

RECOMMENDATION C.2.12:
All Message Based Devices should have A24 Master capability.

RULE C.2.41:
All Message Based A16 Masters **SHALL** be capable of accessing the entire A16 address range.

RULE C.2.42:
All Message Based A24 Masters **SHALL** be capable of accessing the entire address range from 200000_{16} through $DFFFFF_{16}$.

PERMISSION C.2.13:
A Message Based A24 Master **MAY** be capable of accessing the entire A24 address range from 000000_{16} through $FFFFFF_{16}$.

SUGGESTION C.2.3:
Message Based Devices should have D08(EO) Master capability.

OBSERVATION C.2.49:
D08(EO) Master capability facilitates compatibility with a broader range of Non-VXIbus devices than would otherwise be possible.

PERMISSION C.2.14:
Message Based Devices **MAY** implement A32 and/or D32 capabilities.

OBSERVATION C.2.50:
A32 Master capability is required for control of any A32 Register Based or Memory devices or any A32 Non-VXIbus devices.

RULE C.2.43:
All Message Based A32 Masters **SHALL** be capable of accessing the entire A32 address range from 20000000_{16} through $DFFFFFFF_{16}$.

PERMISSION C.2.15:
A Message Based A32 Master **MAY** be capable of accessing the entire A32 address range from 00000000_{16} through $FFFFFFFF_{16}$.

OBSERVATION C.2.51:
VXIbus protocols support the inclusion of A32 or A24 RAM in Message Based Devices.

The preceding rules, etc. are summarized in the following table.

	A16	A24	A32	D08(EO)	D16	D32
MASTER	S/R	REC	MAY	SUG	S/R	MAY
SLAVE	**SHALL**	MAY	MAY	MAY	**SHALL**	MAY

SUG - Suggested.
REC - Recommended.
S/R - **SHALL** for Commanders. Recommended for others.

C.2.4.2 Priority Interrupts

RECOMMENDATION C.2.13:
A Message Based Device should implement either VMEbus Master or Interrupter capability.

Interrupt Handler capability is entirely optional.

OBSERVATION C.2.52:
The *Asynchronous Mode Control* command may be used to select and enable the transmission mode (interrupt or signal) of the different classes (response and event) of interrupts.

RULE C.2.44:
Message Based devices with Interrupter capability **SHALL** implement Release On Acknowledge (ROAK).

There are three Word Serial commands that allow a device's Commander to query and modify the interrupt lines selected by a Message Based Device with Interrupter capability. These commands are:

Read Interrupters - This command is used to read the number of interrupt lines that the device can drive simultaneously, i.e. the number of Interrupters in the device.

Read Interrupter Line - This command is used to read which VMEbus interrupt line is driven by a particular Interrupter in the device.

Assign Interrupt Line - This command is used to assign one of the device's Interrupters to a particular VMEbus interrupt line.

Within a device, multiple Interrupters are identified in some device-dependent consecutive order, starting with #1. They are designated *Interrupter #1, Interrupter #2,* etc.

RECOMMENDATION C.2.14:
A Message Based Device with Interrupter capability should implement Programmable Interrupter (PI) capability, which provides a mechanism to program and read back each interrupt line connection to that device's Interrupter(s). This is achieved by supporting the *Read Interrupters*, *Read Interrupter Line*, and *Assign Interrupter Line* commands.

PERMISSION C.2.16:
Message Based Commanders **MAY** implement Interrupt Handler capability.

There are three Word Serial commands that allow a device's Commander to query and modify the interrupt lines selected by a Message Based Device with Interrupt Handler capability. These commands are:

Read Handlers - This command is used to read the number of interrupt lines that the device can handle simultaneously, i.e. the number of Interrupt Handlers in the device.

Read Handler Line - This command is used to read which VMEbus interrupt line is handled by a particular Interrupt Handler in the device.

Assign Handler Line - This command is used to assign one of the device's Interrupt Handlers to a particular VMEbus interrupt line.

Within a device, multiple Interrupt Handlers are identified in some device-dependent consecutive order, starting with #1. They are designated *Handler #1*, *Handler #2*, etc.

RECOMMENDATION C.2.15:
A Message Based Device with Interrupt Handler capability should implement Programmable Handler (PH) capability, which provides a mechanism to program and read back each interrupt line connection to that device's Interrupt Handler(s). This is achieved by supporting the *Read Handlers, Read Handler Line*, and *Assign Handler Line* commands.

RULE C.2.45:
IF a Message Based Device implements Interrupter capability
THEN it **SHALL** respond to an interrupt acknowledge with the 16-bit Status/ID word defined in the following tables.

There are two different formats for the STATUS/ID word, distinguished by the value of its bit #15.

Bit #	31 - 16	15	14 - 8	7 - 0
Contents	D32 Extension	0	Response	Logical Address

Bit #	31 - 16	15	14 - 8	7 - 0
Contents	D32 Extension	1	Event	Logical Address

The fields of the STATUS/ID word are defined below.

- **D32 Extension** The contents of this optional field are device dependent.

OBSERVATION C.2.53:
D08(O) and D16 STATUS/ID generators will always return $FFFF_{16}$ in this field.

- **Response** This field is identical to bits 14 - 8 of the device's Response register, as defined in Section C.2.4.3.1, "Message Based Device Communication Registers".

OBSERVATION C.2.54:
A Message Based Device will not generate any response signals or response interrupts, unless such response signals or response interrupts have been enabled by the *Control Response* command.

OBSERVATION C.2.55:
Devices which are not capable of generating response signals or response interrupts are not required to support the *Control Response* command.

- **Event** This field indicates the event associated with the interrupt. See Section E.4, "Protocol Events", for a listing of the defined events.

OBSERVATION C.2.56:
D08(O) STATUS/ID generators will always return FF_{16} in bits 15 - 8.

- **Logical Address** This field contains the Logical Address of this device.

OBSERVATION C.2.57:
The VXIbus provides alternate mechanisms for transmitting responses and events: interrupts and signals (See the Signal register definition in Section C.2.4.3.1,

"Message Based Device Communication Registers"). Non-master devices always transmit via interrupts, while master devices may send signals. Some devices are capable of receiving only signals, whereas others may be only Interrupt Handlers. The VMEbus provides only seven interrupt lines. In a large system these lines may be overloaded. Such a system may provide a mechanism for converting signals to interrupts, converting interrupts to signals, or even translating from one interrupt line to another.

C.2.4.3 Message Based Device Registers

Message Based Devices have both VXIbus standard configuration registers and operational registers. The location and quantities of these registers are determined from the device's Logical Address, and other information stored by its configuration registers (See Section C.2.1.1.2, "Configuration Registers"). The *Address Space* field of the ID register defines the location of the operational registers as follows.

A16 Only: The operational registers occupy addresses 06_{16} - $3F_{16}$ relative to the device's A16 base address.

A16/A24: The operational registers occupy addresses 08_{16} - $3F_{16}$ relative to the device's A16 base address, and a block of A24 addresses located by the device's Offset register.

A16/A32: The operational registers occupy addresses 08_{16} - $3F_{16}$ relative to the device's A16 base address, and a block of A32 addresses located by the device's Offset register.

OBSERVATION C.2.58:
All Message Based Device A24 and A32 operational registers have device-dependent functionality.

C.2.4.3.1 Message Based Device Communication Registers

Message Based Devices implement a standard set of A16 operational registers, known as *communication registers*. These registers are shown in the following table. The addresses are relative to a device's A16 base address. A block of registers is reserved for use by future optional protocols. It is anticipated that the definition of these registers will be tied to the definition of the reserved bits in the Protocol register.

The communication registers are organized as follows:

Offset	Register
$3F_{16}$	DEVICE DEPENDENT REGISTERS
$1F_{16}$ – 18_{16}	VXIbus RESERVED REGISTERS
14_{16}	A32 Pointer
10_{16}	A24 Pointer
$0E_{16}$	Data Low
$0C_{16}$	Data High
$0A_{16}$	Response/Data Extended
08_{16}	Protocol/Signal Register
00_{16}	CONFIGURATION REGISTERS

RULE C.2.46:
All Message Based Devices **SHALL** implement the Protocol, Response, and Data Low registers.

OBSERVATION C.2.59:
The Signal, Data Extended, Data High, A24 Pointer, and A32 Pointer registers are optional.

The communication registers are defined as follows:

Protocol Register: The read contents of this 16-bit register indicate which protocols the device supports, and indicate additional communication capabilities of the device.

Bit #	15	14	13	12	11	10	9 - 4	3 - 0
Contents	CMDR*	Signal Register*	Master*	Interrupter	FHS*	Shared Memory*	RESERVED	Dev Dep

- **CMDR*** A one in this field indicates that the device has Servant only capability. A zero in this field indicates that the device has both Servant and Commander capability.

- **Signal Register** A zero in this field indicates that the device has a Signal register. See below for a definition of a Signal register.

- **Master*** A zero in this field indicates that the device has VMEbus Master capability.

- **Interrupter** A one in this field indicates that the device has Interrupter capability.

- **FHS*** A zero in this field indicates that this device's data registers support the Fast-Handshake mode (See Section C.3.3.2, "Fast-Handshake Transfers"). A one indicates that only the normal transfer mode is implemented.

- **Shared Memory*** A zero in this field indicates that this device supports the optional shared memory protocol (See Section G, "Shared Memory Protocols") and implements one or both of the A24 Pointer and A32 Pointer registers. A one indicates that the shared memory protocol is not supported.

- **RESERVED** Reserved for future definition, and must default to all ones ($3F_{16}$).

- **Dev Dep** These device-dependent bits may be defined by the device's manufacturer.

Signal Register: A write to this 16-bit register is used for general device-to-device signaling. For efficient system operation the Commander must be able to detect and respond quickly to any Signal register write. A signal consists of the Logical Address of the sender and other signal-specific information.

There are two different formats for the signals, distinguished by the value of its bit #15.

Bit #	15	14 - 8	7 - 0
Contents	0	Response	Logical Address

Bit #	15	14 - 8	7 - 0
Contents	1	Event	Logical Address

The fields of the signal are defined below:

- Response: This field is identical to bits 14 - 8 of the device's Response register, as defined below.

- Event: This field indicates the *event* associated with the signal.

OBSERVATION C.2.60:
Response signals are generally used to respond to a request made by a device. Event signals are generally asynchronous events analogous to interrupts. The *Asynchronous Mode Control* command may be used to select and enable the transmission mode (interrupt or signal) of the different classes of signals and interrupts (response and event).

RULE C.2.47:
IF a Message Based Device implements a functional Signal register,
THEN there **SHALL** be a Write mode Location Monitor at the Signal register address.

RULE C.2.48:
IF a Message Based Device supports D08(EO) writes to its Signal register,
THEN its Signal register D08 Location Monitor **SHALL** detect odd byte accesses only.

RULE C.2.49:
IF a write to a device's Signal register would result in loss of data,
THEN that device **SHALL** respond to such a write by asserting BERR* instead of DTACK*.

OBSERVATION C.2.61:
A loss of data would occur if a signal write were to overwrite a previous signal that had not yet been read by the device's internal microprocessor.

OBSERVATION C.2.62:
A device may delay the DTACK* or BERR* handshake up to 20 μsec (see **RULE B.2.1**) in an effort to prevent data loss.

OBSERVATION C.2.63:
If a write to a device's Signal register results in a bus error, then that signal will not be received by the device. In such a case the sender may choose to retransmit the signal.

PERMISSION C.2.17:
A Message Based Device **MAY** implement a non-functional Signal register. This is indicated by a one in bit 14 of the device's Protocol register. In this case the device may respond with DTACK* to Signal register writes, even though it stores no data from that write cycle.

RECOMMENDATION C.2.16:
If a Commander is intended to support a variety of Servants and configurations, it should implement a functional Signal register and the receipt of signals.

Response Register: A read of this16-bit register returns the status of the device's communication registers and their associated functions.

The Message Based device's Response register indicates critical device status information. Each bit indicates a single status item as follows:

Bit #	15	14	13	12	11	10	9	8	7	6 - 0
Contents	0	RESERVED	DOR	DIR	Err*	Read Ready	Write Ready	FHS Active*	Locked*	Device Dependent

The fields of the Response register are defined below:

- Bit 15 is always zero.

OBSERVATION C.2.64:
Bit 15 is defined to be zero in order to provide consistency with the format of signals and Interrupter STATUS/ID words.

- **RESERVED** This bit is reserved for future VXIbus definition. Its value is always one.

- **DOR** (Data Out Ready) [optional] A one indicates that the device is ready to output data to its Commander. Its operation is described in Section C.3.3.3, "Byte Transfer Protocol".

OBSERVATION C.2.65:
Devices that support the *Byte Request* command are required to implement the *DOR* bit.

- **DIR** (Data In Ready) [optional] A one indicates that the device is ready to accept data from its Commander. Its operation is described in Section C.3.3.3, "Byte Transfer Protocol".

OBSERVATION C.2.66:
Devices that support the *Byte Available* command are required to implement the *DIR* bit.

- **Err*** (Error) [required] A zero indicates that an error has occurred in one of the serial protocols (Word Serial, Longword Serial, or Extended Longword Serial. See Section C.3.3.1, "Serial Protocols".), and has not yet been reported. The nature of the error can be read using the *Read Protocol Error* command. A one indicates that no errors are pending. See Section C.3.3.4, "Error Handling".

OBSERVATION C.2.67:
The *Read Protocol Error* command is used to report the reason for an assertion of the *Err** bit.

- **Read Ready** [Required] A one indicates that its data registers contain data to be read. This bit is set when data from internal operations is available in the Data Low and Data High registers. It is cleared to zero by a read of the Data Low register.

RULE C.2.50:
The *Read Ready* bit **SHALL** be cleared before DTACK* is released during any data transfer cycle that reads the Data Low register.

- **Write Ready** [Required] A one indicates that this device is ready for data transfers to its data registers. This bit is set when the device is ready for a write to any of the data registers. It is cleared by a write to the Data Low register.

RULE C.2.51:
The *Write Ready* bit **SHALL** be cleared before DTACK* is released during any data transfer cycle that writes to the Data Low register.

- **FHS Active*** [optional] A zero indicates that Fast-Handshakes are currently enabled for this device's Servant registers (See Section C.3.3.2, "Fast-Handshake Transfers"). This bit is cleared to zero whenever the Fast-Handshake Mode is entered, and is set to one any time the device exits the Fast-Handshake Mode.

- **Locked*** [optional] A zero indicates that the Commander of this device has locked access to it from other local sources. (e.g., local lockout of IEEE-488)

- **Device dependent** The bits in this field are available to the device designer for device-specific and/or application-specific status information.

OBSERVATION C.2.68:
Transitions of any of bits 14 - 8 of the Response register may be enabled, to generate response interrupts or response signals, as defined in Section C.2.4.2, "Priority Interrupts".

Data Extended Register: A write to this optional register is the most significant word of input data or a command.

PERMISSION C.2.18:
A Message Based Device **MAY** implement a non-functional Data Extended register.

RULE C.2.52:
IF a Message Based Device implements a functional Data Extended register,
THEN there **SHALL** be a Write mode Location Monitor at the Data Extended register address.

RULE C.2.53:
IF a Message Based Device supports D08(EO) writes to its Data Extended register,
THEN its Data Extended register D08 Location Monitor **SHALL** detect odd-byte accesses only.

Data High Register: A write to this optional register is the second least significant word of write data. A read of this register is the most significant word of read data.

PERMISSION C.2.19:
A Message Based Device **MAY** implement a non-functional Data High register.

RULE C.2.54:
IF a Message Based Device implements a functional Data High register,
THEN there **SHALL** be a Write mode Location Monitor at the Data High register's address.

RULE C.2.55:
IF a Message Based Device supports D08(EO) writes to its Data High register,
THEN its Data High register D08 Location Monitor **SHALL** detect odd byte accesses only.

Data Low Register : A write to this register is the least significant word of write data. A write to this register causes the device to execute some action based on the data written to the Data High, Data Low and Data Extended registers. A read of this register is the least significant word of read data.

RULE C.2.56:
There **SHALL** be a Read mode and a Write mode Location Monitor at the Data Low register's address.

RULE C.2.57:
IF a Message Based Device supports D08(EO) accesses of its Data Low register,
THEN its Data Low register D08 Location Monitors **SHALL** detect odd byte accesses only.

OBSERVATION C.2.69:
The Location Monitors at the data registers provide the device with the ability to detect the width of the data or command being sent. 16-bit, 32-bit and 48-bit data transfers can all be independently detected by this information.

A24 Pointer Register This 24-bit register is defined by the optional Shared Memory protocol. See Section G, "Shared Memory Protocols".

A32 Pointer Register This 32-bit register is defined by the optional Shared Memory protocol. See Section G, "Shared Memory Protocols".

VXIbus Reserved Registers These registers are reserved for future VXIbus definition. The effects of reads and writes of these registers are undefined.

C.2.4.4 Message Based Device Operation

The specification of standard communication protocols requires that all Message Based Devices have a uniform and clearly defined behavior when using common resources. For a Message Based Device, three sub-states are added in the PASSED state, defined in Section C.2.1.2, "Device Initialization and Diagnostics". The first, CONFIGURE sub-state, is the state in which a Message Based Device is configured. The second, NORMAL OPERATION sub-state, is the state which is entered when the configuration has been completed. Transitions between these sub-states are initialized by the *Begin Normal Operation*, *End Normal Operation*, and *Abort Normal Operation* commands. The third sub-state, INITIALIZE, is the state of the device while it is executing the *Begin Normal Operation* command.

The sub-states and the transitions between them are described in Figure C.5 as an extension to Figure C.4. All state transitions already defined in Figure C.4 (Section C.2.1.2, "Device Initialization and Diagnostics") are the same.

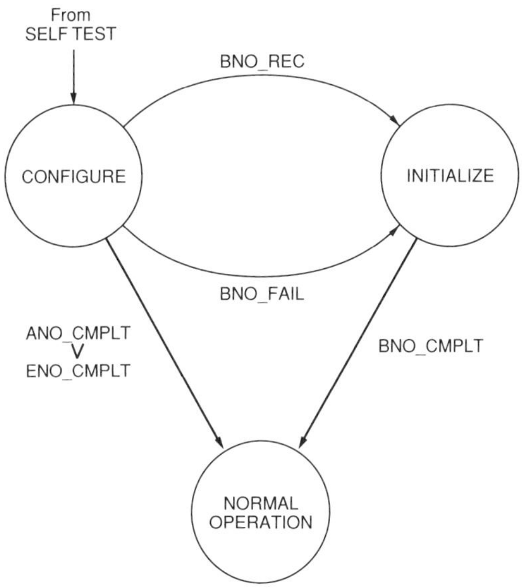

BNO_FAIL: *Begin Normal Operation* command failed
BNO_REC: *Begin Normal Operation* command received
BNO_CMPLT: *Begin Normal Operation* command successfully completed
ENO_CMPLT: *End Normal Operation* command successfully completed
ANO_CMPLT: *Abort Normal Operation* command successfully completed

Figure C.5. Configure State Diagram

OBSERVATION C.2.70:
The CONFIGURE, INITIALIZE, and NORMAL OPERATION sub-states are undefined in all states other than PASSED.

RULE C.2.58:
All Message Based Devices **SHALL** support the Word Serial protocol when in the PASSED state (See Section C.2.1.2, "Device Initialization and Diagnostics").

RULE C.2.59:
While in the CONFIGURE sub-state, a Message Based Device **SHALL** remain deactivated.

OBSERVATION C.2.71:
A deactivated device may not assert any VMEbus IRQ lines, or initiate any data transfer cycles.

C.2.4.4.1 State Dependency of Commands

A Message Based Device responds to different sets of commands, depending on the sub-state it is in: CONFIGURE, INITIALIZE, or NORMAL OPERATION. Some commands are used in all states. Whether the commands in the sets are mandatory or optional is defined in Section C.2.4.4.2, "Required Commands".

The commands responded to in the CONFIGURE sub-state are called *configuration commands*:

- *Assign Handler Line*
- *Assign Interrupter line*
- *Grant Device*
- *Identify Commander*
- *Release Device*
- *Read Servant Area*

The commands responded to in the NORMAL OPERATION sub-state are called *normal operation commands*:

- *Byte Available*
- *Byte Request*
- *Clear Lock*
- *Read STB*
- *Set Lock*
- *Trigger*
- *User Defined*

The commands, responded to in both the CONFIGURE and the NORMAL OPERATION sub-states, are called *state-independent commands*:

- *Abort Normal Operation*
- *Asynchronous Mode Control*

- *Begin Normal Operation*
- *Clear*
- *Control Event*
- *Control Response*
- *End Normal Operation*
- *Read Handler Line*
- *Read Handlers*
- *Read Interrupter Line*
- *Read Interrupters*
- *Read MODID*
- *Read Protocol*
- *Read Protocol Error*
- *Set Lower MODID*
- *Set Upper MODID*

RULE C.2.60:
A Commander **SHALL NOT** send any normal operation commands to a Servant which is in the CONFIGURE sub-state.

RULE C.2.61:
A Commander **SHALL NOT** send any configure commands to a Servant which is in the NORMAL OPERATION sub-state.

OBSERVATION C.2.72:
It is a protocol violation to send any commands to any device while it is in the INITIALIZE sub-state.

C.2.4.4.2 Required Commands

This section defines minimal conformance requirements for Message Based Devices, in terms of their capabilities and sub-states. In this context a phrase of the form "A Message Based Device shall properly respond to..." means that:

- The Message Based Device must support at least the minimum capability implied by the command.

- The Message Based Device must recognize and interpret the command in a manner consistent with the intent of the command.
- The Message Based Device must implement full functional support of the command, according to its capabilities.
- If the command requires a response, the Message Based Device's response must conform to the response formats defined in Section E, "Command and Event Formats".

RULE C.2.62:
All Message Based Devices **SHALL** properly respond to the following commands in both the CONFIGURE and NORMAL OPERATION sub-states:

- *Abort Normal Operation*
- *Begin Normal Operation*
- *Clear*
- *End Normal Operation*
- *Read Protocol*
- *Read Protocol Error*

RULE C.2.63:
All Message Based Devices with VMEbus Master capability **SHALL** properly respond to the following command in the CONFIGURE sub-state:

- *Identify Commander*

OBSERVATION C.2.73:
VMEbus Master capability is indicated by the *Master** bit in the Protocol register.

RULE C.2.64:
All Message Based Commanders **SHALL** properly respond to the following commands in the CONFIGURE sub-state:

- *Read Servant Area*
- *Grant Device*
- *Release Device*

OBSERVATION C.2.74:
Message Based Commander capability is indicated by the CMDR* bit in the Protocol register.

RULE C.2.65:
All Message Based Devices having Programmable Handler capability **SHALL** properly respond to the following commands in both the CONFIGURE and NORMAL OPERATION sub-states:

- *Read Handlers*
- *Read Handler Line*

RULE C.2.66:
All Message Based Devices having Programmable Handler capability **SHALL** properly respond to the following command in the CONFIGURE sub-state:

- *Assign Handler Line*

OBSERVATION C.2.75:
Programmable Handler capability is indicated by the response to the *Read Protocol* command.

RULE C.2.67:
All Message Based Devices having Programmable Interrupter capability **SHALL** properly respond to the following commands in both the CONFIGURE and NORMAL OPERATION sub-states:

- *Read Interrupters*
- *Read Interrupter Line*

RULE C.2.68:
All Message Based Devices having Programmable Interrupter capability **SHALL** properly respond to the following command in the CONFIGURE sub-state:

- *Assign Interrupter Line*

OBSERVATION C.2.76:
Programmable Interrupter capability is indicated by the response to the *Read Protocol* command.

RULE C.2.69:
IF a Message Based Servant implements response interrupts or response signals,
THEN it **SHALL** have the Response Generation (RG) capability.

RULE C.2.70:
All Message Based Devices having Response Generation capability **SHALL** properly respond to the following commands in the CONFIGURE or NORMAL OPERATION sub-state:

- *Asynchronous Mode Control*
- *Control Response*

RULE C.2.71:
A Message Based Device having Response Generation capability **SHALL** enable or disable response generation on each Response register bit transition for which it supports response generation, in response to the *Control Response* command.

OBSERVATION C.2.77:
Response Generation capability can be determined by the Read Protocol command.

RULE C.2.72:
IF a Message Based Servant implements event interrupts or event signals,
THEN it **SHALL** have the Event Generation (EG) capability.

RULE C.2.73:
All Message Based Devices having Event Generation capability **SHALL** properly respond to the following command in the CONFIGURE or NORMAL OPERATION sub-state:

- *Asynchronous Mode Control*
- *Control Event*

RULE C.2.74:
A Message Based Device having Event Generation capability **SHALL** enable or disable event generation for each event it is capable of generating, in response to the *Control Event* command.

OBSERVATION C.2.78:
Event Generation capability can be determined by the Read Protocol command.

C.2.4.5 Message Based Device Configuration

The specification of standard communication protocols requires that all Message Based Devices be configured in a uniform fashion after a Hard Reset, Soft Reset, or after receiving one of the Word Serial *Abort Normal Operation*, *End Normal Operation* or *Clear* commands.

C.2.4.5.1 The Default Configuration

Message Based Devices have a uniform default configuration. The default configuration is assumed when the devices are reset or receive the *Abort Normal Operation* command. The configuration may be modified (from the default) with the configuration commands. The resulting configuration becomes active when the device enters the NORMAL OPERATION sub-state.

RULE C.2.75:
Upon entering the PASSED state, a Message Based Device **SHALL** assume the CONFIGURE sub-state with the default configuration wherein:

1. **IF** the Message Based Device is a Commander,
   THEN its Servant list is empty.

2. Responses and events are globally enabled (See the definition of the *Asynchronous Mode Control* command in Section E.1, "Word Serial Commands").

3. **IF** the Message Based Device has VMEbus Master capability,
   THEN responses and events are sent via signals to Logical Address zero.

4. **IF** the Message Based Device does not have VMEbus Master capability,
   THEN responses and events are sent via interrupts.

5. **IF** the Message Based Device has Programmable Interrupter capability,
   THEN the Interrupter capabilities are not connected to any IRQ lines.

6. **IF** the Message Based Device has Programmable Handler capability,
   THEN the Interrupt Handler capabilities are not connected to any IRQ lines.

7. Generation of all supported events is enabled. (See the definition of the *Control Event* command in Section E.1, "Word Serial Commands".)

8. Generation of B14*, DOR*, DIR*, Err*, RR*, WR*, FHS* responses is disabled. (See the definition of the *Control Response* command in Section E.1, "Word Serial Commands".)

9. The device is not processing any incoming signals or interrupts.

OBSERVATION C.2.79:
Since the device is deactivated in the CONFIGURE sub-state, even though responses and events may be enabled, signals and interrupts are not generated until the device enters the NORMAL OPERATION sub-state.

The default configuration is summarized in Table C.1.

C.2.4.5.2 Modifying The Configuration

Changes can be made to the current configuration in the CONFIGURE sub-state using the configuration commands and state independent commands.

RULE C.2.76:
Upon receipt of the *Grant Device* command, a Commander **SHALL** add the granted device to its Servant list.

TABLE C.1. The Default Configuration

DEFAULT CONFIGURATION	
Identify Commander:	Resource Manager
Asynchronous Mode:	
Events	Enabled (via signals if Bus Master)
Responses	Enabled (via signals if Bus Master)
Control Response:	All Bits Disabled
Control Event:	All Events Enabled
Programmable Handlers	All Disconnected
Programmable Interrupters	All Disconnected
Servant List	Empty

OBSERVATION C.2.80:
Each Commander maintains a list of the Logical Addresses of Servant devices, referred to as the Servant list.

RULE C.2.77:
Upon receipt of the *Release Device* command, a Commander **SHALL** delete the released device from its Servant list.

RULE C.2.78:
IF a Message Based Device supports the *Identify Commander* command,
THEN upon receipt of the *Identify Commander* command, the Message Based Device **SHALL** register the identified Commander as the device's Commander.

RULE C.2.79:
IF a Message Based Device supports the *Assign Handler Line* command,
THEN upon receipt of the *Assign Handler Line* command the Message Based Device **SHALL** connect the IRQ line to the Interrupt Handler capability.

RULE C.2.80:
IF a Message Based Device supports the *Assign Interrupter Line* command,
THEN upon receipt of the *Assign Interrupter Line* command the Message Based Device **SHALL** connect the IRQ line to the Interrupter capability.

C.2.4.6 Initiating Operation

Once configured, a Message Based Device is activated via the *Begin Normal Operation* command. Receipt of the *Begin Normal Operation* command causes a device to initialize its Servants and enter the NORMAL OPERATION sub-state. While in the NORMAL OPERATION sub-state, the device will have its Ready bit set to one.

OBSERVATION C.2.81:
While in the NORMAL OPERATION sub-state the Commander/Servant hierarchy and the interrupt line configuration cannot be changed, according to Section C.2.4.4.1, "State Dependency of Commands".

OBSERVATION C.2.82:
The *Begin Normal Operation* command indicates, via the *Top Level* bit, whether a device is a Top Level device or not. A top level device has no Commander, and in the NORMAL OPERATION sub-state it may operate autonomously with respect to all other devices except a device involved in runtime resource management. A device involved in runtime resource management retains the capability of forcing a top level device into the CONFIGURE sub-state by sending the *End Normal Operation* or the *Abort Normal Operation* command to the top level device.

RULE C.2.81:
Upon the receipt of a *Begin Normal Operation* command, a Commander which is in the CONFIGURE sub-state **SHALL** execute the following process.

1. Enter the INITIALIZE sub-state.

2. Discard all pending and initiated signals and VMEbus interrupts.

3. Determine which Message Based Servants are VMEbus Masters by reading the Protocol register.

4. Send the *Identify Commander* command to all Message Based Servants which are VMEbus Masters.

5. Send the *Begin Normal Operation* command to all Message Based Servants and wait for their responses.

6. Configure all its non-Message Based Servants as needed. This includes allocating IRQ lines.

7. If the above process is successful, enter the NORMAL OPERATION sub-state and set the *Ready* bit of the Status register to one. Otherwise, return to the CONFIGURE sub-state.

8. Return the appropriate response, indicating the state(s) of itself and its hierarchical Servants, in the Data Low register. This response will incorporate any device failure information returned (in response to *Begin Normal Operation*) by the Commander's Servants.

OBSERVATION C.2.83:
In general, a device will enter the NORMAL OPERATION sub-state if it is capable of normal operation.

RULE C.2.82:
Upon the receipt of a *Begin Normal Operation* command, a Commander which is in the NORMAL OPERATION sub-state **SHALL** return the appropriate response, indicating the state(s) of itself and its hierarchical Servants, in the Data Low register.

OBSERVATION C.2.84:
Since a device in the NORMAL OPERATION sub-state has already initialized its Servants, no action other than the response specified in the previous rule is required.

RULE C.2.83:
A Message Based Device which is in either the CONFIGURE or INITIALIZE state **SHALL** have the *Ready* bit of its Status register cleared to zero.

RULE C.2.84:
A Message Based Device which is in the NORMAL OPERATION sub-state **SHALL** have the *Ready* bit of its Status register set to one.

OBSERVATION C.2.85:
Normally, the Resource Manager will configure a Message Based Device with PI capability to drive an IRQ line that its Commander handles.

OBSERVATION C.2.86:
A Servant may be programmed to drive an IRQ line which is not handled by its Commander.

OBSERVATION C.2.87:
It is the responsibility of the system integrator to ensure that the IRQ lines are properly allocated.

RULE C.2.85:
A VXIbus device **SHALL NOT** attempt to control, through the VXIbus registers, any VXIbus device not in its own Servant list.

RULE C.2.86:
A Commander lacking device-specific knowledge **SHALL NOT** access any of its Servants' device-dependent registers, other than the VXIbus defined registers.

OBSERVATION C.2.88:
It is not safe to assume that a Commander may freely write to any device-dependent register of a Servant Device. Such accesses should not (in general) be attempted without good cause.

C.2.4.7 Terminating Operation

Four methods are available for terminating the operation of Message Based Devices. These are *Hard Reset*, *Soft Reset*, *Abort Normal Operation* and *End Normal Operation*. Each of these differ in the level of reconfiguration they impose on the device. Table C.2 outlines the actions imposed by these methods. Also included in the table are the actions imposed by the *Clear* command. Though the *Clear* command, as defined for Message

Based Devices (Section C.2.4.8, "Clearing a Message Based Device"), does not terminate the operation of the received device, it has been included in this table for completeness.

C.2.4.7.1 Hard Reset

Upon receipt of a Hard Reset (assertion of SYSRESET*), a Message Based Device enters, as does any other device type, the HARD RESET state. From that state a device continues its initialization until it ends in one of the possible states, as defined in Section C.2.1.2, "Device Initialization and Diagnostics". A Message Based Device will, if it enters the PASSED state, enter the CONFIGURE sub-state configured with the default configuration.

Hard Reset is used only when it is necessary to promptly terminate operation of all devices in a VXIbus system. A Hard Reset always affects the entire VXIbus system, and will result in the system being configured by the Resource Manager.

C.2.4.7.2 Soft Reset

Upon receipt of a Soft Reset (setting and clearing the *Reset* bit), a Message Based Device enters (as does any other device type) the SOFT RESET state. From that state a device passes through the SELF TEST state, until it ends in one of the possible states, as defined in Section C.2.1.2 "Device Initialization and Diagnostics". A Message Based Device will, if it enters the PASSED state, enter the CONFIGURE sub-state configured with the default configuration.

A Soft Reset affects only a single device, and is used to deactivate a device. A deactivated device is prevented from using the VXIbus.

RECOMMENDATION C.2.17:
In general, a Commander should soft reset a Message Based Servant only if all VXIbus termination commands fail.

OBSERVATION C.2.89:
A Message Based Device that has been Soft Reset will have all of its configuration parameters reset (to the Default Configuration). Its state may also be inconsistent with the remainder of the system, due to lost signals or abandoned communication links with other VXIbus devices. It is important that such a device not be activated until its configuration is restored, and all inconsistencies have been resolved. In general, the Resource Manager is the only entity capable of doing so.

RULE C.2.87:
When a Commander soft resets a device it **SHALL** ensure that the operation of all Servants of that device is also terminated.

OBSERVATION C.2.90:
In general, the Resource Manager is the only device with full knowledge of Servant hierarchies.

TABLE C2. Message Based Device Termination and Clear Actions

	Hard Reset	Soft Reset	ANO	ENO	Clear
Generated By	SYSRESET*	Reset Bit	*Abort Normal Operation* Command	*End Normal Operation* Command	*Clear* Command
Action Defined	Upon Entry to PASSED State	Upon Entry to PASSED State	After Command Processed	After Command Processed	After Command Processed
All Devices: Sub-State	CONFIGURE	CONFIGURE	CONFIGURE	CONFIGURE	Unchanged
Configuration	Default	Default	Default	Unchanged	Unchanged
Status Register: A24/A32 Active	0	Unchanged	Unchanged	Unchanged	Unchanged
Offset Register:	Initial	Unchanged	Unchanged	Unchanged	Unchanged
Logical Address:	Initial	Unchanged	Unchanged	Unchanged	Unchanged
Response Register: *Err* bit*	1	1	1	1	1
Read Ready	0	0	Valid	Valid	0
*FHS Active**	1	1	1	1	Unchanged
*Locked**	1	1	1	1	Unchanged
Bus Masters: Data Transfers	No	No	No	No	Unchanged
Interrupters: IRQ Assertions	No	No	No	No	Unchanged

RULE C.2.88:
Before a Commander sends *Begin Normal Operation* to a Servant (after having soft reset that Servant) the Commander **SHALL** guarantee that the device and the system are in a consistent configuration and state.

OBSERVATION C.2.91:
When a Commander receives *Begin Normal Operation* while it is in the CONFIGURE sub-state) the Commander can assume that all its Servants are in a state and configuration consistent with the system.

C.2.4.7.3 Abort Normal Operation

An *Abort Normal Operation* affects only a single device, and is used by a Commander (or by the Resource Manager) to deactivate a device, i.e. prevent it from using the VXIbus. After an *Abort Normal Operation*, a Message Based Device will enter the CONFIGURE sub-state with the default configuration. In contrast to the soft reset, the device does not pass through the SELF TEST state.

RECOMMENDATION C.2.18:
In general, a Commander should send the *Abort Normal Operation* command to one of its Servants only after an attempt to send *End Normal Operation* has already failed.

RULE C.2.89:
Upon the receipt of the *Abort Normal Operation* command in the NORMAL OPERAT!ON sub-state, a Message Based Device **SHALL** respond in the following manner:

1. Abort operation as fast as possible.
2. Exit the NORMAL OPERATION sub-state to the CONFIGURE sub-state, clearing the *Ready* bit of its Status register to zero.
3. Stop processing incoming signals and interrupts.
4. Restore the default configuration. (The Logical Address and programmed offset is kept.)
5. Return the appropriate code in the Data Low register.

RULE C.2.90:
Upon the receipt of the *Abort Normal Operation* command in the CONFIGURE sub-state, a Message Based Device **SHALL** respond in the following manner:

1. Restore the default configuration. (The Logical Address and programmed offset is kept.)
2. Return the appropriate code in the Data Low register.

OBSERVATION C.2.92:
As a consequence of entering (or reentering) the CONFIGURE sub-state, the response signal or interrupt generation is disabled, and all pending bus requests and interrupts are unasserted.

OBSERVATION C.2.93:
Abort Normal Operation is not propagated through the hierarchy.

OBSERVATION C.2.94:
A device that has been Aborted will have all of its configuration parameters reset to the Default Configuration. Its state may also be inconsistent with the remainder of the system, due to lost signals or abandoned communication links with other VXIbus devices. It is important that such a device not be activated until its configuration is restored, and all inconsistencies have been resolved. In general, the Resource Manager is the only entity capable of doing this.

RULE C.2.91:
When a Commander terminates the operation of a device with the *Abort Normal Operation* command, it **SHALL** ensure that the operation of all Servants of that device is also terminated.

OBSERVATION C.2.95:
In general, the Resource Manager is the only device with full knowledge of Servant hierarchies.

RULE C.2.92:
Before a Commander sends *Begin Normal Operation* to a Servant, after having aborted that Servant's operation, the Commander **SHALL** guarantee that the device and the system are in a consistent configuration and state.

OBSERVATION C.2.96:
When a Commander receives *Begin Normal Operation* while it is in the CONFIGURE sub-state, the Commander can assume that all its Servants are in a state and configuration consistent with the system.

C.2.4.7.4 End Normal Operation

An *End Normal Operation* command affects a hierarchy of devices, and is used by a Commander or by the Resource Manager to bring the devices into a state from which they can be restarted. After a successful *End Normal Operation* command, the devices in the hierarchy are deactivated.

RULE C.2.93:
Upon the receipt of the *End Normal Operation* command in the NORMAL OPERATION sub-state, a Message Based Device **SHALL** respond in the following manner:

1. Allow operation to complete in a device-specific-consistent manner. All operation shall eventually end, but there is no time limit.

2. Bring any Register Based or Extended Device Servants into a deactivated state.

3. Send the *End Normal Operation* command to all its Message Based Servants, and wait for their responses.

4. Stop processing incoming signals and interrupts.

5. Exit the NORMAL OPERATION sub-state to the CONFIGURE sub-state, clearing the *Ready* bit of the Status register to zero. The current configuration is kept.

6. Return the appropriate response code in the Data Low register. This response will incorporate any device failure information returned (in response to *End Normal Operation*) by the Commander's Servants.

RULE C.2.94:
Upon the receipt of the *End Normal Operation* command in the CONFIGURE sub-state, a Message Based Device **SHALL** return the appropriate response code in the Data Low register.

OBSERVATION C.2.97:
Since a device in the CONFIGURE sub-state has already been deactivated, no action other than the response specified in the previous rule is required.

OBSERVATION C.2.98:
As a consequence of entering the CONFIGURE sub-state, the device is deactivated i.e. prevented from using the VXIbus.

OBSERVATION C.2.99:
Those state parts (e.g. higher level configurations) which are not covered by the rules in this section can, upon reception of *End Normal Operation,* be changed in any manner that is device specific, or may be standardized in the future.

OBSERVATION C.2.100:
The *End Normal Operation* command is propagated from Commander to Servant until it has reached the bottom of the hierarchy.

OBSERVATION C.2.101:
It is device dependent if (or how) a device resumes operation, on receipt of *Begin Normal Operation* after *End Normal Operation.*

C.2.4.8 Clearing A Message Based Device

A *Clear* command is used only to remove pending output data from a Message Based Device's Data Low, High and Extended registers. Just as in the case of *Abort Normal Operation* and *End Normal Operation*, it may have an additional meaning to higher level protocols.

RULE C.2.95:
Upon the receipt of the *Clear* command, a Message Based Device **SHALL**:

- Remove any data output in the Data Low, Data High and Data Extended registers.

- Remove any buffered data output to the Data Low, Data High and Data Extended registers.

- Discard output data from any output-data-generating operations previously started.
- Become ready to receive a new command.
- Clear the *Read Ready* bit of its Response register to zero.
- De-assert the *Err** bit of the Response register, and cancel any pending *Read Protocol Error* command responses.

RULE C.2.96:
The *Clear* command **SHALL NOT** affect:

- The CONFIGURE/NORMAL OPERATION sub-state.
- The PI or PH settings.
- The Servant list.
- The Signal register.
- Assertion of IRQ lines.
- Communication links with other VXIbus devices.

OBSERVATION C.2.102:
The effects of *Clear* on a VXIbus instrument are documented in Section D.1.1.3, "Clearing a VXIbus Instrument".

C.2.5 Extended Devices

An Extended device is a VXIbus device with provisions for defining new subclasses of VXIbus devices for future applications. Each Extended Device has a subclass register, in addition to its configuration registers. The subclass register allows for the definition of both standard and manufacturer-specific Extended Device subclasses.

OBSERVATION C.2.103:
An Extended Device implements the standard set of VXIbus configuration registers.

RULE C.2.97:
An Extended Device **SHALL** implement a subclass register.

The Extended Device register map is illustrated in the following table:

Address	Registers
$3F_{16}$ – 20_{16}	SUBCLASS DEPENDENT REGISTERS
$1E_{16}$	Subclass Register
$1D_{16}$ – 08_{16}	SUBCLASS DEPENDENT REGISTERS
00_{16}	CONFIGURATION REGISTERS

The registers are defined as follows:

Subclass Register: The read contents of this 16-bit register indicate the subclass of the device. This register has two alternate formats, depending on the value of bit 15:

Bit #	15	14 - 0
Contents	1	Reserved Subclass

Bit #	15	14 - 12	11 - 0
Contents	0	Manufacturer Subclass	Manufacturer ID

Bit 15 is used to differentiate between VXIbus-defined subclasses and manufacturer-defined subclasses. The other fields are defined as follows:

- **Reserved Subclass** These bits identify the device's subclass. All values of this field are reserved for future VXIbus definition.

- **Manufacturer Subclass** These bits indicate the device's subclass, as defined by a particular VXIbus manufacturer. The values of this field are defined by the manufacturer indicated by the adjacent *Manufacturer ID* field.

- **Manufacturer ID** These bits identify the manufacturer responsible for the definition of this particular subclass. The format is the same as in the *Manufacturer ID* field of the ID register (See Section C.2.1.1.2, "Configuration Registers").

OBSERVATION C.2.104:
Each VXIbus manufacturer may define up to 8 Extended Device Subclasses. Each of these subclasses has a unique Manufacturer Subclass/ID combination.

The effects of writes to the Subclass Register location are subclass dependent.

Subclass Dependent Registers These registers will be defined according to a device's subclass requirements as a part of that subclass's definition.

C.3 DEVICE COMMUNICATION PROTOCOLS:

This section describes the protocols necessary for proper communication between VXIbus devices.

C.3.1 Communication Elements

C.3.1.1 Register Based Servants

Register Based Servant describes the communication element of a Register Based device. A Register Based Servant responds to no defined protocols, since the device's registers are entirely card dependent.

C.3.1.2 Message Based Servants

These devices are generally capable of independently executing complex commands, and may also control other devices in a hierarchical instrumentation system. Message Based Servants communicate using the VXIbus Message Based device protocols.

C.3.1.3 Message Based Commanders

A Message Based Commander is the interface for a Message Based Device exercising control over other devices. Message Based Commanders communicate using the VXIbus Message Based Device protocols.

C.3.2 Register Based Servant Control

The protocols for control of Register Based Devices are completely device dependent. The designer of a Register Based Device is free to specify any register interactions and control protocols required for proper operation of that device.

C.3.3 Message Based Servant Control

The protocols for Commander-Servant communication involve the Servant's Protocol, Response, and Data registers. They may optionally use the Commander's Signal register, or the VMEbus interrupts. The simplest communications use the data and Response registers to transfer data in a Word Serial fashion. This mode is defined as the *Word Serial protocol.* All Message Based Devices implement this protocol. The Word Serial protocol is the most basic method of communication defined for Message

Based Devices. It is simple to implement (in both hardware and software), yet provides the required communication capability to achieve system objectives.

Communication between Message Based Devices is established with Word Serial protocol. Then (upon notification of device capabilities) it may progress to more sophisticated protocols. These may include higher performance serial protocols, as well as shared-memory protocols. A Commander uses the *read protocol* command to determine which higher performance protocols are supported by a Servant.

RULE C.3.1:
All Message Based Devices **SHALL** implement the *read protocol* command.

C.3.3.1 Serial Protocols

The term *Serial Protocol* is used to apply to any communication protocol in which data is transferred serially to (or from) a fixed location; in this case the read or write data register of the Servant.

The serial protocols are based upon a generalized model of a full duplex UART (Universal Asynchronous Receiver/Transmitter). Each implementation uses bidirectional data registers and a Response register. The data registers are categorized as full duplex, because reads and writes are totally independent. Each write to the data registers is interpreted as a command, unless a previous command has defined it as data. Commands may contain embedded data, or may require supporting data to be transferred in succeeding writes. Such command/data sequences are generally not interruptible.

Progress of a data transfer is paced by bits in the Response register, which indicate whether the write-data registers are empty, and whether the read-data registers are full.

Data may be written into the write-data registers only when the *Write Ready* bit in the Response register is set to one. When data is placed in the write-data registers, the *Write Ready* bit is cleared to zero, until the data is accepted by the Servant.

Valid data may be read from the read data registers only when the *Read Ready* bit in the Response register is set to one. When the data is read from the read-data registers, the *Read Ready* bit is cleared to zero, until the Servant places another word in the read data registers.

In most cases, a Servant will source data to its read-data registers only in response to a command. It is up to its Commander to read such data before issuing a command requiring the output of more data. Thus the Servant is not required to maintain an output queuing structure.

RULE C.3.2:
A Commander **SHALL NOT** send any command requiring a Servant to place data in its data registers until the Commander has read (from the data registers) all data generated by previous commands (and the *Read Ready* bit is zero).

OBSERVATION C.3.1:
The preceding rule does NOT restrict the implementation of output queuing structures within higher level protocols.

Three forms of serial communications are defined in this section: Word Serial (16 bit), Longword Serial (32 bit), and Extended Longword Serial (48 bit). Support for the Longword and Extended LongWord Serial protocols is optional. Data transfers using any of these three protocols may be freely intermixed with data transfers of the other two protocols.

OBSERVATION C.3.2:
Longword and Extended Longword transfers to a Servant that does not support these protocols is meaningless, and will have unpredictable results.

OBSERVATION C.3.3:
Every Message Based Device has write-mode Location Monitors at each of its supported data registers (Data Low, Data High, and Data Extended). These Location Monitors indicate to the device which protocol is being invoked by its Commander for each transfer.

OBSERVATION C.3.4:
The command sets for all forms of serial communication are orthogonal (mutually exclusive). The command sets are described in Section E, "Command and Event Formats".

C.3.3.1.1 Word Serial Transfers

The Word Serial transfer is the minimum usable data transfer protocol. The data path width is 16 bits (one word).

Data is transferred by reads or writes of the Data Low register. By default, all writes are interpreted as commands. Each transfer affects the state of the appropriate bit (*Read Ready* or *Write Ready*) of the Response register.

RULE C.3.3:
IF a Word Serial data transfer requires more than a single bus cycle,
THEN the least significant byte must be transferred last, to ensure validity of the associated *Read Ready* or *Write Ready* bit.

RULE C.3.4:
All Message Based Devices **SHALL** always respond to the Word Serial protocol.

C.3.3.1.2 Longword Serial Transfers

The Longword Serial transfer is a higher performance protocol than the Word Serial transfer, having a 32-bit-wide data path.

Data is transferred by reads or writes of the Data High and Data Low registers. By default, all writes are interpreted as commands. Each Longword Serial transfer affects the state of the appropriate bit (*Read Ready* or *Write Ready*) of the Response register.

RULE C.3.5:
IF a Longword Serial data transfer requires more than a single bus cycle,
THEN the least significant word or byte must be transferred last, to ensure validity of the associated *Read Ready* or *Write Ready* bit.

OBSERVATION C.3.5:
Support of the Longword Serial communication protocol is optional for Message Based Devices. This capability is indicated in the response to the *read protocol* command (see Section E.1, "Word Serial Commands").

C.3.3.1.3 Extended Long Word Serial Transfers

The Extended Longword Serial transfer is a higher performance protocol than the Longword Serial, having a 48-bit-wide data path. This is a write-only protocol.

Data is transferred by writes to the Data Extended, Data High, and Data Low registers. By default, all writes are interpreted as commands. Each Extended Longword Serial transfer affects the state of the *Write Ready* bit of the Response register.

RULE C.3.6:
IF an Extended Longword Serial data transfer requires more than a single bus cycle,
THEN the least significant word or byte must be transferred last, to ensure validity of the associated *Read Ready* or *Write Ready* bit.

PERMISSION C.3.1:
Support of the Extended Longword Serial communication protocol is optional for Message Based Devices. This capability is indicated in the response to the read protocol command (see Section E.1, "Word Serial Commands").

C.3.3.2 Fast Handshake Transfers

This section describes a faster version of the Word Serial protocol.

C.3.3.2.1 Handshake Types

The Word Serial protocols may use either of two handshake modes for data transfer, the **Normal Transfer** mode or the **Fast-Handshake** mode.

The Normal-Transfer mode uses the Servant's *Read Ready* and *Write Ready* bits to synchronize data transfers, as described previously.

The Fast-Handshake mode uses the Servant's DTACK* and BERR* lines to enforce proper synchronization. In this mode, a Servant may hold off a VMEbus transfer for up to 20 μsecs, while waiting for a read or write ready condition. If the 20 μsec time limit expires before the appropriate readiness condition is true, the Servant will assert BERR*. Otherwise the Servant will assert DTACK*, and complete the transfer normally.

C.3.3.2.2 Servant Data Transfer Modes

RULE C.3.7:
A Message Based Servant **SHALL** always support the Normal-Transfer mode, even when in the Fast-Handshake mode.

OBSERVATION C.3.6:
In the Fast-Handshake mode, a device is always in the *Read Ready* and *Write Ready* states. However, some cycles may require holding up the backplane for up to 20μsecs to prevent loss of data.

PERMISSION C.3.2:
When in Fast-Handshake mode, a Servant **MAY** set both the *Read Ready* and the *Write Ready* bits in the Response register continuously to one.

PERMISSION C.3.3:
When in Fast-Handshake mode, a Servant **MAY** control the *Read Ready* and the *Write Ready* bits of the Response register in a manner that allows normal mode transfers to execute without long VMEbus data transfer cycles.

A Servant indicates its support of the Fast-Handshake mode via the *FHS** bit in its protocol register. The *FHS Active** bit of its Response register indicates the current status of the Servant's Fast-Handshake mode. The meanings of these bits are summarized in the following table.

BIT	Value	
	0	1
FHS*	Capable	Incapable
FHS Active*	Active	Inactive

PERMISSION C.3.4:
A Servant **MAY** initiate the Fast-Handshake mode at any time by clearing the *FHS Active** bit. This may be in response to an explicit request by its Commander, an implicit requirement of its application, or any other device-dependent or system-dependent condition.

PERMISSION C.3.5:
A Servant **MAY** have its Fast-Handshake mode always enabled.

RULE C.3.8:
A Servant **SHALL** assert BERR* if it cannot complete a Fast-Handshake data transfer cycle in 20 μsecs.

A Servant terminates the Fast-Handshake mode by asserting BERR* during a data transfer cycle. Under this condition, the Servant updates neither its read nor write queue. It sets its *FHS Active** bit to one, and operates in the Normal-Transfer mode until it is ready to resume the Fast-Handshake mode.

C.3.3.2.3 Commander Data Transfer Modes

The Fast-Handshake mode is optional for Commanders, since the Normal-Transfer mode always functions properly. This is true even if its Servant is in the Fast-Handshake mode, because its *Read Ready* and *Write Ready* status bits are always correct.

A Commander that supports the Fast-Handshake mode may execute Fast-Handshake transfers only with Servants that are capable of Fast-Handshake transfers, and which have the Fast-Handshake mode active.

RULE C.3.9:
A Commander **SHALL NOT** attempt Fast-Handshake mode operations with a Servant having the FHS* bit of its protocol register set to one.

RULE C.3.10:
A Commander **SHALL NOT** attempt Fast-Handshake operations with a Servant while the Servant's *FHS Active** bit is set to one.

A Commander may begin Fast-Handshake transfers with a Fast-Handshake-capable Servant any time it detects the Servant's *FHS Active** bit to be zero. It may exit the fast handshake mode at any time. However, it must exit the Fast-Handshake mode after any data transfer attempt that results in a bus error. The Fast-Handshake mode must not be resumed until the Servant again clears its *FHS* Active* bit to zero.

RULE C.3.11:
IF a Word Serial data transfer attempt to a Fast-Handshake-capable Servant results in the assertion of BERR*,
THEN the Commander **SHALL NOT** attempt any Fast-Handshake transfers with the same Servant until that Servant clears its *FHS Active** bit to zero.

OBSERVATION C.3.7:
No data is transferred during a cycle that causes a bus error. The failed transfer should be retried.

OBSERVATION C.3.8:
Once a Commander detects a Servant in the Fast-Handshake-active condition, it does not need to check the Servant's *Read Ready*, *Write Ready*, *Err**, or *FHS Active** bits again until the next bus error.

C.3.3.3 Byte Transfer Protocol

The **Byte Transfer Protocol** is a mechanism for transferring 8-bit data between Commanders and their Servants. It uses the Word Serial *Byte Available* and *Byte*

Request commands. The *Byte Available* command is used to transfer data from a Commander to its Servant. The *Byte Request* command is used to transfer data from a Servant to its Commander. The data flow is regulated by the use of the *DIR* and *DOR* bits in the Servant's Response register.

RULE C.3.12:
When a Message Based Device is not ready to process a *Byte Available* command, it **SHALL** clear the *DIR* bit in its Response register to zero.

OBSERVATION C.3.9:
When a Message Based Device is ready to process a *Byte Available* command, it will normally set the *DIR* bit in its Response register to one.

PERMISSION C.3.6:
A Message Based Device **MAY** clear the *DIR* bit when it is unable to accept other data input commands.

RULE C.3.13:
When a Message Based Device is not ready to process a *Byte Request* command, it **SHALL** clear the *DOR* bit in its Response register to zero.

OBSERVATION C.3.10:
When a Message Based Device is ready to process a *Byte Request* command, it will normally set the *DOR* bit in its Response register to one.

PERMISSION C.3.7:
A Message Based Device **MAY** clear the *DOR* bit when it is unable to accept other data output commands.

RULE C.3.14:
IF a command to a Message Based Device, not in the Fast-Handshake mode (See Section C.3.3.2, "Fast-Handshake Transfers"), causes the de-assertion of the *DIR* bit,
THEN the *Write Ready* bit **SHALL NOT** be set to one until the *DIR* bit is cleared to zero.

RULE C.3.15:
IF a command to a Message Based Device, not in the Fast-Handshake mode (See Section C.3.3.2, "Fast-Handshake Transfers"), causes the de-assertion of the *DOR* bit,
THEN the *Write Ready* bit **SHALL NOT** be set to one until the *DOR* bit is cleared to zero.

RULE C.3.16:
A Commander **SHALL NOT** send a *Byte Available* command to any Servant which has its *DIR* bit cleared to zero.

RULE C.3.17:
A Commander **SHALL NOT** send a *Byte Request* command to any Servant which has its *DOR* bit cleared to zero.

C.3.3.4 Error Handling

VXIbus Message Based Devices implement a uniform method of reporting Word Serial protocol errors. It is a multi-level approach wherein all Word Serial protocol conformance errors are reported via the *Err** bit in the Response register and the *Read Protocol Error* command. Higher level errors are reported at higher levels, often as status responses to commands.

There are six defined Word Serial protocol errors:

- **Unsupported Command** This error is detected when a Servant receives a command which it does not support.
- **Multiple Query** This error occurs when a Servant receives a command requiring it to output a response to its Data Low register, and is unable to respond because of an unread response to a previous command.
- **DIR Violation** This error occurs when a Servant receives a DIR synchronized command (such as *Byte Available*) when it is unable to process that command because the *DIR* bit is zero.
- **DOR Violation** This error occurs when a Servant receives a DOR synchronized command (such as *Byte Request*) when it is unable to process that command because the *DOR* bit is zero.
- **Write Ready Violation** This error occurs when data is written to a Servant while its *Write Ready* bit is zero.
- **Read Ready Violation** This error occurs when data is read from a Servant while its *Read Ready* bit is zero.

RULE C.3.18:
A Message Based Device **SHALL** detect the *Unsupported Command, Multiple Query, DIR Violation,* and *DOR Violation* errors.

RULE C.3.19:
IF a command to a Message Based Device, not in the Fast-Handshake mode (See Section C.3.3.2, "Fast-Handshake Transfers"), causes the detection of one of the defined Word Serial protocol errors,
THEN the *Write Ready* bit **SHALL NOT** be set to one until both the *Err** and *Read Ready* bits are cleared to zero.

OBSERVATION C.3.11:
Since the *Err** bit is required to be valid after each transfer, there is no need to queue Word Serial protocol errors on the part of a servant.

C.3.3.5 Device Failures

When a device experiences a catastrophic failure, it enters the FAILED state, asserting SYSFAIL* and clearing its *Passed* bit to zero.

RULE C.3.20:
IF a device is in the FAILED state,
THEN that device's Commander **SHALL** set the *Sysfail Inhibit* bit of the Control register on the failed device to one.

PERMISSION C.3.8:
When a device fails, the device's Commander **MAY** also set the *Reset* bit of the Control register to one on the failed device.

OBSERVATION C.3.12:
A Commander may detect that a Servant has failed by monitoring the SYSFAIL* signal, or by polling the *Passed* bit in the Status register of a device.

RECOMMENDATION C.3.1:
When a top-level device fails, the Resource Manager should set the *Sysfail Inhibit* bit of the Control register to one on the failed device.

C.4 SYSTEM RESOURCES

The VXIbus architecture provides for a set of common system resources. These include Slot 0 services and system configuration management. This section describes how these services are provided. The configuration management services are provided by a central Resource Manager, and Slot 0 services are either provided by the Resource Manager or by Slot 0 devices.

C.4.1 Resource Manager

The VXIbus Resource Manager is a device at Logical Address zero that performs the following functions at system power-on.

1. Identify all VXIbus devices in the system.
2. Manage the system self test and diagnostic sequence.
3. Configure the system's A24 and A32 address maps.
4. Configure the system's Commander/Servant hierarchies.
5. Allocate the VMEbus IRQ lines.
6. Initiate normal system operation.

The logical addresses of VXIbus devices can be statically set or dynamically assigned. In this section, the Resource Manager's functions are described for the case of statically-set logical addresses. The functions for dynamic assignment of logical addresses are described in Section F, "Dynamic Configuration".

RULE C.4.1:
A Resource Manager **SHALL** be a Message Based Device with commander capability.

PERMISSION C.4.1:
The functions of a Resource Manager **MAY** be combined with those of another Message Based Commander.

OBSERVATION C.4.1:
If a Resource Manager is combined with other Message Based functionality, the Resource Manager configures its Message Based functions as it would any other Message Based Device. In this case it uses device-specific internal communications, instead of Word Serial commands.

RULE C.4.2:
IF a Resource Manager device is not configured as a top-level Commander,
THEN it **SHALL** behave as a normal Message Based Device toward its Commander and Servants.

RULE C.4.3:
A Resource Manager **SHALL** provide a device-specific mechanism for reporting system configuration errors.

PERMISSION C.4.2:
A Resource Manager **MAY** provide Slot 0 services, in addition to its other duties.

RULE C.4.4:
A Resource Manager which lacks device-specific knowledge of a device **SHALL** write a one to all the device-dependent bits in the Control register, whenever it writes to the Control register of that device.

OBSERVATION C.4.2:
A Resource Manager will normally write to the Control register of a device to affect the *A24/A32 Enable, Sysfail Inhibit,* or *Reset* bits.

C.4.1.1 Device Identification

The Resource Manager identifies the available VXIbus devices according to the following procedure.

RULE C.4.5:
All Resource Managers **SHALL** implement the following device identification procedure.

1. After the de-assertion of SYSRESET*, wait until either SYSFAIL* is de-asserted, or five seconds have elapsed (whichever occurs first) before accessing any other VXIbus device's A16 configuration registers.

2. Attempt a read of the Status register at each of the 256 defined configuration register locations. If an attempt is successful, the corresponding device is present. If a bus error results, the device is absent.

OBSERVATION C.4.3:
Local VMEbus operations (that do not access any of the registers of VXIbus devices) are permitted before the de-assertion of SYSFAIL*, or the five-second timeout.

C.4.1.2 System Self Test Management

RULE C.4.6:
All Resource Managers **SHALL** implement the following self test management procedure.

1. Wait until all self tests have completed.

2. Force all failed devices into the SOFT RESET state with SYSFAIL* assertion inhibited (see Section C.2.1.2, "Device Initialization and Diagnostics").

OBSERVATION C.4.4:
A device has completed its self test when its *Passed* bit is one, or when at least five seconds have elapsed after the deassertion of SYSRESET*.

OBSERVATION C.4.5:
The Resource Manager forces a device into its SOFT RESET state by writing a one to the device's *Reset* bit.

OBSERVATION C.4.6:
The Resource Manager causes a device to inhibit SYSFAIL* assertion by writing a one to the device's *Sysfail Inhibit* bit.

PERMISSION C.4.3:
A Resource Manager **MAY** perform diagnostic tests of a failed device by some device-dependent method.

C.4.1.3 Address Map Configuration

RULE C.4.7:
All Resource Managers **SHALL** implement A24 and A32 address map configuration.

In order to configure the address maps, the Resource Manager reads the address space requirements of each A16/A24 and A16/A32 device, and assigns a base address offset to that device. The offsets are calculated such that no two device address spaces overlap.

OBSERVATION C.4.7:
The algorithm for determining address offsets is not specified by the VXIbus standard. A Resource Manager may assign offsets according to different criteria, such as device class and memory attributes.

C.4.1.4 Commander/Servant Hierarchies

The Resource Manager establishes a system-wide control hierarchy. This structure has the form of one or more inverted trees. The hierarchical relationships are described by the terms Commander and Servant. A Commander is any device in the hierarchy with one or more associated lower-level devices. A Servant is any device in the subtree of a Commander. A device may be both a Commander and a Servant in a multiple-level hierarchy. A Commander has exclusive control of its immediate Servants' communication and control registers. A top-level Commander has no Commander, but may have Servants. All Commanders are Message Based Devices.

A Resource Manager establishes the system hierarchy in the following manner.

1. Find all Commanders by checking the *CMDR* bit in the protocol register of each Message Based Device.

2. Read the Servant area size from each Commander, using the *Read Servant Area* query.

3. Determine the Commander/Servant hierarchies.

4. Assign Servants to Commanders, using the *grant device* command.

RULE C.4.8:
A Resource Manager **SHALL** implement some means of building a Commander/Servant hierarchy.

RECOMMENDATION C.4.1:
A Resource Manager should implement the default Commander/Servant mapping algorithm described in the following section.

OBSERVATION C.4.8:
Each Commander maintains a list of its Servants, received via the *grant device* command. It maintains this list until the next activation of HARD RESET or SOFT RESET, or the receipt of the *Abort Normal Operation* command.

RULE C.4.9:
A Resource Manager **SHALL NOT** grant a failed Servant device to any Commander (see Section C.2.1.2.2, "Self Test Operation").

C.4.1.4.1 Commander/Servant Mapping

The default Commander/Servant mapping algorithm allows for a virtually unlimited number of hierarchical levels. It is based on each VXIbus device's Logical Addresses, and each Commander's Servant area size. The Servant area size is an 8 bit (0-255) value selected either by the user, or at the factory. It is stored in a non-volatile fashion in each Commander device.

A Commander's Servant area starts at the next consecutive Logical Address after the Commander's logical address, and includes the number of contiguous Logical Addresses given by the Commander's Servant area size.

If a device is in the Servant area of a given Commander, and it is not in the Servant area of any other device within that Commander's Servant area, then it is a Servant of that Commander, and may be assigned to that Commander by the Resource Manager.

OBSERVATION C.4.9:
All of a device's default Servants are in its own Servant area.

OBSERVATION C.4.10:
A device's Servant area may include the Servant areas of its Servants.

OBSERVATION C.4.11:
None of a device's Servants are located in any of the Servant areas of its Servants.

C.4.1.5 Allocating IRQ Lines

The Resource Manager is responsible for allocating the VMEbus IRQ lines among the various Interrupt Handlers and Interrupters in the system. Each IRQ line may be assigned to only one handler, but it may be assigned to several Interrupters. Provision must be made for devices with jumper-selectable IRQ line assignments, as well as for programmable Interrupters and Interrupt Handlers. This requires that some interrupt configuration information be supplied by the system user.

RULE C.4.10:
A Resource Manager **SHALL** provide a means for a user to supply interrupt configuration information relating Interrupt Handlers and Interrupters to any or all interrupt lines.

OBSERVATION C.4.12:
The form of the interrupt configuration information is device dependent. It is anticipated that the information will typically be loaded from a system console, downloaded from a host computer, or retrieved from some form of non-volatile storage.

RULE C.4.11:
A Resource Manager **SHALL** allocate IRQ lines to Interrupt Handlers according to the following procedure.

1. Determine which Message Based Devices support Programmable Interrupt Handler (PH) capability, using the *Read Protocol* command.

2. Allocate the IRQ lines as follows:

   a. First use the supplied interrupt allocation information.

   b. Then allocate one of the remaining IRQ lines to each PH-capable Commander, in any desired order.

   c. Finally allocate any remaining IRQ lines to any Servant-only PH-capable Message Based Devices, and to any PH Commanders, with multiple handler capability.

3. For each Programmable Handler assigned an IRQ line in step 2, program its IRQ allocation, using the *Read Handlers, Read Handler Line*, and *Assign Handler Line* commands.

RULE C.4.12:
A Resource Manager **SHALL** allocate IRQ lines to Interrupters according to the following procedure.

1. Determine which Message Based Servants have Programmable Interrupter (PI) capability, using the *Read Protocol* command.

2. Allocate the IRQ lines as follows:

   a. First use the supplied interrupt allocation information.

   b. Then allocate an IRQ line to each PI-capable Servant, based on its Commander's IRQ line allocation.

3. For each Programmable Interrupter assigned an IRQ line in step 2, program its IRQ allocation, using the *Read Interrupters, Read Interrupter Line* and the *Assign Interrupter Line* commands.

C.4.1.6 Initiating Normal Operation

After allocating IRQ lines, a Resource Manager may provide some system-dependent start-up services. It then sends the *Begin Normal Operation* command (with the *Top Level* bit set to one, to all top level Commanders, in order of increasing logical address. At this point the Resource Manager's power-on sequence is complete, and it may enter its run-time operating mode.

Each top level Commander is then required to initialize its Servants. Each Commander must send the *Begin Normal Operation* command (with the *Top Level* bit cleared to zero) to all of its Message Based Servants. Therefore, the *Begin Normal Operation* command propagates down the hierarchy from Commanders to Servants, until all Message Based Devices have received it.

C.4.2 Runtime Resource Management

In Section C.4.1, "Resource Manager", and in Section F, "Dynamic Configuration", resource management is defined only for default configuration. The behavior of a Resource Manager (which provides ongoing services for a device in NORMAL OPERATION sub-state) has not been defined. Ongoing resource management services can be used to support shared resource utilization, or other system operations. In the following discussion, the term "runtime resource management" is used for such ongoing resource management.

PERMISSION C.4.4:
A Resource Manager which handles default system configuration (as described in Section C.4.1, "Resource Manager" and in section F, "Dynamic Configuration") **MAY** handle runtime resource management.

OBSERVATION C.4.13:
It is a violation of the Word Serial protocol for a device involved in runtime resource management to send any Word Serial command (including *End Normal Operation* and *Abort Normal Operation*) to any device, while that device's Commander is in the NORMAL OPERATION sub-state.

C.4.3 VXIbus Subsystem Slot 0

A VXIbus Slot 0 device provides common resources to its VXIbus subsystem slots 1-12. On P2, the resources provided are CLK10 and MODID. On P3, the Slot 0 device provides CLK100. It may also provide SYNC100, STARX, and STARY signal distribution on P3.

OBSERVATION C.4.14:
VXIbus Slot 0 devices exist only on B, C and D form factors.

RULE C.4.13:
All Slot 0 devices **SHALL** supply CLK10.

RULE C.4.14:
All Slot 0 devices **SHALL** support MODID.

RULE C.4.15:
All P3 Slot 0 devices **SHALL** provide CLK100.

OBSERVATION C.4.15:
Slot 0 devices are not required to support SYNC100, STARX, and STARY.

RECOMMENDATION C.4.2:
If only one STAR signal set is provided, it should be STARX.

RECOMMENDATION C.4.3:
IF a Slot 0 device supports either STAR signal set,
THEN it should support a contiguous set of slots, including slot 1.

OBSERVATION C.4.16:
A Slot 0 device is only guaranteed to provide CLK10 and MODID support. It may also provide other services such as SYNC100 arbitration, trigger line buffering, and STARX and STARY routing.

A Slot 0 device may be identified by its *Model Code*, which has a value in the range 0 - 255 (0_{16} - FF_{16}).

RULE C.4.16:
All VXIbus Slot 0 devices **SHALL** have model codes in the range 0_{16} - FF_{16}.

RULE C.4.17:
A VXIbus device which is not enabled for Slot 0 functionality **SHALL NOT** have a model code in the range 0_{16} - FF_{16}. A Slot 0 device with its Slot 0 functions disabled (via switches, jumpers, cut traces, etc) **SHALL** also have its *Model Code* modified to some value outside the range 0_{16} - $FF00_{16}$.

RECOMMENDATION C.4.4:
Some provisions should be made to disable the Slot 0 capability and the VMEbus system controller capability on a Slot 0 module, in order to allow that module to be placed in any slot position.

RULE C.4.18:
All MODID drivers **SHALL** be disabled by both hard and soft resets (See Section C.2.1.2.2, "Self Test Operation").

C.4.3.1 Register Based Slot 0 Devices

A Register Based Slot 0 Device is a standard VXIbus Register Based Device that supports Slot 0 functionality. Its *Model Code* (in its Device Type register - See Section C.2.1.1.2, "Configuration Registers") is in the range 0 - 255 (0_{16} - FF_{16}), and it has a MODID register at its A16 base address + 8_{16}. It may implement additional Slot 0 related registers.

RULE C.4.19:
A Register Based VXIbus Slot 0 Device **SHALL** implement a MODID register, as defined in the following section (C.4.3.1.1) at its A16 base address + 8_{16}.

PERMISSION C.4.5:
The remaining device-dependent registers **MAY** be used to implement other Slot 0 related services.

C.4.3.1.1 MODID Register

The MODID register indicates the status of (and provides control of) the associated MODID lines. It conforms to the following format.

Bit #	15 - 14	13	12 - 0
Contents	RESERVED	Output Enable	MODID00-12

The fields of the MODID register are defined below.

- **RESERVED** This field is reserved for future VXIbus definition. A write to this field has no effect. A read of this field returns all ones (3_{16}).

- **Output Enable** Writing a one to this bit enables the Slot 0 MODID drivers. Writing a zero to this bit disables the MODID drivers. A one read from this bit indicates that the MODID drivers are enabled. A zero indicates that the drivers are disabled.

OBSERVATION C.4.17:
All MODID drivers are disabled by device resets (The *Output Enable* bit is cleared to zero).

- MODID 00-12: Writing a one to any of these bits drives the corresponding MODID line high, if the *Output Enable* bit is set. Writing a zero to any of these bits drives the corresponding line low, if the *Output Enable* bit is set. When read, each of these bits indicates the actual level of the corresponding MODID line (1 = HI, 0 = LOW).

OBSERVATION C.4.18:
After system power up, the bit values of MODID00-12 in the MODID write register might be undefined.

C.4.3.2 Message Based Slot 0 Devices

A Message Based Slot 0 Device is a standard VXIbus Message Based Device that supports Slot 0 functionality. Its *Model Code* (in its Device Type register - See Section C.2.1.1.2, "Configuration Registers") is in the range 0 - 255, and it supports the MODID-related commands. It may implement additional Slot 0 related commands.

RULE C.4.20:
A Message Based VXIbus Slot 0 Device **SHALL** implement the *Read MODID, Set Lower MODID,* and *Set Upper MODID* commands, as defined in Section E.1, "Word Serial Commands".

PERMISSION C.4.6:
A Message Based Slot 0 Device **MAY** implement commands to provide other Slot 0 related services.

The MODID-related commands are used to indicate the status of the MODID lines, and to control the levels of the MODID lines.

OBSERVATION C.4.19:
All MODID drivers are disabled by device resets.

C.4.3.3 Other Slot 0 Devices

Since VXIbus only requires support of MODID and CLK10 lines from a Slot 0 device, the Resource Manager may provide the Slot 0 services, in addition to its other functions. This arrangement hides the Slot 0 functionality inside the Resource Manager, and therefore allows another device to provide Slot 0 functionality, without using a Logical Address for a minimal Slot 0 implementation.

SECTION D

VXIbus Device Implementations

This section describes the design rules for certain devices which facilitate manufacturer-to-manufacturer interchangeability. The protocols are defined at a higher level than defined by the system architecture. These protocols, along with the common command formats, are specified in the following sections.

One common device is a VXIbus-to-IEEE-488-bus [2] interface. This device allows VXIbus devices to be controlled via the IEEE 488 bus, as if they are IEEE 488 devices. The following sections define such an interface, its VXIbus protocols, and the requirements for an instrument to interface to it. The protocols are intended to be extensible to other interfaces and Commanders.

D.1 VXIbus INSTRUMENTS

This section specifies the interfacing requirements for *VXIbus Instruments*. These are devices which may be controlled with the VXIbus Instrument Protocol, implemented by a *VXIbus Interface Device*.

D.1.1 VXIbus Instrument Protocols

RULE D.1.1:
A VXIbus instrument **SHALL** be a Message Based Device, and **SHALL** provide the VXIbus word-serial communication protocol.

VXIbus instruments communicate with the core set of commands and events described in the following sections.

OBSERVATION D.1.1:
The following commands are considered to be Instrument Protocol commands:

- *Byte Available*
- *Byte Request*
- *Clear*
- *Trigger*
- *Set Lock*
- *Clear Lock*
- *Read STB*

[2] See References at the end of Appendix IV (page VXI-246)

D.1.1.1 Data Transfer From Commanders To Instruments

The *Byte Available* command is used to send a data byte (DAB) to a VXIbus instrument. The command includes an END bit, which may be used as a message terminator.

RULE D.1.2:
All VXIbus instruments **SHALL** support the *Byte Available* command, in conformance to the Byte Transfer Protocol.

OBSERVATION D.1.2:
The Byte Transfer Protocol provides for an instrument to hold off incoming Byte Available commands, while still allowing other commands such as *Clear* and *Read STB*.

OBSERVATION D.1.3:
When the *Write Ready* bit becomes asserted, the state of *DIR* allows a Commander to determine whether or not to send a *Byte Available* or *Trigger* command to the Servant. If *DIR* is set, then the Commander may send a *Byte Available* or *Trigger* command.

D.1.1.2 DataTransfer from Instruments To Commanders

Data transfer from an instrument to its Commander is initiated when the Commander sends a *Byte Request* command to the VXIbus instrument. The instrument responds by making the data byte available in its data registers. The returned data includes an END bit, which may be used as a message terminator.

RULE D.1.3:
All VXIbus instruments **SHALL** support the *Byte Request* command, in conformance to the Byte Transfer Protocol.

OBSERVATION D.1.4:
The Byte Transfer Protocol allows an instrument to indicate the presence of data in its output buffer (and thus its readiness for a *Byte Request* command).

OBSERVATION D.1.5:
When the *Write Ready* bit becomes asserted, the state of *DOR* allows a Commander to determine whether or not to send a *Byte Request* command to the Servant. If *DOR* is set, then the Commander can send the *Byte Request* command.

D.1.1.3 Clearing A VXIbus Instrument

A VXIbus instrument which receives a *Clear* command will respond according to the following rules (in addition to Rules C.2.75 and C.2.76).

RULE D.1.4:
Upon receipt of the *Clear* command, the VXIbus instrument **SHALL** perform the following actions:

- clear its input buffer (set *DIR*)
- clear its output queue (clear *DOR*, if appropriate)
- become ready to process new commands

Further clarification of the proper operation of the *Clear* command may be found in the IEEE 488.2 Standard [3], Section 5.8, "Device Clear Requirements".

OBSERVATION D.1.6:
The VXIbus *Clear* command may affect current instrument operation for VXIbus instruments, while for VXIbus IEEE 488.2 instruments it will not.

RULE D.1.5:
All VXIbus instruments **SHALL** support the *Clear* command.

D.1.1.4 Triggering An Instrument

RULE D.1.6:
All VXIbus instruments which support a trigger function **SHALL** be capable of being triggered via the *Trigger* command.

PERMISSION D.1.1:
VXIbus Instruments may support other methods of triggering (via P2 or P3 hardware trigger buses).

RULE D.1.7:
When a VXIbus instrument is not ready to process a *Trigger* command, it **SHALL** clear the *DIR* bit in its Response register to zero.

OBSERVATION D.1.7:
When a VXIbus instrument is ready to process a *Trigger* command, it will normally set the *DIR* bit in its Response register to one.

D.1.1.5 Local Lockout

A Commander can issue the *Set Lock* command to its Servant if it wishes the Servant to ignore commands from local sources. The *Clear Lock* command is used by a Commander to cause a device to exit the Lock state.

RULE D.1.8:
IF an instrument supports the *Set Lock* command,
THEN upon receipt of the *Set Lock* command it **SHALL** clear the *Locked** bit in its Response register to zero.

[3] See References at the end of Appendix IV (page VXI-246)

RULE D.1.9:
IF an instrument supports the *Clear Lock* command,
THEN upon receipt of the *Clear lock* command, it **SHALL** set the *Locked** bit in its Response register to one.

RULE D.1.10:
IF an instrument supports the *Set Lock* command,
THEN local control of the device **SHALL** be disabled, upon the receipt of a *Set Lock* command.

RULE D.1.11:
IF an instrument has had local control disabled by the receipt of a *Set Lock* command,
THEN local control of that device **SHALL NOT** be re-enabled until the receipt of a *Clear Lock* command.

PERMISSION D.1.2:
Local operations that do not affect device operations are allowed while the device is in the locked state, as per IEEE 488.2.

PERMISSION D.1.3:
VXIbus instruments **MAY** implement the *Set Lock* and *Clear Lock* commands.

RECOMMENDATION D.1.1:
If an instrument implements either the *Set Lock* or the *Clear Lock* command, then it should implement both commands.

D.1.1.6 SRQ Operation

When a device needs to request service on an interface, it sends a *request true* event to its Commander. This event can be sent via the event form of the interrupt response word, or as a signal. If the need for service is removed, a device which has previously sent a *request true* event must send the *request false* event. This event is sent in an analogous manner to the *request true* event.

PERMISSION D.1.4:
VXIbus instruments **MAY** implement the *request true* and *request false* events.

RULE D.1.12:
IF a VXIbus Instrument has VMEbus Master capability
THEN the *request true* and *request false* events **SHALL** be sent using signals, unless modified by the *Asynchronous Mode Control* command.

RULE D.1.13:
IF a VXIbus Instrument does not have VMEbus Master capability
THEN the *request true* and *request false* events **SHALL** be sent using interrupts, unless modified by the *Asynchronous Mode Control* command.

D.1.1.7 SPOLL Operation

PERMISSION D.1.5:
VXIbus instruments **MAY** implement the *Read STB* command.

RULE D.1.14:
A VXIbus instrument which implement the *Read STB* command **SHALL** send (as bit 6 of its response) a bit that is equivalent to the MSS bit of the STB? response in IEEE Standard 488.2-1987.

OBSERVATION D.1.8:
This bit indicates whether or not the device is currently requesting service. With an interface other than IEEE 488.1, the RQS function may not be available. In this case, the device indicates it's current needs to the requester via this bit.

D.1.1.8 Error Reporting

Error reporting is divided into two parts: Word Serial protocol errors, and errors resulting from messages sent via the *Byte Available* command. Word Serial protocol error reporting is described in Section C.3.3.4, "Error Handling". Errors that result from higher-level messages (received via the *Byte Available* command) are reported via the *request true* event.

RULE D.1.15:
IF a device reports errors via signals or interrupts, as a result of messages received via the *Byte Available* command,
THEN that device **SHALL** use the *request true* event to notify its Commander of the error.

RECOMMENDATION D.1.2:
A VXIbus Instrument should follow the IEEE 488.2 error reporting conventions (Reference: IEEE 488.2, Sections 11.3 and 11.5) including the enabling and disabling of the *request true* event generation.

D.1.1.9 Initialization

VXIbus instruments require no initialization beyond that required for all Message Based Instruments. Once a VXIbus system is configured, the instruments are ready to process Instrument Protocol commands.

RULE D.1.16:
After the receipt of the *Begin Normal Operation* command, a VXIbus Instrument **SHALL** become ready to accept Instrument protocol commands.

OBSERVATION D.1.9:
Normally, when a device is ready to accept Instrument protocol commands, it will have the following indications: *DIR* =1, *DOR* =0, *Locked* =1, *Read Ready* =0, *Write Ready* =1, and *Err** =1.

OBSERVATION D.1.10:
A device which implements a self-initiated trigger, may have data ready for output when it is ready to accept Instrument protocol commands. It will indicate this with *DOR* =1.

D.1.2 VXIbus IEEE-488.2 Instrument Protocols

This section describes additions to the VXIbus Instrument Protocols that apply to VXIbus IEEE-488.2 Instruments.

RULE D.1.17:
All VXIbus IEEE-488.2 Instruments **SHALL** comply with all the rules for VXIbus Instruments.

D.1.2.1 Clearing A VXIbus IEEE-488.2 Instrument

RULE D.1.18:
The *Clear* command **SHALL NOT** affect current instrument operation for a VXIbus IEEE-488.2 Instrument.

RULE D.1.19:
VXIbus IEEE-488.2 Instruments **SHALL** implement the *Clear* command according to the specifications in the IEEE 488.2 Standard, Section 5.8, "Device Clear Requirements".

D.1.2.2 Triggering An IEEE-488.2 Instrument

OBSERVATION D.1.11:
VXIbus IEEE 488.2 instruments **MAY** support either DT0 or DT1 IEEE 488.1 subsets.

D.1.2.3 Local Lockout

OBSERVATION D.1.12:
VXIbus IEEE 488.2 instruments **MAY** support either RL0 or RL1 IEEE 488.1 subsets.

D.1.2.4 SRQ Operation

RULE D.1.20:
All VXIbus IEEE 488.2 instruments **SHALL** implement the *request true* and *request false* events.

D.1.2.5 SPOLL Operation

RULE D.1.21:
All VXIbus IEEE 488.2 instruments **SHALL** implement all the required 488.2 status reporting registers and protocols and the *Read STB* command.

OBSERVATION D.1.13:
The response to a *Read STB* command will have a bit 6, that is equivalent to the MSS bit of the STB? response in IEEE Standard 488.2-1987.

D.2 488-VXIbus INTERFACE

The 488-VXIbus interface device converts IEEE 488 protocols to VXIbus instrument control protocols, allowing IEEE 488 external controllers to control VXIbus instruments in a manner analogous to a typical rack and stack instrument system. The primary function of the 488-VXIbus Interface device is to route the IEEE 488 messages to the appropriate VXIbus instrument, and to return the results from the instrument to the external IEEE 488 controller. A VXIbus system may or may not provide one or more 488-VXIbus Interface devices.

RULE D.2.1:
A 488-VXIbus Interface **SHALL** be a Message Based Commander.

RULE D.2.2:
A 488-VXIbus Interface device **SHALL** support the receipt of signals.

RULE D.2.3:
A 488-VXIbus Interface device **SHALL** include an Interrupt Handler.

PERMISSION D.2.1:
When a 488-VXIbus Interface device detects that a Servant has not recognized a command (error response from the Servant) the interface action is left to the device designer.

D.2.1 IEEE 488 Address Mapping

The mapping of an IEEE 488 address to a VXIbus Logical Address is a function which is handled by the 488-VXIbus Interface device. The particulars of this mapping are interface device specific. Several potential mappings are described below:

- The 488-VXIbus Interface device supports primary addressing of 31 potential addresses. The 31 IEEE 488 addresses will be associated one-for-one with 31 Logical Addresses of the VXIbus instruments.

- The 488-VXIbus Interface device supports secondary addressing of 31 potential secondary addresses. The 488-VXIbus Interface device will provide a means of selecting the primary address. The 31 IEEE 488 secondary addresses will be associated one-for-one with 31 Logical Addresses of the VXIbus instruments.

- The 488-VXIbus Interface device supports a single primary IEEE 488 address, through which all VXIbus Instruments are addressed. The particular language constructs which implement routing are up to the interface device designer.

OBSERVATION D.2.1:
488-VXIbus Interface devices can use these (or any other) address mapping scheme.

OBSERVATION D.2.2:
488-VXIbus Interface devices which implement different address mapping schemes might require address changes for portability. The application which controls VXIbus instruments must consider the address mapping.

D.2.2 488-VXIbus Interface Device to IEEE 488 Bus Functions.

A 488-VXIbus interface device will provide a set of capabilities which enables a VXIbus instrument to comply with IEEE 488.2. The actual choice of whether or not a VXIbus instrument will implement IEEE 488.2 is left to the instrument designer.

RULE D.2.4:
All 488-VXIbus Interface devices **SHALL** implement the following IEEE 488.1 subsets, in an IEEE 488.2 compatible fashion, for each VXIbus Instrument it controls:

- SH1 Source Handshake.
- AH1 Acceptor Handshake.
- T6 Addressing Talker or TE6 Extended Addressing Talker - (No Talk-Only Mode).
- L4 Listener or LE4 Extended Listener - (No Listen Only Mode).
- SR1 Service Request.
- DC1 Device Clear.
- DT1 Device Trigger.

OBSERVATION D.2.3:
The preceding rule guarantees that a VXIbus Instrument that is compatible with IEEE 488.2 will be IEEE 488.2 compatible when accessed through the 488-VXIbus interface device.

PERMISSION D.2.2:
RL1 Remote/Local capability **MAY** be implemented by the 488-VXIbus Interface device.

PERMISSION D.2.3:
Controller capability **MAY** be implemented by the 488-VXIbus Interface device.

RULE D.2.5:
IF a 488-VXIbus Interface device supports controller capability,
THEN the controller **SHALL** follow the IEEE 488.2 implementation.

RECOMMENDATION D.2.1:
488-VXIbus Interface devices should implement E2 three-state electrical drivers.

RULE D.2.6:
IF a 488-VXIbus Interface device implements parallel poll capability,
THEN it **SHALL** be implemented according to IEEE 488.2 rules (PP1).

RULE D.2.7:
The 488-VXIbus Interface device **SHALL** preserve the order of received messages, including data bytes, END messages, Group Execute Trigger, Device Clear, and Selective Device Clear.

D.2.3 VXIbus Instrument Protocol

RULE D.2.8:
VXIbus Instruments and VXIbus Interface devices SHALL be logically partitioned according to Figure D.1.

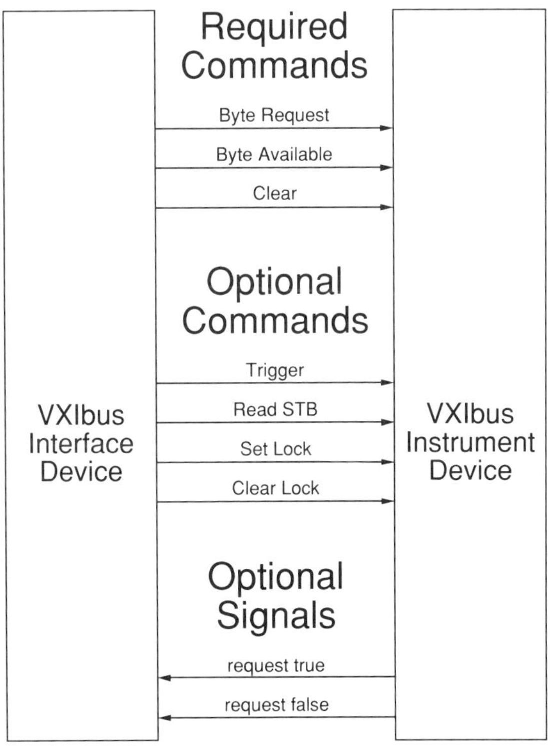

Figure D.1. VXIbus Instrument Protocol

RULE D.2.9:
IEEE 488.2 compatible VXIbus Instruments **SHALL** be logically partitioned according to Figure D.2.

D.2.3.1 Data Transfer From Interface Device To VXIbus Instrument

RULE D.2.10:
IF EOI is asserted on receipt of a data byte from the external interface,
THEN the interface device **SHALL** send the *Byte Available* command with END asserted to all Servants which are addressed to listen (LACS).
ELSE the interface device **SHALL** send the *Byte Available* command with END unasserted to all Servants which are addressed to listen (LACS).

OBSERVATION D.2.4:
A VXIbus Interface device cannot send END for message terminators that do not have EOI asserted.

OBSERVATION D.2.5:
A VXIbus Interface device cannot send a *Byte Available* command to an instrument which has its *DIR* bit cleared to zero.

D.2.3.2 Data Transfer From VXIbus Instrument To VXIbus Interface Device

RULE D.2.11:
When an interface device receives a byte request from the external interface, it **SHALL** perform the following actions on the VXIbus:

- The interface device sends the *Byte Request* command to the Servant device that is currently addressed to talk (TACS).
- The interface device then reads the result of the *Byte Request* command from the appropriate Servant device, using the Word Serial protocol.
- The interface places the DAB portion of the response onto the external interface, and simultaneously drives the EOI line to the appropriate level...
  IF the END bit in the response is asserted
  THEN simultaneously, the interface **SHALL** assert the EOI line
  ELSE the interface **SHALL NOT** assert the EOI line.

OBSERVATION D.2.6:
A VXIbus Interface device cannot send a *Byte Request* command to an instrument which has its *DOR* bit cleared to zero.

D.2.3.3 Device Clear Operation

This section describes the operation of a 488-VXIbus Interface device, when it receives an IEEE 488 device clear message.

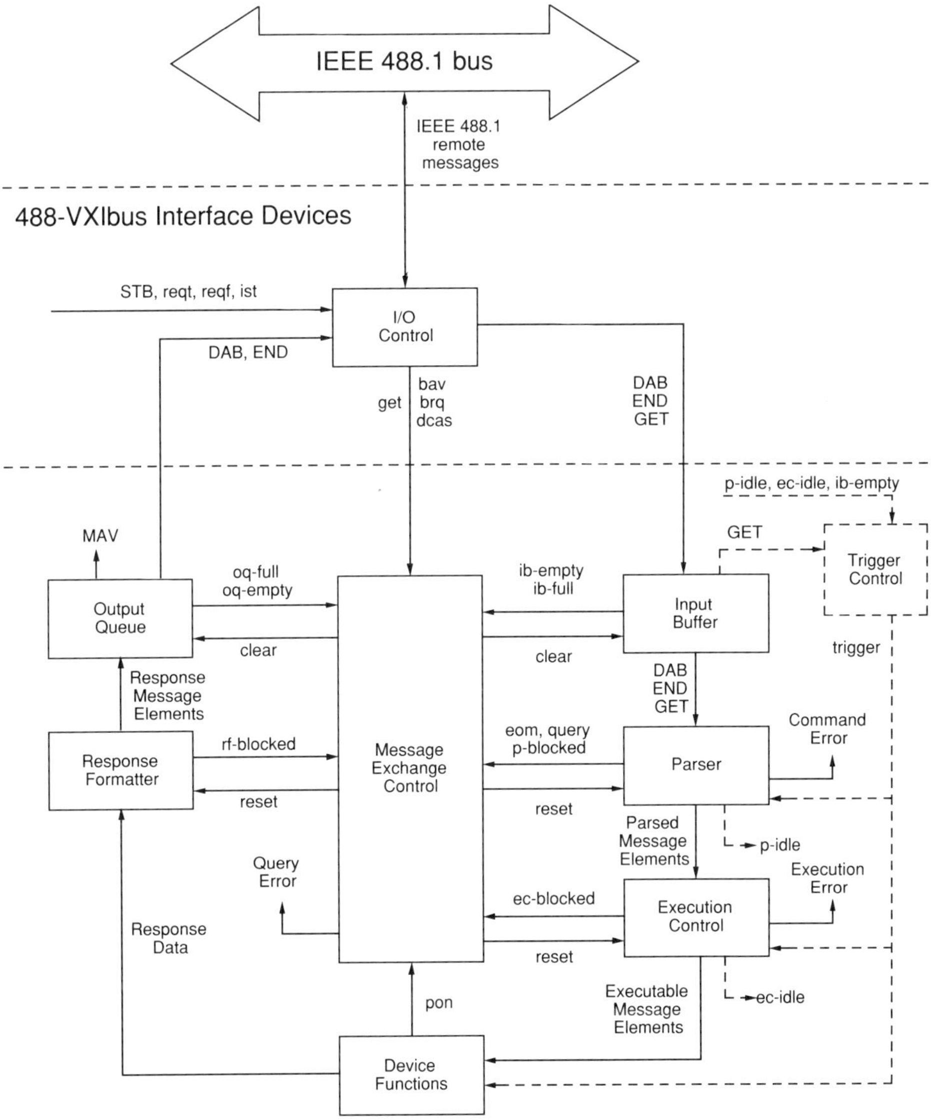

Figure D.2. Message Exchange Control Interface Functional Blocks

RULE D.2.12:
When the 488-VXIbus Interface device receives the 488.1 DCL command, it **SHALL** send the *Clear* command to all VXIbus instruments that it controls.

RULE D.2.13:
When the 488-VXIbus Interface device receives the 488.1 SDC command, it **SHALL** send the *Clear* command to all VXIbus instruments that it controls, and that are addressed to listen.

RULE D.2.14:
The 488-VXIbus Interface device **SHALL** hold off NDAC on the IEEE 488 interface until the *Clear* command has been sent to all affected devices.

RULE D.2.15:
The 488-VXIbus interface **SHALL NOT** perform a soft reset of any VXIbus instruments in response to any 488.1 defined commands or messages, such as DCL, SDC, or IFC.

D.2.3.4 Trigger Operation

This section describes the operation of a 488-VXIbus Interface device when it receives an IEEE 488 trigger message.

RULE D.2.16:
Upon receipt of the 488.1 GET message, the 488-VXIbus Interface device **SHALL** send the Trigger command to all VXIbus instruments that:

- are addressed to listen,
- AND are controlled by that particular interface device,
- AND implement the Trigger command,
- AND are not configured for triggering via an alternate trigger implementation.

OBSERVATION D.2.7:
The fourth condition in the above rule allows Servants that are configured for triggering via an alternate trigger implementation to be triggered via that implementation, instead of with a Trigger command.

RULE D.2.17:
The 488-VXIbus Interface device **SHALL** hold off NDAC on the IEEE 488 interface until the *Trigger* command has been sent to all affected devices.

OBSERVATION D.2.8:
There will be a time delay between the receipt of the *Trigger* command by the various VXIbus instruments in the VXIbus system.

OBSERVATION D.2.9:
The possible inclusion of P1-only VXIbus devices in the system mandates a Trigger command, rather than use of a trigger line.

RULE D.2.18:
A VXIbus Interface device **SHALL NOT** send a *Trigger* command to an instrument that has its *DIR* bit cleared to zero.

D.2.3.5 IEEE 488 Remote Local

This section discusses the implementation of the IEEE-488 RL1 functions within the context of a VXIbus system. Local operations are defined to be operations that are initiated from any non-VXIbus device interface.

D.2.3.5.1 Locking Devices

The 488-VXIbus Interface will issue the *Set Lock* and *Clear Lock* commands according to the remote/local state of the associated device (as maintained in a *state table*) according to the rules below:

RULE D.2.19:
IF A 488-VXIbus Interface Device implements RL1,
THEN it **SHALL** issue the *Set Lock* command to a device which implements the *Set Lock* command when the interface determines that the device should enter the RWLS state of the IEEE-488.1.

RULE D.2.20:
IF A 488-VXIbus Interface Device implements RL1,
THEN it **SHALL** issue the *Clear Lock* command to a device which implements the *Clear Lock* command when the interface determines that the device should exit the RWLS state of the IEEE-488.1.

D.2.3.6 SRQ Operation

This section describes the 488-VXIbus Interface device requirements for support of the IEEE 488.2 SRQ capability. 488.2 simplifies the implementation of service requests by requiring two synchronization messages, which control the assertion of the 488.1 rsv message. Two messages are sourced by the device: the reqt message is used to assert a 488.1 rsv (requires service message) and the reqf message is used to asynchronously deassert rsv. In a VXIbus system these messages are sent using the *request true* and *request false* events. The 488.2 synchronization protocols allow the interface device to:

- Assert SRQ if a previously "enabled" condition occurs.
- Keep SRQ asserted until the controller has recognized the service request and has serial polled the device, or the instrument has removed the request (by sending a *request false* event).

- Release SRQ when serial polled, so that the controller can detect an SRQ from another instrument.

- Assert SRQ again if another condition occurs, whether or not the controller has cleared the first condition.

For further clarification of SRQ and serial poll operation of the 488-VXIbus Interface device, see IEEE 488.2, Section 11.3.3, "Service Request Generation".

RULE D.2.21:
A 488-VXIbus Interface device, upon receipt of a *request true* event from a VXIbus instrument, **SHALL** set the 488.1 rsv message true.

RULE D.2.22:
A 488-VXIbus Interface device, upon receipt of a *request false* event from a VXIbus instrument, **SHALL** set the 488.1 rsv message false only if no *request true* has been received from any other VXIbus Instrument.

OBSERVATION D.2.10:
The VXIbus interface device has to keep track of requests for each of its Servants in order to properly control the SRQ line on the IEEE 488.1 interface.

RULE D.2.23:
A 488-VXIbus Interface device **SHALL** implement the IEEE 488.2 SRQ protocols.

OBSERVATION D.2.11:
Request true and *request false* events may be sent via either interrupts or signals. The 488-VXIbus Interface device is required to accept these events, regardless of method of transmission.

D.2.3.7 SPOLL Operation

This section discusses the implementation of the IEEE 488 Serial Poll function within the context of a VXIbus system. The IEEE-488-specific functions (such as whether the device is in SPAS) are managed by the IEEE 488 VXIbus Interface device. When the IEEE 488-VXIbus interface device is ready to source the serial poll response to the 488 interface, it sends a *Read STB* command to the device for which the serial poll was requested. It then reads (using the word serial protocol) the status byte of the polled instrument, and makes this byte available to the interface as the serial poll response.

RULE D.2.24:
IF a VXIbus instrument supports the *Read STB* command,
THEN the IEEE 488-VXIbus Interface device **SHALL** respond to a serial poll of that instrument with the status byte obtained by the *Read STB* command.

RULE D.2.25:

IF a VXIbus instrument doesn't support the *Read STB* command,

THEN the IEEE 488-VXIbus Interface device **SHALL** respond to a serial poll of that instrument with bit 6 of the serial poll response byte properly indicating the rsv state of the instrument. Bit 6 will be set to one if the instrument is requesting service. Otherwise it is cleared to zero.

RULE D.2.26:

An IEEE 488-VXIbus Interface device **SHALL** implement SPOLL in the manner specified by IEEE 488.2.

SECTION E

Command And Event Formats

This section summarizes the required commands, the optional commands, and the events that support the instrument protocols.

E.1 WORD SERIAL COMMANDS

The following commands are the standard set of VXIbus device commands, that are sent with the word-serial protocol.

RULE E.1.1:
Message Based Devices **SHALL** implement the appropriate subsets of the following command set, as specified in Table E.1 and E.2.

TABLE E.1. General Command Requirements

	Primary Classification			CMDR	MASTER	Slot 0	Comments
	All MBD	I	I4				
Abort Normal Operation	R	R	R	R	R	R	
Begin Normal Operation	R	R	R	R	R	R	
Byte Available		R	R				Requires *DIR* bit
Byte Request		R	R				Requires *DOR* bit
Clear	R	R	R	R	R	R	
Clear Lock							Required with RL 1 Requires *Locked** bit
End Normal Operation	R	R	R	R	R	R	
Grant Device				R			
Identify Commander				R	R		
Read MODID						R	
Read Protocol	R	R	R	R	R	R	
Read Protocol Error	R	R	R	R	R	R	
Read STB			R				
Read Servant Area				R			
Release Device				R			
Set Lock							Required with RL 1 Requires *Locked** bit
Set Lower MODID						R	
Set Upper MODID						R	
Trigger							Optional for all devices

R - Required for this type of device.

TABLE E.2. Asynchronous Command Requirements

	Programmable Handler	Programmable Interrupter	Response Generation	Event Generation	Comments
Assign Handler Line	R				
Assign Interrupter Line		R			
Asynchronous Mode Control			R	R	
Control Event				R	
Control Response			R		
Read Handler Line	R				
Read Handlers	R				
Read Interrupter Line		R			
Read Interrupters		R			

R - Required for this type of device.

The following is a list of commands:

Abort Normal Operation The *Abort Normal Operation* command causes a device to cease normal operation. Upon receipt of this command the device returns to its default configuration, aborting all operations. The aborted state is defined as follows:

- All VMEbus Master activity is halted.
- Pending interrupts are unasserted.
- No new bus requests or interrupts may be asserted.
- The device is in a generally inactive state, and is ready to accept commands.

The syntax of the *Abort Normal Operation* command is defined in the following table.

15	14	13	12	11	10	9	8	7	6	5	4	3	2	1	0
1	1	0	0	1	0	0	0	1	1	1	1	1	1	1	1

When the abort operation has completed (and the device is in the aborted state) response data is placed in the Data Low register in the following format:

15	14	13	12	11	10	9	8	7	6	5	4	3	2	1	0
1	1	1	1	1	1	1	1	1	1	1	1	1	1	1	0

Assign Handler Line The *Assign Handler Line* command assigns a VMEbus IRQ line to a particular Interrupt Handler in the Servant device. The syntax of this command is defined in the following table.

15	14	13	12	11	10	9	8	7	6	5	4	3	2	1	0
1	0	1	0	1	0	0	1	X	Hand_ID			X	Line		

- **X** Don't care. The value written to this bit has no effect.

- **Hand_ID** This is a unique identifier of the particular Interrupt Handler being assigned. It has a range of 1 to 7. Within a device, multiple handlers are identified in consecutive order, starting at 1.

- **Line** This is the VMEbus IRQ line number. A value of zero (000) indicates that the handler is to be disconnected.

When the assignment operation has completed, response data is placed in the Data Low register in the following format:

15	14	13	12	11	10	9	8	7	6	5	4	3	2	1	0
Status				1	1	1	1	1	1	1	1	1	1	1	0

- **Status** This field indicates the execution status of the command. It may have the following values:

  - F_{16} The command successfully completed.

  - 7_{16} Command failed - The handler referenced in the Hand_ID field is unknown to this device.

OBSERVATION E.1.1:
There is no implicit relationship between the Handler ID (Hand_ID) and the functionality of the associated handler.

Assign Interrupter Line The *Assign Interrupter Line* command is used to assign a VMEbus IRQ line to a particular Interrupter in the Servant device. The syntax of this command is defined in the following table.

15	14	13	12	11	10	9	8	7	6	5	4	3	2	1	0
1	0	1	0	1	0	1	0	X	Int_ID			X	Line		

- **X** Don't care. The value written to this bit has no effect.

- **Int_ID** This is a unique identifier of the particular Interrupter being assigned. It has a range of 1 to 7. Within a device, multiple Interrupters are identified in consecutive order, starting at 1.

- **Line** This is the VMEbus IRQ line number. A value of zero indicates that the Interrupter is to be disconnected.

When the assignment operation has completed, response data is placed in the Data Low register in the following format:

15	14	13	12	11	10	9	8	7	6	5	4	3	2	1	0
Status				1	1	1	1	1	1	1	1	1	1	1	0

- **Status** This field indicates the execution status of the command. It may have the following values:

  - F_{16} The command successfully completed.

  - 7_{16} Command failed - The handler referenced in the Int_ID field is unknown to this device.

OBSERVATION E.1.2:
There is no implicit relationship between the Interrupter ID (Int_ID) and the functionality of the associated Interrupter.

Asynchronous Mode Control The *Asynchronous Mode Control* command is used by a Commander to direct the path of events and responses. The syntax of this command is defined in the following table.

15	14	13	12	11	10	9	8	7	6	5	4	3	2	1	0
1	0	1	0	1	0	0	0	X				Resp. En*	Event En*	Resp. Mode	Event Mode

The bits have the following meanings:

- **X** Don't care. The value written to this bit has no effect.

- **Resp. En*** A zero enables generation of responses. A one disables generation of responses.

- **Event En*** A zero enables generation of events. A one disables generation of events.

- **Resp. Mode** A one indicates that responses should be sent as signals. A zero indicates that responses should be sent as interrupts.

- **Event Mode** A one indicates that events should be sent as signals. A zero indicates that events should be sent as interrupts.

The result data is placed in the Data Low register with the following format. This result is a confirmation or denial of the command.

15–12	11	10	9	8	7	6	5	4	3	2	1	0
Status	1	1	1	1	1	1	1	1	Resp. En*	Event En*	Resp. Mode	Event Mode

The bits have the following meanings:

- **Status** This field indicates the execution status of the command. It may have the following values:

  - **F_{16}** The command successfully completed, as expected.

  - **7_{16}** Command failed - A requested option is not supported.

- **Resp. En*** A zero indicates that the generation of responses is enabled. A one indicates that the generation of responses is disabled.

- **Event En*** A zero indicates that the generation of events is enabled. A one indicates that the generation of events is disabled.

- **Resp. Mode** A one indicates that responses are being sent as signals. A zero indicates that responses are being sent as interrupts.

- **Event Mode** A one indicates that events are being sent as signals. A zero indicates that events are being sent as interrupts.

Begin Normal Operation The *Begin Normal Operation* command notifies the device that it can begin normal operation now. A Commander may not initiate any VMEbus operations after power-up, until it receives the *Begin Normal Operation* command. It then sends the Begin Normal Operation command to all of its current Message Based Servants, and begins normal execution of commands and/or application programs. The *Top_Level* field of the *Begin Normal Operation* command is provided to inform a device whether or not it is a top level Commander. The syntax of this command is defined in the following table.

15	14	13	12	11	10	9	8	7	6	5	4	3	2	1	0
1	1	1	1	1	1	0	Top Level	1	1	1	1	1	1	1	1

- **Top Level** A one in this field indicates that the device is a top level Commander. A zero indicates that it is a Servant to another device.

When the begin operation has completed, response data is placed in the Data Low register in the following format:

15	14	13	12	11	10	9	8	7	6	5	4	3	2	1	0
Status				State				Logical Address							

- **Status** This field indicates the execution status of the command. It may have the following values:

  - F_{16} The Begin Normal Operation command has been successfully executed. The value FE_{16} is reported in the Logical Address field.

  - 7_{16} The Commander device is not able to command a Servant. The Servant's Logical Address is reported in the Logical Address field.

  - 6_{16} The Commander device is not able to configure a Servant device. The Servant's Logical Address is reported in the Logical Address field.

  - 5_{16} The Commander device timed out when waiting for a response from a Servant device. The Servant's Logical Address is reported in the Logical Address field.

  - 4_{16} The device is not able to successfully initialize itself. The value FE_{16} is reported in the Logical Address field.

  - 3_{16} The device is not able to be a top level Commander. The value FE_{16} is reported in the Logical Address field.

  - 2_{16} The device must be a top level Commander. The value FE_{16} is reported in the Logical Address field.

  - 1_{16} An undefined error was caused. The value FE_{16} is reported in the Logical Address field.

- **State** This field indicates the state of this device and its sub-tree. It may have the following values:

  - 3_{16} This device (and all of its Servant tree) are in the CONFIGURE sub-state.

  - 7_{16} This device is in the CONFIGURE sub-state. However at least one device in its Servant tree is in the NORMAL OPERATION sub-state.

  - B_{16} This device is in the NORMAL OPERATION sub-state. However at least one device in its Servant tree is in the CONFIGURE sub-state.

  - F_{16} This device and all of its Servant tree are in the NORMAL OPERATION sub-state.

- **Logical Address** This field contains the Logical Address corresponding to the status field values.

Byte Available The *Byte Available* command is used by a Commander to send a byte of data to a Servant device. The END field identifies that this is the last byte of the message. The syntax of this command is defined in the following table.

15	14	13	12	11	10	9	8	7	6	5	4	3	2	1	0
1	0	1	1	1	1	0	END	Datum							

Byte Request The *Byte Request* command is used by a Commander to read a byte of data from a Servant device. The syntax of this command is defined in the following table.

15	14	13	12	11	10	9	8	7	6	5	4	3	2	1	0
1	1	0	1	1	1	1	0	1	1	1	1	1	1	1	1

The result data is placed in the Data Low register with the following format. The END field is used to indicate the last byte of the message.

15	14	13	12	11	10	9	8	7	6	5	4	3	2	1	0
1	1	1	1	1	1	1	END	Datum							

Clear The *Clear* command is used by a Commander to cause a Servant device to clear the VXIbus interface, and any pending operations. Any initiated operations in the receiving device are undisturbed. The syntax of this command is defined in the following table.

15	14	13	12	11	10	9	8	7	6	5	4	3	2	1	0
1	1	1	1	1	1	1	1	1	1	1	1	1	1	1	1

Clear Lock The *Clear* Lock command is used by a Commander to cause a Servant device to exit the locked state. When a device receives this command, it will set its Locked* bit. The syntax of this command is defined in the following table.

15	14	13	12	11	10	9	8	4	3	2	1	0			
1	1	1	0	1	1	1	1	1	1	1	1	1	1	1	1

Control Event The *Control Event* command is used by a Commander to selectively enable the generation of events by a Servant. A one in the enable field enables the generation of the specific event. A zero in the enable field disables the generation of the specific event. The syntax of this command is defined in the following table.

15	14	13	12	11	10	9	8	7	6	5	4	3	2	1	0
1	0	1	0	1	1	1	1	Enable	Event						

- **Event** These bits (6 - 0) are the identifying bits (14 - 8) of the event being enabled or disabled. See Section E.4, "Protocol Events", for further information.

The device returns the following data in the Data Low register:

15	14	13	12	11	10	9	8	7	6	5	4	3	2	1	0
Status				1	1	1	1	1	1	1	1	1	1	1	0

- **Status** This field indicates the execution status of the command. It may have the following values:

  - F_{16} The command successfully completed.

  - 7_{16} Command failed - The event referenced is not generated by this device.

Control Response The *Control Response* command is required for devices which implement response signals or response interrupts. It is used to enable response signals or response interrupts on certain response register bit transitions. The syntax of this command is defined in the following table.

15	14	13	12	11	10	9	8	7	6	5	4	3	2	1	0
1	0	0	0	1	1	1	1	X	B14*	DOR*	DIR*	Err*	RR*	WR*	FHS*

The bits have the following meanings:

- **X** Don't care. The value written to this bit has no effect.

- **B14*** A zero enables a signal/interrupt on transitions of bit 14 of the Response register. A one disables this capability. Since bit 14 of the Response register is reserved (always one), the value of this bit may be ignored by the Servant.

- **DOR*** A zero enables a signal/interrupt on a zero-to-one transition of the *DOR* bit. A one disables this capability.

- **DIR*** A zero enables a signal/interrupt on a zero-to-one transition of the *DIR* bit. A one disables this capability.

- **Err*** A zero enables a signal/interrupt on a one-to-zero transition of the *Err** bit. A one disables this capability.

- **RR*** A zero enables a signal/interrupt on a zero-to-one transition of the *Read Ready* bit. A one disables this capability.
- **WR*** A zero enables a signal/interrupt on a zero-to-one transition of the *Write Ready* bit. A one disables this capability.
- **FHS*** A zero enables a signal/interrupt on a one-to-zero transition of the *FHS Active** bit. A one disables this capability.

The result data is placed in the Data Low register with the following format. This result is a confirmation/denial of the command.

15	14	13	12	11	10	9	8	7	6	5	4	3	2	1	0
Status				1	1	1	1	1	B14*	DOR*	DIR*	Err*	RR*	WR*	FHS*

The bits have the following meanings:

- **Status** This field indicates the execution status of the command. It may have the following values:

  - F_{16} The command successfully completed, as expected.

  - 7_{16} Command failed - an unsupported bit transition was requested.

- **B14*** A zero indicates that interrupt/signal generation on transitions of bit 14 of the Response register is enabled. A one indicates that this capability is disabled. Since bit 14 of the Response register is reserved (and should never change) this bit should always be set to one.
- **DOR*** A zero indicates that interrupt/signal generation on a zero-to-one transition of the *DOR* bit is enabled. A one indicates that this capability is disabled.
- **DIR*** A zero indicates that interrupt/signal generation on a zero-to-one transition of the *DIR* bit is enabled. A one indicates that this capability is disabled.
- **Err*** A zero indicates that interrupt/signal generation on a one-to-zero transition of the *Err** bit is enabled. A one indicates that this capability is disabled.
- **RR***: A zero indicates that interrupt/signal generation on a zero-to-one transition of the *Read Ready* bit is enabled. A one indicates that this capability is disabled.
- **WR*** A zero indicates that interrupt/signal generation on a zero-to-one transition of the *Write Ready* bit is enabled. A one indicates that this capability is disabled.
- **FHS*** A zero indicates that interrupt/signal generation on a one-to-zero transition of the *FHS* Active* bit is enabled. A one indicates that this capability is disabled.

OBSERVATION E.1.3:
A Servant might not implement response signals or response interrupts for some of the defined bits. Such a Servant will always return a one in the corresponding bit position(s), in its response to the *Control Response* command.

End Normal Operation The *End Normal Operation* command is used to cause a Commander to cease normal operation in an orderly manner. Upon receipt of this command, the Commander issues the *End Normal Operation* command to all of its Message Based Servants. The Commander is responsible for halting all its Servants. The "ended" state is defined as follows: All VMEbus Master activity is halted. Pending interrupts are unasserted. No new bus requests or interrupts may be asserted. The device is in a generally inactive state, and is ready to accept commands.

The syntax of the *End Normal Operation* command is defined in the following table.

15	14	13	12	11	10	9	8	7	6	5	4	3	2	1	0
1	1	0	0	1	0	0	1	1	1	1	1	1	1	1	1

When the "ending" operation has completed, response data is placed in the Data Low register in the following format:

15	14	13	12	11	10	9	8	7	6	5	4	3	2	1	0
Status				State				Logical Address							

- **Status:** This field indicates the execution status of the command. It may have the following values:

- F_{16} The End Normal Operation command has been successfully executed. The value FE16 is reported in the Logical Address field.

- 7_{16} The device was already in the CONFIGURE sub-state. The value FE_{16} is reported in the Logical Address field.

- 6_{16} The Commander device timed out when waiting for the response from a Servant device. The Servant's Logical Address is reported in the Logical Address field.

- 5_{16} The device is not able to end its operation in a consistent manner. The value FE_{16} is reported in the Logical Address field.

- 4_{16} The Commander device is not able to deactivate a Servant. The Servant's Logical Address is reported in the Logical Address field.

- 3_{16} An undefined error was caused. The value FE_{16} is reported in the Logical Address field.

- **State** This field indicates the state of this device and its sub-tree. It may have the following values:

  - F_{16} This device and all of its Servant tree are in the CONFIGURE sub-state.

  - B_{16} This device is in the CONFIGURE sub-state, however at least one device in its Servant tree is in the NORMAL OPERATION sub-state.

  - 9_{16} This device is in the CONFIGURE sub-state. However the sub-states of the devices in its Servant tree are unknown.

  - 7_{16} This device is in the NORMAL OPERATION sub-state. However at least one device in its Servant tree is in the CONFIGURE sub-state.

  - 3_{16} This device (and all of its Servant tree) are in the NORMAL OPERATION sub-state.

- Logical Address: This field contains the Logical Address corresponding to the status field values.

OBSERVATION E.1.4:
The *End Normal Operation* command (unlike the *Abort Normal Operation* command) allows for the completion of some pending operations before "ending".

Grant Device The *Grant Device* command is used by the Resource Manager to inform a Commander of the Logical Address of its Servants. The syntax of this command is defined in the following table.

15	14	13	12	11	10	9	8	7	6	5	4	3	2	1	0
1	0	1	1	1	1	1	1	Servant's Logical Address							

Identify Commander The *Identify Commander* command is used by a Commander to tell a Servant the Commander's Logical Address. This command allows the Servant to reply with signals rather than interrupts. The syntax of this command is defined in the following table.

15	14	13	12	11	10	9	8	7	6	5	4	3	2	1	0
1	0	1	1	1	1	1	0	Commander's Logical Address							

Read Handler Line The *Read Handler Line* command is used to determine which VMEbus IRQ line a particular Interrupt Handler in the Servant device is connected to. The syntax of this command is defined in the following table.

15	14	13	12	11	10	9	8	7	6	5	4	3	2	1	0
1	0	0	0	1	1	0	0	X					Hand_ID		

- **X** Don't care. The value written to this field has no effect.

- **Hand_ID** This is a unique identifier of the particular Interrupt Handler being queried. It has a range of 1 to 7. Within a device, multiple Interrupters are identified in consecutive order, starting at 1.

The VMEbus IRQ line number is placed in the Data Low register with the following format.

15	14	13	12	11	10	9	8	7	6	5	4	3	2	1	0
Status				1	1	1	1	1	1	1	1	1	Line		

- **Status** This field indicates the execution status of the command. It may have the following values:

  - **F_{16}** The command successfully completed.

  - **7_{16}** Command failed - The handler referenced in the Hand-ID field is unknown to this device.

- **Line** This is the VMEbus IRQ line number currently assigned. A value of zero (000) indicates that the handler is disconnected.

Read Handlers The *Read Handlers* command is used to determine the number of Interrupt Handlers within a Servant device. The syntax of this command is defined in the following table.

15	14	13	12	11	10	9	8	7	6	5	4	3	2	1	0
1	1	0	0	0	1	1	1	1	1	1	1	1	1	1	1

The number of Interrupt Handlers is placed in the Data Low register with the following format.

15	14	13	12	11	10	9	8	7	6	5	4	3	2	1	0
1	1	1	1	1	1	1	1	1	1	1	1	1	Hand_no		

- Hand_no: This is the number of Interrupt Handlers within this device.

Read Interrupter Line The *Read Interrupter Line* command is used to determine which VMEbus IRQ line a particular Interrupter in the Servant device is connected to. The syntax of this command is defined in the following table.

15	14	13	12	11	10	9	8	7	6	5	4	3	2	1	0
1	0	0	0	1	1	0	1	X					Int_ID		

- **X** Don't care. The value written to this field has no effect.

- **Int_ID** This is a unique identifier of the particular Interrupter being queried. It has a range of 1 to 7. Within a device, multiple Interrupters are identified in consecutive order, starting at 1.

The VMEbus IRQ line number is placed in the Data Low register with the following format.

15	14	13	12	11	10	9	8	7	6	5	4	3	2	1	0
Status				1	1	1	1	1	1	1	1	1	Line		

- **Status** This field indicates the execution status of' the command. It may have the following values:

  - F_{16} The command successfully completed.

  - 7_{16} Command failed - The Interrupter referenced in the Int_ID field is unknown to this device.

- **Line** This is the VMEbus IRQ line number currently assigned. A value of zero (000) indicates that the Interrupter is disconnected.

Read Interrupters *The Read Interrupters* command is used to determine the number of Interrupters within a Servant device. The syntax of this command is defined in the following table.

15	14	13	12	11	10	9	8	7	6	5	4	3	2	1	0
1	1	0	0	1	0	1	0	1	1	1	1	1	1	1	1

The number of Interrupters is placed in the Data Low register with the following format.

15	14	13	12	11	10	9	8	7	6	5	4	3	2	1	0
1	1	1	1	1	1	1	1	1	1	1	1	1	Int_no		

- **Int_no** This is the number of Interrupters within this device.

Read MODID *The Read MODID* command is used by a Commander to determine the state of the MODID lines controlled by a given Slot 0 device. The syntax of this command is defined in the following table.

15	14	13	12	11	10	9	8	7	6	5	4	3	2	1	0
1	1	0	0	1	1	0	0	1	1	1	1	1	1	1	1

The resulting MODID status word is placed in the Data Low register with the following format.

15	14	13	12	11	10	9	8	7	6	5	4	3	2	1	0
1	1	Output Enable	MODID												

- **Output Enable** A one in this position indicates that the Slot 0 device is driving the MODID lines.

- **MODID** This field contains the state of the 13 MODID lines.

Read Protocol The *Read Protocol* command is used by a Commander to find out what protocols (in addition to the word-serial protocol) the Servant supports. The syntax of this command is defined in the following table.

15	14	13	12	11	10	9	8	7	6	5	4	3	2	1	0
1	1	0	1	1	1	1	1	1	1	1	1	1	1	1	1

The protocol support word is placed in the Data Low register with the following format.

15	14 13 12 11	10	9	8	7	6	5	4	3	2	1	0
1	Device Dependent	RSVD	RG*	EG*	0	PI*	PH*	TRG*	I4*	I*	ELW*	LW*

- **Device Dependent** These bits may be defined by the device's manufacturer.

- **RSVD** This bit is always one.

- **RG*** A zero in this position indicates that this device is capable of response generation.

- **EG*** A zero in this position indicates that this device is capable of event generation.

- **PI*** A zero in this position indicates that this device supports the *Read Interrupters, Read Interrupter Line*, and *Assign Interrupter Line* commands.

- **PH*** A zero in this position indicates that this device supports the *Read Handlers, Read Handler Line,* and *Assign Handler Line* commands.

- **TRG*** A zero in this position indicates that this device supports the Word Serial *Trigger* command.

- **I4*** A zero in this position indicates that this device supports the VXIbus 488.2 Instrument protocol.

- **I*** A zero in this position indicates that this device supports the VXIbus Instrument protocol.
- **ELW*** A zero in this position indicates that this device supports the Extended Longword Serial Protocol.
- **LW*** A zero in this position indicates that this device supports the Longword Serial Protocol.

Read Protocol Error The *Read Protocol Error* command is used by a Commander to tell a Servant device to report its most recent error code. When the error code has been reported by the Servant, the Err* bit is reset before Read Ready is asserted on the error code output. The syntax of this command is defined in the following table.

15	14	13	12	11	10	9	8	7	6	5	4	3	2	1	0
1	1	0	0	1	1	0	1	1	1	1	1	1	1	1	1

The error codes are placed in the Data Low register with the following formats.

No error:

15	14	13	12	11	10	9	8	7	6	5	4	3	2	1	0
1	1	1	1	1	1	1	1	1	1	1	1	1	1	1	1

Multiple Queries The device was requested to overwrite previous unread response data.

15	14	13	12	11	10	9	8	7	6	5	4	3	2	1	0
1	1	1	1	1	1	1	1	1	1	1	1	1	1	0	1

Unsupported Command This device has received a command that it does not implement.

15	14	13	12	11	10	9	8	7	6	5	4	3	2	1	0
1	1	1	1	1	1	1	1	1	1	1	1	1	1	0	0

DIR Violation This device has received a command that violates the *DIR* handshake.

15	14	13	12	11	10	9	8	7	6	5	4	3	2	1	0
1	1	1	1	1	1	1	1	1	1	1	1	1	0	1	1

DOR Violation This device has received a command that violates the *DOR* handshake.

15	14	13	12	11	10	9	8	7	6	5	4	3	2	1	0
1	1	1	1	1	1	1	1	1	1	1	1	1	0	1	0

Read Ready Violation This device has received a command that violates the *Read Ready* handshake.

15	14	13	12	11	10	9	8	7	6	5	4	3	2	1	0
1	1	1	1	1	1	1	1	1	1	1	1	1	0	0	1

Write Ready Violation This device has received a command that violates the *Write Ready* handshake.

15	14	13	12	11	10	9	8	7	6	5	4	3	2	1	0
1	1	1	1	1	1	1	1	1	1	1	1	1	0	0	0

All other error codes are reserved for future use by VXIbus.

Read STB The *Read STB* command is used by a Commander to read the status word from a Servant device. The syntax of this command is defined in the following table.

15	14	13	12	11	10	9	8	7	6	5	4	3	2	1	0
1	1	0	0	1	1	1	1	1	1	1	1	1	1	1	1

The status byte is placed in the Data Low register with the following format.

15	14	13	12	11	10	9	8	7–0
1	1	1	1	1	1	1	1	Status Byte

Read Servant Area: The *Read Servant Area* command is used by the Resource Manager to read the Servant area size of Commanders in the system. The syntax of this command is defined in the following table.

15	14	13	12	11	10	9	8	7	6	5	4	3	2	1	0
1	1	0	0	1	1	1	0	1	1	1	1	1	1	1	1

The Servant area size is placed in the Data Low register with the following format.

15	14	13	12	11	10	9	8	7–0
1	1	1	1	1	1	1	1	Servant Area Size

Release Device: The *Release Device* command causes a Commander to relinquish control of a specified Servant device. After releasing the Servant, the Commander will not attempt any further control of that Servant. The syntax of this command is defined in the following table.

15	14	13	12	11	10	9	8	7	6	5	4	3	2	1	0
1	0	0	0	1	1	1	0	Servant Logical Address							

The command termination status is placed in the Data Low register with the following format.

15	14	13	12	11	10	9	8	7	6	5	4	3	2	1	0
Status				1	1	1	1	1	1	1	1	1	1	1	0

- **Status** This field indicates the execution status of the command. It may have the following values:

  - F_{16} The command successfully completed.

  - 7_{16} Command failed - The requested device is not in the Servant list.

Set Lock The *Set Lock* command is used by a Commander to cause a Servant device to enter the locked state. When a device receives this command it will clear its *Locked** bit. The syntax of this command is defined in the following table.

15	14	13	12	11	10	9	8	7	6	5	4	3	2	1	0
1	1	1	0	1	1	1	0	1	1	1	1	1	1	1	1

Set Lower MODID The *Set Lower MODID* command is used by a Commander to change the state of the MODID lines controlled by a given Slot 0 device. This command only affects the state of MODID0 through MODID6 (the lower 7 lines). The enable field, (when set to one) causes the Slot 0 device to enable its MODID drivers. (The function of this field is analogous to the output enable bit in the MODID register of a register based Slot 0 device.) The syntax of this command is defined in the following table.

15	14	13	12	11	10	9	8	7	6	5	4	3	2	1	0
1	0	1	0	1	1	1	0	Enable	MODID6 - MODID0						

Upon the completion of the MODID operation, the device places the following response in the Data Low register:

15–12	11	10	9	8	7	6	5	4	3	2	1	0
Status	1	1	1	1	1	1	1	1	1	1	1	0

- **Status** This field indicates the execution status of the command. It may have the following values:

  - **- F_{16}:** The command successfully completed.

  - **- 7_{16}:** Command failed - The device is not installed in Slot 0.

Set Upper MODID The *Set Upper MODID* command is used by a Commander to change the state of the MODID lines controlled by a given Slot 0 device. This command only affects the state of MODID7 through MODID12 (the upper 6 lines). The enable field (when set to one) causes the Slot 0 device to enable its MODID drivers. (The function of this field is analogous to the output enable bit in the MODID register of a register based Slot 0 device.) The syntax of this command is defined in the following table.

15	14	13	12	11	10	9	8	7	6	5–0
1	0	1	0	1	1	0	1	Enable	X	MODID12 - MODID7

Upon the completion of the MODID operation, the device places the following response in the Data Low register:

15–12	11	10	9	8	7	6	5	4	3	2	1	0
Status	1	1	1	1	1	1	1	1	1	1	1	0

- **Status** This field indicates the execution status of the command. It may have the following values:

  - **- F_{16}** The command successfully completed.

  - **- 7_{16}** Command failed - The device is not installed in Slot 0.

Trigger The *Trigger* command is used by a Commander to cause a Servant device to trigger any previously armed operation. The syntax of this command is defined in the following table.

Bit #

15	14	13	12	11	10	9	8	7	6	5	4	3	2	1	0
1	1	1	0	1	1	0	1	1	1	1	1	1	1	1	1

User Defined The blocks of commands with the following bit patterns are reserved for device-specific actions.

15	14 13 12 11 10 9 8 7 6 5 4 3 2 1 0
0	User Defined Command

OBSERVATION E.1.5:
Commands should indicate the presence of an error with a response which has bit 15 cleared to zero.

All other commands are reserved for future use by VXIbus.

E.2 LONGWORD SERIAL COMMANDS

The following commands are the standard set of VXIbus device commands that are sent with the Longword serial protocol.

User Defined The blocks of commands with the following bit patterns are reserved for device-specific actions.

31	30	29 - 0
1	0	User Defined Command

31	30 - 0
0	User Defined Command

All other commands are reserved for future use by VXIbus.

E.3 EXTENDED LONGWORD SERIAL COMMANDS

The following commands are the standard set of VXIbus device commands that are sent with the Extended Longword Serial protocol.

User Defined The blocks of commands with the following bit patterns are reserved for device-specific actions.

47	46	45 - 0
1	0	User Defined Command

47	46 - 0
0	User Defined Command

All other commands are reserved for future use by VXIbus.

E.4 PROTOCOL EVENTS

If a Message Based VXIbus Servant interrupts (or signals) its Commander, it may pass a response (or an event) in its Status/ID word. If the Status/ID word has bit 15 set to one, then it is defined as an event. If bit 15 is zero, then it is defined as a response. The lower 8 bits are the Logical Address of the device. This section only covers events. For information on signals see the discussions of interrupts and signals in Section C.2.4, "Message Based Devices".

No Cause Given This event is sent by a device which has only one reason to signal its Commander. Usually this event simply means "operation complete". The syntax of this event is defined in the following table.

15	14	13	12	11	10	9	8	7	6	5	4	3	2	1	0
1	1	1	1	1	1	1	1	Sender's Logical Address							

Request True This event is sent by a device when the device requires service from its Commander. The syntax of this event is defined in the following table.

15	14	13	12	11	10	9	8	7	6	5	4	3	2	1	0
1	1	1	1	1	1	0	1	Sender's Logical Address							

Request False This event is sent by a device when the device no longer requires service from its Commander. The syntax of this event is defined in the following table.

15	14	13	12	11	10	9	8	7	6	5	4	3	2	1	0
1	1	1	1	1	1	0	0	Sender's Logical Address							

User Defined This block of events is reserved for user-defined events. The syntax of these events is defined in the following table.

15	14	13	12	11	10	9	8	7	6	5	4	3	2	1	0
1	0	User Defined						Sender's Logical Address							

All other events are reserved for future use by VXIbus.

SECTION F

Dynamic Configuration

Dynamic Configuration is an optional, alternative method of assigning Logical Addresses to VXIbus devices, based upon their slot positions. It allows each slot to contain one or more devices. Provision is made for the different devices occupying a slot to share address decoding hardware. In addition, Dynamic Configuration requires hardware support in the devices being configured, as well as software support in the Resource Manager.

The following sections present an overview of Dynamic Configuration, including definition of terms, hardware and software requirements, and a description of the system configuration algorithm.

F.1 DEFINITIONS

In the following pages, the abbreviation DC means *Dynamic Configuration* and the abbreviation SC means *Static Configuration*.

- A DC device is a device whose Logical Address can be programmed via the VMEbus, by a DC Resource Manager.
- An SC device is any device whose Logical Address is fixed manually, and cannot be changed, except through manually configuring the device.
- A DC Resource Manager is a Resource Manager which supports Dynamic Configuration.
- An SC Resource Manager is any Resource Manager which does not support Dynamic Configuration, i.e. a Resource Manager as defined in Section C.4.1, "Resource Manager".
- A DC system is a VXIbus system with a DC Resource Manager.
- An SC system is a VXIbus system with an SC Resource Manager.

F.2 DC DEVICE REQUIREMENTS

DC devices have the standard VXIbus configuration registers, with some additional requirements. Each DC device has a Logical Address Register, which is used for setting the device's Logical Address. In addition, the device has certain information encoded in its Offset register at power-on. DC devices are also required to support the MODID line.

F.2.1 Logical Address Register

RULE F.2.1:
Each DC device **SHALL** have a writable Logical Address register, located at address offset 0 in its configuration register area, and conforming to the following format:

15 - 8	7 - 0
Undefined	Logical Address

The fields of the Logical Address register are defined as follows:

- **Undefined** The values written to these bits have no effect.
- **Logical Address** The value written to this field is the device's new Logical Address.

RULE F.2.2:
A DC device **SHALL** change its Logical Address before DTACK* is released, during the write cycle which transfers the new Logical Address.

RULE F.2.3:
Any DC device which has moved to a new Logical Address **SHALL NOT** respond to any further accesses at its old Logical Address.

OBSERVATION F.2.1:
Like all VXIbus devices, DC devices have non-volatile Logical Address selectors.

F.2.2 DC Device Logical Address Assignment

A fundamental requirement of DC devices is that (at power-up time) they default to Logical Address 255 (FF_{16}). The Resource Manager will expect to find all DC devices at that Logical Address.

RULE F.2.4:
A DC device **SHALL** have its Logical Address selector configured to Logical Address 255 (FF_{16}) to have its DC feature enabled.

OBSERVATION F.2.2:
Setting the Logical Address of a DC device to 255 (FF_{16}) will cause a DC device to participate in the dynamic Logical Address assignment process. Setting the DC device Logical Address to any other Logical Address will fix the DC device at that Logical Address.

RULE F.2.5:
SC-only devices **SHALL NOT** be set at Logical Address 255 (FF_{16}) in DC systems.

F.2.3 Offset Register

Multiple device modules may share address-decoding hardware. This can result in a significant hardware reduction. In such a case, the devices will share one or more Logical Address bits. A set of such devices is defined to be *address-blocked.* These devices will be configured to a block of contiguous Logical Addresses. The Offset register is used to indicate the number of devices sharing the addressing hardware.

RULE F.2.6:
After SYSRESET* is de-asserted (and prior to entering the SELF TEST state) a DC device **SHALL** load its Offset register in the following format:

15 - 8	7 - 0
Undefined	Devices

The fields of the Offset register are defined as follows:

- **Undefined** The contents of this field are undefined, and are not used.
- **Devices** The value in this field is the number of devices sharing the common address-decoding hardware. Values of 0, 1, and 255 all indicate a single device.

The devices will have the least significant bits of their Logical Addresses hard wired to predetermined values, and will share the decoding hardware for the most significant bits.

RULE F.2.7:
A set of address-blocked DC devices **SHALL** share address bits in accordance with the following table:

Devices	Shared Bits	Hard-coded bits
2	7- 1	0
3 - 4	7 - 2	1 - 0
5 - 8	7 - 3	2 - 0
9 - 16	7 - 4	3 - 0
17 - 32	7 - 5	4 - 0
33 - 64	7 - 6	5 - 0
65 - 128	7	6 - 0
129 - 254	-	7 - 0

RULE F.2.8:
A set of address-blocked DC devices **SHALL** implement the lowest-valued block of contiguous Logical Addresses possible within its hard-coded bits.

OBSERVATION F.2.3:
The above rule guarantees that a set of address-blocked DC devices occupy the lowest-valued addresses available with its block of available addresses.

OBSERVATION F.2.4:
A write to the Logical Address register of an address-blocked module sets the Logical Addresses of all the related devices.

F.2.4 MODID Support

A DC device, when at Logical Address 255 (FF_{16}), uses the MODID line as a device select qualifier. The device responds to data transfers when MODID is asserted. After the device has been moved to another address, the device responds at that address, independent of the state of the MODID line.

RULE F.2.9:
A DC device **SHALL** use its MODID line as a device-select qualifier for Logical Address 255 (FF_{16}).

RULE F.2.10:
A DC device **SHALL NOT** use MODID as a device select qualifier for any Logical Address other than 255 (FF_{16}).

RULE F.2.11:
DC modules which contain multiple DC entities with independent address decoding hardware **SHALL** respond to sequential accesses to Logical Address 255 (FF_{16}), qualified with that module's MODID line, one entity at a time, until all entities have been moved to new Logical Addresses.

OBSERVATION F.2.5:
The entities referred to in the preceding rule may be single devices, or a set of devices that share address-decoding hardware.

RULE F.2.12:
A multiple-board DC module **SHALL** respond to exactly one MODID line, specified by the device designer.

F.3 DC SYSTEM REQUIREMENTS

The DC system Resource Manager must include the software necessary to implement the three-part configuration algorithm defined in the following section. Additionally, some limitations are imposed on Resource Manager and Slot 0 devices.

RULE F.3.1:
A DC system **SHALL** include a DC Resource Manager.

RULE F.3.2:
DC Resource Managers **SHALL** perform the system configuration responsibilities outlined in Section C.4.1, "Resource Manager", except for the Device Identification and Commander/Servant Hierarchy steps alternatively defined in the following paragraphs.

OBSERVATION F.3.1:
To meet the requirements of Section C.4.1, the DC Resource Manager must reside at Logical Address 0. Therefore, the Resource Manager must be an SC device, and cannot be dynamically configured.

A DC Resource Manager uses the Slot-0-controlled MODID lines as board select qualifiers when assigning Logical Addresses. To access the Slot 0 devices, the Resource Manager must know the Logical Addresses of the system's Slot 0 devices.

RULE F.3.3:
Slot 0 devices **SHALL** be SC devices; i.e. Slot 0 devices cannot be dynamically configured.

OBSERVATION F.3.2:
SC devices can be included in a DC system.

F.3.1 System Configuration Algorithm

To configure a Dynamic Configuration system, the DC system Resource Manager performs the following three major functions:

- SC Device Identification
- DC Device Logical Address Assignment

The previous two functions modify the "Device Identification" step of the initialization process outlined in Section C.4.1, "Resource Manager".

- Hierarchy Construction

This function modifies the "Assign Commander/Servant Hierarchies" step of the initialization process outlined in Section C.4.1, "Resource Manager".

F.3.1.1 SC Device Identification

In any Dynamic Configuration system, there will always be at least one (and possibly several) SC devices present: the DC Resource Manager and the system Slot-0 device(s). Also, users may wish to install other types of SC devices in DC systems. Since the Logical Addresses of these SC devices are fixed, the DC Resource Manager must first identify all SC devices, as specified in Section C.4.1.1, "Device Identification". Having located the SC devices, the DC Resource Manager will not attempt to assign their Logical Addresses to any DC devices.

OBSERVATION F.3.3:
If an SC device is found at Logical Address 255 (FF_{16}), no DC devices can be configured.

F.3.1.2 DC Device Logical Address Assignment

After identifying the SC devices, the DC Resource Manager must assign Logical Addresses to all of the DC devices in the system. This requires that the DC Resource Manager:

- Find every DC device by asserting each MODID line and reading a configuration register at Logical Address 255 (FF_{16}).

AND

- Write a new Logical Address to each DC device's Logical Address register.

OBSERVATION F.3.4:
No single order of asserting MODID lines is mandated. The order in which a DC Resource Manager asserts the MODID lines is at its manufacturer's discretion. Consequently, the order in which Logical Addresses are assigned to DC devices might vary from system to system.

PERMISSION F.3.1:
In assigning Logical Addresses, the DC Resource Manager **MAY** skip the access of slots that are known to contain SC devices.

A DC resource manager can implement any procedure that assigns Logical Addresses to every DC device in the system, such as the following:

For every Slot (1 - 12) in the system do the following steps (1 - 3):

1. **Assert the selected slot's MODID line.**

2. **Repeat the following steps (a - c) until a bus error occurs.**

   a. Read the Offset register at Logical Address 255 (FF_{16}).

   b. **IF** the Offset register indicates a single device,
   THEN determine the Logical Address to be assigned to that device.
   ELSE a set of address-blocked devices is being accessed, so determine the Logical Addresses to be assigned to these devices. (The lowest address of the block is written in the next step.)

   c. Write the new Logical Address to the Logical Address register at Logical Address 255 (FF_{16}).

   OBSERVATION F.3.5:
   An address-blocked group of devices will modify only the *shared* bits of their Logical Addresses. The hard-coded bits will cause the devices to occupy the lowest available Logical Addresses within the block defined by the shared bits. As a consequence, the first device will have Logical Address B x 2^H, where *B* is the value assigned to the shared Logical Address bits, and *H* is the number of hard-coded Logical Address bits. The last device will have

Logical Address (B x 2^H)+(D-1), where D is is the number of address-blocked devices in the group. The Logical Addresses (B x 2^H)+D through ([B+1] x 2^H)-1 will be unused, and will be available for assignment to other devices.

RECOMMENDATION F.3.1:
A DC resource manager should verify that each DC device is at its new Logical Address by reading a configuration register at the new Logical Address.

3. De-assert the MODID line.

F.3.1.3 Hierarchy Construction

DC Resource Managers are not required to respect the system hierarchy configuration encoded in the Servant Area Size of the system devices, as defined in Section C.4.1.4, "Commander/Servant Hierarchies".

RULE F.3.4:
IF the DC Resource Manager does not respect the hierarchy encoded in the Servant Area Size of the system devices,
THEN the DC Resource Manager **SHALL** provide some means for building an initial hierarchy.

SECTION G

Shared Memory Protocols

Shared Memory protocols are intended to supplement the VXI word-serial protocols. These protocols will allow a large amount of information to be transferred with less overhead than would be required for the Word Serial protocols. In addition, they can be used by devices to communicate outside of the commander/servant relationship. The protocols will be implemented in a layered approach, similar to many of the networking protocols used in computer systems.

APPENDIX I

VXIbus Register Overview

This section is provided to improve the overview of the VXIbus configuration registers. The following figures contain register maps for the various device types and register content overviews.

Signature explanation:

[ ] Optional register or field

Register with location monitor

( ) Required for particular type of device.

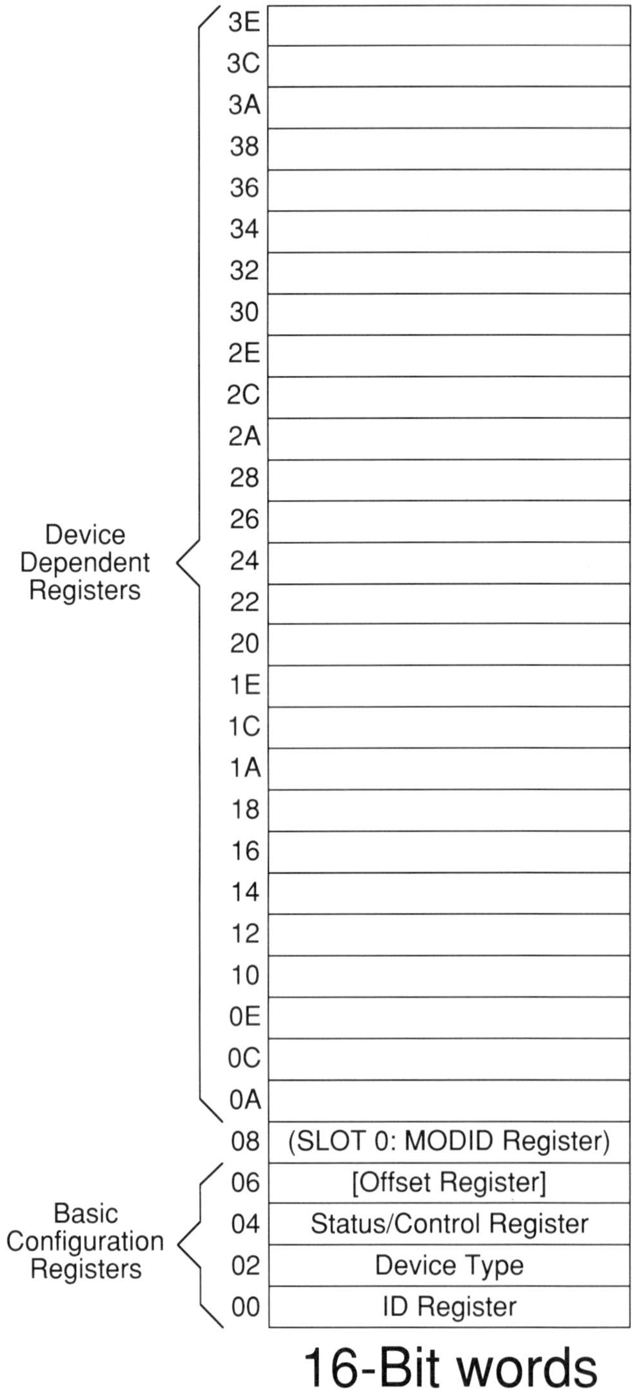

Figure I.1. Register Map for Register Based Devices

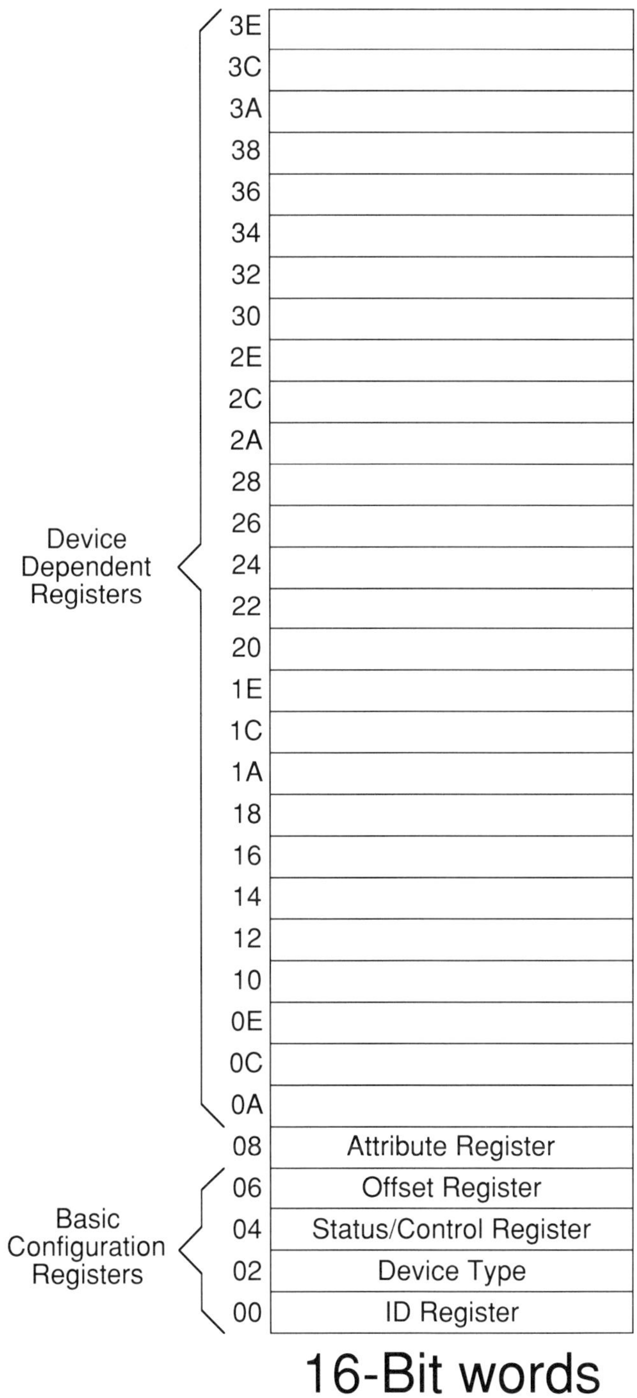

Figure I.2. Register Map for Memory Devices

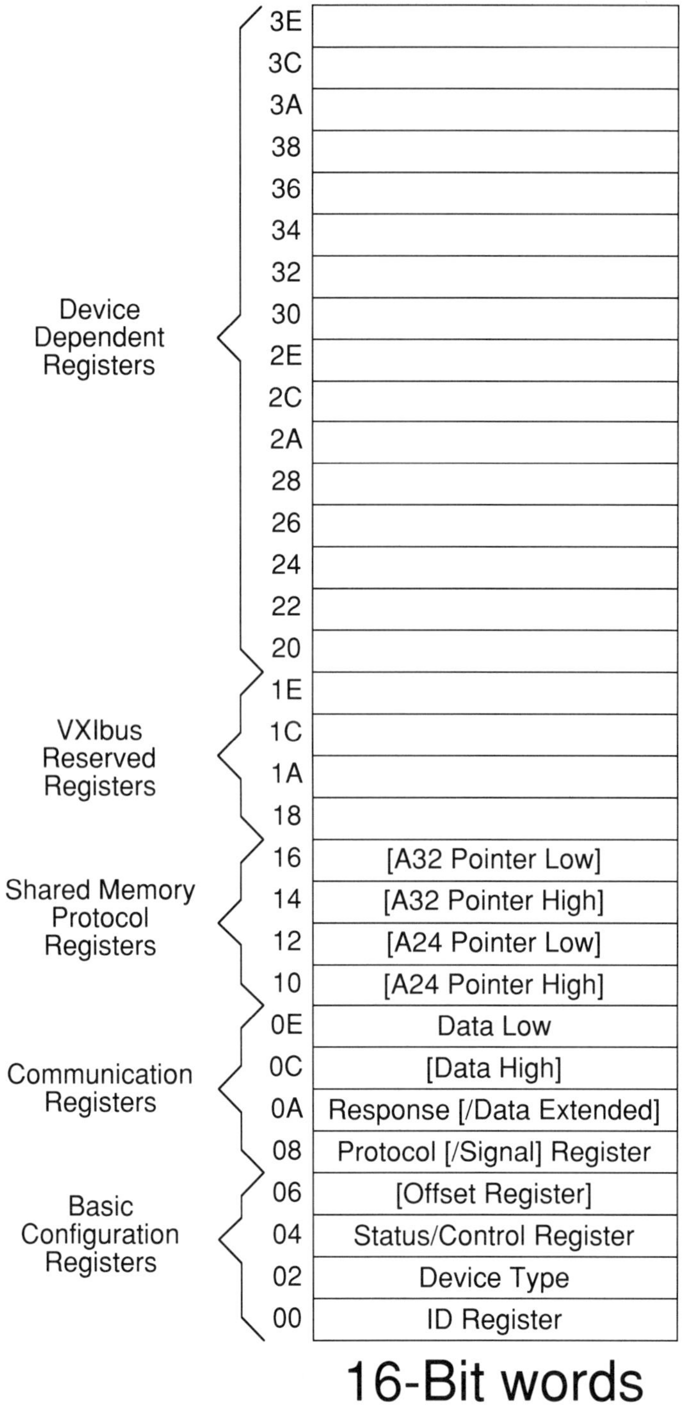

Figure I.3. Register Map for Message Based Devices

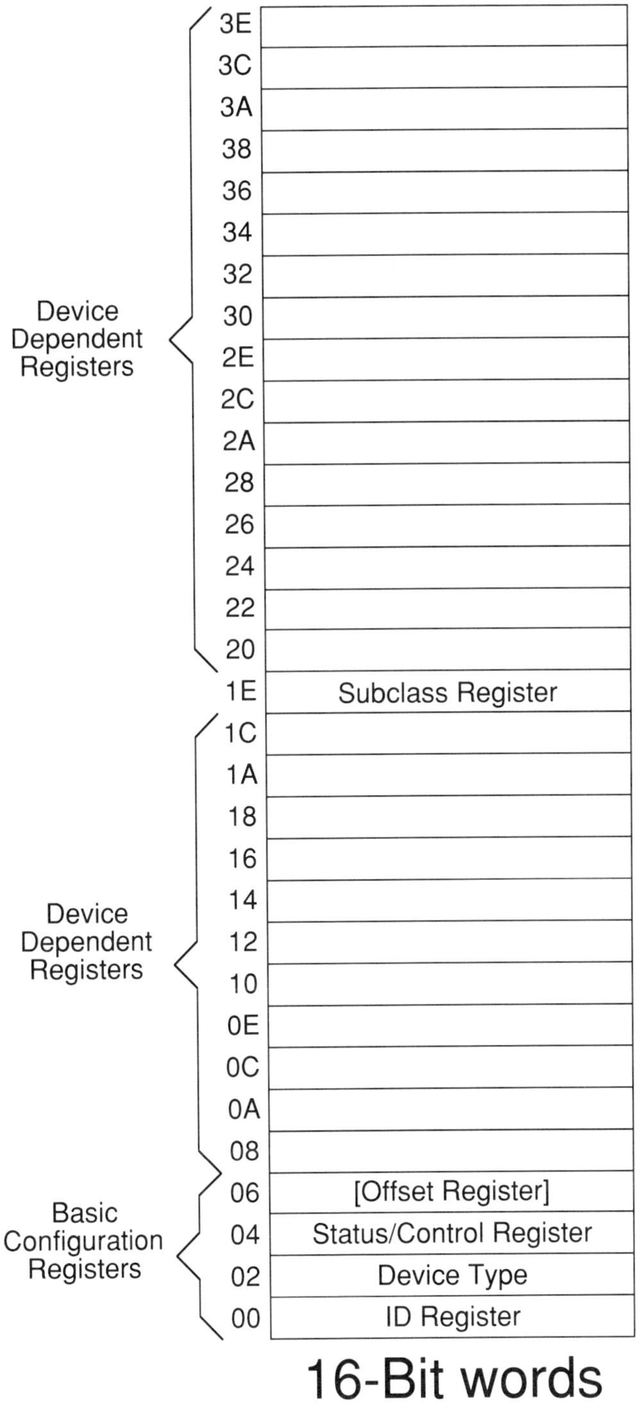

Figure I.4. Register Map for Extended Devices

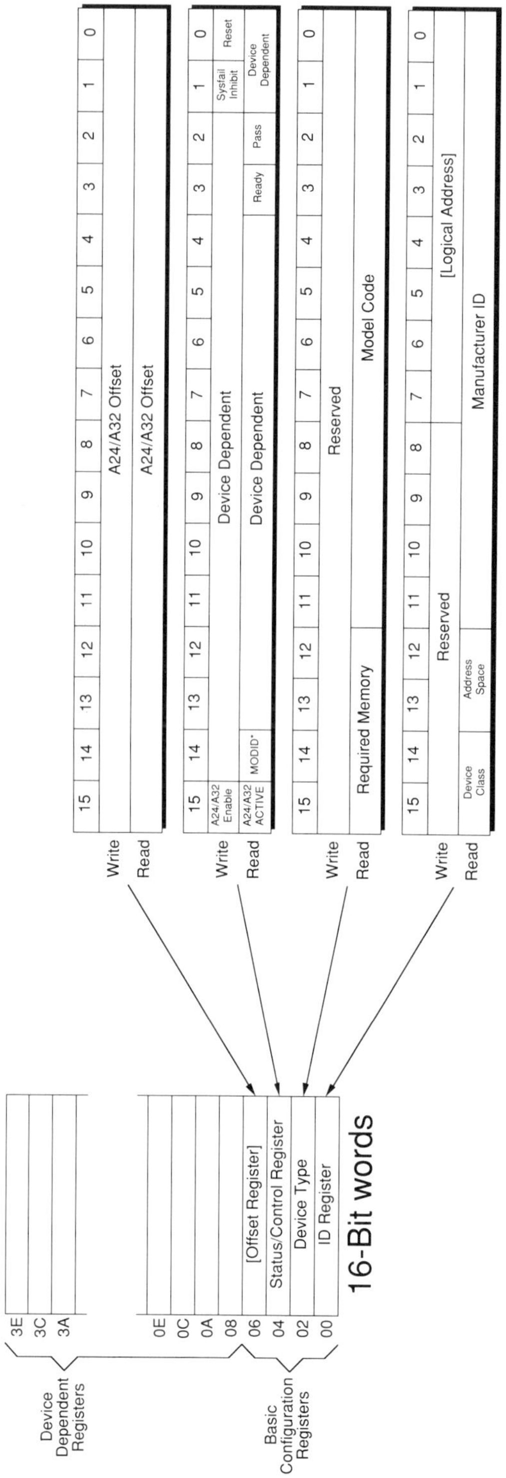

Figure I.5. Register Overview for Basic Configuration Registers

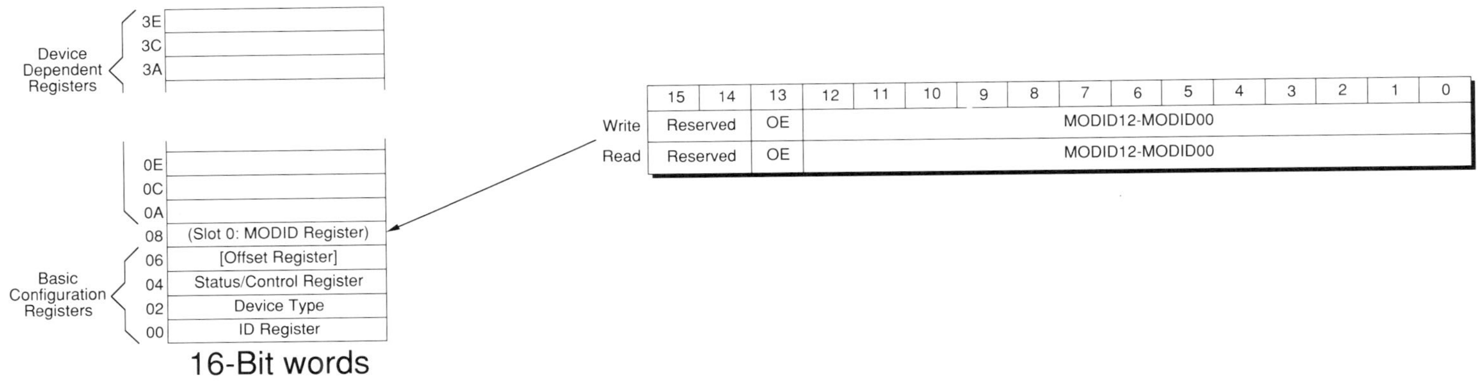

Figure I.6. Register Overview for MODID Register in Slot 0 Devices

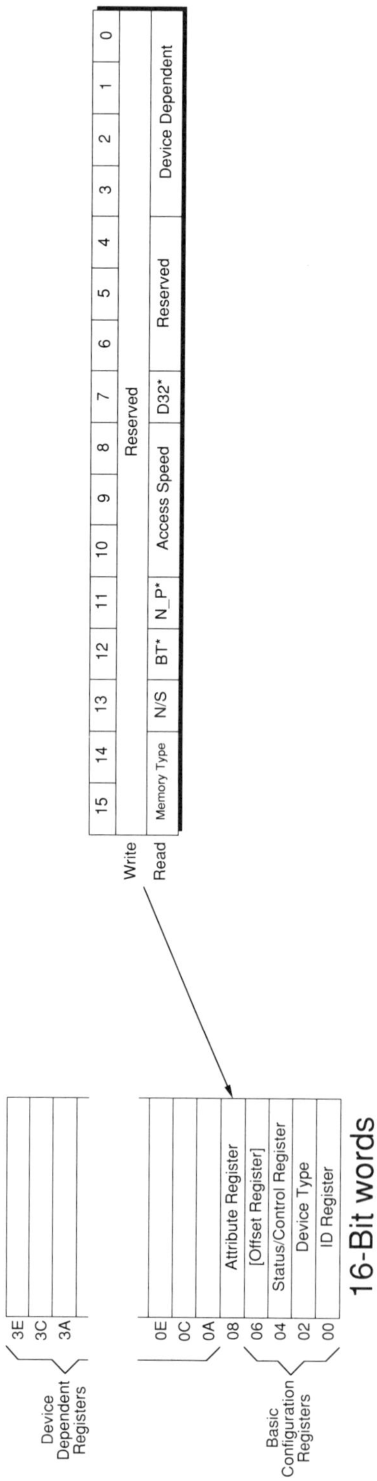

Figure I.7. Register Overview for Attribute Register in Memory Devices

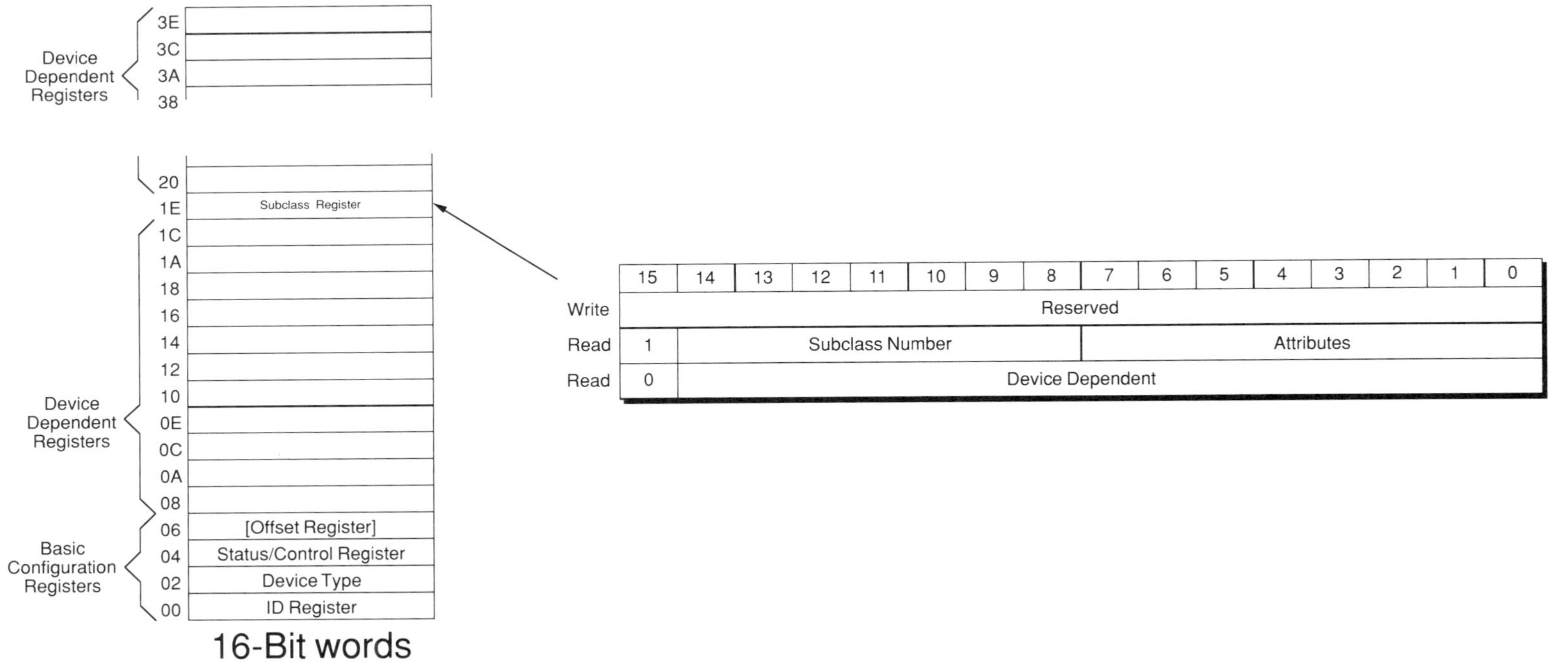

Figure I.8. Register Overview for Subclass Register in Extended Devices

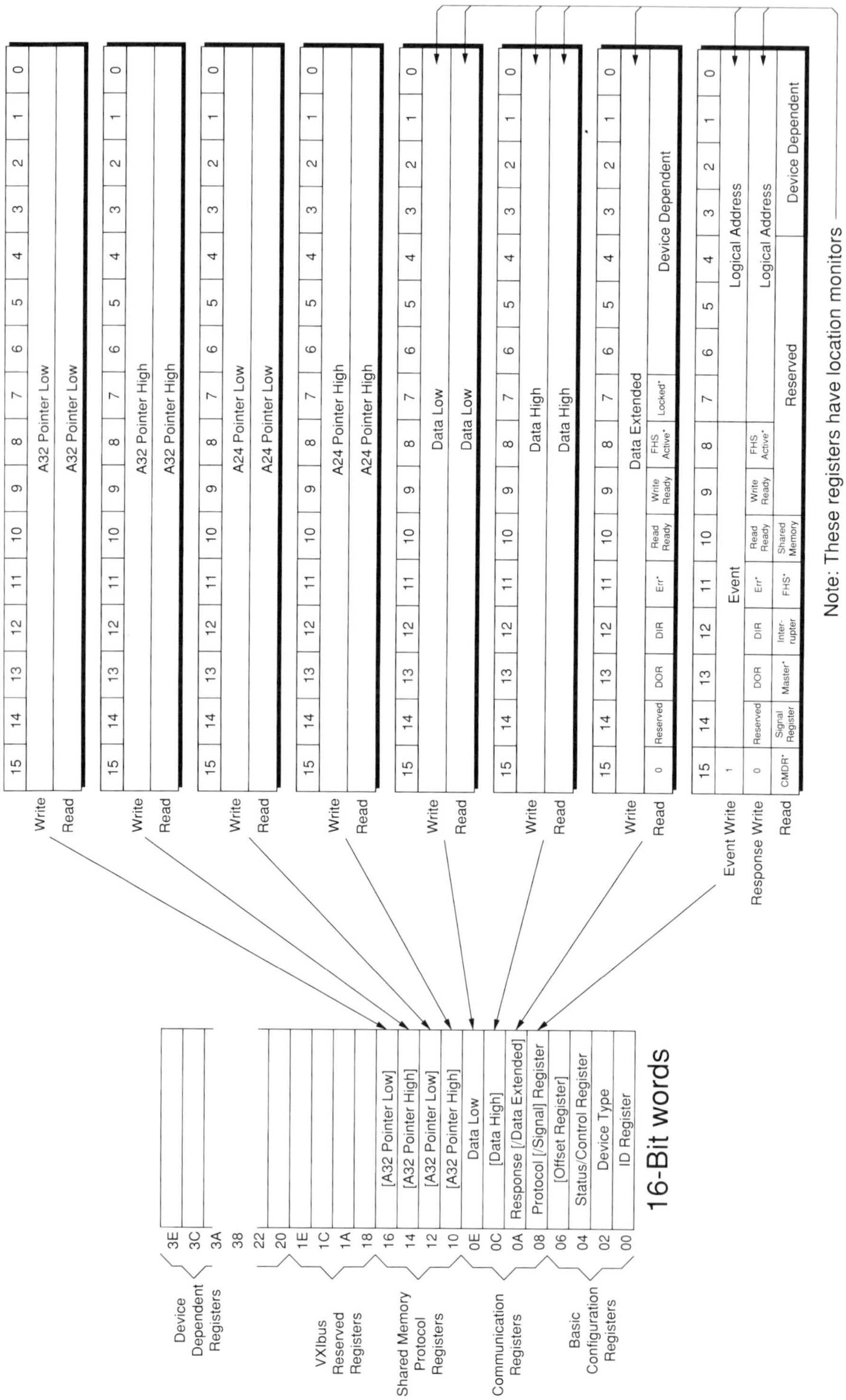

Figure I-9. Register Overview for Communication Registers

APPENDIX II

Suggested Backplane Design

II.1 SUGGESTED BACKPLANE STRUCTURE

A reasonable backplane layer assignment that minimizes noise and maximizes high frequency performance follows:

TOP (CONNECTOR SIDE)

1	GROUND PLANE
2	SIGNAL LAYER (TTL)
3	+5 VOLT PLANE
4	GROUND PLANE
5	-5.2 VOLT PLANE
6	-2 VOLT PLANE
7	SIGNAL LAYER (ECL)
8	GROUND PLANE

BOTTOM (SOLDER SIDE)

By making layers 1 and 8 ground planes, EMI is reduced because of the shielding of the internal traces. Layer 1 also provides the ground layer for attachment of the EMI gasketing to the instrument modules. The thickness between planes 3 and 4, planes 4 and 5, and planes 6 and 7 should be as small as possible (.005 inches is reasonable). This provides high distributed capacitance between those layers, and allows high *di/dt* in the voltage planes. One signal layer also exists between the ground layer and the + 5 volt layer. Since -2 Volts is the signal return for ECL signals, layer 7 should contain as many of the critical ECL signals as possible.

II.2 BACKPLANE LAYOUT

While noise in any logic system can be due to electrostatic coupling, the primary means of coupling is usually due to electromagnetic coupling. The key to designing a low noise system, where electromagnetic coupling is the primary noise source, is the control of current loop areas. The width of the loop is the spacing between the signal current path and the return current path. The main reason for a ground plane is not to provide a shield, but to allow the return current (which flows through the ground plane) to flow as close as possible to the signal current path. This creates the smallest possible loop area. A designer should always examine the route that any return current might take.

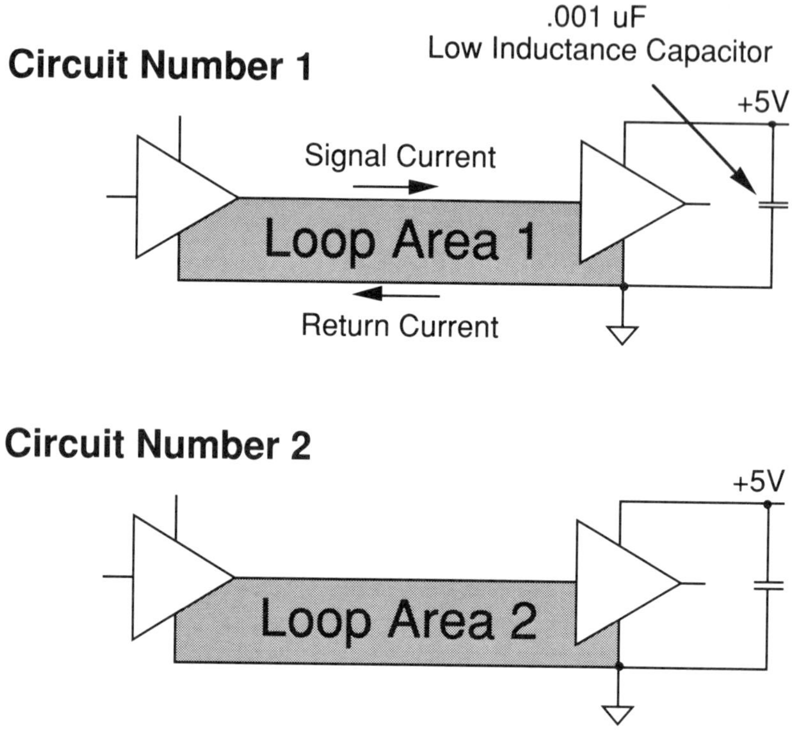

Figure II.1. Typical Logic Circuits

In Figure II.1, if Circuit number 1 is driven, then there must be a return current. This return current forms a loop, called loop area 1. Circuit number 2 is assumed to be quiet. However there is also a loop area in this circuit, which consists of the signal current path, and the return current path. The noise induced in the signal line of Circuit number 2 is directly proportional to the loop areas 1 and 2. If these two loop areas are large enough, it is possible to generate a voltage in Circuit number 2 large enough to create a threshold crossing.

Please note that there is another "hidden" loop in Figure II.1. This is the power supply current path. When the receiving gate in Circuit number 1 changes state, a large amount of current is drawn from the supply voltage line. However, if a voltage plane is used, the loop area due to this current will be small. If a plane isn't used (or if the plane is discontinuous), then the loop area of the power supply return current can be quite large. (In fact, it might encompass the entire board.) If this area is not controlled, it can easily become the dominant noise source. In fact, when ground and voltage planes are not used, this power supply loop is almost always the dominant noise source. If power and ground planes are used, the need for external capacitors is reduced because of the high distributed capacitance.

IT IS HIGHLY RECOMMENDED THAT GROUND AND VOLTAGE PLANES ALWAYS BE USED (PARTICULARLY ON THE BACKPLANE) AND THAT A LOW-INDUCTANCE, LOW-ESR .001μF CAPACITOR BE PLACED AS CLOSE AS POSSIBLE TO THE POWER AND GROUND PINS OF GATES.

Figure II.2 shows a pair of gates that exist on a PC board with a ground plane, but with a cut made in the ground plane.

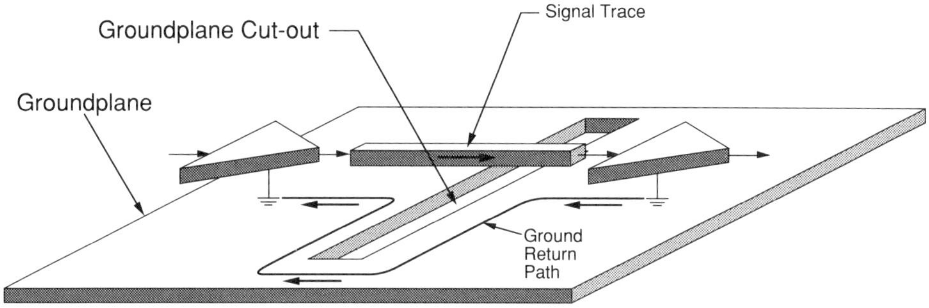

Figure II.2. Loop Area With Cut in Ground Plane

In Figure II.2 the signal is routed on a layer other than the ground plane. If the *ground plane cut* had not been made, the return current would flow adjacent to the signal run, but in the ground plane. With the *cut,* the return current is forced away from the signal current, creating a large loop area. The layout shown in Figure II.2 will typically create large amounts of noise. It will also typically pick up large amounts of noise.

FOR BEST RESULTS DON'T MAKE ANY CUTS IN A GROUND PLANE OR VOLTAGE PLANE FOR ANY REASON.

In Figure II.3 two connectors are shown. In connector 2, the plane is not routed between pins. This results in large loop areas. In fact, this design is susceptible to simultaneous switching of signal lines, creating *pattern sensitive noise*. This noise can be high enough to cross a threshold for certain bit patterns.

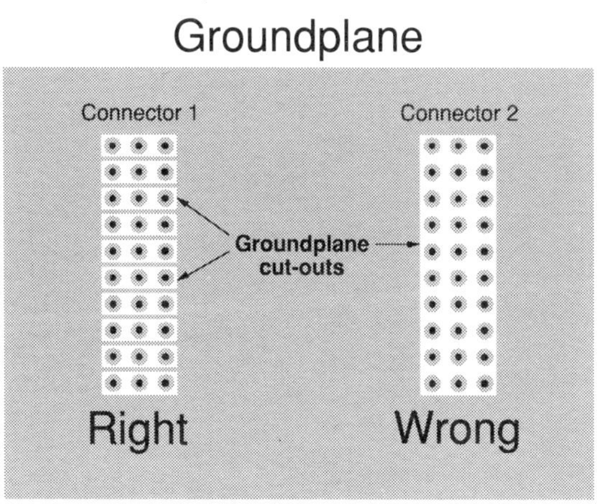

Figure II.3. Backplane Cutouts for Connectors

IT IS HIGHLY RECOMMENDED THAT VOLTAGE AND GROUND PLANES BE CONNECTED BETWEEN THE PINS OF ALL BACKPLANE CONNECTORS.

APPENDIX III

Support Of Revisions 1.2 Devices

Some Word Serial commands that were *optional* in Revision 1.2 of the VXIbus specification are *required* in Revision 1.3. Revision 1.2 also permitted some activities that are prohibited in Revision 1.3. The following is a minimal set of exceptions to the present specification which, if implemented by the Commander or Resource Manager, should provide compatible support of a Revision 1.2 servant-only device. A bit in the response to the *Read Protocol* command identifies Revision 1.2 devices.

RECOMMENDATION III.0.1:
A Resource Manager should implement the following operational changes when communicating with a Revision 1.2 Message Based device.

- Identify any Message Based device that responds to the *Read Protocol* command with bit 15 cleared to zero as a Revision 1.2 device.
- Identify any Revision 1.2 devices that do not support the *Err** bit. This is indicated by a 1 in bit 7 of the response to the *Read Protocol* command.
- If the device does not support the *Err** bit, the device will report unsupported commands with the *unrecognized command* event, which was defined for Revision 1.2 Message Based devices. This event will be returned as an event signal if the device is a VMEbus Master, or as an event interrupt if the device is not a Master. The Resource Manager should provide an event handler for these *unrecognized command* events, and use the unsupported command information as necessary.
- A Revision 1.2 device will not return a response to the *Begin Normal Operation* command. Therefore a Resource Manager or Commander should not expect one.
- Revision 1.2 did not prohibit interrupt generation by a device when not in the NORMAL OPERATION sub-state. The Resource Manager should not send the *Read STB, Read Protocol Error, End Normal Operation,* or *Abort Normal Operation* commands (all optional in Revision 1.2) to a device prior to a *Begin Normal Operation* command, to avoid unwanted interrupts prior to the NORMAL OPERATION sub-state.
- The *End Normal Operation* and *Abort Normal Operation* commands were optional commands for servant-only devices in Revision 1.2. Resource Managers or Commanders should either not send these commands to Revision 1.2 devices, or should provide for handling of a possible *unrecognized command* event interrupt or signal. Since these termination commands may not be supported by the device, proper termination of the device may be application specific.

- The *Asynchronous Mode Control, Control Response, Control Event, Abort Normal Operation, End Normal Operation*, and *Read STB* commands, if supported by a Revision 1.2 device, did not require bits 15-8 of the *successfully completed* response to these commands to be set to one. The Resource Manager or Commander should mask bits 15-8 when interpreting the response from these commands. The *Control Event* command did not specify a response in Revision 1.2. Therefore, the Resource Manager or Commander should not expect one. The EG* and RG* fields of the *Read Protocol* command were not defined in Revision 1.2. Lack of support for the *Asynchronous Mode Control, Control Response, Control Event, Abort Normal Operation, End Normal Operation*, and *Read STB* commands is indicated by an unsupported command error.

- If Programmable Handler (PH) or Programmable Interrupter (PI) capability is supported by a Revision 1.2 device, the responses to these commands differed from the present specification. The *Assign Handler Line* and *Assign Interrupter Line* commands did not permit a response, and the *Read Handlers, Read Handler Line, Read Interrupters*, and *Read Interrupter Line* commands did not require bits 15-3 of the response to be set to one. The Resource Manager or Commander should not expect a response to the *Assign Handler Line*, or the *Assign Interrupter Line* commands, and should mask bits 15-3 when intrepreting the response to the *Read Handlers, Read Handler Line, Read Interrupters,* and *Read Interrupter Line* commands.

- *DOR* and *DIR* support was not required of Revision 1.2 devices. Devices which did not support *DIR* and *DOR* held the associated bits high. A Resource Manager or Commander may not assume that the *DOR* and *DIR* bits in a Revision 1.2 device will accurately reflect the device's state.

APPENDIX IV

Glossary Of Terms

CONFIGURATION REGISTERS: A device's A16 registers, which are required for the system configuration process.

BACKPLANE: An assembly, typically a printed circuit board, with 96-pin connectors and signal paths that bus those connector pins. VXIbus systems will have two sets of bused connectors (called the J1 and J2 backplanes) or will have three sets of bused connectors (called the J1, J2, and J3 backplane).

BOARD: A blank printed circuit board.

BOARD ASSEMBLY: A board and its associated electrical components and connectors.

CHASSIS SHIELD: A shield that resides between two modules, and attaches to the mainframe.

COMMAND: Any communication from a commander to a message-based servant, consisting of a write to the servant's Data Low register, possibly preceded by a write to the servant's Data High register, or a write to both its Data High and its Data Extended register.

COMMANDER: A message based device that is also a bus master, and can control one or more servants.

DEVICE: A component of a VXIbus system. Normally, a device will consist of one VXIbus board. However, multiple-slot devices and multiple-device modules, are permitted. Some examples of devices are computers, multimeters, multiplexers, oscillators, operator interfaces, and counters.

DYNAMIC CONFIGURATION: A method of automatically assigning Logical Addresses to VXIbus devices when the system is powered up, or possibly at other times. It allows for each slot to contain one or more devices, as well as different devices within a slot to share address decoding hardware.

EVENTS: Signals or interrupts generated by a device to notify another device of an asynchronous event. The contents of events are device dependent.

FAST HANDSHAKE: A high speed mode of operation which uses the same communication registers as the Word Serial protocol, and allows data transfers without the need for polling after each transfer.

EXTENDED DEVICE: A device that has VXIbus configuration registers as well as a subclass register. This category is intended to allow for definition of additional device types.

EXTENDED LONGWORD SERIAL: A form of Word Serial communication that allows 48-bit data transfers between commanders and servants.

HYBRID DEVICE: A VMEbus-compatible device that includes application-specific subsets of VXIbus protocols.

LOCATION MONITOR: A function that monitors data transfers over the VMEbus, in order to detect accesses to the locations it has been assigned to watch. When an access occurs to one of these assigned locations, the LOCATION MONITOR generates a local signal.

LOGICAL ADDRESS: An 8-bit number which uniquely identifies each VXIbus Device in a system. It defines a device's A16 register addresses, and indicates Commander/Servant relationships.

LONGWORD SERIAL: A form of Word Serial communication that allows 32-bit data transfers between commanders and servants.

MAINFRAME: A rigid framework that provides mechanical support for modules inserted into the backplane, ensuring that connectors mate properly and that adjacent modules do not contact each other. It also provides cooling airflow, and ensures that modules do not disengage from the backplane, due to vibration or shock.

MEMORY DEVICE: A device which contains only memory, and implements configuration registers.

MESSAGE BASED DEVICE: An intelligent device which includes the defined VXIbus registers and communication protocols.

MODULE: Typically consists of a board assembly and its associated mechanical parts, front panel, optional shields, etc. A module contains everything required to occupy a slot in a mainframe. A module might occupy one or more slots.

OBSERVATION: Observations spell out implications of rules and bring attention to things that might otherwise be overlooked. They also give the rationale behind certain rules, so that the reader understands why the rule must be followed.

OPERATIONAL REGISTER: An operational register is any register on a device that is not required for the system configuration process.

PERMISSION: Permissions are included to clarify the areas of the specification that are not specifically prohibited. Permissions reassure the reader that a certain approach is acceptable, and will cause no problems. The word **MAY** is reserved for indicating permissions.

RECOMMENDATION: Recommendations consist of *advice* to implementors, which will affect the usability of the final device. Discussions of particular hardware to enhance throughput would fall under a recommendation. Recommendations should be followed to avoid problems, and to obtain optimum performance.

REGISTER BASED DEVICE: A servant-only device which supports VXIbus configuration registers. Register-based devices are typically controlled by message-based devices, via device-dependent register reads and writes.

RESOURCE MANAGER: A message-based commander located at Logical Address 0, which provides configuration management services such as address map configuration, commander/servant mappings, self test, and diagnostics management.

RESPONSES: Signals or interrupts generated by a device to notify another device of an asynchronous event. Responses contain the information in the sender's Response register.

RULE: Rules **MUST** be followed to ensure compatibility between modules in the system. A rule is characterized by the use of the words **SHALL** and **SHALL NOT**. These words are not used for any other purpose other than stating rules.

SERVANT: A device that is controlled by a commander. There are message-based and register-based servants.

SERVANT POINTER: An 8-bit number which is the Logical Address of the lowest-numbered servant associated with this commander.

SHARED MEMORY PROTOCOL: A communication protocol designed to allow message based devices to communicate by sharing an area of memory accessible to both. The protocol specifies connection and operation sequences to be followed by both devices.

SIGNAL: Any communication between message-based devices, consisting of a write to a Signal register.

SLOT: A slot is a position where a module can be inserted into a VXIbus backplane. Each slot provides the 96-pin J connectors that interface to the P connectors on the plug-in modules. A slot might provide one, two, or three J connectors, called J1, J2, and J3.

SUGGESTION: A suggestion contains advice which is helpful but not vital. The reader is encouraged to consider the advice before discarding it. Suggestions are included to help the novice designer with areas of design that can be problematic.

SYSTEM: A system consists of one or more mainframes that are connected, all sharing a common resource manager. Each device in a system has a unique Logical Address.

VXIbus 488 INSTRUMENT: A VXIbus 488 INSTRUMENT is defined to be a message-based device with communication capability to support the 488-VXIbus Interface.

VXIbus SUBSYSTEM: A VXIbus Subsystem consists of a central timing module (referred to as a Slot 0 module) with up to 12 additional adjacent VXIbus modules. The VXIbus Subsystem bus defines the lines on the P2 and P3 connectors.

WORD SERIAL: The simplest communication protocol that must be supported by VXIbus Message Based devices. It uses the A16 communication registers to transfer data, using a simple polling handshake method.

488-VXIbus INTERFACE DEVICE: A 488-VXIbus INTERFACE device is a message-based device which provides a communication path between an IEEE 488 bus and a VXIbus system.

References

1. THE VMEbus SPECIFICATION Revision C.1 Micrology pbt, Inc. 2618 S. Shannon Dr., Tempe, AZ 85282 602-966-5936

2. ANSI/IEEE Std 488.1-1987, IEEE Standard Digital Interface for Programmable Instrumentation, The IEEE, Inc. 345 East 47th St., New York, NY 800-678-4333

3. IEEE Std 488.2-1987, Codes, Formats, Protocols, and Common Commands For use with ANSI/IEEE Std. 488.1-1987, IEEE Standard Digital Interface for Programmable Instrumentation, The IEEE, Inc. 345 East 47th St. New York, NY 800-678-4333